DATE DUE

DEMCO, INC. 38-2931

JUL 1 9 2005

ANTICANCER AGENTS from NATURAL PRODUCTS

ANTICANCER AGENTS from NATURAL PRODUCTS

Edited by
Gordon M. Cragg
David G.I. Kingston
David J. Newman

Taylor & Francis
Taylor & Francis Group

Boca Raton   London   New York   Singapore

A CRC title, part of the Taylor & Francis imprint, a member of the
Taylor & Francis Group, the academic division of T&F Informa plc.

Published in 2005 by
CRC Press
Taylor & Francis Group
6000 Broken Sound Parkway NW, Suite 300
Boca Raton, FL 33487-2742

© 2005 by Taylor & Francis Group, LLC
CRC Press is an imprint of Taylor & Francis Group

No claim to original U.S. Government works
Printed in the United States of America on acid-free paper
10 9 8 7 6 5 4 3 2 1

International Standard Book Number-10: 0-8493-1863-7 (Hardcover)
International Standard Book Number-13: 978-0-8493-1863-4 (Hardcover)
Library of Congress Card Number 2004065568

Library of Congress Cataloging-in-Publication Data

Kingston, David.
    Anticancer agents from natural products / David Kingston, Gordon Cragg, David Newman.
      p. cm.
    ISBN 0-8493-1863-7
    1. Antineoplastic agents. 2. Pharmacognosy. 3. Natural products. I. Cragg, Gordon M. L. II. Newman, David J. III. Title.

RS431.A64K545 2005
616.99'4061--dc22
                                                    2004065568

Taylor & Francis Group
is the Academic Division of T&F Informa plc.

Visit the Taylor & Francis Web site at
http://www.taylorandfrancis.com

and the CRC Press Web site at
http://www.crcpress.com

Preface

Natural products have made an enormous contribution to cancer chemotherapy, and over half of the current anticancer agents in clinical use are natural products or are derived from natural products. In spite of this fact, no book published in recent years has brought together the disparate information on anticancer natural products that is currently scattered throughout the chemical, biological, and medical literature. The most recent book to do this was *Anticancer Agents Based on Natural Product Models*, edited by John Cassady and John Douros, but this text was published in 1980 and is now sadly out of date. The present book covers the current clinically used anticancer agents that are either natural products or are clearly derived from natural product leads. In addition, a number of drug candidates that are in clinical development are also covered, albeit more briefly, as many of these will be clinically used drugs in the future.

It is expected that this volume will appeal to several classes of reader. It will be of interest to natural products chemists, medicinal chemists, and pharmacognosists as an important reference work in their area of interest. It will appeal to synthetic organic chemists as a source of information on challenging synthetic targets. It will also appeal to oncologists as a source of background information on the drugs they use, although it is not intended as a primary clinical text.

Some of the key features of the book include up-to-date coverage of a field that is scattered among many different journals and review articles, inclusion of information on drugs in clinical development, and authorship by leading scientists on each drug; in some cases the drug developer or a close associate is the chapter author. These features should give the book real value for scientists looking for information on the next generation of anticancer drugs.

The editors express their appreciation to the staff of Taylor & Francis for their excellent help, especially Randi Cohen, Erika Dery, Lindsey Hofmeister, and Jay Margolis. The volume is dedicated to the memory of those scientists who have blazed the trail to successful natural products-based anticancer drug development, including Drs. John Douros, John Faulkner, Jonathan Hartwell, Morris Kupchan, Paul Scheuer, Matthew Suffness, and Monroe Wall. May their tribe increase!

Gordon M. Cragg, David G. I. Kingston, and David J. Newman
Frederick, MD, and Blacksburg, VA

Editors

Gordon M. Cragg was born in Cape Town, South Africa, and obtained his undergraduate training in chemistry at Rhodes University before proceeding to Oxford University, where he obtained his D. Phil. in organic chemistry in 1963. After 2 years of postdoctoral research in natural products chemistry at the University of California, Los Angeles, he returned to South Africa to join the Council for Scientific and Industrial Research. In 1966, he was appointed to the staff of the Department of Chemistry at the University of South Africa and transferred to the University of Cape Town in 1972. In 1979, he returned to the United States to join the Cancer Research Institute at Arizona State University, working with Professor G. Robert Pettit on the isolation of potential anticancer agents from plant and marine invertebrate sources. In 1985, he moved to the National Cancer Institute in Bethesda, Maryland, and was appointed Chief of the Natural Products Branch in 1989. His major interests lie in the discovery of novel natural product agents for the treatment of cancer and AIDS. He has been awarded National Institutes of Health Merit Awards for his contributions to the development of the drug, taxol (1991), leadership in establishing international collaborative research in biodiversity and natural products drug discovery (2004), and contributions to developing and teaching NIH technology transfer courses (2004). In 1998–1999 he served as President of the American Society of Pharmacognosy, and was elected to honorary membership of the society in 2003. He has established collaborations between the National Cancer Institute and organizations in many countries promoting drug discovery from their natural resources. He has given over 100 invited talks at conferences in many countries and has published more than 140 papers related to these interests.

David G. I. Kingston was born in London, England, and obtained both his undergraduate and graduate training in chemistry at Cambridge University. His graduate supervisors were Lord Todd and Dr. D. W. Cameron, and his Ph.D. research was on the chemistry of the aphid pigments; he completed his Ph.D. degree in 1963. He then did 3 years of postdoctoral research; one in the Division of Biochemistry at MIT under Professor J. M. Buchanan and two back at Cambridge, where he was a Research Fellow of Queens College and worked with Lord Todd, Franz Sondheimer, and Dudley Williams. He moved to the State University of New York at Albany in 1966 and then to Virginia Polytechnic Institute and State University in 1971, where he currently holds the rank of University Distinguished Professor. He served as President of the American Society of Pharmacognosy in 1988–1989. His research interests are on the isolation and structure elucidation of novel natural products, especially those with anticancer activity, and on the chemistry and mechanism of action of tubulin-binding natural anticancer agents such as taxol, epothilone, and discodermolide, and he currently serves as the principal investigator of the Madagascar International Cooperative Biodiversity Group. He received the Research Achievement Award of the American Society of Pharmacognosy in 1999 and was named Virginia Scientist of the Year in 2002. He has published over 280 papers and holds 14 patents. He is also an Elder in his church, the Blacksburg Christian Fellowship.

David J. Newman was born in Grays, Essex. Initially he trained as a chemical analyst (Grad. RIC), followed by his being awarded an M.Sc. in organic chemistry (University of Liverpool), and then after time in the U.K. chemical industry, he obtained a D. Phil. in microbial chemistry from the University of Sussex in 1968. Following two years of postdoctoral studies on the structure of electron transport proteins at the University of Georgia, he worked for Smith Kline and French in

Philadelphia, Pennsylvania, as a biological chemist predominantly in the area of antibiotic discovery. During this time, he obtained an M.S. in information sciences in 1977 from Drexel University, Philadelphia. He has worked for a number of United States-based pharmaceutical companies in natural products-based discovery programs in antiinfective and cancer treatments and joined the Natural Products Branch of the National Cancer Institute in 1991. He is responsible for the marine and microbial collection programs of the National Cancer Institute and, in concert with Gordon Cragg, for the National Cancer Institute's Open and Active Repository programs. In 2003 he was awarded the National Institutes of Health Merit Award for his contributions to the development of potential anticancer agents from marine and microbial sources. His scientific interests are in the discovery and history of novel natural products as drug leads in the antiinfective and cancer areas and in the application of information technologies to drug discovery. In conjunction with Gordon Cragg, he has established collaborations between the National Cancer Institute and organizations in many countries promoting drug discovery from their natural resources. He has published over 60 papers, presented over 60 abstracts, holds 17 patents that are related to these interests, is both a UK Chartered Chemist and a UK Chartered Biologist, and is also an adjunct full professor at the Center of Marine Biotechnology, University of Maryland.

Contributors

Dr. Rima Al-awar
Eli Lilly and Company
Lilly Corporate Center
Discovery Chemistry Research
  and Technology
Drop Code 2810
Indianapolis, IN 46285
AL-AWAR_RIMA_S@LILLY.COM

Professor Raymond J. Andersen
Dept of Chemistry
University of British Columbia
Vancouver, BC
Canada V6T 1Z1
randersn@interchange.ubc.ca

Dr. Federico Arcamone
Via Quattro Novembre 26
Nerviano
20014
Milano, Italy
farcamone@virgilio.it

Dr. Kathryn A. Bixby
Department of Bacteriology
University of Wisconsin-Madison
777 Highland Avenue
Madison, WI 53705-2222

Dr. Don Borders
13 Heatherhill Lane
Suffern, NY 10901
biosource@prodigy.net

Dr. David J. Chaplin
Oxigene Inc.
230 Third Avenue
Waltham, MA 02451

Dr. Gordon M. Cragg
Natural Products Branch
Developmental Therapeutics Program
Division of Cancer Treatment and Diagnosis
National Cancer Institute
NCI-Frederick
Fairview Center, Room 206
P.O. Box B
Frederick, MD 21702-1201
craggg@mail.nih.gov

Dr. Carmen Cuevas
PharmaMar
Av. de los Reyes, 1, P.I. La Mina
28770 Colmenar Viejo
Madrid, Spain
ccuevas@pharmamar.com

Dr. Klaus Edvardsen
Department of Cell and Molecular Biology
Section for Tumor Immunology
University of Lund
I12, 221 84
Lund, Sweden

Dr. Jacques Fahy
Division de Chimie Médicinale 5
Centre de Recherches Pierre F.
17, avenue Jean Moulin
81106 CASTRES, France
Jacques.fahy@pierre-fabre.com

Dr. Glynn T. Faircloth
PharmaMar USA, Inc.
320 Putnam Avenue
Cambridge, MA 02139
gfaircloth@pharmamarusa.com

Dr. Erik Flahive
Pfizer Global Research and
  Development - La Jolla Labs
10578 Science Center Drive
San Diego, CA 92109
Erik.flahive@pfizer.com
eflahive@earthlink.net

Professor Heinz G. Floss
University of Washington
Department of Chemistry
Campus Box 351700
Seattle, WA 98195-1700
floss@chem.washington.edu

Dr. (Mrs) Françoise Guéritte
I.C.S.N./C.N.R.S.
Avenue de la Terrasse
91198 Gif-sur-Yvette, France
Francoise.Gueritte@icsn.cnrs-gif.fr

Dr. Sarath P. Gunasekera
Harbor Branch Oceanographic Institution
5600 US 1 North
Fort Pierce, FL 34946
sgunaseker@hboi.edu

Dr. Philip R. Hamann
Chemical and Screening Sciences
Wyeth Research Laboratories
401 N. Middletown Road
Pearl River, NY 10965
HAMANNP@wyeth.com

Professor Sidney M. Hecht
Department of Chemistry
University of Virginia
McCormick Road, PO Box 400319
Charlottesville, VA 22904-4319
sidhecht@virginia.edu

Dr. Rubén Henríquez
PharmaMar
Av. de los Reyes, 1, P.I. La Mina
28770 Colmenar Viejo
Madrid, Spain
rhenriquez@pharmamar.com

Dr. Gerhard Höfle
Bereich Naturstoffe
GBF – Gesellschaft für Biotechnologische
  Forschung
Mascheroder Weg 1
D-38124
Braunschweig, Germany
gho@gbf.de

Professor Hideji Itokawa
Department of Medicinal Chemistry, School of
  Pharmacy
Beard Hall CB7360
University of North Carolina
Chapel Hill, NC 27599-7360

Dr. Christopher Jelinek
Department of Chemistry and Biochemistry
P.O. Box 97348
Baylor University
Waco, Texas 76798-7348

David G. I. Kingston
Department of Chemistry, M/C 0212
Virginia Polytechnic Institute and State
  University
Blacksburg, VA 24061
DKingston@vt.edu

Professor Yoshita Kishi
Dept. of Chemistry
Harvard University
Cambridge, MA 02138
Kishi@chemistry.harvard.edu

Dr. Rohtash Kumar
Department of Chemistry
University of Alberta
Edmonton
Alberta T6G 2G2
Canada

Professor Helmut Lackner
Institut für
Universität Göttingen
Tammannstraße 2
D-37077
Göttingen, Germany
hlackne@gwdg.de

Profesor K. H. Lee
Department of Medicinal Chemistry, School of
  Pharmacy
Beard Hall CB7360
University of North Carolina
Chapel Hill, NC 27599-7360
khlee@unc.edu

Dr. Bruce A. Littlefield
Eisai Research Institute
One Corporate Drive
Andover, MA 01810
bruce_littlefield@eri.eisai.com

Professor J. W. Lown
Department of Chemistry
University of Alberta
Edmonton
Alberta T6G 2G2
Canada
Lynne.lechelt@ualberta.ca

Dr. Anthony Mauger
10206 Frederick Ave
Kensington, MD 20891-3304
maugerai@yahoo.com

Dr. David J. Newman
Natural Products Branch
Developmental Therapeutics Program
Division of Cancer Treatment and Diagnosis
National Cancer Institute
NCI-Frederick
Fairview Center, Room 206
P.O. Box B
Frederick, MD 21702-1201
newmand@mail.nih.gov

Professor George R. Pettit
Cancer Research Institute
Arizona State University, Main Campus
P.O. Box 872404
Tempe, AZ 85287-2404

Professor Kevin G. Pinney
Department of Chemistry and Biochemistry
P.O. Box 97348
Baylor University
Waco, TX 76798-7348
Kevin_Pinney@baylor.edu

Dr. Michelle Prudhomme
Université Blaise Pascal
Laboratoire de Synthèse et Etude de Systèmes
  à Intérêt Biologique
UMR 6504 du CNRS
URA 485 du CNRS
63177
Aubière, France
mprud@chimtp.univ-bpclermont.fr

Dr. Nicholas J. Rahier
Department of Chemistry
University of Virginia
McCormick Road, PO Box 400319
Charlottesville, VA 22904-4319

Professor Hans Reichenbach
Bereich Naturstoffe
Gesellschaft für Biotechnologische Forschung
Mascheroder Weg 1
D-38124
Braunschweig, Germany
hre@gbf.de

Dr. William Remers
AmpliMed Corporation
4280 N. Campbell Avenue
Tucson, AZ 85718
remers@pharmacy.arizona.edu

Professor Michel Roberge
Department of Biochemistry and Molecular
  Biology
University of British Columbia
Vancouver, BC
Canada V6T 1Z1
michelr@interchange.ubc.ca

Professor Ben Shen
Division of Pharmaceutical Sciences
School of Pharmacy
University of Wisconsin-Madison
777 Highland Avenue
Madison, WI 53705-2222
bshen@pharmacy.wisc.edu

Dr. Chuan Sih
Eli Lilly and Company
Lilly Corporate Center
Discovery Chemistry Research and Technology
Indianapolis, IN 46285
SHIH-CHUAN@LILLY.COM

Dr. Kenneth Snader
1346 34th Avenue
Vero Beach, FL 32960
snaderk@yahoo.com

Dr. Jay Srirangam
Chemical Research & Development
Pfizer Global R&D - La Jolla Labs
10578 Science Center Drive
San Diego, CA 92109
Jay.srirangam@pfizer.com

Dr. Craig J. Thomas
Department of Chemistry
University of Virginia
McCormick Road, PO Box 400319
Charlottesville, VA 22904-4319

Dr. Michael G. Thomas
Department of Bacteriology
University of Wisconsin-Madison
777 Highland Avenue
Madison, WI 53705-2222

Dr. Janis Upeslacis
Chemical and Screening Sciences
Wyeth Research Laboratories
401 N. Middletown Road
Pearl River, NY 10965
UPESLAJ@wyeth.com

Dr. Amy E. Wright
Harbor Branch Oceanographic Institution
5600 US 1 North
Fort Pierce, FL 34946
wright@hboi.edu

Ms. Zhiyan Xiao
Department of Medicinal Chemistry, School of
  Pharmacy
Beard Hall CB7360
University of North Carolina
Chapel Hill, NC 27599-7360

Dr. Melvin Yu
Eisai Research Institute
4 Corporate Drive
Andover, MA 01810
Melvin_Yu@eri.eisai.com

Dr. Tin-Wein Yu.
Department of Biological Sciences
Louisiana State University
Baton Rouge, LA 70803-1715
yu@lsu.edu

Contents

Chapter 1   Introduction..1
Gordon M. Cragg, David G. I. Kingston, and David J. Newman

Chapter 2   Camptothecin and Its Analogs 5
Nicolas J. Rahier, Craig J. Thomas, and Sidney M. Hecht

Chapter 3   The Discovery and Development of the Combretastatins.......................23
Kevin G. Pinney, Christopher Jelinek, Klaus Edvardsen, David J. Chaplin, and George R. Pettit

Chapter 4   Homoharringtonine and Related Compounds..47
Hideji Itokawa, Xihong Wang, and Kuo-Hsiung Lee

Chapter 5   Podophyllotoxins and Analogs...71
Kuo-Hsiung Lee and Zhiyan Xiao

Chapter 6   Taxol and Its Analogs..89
David G. I. Kingston

Chapter 7   The Vinca Alkaloids 123
Françoise Guéritte and Jacques Fahy

Chapter 8   The Bryostatins..137
David J. Newman

Chapter 9   The Isolation, Characterization, and Development of a Novel Class
                of Potent Antimitotic Macrocyclic Depsipeptides: The Cryptophycins..................151
Rima S. Al-awar and Chuan Shih

Chapter 10 Chemistry and Biology of the Discodermolides, Potent Mitotic Spindle
                Poisons...171
Sarath P. Gunasekera and Amy E. Wright

Chapter 11 The Dolastatins: Novel Antitumor Agents from *Dolabella auricularia* 191
Erik Flahive and Jayaram Srirangam

Chapter 12 Ecteinascidin 743 (ET-743; Yondelis™), Aplidin, and Kahalalide F 215

Rubén Henríquez, Glynn Faircloth, and Carmen Cuevas

Chapter 13 Discovery of E7389, a Fully Synthetic Macrocyclic Ketone Analog
of Halichondrin B...241

Melvin J. Yu, Yoshito Kishi, and Bruce A. Littlefield

Chapter 14 HTI-286, A Synthetic Analog of the Antimitotic Natural Product
Hemiasterlin..267

Raymond J. Andersen and Michel Roberge

Chapter 15 The Actinomycins...281

Anthony B. Mauger and Helmut Lackner

Chapter 16 Anthracyclines 299

Federico Maria Arcamone

Chapter 17 Ansamitocins (Maytansinoids)..321

Tin-Wein Yu and Heinz G. Floss

Chapter 18 Benzoquinone Ansamycins 339

Kenneth M. Snader

Chapter 19 Bleomycin Group Antitumor Agents 357

Sidney M. Hecht

Chapter 20 Biochemical and Biological Evaluation of (+)-CC-1065 Analogs and
Conjugates with Polyamides...383

Rohtash Kumar and J. William Lown

Chapter 21 Epothilone, a Myxobacterial Metabolite with Promising Antitumor Activity........413

Gerhard Höfle and Hans Reichenbach

Chapter 22 Enediynes..451

Philip R. Hamann, Janis Upeslacis, and Donald B. Borders

Chapter 23 The Mitomycins 475

William A. Remers

Chapter 24 Staurosporines and Structurally Related Indolocarbazoles as
Antitumor Agents..499

Michelle Prudhomme

Chapter 25 Combinatorial Biosynthesis of Anticancer Natural Products....................519

Michael G. Thomas, Kathryn A. Bixby, and Ben Shen

Chapter 26 Developments and Future Trends in Anticancer Natural Products
Drug Discovery..553

David J. Newman and Gordon M. Cragg

Index 573

1 Introduction

Gordon M. Cragg, David G. I. Kingston, and David J. Newman

CONTENTS

References 2

The search for new lead compounds is a crucial element of modern pharmaceutical research. Natural products provided the only source of pharmaceuticals for thousands of years, and natural products have made enormous contributions to human health through compounds such as quinine, morphine, aspirin (a natural product analog), digitoxin, and many others. The potential of using natural products as anticancer agents was recognized in the 1950s by the U.S. National Cancer Institute (NCI) under the leadership of the late Dr. Jonathan Hartwell, and the NCI has since made major contributions to the discovery of new naturally occurring anticancer agents through its contract and grant support, including an important program of plant and marine collections. Many, although not all, of the compound classes described in this text owe their origin in whole or in part to NCI support. In spite of the success of the natural-products approach to anticancer drug discovery, as exemplified by the following chapters, in recent years their importance as a source of molecular diversity for drug discovery research and development has been overshadowed by various newer approaches currently in favor. These approaches include chemical ones which make heavy use of combinatorial chemistry, and biological ones such as manipulation of biosynthetic pathways of microbial metabolites through combinational biosynthetic techniques. It is thus worthwhile to review briefly the major reasons why the natural products are so important.

First, there is a strong biological and ecological rationale for plants and marine invertebrates to produce novel bioactive secondary metabolites. The importance of plants and marine organisms as a source of novel compounds is probably related in large measure to the fact that they are not mobile and hence must defend themselves by deterring or killing predators, whether insects, fish, microorganisms, or animals. Plants and marine organisms have thus evolved a complex chemical defense system, which can involve the production of a large number of chemically diverse compounds. It has been proposed that all natural products have evolved to bind to specific receptors,[1] and evidence for the advantage that natural products give an organism over predators has been found in ecological studies in the marine environment related to predator deterrence.[2] As far as microbial species are concerned, secondary metabolites from the prokaryota and also from certain phyla in the eukaryota are actually synthesized in a combinatorial manner.[3] This is the case because the genes that are ultimately responsible for the chemical structures obtained from these microbes, plants, and marine invertebrates can frequently be "shuffled" between taxa, and these shuffled genes produce diverse secondary metabolites, some of which are of use to man.

Second, natural products have historically provided many major new drugs. Several recent reviews have provided data that document the importance of natural products as a source of bioactive compounds,[3–6] and the conclusions can be summarized by a recent review that states, "In retrospect, the use of natural products has been the single most successful strategy in the discovery of modern medicines."[7]

Third, natural products provide drugs that would be inaccessible by other routes. A large part of the reason for the importance of natural products in drug discovery is that natural products provide drugs that would be inaccessible by other routes. Thus, compounds such as paclitaxel (Taxol®) or halichondrin would never be prepared by standard "medicinal chemistry" approaches to drug discovery, even including the newer methods of combinatorial chemistry. Likewise, the new approach of combinatorial biosynthesis, although important, is unlikely in the near future to yield new compounds that have the complexity of bryostatin or camptothecin. The claim that synthetic compounds cannot match natural products for their structural diversity is borne out by a statistical study by Henkel and his collaborators, in which over 200,000 compounds from various databases were compared in various ways for their molecular diversity. The authors concluded "about 40% of the natural products are not represented by synthetic compounds" and that "the potential for new natural products is not exhausted and natural products still represent an important source for the lead-finding process."[8]

Fourth, natural products can provide templates for future drug design. In many cases, the isolated natural product may not be an effective drug for any of several possible reasons, but it may nevertheless have a novel pharmacophore. Certainly, natural products, with their complex three-dimensional structures, are well suited as ligands for such targets as protein–protein interactions and for the selectivity needed to differentiate between the various protein kinases, to cite just two examples. Once a novel natural product chemotype has been discovered, it can be developed by medicinal chemistry or combinatorial chemistry into improved agents, as described in several chapters in this book. Examples of this are the drugs etoposide and teniposide, derived from the lead compound podophyllotoxin (Chapter 5); numerous analogs derived from Taxol (Chapter 6); topotecan, derived from camptothecin (Chapter 2); and the synthetic clinical candidates E7389 and HTI-286, developed from the marine leads halichondrin B and hemiasterlin, respectively (Chapters 14 and 15).

In summary, the approach to drug discovery from natural sources has a historical justification (it has yielded many important new pharmaceuticals), a biochemical rationale (the position of marine organisms and plants in the ecosystem demands that they produce defense substances, many of which have a novel phenotype, and microbial species can produce new secondary metabolites by gene shuffling), and a chemical rationale (natural products provide templates for drug design).

This book covers the major classes of clinically used anticancer natural products. The chapters are grouped by the source organism (plant, marine, or microbial), although it is recognized that the distinction is not always clear-cut, and some so-called marine natural products, for example, may in fact be produced by symbiotic microbial species (see, e.g., the discussion of bryostatin in Chapter 8). Within each chapter, the coverage normally includes the history of the drug and a discussion of its mechanism in action, its medicinal chemistry, its synthesis, and its clinical applications. The book concludes with a chapter on the increasingly important biosynthetic approaches to "unnatural natural products" and with a final chapter looking ahead to future developments in anticancer natural products drug discovery. Because of the wealth of information available, the coverage in all of these chapters is of necessity selective rather than comprehensive, but the authors and editors have attempted to provide the most important recent results in each area.

The editors hope that this volume will not only serve as a convenient summary of the current status of research and development of some of the most effective anticancer agents available today, but will also serve as an inspiration and a challenge to a new generation of scientists to engage in developing new and even better drugs from Nature's bounty.

REFERENCES

1. Williams, D.H. et al., Why are secondary metabolites (natural products) biosynthesized? *J. Nat. Prod.* 52, 1189, 1989.

2. Paul, V.J. *Ecological Roles of Marine Natural Products*. Cornell University Press, Ithaca, NY, 1992.
3. Kingston, D.G.I. and Newman, D.J. Mother Nature's combinatorial libraries: Their influence on the synthesis of drugs. *Curr. Opin. Drug Disc. Dev.* 5, 304, 2002.
4. Shu, Y.-Z. Recent natural products based drug development: A pharmaceutical industry perspective. *J. Nat. Prod.* 61, 1053, 1998.
5. Newman, D.J., Cragg, G.M., and Snader, K.M. Natural products as sources of new drugs over the period 1981–2002. *J. Nat. Prod.* 66, 1022, 2003.
6. Newman, D.J., Cragg, G.M., and Snader, K.M. The influence of natural products upon drug discovery. *Nat. Prod. Rep.* 17, 215, 2000.
7. Tulp, M. and Bohlin, L. (2002) Functional versus chemical diversity: Is biodiversity important for drug discovery? *Trends Pharm. Sci.* 23, 225, 2002.
8. Henkel, T., et al., Statistical investigation into the structural complementarity of natural products and synthetic compounds. *Angew. Chem. Int. Ed.*, 38, 643, 1999.

2 Camptothecin and Its Analogs

Nicolas J. Rahier, Craig J. Thomas, and Sidney M. Hecht

CONTENTS

I. Introduction 5
II. Mechanism of Action..6
    A. Poisoning of Topoisomerase I 6
    B. Other Biochemical Effects of CPT 8
III. Synthetic Studies...9
    A. Synthesis of Racemic CPT...9
    B. Asymmetric Synthesis 9
    C. Semisynthetic Methods...11
IV. Medicinal Chemistry...12
    A. A/B/C/D Ring Analogs...12
    B. The E-Ring Lactone 13
    C. Water Soluble/Insoluble Analogs 14
    D. Conjugates..14
V. Development of Clinically Useful Camptothecin Analogs 15
    A. Topotecan...15
    B. Irinotecan 15
    C. Analogs Currently under Evaluation..16
VI. Conclusions 17
VII. Acknowledgments 17
References 17

I. INTRODUCTION

The antineoplastic agent camptothecin (CPT) (**1**) (Figure 2.1) was isolated from extracts of *Camptotheca acuminata* by Wani and Wall.[1] Preclinical studies revealed that CPT had remarkable activity against L1210 leukemia. The marginal water solubility of CPT encouraged researchers to initiate clinical investigation of the drug with the water-soluble sodium salt (**2**). Unfortunately, poor tumor suppression and numerous side effects were associated with treatment by **2**; consequently, the trials were suspended.[2] Interest in CPT was reestablished with the discovery that topoisomerase I (topo I) was the principal cellular target of the drug.[3] Since this discovery, CPT has been the focus of numerous studies; to date, two analogs have been approved for clinical use — the semisynthetic, water-soluble analogs topotecan (Hycamtin) (**3**) and irinotecan (Camptosar) (**4**) (Figure 2.1). This chapter summarizes studies of the mechanism of action of CPT, as well as several of the synthetic studies, and provides an account of the current structure-activity studies. Also discussed briefly is the clinical development of topotecan, irinotecan, and several analogs that are currently in various stages of clinical trials.

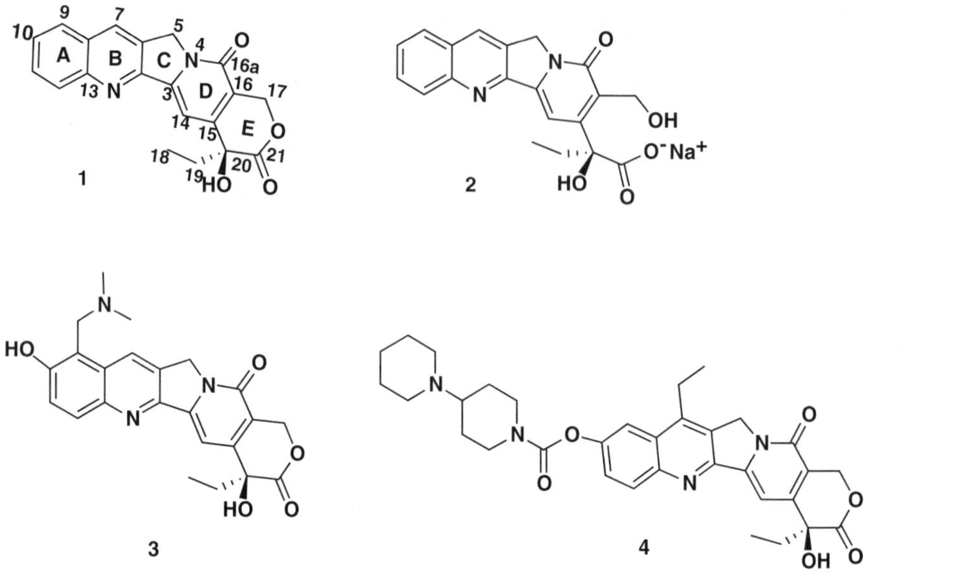

FIGURE 2.1 Structures of camptothecin (**1**), the water-soluble sodium salt (**2**), topotecan (**3**), and irinotecan (**4**).

II. MECHANISM OF ACTION

A. POISONING OF TOPOISOMERASE I

In 1985, it was reported that CPT stabilized the DNA–topoisomerase I covalent binary complex.[3] It had previously been shown that CPT is capable of inhibiting DNA synthesis, thereby causing cell death during the S-phase of the cell cycle.[4] The S-phase specific cytotoxicity is directly correlated to the occurrence of irreversible DNA cleavage when the replication fork encounters the covalent DNA–enzyme binary complex (Figure 2.2).[5] It has been demonstrated that deletion of the gene for topo I from *Saccharomyces cerevisiae* results in viable cells that are fully resistant to CPT.[6] Further, a number of CPT-resistant cell lines have been identified, each containing mutations within topo I.[7] These studies clearly support a mechanism of action for CPT involving topo I–mediated DNA cleavage.

The topoisomerases (type I and type II) are a class of enzymes that mediate the relaxation of chromosomal DNA prior to DNA replication and transcription.[8] Mechanistically, topo II effects this relaxation via transient double-strand cleavage, DNA strand passage, and religation of the phosphodiester backbone. This process requires ATP and alters the DNA linking number by multiples of two. The topo I mechanism involves energy-independent single-strand DNA cleavage, followed by strand passage and religation.[9] The mechanism of topo I–mediated DNA relaxation is known to involve an active site tyrosine that cleaves DNA by nucleophilic attack of the active site tyrosine phenolic OH group on the phosphodiester backbone (Figure 2.3). The resulting DNA–topo I intermediate is a covalent binary complex. The cleaved DNA strand can be passaged around the unbroken DNA strand; the intact duplex is reformed on religation of the phosphodiester bond, with the concomitant release of topo I.

Although topo I may be capable of DNA cleavage at a number of sites, it exhibits a strong preference for the nucleoside thymidine as the nucleobase directly upstream (the -I position) (cf. Figure 2.3). Stabilization of the covalent binary complex by CPT has been noted to involve an additional preference for guanosine at the +1 position.

Most of the agents that poison type I and type II topoisomerases are characterized by their ability to inhibit the religation step during DNA relaxation (Figure 2.3). Topo II poisons include a

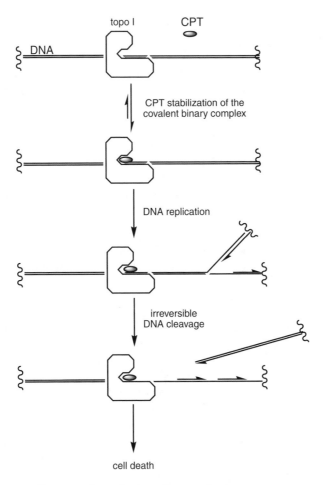

FIGURE 2.2 S-phase specific cytotoxic mechanism of camptothecin.

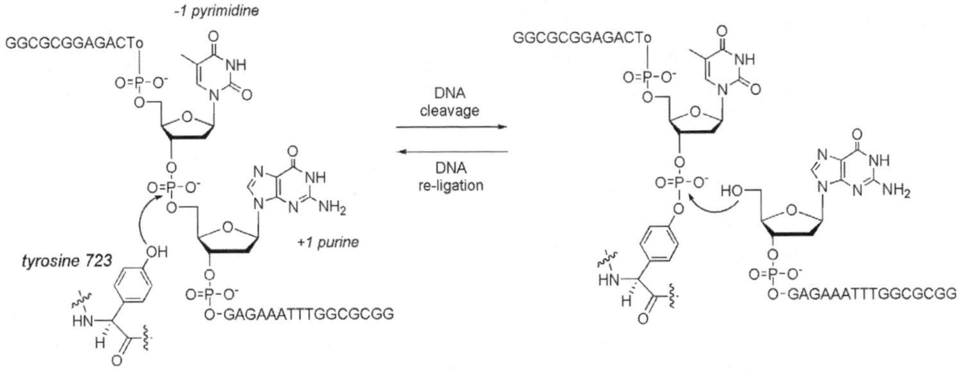

FIGURE 2.3 Mechanism of DNA relaxation by human topoisomerase I.

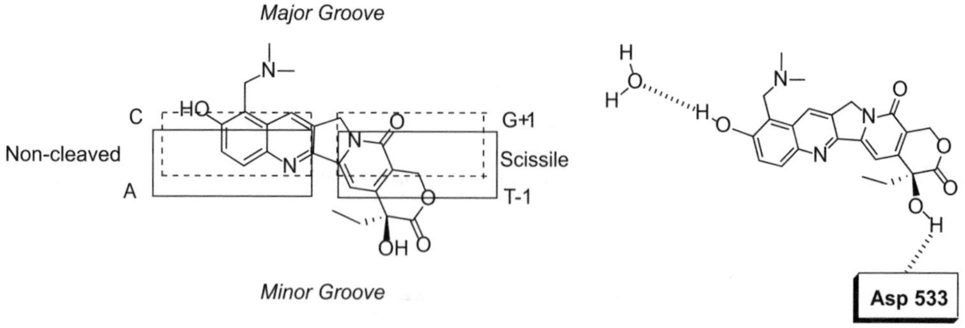

FIGURE 2.4 Two-dimensional representation of the x-ray crystal structure of topotecan within the covalent binary complex.

number of well-characterized clinical agents such as amsacrine and etopside.[10] Several inhibitors of topo I have also been identified; however, CPT remains the most widely studied of this class of medicinal agents.

The development of a precise understanding of the way in which CPT stabilizes the DNA–topo I covalent binary complex, and thereby inhibits the religation of duplex DNA, is an important current goal. CPT has no binding affinity for topo I, and only the positively charged CPT analog topotecan (**3**) has shown any DNA binding.[11,12] However, the ability of CPT to bind to the covalent binary complex formed between topo I and DNA is sufficient to inhibit religation. There is compelling evidence that CPT is capable of interacting with the covalent binary complex at or near the interface between DNA and topo I. Hertzberg and coworkers have established that a CPT analog containing a bromoacetamide at carbon 10 was, on prolonged exposure, capable of forming a drug–enzyme crosslink.[13] Pommier and coworkers later showed that a 7-chloromethylated CPT analog alkylated N3 of the guanine at the +1 cleavage site of DNA.[14]

A number of computational models have been formulated depicting the interaction between CPT and the covalent binary complex.[15–17] These models posit different energy-minimized interactions between CPT, the DNA 5-TpG-3 base pair, and selected topo I amino acid residues known to be important based on biochemical studies or their proximity to the putative CPT binding site.[18–21] The analysis of several CPT-resistant cell lines has provided important information regarding the amino acid residues that play a role in CPT binding.[7] For example, mutational analysis of Asp 533, Arg364, Asn722, and Lys532 has revealed that each of these amino acid residues likely play a role in CPT binding.[19–21]

Staker et al. have recently reported the X-ray crystal structure of a ternary complex formed between DNA, topo I, and topotecan (**3**).[22] The reported complex used a DNA oligonucleotide containing a 5-bridging phosphothioate to facilitate crystal formation and indicated that topotecan bound the covalent binary complex in an intercalative fashion (Figure 2.4). The only direct drug–enzyme hydrogen bond interaction was between Asp533 and the 20-hydroxyl group of topotecan. In addition, one hydrogen bond was observed with a water molecule. Carbons 7, 9, and 10 of CPT were positioned in a manner that situated them in the vicinity of the major groove.

B. OTHER BIOCHEMICAL EFFECTS OF CPT

It is generally accepted that the basis for CPT-induced cytotoxicity is contingent on CPT acting as a topo I poison (as opposed to an inhibitor of enzyme activity, per se).[3–7] However, the antitumor selectivity of CPT is somewhat surprising given that topo I is an enzyme found in all cell types. Elevated levels of topo I are present in tumors of the colon, ovary, and prostate, which may explain the therapeutic index of CPT.[23] Deficiencies in DNA repair capabilities in some cancer cells may provide another possible basis for cancer cell selectivity. The selective inhibition of all dividing cell populations represents another possible source of antitumor selectivity. Further, other effects of CPT exposure have been noted and merit discussion.

Kauh and Bjornsti, using a genetic screen, have identified six dominant suppressors of camptothecin toxicity at a single genetic locus (*SCT1*).[24] Mutant *SCT1* cells were shown to express wild-type topo I, indicating additional factors in the overall cellular response to CPT. One report indicated that irinotecan, but not topotecan, inhibits acetylcholinesterase activity.[25] Additional reports indicate that CPT activates the transcription factor NFκB, which has been implicated in numerous activities *in vivo*.[26] Importantly, activation of NFκB has been implicated in the overproduction of interferons, triggering several cellular responses.[26–29]

In addition to its role in relaxation of supercoiled DNA, topo I is able to regulate transcription,[30] recognize and cleave mismatched nucleotides at intrinsic cleavage sites,[31] and associate with numerous proteins *in vivo*. Tazi and coworkers have reported that topo I is capable of influencing gene splicing by acting as a phosphorylating enzyme for SR proteins.[32] This kinase activity is inhibited by CPT despite the fact that it has been shown to be unconnected to its DNA relaxation activity through mutational studies.[33] Analysis of a family of topoisomerase I related function proteins (*TRFp*) demonstrate that one member of this family, *TRF4p*, plays a critical role in mitotic chromosome condensation during the S phase of the cell cycle.[34] *TRF4p* is associated with the DNA binding protein *Smc1p* during chromosomal condensation, and reports detailing mutations to *TRF4p* have produced cell lines with unexpected hypersensitivity to CPT.[35]

III. SYNTHETIC STUDIES

Given the finite supplies of natural CPT and the need to develop additional analogs, much effort has focused on practical synthetic routes to CPT and its analogs. The following are descriptions of several of the synthetic approaches to CPT and of the semisynthesis of topotecan (**3**) and irinotecan (**4**).

A. SYNTHESIS OF RACEMIC CPT

Following the initial publication of the structure of CPT, a number of synthetic strategies for the preparation of CPT were reported. The first total synthesis of (*R,S*)-CPT was described by Stork and Schultz in 1971.[36] Numerous successful synthetic approaches to 20(*R,S*)-CPT have since been published and reviewed.[37,38]

Given the early availability of methods for racemic synthesis, relative to asymmetric approaches, research also focused on the chiral resolution of the racemates or key synthetic intermediates. In 1975, Corey and coworkers were the first to report the successful resolution of a chiral intermediate, leading to the preparation of 20(*S*)-CPT.[39] Wani et al. and Teresawa et al. have also reported the successful resolution of the intermediates in the synthesis of CPT, using (*R*)-(+)-α-methyl-benzylamine.[40,41]

B. ASYMMETRIC SYNTHESIS

Ejima et al. reported the first stereocontrolled synthesis of CPT via a diastereoselective ethylation process using *N*-tosyl-(*R*)-proline (**6**) (Scheme 2.1).[42] Indolizine **5**, the CD-ring precursor of the parent alkaloid, was employed in an approach to construction of ring E. Bromination of **5** followed by treatment with *N*-tosyl-(*R*)-proline (**6**) in the presence of base afforded compound **7**. A diastereomeric mixture of (*S,R*)-**8** and (*R,R*)-**9** was quantitatively prepared by facial differentiated ethylation of **7** in 82% de; selective recrystallization provided pure **8** in 56% yield. Following Raney Ni catalyzed reduction of **8** and subsequent treatment with NaNO$_2$, the optically pure ester **10** was obtained in 74% yield. The triester **10** was hydrolyzed using LiOH and lactonized to provide hydroxy lactone **11** having the proper 20(*S*)-configuration in 90% yield. Hydrolysis of the ketal

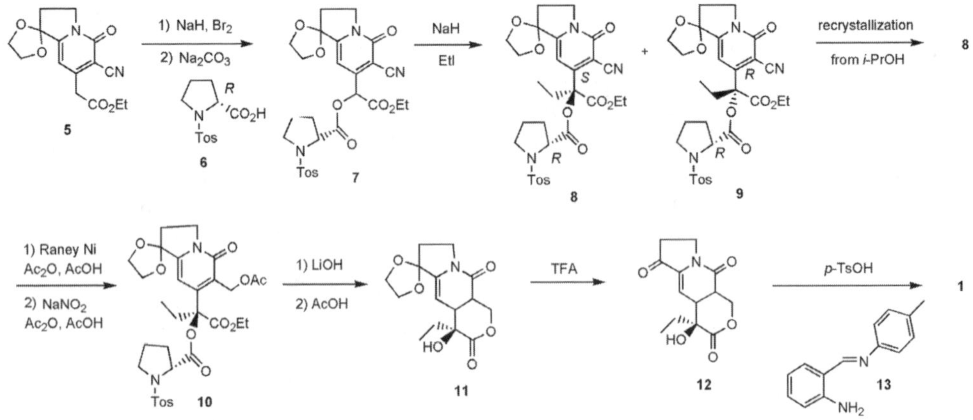

SCHEME 2.1

functionality gave optically pure key intermediate **12**, which was converted to 20(*S*)-CPT by Friedländer reaction with 2-aminobenzaldehyde derivative **13** in 84% yield.

Comins and coworkers have successfully employed chiral auxiliaries to establish the correct stereochemistry for the 20(*S*)-hydroxyl group (Scheme 2.2).[43] Refinement of their method has culminated in the asymmetric synthesis of 20(*S*)-CPT in only six steps, using the α-ketobutyric ester derived from (-)-trans-2-(α-cumyl)cyclohexanol [(-)-TCC] (**17**) as chiral auxiliary (Scheme 2.2).[44] Treatment of commercially available 2-methoxypyridine (**14**) with mesityllithium, followed by addition of *N*-formyl-*N*,*N*′,*N*′-trimethylethylenediamine, effected the alkylation of the aromatic ring at C3. Addition of *n*-BuLi followed by iodine and workup with aqueous NaBH$_4$/CeCl$_3$ provided alcohol **15** in 46% yield via a one-pot process. Conversion of **15** directly to 1,3-dioxane **16** using NaI/TMSCl/paraformaldehyde was accomplished in 87% yield. The DE ring precursor **18** was fashioned via another one-pot process involving lithium–halogen exchange effected with *n*-BuLi, followed by addition of chiral auxiliary **17**. Addition of HCl effected protonation, acetal hydrolysis, and lactonization to give intermediate **18** in 60% yield (93% ee). Coupling of **18** with the quinoline intermediate **19** was accomplished via displacement of the primary iodide to provide enantiopure **20** in 81% yield. The C-ring was closed using a Heck reaction through treatment of **20** with Pd(II) and potassium acetate to provide 20(*S*)-CPT in 64% yield.

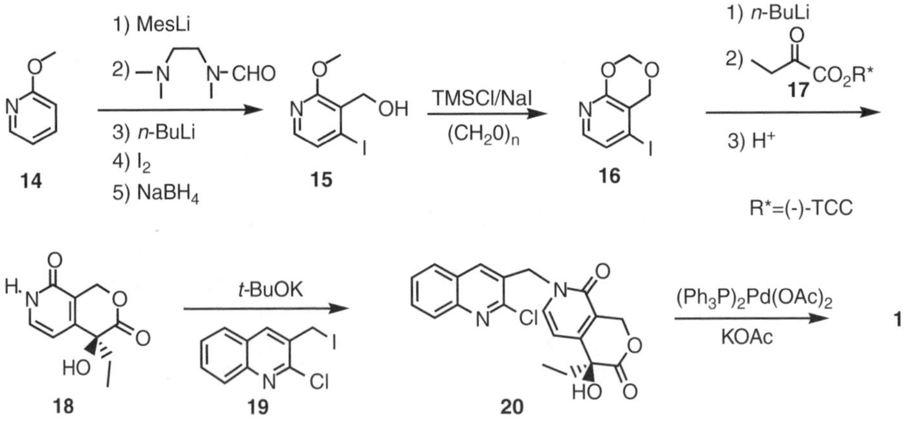

SCHEME 2.2

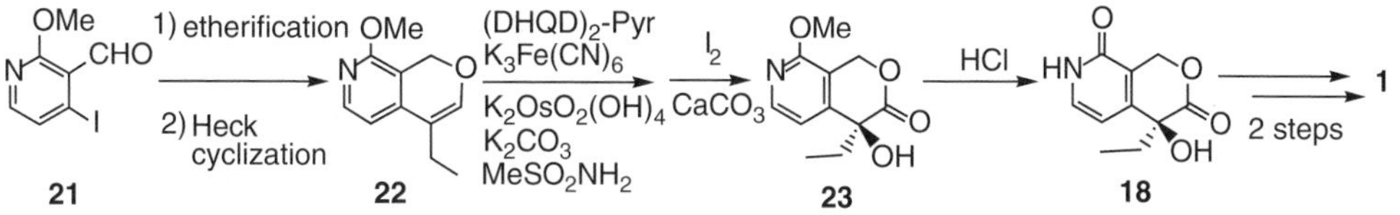

SCHEME 2.3

Two other research groups have employed chiral auxiliaries to establish the *S*-configuration at the C20 position of CPT. Tagami et al. used a Davis reagent, (2*R*, 8a*S*)-(+)-(camphorylsulfonyl)oxaziridine), to asymmetrically hydroxylate 20-deoxycamptothecin.[45] In 2002, Bennasar et al. made use of (2*R*,5*R*)-2-*tert*-butyl-5-ethyl-1,3-dioxolan-4-one to establish C20 asymmetry and synthesize 20(*S*)-CPT.[46]

To control absolute stereochemistry at C20, Ciufolini and Roschangar reported a synthesis of 20(*S*)-CPT that made use of an aldehyde intermediate obtained by an enzymatic desymmetrization of a corresponding malonate.[47] In 1998, Imura and coworkers described the first asymmetric synthesis of a key chiral intermediate, using enzyme-catalyzed resolution.[48]

Fang et al. applied the first chiral catalytic method to prepare 20(*S*)-CPT (Scheme 2.3).[49] Reductive etherification of aldehyde **21** followed by intramolecular Heck reaction gave the cyclic enol ether **22** in 45% yield (Scheme 2.3). Sharpless asymmetric dihydroxylation of **22** using (DHQD)$_2$-Pyr as the chiral catalyst followed by oxidation with iodine and CaCO$_3$ was performed to synthesize the DE ring precursor **23** with 94% ee (90% yield). Treatment of **23** with HCl provided the enatiomerically pure pyridone **18** in 74% yield. The authors completed the synthesis of 20(*S*)-CPT using the Comins route (cf Scheme 2.2).[43,44]

Jew et al. made further use of the catalytic asymmetric Sharpless dihydroxylation using (DHQD)$_2$-Pyr as chiral catalyst, resulting in stereocontrolled oxidation of carbon 20 in greater than 90% ee in the total synthesis of 20(*S*)-CPT.[50] In 2002, Blagg and Boger established the configuration of the C20 (*S*) tertiary alcohol through a Sharpless asymmetric dihydroxylation reaction, using a 3,4,5-trimethoxyphenyl-derived DHQ dimer ligand in 86% ee.[51]

Curran and coworkers have reported a synthesis of 20(*S*)-CPT based on a 4+1 radical cascade annulation (Scheme 2.4).[52] Lactone **24** was obtained in a fashion similar to the synthesis developed by Fang et al.[49] Exchange of the TMS group in **24** for iodine, followed by demethylation, provided **25** in 33% yield (Scheme 2.4). *N*-Propargylation of lactone **25** provided alkyne **26** in 88% yield followed by isonitrile treatment under irradiation to provide 20(*S*)-CPT in 63% yield. Curran and coworkers have subsequently reported improvements in the synthesis of the enantiopure DE ring precursor **25**, using a samarium catalyst.[53] Radical methods of this type have been shown to tolerate A- and B-ring substituents and have, therefore, been used for the synthesis of numerous CPT analogs.[52]

C. SEMISYNTHETIC METHODS

In addition to these total syntheses, many synthetic efforts have been focused on producing analogs of CPT. The majority of these efforts have involved the semisynthetic manipulation of CPT. The approaches to irinotecan and topotecan are described.

Sawada et al. first reported the synthesis of irinotecan (**4**) in 1991 (Scheme 2.5).[54] Hydrogen peroxide was added to a solution of 20(*S*)-CPT in aqueous sulfuric acid in the presence of ferrous

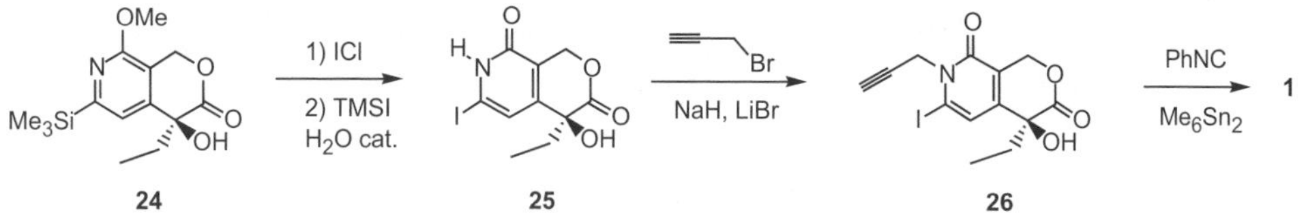

SCHEME 2.4

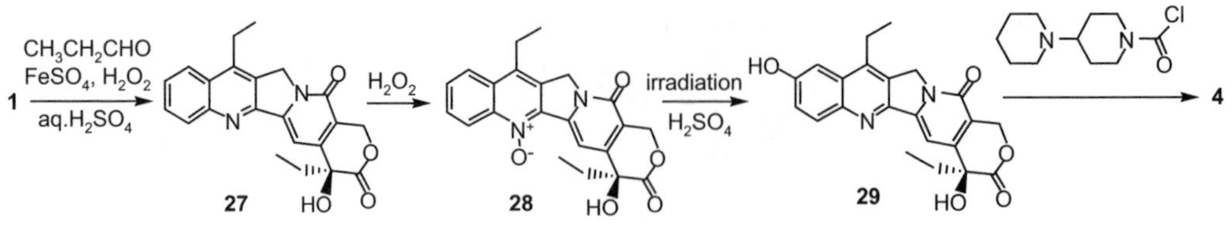

SCHEME 2.5

sulfate and propionaldehyde to afford 7-ethyl CPT (**27**) in 77% yield (Scheme 2.5). 7-Ethyl CPT (**27**) was converted to the corresponding N-oxide **28** using hydrogen peroxide in acetic acid. Irradiation of **28** in acidic media furnished the active metabolite of irinotecan (SN-38) (**29**) in 49% yield. Treatment of SN-38 (**29**) with 4-(1-piperidino)-1-(piperidino)-chloroformate then provided irinotecan (**4**) in 80% yield.

The second currently marketed CPT derivative, topotecan, was synthesized in 1991 by Kingsbury et al. in two steps starting from 20(S)-CPT (Scheme 2.6).[55] Conversion of 20(S)-CPT to 10-hydroxy CPT (**30**) was accomplished through a reduction-oxidation sequence in 71% yield (Scheme 2.6). Treatment of **30** with dimethylamine in aqueous formaldehyde and acetic acid provided topotecan (**3**) in 62% yield.

IV. MEDICINAL CHEMISTRY

Synthetic and semisynthetic analogs of CPT have provided the means to study the relationship between CPT structure and function. A clear understanding of the mechanism by which CPT inhibits DNA religation and insight into the exact roles of the various structural elements of CPT are essential to preparing novel, therapeutically important derivatives. At present, much is known regarding the SAR profile of CPT, and the following text is a summary description of innovative and important derivatives.[38] This discussion focuses on the significance of the individual rings within the pentacyclic ring system, as well as the lone stereocenter. Further, we will consider analogs with substitutions, additions, and deletions within each of the five rings of CPT and specific derivatives that increase water solubility and, conversely, lipophilicity; and, finally, we will present an account of several CPT conjugates.

A. A/B/C/D Ring Analogs

Alterations to the aromatic core of CPT are normally confined to substitutions on the A/B quinoline ring system. There is general agreement that the planarity of the four ring aromatic core is a requirement for topo I inhibition, as illustrated by studies that alter both the quinoline ring and the C/D ring systems.[56-58] Additions to the C/D ring system are restricted by the lack of accessible carbons (limited to C5 and C14), and analogs with additions at those sites have generally displayed diminished activity. Exceptions include the C5 substituted 2-fluoroethyl ester analog and 5-ethylidene CPT, which displayed various degrees of *in vitro* and *in vivo* activity.[59,60]

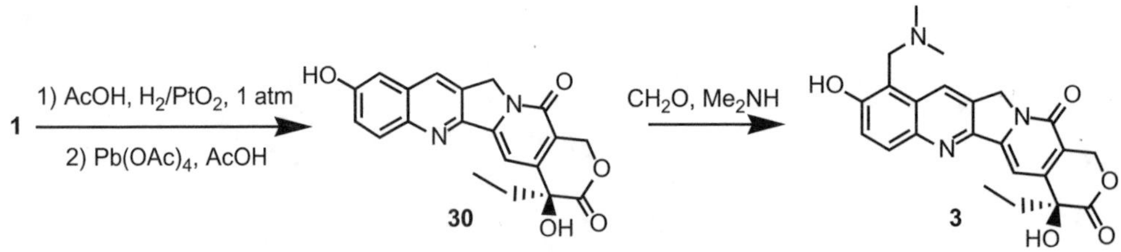

SCHEME 2.6

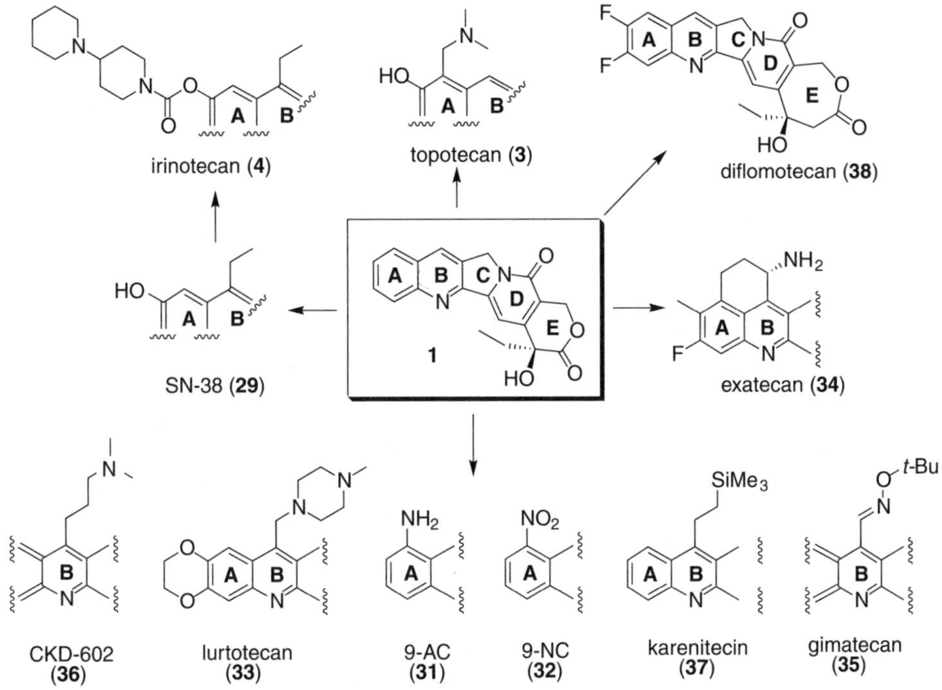

FIGURE 2.5 Structure of selected camptothecin analogues.

The vast majority of studies involving CPT modification have focused on additions to the quinoline ring system.[38,61-63] Studies by Wani, Wall, and coworkers described a number of CPT analogs that established several broadly applicable principles regarding CPT structure and function.[64-67] Among these principles were that substitutions at carbons 11 and 12 were generally unfavorable, albeit with some exceptions, whereas substitutions at positions 7, 9, and 10 were generally accepted without significant loss of function. In addition, a study by Kingsbury and coworkers illustrated the accuracy of these observations.[55] Included within these studies were the first descriptions of the clinically significant derivatives 9-aminoCPT (**31**), 9-nitroCPT (**32**) and topotecan (**3**) (Figure 2.5).

B. THE E-RING LACTONE

Although additions to the quinoline ring have been useful for preparing functional CPT analogs, alterations of the E-ring lactone have proven to be less successful. The two functional groups of the E-ring, that is, the 20(S)-hydroxyl group and the lactone, are both highly sensitive to alterations. The 20(R) enantiomer, which is accessible through synthetic methods, is inactive as both a topo I poison and an antitumor agent.[68,69] The inactivity of the 20(R)-CPT argues for a remarkably specific interaction between the 20-hydroxyl group and the topo I–DNA covalent binary complex. Replacement of the 20(R) hydroxyl group with an amine results in a CPT analog with largely diminished activity, whereas 20-halogeno derivatives maintained substantial potency.[69] The deoxy species is capable of inhibiting topo I–induced rearrangement of DNA but is inactive as a topo I poison and produces no cytotoxicity.[69] A notable difference between these 20-substituents involves their ability to participate in hydrogen-bonding interactions. In the aggregate, these studies suggest that the role of the 20-hydroxyl group may be that of a specific hydrogen bond acceptor, rather than a donor. Alternatively, a 20-substituent may be required to maintain a conformation of ring E conducive to ternary complex stabilization.

The E-ring lactone is also important for maintaining the potency of CPT as a topo I poison and cytotoxic agent. The apparent inactivity of the carboxylate form of CPT (**2**) led researchers to consider the more hydrolytically stable lactam derivative.[70] Surprisingly, although the CPT-lactam did, in fact, enhance the stability of the E-ring in water-based assays, it was ineffectual as a topo I poison. Likewise, the carbinol lactam, thiolactone, and the imide derivatives were also essentially inactive.[70]

Unsurprisingly, the α-hydroxylactone ring exists in an equilibrium that favors the inactive carboxylate form at physiological pH.[70] Further, there is a strong preference of human serum albumin (HSA) to bind the carboxylate form of CPT, which could plausibly constitute an obstacle in delivering the active form of the drug.[71] Although the lactam derivative was unsuccessful at overcoming this difficulty, the replacement of the α-hydroxylactone ring with a β-hydroxylactone ring resulted in a class of CPT analogs, designated homo-CPT, with increased plasma stability and good inhibitory activity.[72]

In spite of the obvious importance of the E-ring lactone to the topoisomerase I inhibitory activity, Cagir et al. have recently reported that the naturally occurring pyrroloquinazolinoquinoline alkaloid luotonin A functions as a topoisomerase I poison in much the same fashion as CPT.[73] Luotonin A is identical with CPT in rings A–C and differs mainly in the E-ring which is an unsubstituted benzene ring in luotonin A.[73]

C. WATER SOLUBLE/INSOLUBLE ANALOGS

The success of topotecan and irinotecan in the clinic has prompted the evaluation of other water-soluble analogs of CPT. Numerous examples exist, and studies highlight both the biochemical activity of these derivatives and the increased stability of the E-ring lactone.[38] Among the most potent of these analogs is 10,11-(methylenedioxy)-7-((N-methylpiperazino)methyl)CPT (lurtotecan) (**33**) (Figure 2.5), which is reported to be a more effective cytotoxic agent than topotecan.[74] Exactecan (**34**) (Figure 2.5) is another highly water soluble analog that has received significant attention. The potency of exatecan has been reported to be up to 28-fold greater than that of topotecan toward various human malignant cells.[75]

The required integrity of the E-ring lactone in comparison with the preference of HSA to bind the carboxylate form of CPT has also prompted researchers to examine analogs with increased lipophilicity. Burke and coworkers, using silicon-containing derivatives of CPT, have prepared a number of highly active analogs, designated silatecans.[76] Similarly, Zunino and coworkers have used aminomethyl, iminomethyl, and oxyiminomethyl substitutions as an alternative method to increase the lipophililic nature of CPT.[77] Two of the most successful of these analogs are 7-(*tert*-butyldimethylsilyl)-10-hydroxyCPT (DB-67) and 7-(*tert*-butoxyiminomethyl)CPT (Gimatecan)(**35**) (Figure 2.5), which have both shown high levels of cytotoxicity.[77,78] Curran, Burke, and coworkers have also demonstrated the success of a silatecan/homoCPT derivative (Du1441) that has demonstrated the highest level of lactone stability yet reported.[79] The success of these water-insoluble derivatives, in conjunction with liposomal core loading of the drug, constitutes a powerful tool for the delivery of CPT in the E-ring lactone form.

D. CONJUGATES

The development of CPT conjugates has been successful in affording CPT analogs having novel properties.[38] Two strategies have been employed for the preparation of these conjugates. The first uses the 20(S)-hydroxyl group as a conjugation point, and the second relies on semisynthetic functional groups appended to the quinoline ring (e.g. amino, hydroxyl, and carboxylic acid groups). Several studies have produced derivatives of potential therapeutic importance.

Firestone and coworkers have reported the elaboration of an immunoconjugate derived from CPT and the tumor-specific antibody BR96.[80] The conjugation was achieved using a carbamate linkage that can be cleaved by cathespin B. The conjugate was designed to eliminate dose-limiting side effects, and preliminary *in vitro* activity was reported to be superior to that of CPT alone.

The CPT-induced sequence-specific cleavage of DNA is limited to TpG sequences and, accordingly, lacks the wherewithal to be used in an antisense-type approach of targeting specific genetic sequences. Conjugates involving triplex-forming oligonucleotides (TFOs) covalently attached to campothecin through amide linkages with various aliphatic spacers have extended the sequence specificity of CPT while maintaining the potency of the drug.[81] Wang and Dervan have extended the generality of this strategy by using the minor groove binding hairpin polyamide within CPT conjugates.[82]

Greenwald and coworkers have studied the use of CPT-PEG prodrugs as a mechanism for delivery of CPT solid tumors.[83] Specifically, the preparation of glycine- and alanine-based linkers in PEG-β-CPT was demonstrated as a successful strategy at the levels of biodistribution and *in vivo* potency.[83] Recently, Greenwald et al. have reported the preparation and high *in vivo* activity of two PEG-CPT derivatives conjugated through the open carboxylate form of CPT.[84]

V. DEVELOPMENT OF CLINICALLY USEFUL CAMPTOTHECIN ANALOGS

CPT is not used as an anticancer agent because of its poor water solubility and numerous toxicities. Two CPT derivatives, topotecan (**3**) and irinotecan (**4**), are used in several countries for the treatment of solid tumors. Further, there are a number of new CPT analogs in various stages of clinical trials as well as expanding clinical development of CPT analogs involving combination regimens. The clinical applications of the CPT analogs have been discussed in several excellent reviews, and a brief summary is detailed here.[85,86]

A. TOPOTECAN

Topotecan (**3**) (Hycamtin, GlaxoSmithKline) is a semisynthetic derivative of CPT with a basic *N*,*N*-dimethylaminomethyl functional group at C9 that confers improved water solubility to the molecule. Topotecan was approved by the U.S. Food and Drug Administration in 1996 and is presently used as second-line therapy for advanced ovarian cancer in patients, following unsuccessful treatment with platinum-based therapeutics or paclitaxel-containing chemotherapy regimens. It is also approved as a therapeutic option for recurrent small-cell lung cancer.[85] Topotecan is most commonly administered as an intravenous infusion. The biological half-life of topotecan in humans is much shorter than that of other CPT derivatives, and as a consequence, drug accumulation does not occur. The fraction of topotecan bound to plasma proteins is also much lower than that of other CPT derivatives.

Neutropenia has proven to be the most encountered dose-limiting toxicity; however, a degree of thrombocytopenia is often associated with topotecan treatment as well. In addition to its action on ovarian and small-cell lung cancers, topotecan has shown activity against hematological malignancies. Topotecan combination regimens with paclitaxel, etoposide, cisplatin, cytarabine, and cyclophosphamide, as well as with other treatment modalities such as radiation therapy, are in development.[86]

B. IRINOTECAN

Irinotecan (**4**) (Camptosar, Pfizer) is a water-soluble prodrug designed to facilitate parenteral administration of the potent 7-ethyl-10-hydroxycamptothecin analog SN 38 (**29**). It contains a dibasic bispiperidine moiety, linked through a carbonyl group to a 10-hydroxy functionality. Enzymatic cleavage of the bispiperidine functionality by carboxylesterases predominantly located in the liver affords SN-38, the biologically active compound, which is 1000-fold more potent as an inhibitor of purified topoisomerase I *in vitro* than irinotecan. Irinotecan was approved in 2000 by the U.S. Food and Drug Administration for use in the treatment of advanced colorectal cancer, both as first-line therapy in combination with 5-fluorouracil and as salvage treatment in 5-fluorouracil refractory disease. Irinotecan is most commonly administered as an intravenous infusion.[85] The biological half-life of the lactone form of SN-38 exceeds that of topotecan, which represents a potential pharmacological advantage. Further, in comparison with other CPT analogs, a relatively large percentage of the intact lactone form of both irinotecan and SN-38 persists in the plasma of patients after drug administration, attributable to the preferential binding of irinotecan to serum albumin in its lactone form.

The principal dose-limiting toxicity of irinotecan is delayed diarrhea, with or without neutropenia. Promising antitumor activity has also been observed against small-cell and non-small-cell lung cancer, ovarian cancer, and cervical cancer and in recent clinical trials involving patients with malignant gliomas. Studies to evaluate additional irinotecan drug combinations with taxanes, anthracyclines, vinca alkaloids, or alkylating agents are in progress.

C. ANALOGS CURRENTLY UNDER EVALUATION

Exatecan mesylate dihydrate (DX-8951f) (**34**) (Daiichi Pharmaceuticals) is a synthetic hexacyclic water-soluble analog of CPT that inhibits the growth of human tumor cell lines *in vitro* and that of tumor xenografts *in vivo*, including tumors resistant to topotecan and irinotecan.[85] Reports of phase II clinical trials indicate that exatecan may have significant activity against small-cell and non-small-cell lung cancer, hepatocellular cancer, colorectal cancer, and pancreatic carcinoma.[85-88]

CKD-602 (**36**) (Chong Kun Dang Pharmaceuticals) is a potent topo I inhibitor that overcomes the poor aqueous solubility and toxicity profile of CPT.[87] This compound shows enhanced activity against a series of cell lines *in vitro* and against L1210 *in vivo*.[88] Furthermore, partial responses were observed in patients with stomach and ovarian cancer. CKD-602 is presently in phase II clinical trials.[87]

NX211 (NeXstar Pharmaceuticals) is liposomal formulation of lurtotecan (**33**) that prolonged the systemic duration of lurtotecan and increased its therapeutic index. In phase I clinical trials, tumor responses were observed in patients with ovarian cancer.[85-88]

9-AC (IDEC-132) (**31**) (National Cancer Institute) is a semisynthetic CPT analog with potent antitumor activity against a wide spectrum of human tumor xenograft models.[85] Phase II studies using intravenous infusion have been conducted in patients with various type of malignancies with disappointing results. A solid oral dosage form was developed by incorporating 9-AC into poly(ethylene)glycol-1000, which yielded bioavailabilities of 49% in cancer patients. Thus, the oral bioavailability of 9-AC in cancer patients is far superior to that observed with any other CPT derivative to date. Phase II studies using this formulation have been initiated in Europe.[85]

9-NC (Rubitecan) (**32**) (SuperGen) is a potent but poorly soluble CPT analog developed exclusively for oral administration.[85] This compound is metabolically converted *in vivo* into 9-AC. Rubitecan is more robust and relatively inexpensive to prepare compared to 9-AC.[86] It is in phase III clinical trials in pancreatic cancer as well as phase II clinical testing for the treatment of 11 additional tumor types.[87] An aerosol delivery of liposomal 9-NC has been employed in phase I clinical studies.[86]

Karenitecin (BNP-1350) (**37**) (Bionumerik Pharmaceuticals) is a semisynthetic, lipophilic compound that exhibits more potent cytotoxic activity than CPT both *in vitro* and *in vivo*, with enhanced oral bioavailability. The *in vitro* assays for antiproliferative capacity in colon cancer cell lines indicated that Karenitecin was similar in effectiveness to SN-38. Karenitecin has completed phase I clinical trials for pancreatic and colorectal cancer, and it is undergoing phase II clinical trials.[87]

Gimatecan (**35**) (ST1481) (Sigma-Tau and Italian National Cancer Institute) is a semisynthetic, lipophilic 7-oxyiminomethylCPT. Gimatecan showed cytotoxicity against a panel of human tumor xenografts and is undergoing phase I clinical trials.[88]

Diflomotecan (BN-80915) (**38**) (Beaufour Ipsen) is a homoCPT analog that demonstrates enhanced E-ring stability. Diflomotecan exhibited high antiproliferative activity in a panel of tumor cell lines, including those exhibiting multidrug resistance, and was found to be active at very low doses in a variety of human tumor xenografts. BN-80915 is currently undergoing phase I clinical evaluation in Europe.[86–89]

In addition, efforts to optimize delivery strategies for CPT derivatives such as polymer conjugation, nanoparticule encapsulation, and liposomal core loading are under way.[90]

VI. CONCLUSIONS

Camptothecin and its analogs represent an exciting class of antineoplastic agents targeted at numerous types of cancers. At present there exists considerable interest in CPT, and studies continue to offer meaningful insights into the drug nearly 40 years after Wani and Wall first reported its pentacyclic structure. Today the biochemical basis of CPT activity is better defined, allowing scientists to focus their investigations more effectively. Recent advances in the synthesis of CPT and its analogs are such that the construction of new analogs is achievable on a scale not possible even a decade ago. Development of novel analogs that optimize and exploit important structural features and further expand the therapeutic potential of CPT is being reported.

VII. ACKNOWLEDGMENTS

We thank Dr. Cyrille Gineste, University of Virginia, for helpful discussions during the preparation of this review. Work in our laboratory has been supported by National Institutes of Health research grant CA78415, awarded by the National Cancer Institute, and by American Cancer Society fellowship PF-02-090-01-CDD (to C.J.T.).

REFERENCES

1. Wall, M.E. et al., Plant antitumor agents. I. The isolation and structure of camptothecin, a novel alkaloidal leukemia and tumor inhibitor from *Camptotheca acuminata*, *J. Am. Chem. Soc.*, 88, 3888, 1966.
2. Moertel, C.G. et al., Phase II study of camptothecin (NSC-100880) in the treatment of advanced gastrointestinal cancer, *Cancer Chemother. Rep.*, 56, 95, 1972.
3. Hsiang, Y.-H. et al., Camptothecin induces protein-linked DNA breaks via mammalian DNA topoisomerase I, *J. Biol. Chem.*, 260, 14873, 1985.
4. Gallo, R.C., Whang-Peng, J., and Adamson, R.H., Studies on the antitumor activity, mechanism of action, and cell cycle effects of camptothecin, *J. Natl. Cancer Inst.*, 46, 789, 1971.
5. Hsiang, Y.-H., Lihou, M.G., and Liu, L.F., Arrest of replication forks by drug-stabilized topoisomerase I-DNA cleavable complex as a mechanism of cell killing by camptothecin, *Cancer Res.*, 49, 5077, 1989.
6. Nitiss, J. and Wang, J.C., DNA topoisomerase-targeting antitumor drugs can be studied in yeast, *Proc. Natl. Acad. Sci. USA*, 85, 7501, 1988.
7. Pommier, Y. et al., Topoisomerase I inhibitors: selectivity and cellular resistance, *Drug Resistance Updates*, 2, 307, 1999.
8. Champoux, J.J., DNA topoisomerases: structure, function, and mechanism, *Annu. Rev. Biochem.*, 70, 369, 2002.
9. Pommier, Y. et al., Mechanism of action of eukaryotic DNA topoisomerase I and drugs targeted to the enzyme, *Biochim. Biophys. Acta*, 1400, 83, 1998.
10. Sengupta, S.K., Topoisomerase II inhibitors, in *Cancer Chemotherapeutic Agents*, Foye, W.O., Ed., American Chemical Society, Washington D.C., 1995, pp. 205–292.
11. Hertzberg, R.P., Caranfa, M.J., and Hecht, S.M., On the mechanism of topoisomerase I inhibition by camptothecin: evidence for binding to an enzyme-DNA complex, *Biochemistry*, 28, 4629, 1989.

12. Yang, D. et al., DNA interactions of two clinical camptothecin drugs stabilize their active lactone forms, *J. Am. Chem. Soc.*, 120, 2979, 1998.

13. Hertzberg, R.P. et al., Irreversible trapping of the DNA-topoisomerase I covalent complex, *J. Biol. Chem.*, 265, 19287, 1990.

14. Pommier, Y. et al., Interaction of an alkylating camptothecin derivative with a DNA base at topoisomerase I-DNA cleavage sites, *Proc. Natl. Acad. Sci. USA*, 92, 8861, 1995.

15. Fan, Y. et al., Molecular modeling studies of the DNA-topoisomerase I ternary cleavable complex with camptothecin, *J. Med. Chem.*, 41, 2216, 1998.

16. Kerrigan, J.E. and Pilch, D.S., A structural model for the ternary cleavable complex formed between human topoisomerase I, DNA, and camptothecin, *Biochemistry*, 40, 9792, 2001.

17. Laco, G.S. et al., Human topoisomerase I inhibition: docking camptothecin and derivatives into a structure-based active site model, *Biochemistry*, 41, 1428, 2002.

18. Redinbo, M.R. et al., Crystal structures of human topoisomerase I in covalent and noncovalent complexes with DNA, *Science*, 279, 1504, 1998.

19. Yoshimasa, U. et al., Characterization of a novel topoisomerase I mutation from a camptothecin-resistant human prostate cancer cell line, *Cancer Res.*, 61, 1964, 2001.

20. Fujimori, A. et al., Mutation at the catalytic site of topoisomerase I in CEM/C2, a human leukemia cell line resistant to camptothecin, *Cancer Res.*, 55, 1339, 1995.

21. Jensen, A.D., and Svejstrup, J.Q., Purification and characterization of human topoisomerase I mutants, *Eur. J. Biochem.*, 236, 389, 1996.

22. Staker, B.L. et al., The mechanism of topoisomerase I poisoning by a camptothecin analog, *Proc. Natl. Acad. Sci. USA*, 99, 15387, 2002.

23. Husain, I. et al., Elevation of topoisomerase I messenger RNA, protein, and catalytic activity in human tumors: demonstration of tumor-type specificity and implications for cancer chemotherapy, *Cancer Res.*, 54, 539, 1994.

24. Kauh, E.A. and Bjornsti, M.-A., SCT1 mutants suppress the camptothecin sensitivity of yeast cells expressing wild-type DNA topoisomerase I, *Proc. Natl. Acad. Sci. USA*, 92, 6299, 1995.

25. Kawato, Y. et al., Inhibitory activity of camptothecin derivatives against acetylcholinesterase in dogs and their binding activity to acetylcholine receptors in rats, *J. Pharm. Pharmacol.*, 45, 444, 1993.

26. Siddoo-Atwal, C., Haas, A.L., and Rosin, M.P., Elevation of interferon beta-inducible proteins in ataxia telangiectasia cells, *Cancer Res.*, 56, 443, 1996.

27. DeSai, S.D. et al., Ubiquitin, SUMO-1, and UCRP in camptothecin sensitivity and resistance, *Ann. N. Y. Acad. Sci.*, 922, 306, 2000.

28. Ohwada, S. et al., Interferon potentiates antiproliferative activity of CPT-11 against human colon cancer xenografts, *Cancer Lett.*, 110, 149, 1996.

29. Ibuki, Y., Mizuno, S., and Goto, R., γ-Irradiation-induced DNA damage enhances NO production via NF-κB activation in RAW264.7 cells, *Biochim. Biophys. Acta*, 1593, 159, 2003.

30. Kretzschmar, M., Meistererst, M., and Roeder, R.G., Identification of human DNA topoisomerase I as a cofactor for activator-dependent transcription by RNA polymerase II, *Proc. Natl. Acad. Sci. USA*, 90, 11508, 1993.

31. Yeh, Y.C. et al., Mammalian topoisomerase I has base mismatch nicking activity, *J. Biol. Chem.*, 269, 15498, 1994.

32. Rossi, F. et al., Specific phosphorylation of SR proteins by mammalian DNA topoisomerase I, *Nature*, 381, 80, 1996.

33. Rossi, F. et al., The C-terminal domain but not the tyrosine 723 of human DNA topoisomerase I active site contributes to kinase activity, *Nucleic Acids Res.*, 26, 2963, 1998.

34. Castano, I.B. et al., Mitotic chromosome condensation in the rDNA requires TRF4 and DNA topoisomerase I in *Saccharomyces cerevisiae*, *Genes Dev.*, 10, 2564, 1996.

35. Walowsky, C. et al., The topoisomerase-related function gene TRF4 affects cellular sensitivity to the antitumor agent camptothecin, *J. Biol. Chem.*, 274, 7302, 1999.

36. Stork, G. and Schultz, A.G., The total synthesis of dl-camptothecin, *J. Am. Chem. Soc.*, 93, 4074, 1971.

37. Jew, S.-S. et al., Synthesis and antitumor activity of camptothotecin analogues, *Korean J. Med. Chem.*, 6, 263, 1996.

38. Thomas, C.J., Rahier, N.J., and Hecht, S.M., Camptothecin: current perspectives, *Bioorg. Med. Chem.*, 12, 1585, 2004.
39. Corey, E.J., Crouse, D.N., and Anderson, J.E., A total synthesis of natural 20(S)-camptothecin, *J. Org. Chem.*, 40, 2140, 1975.
40. Wani, M.C., Nicholas, A.W., and Wall, M.E., Plant antitumor agents 28. Resolution of a key tricyclic synthon, 5′(RS)-1,5-dioxo-5′-ethyl-5′-hydroxy-2′H,5′H,6′H-6′-oxopyrano[3′,4′-f]Δ6,8-tetrahydro-indolizine: total synthesis and antitumor activity of 20(S)- and 20(R)-camptothecin, *J. Med. Chem.*, 30, 2317, 1987.
41. Terasawa, H. et al., Antitumor agents III. A novel procedure for inversion of the configuration of a tertiary alcohol related to camptothecin, *Chem. Pharm. Bull.*, 37, 3382, 1989.
42. Ejima, A. et al., Antitumour agents. Part 2. Asymmetric synthesis of (S)-camptothecin, *J. Chem. Soc., Perkin Trans. 1*, 27, 1990.
43. Comins, D.L., Hong, H., and Jianhua, G., Asymmetric synthesis of camptothecin alkaloids: A nine-step synthesis of (S)-camptothecin, *Tetrahedron Lett.*, 35, 5331, 1994.
44. Comins, D.L. and Nolan, J.M., A practical six-step synthesis of (S)-camptothecin, *Org. Lett.*, 3, 4255, 2001.
45. Tagami, K. et al., Asymmetric synthesis of (+)-camptothecin and (+)-7-ethyl-10-methoxycamptothecin, *Heterocycles*, 53, 771, 2000.
46. Bennasar, M.-L. et al., Addition of ester enolates to N-alkyl-2-fluoropyridinium salts: total synthesis of (±)-20-deoxycamptothecin and (+)-camptothecin, *J. Org. Chem.*, 67, 7465, 2002.
47. Ciufolini, M.A. and Roschangar, F., Practical synthesis of (20 S)-(+)-camptothecin: the progenitor of a promising group of anticancer agents, *Targets Heterocyclic Syst.*, 4, 25, 2000.
48. Imura, A., Itoh, M., and Miyadera, A., Enantioselective synthesis of 20(S)-camptothecin using an enzyme-catalyzed resolution, *Tetrahedron: Asymmetry*, 9, 2285, 1998.
49. Fang, F.G., Xie, S., and Lowery, M.W., Catalytic enantioselective synthesis of 20(S)-camptothecin: a practical application of the Sharpless asymmetric dihydroxylation reaction, *J. Org. Chem.*, 59, 6142, 1994.
50. Jew. S.-S. et al., Enantioselective synthesis of 20(S)-camptothecin using Sharpless catalytic asymmetric dihydroxylation, *Tetrahedron: Asymmetry*, 6, 1245, 1995.
51. Blagg, B.S.J. and Boger, D.L., Total synthesis of (+)-camptothecin, *Tetrahedron*, 58, 6343, 2002.
52. Curran, D.P. et al., The cascade radical annulation approach to new analogues of camptothecins. Combinatorial synthesis of silatecans and homosilatecans, *Ann. N. Y. Acad. Sci.*, 922, 112, 2000.
53. Yabu, K. et al., Studies toward practical synthesis of (20S)-camptothecin family through catalytic enantioselective cyanosilylation of ketones: improved catalyst efficiency by ligand-tuning, *Tetrahedron Lett.*, 43, 2923, 2002.
54. Sawada, S. et al., Synthesis and antitumor activity of 20(S)-camptothecin derivatives: carbamate-linked, water-soluble derivatives of 7-ethyl-10-hydroxycamptothecin, *Chem. Pharm. Bull.*, 39, 1446, 1991.
55. Kingsbury, W.D. et al., Synthesis of water-soluble (aminoalkyl)camptothecin analogs: inhibition of topoisomerase I and antitumor activity, *J. Med. Chem.*, 34, 98, 1991.
56. Lackey, K. et al., Rigid analogues of camptothecin as DNA topoisomerase I inhibitors, *J. Med. Chem.*, 38, 906, 1995.
57. Ihara, M. et al., Studies on the synthesis of heterocyclic compounds and natural products. 999. Double enamine annelation of 3,4-dihydro-1-methyl-β-carboline and isoquinoline derivatives with 6-methyl-2-pyrone-3,5-dicarboxylates and its application for the synthesis of (+/-)-camptothecin, *J. Org. Chem.*, 48, 3150, 1983.
58. Kurihara, T. et al., Synthesis of C-nor-4,6-secocamptothecin and related compounds, *J. Heterocycl. Chem.*, 30, 643, 1993.
59. Subrahmanyam, D. et al., Novel C-ring analogues of 20(S)-camptothecin Part 2. Synthesis and *in vitro* cytotoxicity of 5-C-substituted 20(S)-camptothecin analogues, *Bioorg. Med. Chem. Lett.*, 9, 1633, 1999.
60. Sugimori, M. et al., Antitumor agents. VI. Synthesis and antitumor activity of ring A-, ring B-, ring C-modified derivatives of camptothecin, *Heterocycles*, 38, 81, 1993.
61. *Camptothecins: New Anticancer Agents*, Potmesil, M. and Pinedo, H., Eds., CRC Press, Boca Raton, 1995.

62. Kawato, Y. and Terasawa, H., Recent advances in the medicinal chemistry and pharmacology of camptothecin, in *Progress in Medicinal Chemistry*, Ellis, G.P., and Luscombe, D.K., Eds., Elsevier, London, 1997, Vol 34, pp. 70–100.

63. Wall, M.E. and Wani, M.C., Recent advances in the medicinal chemistry and pharmacology of camptothecin, in *Cancer Chemotherapeutic Agents*, Foye, W.O., Ed., American Chemical Society, Washington D.C., 1995, pp. 293–310.

64. Wani, M.C. et al., Plant antitumor agents 18. Synthesis and biological activity of camptothecin analogues, *J. Med. Chem.*, 23, 554, 1980.

65. Wani, M.C., Nicholas, A.W., and Wall, M.E., Plant antitumor agents 23. Synthesis and antileukemic activity of camptothecin analogs, *J. Med. Chem.*, 29, 2358, 1986.

66. Wani, M.C. et al., Plant Antitumor Agents 25. Total synthesis and antileukemic activity of ring A substituted camptothecin analogues. Structure-activity correlations, *J. Med Chem.*, 30, 1774, 1987.

67. Wall, M.E. et al., Plant antitumor agents 30. Synthesis and structure activity of novel camptothecin analogues, *J. Med. Chem.*, 36, 2689, 1993.

68. Jaxel, C. et al., Structure-activity study of the actions of camptothecin derivatives on mammalian topoisomerase I: evidence for a specific receptor site and a relation to antitumor activity, *Cancer Res.*, 49, 1465, 1989.

69. Wang, X., Zhou, X., and Hecht, S.M., Role of the 20-hydroxyl group in camptothecin binding by the topoisomerase I-DNA binary complex, *Biochemistry*, 38, 4374, 1999.

70. Hertzberg, R.P. et al., Modification of the hydroxy lactone ring of camptothecin: Inhibition of mammalian topoisomerase I and biological activity, *J. Med. Chem.*, 32, 715, 1989.

71. Burke, T.G. and Mi, Z., Ethyl substitution at the 7 position extends the half-life of 10-hydroxycamptothecin in the presence of human serum albumin, *J. Med. Chem.*, 36, 2580, 1993.

72. Lavergne, O. et al., BN80245: An E-ring modified camptothecin with potent antiproliferative and topoisomerase I inhibitory activities, *Bioorg. Med. Chem. Lett.*, 7, 2235, 1997.

73. Cagir, A. et al., Luotonin A. A naturally occurring DNA topoisomerase I poison, *J. Am. Chem. Soc.*, 125, 13628, 2003.

74. Luzzio, M.J. et al., Synthesis and antitumor activity of novel water soluble derivatives of camptothecin as specific inhibitors of topoisomerase I, *J. Med. Chem.*, 38, 395, 1995.

75. van Hattum, A.H. et al., The activity profile of the hexacyclic camptothecin derivative DX-8951f in experimental human colon cancer and ovarian cancer, *Biochem. Pharmacol.*, 64, 1267, 2002.

76. Josien, H. et al., 7-Silylcamptothecins (silatecans): a new family of camptothecin antitumor agents, *Bioorg. Med. Chem. Lett.*, 7, 3189, 1997.

77. Dallavalle, S. et al., Novel 7-oxyiminomethyl derivatives of camptothecin with potent *in vitro* and *in vivo* antitumor activity, *J. Med. Chem.*, 44, 3264, 2001.

78. Pollack, I.F. et al., Potent topoisomerase I inhibition by novel silatecans eliminates glioma proliferation *in vitro* and *in vivo*, *Cancer Res.*, 59, 4898, 1999.

79. Bom, D. et al., Novel A, B, E-ring-modified camptothecins displaying high lipophilicity and markedly improved human blood stabilities, *J. Med. Chem.*, 42, 3018, 1999.

80. Walker, M. A. et al., Synthesis of an immunoconjugate of camptothecin, *Bioorg. Med. Chem. Lett.*, 12, 217, 2002.

81. Arimondo, P.B. et al., Design and optimization of camptothecin conjugates of triple helix-forming oligonucleotides for sequence-specific DNA cleavage by topoisomerase I, *J. Biol. Chem.*, 277, 3132, 2002.

82. Wang, C.C.C. and Dervan, P.B., Sequence-specific trapping of topoisomerase I by DNA binding polyamide-camptothecin conjugates, *J. Am. Chem. Soc.*, 123, 8657, 2001.

83. Conover, C.D. et al., Camptothecin delivery systems: enhanced efficacy and tumor accumulation of camptothecin following its conjugation to polyethylene glycol via a glycine linker, *Cancer Chemother. Pharmacol.*, 42, 407, 1998.

84. Greenwald, R.B., Zhao, H., and Xia, J., Tripartate poly(ethylene glycol) prodrugs of the open lactone form of camptothecin, *Bioorg. Med. Chem.*, 11, 2635, 2003.

85. Garcia-Carbonero, R. and Supko, J.G., Current perspectives on the clinical experience, pharmacology, and continued development of the camptothecins, *Clinical Cancer Res.*, 8, 641, 2002.

86. Ulukan, H. and Swaan, P.W., Camptothecins. A review of their chemotherapeutic potential, *Drugs*, 62, 2039, 2002.

87. Kim, D.-K. and Lee, N., Recent advances in topoisomerase I-targeting agents, camptothecin analogues, *Mini Rev. Med. Chem.*, 2, 611, 2002.
88. Dallavalle, S. et al., Perspectives in camptothecin development, *Expert Opin. Ther. Patents*, 12, 837, 2002.
89. Bailly, C., Homocamptothecins: Potent topoisomerase I inhibitors and promising anticancer drugs, *Critical Rev. Oncology/Hematology*, 45, 91, 2003.
90. Hatefi, A. and Amsden, B., Camptothecin delivery methods, *Pharm. Res.*, 19, 1389, 2002.

3 The Discovery and Development of the Combretastatins

Kevin G. Pinney, Christopher Jelinek, Klaus Edvardsen, David J. Chaplin, and George R. Pettit

CONTENTS

I. Introduction, History, and Structures of Combretastatins 23
II. Biochemical and Biological Mechanism of Action 25
III. Therapeutic Intervention through Vascular Targeting 27
IV. Synthesis of Combretastatins...27
V. Combretastatin Derivatives and Synthetic Analogs Inspired by the Combretastatins 33
VI. Preclinical Studies..34
    A. Effects of CA4P on Tumor Blood Flow 34
    B. Tumor Response to CA4P as a Single Agent..35
    C. Treatment Response to CA4P in Combination Therapy 37
    D. Preclinical Studies with CA1P (OXi4503) 37
VII. Clinical Experience..38
VIII. Coda 39
IX. Acknowledgments 40
References 40

I. INTRODUCTION, HISTORY, AND STRUCTURES OF COMBRETASTATINS

The Combretaceae plant family (comprising 600 or more species) of shrubs and trees is divided among 20 genera, of which the *Combretum* genus (250 species) of tropical and deciduous trees encompasses the largest number.[1] Some 24 species of *Combretum* are well-known in African folk medicine for applications and problems ranging from heart and worm remedies to wound dressings, treatment for the mentally ill, and scorpion stings.[2] Only the Indian *Combretum latifolium* appears to have been recorded as a folk medical treatment for cancer.[3] However, over 30 years ago, as part of the U.S. National Cancer Institute's (NCI's) worldwide exploratory survey of terrestrial plants, both *Combretum molle*[4] and *Combretum caffrum* (Eckl. and Zeyh.) Kuntze were found to provide extracts significantly active against the murine P-388 lymphocytic leukemia (PS system). *C. caffrum* is a deciduous tree (growing to 15 m high) in Africa, found principally in the Eastern Cape and Transkei to Natal. This willow-like tree ("bushwillow") is a common sight overhanging stream beds. In autumn, these trees become quite prominent, with displays of reddish-brown fruit and leaves that turn bright red before falling.[5] The powdered root bark has been used by the Zulu of South Africa as a charm to harm an enemy.[2]

In a broader context, the Combretaceae family is well represented in traditional medical practices of, especially, Africa and India.[6] Illustrative is a study of Combretaceae species used in Somalia.[7] These species range from *Combretum hereroense* (young shoots used for respiratory infection) to *Terminalia brevipes* (root bark employed for hepatitis and malaria) to *Commelina forskoolii* (juice for treatment of uterine cancer). Nine other species including *Anogeissus leiocorpus* (fruit and roots as an anthelmintic treatment), *Guiera senegalensis* (fruit and leaves for leprosy and dysentery), and *Quisqualis indica* (leaves for vermifuge) are more widely used in Africa.[8] Other Combretaceae species, such as *C. coccineum* and *C. constrictum*,[9] are well-known medicinal plants in Indian Ocean Island areas. Even better known is the medicinal herb *C. micronthum*, used as an opium antidote for addiction and for other indications from blackwater fever to nausea.[10] Importantly, seven species of the genus *Terminalia*[11] and one *Guiera* species have a long history in African and Asian primitive medical treatments for cancer.[3] In Zimbabwe, *Combretum molle* and three other *Combretum* species have been summarized along with two *Terminalia* species as plant extract remedies for diarrhea.[12]

The early events leading to the discovery of the combretastatins began with the first (1973) collection of *C. caffrum* in Africa for the NCI in the former Rhodesia (now Zimbabwe). Several scale-up recollections of *C. caffrum* were completed, and chemical investigations of constituents were pursued by an NCI collaborative research group, but by 1979 that research had led to a dead end in respect to anticancer constituents.

In 1979 one of us (G.R.P.) and colleagues continued with a 1979 *C. caffrum* collection in collaboration with the NCI Natural Products Branch. As summarized in the sequel to this introduction, this research subsequently led to the discovery of 20 cancer cell growth inhibitory stilbenes, bibenzyls, dihydrophenanthrenes, and phenanthrenes from *C. caffrum*. Some of those discoveries led to a number of important advances that range from the first isolation of a cancer cell growth inhibitor, (-)-combretastatin, isolated by using the NCI astrocytoma bioassay, to the first well-established vascular-targeting anticancer drug — sodium combretastatin A-4 phosphate (CA4P), which has entered human cancer clinical trials.

Several of the natural products isolated from *C. caffrum* are characterized by their remarkable biological activity as inhibitors of tubulin polymerization, potent *in vitro* inhibition against human cancer cell lines, and demonstrated *in vivo* efficacy as vascular targeting agents (VTAs; in suitable prodrug formulation) and antiangiogenesis agents. The combretastatins are especially noteworthy for their robust activity in biological systems, while maintaining remarkable simplicity in terms of chemical structure. Combretastatin A-4 (CA4) and combretastatin A-1 (CA1), along with their corresponding phosphate prodrug salts (CA4P and CA1P, respectively), have emerged as compounds of profound therapeutic interest (Figure 3.1). Several excellent reviews of the combretastatins and novel analogs inspired by the combretastatins have been published.[13–24] In addition, numerous patents have been issued, or are in the application process, for a wide variety of combretastatins and their analogs.[25–34]

The discovery and development of small-molecule (i.e., compounds with molecular weights in the range of 300–800 g/mol) inhibitors of tubulin assembly has been (and continues to be) a challenging and fruitful goal that, to a large degree, began with colchicine (Figure 3.1) in the 1930s and 1940s. Representative naturally occurring antimitotic ligands[19] in addition to colchicine include paclitaxel,[35] epothilone A,[36] vinblastine,[37] dolastatin 10,[38,39] and CA4.[40–43] In addition, there are a variety of synthetic compounds that also demonstrate efficient inhibition of tubulin polymerization.[42,43] Because of the poor solubility of CA1 and CA4, sodium phosphate prodrugs CA1P[44] and CA4P[45] were developed in the mid-1990s.

Isolation of the first constituent of *C. caffrum*, (-)-combretastatin (Figure 3.2), as noted above, was achieved in 1982 by Pettit and coworkers.[46] By 1987, the absolute stereochemistry, attributed to the single stereogenic center in the molecule, was determined to be the *R*-configuration.[47] The combretastatins are currently divided into four major groups on the basis of their structural characteristics. These include the A-Series (cis-stilbenes), B-series (diaryl-ethylenes), C-series (quinone), and D-series (macrocyclic lactones). Representative members of each of these groups are depicted in Figure 3.2.

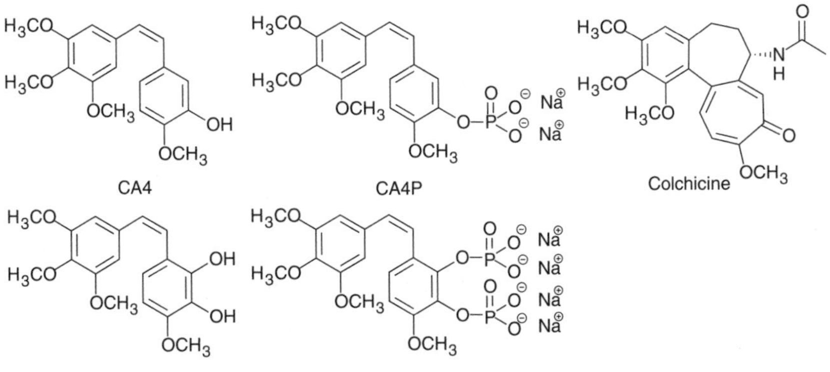

FIGURE 3.1 Structures of combretastatin A-4, combretastatin A-4 phosphate, combretastatin A-1, combretastatin A-1 phosphate, and colchicine.

The combretastatin A and B series contain two phenyl rings tilted at 50–60° to each other, and are linked by a two-carbon bridge.[15] The bridge is either unsaturated, as in the case of stilbenes, or saturated. The substituents differ in their respective locations on the phenyl rings, and they usually contain methoxy and hydroxy functionalities. The combretastatin A series is made up of six constituents, combretastatin A-1[48] (CA1), CA2,[49] CA3,[49] CA4,[50] CA5,[50] and CA6,[50] which retain a *cis*-stilbene moiety. The B series contains four compounds — CB1,[48] CB2,[49] CB3,[51] and CB4[51] — with three similar structures not named according to the prescribed nomenclature.[51] These compounds are similar to the A series except for a saturated carbon–carbon bridge between the two aryl rings. There is one compound in the C series, designated combretastatin C1[52] (CC1), that contains a three-ring quinone structure. Two unusual macrocyclic lactones make up the D series, CD1[53] and CD2.[54] Finally, there are four phenanthrene-like compounds that are currently not named other than by their IUPAC designation.[16]

II. BIOCHEMICAL AND BIOLOGICAL MECHANISM OF ACTION

Antimitotic anticancer drugs, which disrupt the mitotic spindle, have long been a major component in the arsenal of "weaponry" available to wage the battle against cancer. The tubulin–microtubule protein system is the biological target for the majority of these compounds.[17,19] Tubulin is a cytoskeletal protein ubiquitous to eukaryotic cells. In addition to providing cells with shape and serving as a cellular mobility (transportation) system, it also plays a key role during mitosis.[55] During metaphase of the cell cycle, the tubulin heterodimer (composed of both alpha and beta subunits) undergoes assembly (also referred to as polymerization) to form microtubules, which, collectively, are essential components of the mitotic spindle. The assembly process, which is reversible through a disassembly mechanism, localizes in the microtubule organizing centers (centrosomes) of dividing cells and provides a venue by which the microtubules are able to interdigitate between the separating chromosomes, thereby pushing them apart. Certain compounds such as vinblastine, vincristine, colchicine, and podophyllotoxin, function biologically by binding to tubulin heterodimers and causing conformational changes and a steric bias that inhibits the assembly of tubulin into microtubules.[17,19] Other compounds, such as paclitaxel and the epothilones, bind to the microtubule itself and inhibit the disassembly process into the tubulin protein.[17,19] In both cases, cellular division is disrupted by these compounds, and although healthy cells are also affected, it is the tumor cells that are more prominently targeted because they are characterized by rapid proliferation, which is tubulin dependent.

Tubulin is bound by two molecules of guanosine triphosphate (GTP) with a cross-sectional diameter of approximately 24 nm. A tertiary structure of the α,β–tubulin heterodimer (as contained

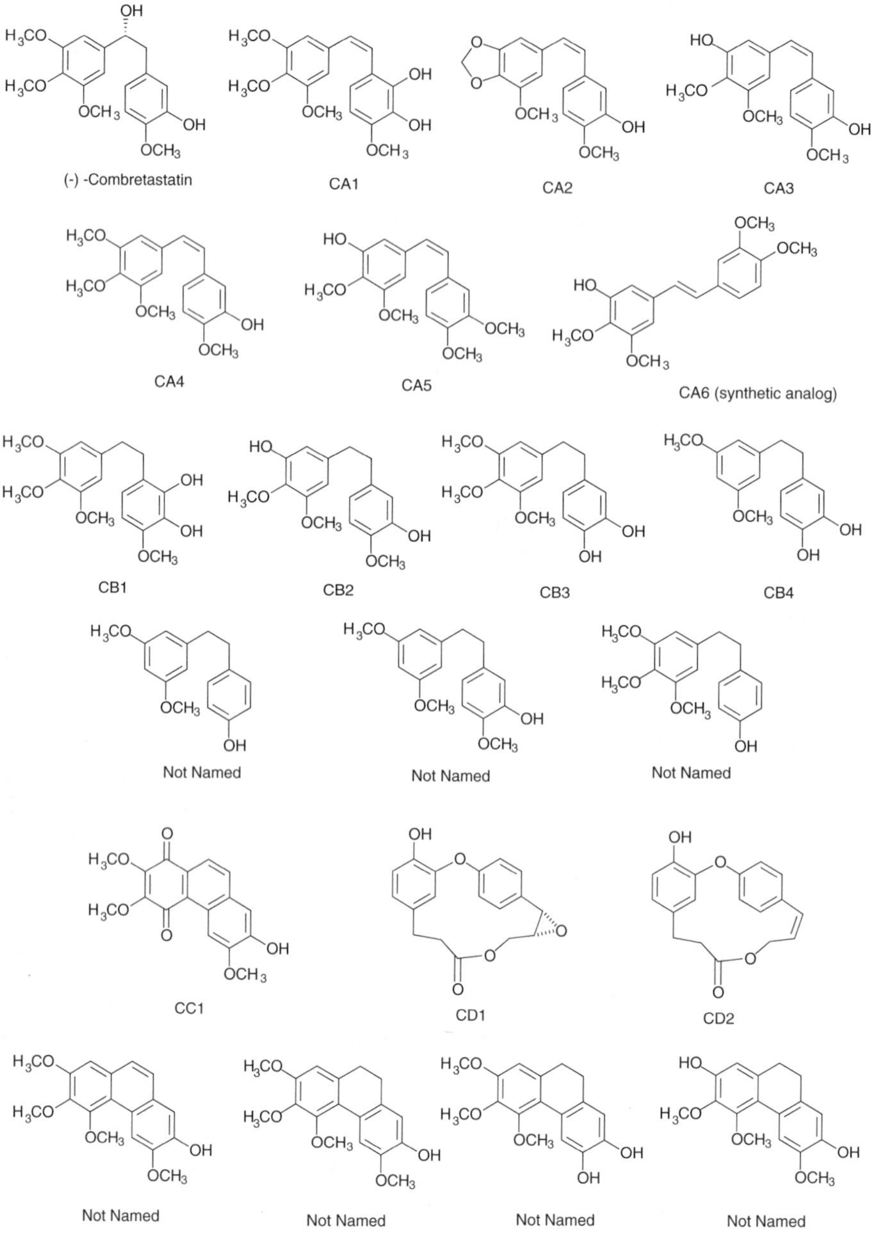

FIGURE 3.2 Natural products isolated from *Combretum caffrum*.

in the microtubule stabilized by taxol) was reported in 1998 by Downing and coworkers at a resolution of 3.7 Å, using electron crystallography.[56] This accomplishment marks the culmination of decades of work directed toward the elucidation of this structure and should contribute to the identification of small-molecule binding sites, such as the colchicine site, through techniques such as photoaffinity and chemical affinity labeling.[57–70] Small molecules can bind in at least three known locations or sites on the tubulin system: the colchicine site, the vinca-alkaloid site, and the taxoid site. These sites are named because of the binding of colchicine, vinblastine, and taxol, respectively, to certain distinct binding pockets on the tubulin dimer itself, or on the microtubule in the case of taxol.[71] Ligands that interact at the taxoid site (such as paclitaxel and the epothilones) stabilize the microtubule, whereas ligands interacting at the vinca domain (such as vinblastine) or the colchicine site (such as colchicine and CA4) disrupt the formation of microtubules. The combretastatin A series is historically known for its remarkable biological activity in terms of inhibition of tubulin assembly and *in vitro* cytotoxicity against human cancer cell lines. Combretastatin A1 and CA4 are among the most active of the natural products isolated from *C. caffrum* in terms of *in vitro* activity[16] (Table 3.1).

III. THERAPEUTIC INTERVENTION THROUGH VASCULAR TARGETING

Although antimitotic agents attack the tumor cells directly and disrupt mitosis, some molecules have been shown to exhibit an antivascular and antiangiogenic effect as well. Often, the two classes overlap, as in the case of the combretastatins, where certain compounds that cause direct vascular damage can also prevent angiogenesis. Vascular targeting agents (VTAs) function by destroying existing tumor vasculature through a mechanism that induces morphological changes within endothelial cells, thereby disrupting shape, transport, and motility. In contrast, antiangiogenic agents inhibit the formation of new blood vessels. In either case, this disruption of vasculature essentially starves the tumor. Recently, it has been discovered that certain of these tubulin binding compounds, in prodrug form, demonstrate remarkable selectivity for the destruction of tumor vasculature.[72–75] Solid tumor survival is dependent on blood, oxygen, and nutrient supply from surrounding vessels.[76] Growing tumors receive their nutrients through existing vasculature in the surrounding area as well as through newly formed neovascularization (angiogenesis). In prodrug form, these VTAs are, for the most part, protected from binding to tubulin. However, once they undergo enzymatic cleavage to their corresponding parent free phenol or amine analog, they function as potent inhibitors of tubulin assembly. In the microvessels feeding tumor cells, these compounds bind to endothelial cell tubulin and cause rapid morphology changes in the endothelial cells lining the microvessels.[77–79] These morphology changes (known as rounding up of the cells) disrupt blood flow, which ultimately renders both new microvessels as well as mature and established vessels nonfunctional for blood delivery, causing hypoxia and necrosis of the tumor. The reason for the selectivity of the VTAs for the microvessels of tumors is largely unknown. However, it may reflect, in part, variability in the cytoskeletal make-up of rapidly proliferating endothelial cells inherent to microvessels feeding tumor cells versus the normal proliferating endothelial cells of microvessels serving healthy cells.[77] Because one single microvessel can feed hundreds or thousands of tumor cells, this method of vascular targeting has the potential to be extremely effective in destroying tumor cells without targeting them directly (as it is the microvessels that are immediately affected).

CA4P and other VTAs are useful as chemotherapeutic agents for cancer treatment as well as for therapeutic intervention in ocular diseases (such as wet, age-related macular degeneration).[80] At present, there are at least three compounds in human clinical development as potential VTAs for cancer chemotherapy that use a tubulin-based biological mechanism. Although none of these compounds has achieved NDA (New Drug Application) approval for marketing by the U.S. Food and Drug Administration, CA4P is the most advanced in clinical trials. In June 2003, CA4P received orphan drug approval by the U.S. Food and Drug Administration for the thyroid cancer group, and a month later it received approval for a "fast-track" phase II clinical trial against the usually quite lethal (median survival time is 4 months) anaplastic thyroid cancer.

IV. SYNTHESIS OF COMBRETASTATINS

Isolation of CA1 and CA4, as with the other constituents, began with a methylene chloride-methanol extract, initially from the macerated branches, leaves, and fruit, and subsequently from the stem wood of *C. caffrum*.[48,50,81] These extracts were then separated by bioassay (PS system) guided techniques. For example, with respect to CA4, a trace fraction was isolated from 77 kg of dry stem wood. Further separation led to 26.4 mg of a fraction that at first seemed to be a pure compound, but high-field ¹H

TABLE 3.1
In Vitro Activities of Combretastatins

Name	Structure	Tubulin[a]	P388[b]	L1210[c]	Name	Structure	Tubulin[a]	P388[b]	L1210[c]
CA1	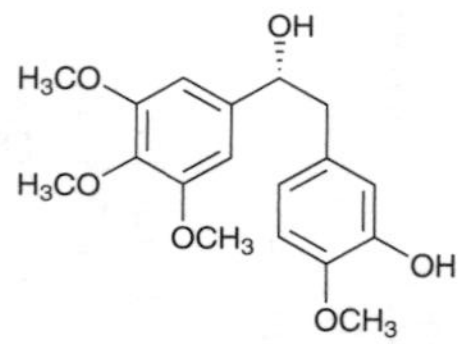 (structure)	5–7.5	0.011[d]	0.05	CB4	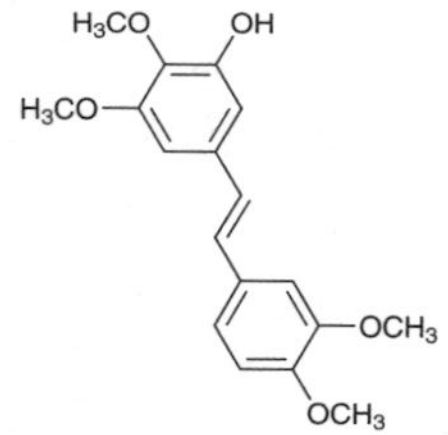 (structure)	25–30	1.7[h]	1
CA2	(structure)	2–3	0.99[e]	0.6	—	(structure)	10–15	NA	1
CA3	(structure)	4–5	0.027[f]	0.1	—	(structure)	>100	NA	4
	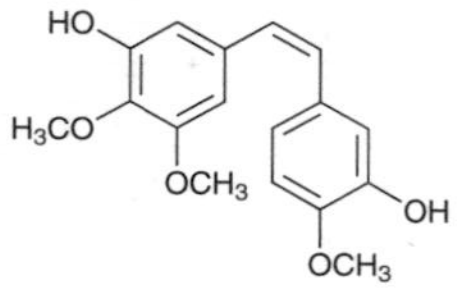 (structure)	4–5	0.026[f]	0.04	—	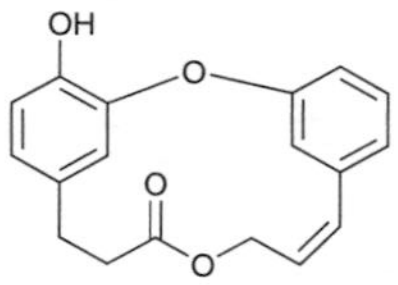 (structure)	>100	NA	50

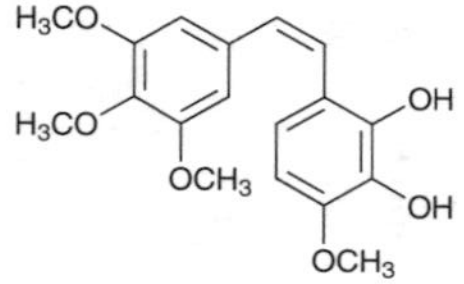

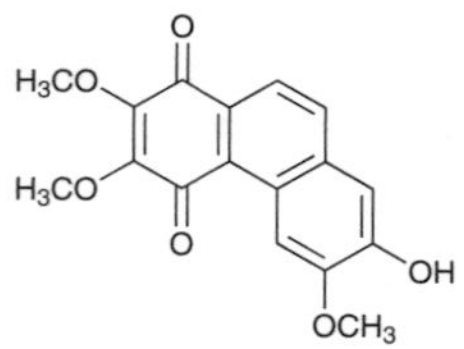

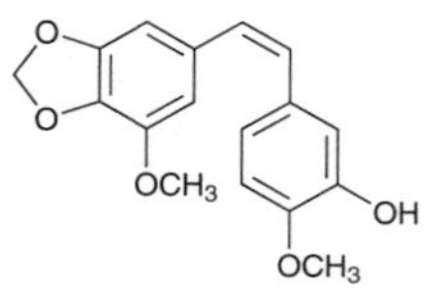

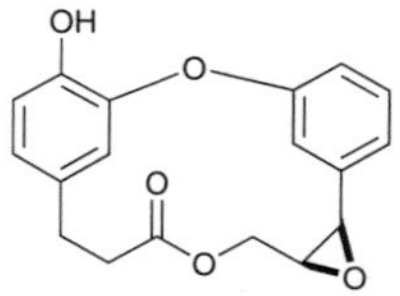

Compound			
CA4	0.007	0.0034	1.9[g] / 2–3
trans CA4	0.02	0.050	7.5–10
CA5	2	0.9	75–100
CA6	30	18	>100
CB1	2	1.7[e]	3[e] / 4–5
CC1	NA	2.2[i]	NA
CD1	NA	3.3[j]	NA
CD2	NA	5.2[k]	51[l]
—	100	NA	1
—	>100	NA	3

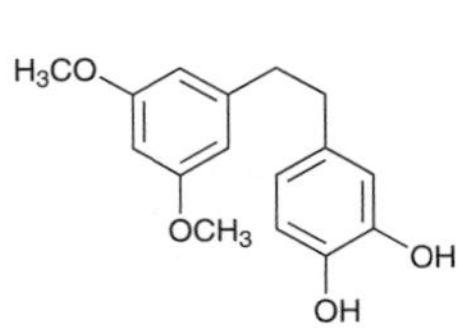

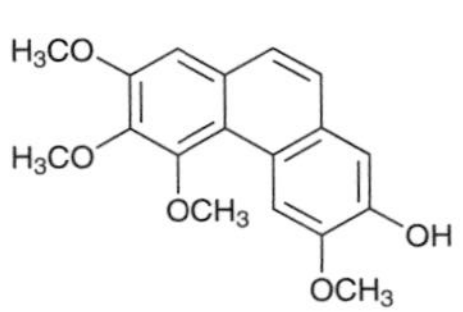

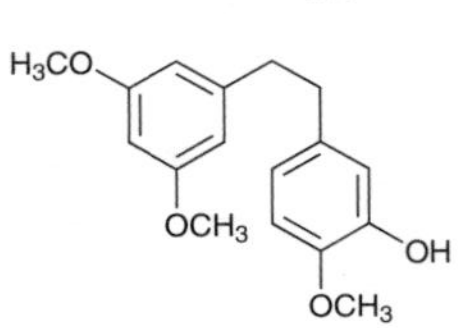

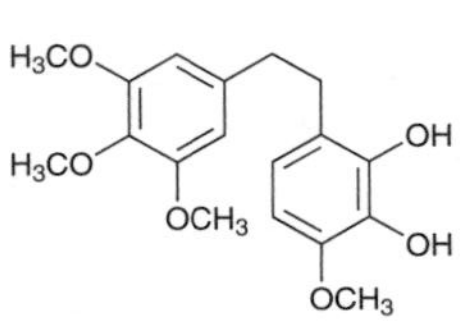

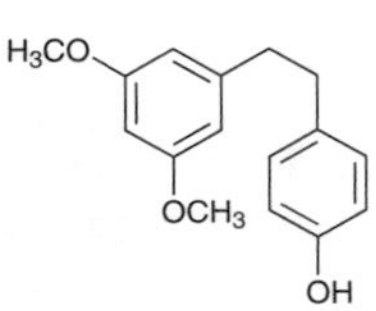

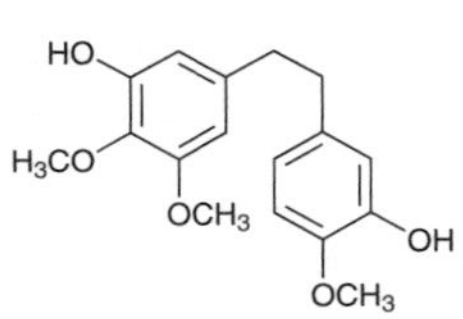

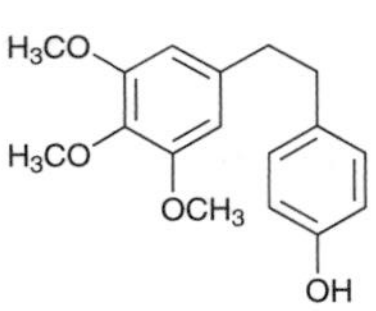

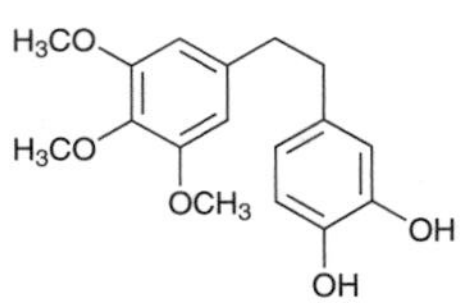

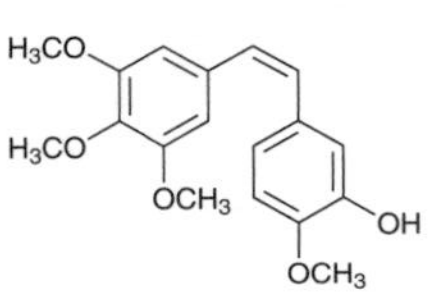

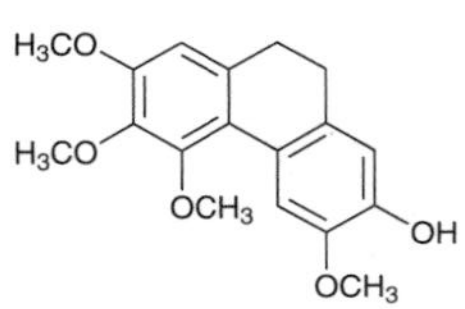

TABLE 3.1
In Vitro Activities of Combretastatins

Name	Structure	Tubulin[a]	P388[b]	L1210[c]	Name	Structure	Tubulin[a]	P388[b]	L1210[c]
CB2	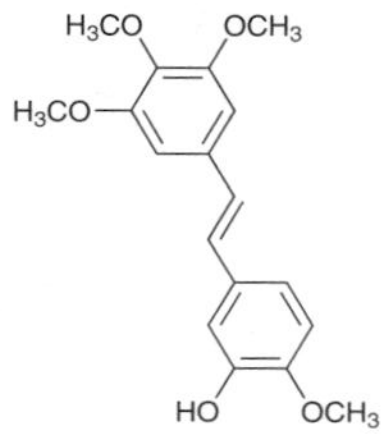	40	0.32[f]	3	—	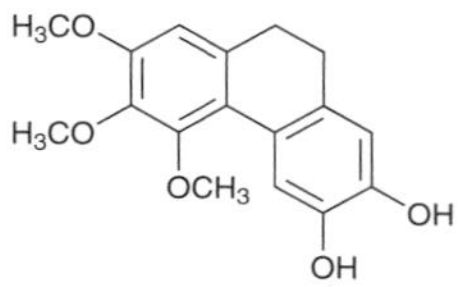	>100	NA	20
CB3	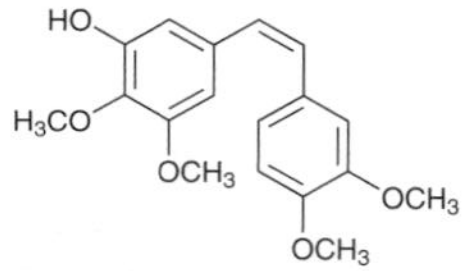	>100	0.4[h]	3	—	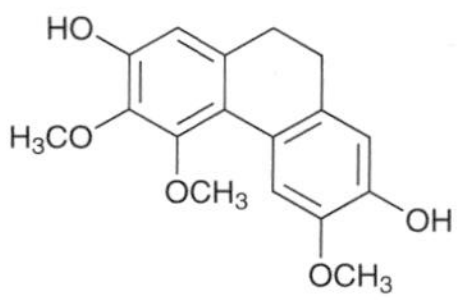	>100	NA	20

Note: The compound is (–)-combretastatin.

[a] Indicates tubulin polymerization inhibition (IC_{50} micromolar) (data taken from ref. 16, except as noted).
[b] Indicates cytotoxicity for the PS cell line (P388 leukemia) (ED_{50} in µg/mL) (data taken from ref. 50, except as noted).
[c] Indicates cytotoxicity for the L1210 cell line (murine leukemia) (ED_{50} micromolar) (data taken from ref. 16, except as noted).
[d] Data taken from ref. 46.
[e] Data taken from ref. 48.
[f] Data taken from ref. 49.
[g] Data taken from ref.13.
[h] Data taken from ref.51.
[i] Data taken from ref.52.
[j] Data taken from ref. 53.
[k] Data taken from ref. 54.
[l] Data taken from ref. 83.

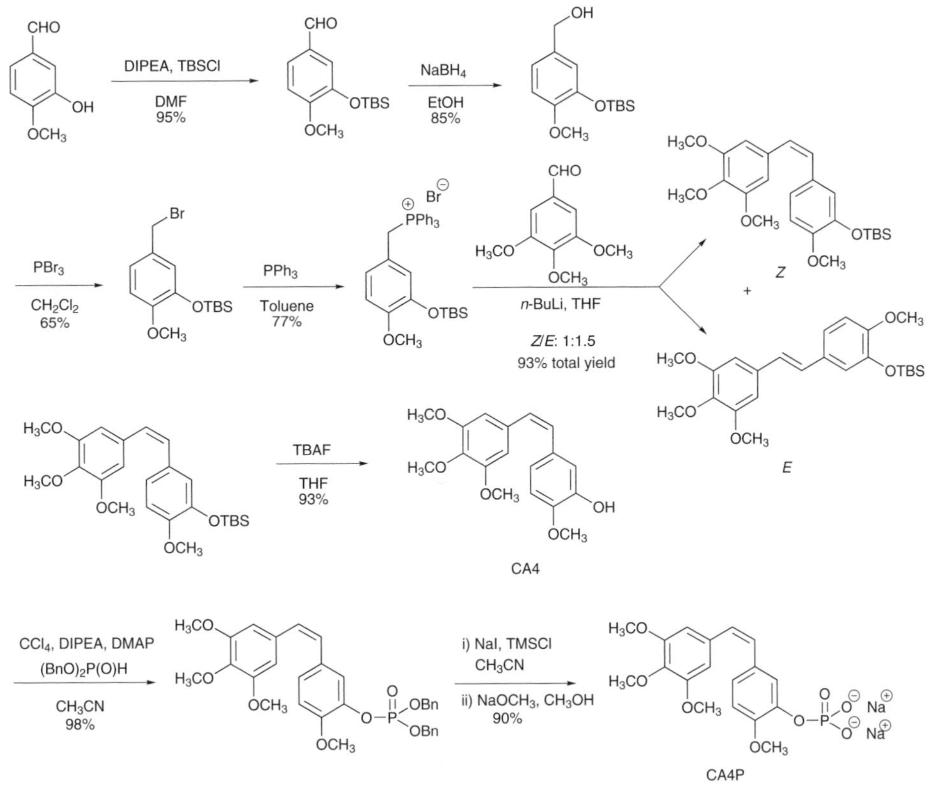

SCHEME 3.1 Synthesis of combretastatin A-4 and its disodium phosphate prodrug.

and ^{13}C-NMR (nuclear magnetic resonance) indicated a mixture of three compounds. Conversion to a *t*-butyldimethylsilyl ether derivative and further separation afforded about 9 mg of CA4 silyl ether. Following cleavage of the protecting group from a 1.7-mg sample, a 1.15-mg sample of pure CA4 was obtained. By analogous techniques, CA5 and CA6 were obtained as the other two components of this complex mixture. The structure and configuration of CA4 was established by ^{1}H-NMR spectra,[50] which gave evidence of four methoxy groups and seven protons in the aromatic region. Two of the methoxy groups were equivalent, and three protons were displayed at δ 6.734 (d, *J* = 8.4 Hz), 6.80 (dd, *J* = 8.42, 2.04 Hz), and 6.92 (d, *J* = 2.04 Hz), indicating ortho–ortho and ortho–meta coupling, and one singlet at δ 6.53 represented the two symmetrical aromatic protons on the A ring of CA4. The cis (Z) stereochemistry of the stilbene was indicated by the coupling constant (*J* = 12.16 Hz) for each of the two doublets at d 6.471 and d 6.412. The *trans*(*E*) derivative of CA4 showed a vinyl coupling of *J* = 16.30 Hz. The synthesis of CA4,[50] (Scheme 3.1) was achieved through a Wittig reaction between the silyl-protected phosphonium bromide from isovanillin and 3,4,5-trimethoxybenzaldehyde. The *Z* geometry was confirmed by an x-ray crystal structure determination. The *Z* isomer was then desilylated with tetrabutylammonium fluoride to afford CA4.

Similarly, the structure and configuration of CA1[48] was confirmed to be *Z* with an AB system of doublets at δ 6.453 (*J* = 12.2 Hz) and 6.523 (*J* = 12.2 Hz) and a two-proton signal at δ 5.438, representing the presence of phenolic groups. The synthesis of CA1[48] (Scheme 3.2) was also achieved through a Wittig reaction; however, availability and inefficiency of certain starting materials required that 2,3,4-trihydroxybenzaldehyde be used as a starting substance. After selective methylation at the para position and silylation, the benzaldehyde was allowed to react with the phosphonium bromide of 3,4,5-trimethoxybenzyl alcohol to afford the protected *Z* isomer. After deprotection using tetrabutylammonium fluoride, CA1 was obtained.

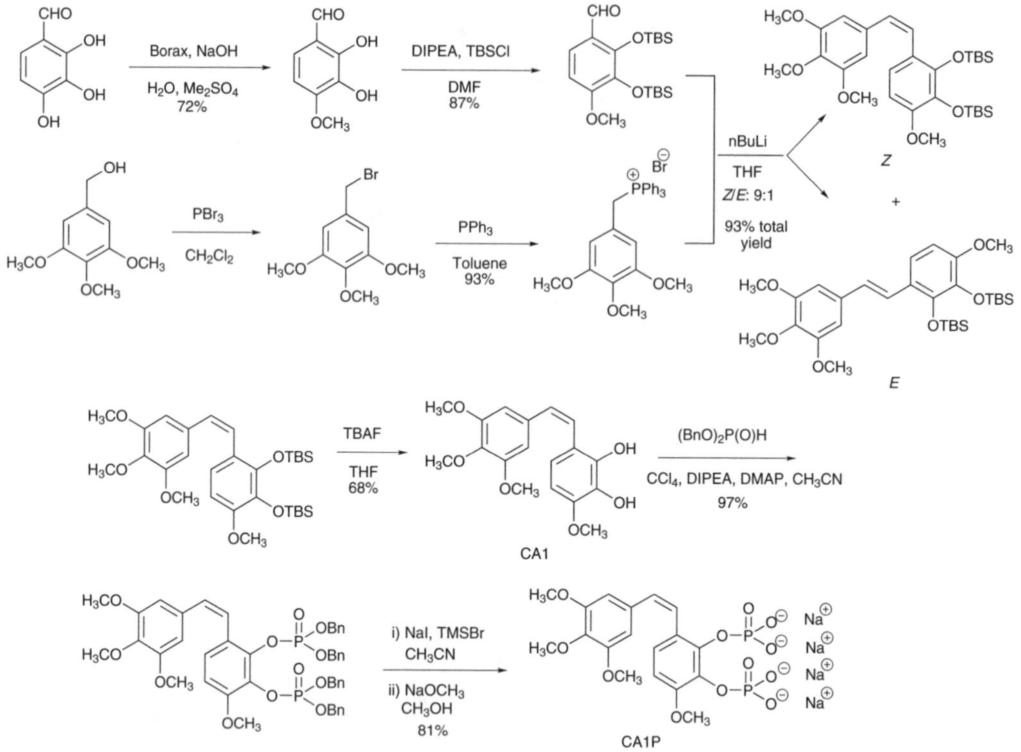

SCHEME 3.2 Synthesis of combretastatin A-1 and its prodrug combretastatin A-1 phosphate.

The sodium phosphate salt of CA4[82] and disodium phosphate salt of CA1[44] were both synthesized via an efficient sequence using a dibenzylphosphite-carbon tetrachloride procedure, followed by debenzylation with sodium iodide and chlorotrimethylsilane.

In addition to the Wittig reaction used by Pettit and coworkers, other methods have been developed to synthesize CA4 and CA1, including a Perkin condensation of 3,4,5-trimethoxyphenylacetic acid and 3-hydroxy-4-methoxybenzaldehyde,[85] a Suzuki cross-coupling of an ethenyl bromide and 3,4,5-trimethoxyphenyl boronic acid,[85] a biooxidation of arenes to catechols using *Escherichia coli* JM109 followed by a Pd-catalyzed Suzuki coupling,[86] and a modified Suzuki reaction through a hydroboration/protonation of a diaryl-alkyne.[87]

Further study of *C. caffrum* constituents led to the isolation, characterization, and synthesis of B, C, and D series members. In general, B series constituents were synthesized by methods analogous to A series syntheses followed by hydrogenation.[48,49,51] For example, for CB1,[48] a mixture of CA1 in methanol and 5% Pd/C was treated with a positive pressure of H_2 at ambient temperature overnight to afford CB1.

Synthesis of CC1[52] (Scheme 3.3) was achieved through photochemical cyclization of the corresponding stilbene. The improved synthesis of CC1 included selective demethylation of 2,3,4-trimethoxybenzaldehyde with aluminum chloride followed by silylation. Subsequent coupling of the aldehyde with the corresponding ylide and photochemical cyclization, using iodine and a mercury lamp, provided the phenanthrene product. Desilylation using tetrabutylammonium fluoride afforded the phenol, which, on Fremy's salt oxidation, yielded the quinone CC1.

The unusual macrocyclic lactones, CD1 and CD2, have been synthesized by several groups.[83,88–90] Boger's synthesis using an intramolecular Ullmann macrocyclization[83] to afford CD2 is presented here (Scheme 3.4). Esterification of 3-(3-hydroxy-4-methoxyphenyl)propanoic acid with (Z)-3-(4-iodophenyl)-2-propenol under Mitsunobu conditions provided the bridged diaryl ester.

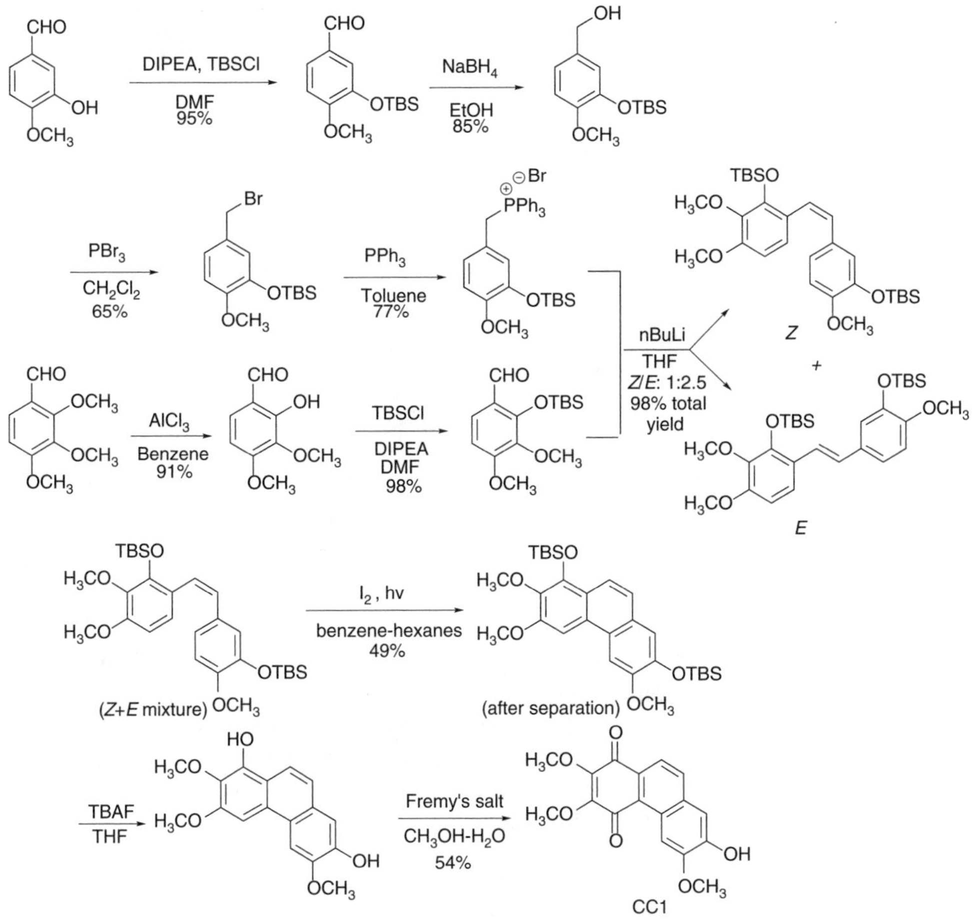

SCHEME 3.3 Synthesis of combretastatin C1.

Upon intramolecular Ullman macrocyclization with CuCH$_3$ and pyridine, followed by aryl methyl ether deprotection, CD2 was obtained.

V. COMBRETASTATIN DERIVATIVES AND SYNTHETIC ANALOGS INSPIRED BY THE COMBRETASTATINS

Many compounds (Figs. 3.3 and 3.4) have been synthesized to mimic the structure and efficacy of the combretastatins, especially CA4 and CA1.[13,16,91] A wide variety of combretastatin analogs with functional group modification have been prepared with the A and B aryl rings and the ethylene bridge intact. A sampling of these compounds includes 3-nitro,[64,92] 3-amino,[64,92] 3-amino acid salts,[92,93] 2-nitrogen substituted derivatives,[94] 3-azido,[64] 3,4-methylenedioxy-3-amino derivatives,[95] 3-fluoro,[96] 4-methyl,[97] and 3,4,5-trimethyl[98] combretastatin analogs (Figure 3.3). In addition, many combretastatin analogs feature significant structural variation from the basic stilbenoid molecular core structure. A representative group (not intended to be inclusive) includes phenstatin,[99] hydroxyphenstatin,[100] heterocombretastatins,[101] combretatropones,[102,103] *aza*-combretastatins,[104] bridge-modified vicinal diols,[105,106] combretadioxolane analogs,[107–109] chalcone derivatives,[110] disubstituted imidazole analogs,[111] pyridone analogs,[112] and sulfonamide analogs[113,114] (Figure 3.3). In addition, 3-aroyl-2-aryl indoles,[72,115,116] 2,3-diaryl indoles,[116] 3-formyl indoles,[117] benzo[b]thiophene,[43,118-123] and benzofuran analogs,[116,124,125] along with dihydronaphthalenes,[119,126,127] dihydroxyindolo[2,1-α]isoquinolines,[128]

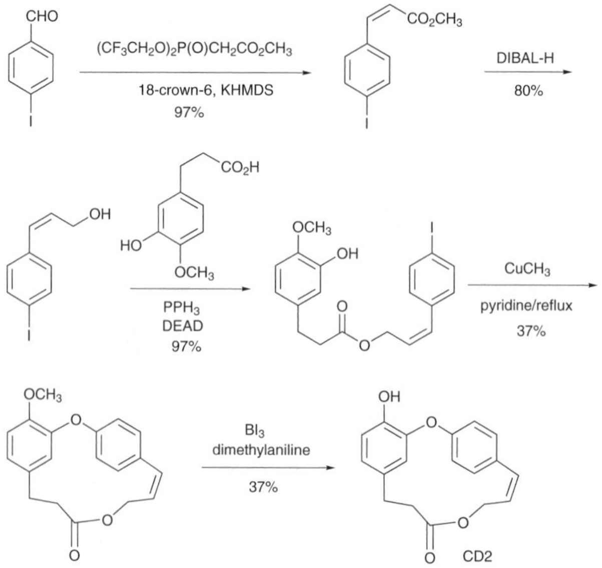

SCHEME 3.4 Synthesis of CD2.

and indolyloxazoline analogs,[129,130] have been prepared (Figure 3.4). The relative simplicity of the basic stilbenoid molecular architecture characteristic of CA4 and CA1, along with the ease of synthesis and the versatility afforded by the variety of synthetic routes available for the formation of stilbenoids, has resulted in extensive structure activity relationship studies around the combretastatins. These studies have led to the discovery and development of a large number of structurally distinct combretastatin analogs. In fact, the literature, with regard to the combretastatins, has exploded in the last 10 years. For example, in 1994, a literature search for CA4 would turn up approximately 30 "hits," and a search today (2004) yields over 600 papers and professional presentations.

Three small-molecule VTAs that are currently in human clinical development function through a biological mechanism ultimately involving the tubulin–protein system (Figure 3.5). These drugs include CA4P (discovered by George R. Pettit, Arizona State University, sponsored by Oxigene, Inc.; see appropriate references throughout), AVE8062[92,93,131] (discovered by Ajinomoto, Inc. [AC7700] and sponsored by Aventis), and ZD-6126[132–135] (licensed from Angiogene Pharmaceuticals UK and sponsored by AstraZeneca). The similarity of CA4P and AVE8062 to the natural product CA4 is obvious. Likewise, the six, seven, six ring fusion in ZD-6126 is highly reminiscent of the six, seven, seven (tropolone) ring system of the natural product colchicine.

VI. PRECLINICAL STUDIES

The recent interest in the anticancer potential of the combretastatins has gained momentum from the observation that CA4P can induce rapid and selective blood flow shutdown within tumor tissue.[76,136,137] As the development and activity of CA4P as a VTA has been the subject of several recent reviews[15,22,74,138] only a brief summary of the preclinical data will be provided here.

A. EFFECTS OF CA4P ON TUMOR BLOOD FLOW

A number of studies have demonstrated the rapid and selective effects of CA4P on tumor vascular function.[41,137,139–148] These studies have established that blood flow reductions occur in transplanted and spontaneous rodent tumors and in xenografted human tumors. Blood flow effects can be measured within 10–20 minutes of CA4P administration, a timeframe that closely mimics the *in vitro* effects on endothelial cell shape. These blood flow effects persist for hours, although in some

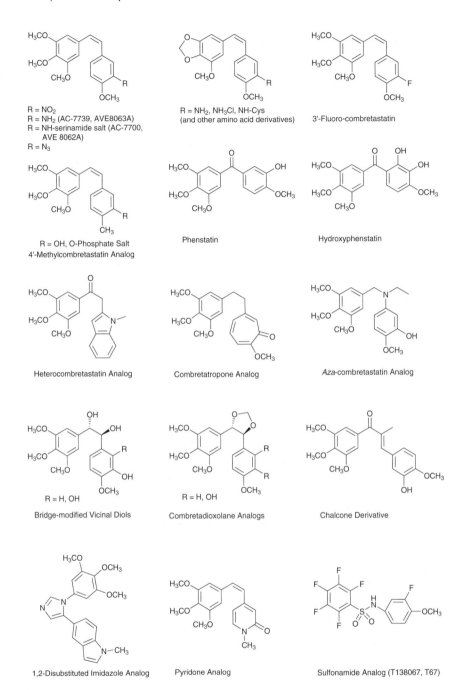

FIGURE 3.3 Combretastatin derivatives.

tumors restoration of flow can be observed 24 hours following treatment. Because it is known that dramatic tumor cell kill is induced within a few hours of cutting off the blood supply,[149] even short periods of blood flow reduction are expected to induce a significant reduction in tumor cell burden.

B. TUMOR RESPONSE TO CA4P AS A SINGLE AGENT

Measurements of clonogenic tumor cell survival in tumors excised following CA4P treatment have shown that extensive tumor cell kill is induced.[41,150] Indeed, these studies show that 24 hours after

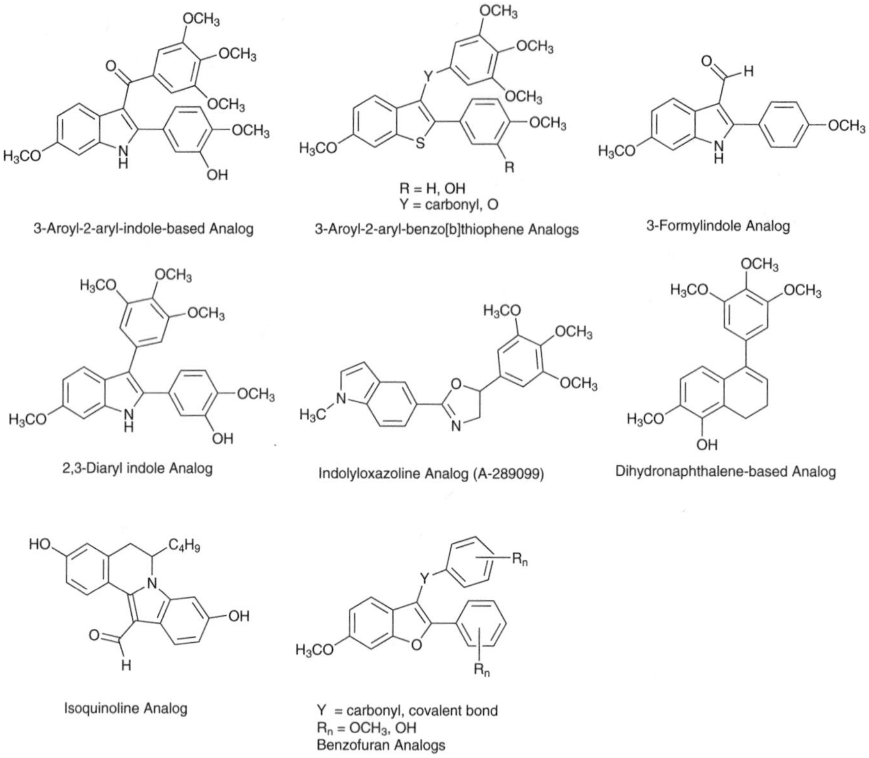

FIGURE 3.4 Synthetic analogs inspired by the combretastatins.

a dose of 100 mg/kg, tumor cell survival is reduced by 90%–99% in a number of experimental tumors. Despite the extensive cell kill observed following the vascular shutdown achieved with CA4P, significant growth retardation is very rarely seen. This lack of retardation has been attributed to rapid regrowth from the rim of viable cells surviving at the tumor periphery.[41] Of interest is that treatment with conventional cytotoxic chemotherapy or radiation at doses that induce similar or

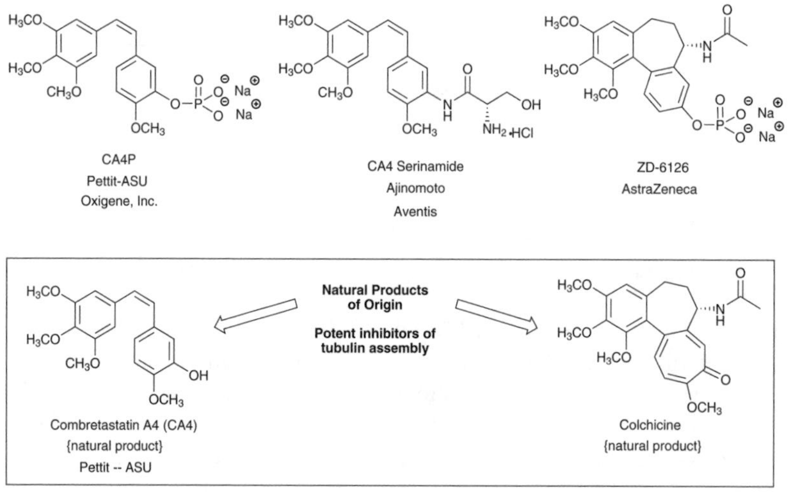

FIGURE 3.5 Small-molecule vascular targeting agents in clinical development.

lower levels of cell kill can induce significant retardation of tumor growth.[74] One possible explanation for this difference is that the cells in the periphery have access to a more nutritive blood flow than those in the center, which are dependent on the relatively poor and chaotic vascular network present within the tumor mass. This has several consequences, one of which is that the hypoxic and nutrient-deprived cells that are more prevalent in the tumor core will be more resistant to conventional treatments such as radiation and cytotoxic approaches, which are known to be more toxic to well-oxygenated and rapidly proliferating cell populations.[151] Therefore, if this is the case, treatments such as radiation and cisplatin, in contrast to CA4P, will preferentially kill cells in the periphery and spare those in the center of the tumor. Thus, if the cells that have survived treatment are already proliferating and adjacent to a good supply of oxygen and nutrients, they will initiate rapid regrowth, whereas, if the surviving cells are less well supplied, there will be a lag phase before proliferation is initiated. If this hypothesis is correct, CA4P, along with other VTAs, offers great potential to enhance the effectiveness of current treatment approaches by the complementary cell killing of different cell populations.

C. Treatment Response to CA4P in Combination Therapy

Several studies have already indicated the therapeutic potential of combining CA4P with radiation and a variety of chemotherapy agents.[41,142,152] In addition, CA4P has been shown to enhance other therapeutic approaches such as hyperthermia, antibody-based approaches, and even antiangiogenic agents.[146,153–155]

In a study by Siemann and colleagues, the tumor cell kill induced by cisplatin or cyclophosphamide was enhanced by a factor of 10–500 by post-treatment administration of CA4P.[150] Also, no increase in the bone marrow stem-cell toxicity associated with these anticancer drugs was observed, giving rise to a real therapeutic gain. Numerous other studies have shown significant enhancement of the effects of conventional chemotherapeutic agents on tumor response, including cisplatin,[41] fluorouracil,[156] doxorubicin,[157] chlorambucil,[15] melphalan,[15] CPT-11,[158] and Taxol®.[41]

It has been shown that CA4P, when administered shortly following radiation treatment, can significantly enhance tumor response.[77,142,153,154,159] In contrast, no enhancement of radiation damage has been observed in normal tissue.[160]

Combining VTAs with hyperthermia is another potentially useful approach. Clear evidence exists that blood flow to the tumor is important in determining the response to heat.[161] Furthermore, a number of studies have now shown that decreasing tumor blood flow by CA4P treatment can provide an effective means of improving tumor response to hyperthermia.[154,162]

One of the most dramatic enhancements of activity by CA4P in preclinical models was reported by Pedley and colleagues.[155] In a study evaluating the efficacy of a radiolabeled anti-CEA antibody against an established human colon cancer in mice, the researchers found that administration of CA4P could improve long-term tumor control from 0% to 85%.

D. Preclinical Studies with CA1P (OXi04503)

The preclinical activity of CA1P has only recently been evaluated.[163–166] However, the data show that not only is it a much more potent agent than CA4P, but it can also induce tumor growth delays and regressions when used as a single agent (something not usually observed with CA4P). This enhanced activity was unexpected based on the *in vitro* data for tubulin binding and for inhibition of cell proliferation for the active parent drug, CA1. Because regressions are observed, it indicates that CA1P, in addition to the vascular effects, is also directly attacking the remaining viable cells at the rim of the tumor mass. One possible explanation for this unexpected activity is that it is metabolized *in vivo* to a reactive and cytotoxic *o*-quinone. Supportive evidence for this has been recently provided in a paper by Kirwan and colleagues, who show that CA1 is metabolized by tumor tissue to an agent that covalently binds to the cellular contents of the tumor.[167]

TABLE 3.2
Completed and Ongoing Clinical Trials with Combretastatin A-4 Phosphate (OXi 2021) in Oncology

Compound	Indication/Use	Clinical Phase	Notes
Combretastatin A-4P	Solid tumors; single agent	Phase I	Three trials completed
Combretastatin A-4P	Solid tumors; combination with carboplatin	Phase Ib	Patient enrollment complete
Combretastatin A-4P	Head and neck/lung/prostate; combination with radiation	Phase I/II	Ongoing
Combretastatin A-4P	Anaplastic thyroid cancer	Phase II	Ongoing
Combretastatin A-4P	Anaplastic thyroid cancer in combination with chemotherapy and radiotherapy	Phase I/II	Ongoing
Combretastatin A-4P	Advanced ovarian cancer; combination with carboplatin and paclitaxel	Phase I/II	Ongoing
Combretastatin A-4P	Advanced colorectal cancer, combination with iodine-labeled antibody A5B7	Phase I/II	Ongoing

VII. CLINICAL EXPERIENCE[80]

Under the sponsorship of Oxigene, Inc., CA4P and CA1P have moved into clinical trials. To date, there have been nine clinical trials involving CA4P: three completed and six ongoing (Table 3.2), and aspects of the clinical experience with CA4P have been published by Young and Chaplin.[80]

Ninety-six patients with advanced malignancies were enrolled in three phase I clinical trials, beginning in 1998. These trials explored the safety and pharmacokinetics of three different intravenous administration schedules of CA4P as a single agent (i.e., single dose schedule every 21 d ["1x/21"] study); weekly for three consecutive weeks every 28 d ("3x/28" study); and daily for five consecutive days every 21 d ("dx5" study).[168–172] Doses as high as 114 mg/m^2 (weekly schedule) have been studied. Dose-limiting toxicities have been identified in all three trials, resulting in a maximum tolerated dose in the range of 60–68 mg/m^2. Dose-limiting toxicities have included dyspnea, myocardial ischemia, reversible neurological events, and tumor pain.

A major objective of these studies, in addition to establishing the maximum tolerated dose, was to evaluate the effects on tumor blood flow. Significant reductions in tumor blood flow were seen in all three clinical trials.[168–172] Blood flow reductions were observed at all dose levels evaluated between 52 and 114 mg/m^2, with no clear dose response being discernable. Additional studies have shown blood flow reductions down to 30 mg/m^2 (Oxigene, Inc., unpublished data). Thus, the data on humans confirm the results in animals that tumor blood flow reductions can be seen at doses at or below the maximum tolerated dose. These findings have provided the impetus to move CA4P into later-stage clinical trials.

The preclinical data to date for CA4P indicate that although the compound elicits significant reductions in blood flow, resulting in histologically demonstrable tumor necrosis, it appears to leave a rim of viable tumor cells at the periphery. Replication and growth of these few remaining tumor cells results in an insignificant tumor growth delay in a variety of preclinical models. Because of this phenomenon, the clinical focus was directed at combining CA4P with other treatment therapies and modalities that are more effective at the outer tumor regions. Therefore, the current clinical development of CA4P is mainly focused on combinations with several common chemotherapeutics agents, with fractionated external beam radiotherapy, and with an antibody-based radioimmunotherapeutic agent.

A phase Ib, open-label, single-center, dose-escalation clinical trial of CA4P in combination with carboplatin has recently been completed in cancer patients with nonhematological malignancies that have progressed on standard therapy or for whom no life-prolonging treatment is known. The rationale for combining CA4P and carboplatin stems from their different and potentially complementary mechanisms of action and their nonoverlapping toxicity profiles.

Additional active clinical trials include a (phase II) single-agent study to evaluate the safety and efficacy of CA4P in anaplastic thyroid carcinoma on a weekly administration schedule as a single agent. Although clinical effectiveness of CA4P as a single agent was not anticipated, an anaplastic thyroid carcinoma patient experienced a complete pathological response with CA4P alone, and thus, a clinical trial to evaluate the potential of CA4P in this aggressive disease was initiated. Another (phase I/II) clinical trial is also ongoing to evaluate the safety of CA4P in combination with radiation therapy in a variety of solid tumor types, including head and neck, prostate, and non–small cell lung cancer. The fourth active trial is a (phase I/II) safety and tolerability trial of CA4P in combination with carboplatin and paclitaxel in advanced ovarian cancer. A fifth study (phase I/II) recently initiated is CA4P in combination with the ^{131}I-radiolabeled anti-CEA antibody A5B7 for the treatment of colorectal carcinoma. A sixth study (phase I/II) soon to be initiated is CA4P in combination with doxorubicin, cisplatin, and radiotherapy for the treatment of newly diagnosed anaplastic thyroid carcinoma.

Combretastatin A-1P (Oxi-4503) has also shown excellent activity in preclinical studies. Oxi-4503 appears to be able to cause tumor regressions with single-agent activity, whereas CA4P's *in vivo* activity is best with combination studies (i.e., phase Ib with carboplatin). In addition, CA1P has the ability to destroy vessels in all regions of the tumor rather than just at the periphery of the tumor, as is the case with CA4P.[166]

VIII. CODA

In the classic 1962 compilation and discussion of "Medicinal and Poisonous Plants of Southern and Eastern Africa," Watt and Breyer-Brandwijk described the root bark of *C. caffrum* (aka *salicifolium* E. Mey.) as the Zulu charm for harming an enemy. "Some of the powdered material is heated with water in a piece of broken earthenware pot, and the points of two assegais are dipped into the boiling solution. The assegai points are brought near the mouth and the operator licks them and spits in the direction of his enemy whose name is shouted out. This is repeated several times, the weapons being thrust in the same direction and finally into the ground." Commencing with the discovery of the combretastatins, the question has arisen many times as to whether or not some of the indigenous people of Africa and elsewhere had used extracts of *C. caffrum* for medicinal purposes that included treating cancer. The scientific/medical literature appears to be silent on that issue. In that respect, one of us (GRP) was pleased to be contacted in November 1999 by a South African chemist, Case van Hattem. In the helpful correspondence that ensued, van Hattem provided some interesting information. He reports that, "I have spoken with the chairman of the Transkei Traditional Healers Association, and he confirms that they, 'the Xhosa people,' have been using *C. caffrum* for decades and longer for the treatment of cancer amongst other ailments." The chairman also cautioned that extracts of *C. caffrum* should be used in combination with other plants to be effective, presumably to moderate the toxicity. Another useful contribution made by van Hattem was the observation that, "The tribe in the area of *C. caffrum* are Xhosa and not Zulu. There is a gradual change from *C. caffrum* in the Xhosa area to *C. erythrophyllum* in the Zulu area as one travels northwards from East London through the Transkei." Furthermore, the Xhosa and Zulu do not differentiate between these two botanical species, and in his discussions with Professor A. E. van Wyk, curator of the Herbarium, Department of Botany, University of Pretoria, the conclusion is that *C. caffrum* and *C. erythrophyllum* are nearly identical.

Until Mr. Hattem's communications in 1999, none of us involved in the sampling of *C. caffrum* over 30 years ago, and subsequently, in discovery of the combretastatins, had any knowledge of this plant being used in traditional medical treatments for cancer. From a humanitarian viewpoint, this historical uncertainty is essentially moot. This writer (GRP) thinks the real benefit to the people of Zimbabwe and South Africa will be derived from the eventual worldwide success of the CA4 and CA1 prodrugs in treating human cancer. Those drugs should become two of the least expensive anticancer drugs of all time, greatly benefiting both African countries and their citizens. Indeed, if it were not for the NCI and Arizona State University's Cancer Research Institute, the discovery of CA4 prodrug and its development to the present phase I and II human cancer clinical trials might have been delayed many decades or centuries into the future. Hence, there needs to be a worldwide realization that it is in everyone's best interest if the collection of plants and animals for medicinal purposes is facilitated by clear and transparent bioprospecting laws. Both the source country and patients worldwide will then benefit from the investment by governments, universities, and the private sector in the discovery and development of new anticancer and other drugs.

IX. ACKNOWLEDGMENTS

We extend our appreciation and special thanks to each of the colleagues named as coauthors on publications involving the discovery and development of the combretastatins. In addition, we express appreciation for helpful communications with Professor A. E. van Wyk and Mr. Case van Hattem.

REFERENCES

1. Willis, J.C. and Arey Shaw, H.K., *A Dictionary of the Flowering Plants & Ferns*, 8th ed., Cambridge University Press, London, 1973.
2. Watt, J.M. and Breyer-Brandwijk, M.G., *The Medicinal and Poisonous Plants of Southern and Eastern Africa*, 2nd ed., E. and S. Livingston, London, 1962, 194.
3. Hartwell, J.L., *Plants Used Against Cancer*, Quarterman Publications, Laurence, MA, 1982, 108.
4. Wall, M.E. et al., Plant antitumor agents. 3. A convenient separation of tannins from other plant constituents, *J. Pharm. Sci.*, 58, 839, 1969.
5. Palmer, E. and Pitman, N, *Trees of Southern Africa*, A.A. Balkema, Cape Town, 1972, Vol. 3.
6. Kokwaro, O., *Medicinal Plants of East Africa*, East African Literature Bureau, Nairobi, 1976.
7. Samuelsson, G. et al., Inventory of plants used in traditional medicine in Somalia. II. Plants of the families combretaceae to labiatae, *J. Ethnopharm.*, 37, 47, 1992.
8. Iwu, M.M., *Handbook of African Medicinal Plants*, CRC Press, Boca Raton, FL, 1993.
9. Gurib-Fakim, A. and Brendler, T., *Medicinal and Aromatic Plants of Indian Ocean Islands: Madagascar, Comoros, Seychelles, and Mascarenes*, Medpharm Scientific Publishers, Stuttgart, Germany, 2004, 194, 463.
10. Duke, J., *Medicinal Herbs*, 2nd ed., CRC Press, New York, 2002, 539.
11. Pettit, G.R. et al., Antineoplastic agents 338. The cancer cell growth inhibitory. Constituents of terminalia arjuna (combretaceae), *J. Ethnopharm.*, 53, 57, 1996.
12. Chinemana, F. et al., Indigenous plant remedies in Zimbabwe, *J. Ethnopharm.*, 14, 159, 1985.
13. Cushman, M. et al., Synthesis and evaluation of stilbene and dihydrostilbene derivatives as potential anticancer agents that inhibit tubulin polymerization, *J. Med. Chem.*, 34, 2579, 1991.
14. Sackett, D.L., Podophyllotoxin, steganacin, and combretastatin: natural products that bind at the colchicine site of tubulin, *Pharm. Ther.*, 59, 163, 1993.
15. Tozer, G.M. et al., The biology of the combretastatins as tumour vascular targeting agents. *Int. J. Exp. Path.*, 83, 21, 2002.
16. Lin, C.M. et al., Interactions of tubulin with potent natural and synthetic analogs of the antimitotic agent combretastatin: a structure-activity study, *Mol. Pharm.*, 34, 200, 1988.
17. Li, Q. and Sham, H.L., Discovery and development of antimitotic agents that inhibit tubulin polymerization for the treatment of cancer, *Exp. Opin. Ther. Pat.*, 12, 1663, 2002.
18. Rowinsky, E.K. et al., Horizons in cancer therapeutics: from bench to bedside, *Continuing Education*, 3, 1, 2002.
19. Hamel, E., Antimitotic natural products and their interactions with tubulin, *Med. Res. Rev.*, 16, 207, 1996.
20. Thorpe, P.E., Vascular targeting agents as cancer therapeutics, *Clin. Cancer Res.*, 10, 415, 2004.

21. Griggs, J., Metcalfe, J.C., and Hesketh, R., Targeting tumour vasculature: the development of combretastatin A4, *Lancet Onc.*, 2, 82, 2001.
22. Bibby, M.C., Combretastatin anticancer drugs, *Drugs of the Future*, 27, 475, 2002.
23. Cirla, A. and Mann, J., Combretastatins: from natural products to drug discovery, *Nat. Prod. Rep.*, 20, 558, 2003.
24. Nam, N-H., Combretastatin A-4 analogues as antimitotic antitumor agents, *Curr. Med. Chem.*, 10, 1697, 2003.
25. Pettit, G.R. and Lippert, J.W., Preparation of combretastatin A-1 phosphate and combretastatin B-1 phosphate prodrugs with increased solubility, PCT Int. Application WO2001081355 A1, 2001.
26. Pettit, G.R. and Minardi, M.D., Preparation of combretastatin A3 diphosphate prodrugs for the treatment of cancer, PCT Int. Application WO2002102766 A2, 2002.
27. Pettit, G.R. and Singh, S.B, Combretastatin A-4, US Patent 4,996,237, 1991.
28. Pettit, G.R. and Singh, S.B., Isolation, structural elucidation and synthesis of novel antineoplastic substances denominated "combretastatins," US Patent 5,569,786, 1996.
29. Pettit, G. R., Combretastatin A-4 prodrug, US Patent 5,561,122, 1996.
30. Pinney, K. G. et al., Description of anti-mitotic agents which inhibit tubulin polymerization, US Patent 6,350,777, 2000.
31. Pettit, G.R. and Grealish, M.P., Synthesis of hydroxyphenstatin and the prodrugs thereof as anticancer and antimicrobial agents, PCT Int. Application WO2001081288 A1, 2001.
32. Pettit, G.R. and Brian, T., Synthesis and formulation of phenstatin and related prodrugs for use as antitumor agents, PCT Int. Application WO9934788 A1, 1999.
33. Pettit, G.R. and Rhodes, M.R., Preparation and formulation of combretastatin A4 prodrugs and their trans-isomers for use as antitumor agents, PCT Int. Application WO9935150 A1, 1999.
34. Pettit, G.R. and Moser, B.R., Preparation of combretastatin A-2 prodrugs as antitumor and antimicrobial agents, PCT Int. Application WO2003059855 A1, 2003.
35. Kingston, D.G.I., Samaranayake, G., and Ivey, C.A., The chemistry of taxol, a clinically useful anticancer agent, *J. Nat. Prod.*, 53, 1, 1990.
36. Nicolaou, K.C. et al., Synthesis of epothilones A and B in solid and solution phase, *Nature*, 387, 268, 1997.
37. Owellen, R.J. et al., Inhibition of tubulin-microtubule polymerization by drugs of the vinca alkaloid class, *Cancer Res.*, 36, 1499, 1976.
38. Pettit, G.R. et al., The isolation and structure of a remarkable marine animal antineoplastic constituent: dolastatin 10, *J. Am. Chem. Soc.*, 109, 6883, 1987.
39. Pettit, G.R. et al., Antineoplastic agents 365. Dolastatin 10 SAR probes, *Anticancer Drug Des.*, 13, 243, 1998.
40. Lin, C.M. et al., Antimitotic natural products combretastatin A-4 and combretastatin A-2: studies on the mechanism of their inhibition of the binding of colchicine to tubulin, *Biochemistry*, 28, 6984, 1989.
41. Chaplin, D.J., Pettit, G.R., and Hill, S.A., Anti-vascular approaches to solid tumor therapy: evaluation of combretastatin A4 phosphate, *Anticancer Res.*, 19, 189, 1999.
42. Jordon, A. et al., Tubulin as a target for anticancer drugs: agents which interact with the mitotic spindle, *Med. Res. Rev.*, 18, 259, 1998.
43. Pinney, K.G. et al., A new anti-tubulin agent containing the benzo[b]thiophene ring system, *Bioorg. Med. Chem. Lett.*, 9, 1081, 1999.
44. Pettit, G.R. and Lippert, J.W., Antineoplastic agents 429. Syntheses of the combretastatin A-1 and combretastatin B-1 prodrugs, *Anticancer Drug Des.*, 15, 203, 2000.
45. Pettit, G.R. et al., Antineoplastic agents 322. Synthesis of combretastatin A-4 prodrugs, *Anticancer Drug Des.*, 10, 299, 1995.
46. Pettit, G.R. et al., Isolation and structure of combretastatin, *Can. J. Chem.*, 60, 1374, 1982.
47. Pettit, G.R., Cragg, G.M., and Singh, S.B., Antineoplastic agents, 122. Constituents of *Combretum caffrum*, *J. Nat. Prod.*, 50, 386, 1987.
48. Pettit, G.R. et al., Isolation, structure, and synthesis of combretastatins A-1 and B-1, potent new inhibitors of microtubule assembly, derived from *Combretum caffrum*, *J. Nat. Prod.*, 50, 119, 1987.
49. Pettit, G.R. and Singh, S.B., Isolation, structure, and synthesis of combretastatin A-2, A-3, and B-2, *Can. J. Chem.*, 65, 2390, 1987.
50. Pettit, G.R. et al., Antineoplastic agents. 291. Isolation and synthesis of combretastatins A-4, A-5, and A-6, *J. Med. Chem.*, 38, 1666, 1995.

51. Pettit, G.R., Singh, S.B., and Schmidt, J.M., Isolation, structure, synthesis, and antimitotic properties of combretastatins B-3 and B-4 from *Combretum caffrum*, *J. Nat. Prod.*, 51, 517, 1988.

52. Singh, S.B. and Pettit, G.R., Isolation, structure, and synthesis of combretastatin C-1, *J. Org. Chem.*, 54, 4105, 1989.

53. Pettit, G.R., Singh, S.B., and Niven, M.L., Isolation and structure of combretastatin D-1: a cell growth inhibitory macrocyclic lactone from *Combretum caffrum*, *J. Am. Chem. Soc.*, 110, 8539, 1988.

54. Singh, S.B. and Pettit, G.R., Antineoplastic agents. 206. Structure of the cytostatic macrocyclic lactone combretastatin D-2, *J. Org. Chem.*, 55, 2797, 1990.

55. White, A., *Principles of Biochemistry*, 59th ed., McGraw-Hill, New York, 1978.

56. Nogales, E., Wolf, S.G., and Downing, K.H., Structure of the α,β-tubulin dimer by electron crystallography, *Nature*, 391, 199, 1998.

57. Rao, S., Horwitz, S.B., and Ringel, I., Direct photoaffinity labeling of tubulin with taxol, *J. Natl. Cancer Inst.*, 84, 785, 1992.

58. Staretz, M.E. and Hastie, S.B., Synthesis, photochemical reactions, and tubulin binding of novel photoaffinity labeling derivatives of colchicine, *J. Org. Chem.*, 58, 1589, 1993.

59. Hahn, K.M., Hastie, S.B., and Sundberg, R.J., Synthesis and evaluation of 2-diazo-3,3,3-trifluopropanoyl derivatives of colchicine and podophyllotoxin as photoaffinity labels: reactivity, photochemistry, and tubulin binding, *Photochem. Photobiol.*, 55, 17, 1992.

60. Wolff, J. et al., Direct photoaffinity labeling of tubulin with colchicine, *Proc. Natl. Acad. Sci. USA*, 88, 2820, 1991.

61. Floyd, L.J., Barnes, L.D., and Williams, R.F., Photoaffinity labeling of tubulin with (2-nitro-4-azidophenyl)deacetylcolchicine: direct evidence for two colchicine binding sites, *Biochemistry*, 28, 8515, 1989.

62. Bai, R. et al., Identification of cysteine 354 of beta-tubulin as part of the binding site for the A ring of colchicine, *J. Biol. Chem.*, 271, 12639, 1996.

63. Olszewski, J. D. et al., Potential photoaffinity labels for tubulin. Synthesis and evaluation of diazocyclohexadienone and azide analogs of colchicine, combretastatin, and 3,4,5-trimethoxybiphenyl, *J. Org. Chem.*, 59, 4285, 1994.

64. Pinney, K.G. et al., Synthesis and biological evaluation of aryl azide derivatives of combretastatin A-4 as molecular probes for tubulin, *Bioorg. Med. Chem.*, 8, 2417, 2000.

65. Sawada, T. et al., A fluorescent probe and a photoaffinity labeling reagent to study the binding site of maytansine and rhizoxin on tubulin, *Bioconj. Chem.*, 4, 284, 1993.

66. Sawada, T. et al., Identification of the fragment photoaffinity-labeled with azidodansyl-rhizoxin as Met-363-Lys-379 on beta-tubulin, *Biochem. Pharm.*, 45, 1387, 1993.

67. Sawada, T. et al., Fluorescent and photoaffinity labeling derivatives of rhizoxin, *Biochem. Biophys. Res. Comm.*, 178, 558, 1991.

68. Chavan, A.J. et al., Forskolin photoaffinity probes for the evaluation of tubulin binding sites, *Bioconj. Chem.*, 4, 268, 1993.

69. Safa, A.R., Hamel, E., and Felsted, R.L., Photoaffinity labeling of tubulin subunits with a photoactive analogue of vinblastine, *Biochemistry*, 26, 97, 1987.

70. Williams, R.F. et al., A photoaffinity derivative of colchicine: 6-(4-azido-2-nitrophenylamino)hexanoyldeacetylcolchicine; photolabeling and location of the colchicine-binding site on the alpha-subunit of tubulin, *J. Biol. Chem.*, 260, 13794, 1985.

71. Nogales, E., Structural insights into microtubule function, *Ann. Rev. Biochem.*, 69, 277, 2000.

72. Hadimani, M. et al., 2-(3-tert-Butyldimethylsiloxy-4-methoxyphenyl)-6-methoxy-3-(3,4,5-trimethoxybenzoyl)indole, *Acta Cryst., Sec. C Crystal Struct. Comm.*, C58, 330, 2002.

73. Siemann, D.W., Vascular targeting agents: an introduction, *Int. J. Radiat. Oncol. Biol. Phys.*, 54, 1472, 2002.

74. Chaplin, D.J. and Hill, S.A., The development of combretastatin A4 phosphate as a vascular targeting agent, *Int. J. Radiat. Oncol. Biol. Phys.*, 54, 1491, 2002.

75. Thorpe, P.E., Chaplin, D.J., and Blakeley, D.C., The first international conference on vascular targeting: meeting overview, *Cancer Res.*, 63, 1144, 2003.

76. Chaplin, D.J. and Dougherty, G.J., Tumour vasculature as a target for cancer therapy, *Br. J. Cancer*, 80, 57, 1999.

77. Kanthou, C. and Tozer, G.M., The tumor vascular targeting agent combretastatin A-4-phosphate induces reorganization of the actin cytoskeleton and early membrane blebbing in human endothelial cells, *Blood*, 99, 2060, 2002.

78. Tozer, G.M. et al., Mechanisms associated with tumor vascular shut-down induced by combretastatin A-4 phosphate: intravital microscopy and measurement of vascular permeability, *Cancer Res.*, 61, 6413, 2001.

79. Galbraith, S.M. et al., Effects of combretastatin A4 phosphate on endothelial cell morphology *in vitro* and relationship to tumor vascular targeting activity *in vivo*, *Anticancer Res.*, 21, 93, 2001.

80. Young, S. and Chaplin, D.J., Combretastin A4 phosphate: background and current clinical status, *Expert Opin. Investig. Drugs*, 13, 1171, 2004.

81. Pettit, G.R. et al., Isolation and structure of the strong cell growth and tubulin inhibitor combretastatin A-4, *Experientia*, 45, 209, 1989.

82. Pettit, G.R. and Rhodes, M.R., Antineoplastic agents 389. New syntheses of the combretastatin A-4 prodrug, *Anticancer Drug Des.*, 13, 183, 1998.

83. Boger, D.L., Sakya, S.M., and Yohannes, D., Total synthesis of combretastatin D-2: intramolecular Ullmann macrocyclization reaction, *J. Org. Chem.*, 56, 4204, 1991.

84. Deleted from text.

85. Gaukroger, K. et al., Novel syntheses of cis and trans isomers of combretastatin A-4, *J. Org. Chem.*, 66, 8135, 2001.

86. Bui, V.P. et al., Direct biooxidation of arenes to corresponding catechols with *E. coli* JM109 (PDTG602). Application to synthesis of combretastatins A-1 and B-1, *Tetrahedron Lett.*, 43, 2839, 2002.

87. Lawrence, N.J. et al., The synthesis of (E) and (Z)-combretastatins A-4 and a phenanthrene from *Combretum caffrum*, *Synthesis*, 9, 1656, 1999.

88. Deshpande, V.H. and Gokhale, N.J., Synthesis of combretastatin D-2, *Tetrahedron Lett.*, 33, 4213, 1992.

89. Couladouros, E.A. and Soufli, I.C., Total synthesis of natural (-)-combretastatin D-1, *Tetrahedron Lett.*, 36, 9369, 1995.

90. Couladouros, E.A. and Soufli, I.C., Synthesis of combretastatin D-2. An efficient route to caffrane macrolactones, *Tetrahedron Lett.*, 35, 4409, 1994.

91. Chaplin, D.J. et al., Functionalized stilbene derivatives as improved vascular targeting agents, US Patent 20030149003 A1, 2003.

92. Ohsumi, K. et al., Novel combretastatin analogues effective against murine solid tumors: design and structure-activity relationships, *J. Med. Chem.*, 41, 3022, 1998.

93. Ohsumi, K. et al., Synthesis and antitumor activities of amino acid prodrugs of amino-combretastatins, *Anticancer Drug Des.*, 14, 539, 1999.

94. Siles, R., Hadimani, M., and Pinney, K.G., Cancer chemotherapy: prodrug strategies for enhanced bioavailability and selectivity of vascular targeting agents, 58th Southwest Regional Meeting of The American Chemical Society (abstract 236), Austin, TX, November 4, 2002.

95. Pettit, G.R. et al., Antineoplastic agents. 487. Synthesis and biological evaluation of the antineoplastic agent 3,4-methylenedioxy-5,4-dimethoxy-3-amino-Z-stilbene and derived amino acid amides, *J. Med. Chem.*, 46, 525, 2003.

96. Lawrence, N.J. et al., Synthesis and anticancer activity of fluorinated analogues of combretastatin A-4, *J. Fluorine Chem.*, 123, 101, 2003.

97. Davis, P.D., Compositions with vascular damaging activity, PCT Int. Application WO0214329 A1, 2002.

98. Gaukroger, K., Structural requirements for the interaction of combretastatins with tubulin: how important is the trimethoxy unit? *Org. Biomol. Chem.*, 1, 3033, 2003.

99. Pettit, G.R. et al., Antineoplastic agents. 379. Synthesis of phenstatin phosphate, *J. Med. Chem.*, 41, 1688, 1998.

100. Pettit, G.R. et al., Antineoplastic agents. 443. Synthesis of the cancer cell growth inhibitor hydroxyphenstatin and its sodium diphosphate prodrug, *J. Med. Chem.*, 43, 2731, 2000.

101. Medarde, M. et al., Synthesis and antineoplastic activity of combretastatin analogues: heterocombretastatins, *European J. Med. Chem.*, 33, 71, 1998.

102. Andres, C.J. et al., "Combretatropones" — hybrids of combretastatin and colchicine. Synthesis and biochemical evaluation, *Bioorg. Med. Chem. Lett.*, 3, 565, 1993.

103. Janik, M.E. and Bane, S.L., Synthesis and antimicrotubule activity of combretatropone derivatives, *Bioorg. Med. Chem.*, 10, 1895, 2002.

104. Shirai, R. et al., Synthesis and anti-tubulin activity of *aza*-combretastatins, *Bioorg. Med. Chem. Lett.*, 4, 699, 1994.

105. Pettit, G.R. et al., Antineoplastic agents. 410. Asymmetric hydroxylation of trans-combretastatin A-4, *J. Med. Chem.*, 42, 1459, 1999.

106. Pettit, G.R. et al., Antineoplastic agents 440. Asymmetric synthesis and evaluation of the combretastatin A-1 SAR probes (1S, 2S)-and (1R, 2R)-1,2-dihydroxy-1-(2,3-dihydroxy-4-methoxyphenyl)-2-(3,4,5-trimethoxyphenyl)-ethane, *J. Nat. Prod.*, 63, 969, 2000.

107. Shirai, R. et al., Asymmetric synthesis of antimitotic combretadioxolane with potent antitumor activity against multi-drug resistant cells, *Bioorg. Med. Chem. Lett.*, 8, 1997, 1998.

108. Pettit, G.R. et al., Antineoplastic agents 442. Synthesis and biological activities of dioxostatin, *Anticancer Drug Des.*, 15, 361, 2000.

109. Shirai, R., Okabe, T., and Iwasaki, S., Synthesis of comformationally restricted combretastatins, *Heterocycles*, 46, 145, 1997.

110. Ducki, S. et al., Potent antimitotic and cell growth inhibitory properties of substituted chalcones, *Bioorg. Med. Chem. Lett.*, 8, 1051, 1998.

111. Wang, L. et al., Potent, orally active heterocycle-based combretastatin A-4 analogues: synthesis, structure-activity relationship, pharmacokinetics, and *in vivo* antitumor activity evaluation, *J. Med. Chem.*, 45, 1697, 2002.

112. Hatanaka, T. et al., Novel B ring modified combretastatin analogues: syntheses and antineoplastic activity, *Bioorg. Med. Chem. Lett.*, 8, 3371, 1998.

113. Medina, J.C. et al., Pentafluorobenzenesulfonamides and analogs, US Patent 5880151, 1999.

114. Flygare, J.A. et al., Pentafluorobenzenesulfonamides and analogs, US Patent 6121304, 2000.

115. Pinney, K.G., Wang, F., and Del Pilar Mejia, M., Preparation of indole-containing and combretastatin-related anti-mitotic and anti-tubulin polymerization agents, PCT Int. Application WO2001019794 A2, 2001.

116. Flynn, B.L., Hamel, E., and Jung, M.K., One-pot synthesis of benzo[b]furan and indole inhibitors of tubulin polymerization, *J. Med. Chem.*, 45, 2670, 2002.

117. Gastpar, R. et al., Methoxy-substituted 3-formyl-2-phenylindoles inhibit tubulin polymerization, *J. Med. Chem.*, 41, 4965, 1998.

118. Chen, Z. et al., Preparation of new anti-tubulin ligands through a dual-mode, addition-elimination reaction to a bromo-substituted,-unsaturated sulfoxide, *J. Org. Chem.*, 65, 8811, 2000.

119. Pinney, K.G. et al., Preparation of trimethoxyphenyl-containing tubulin binding ligands and corresponding prodrug constructs as inhibitors of tubulin polymerization and antimitotic agents, PCT Int. Application WO2001068654, 2001.

120. Pinney, K.G. et al., Anti-mitotic agents which inhibit tubulin polymerization, US Patent 6162930, 2000.

121. Pinney, K.G. et al., Anti-mitotic agents which inhibit tubulin polymerization, Australian Patent 732917, 2001.

122. Flynn, B.L. et al., The synthesis and tubulin binding activity of thiophene-based analogues of combretastatin A-4, *Bioorg. Med. Chem. Lett.*, 11, 2341, 2001.

123. Flynn, B.L., Verdier-Pinard, P., and Hamel, E., A novel palladium-mediated coupling approach to 2,3-disubstituted benzo[b]thiophenes and its application to the synthesis of tubulin binding agents, *Org. Lett.*, 3, 651, 2001.

124. Pinney, K.G. et al., Preparation of aryl and arylcarbonylbenzothiophenes, -benzofurans, -indenes, and indoles as tubulin binding ligands and corresponding prodrug constructs thereof useful as antitumor agents, US Patent 2004044059, 2004.

125. Flynn, B.L. and Hamel, E., Synthesis of indoles, benzofurans, and related compounds as potential tubulin binding agents, PCT Int. Application WO2002060872 A1, 2002.

126. Pinney, K.G. et al., Tubulin binding ligands and corresponding prodrug constructs, US Patent 6593374, 2003.

127. Ghatak, A. et al., Synthesis of methoxy and hydroxy containing tetralones: versatile intermediates for the preparation of biologically relevant molecules, *Tetrahedron Lett.*, 44, 4145, 2003.

128. Goldbrunner, M. et al., Inhibition of tubulin polymerization by 5,6-dihydroindolo[2,1-α]isoquinoline derivatives, *J. Med. Chem.*, 40, 3524, 1997.

129. Li, Q. et al., Synthesis and biological evaluation of 2-indolyloxazolines as a new class of tubulin polymerization inhibitors. Discovery of A-289099 as an orally active antitumor agent, *Bioorg. Med. Chem. Lett.*, 12, 465, 2002.

130. Tahir, S.K. et al., Biological activity of A-289099: an orally active tubulin-binding indolyloxazoline derivative, *Mol. Cancer Ther.*, 2, 227, 2003.

131. Hori, K. and Saito, S., Microvascular mechanisms by which the combretastatin A-4 derivative AC7700 (AVE8062) induces tumour blood flow stasis, *Br. J. Cancer*, 89, 1334, 2003.

132. Blakey, D.C. et al., ZD6126: a novel small molecule vascular targeting agent, *Int. J. Radiat. Onc. Biol. Phys.*, 54, 1497, 2002.

133. Siemann, D.W. and Rojiani, A.M., Antitumor efficacy of conventional anticancer drugs is enhanced by the vascular targeting agent ZD6126, *Int. J. Radiat. Oncol. Biol. Phys.*, 54, 1512, 2002.

134. Siemann, D.W. and Rojiani, A.M., Enhancement of radiation therapy by the novel vascular targeting agent ZD6126, *Int. J. Radiat. Oncol. Biol. Phys.*, 53, 164, 2002.

135. Micheletti, G. et al., Vascular-targeting activity of ZD6126, a novel tubulin-binding agent, *Cancer Res.*, 63, 1534, 2003.

136. Chaplin, D.J. et al., Antivascular approaches to solid tumour therapy: evaluation of tubulin binding agents, *Br. J. Cancer*, 27, S86, 1996.

137. Dark, G.D. et al., Combretastatin A-4, an agent that displays potent and selective toxicity toward tumor vasculature, *Cancer Res.*, 57, 1829, 1997.

138. West, C.M.L. and Price, P., Combretastatin A4 phosphate, *Anticancer Drugs*, 15, 179, 2004.

139. McGown, A.T. and Fox, B.W., Structural and biochemical comparison of the anti-mitotic agents colchicine, combretastatin A4 and amphethinile, *Anticancer Drug Des.*, 3, 249, 1989.

140. Hamel, E. and Lin, C.M., Interactions of combretastatin, a new plant-derived antimitotic agent, with tubulin, *Biochem. Pharm.*, 32, 3864, 1983.

141. Horsman, M.R. et al., The effect of combretastatin A-4 disodium phosphate in a C3H mouse mammary carcinoma and a variety of murine spontaneous tumors, *Int. J. Radiat. Oncol. Biol. Phys.*, 42, 895, 1998.

142. Li, L., Rojiani, A., and Siemann, D.W., Targeting the tumor vasculature with combretastatin A-4 disodium phosphate: effects on radiation therapy, *Int. J. Radiat. Oncol. Biol. Phys.*, 42, 899, 1998.

143. Tozer, G.M., et al., Combretastatin A-4 phosphate as a tumor vascular-targeting agent: early effects in tumors and normal tissues, *Cancer Res.*, 57, 1829, 1997.

144. Grosios, K. et al., *In vivo* and *in vitro* evaluation of combretastatin A-4 and its sodium phosphate prodrug, *Br. J. Cancer*, 81, 1318, 1999.

145. Beauregard, D.A. et al., Magnetic resonance imaging and spectroscopy of combretastatin A4 prodrug-induced disruption of tumour perfusion and energetic status, *Br. J. Cancer*, 77, 1761, 1998.

146. Eikesdal, H.P. et al., The new tubulin-inhibitor combretastatin A-4 enhances thermal damage in the BT4An rat glioma, *Int. J. Radiat. Oncol. Biol. Phys.*, 46, 645, 2000.

147. Maxwell, R.J. et al., Effects of combretastatin on murine tumours monitored by ^{31}P MRS, ^{1}H MRS and ^{1}H MRI, *Int. J. Radiat. Oncol. Biol. Phys.*, 42, 891, 1998.

148. Ahmed, B. et al., Vascular targeting effect of combretastatin A-4 phosphate dominates the inherent angiogenesis inhibitory activity, *Int. J. Cancer*, 105, 20, 2003.

149. Chaplin, D.J. and Horsman, M.R., The influence of tumour temperature on ischemia-induced cell death: potential implications for the evaluation of vascular mediated therapies, *Radiother. Oncol.*, 30, 59, 1994.

150. Siemann, D.W. et al., Vascular targeting agents enhance chemotherapeutic agent activities in solid tumor therapy, *Int. J. Cancer*, 99, 1, 2002.

151. Chaplin, D.J. et al., Therapeutic significance of microenvironmental factors, Medical *Radiology, (Radiation Oncology): Blood Perfusion and Microenvironment of Human Tumors*, Springer, Berlin, 1998, 133.

152. Horsman, M.R. et al., Combretastatins: novel vascular targeting drugs for improving anti-cancer therapy. Combretastatins and conventional therapy, *Adv. Exp. Med. Biol.*, 476, 311, 2000.

153. Landuyt, W. et al., *In vivo* antitumor effect of vascular targeting combined with either ionizing radiation or anti-angiogenesis treatment, *Int. J. Radiat. Oncol. Biol. Phys.*, 49, 443, 2001.

154. Horsman, M.R. and Murata, R., Combination of vascular targeting agents with thermal or radiation therapy, *Int. J. Radiation Oncol. Biol. Phys.*, 54, 1518, 2002.

155. Pedley, R.B. et al., Eradication of colorectal xenografts by combined radioimmunotherapy and combretastatin A-4 3-O-phosphate, *Cancer Res.*, 61, 4716, 2001.

156. Grosios, K. et al., Combination chemotherapy with combretastatin A-4 phosphate and 5-fluorouracil in an experimental murine colon adenocarcinoma, *Anticancer Res.*, 20, 229, 2000.

157. Nelkin, B.D. and Ball, D.W., Combretastatin A-4 and doxorubicin combination treatment is effective in a preclinical model of human medullary thyroid carcinoma, *Oncol. Rep.*, 8, 157, 2001.

158. Wildiers, H. et al., Combretastatin A-4 phosphate enhances CPT-11 activity independently of the administration sequence, *Eur. J. Cancer*, 40, 284, 2004.

159. Murata, R. et al., Interaction between combretastatin A-4 disodium phosphate and radiation in murine tumors, *Radiother. Oncol.*, 60, 155, 2001.

160. Murata, R., Overgaard, J., and Horsman, M.R., Combretastatin A-4 disodium phosphate: a vascular targeting agent that improves that improves the anti-tumor effects of hyperthermia, radiation, and mild thermoradiotherapy, *Int. J. Radiat. Oncol. Biol. Phys.*, 51, 1018, 2001.

161. Horsman, M.R., Christensen, K.L., and Overgaard, J., Hydralazine-induced enhancement of hyperthermic damage in a C3H mammary carcinoma *in vivo*, *Int. J. Hyperthermia*, 5, 123, 1989.

162. Eikesdal, H.P. et al., Combretastatin A-4 and hyperthermia; a potent combination for the treatment of solid tumors, *Radiother. Oncol.*, 60, 147, 2001.

163. Holwell, S.E. et al., Anti-tumor and anti-vascular effects of the novel tubulin-binding agent combretastatin A-1 phosphate, *Anticancer Res.*, 22, 3933, 2002.

164. Hill, S.A. et al., Preclinical evaluation of the antitumour activity of the novel vascular targeting agent OXi 4503, *Anticancer Res.*, 22, 1453, 2002.

165. Shnyder, S.D. et al., Combretastatin A-1 phosphate potentiates the antitumour activity of cisplatin in a murine adenocarcinoma model, *Anticancer Res.*, 23, 1619, 2003.

166. Hua, J. et al., Oxi4503, a novel vascular targeting agent: effects on blood flow and antitumor activity in comparison to combretastatin A-4 phosphate, *Anticancer Res.*, 23, 1433, 2003.

167. Kirwan, I.G. et al., Comparative preclinical pharmacokinetic and metabolic studies of the combretastatin prodrugs combretastatin A4 phosphate and A1 phosphate, *Clin. Cancer Res.*, 10, 1446, 2004.

168. Dowlati, A. et al., A phase 1b pharmacokinetic and translational study of the novel vascular targeting agent combretastatin A-4 phosphate on a single-dose intravenous schedule in patients with advanced cancer, *Cancer Res.*, 62, 3408, 2002.

169. Stevenson, J.P. et al., Phase I trial of the antivascular agent combretastatin A4 phosphate on a 5-day schedule to patients with cancer: magnetic resonance imaging evidence for altered tumor blood flow, *J. Clin. Oncol.*, 21, 4428, 2003.

170. Rustin, G.J. et al., Phase I clinical trial of weekly combretastatin A4 phosphate: clinical and pharmacokinetic results, *J. Clin. Oncol.*, 21, 2815, 2003.

171. Galbraith, S.M. et al., Combretastatin A4 phosphate has tumor antivascular activity in rat and man as demonstrated by dynamic magnetic resonance imaging, *J. Clin. Oncol.*, 21, 2831, 2003.

172. Anderson, H.L. et al., Assessment of pharmacodynamic vascular response in a phase I trial of combretastatin A4 phosphate, *J. Clin. Oncol.*, 21, 2823, 2003.

4 Homoharringtonine and Related Compounds

Hideji Itokawa, Xihong Wang, and Kuo-Hsiung Lee

CONTENTS

I. Introduction 47
II. Structures of the Drug and Related Compounds 48
    A. Isolation of New Cephalotaxus Alkaloids..48
    B. Modification of the Skeleton of Homoharringtonine (HHT)
        through Unusual Rearrangements 50
III. Synthesis of the Drug 50
    A. Synthesis of Acyl Groups...50
    B. Total Synthesis..53
IV. Mechanism of Action..56
V. Medicinal Chemistry of the Drug...58
VI. Development of the Clinical Drug from the Natural Products Lead............................63
VII. Clinical Applications...63
    A. Chinese Clinical Trials Treating Leukemias with Cephalotaxine Esters 63
    B. Phase I Studies in the United States 63
    C. Phase II Studies in Solid Tumors...63
    D. Phase I–II Studies in Patients with Acute Leukemia...64
    E. Phase II Studies in Myelodysplastic Syndrome (MDS)...64
    F. Studies in Acute Promyelocytic Leukemia (APL)...64
    G. Phase II Studies in Chronic Myeloid Leukemia (CML) Patients 64
VIII. Conclusions 65
Acknowledgments 65
References 65

I. INTRODUCTION

The genus *Cephalotaxus* contains coniferous evergreen trees and shrubs that are indigenous to Asia. Historically, the bark has long been used in China as traditional medicine for a variety of indications. Through screening assays and preliminary clinical trials, Chinese investigators discovered that the total alkaloid fraction from *Cephalotaxus fortunei Hook. F* possessed definite antitumor activity.[1] Homoharringtonine (**1**, HHT) (cephalotaxine 4-methyl-2-hydroxy-4-methylpentyl butanedioate), one of the most active *Cephalotaxus* alkaloids, was obtained by alcoholic extraction from the Chinese evergreen tree *Cephalotaxus harringtonia K. Koch var. harringtonia*.[2] Other *Cephalotaxus* alkaloids have been isolated from various *Cephalotaxus spp.*[3–5]

The parent compound cephalotaxine (**2**) was first isolated by Paudler et al.[3] from two *Cephalotaxus* species. Their initial structural proposal was later revised, and Abraham et al.[6] proposed the correct structure and relative stereochemistry of **2** through x-ray crystallographic analysis of cephalotaxine

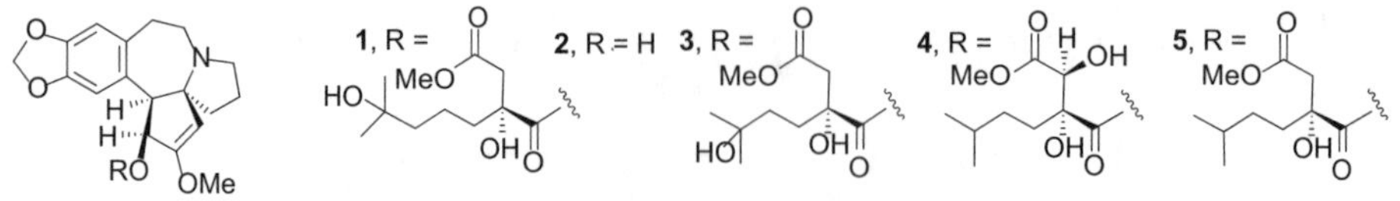

STRUCTURES 1–5

methiodide. The parent compound and its esters represent a group of alkaloids with a unique ring system.[7–12] Although cephalotaxine is devoid of antitumor activity, ester alkaloids derived from cephalotaxine are of particular anticancer interest. The most active compounds in this series are HHT (**1**) and harringtonine (**3**),[8] isolated by Powell et al.[9] In addition to **1** and **3**, the active ester alkaloids isoharringtonine (**4**) and deoxyharringtonine (**5**) have also been isolated from *C. harringtonia*.[8]

HHT (**1**) has been investigated in phase II clinical trials as an anticancer agent at the National Cancer Institute, Bethesda, Maryland. However, its side effects still remain an issue. Work has thus continued on the study of new components from this plant species and on the development of new analogs on the basis of the structure–activity relationships of these alkaloids. This chapter gives an overview of this work. The HHT alkaloids and related cephalotaxine esters have been reviewed previously by Cassady and Douros[13] and by Kantarijian et al.,[14] who selected HHT as the best compound to develop as an anticancer agent.

II. STRUCTURES OF THE DRUG AND RELATED COMPOUNDS

A. ISOLATION OF NEW *CEPHALOTAXUS* ALKALOIDS

In the last several years, more than 20 *Cephalotaxus* alkaloids have been isolated from various *Cephalotaxus* spp.[15–20] *Cephalotaxus* alkaloids isolated by Takano et al.[15–20] were classified into three types based on the side chain structure. One group contains compounds having a carboxyl group at the end of side chain (e.g., 5'-des-*O*-methylharringtonine [**6**], 5'-des-*O*-methylhomoharringtonine [**7**], 5'-des-*O*-methylisoharringtonine [**8**] and 3'*S*-hydroxy-5'-des-*O*-methylharringtonine [**9**]). A second group contains compounds that vary by the number of methylene groups in the side chain (e.g., nordeoxyharringtonine [**10**], homodeoxyharringtonine [**11**], and bishomodeoxyharringtonine [**12**]). The third group of compounds contains terminal aromatic rings in the side chain (e.g., neoharringtonine [**13**], homoneoharringtonine [**14**] and 3*S*-hydroxyneoharringtonine [**15**]).[15–20]

Cephalezomines C-F (**16–19**) were isolated by Morita et al.,[21] and four new *Cephalotaxus* alkaloids, cephalotaxine α–N-oxide (**20**), cephalotaxine β-N-oxide (**21**), 11-β-hydroxycephalotaxine β-N-oxide (**22**), and isocephalotaxine (**23**), together with several known alkaloids, were isolated recently from an EtOAc extract of the fruits of *C. fortunei*.[22] These latter four compounds displayed

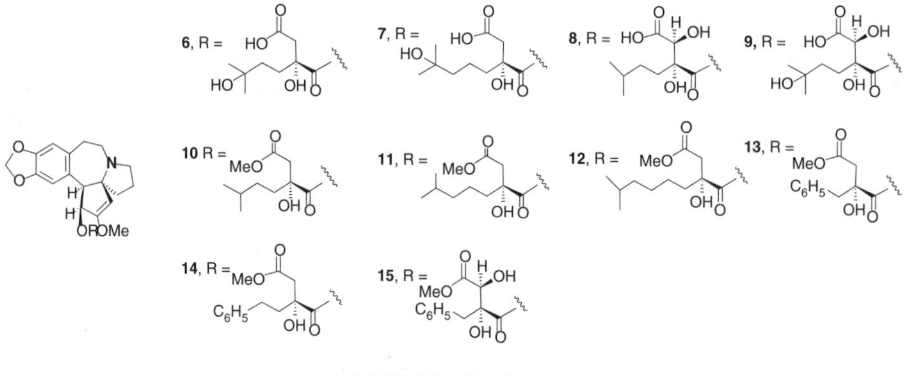

STRUCTURES 6–15

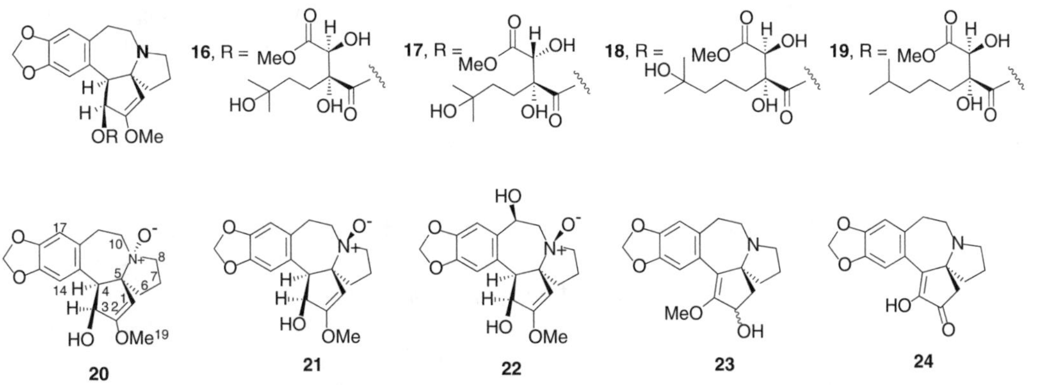

STRUCTURES 16–24

weak cytotoxicity against KB cells with IC_{50} values of 30, 14, 31, and 15 µg/mL, respectively.[22] Desmethylcephalotaxinone (**24**) was obtained from *C. harringtonia* species.[12]

Other related compounds have a hydroxyl group at carbon-11 (e.g., 11α-hydroxyhomodeoxyharringtonine [**25**], 11β-hydroxyhomodeoxyharringtonine [**26**], and 11β-hydroxydeoxyharringtonine [**27**]).[19] In addition, the ester alkaloid drupangtonine (**28**) has a drupacine (**32**) skeleton.[15] Two additional alkaloids, cephalezomines A (**29**) and B (**30**), were obtained from *C. harringtonia* f. *drupacea*.[21]

The Japanese species *C. harringtonia* (Knight) Koch. f. *drupacea* (Sieb. & Zucc.) Kitamura [= *C. harringtonia var. drupacea* (Sieb. & Zucc.) = *C. drupacea* Sieb. & Zucc.] was reported to contain the characteristic *Cephalotaxus* alkaloids HHT (**1**), cephalotaxine (**2**), harringtonine (**3**), isoharringtonine (**4**), deoxyharringtonine (**5**), and drupacine (**32**). 11-Hydroxycephalotaxine (**31**) and drupacine (**32**) were isolated from *C. harringtonia var. drupacea*,[11] and compound **31** was converted to **32** under acidic conditions. The dimeric cephalotaxidine (**33**), obtained by Takano et al.,[18] and cephalocyclidin A (**34**),[23] are examples of unusual *Cephalotaxus* alkaloids. Epischellhammericine,[24] and 3-epischellhammericine[25] were also obtained, but they belong to the homoerythrina alkaloid class.[26]

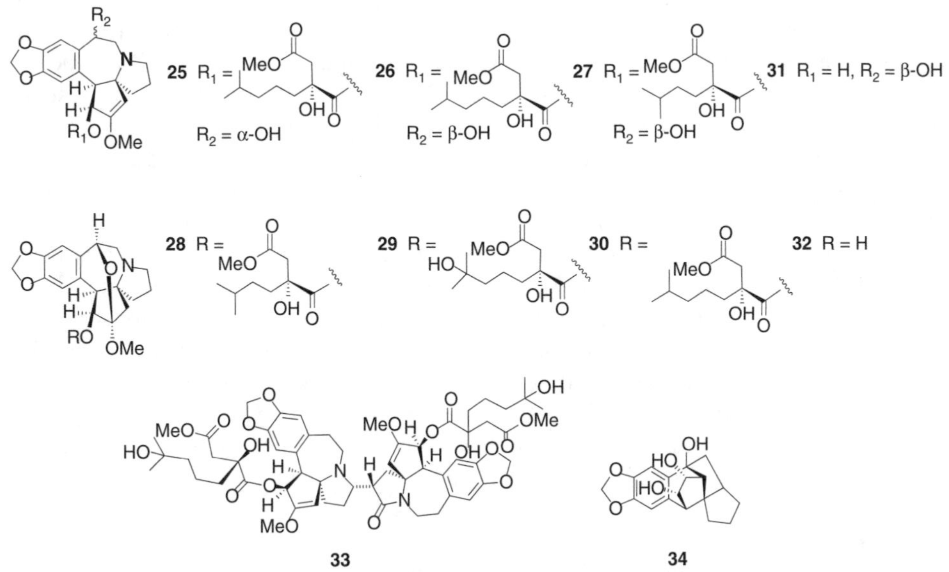

STRUCTURES 25–34

Arora et al.[27,28] determined the absolute stereochemistry of cephalotaxine through x-ray crystallography. Working with the *p*-bromo-benzoate of **2**, the researchers ascertained that natural (-)-cephalotaxine has an 3*S*,4*S*,5*R* configuration.

B. Modification of the Skeleton of Homoharringtonine (HHT) through Unusual Rearrangements[29]

Early biological evaluation indicated that the acyl group of **1** is very important for the activity of *Cephalotaxus*-type alkaloids. Numerous minor alkaloids possessing different ester groups have been isolated from *Cephalotaxus* plants, and previous efforts toward the modification of *Cephalotaxus* alkaloids addressed the replacement of the parent ester moiety of **1** by various acyl groups.

Several skeleton-modified analogs of HHT (**1**), which is the most abundant ester-type *Cephalotaxus* alkaloid showing potent antitumor activity, have been designed and synthesized. In addition, because the nitrogen lone pair might play an important role in expressing the activity, this region was selected for modification.

Oxidation of **1** with hydrogen peroxide in methanol gave β-*N*-oxide **35** and α-*N*-oxide **36** in 26% and 36% yield, respectively (Figure 4.1). When a 1,2-dimethoxyethane solution of β-*N*-oxide **35** was heated in a sealed tube at 105°C for 2 h, compound **37** and two unexpected compounds, **38** and **39**, were obtained in yields of 37%, 44%, and 7.7%, respectively. Heating α-*N*-oxide **36** under the same conditions also gave compounds **37** (32%), **38** (36%), and **39** (7.6%). Zinc and acetic acid reduction of **38** and **39** gave ring contracted HHT analogs **40** and **41** in yields of 96% and 67%, respectively. Their stereostructures were confirmed from NOESY spectral data.

III. SYNTHESIS OF THE DRUG

A. Synthesis of Acyl Groups

Alkaline hydrolysis of **1** or **3–5** yields cephalotaxine **2** plus a carboxylic acid moiety. Initially, the structures of the resulting dicarboxylic acids were identified by nuclear magnetic resonance and mass spectral studies.[9,10] The structure of **42**, the dimethyl ester of the acid moiety from **5**, was also verified by synthesis. Thus, methyl isopentyl ketone (**43**) was condensed with diethyl carbonate (**44**) to give ethyl 6-methyl-3-oxo-heptanoate (**45**), which was converted to a cyanohydrin (**46**). Acid catalyzed methanolysis of **46** provided a racemic dimethyl ester (**42**) identical with the product from **5**, except for its chiral properties (Figure 4.2).[10] This confirmatory synthesis supported structures assigned to the corresponding acid moieties of **1**, **3**, and **4**.

However, characterization of these dicarboxylic acids did not establish which of their two available carboxyl groups was attached to the cephalotaxine moiety. This feature was revealed through an attempted synthesis of **5**. When cephalotaxine was acylated with the requisite half ester of **42**, which had the primary carboxyl group in the acid chloride form, the resulting product was an isomer of **5**, ψ-deoxyharringtonine (**47**). The nuclear magnetic resonance spectrum of **47** was distinctly different from that of deoxyharringtonine (**5**), thus providing evidence for structure **5**, in which the ester side chain is attached through the tertiary carboxyl group. From close parallels of the nuclear magnetic resonance spectra, Mikolajczak and associates[10] concluded that the acyl moiety was constituted similarly in **1** and **3–5**. This conclusion was reinforced by additional synthesis of the dicarboxylic acid moieties. Auerbach and coworkers[30] carried out an alternative synthesis of **42** (Figure 4.3). Methyl itaconate (**48**) was epoxidized with buffered peroxytrifluoroacetic acid, and the resulting epoxide (**49**) was condensed with a reagent prepared from isobutyl lithium and cuprous iodide. The resulting product was **42**.

Ipaktchi and Weinreb[31] synthesized two diastereomeric compounds corresponding to the acid moiety of isoharringtonine (**4**). Isoamylacetoacetate (**50**) was converted through a haloform-type reaction to isoamylfumaric acid (**51**), which, in turn, was dehydrated to an anhydride. This anhydride

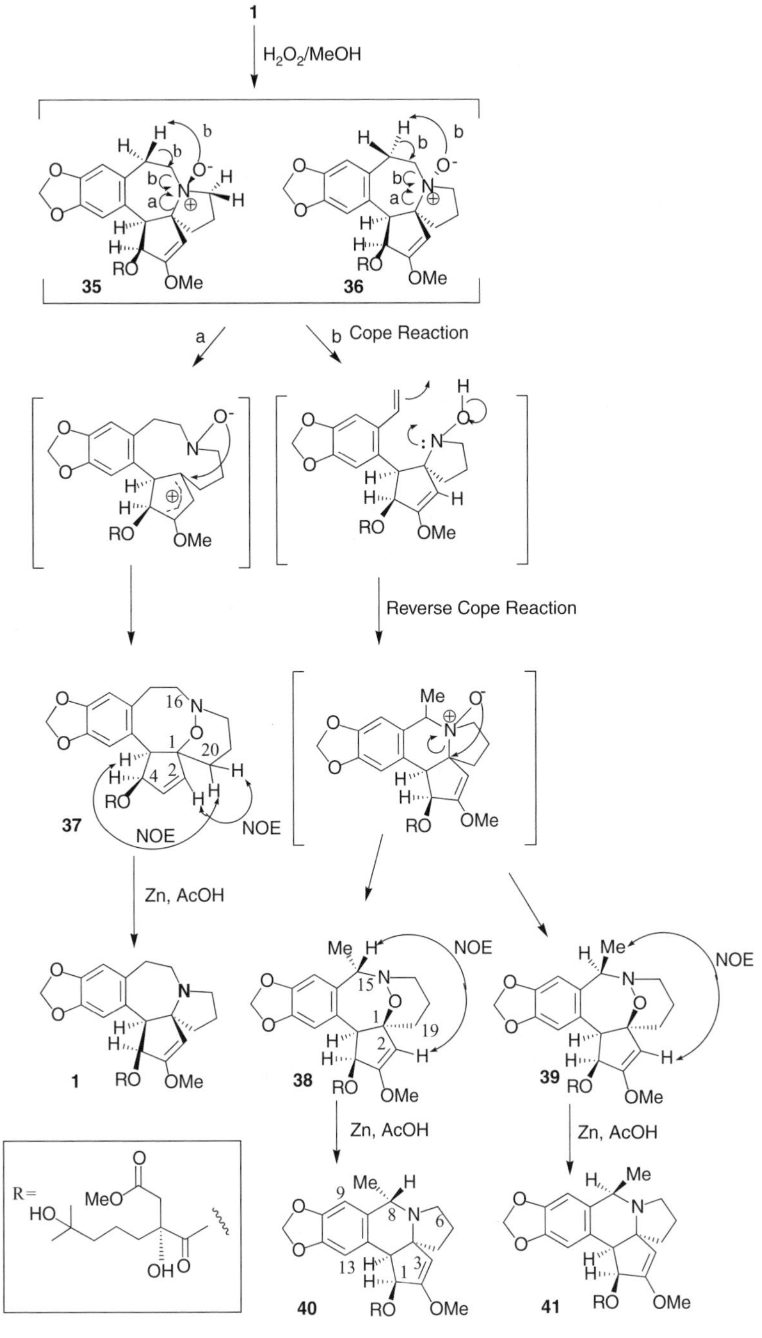

FIGURE 4.1 Modification of the skeleton of HHT through unusual rearrangements.

was converted to a dimethyl maleate derivative, which was reacted with osmium tetroxide-hydrogen peroxide in *t*-butanol to give a diol (**52**). This diol was identical with the dimethyl ester from **4**, except for optical rotation. Thus, the relative configuration of **52** was established as erythro, although the absolute configuration had yet to be determined (Figure 4.4).

Kelly and coworkers[32] synthesized the diacid from hydrolysis of harringtonine (**3**) through an α-keto ester intermediate (**54**) (Figure 4.4). Benzylated hydroxyacetylene **53** was treated with

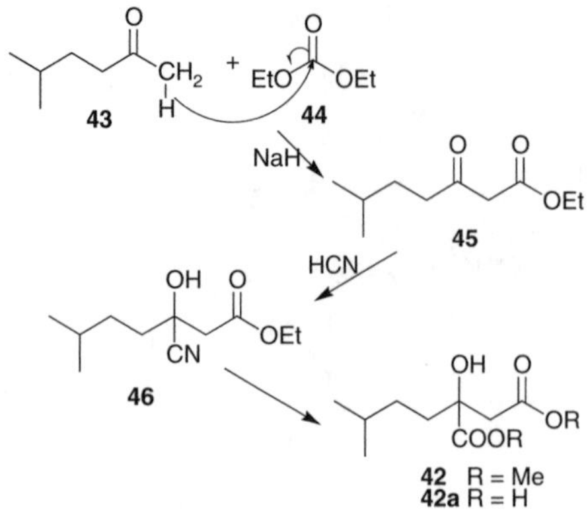

FIGURE 4.2 Synthesis of deoxyharringtonine side chain.

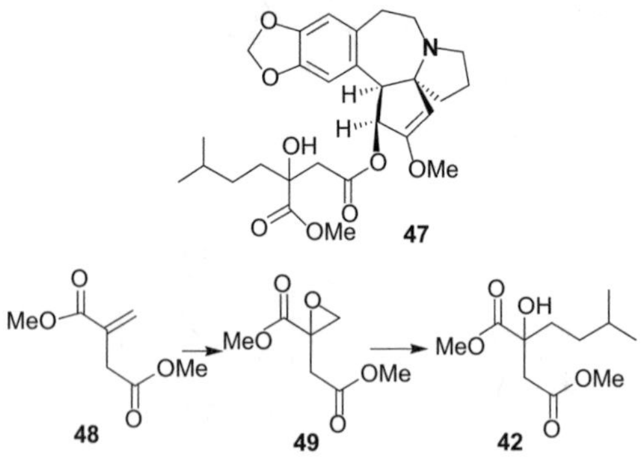

FIGURE 4.3 Auerbach synthesis of deoxyharringtonine side chain.

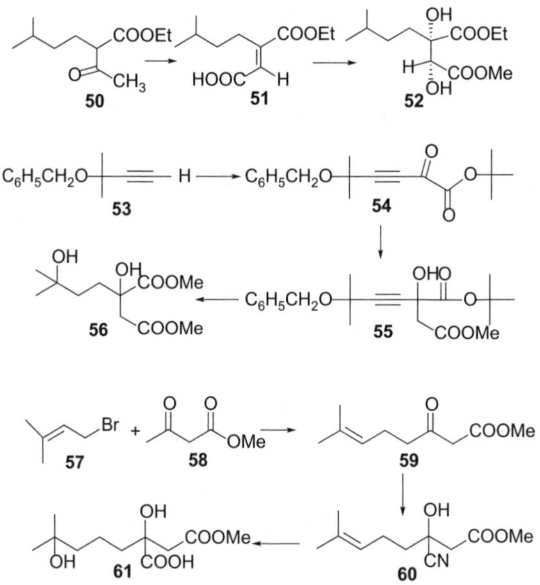

FIGURE 4.4 Additional synthesis of side chain acids.

butyllithium, and the resulting lithio derivative was condensed with ethyl *t*-butyl oxalate to provide the α-keto ester **54**. At this stage, the carbomethoxymethyl side chain was introduced by condensing **54** with lithium methyl acetate (LiCH₂CO₂Me). Debutylation of the product **55** was effected with trifluoroacetic acid. After methylation with diazomethane and debenzylation by hydrogenolysis, the dimethyl ester (**56**) of the desired diacid was obtained.

Utawanit[33] synthesized the diacid resulting from hydrolysis of HHT (**1**). 1-Bromo-3-methyl-2-butene (**57**) was condensed with methyl acetoacetate (**58**) to give (**59**) (Figure 4.4). This unsaturated ester (**59**) was converted to the corresponding cyanohydrin (**60**), and subsequent acid hydrolysis gave the racemic form of the desired diacid (**61**). Hydration of the double bond was accompanied by hydrolysis of the nitrite group.

B. TOTAL SYNTHESIS

The unique structure and the therapeutic potential of this group of alkaloids have stimulated many syntheses of cephalotaxine (CET, **2**),[34–37] as well as numerous studies on the construction of the pentacyclic ring system.[38–43] The unique heterocyclic ABCDE ring system of CET has continued to be a proving ground for new synthetic strategies and methods. Excellent review articles on this subject have been published by Huang and Xue,[44] Jalil Miah et al.,[34] and most recently, by Li and Wang.[45]

Total synthesis of CET was accomplished on the basis of a conceptually novel strategy that featured key transannular reductive skeletal rearrangements to construct the pentacyclic ring skeleton. The synthetic potential of the designated Clemments–Clemo–Prelog–Leonard reductive rearrangement was demonstrated for the first time in a facile synthesis of the benzazepine subunit of CET. A novel endocyclic enamine (cyclopentenone) annulation was discovered as an unusual azo-Nazarov-type cyclization.

A key pentacyclic precursor, ketone **62,** was based on a transannular skeletal rearrangement (Figure 4.5, path a) of the cyclic enone **63**.[46] This pathway parallels the biogenesis of CET that was postulated by Parry et al. (Figure 4.6).[47] Enone **63** could in turn be derived from cyclic ketone **65**. A more conventional strategy is outlined in Figure 4.5, path b, which involves a transannular rearrangement of ketone **65** to benzazepine precursor **64**.

The synthesis of **62** is shown in Figure 4.7.[46] β-(3, 4-Methylenedioxy)-phenethylamine (**66**) was converted to **67** in 62% overall yield by the following sequence: (1) Bischler–Napiralsky cyclization of corresponding oxalamide, (2) catalytic hydrogenation, and (3) N-alkylation with ethyl 4-bromobutyrate. Reaction of diester **67** with allyl bromide in DMSO gave the allylation product

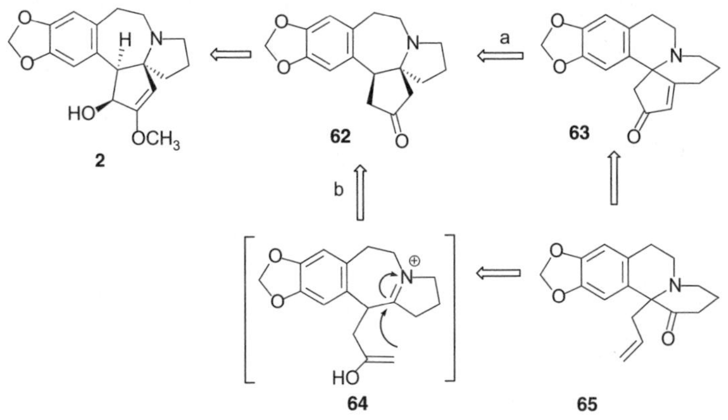

FIGURE 4.5 Retrosynthetic strategic plan for synthesis of **2**.

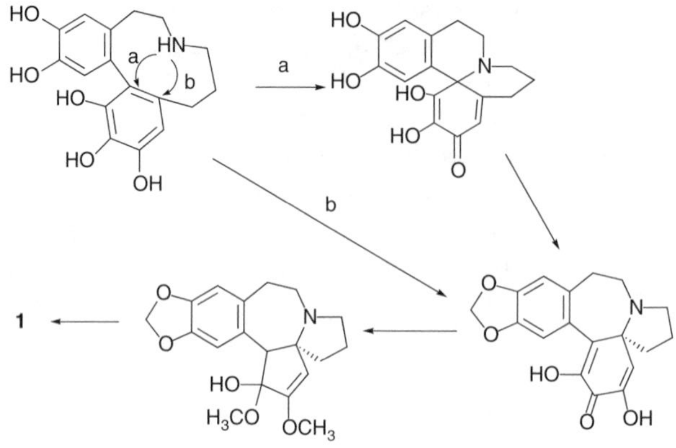

FIGURE 4.6 Parry pathway for biogenesis of **1**.

69, presumably via a facile 2,3-sigmatropic rearrangement of the corresponding N-ylide of ammonium salt **68**. Dieckman cyclization of diester **69** and subsequent decarboxylation produced the cyclic ketone **65**. Pd/Cu catalyzed oxidation furnished the diketone **70**, which was cyclized to form the pentacyclic enone **63**. On short exposure to zinc dust in hot glacial acetic acid, enone **63** was transformed into the desired pentacyclic ketone **62**. This remarkably facile transformation can be rationalized as a transannular reductive rearrangement via a transient bridged aziridinium intermediate;[48] this type of skeletal rearrangement can also be found in Buchi's synthesis of Iboga alkaloids.[49]

Subjecting **62** to Moriarity oxidation gave a hydroxy dimethylketal as the sole product,[50] which was deduced to be **71** by spectroscopic analysis.[51] Deketalization of **71** in THF followed by autoxidative dehydrogenation of the resulting crude hydroxy ketone **72** (epidesmethylcephalotaxine) afforded desmethylcephalotaxinone (**73**).[52] The synthetic product was identical in every respect with a sample prepared from natural (-)-CET by hydrolysis and autoxidative dehydrogenation, as for **71→73**.

Because desmethylcephalotaxinone (**73**) has been converted to natural (-)-CET via methylation, optical resolution with L-tartaric acid, and borohydride reduction,[53] the sequence outlined in Figure 4.7

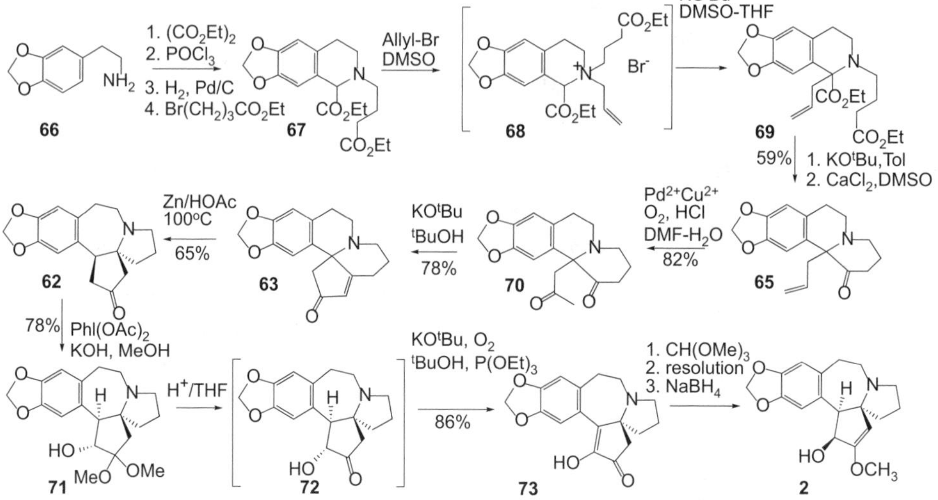

FIGURE 4.7 Total synthesis of cephalotaxine.

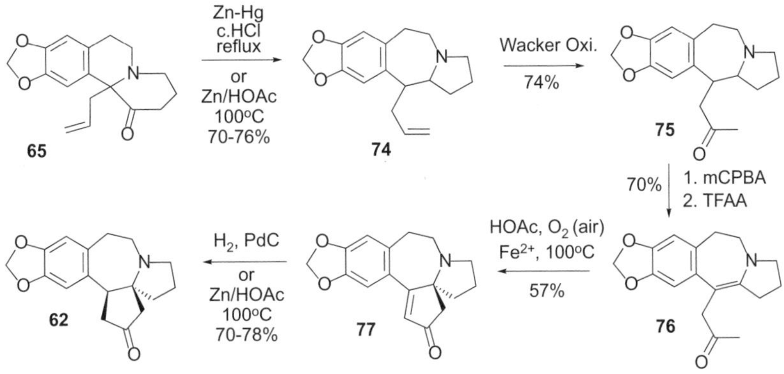

FIGURE 4.8 Enamine cyclopentenone annulation pathway.

constitutes a total synthesis of CET with an overall yield of about 12% to **73** from diester **67** through an eight-stage sequence.

The synthesis described above evolved from an initial strategic plan (Figure 4.5, plan b) that was based on an intramolecular Mannich cyclization of the iminium intermediate for the E-ring formation.[54] The precursor to **64** would be generated from cyclic ketone **65** by a Clemmensen reductive rearrangement. The anomalous Clemmensen reaction of a cyclic α-amino ketone was first noticed by Clemo et al. in 1931,[55,56] clarified by Prelog in 1939,[57] and later studied systematically by Leonard and coworkers in 1949.[58,59] This unique Clemmensen–Clemo–Prelog–Leonard reductive rearrangement was applied to constructing the benzazepine-bearing ABCD-ring system of CET. Clemmensen reduction (Zn-Hg, conc. HCl, reflux) of α-amino ketone **65** led to allyl benzazepine derivative **74** (Figure 4.8). Clemmensen reduction conditions, with zinc dust in hot glacial acetic acid, were also equally effective for this reductive rearrangement.[60] This facile rearrangement can be generalized as an acid-catalyzed transannular interaction of nitrogen with the protonated carbonyl group, leading to a transient aziridinium intermediate that would facilitate reductive cleavage of the benzylic C-N bond.[61–64] The electron-rich aromatic system certainly contributed to this facile process.

Acidic Wacker oxidation of **74** gave methyl ketone **75** (Figure 4.8). However, the attempted generation of iminium intermediate **64** from **75** using Hg(OAc)$_2$ resulted in a slow decomposition of **75** under a variety of reaction conditions.[65,66] In the hope that an alternative Polonski–Potier protocol might be more selective, the corresponding amine oxide of **75,** prepared by m-CPBA oxidation, was treated with excess trifluoroacetic anhydride in CH$_2$Cl$_2$. Enamine ketone **76** was obtained as the sole product, apparently formed by isomerization of the initially generated iminium salt **64**.[67] Compound **76** had been previously synthesized by an alternative enamine alkylation of the so-called Dolby–Weinreb enamine.[34] It was thought[54,68] that cyclization to **62** (formally an endocyclic enamine annulation) could not be realized under various conditions, although Dolby et al.[46] reported that acid-catalyzed cyclization of **76** led to a rearranged product (saturated enone **63**) in low yield.

Although keto enamine **76** appeared not to be a promising intermediate, various acidic treatments of **76** were examined. A hot mixture of 40% acetic acid and **76** in an open flask gradually generated a sole isolatable product, whose structure was verified as the pentacyclic enone **77** (Figure 4.8). Further experimentation improved the yield in this cyclization to 57%. This cyclization appears to be an acid-catalyzed O$_2$-dependent process, reasoned to be an unusual azo-Nazarov-type cyclization initiated by an acid-catalyzed autooxidation. Related acid-catalyzed, oxidative cyclization can be found in Woodward's synthesis of chlorophyll-a[69] and in an oxy-Nazarov cyclization[40] in Weinreb's pioneering CET synthesis.[52] Because endocyclic enamine **76** is a direct enamine alkylation product, this unusual cyclization can be regarded as a novel endocyclic enamine cyclopentenone annulation.[70] Such processes could have broad applications to the synthesis of complex polycyclic natural alkaloids and other heterocycles.

Catalytic hydrogenation of pentacyclic enone **77** gave the pentacyclic ketone **62** in 76% yield. Interestingly, the action of zinc dust in hot HOAc on **77** smoothly produced ketone **62** without causing skeletal rearrangement (Figure 4.8), which further supports the mechanistic rationale for the transannular reductive rearrangement of enone **63** to pentacyclic ketone **62**.

In conclusion, a short, practical total synthesis of CET has been developed based on a computationally novel strategy featuring key transannular reductive rearrangements to establish the pentacyclic ring skeleton. The efficiency and practicality of the synthesis stem from its brevity (nine stages from diester **67**), readily available starting materials, simple reactions, inexpensive reagents, and good overall yield. Finally, this synthesis of CET may be regarded as a biomimetic synthesis[71,72] in terms of pentacyclic ring construction.[47] Further work on the asymmetric strategy is under way.[45]

IV. MECHANISM OF ACTION

HHT and its analogs are dose- and time-dependent inhibitors of protein synthesis. Effects on DNA may also be important. HHT inhibited 50% of tritiated leucine uptake in Hela cells and globin synthesis in reticulocyte lysates at doses of 0.04 and 0.10 μM, respectively. These doses were significantly lower compared with those necessary for other esters to achieve the same results.[73] Degradation of polyribosomes, release of completed protein chains, and inhibition of initiation of protein synthesis were observed. HHT and analogs did not affect mRNA binding to the donor site of the ribosome. The alkaloids blocked poly (U)-directed polyphenylalanine synthesis, peptide bond formation, and enzymatic and nonenzymatic binding of the Phe-tRNA to ribosomes.[74] These results indicate that HHT inhibited the elongation phase of translation by preventing substrate binding to the acceptor site on the 60s ribosome subunit and, therefore, blocked aminoacyl-tRNA binding and peptide bond formation. Tujebajeva et al. confirmed these observations in 1989.[75] In poly (U)-programmed human placenta 80-S ribosomes, he demonstrated that neither nonenzymatic binding nor eEF-1-dependent Phe-tRNA binding at A and P. sites were hindered by HHT, whereas diphenylalanine synthesis and puromycin reactions were strongly inhibited by HHT. Thus, HHT was found to be a selective inhibitor of transpeptidation during the elongation cycle. More recent studies in chronic myeloid leukemia (CML) cells indicated both differentiation and induction of apoptosis as potential downstream mechanisms for HHT effects.[76–79]

The first studies regarding the biological activity of HHT were investigated simultaneously with structural studies.[7–9,11,12,26] Cytokinetic studies showed that HHT cytotoxicity was cell-cycle specific, affecting mostly cells in the G_1 and G_2 phases, as expected for a drug that primarily inhibits protein synthesis.[80] Although this finding suggested a possible role of HHT against slowly growing tumors, clonogenic assay studies demonstrated greater activity against rapidly growing tumors.[81] Prolonged drug exposure was required for a maximal antitumor effect *in vitro*, and *in vivo* studies showed recovery of protein synthesis within 24 h of injection.[74,82–84] These data may be relevant to the interpretation of the results with HHT in solid tumors, which mostly used short infusion and exposure schedules.

At dose ranges of 0.5 mg to 4.0 mg/kg, HHT had remarkable *in vivo* effects on L1210 cells in mice. The inhibition of neoplastic proliferation was measured by mitotic index of leukemic cells and by survival of treated animals, which was 142% of control mice. In mice bearing P388 lymphocytic leukemia, survival prolongation of treated animals was >300% that of nontreated mice and indicated the superior antineoplastic effect of HHT compared with isodoses of three other cephalotaxine alkaloids.

In vitro and *in vivo* studies of HHT and analog esters against L1210 leukemia sublines demonstrated a proportional relation between cytotoxicity and protein synthesis inhibition. Intracellular retention of radiolabeled harringtonine was found to correlate with protein synthesis, and the majority of the drug was tightly bound to the microsomal structures. Resistance in two sublines corrrelated with rapid drug egress compared with wild-type lines.[85]

HHT had moderate activity against CD8F1 mammary carcinoma and marginal effect against B16 intraperitoneal melanoma but was ineffective against Lewis lung carcinoma and human colon, lung, and mammary tumor xenografts in immunodeficient mice.[86] It caused extensive necrosis of subcutaneous colon 38 tumors in mice and appeared to delay tumor growth significantly,[87] and it had variable efficacy against different solid tumors and leukemias, with a wide range of cytotoxic effects, mainly directed against rapidly proliferating cells, irrespective of their lymphoid or myeloid origin.

Among 80 tumors of 13 histologic subtypes tested, HHT was active in 13 (16%). When it was tested *in vitro* against 10 human leukemia and lymphoma cell lines,[81] it demonstrated a 70-fold difference in growth inhibition between the most sensitive and the most resistant lines. The most sensitive lines were HL-60 (presumed acute promyeocytic) and RPMI-8402 (T-cell acute lymphocytic); moderately sensitive lines included DND-39A (B-cell lymphoma), ML-2 (acute myeloid), MOLT-3 (T-cell acute lymphocytic), and KG-1 (erythroleukemia); and less sensitive lines were Daudi (B-cell lymphoma), NALL-1 (null cell acute lymphocytic), BALM-2 (B-cell acute lymphocytic), and DND-41 (T-cell acute lymphocytic). Tumors with high growth fractions appeared to be more sensitive to HHT.

In contrast to leukemia cell lines, six human neuroblastoma cell lines were found to be sensitive only after exposure to >300 ng/ml of HHT for 7 d.[88] The high HHT concentrations needed to suppress tumor growth (>10,000 ng/mL in four of six cell lines) indicated that doses used in the majority of clinical trials would be inactive against neuroblastoma but might be effective against leukemia. However, *in vivo* studies showed 53% inhibition of neuroblastoma in mice treated with 3.5 mg/kg HHT intraperitoneally 2 d a week for 3 weeks and 41% inhibition of neuroblastoma in mice treated with 0.7 mg/kg daily for 3 d, followed by 2.1 mg/kg daily for 7 d. Survival of treated mice was prolonged significantly.[89]

Harringtonine induced monocytic differentiation of HL-60 cells after 7 d of liquid culture; morphologic changes typical of granulocytic or monocytic differentiation were detected in four of eight human acute myeloid leukemia (AML) primary cell cultures performed for days in the absence of growth factors.[90] After maturation and proliferation arrest, when cultures were prolonged for 10 d, morphologic features of apoptosis were present in recovered cells (i.e., pyknosis of nuclei, karrhexis, and shrinkage of cytoplasm). These results were supported by Zhou et al., who found that HHT at low concentrations of 2–20 ng/ml induced 28% of HL-60 cells to differentiate to macrophage-like cells.[91] Visani et al. (1997)[76] reported that HHT exerted more cytotoxicity against CML chronic phase cells, *in vitro* on semisolid cultures, compared with normal bone marrow at 50 ng/ml ($P = 0.02$) and 200 ng/mL ($P = 0.01$). They also demonstrated synergistic effects for HHT combinations with interferon-α (IFN-α), cytarabine, or both. For CML blastic phase cells, only the triple combination was effective. Induction of apoptosis by HHT was dose dependent when tested at 100 ng/mL, 200 ng/mL, and 1000 ng/mL in both chronic phase and normal cells, but no apoptosis was observed for CML blastic phase cells. At 2×10^{-7} mol/L and 10^{-7} mol/L, harringtonine could induce apoptosis in HL-60 cells after exposure for 4 h.[78] In agarose gel electrophoresis, DNA extracted from HL-60 cells treated with harringtonine and HHT showed typical internucleosomal DNA degradation (i.e., DNA ladder, nuclear chromosome segmentation and condensation, and cytoplasmic vacuolation). This HHT effect was concentration and time dependent.

Exposure to HHT reversibly inhibited glycoprotein synthesis and stimulated lipid-linked oligosaccharide formation *in vitro* in bladder carcinoma cells, whereas prolonged treatment (>8 h) resulted in a generalized suppression of both glycoprotein biosynthesis and lipid-linked oligosaccharide formation. Kinetic studies indicated that the time course for the decrease in glycoprotein biosynthesis and the accumulation of dolichol-linked oligosaccharide paralleled the HHT-induced decrease in protein synthesis.[92]

HHT inhibited colony formation in a variety of myeloid (50% inhibition at concentration ranges of 7–12 ng/mL) and lymphocytic (ranges of 4–7 ng/mL) cells, as well as in fresh myeloid leukemic cells from patients (50% inhibition at concentration ranges of 2–25 ng/mL) in a dose-dependent manner. Pulse exposure studies showed 50% inhibition of colony formation of HL.60 cells (10–20 ng/ml) after 45 h of exposure, followed by complete inhibition at 72 h of exposure. Radioactive precursor studies in the HL-60 experiment verified inhibition of protein synthesis by HHT, rather than inhibition of RNA and DNA synthesis.[91]

In vitro and *in vivo* studies indicated that although P388 cells resistant to doxorubicin and vincristine also showed cross-resistance to HHT, cytarabine-resistant P388 cell lines showed a surprising collateral sensitivity to HHT.[93] These observations have potential implications related to multidrug-resistant (MDR) mechanisms of HHT resistance and to therapeutic strategies including combinations of HHT and MDR-blocking agents, as well as HHT plus cytarabine. The wide differences in apoptotic and cytotoxic effects in different cell lines may be a result of differences in expression of P170 glycoprotein (e.g., in blastic versus chronic phases of CML).

Resistance to HHT appears to be MDR related (similarly to anthracyclines, vinca alkaloids, and taxol).[94] Cell lines with high MDR expression were 15 times more resistant to HHT effects.[95] Two leukemic cell lines (K562-HHT and L1210-HHT), rendered 17-fold and 13-fold, respectively, resistant to HHT, demonstrated cross-resistance to doxorubicine, vincristine, and daunorubicine and had increased expression of the MDRI gene. Expression of MDRI P170 in HHT-resistant sublines of K562 also was increased. These observations may explain the reduced antileukemic activity of HHT in refractory AML and in the accelerated–blastic phases of CML compared with chronic phase CML.

To elucidate the mechanism of resistance to HHT in leukemic cells, five sublines of human myeloid leukemia K562 cells, which demonstrated progressive resistance to different concentrations of HHT, were established. These sublines were cross-resistant with daunorubicin, vincristine, etoposide, and mitoxanthrone. Immunofluorescence with monoclonal anti-P-glycoprotein antibody MRK16 and Northern blot analysis demonstrated that resistance to HHT was related to the sequential emergence of MRP gene and MDR-1 gene overexpression. In the highly resistant cell sublines, MDRI gene overexpression predominated.[96] Another harringtonine-resistant HL-60 cell line, HR20, showed cross-resistance with HHT, doxorubicin, daunorubicin, vincristine, and colchicines. The growth-doubling time and cell numbers in the G_1 phase were increased, and the cell line overexpressed mdr-1 gene and P-glycoprotein.[97]

HHT was investigated *in vitro* in combination with cytotoxic agents to detect potential synergistic interactions. With the exception of consistent significant synergistic effects with cytarabine[14] and modest synergy with 5-fluorouracil and hexamethylene bisacetamide,[97,98] no other synergistic interactions were observed.[99]

V. MEDICINAL CHEMISTRY OF THE DRUG

Different activities among naturally occurring *Cephalotaxus* alkaloids and the antileukemic esters of cephalotaxine are significant issues for drug development. Smith et al.[100] have reviewed structure–activity relationships among *Cephalotaxus* alkaloids and their derivatives.

HHT (**1**) and harringtonine (**3**) show comparable levels of activity in P388 and L1210 experimental leukemia systems (Table 4.1 and Table 4.2). Insertion of an additional methylene group in the terminal portion of the acyl side chain of **3** has little effect on activity. In contrast, removal of the hydroxyl group from the penultimate carbon of the acyl moiety [to give deoxyharringtonine (**5**)] reduces activity by about one-half. Shifting this same hydroxyl to give isoharringtonine (**4**) lowers P-388 activity by nearly an order of magnitude (Table 4.1).

As indicated previously, natural (-)-cephalotaxine (**2**) is devoid of antileukemic activity. The corresponding dimethyl ester (**42**) and its isomeric acid (**42a**), provided by hydrolysis of **5**, are likewise inactive. Thus, an ester linkage is evidently a required, but not sufficient, structural element for activity. In addition, although **56**, the dimethyl ester of the dicarboxylic acid derived from

TABLE 4.1
Activity of Selected *Cephalotaxus* Alkaloids against P-388 Lymphocytic Leukemia

Compound	Dose (mg/kg)	Survivors	Animal weight difference (*T-C*)	Survival time, *T/C* (d)	*T/C* (%)
Homoharringtonine (**1**)	2.00	6/6	−3.8	7.5/9.0	
	1.00	6/6	−2.8	30.5/9.0	338
	0.50	6/6	−1.8	24.5/9.0	272
	0.25	6/6	−2.2	22.0/9.0	244
Harringtonine (**3**)	4.00	2/6	−5.5	5.0/9.0	
	2.00	6/6	−3.3	18.5/9.0	205
	1.00	6/6	−2.3	36.5/9.0	405
	0.50	6/6	−1.0	26.5/9.0	294
Isoharringtonine (**4**)	15.00	6/6	−4.3	9.5/9.0	105
	7.50	6/6	−3.0	24.5/9.0	272
	3.75	6/6	−2.8	15.5/9.0	172
	1.87	6/6	−1.3	13.5/9.0	150
Deoxyharringtonine (**5**)	4.00	6/6	−3.4	14.0/10.0	140
	2.00	6/6	−3.3	18.0/10.0	180
	1.00	6/6	−2.4	15.5/10.0	155
	0.50	6/6	−1.2	14.5/10.0	145

Note: Data presented are representative of results from several assays with different samples of each alkaloid. Materials are considered active if the survival time of animals treated (*T*) with them is ≥ 125% of that of the controls (*C*) (i.e., *T/C* ≥ 125%). Values are quoted from ref. 8.

harringtonine, has not been assayed in the usual tumor systems (P-388, KB), work with a Hela cell system indicated that this acid has no inhibitory activity when tested as a separate entity.[73]

Various research groups have prepared numerous cephalotaxine derivatives that incorporate "unnatural" acyl groups or other structural variations. Some seemingly subtle structural alterations abolish activity, whereas certain active esters of cephalotaxine have been prepared whose acyl groups bear no resemblance to those of the natural alkaloids.

The following paragraph contains examples of modifications leading to abolished or reduced activity. The partial synthesis of deoxyharringtonine (**87**) provided a diastereomer (**88**) differing only in the configuration at C2′ of the acyl side chain.[101] Compound **88** appears to be less active than **87**, although further testing is needed to establish this point definitively.[102] Attaching the "wrong" carboxyl group of the acyl moiety to **2** gives an inactive isomer (**85**) of deoxyharringtonine. A "rearranged" isomer (**5a**) of deoxyharringtonine, in which the acyloxy group is shifted from C1 to C3, likewise has little or no antileukemic activity.[103] Removal of the hydroxyl function from C2′ of the acyl side chain of deoxyharringtonine (**87**) produces **106** and thereby abolishes activity. When cephalotaxine was acylated with an acid having a hydroxyl group α to a tertiary carboxyl, although the resulting ester (**107**) incorporated some steric features of the active esters, it showed no more than marginal activity.[104]

A structurally diverse series of cephalotaxine esters has been prepared with acyl groups that contain conjugated double bonds, aromatic rings (both benzenoid and heterocyclic), chloro, nitro, hydroxyl, and sulfonate groups (**78**–**107**) (Figure 4.9). Esters showing at least marginal activity are listed in Table 4.3; others are inactive.[102] It has been noted that some seemingly minor alterations in the acyl moieties of the more potent antileukemic esters of cephalotaxine markedly diminish their activity. Accordingly, the structures of some active esters in Table 4.3 may occasion some surprise. No apparent rationale emerges to provide a logical statement of structure–activity relationships for

TABLE 4.2
Activity of Selected *Cephalotaxus* Alkaloids against P-1210 Lymphocytic Leukemia

Compound	Dose (mg/kg)	Survivors	Weight difference (*T–C*)	Survival time, *T/C* (d)	*T/C* (%)
Homoharringtonine (**1**)	2.00	6/6	–3.0	9.2/9.1	101
	1.00	6/6	–1.4	13.0/9.1	142
	0.50	6/6	–0.8	11.0/9.1	120
	0.25	6/6	–0.5	11.2/9.1	123
Cephalotaxine (**2**)	220.00	6/6	–1.2	9.8/9.6	102
	110.00	6/6	0.2	10.3/9.6	107
	55.00	6/6	0.5	9.5/9.6	98
Cephalotaxine acetate (**2a**)	100.00	6/6	–0.6	9.8/9.6	102
	50.00	6/6	0.0	10.2/9.6	106
	25.00	6/6	–0.3	10.2/9.6	106
Harringtonine (**3**)	4.00	1/6	–4.1	0.0/9.1	—
	2.00	6/6	–2.4	12.5/9.1	137
	1.00	6/6	–1.2	12.3/9.1	135
	0.50	6/6	–1.0	12.0/9.1	131
Isoharringtonine (**4**)	15.00	6/6	–3.5	10.0/9.1	109
	7.50	6/6	–1.3	11.5/9.1	126
	3.75	6/6	–0.5	11.3/9.1	124
	1.87	6/6	–1.1	10.0/9.1	109

Note: Data presented are representative of results from several assays with different samples of each akanoid. Materials are considered active if the survival time of animals treated (*T*) with them is ≥ 125% of that of the controls (*C*) (i.e., *T/C* ≥ 125%). Values are quoted from ref. 8.

the series of compounds in Table 4.3. One particularly intriguing observation is that methyl cephalotaxyl itaconate (**83**) prepared from the natural (-)-isomer of **2** is active (and exceptionally nontoxic), whereas its optical antipode (**84**) is inactive. However, until additional representatives of the (+)-cephalotaxine ester series have been prepared and tested, generalizations about their activity will be premature. In terms of dose requirements for effective tumor inhibition, harringtonine and HHT are at least an order of magnitude more active than any of the esters in Table 4.3.

This work has delineated a number of structural variations in the acyl moiety that greatly reduce or abolish activity. Indeed, with these examples alone, one might infer that there are rather rigid structural requirements for activity in the harringtonine series. Nevertheless, activity appears in several cephalotaxine esters (e.g., **105**, the trichloroethoxycarbonyl derivative) having acyl groups with no apparent relationship to that of harringtonine. It should also be noted that most of the cephalotaxine esters described by Mikolajczak et al. in 1977,[102] including **78–107**, can be prepared without the severe steric constraints that hamper conversion of cephalotaxine to harringtonine. Obviously, possible variations in the acyl group of harringtonine are endless, and the search for harringtonine analogs with improved properties continues.

Except in the case of the "rearranged" ester **5a**, structure–activity relationships in the harringtonine series have not been extended to the cephalotaxine moiety. Many structural variations in the cephalotaxine ring system await physiological investigation, with some of the more obvious variations being (a) transformation of the methylenedioxy group at C18 into dimethoxy, dihydroxy, or other functions; (b) aromatic substitutions at C14 or C17; (c) inversion of the oxygen function at C3 to give an epimeric series of esters; (d) quaternarization of the nitrogen (e.g., to form a methiodide); (e) substitution or oxidation at one or more of the methylene groups — C6, C7, C8,

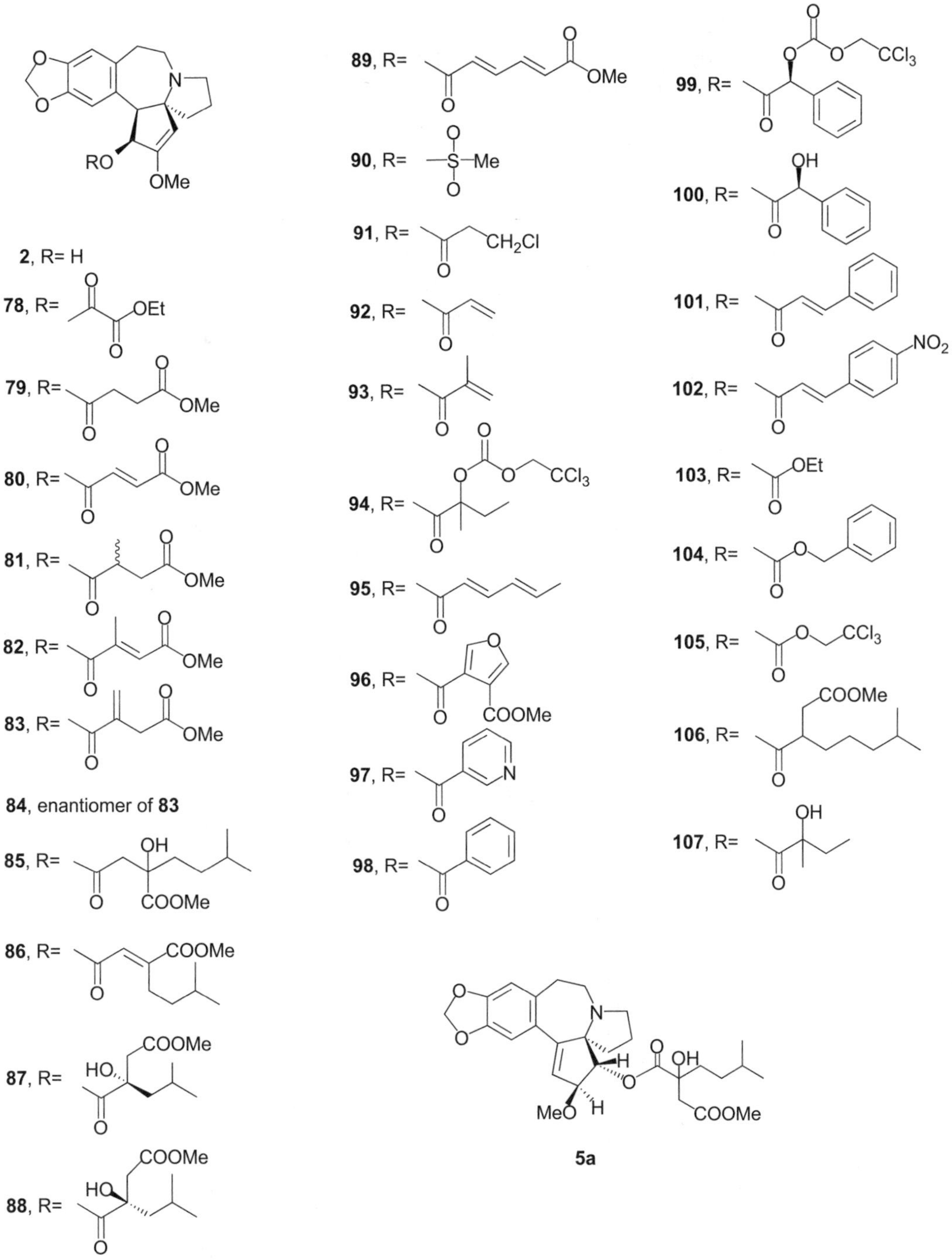

FIGURE 4.9 Various synthetic analogs.

TABLE 4.3
Activity of Various Cephalotaxine Esters against P-388
Lymphocytic Leukemia

Compound	Vehicle[a]	Dose (mg/kg/inj)	Weight Difference (T-C)	T/C (%)
78	D	20	0.1	135
	D	20	−0.2	211
	D	13	−0.8	154
	T	20	−0.1	129
80	B	80	−0.7	145
	B	40	−0.9	134
	B	20	−1.0	125
	B	10	−1.2	136
	D	4.4	−1.5	147
	D	1.9	0.5	134
83	D	365	−3.0	198
	D	240	−1.4	169
	D	160	−0.1	183
	D	160	−1.1	167
	D	80	−0.1	183
	D	40	−1.3	135
86	C	20	0.9	131
87	D	5.9	−0.9	184
	A	4	0.0	174
	A	2	0.8	126
88	D	9	−2.2	131
	A	4	0.5	150
95	D	80	0.9	150
	D	40	−1.8	125
	D	20	0.9	130
99	A	320	−2.3	136
	A	160	1.2	154
	A	80	−1.0	138
105	D	320	−1.0	172
	D	160	−0.9	162
	D	160	−0.3	155
	D	80	−0.9	183
	D	80	−1.3	160
	D	40	−0.4	140
	D	40	0.9	183
	D	20	−2.7	128
	D	20	−1.5	160
	D	20	−1.0	195
	D	13	−0.5	138
	D	8.8	−0.3	170

[a] A: saline; B: water+alcohol+acetone; C: water+acetone; D: water+alcohol, T: saline+Tween 80. Values are quoted from ref. 100.

C10, or C11; and (f) unique skeletal alterations via total synthesis procedures including photo-rearrangements or other approaches.

VI. DEVELOPMENT OF THE CLINICAL DRUG FROM THE NATURAL PRODUCTS LEAD

As previously noted, HHT (**1**) is the most promising lead compound among the *Cephalotaxus* alkaloids and is active against a number of murine tumors, including B26 melanoma, CD8F1 mammary carcinoma, L1210 leukemia, and especially colon 38 tumor and P388 leukemia.[105] Its concentration in plant extracts is much higher than those of the other alkaloids, and research with **1** has predominated because of its more ready supply and high *in vitro* activity.[14,105] Even though many other *Cephalotaxus* alkaloids have been isolated from the *Cephalotaxus* genus and numerous semisynthetic derivatives prepared, HHT (**1**) and harringtonine (**3**) remain the most active compounds and have entered clinical trials for the treatment of leukemia, as described in the next section.

VII. CLINICAL APPLICATIONS

A. CHINESE CLINICAL TRIALS TREATING LEUKEMIAS WITH CEPHALOTAXINE ESTERS

HHT has been used in China since 1974 for the treatment of leukemia. Many Chinese patients were treated with mixtures of harringtonine and HHT, using a variety of doses and treatment schedules. The greatest benefit was seen in those patients with AML.[105] In one initial study, the mixture was used at a dose of 2 mg daily for 14 d by intramuscular injection, and the dose subsequently was increased to 4 mg daily intravenously for 22–86 d, using a ratio of 2:1 harringtonine:HHT. This preparation was also evaluated at a dose of 5–6 mg daily for 4–6 d every 2–3 weeks. These studies indicated that the maximum tolerated, therapeutically effective dose was 4 mg daily for 14 d. Major dose-limiting toxic effects were cardiovascular disturbances, including hypotension, ventricular tachycardia, and myelosuppression.[14]

B. PHASE I STUDIES IN THE UNITED STATES

The success of the Chinese studies prompted the initiation of clinical studies in the United States, using highly purified HHT. Initial phase I trials in patients with solid tumors employed an infusion of HHT, ranging from 10 min to 3 h.[2] Hypotension accompanied by tachycardia proved to be dose limiting in several trials. Continuous infusion ameliorated this tarchycardia and hypotension, and it was recommended for further trials. On the completion of the phase I clinical trials in patients with solid tumors, clinical studies of HHT leukemia were initiated in 49 adult patients with leukemia, including 36 patients with acute nonlymphocytic leukemia (ANLL). Myelosuppression was the major side effect, and the cardiovascular side effects were minimal.[14]

C. PHASE II STUDIES IN SOLID TUMORS

Ajani et al.[106] conducted five phase II trials of HHT in 80 patients with advanced solid tumors including malignant melanoma, sarcoma, head and neck carcinoma, breast carcinoma, and colorectal carcinoma. Among the 74 evaluable patients, there were no complete (CR) or partial remissions. Runge-Morris et al.[107] treated 18 patients with advanced head and neck squamous cell carcinoma with HHT (4 mg/m^2) by continuous infusion daily for 5 d every 4 weeks. Hypotension and myelosuppression were the major severe toxicities; none of the 14 evaluable patients demonstrated an objective response. HHT also was given to 15 patients with recurrent or progressive malignant glioma at a dose of 3-4 mg/m^2 by continuous intravenous infusion daily for 5 d every 3–4 weeks. No objective tumor regressions occurred.[108] Witte et al.[109,110] treated previously untreated patients with advanced colorectal carcinoma and renal carcinoma with HHT. No objective responses were observed. Other phase II studies in solid tumors did not demonstrate antitumor efficacy.[111] The above-mentioned phase II studies indicated a lack of efficacy for HHT in solid tumors at the dose schedules used.

D. Phase I–II Studies in Patients with Acute Leukemia

Stewart and Krakoff[105] evaluated HHT in a phase I clinical trial using a five-times-daily schedule with both bolus administration and continuous infusion of the drug. Their results were consistent with preclinical laboratory results, showing that the antitumor effects of HHT are more dependent on exposure time than on concentration.[81] Kantarjian et al.[112] investigated HHT in a lower-dose, longer infusion schedule in patients with refractory-recurrent acute leukemia in an attempt to demonstrate efficacy without hypotensive events. HHT was administered by continuous infusion at a dose of 2.5 mg/m² daily for 15–21 d to 13 patients and at a dose of 3.0 mg/m² daily for 15 d to 18 patients. Only one patient (3%) achieved a CR. The authors concluded that this schedule of HHT had low antileukemic efficacy.

Feldman et al.[113] used HHT, 5 mg/m², by continuous infusion daily for 9 d in patients with refractory-recurrent acute leukemia or blastic phase CML. CRs were achieved in 7 of 43 patients (16%) with recurrent AML, in 0 of 11 patients with AML that primarily was resistant to anthra-cycline-cytarabine combinations, and in 2 of 3 patients whose disease was resistant to low-dose cytarabine. Side effects included significant hypotensive events, fluid retention, weight gain (29%), and hyperglycemia (63%).[113] Ekert et al.[114,115] treated 25 Australian patients with AML using either HT or HHT, and two CRs (8%) were obtained. Sullivan and Leydan[116] also reported that an elderly patient with newly diagnosed AML achieved a response to single-agent HT.

Chinese investigators have suggested that the maximum tolerated dose in pediatric leukemia patients is 7 mg/m² per day infused over 10 d.[117] Two phase I trials in pediatric leukemia demonstrated that children could tolerate higher doses of HHT. The maximum tolerated dose was 8.5 mg/m² daily for 10 d, and the dose-limiting toxicity was myalgia.[117–119]

E. Phase II Studies in Myelodysplastic Syndrome (MDS)

Twenty-eight patients with MDS or with MDS that evolved into AML were treated with HHT, 5 mg/m² daily by continuous infusion for 9 d. Seven patients (25%) achieved a CR. In this study, induction death was high (13 of 28 patients), mainly because of myelosuppression-related infections. Myelosuppression was prolonged and severe, but other toxicities were mild.[120] The dose schedule of HHT used in this study of elderly patients was effective but toxic. It was postulated that lower dose schedules of HHT might demonstrate similar efficacy and lower toxicity in patients with MDS.[14]

F. Studies in Acute Promyelocytic Leukemia (APL)

Ye et al. (1988) treated 10 patients with APL with harringtonine, 1–3 mg intravenously over 4–5 h for 1–3 courses, each lasting for 13–81 d.[121] Interruptions between courses lasted for 5–11 d. A CR was achieved in seven patients (70%). The major side effects were pancytopenia and bone marrow hypoplasia or aplasia. Feldman et al.[122] reported a 60% remission rate in patients with refractory-recurrent APL who received HHT. This result indicates that HHT could be an interesting agent in patients with relapsed APL.[123]

G. Phase II Studies in Chronic Myeloid Leukemia (CML) Patients

Results of the lower-dose, longer-exposure schedule of HHT indicated effective antiproliferative activity and an acceptable toxicity profile, which could be potentially useful in a chronic myeloproliferative disorder such as CML.[14] O'Brien et al.[124] conducted a study in patients with late chronic phase CML. Patients received induction with HHT, 2.5 mg/m^2, daily by continuous infusion for 14 d every 4 weeks until complete hematologic response (CHR), followed by maintenance therapy with HHT, 2.5 mg/m^2, daily for 7 d every month. Among the 71 patients treated, 72% achieved a CHR, and 31% achieved cytogenetic responses. The investigators concluded that HHT was effective as an initial single-agent therapy in patients with chronic phase CML.

VIII. CONCLUSIONS

Because of their unique structures and strong antileukemic activity, *Cephalotaxus* alkaloids have attracted a great deal of attention. Many alkaloids were isolated from natural sources and derived by synthetic methodology. Furthermore, much work will be followed from the viewpoint of medicinal chemistry.

There is clear evidence that HHT (**1**) and harringtonine (**3**) have an antileukemic effect, but the dose and duration of administration of the drug can determine the efficacy and toxicity. Combination trials of HHT with other antineoplastic agents are now being investigated, and continuous infusion is recommended for further studies.

ACKNOWLEDGMENTS

We thank Professor K. Takeya, Professor T. Nagasaka, Dr. Y. Hitotsuyanagi, and Dr. S. Morris-Natschke for their valuable suggestions. Thanks are also due to partial support from National Institutes of Health grant CA17625 (K.H.L.).

REFERENCES

1. Huang, C.C. et al., Cytotoxicity and sister chromatid exchanges induced *in vitro* by six anticancer drugs developed in the People's Republic of China, *J. Natl. Cancer Inst.*, 71, 841, 1983.
2. Grem, J.L. et al., Cephalotaxine esters: anti-leukemic advance or therapeutic failure? *J. Natl. Cancer Inst.*, 80, 1095, 1988.
3. Paudler, W.W., Kerley, G.I., and McKay, J., The alkaloids of *Cephalotaxus drupacea* and *Cephalotaxus fortunei*, *J. Org. Chem.*, 28, 2194, 1963.
4. Perdue, R.E., Spetzman, L.A., and Powell, R.G., *Am. Hortic. Mag.*, 49, 129, 1970.
5. Spencer, G.F., Platner, R.D., and Powell, R.G., Quantitative gas chromatography and gas chromatography-mass spectrometry of *Cephalotaxus* alkaloids, *J. Chromatogr.*, 120, 335, 1976.
6. Abraham, D.J., Rosenstein, R.D., and McGandy, E.L., Single crystal X-ray structures of chemotherapeutic agents II, the structure of cephalotaxine methiodide, *Tetrahedron Lett.*, 10, 4085, 1969.
7. Powell, R.G. et al., Structure of cephalotaxine and related alkaloids, *Tetrahedron Lett.*, No. 46, 4081, 1969.
8. Powell, R.G., Weisleder, D., and Smith, C.R., Antitumor alkaloids from *Cephalotaxus harringtonia*: structure and activity, *J. Pharm. Sci.*, 61, 1227, 1972.
9. Powell, et al., Structures of harringtonine, isoharringtonine, and homoharringtonine, *Tetrahedron Lett.*, 825, 1970.
10. Mikolajczak, K.L., Powell, R.G., and Smith, C.R., Deoxyharringtonine, a new antitumor alkaloid from *Cephalotaxus*. Structure and synthetic studies, *Tetrahedron*, 28, 1995, 1972.
11. Powell, R.G. et al., Alkaloids of *Cephalotaxus harringtonia var. drupacea*. 11-Hydroxycephalotaxine and drupacine, *J. Org. Chem.*, 39, 676, 1974.
12. Powell, R.G. et al., Desmethylcephalotaxine and its correlation with cephalotaxine, *Phytochem.*, 12, 2987, 1973.
13. Cassady, J.M. and Douros, J.D., *Anticancer Agents based on Natural Product Models, Medicinal Chemistry*, Vol. 16, Academic Press, New York, 1980, 341.

14. Kantarjian, H.M. et al., Homoharringtonine in hematologic and solid malignancies (history, current research, and future directions), *Cancer*, 92, 1591, 2001.

15. Takano, I. et al., Drupangtonine, a novel antileukemic alkaloid from *Cepalotaxus harringtonia var. Drupacea*, *Bioorg. Med. Chem. Lett.*, 6, 1689, 1996.

16. Takano, I. et al., Alkaloids from *Cephalotaxus harringtonia*, *Phytochem.*, 43, 299, 1996.

17. Takano, I. et al., New *Cephalotaxus* alkaloids from *Cephalotaxus harringtonia* var. *drupacea*, *J. Nat. Prod.*, 59, 965, 1996.

18. Takano, I. et al., Cephalotaxidine, a novel dimeric alkaloid from *Cepharotaxus harringtonia* var. *drupacea*, *Tetrahedron Lett.*, 37, 7053, 1996.

19. Takano, I. et al., New oxygenated *Cephalotaxus* alkaloids from *Cephalotaxus harringtonia* var. *drupacea*, *J. Nat. Prod.*, 59, 1192, 1996.

20. Takano, I., et al., Ester-type *Cephalotaxus* alkaloids from *Cephalotaxus harringtonia* var. *drupacea*, *Phytochem.*, 44, 735, 1997.

21. Morita, H. et al., Cephalezomines A-F, potent cytotoxic alkaloids from *Cephalotaxus harringtonia var. nana*, *Tetrahedron*, 56, 2929, 2000.

22. Bocar, M., Jossang, A., and Bodo, B., New alkaloids from *Cephalotaxus fortunei*, *J. Nat. Prod.*, 66, 152, 2003.

23. Kobayashi, J. et al., Cephalocyclidin A, a novel pentacyclic alkaloid from *Cephalotaxus harringtonia* var. nana, *J. Org. Chem.*, 67, 2283, 2002.

24. Powell, R.G., Structures of homoerythrina alkaloids from *Cephalotaxus harringtonia*, *Phytochem.*, 11, 1467, 1972.

25. Furukawa, H. et al., Alkaloids of *Cephalotaxus wilsoniana Hay.* in Taiwan, *Yakugaku Zasshi*, 96, 1373, 1976.

26. Powell, R.G. et al., Alkaloids of *Cephalotaxus* wilsonia, *Phytochem.*, 11, 3317, 1972.

27. Arora, S.K. et al., Crystal and molecular structure of cephalotaxine p-bromobenzoate, *J. Org. Chem.*, 39, 1269, 1974.

28. Arora, S.K. et al., Crystal and molecular structure of cephalotaxine, *J. Org. Chem.*, 41, 551, 1976.

29. Takano, I. et al., Modification of the skeleton of homoharringtonine through unusual rearrangements, *J. Org. Chem.*, 62, 8251, 1997.

30. Auerbach, J., Ipaktchi, T., and Weinreb, S.M., Synthesis of the diacid side chain of deoxyharringtonine, *Tetrahedron Lett.*, 4561, 1973.

31. Ipaktchi, T. and Weinreb, S.M., Relative configuration of the diacid side chain of isoharringtonine., *Tetrahedron Lett.*, 3895, 1973.

32. Kelly, T. et al., Regiospecific synthesis of the acyl portion of harringtonine, *Tetrahedron Lett.*, 36, 3501, 1973.

33. Utawanit, T., Ph.D. Thesis, University of Illinois at Urbana-Champaign, Urbana Illinois, 1975.

34. Jalil Miah, M.A., Hudlicky, T., and Reed, J.W., in *The Alkaloids*; Vol. 51, Brossi, A., Ed, Academic Press, New York, 1998, 199.

35. Suga, S., Watanabe, M., and Yoshida, J., Electroauxiliary-assisted sequential introduction of two carbon nucleophiles on the same α-carbon of nitrogen: Application to the synthesis of spiro compounds, *J. Amer. Chem. Soc.*, 124, 14824, 2002.

36. Koseki, Y. et al., A formal total synthesis of (+)-cephalotaxine using sequential N-acylimnium ion reactions, *Org. Lett.*, 43, 6011, 2002.

37. Tietze, L.F. and Schirok, H., Enantioselective highly efficient synthesis of (-)-cephalotaxine using two palladium-catalyzed transformation, *J. Am. Chem. Soc.*, 121, 10264, 1999.

38. Worden, S.M., Mapitse, R., and Hayes, C., Towards a total synthesis of (-)-cephalotaxine: construction of the BCDE-tetracyclic core, *Tetrahedron Lett.*, 43, 6011, 2002.

39. Bocker-Milburn, K.I. et al., Formal intramolecular [5 + 2] photocycloaddition reactions of maleimides: a novel approach to the CDE ring skeleton of (-)-cephalotaxine. *Org. Lett.*, 3, 3005, 2001.

40. Kim, S.H. and Cha, J.K., Synthetic studies toward cephalotaxine: Functionalization of tertiary N-acylhemiaminals by Nazarov cyclization, *Synthesis*, 2113, 2000.

41. Beall, L.S. and Padwa, A., An approach to the cephalotaxine ring skeleton using an ammonium ylide/steveno[1,2]-rearrangement, *Tetrahedron Lett.*, 39, 4159, 1998.

42. Molander, G.A. and Hierseman, M., Intramolecular 1,3-dipolar cycloaddition as a total for the preparation of azaspirocyclic keto aziridines. Synthesis of intermediates for the total synthesis of (±)cephalotaxine, *Tetrahedron Lett.*, 38, 4347, 1997.

43. De Oliveira, E.R., Dumas, P., and D'Angela, J., A simple, efficient access to functionalized pyrrolobenzazapines related to the ABC core of cephalotaxine, *Tetrahedron Lett.*, 38, 3723, 1997.

44. Huang, L. and Xue, Z., in *The Alkaloids*, Vol. 23, Brossi, A., Ed, Academic Press, New York, 1984, 157.

45. Li, W.D.Z. and Wang, Y.Q., A novel and efficient total synthesis of cephalotaxine, *Org. Lett.*, 5, 2931, 2003.

46. Dolby, L.I., Nelson, S.I., and Senkovich, D., Synthesis of cephalotaxine, *J. Org. Chem.*, 37, 3691, 1972.

47. Parry, R.I. et al., Biosynthesis of the *Cephalotaxus* alkaloids. Investigations of the early and late stages of cephalotaxine biosynthesis, *J. Am. Chem. Soc.*, 102, 1099, 1980.

48. Nagata, W., Lectures in heterocyclic chemistry, Castle, R.N., Elslager, E.F., Eds., *J. Heterocyclic Chem.*, 1, 529, 1972.

49. Büchi, G. et al., The total synthesis of Iboga alkaloids, *J. Am. Chem. Soc.*, 88, 3099, 1966.

50. Moriarity, R.M., Hu, H., and Gupta, S.C., Direct α-hydroxylation of ketones using iodosobenzene, *Tetrahedron Lett.*, 22, 1283, 1981.

51. Yasuda, S., Yamada, T., and Hanaoka, M., A novel and stereoselective synthesis of (±)-cephalotaxine and its analogue, *Tetrahedron Lett.*, 27, 2023, 1986.

52. Weinreb, S.M. and Auerbach, I., Total synthesis of the *Cephalotaxus* alkaloids cephalotaxine, cephalotaxinone, and demethylcephalotaxinone, *J. Am. Chem. Soc.*, 97, 2503, 1975.

53. Zhong, S. et al. Total synthesis of (-)-cephalotaxine. *Zhongguo Yaowu Huaxue Zazhi*, 4, 84, 1994.

54. Overman, L.E. and Ricca, D.J., in *Comprehensive Organic Synthesis*, Vol. 3, Trost, B.M. and Flemming, I., Eds., Pergamon Press, Oxford, 1991, 1007.

55. Clemo, G.R. and Ramage, G.R., The lupine alkaloids. Part IV. The synthesis of octahydropyridocoline, *J. Chem. Soc.*, 437, 1931.

56. Clemo, G.R., Raper, R. and Vipond, H.J., The Clemmensen reduction of certain α-amino-ketones and its bearing on the reduction of 1-keto-octahydropyridocoline, *J. Chem. Soc.*, 2095, 1949.

57. Prelog, V. and Seiwerth, R., Constitution of the so-called norlupinane B, *Chem. Ber.*, 72, 1638, 1939.

58. Leonard, N.J. and Ruyle, W.V., Rearrangement of α-aminoketones during Clemmensen reduction. II. Contraction of a six-membered ring in the monocyclic series, *J. Am. Chem. Soc.*, 71, 3094, 1949.

59. Leonard, N.J. and Barthel, E., Jr., Rearrangement of α-aminoketones during Clemmensen reduction. III. Contraction of a seven-membered ring in the monocyclic series, *J. Am. Chem. Soc.*, 71, 3098, 1949.

60. Vedejs, E., Clemmensen reduction of ketones in anhydrous organic solvents, *Org. React.*, 22, 401, 1975.

61. Leonard, N.J., *J. Rec. Chem. Prog.*, 17, 243, 1954.

62. Gaskel, A.J. and Joule, I.A., The zinc-acetic acid reduction of reserpine and other tetrahydro--carbolone alkaloids, *Tetrahedron*, 24, 5115, 1968.

63. Noe, E. et al., Synthesis of the new (cyclopenta[b]pyrolo[1,2-d]azepine[4,5-b]indole ring system, *Tetrahedron Lett.*, 37, 5701, 1996.

64. Ait-Mohand, S. et al., Rearrangements and cyclizations of 2-chloropropenyl-appended indolo[2,3-a]quinolizidine derivatives, *Eur. J. Org. Chem.*, 3429, 1999.

65. Leonard, N.J. et al., Unsaturated amines. III. Introduction of α,β-unsaturation by means of mercuric acatate:[1(10)-]dehydroquinolizidine, *J. Am. Chem. Soc.*, 77, 439, 1955.

66. Leonard, N.J., Fulmax, R.W. and Hay, A.S., Unsaturated amines. VII. Introduction of α, β-unsaturation by means of mercuric acetate: methylquinolizidines, *J. Am. Chem. Soc.*, 78, 3457, 1956.

67. Grieson, D.S. and Husson, H.P., in *Comprehensive Organic Synthesis*; Vol. 6, Trost B.M., Flemming, I. Eds, Pergamon Press, Oxford, 1991, 910.

68. Weinstein, B. and Craig, A.R., Synthetic approach to the cephalotaxine skeleton, *J. Org. Chem.*, 41, 875, 1976.

69. Woodward, R.B. et al., The total synthesis of chlorophyll-a, *Tetrahedron*, 46, 7599, 1990.

70. Stevens, R.V., General methods of alkaloid synthesis, *Acc. Chem. Res.*, 10, 193, 1977.

71. Marino, J.P. and Samanen, J.M., Biogenetic-type approach to homoerythrina alkaloids, *J. Org. Chem.*, 41, 179, 1976.

72. Kupchan, S.M., Dhingra, O.P. and Kim, C.K., New biogenetic-type approach to *Cephalotaxus* alkaloids and the mechanism of Schelhammer-type homoerythrinadienone formation *in vitro*, *J. Org. Chem.*, 43, 4461, 1978.

73. Huang, M.T., Harringtonine, an inhibitor of initiation of protein biosynthesis, *Molec. Pharmacol.*, 11, 511, 1975.

74. Fresno, M., Jimenez, A., and Vazquez, D., Inhibition of translation in eukaryotic systems by harringtonine, *Eur. J. Biochem.*, 72, 323, 1977.

75. Tujebajeva, R.M. et al, Alkaloid homoharringtonine inhibits polypeptide chain elongation on human ribosomes on the step of peptide bone formation, *FEBS Lett.*, 257, 254, 1989.

76. Visani, G. et al., Effect of homoharringtonine alone and in combination with alpha interferine and cystosine arabinoside on *"in vitro"* growth and induction of apoptosis in chronic myeloid leukemia and normal hematopoitic progenitors, *Leukemia*, 11, 624, 1997.

77. Kuliczkowski, K., Influence of harringtonine on human leukemia cell differentiation, *Arch. Immunol. Ther. Exp. (Warsz)*, 37, 69, 1989.

78. Li, L. et al., Induction of apoptosis by harringtonine and homoharringtonine in HL-60 cells, *Yao Hsueh Hsueh Pao*, 29, 667, 1994.

79. O'Brien, S., et al., Homoharringtonine induces apoptosis in chronic myelogenous leukemia cells, *Blood*, 82(Suppl.), 555a. (2203), 1993.

80. Baaske, D.M. and Heinstein, P., Cytotoxicity and cell cycle specificity of homoharringtonine, *Antimicrob. Agents Chemother.*, 12, 298, 1977.

81. Takemura, Y. et al, Biological and pharmacologic effects of harringtonine on human leukemia-lymphoma cells. *Cancer Chemother. Pharmacol.*, 14, 206, 1985.

82. Han, R. and Ji, X.J., Chemical, pharmacological and clinical studies on the antitumor active principle of *Cephalotaxus hainanesis Li* (in Chinese), *Zhonghua Zhongliu Zazhi*, 1, 176, 1979. [NIH Library translation (NIH Pub No. 87-2361)].

83. Wang, Y., Pan, Z., and Han, R., The cytokinetic effects of harringtonine on leukemia L1210 cells, II. Studies by microscopic photometer, *Zhonghua Zhognliu Zachi*, 2, 247, 1980.

84. Xu, Y. et al., The effect of harringtonine and its allied alkaloids on the incorporation of labeled amino acids into proteins of cells of transplantable leukemias L615 and P388, *Acta Pharm. Sinica*, 16, 661, 1981.

85. Chou, T.C. et al., Uptake, initial effects, and chemotherapeutic efficacy of harringtonine in murine leukemia cells sensitive and resistant to vincristine and other chemotherapeutic agents, *Cancer Res.*, 43, 3074, 1983.

86. O'Dwyer, P.J. et al., Homoharringtonine perspective on an active new natural product, *J. Clin. Oncol.*, 4, 1563, 1986.

87. Baguley, B.C. et al., Comparison of the effects of flavone acetic acid, fostriecin, homoharringtonine and tumor necrosis factor alpha on colon38 tumours in mice, *Eur. J. Cancer Clin. Oncol.*, 25, 263, 1989.

88. Tebbi, C.K., Chervinsky, D., and Murphy, T., Effects of homoharringtonine on human neuroblastoma cell lines, *Proc. Am. Assoc. Cancer Res.*, 28, 425, 1987.

89. Tebbi, C.K. and Chervinsky, D., The effects of homoharringtonine (HHT) on mouse C-1300 neuroblastoma *in vitro* and *in vivo*, *Proc. Am. Assoc. Cancer Res.*, 27, 274, 1986.

90. Boyd, A.W. and Sullivan, J.R., Leukemic cell differentiation *in vivo* and *in vitro*: arrest of proliferation parallels the differentiation induced by the antileukemic drug harringtonine, *Blood*, 63, 384, 1984.

91. Zhou, J.Y. et al., Effect of homoharringtonine on proliferation and differentiation of human leukemic cells *in vitro*, *Cancer Res.*, 50, 2031, 1990.

92. Ling, Y.H., Tseng, M.T., and Harty, J.L., Effects of homoharringtonine on protein glycosylation in human bladder carcinoma cell T-24, *Cancer Res.*, 49, 76, 1989.

93. Wilkoff, L.J. et al., Effect of homoharringtonine on the viability of murine leukemia P388 cells resistant to either adriamycin, vincristine, or 1-β-D-arabinofuranosycytosine, *Cancer Chemother. Pharmacol.*, 23, 145, 1989.

94. Tebbi, C.K., Chervinsky, D., and Baker, R.M., Modulation of drug resistance in homoharringtonine-resistant C-1300 neuroblastoma cells with cyclosporine A and dipyridamole, *J. Cell. Physiol.*, 148, 464, 1991.

95. Russo, D. et al., MDR-related P170-glycoprotein modulates cytotoxic activity of homoharringtonine, *Leukemia*, 9, 513, 1995.

96. Zhou, D.C. et al., Sequential emergence of MRP-and MDRI-gene over-expression as well as MDRI-gene translocation in homoharringtonine-selected K562 human leukemia cell lines, *Int. J. Cancer*, 65, 365, 1996.

97. Laster, W.R. et al., Therapeutic synergism (TS) of homoharringtonine (H) plus 5-fluorouracil (FU) against leukemia P388 (P388/o) and ARA-C-resistant P388 (P388/ARA-C), *Proc. Am. Assoc. Cancer Res.*, 23, 199, 1982.

98. Fanucchi, M.P., Kong, X.R., and Chou, T.C., Hexamethylene bisacetamide (HMBA) does not enhance the cytotoxic effects of adriamycin (ADR), 1β-D-arabinofuranosylcytosine (ARA-C) and harringtonine (HT) in HL-60 cells [abstract 1492], *Proc. Am. Assoc. Cancer Res.*, 27, 376, 1986.

99. Okano, T. et al., Effects of harringtonine in combination with acivicin, adriamycin, L-asparaginase, cytosine arabinoside, dexamethasone, fluorouracil or methotrexate on human acute myelogenous leukemia cell line KG-1, *Invest. New Drugs*, 1, 145, 1983.

100. Smith, C.R., Jr., Mikolajczak, K.L., and Powell, R.G., Harringtonine and related cephalotaxine esters, in *Anticancer Agents Based on Natural Product Models*, Cassady, J.M. and Douros, J.D. Eds., Academic Press New York, 1980, pp. 407-414.

101. Mikolajczak, K.L. et al., Synthesis of deoxyharringtonine, *Tetrahedron Lett.*, 283, 1974.

102. Mikolajczak, K.L., Smith, C.R., and Weisleder, D., Synthesis of cephalotaxine esters and correlation of their structures with antitumor activity, *J. Med. Chem.*, 20, 328, 1977.

103. Mikolajczak, K.L., Powell, R.G., and K.L., Smith, C., Preparation and antitumor activity of a rearranged ester of cephalotaxine, *J. Med. Chem.*, 18, 63, 1975.

104. Mikolajczak, K.L., Smith, C.R., and Powell, R.G., Partial synthesis of harringtonine analogs, *J. Pharm. Sci.*, 63, 1280, 1974.

105. Stewart, J.A. and Krakoff, I.H., Homoharringtonine: a phase I evaluation, *Investigational New Drugs*, 3, 279, 1985.

106. Ajani, J.A., et al., Phase II studies of homoharringtonine in patients with advanced malignant melanoma; sarcoma; and head and neck, breast, and colorectal carcinomas., *Cancer Treat. Rep.*, 70, 375, 1986.

107. Runge-Morris, M.A. et al., Evaluation of homoharringtonine efficacy in the treatment of squamous cell carcinoma of the head and neck: A phase II Illinois Cancer Council Study, *Invest. New Drugs*, 7, 269, 1989.

108. Feun, L.G. et al., Phase II study of homoharringtonine in patients with recurrent primary malignant central nervous system tumors, *J. Neurooncol*, 9, 159, 1990.

109. Witte, R.S. et al., A phase II trial of amonafide, caracemide, and homoharringtonine in the treatment of patients with advanced renal cell cancer, *Invest. New Drugs*, 14, 409, 1996.

110. Witte, R.S., A phase II trial of homoharringtonine and caracemide in the treatment of patients with advanced large bowel cancer, *Invest. New Drugs*, 17, 173, 1999.

111. Kavanagh, J.J. et al., Intermittent IV homoharringtonine for the treatment of refractory epithelial carcinoma of the ovary: a phase II trial, *Cancer Treat. Rep.*, 68, 1503, 1984.

112. Kantarjian, H.M. et al., Phase II study of low-dose continuous infusion homoharringtonine in refractory acute myelogenous leukemia, *Cancer*, 63, 813, 1989.

113. Feldman, E.J. et al., Homoharringtonine is safe and effective for patients with acute myelogenous leukemia, *Leukemia*, 6, 1185, 1992.

114. Ekert, H. and Richards, M., Experience with homoharringtonine in one patient with acute myeloid leukemia, *Proc. Clin. Oncol. Soc. Aust.*, 8, 152, 1980.

115. Ekert, H., et al., Treatment of acute myeloid leukemia with the harringtonines. *Proc. Clin. Oncol. Soc. Aust.*, 9, 122, 1982.

116. Sullivan, J. and Leyden, M., Long survival in an elderly patient with acute myeloid leukemia after treatment with harringtonine, *Med. J. Aust.*, 142, 693. 1985.

117. Bell, B.A., Chang M.N., and Weinstein H.J., A Phase II study of homoharringtonine for the treatment of children with refractory or recurrent acute myelogenous leukemia: a pediatric oncology group study, *Med. Pediatric. Oncol.*, 37, 103, 2001.

118. Bell, B.A. et al., Phase II study of homoharringtonine (HHT) for the treatment of children with refractory nonlymphoblastic leukemia (ANLL), *Proc. Am. Soc. Clin. Oncol.*, 13, A1060, 1994.

119. Tan, C.T.C. et al., Phase I trial of homoharringtonine in children with refractory leukemia, *Cancer Treat. Rep.*, 71, 1245, 1987.
120. Feldman, E.J. et al., Homoharringtonine in patients with myelodysplastic syndrome (MDS) and MDS evolving to acute myeloid leukemia, *Leukemia*, 10, 40, 1996.
121. Ye, J.S. et al., Small-dose harringtonine induces complete remission in patients with acute promyelocytic leukemia, *Leukemia*, 2, 427, 1988.
122. Feldman, E.J. et al., Acute promyelocytic leukemia: A 5-year experience with new antileukemic agents and a new approach to preventing fatal hemorrhage, *Acta Haematol*, 82, 117, 1989.
123. Cortes, J. and Kantarjian, H.M., Promising approaches in acute leukemia, *Invest. New Drugs*, 18, 57, 2000.
124. O'Brien, S. et al., Homoharringtonine therapy induces responses in patients with chronic myelogenous leukemia in late chronic phase, *Blood*, 86, 3322, 1995.

5 Podophyllotoxins and Analogs[1]

Kuo-Hsiung Lee and Zhiyan Xiao

CONTENTS

I. Introduction 71
II. History 72
III. Structure 72
IV. Development of Etoposide and Teniposide 73
V. Mechanisms of Action 74
    A. Inhibition of Tubulin Polymerization 74
    B. Inhibition of DNA Topoisomerase II 74
    C. Other Antineoplastic Mechanisms 75
VI. Structure–Activity Relationships 75
    A. Molecular Area-Oriented Analog Syntheses 76
        1. Ring A 76
        2. Ring B 76
        3. Ring C 76
        4. Ring D 77
        5. Ring E 78
    B. SAR Models 80
        1. Composite Pharmacophore Model 80
        2. Comparative Molecular Field Analysis Model 80
        3. K-Nearest Neighbor QSAR Model 81
    C. Representative Analogs 81
        1. Etopophos 81
        2. NK 611 81
        3. GL-331 82
        4. TOP-53 82
VII. Syntheses 82
VIII. Clinical Applications 84
IX. Future Perspectives 85
Acknowledgments 85
References 85

I. INTRODUCTION

Podophyllotoxins are important natural products in the armamentarium of antineoplastic agents. The biological assessment of podophyllotoxin (**1**) was followed by discovery of its mode of action and culminated in the synthesis of the anticancer drugs etoposide (**2**) and teniposide (**3**). The long journey from podophyllotoxin to etoposide and teniposide illustrates the fascinating development of clinically useful anticancer drugs from natural product prototypes through chemical modification. It is particularly distinctive that structural variation of podophyllotoxin caused a radical change in

---

[1] Antitumor Agents 240.

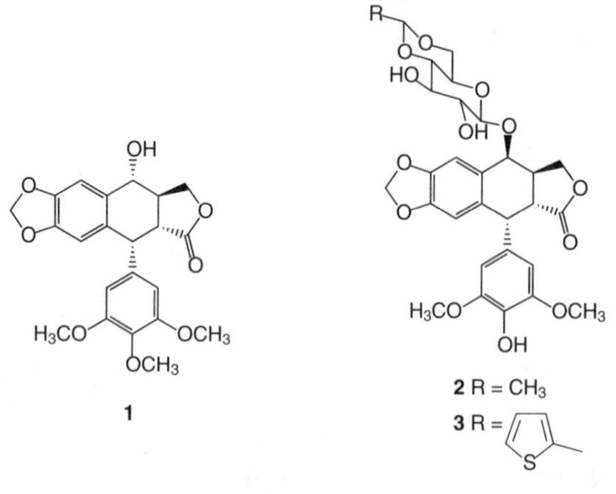

STRUCTURES 1–3

the mechanism of action. Today, several new podophyllotoxin analogs have emerged as potential anticancer drugs. Some recent literature contributions have provided comprehensive updates on various aspects of this compound class.[1–7] In this chapter, we highlight recent developments and emphasize critical features of these analogs.

II. HISTORY

The genus *Podophyllum* (Podophyllaceae), including American *P. peltatum* L. and Indian or Tibetan *P. emodi* Wall (syn. *P. hexandrum* Royle), has been used for centuries for its medicinal properties. *Podophyllum* plants have long been valued for their cathartic and cholagogic properties (increasing the flow of bile into the intestine) by the indigenous populations of North America and the Himalayas. Podophyllin, the alcoholic extract of *Podophyllum* rhizome, was listed in the first American Pharmacopoeia (1820), and was later introduced into European Pharmacopoeias. However, it was removed from the U.S. Pharmacopoeia in 1942 because of its undesirable toxicity.[3]

In the same year, Kaplan demonstrated the curative effects of podophyllin on the benign tumor *Condylomata acuminata*,[8] which rekindled interest in *Podophyllum* plants and stimulated intensive studies on the action mode and chemical constituents of podophyllin. Podophyllin was first reported in 1946 to show toxic effects against dividing cells in a manner similar to that of colchicine, the classic antimicrotubule agent.[9] Later, its major constituent podophyllotoxin was found to inhibit assembly of the mitotic spindle.[10] Other podophyllotoxin compounds were also reported to induce cell cycle arrest at mitosis.[11] Both podophyllin and podophyllotoxin exhibited destructive effects on mice tumors,[12] which raised hopes for their use in the treatment of malignant tumors. However, the zeal in developing podophyllotoxin as cancer chemotherapy was tempered by its unacceptable side effects, particularly gastrointestinal toxicity.[5] Although the direct therapeutic application of podophyllotoxin failed, the pioneer work provided a natural prototype and eventually led to the serendipitous discovery of the semisynthetic anticancer drugs etoposide and teniposide.

III. STRUCTURE

Podophyllotoxin (**1**), the major constituent of podophyllin, was first isolated in 1880.[13] Its correct structure was resolved chemically in 1951[14] and was later confirmed by total synthesis.[15] Similar to other *Podophyllum* constituents, **1** is an aryltetralinlactone cyclolignan. Its skeleton is a flat, rigid, five-ring system. The methylenedioxy ring A, tetrahydronaphthalene rings B and C, and lactone ring D make a four-ring pseudoplane to which the pendent aryl ring E is attached pseudoaxially at C_1

TABLE 5.1
Chemical Structures of Major Podophyllotoxin Analogs

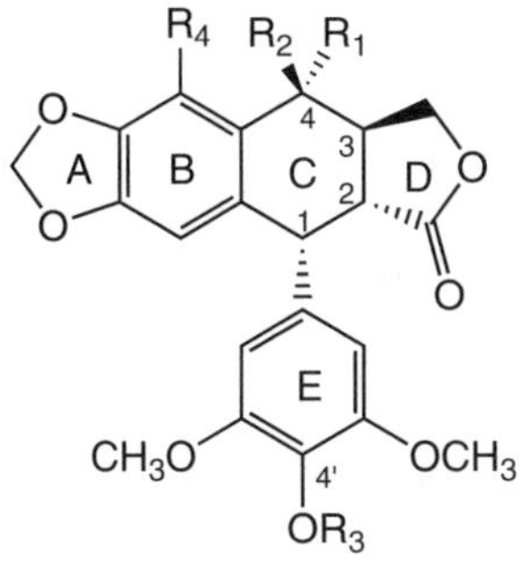

	R_1	R_2	R_3	R_4
Podophyllotoxin	OH	H	CH_3	H
Deoxypodophyllotoxin	H	H	CH_3	H
4′-Demethylpodophyllotoxin	OH	H	H	H
4′-Demethylepipodophyllotoxin	H	OH	H	H
α-Peltatin	H	H	H	OH
β-Peltatin	H	H	CH_3	OH

(Table 5.1). The configurations at the four asymmetric centers (C_1, C_2, C_3, and C_4) and the highly strained *trans*-lactone D ring characterize unique structures of this compound class. Podophyllotoxins are widespread in the plant kingdom and are not limited to the Podophyllaceae family. Key chemical structures include podophyllotoxin, deoxypodophyllotoxin, 4′-demethylpodophyllotoxin, 4′-demethylepipodophyllotoxin (DMEP), α-peltatin, β-peltatin, and their corresponding glycosides (Table 5.1). Most clinically relevant antineoplastic podophyllotoxins are DMEPs.

IV. DEVELOPMENT OF ETOPOSIDE AND TENIPOSIDE

The complex path from *Podophyllum* plants to podophyllotoxin and eventually to etoposide and teniposide involved more than a century of study and resulted in successful development of clinically useful drugs from natural sources. The investigators themselves described the story retrospectively.[16]

In the early 1950s, scientists in Sandoz, Ltd. assumed that, in analogy to cardiac glycosides, podophyllotoxin glycosides might exhibit pharmacological profiles superior to those of the aglycone. This assumption stimulated extensive efforts to acquire both natural and synthetic *Podophyllum* glycosides and led in 1963 to the development and commercialization of SP-G, the condensation product of the crude *Podophyllum* glycoside fraction with benzaldehyde. A highly active "antileukemia factor" was later isolated from SP-G as a minor component (<0.25%). At low doses, this component significantly inhibited cell proliferation *in vitro* and considerably prolonged the survival time of leukemic mice. It was identified as 4′-*O*-demethyl-epipodophyllotoxin benzylidene β-D-glucoside (DEPBG, **4**) and had the unique structural features of a free phenolic hydroxyl group at $C_4′$ and an epi configuration at C_4. Subsequent synthetic efforts to condense 4′-*O*-demethyl-epipodophyllotoxin glucoside (DEPG, **5**) with various aldehydes and ketones led in the late 1960s to the discovery of etoposide (**2**) and teniposide (**3**).[16] The FDA approved etoposide for the treatment of testicular cancer in 1983, and teniposide was brought into the U.S. market in 1992. These drugs are currently used against a variety of cancers, including small cell lung cancer, testicular cancer, lymphoma, leukemia, and Kaposi's sarcoma.[17]

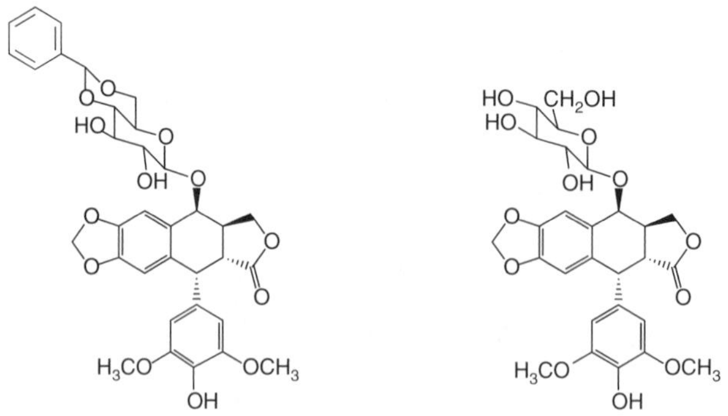

STRUCTURES 4–5

V. MECHANISMS OF ACTION

Primary molecular mechanisms underlying the antineoplastic activities of podophyllotoxin analogs include preventing the assembly of tubulin into microtubules or inhibiting the catalytic activity of DNA topoisomerase II, although other known and in some cases ambiguous mechanisms are also involved.

A. INHIBITION OF TUBULIN POLYMERIZATION

As early as 1947, podophyllotoxin was reported to inhibit assembly of the mitotic spindle and induce arrest of the cell cycle at mitosis.[10] It was later found to bind tubulin, the fundamental monomeric protein subunit of microtubules, at the same binding site as colchicine. It binds to tubulin as strongly as colchicine does, but the binding is more rapid and is, in contrast to that of colchicine, reversible.[18]

The microtubule network forms a vital part of the cytoskeleton in eukaryotic cells and plays an important role in mitosis. Podophyllotoxin reversibly binds to tubulin, disturbs the dynamic equilibrium between the assembly and disassembly of microtubules, and eventually causes mitotic arrest. Most *Podophyllum* compounds, including the acetal products of podophyllotoxin glucosides and peltatins, share this mechanism of action. On treatment with these compounds, cells can still enter mitosis and undergo a normal pro-phase, but the separation of chromosomes is blocked as a result of the inhibition of mitotic spindle formation. Characterizing the cytotoxic properties of these compounds, cells with clumped chromosomes accumulate in metaphase.

B. INHIBITION OF DNA TOPOISOMERASE II

Unlike podophyllotoxin, etoposide prevents cells from entering mitosis rather than trapping cells in metaphase. Time course analysis with etoposide in tissue culture shows that disappearance of mitoses begins less than 1 h after drug addition.[19] This observation implies that the compound acts in late S or G_2 phase of the cell cycle, which is the checkpoint for DNA damage or DNA replication, rather than spindle assembly. Etoposide causes few effects on tubulin polymerization. However, fragmentation of DNA in HeLa cells was observed on treatment with etoposide.[20] It was not recognized until the 1980s, however, that the ability of etoposide to induce DNA breaks was mediated by DNA topoisomerase II (topo II).[21]

DNA topoisomerases are ubiquitous enzymes that control the topological state of DNA. Topo II catalyzes the cleavage and religation of double-strand DNA. Etoposide stabilizes the covalent DNA-enzyme cleavable complex, inhibits the catalytic activity of topo II, and induces topo II–mediated

DNA breakage. These actions convert the essential enzyme into a cellular poison, trigger cascade reactions, and eventually lead to cell death.[22] The topo II inhibition mechanism is shared by etoposide, teniposide, and other therapeutically important DMEPs.

The structural preferences of topo II inhibitors over antimicrotubule agents have been roughly identified as (1) 4′-demethylation, (2) 4β-configuration, and (3) 4β-bulky substitution.[3] Because of the severe toxic effects of antimicrotubule agents acting on the colchicine-binding site, topo II inhibitory compounds are of more clinical relevance.

C. OTHER ANTINEOPLASTIC MECHANISMS

Peroxidases or cytochrome P450 can generate catechol, *ortho*-quinone, and phenoxyradicals from epipodophyllotoxins. These radical species can covalently bind to DNA and cause DNA damage by forming chemical adducts.[23] DNA strand cleavage does not occur with etoposide alone; however, DNA cleavage can be induced in the presence of metal ions, such as Cu^{2+} and Fe^{3+}, or ultraviolet irradiation, and can be inhibited with known radical scavengers.[24] These facts support a role of radical formation for DNA cleavage induced by epipodophyllotoxins.

Etoposide and GL-331 (Figure 5.1; cf. Sections VI.A.3 and VI.C.3) also induce cancer cell death with signs of apoptosis. Cellular protein tyrosine phosphatase (PTP) activity was increased significantly after GL-331 treatment, and in addition, GL-331-induced internucleosomal cleavage was efficiently prevented by two PTP inhibitors but not by an inhibitor of serine/threonine phosphatase.[25] Abnormal activation of cyclin B-associated CDC 2 kinase has been noted after treatment with etoposide or GL-331 in various cancer cells. GL-331 treatment in NPC-TW01 cells also increased CDC 25A phosphatase activity and facilitated the association of CDC 25A with Raf-1. Apoptotic DNA fragmentation induced by GL-331 was inhibited by treatment with cyclin B1-specific antisense oligonucleotides.[26] These results indicate that PTP, CDC2 kinase, and CDC25 phosphatase might be involved in the induced apoptosis.

VI. STRUCTURE–ACTIVITY RELATIONSHIPS

The impressive antitumor potency and clinical efficacy of **2** and **3** have prompted extensive molecular modifications of the podophyllotoxin prototype, and numerous podophyllotoxin analogs have been synthesized and evaluated since the 1950s. As highlights of such global efforts, several synthetic analogs, including Etopophos (Bristol-Myers), NK 611 (Nippon-Kayaku), GL-331 (NPL at UNC), and TOP-53 (Taiho) (Figure 5.1), have been produced as either clinical drugs or novel clinical trial candidates for various cancers.

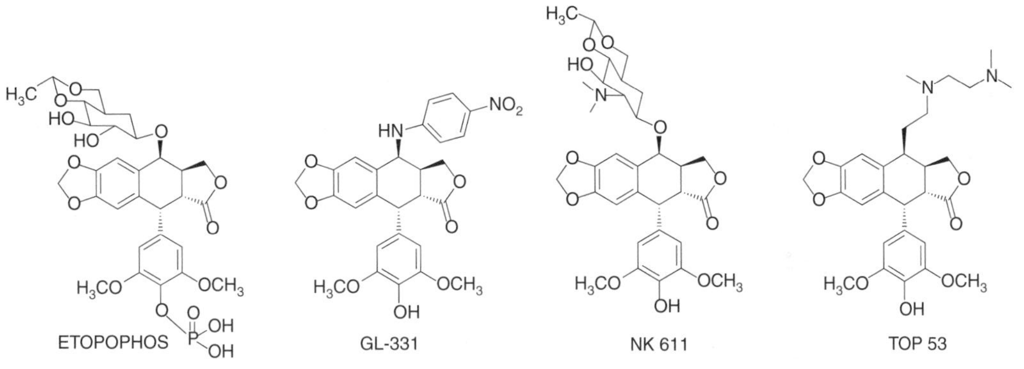

FIGURE 5.1 Representative podophyllotoxin analogs.

TABLE 5.2
Molecular Modifications of Podophyllotoxin

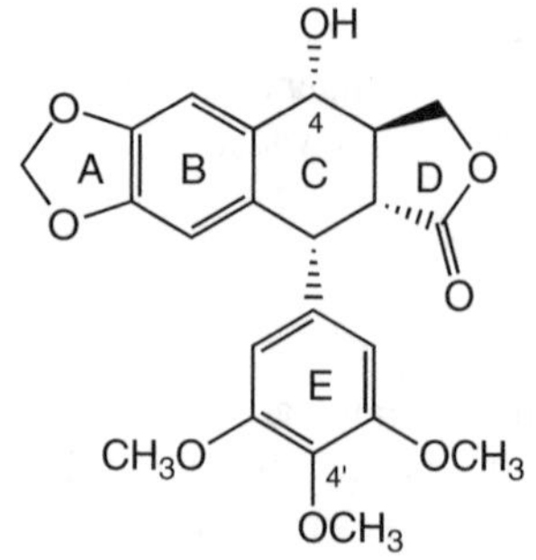

Modifications

A ring	Removal of the methylenedioxy group to give hydroxy, methoxy, or other oxygenated substituents; replacement of the methylenedioxy ring with heteroaromatic ring systems
B ring	5-Oxygenation to give hydroxy or alkoxy groups
C ring	Extensive C_4 modifications, including C_4 sugar- and nonsugar- (with O-, S-, N- and C-linkages) substituted derivatives; aromatization of C ring
D ring	Conversion of the γ-lactone to lactam, cyclopentanone, cyclopentane, sulfide, sulfoxide, sulfone, cyclic ether, and homolactone; substitution of the hydrogen at C_2 to halogen atoms, methyl or hydroxyl; replacement of C_2 to nitrogen
E ring	Demethylations; oxidation to the O-quinone; esterification of the $C_{4'}$ hydroxyl

A. Molecular Area-Oriented Analog Syntheses

Previous reviews[1,6] have thoroughly discussed the extensive molecular modifications of podophyllotoxin, which are summarized in Table 5.2. We highlight herein only the chemical modifications most relevant to the structure–activity relationships (SARs).

1. Ring A

The A ring has been modified via two different approaches: opening the methylenedioxy bridge and selectively functionalizing the two phenols,[27] and replacing the methylenedioxy ring with heteroaromatic ring systems.[28] Neither approach has provided derivatives with potent cytotoxicity or topo II inhibitory activity.

2. Ring B

B-ring modification is mainly related to two natural products, α-peltatin and β-peltatin (Table 5.1). Similar to podophyllotoxin, these derivatives act as antimitotic agents and interact only weakly with topo II. Many α-peltatin derivatives were synthesized but were much less active than etoposide against topo II and P388 leukemia.[29,30]

3. Ring C

Modification of the C ring has been quite extensive and focused almost exclusively on the C_4 position. 4-Epimerization of podophyllotoxin shifts the molecular target from tubulin to topo II,[1] which is the clinically relevant target; therefore, most C_4 modifications use epipodophyllotoxins.

Etoposide and teniposide were developed through C_4 modification in the late 1960s.[31,32] Subsequently, many other sugar substituted derivatives were prepared. Some of the clinically useful

TABLE 5.3
Podophyllotoxin Analogs with
Nitrogen-Substituted Sugars

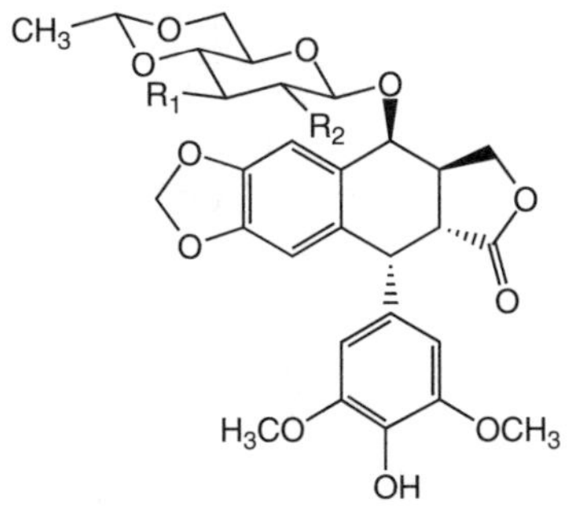

No.	R_1	R_2	L1210 (Max % T/C)
1	—	—	131
2	OH	OH	184
6	OH	NH_2	272
7	NH_2	OH	361
8	OH	NHMe	113
9 (NK 611)	OH	NMe_2	438

compounds incorporated nitrogen into the sugar moiety to maintain a favorable activity profile and overcome one of etoposide's main drawbacks — poor water solubility. These compounds increased survival time in mice with leukemia L-1210 (Table 5.3).[33] One of these compounds, NK 611 (**9**), was brought to clinical trials.

The preparation of nonglycosidic C_4-substituted epipodophyllotoxins began by making various changes to the hydroxyl moiety, including to esters, carbonates, carbamates, ethers, and thioethers.[1] The introduction of nitrogen at the 4-position provided 4β-alkylamino,[34] arylamino,[35] benzylamino,[36] and halogenated anilino[37] derivatives. In comparison with etoposide, most nitrogen-analogs exhibited superior potency in cellular protein–DNA complex formation and topo II inhibitory assays, as well as comparable cytotoxicity. Table 5.4 lists some of the most active analogs, among which GL-331 (**10**) was selected for clinical evaluation. Notably, many 4β-nitrogen substituted derivatives retained cytotoxicity against **2**-resistant KB variants (KB = human epidermoid carcinoma of the nasopharynx) with decreased cellular uptake of **2**, decreased expression of topo II, or overexpression of multidrug-resistant protein (MDR1).[38] These results implied that different C_4 substituents might play a significant role in the biochemical determinants of cellular drug uptake in **2**-resistant cell lines.

Another modification leading to potent compounds was alkylation at the 4β-position. Carbon chains containing hydroxyl, amino, or amido groups were introduced.[39] Many of the derivatives exhibited potent cytotoxicity and topo II inhibition. Selected data for representative 4β-alkylated analogs are shown in Table 5.5. TOP-53 (**16**) is currently in Phase I clinical trial.

4. Ring D

The *trans*-fused γ-lactone D ring in etoposide and its derivatives can be converted to the *cis*-fused lactone (*picro* form) under basic conditions and can also be extensively metabolized *in vivo* to open-ring hydroxy acids. All three metabolic species are inactive.[1]

To overcome the problem of metabolic inactivation, podophyllotoxin derivatives with differently modified D rings, including lactam,[40] cyclopentanone,[41] cyclopentane,[41] sulfide,[41] sulfoxide,[41] sulfone,[41]

TABLE 5.4

4β-Arylamino Derivatives of Podophyllotoxin

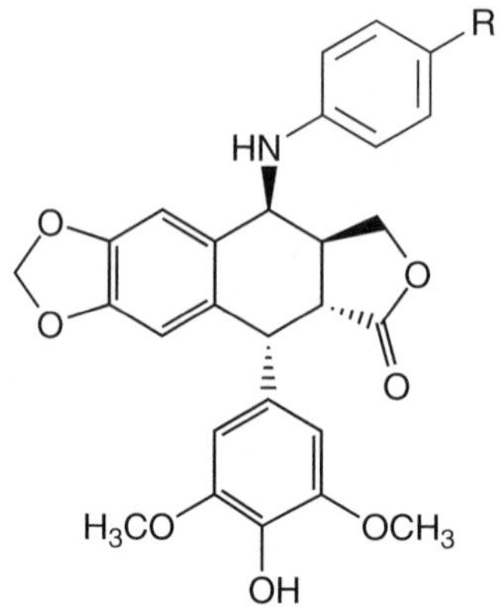

No.	R	KB cells (IC50 µM)	Topo II (ID50 µM)	% Protein–DNA Complex Formation
Etoposide	—	0.2	50	100
10 (GL-331)	NO$_2$	0.49	10	323
11	NH$_2$·HCl	0.8	5	330
12	F	0.24	5	213
13	CN	0.64	10	211
14	COOEt	0.84	5	207
15	3,4-O(CH$_2$)$_2$O	0.68	10	279

cyclic ether,[42] 2-aza,[43] homolactone,[44] and C$_2$ substituted[45,46] derivatives, have been prepared. Many of these derivatives were less active than their D ring intact congeners, especially in topo II inhibition assays.

Tetrahydrofuran D ring derivatives were designed to simultaneously eliminate hydrolysis and epimerization of the γ-lactone ring by replacing the lactone carbonyl with a methylene group (Table 5.6). Compounds **20** and **21** were found to be comparable to **2** in inhibiting topo II and causing DNA breakage but inferior to their parent compounds with an intact D ring.[42]

5. Ring E

4′-Demethylation increases the etoposide-like activity (e.g., topo II inhibition) and is considered one of the structural fingerprints for topo II inhibitors in this compound class. Extensive E ring modification, including 3′-nitrogen or 3′,4′-dinitrogen substituted, 4′-acyl, 4′-coupled dimeric, and phosphorous-containing derivatives, was performed by the Bristol Myers Company. Many of these derivatives displayed significant cytotoxicity. The most pronounced compound was the 4′-phosphate ester of etoposide (Etopophos).[47] This compound is a water-soluble prodrug of etoposide and is currently used clinically by intravenous administration.

Introduction of a chlorine atom in the 2′-position was reported to stabilize the resulting compound to C$_2$ epimerization[48]; however, it significantly decreased the activity in both cytotoxicity and topo II inhibition assays.[49] To conclude the molecular area-oriented SAR discussion, Figure 5.2 summarizes the structural features that are critical for the antineoplastic activity of podophyllotoxin analogs.[7] Among these features, a 4β-configuration and a 4′-hydroxyl group are considered to be structural determinants for topo II inhibition, the primary mechanism of action of therapeutically useful analogs.

TABLE 5.5
4β-Alkylated Derivatives of Podophyllotoxin

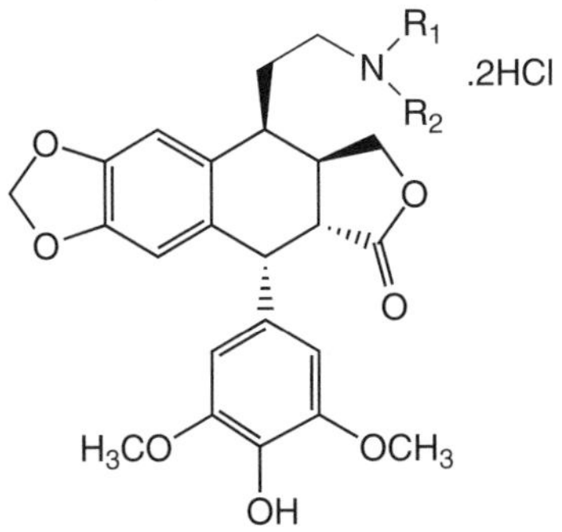

No.	R_1	R_2	Topo II (IC50 μ*M)	Cytotoxicity (ED50 μg/mL)[a]				
				SBC-3	A-549	HLE	G-402	COLO320DM
1	—	—	59.2	1.6	2.9	1.5	4.8	14
16	CH_3	$(CH_2)_2N(CH_3)_2$	32.5	0.41	0.82	0.23	0.33	2.0
17	CH_3	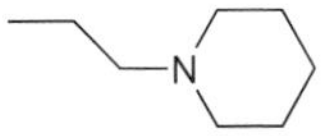	60.9	0.54	1.8	0.36	0.35	3.2
18		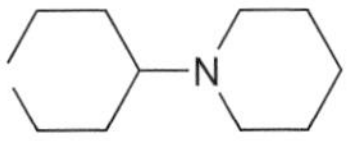	29.8	0.16	1.0	0.31	0.18	2.1
19		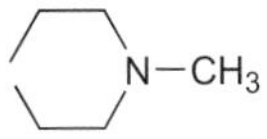	33.6	0.28	0.76	0.13	0.25	0.99

[a] SBC-3, small cell lung cancer; A-549, non–small cell lung cancer; HLE, hepatoma; G-402, renal cancer; COLO320DM, colon carcinoma.

TABLE 5.6
Tetrahydrofuran D-Ring Derivatives of Podophyllotoxin

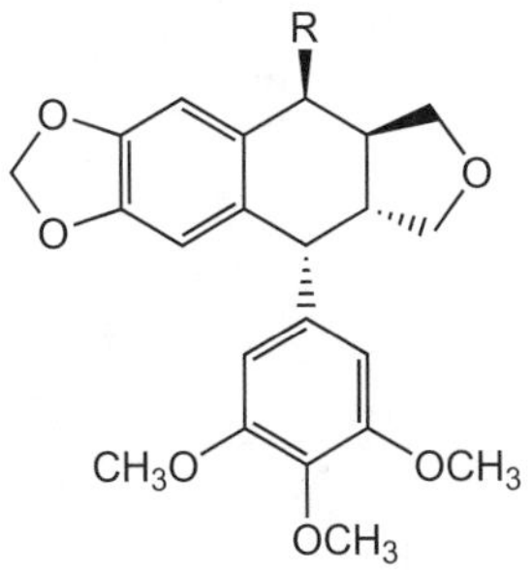

20 R = HNC_6H_5

21 R = HNC_6H_4F

	Topo II Inhibition, ID_{50} (μM)	% Protein–DNA Complex Formation
2	50	100
20	50	139
21	50	125

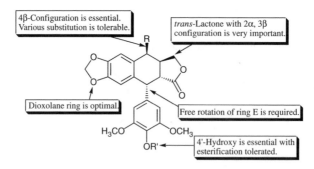

FIGURE 5.2 Summary of structural features critical for the antineoplastic activity of podophyllotoxin derivatives.

B. SAR MODELS

As the number of epipodophyllotoxin derivatives increases, informative SAR must be assembled in a readily usable format. To date, three different approaches have been applied to generate SAR models for this compound class.

1. Composite Pharmacophore Model

The first model is the composite pharmacophore model proposed by MacDonald.[50] This model was derived from the superimposition of several topo II inhibitors of the epipodophyllotoxin, anthracycline, and aminoacridine classes. As shown in Figure 5.3, the model defined three pharmacophoric domains: the DNA intercalating moiety, which is a planar, polycyclic molecular surface (**A**); the minor groove binding site, which is a pendant, *para*-hydroxy- or sulfonylamino-phenyl ring (**B**); and a variable molecular region, which can accommodate considerable structural diversity (**C**). Although later modifications, particularly various C_4 modifications, have produced SAR consistent with this model, the preparation and evaluation of podophenazine derivatives[28] failed to support a crucial role for molecular area **A** in topo II inhibition. This generalized SAR model for diverse topo II inhibitors implied a common interaction site on the enzyme; nevertheless, no experimental evidence validated the existence of such a putative site.

2. Comparative Molecular Field Analysis Model

More recently, the comparative molecular field analysis (CoMFA) technique was applied to 102 epipodophyllotoxin derivatives to generate quantitative SAR (QSAR) models.[51] The steric and

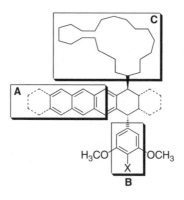

FIGURE 5.3 MacDonald's pharmacophore model.

electrostatic contour plots of the final CoMFA model indicated that diverse and bulky substitutions could be accommodated at C_4, which agreed with MacDonald's model. The CoMFA model also suggested that the C_4 substituent would interact with the DNA minor groove. On the basis of this model, several etoposide analogs bearing minor groove binding moieties at C_4 have been synthesized.[52] Some of these compounds were more active than etoposide in both cytotoxicity and topo II inhibition assays.

3. K-Nearest Neighbor QSAR Model

The application of CoMFA is often undermined by poor reproducibility resulting from inconsistent three-dimensional alignment. A k-nearest neighbor quantitative structure activity relationship (kNN-QSAR) method using alignment-free two-dimensional topological descriptors was applied to 157 epipodophyllotoxins.[58] QSAR models were generated using molecular connectivity indices (MCI) and molecular operating environment (MOE) descriptors and were characterized by the values of the internal leave-one-out cross-validated R^2 (q^2) for the training set and external predictive R^2 for the test set. As compared with those values obtained with CoMFA method, the kNN QSAR models afforded higher values of q^2 and predictive R^2. Because of the physicochemical ambiguity of topological descriptors, the kNN-QSAR models cannot direct chemical modifications on specific molecular areas. However, their high predictive ability should guide the rational design of novel derivatives and focused libraries based on the epipodophyllotoxin skeleton, as well as facilitate the search for bioactive structures from large databases.

C. REPRESENTATIVE ANALOGS

Despite their impressive clinical efficacy, therapeutic application of etoposide and teniposide is often impeded by problems such as poor water solubility, acquired drug resistance, and metabolic inactivation. Several novel podophyllotoxin analogs, including Etopophos (etoposide phosphate), NK 611, GL-331, and TOP-53, have resulted from the wide range of research programs geared to overcome the above problems.

1. Etopophos

The poor water solubility of **2** and **3** presents problems in drug administration and formulation. Accordingly, Etopophos, a water-soluble prodrug of etoposide, was developed.[47] Etopophos is less toxic and more active than etoposide in *in vivo* tumor models. It can be efficiently converted *in vivo* by endogenous phosphatase to the active drug etoposide and exhibits pharmacological and pharmacokinetic profiles similar to those of etoposide. Notably, the prodrug approach increases the *in vivo* bioavailability from 0.04% to over 50% and also results in more predictable oral bioavailability.[53] The improved water solubility and bioavailability made Etopophos preferable to etoposide for routine clinical use, and in 1996, the FDA approved its intravenous use.

2. NK 611

To tackle the problem of poor water solubility, another etoposide derivative, NK 611, was developed.[33] Introduction of a 2-dimethylamino group led to a 120-fold increase in water solubility. Compared with **2**, NK 611 showed similar antitumor activity against human tumor xenografts but more potent topoisomerase II inhibitory effects and cytotoxic activity against various human cancer lines including lung, gastrointestinal, ovarian, testicular, breast, and head and neck cancers and leukemias. Clinical tests of intravenous and oral formulations of NK 611 indicate that it has better bioavailability than etoposide. However, evidence for cross-resistance between etoposide and NK 611 was also found.[7]

3. GL-331

GL-331 is a 4β-arylamino analog of etoposide, with the sugar moiety in **2** replaced by a *p*-nitro anilino group. GL-331 is more active than **2** in causing DNA double-strand breakage and G2-phase arrest.[35] It is also more potent against tumor cells both *in vitro* and *in vivo* and, remarkably, overcomes multidrug resistance in many cancer cell lines. Formulated GL-331 shows desirable stability and biocompatibility, as well as similar pharmacokinetic profiles, to those of **2**. Initial results from phase I clinical trials in non-small and small cell lung, colon, and head and neck cancers showed marked antitumor efficacy. Side effects were minimal, with cytopenias being the major toxicity. GL-331 is not active against stomach cancer. Phase II clinical evaluation against several forms of cancer, especially etoposide-resistant malignancies, is being planned.[54]

4. TOP-53

TOP-53 was selected from a series of 4β-alkylated etoposide analogs in which the glycoside group of etoposide was replaced with a carbon chain containing hydroxyl, amino, or amido groups.[39] As compared with **2**, TOP-53 was a more potent inhibitor of topo II. It showed nearly wild-type potency against a mutant yeast type II enzyme highly resistant to **2**, implying therapeutic potential for drug-resistant cancers. TOP-53 exhibited strong activity against a wide variety of tumor cells, with especially high activity against non-small cell lung cancer in both tumor cells and animal tumor models.[55] This compound is currently in phase I clinical trials.[7]

VII. SYNTHESES

Gensler et al. accomplished the first total synthesis of podophyllotoxin in the 1960s.[15] Major challenges for total syntheses of podophyllotoxin analogs include the presence of four contiguous stereocenters and a base-sensitive *trans*-lactone ring. To date, four general approaches to the syntheses of podophyllotoxin derivatives have been developed. Key steps in these routes involve the elaboration of an χ-oxo ester, the lactonization of a dihydroxy acid, the cyclization of a conjugate addition product, or the use of a Diels–Alder reaction to construct the aryltetralin unit.[5] Although several synthetic approaches modified from the above routes provided excellent enantiopurities, the low overall yields disqualified total syntheses as an alternative for naturally produced materials. Therefore, natural (–)-podophyllotoxin remains the major source for the preparation of clinically useful analogs.

DMEP was synthesized from podophyllotoxin in the late 1960s via simultaneous 4-epimerization and selective 4′-demethylation using hydrogen bromide.[56] After 4′-benzyloxycarbonyl protection to give intermediate **22**, the 4β-hydroxyl group was then glycosylated by reaction with tetra-*O*-acetyl-α-D-glucopyranosyl bromide and subsequent ZnCl$_2$-catalyzed methanolysis.[31] The resulting glycosidic derivative **23** was treated with aldehydes or ketones in the presence of acid catalysts to yield the corresponding cyclic acetals or ketals, including etoposide and teniposide (Scheme 5.1).[32]

An efficient preparation of Etopophos is shown in Scheme 5.2.[47] The benzyl phosphate protection of the 4′-phenol and glycosylation of C$_4$ with an acetylated sugar characterize this preparation. The overall yield (starting from podophyllotoxin) was over 19%.

The analog NK 611 was prepared from the aminosugar intermediate **24**, which was itself prepared following the procedure used to synthesize etoposide and teniposide (cf. Scheme 5.1). NK 611 was then produced by reductive methylation with aqueous formaldehyde in the presence of NaBH$_3$CN (Scheme 5.3).[33]

In sharp contrast to the preparation of the other clinically useful derivatives, a two-step, one-pot procedure was employed in the manufacture of GL-331. GL-331 was obtained from podophyllotoxin through subsequent 4-bromination and 4′-demethylation with hydrogen bromide (HBr, g) and nucleophilic displacement with the appropriate amine in the presence of BaCO$_3$ (Scheme 5.4).[35]

The preparation of the 4β-alkyl derivative TOP-53 is shown in Scheme 5.5.[39] The 4′-hydroxyl of DMEP was protected by a benzyloxycarbonyl group before regio- and stereospecific introduction

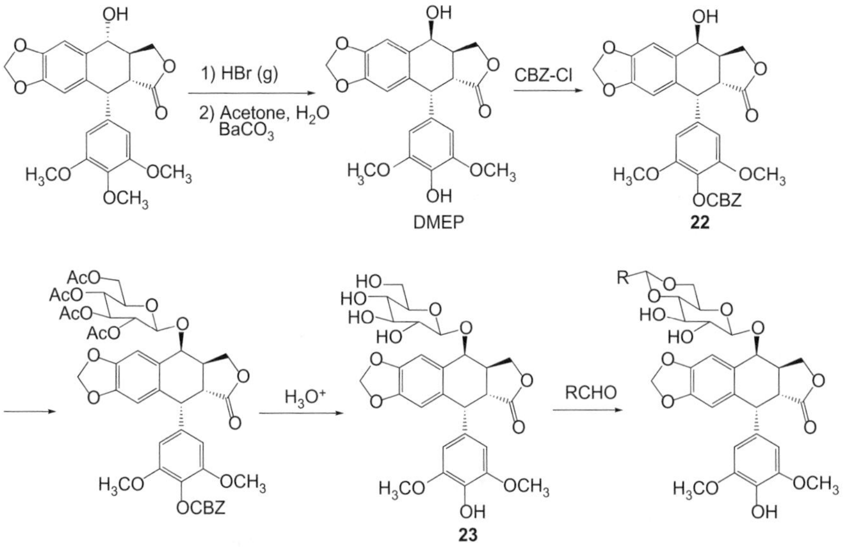

Etoposide (R = CH_3)
Teniposide (R = C_4H_3S)

SCHEME 5.1 Synthesis of etoposide and teniposide.

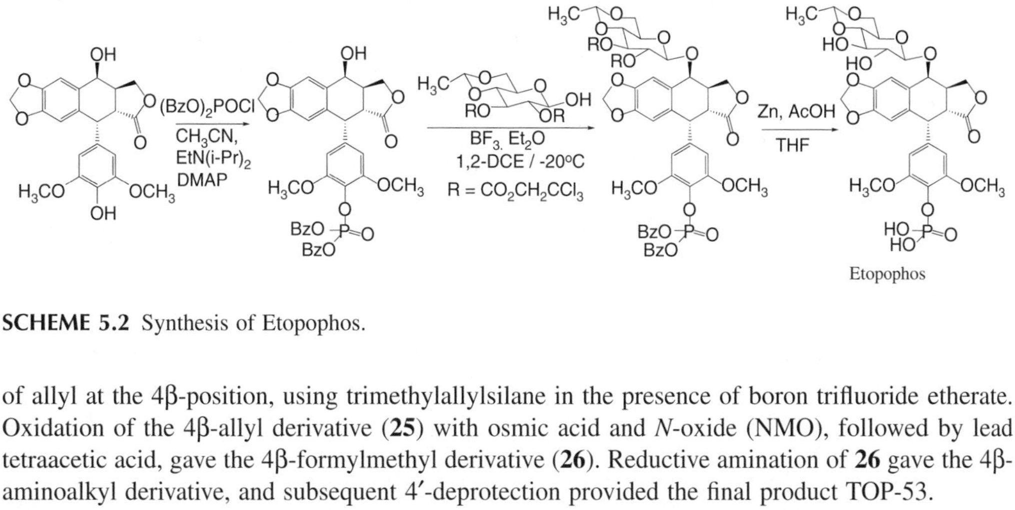

Etopophos

SCHEME 5.2 Synthesis of Etopophos.

of allyl at the 4β-position, using trimethylallylsilane in the presence of boron trifluoride etherate. Oxidation of the 4β-allyl derivative (**25**) with osmic acid and *N*-oxide (NMO), followed by lead tetraacetic acid, gave the 4β-formylmethyl derivative (**26**). Reductive amination of **26** gave the 4β-aminoalkyl derivative, and subsequent 4′-deprotection provided the final product TOP-53.

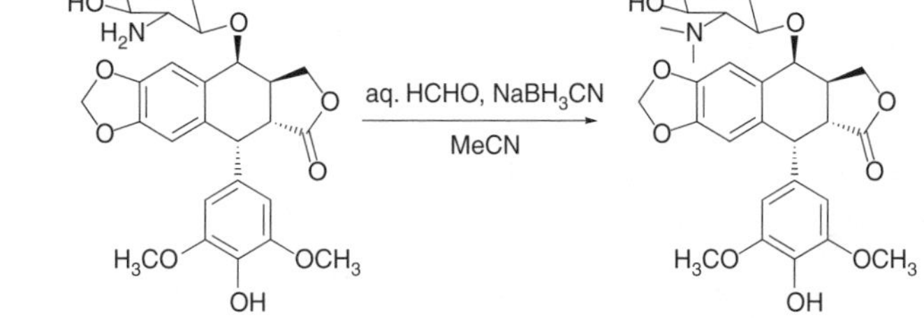

24

NK-611

SCHEME 5.3 Synthesis of NK 611.

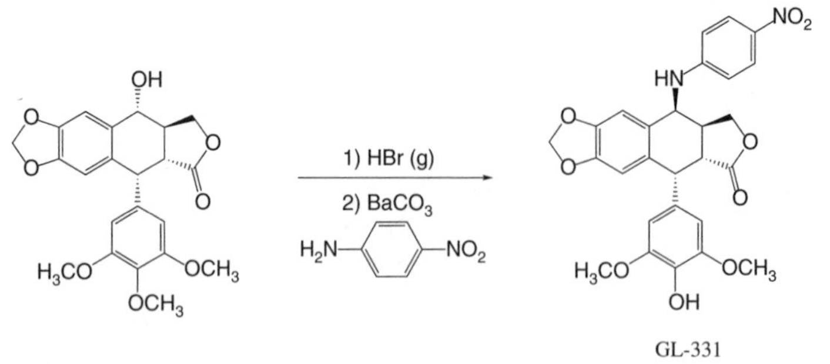

SCHEME 5.4 Synthesis of GL-331.

VIII. CLINICAL APPLICATIONS

Etoposide shows broad-spectrum antitumor activity against a variety of cancers. It is effective against ascitic tumors, testicular tumors, ovarian and gestational carcinomas, different types of lung cancer (small cell, squamous, adenocarcinoma, Lewis carcinoma), leukemias (monocytic and refractory), malignant and recurrent lymphomas, urogenital tumors, sarcomas, melanomas, and experimentally induced colon cancer.[6]

Etoposide is widely used as an antineoplastic agent, especially in combination chemotherapeutic regimens. The combination of etoposide with *cis*-platin is a most effective protocol and shows highly productive first-line therapeutic synergism against small cell cancer and as a second-line or salvage therapy for CAV (cyclophosphamide, Adriamycin [doxorubicin], and vincristine)–resistant or CAV-failure small cell lung cancer. It is also effective for the treatment of testicular cancer and non–small cell lung cancer. When incorporated into other multidrug treatment protocols, etoposide is also used for the treatment of non-Hodgkin lymphomas resistant to other agents, different types of lymphomas, refractory childhood leukemia, hepatocellular metastasis, refractory acute lymphoblastic leukemia, and other types of cancers.[6,57]

As a water-soluble prodrug of etoposide, Etopophos exhibits pharmacological and pharmacokinetic profiles similar to those of etoposide and is preferable for routine clinical use, especially for intravenous administration.

Teniposide, which is more potent than etoposide, shows good antineoplastic activity against different types of cancer, including lymphoblastic, acute lymphocytic leukemia and other experimentally induced leukemias, infantile non-Hodgkin lymphoblastic lymphoma, multiple myeloma, ascitic tumors, malignant brain tumors, colorectal and refractory or recurrent testicular carcinomas, and small cell and non–small cell lung cancer. Combination chemotherapy incorporating teniposide has also been applied to various cancers. It is used in combination with *cis*-platin against neuroblastoma, with cytarabine (ara-C), against acute lymphoblastic leukemia, and with carboplatin against small cell lung cancer.[6,57]

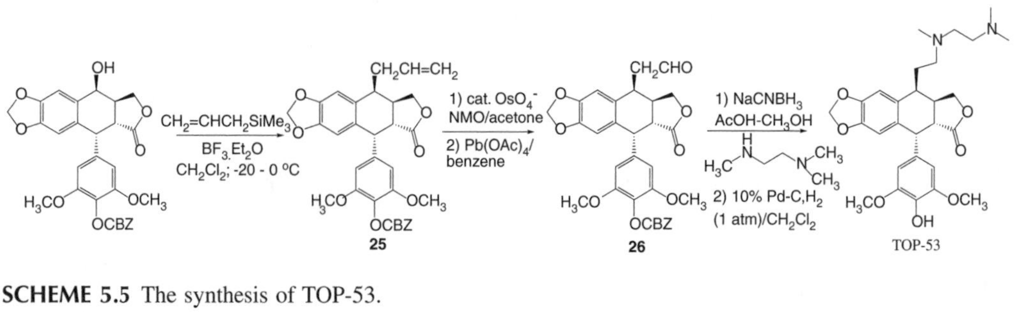

SCHEME 5.5 The synthesis of TOP-53.

IX. FUTURE PERSPECTIVES

Etoposide and teniposide have been used in cancer chemotherapy for over two decades. However, their clinical application is often impeded by problems of acquired drug resistance and poor water solubility. The recent development of anticancer candidates, including NK 611, GL-331, and TOP-53, indicates that molecular modification at position 4 of etoposide can effectively yield novel therapeutic analogs with better pharmacological and pharmacokinetic profiles. The availability of reliable and interpretable QSAR models will significantly benefit continued chemical efforts to produce improved analogs. Readily usable QSAR information must be extracted from accumulated structural and biological data.

Although antineoplastic properties are arguably the most pronounced pharmacological effect of podophyllotoxin analogs, their antiviral, anti-inflammatory, and immunosuppressive activities are drawing more and more attention. We anticipate podophyllotoxin analogs with therapeutic utility other than cancer chemotherapy to be produced by exploring other biological activities.

ACKNOWLEDGMENTS

The research on podophyllotoxin analogs performed in the Natural Products Laboratory UNC-Chapel Hill was supported by National Institutes of Health grant CA 17625, awarded to K. H. Lee.

REFERENCES

1. Zhang, Y.L. and Lee, K.H., Recent progress in the development of novel antitumor etoposide analogs, *Chin. Pharm. J.*, 46, 319, 1994.
2. Bohlin, L. and Rosén, B. Podophyllotoxin derivatives: drug discovery and development, *Drug Discovery Today*, 8, 343, 1996.
3. Imbert, T.F., Discovery of podophyllotoxins, *Biochimie*, 80, 207, 1998.
4. Damayanthi, Y. and Lown, J.W. Podophyllotoxins: Current status and recent developments. *Curr. Med. Chem.*, 5, 205, 1998.
5. Canel, C. et al., Podophyllotoxin, *Phytochemistry*, 54, 115, 2000.
6. Gordaliza, M. et al., Antitumor properties of podophyllotoxin and related compounds, *Curr. Pharm. Design*, 6, 1811, 2000.
7. Moraes, R.M., Dayan, F.E., and Canel, C., The lignans of *Podophyllum*, *Studies in Natural Products Chemistry*, 26 (Bioactive Natural Products [Part G]), 149, 2002.
8. Kaplan I.W., *Condylomata acuminata*, *New Orleans Med. Surg. J.*, 94, 388, 1942.
9. King, L.S. and Sullivan, M., The similarity of the effect of podophyllin and colchicine and their uses in the treatment of *Condylomata acuminata*, *Science*, 104, 244, 1946.
10. Sullivan, B.J. and Weshler, H.I., The cytological effects of podophyllotoxin, *Science*, 105, 433, 1947.
11. Seidlova-Massinova, V., Malinsky, J., and Santavy, F., The biological effect of some podophyllotoxin compounds and their dependence on chemical structure, *J. Natl. Cancer Inst.*, 18, 359, 1957.
12. Hartwell, J.L. and Shear, M.J., Chemotherapy of cancer: classes of compounds under investigation, and active components of podophyllin. *Cancer Res.*, 7, 716, 1947.
13. Podwyssotzki, V., Pharmakologische studien über *Podophyllum peltatum*, *Arch. Exp. Pathol. Pharmakol.*, 13, 29, 1880.
14. Hartwell, J.L. and Schrecker, A.W., Components of podophyllin. V. The constitution of podophyllotoxin, *J. Am. Chem. Soc.*, 73, 2909, 1951.
15. Gensler, W.J. and Gatsonis C.D., Synthesis of podophyllotoxin, *J. Am. Chem. Soc.*, 84, 1748, 1962.
16. Stähelin, H. and von Wartburg, A., The chemical and biological route from podophyllotoxin glucoside to etoposide: ninth Cain memorial award lecture, *Cancer Res.*, 51, 5, 1991.
17. O'Dwyer, P.J. et al., Etoposide (VP-16-213): current status of an anticancer drug, *N. Engl. J. Med.*, 312, 692, 1985.

18. Cortese, F., Bhattacharyya, B., and Wolf, J., Podophyllotoxin as a probe for the colchicine binding site of tubulin, *J. Biol. Chem.*, 252, 1134, 1977.
19. Stähelin, H.F., 4-Demethyl-epipodophyllotoxin thenylidene glucoside (VM-26), a podophyllum compound with a new mechanism of action, *Eur. J. Cancer*, 6, 303, 1970.
20. Loike, J.D. and Horwitz, S.B., Effect of VP-16-213 on the intracellular degradation of DNA in HeLa cells, *Biochemistry*, 15, 5443, 1976.
21. Ross, W. et al., Role of topoisomerase II in mediating epipodophyllotoxin-induced DNA cleavage, *Cancer Res.*, 44, 5857, 1984.
22. Berger J.M. and Wang J.C., Recent developments in DNA topoisomerase II structure and mechanism, *Curr. Opin. Struct. Biol.*, 6, 84, 1996.
23. van Maanen, J.M.S. et al., Effects of the *ortho*-quinone and catechol of the antitumor drug VP-16-213 on the biological activity of single-stranded and double stranded ΦX174 DNA, *Biochem. Pharmacol.*, 37, 3579, 1988.
24. Sakurai, H. et al., Metal- and photo-induced cleavage of DNA by podophyllotoxin, etoposide, and their related compounds, *Mol. Pharmacol.*, 40, 965, 1991.
25. Huang, T.S. et al., Protein tyrosine phosphatase activities are involved in apoptotic cancer cell death induced by GL-331, a new homolog of etoposide, *Cancer Lett.*, 110, 77, 1996.
26. Huang, T.S. et al., Activation of CDC 25 phosphatase and CDC2 kinase involved in GL-331- induced apoptosis, *Cancer Res.*, 57, 2974, 1997.
27. Wang, Z.Q. et al., Antitumor agents 124. New 4β-substituted aniline derivatives of 6,7-*O,O*-demethylpodophyllotoxin and related compounds as potent inhibitors of human DNA topoisomerase II, *J. Med. Chem.*, 35, 871, 1992.
28. Cho, S.J. et al., Antitumor agents 164. Podophenazine, 2″,3″-dichloropodophenazine, benzopodophenazine, and their 4β-*p*-nitroaniline derivatives as novel DNA topoisomerase II inhibitors, *J. Med. Chem.*, 39, 1396, 1996.
29. Thurston, L.S. et al., Antitumor agents 78. Inhibition of human DNA topoisomerase II by podophyllotoxin and α-peltatin analogues, *J. Med. Chem.*, 29, 1547, 1986.
30. Saito, H. et al., Studies on lignan lactone antitumor agents IV. Synthesis of glycosidic lignan variants related to α-peltatin, *Bull. Chem. Soc. Jpn.*, 61, 1259, 1988.
31. Kuhn, M. and von Wartburg, A., On a new glycoside synthesis process II. Glycosides of 4′-demethylepipodophyllotoxins, *Helv. Chem. Acta*, 52, 948, 1969.
32. Keller-Juslen, C. et al., Synthesis and antimitotic activity of glycosidic lignan derivatives related to podophyllotoxin, *J. Med. Chem.*, 14, 936, 1971.
33. Saito, H. et al., Studies on lignan lactone antitumor agents. II. Synthesis of N-alkylamino- and 2,6-dideoxy-2-aminoglycosidic lignan variants related to podophyllotoxin, *Chem. Pharm. Bull.*, 34, 3741, 1986.
34. Lee, K.H. et al., Antitumor agents 107. New cytotoxic 4-alkylamino analogues of 4′-demethylepipodophyllotoxin as inhibitors of human DNA topoisomerase II, *J. Nat. Prod.*, 52, 606, 1989.
35. Wang, Z.Q. et al., Antitumor agents 113. New 4β-arylamino derivatives of 4′-*O*-demethylepipodophyllotoxin and related compounds as potent inhibitors of human DNA topoisomerase II, *J. Med. Chem.*, 33, 2660, 1990.
36. Zhou, X.M. et al., Antitumor agents 120. New 4-substituted benzylamine and benzyl ether derivatives of 4′-*O*-demethylepipodophyllotoxin as potent inhibitors of human DNA topoisomerase II, *J. Med. Chem.*, 34, 3346, 1991.
37. Lee, K.H. et al., Antitumor agents 111. New 4-hydroxylated and 4-halogenated anilino derivatives of 4-demethylepipodophyllotoxin as potent inhibitors of human DNA topoisomerase II, *J. Med. Chem.*, 33, 1364, 1990.
38. Chang, J.Y. et al., Effect of 4β-arylamino derivatives of 4′-*O*-demethylpodophyllotoxin on human DNA topoisomerase II, tubulin polymerization, KB cells, and their resistant variants, *Cancer Res.*, 51, 1755, 1991.
39. Terada, T. et al., Antitumor agents 3. Synthesis and biological activity of 4β-alkyl derivatives containing hydroxyl, amino, and amido groups of 4′-*O*-demethyl-4-deoxypodophyllotoxin as antitumor agents, *J. Med. Chem.*, 36, 1689, 1993.
40. Kadow, J.F., Vyas, D.M., and Doyle, T.W., Synthesis of etoposide lactam via a Mitsunobu reaction sequence, *Tetra. Lett.*, 30, 3299, 1989.

41. Gensler, W.J., Murthy, C.D., and Trammell, M.H., Nonenolizable podophyllotoxin derivatives, *J. Med. Chem.*, 20, 635, 1977.

42. Zhou, X.M. et al., Antitumor agents 144. New γ-lactone ring-modified arylamino etoposide analogs as inhibitors of human DNA topoisomerase II, *J. Med. Chem.*, 37, 287, 1994.

43. Tomioka, K., Kubota, Y., and Koga, K., Synthesis and antitumor activity of podophyllotoxin aza-analogues, *Tetra. Lett.*, 30, 2953, 1989.

44. Roulland, E. et al., Synthesis of picropodophyllotoxin homolactone, *Tetra. Lett.*, 41, 6769, 2000.

45. Glinski-Oomen, M.B., Freed, J.C., and Drust, T., Preparation of 2-substituted podophyllotoxin derivatives, *J. Org. Chem.*, 52, 2749, 1987.

46. VanVliet, D.S. et al., Antitumor agents: 207. Design, synthesis, and biological testing of 4β–anilino-2-fluoro-4′-demethylpodophyllotoxin analogues as cytotoxic and antiviral agents, *J. Med. Chem.*, 44, 1422, 2001.

47. Saulnier, M.G. et al., Synthesis of etoposide phosphate, BMY-40481: a water-soluble clinically active prodrug of etoposide, *Bioorg. Med. Chem. Lett.*, 4, 2567, 1994.

48. Ayres, D.C. and Lim, C.K., Lignans and related phenols. Part XIII. Halogenated derivatives of podophyllotoxin, *J. Chem. Soc., Perkin I*, 111, 1350, 1972.

49. Hu, H. et al., Antitumor agents 123. Synthesis and human DNA topoisomerase II inhibitory activity of 2′-chloro derivatives of etoposide and 4β-(arylamino)-4′-*O*-demethylpodophyllotoxins, *J. Med. Chem.*, 35, 866, 1992.

50. MacDonald, T.L. et al., On the mechanism of interaction of DNA topoisomerase II with chemotherapeutic agents. In *DNA Topoisomerase in Cancer*, Potmesil, M., Kohn, K.W., Eds., Oxford University Press, New York, 1991, 119.

51. Cho, S.J. et al., Antitumor agents 163. Three-dimensional quantitative structure-activity relationship study of 4′-*O*-demethylepipodophyllotoxin analogs using the modified CoMFA/q^2-GRS approach, *J. Med. Chem.*, 39, 1383, 1996.

52. Ji, Z. et al., Antitumor agents 177. Design, syntheses, and biological evaluation of novel etoposide analogs bearing pyrrolecarboxamidino group as DNA topoisomerase II inhibitors, *Bioorg. Med. Chem. Lett.*, 7, 607, 1997.

53. Hande, K.R., Etoposide: Four decades of development of a topoisomerase II inhibitor, *Eur. J. Cancer*, 34, 1514, 1998.

54. Lee, K.H., Antitumor agents 197. Novel antitumor agents from higher plants, *Med. Res. Rev.*, 19, 569, 1999.

55. Byl, J.A.W., DNA topoisomerase II as the target for the anticancer drug TOP-53: mechanistic basis for drug action, *Biochemistry*, 40, 712, 2001.

56. Kuhn, M., Keller-Juslen, C., and von Wartburg, A., Partial synthesen von 4′-demethylepipodophyllotoxins, *Helv. Chem. Acta*, 52, 944, 1969

57. Ayres, D.C. and Loike J.D., Lignans. Chemical, biological and clinical properties, Cambridge University Press, Cambridge, 1990, 113.

58. Xiao, Z. et al., Antitumor agents 213. Modeling of epipodophyllotoxin derivatives using variable selection *k* Nearest Neighbor QSAR method. *J. Med. Chem.*, 45, 2294, 2002.

6 Taxol and Its Analogs

David G. I. Kingston

CONTENTS

I. Introduction 89
II. History 90
    A. Discovery 90
    B. Preclinical Development...90
    C. Clinical Development 91
    D. The Taxol Supply Crisis..92
    E. Taxotere...92
III. Biosynthesis and Bioproduction of Taxol 93
    A. Taxol Biosynthesis...93
        1. The Diterpenoid Ring System...93
        2. The Side Chain 94
    B. Taxol Bioproduction 94
IV. Taxol's Mechanism of Action..95
V. Medicinal Chemistry of Taxol 96
    A. Ring A Modifications 96
    B. Ring B Modifications 97
    C. Ring C Modifications 98
    D. Ring D Modifications 100
    E. Side Chain Analogs 101
    F. Prodrugs of Taxol 102
    G. Targeted Analogs of Taxol...102
    H. Summary of Taxol's SAR 104
VI. Synthetic Studies...105
    A. Semisynthetic Methods..105
    B. Total Synthesis...107
VII. The Taxol–Tubulin Interaction..109
VIII. Other Natural Products with a Similar Mechanism of Action 111
IX. Clinical Applications of Taxol and Docetaxel...112
X. Taxol Analogs in Clinical Trials 113
XI. Conclusion...113
Acknowledgments 113
References 114

I. INTRODUCTION

Arguably no naturally occurring anticancer agent has had a bigger effect on cancer treatment than Taxol® (**I.1**), now known as paclitaxel. Although it is now recognized as one of the most important drugs available for the treatment of breast and ovarian cancers, and although it has

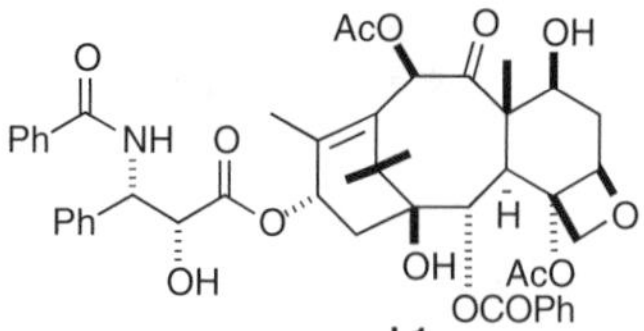

SCHEME 6.1

spawned several analogs that are now in clinical trials, taxol itself almost never became a drug at all. This review briefly covers the early history of the discovery and development of taxol and then goes on to describe its medicinal chemistry, its synthesis, its interaction with tubulin, and its relationship to compounds such as the epothilones and discodermolide, which have similar mechanisms of action.

II. HISTORY

A. DISCOVERY

The story of taxol began on August 21, 1962, when a team of botanists led by Dr. Arthur Barclay from the U.S. Department of Agriculture, working under a contract from the National Cancer Institute (NCI), collected a sample of *Taxus brevifolia* Nutt. in the Gifford Pinchot National Forest in Washington state. As was the procedure at that time, the sample was extracted by the Wisconsin Alumni Research Foundation contract laboratory, and the extract was tested for cytotoxicity to KB (human epidermoid carcinoma of the nasopharynx) cells by Microbial Associates in Bethesda, Maryland. A positive response to the stem and bark extract led to the assignment of the extract to Dr. Monroe Wall, who had recently moved to the newly established Research Triangle Institute in North Carolina. Work in the Wall laboratory was also carried out under a contract from the NCI, but it proceeded slowly at first. This was in part because of the complexity of the structural problem and in part because Dr. Wall also had another exciting compound, camptothecin, under investigation, and this study consumed much of his resources. It was thus not until 1971 that Dr. Wall, together with his collaborators Dr. Mansukh Wani and Dr. Andrew McPhail, announced the structure of the major active constituent of *T. brevifolia* as taxol (**I.1**).[1] Unknown to the discoverers, the name taxol had been trademarked by a French company for an unrelated laxative product. This trademark was later acquired by Bristol-Myers Squibb (BMS), who then applied it to their formulation of the drug. The generic name paclitaxel was assigned to the chemical compound of structure I.1 (Scheme 6.1). Because this review is partly historical in nature, the name taxol is retained for compound I.1. No infringement of the BMS trademark is implied. Interestingly, it has been pointed out[2] that the selection of *T. brevifolia* as the source of taxol was very fortunate, as this species only contains low amounts of the toxic alkaloids taxine A and B. Had a different yew species been investigated, it is probable that Dr. Wall's fractionation would have led to the isolation of the taxines as the cytotoxic constituents, and the smaller amounts of taxol present in these other species might have gone undetected.

B. PRECLINICAL DEVELOPMENT

The publication of taxol's structure and activity excited interest in the natural products community, but not in the pharmaceutical industry. The reasons for this are not hard to discern, as taxol presented a number of seemingly intractable problems to a would-be developer. In the first place it occurred only in low yield in the bark of *T. brevifolia*, which was its best source, and this bark was relatively thin; it would thus take a Herculean effort to obtain enough taxol for preclinical studies, let alone

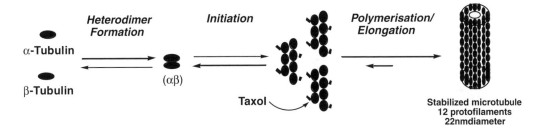

FIGURE 6.1 Tubulin polymerization by taxol.

clinical trials and clinical use. Complicating this situation was the fact that taxol's structure was very complex, so it could not be synthesized readily, and it was only very sparingly soluble in water, so its formulation would be difficult. Finally, it only showed activity against KB cell culture and various leukemias and one carcinosarcoma, so it was not clear that it would be a successful drug against solid tumors.

Given these handicaps, it is fortunate that taxol became more than a laboratory curiosity. That it did so is in no small part a result of the perseverance of Dr. Wall and Dr. Mathew Suffness at the NCI. Modest amounts of taxol were isolated in the early 1970s to allow for additional animal testing, and a turning point was reached when it was found that taxol showed excellent activity against various human solid tumor xenografts in nude mice, including the CX-1 colon and MX-1 breast xenografts. These results were encouraging enough that in 1977 the NCI decided to begin preclinical formulation and toxicity studies of taxol, with a view to eventual clinical trials.

A second major turning point in taxol's development was the discovery by Dr. Susan Horwitz of its unique activity as a promoter of tubulin polymerization (Figure 6.1).[3] Although clinically used drugs such as vinblastine (Velban) and vincristine (Oncovin) were known to act as inhibitors of tubulin polymerization, taxol was the first compound shown to act in the opposite direction. This discovery demonstrated that taxol was not "just another cytotoxic agent" and awakened a renewed interest in the compound — not least in the biological community — which turned out to be an important factor in keeping interest in taxol alive when it encountered problems during its initial clinical trials.

C. CLINICAL DEVELOPMENT

Taxol entered phase 1 clinical trials in 1984, but unfortunately it gave some allergic reactions, most probably related to its formulation as an emulsion with Cremophor EL®, a polyethoxylated castor oil. These reactions were responsible for at least one death.[4] These problems would have spelled the demise of most drug candidates, but taxol's unique mechanism of action was enough to encourage the clinicians to persevere, and the problems were overcome by lengthening the infusion period and premedicating patients. Ironically, these changes may have contributed to taxol's eventual success, as there is evidence that the cytotoxicity and *in vivo* activity of taxol are increased on prolonged exposure to the drug.[5] In any event, phase II trials, using the lengthened infusion period, were initiated in 1985, and these proved to be very successful. The first definitive clinical results were obtained in 1989 against drug-refractory ovarian cancer,[6] and 2 yr later, excellent clinical results were reported against breast cancer.[7] The drug was approved by the U.S. Food and Drug Administration for the treatment of refractory ovarian cancer in 1992 and for treatment of refractory or anthracycline-resistant breast cancer in 1994.

The discovery and preclinical work as well as the initial clinical studies on taxol were sponsored by the NCI. Once it became clear that the compound had real therapeutic value, but that its further development would require more resources than NCI could allocate, its development was offered to the pharmaceutical industry under a Cooperative Research and Development Agreement

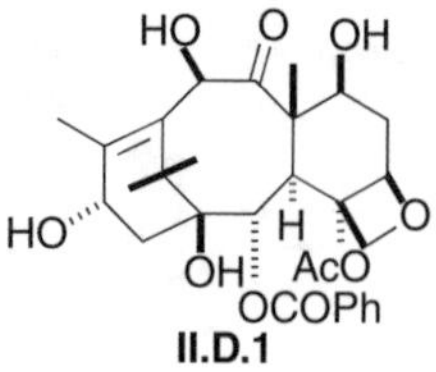

II.D.1

SCHEME 6.2

(CRADA). An agreement was signed with BMS in 1989, and because taxol was never patented, the CRADA gave BMS a 7-yr period of exclusivity in return for its development investment.

D. THE TAXOL SUPPLY CRISIS

The therapeutic efficacy of taxol became generally known by 1990, but unfortunately the supply of the drug from its natural source of *T. brevifolia* bark was not immediately adequate to the task of supplying a dramatically expanded demand. The situation was further complicated by the fact that *T. brevifolia* was primarily obtained from the old-growth forests of the Pacific Northwest of the United States, which forests were the natural habitat of the endangered spotted owl. The prospect of large-scale harvesting of *T. brevifolia* for taxol production thus raised serious environmental concerns.[8]

The initial approach to solving this vexing problem was to mount a large-scale collection and extraction program for *T. brevifolia* bark, which was successfully carried out by Hauser Chemical Research of Boulder, Colorado, under contract from BMS. This costly and labor-intensive approach was then replaced by a semisynthetic process from 10-deacetylbaccatin III (**II.D.1**), which Potier and his coworkers had found to be abundant in the European yew, *Taxus baccata*.[9,10] Several groups developed methods to convert 10-deacetylbaccatin III to taxol, including those of Potier[10] and Holton;[11] the latter group's method was selected by BMS for their commercial preparation of taxol.

E. TAXOTERE

This historical section would not be complete without a brief account of the discovery of docetaxel (Taxotere), the only other taxoid drug currently in clinical use. The Potier group at the Centre National de la Recherche Scientifique in Paris became interested in taxol in the early 1980s and carried out a series of isolation and semisynthetic studies. As noted above, an important initial finding was that the taxol precursor 10-deacetylbaccatin III (**II.D.1**) could be obtained in good yield from the needles of *T. baccata* — the English (or European) yew.[9,10] Building on the availability of this compound, the group then developed various approaches to the semisynthesis of taxol from 10-deacetylbaccatin III. One of the first syntheses involved hydroxyamination of a cinnamoyl substituted baccatin III derivative to give a mixture of stereo- and regioisomeric hydroxyamines.[12] Because the BOC group was being used as an amine protective group, one of these products was a protected version of compound **II.E.1**, and so this intermediate was converted to **II.E.1** and tested for biological activity. It turned out to have excellent activity — better than taxol in some assays — and it was

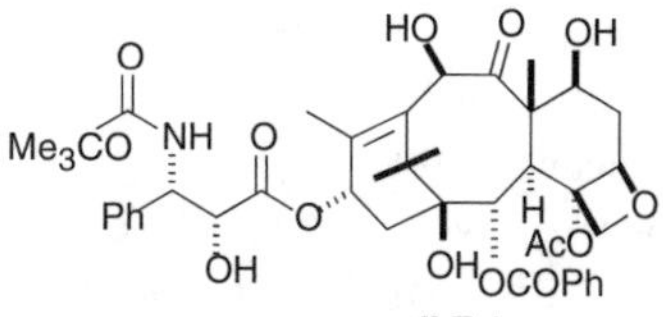

II.E.1

SCHEME 6.3

thus developed as a parallel drug to taxol. Taxotere, as the compound was named, entered phase I clinical trials in 1990[13] and was approved for treatment of advanced breast cancer in 1996 and for non–small cell lung cancer in 1999; its generic name is docetaxel.[14]

III. BIOSYNTHESIS AND BIOPRODUCTION OF TAXOL

Interest in the biosynthesis of taxol has been driven in part by a need to develop a renewable and environmentally friendly source of this important compound. Significant work has been carried out both on the scientific study of the biosynthetic pathways to taxol and on the important practical application of plant tissue culture methods to the commercial production of the compound. This latter area of work has progressed significantly since a recent review,[15] and taxol is now produced on a commercial scale by plant tissue culture by BMS at its plant in Ireland.[16]

A. TAXOL BIOSYNTHESIS

1. The Diterpenoid Ring System

Taxol is a diterpenoid, and until a few years ago all diterpenoids were thought to arise from geranylgeranyl diphosphate, which was biosynthesized from mevalonic acid. In recent years, however, a nonmevalonoid pathway to terpenoids has been uncovered,[17] and it has been shown that at least some taxane diterpenoids are biosynthesized by this pathway. Thus, a feeding experiment with $[U-^{13}C_6]$ glucose showed the incorporation of an intact 3-carbon precursor, eliminating the acetate-derived mevalonic acid as a precursor.[18]

The first committed step in the biosynthesis of the diterpenoid ring system of taxol is the cyclization of geranylgeranyl diphosphate (**III.1**) to taxa-4(5),11(12)-diene **III.5**.[19] This step was thought initially to be the slow step, based on the low levels of **III.5** found in *T. brevifolia* bark and the fact that the cyclization step is slow relative to subsequent oxygenations. Based on the presumed rate-determining nature of the cyclization of **III.1** to **III.5**, the enzyme responsible for the cyclization, taxadiene synthase, was investigated in detail. The enzyme was purified from *T. brevifolia* bark and characterized as a 79-kDa protein.[20] Genetic studies led to the isolation and expression of a cDNA fragment, which was expressed in *Escherichia coli* to produce the same protein.[21] A study of the activity of taxadiene synthase in *T. canadensis* cell cultures showed that it was greater than that needed for taxol production *in vivo*, indicating that the true rate-limiting step of taxol biosynthesis occurs further down the pathway.[22]

The mechanism of the cyclization of **III.1** to **III.5** has been shown to involve an intramolecular hydride ion transfer (presumably via an enzyme-bound intermediate) from the putative verticillyl ion **III.2** to the cation **III.3**, which then undergoes intramolecular cyclization to the tricyclic cation **III.4** and finally deprotonation to **III.5**.[23]

Because the taxane ring system is oxygenated at eight positions in taxol, the conversion of **III.5** to taxol requires a series of oxidation reactions. A survey of the oxygenation patterns of known taxoids[24] indicated the order of the oxygenation to be C5 and C10, then C2, C9, and C13, then C7 and C1, followed finally by epoxidation of the C4–C20 double bond and ring expansion to the oxetane ring. The first oxygenation step was shown to be formation of the C5 alcohol **III.6**;[25] the low levels of **III.6** found in *T. brevifolia* bark indicate that the hydroxylation of **III.5** to **III.6** may be the slow step of taxol biosynthesis. Conversion of **III.6** to the acetate **III.7** was demonstrated with a soluble enzyme preparation from *T. canadensis*, which is most probably the next step in the biosynthesis.[26]

It has proved to be difficult to elucidate the order of the hydroxylation reactions downstream of **III.7**, but a probable sequence is hydroxylation to the 10β-ol **III.8** or the 13β-ol **III.9**,[27] and thence to the tetrol **III.10**; the intermediacy of 2α,10β-diol derivatives has also been proposed.[28] The conversion of taxusin (**III.10**, R = Ac) to its 7β-hydroxy derivative **III.11** has also recently

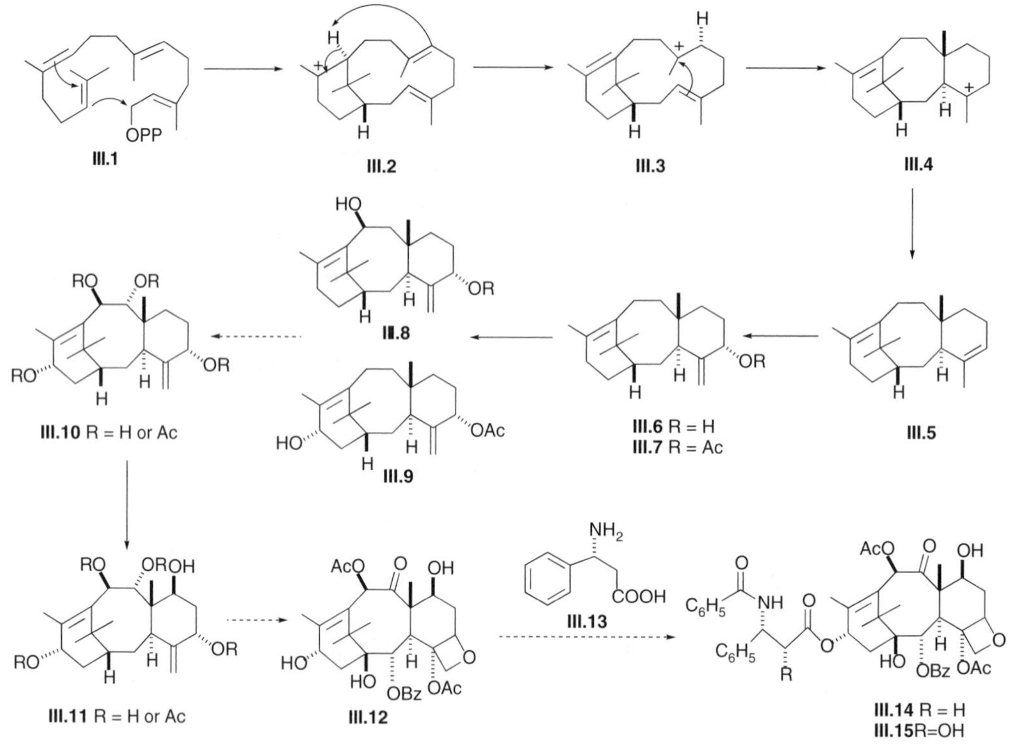

SCHEME 6.4

been demonstrated.[29] Substrate specificity studies with a set of available P450 hydroxylases indicate that there may well be more than one biosynthetic pathway to taxol in the yew, with the various pathways most probably converging on baccatin III (**III.12**).[30] Recently, cDNAs encoding two new P450 taxane hydroxylases and candidate genes for most of the steps of the taxol biosynthesis pathway have been identified.[31]

2. The Side Chain

The characteristic N-benzoylphenyl isoserine side chain of taxol has been shown to arise from phenylalanine, which was efficiently incorporated into taxol.[32,33] The intact side chain was not incorporated into taxol, but the N-debenzoyl side chain was incorporated — although not as efficiently as β-phenylalanine.[34] It is thus probable that hydroxylation of the side chain occurs after attachment to baccatin III, and that the major pathway is thus from α-phenylalanine to β-phenylalanine **III.13**. β-phenylalanine probably couples to baccatin III to give the 2-deoxytaxol **III.14**, and this finally undergoes hydroxylation to taxol (**III.15**).

The overall pathway shown in Scheme 6.1 may be regarded as a plausible pathway to taxol, but the detailed steps, and especially the order of the various oxidation steps, have not yet been completely established. Steps that have been demonstrated are shown with solid arrows, and those that remain undefined are shown with dotted arrows.

B. TAXOL BIOPRODUCTION

The original source of taxol was the bark of the western yew, *T. brevifolia*, but this source is a limited and nonrenewable one, as the removal of the bark destroys the tree. In addition to this, *T. brevifolia* is concentrated primarily in the old-growth forests of the Pacific Northwest of the United

States, and these forests are also home to the endangered spotted owl. The initial large-scale isolations of taxol for the licensee, BMS, were carried out by Hauser Chemical Research, Inc., and met with considerable controversy.[9] In spite of this, large quantities of the drug were produced in this way until 1994, when BMS ended its contract with Hauser.

The second phase of large-scale production began in 1993, with the adoption of a semisynthetic process for the synthesis of taxol from 10-deacetyl baccatin III (10-DAB), a compound that is available in yew needles in relatively large amounts.[9,10] The conversion of 10-DAB to taxol involved a coupling reaction with a synthetic β-lactam intermediate; the details of this synthesis will be described later. Because 10-DAB is available from yew needles, which could in principle be harvested in a renewable way, this semisynthetic approach ensured a stable supply of taxol.

The third phase of taxol's bioproduction is that of plant tissue culture. An enormous effort was expended on the development of viable methods of taxol production by plant tissue culture methods, and this subject has been reviewed elsewhere.[16,35,36] Production of taxol is enhanced by elicitation with methyl jasmonate and other elicitors, and yields of up to 115 mg/L in 2-week cell cultures of *Taxus media* grown in flasks have been reported.[37] These yields are close to those needed for commercial production, and undoubtedly unpublished proprietary improvements have been made to facilitate large-scale production. The commercial production of taxol by cell culture methods has been reported in 2003 by the Korean company Samyang Genex,[38] who name their product Genexol®. BMS also announced in 2003 that it is producing taxol by plant cell culture methods. The details of this process have not been released, but it is known that production is based on cultures of *T. chinensis* in Germany, and that final purification is carried out at the company's plant in Swords, Ireland, based on a method developed in collaboration with Phyton, Inc.[16] The method is reported to be much more environmentally friendly than the previous semisynthetic method, requiring only five organic solvents instead of the 13 solvents needed for the earlier method.

The production of taxol by the endophytic fungus *Taxomyces andreanae* has also been reported,[39] but the yield was extremely low (approximately 50 ng/L). More recently, the fungus *Periconia sp.* was found to produce taxol at the 800 ng/L level when stimulated with benzoic acid.[40] However, at this point the maximum yields of taxol by the fungal route lag far behind those of the plant tissue culture route, and it has been suggested that the best prospects for microbial taxol production may require the discovery of endophytes that make large quantities of a precursor taxane that could be converted chemically to taxol.[41]

In summary, the commercial production of taxol has now moved to plant tissue culture methods, but semisynthesis from 10-DAB or other precursors will continue to be an important option for the commercial synthesis of the clinical candidates with modified structures described in section X.

IV. TAXOL'S MECHANISM OF ACTION

The discovery that taxol was a tubulin polymerization promoter[3] was key to maintaining interest in it during its lengthy development, as at the time this was an unprecedented mechanism of action for an anticancer drug. Since that time, other natural products have been shown to have the same (or a very similar) mechanism, and at least two of these products, the epothilones and discodermolide, are in clinical trials as anticancer agents. Each of these compounds is described in a separate chapter in this book and will thus not be discussed here.

The primary mechanism of action of taxol has been investigated extensively. It promotes the polymerization of tubulin heterodimers to microtubules,[42] binding with a stoichiometry of approximately 1 M taxol to 1 M tubulin dimmer. At high concentrations taxol both stabilizes microtubules and increases the total polymer mass,[42] but these concentrations are higher than those needed to inhibit microtubule functions.[43] At clinically relevant concentrations, taxol suppresses dynamic changes in microtubules, leading to mitotic arrest.[44] Taxol interferes with the formation of the mitotic spindle, which causes the chromosomes not to segregate.[45]

Taxol has been shown to have many biological effects in addition to its ability to stabilize microtubules. Probably its most important activity after its microtubule assembly activity is the inactivation of the antiapoptotic protein Bcl-2. It does this by inducing phosphorylation,[46] which leads to inactivation.[47] The phosphorylation of Bcl-2 may occur through activation of Raf-1 kinase, and it has been proposed that Raf-1 is activated following drug-induced disruptions of microtubules.[48] Taxol also binds directly to Bcl-2.[49]

In spite of taxol's effect on Bcl-2 and other effects such as induction of the production of cytokines,[49] it has been proposed that all the significant effects of taxol are directly traceable to its microtubule-binding activity; as a recent critical review states, "Unless convincingly proven otherwise, all relevant paclitaxel effects should be assumed to result from its microtubule-binding activity."[50] The actual effects resulting from microtubule binding are, however, rather complex. As another recent review states, "Depending on the phase of the cell cycle, taxanes can affect spindle formations, chromosome segregation, or completions of mitosis, thus activating the mitotic or the DNA-damage checkpoints and blocking cell-cycle progression. This complex scenario, with different cell cycle responses related to the specific microtubule function affected in each phase, is reflected in the variety of pathways described to result in apoptosis upon taxane treatment. "[51]

Because the most important biological effects of taxol can be attributed to its microtubule stabilization, the nature of its binding to microtubules becomes of great importance; this subject will be discussed in section VII of this chapter.

V. MEDICINAL CHEMISTRY OF TAXOL

The chemistry and medicinal chemistry of taxol have been extensively investigated, and it is not possible to discuss every aspect of this subject in the space allotted to this chapter. The emphasis of this section will thus be on work that has led to important new understandings or to compounds with improved activity. Readers interested in broader aspects of the subject are referred to any of several recent reviews.[52–57]

A. RING A MODIFICATIONS

The C11(12) double bond in the A-ring of taxol is unreactive to hydrogenation; exhaustive hydrogenation of baccatin III gives a hexahydroproduct with a cyclohexylcarbonyl group replacing the C2 benzoyl group and an intact C11(12) double bond.[58] It can, however, be epoxidized, and compound **V.A.1** was prepared by epoxidation of 10-deacetoxytaxol. **V.A.1** is more active than taxol in a tubulin-assembly assay but is less cytotoxic to B16 melanoma cells.[59] A similar result was observed for the derivatives **V.A.2** and **V.A.3**.[60] The enol **V.A.4** is surprisingly stable and can be converted to the analog **V.A.5**, which is slightly more cytotoxic than taxol.[61]

The C18 analogs **V.A.6** (X = Me, N3, OAC, CN, or Br) could be prepared by allylic homination of 7,13-di(triethylsilyl) baccatin III followed by nucleophilic displacement and side chain attachment, but all were less cytotoxic than taxol.[62] A-nortaxol analogs such as **V.A.7** were prepared by treatment of taxol with Lewis acids;[63] compounds of this type are much less cytotoxic than taxol,

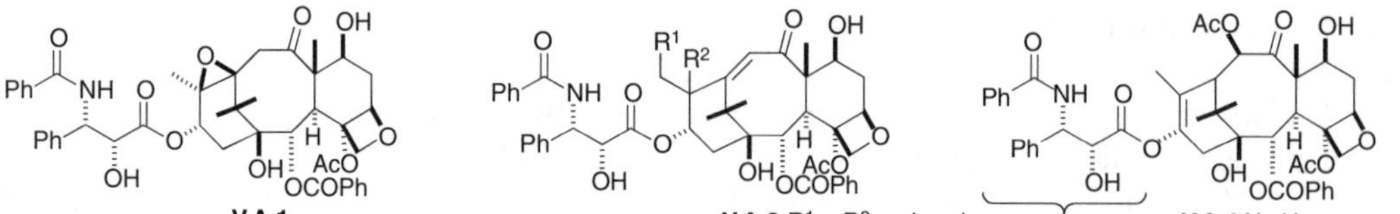

SCHEME 6.5

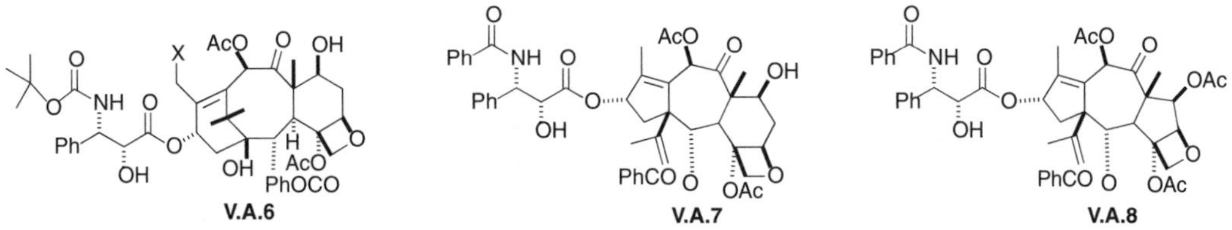

SCHEME 6.6

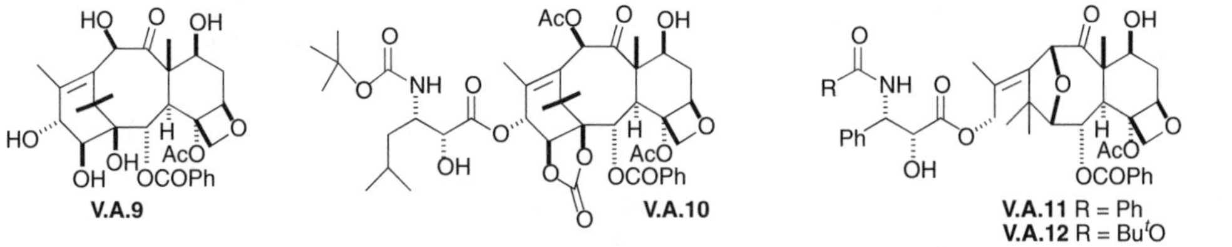

SCHEME 6.7

although they do retain some tubulin-assembly activity. Similarly, compound **V.A.8** with contracted A and C rings is noncytotoxic, but still shows tubulin-assembly promotion activity.

The hydroxylated baccatin III derivative 14β-hydroxy-10-deacetylbaccatin III (**V.A.9**) was isolated from *Taxus wallichiana* leaves[64] and has been used as the precursor for a series of 14β-hydroxytaxol analogs.[65] The best of these turned out to be **V.A.10**, which is in phase II clinical trials under the name ortataxel.[56] Other compounds prepared from **V.A.9** include the A-seco analogs **V.A.10** and **V.A.11**, which were as active as taxol in the resistant cell line MCF7-R but were from 20 to 40 times less cytotoxic against several normal cell lines.[66]

B. RING B MODIFICATIONS

Taxol can be selectively deacetylated at C10 by treatment with hydrazine[67] or sodium bicarbonate and hydrogen peroxide,[68] and it can be deoxygenated with samarium diiodide to give 10-deacetoxyltaxol **V.B.1**. Similar treatment of docetaxel gives the docetaxel analog **V.B.2**.[69,70] Both 10-deacetoxyl[71] and its docetaxel analog[71] had similar cytotoxicity to taxol. Acylation of 10-deacetyltaxol has yielded a number of 10-acyl analogs with improved bioactivities,[72] although the best compounds (**V.B.3** and **V.B.4**) also had modified side chains; these compounds were two orders of magnitude more active than taxol.

The C9 position is occupied by a carbonyl group in taxol, but (9R)-13-acetyl-9-dihydrobaccatin III was isolated from *T. canadensis* and converted to 9-dihydrotaxol **V.B.5**.[73] Various analogs of **V.B.5** were also prepared.[74,75] Deoxygenation of (9R)-13-acetyl-dihydrobaccatin III via the Barton procedure, followed by side chain attachment, gave the 9-deoxytaxol derivative **V.B.6**.[76]

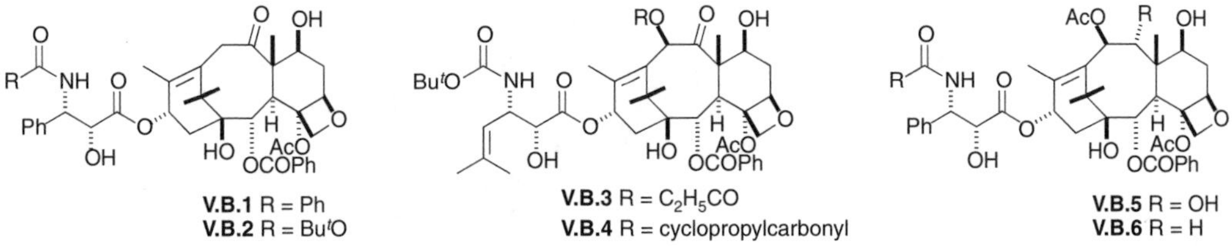

SCHEME 6.8

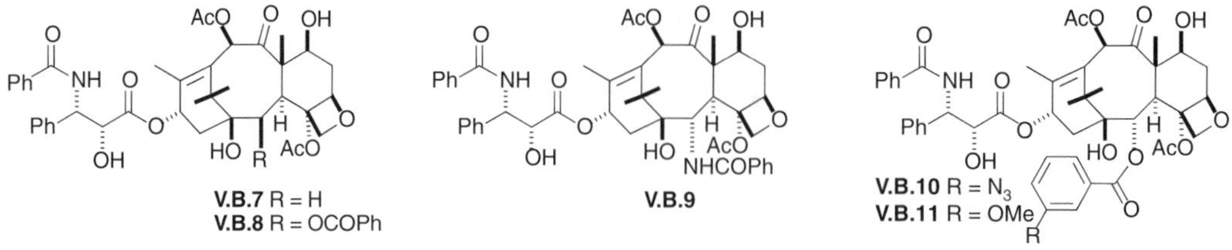

V.B.7 R = H
V.B.8 R = OCOPh

V.B.9

V.B.10 R = N$_3$
V.B.11 R = OMe

SCHEME 6.9

Modifications at the C2 position have yielded interesting results. The C2 benzoate group or a similar group is necessary for activity, and both 2-debenzoyloxy taxol[77] (**V.B.7**) and 1-benzoyl-2-debenzoyloxytaxol are both inactive.[78] 2-epi-Taxol (**V.B.8**) is also inactive,[79] as is 1-benzoyl-2-debenzoyloxytaxol. On the other hand, 2-debenzoyloxy-2α-benzoylamidotaxol (**V.B.9**) retains cytotoxicity against three cell lines, with diminished activity as compared to taxol in two cell lines but with slightly increased activity in the third.[80] Taxol analogs with substituted benzoyl groups at C2 can, however, be more active than taxol, although there is an interesting substituent effect on activity, in that *para*-substituted benzoate analogs are uniformly less active than taxol, whereas *ortho*- and *meta*-substituted analogs can be more active than taxol. Two of the best analogs are the *m*-azidobenzoyl derivative **V.B.10** and the *m*-methoxybenzoyl derivative **V. B.11**.[81,82] Analogs with 2-heteroaryl groups were also investigated, and thiophene carbonyl and furan carbonyl groups gave comparable activity to taxol.[82–84] Most nonaromatic C2 esters at C2 gave analogs with lower activity, but the 3,3-dimethylacrylic ester showed better cytotoxicity against several cell lines.[85]

Direct deoxygenation of taxol at the C1 position has not proved possible, but a small series of 1-deoxytaxol analogs was prepared by modifications of baccatin VI.[86] The analog **V.B.12** was approximately half as active as taxol in a tubulin-assembly assay.

C. RING C MODIFICATIONS

The C7 hydroxy group is the most readily modified group in the diterpenoid ring system of taxol, and extensive modifications of it have been made. As the hydroxyl group of a β-hydroxyketone it is readily epimerized, and it can also be readily and selectively derivatized if the C2 hydroxyl group on the side chain is suitably protected. Some of the various acyl derivatives that have been prepared have had improved properties compared with taxol, and the DHA derivative **V.C.1**[87] and 7-hexanoyltaxol **V.C.2**[88] are both currently in phase II clinical trial. Some simple ether derivatives also have had good activities; thus, thioethers **V.C.3** and **V.C.4** both had improved cytotoxicity and tubulin-assembly activity compared with taxol, and **V.C.3** is currently in phase II clinical trials.[89–91]

Oxidation of the C7 hydroxyl group leads to the corresponding ketone, which is unstable in base and undergoes ring-opening to the αβ-unsaturated ketone **V.C.5**; this compound is devoid of activity.[92] Deoxygenation at C7 has been achieved by several groups, and both 7-deoxytaxol **V.C.6**[93,94] and 10-acetyl-7-deoxydocetaxel **V.C.7**[95] have been prepared; these compounds are slightly less active than taxol in the tubulin assembly assay but are comparably cytotoxic to HCT 116 cells.

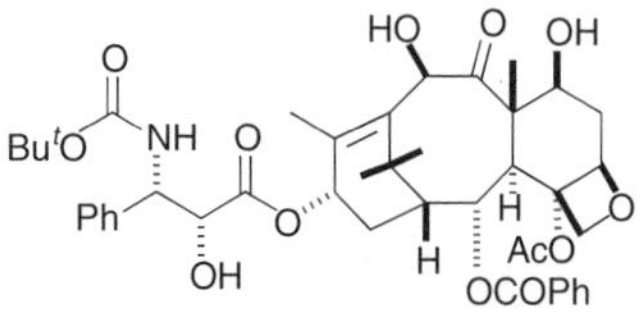

V.B.12

SCHEME 6.10

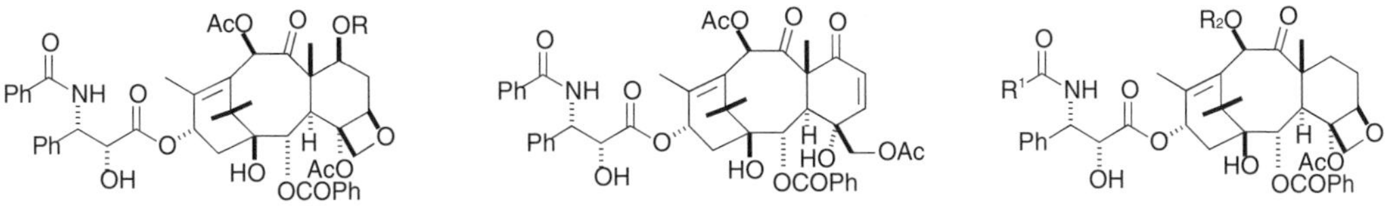

V.C.1 R = COCH$_2$(CH$_2$CH=CH)$_6$CH$_2$CH$_3$
V.C.2 R = CO(CH$_2$)$_5$CH$_3$
V.C.3 R = OCH$_2$SCH$_3$
V.C.4 R = SCH$_2$OCH$_3$

V.C.5

V.C.6 R^1 = C$_6$H$_5$; R^2 = CHCO
V.C.7 R^1 = Bu^{t}O, R^2 = H$_3$

SCHEME 6.11

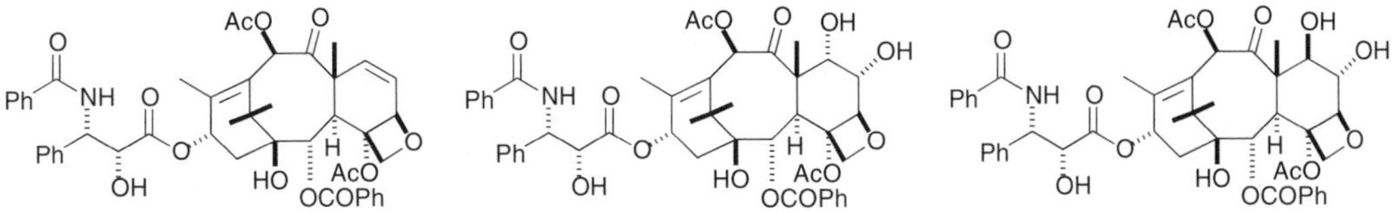

V.C.8   **V.C.9**   **V.C.10**

SCHEME 6.12

Dehydration of taxol was achieved by treatment of the 7-triflate derivative with a nonnucleophilic base to give 6,7-dehydrotaxol **V.C.8**,[96,97] which was comparable to taxol in its tubulin assembly and cytotoxic activity. Compound **V.C.8** could be converted to the corresponding diol **V.C.9** by osmylation, and this could be epimerized (albeit in low yield) to **V.C.10**, the major human metabolite of taxol.[98] An improved synthesis of compound **V.C.10** was reported by workers at BMS.[99]

Another interesting product was obtained by reaction of **V.C.9** with lead tetraacetate to give **V.C.11**, but this compound was much less active than taxol.[100] Two modifications of the C-ring of taxol have, however, led to compounds that have advanced to clinical trial or clinical candidacy. The first of these is the formation of a cyclopropane ring between C7 and C18. The cyclopropataxol **V.C.12** and its docetaxel analog **V.C.13** were discovered independently by workers at BMS[101] and Rhone-Poulenc,[102] and the analog **V.C.14** is in phase II clinical trials as RPR-109881A.

The second series of compounds is a surprising one, as it involves cleavage of ring C to give products that no longer maintain the rigid inverted-cup geometry of taxol. The compounds were discovered serendipitously by Appendino,[103] but they showed good activity in the NCI 60-cell line panel and limited toxicity, and the C-seco analog **V.C.15** has been selected as an antiangiogenic candidate for extended treatments.[104] Various acyl derivatives of **V.C.15** have been prepared, and these have similar activity to the parent compound, so *in vivo* oxidation and re-aldolization to the corresponding taxane derivative is excluded as the reason for the observed activity.[105]

Taxol analogs with modifications at C4 have also been prepared. A C4 acyl group is important for taxol's activity, as removal of the C4 acetate leads to an inactive product.[106] Reacylation of C4

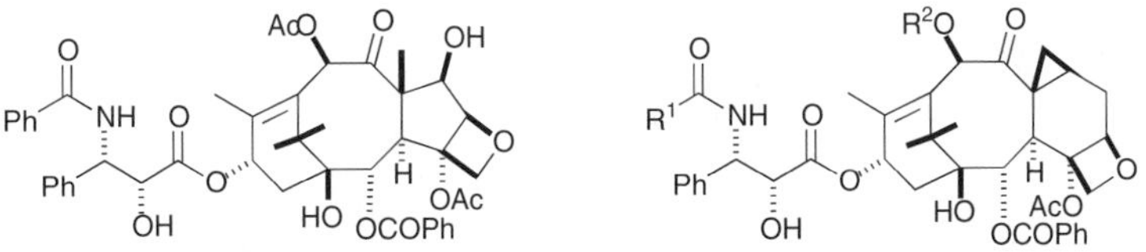

V.C.11

V.C.12 R^{11} = C$_6$H$^t_5$, R^{22} = CH$_3$CO
V.C.13 R^1 = Bu^{t}O, R^2 = H
V.C.14 R = BuO, R = CH$_3$CO

SCHEME 6.13

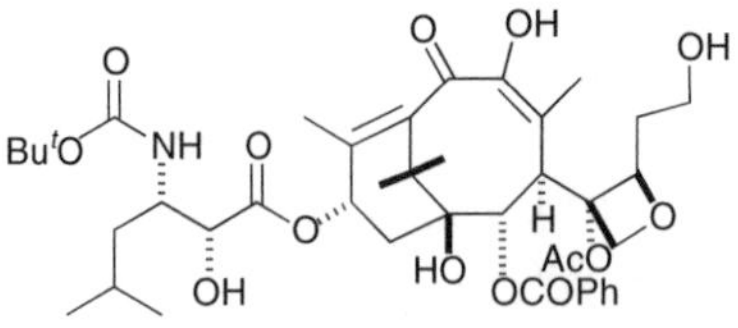

V.C.15 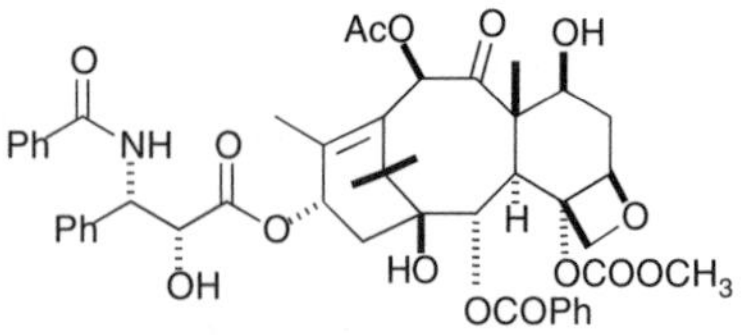   V.C.16

SCHEME 6.14

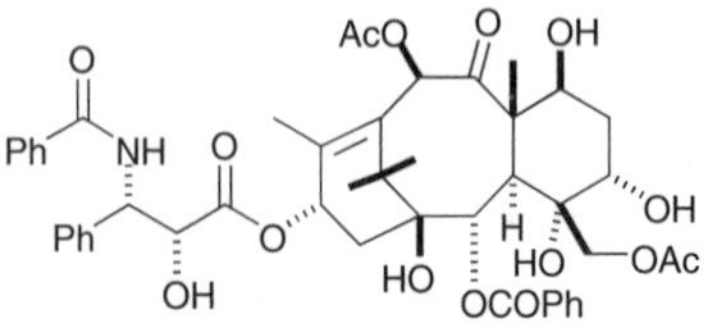

V.D.1

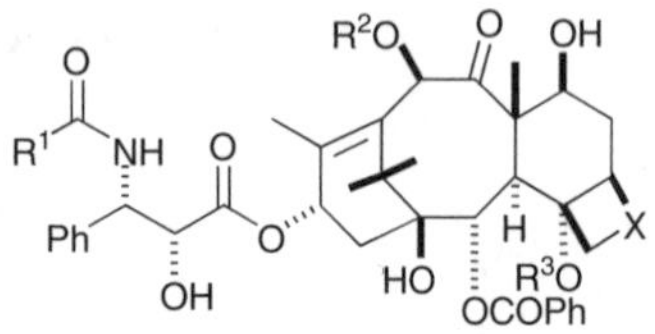

V.D.2 R^1 = Bu^{t}O, R^2 = H, R^3 = COCH$_3$, X = NH
V.D.3 R^1 = Ph, R^2 = OCOCH$_3$, R^3 = COOCH$_3$, X = S

SCHEME 6.15

with various acyl groups gave some products with improved activity, and the C4 carbonate **V.C.16** is in Phase II clinical trials.[107,108] Various analogs with C2 and C4 acyl modifications have also been made, and some of these compounds showed enhanced activity against resistant cell lines.[109]

D. Ring D Modifications

The oxetane ring is an unusual feature of the taxol structure, and it plays a significant role in taxol's unique activity. Thus, opening of the oxetane ring to a compound such as **V.C.5** destroys taxol's activity, as does ring opening by electrophilic reagents to give compound **V.D.1**.[63] Taxol analogs in which the oxetane oxygen atom is replaced with nitrogen (**V.D.2**)[110] or sulfur (**V.D.3**)[111] are much less active than taxol.

   The importance of the oxetane ring cannot be because of its chemical reactivity, as taxol binds noncovalently to tubulin. It is thus probable that it serves as a conformational lock, holding the diterpenoid ring system in its characteristic inverted-cup conformation, and also (in conjunction with the C4 acetate) as a hydrogen bond acceptor. These conclusions were put on an quantitative basis by Snyder and his colleagues,[112] who predicted that other functional groups would be capable of assisting the binding of taxol to tubulin as effectively as the oxetane ring. This prediction was borne out by the synthesis of the cyclopropane derivative **V.D.4**, which is almost as active as taxol in a tubulin assembly assay.[113] Compound **V.D.4** contains both the necessary C4 acetate and rigidifying cyclopropane ring, and both features appear to be necessary for activity. Compound **V.D.5**, which contains only the C4 acetate function, is much less active than taxol.[114] Similarly, compound **V.D.6**, with both C4 and C20 acetate functions, is also inactive as a tubulin assembly agent.[115]

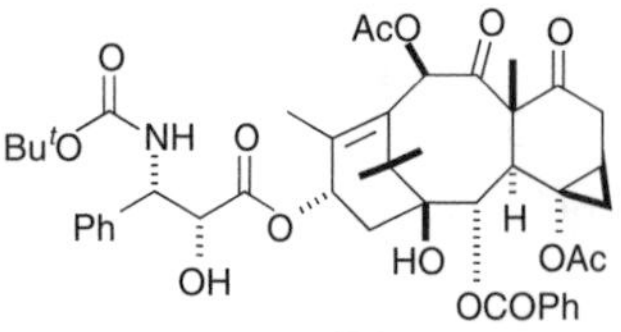

V.D.4 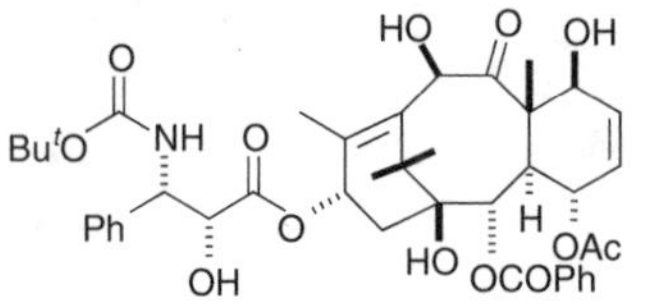   V.D.5 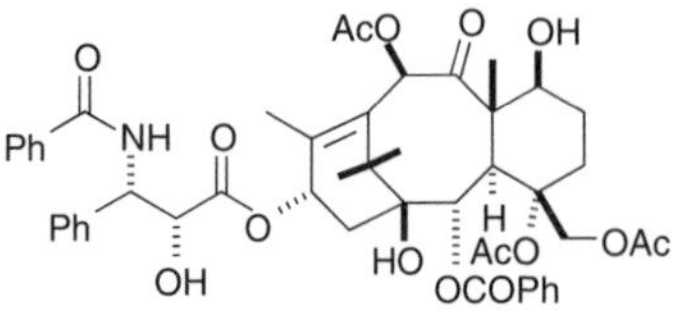   V.D.6

SCHEME 6.16

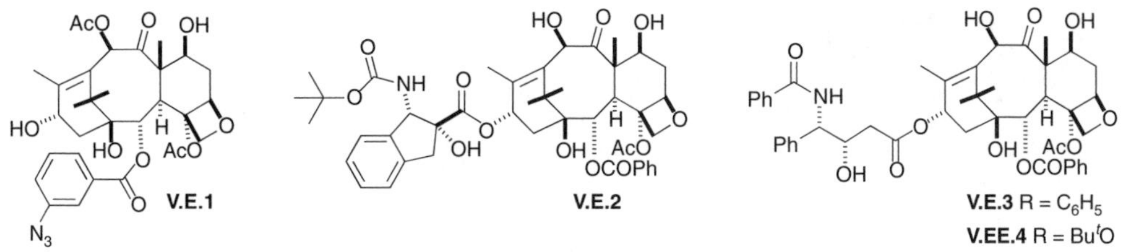

SCHEME 6.17

E. Side Chain Analogs

The side chain of taxol is in many ways the most readily modified part of the molecule, as it can be prepared synthetically and linked to baccatin III by one of the various methods described in section VI.A. It has been found that the 3′-phenyl and N-benzoyl groups of taxol are not specifically required for activity, and several of the taxol analogs in clinical development or in clinical trials have modified side chains.[116] Space does not permit a detailed discussion of all the side chain variations that have been investigated; more complete details can be found in the reviews previously mentioned.[52-56]

The side chain itself is necessary for the full activity of taxol, and all the compounds in clinical trial have a substituted isoserine side chain of some nature. However, suitably modified baccatin III derivatives can retain significant activity, as is the case with 2-(*m*-azidobenzoyl) baccatin III (**V.E.1**).[117]

Not only is the side chain necessary for full activity, its regio- and stereochemistry are also indispensable. Analogs with simplified side chains, such as N-benzoylisoserine or phenyllactic acid,[118] are significantly less active, and analogs with different stereochemistries than the 2*R*, 3*S* stereochemistry of taxol are also less active than taxol.[119] Some analogs with conformationally restricted side chains have been prepared, and compound **V.E.2** is only moderately less active than taxol.[120] Extension of the side chain by one carbon, as in **V.E.3** and **V.E.4**, caused a 30-fold loss of tubulin assembly activity,[121] whereas transposition of the hydroxyl and amino substituents also caused a reduction in activity by a factor of about ten.[119]

Modifications of the 3-phenyl and N-benzoyl substituents have been more fruitful. Replacement of the 3′-phenyl group with various substituted phenyl groups is largely unproductive, but replacement with moderately sized alkyl groups such as isobutyl or isobutenyl gives compounds with improved activity,[122] and as noted earlier, the analog **V.A.10** with this substitution is in clinical trials. Replacement of the 3′-phenyl with a 3′-(2-furyl) group also gives improved activity,[72] and the compound MAC-321 (also designated TL-00139, **V.E.5**) is in clinical trials.[123] The 3-(fluoro-pyridyl) group also imparts improved activity, and the analog **V.E.6** is in clinical trials,[124] and a final example is the 3′-(*t*-butyl) analog **V.E.7**, which is as active in mice when given orally as taxol is when given intravenously.[125]

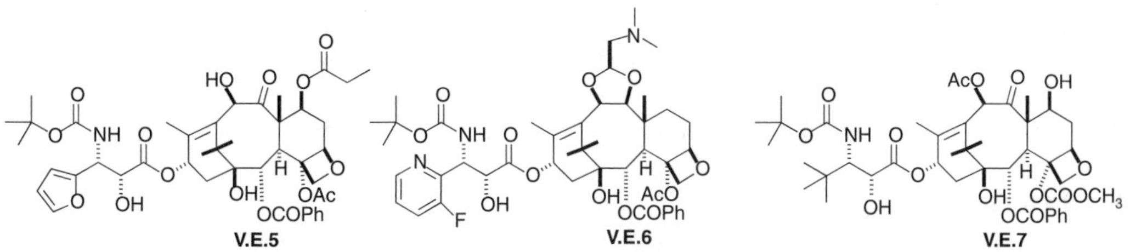

SCHEME 6.18

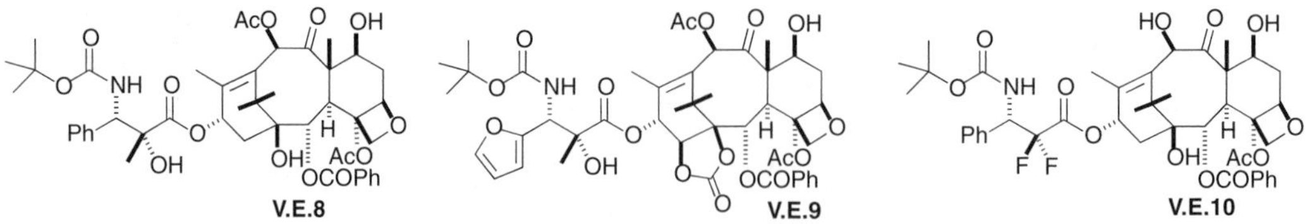

SCHEME 6.19

Various modifications of the 3'-N substituent have been studied. An acyl substituent at this position is required for activity, and the t-butoxycarbonyl group of docetaxel (Taxotere, **II.E.1**) has been proven an effective alternative acyl group to the benzoyl group of taxol. All the taxol analog clinical candidates surveyed in a recent review[116] have a benzoyl group, a t-butoxycarbonyl group, or in one case, a substituted benzoyl group as the 3-N substituent.

The 2'-hydroxyl group was early on shown to be necessary for activity,[126] and successful modifications (other than acylation to give a prodrug, discussed in the next sections) have been rare. A few cases of important activity have been discovered, however. Several 2'-methyl taxol derivatives such as **V.E.8** have been prepared and have been found to have improved cytotoxicity as compared with taxol.[127–129] A recent paper describes the synthesis of the analog **V.E.9**, which is more active than taxol in both normal A2780 cells and taxol-resistant A2780 cells.[130] Replacement of the 2-hydroxyl group with a thiol group gave an inactive compound,[131] but the 2,2-difluoro analog **V.E.10** is somewhat more active than taxol toward P388 cells.[132]

F. PRODRUGS OF TAXOL

The low water-solubility of taxol was a major problem in its early development, so the preparation of water-soluble prodrugs was an important part of the early work on the drug. The development of more active and more water-soluble analogs of taxol described in the preceding sections and summarized in section IX below has rendered much of the prodrug work obsolete. At present the only prodrug that is included in a recent survey of taxanes in development[116] is T-3782 (**V.F.1**; compound **4a** in Yamaguchi et al.[133]). This section is thus abbreviated and will only mention a few of the variations that have been prepared. A more detailed treatment can be found in the reviews previously mentioned.[52,134]

Initial studies focused on simple ester derivatives at the 2-position, as these are readily hydro-lyzed to taxol. Various succinate and glutarate derivatives were prepared,[126,135] as wall as sulfonic acid salts[136] and amino acid derivatives.[126,137] The prodrug **V.F.1** noted above is of this general type, but with modified side chain substituents as well as the amino acid linked through a glycolate spacer.

Phosphate prodrugs have proved attractive because of phosphatases present in cells, and inge-nious use was made of this fact in the synthesis of prodrugs such as **V.F.2**; this compound is stable in water but releases taxol by internal lactonization after dephosphorylation by phosphatases.[138] Regrettably, **V.F.2** was found to bind to plasma proteins and was deemed to be unsuitable for use as a prodrug, although it did show good *in vitro* activity against the murine M109 tumor.[138]

A second ingenious solution to the water-solubility problem was developed by noting that the O-benzoyl to N-benzoyl migration of 2'-benzoyl-3'N-debenzoyltaxol (**V.F.3**) is slow at pH 4.0. This compound is much more water soluble than taxol at this pH and can thus be used as a prodrug, with conversion to taxol occurring relatively quickly (t_f = 15 min) under physiological conditions.[139] The same strategy has recently been applied to generate **V.F.4**, a prodrug of canadensol.[140]

G. TARGETED ANALOGS OF TAXOL

Although the need for prodrugs of taxol has been alleviated by the development of more potent water-soluble analogs, there is still a need to develop analogs that can be delivered more selectively

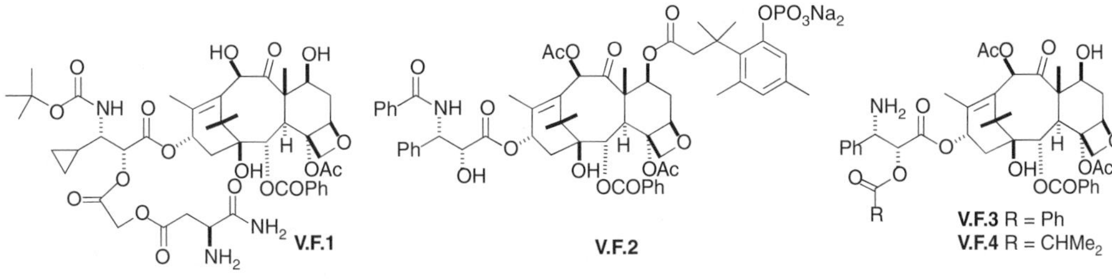

SCHEME 6.20

to the tumor cells. Some of the methods for doing this involve new pharmaceutical formulations and will be discussed in section IX. In other cases, targeting agents have been covalently linked to taxol, and these modifications are described here.

Several targeting analogs make use of monoclonal antibodies (mAbs) to target taxol to a tumor site. In the antibody-directed enzyme prodrug therapy (ADEPT) approach,[141] specific enzymes are delivered to tumor sites by mAbs, and these enzymes are then available to activate a "protected" taxol derivative selectively cleaved by this enzyme, resulting in release of taxol. The taxol analog **V.G.1**, for example, can be converted to taxol by hydrolysis of the glucuronide with a glucuronidase, decarboxylation, and internal lactam formation.[142] A *p*-nitrophenol linker has also been used with similar results,[143] and docetaxel has also been targeted by this approach.[144]

A second strategy makes use of the fact that the protease plasmin plays an important role in tumor invasion and metastasis and is thus present in tumors. Taxol prodrugs were thus prepared in which plasmin substrate peptides were linked through a spacer to the 2-position of taxol. One of the best such compounds is **V.G.2**: This compound is one of the least toxic taxol prodrugs reported, and yet it is converted by plasmin to taxol with a half life of 42 min.[145]

Another tumor-associated enzyme is neuraminidase, and the taxol-sialic acid hybrid **V.G.3** was prepared as a neuraminidase-cleavable prodrug.[146] Compound **V.G.3** was somewhat less cytotoxic than taxol, but it was much more water-soluble. No information on its cleavage by neuraminidase was provided.

A taxol-hyaluronic acid conjugate has been prepared in which the polymeric hyaluronic acid is linked to taxol through a linker at the 2-position.[147] The conjugates showed selective toxicities toward breast, colon, and ovarian cancer cell lines that are known to overexpress hyaluronic acid receptors, but were nontoxic to a mouse fibroblast cell line.

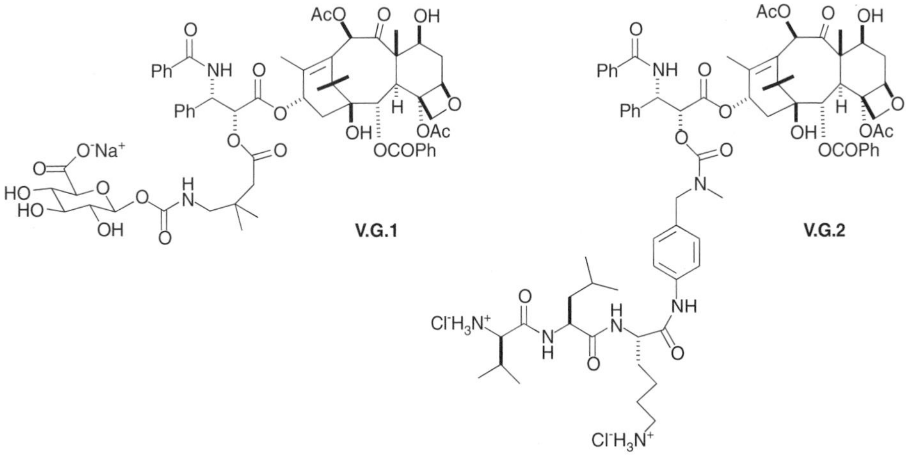

SCHEME 6.21

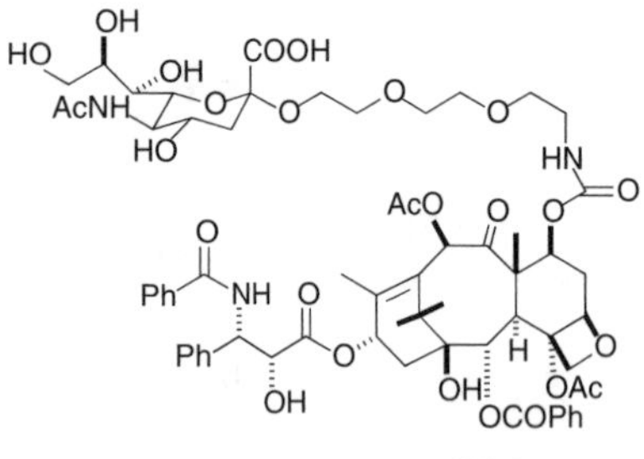

SCHEME 6.22

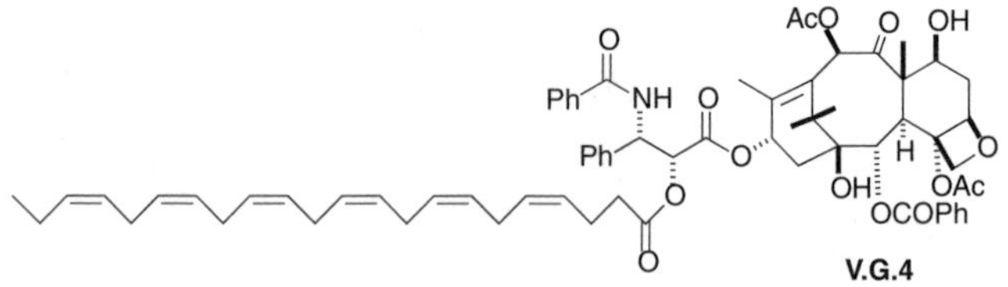

A different approach was taken in the synthesis of the docosahexaenoic acid conjugate **V.G.4**, based on the observation that tumors rapidly take up some fatty acids from arterial blood. This compound was evaluated in a phase I clinical trial and was found to be well tolerated by the patients. It also proved to be curative for mice with the M109 tumor, in contrast to taxol, which was not curative in this system.[148,149] The C7 analog **V.C.1** mentioned previously is also a targeted analog of this type.

Folic acid is another compound that is taken up selectively by cancer cells, and taxol-folic acid conjugates have been prepared and evaluated.[150] Compound **V.G.5** was found to increase the life span of mice with the M109 tumor, but it was disappointingly less effective than taxol itself.

It has been found that short oligomers of arginine exhibit superior membrane translocation activity, and a series of arginine-based molecular transporters of taxol has been prepared.[151] The general structure of these compounds is **V.G.6**; they are much more water soluble than taxol and release free taxol with half-lives of minutes to hours, depending on the linker structure and the pH of the medium.

The final and most direct strategy is that of linking taxol directly to a mAb. This approach has been adopted by more than one group, with promising results. Thus, conjugation of taxol to the anti-p75 mAb MC192 was achieved through a 2-glutaric acid linker, and the resulting conjugate had improved cytotoxicity and improved selectivity as compared to taxol.[152] In another approach, several taxane–antibody immunoconjugates have been prepared by Ojima.[153] As one example, the conjugate **V.G.7** was prepared using an antibody that recognized the human epidermal growth factor (EGFR) expressed in human squamous cells. The conjugate was shown to have excellent activity *in vivo* against EGFR — expressing A431 tumor xenografts in severe combined immunodeficiency (SCID) mice.[153]

In summary, the targeting of taxol and other taxanes through the approaches described above offers a way to improve therapeutic effect and reduce toxicity, and it seems probable that some targeted analogs will be introduced into clinical trials within the next few years.

H. SUMMARY OF TAXOL'S SAR

A summary of the SARs of taxol describes in the preceding sections is provided in structural form in Figure 6.2.

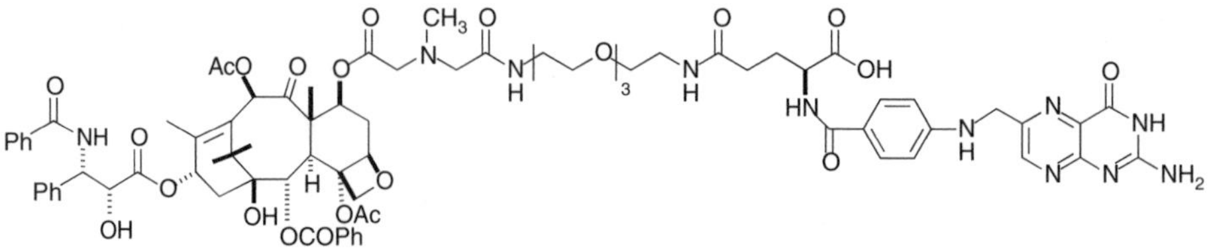

SCHEME 6.23

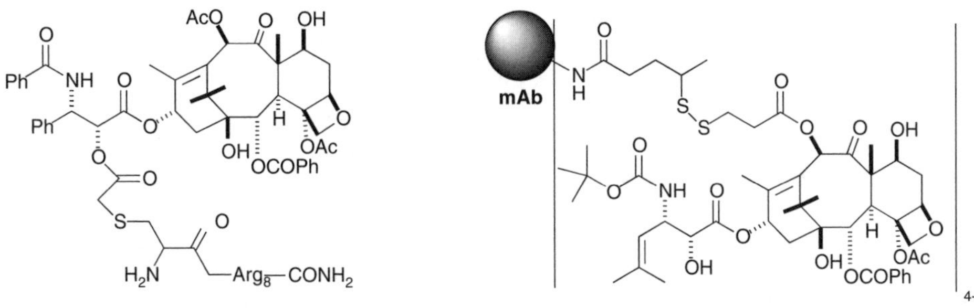

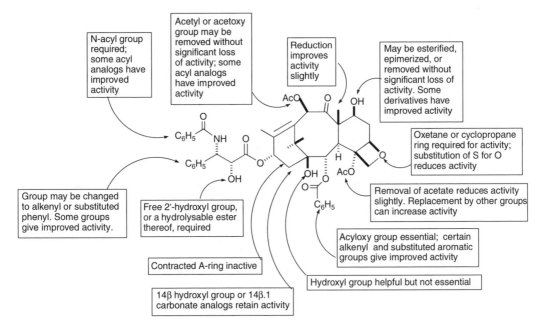

V.G.6   **V.G.7**

SCHEME 6.24

VI. SYNTHETIC STUDIES

A. Semisynthetic Methods

Much of the work described in the preceding sections has only been possible because synthetic methods are available for the attachment of the taxol side chain to baccatin III. In addition, until recently, taxol itself was manufactured on a large scale by semisynthesis from baccatin III, lending a large incentive for the discovery of improved or alternate methods for its synthesis.

All syntheses of taxol and its analogs start with 10-DAB (**II.D.1**), which must be protected at C7 and C10 before acylation at the C13 position. If a 10-acyl compound such as taxol is the desired product, 10-DAB is normally protected as its 7-triethylsilyl ether and then selectively acylated at C10 before reaction with a suitable side chain precursor.[154]

Many different approaches have been used to couple the side chain to protected baccatin III, and only the most important will be described here. The most important method from a commercial point of view is that which is based on β-lactam chemistry, as this was the method used by BMS in their semisynthesis of taxol until it was replaced by the plant tissue culture production method.

FIGURE 6.2 The Structure-Activity Relationships of Taxol.

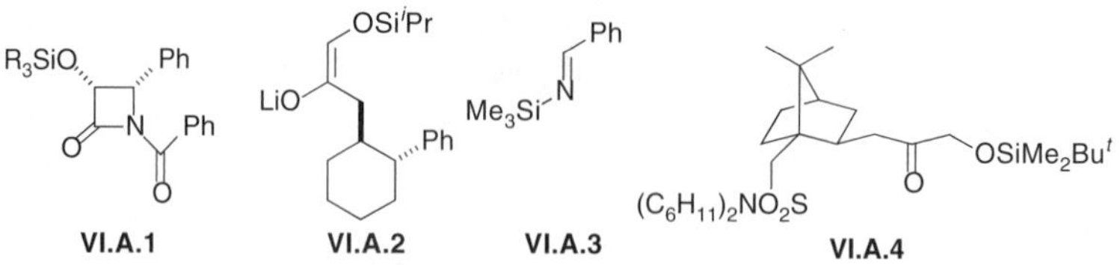

VI.A.1 VI.A.2 VI.A.3 VI.A.4

SCHEME 6.25

The chemistry was developed independently by Holton[11] and Ojima[155] and is a simple and efficient route to taxol.

The key nonracemic β-lactam **VI.A.1** can be prepared by various methods, all of which involve a [2+2] cycloaddition reaction. Two different types of approach can be used, depending on whether the chiral auxiliary is associated with the enolate or the imine half of the cycloaddition reaction. Methods that use a chiral enolate include the use of a 2-phenylcyclohexanol enolate (**VI.A.2** + **VI.A.3**)[156] and of a chiral enolate of Oppolzer's auxiliary (**VI.A.4**).[157] A chiral oxazolidinone enolate has also been used, although this approach requires several steps to remove the chiral auxiliary when it is no longer needed.[158,159]

Chiral imines have also yielded diastereoselective 2+2 cycloadditions. The imine **VI.A.5** gave 84% diastereomeric excess of the desired β-lactam, although several steps were required to remove the chiral auxiliary.[160] The imine **VI.A.6** gave a 50%–60% diastereomeric excess of β-lactam, but this had to be hydrolyzed to the open-chain amino acid to remove the chiral auxiliary by hydrogenolysis.[161] A better diastereomeric excess was achieved with the imine **VI.A.7**.[162] The β-lactam can also be prepared selectively from an α-bromoamide precursor; thus treatment of the intermediate **VI.A.8** with NaH or TBAF gave β-lactam in good yield without significant loss of optical purity.[163] An alternative and experimentally simpler approach is to prepare the acetate of racemic β-lactam **VI.A.9** by simple 2+2 cycloaddition and then resolve it with a commercially available lipase.[164] Ironically, all this effort to prepare enantiopure β-lactams may be unnecessary, as Ojima has shown that kinetic resolution can be obtained in coupling racemic β-lactams with baccatins.[165]

Coupling of the protected β-lactam, however prepared, with protected baccatin III is an efficient and high-yield reaction. Coupling is normally carried out in the presence of NaH or LiHDMS, and yields greater than 90% are routinely achieved. Thus, coupling of β-lactam **VI.A.1** with the protected baccatin III derivative **VI.A.10** gives the protected taxol derivative **VI.A.11**, which can be deprotected to give taxol **I.1**.

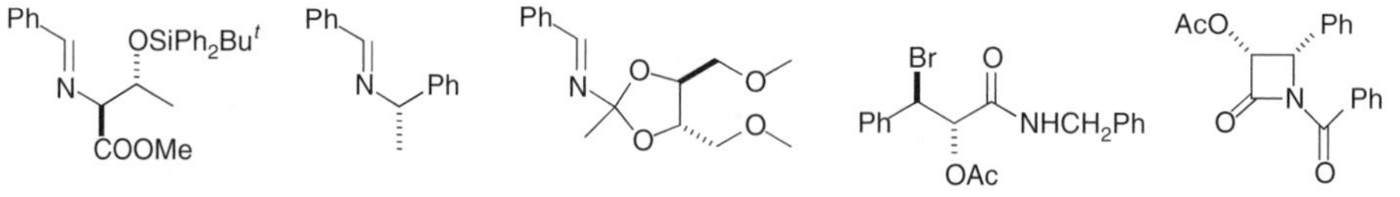

VI.A.5 VI.A.6 VI.A.7 VI.A.8 VI.A.9

SCHEME 6.26

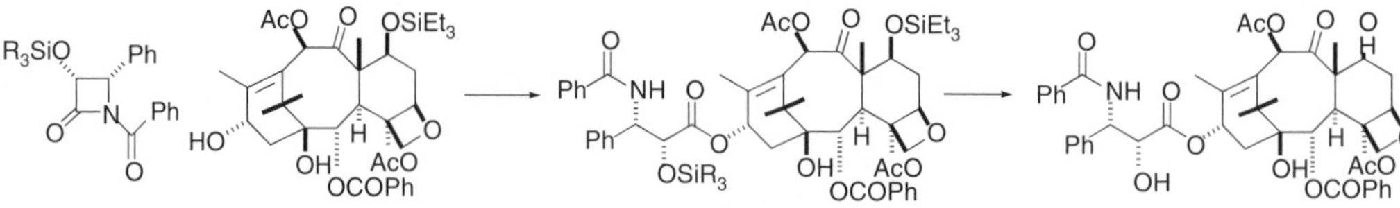

VI.A.1 VI.A.10 VI.A.11 I.1

SCHEME 6.27

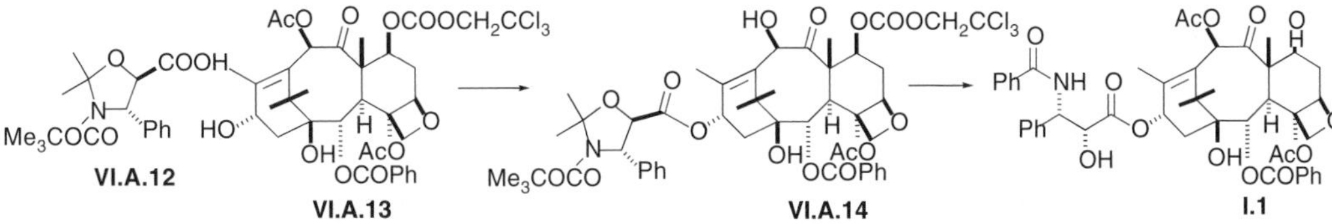

SCHEME 6.28

Most of the other approaches to taxol from baccatin III also involve cyclic intermediates, as these minimize the steric constraints inherent in acylating the very hindered C13 hydroxyl group of baccatin III. Groups that have been used successfully include the oxazolidine group, such as **VI.A.12**, which has been coupled with the protected baccatin III **VI.A.13**; the coupled product **VI.A.14** must then be deprotected by treatment with formic acid and benzoylated to give taxol.

Oxazolines such as **VI.A.15** can also be coupled in high yield; hydrolysis of the intermediate oxazoline **VI.A.16** gives taxol **I.1**. As is also the case with the β-lactam approach, this route is atom economical in that the protecting group is part of the final product.[166,167]

Many other methods have been developed for the synthesis of taxol and related compounds from baccatin III. These include coupling of oxazolidine thioesters,[168] coupling with cis-glycidic acids followed by appropriate manipulations,[169] and coupling with a β-ketoester and subsequent functionalizations.[170]

B. Total Synthesis

The structure of taxol, with its complex stereochemistry and unique ring structure, represents one of the most challenging natural product synthetic targets. In spite of this complexity, six independent total syntheses have been achieved, together with numerous partial syntheses and unique synthetic approaches. Because of space limitations, and also because this subject has been recently reviewed elsewhere,[53,171] only a general summary of the synthetic routes will be presented here. Also, because baccatin III can be converted to taxol by any of the routes described in the preceding section, a synthesis of baccatin III constitutes a formal synthesis of taxol.

The first two syntheses were published essentially simultaneously by Holton[172,173] and Nicolaou.[174] The Holton synthesis started with the natural product β-patchoulene, which eventually gave the A and B rings, which were subsequently elaborated into the final product. The strategy was thus a linear one, of the form AB→ABC→ABCD. β-Patchoulene was converted to the protected diol **VI.B.1** and thence through a clever ring opening of its epoxide to the bicyclic AB ring system **VI.B.2**. This was then converted to the carbonate **VI.B.3**, and thence to the ABC ring system **VI.B.4**. Final elaboration of ring D and appropriate functional group manipulations gave baccatin III.[173] This synthesis gave the unnatural enantiomer (+)-taxol if the commercially available β-patchoulene was used, but the natural (−)-taxol enantiomer could be prepared if (−)-borneol was used as starting material, although this required additional synthetic manipulations.

The Nicolaou synthesis[174] used the synthetic design A+C→A–C→ABC→ABCD. The preformed A (**VI.B.5**) and B (**VI.B.6**) rings were coupled by a Shapiro reaction to give the A–C

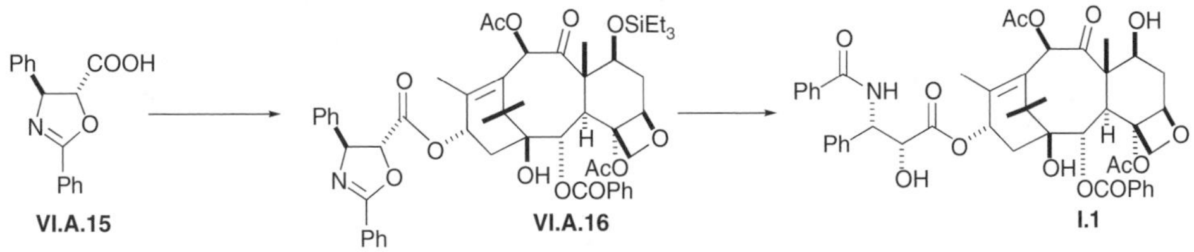

SCHEME 6.29

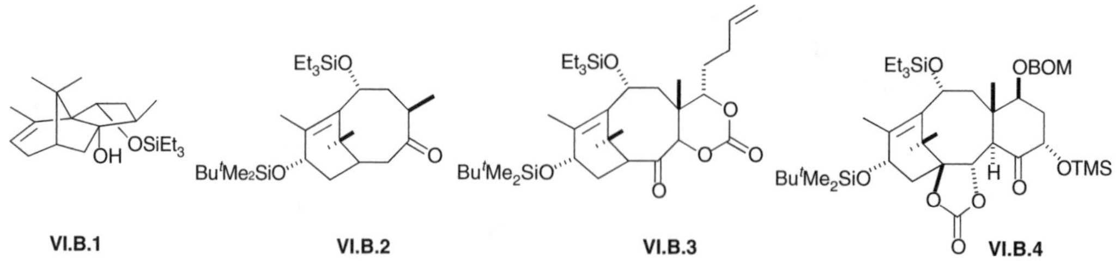

VI.B.1 **VI.B.2** **VI.B.3** **VI.B.4**

SCHEME 6.30

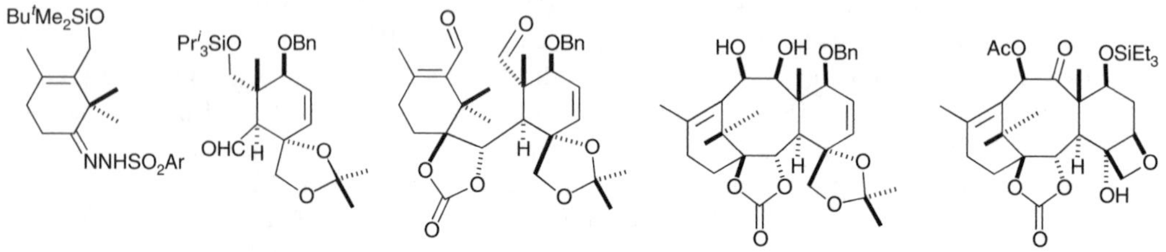

VI.B.5 **VI.B.6** **VI.B.7** **VI.B.8** **VI.B.9**

SCHEME 6.31

intermediate **VI.B.7.** This was then converted into the ABC system (**VI.B.8**) by a McMurry coupling; this reaction proved to be the Achilles' heel of the synthesis, with a yield of only 23%. Compound **VI.B.8** was then elaborated to baccatin III.

The Danishefsky synthesis[175] is of the form CD + A→A–CD→ABCD and is the only synthesis in which the oxetane ring is formed at an early stage and carried through the entire synthetic sequence. This was possible because the C4 hydroxyl group was protected as its benzyl ether, thus avoiding complications from neighboring group participation by the C4 acetate. Coupling of the A-ring synthon **VI.B.10** to the CD-synthon was achieved by lithiation of **VI.B.10** and 1,2-addition to the aldehyde group of **VI.B.11**. Cyclization to the ABCD system **VI.B.13** was brought about by a Heck reaction. Functional group manipulations converted **VI.B.13** to **VI.B.14** and the latter compound to baccatin III.

The Wender synthesis[176,177] is similar to the Holton synthesis in being a linear synthesis of the form A→AB→ABC→ABCD. The starting material was verbenone, which was converted by some clever chemistry into **VI.B.15**. Oxidative cleavage of this compound was carried out by a method related to the Holton conversion of **VI.B.1** to **VI.B.2** and gave **VI.B.16**. This compound was then converted to **VI.B.17** through elaboration of the C3 substituent, and aldol condensation of this formed the C ring. The synthesis was completed through intermediate **VI.B.18** and formation of the oxetane ring. The overall synthesis, at 37 steps from verbenone, is the shortest recorded synthesis of taxol to date.

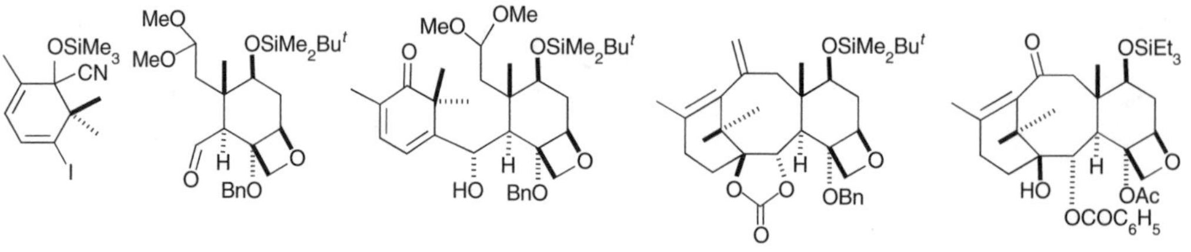

VI.B.10 **VI.B.11** **VI.B.12** **VI.B.13** **VI.B.14**

SCHEME 6.32

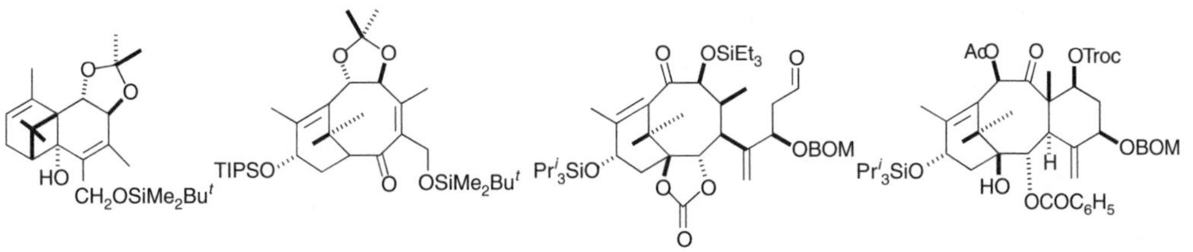

VI.B.15 VI.B.16 VI.B.17 VI.B.18

SCHEME 6.33

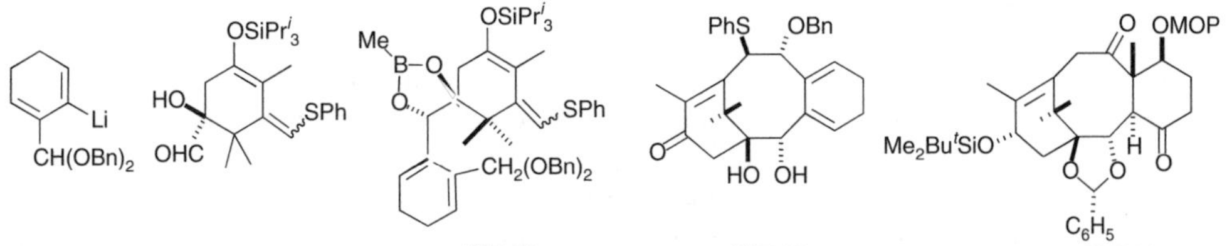

VI.B.19 VI.B.20 VI.B.21 VI.B.22 VI.B.23

SCHEME 6.34

The Kuwajima synthesis was reported in a preliminary communication in 1998, with the full report appearing in 2000.[178] It is of the form A + C→A–C→ABC→ABCD, and uses a 1,2-addition of the organolithium **VI.B.19** to the aldehyde group of **VI.B.20** to form the key A–C bond in **VI.B.21**. The second ring closure to form the B ring in **VI.B.22** was brought about by Lewis acid catalysis, with subsequent conversion to baccatin III occurring via the diketone **VI.B.23**.

The Mukaiyama synthesis took the unusual pathway of forming the B ring first, and is of the form B→BC→ABC→ABCD.[179] Ring B was formed by cyclization of the linear precursor **VI.B.24** with SmI_2, and ring C was added through conjugate addition of a lithium cuprate reagent to the αβ-unsaturated ketone of **VI.B.25**. Ring A was added by addition of an organolithium reagent to the ketone of **VI.B.26**, and ring D was formed by cyclization of the diol precursor **VI.B.27**.

As noted earlier, all of these syntheses stand out as major synthetic achievements, but none of them is new knowledge of taxol's SARs capable of providing an economical approach to the large-scale synthesis of taxol. Fortunately, the plant tissue culture methods described earlier have provided an alternate and environmentally friendly way of producing taxol, but the synthetic efforts have nevertheless been very useful as platforms for the development of new chemistry.

VII. THE TAXOL–TUBULIN INTERACTION

As noted in section IV, the primary mechanism of action of taxol is its ability to stabilize microtubules and thus to disturb the equilibrium between tubulin and microtubules that is so essential

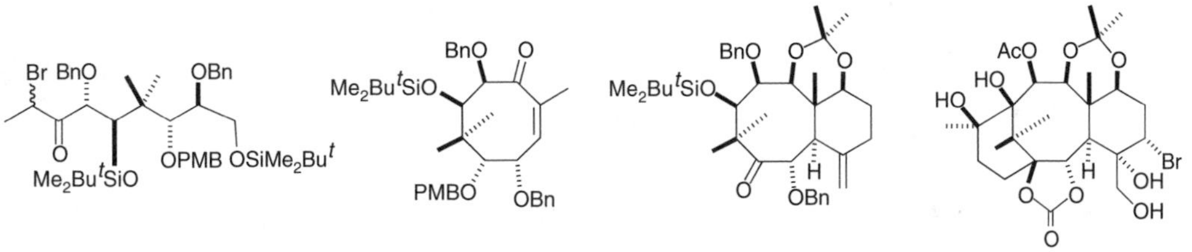

VI.B.24 VI..B.25 VI.B.26 VI.B.27

SCHEME 6.35

for normal cellular function. The nature of interactions between taxol and tubulin has thus been a subject of intense interest, and many different tools have been used to study it.

The location of the taxol binding site within the tubulin molecule was shown to be on β-tubulin by photoaffinity labeling studies with 3-N-debenzoyl-3-N-(p-azidobenzoyl) taxol ([³H]-**VII.1**), which photolabeled the N-terminal proton of β-tubulin.[180] In contrast, 2-debenzoyl-2-(m-azidobenzoyl) taxol (**VII.2**) labeled amino acids 217-231 of β-tubulin,[181] and taxol with a photoaffinity label at C7 bound to Arg²⁸² in β-tubulin.[182]

Although β-tubulin is the tubulin molecule that is labeled most extensively by photoaffinity-labeled taxol, some analogs do label α-tubulin as well as (or instead of) β-tubulin. Thus [³H]-**VII.1** does bind to α-tubulin to a modest extent,[183] and [³H]-**VII.3** labeled amino acids 281–304 of α-tubulin.[184] These results thus indicate that taxol most probably occupies a binding site located between α- and β-tubulin, but largely on β-tubulin.

In ideal circumstances the binding site of taxol on tubulin would be elucidated by x-ray crystallography, but it has not been possible to obtain crystals of the polymeric complex. It has proved possible, however, to obtain electron diffraction data on the αβ-tubulin-taxol complex as zinc-induced sheets, and this has proved enormously important in understanding the taxol–tubulin interaction.[185,186] In this structure, taxol occupies a hydrophobic cleft on β-tubulin, and the binding of taxol in this cleft converts it into a hydrophilic surface.[187] This binding location is consistent with the photoaffinity results reported above, showing labeling of β-tubulin by taxol; the labeling of α-tubulin is difficult to explain.

Although the electron crystallographic results show the location of taxol on tubulin, the actual binding conformation of taxol cannot be discerned at the resolution obtained, and indirect methods thus must be used to make this determination. A knowledge of the binding conformation of taxol is important, in that it would help to explain how compounds as structurally diverse as the epothilones, discodermolide, and eleutherobin could all have the same mechanism of action, and it could also provide a guiding model for the synthesis of simplified compounds that retained tubulin-binding ability.

Information on the binding conformations of taxol has been obtained by two studies using solid-state nuclear magnetic resonance (NMR) to probe the conformation of tubulin-bound taxol. In the first study, an F–F distance of 6.5 Å was determined for the compound **VII.4**,[188] and in the second, F–¹³C distances of 9.8 and 10.3 Å were determined for the distances to the C3 amide carbonyl and the C3 methine carbons, respectively, of **VII.5**.[189]

Studies of the NMR spectra of taxol in solution indicated the presence of two major conformers — a nonpolar conformer with the C2 benzoyl group and the C3 phenyl side chain exhibiting hydrophobic collapse[190–192] and a polar form in which the C2 benzoyl group and C3 benzamido side chain are in close proximity.[193–195] However, NMR deconvolution analysis in chloroform also revealed the presence of a third conformer, designated T-taxol, in which the C2 benzoyl group bisects the angle between the C3 phenyl and C3 benzamido side chains.[196] The T-taxol conformation is consistent with the solid-state NMR studies previously referred to and also with the unexpected activity of 2-m-azidobaccatin III.[197] It also docks well into the hydrophobic cleft in β-tubulin described earlier.[187]

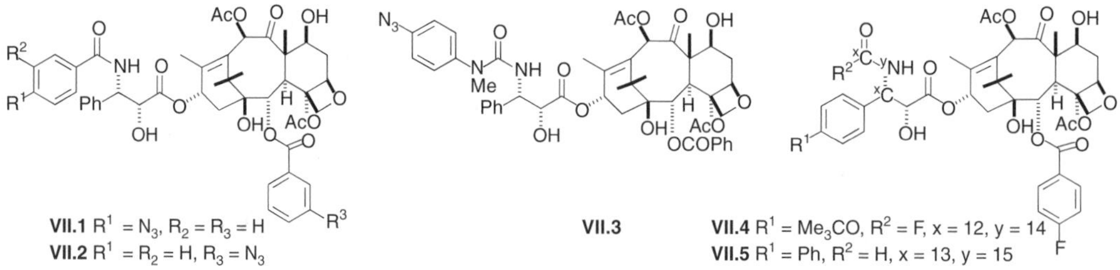

VII.1 R¹ = N₃, R₂ = R₃ = H
VII.2 R¹ = R₂ = H, R₃ = N₃

VII.3

VII.4 R¹ = Me₃CO, R² = F, x = 12, y = 14
VII.5 R¹ = Ph, R² = H, x = 13, y = 15

SCHEME 6.36

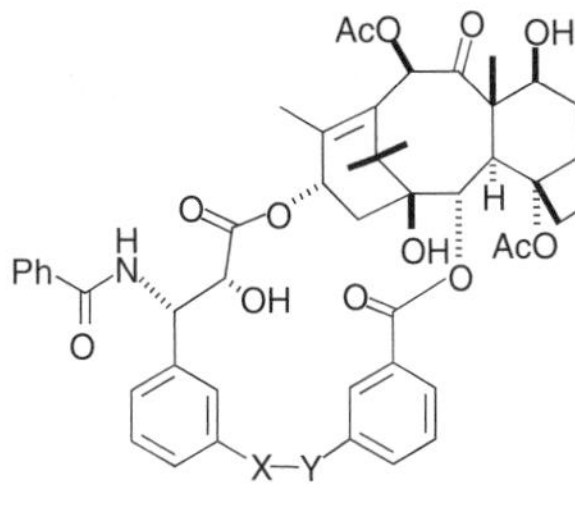

SCHEME 6.37

Tests of the various models have been made using the synthesis of conformationally constrained taxols. A test of the hydrophobically collapsed model was made by the synthesis of compounds **VII.7** and **VII.8**, and both of these compounds were completely inactive in a tubulin assembly assay, indicating that taxol does not adopt a hydrophobically collapsed conformation in the bound state.[198] A second series of compounds was based on the proposal of a common pharmacophore for taxol, the epothilones, discodermolide, and eleutherobin.[199] The analog **VII.9** was prepared based on this model, and it did show cytotoxic activity against the MDA435/LCC6-WT cell line, but it was over 100-fold less active than taxol. It also was active as a tubulin-assembly agent, with 37% of the activity of taxol. A third approach to an active macrocyclic taxol analog is exemplified by the compound series **VII.10**, where n and n′ range from 2 to 4.[200] The best of these compounds was sevenfold less active than taxol as a tubulin-assembly agent but was over three orders of magnitude less cytotoxic.

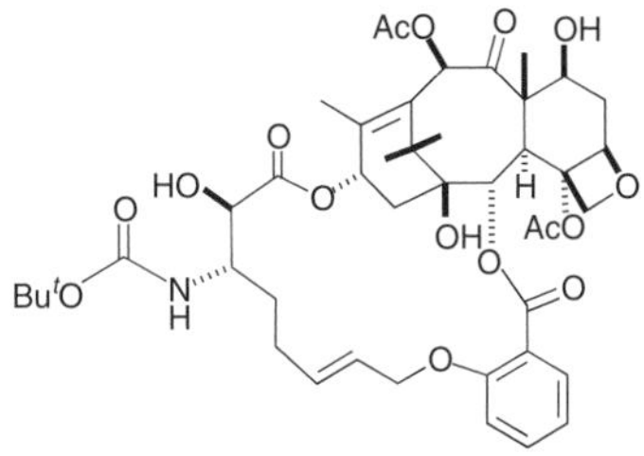

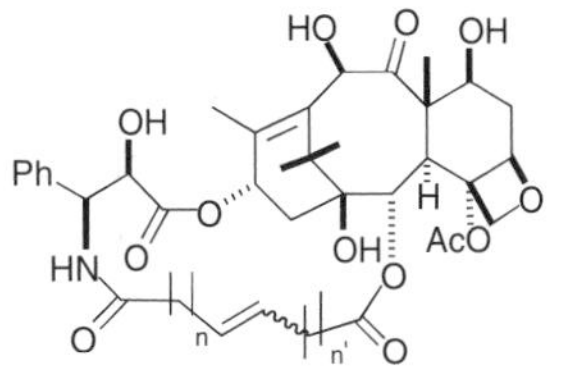

All of the bridged analogs described above are linked from the side chain to the C2 positions. A different series has been prepared on the basis of the T-taxol design concept, and this series has been spectacularly successful in providing bridged analogs that not only equal but surpass taxol in its activity.[201] A key compound is the bridged taxol **VII.11**; this compound is about 20-fold more cytotoxic than taxol to A2780 cells and has a critical concentration for tubulin polymerizations over threefold lower than taxol's. The compound maps well onto the T-taxol structure and fits the taxol-binding pocket on β-tubulin as well as taxol itself. This structure thus represents the best available model of the tubulin-bound conformation of taxol and presents an attractive target for the synthesis of simplified analogs that retain the active T-conformations of taxol.

VIII. OTHER NATURAL PRODUCTS WITH A SIMILAR MECHANISM OF ACTION

In recent years several natural products with a similar mechanism of action to taxol's have been discovered, and some of these have been developed as clinical candidates or are in clinical trial. These include the epothilones A (**VIII.1**) and B (**VIII.2**),[202,203] discodermolide **VIII.3**,[204] and

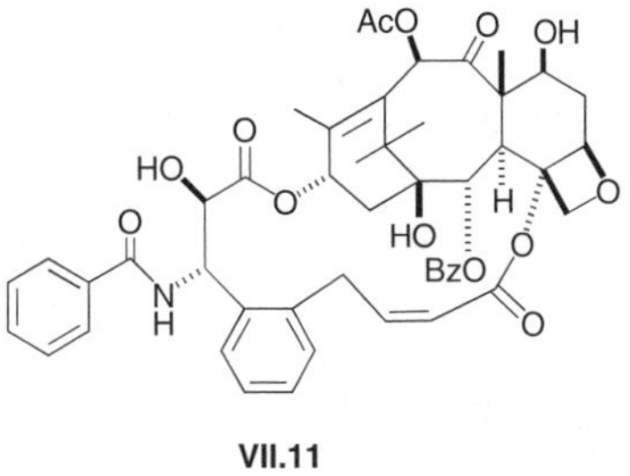

VII.11

SCHEME 6.38

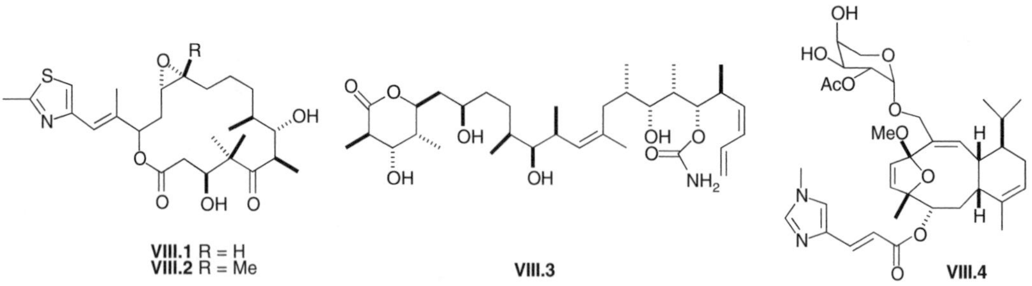

VIII.1 R = H
VIII.2 R = Me

VIII.3

VIII.4

SCHEME 6.39

eleutherobin **VIII.4**.[205] These epothilones and discodermolide are discussed in more detail in separate chapters in this volume. Other compounds with similar mechanisms of action include rhazinilam,[206] which inhibits the disassembly of microtubules by a different mechanism than taxol, laulimalide and isolaulimalide,[207] WS9885B,[208] polyisoprenylated benzophenones related to guttiferone E,[209] jatrophane esters,[210] dictyostatin,[211] and even some steroid derivatives[212] and dicoumarol.[213] Several reviews covering this area have appeared.[214–216]

The tubulin-bound conformations of these agents has proved a fruitful field for modeling and computational approaches. Several authors have advanced models based on presumed common pharmacophores,[217–221] and the field has also been reviewed.[222]

IX. CLINICAL APPLICATIONS OF TAXOL AND DOCETAXEL

The preceding discussions have provided a brief and somewhat selective overview of the chemistry and biology of taxol. For most nonscientists, however, the key question is, "How well does it work: What are the clinical benefits of taxol and its sister drug docetaxel to the cancer patient?" It is difficult to give an unqualified answer to this question, in part because treatment modalities are continuing to improve. The use of taxol in cancer therapy has recently been reviewed,[223] and several reviews have also covered the use of taxol or docetaxel for the treatment of specific cancers.[224–228] At present, the taxane drugs taxol and docetaxel are used for the treatment of breast, lung, and ovarian carcinomas and also for AIDS-related Kaposi's sarcoma.

The taxanes have proven to be among the most active agents in treating advanced metastatic breast cancer,[225] but their benefits in treating early-stage breast cancer have only recently been evaluated. Two recent reviews have now documented their value, however. One reviewer states, "The evidence is now clear that taxanes added to standard adjuvant regimens … can improve outcomes for patients with breast cancer,"[229] and a second systematic review of taxane versus nontaxane regimens for treatment of early breast cancer concludes, "The results of this systematic review support the use of taxanes as adjuvant chemotherapy for women with early breast cancer and involved lymph nodes." In the studies summarized in this review, the 5-yr relapse-free survival rate was 74% for the control (nontaxane) group versus 79% for the taxane-treated group.[230]

In the lung cancer area docetaxel is the drug of choice for the treatment of advanced non–small cell lung cancer that is refractory to primary chemotherapy.[226] A study to compare the efficacy of three different treatment regimens (cisplatin and gemcitabine, cisplatin and docetaxel, or carboplatin and taxol) found that the response rates and survival did not differ significantly between the regimens.[224]

The significance of taxol as an anticancer drug was first discovered for ovarian cancer, and this remains an important target. A European–Canadian study confirmed early findings that the combination of taxol–cisplatin is superior to the previously used regimen of cyclophosphamide–cisplatin,[231] and that this is now the standard of care for women with advanced ovarian cancer.

Although the taxanes are certainly not the wonder drugs they were at one time thought to be, they have nevertheless brought significant benefits to many cancer patients. The new analogs referred to in the following section promise real improvements in efficacy over taxol and docetaxel, and improvements in drug delivery will also play a major role in improving treatment. In these connections, the recent announcement of the successful phase III trial of the albumin nanoparticle-based formulations of taxol ABI-007[232] shows that improved formulations can have a dramatic effect. The overall response rate for ABI-007 in patients with metastatic breast cancer was 33%, compared with 19% for taxol. It is thus clear that taxol and its analogs will continue to play an important role in cancer chemotherapy well into the twenty-first century.

X. TAXOL ANALOGS IN CLINICAL TRIALS

As described in section V, numerous taxol analogs have been prepared, and several have entered preclinical development or clinical trial. An extensive listing of taxol analogs in development is contained in an excellent recent review,[116] and readers are referred to this review for a listing of taxol analogs in preclinical development.

Six taxol analogs are currently in phase II clinical trial. They are the simple C7 acyl derivatives 7-DHA-taxol (Taxoprexin, **V.C.1**),[88] 7-hexanoyltaxol (**V.C.2**),[88] the C7 thioether **V.C.3** (BMS-184476),[89–91] the cyclopropyl derivative **V.C.14** (RPR-109881A),[233] the C4 carbonate **V.C.16** (BMS-188797),[107] and the 7,10-dimethyl ether **X.1** (TXD258).[234] Four additional taxol analogs are in phase I clinical trial. These are ortataxel (**V.A.10**),[56] TL-00139 (**V.E.5**),[123] DJ-927 (**V.E.6**),[124] and BMS-275183 (**X.2**). A further 23 taxol analogs are in preclinical development; their structures are given in the review cited above.

XI. CONCLUSION

The botanists who collected the bark of *T. brevifolia* in 1962 could not have foreseen the enormous consequences that would flow from that simple act. Providentially, their action initiated a chain of events that has led to the present situation, in which the one lead compound of taxol has not only yielded two currently approved clinical agents but led to the development of 33 compounds that are either in clinical trial or advanced preclinical development. It is certain that some of these compounds will advance to clinical use, and taxol will thus be joined by a number of related compounds in providing clinical anticancer agents. In addition, the mechanism of action of taxol, which was unique when it was first discovered, has led to the discovery of several new natural and synthetic compounds with the same mechanism, as described in section VIII. In a real sense, taxol can thus be viewed as the progenitor of an even larger number of potential clinical agents. It is almost a certainty that some of them will turn out to have significantly better clinical profiles than taxol itself, with significant life extensions and even cures in some cases. That botanical collection in 1962, coupled with the skill and expertise of numerous scientists, has thus made a huge contribution to cancer treatment and must surely rank as the most significant such collections ever made!

ACKNOWLEDGMENTS

John Donne wrote that "No man is an island," and this is certainly true of all scientists who have worked on taxol, as we all build on the knowledge and results of others in the area. The author is particularly grateful to his colleagues at the National Cancer Institute, and especially to the late Matthew Suffness for his early support of work on taxol when few others believed that it would ever become a clinically used drug. More recently, Gordon Cragg and David Newman, the coeditors of this volume, and Yali Hallock have provided continued support and encouragement.

The author's work reported herein was supported by grants from BMS and the National Cancer Institute, most recently under grant CA 69571, and such support is gratefully acknowledged. The author also acknowledges with gratitude the collaboration of his colleagues Susan Bane (SUNY Binghamton), Jake Schaefer (Washington University), and Jim Snyder (Emory University) and of the talented graduate students and research associates whose names are recorded in the publications from his group.

REFERENCES

1. Wani, M.C. et al., Plant antitumor agents. VI. The isolation and structure of taxol, a novel antileukemic and antitumor agent from *Taxus brevifolia*, *J. Am. Chem. Soc.*, 93, 2325, 1971.
2. Itokawa, H. Introduction, in *Taxus: The Genus Taxus*, Itokawa, H. and Lee, K.-H., Eds. Taylor & Francis, London, 2003, 1.
3. Schiff, P.B., Fant, J., and Horwitz, S.B., Promotion of microtubule assembly *in vitro* by taxol, *Nature*, 277, 665, 1979.
4. Weiss, R.B. et al., Hypersensitivity reactions from taxol, *J. Clin. Oncol.*, 8, 1263, 1990.
5. Rose, W.C. Preclinical Antitumor Activity of Taxanes, in *Taxol, Science and Applications,* Suffness, M. Ed., CRC Press, Inc., Boca Raton, FL, 1995, chap. 8.
6. McGuire, W.P. et al., Taxol: a unique antineoplastic agent with significant activity in advanced ovarian epithelial neoplasms, *Ann. Intern. Med.*, 111, 273, 1989.
7. Holmes, F.A. et al., Phase II trial of taxol, an active drug in the treatment of metastatic breast cancer, *J. Nat. Cancer Inst.*, 83, 1797, 1991.
8. Chase, M., Cancer drug may save many human lives — at cost of rare trees, *The Wall Street Journal*, 217 (69), A1, A8, 1991.
9. Chauvière G. et al., Analyse structurale et etude biochimique de produits isoles de l'if: *Taxus baccata* L. (Taxacees), *C. R. Acad. Sci. Paris, Ser. II*, 293, 501, 1981.
10. Denis, J.-N. et al., A highly efficient, practical approach to natural taxol, *J. Am. Chem. Soc.*, 110, 5917, 1988.
11. Holton, R.A., Biediger, R.J., and Boatman, P.D. Semisynthesis of taxol and taxotere. In *Taxol, Science and Applications,* Suffness, M., Ed., CRC Press, Inc., Boca Raton, FL, 1995, chap. 5.
12. Guèritte-Voegelein, F. et al., Chemical studies of 10-deacetyl baccatin III. Hemisynthesis of taxol derivatives, *Tetrahedron*, 42, 4451, 1986.
13. Bissery, M.-C. et al., Docetaxel (Taxotere): a review of preclinical and clinical experience. Part I: preclinical experience, *Anti-Cancer Drugs,* 6, 339, 1995.
14. Guèritte, F. General and recent aspects of the chemistry and structure-activity relationships of taxoids, *Curr. Pharm. Design,* 7, 1229, 2001.
15. Takeya, K.; Plant tissue culture of taxoids, in *Taxus: The Genus Taxus*, Itokawa, H. and Lee, K.-H., Eds., Taylor and Francis, London, 2003, chap. 5.
16. Anonymous. Sustaining a responsible pipeline of new products. Bristol-Myers Squibb 2003 Sustainability Report http://www.bms.com/static/ehs/report/data/s03p25.html, Accessed May 13, 2004.
17. Rohmer, M. et al., Isoprenoid biosynthesis in bacteria: a novel pathway for the early steps leading to isopentenyl diphosphate, *Biochem. J.,* 295, 517, 1993.
18. Eisenreich, W. et al., Studies on the biosynthesis of taxol: the taxane carbon skeleton is not of mevalonoid origin, *Proc. Natl. Acad. Sci. USA,* 93, 6431, 1996.
19. Koepp, A.E. et al., Cyclization of geranylgeranyl diphosphate to taxa-4(5),11(12)-diene is the committed step of taxol biosynthesis in Pacific yew, *J. Biol. Chem.,* 270, 8686, 1995.
20. Hezari, M., Lewis, N.G., and Croteau, R., Purification and characterization of taxa-4(5),11(12)-diene synthase from Pacific yew (*Taxus brevifolia)* that catalyzes the first committed step of taxol biosynthesis, *Arch. Biochem. Biophys.,* 322, 437, 1995.
21. Wildung, M.R. and Croteau, R., A cDNA clone for taxadiene synthase, the diterpene cyclase that catalyzes the committed step of taxol biosynthesis, *J. Biol. Chem.,* 271, 9201, 1996.
22. Hezari, M. et al., Taxol production and taxadiene synthase activity in *Taxus canadensis* cell suspension cultures, *Arch. Biochem. Biophys.,* 337, 185, 1997.

23. Lin, X. et al., Mechanism of taxadiene synthase, a diterpene cyclase that catalyzes the first step of taxol biosynthesis in Pacific yew, *Biochemistry,* 35, 2968, 1996.

24. Floss, H.G. and Mocek, U., Biosynthesis of taxol, in *Taxol: Science and Applications*, Suffness, M., Ed., CRC Press, Inc., Boca Raton, FL, 1995, chap. 7.

25. Hefner, J. et al., Cytochrome P450-catalyzed hydroxylation of taxa-4(5),11(12)-diene to taxa-4(20),11(12)-dien-5α-ol: the first oxygenation step in taxol biosynthesis, *Chem. Biol.*, 3, 479, 1996.

26. Walker, K. et al., Partial purification and characterization of acetyl coenzyme A: taxa-4(20),11(12)-dien-5α-ol O-acetyl transferase that catalyzes the first acylation step of taxol biosynthesis, *Arch. Biochem. Biophys.*, 364, 273, 1999.

27. Wheeler, A.L. et al., Taxol biosynthesis: differential transformations of taxadien-5α-ol and its acetate ester by cytochrome P450 hydroxylases from *Taxus* suspension cells, *Arch. Biochem. Biophys.*, 390, 265, 2001.

28. Horiguchi, T. et al., Studies on taxol biosynthesis. Preparation of 5α-acetoxytaxa-4(20),11-dien-2α,10β-diol derivatives by deoxygenation of a taxadiene tetra-acetate obtained from Japanese yew, *Tetrahedron,* 59, 267, 2003.

29. Chau, M. et al., Taxol biosynthesis: molecular cloning and characterization of a cytochrome P450 taxoid 7β-hydroxylase, *Chem. Biol.*, 11, 663, 2004.

30. Jennewein, S. et al., Taxol biosynthesis: taxane 13α-hydroxylase is a cytochrome P450-dependent monooxygenase, *Proc. Natl. Acad. Sci. USA*, 98, 13595, 2001.

31. Jennewein, S. et al., Random sequencing of an induced *Taxus* cell cDNA library for identification of clones involved in taxol biosynthesis, *Proc. Natl. Acad. Sci. USA,* 101, 9149, 2004.

32. Fleming, P.E., Mocek, U., and Floss, H.G., Biosynthesis of taxoids. Mode of formation of the taxol side chain, *J. Am. Chem. Soc.*, 115, 805, 1993.

33. Fleming, P.E. et al., Biosynthetic studies on taxol, *Pure Appl. Chem.*, 66, 2045, 1994.

34. Fleming, P.E. et al., Biosynthesis of taxoids. Mode of attachment of the taxol side chain, *J. Am. Chem. Soc.*, 116, 4137, 1994.

35. Zhong, J.-J., Plant cell culture for production of paclitaxel and other taxanes, *J. Biosci. Bioeng.*, 94, 591, 2002.

36. Tabata, H., Paclitaxel production by plant-cell-culture technology. *Adv. Biochem. Engin./Biotechnol.,* 7, 1, 2004.

37. Yukimune, Y. et al., Methyl jasmonate-induced overproduction of paclitaxol and baccatin III in *Taxus* cell suspension cultures, *Nat. Biotechnol.*, 14, 1129, 1996.

38. Genexol. Genexol: The World's Best Paclitaxel. http://www.genexol.com/eng/app//manufacture/default.asp. Accessed Feb. 21, 2005.

39. Stierle, A., Strobel, G., and Stierle, D., Taxol and taxane production by *Taxomyces andreanae*, an endophytic fungus of Pacific yew, *Science*, 260, 214, 1993.

40. Li, J.Y. et al., The induction of taxol production in the endophytic fungus — *Periconia* sp. from *Torreya grandifolia*, *J. Ind. Micro. Biotechnol.*, 20, 259, 1998.

41. Strobel, G. et al., Natural products from endophytic microorganisms, *J. Nat. Prod.*, 67, 257, 2004.

42. Schiff, P. B. and Horwitz, S. B., Taxol stabilizes microtubules in mouse fibroblast cells, *Proc. Natl. Acad. Sci. USA,* 77, 1561, 1980

43. Jordan, M.A. and Wilson, L. Microtubules and actin filaments: dynamic targets for cancer chemotherapy, *Curr. Opin. Cell Biol.,* 10, 123, 1998.

44. Jordan, M.A. et al., Mechanism of mitotic block and inhibition of cell proliferation by taxol at low concentrations, *Proc. Natl. Acad. Sci. USA*, 90, 9552, 1993.

45. Long, B.H. and Fairchild, C.R., Paclitaxel inhibits progression of mitotic cells to G_1 phase by interference with spindle formation without affecting other microtubule functions during anaphase and telephase, *Cancer Res.*, 54, 4355, 1994.

46. Haldar, S., Chintapalli, J., and Croce, C. M., Paclitaxel induces Bcl-2 phosphorylation and death of prostate cancer cells, *Cancer Res.*, 56, 1253, 1996.

47. Blagosklonny, M.V. et al., Raf-1/bcl-2 phosphorylation: a step from microtubule damage to cell death, *Cancer Res.*, 57, 130, 1997.

48. Rodi, D.J. et al., Screening of a library of phage-displayed peptides identifies human bcl-2 as a taxol-binding protein, *J. Mol. Biol.*, 285, 197, 1999.

49. Carboni, J. M., Singh, C. and Tepper, M. A., Taxol and lipopolysaccharide activation of a murine macrophage cell line and induction of similar tyrosine phosphoproteins, *J. Natl. Cancer. Inst. Monogr.*, 15, 95, 1993.
50. Blagosklonny, M.V. and Fojo, T., Molecular effects of paclitaxel: Myths and reality (a critical review), *Int. J. Cancer*, 83, 151, 1999.
51. Abal, M., Andreu, J.M., and Barasoain, I., Taxanes: Microtubule and centrosome targets, and cell cycle dependent mechanisms of action, *Curr. Cancer Drug Targets*, 3, 193, 2003.
52. Wang, X., Itokawa, H., and Lee, K.-H., Structure-activity relationships of taxoids, in *Taxus: The Genus Taxus*, Itokawa, H. and Lee, K.-H., Eds. Taylor & Francis, London, 2003, 298.
53. Kingston, D. G. I. et al., The chemistry of taxol and related taxoids, in *Progress in the Chemistry of Organic Natural Products Vol. 84*, Herz, W. Falk, H., and Kirby, G. W., Eds., Springer, Vienna, 2002, 53–224.
54. Kingston, D. G. I., Taxol, a molecule for all seasons, *Chem. Commun.*, 867, 2001.
55. Ojima, I. et al., New generation taxoids and hybrids of microtubule-stabilizing anticancer agents, in *ACS Symposium Series 796*, I. Ojima, Altmann K.-H., Eds., American Chemical Society, Washington, DC, 2001, chap. 4.
56. Ojima, I. et al., Medicinal chemistry and chemical biology of new generation taxane antitumor agents, *Life*, 53, 269, 2002.
57. Guèritte, F., General and recent aspects of the chemistry and structure-activity relationships of taxoids, *Curr. Pharmaceutical Design*, 7, 1229, 2001.
58. Samaranayake, G., Neidigh, K.A., and Kingston, D.G.I., Modified taxols. 8. Deacylation and reacylation of baccatin III, *J. Nat. Prod.*, 56, 884, 1993.
59. Harriman, G.C.B. et al., The chemistry of the taxane diterpene: stereoselective synthesis of 10-deacetoxy-11,12-epoxypaclitaxel, *Tetrahedron Lett.*, 36, 8909, 1995.
60. Chen, S.-H. et al., Taxol structure-activity relationships: synthesis and biological evaluation of 10-deoxytaxol, *J. Org. Chem.*, 58, 2927, 1993.
61. Kelly, R.C. et al., 12-13-Isobaccatin III. Taxane enol esters (12,13-isotaxanes), *J. Am. Chem. Soc.*, 118, 919, 1996.
62. Uoto, K. et al., First synthesis and cytotoxic activity of novel docetaxel analogs modified at the C-18-position, *Bioorg. Med. Chem. Lett.*, 7, 2991, 1997.
63. Samaranayake, G. et al., Modified taxols. 5. Reaction of taxol with electrophilic reagents, and preparation of a rearranged taxol derivative with tubulin assembly activity, *J. Org. Chem.*, 56, 5114, 1991.
64. Appendino, G. et al., 14β-Hydroxy-10-deacetylbaccatin III, a new taxane from Himalayan yew (Taxus wallichiana Zucc.), *J. Chem. Soc., Perkin Trans.*, 1, 21, 2925, 1992.
65. Ojima, I. et al., Structure-activity relationships of new taxoids derived from 14β-hydroxy-10-deacetyl-baccatin III, *J. Med. Chem.*, 37, 1408, 1994.
66. Ojima, I. et al., Syntheses and structure-activity relationships of novel nor-seco taxoids, *J. Org. Chem.*, 63, 1637, 1998.
67. Johnson, R.A. et al., Taxol chemistry. 7-O-triflates as precursors to olefins and cyclopropanes, *Tetrahedron Lett.*, 35, 7893, 1994.
68. Zheng, Q.Y. et al., Deacetylation of paclitaxel and other taxanes, *Tetrahedron Lett.*, 36, 2001, 1995.
69. Holton, R.A., Somoza, C., and Chai, K.-B., A simple synthesis of 10-deacetoxytaxol derivatives, *Tetrahedron Lett.*, 35, 1665, 1994.
70. Georg, G.I. and Cheruvallath, Z.S., Samarium diiodide-mediated deoxygenation of taxol: a one-step synthesis of 10-deacetoxytaxol, *J. Org. Chem.*, 59, 4015, 1994.
71. Chaudhary, A.G. and Kingston, D.G.I., Synthesis of 10-deacetoxytaxol and 10-deoxytaxotere, *Tetrahedron Lett.*, 34, 4921, 1993.
72. Ojima, I. et al., Syntheses and structure-activity relationships of the second-generation antitumor taxoids: exceptional activity against drug-resistant cancer cells, *J. Med. Chem.*, 39, 3889, 1996.
73. Klein, L.L., Synthesis of 9-dihydrotaxol: a novel bioactive taxane, *Tetrahedron Lett.*, 34, 2047, 1993.
74. Li, L. et al., Synthesis and biological evaluation of C-3-modified analogs of 9(R)-dihydrotaxol, *J. Med. Chem.*, 37, 2655, 1994.
75. Klein, L.L. et al., Antitumor activity of 9(R)-dihydrotaxane analogs, *J. Med. Chem.*, 38, 1482, 1995.
76. Klein, L. L., Yeung, C. M., and Li, L., 9-Deoxotaxanes, U.S. patent 5,440,056, August 8, 1995.

77. Chen, S.-H., Wei, J.-M., and Farina, V., Taxol structure-activity relationships: Synthesis and biological evaluation of 2-deoxytaxol, *Tetrahedron Lett.*, 34, 3205, 1993.

78. Chaudhary, A.G., Chordia, M.D., and Kingston, D.G.I., A novel benzoyl group migration: synthesis and biological evaluation of 1-benzoyl-2-des(benzoyloxy)paclitaxel, *J. Org. Chem.*, 60, 3260, 1995.

79. Chordia, M.D. and Kingston, D.G.I., Synthesis and biological evaluation of 2-*epi*-paclitaxel, *J. Org. Chem.*, 61, 799, 1996.

80. Fang, W.-S., Fang, Q.-C., and Liang, X.-T., Synthesis of the 2α-benzoylamido analogue of docetaxel, *Tetrahedron Lett.*, 42, 1331, 2001.

81. Chaudhary, A.G. et al., Unexpectedly facile hydrolysis of the 2-benzoate group of taxol and synthesis of analogs with increased activities, *J. Am. Chem. Soc.*, 116, 4097, 1994.

82. Kingston, D.G.I. et al., Synthesis and biological evaluation of 2-acyl analogues of paclitaxel (taxol), *J. Med. Chem.*, 41, 3715, 1998.

83. Nicolaou, K.C. et al., Chemical synthesis and biological evaluation of C-2 taxoids, *J. Am. Chem. Soc.*, 117, 2409, 1995.

84. Georg, G.I. et al., Synthesis of 2-O-heteroaroyl taxanes: Evaluation of microtubule assembly promotion and cytotoxicity, *Bioorg. Med. Chem. Lett.*, 5, 115, 1995.

85. Ojima, I. et al., Syntheses and structure-activity relationships of nonaromatic taxoids: Effects of alkyl and alkenyl ester groups on cytotoxicity, *J. Med. Chem.*, 40, 279, 1997.

86. Kingston, D.G.I., Chordia, M.D., and Jagtap, P.G., Synthesis and biological evaluation of 1-deoxy-paclitaxel analogues, *J. Org. Chem.*, 64, 1814, 1999.

87. Bradley, M. O. et al., Preparation of conjugates of all-cis-docosahexaenoic acid and paclitaxel as antitumor agents. PCT Int. Appl. WO 9,744,336, 1997.

88. Ali, S. et al., Hydrolyzable hydrophobic taxanes: synthesis and anti-cancer activities, *Anti-Cancer Drugs,* 12, 117, 2001.

89. Altstadt, T.J. et al., Synthesis and antitumor activity of novel C-7 paclitaxel ethers: discovery of BMS-184476, *J. Med. Chem.*, 44, 4577, 2001.

90. Ojima, I. and Geney, R., BMS-184476 (Bristol-Myers Squibb), *Curr. Opin. Invest. Drugs,* 4, 732, 2003.

91. Plummer, R. et al., Phase I and pharmacokinetic study of the new taxane analog BMS-184476 given weekly in patients with advanced malignancies, *Clin. Cancer Res.*, 8, 2788, 2002.

92. Magri, N.F. and Kingston, D.G.I., Modified taxols. 2. Oxidation products of taxol, *J. Org. Chem.*, 51, 797, 1986.

93. Chaudhary, A.G., Rimoldi, J.M., and Kingston, D.G.I., Modified taxols. 10. Preparation of 7-deoxytaxol, a highly bioactive taxol derivative, and interconversions of taxol and 7-*epi*-taxol, *J. Org. Chem.*, 58, 3798, 1993.

94. Chen, S.-H. et al., Synthesis of 7-deoxy- and 7,10-dideoxytaxol via radical intermediates, *J. Org. Chem.*, 58, 5028, 1993.

95. Poujol, H. et al., Taxoides: 7-deshydroxy-10-acetyldocetaxel et nouveaux analogues prepares a partir des alcaloides de l'if, *Tetrahedron*, 53, 12575, 1997.

96. Johnson, R.A. et al., Taxol chemistry. 7-O-triflates as precursors to olefins and cyclopropanes, *Tetrahedron Lett.*, 35, 7893, 1994.

97. Liang, X. et al., Synthesis and biological evaluation of paclitaxel analogs modified in ring C, *Tetrahedron Lett.*, 36, 2901, 1995.

98. Yuan, H. and Kingston, D.G.I., Synthesis of 6α-hydroxypaclitaxel, the major human metabolite of paclitaxel, *Tetrahedron Lett.*, 39, 4967, 1998.

99. Wittman, M.D., Kadow, J.F., and Vyas, D.M., Stereospecific synthesis of the major human metabolite of paclitaxel, *Tetrahedron Lett.*, 41, 4729, 2000.

100. Liang, X. et al., Paclitaxel analogs modified in ring C: Synthesis and biological evaluation, *Tetrahedron*, 53, 3441, 1997.

101. Chen, S.-H. et al., Serendipitous synthesis of a cyclopropane-containing taxol analog via anchimeric participation of an unactivated angular methyl group, *J. Org. Chem.*, 58, 4520, 1993.

102. Bouchard, H. et al., Improved access to 19-nor-7β,8β-methylene-taxoids and formation of a 7-membered C-ring analog of docetaxel by electrochemistry, *Tetrahedron Lett.*, 35, 9713, 1994.

103. Appendino, G. et al., Synthesis and evaluation of C-*seco* paclitaxel analogues, *Tetrahedron Lett.*, 38, 4273, 1997.

104. Pratesi, G. et al., IDN 5390: an oral taxane candidate for protracted treatment schedules, *Br. J. Cancer,* 88, 965, 2003.

105. Appendino, G. et al., Structure-activity relationships of ring C-secotaxoids. 1. Acylative modifications, *J. Nat. Prod.*, 67, 184, 2004.

106. Neidigh, K.A. et al., Synthesis and biological evaluation of 4-deacetylpaclitaxel, *Tetrahedron Lett.*, 35, 6839, 1994.

107. Chen, S.-H., Discovery of a novel C-4 modified 2nd generation paclitaxel analog BMS-188797, *Frontiers Biotechnol. Pharmaceut.*, 3, 157, 2002.

108. Mastalerz, H. et al., The discovery of BMS-275183: an orally efficacious novel taxane, *Bioorg. Med. Chem.* 11, 4315, 2003.

109. Chordia, M.D. et al., Synthesis and bioactivity of 2,4-diacyl analogues of paclitaxel, *Bioorg. Med. Chem.*, 9, 171, 2001.

110. Marder-Karsenti, R. et al., Synthesis and biological evaluation of D-ring-modified taxanes: 5(20)-azadocetaxel analogs, *J. Org. Chem.*, 62, 6631, 1997.

111. Gunatilaka, A.A.L. et al., Synthesis and biological evaluation of novel paclitaxel (taxol) D-ring modified analogues, *J. Org. Chem.*, 64, 2694, 1999.

112. Wang, M. et al., The oxetane ring in taxol, *J. Org. Chem.*, 65, 1059, 2000.

113. Dubois, J. et al., Synthesis of 5(20)deoxydocetaxel, a new active docetaxel analogue, *Tetrahedron Lett.*, 41, 3331, 2000.

114. Deka, V. et al., Deletion of the oxetane ring in docetaxel analogues: synthesis and biological evaluation, *Org. Lett.*, 5, 5031, 2003.

115. Barboni, L. et al., Novel D-seco paclitaxel analogues: synthesis, biological evaluation, and model testing, *J. Org. Chem.*, 66, 3321, 2001.

116. Cragg, G.M. and Newman, D.J., A tale of two tumor targets: topoisomerase I and tubulin. The Wall and Wani contribution to cancer chemotherapy, *J. Nat. Prod.*, 67, 232, 2004.

117. He, L. et al., A common pharmacophore for taxol and the epothilones based on the biological activity of a taxane molecule lacking a C-13 side chain, *Biochemistry*, 39, 3972, 2000.

118. Swindell, C.S. et al., Biologically active taxol analogues with deleted A-ring side chain substituents and variable C-2 configurations, *J. Med. Chem.*, 34, 1176, 1991.

119. Guèritte-Voegelein, F. et al., Relationships between the structure of taxol analogues and their antimitotic activity, *J. Med. Chem.*, 34, 992, 1991.

120. Barboni, L. et al., Synthesis and NMR-driven conformational analysis of taxol analogues conformationally constrained on the C13 side chain, *J. Med. Chem.*, 44, 1576, 2001.

121. Jayasinghe, L.R. et al., Structure-activity studies of antitumor taxanes: synthesis of novel C-13 side chain homologated taxol and taxotere analogs, *J. Med. Chem.*, 37, 2981, 1994.

122. Georg, G.I. et al., Synthesis, conformational analysis, and biological evaluation of heteroaromatic taxanes, *J. Org. Chem.*, 61, 2664, 1996.

123. Sampath, D. et al., MAC-321, a novel taxane with greater efficacy than paclitaxel and docetaxel *in vitro* and *in vivo*, *Mol. Cancer Ther.*, 873, 2003.

124. Shionoya, M. et al., DJ-927, a novel oral taxane, overcomes P-glycoprotein-mediated multidrug resistance *in vitro* and *in vivo*, *Cancer Sci.* 94, 459, 2003.

125. Rose, W.C. et al., Preclinical pharmacology of BMS-275183, an orally active taxane, *Clin. Cancer Res.*, 7, 2016, 2001.

126. Magri, N.F. and Kingston, D.G.I., Modified taxols. 4. Synthesis and biological activity of taxols modified in the side chain, *J. Nat. Prod.,* 51, 298, 1988.

127. Denis, J.-N. et al., Docetaxel (taxotere) derivatives: novel NbCl₃-based stereoselective approach to 2-methyldocetaxel, *J. Chem. Soc. Perkin Trans.* 1, 1811, 1995.

128. Kant, J. et al., Diastereoselective addition of Grignard reagents to azetidine-2,3-dione: Synthesis of novel taxol analogues, *Tetrahedron Lett.*, 37, 6495, 1996.

129. Ojima, I., Wang, T., and Delaloge, F., Extremely stereoselective alkylation of 3-siloxy-β-lactams and its applications to the asymmetric syntheses of novel 2-alkylisoserines, their dipeptides, and taxoids, *Tetrhedron Lett.*, 39, 3663, 1998.

130. Battaglia, A. et al., Synthesis and biological evaluation of 2-methyl taxoids derived from baccatin III and 14β-OH-baccatin II 1,14-carbonate, *J. Med. Chem.*, 46, 4822, 2003.

131. Qi, X. et al., Synthesis of novel thiol surrogate of taxol: 2-deoxy-2-mercaptopaclitaxel, *Tetrahedron,* 60, 3599, 2004.

132. Uoto, K. et al., Synthesis and structure-activity relationships of novel 2,2-difluoro analogues of docetaxel, *Chem. Pharm. Bull.,* 45, 1793, 1997.

133. Yamaguchi, T. et al., Synthesis of taxoids 5. Synthesis and evaluation of novel water-soluble prodrugs of a 3-desphenyl-3-cyclopropyl analogue of docetaxel, *Bioorg. Med. Chem. Lett.,* 9, 1639, 1999.

134. Georg, G.I. et al., The medicinal chemistry of taxol, in *Taxol: Science and Applications,* Suffness, M., Ed., CRC Press, Inc., Boca Raton, FL, 1995, chap. 13.

135. Deutsch, H.M. et al., Synthesis of congeners and prodrugs. 3. Water-soluble prodrugs of taxol with potent antitumor activity, *J. Med. Chem.,* 32, 788, 1989.

136. Zhao, Z., Kingston, D.G.I., and Crosswell, A.R., Modified taxols, 6. Preparation of water-soluble prodrugs of taxol, *J. Nat. Prod.,* 54, 1607, 1991.

137. Mathew, A.E. et al., Synthesis and evaluation of some water-soluble prodrugs and derivatives of taxol with antitumor activity, *J. Med. Chem.,* 35, 145, 1992.

138. Ueda, Y. et al., Novel water soluble phosphate prodrugs of taxol possessing *in vivo* antitumor activity, *Bioorg. Med. Chem. Lett.,* 3, 1761, 1993.

139. Hayashi, Y. et al., A novel approach of water-soluble paclitaxel prodrug with no auxiliary and no byproduct: design and synthesis of isotaxel, *J. Med. Chem.,* 46, 3782, 2003.

140. Skwarczynski, M. et al., O-N intramolecular acyl migration strategy in water-soluble prodrugs of taxoids, *Bioorg. Med. Chem.,* 13, 4441, 2003.

141. Leenders, R.G.G. et al., β-glucuronyl carbamate based pro-moieties designed for prodrugs in ADEPT, *Tetrahedron Lett.,* 36, 1701, 1995.

142. de Bont, D.B.A. et al., Synthesis and biological activity of β-glucuronyl carbamate-based prodrugs of paclitaxel as potential candidates for ADEPT, *Bioorg. Med. Chem.* 5, 405, 1997.

143. Schmidt, F. et al., Cancer chemotherapy: a paclitaxel prodrug for ADEPT (antibody-directed enzyme prodrug therapy), *Eur. J. Org. Chem.,* 2129, 2001.

144. Bouvier, E. et al., First enzymatically activated Taxotere prodrugs designed for ADEPT and PMT, *Biorg. Med. Chem.,* 12, 969, 2004.

145. de Groot, F.M.H. et al., Synthesis and biological evaluation of 2-carbamate-linked and 2-carbonate-linked prodrugs of paclitaxel: selective activation by the tumor-associated protease plasmin, *J. Med. Chem.,* 43, 3093, 2000.

146. Takahashi, T., Tsukamoto, H., and Yamada, H., Design and synthesis of a water-soluble taxol analog: taxol-sialyl conjugate, *Bioorg. Med. Chem. Lett.,* 8, 113, 1998.

147. Luo, Y. and Prestwich, G.D., Synthesis and selective cytotoxicity of a hyaluronic acid-antitumor bioconjugate, *Bioconjugate Chem.,* 10, 755, 1999.

148. Bradley, M.O. et al., Tumor targeting by conjugation of DHA to paclitaxel, *J. Controlled Release,* 74, 233, 2001.

149. Bradley, M.O. et al., Tumor targeting by covalent conjugation of a natural fatty acid to paclitaxel, *Clin. Cancer Res.,* 7, 3229, 2001.

150. Lee, J.W. et al., Synthesis and evaluation of taxol-folic acid conjugates as targeted antineoplastics, *Bioorg. Med. Chem.,* 10, 2397, 2002.

151. Kirschberg, T.A. et al., Arginin-based molecular transporters: the synthesis and chemical evaluation of releasable taxol-transporter conjugates, *Org. Lett.,* 5, 3459, 2003.

152. Guillemard, V. and Saragovi, H.U., Taxane-antibody conjugates afford potent cytotoxicity, enhanced solubility, and tumor target selectivity, *Cancer Res.,* 61, 694, 2001.

153. Ojima, I. et al., Tumor-specific novel taxoid-monoclonal antibody conjugates, *J. Med. Chem.,* 45, 5620, 2002.

154. Wuts, P.G.M., Semisynthesis of taxol, *Curr. Opin. Drug Disc. Dev.,* 1, 329, 1998.

155. Ojima, I. et al., New and efficient approaches to the semisynthesis of taxol and its C-13 side chain analogs by means of β-lactam synthon method, *Tetrahedron,* 48, 6985, 1992.

156. Ojima, I. et al., Efficient and practical asymmetric synthesis of the taxol C-13 side chain, N-benzoyl-(2R,3S)-3-phenylisoserine, and its analogues via chiral 3-hydroxy-4-aryl-β-lactams through chiral ester enolate-imine cycl, *J. Org. Chem.,* 56, 1681, 1991.

157. Georg, G.I. et al., An efficient semisynthesis of taxol from (3*R*,4*S*)-*N*-benzoyl-3-[(*t*-butyldimethylsi-lyl)oxy]-4-phenyl-2-azetidinone and 7-(triethylsilyl)baccatin III, *Bioorg. Med. Chem. Lett.*, 3, 2467, 1993.

158. Palomo, C. et al., Asymmetric synthesis of α-keto β-lactams via [2+2] cycloaddition reaction — a concise approach to optically active α-hydroxy β-lactams and β-alkyl(aryl)isoserines, *Tetrahedron Lett.*, 34, 6325, 1993.

159. Holton, R.A. and Liu, J.H., A novel asymmetric synthesis of cis-3-hydroxy-4-aryl azetidin-2-ones, *Bioorg. Med. Chem. Lett.*, 3, 2475, 1993.

160. Farina, V., Hauck, S.I., and Walker, D.G., A simple chiral synthesis of the taxol side chain, *Synlett*, 1, 761, 1992.

161. Bourzat, J.D. and Commercon, A., A practical access to chiral phenylisoserinates, preparation of taxotere analogs, *Tetrahedron Lett.*, 34, 6049, 1993.

162. Shimizu, M., Ishida, T., and Fujisawa, T., Highly stereocontrolled construction of 3-alkoxyazetidin-2-ones using ester enolate-imine condensation, *Chem. Lett.*, 1403, 1994.

163. Song, C.E. et al., A new synthetic route to (3*R*,4*S*)-3-hydroxy-4-phenylazetidin-2-one as a taxol side chain precursor, *Tetrahedron: Asymmetry*, 9, 983, 1998.

164. Brieva, R., Crich, J.Z., and Sih, C.J., Chemoenzymatic synthesis of the C-13 side chain of taxol: optically-active 3-hydroxy-4-phenyl β-lactam derivatives, *J. Org. Chem.*, 58, 1068, 1993.

165. Lin, S. et al., Synthesis of highly potent second-generation taxoids through effective kinetic resolution coupling of racemic β-lactams with baccatins, *Chirality*, 12, 431, 2000.

166. Kingston, D.G.I. et al., Synthesis of taxol from baccatin III via an oxazoline intermediate, *Tetrahedron Lett.*, 35, 4483, 1994.

167. Gennari, C. et al., Taxol semisynthesis: a highly enantio- and diastereoselective synthesis of the side chain and a new method for ester formation at C-13 using thioesters, *J. Org. Chem.*, 62, 4746, 1997.

168. Gennari, C. et al., Semisynthesis of taxol: A highly enantio- and diastereoselective synthesis of the side chain and a new method for ester formation at C13 using thioesters, *Angew. Chem. Int. Ed. Engl.*, 35, 1723, 1996.

169. Yamaguchi, T. et al., Synthesis of taxoids 4. Novel and versatile methods for preparation of new taxoids by employing *cis*- or *trans*-phenyl glycidic acid, *Tetrahedron*, 55, 1005, 1999.

170. Mandai, T. et al., A semisynthesis of paclitaxel via a 10-deacetylbaccatin III derivative bearing a β-keto ester appendage, *Tetrahedron Lett.*, 41, 243, 2000.

171. Xiao, Z., Itokawa, H., Lee, and K.-H., Total synthesis of taxoids, in *Taxus: The Genus Taxus,* Itokawa, H. and Lee, K.-H., Eds., Taylor & Francis, London, 2003, chap. 9.

172. Holton, R.A. et al., First total synthesis of taxol. 1. Functionalization of the B ring, *J. Am. Chem. Soc.*, 116, 1597, 1994.

173. Holton, R. A. et al., First total synthesis of taxol. 2. Completion of the C and D rings, *J. Am. Chem. Soc.,* 116, 1599, 1994.

174. Nicolaou, K.C. et al., Total synthesis of taxol, *Nature*, 367, 630, 1994.

175. Danishefsky, S.J. et al., Total synthesis of baccatin III and taxol, *J. Am. Chem. Soc.*, 118, 2843, 1996.

176. Wender, P.A. et al., The pinene path to taxanes. 5. Stereocontrolled synthesis of a versatile taxane precursor, *J. Am. Chem. Soc.*, 119, 2755, 1997.

177. Wender, P.A. et al., The pinene path to taxanes. 6. A concise stereocontrolled synthesis of taxol, *J. Am. Chem. Soc.*, 119, 2757, 1997.

178. Kusama, H. et al., Enantioselective total synthesis of (–)-taxol, *J. Am. Chem. Soc.*, 122, 3811, 2000.

179. Mukaiyama, T. et al., Asymmetric total synthesis of taxol, *Chem. Eur. J.*, 5, 121, 1999.

180. Rao, S. et al., 3-(*p*-azidobenzamido)taxol photolabels the N-terminal 31 amino acids of β-tubulin, *J. Biol. Chem.*, 269, 3132, 1994.

181. Rao, S. et al., Characterization of the taxol binding site on the microtubule, *J. Biol. Chem.*, 270, 20235, 1995.

182. Rao, S. et al., Characterization of the taxol binding site on the microtubule, *J. Biol. Chem.*, 274, 37990, 1999.

183. Dasgupta, D. et al., Synthesis of a photoaffinity taxol analogue and its use in labeling tubulin, *J. Med. Chem.*, 37, 2976, 1994.

184. Loeb, C. et al., [³H](azidophenyl)ureido taxoid photolabels peptide amino acids 281–304 of β-tubulin, *Biochemistry*, 36, 3820, 1997.
185. Lowe, J. et al., Refined structure of β-tubulin at 3.5 Å resolution, *J. Mol. Biol.*, 313, 1045, 2001.
186. Nogales, E., Wolf, S.G., and Downing, K.H., Structure of the αβ-tubulin dimer by electron crystallography, *Nature*, 39, 199, 1998.
187. Snyder, J.P. et al., The binding conformation of taxol in β tubulin: a model based on the electron crystallographic density, *Proc. Natl. Acad. Sci. USA*, 98, 5312, 2001.
188. Ojima, I., Inoue, T., and Chakravarty, S., Enantiopure fluorine-containing taxoids: potent anticancer agents and versatile probes for biomedical problems, *J. Fluorine Chem.*, 97, 3, 1999.
189. Li, Y. et al., Conformation of microtubule-bound paclitaxel determined by fluorescence spectroscopy and REDOR NMR, *Biochemistry*, 39, 281, 2000.
190. Dubois, J. et al., Conformation of taxotere and analogues determined by NMR spectroscopy and molecular modeling studies, *Tetrahedron*, 49, 6533, 1993
191. Williams, H.J. et al., NMR and molecular modeling study of the conformations of taxol and of its side chain methylester in aqueous and non-aqueous solution, *Tetrahedron*, 49, 6545, 1993.
192. Cachau, R.E. et al., Solution structure of taxol determined using a novel feedback-scaling procedure for NOE-restrained molecular dynamics, *Supercomputer Appl. High Performance Comput.*, 8, 24, 1994.
193. Vander Velde, D.G. et al., "Hydrophobic collapse" of taxol and taxotere solution conformations in mixtures of water and organic solvent, *J. Am. Chem. Soc.*, 115, 11650, 1993.
194. Paloma, L.G. et al., Conformation of a water-soluble derivative of taxol in water by 2D-NMR spectroscopy, *Chem. Biol.*, 1, 107, 1994.
195. Ojima, I. et al., A novel approach to the study of solution structures and dynamic behavior of paclitaxel and docetaxel using fluorine-containing analogs as probes, *J. Am. Chem. Soc.*, 119, 5519, 1997.
196. Snyder, J.P. et al., The conformations of taxol in chloroform, *J. Am. Chem. Soc.*, 122, 724, 2000.
197. He, L. et al., A common pharmacophore for taxol and the epothilones based on the biological activity of a taxane molecule lacking a C-13 side chain, *Biochemistry*, 39, 3972, 2000.
198. Boge, T.C. et al., Conformationally restricted paclitaxel analogues: Macrocyclic mimics of the "hydrophobic collapse" conformation, *Bioorg. Med. Chem. Lett.*, 9, 3047, 1999.
199. Ojima, I. et al., A common pharmacophore for cytotoxic natural products that stabilize microtubules, *Proc. Natl. Acad. Sci. USA*, 96, 4256, 1999.
200. Querolle, O. et al., Synthesis of novel macrocyclic docetaxel analogues. Influence of their macrocyclic ring size on tubulin activity, *J. Med. Chem.*, 46, 3623, 2003.
201. Ganesh, T. et al., The bioactive taxol conformation of β-tubulin: experimental evidence from highly active constrained analogs, *Proc. Natl. Acad. Sci. USA*, 101, 10006, 2004.
202. Gerth, K. et al., Antibiotics from gliding bacteria. 74. Epothilons A and B: antifungal and cytotoxic compounds from *Sorangium cellulosum* (Myxobacteria): production, physico-chemical and biological properties, *J. Antibiot.*, 49, 560, 1996.
203. Bollag, D.M. et al., Epothilones, a new class of microtubule-stabilizing agents with a taxol-like mechanism of action, *Cancer Res.*, 55, 2325, 1995.
204. ter Haar, E. et al., Discodermolide, a cytotoxic marine agent that stabilizes microtubules more potently than taxol, *Biochemistry*, 35, 243, 1996.
205. Lindel, T. et al., Eleutherobin, a new cytotoxin that mimics paclitaxel (taxol) by stabilizing microtubules, *J. Am. Chem. Soc.*, 119, 8744, 1997.
206. Dupont, C. et al., D-ring substituted rhazinilam analogues: semisynthesis and evaluation of antitubulin activity, *Bioorg. Med. Chem.*, 7, 2961, 1999.
207. Mooberry, S.L. et al., Laulimalide and isolaulimalide, new paclitaxel-like microtubule-stabilizing agents, *Cancer Res.*, 59, 653, 1999.
208. Vanderwal, C.D. et al., Postulated biogenesis of WS9885B and progress toward an enantioselective synthesis, *Org. Lett.*, 1, 645, 1999.
209. Roux, D. et al., Structure-activity relationship of polyisoprenyl benzophenones from *Garcinia pyrifera* on the tubulin/microtubule system, *J. Nat. Prod.*, 63, 1070, 2000.
210. Miglietta, A. et al., Biological properties of jatrophane polyesters, new microtubule-interacting agents, *Cancer Chemother. Pharmacol.*, 51, 67, 2003.

211. Isbrucker, R.A. et al., Tubulin polymerizing activity of dictyostatin-1, a polyketide of marine sponge origin, *Biochem. Pharmacol.*, 66, 75, 2003.

212. Wang, Z. et al., Synthesis of B-ring homologated estradiol analogues that modulate tubulin polymerization and microtubule stability, *J. Med. Chem.*, 43, 2419, 2000.

213. Madari, H. et al., Dicoumarol: a unique microtubule stabalizing natural product that is synergistic with taxol, *Cancer Res.*, 63, 1214, 2003.

214. Fojo, T. and Giannakakou, P., Taxol and other microtubule-interactive agents, *Curr. Opin. Oncol. Endocrine Metabol. Investig. Drugs*, 2, 293, 2000.

215. Altmann, K.-H., Microtubule-stabilizing agents: a growing class of important anticancer drugs, *Curr. Opin. Chem. Biol.*, 5, 424, 2001.

216. Myles, D.C., Emerging microtubule stabilizing agents for cancer chemotherapy, *Ann. Rep. Med. Chem.*, 37, 125, 2002.

217. Jimenez-Barbero, J., Amat-Guerri, F., and Snyder, J.P., The solid state, solution and tubulin-bound conformations of agents that promote microtubule stabilization, *Curr. Med. Chem. Anti-Cancer Agents*, 2, 91, 2002.

218. Manetti, F. et al., 3D QSAR studies of the interaction between β-tubulin and microtubule stabilizing antimitotic agents (msaa). A combined pharmacophore generation and pseudoreceptor modeling approach applied to taxanes and epothilones, *Il Farmaco*, 58, 357, 2003.

219. Winkler, J.D. and Axelsen, P.H., A model for the taxol (paclitaxel)/epothilone pharmacophore, *Bioorg. Med. Chem. Lett.*, 6, 2963, 1996.

220. Giannakakou, P. et al., A common pharmacophore for epothilone and taxanes: Molecular basis for drug resistance conferred by tubulin mutations in human cancer cells, *Proc. Natl. Acad. Sci. USA*, 97, 2904, 2000.

221. Wang, M. et al., A unified and quantitative receptor model for the microtubule binding of paclitaxel and epothilone, *Org. Lett.*, 1, 43, 1999.

222. He, L., Orr, G.A., and Horwitz, S.B., Novel molecules that interact with microtubules and have functional activity similar to taxol, *Drug Discov. Today*, 6, 1153, 2001.

223. Mekhail, T. and Markman, M., Paclitaxel in cancer therapy, *Expert Opin. Pharmacother.*, 3, 755, 2002.

224. Levin, M., The role of taxanes in breast cancer treatment, *Drugs Today*, 37, 57, 2001.

225. Rowinsky, E.K. The development and clinical utility of the taxane class of antimicrotubule chemotherapy agents. *Ann. Rev. Med.* 48, 353, 1997.

226. Kris, M.G. and Manegold, C., Docetaxel (taxotere) in the treatment of non–small cell lung cancer: an international update, *Sem. Oncol.*, 28, 1, 1, 2001.

227. Michaud, L.B., Valero, V., and Hortobagyi, G., Risks and benefits of taxanes in breast and ovarian cancer, *Drug Safety*, 23, 401, 2000.

228. Calderoni, A. and Cerny, T., Taxanes in lung cancer: a review with focus on the European experience, *Crit. Rev. Oncol./Hematol.*, 38, 105, 2001.

229. Hudis, C., The use of taxanes in early breast cancer, *EJC Suppl.*, 1, 1, 2003.

230. Nowak, A.K. et al., Systematic review of taxane-containing versus non-taxane-containing regimens for adjuvant and neoadjuvant treatment of early breast cancer, *Lancet Oncol.*, 5, 372, 2004.

231. Piccart, M.J. et al., Randomized intergroup trial of cisplatin-paclitaxel versus cisplatin-cyclophosphamide in women with advanced epithelial ovarian cancer: three-year results, *J. Natl. Cancer Inst.*, 92, 699, 2000.

232. Garber, K., Improved paclitaxel formulation hints at new chemotherapy approach, *J. Natl. Cancer Inst.*, 96, 90, 2004.

233. Ojima, I. and Geney, R., 109881 (Aventis), *Curr. Opin. Investig. Drugs*, 4, 737, 2003.

234. Cisternino, S., et al., Nonlinear accumulation in the brain of the new taxoid TXD258 following saturation of P-glycoprotein at the blood-brain barrier in mice and rats. *Br. J. Pharmacol.* 138, 1367, 2003.

7 The Vinca Alkaloids[‡]

Françoise Guéritte and Jacques Fahy

CONTENTS

I. Introduction 123
II. Discovery of the Antitumor Vinca Alkaloids 123
III. Mechanism of Action...124
IV. Semisynthesis and Total Synthesis of Vinca Alkaloids..125
V. Medicinal Chemistry...128
VI. Clinical Applications...131
VII. Conclusions 132
References 132

I. INTRODUCTION

Vinca alkaloids are a family of indole–indoline dimeric compounds coming from the genus Apocynaceae, and they represent one of the most important classes of anticancer agents. Vinca alkaloids became clinically useful in cancer chemotherapy after their discovery in the late 1950s. More than 40 years later, the two natural drugs vinblastine and vincristine are still widely used in cancer chemotherapy, and semisynthetic analogs, such as vindesine and vinorelbine, have been developed after intensive synthetic and structure–activity relationship studies. Other analogs have emerged as potentially potent drugs, and some of these, such as vinflunine, will certainly enrich the family of anticancer vinca alkaloids in the near future. This review summarizes current knowledge about the vinca alkaloids, from their discovery to clinical applications, with an emphasis on the latest analogs that may be of interest in cancer chemotherapy.

II. DISCOVERY OF THE ANTITUMOR VINCA ALKALOIDS[1]

Catharanthus roseus (L.) G. Don (formerly named as *Vinca rosea* L.) is known as the Madagascar periwinkle and belongs to the Apocynaceae family. This perennial plant has dark green leaves and pink or white flowers. It is cultivated in most warm countries as an ornamental in parks and gardens, and many varieties have been developed with various colors. In folkloric medicine, leaves of *C. roseus* were used to treat various diseases, and investigators from different countries of the world have reported hypotensive, hypoglycemic, and purgative properties for this plant.[2] The antitumor properties of *C. roseus* were discovered independently in the 1950s by two teams, one Canadian and one American. In the early 1950s, Clark Noble, one of the first members of the University of Toronto insulin team, sent *C. roseus* leaves used to treat diabetes in Jamaica to his brother, Robert Noble. The challenge for Noble, working in Collip's laboratory at the University of Western Ontario, was to find substances that could affect blood glucose levels. Instead of these properties, Noble found the leaves to have a strong effect on white blood cell counts and bone marrow. Together with the chemist Charles T. Beer, he

---

[‡] This review is dedicated to Professor Pierre Potier, whose contribution in the field of Vinca alkaloids led to the discovery of new antitumor drugs.

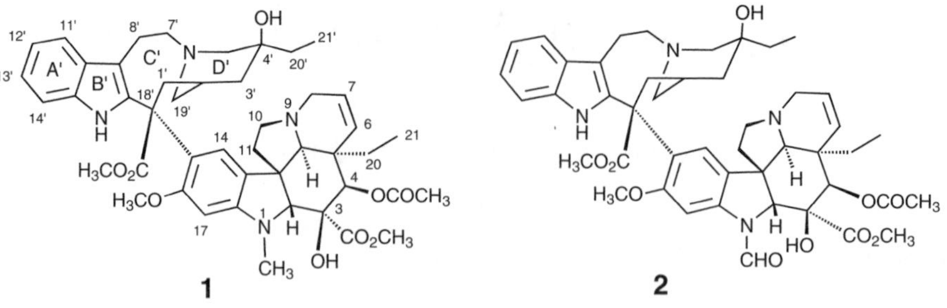

FIGURE 7.1 Vinblastine (**1**) and Vincristine (**2**).

isolated the active principle, which they named vincaleukoblastine; the name was later changed to vinblastine (**1**).[3] Other investigations by Svoboda, Johnson, and collaborators at Lilly Laboratories, based on the reproducible antitumor activity of extracts of *C. roseus* leaves, resulted in the isolation of vinblastine.[4] These studies also led to the isolation of other bioactive alkaloids, including leurocristine, also named vincristine **2**, which was isolated from the leaves of the periwinkle.[5]

Vinblastine **1** and vincristine **2** are isolated in minute amounts from the leaves of *C. roseus* (Figure 7.1). For this reason, the semisynthesis and total synthesis of these binary alkaloids has been the subject of a number of studies, as we will see later on. Since the initial clinical trials in the 1960s, the so-called vinca alkaloids, vinblastine (Velban® or Velbe®) and vincristine (Oncovin®), have been widely used in the treatment of different types of cancer.

The complex structure and stereochemistry of vinblastine **1** and vincristine **2** were elegantly established by Neuss and his colleagues[6] and by Moncrief and Lipscomb.[7] Both vinblastine and vincristine are dimeric compounds possessing an indole moiety (the velbanamine "upper" part) and a vindoline dihydroindole nucleus. They differ by the substituent attached to the nitrogen of the dihydroindole part, with vinblastine bearing an N-methyl group and vincristine possessing an N-formyl function. Two systems have been adopted to number the bisindole-type alkaloids: biogenetic and International Union of Pure and Applied Chemistry numbering.[1] The latter will be used throughout this review. Despite their small structural differences, vinblastine **1** and vincristine **2** differ strongly in their antitumor properties and clinical toxicities. Two other modified natural antitumor vinca alkaloids, vindesine (Eldesine®)[8] **3** and vinorelbine (Navelbine®)[9] **4** have been developed as antitumor compounds, and more recently, a fluorinated structural analog, vinflunine **5**, was shown to possess potent antitumor properties[10,11] and is now in phase III clinical trials in Europe (Figure 7.2). Thus, the discovery of vinblastine was a milestone in the development of cancer chemotherapy, inducing extensive chemical, biological, and clinical studies that finally led to the discovery of new potent antitumor vinca alkaloids.

III. MECHANISM OF ACTION

Soon after the discovery of the *in vivo* antitumor properties of the vinca alkaloids, a number of studies focused on their mode of action. Vinblastine was found to be a cell cycle–dependent

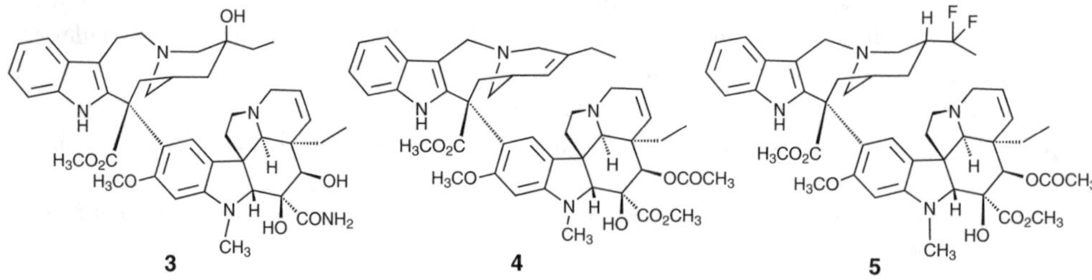

FIGURE 7.2 Vindesine (**3**), Vinorelbine (**4**), and Vinflunine (**5**).

antimitotic agent that interacts with tubulin,[12] a ubiquitous heterodimeric protein present in all eukaryotic cells. Tubulin and its polymerized form, microtubules, play crucial roles in the maintenance of cellular morphology and intracellular transport and in the construction of the mitotic spindle during cell division.[13] The formation of microtubules is a dynamic process involving the assembly of tubulin and the disassembly of the polymers. The antiproliferative activity of the vinca alkaloids was shown to be a result of their interaction with the mitotic spindle.[14] The alkaloids inhibit the assembly of tubulin into microtubules and, consequently, prevent the cells from undergoing division. They bind to β–tubulin at different sites from colchicine and the antitumor taxoids, but a number of other natural products such as Rhizoxin, maytansine, cryptophycins, and dolastatins share the same binding site, interacting with the so-called vinca domain of tubulin.[15-17] *In vitro*, the effects of vinca alkaloids on tubulin are concentration dependent. At low concentrations (submicromolar), they inhibit the formation and function of microtubules from tubulin, whereas spirals are formed at higher concentration.[18,19] Depending on their structure, the effects of vinca alkaloids on microtubule dynamics are different, and these effects have been associated with the differences observed in the alkaloids' efficacies and neurotoxicities. [20] Although the cellular target of the vinca alkaloids has been known since the 1970s, the precise location of the binding site is still unknown. Nevertheless, cross-linking experiments indicate that vinblastine interacts with the β-subunit of tubulin,[21] and a recent synthesized fluorescent analog of vinblastine should give information about the binding site of vinblastine in the near future.[22]

IV. SEMISYNTHESIS AND TOTAL SYNTHESIS OF VINCA ALKALOIDS

The availability of bioactive natural substances is often a major problem when they have to be produced on a commercial basis. Vinblastine **1** and vincristine **2** are isolated in very low yields from the leaves of *C. roseus*. Isolation yields of vincristine are on the order of 0.0003% from the dried leaves. Vinblastine is obtained in higher yield (0.01%) and has thus been used to produce vincristine, with oxidation of the indolinyl N-methyl group of the former leading to the latter. Vindesine **3** is also prepared from vinblastine by hydrazinolysis and subsequent hydrogenolysis of the newly formed N–N bond.[8]

Although cultivation of *C. roseus* provides a renewable source of vinblastine, a number of teams have studied the potential of preparing the antitumor vinca alkaloids by semisynthesis and total synthesis. The structural complexity of these molecules made this a significant synthetic challenge.

The successful strategy of the semisynthesis of vinca alkaloids was based on a biogenetic hypothesis involving vindoline **7** and catharanthine **6** as precursors. These alkaloids are two of the more abundant alkaloids isolated from *C. roseus*, and the idea was that vinblastine **1**, as well as other binary vinca alkaloids, could result from the union of vindoline **7** with an intermediate derived from catharanthine **6**. After a number of attempts,[1] a vinblastine-type alkaloid possessing the C18′S configuration, essential for bioactivity, could be obtained by applying a modification of the Polonovski reaction, also called the Polonovski–Potier reaction, to the N_b-oxide of catharanthine **8** and vindoline **7**. Thus, treatment of catharanthine Nb-oxide **8** and vindoline **7** with trifluoroacetic anhydride led, after reduction by sodium borohydride, to the new bisindole alkaloid anhydrovinblastine (**9**), which possesses the natural 18′S configuration (Scheme 7.1).[23-25]

Semisynthesis of anhydrovinblastine **9** was also realized from catharanthine and vindoline through ferric ion-mediated coupling in acidic aqueous media.[26] Anhydrovinblastine was then found to be a natural vinca alkaloid, first by feeding experiments of both radioactive catharanthine and vindoline to *C. roseus*,[27] and second by its extraction from the plant.[28] More recently, a basic peroxidase with anhydrovinblastine synthase activity was purified from the leaves of *C. roseus* and was shown to produce the dimer from catharanthine and vindoline.[29]

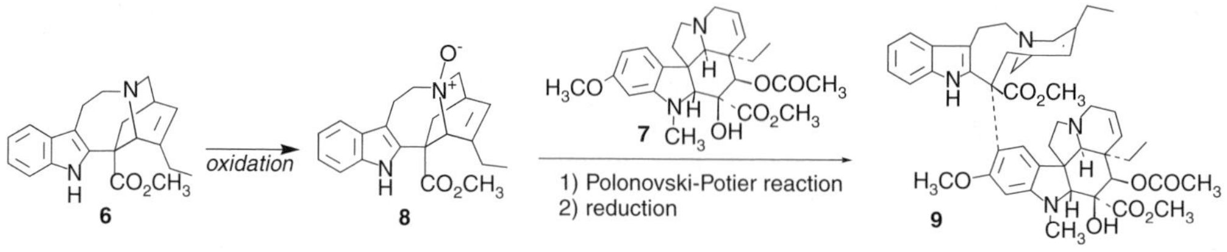

SCHEME 7.1 Semisynthesis of anhydrovinblastine from catharathine.

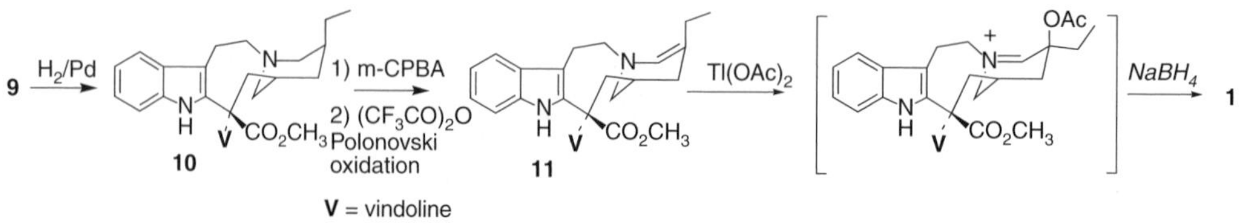

V = vindoline

SCHEME 7.2 Conversion of anhydrovinblastine to vinblastine.

Because of its facile preparation from catharanthine and vindoline, anhydrovinblastine **9** was considered an attractive key intermediate in the synthesis of other binary alkaloids.[1] Thus, vinblastine was obtained from **9** through different steps involving hydrogenation to 4′-deoxyleurosidine **10**, N_b-oxidation and Polonovski reaction leading to enamine **11**, treatment with thallium triacetate, and reduction with hydrolysis of the C4′ acetoxy group (Scheme 7.2).[30]

Another improved way to provide vinblastine from anhydrovinblastine was to subject enamine **11** to $FeCl_3$-promoted oxygenation.[31] Two other strategies have also been used to generate the C18′S–C15 bond of vinblastine **1** from natural vindoline **7** and substrates that could lead to the velbanamine part of the binary indole–indoline alkaloids.[1] The first involved reaction of chloroindolenine **12** with silver tetrafluoroborate and vindoline **7** and led to intermediate **13**, which produced vinblastine **1** after cyclization and deprotection (Scheme 7.3).[32]

In the second strategy (Scheme 7.4), the coupling methodology is based on a nonoxidative cleavage of the tertiary amine **14** that undergoes fragmentation after treatment with $ClCO_2CH_2C_6H_4NO_2$ and vindoline **7**.[33] The indole–indoline dimer **15** was then converted to hydroxy aldehyde **16**, which led to vinblastine **1** after hydrogenolysis and reduction of the iminium species **17**.

A new total synthesis of vinblastine was achieved recently by coupling chloroindolenine **19** with vindoline **7** synthesized from 7-mesyloxyquinoline **18** (Scheme 7.5).[34] The iminium salt intermediate was formed by activation of **19** by trifluoroacetic acid, and electrophilic substitution with vindoline **7** led to dimer **20**, which was cyclized to vinblastine after deprotection of the tertiary alcohol and amine groups.

If anhydrovinblastine **9** is an important intermediate in the semisynthesis of a number of natural binary alkaloids, it is also the precursor of two unnatural anticancer drugs of the *Vinca* family: vinorelbine **4** and vinflunine **5**. The first synthesis of vinorelbine came from the application of the

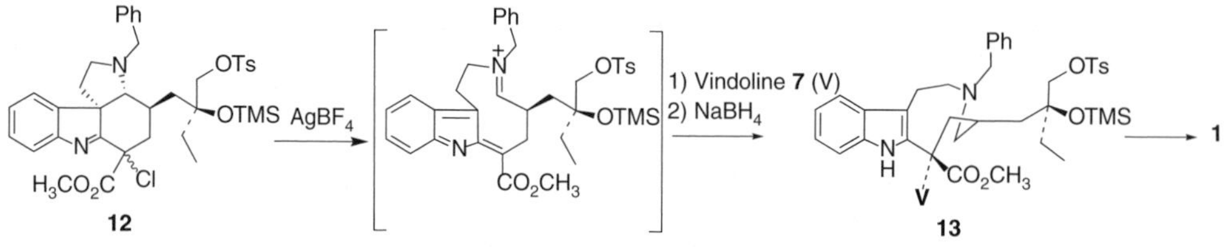

SCHEME 7.3 Conversion of chloroindolenine (**12**) to vinblastine.

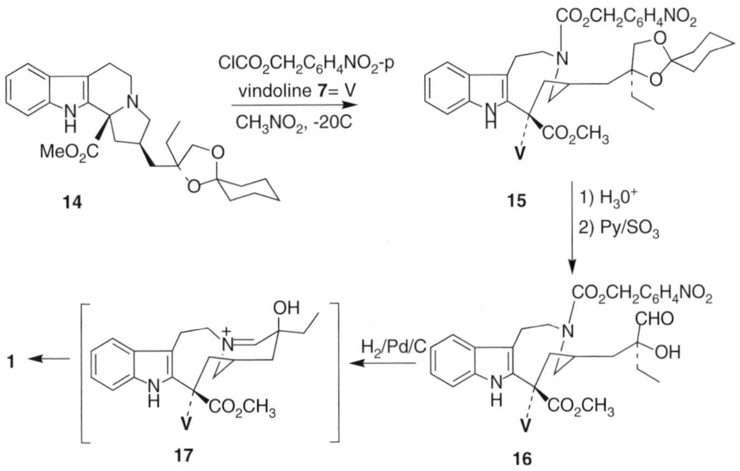

SCHEME 7.4 Conversion of tertiary amine (**14**) to vinblastine.

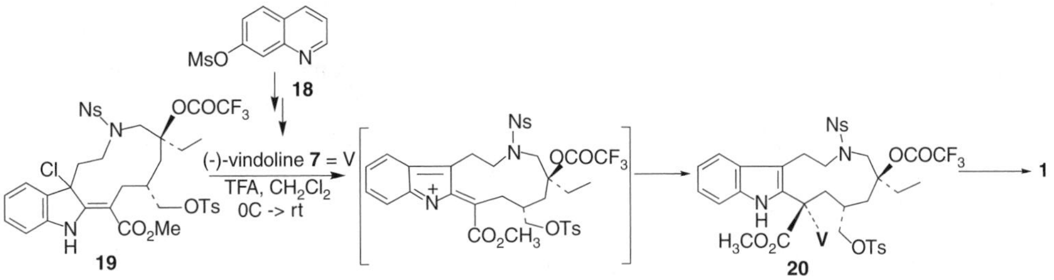

SCHEME 7.5 Conversion of chloroindolenine (**19**) to vinblastine.

Polonovski–Potier reaction to anhydrovinblastine N_b-oxide **21** (Scheme 7.6). Treatment of **21** with trifluoroacetic anhydride led to an unstable quaternary ammonium salt **22**, which undergoes fragmentation to a bisiminium salt **23**. Addition of water and subsequent loss of formaldehyde produces a nucleophilic secondary amine **24** that gives rise to vinorelbine **4** after intramolecular trapping of the iminium group.[9] Improvement in the synthesis of vinorelbine **4** could be obtained by using the chloro- or bromo-indolenines of anhydrovinblastine **25**.[35,36]

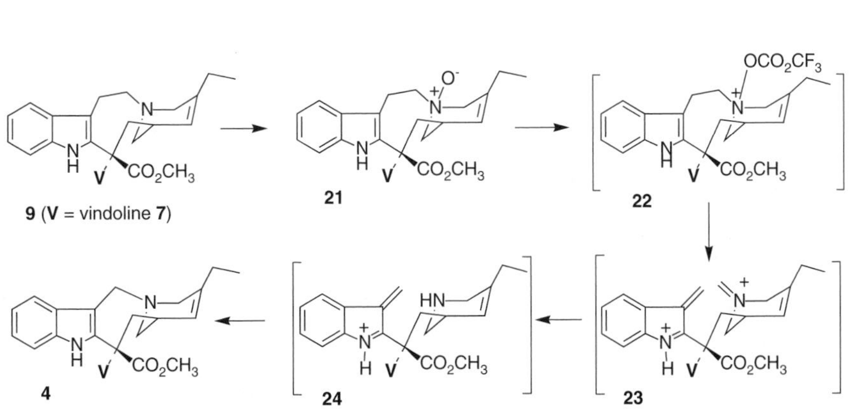

SCHEME 7.6 Conversion of anhydrovinblastine N_b-oxide (**21**) to vinorelbine (**4**).

V. MEDICINAL CHEMISTRY

The medicinal chemistry of the vinca alkaloids has been extensively reviewed by Pearce[37] and Borman and Kuehne.[38] Since the discovery of their antitumor properties, many derivatives have been synthesized in the pharmaceutical industry, with the aim of improving their pharmacological activities to identify new drugs exhibiting a wider spectrum of clinical efficacy. The natural vinblastine **1** extracted from *C. roseus* leaves was used as a starting material because for a long time it was the only compound available in sufficient quantities.

Briefly, most of the new analogs have been obtained by modifications of the vindoline "lower" part, bearing several reactive functions. Many such derivatives, including vinglycinate **26**, vindesine **3**, and vinzolidine **27**, have been synthesized and evaluated by the Eli Lilly group (Figure 7.3). Vinglycinate **26**, with a glycine residue at the vindoline C4 position, was the first vinblastine analog to enter phase I clinical trials in 1967.[39] Subsequently, modifications at the C3 position led to the amido-derivative vindesine **3**,[8] which can be considered a chemical precursor of vinzolidine **27**[37] and vintriptol **28**.[40] Vinepidine **29**, corresponding to 4′-*epi*-4′-deoxyvincristine, was also developed by the Lilly group for its increased tubulin affinity relative to vinblastine **1**.[37] Nevertheless, none of these semisynthetic compounds showed marked benefits in clinical evaluations relative to vinblastine **1** and vincristine **2**.

In the meantime, as previously mentioned, the preparation of anhydrovinblastine **9** from catharanthine **8** and vindoline **7** facilitated the access to vinca alkaloids analogs exhibiting antimitotic properties,[24] leading to vinorelbine **4** and vinflunine **5**. To date, the semisynthetic compounds that have been approved for clinical use as anticancer drugs are vindesine **3** and vinorelbine **4**.

Although most of the work reported has been on the synthesis of analogs modified in the "lower half," two reviews have appeared recently describing the consequences of modifications in the velbanamine "upper part" of the vinblastine skeleton.[41,42]

After 1990, only a few new derivatives have been prepared and evaluated with the aim of discovering novel anticancer drugs with clinical efficacy. These include the aminophosphonate derivative vinfosiltine **30** (Figure 7.3), selected for its unusual high potency both *in vitro* and *in vivo* compared to the classical vinblastine and vincristine.[43] Vinfosiltine was designed on the basis of the similarity between α-amino phosphonic acids and natural amino acids. However, no evidence of marked benefice in phase II clinical trials, relative to other vinca alkaloids, has been obtained, and development of vinfosiltine was discontinued in 1995.[44]

The natural anhydrovinblastine **9**, biogenetic precursor of the dimeric vinca alkaloid, entered phase I clinical trials in 1999 and is claimed to be under phase II investigation in the United States as an agent against non–small cell lung cancer (NSCLC).[45]

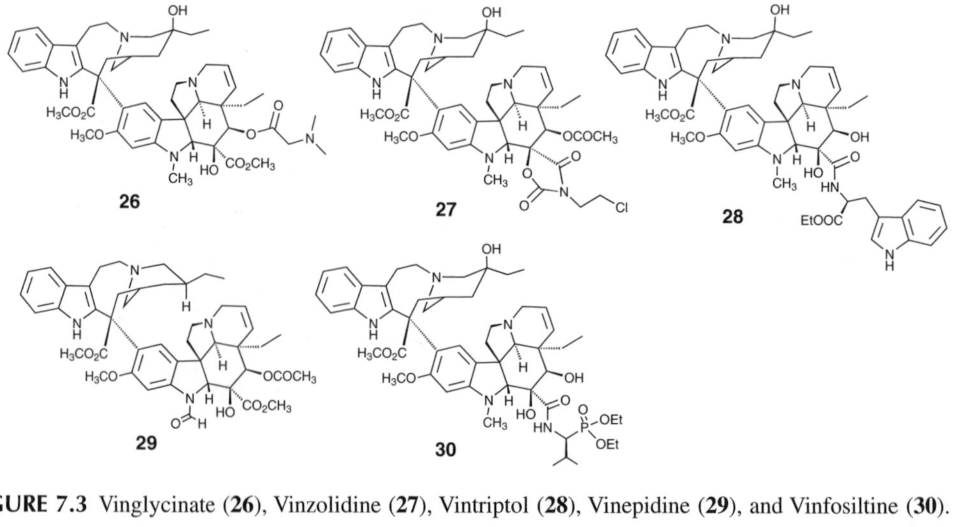

FIGURE 7.3 Vinglycinate (**26**), Vinzolidine (**27**), Vintriptol (**28**), Vinepidine (**29**), and Vinfosiltine (**30**).

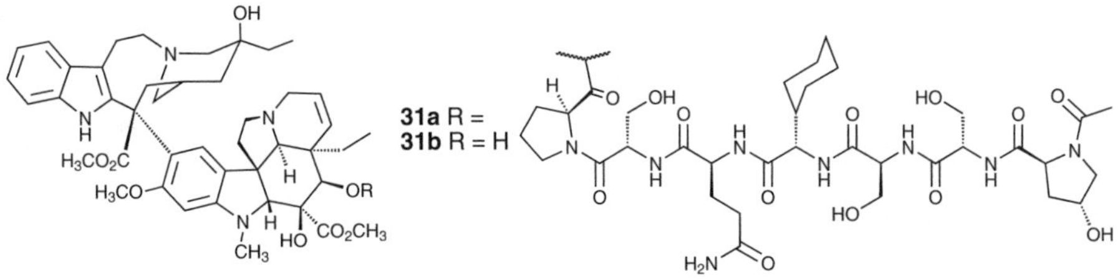

FIGURE 7.4 Vinblastine-peptide derivative (**31a**) and desacetylvinblastine (**31b**).

In a different domain, targeted delivery of vinca alkaloids has also been investigated. Several experiments were conducted during the 1980s with conjugates of vinblastine with monoclonal antibodies to deliver the cytotoxic drug to malignant tissues, but no further clinical development has been reported.[46] More recently, a peptide–vinblastine derivative **31a** targeted at prostate cancer cells has been reported (Figure 7.4).[47] The proteolytic activity of prostate-specific antigen has been exploited to convert the conjugate **31a** to deacetyl-vinblastine **31b** at the site of the tumor, as demonstrated by *in vivo* studies using prostate-specific-antigen-secreting implanted human prostate cancer cells on nude mice. It should be pointed out that the putative effective drug 4-deacetylvinblastine **31b**, resulting from the hydrolysis of **31a**, exhibited no *in vivo* activity under the same experimental conditions.

Kuehne's group has extensively documented the synthesis of vinblastine derivatives, mainly modified at the piperidine ring D′ of the velbenamine part.[32,48] As a result it was clearly shown that subtle modifications at the C4′ position dramatically affect the interaction with tubulin[49] together with cytotoxicity.[50] More recently, two series of *homo*-vinblastine derivatives **32** and **33** (Figure 7.5) have been synthesized by the same group, highlighting the possible isolation of atropoisomers exhibiting *in vitro* activities different from those of the compound bearing the natural conformation. The authors also mentioned the concept of "thermal pro-drug activation" potentially applicable to these compounds.[51,52] Among them, the 7′a-*homo*-vinblastine **32**, including a 10-membered ring C′, was able to inhibit the tubulin polymerization at submicromolar concentrations, similar to vinblastine, but was slightly less cytotoxic than vinblastine when tested against the murine L1210 and S180 cell lines. However, the 18′a-*homo*-derivatives were shown to be inactive under the same experimental conditions. In a more recent publication, a large series of 62 vinblastine congeners was prepared and evaluated for their *in vitro* pharmacological properties.[53] As an example, derivative **34**, including an additional cyclohexane ring fused at C3′–4′ (Figure 7.6) induced a very high level of cytotoxicity toward the L1210 murine leukemia cell line (IC$_{50}$ < 1 p*M* compared with 0.4 n*M* for vinblastine). However, the activity of **34** with RCC-2 cells (rat colon cancer) was slightly lower than that of the parent compound. In a similar way, it was shown that further ring D′ modified compounds such as **35** and **36** were as potent as vinblastine in terms of cytotoxicity (Figure 7.6).[53,54]

Surprisingly, superacid chemistry applied to vinca alkaloids resulted in access to newly modified compounds, specifically at the C4′ and C20′ positions, including vinflunine **5**.[10] These include

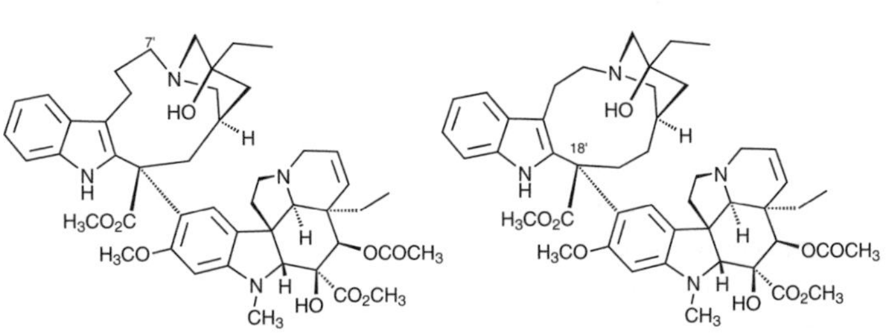

32   **33**

FIGURE 7.5 *Homo*-vinblastine derivatives.

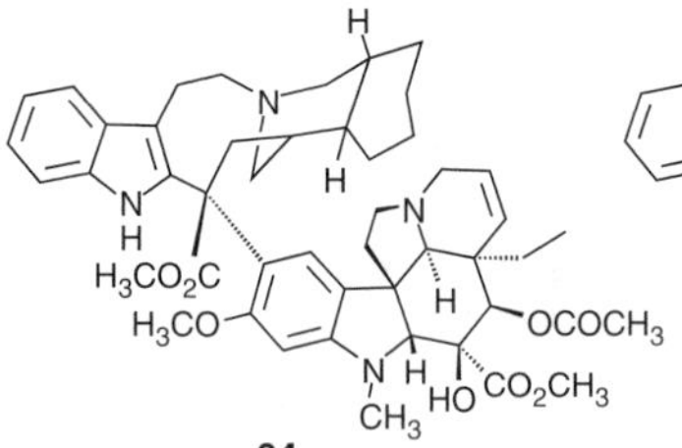

FIGURE 7.6 Ring D′ modified analogs.

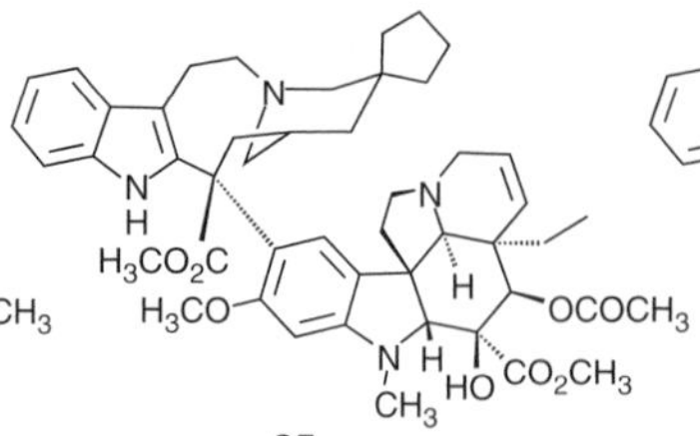

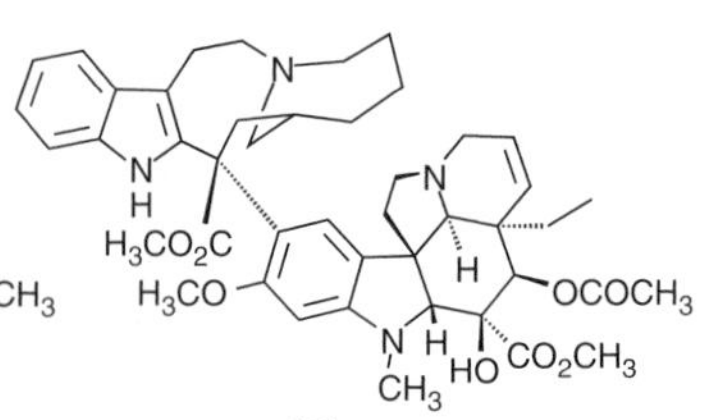

compounds **37a** and **37b**, substituted by halogen atoms (fluorine or chlorine) when the reaction was conducted in the presence of chlorinated solvents, and oxygenated derivatives **37c–e** or **38** (alcohols or ketones) in the presence of hydrogen peroxide (Figure 7.7).[41]

A series of such derivatives has been evaluated for their overall pharmacological properties: tubulin polymerization inhibition, cytotoxicity, and *in vivo* antitumor activity against the P388 murine leukemia model. However, a lack of correlation between *in vitro* and *in vivo* results was revealed, and no clear structure-activity relationships were obtained. In the tubulin polymerization inhibition assay, all the compounds of the study exerted an activity in the micromolar range of IC_{50} close to those of the reference compounds. Clearly, this assay was not discriminatory enough to be useful in the selection of such derivatives. More surprisingly, binding of [³H]-vinflunine to tubulin was undetectable using the standard centrifugal gel filtration, unlike the other vinca alkaloids.[55] This observation could appear somewhat paradoxical considering the overall *in vivo* antitumor activity of vinflunine **5**. However, cytotoxicity values against the L1210 murine leukemia cells correlated with the induction of tubulin spirals for a subset of these compounds.[56] Detailed investigations of pharmacokinetic and metabolism properties did not provide adequate results to understand this unusual profile of activities.[57]

It should be pointed out that alcohol **38**, isolated after reaction in superacidic medium in 85% yield, corresponds to a minor metabolite of vinorelbine **4**.[58] Addition of fluorine atoms in vinorelbine **4** (resulting in vinflunine **5**, for example) could block a metabolic pathway and modify its pharmacodynamic properties. The data available, however, do not explain the unusual pharmacological profile of **5**: Similar to the other vinca alkaloids, vinorelbine is poorly metabolized, and compound **38**, found at very low levels, is not a circulatory metabolite.[59] Should superacids be considered as tools for biomimetic reactions under particular conditions? Or, on the contrary, should certain cytochromes be able to catalyze superacidic reactions? Further experiments need to be undertaken to validate this hypothesis.

Taken together, all the results available in the literature appear to indicate that the strength of interaction with tubulin is not a sufficient criterion for the selection of a vinblastine derivative for further pharmacological studies. A quantitative comparison permitted the establishment of relative binding affinities for tubulin of vincristine > vinblastine > vinorelbine, in parallel with the ability to induce the formation of tubulin spirals, which are considered to be responsible for the neurotoxicity observed as an undesirable side-effect in the clinic.[60]

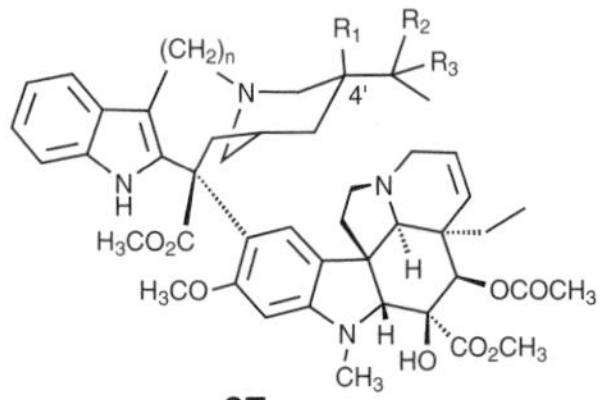

	n	R_1	R_2	R_3	C-4′
37a	1,2	H	H	Cl	*R*
37b	1,2	F	H	Cl	*R* and *S*
37c	1	H	---O---		*R*
37d	1	F	H	OH	*R*
37e	1	OH	H	F	*S*

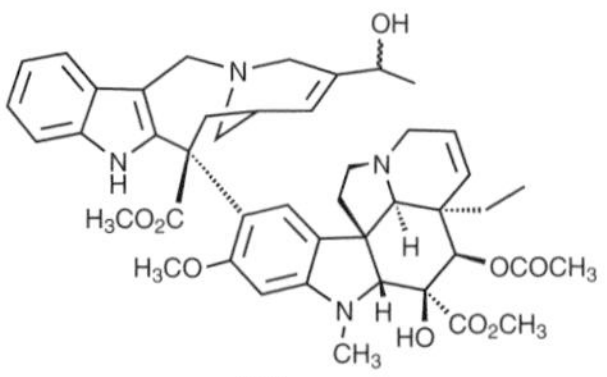

FIGURE 7.7 Superacid-generated analogs.

The cytotoxic activity is also not sufficient on its own: a very potent compound such as vinfosiltine **30** may demonstrate a high *in vivo* antitumor activity, but within a sharp dose range, resulting in a low therapeutic index. On the contrary, vinflunine **5** exhibited markedly lower cytotoxic properties in a panel of several murine and human cell lines compared to the standard vinca alkaloids[55] but was the most active compound in a series of *in vivo* experiments that included murine models[61] and human xenograft models.[11]

Microtubules are dynamic structures that are continuously shortening and growing. Detailed investigations evaluating microtubule dynamics perturbations have demonstrated that vinorelbine **4** and vinflunine **5** have a qualitative mode of action different from vinblastine **1**.[20] Moreover, examinations of the intracellular concentrations of vinorelbine **4** and vinflunine **5** compared with vinblastine **1** indicated the possible presence of sequestered drugs in "intracellular reservoirs."[62]

The effects of vinca alkaloids on centromere dynamics have also been studied, leading to the conclusion that suppression of microtubule dynamics is the primary mechanism of action by which vinca alkaloids block cell mitosis.[63] Overall, all these observations may contribute, in part, to the original *in vivo* efficacy of these new derivatives, but a correlation between tubulin interactions and antitumor activity remains to be established.

Nevertheless, based on the numerous results obtained in the studies mentioned above, it appears clear that chemical modifications in the piperidine ring D′ or at the C4′ position induce dramatic changes in the pharmacological properties of vinca alkaloids. Despite the recent determination of the structure of the α,β-tubulin dimer resolved by electron crystallography[64] and the subsequent studies aimed at improving the crystal resolution, the so-called vinca binding site remains insufficiently known to investigate drug–protein interactions at the molecular level, rendering the rational design of more selective compounds almost impossible. To date, only a hypothetical model in which vinblastine **1** binds at the "plus end" interface of the polar part of the β-tubulin subunit has been proposed.[65,66] Further investigations involving new crystallographic techniques, for example, will certainly allow a better understanding of the mechanism of action of vinca alkaloids in a near future.

VI. CLINICAL APPLICATIONS

The vinca alkaloids have been used in both curative and palliative chemotherapy regimens in clinical oncology for approximately 40 yr. A comprehensive review of their clinical applications can be found in *Cancer: Principles and Practice of Oncology*.[67] The recent availability of vinorelbine has resulted in renewed interest in this class of compounds.[68,69]

Vincristine **1** plays a major role in combination chemotherapy in the treatment of acute lymphoblastic leukemias and lymphomas. Vinblastine **2** is commonly used in combination with other anticancer drugs to treat bladder and breast cancers and is an essential component in the curative regimen for Hodgkin's disease. Vinorelbine **4** has been approved worldwide for treating NSCLC either as a single agent or in combination with cisplatin. It has also been registered for advanced breast cancer in Europe, and an oral formulation is now available. Vinflunine **5** entered phase III clinical evaluations in 2003 in Europe against NSCLC and bladder cancer.[70]

Neurotoxic effects are the main side-effects observed with vincristine **1** and, to a lesser extent, with the other vinca alkaloids. These toxic effects have been associated with the affinity for axonal microtubules[71] and with the ability to induce microtubule spirals, as mentioned before.[56] Neutropenia is the principal dose-limiting toxicity of vinca alkaloids, but recovery occurs after treatment.

The emergence of drug-resistant cells is the major limitation on the clinical usefulness of vinca alkaloids, as for many other anticancer drugs.[72] The best described mechanism involves the amplification of the multidrug resistance protein, resulting in an efflux of the drug out of the cells mediated by the phosphoglycoprotein (Pgp) pump. However, recent studies with vinflunine demonstrated that the level of cross-resistance is much lower than that observed with vinorelbine or vincristine, which has positive implications for its clinical usage.[73] Resistance phenomena have been extensively studied, and numerous reviews covering these aspects appear regularly in the literature.[74,75]

The discovery of the antiangiogenic properties of vinca alkaloids in the early 1990s offered new research areas with potential clinical applications.[72,76] Moreover, recent studies highlighted the definite antiangiogenic activity of vinblastine at noncytotoxic doses,[77] as demonstrated both *in vitro* on endothelial cells and *in vivo* using the chick embryo chorioallantoic assay. In addition, newer data indicate that vinflunine mediates its antitumor activity at least in part via an antivascular mechanism.[78] These observations will certainly lead to new clinical applications of the vinca alkaloids when used at subtherapeutic doses, such as potentiation of standard cytotoxic compounds or potentiation of specific antiangiogenic derivatives.

The association of taxanes with vinca alkaloids such as vinorelbine has revealed evidence of potential synergy in preclinical models,[79] supporting a promising clinical efficacy, particularly in breast cancer.

Finally, "new" targets of vinca alkaloids may be discovered, and their participation in the antitumor and anticancer activities will have to be established. As an example, vinorelbine has been described as a potent calmodulin binder, inhibiting the association of calmodulin to its own target proteins.[80] Further research is underway to investigate the potential role of calmodulin inhibitors such as vinca alkaloids and other tubulin interacting agents in the cellular division.[81]

VII. CONCLUSIONS

Although the anticancer activity linked to clinical efficacy of vinca alkaloids has been recognized for more than three decades, further studies in different domains remain to be undertaken to better understand their mechanism of action. Recent advances in cell biology have identified numerous regulators of cell cycle such as molecular motors or specific kinases involved in the mitotic process. The question might be asked whether a mitotic blocker such as a new-generation vinca alkaloid could target certain of these proteins. Progress in structural biology should permit the determination of the tubulin structure at high resolution, allowing a better investigation of the interactions with vinca alkaloids at the molecular level. Then design of more selective molecules will be accessible with the aim of establishing the importance of the strength of interactions with tubulin for the antimitotic activity.

In conclusion, continued research of novel tubulin interacting agents, including vinca alkaloids, might lead to more specific and more effective anticancer drugs.

REFERENCES

1. Brossi, A. and Suffness, M., Eds. *The Alkaloids*, Vol. 37: *Antitumor Bisindole Alkaloids from* Catharanthus roseus *(L.)*. Academic Press, Inc., San Diego, 1990.
2. Svoboda, G.H. and Blake, D.A., The phytochemistry and pharmacology of *Catharanthus roseus* (L.) G. Don, in *Catharanthus Alkaloids*, Taylor, W.I. and Farnsworth, N.R., Eds, Marcel Dekker, Inc., New York, 1975, chap. 2.
3. Noble, R.L., Beer, C.T., and Cutts, J.H., Role of chance observation in chemotherapy: *Vinca rosea*, *Ann. N. Y. Acad. Sci.*, 76, 882, 1958.
4. Johnson, I.S., Wright, H.F., and Svoboda, G.H., Experimental basis for clinical evaluation of antitumor principles from *Vinca rosea* Linn, *J. Lab. Clin. Med.*, 54, 830, 1959.
5. Svoboda, G.H., Alkaloids of *Vinca rosea* Linn. (*Catharanthus roseus*). 1X: Extraction and characterization of leurosidine and leurocristine, *Llyodia*, 24, 173, 1961.
6. Neuss, N. et al., *Vinca* alkaloids, XXI. The structure of the oncolytic alkaloids vinblastine (VLB) and vincristine (VCR), *J. Am. Chem. Soc.*, 86, 1440, 1964.

7. Moncrief, J.W. and Lipscomb, W.N., Structure of leurocristine methiodide dihydrate by anomalous scattering methods: relation to leurocristine (vincristine) and vincaleukoblastine (vinblastine), *Acta Cryst.*, 21, 322, 1966.

8. Barnett, C.J. et al., Structure-activity relationships of dimeric Catharanthus alkaloids. 1. Deacetylvinblastine amide (vindesine) sulfate, *J. Med. Chem.*, 21, 88, 1978.

9. Mangeney, P. et al., 5′-nor anhydrovinblastine, prototype of a new class of vinblastine derivatives, *Tetrahedron*, 35, 2175, 1979.

10. Fahy, J. et al., *Vinca* alkaloids in superacidic media: a method for creating a new family of antitumor derivatives, *J. Am. Chem. Soc.*, 119, 8576, 1997.

11. Hill, B. et al., Superior *in vivo* experimental antitumor activity of vinflunine, relative to vinorelbine, in a panel of human tumor xenografts, *Eur. J. Cancer*, 35, 512, 1999.

12. Lee, J.C., Harrison, D., and Timasheff, S.N., Interaction of vinblastine with calf brain microtubule protein, *J. Biol. Chem*, 24, 9276, 1975.

13. Dustin P., *Microtubules*, 2nd ed., Springer-Verlag, Berlin, 1984.

14. Jordan, M.A., Thrower, D., and Wilson, L., Mechanism of inhibition of cell proliferation by *Vinca* alkaloids, *Cancer Res.*, 51, 2212, 1991.

15. Hamel, E., Natural products which interact with tubulin in the *Vinca* domain: maytansine, Rhizoxin, phomopsin A, dolastatins 10 and 15 and halichondrin B, *Pharm. Ther.*, 55, 31, 1992.

16. Shih, C. and Teicher, B.A., Cryptophycins: a novel class of potent antimitotic antitumor depsipeptides, *Curr. Pharm. Design*, 7, 1259, 2001.

17. Gupta, S. and Bhattacharyya, B., Antimicrotubular drugs binding to the *Vinca* domain of tubulin, *Mol. Cell. Biochem.*, 253, 41, 2003.

18. Himes, R.H. et al., Action of the *Vinca* alkaloids, vincristine, vinblastine and desacetyl vinblastine amide on microtubules *in vitro*, *Cancer Res.*, 36, 3798, 1976.

19. Zavala, F., Guénard, D., and Potier, P., Interaction of vinblastine analogues with tubulin, *Experientia*, 34, 1497, 1978.

20. Ngan, V.K. et al., Novel actions of the antitumor drugs vinflunine and vinorelbine on microtubules, *Cancer Res.*, 60, 5045, 2000.

21. Rai, S.S. and Wolff, J., Localization of the vinblastine-binding site on beta-tubulin, *J. Biol. Chem.*, 271, 14707, 1996.

22. Chatterjee, S.K. et al., Interaction of tubulin with a new fluorescent analogue of vinblastine, *Biochemistry*, 41, 14010, 2002.

23. Potier, P. et al., Partial synthesis of vinblastine-type alkaloids, *J. Chem. Soc., Chem. Commun.*, 670, 1975.

24. Langlois, N. et al., Application of a modification of the Polonovski reaction to the synthesis of vinblastine-type alkaloids, *J. Am. Chem. Soc.*, 98, 7017, 1976.

25. Kutney, J.-P. et al., Studies on the synthesis of bisindole alkaloids II. The synthesis of 3′-4′-dehydrovinblastine, 4′-deoxovinblastine and related analogues. The biogenetic approach, *Heterocycles*, 3, 639, 1975.

26. Vucovic, J. et al., Production of 3′,4′-anhydrovinblastine: a unique chemical synthesis, *Tetrahedron*, 44, 325, 1988.

27. Scott, A.I., Guéritte, F., and Lee, S.L., Role of anhydrovinblastine in the biosynthesis of the antitumor dimeric indole alkaloids, *J. Am. Chem. Soc.*, 100, 6253, 1978.

28. Goodbody, A.E. et al., Extraction of 3′,4′-anhydrovinblastine from *Catharanthus roseus*, *Phytochemistry*, 27, 1713, 1988.

29. Sottomayor, M. et al., Purification and characterization of α-3′,4′-anhydrovinblastine synthase (peroxidase-like) from *Catharanthus roseus* (L.) G. Don., *FEBS Lett*, 428, 299, 1998.

30. Mangeney, P. et al., Preparation of vinblastine, vincristine and leurosidine antitumour alkaloids, *J. Am. Chem. Soc.*, 101, 2243, 1979.

31. Kutney, J.-P. et al., A highly efficient and commercially important synthesis of the antitumor *Catharanthus* alkaloids vinblastine and leurosidine from catharanthine and vindoline, *Heterocycles*, 27, 1845, 1988.

32. Kuehne, M.E., Matson, P.A., and Bornmann, W.G., Enantioselective syntheses of vinblastine, leurosidine, vincovaline, and 20′-*epi*-vincovaline, *J. Org. Chem.*, 56, 513, 1991.

33. Magnus, P. et al., Nonoxidative coupling methodology for the synthesis of the antitumor bisindole alkaloid vinblastine and a lower-half analogue: solvent effect on the stereochemistry of the crucial C-15′/C-18′ bond, *J. Am. Chem. Soc.*, 114, 10232, 1992.

34. Yokoshima, S. et al., Stereocontrolled total synthesis of (+)-vinblastine, *J. Am. Chem. Soc.*, 124, 2137, 2002.

35. Andriamializoa, R.Z. et al., Composés antitumoraux du groupe de la vinblastine: nouvelle méthode de préparation, *Tetrahedron*, 36, 3053, 1980.

36. Guéritte, F. et al., Composés antitumoraux du groupe de la vinblastine: dérivés de la nor-5′ anhy-drovinblastine, *Eur. J. Med. Chem.*, 18, 419, 1983.

37. Pearce, H.L., Medicinal chemistry of bisindole alkaloids from *Catharanthus*, in *The Alkaloids, Vol. 37: Antitumor bisindole alkaloids from* Catharanthus roseus *(L.)*. Brossi, A. and Suffness, M., Eds. Academic Press, Inc., San Diego, CA, 1990, 145.

38. Borman, L.S. and Kuehne, M.E., Functional hot spot at the C-20′ position of vinblastine, in *The Alkaloids, Vol. 37: Antitumor bisindole alkaloids from* Catharanthus roseus *(L.)*. Brossi, A. and Suffness, M., Eds. Academic Press, Inc., San Diego, CA, 1990, 133.

39. Armstrong, J.G. et al., Initial clinical experience with vinglycinate sulfate, a molecular modification of vinblastine, *Cancer Res.*, 27, 221, 1967.

40. Rao, K.S.P.B. et al., Vinblastin-23-oyl amino acid derivatives: chemistry, physicochemical data, toxicity, and antitumor activities against P388 and L1210 leukemias, *J. Med. Chem.*, 28, 1079, 1985.

41. Fahy, J., Modifications in the "upper" or velbenamine part of the *Vinca* alkaloids have major impli-cations for tubulin interacting activities, *Curr. Pharm. Design*, 7, 1181, 2001.

42. Duflos, A., Kruczynski, A. and Barret, J.-M., Novel aspect of natural and modified *Vinca* alkaloids, *Curr. Med. Chem. Anti-Cancer Agents*, 2, 55, 2002.

43. Lavielle, G. et al., New alpha-amino phosphonic acid derivatives of vinblastine: chemistry and antitumor activity, *J. Med. Chem.*, 34, 1998, 1991.

44. Adenis, A. et al., Phase II study of a new *Vinca* alkaloid derivative, S12363, in advanced breast cancer, *Cancer Chemother. Pharmacol.*, 35, 527, 1995.

45. Ramnath, N. et al., Phase I and pharmacokinetic study of anhydrovinblastine every 3 weeks in patients with refractory solid tumors, *Cancer Chemother. Pharmacol.*, 51, 227, 2003.

46. Laguzza, B.C. et al., New antitumor monoclonal antibody-*Vinca* conjugate LY203725 and related compounds, *J. Med. Chem.*, 32, 548, 1989.

47. Brady, S.F. et al., Design and synthesis of a pro-drug of vinblastine targeted at treatment of prostate cancer with enhanced efficacy and reduced systemic toxicity, *J. Med. Chem.*, 45, 4706, 2002.

48. Kuehne, M.E. et al., Three routes to the critical C16′-C14′ parf relative stereochemistry of vinblastine. Syntheses of 20′-desethyl-20′-deoxyvinblastine and 20′-desethyl-20′-deoxyvincovaline, *J. Org. Chem.*, 52, 4340, 1987.

49. Borman, L.S. et al., Single site-modified congeners of vinblastine dissociate its various anti-microtu-bule actions, *J. Biol. Chem.*, 263, 6945, 1988.

50. Borman, L.S. and Kuehne, M.E., Specific alterations in the biological activities of C-20′-modified vinblastine congeners, *Biochem. Phamacol.*, 38, 715, 1989.

51. Kuehne, M.E. et al., Synthesis of 5′a-homo-vinblastine and congeners designed to establish structural determinants for isolation of atropoisomers, *J. Org. Chem.*, 66, 5303, 2001.

52. Kuehne, M.E. et al., The synthesis of 16′a-homo-leurosidine and 16′a-homo-vinblastine. Generation of atropoisomers, *J. Org. Chem.*, 66, 531 7, 2001.

53. Kuehne, M.E. et al., Synthesis and biological evaluation of vinblastine congeners, *Org. Biomol. Chem.*, 1, 2120, 2003.

54. Parish, C.A. et al., Circular dichroism studies of bisindole *Vinca* alkaloids, *Tetrahedron*, 54, 15739, 1998.

55. Kruczynski, A. et al., Antimitotic and tubulin-interacting properties of vinflunine, a novel fluorinated *Vinca* alkaloid, *Biochem. Pharmacol.*, 55, 635, 1998.

56. Lobert, S. et al., *Vinca* alkaloids-induced tubulin spiral formation correlates with cytotoxicity in the leukemic L1210 cell line, *Biochemistry*, 39, 12053, 2000.

57. Bennouna, J. et al., Phase I and pharmacokinetic study of the new *Vinca* alkaloid vinflunine admin-istered as a 10-min infusion every 3 weeks in patients with advanced solid tumours, *Ann. Onc.*, 14, 630, 2003.

58. Yamaguchi, K. et al., Identification of novel metabolites of vinorelbine in rat, *Xenobiotica*, 28, 281, 1998.

59. Van Heugen, J.C. et al., New sensitive liquid chromatography method coupled with tandem mass spectrometric detection for the clinical analysis of vinorelbine and its metabolites in blood, plasma, urine and faeces, *J. Chromatogr. A.*, 926, 11, 2001.

60. Lobert, S., Vulevic, B., and Correia, J.J., Interaction of *Vinca* alkaloids with tubulin: a comparison of vinblastine, vincristine, and vinorelbine, *Biochemistry*, 35, 6806, 1996.

61. Kruczynski, A. et al., Preclinical *in vivo* antitumor activity of vinflunine, a novel fluorinated *Vinca* alkaloid, *Cancer Chemother. Pharmacol.*, 41, 437, 1998.

62. Ngan, V.K. et al., Mechanism of mitotic block and inhibition of cell proliferation by the semisynthetic *Vinca* alkaloids vinorelbine and its newer derivative vinflunine, *Mol. Pharmacol.*, 60, 225, 2001.

63. Okouneva, T. et al., The effects of vinflunine, vinorelbine and vinblastine on centromere dynamics, *Mol. Cancer Ther.*, 2, 427, 2003.

64. Nogales, E., Wolf, S.G. and Downing, K.H., Structure of the α,β-tubulin dimer by electron crystallography, *Nature*, 391, 199, 1998.

65. Downing, K.H. and Nogales, E., Crystallographic structure of tubulin: implications for dynamics and drug binding, *Cell Struct. Funct.*, 24, 269, 1999.

66. Checchi, P.M. et al., Microtubule-interacting drugs for cancer treatment, *Trends Pharm. Sci.*, 24, 361, 2003.

67. Rowinsky, E.K and Tolcher, A.W., Antimicrotubule agents, in *Cancer: Principles and Practice of Oncology*, 6th ed., DeVita, V.T., Hellman, S., and Rosenberg, S.A., Eds, Lippincott-Raven, Philadelphia, PA, 2001, 431.

68. Johnson, S.A. et al., Vinorelbine: an overview, *Cancer Treat. Rev.*, 22, 127, 1996.

69. Gregory, R.K. and Smith, I.E., Vinorelbine — a clinical review, *Br. J. Cancer*, 82, 1907, 2000.

70. Pierre Fabre Laboratories, Castres, France. Press release September 10, 2003; http://www.pierre-fabre.com.

71. Binet, S. et al., Immunofluorescence study of the action of navelbine, vincristine and vinblastine on mitotic and axonal microtubules, *Int. J. Cancer*, 46, 262, 1990.

72. Hill, S.A. et al., *Vinca* alkaloids: anti-vascular effects in murine tumour, *Eur. J. Cancer*, 29, 1320, 1993.

73. Etiévant, C. et al., Markedly diminished drug-resistance inducing properties of vinflunine (20′,20′-difluoro-3′,4′-dihydrovinorelbine) relative to vinorelbine, identified in murine and human tumour cells *in vivo* and *in vitro*, with clinical implications, *Cancer Chemother. Pharmacol.*, 48, 62, 2001.

74. Dumontet, C. and Sikic, B.I., Mechanisms of action and resistance to antitubulin agents: microtubule dynamics, drug transport, and cell death, *J. Clin. Oncol.*, 17, 1061, 1999.

75. Kavallaris, M., Verrills, N.M., and Hill, B.T., Anticancer therapy with novel tubulin-interacting agents, *Drug Resistance Updates*, 4, 392, 2001.

76. Baguley, B.C. et al., Inhibition of growth of colon 38 adenocarcinoma by vinblastine and colchicine. Evidence for a vascular mechanism, *Eur. J. Cancer*, 27, 482, 1991.

77. Vacca, A. et al., Antiangiogenesis is produced by non-toxic doses of vinblastine, *Blood*, 94, 4143, 1999.

78. Kruczynski, A. and Hill, B.T., Vinflunine, the latest *Vinca* alkaloid in clinical development. A review of its preclinical anticancer properties, *Crit. Rev. Oncol. Hematol.*, 40, 159, 2001.

79. Aapro, M.S. et al., Developments in cytotoxic chemotherapy: advances in treatment utilising vinorelbine, *Crit. Rev. Oncol. Hematol.*, 40, 251, 2001.

80. Molnar, A. et al., Anti-calmodulin potency of indol alkaloids in *in vitro* systems, *Eur. J. Pharm.*, 291, 73, 1995.

81. Moisoi, N. et al., Calmodulin-containing substructures of the centrosomal matrix released by microtubule perturbation, *J. Cell Sci.*, 115, 2367, 2002.

8 The Bryostatins

David J. Newman

CONTENTS

I. Introduction 137
II. Biological Activities..138
III. Chemical Syntheses of Bryostatins 140
IV. Clinical Trials of Bryostatin 141
V. Future Sources of Bryostatins 141
    A. Aquaculture..141
    B. Actual Source of the Bryostatins 145
    C. Chemical Analogs..146
VI. Conclusions 147
References 147

I. INTRODUCTION

The bryostatins (Figure 8.1) are a class of highly oxygenated macrolides originally isolated by the Pettit group under the early National Cancer Institute (NCI) program (1955–1982) designed to discover novel antitumor agents from natural sources. The initial discovery (of bryostatin 3, Figure 8.1) was indirectly reported in 1970.[1] Subsequent developments leading to the report of the isolation and x-ray structure of bryostatin 1 (Figure 8.1) in 1982,[2] and the multiyear program that culminated in the isolation and purification of (currently) 20 bryostatin structures, have been well documented by a variety of authors over the years.[3–9]

Structurally, all of the molecules possess a 20-membered macrolactone ring. Modifying the description by Hale et al.,[9] all bryostatins possess a 20-membered macrolactone in which there are three remotely substituted pyran rings that are linked by a methylene bridge and an (*E*)-disubstituted alkene; all have geminal dimethyls at C_8 and C_{18} and a four-carbon side chain (carbons 4–1) from the A ring to the lactone oxygen, with another four-carbon chain (carbons 24–27) on the other side of the lactone oxygen to the C ring. Most have an exocyclic methyl enoate in their B and C rings, though bryostatin 3, in particular, has a butenolide rather than the C-ring methyl enoate, and bryostatins 16 and 17 have glycals in place of the regular C_{19} and C_{20} hydroxyl moieties. In the reviews by Hale[9] and Mutter,[7] 18 structures (bryostatins 1–18) are listed; the remaining two structures are now thought to be desoxy-bryostatin 4 and desoxy-bryostatin 5 isolated from *Lissodendoryx isodictyalis*.[10]

The levels of all bryostatins so far isolated from natural sources have always been at the minuscule level, with the highest yield being that of bryostatin 10, of which 15 mg was isolated from 1.5 kg of wet animal in a sample collected in the Gulf of Aomori.[11] The actual source of these macrolides is debatable, and the evidence for microbial involvement is presented in a later section.

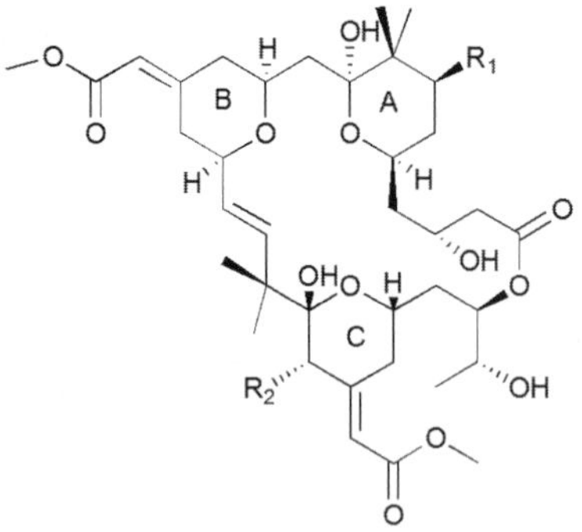

Compound	R₁	R₂
Bryostatin 1	OAc	OCO(CH)$_4$n-Pr
Bryostatin 2	OH	OCO(CH)$_4$n-Pr
Bryostatin 4	OCOC(CH$_3$)$_3$	OCOn-Pr
Bryostatin 5	OCOC(CH$_3$)$_3$	OAc
Bryostatin 6	OCOn-Pr	OAc
Bryostatin 7	OAc	OAc
Bryostatin 8	OCOn-Pr	OCOn-Pr
Bryostatin 9	OAc	OCOn-Pr
Bryostatin 10	OCOC(CH$_3$)$_3$	H
Bryostatin 11	OAc	H
Bryostatin 12	OCO(CH)$_4$n-Pr	OCOn-Pr
Bryostatin 13	OCOn-Pr	H
Bryostatin 14	OCOC(CH$_3$)$_3$	OH
Bryostatin 15	OAc	OCO(CH)$_4$CH(OH)Et

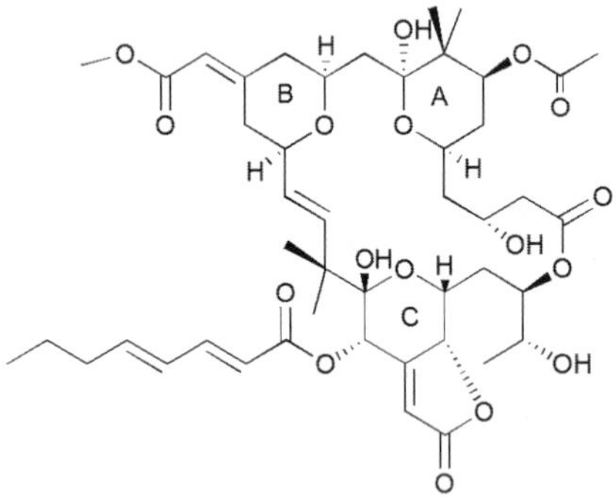

Bryostatin 3

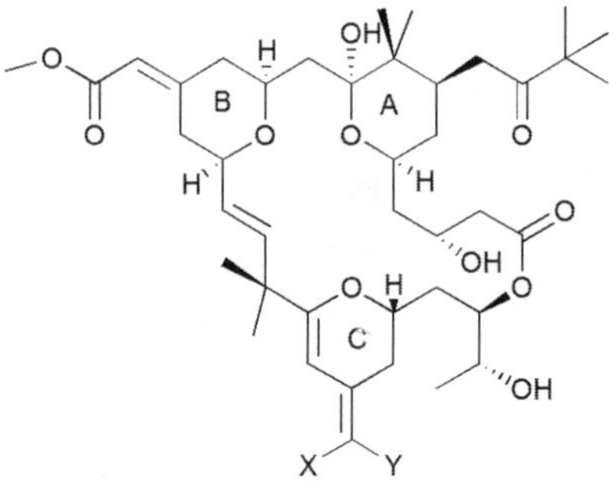

Bryostatin 18

	X	Y
Bryostatin 16	H	C(O)OCH$_3$
Bryostatin 17	C(O)OCH$_3$	H

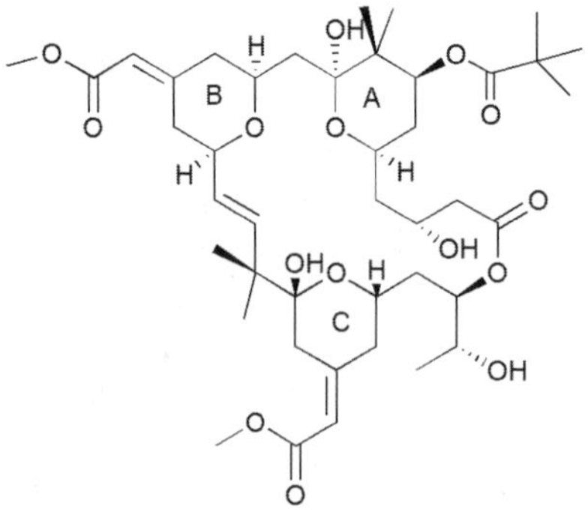

FIGURE 8.1 Structures of the bryostatins.

II. BIOLOGICAL ACTIVITIES

Almost all of the data presented in the following section have been derived from studies with bryostatin 1. However, the basic biology of all of the bryostatins is fundamentally the same; they simply differ in potencies in given systems.

Bryostatin 1 exhibited some extremely interesting biological activities from the beginning. Initially, it demonstrated quite variable activities in the then-current *in vivo* assays that were used at NCI, which were using the P388 and L1210 murine leukemia lines in normal mice. There was very significant variability in the results from batch to batch, but ultimately, using material isolated from *Bugula neritina* collected in the Gulf of California and from bulked fractions from other collection sites, Pettit et al. were able to publish the structure and some biological activities of bryostatin 1 in 1982.[2] During the initial biological workup of bryostatin fractions by NCI it was realized that the response to P388, L1210, and KB cells mimicked the responses shown by extracts from plants of the families *Euphorbiaceae* and *Thymelaceae*. The possibility of isolating a phorbol-like structure from *Bugula* seemed remote, however, as there had never been reports of either phorbol or daphnane-type structures being isolated from marine organisms.[3]

Over the last 20 years, bryostatin 1 has demonstrated a very wide range of biological activities, including immune stimulation, differentiation of transformed cells, and enhancement of cytotoxicity of other agents. Initial experiments by Blumberg's group at NCI and their collaborators in subsequent years demonstrated that bryostatin's anticancer activity was probably based on its interactions with, and subsequent modulation of, protein kinase C isozymes (PKCs) in cells.[12,13] PKC kinases transfer the terminal phosphate group from adenosine triphosphate following binding of diacylglycerol and, frequently, phosphatidyl serine to the PKC isozyme. As a result of these interactions, PKCs are frequently concentrated at the cytosolic surfaces of cell membranes and are active at low physiological concentrations of calcium ions, thus modulating the inositol triphosphate (IP_3) cascade within the cell.

PKCs can be classified into three major groupings: conventional ($\alpha,\beta I/\beta III,\gamma$), where Ca^{2+} is required for activation; novel ($\delta,\epsilon,\theta,\eta/L$), which are independent of Ca^{2+}; and nontypical ($\xi,\lambda/\iota$), where no binding by phorbols is known. The first two groupings have regulatory and catalytic domains, and the regulatory domain in particular has two cysteine-rich domains known as CRD1 and CRD2. Using binding-displacement studies it was shown that bryostatins, phorbol esters (PEs), and diacylglycerol (DAG) all compete for the same binding site or sites on PKC, and through suitable modeling studies, an early pharmacophoric model for DAG and PE was successfully applied to bryostatin binding parameters and to other PKC activators,[14–16] which has subsequently led to the derivation of simpler analogs of the bryostatins by Wender's group (see following).

The main binding sites in the bryostatins included the C_1, C_{19}, and C_{26} oxygen atoms (for convenience, all subsequent comments on chemical analogs will use the bryostatin 1 numeration). This was proven by using chemically modified bryostatins and by following their binding affinities with PKC isoforms (see table 1 in Mutter and Wills for relative binding affinities[7]), but in spite of studies with these agents and other related compounds following cell line responses, no complete explanation as to why the biological responses between bryostatin and PE are so different has emerged to date.

When the specific effects of bryostatin 1 on PKC in cell lines were studied in detail, it was found that although bryostatin down-regulates PKC, the expression of PKC was not directly affected. From both x-ray studies of phorbol 13-acetate bound to murine PKC-δ and the nuclear magnetic resonance solution structure of a PKC-α CRD2 construct, it was determined that the phorbol esters sit in a polar groove that exists at the tip of the CRD. In addition, the binding of such an "activator" does not appear to effect any significant conformational change in the binding domain. Thus, the "activator" sits over the inside polar surface of the groove, creating a continuous hydrophobic surface over approximately one-third of the complexed protein. This hydrophobicity increase of the PKC-δ-phorbol ester complex probably promotes its insertion into the plasma membrane, where it can then engage in signaling related to tumorigenesis,[17] and it is likely that when bryostatin 1 binds to the CRDs of PKCs, similar changes occur, though specific differences may well occur with different PKCs.

Thus, with PKC-δ, where there appears to be a "stabilizing effect" that prevents insertion into the membrane and then its subsequent degradation via the ubiquitin–proteosome pathway, it is possible that the conformational changes are different than those that occur with PKC-α and PKC-δ, where they undergo down-regulation via the degradation pathway. For further details as to the potential route of degradation of ubiquitinylated PKCs, the reader should consult Scheme 2 in the review by Hale et al.[9] Further evidence for specific binding site/overlaps is given in the recent publication from Hale's group, in which they demonstrated, using solution nuclear magnetic resonance techniques that bryostatin 1 (or, more precisely, an analog with specific structural features), does overlap its binding sites with those of phorbol 13-acetate and phorbol-12,13-dibutyrate in the CRD2 site of human PKC-α.[18] A recent review of PKC–drug interactions as potential antitumor therapies should be consulted for further information on PKCs themselves and on other compounds that may well function via these pathways.[19]

In addition to data on the interaction with PKCs, a very significant amount of evidence has accumulated that suggests that bryostatin 1 and, by inference, other similar compounds, but perhaps not the 20-deoxy class such as bryostatin 13, can function as very potent immunostimulants. Thus, resting T cells and neutrophils are activated both *in vitro* and *in vivo*,[20–23] and in clinical trials, it has been shown to raise circulating levels of tumor necrosis factor-α (TNF-α) which is normally produced by the body following immunostimulation.[24] In *in vitro* studies with the murine macrophage line, ANA-1, treatment with bryostatin 1 significantly increased TNF-α mRNA expression and also exhibited synergy with interferon-γ in the production of nitrite and the subsequent expression of the inducible nitric oxide synthetase gene. This gene catalyzes the *in vivo* production of NO from L-arginine, and NO is known to produce strong tumoricidal effects on murine macrophages, probably via induction of the apoptotic cascade.[25] Thus, bryostatins may well be exerting some of their effects via immunomodulation pathways.

In contrast, bryostatin 13, a 20-deoxy-bryostatin, does not stimulate colony formation in bone marrow progenitor cells, whereas bryostatins 1, 3, 8, and 9 do. In addition, bryostatin 13 is claimed to be more potent as an antitumor agent than the other four.[25] Thus, there may well be a component of immunostimulation in the antitumor activities of the C_{20}-O-acyl bryostatins that is not present in the 20-deoxy class.

The recent report by Battle and Frank has demonstrated that one potential mechanism of bryostatin 1–mediated differentiation in human CLL cells is activation of the signal transducer and activator of transcription (STAT). Thus, in cells taken from clinical trials patients, bryostatin 1 appears to activate STAT1 in a PKC-dependent manner by induction of an IFN-γ autocrine loop; this leads to activation of the JAK-STAT1 signaling pathway and to the ultimate differentiation of the cells.[26]

There are also reports that imply that bryostatin 1, under certain conditions, might have some tumor-inducing capability, though this is an extrapolation from cell line studies. In studies in which it was demonstrated that bryostatin 1 can selectively target PKC-II isozymes in K562 (human erythroleukemia) and HL60 (human promyelocytic) cell lines, such treatment led to membrane translocation and lamin B phosphorylation at specific sites on the lamin proteins. Such translocation/phosphorylation causes breakdown of the nuclear envelope during mitosis, and in these cells, such an effect appears to be associated with enhanced proliferation.[27] Thus, care might have to be taken in choosing patients for bryostatin treatment (see later section on clinical trials).

III. CHEMICAL SYNTHESES OF BRYOSTATINS

Since the publication of the first structure by Pettit in 1982, these molecules have been the target of many synthetic chemistry groups. Many partial syntheses have been published in which specific portions of the molecule have been made, but to date, only three of the bryostatins have been synthesized. The first was the enantioselective total synthesis of bryostatin 7 in 1990 by Masamune et al.,[28] the second was by Evans et al. on the enantiomeric total synthesis of bryostatin 2 in 1999,[29] and the third was the synthesis of bryostatin 3 by Nishiyama and Yamamura in 2000.[30] In addition to these papers, three excellent review articles covering through 2002, on the syntheses of these three and other partial bryostatin structures including bryostatin 1, have been published and should be consulted for specific details of reaction schemes and comparisons of routes.[7,9,31]

IV. CLINICAL TRIALS OF BRYOSTATIN

Although a number of bryostatins have been tested in animals and many have had *in vitro* assessments in many cell lines, only bryostatin 1 has entered human clinical trials. Part of the reason is that this is the material that could be obtained in quantities large enough to be able to isolate and produce the compound under cGMP conditions. The history of the initial attempts at large-scale isolations and the methods ultimately used are given in the 1996 review by Newman and are also covered to some extent in the 1991 and 1996 reviews by Pettit. These reviews should be consulted for the details as to sources and methods.[4–6] As a result of these endeavors, enough cGMP-grade material was produced from wild collections to provide a source for most of the clinical trials shown in Tables 8.1 and 8.2. The trials in the United Kingdom in the early 1990s were not performed using the batches made under NCI's auspices but, instead, used batches provided by Pettit; however, subsequent trials have all used the NCI-sourced materials.

To date, there have been reports in the literature from seven phase I and 16 phase II trials in which bryostatin 1 was used as monotherapy (Table 8.1). There have also been reports of three phase I trials in which bryostatin 1 was used in conjunction with vinblastine, cytosine arabinoside (AraC), and fludarabine, respectively, and one phase II trial in which paclitaxel was used with bryostatin 1 (Table 8.2).

Initially, bryostatin 1 demonstrated some partial responses in phase I trials, with myalgia being the dose-limiting toxicity. Subsequent phase I and phase II trials with bryostatin as monotherapy, using a variety of treatment regimens in patients with carcinomas ranging from solid tumors (melanomas, renal and ovarian) to varied leukemias, did not give consistent patterns of responses whether measured as stable disease (SD), or partial (PR) or complete responses (CR). There would be an occasional CR or PR and some SDs, but no consistent responses over the patient population on the trial in the case of the phase IIs. However, when one inspects the patient populations from the aspect of the types of carcinomas, then from the knowledge that is now accruing as to dose-limiting toxicities and methods/timing of administration, there are potential reasons as to why monotherapy by bryostatin is not optimal except perhaps in a very selective cohort of leukemia patients (see earlier comments on choice of patients because of potential tumorigenesis in some carcinoma lines).

However, when bryostatin is combined with another cytotoxin, such as the vinca alkaloids or nucleosides, and the carcinomas are leukemic in nature, then the response rates, even in phase I trials, begin to demonstrate that such mixed treatments may well be worth further investigation (cf. Table 8.2).

Thus, with the combination of high levels of cytosine arabinoside (AraC) and low levels of bryostatin in patients with leukemias, in a population that included patients who had failed high dose AraC or "HiDaC" therapy, 5 of 23 patients presented with complete responses in a phase I trial.[32] Similarly, patients with chronic lymphocytic leukemia (CLL) and Non-Hodgkins lymphoma (NHL) treated with fludarabine and bryostatin, were reported to show close to 50% "objective responses" in the trial report.[33] Finally, in a phase II trial of bryostatin and paclitaxel, 7 of 11 patients with non-small cell lung cancer demonstrated positive responses (PR/SD), but no CRs.[34]

At present (January 2005), there are three phase I and two phase II trials underway (see data from the NCI clinical trials Web site, http://clinicaltrials.nci.nih.gov), and in every case, these are combination studies with biologicals such as interleukin 2 or GM-CSF, nucleoside derivatives such as cladribine, or vincristine. These combinations are being tested against leukemias and lymphomas and against ovarian and prostate carcinomas. It is hoped, similar results to those demonstrated in Table 8.2 will be reported in due course.

V. FUTURE SOURCES OF BRYOSTATINS

A. AQUACULTURE

The work reported to date has been from wild collections, made predominately in either the Gulf of California or in the Pacific Ocean off Palos Verdes, California. However, it became obvious,

TABLE 8.1
Published Data Summary of Bryostatin-1 Monotherapy

Year	Phase	Schedule	Dose Range μg/M²	Tumor Type(s)	No. Patients	CR	PR	SD	Side Effects	Reference
1993	I	1 h infusion; repeated at 14 d; 3 cycles	5–65	Varied	19	0	0	0	Myalgia; phlebitis	Prendiville et al.[51]
1993	I	1 h infusion; 7-, 14-, or 21 d cycle; varied number of cycles	25–50	Varied	35	0	2	0	Myalgia	Philip et al.[24]
1995	I	24 h infusion; repeated at 7 d; intervals for a maximum of 8 cycles	25–50	Varied	19	0	4	0	Myalgia; phlebitis	Jayson et al.[52]
1998	I	24–72 h infusions; repeated at 14 d for 4–42 cycles	12–180	NHL, CLL	29	0	0	11	Myalgia	Varterasian et al.[53]
1998	Ib	0.5–24 h infusion; repeated at 8 and 15 d, then repeated at 28 d; varied number of cycles	12.5–25	Varied	12	0	0	0	Myalgia	Grant et al.[54]
1999	I	1 h infusion; repeated at 8 and 15 d, then repeated at 28 d; varied number of cycles	20–57	Varied	22	0	0	4	Myalgia; photophobia	Weitman et al.[55]
2002	I	4–14 d infusions; repeated at 14–21 d intervals; 2–3 cycles	8–24	Varied	37	0	0	6	Myalgia; fatigue	Marshall et al.[56]
1998	II	1 h infusion; repeated at 8 and 15 d, then repeated at 28 d; 1–6 cycles	25	Melanoma	15	0	0	1	Myalgia	Propper et al.[57]
1999	II	1 h infusion; repeated at 8 and 15 d, then repeated at 28 d; 0.67–8 cycles	25	Melanoma	17	0	0	1	Myalgia	Gonzalez et al.[58]
2000	II	72 h infusion; repeated at 14 d; 4–14 cycles.	120	NHL, CLL	25	1	2	2	Myalgia	Varterasian et al.[59]
2000	II	1 h infusion; repeated at 8 and 15 d, then repeated at 28 d; 1–14 cycles	25	RCC	30	1	1	0	Anemia; myalgia	Pagliaro et al.[60]
2001	II	24 h infusion; repeated at 8 and 15 d, then repeated at 28 d; 2–4 cycles	25–35	CRC	28	0	0	0	Myalgia	Zonder et al.[61]

Year	Phase	Schedule	Dose	Tumor					Adverse effect	Reference
2001	II	24 h infusion; repeated at 8 and 15 d, then repeated at 28 d; 72 h infusion; repeated at 15 d, then repeated at 28 d; 2–4 cycles	25–40	Melanoma	49	0	1	14	Myalgia	Bedikian et al.[62]
2001	II	24 h infusion; repeated weekly; 1–9 cycles	25	NHL	14	0	0	1	Myalgia	Blackhall et al.[63]
2001	II	72 h infusion; at 14 d intervals; 2–8 cycles	9	MM	9	0	0	0	Myalgia	Varterasian et al.[64]
2001	II	72 h infusion; at 14 d intervals; 3 cycles	120	Sarcoma; HNC	24	0	0	6	Myalgia; hyponatremia	Brockstein et al.[65]
2002	II	24 h infusion; repeated at 8 and 15 d, then repeated at 28 d; 1–5 cycles	25	HNC	14	0	0	1	Myalgia (grade 1)	Pfister et al.[66]
2002	II	24 h infusion (low dose); every 7 d; 72 h infusion (high dose); every 14 d; 1–31 cycles	25–120	Melanoma	30	0	0	7	Myalgia	Tozer et al.[67]
2003	II	1 h infusion; repeated at 8 and 15 d, then repeated at 28 d; 1–8 cycles.	35–40	RCC	32	0	2	15	Myalgia	Haas et al.[68]
2003	II	24 h infusion (low dose); 1, 8 and 15 d, repeated every 28 d; 72 h infusion (high dose); every 14 d; 1–9 cycles.	25–120	Ovarian	54	0	1[a]	19	Myalgia	Armstrong et al.[69]
2003	II	24 h infusion; repeated every 7 d; 1–9 cycles.	19–25	Ovarian	17	0	0	4[a]	Myalgia	Clamp et al.[70]
2003	II	24 h infusion; 8 and 15 d, then repeated at 28 d; 1–9 cycles.	25	RCC	13	0	0	3[a]	Myalgia	Madhusudan et al.[71]
2003	II	24 h infusion (low dose); repeated at 8 and 15 d, then repeated at 28 d; 72 h infusion (high dose); every 14 d; varied cycles.	25–120	Cervical	65	0	2	20	Myalgia	Armstrong et al.[72]

[a] Some question as to response level

TABLE 8.2
Combination Therapy with Cytotoxic Drugs

Year	Phase	Schedule	Dose Range	Tumor Type	No. Patients	CR	PR	SD	Side Effects	Reference
2001	I	24 h infusion and bolus of vincristine; dose escalation of bryostatin; 1–5 cycles	12.5–62.5 µg.M² bryostatin; 1.4 mg.M² V	B-cell cancer	25	1	2	4	Myalgia; neuropathy	Dowalti et al.[73]
2002	I	24 h infusion; days 1 and 11; AraC on days 2, 3, 9, 10. bryostatin dose escalation, fixed AraC; 1–6 cycles	12.5–50 µg.M² bryostatin; 1–3 mg.M² AraC	Leukemia	23	5	1[a]	0	Myalgia; neutropenia	Cragg et al.[32]
2002	I	24 h infusion; FAra for days 2–6. Repeat at 28 d; or reverse addition order; 6–9+ cycles	16–50 µg.M² bryostatin; 12.5–25 mg/M² FAra	CLL; NHL	53	?	?	?	Neutropenia	Roberts et al.[33]
2003	II	1 h infusion of paclitaxel on 1, 8, and 15 d; 24 h infusion of bryostatin on 2, 9, 16 d; repeated on 28 d cycle; 1–4 cycles	40–50 µg.M² bryostatin; 90 mg.M² paclitaxel	NSCLC	11	0	2[a]	5	Myalgia	Winegarden et al.[34]

[a] Some question as to response level
? "non-defined objective responses"

even as the initial trials were beginning in the early 1990s, that wild collections would not suffice and that production via chemical synthesis would probably not be viable. Thus, in the early 1990s, NCI, using the Small Business Innovation Research program, established a phase I and a subsequent phase II aquaculture program with a small company, CalBioMarine, with the aim of investigating the possibilities for both on-land and in-sea aquaculture for production of *B. neritina* under conditions that were not at the vagaries of nature. This series of projects was successful, culminating in the proof by CalBioMarine that the organism could be grown under both conditions and also that bryostatins could be isolated from the aquacultured animal in adequate quantities and at costs less than those incurred in wild collections.[35]

Concomitantly with the phase II aquaculture experiments, NCI also used the Small Business Innovation Research mechanism to investigate, with the Massachusetts company Aphios, the potential for supercritical extraction of bryostatins from sources of *B. neritina*, both from wild collections and from some of the material from aquaculture. This was successful, and a demonstration of the potential for such methods was shown in the review by Newman,[6] in which a comparison was made of the methods used for the earlier cGMP purification versus the Aphios technique. The method used for cGMP production involved a four-stage extraction–concentration process, and then a six-step process repeated 15 times to produce 18 g of bryostatin 1 from 13 metric tonnes of *B. neritina*. The overall process was multiweek in duration and used massive amounts of solvents. In contrast, the supercritical technique used basically carbon dioxide as the extraction medium on the wet animal mass, followed by supercritical chromatography, and the process was reduced to six simple operations performed within days, yet yielding material of similar purity.

B. ACTUAL SOURCE OF THE BRYOSTATINS

One interesting observation arising from the search for bryostatin sources was that, despite the ubiquitous occurrence of the nominal producing organism, the number of *B. neritina* colonies actually producing detectable bryostatin 1–3 levels was very low and geographically spread. One possible solution to this question came from the work of Haygood and her collaborators at the Scripps Institution of Oceanography, which has shown that the bryozoan is actually the host to a symbiotic organism that may well be the actual producer of the compound. In an elegant series of experiments, she and her colleagues have demonstrated the presence of a putative (polyketide synthase) PKS-I gene fragment in colonies that produce bryostatin, and that the fragment is absent in nonproducers.[36] In addition, they demonstrated that there are subdivisions within *B. neritina* samples taken from the same sites, but at different depths. Thus, at depths greater than 9 m, (the D or deep type), bryostatins 1–3 and minor components are found (these are also known as producers of chemotype O for "octa-2,4-dienoatic chain"), whereas at less than 9 m (S or shallow type), only the minor derivatives are seen (chemotype M). The symbiotic organisms (*Candidatus* Endobugula neritina) isolated from each type differ in the mitochondrial protein CO I sequences by 8%, giving rise to the possibility that the bryozoans are also different taxonomically.[37] Later reports demonstrated that the possibility of transferring this particular PKS fragment to other, more amenable microbes was being investigated with the goal of producing bryostatin by fermentative means. Recently Haygood[38] suggested that the PKS system resembles that reported by Piel[39] for the *Paederus* beetle's pseudomonal symbiont PKS that produces pederine, in that there are no acyltransferase (AT) domains in the clusters. Further work is ongoing, using "remote" AT domains from another organism. Fuller details of the work have recently been published by Haygood and her collaborators.[40] Further work by Haygood and collaborators has demonstrated that closely related symbiotic microbes are also found in *Bugula simplex*, and that these organisms both form a monophyletic grouping with *E. sertula* and appear to have PKS sequences that mimic the bryostatin PKS cluster.[41] If successful, cultivation of the organism, or a surrogate with the bryostatin PKS system expressed, could probably solve any production problems if bryostatin becomes a viable drug.

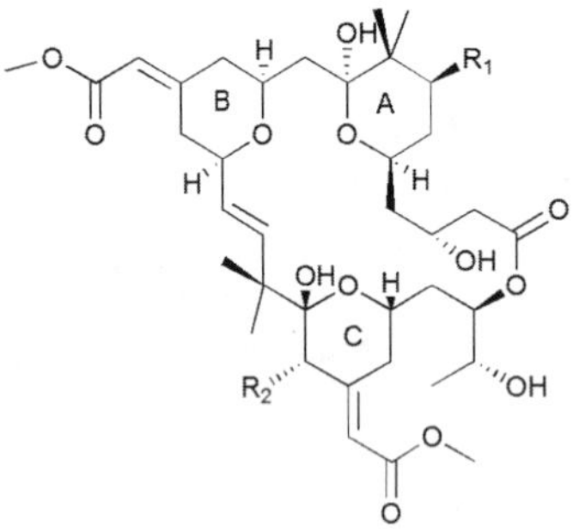

Base Structure for Bryostatins

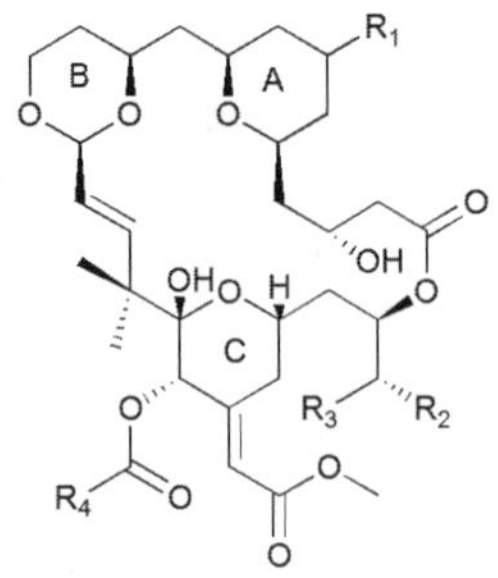

1a, 1b, 1c

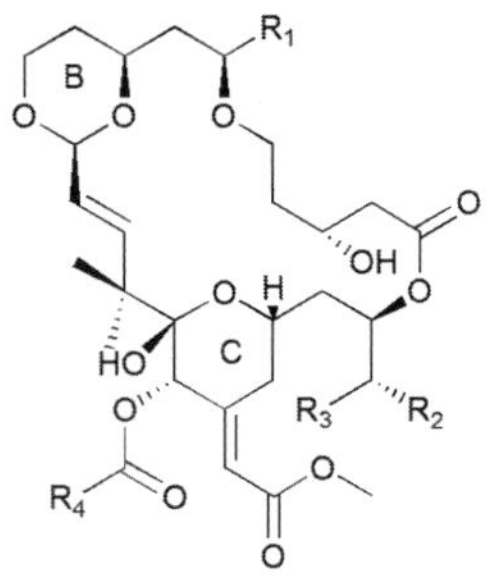

2a, 2b

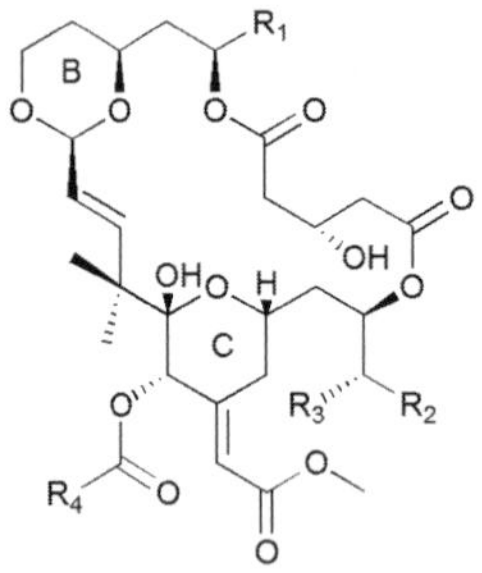

Compound	R₁	R₂	R₃	R₄	Ki (nM)
1a	H	OH	CH₃	C₇H₁₅	3.4
1b	H	OH	H	C₇H₁₅	0.25
1c	H	OAc	CH₃	C₇H₁₅	10000.0
2a	t-Bu	OH	CH₃	C₇H₁₅	8.3
2b	H	OH	CH₃	C₇H₁₅	47.0
3	C(CH₃)₂C₂H₅	OH	CH₃	C₇H₁₃	8.42

3

FIGURE 8.2 Structures of bryostatin analogs.

C. CHEMICAL ANALOGS

At present, the total synthesis of bryostatin 1 is not a feasible process for the production of this agent. However, synthesis of a simpler analog with comparable activity might well be a viable option. In 1986, Wender et al. analyzed the potential binding site of the phorbol esters on PKC as a guide to the design of simpler analogs of these agents.[14] In 1988, this work was expanded by modeling bryostatin 1 onto the same binding site as a result of the initial results indicating that bryostatin 1 interacted with PKC.[15] Subsequently, the modeling work was refined to produce three analogs that would maintain the putative binding sites at the oxygen atoms at C_1 (ketone), C_{19} (hydroxyl), and C_{26} (hydroxyl) in the original molecule. These requirements gave rise to structures **1a**, **2a**, and **2b** (Figure 8.2), which maintained the recognition features but removed a significant amount of the peripheral substituents. These molecules demonstrated nanomolar binding constants when measured in displacement assays of tritiated phorbol esters, with the figures being in the same general range as bryostatin 1, and **1a** and **2a** had activities in *in vitro* cell line assays close to those demonstrated by bryostatin 1 itself.[42–45] Following on from these examples, modifications were made to the base structure **1a** to introduce a second lactone **3** (Figure 8.2) which had 8 nM binding affinity and also inhibited P388 with an ED_{50} of 113 nM.[46] Concomitantly, modifications were made to the base analog **1a**, where different fatty acid esters were made (structures not shown). These, too, exhibited binding affinities for PKC isozymes in the 7–232 nM range, depending on the fatty acid used.[47]

To show the versatility of the base structure, recently Wender et al. published a simple modification in which, by removal of a methyl group in the C_{26} side chain from compound **1a** to produce

1b (Figure 8.2), the binding affinity to PKC was increased to the picomolar level[48] and the compound demonstrated greater potency than bryostatin 1 in *in vitro* cell line assays. Improved syntheses of the molecule that could permit further refinements of the model, and that have the potential for greater overall yields, have been published.[49, 50]

VI. CONCLUSIONS

The unique structural motifs that are part of the bryostatin molecule have now undergone a significant amount of testing in humans. Monotherapy does not appear to be an optimal use, but the potential for usage as part of a combination with other agents is now under intense study, with some promising results — particularly in leukemias — being reported at phase I levels.

The problems of supply have been addressed both biologically and chemically. Although aquaculture-based production is feasible, the isolation of a putative producing microbial symbiont from larvae of animals that do produce bryostatins appears to have greater potential, though it has not as yet been grown. The production by total synthesis of simpler "bryologs" exhibiting similar or better levels of biological activity in both cell lines and against PKC isozymes holds promise for the development of more effective analogs.

It is hoped, within the next few years the promise of this class of molecule (naturally occurring or synthetic) will be realized and a new series of antitumor drugs will reach the general patient population.

REFERENCES

1. Pettit, G. R. et al., Antineoplastic components of marine animals, *Nature*, 227, 962, 1970.
2. Pettit, G. R. et al., Isolation and structure of bryostatin 1, *J. Am. Chem. Soc.*, 104, 6846, 1982.
3. Suffness, M., Newman, D. J., and Snader, K. M., Discovery and development of antineoplastic agents from natural sources, in *Bioorganic Marine Chemistry*, Vol. 3, Scheuer, P. J., Ed., Springer-Verlag, Berlin, 1989, pp 131–168.
4. Pettit, G. R., The bryostatins, in *Progress in the Chemistry of Organic Natural Products*, Vol. 57, Hertz, W., Kirby, G. W., Steglich, W. and Tamm, C., Eds., Springer-Verlag, New York, 1991, pp 135–195.
5. Pettit, G. R., Progress in the discovery of biosynthetic anticancer drugs, *J. Nat. Prod.*, 59, 812, 1996.
6. Newman, D. J., Bryostatin — from bryozan to cancer drug, in *Bryozoans in Space and Time*, Gordon, D. P., Smith, A. M. and Grant-Mackie, J. A., Eds., NIWA, Wellington, 1996, pp 9–17.
7. Mutter, R. and Wills, M., Chemistry and clinical biology of the bryostatins, *Bioorg. Med. Chem.*, 8, 1841, 2000.
8. Pettit, G. R., Herald, C. L., and Hogan, F., Biosynthetic products for anticancer drug design and treatment: the bryostatins, in *Anticancer Drug Development*, Baguley, B. C., and Kerr, D. J., Eds., Academic Press, San Diego, CA, 2002, pp 203–235.
9. Hale, K. J. et al., The chemistry and biology of the bryostatin antitumour macrolides, *Nat. Prod. Rep.*, 19, 413, 2002.
10. Pettit, G. R. et al., Relationship of *Bugula neritina* (*Bryozoa*) antineoplastic constituents to the yellow sponge *Lyssodendoryx isodictyalis*, *Pure Appl. Chem.*, 58, 415, 1986.
11. Kamano, Y. et al., An improved source of bryostatin 10, *Bugula neritina* from the Gulf of Aomori, Japan, *J. Nat. Prod.*, 58, 1868, 1995.
12. Hennings, H. et al., Bryostatin 1, an activator of protein kinase C, inhibits tumor promotion by phorbol esters in SENCAR mouse skin, *Carcinogenesis*, 8, 1343, 1987.
13. Blumberg, P. M. et al., The protein kinase C pathway in tumor promotion, *Prog. Clin. Biol. Res.*, 298, 210, 1989.
14. Wender, P. A. et al., Analysis of the phorbol ester pharmacophore on protein kinase C as a guide to the rational design of new classes of analogs, *Proc. Natl. Acad. Sci. USA*, 83, 4214, 1986.

15. Wender, P. A. et al., Modeling of the bryostatins to the phorbol ester pharmacophore on protein kinase C, *Proc. Natl. Acad. Sci. USA*, 85, 7917, 1988.

16. Lee, J. et al., Protein kinase C ligands based on tetrahydrofuran templates containing a new set of phorbol ester pharmacophores, *J. Med. Chem.*, 42, 4129, 1999.

17. Zhang, G. G. et al., Crystal-structure of the CYS2 activator-binding domain of protein kinase C-delta in complex with phorbol ester, *Cell*, 81, 917, 1995.

18. Hale, K. J. et al., Synthesis of a simplified bryostatin C-ring analog that binds to the CRD2 of human PKC-alpha and construction of a novel BC-analog by an unusual Julia olefination process, *Org. Lett.*, 5, 499, 2003.

19. Goekjian, P. G. and Jirousek, M. R., Protein kinase C inhibitors as novel anticancer drugs, *Expert Opin. Investig. Drugs*, 10, 2117, 2001.

20. May, W. S. et al., Antineoplastic bryostatins are multipotential stimulators of human hematopoietic progenitor cells, *Proc. Natl. Acad. Sci. USA*, 84, 8483, 1987.

21. Hess, A. D. et al., Activation of human T lymphocytes by bryostatin, *J. Immunol.*, 141, 3263, 1988.

22. Berkow, R. L. et al., *In vivo* administration of the anticancer agent bryostatin 1 activates platelets and neutrophils and modulates protein kinase C activity, *Cancer Res.*, 53, 2810, 1993.

23. Esa, A. H. et al., Bryostatins trigger human polymorphonuclear neutrophil and monocyte oxidative metabolism: association with *in vitro* antineoplastic activity, *Res. Immunol.*, 146, 351, 1995.

24. Philip, P. A. et al., Phase I study of bryostatin 1: assessment of interleukin 6 and tumor necrosis factor alpha induction in vivo. The Cancer Research Campaign Phase I Committee, *J. Natl. Cancer Inst.*, 85, 1812, 1993.

25. Taylor, L. S. et al., Bryostatin and IFN-gamma synergize for the expression of the inducible nitric oxide synthetase gene and for nitric oxide production in murine macrophages, *Cancer Res.*, 57, 2468, 1997.

26. Battle, T. E. and Frank, D. A., STAT1 mediates differentiation of chronic lymphocytic leukemia cells in response to Bryostatin 1, *Blood*, 102, 3016, 2003.

27. Hocevar, B., Burns, D. J., and Fields, A. P., Identification of protein kinase C (PKC) phosphorylation sites on human lamin B, *J. Biol. Chem.*, 268, 7545, 1993.

28. Kageyama, M. et al., Synthesis of bryostatin 7, *J. Am. Chem. Soc.*, 112, 7407, 1990.

29. Evans, D. A. et al., Total synthesis of bryostatin 2, *J. Am. Chem. Soc.*, 121, 7540, 1999.

30. Ohmori, K. et al., Total synthesis of bryostatin 3, *Angew. Chem. Int. Ed.*, 39, 2290, 2000.

31. Norcross, R. D. and Paterson, I., Total synthesis of bioactive marine macrolides, *Chem. Rev.*, 95, 2041, 1995.

32. Cragg, L. et al., Phase I trial and correlative laboratory studies of bryostatin 1 (NSC 339555) and high-dose 1-β-D-arabinofuranosylcytosine in patients with refractory acute leukemia, *Clinical Cancer Research*, 8, 2123, 2002.

33. Roberts, J. D. et al., Phase I study of bryostatin-1 and fludarabine in patients with chronic lymphocytic leukemia and indolent non-Hodgkin's lymphoma, *Clinical Lymphoma*, 3, 184, 2002.

34. Winegarden, J. D. et al., A phase II study of bryostatin-1 and paclitaxel in patients with advanced non-small cell lung cancer, *Lung Cancer*, 39, 191, 2003.

35. Mendola, D., Aquaculture of three phyla of marine invertebrates to yield bioactive metabolites: process development and economics, *Biomol. Engin.*, 20, 441, 2003.

36. Davidson, S. K. et al., Evidence for the biosynthesis of bryostatins by the bacterial symbiont "Candidatus Endobugula sertula" of the bryozoan *Bugula neritina*, *Appl. Environ. Microbiol.*, 67, 4531, 2001.

37. Davidson, S. K. and Haygood, M. G., Identification of sibling species of the bryozoan *Bugula neritina* that produce different anticancer bryostatins and harbor distinct strains of the bacterial symbiont "*Candidatus* Endobugula sertula," *Biol. Bull.*, 196, 273, 1999.

38. Haygood, M. G., The role of a bacterial symbiont in the biosynthesis of bryostatins in the marine bryozoan *Bugula neritina*, *Abs. Pap. 6th Int. Mar. Biotech. Conf.*, Abs. S5, 2003.

39. Piel, J., A polyketide synthase-peptide synthase gene cluster from an uncultured bacterial symbiont of *Paederus* beetles, *Proc. Natl. Acad. Sci. USA*, 99, 14002, 2002.

40. Hildebrand, M. et al., Approaches to identify, clone and express symbiont bioactive metabolite genes, *Nat. Prod. Rep.*, 21, 122, 2004.

41. Lim, G. E. and Haygood, M. G., *Candidatus* Endobugula glebosa, a specific bacterial symbiont of the marine bryozoan *Bugula simplex*, *Appl. Environ. Microbiol.*, 70, 4921, 2004.

42. Wender, P. A. et al., Synthesis and biological evaluation of fully synthetic bryostatin analogs, *Tetrahedron Lett.*, 39, 8625, 1998.

43. Wender, P. A. et al., Synthesis of the first members of a new class of biologically active bryostatin analogs, *J. Am. Chem. Soc.*, 120, 4534, 1998.

44. Wender, P. A. et al., The design, computer modeling, solution structure, and biological evaluation of synthetic analogs of bryostatin 1, *Proc. Natl. Acad. Sci. USA*, 95, 6624, 1998.

45. Wender, P. A. et al., The rational design of potential chemotherapeutic agents: synthesis of bryostatin analogs, *Med. Res. Revs.*, 19, 388, 1999.

46. Wender, P. A. and Lippa, B., Synthesis and biological evaluation of bryostatin analogs: the role of the A-ring, *Tetrahedron Lett.*, 41, 1007, 2000.

47. Wender, P. A. and Hinkle, K. W., Synthesis and biological evaluation of a new class of bryostatin analogues: the role of the C20 substituent in protein kinase C binding, *Tetrahedron Lett.*, 41, 6725, 2000.

48. Wender, P. A. et al., The practical synthesis of a novel and highly potent analog of bryostatin, *J. Am. Chem. Soc.*, 124, 13648, 2002.

49. Wender, P. A., Mayweg, A. V. W., and VanDeusen, C. L., A concise, selective synthesis of the polyketide spacer domain of a potent bryostatin analog, *Org. Lett.*, 5, 277, 2003.

50. Wender, P. A., Koehler, M. T. F., and Sendzik, M., A new synthetic approach to the C ring of known as well as novel bryostatin analogs, *Org. Lett.*, 5, 4549, 2003.

51. Prendiville, J. et al., A phase I study of intravenous bryostatin 1 in patients with advanced cancer, *Br. J. Cancer*, 68, 418, 1993.

52. Jayson, G. C. et al., A phase I trial of bryostatin 1 in patients with advanced malignancy using a 24 hour intravenous infusion, *Br. J. Cancer*, 72, 461, 1995.

53. Varterasian, M. L. et al., Phase I study of bryostatin 1 in patients with relapsed non-Hodgkin's lymphoma and chronic lymphocytic leukemia, *J. Clin. Oncol.*, 16, 56, 1998.

54. Grant, S. et al., Phase Ib trial of bryostatin 1 in patients with refractory malignancies, *Clin. Cancer Res.*, 4, 611, 1998.

55. Weitman, S. et al., A phase I trial of bryostatin-1 in children with refractory solid tumors: a pediatric oncology group study, *Clin. Cancer Res.*, 5, 2344, 1999.

56. Marshall, J. L. et al., Phase I study of prolonged infusion Bryostatin-1 in patients with advanced malignancies, *Cancer Biol. Ther.*, 1, 409, 2002.

57. Propper, D. J. et al., A phase II study of bryostatin 1 in metastatic malignant melanoma, *Br. J. Cancer*, 78, 1337, 1998.

58. Gonzalez, R. et al., Treatment of patients with metastatic melanoma with bryostatin-1 — a phase II study, *Melanoma Research*, 9, 599, 1999.

59. Varterasian, M. L. et al., Phase II trial of bryostatin 1 in patients with relapsed low-grade non-Hodgkin's lymphoma and chronic lymphocytic leukemia, *Clin. Cancer Res.*, 6, 825, 2000.

60. Pagliaro, L. et al., A phase II trial of bryostatin-1 for patients with metastatic renal cell carcinoma, *Cancer*, 89, 615, 2000.

61. Zonder, J. A. et al., A phase II trial of bryostatin 1 in the treatment of metastatic colorectal cancer, *Clin. Cancer Res.*, 7, 38, 2001.

62. Bedikian, A. Y. et al., Phase II evaluation of bryostatin-1 in metastatic melanoma, *Melanoma Res.*, 11, 183, 2001.

63. Blackhall, F. H. et al., A phase II trial of bryostatin 1 in patients with non-Hodgkin's lymphoma, *Br. J. Cancer*, 84, 465, 2001.

64. Varterasian, M. L. et al., Phase II study of bryostatin 1 in patients with relapsed multiple myeloma, *Investig. New Drugs*, 19, 245, 2001.

65. Brockstein, B. et al., Phase II studies of bryostatin-1 in patients with advanced sarcoma and advanced head and neck cancer, *Investig. New Drugs*, 19, 249, 2001.

66. Pfister, D. G. et al., A phase II trial of bryostatin-1 in patients with metastatic or recurrent squamous cell carcinoma of the head and neck, *Investig. New Drugs*, 20, 123, 2002.

67. Tozer, R. G. et al., A randomized phase II study of two schedules of bryostatin-1 (NSC339555) in patients with advanced malignant melanoma, *Investig. New Drugs*, 20, 407, 2002.

68. Haas, N. B. et al., Weekly bryostatin-1 in metastatic renal cell carcinoma: A phase II study, *Clin. Cancer Res.*, 9, 109, 2003.

69. Armstrong, D. K. et al., A randomized phase II evaluation of bryostatin-1 (NSC #339555) in recurrent or persistent platinum-sensitive ovarian cancer: a Gynecologic Oncology Group Study, *Investig. New Drugs*, 21, 373, 2003.

70. Clamp, A. R. et al., A phase II trial of bryostatin-1 administered by weekly 24-hour infusion in recurrent epithelial ovarian carcinoma, *Br. J. Cancer*, 89, 1152, 2003.

71. Madhusudan, S. et al., A multicentre phase II trial of bryostatin-1 in patients with advanced renal cancer, *Br. J. Cancer*, 89, 1418, 2003.

72. Armstrong, D. K. et al., A randomized phase II evaluation of bryostatin-1 (NSC #339555) in persistent or recurrent squamous cell carcinoma of the cervix: a Gynecologic Oncology Group Study, *Investig. New Drugs*, 21, 453, 2003.

73. Dowalti, A. et al., Phase I trial of combination bryostatin-1 and vincristine in B-cell malignancies: final report, *Clin. Cancer Res.*, 7 supplement 3814s, Abs 800, 2001.

9 The Isolation, Characterization, and Development of a Novel Class of Potent Antimitotic Macrocyclic Depsipeptides: The Cryptophycins

Rima S. Al-awar and Chuan Shih

CONTENTS

I. Introduction 151
II. Isolation and Characterization of the Cryptophycins 152
III. Mechanism of Action of the Cryptophycins 153
    A.   Cellular Mechanism of Action of Cryptophycin 52 (LY355703) 154
    B.   Effects of Cryptophycin 52 (LY355703) on Microtubule Polymerization
       In Vitro 154
IV. Synthesis of Cryptophycin 52 (LY355703)...156
    A.   Retrosynthetic Analysis and Final Assembly of the Fragments.............156
    B.   Preparation of Fragments (A–D)...158
V. SARs of the Cryptophycins 158
    A.   SAR of the β-Epoxide Region of Fragment A 160
    B.   SAR of the Phenyl Group of Fragment A 160
    C.   SAR of the D-Chlorotyrosine of Fragment B 161
    D.   SAR of the C/D Ester Bond and the C6, C7 Positions................................162
VI. Comparison of Cryptophycins 52, 55, and 55-glycinate 164
VII. Cryptophycin 52 (LY355703)...165
    A.   *In Vivo* Antitumor Activity of Cryptophycin 52 (LY355703) 165
    B.   Clinical Evaluation of Cryptophycin 52 (LY355703) 166
VIII. Conclusions 166
Acknowledgments..166
References 167

I. INTRODUCTION

The cryptophycins are potent antitumor depsipeptides first isolated from terrestrial blue-green algae. Initial cellular mechanism of action studies showed them to be associated with the inhibition of mitotic spindle function. This characterization puts the cryptophycins into one of the clinically very important classes of anticancer agents, such as the vinca alkaloids (vincristine, vinblastine, vinorelbine) and the taxanes (taxol and taxotere), that act primarily on the cellular microtubule structure and function. Extensive preclinical studies have demonstrated that the cryptophycins are highly active against a broad spectrum of murine solid tumors and human tumor xenografts *in vivo*. One very unique pharmacological property of the cryptophycins is that they do not serve as substrates for either Pgp or the MRP multidrug-resistant efflux pumps and are highly active against the resistant tumors that express multiple-drug-resistant phenotypes both *in vitro* and *in vivo*. These results indicate that the cryptophycins may be highly active against human solid tumors that are resistant to the taxanes or the vincas and may offer alternative efficacy or survival advantages over the current therapies. Here we summarize the collaborative structure–activity relationship (SAR) studies conducted among three institutions, Lilly Research Laboratories, University of Hawaii, and Wayne State University, in their identification of cryptophycin 52 (LY355703) as the first-generation cryptophycin to undergo clinical evaluation.

II. ISOLATION AND CHARACTERIZATION OF THE CRYPTOPHYCINS

The cryptophycins are peptolides with a 16-membered macrolide ring structure. The macrocyclic ring is composed of two ester linkages (Fragment A/D and Fragment C/D), two amide linkages (Fragment A/B and Fragment B/C), and seven asymmetric centers (Figure 9.1). Cryptophycin A or 1 was first isolated from the *Nostoc* sp. ATCC 53798[1] by researchers at Merck in the early 1990s and was found to be very active against strains of *Cryptococcus*. The development of cryptophycin A as a potential antifungal agent was unsuccessful mainly because of the toxicity of the agent. During the same period of time, in an effort to identify novel antitumor agents from blue-green algae (cyanobacteria), researchers at the University of Hawaii led by Professor Richard Moore discovered that the lipophilic extract of *Nostoc* sp. GSV224[2] was highly cytotoxic to human nasopharyngeal carcinoma (KB) and human colorectal adenocarcinoma (LoVo) cells. More importantly, this lipophylic extract was found to be highly selective toward solid tumor cells versus leukemia cells and normal fibroblasts in the disk diffusion soft agar colony formation assay developed by Professor Thomas Corbett at Wayne State University.[3] To follow up on this activity, a total of 26 naturally occurring secondary metabolites were isolated from the *Nostoc* sp. GSV224, with cryptophycin 1 accounting for most of the cytotoxic activity in the crude extract. Through rigorous structure elucidation and determination efforts, the University of Hawaii team established

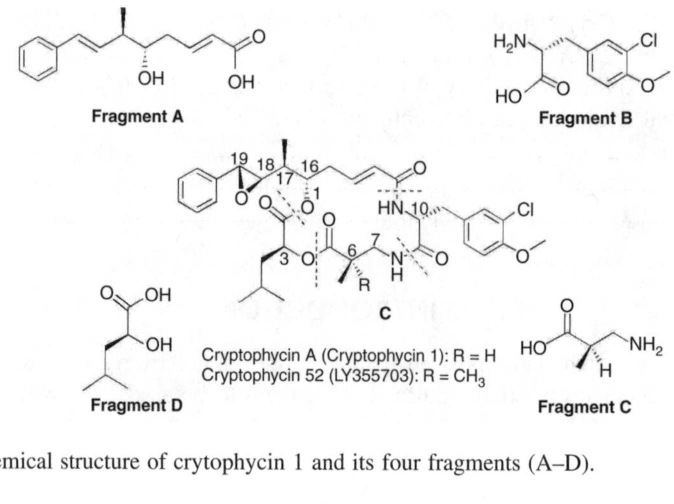

FIGURE 9.1 Chemical structure of crytophycin 1 and its four fragments (A–D).

the absolute stereochemistry all the asymmetric centers of cryptophycin 1 and confirmed that cryptophycin 1 was identical to cryptophycin A, reported on earlier by the Merck group.[4]

The antitumor activity of cryptophycin 1 was then rapidly followed up *in vivo* by the Wayne State University group (Professor Corbett), using various solid-tumor models developed either in syngeneic or severe combined immunodeficiency (SCID) mice. It was quickly determined that cryptophycin 1 was highly active against a number of murine solid tumors and human tumor xenografts when given by intravenous injection. In contrast, the compound was found to be significantly less active when administered by intraperitoneal or oral routes. The lack of activity via these routes was believed to mainly be a result of poorer metabolic stability. Excellent tumor growth inhibition and antitumor activity were observed with cryptophycin 1 in a number of solid tumors, but more importantly, it was highly active against the murine mammary 16/C tumor that is resistant to paclitaxel because of the overexpression of the multiple-drug-resistant protein Pgp.[5] This renders the cryptophycins some of the very few new antitubulin agents that do not serve as substrates for the multiple-drug-resistant protein and, hence, may offer significant advantage for the treatment of resistant human solid tumors.

The discovery of the exciting antitumor activity of cryptophycin 1 prompted the Lilly/Hawaii/Wayne State team to develop a unified strategy not only for producing a large number of analogs for the SAR and the selection of a clinical candidate from the series but also for developing technologies that can be used for the potential commercial-scale production of the cryptophycin class of compounds. Critical analyses revealed that a large-scale bioproduction of the cryptophycins from cultured blue-green algae was not only technically very challenging but also economically unattractive. For example, cryptophycin 1 can only be obtained in an estimated 0.4% yield from the dry algae mass after an extensive high-performance liquid chromatography purification process. The bioproduction approach also offered limited chemical diversity for the SAR studies. A total synthetic approach was eventually elected as the major strategy for the chemical investigation of the cryptophycin series of compounds. It offers several advantages over the bioproduction approach, including chemical diversity, synthetic maneuverability, enabling technologies, and last but not least, speed for advancing the project. The majority of the SARs described below were conducted using materials prepared by total synthetic approaches developed jointly between the Lilly Research Laboratories and the University of Hawaii teams. The clinical candidate compound, cryptophycin 52 (LY355703), was prepared via a convergent synthetic approach that was originally developed by Professors Moore and Tius' groups at the University of Hawaii and was later modified by the Chemical Process Research and Development group at Lilly Research Laboratories, led by Dr. Michael Martinelli.[6] The 30-step convergent synthetic sequence of the cryptophycins illustrates both the complexity and the diversity of such an approach for the preparation of this novel chemical series.

III. MECHANISM OF ACTION OF THE CRYPTOPHYCINS

The cryptophycins are potent antiproliferative and cytotoxic agents that act against a variety of human tumor cell lines derived from both hematopoietic and solid tumors. The mechanism of action behind the potent cytotoxic effects induced by the cryptophycins appears to be associated with inhibition of mitotic spindle function of the cells.[7] This inhibition of mitosis is accompanied by accumulation of cells at the metaphase–anaphase transition, followed by spindle disorganization and fragmentation and, finally, apoptosis or cell death.[8] Biochemically, it was found that at low concentrations, the cryptophycins interact with microtubules to significantly dampen their dynamic characteristics. At higher concentrations, the process of microtubule polymerization was effectively inhibited by the cryptophycins. Compared with other antitubulin agents such as the vincas and the taxanes, the cryptophycins have a major pharmacological advantage in that they are insensitive to the ABC transporters Pgp and MRP, which are implicated in multidrug resistance.[9] The majority of the cellular and biochemical mechanism of action studies were conducted using the synthetic 6,6-gemdimethyl analog (cryptophycin 52) of the natural product cryptophycin 1.

A. Cellular Mechanism of Action of Cryptophycin 52 (LY355703)

Flow cytometry studies showed that cryptophycin 52 can potently cause the accumulation of cells (CCRF-CEM cells, from 15% to 60%; HT-29 cells, from 24% to 88%) in the G2/M phase of the cell cycle, and this effect was found to be both time and concentration dependent. Microscopic examination of cells treated with cryptophycin 52 showed the cells initially accumulating in mitosis and subsequently undergoing apoptosis. Apoptosis was also evaluated by flow cytometry studies by following the end-labeling of fragmented DNA using bromodeoxyuridine (BrdU). It was found that when CCRF-CEM cells were treated with 300 pM of cryptophycin 52, 20% of cells were in interphase, 40% were in the mitotic phase, and 40% were in the apoptotic population (at the 24-h time point), which indicated that the cell death mechanism was mediated through the antimitotic effect induced by the cryptophycins.

Further studies of cryptophycin 52 in HeLa and HT-29 cells using confocal microscopy showed that cryptophycin 52 caused the mitotic spindle to become fragmented, and multiple microtubule foci were found within a ball-like cluster of chromosomes (HeLa cells at 100 pM, approximately equal to the IC$_{50}$). At 300 pM of cryptophycin 52, only small residual fragments of spindle microtubules are evident in the chromosome cluster of treated HeLa cells (Figure 9.2). Similar effects that caused spindle fragmentation and multiple microtubule foci were also observed when HT-29 cells were treated with cryptophycin 52 at 300 pM and 1 nM. Abnormal spindle structures appeared within shorter exposure time when a higher concentration (1 nM) of cryptophycin 52 was used. Microtubule bundles were evident in 1-nM compound-treated cells as early as 2 h posttreatment. These bundles are similar to microtubule bundles described in paclitaxel (Taxol®)-treated cells, but not vinblastine- or vincristine-treated cells. The microscopy data clearly supported the notion that microtubules and mitotic spindles are the primary targets of action for cryptophycin 52 and its derivatives. This was further investigated by studying the effect of crytophycin 52 on the dynamics of microtubule polymerization using bovine brain microtubules.[10]

B. Effects of Cryptophycin 52 (LY355703) on Microtubule Polymerization In Vitro

Using a bovine brain microtubule protein preparation that consisted of 70% tubulin and 30% microtubule-associated proteins (MAPs), it was found that cryptophycin 52 potently inhibited the microtubule polymerization (measured by using radiolabeled GTP/GDP bound to tubulin as a sensitive measure of polymerization) with an IC$_{50}$ of 0.49 μM. This effect was comparable to those exerted by vinblastine (0.35 μM) or vincristine (0.34 μM) and was distinct from paclitaxel, which enhanced (or stabilized) the polymerization of tubulin at concentrations >0.3 μM. Microtubules are not simple equilibrium polymers. The hydrolysis of GTP to GDP and inorganic phosphate as tubulin adds to a growing microtubule end creates highly dynamic polymers whose ends grow and shorten stochastically because of the gain and loss of a stabilizing "cap." The dynamics are extremely rapid both *in vitro* and in cells, especially during mitosis. Considerable recent evidence strongly indicates that not just the presence of microtubules but also the dynamics of the microtubules are critical for mitotic progression.[11] Many of the important antimitotic antitumor drugs such as the vinca alkaloids and paclitaxel exert their powerful antiproliferative actions by suppressing the dynamics of spindle microtubules. Importantly, these antimitotic drugs perturb microtubule dynamics by binding to the ends and along the surface of microtubules at concentrations that are far lower than those required to inhibit microtubule polymerization or induce excess microtubule formation.

The effects of cryptophycin 52 on the growing and shortening dynamics of individual microtubules at steady state in real time using video microscopy were thus examined. MAP-free bovine brain tubulin was assembled to polymer mass steady state at the ends of sea urchin

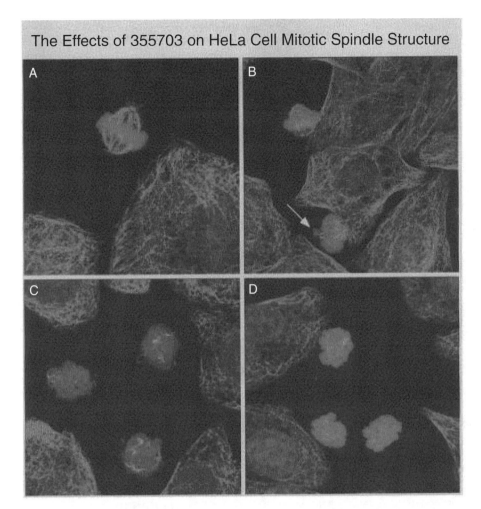

FIGURE 9.2 (See color insert following page 304.) Fluorescence confocal micrographs of HeLa cells. Panel A shows a metaphase mitotic spindle in an untreated control cell. The metaphase spindle is bipolar, with the chromosomes located in a compact metaphase plate at the midpoint between the spindle poles. Panel B shows cells treated with 30 pM of LY355703 (approx. IC$_{50}$) for 8–10 h. The mitotic spindles look relatively normal. The most obvious abnormality involves the displacement of chromosomes from the compact metaphase plate. In the lower spindle, a chromosome (red) can be seen near the spindle pole (arrow). Panel C (8–10 h; 100 pM of LY355703) and D (8–10 h; 300 pM of LY355703) show increased fragmentation of the spindle microtubules and disorganization of the chromosome mass into ball-like clusters.

flagellar axonemal seeds, and the individual dynamics parameters and the changes in length of the microtubules were recorded for approximately 30–45 min. Through such studies, it was found that cryptophycin 52 powerfully suppressed microtubule dynamics at a concentration of 50 nM and at 500 nM, completely shutting down the dynamics of microtubules (Figure 9.3). The studies also indicate that cryptophycin 52 suppresses microtubule shortening rates considerably more powerfully than microtubule growing rates. This is a sharp contrast to vinblastine, which inhibits both the growing and the shortening of microtubules to similar extents. Overall, the relatively weaker effects of cryptophycin 52 on growing as compared with shortening resembles the action of paclitaxel rather than vinblastine. Thus, cryptophycin 52 has characteristics of both the vinca alkaloids and paclitaxel when it comes to influencing microtubule dynamics.[10]

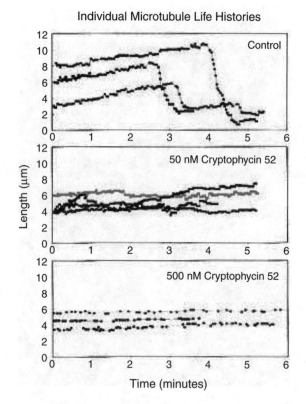

FIGURE 9.3 (See color insert following page 304.) Microtubule life histories.

IV. SYNTHESIS OF CRYPTOPHYCIN 52 (LY355703)

A. Retrosynthetic Analysis and Final Assembly of the Fragments

The retrosynthetic analysis of cryptophycin 52 shown in Figure 9.1 represents a straightforward disconnection of the depsipeptide into four corresponding fragments (labeled A–D) through the two ester and two amide linkages. The molecule contains six asymmetric centers, with four contiguous ones present in fragment A, thus making it the most synthetically challenging fragment. For the latter four, the Sharpless asymmetric epoxidation was used as the essential step in controlling two of the four asymmetric centers (*16S* and *17S*), and the final diastereoselective epoxidation of the styrene olefin provided the other two asymmetric centers (*18R* and *19R*) of fragment A. The two distal asymmetric centers (*3S* and *10R*), in contrast, were obtained from optically pure amino acids (D-tyrosine for fragment B, and L-leucine for fragment D, respectively).

In general, the cryptophycin nucleus can be assembled efficiently in a convergent fashion by first combining fragments A and B and fragments C and D to give the protected AB and CD fragments. The TBS group of the AB fragments was then removed (aq HF, CH_3CN, 95%) to give the secondary alcohol, which was then coupled with the carboxylic acid of the *N*-BOC C′D fragment (DCC, DMAP, CH_2Cl_2, 95%) to give the fully protected *seco*-ABCD compound (Figure 9.4).[6]

Treatment of the *seco* compound with TFA (removal of the BOC group) followed by 2-hydroxypyridine effectively catalyzed the macrolactamization to give the cyclized 16-membered depsipeptide in 62% yield (Figure 9.5).[12] Diastereoselective epoxidation of the styrylolefin (*m*-CPBA, CH_2Cl_2, 95%) gave a 2:1 ratio of epoxides, with the desirable β-epoxide as the major product. The diastereomeric epoxide mixtures can be separated by reverse-phase high-performance

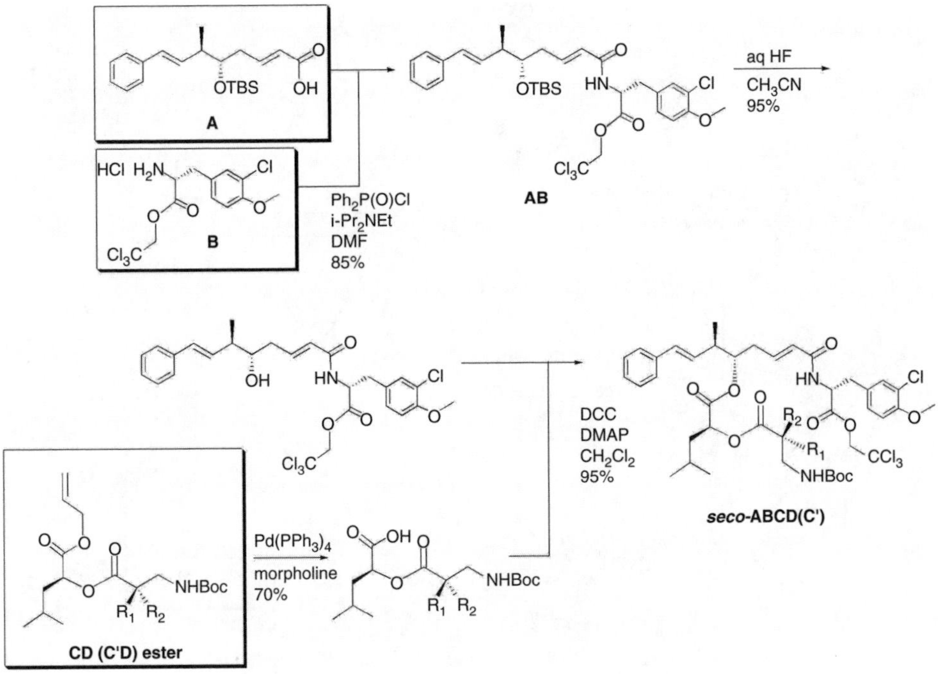

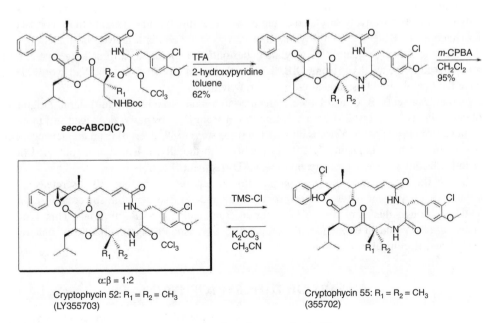

FIGURE 9.4 Convergent synthetic approach to the *seco* ABCD compound.

liquid chromatography to provide the more active β-epoxide (cryptophycin 52) for biological evaluation. Alternatively, the 2:1 epoxide mixture can be treated with trimethylsilylchloride to give the corresponding chlorohydrins, which then can be readily separated by flash silica gel chromatography. Treatment of the pure β-chlorohydrin (cryptophycin 55, 355702) with potassium carbonate in acetonitrile converted it back to cryptophycin 52 in high yields.

FIGURE 9.5 Macrocyclization and the preparation of cryptophycin 52.

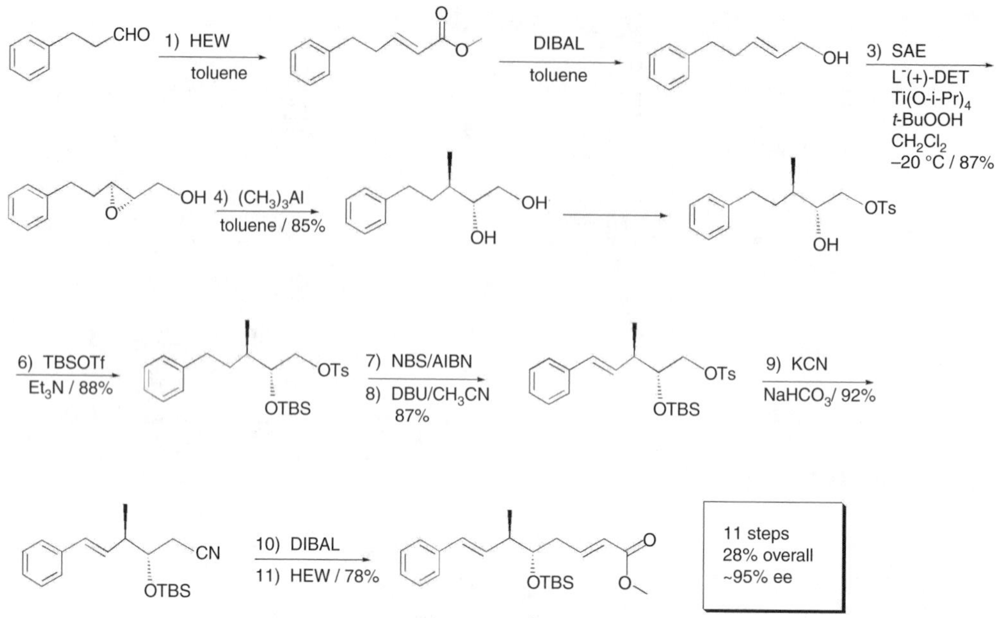

FIGURE 9.6 Sharpless asymmetric epoxidation route to the preparation of Fragment A.

B. Preparation of Fragments (A–D)

Fragment A, the synthetically most challenging fragment, with its four contiguous stereogenic centers, was prepared in 11 total linear steps to give methyl(5S,6R)-5-[(tert-butyldimethylsilyl)oxy]-6-methyl-8-phenylocta-2(E),7(E)-dienoate (**1**, 28% overall yield, 95% ee). In this sequence, the Sharpless asymmetric epoxidation of the allylic alcohol was used as a key reaction to install two of the four asymmetric centers with ~95% ee (Figure 9.6). Regio- and stereoselective opening of the resulting epoxide with trimethylaluminum installed the desired methyl group and resulted in the diols, where the primary one was selectively tosylated in the presence of dibutyl tin oxide and the secondary one was protected as a TBS ether.[13] The styrene moiety was installed via a dehydrobromination, using DBU as the base. Displacement of the tosylate under buffered conditions afforded the crucial nitrile that was reduced to the aldehyde using DIBAL and subsequent Horner-Emmons-Wadsworth olefination gave the final desired unsaturated ester in good yield.[14]

The other fragments, B, C, and D, were prepared in a more straightforward fashion. Fragment B was obtained in six steps (40% overall yield, 99% ee) from D-tyrosine, as described in Figure 9.7. Fragment C′ (for cryptophycin 52) was obtained in three steps (58% overall) from ethylcyanoacetate, which was then coupled (CDI in THF, 91%) readily with the allyl-L-leucic acid (obtained in two steps from L-leucine) to give the corresponding C′-D fragment (Figure 9.8).

Although the total synthesis of cryptophycin 52 required more than 30 discrete synthetic operations, the convergent nature of the process rendered it quite versatile for creating multiple permeations through the combination of different fragments for SAR purposes. This synthetic sequence was also practical enough for the preparation of reasonably large amounts of material for the phase I and II clinical evaluations.

V. SARS OF THE CRYPTOPHYCINS

Over 500 cryptophycin analogs were prepared via either a total synthetic approach or a semisynthetic one, with modifications of the natural cryptophycin 1 or cryptophycin 52, particularly for

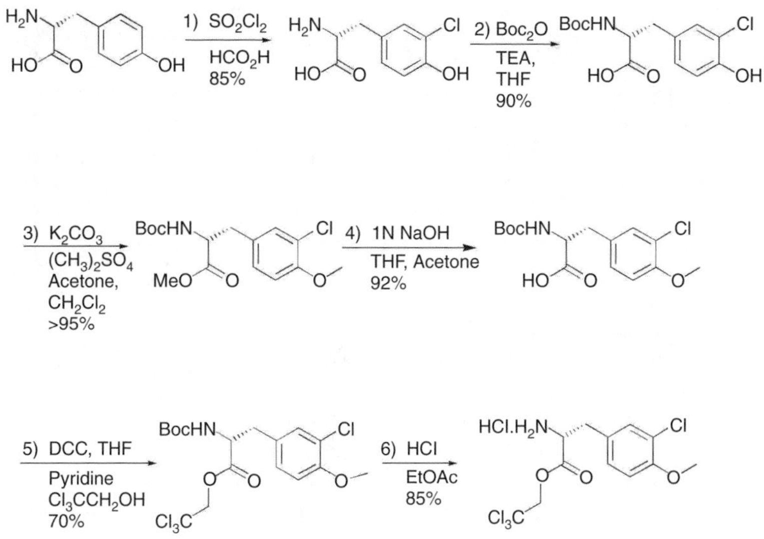

FIGURE 9.7 Preparation of Fragment B.

fragment A analogs. Because the cryptophycins exhibited very potent cytotoxicity toward most of the human malignant carcinoma cell lines tested in culture, a 72-h MTT cell-based assay was used to evaluate the newly synthesized analogs. The IC_{50}s for cryptophycins 1 and 52, for example, were in the range of 20–100 pM against several human tumor cell lines (KB, CCRF-CEM, HT-29, SKOV3, LoVo, and GC3). These *in vitro* cell-based assays provided the required sensitivity and broad range (10 pM ~> 20 μM) of activity essential for differentiating and establishing a firm structure–activity trend for a large number of analogs. In these studies, most of the cytotoxicity data were collected using three cultured human tumor cell lines: KB (Hawaii) and GC3 human colon carcinoma and CCRF-CEM human leukemia (Lilly). The cytotoxicity data cited in the SAR discussion in this section were mostly derived from the GC3 human colon carcinoma cell line unless stated otherwise. The *in vitro* cytotoxicity was then used to prioritize the compounds for the initial *in vivo* antitumor screening, using the Panc 03 (murine pancreatic adenocarcinoma) model, which was then followed by a panel of other murine tumor models and human tumor xenografts (Mam 16C/ADR, Mam 17/ADR, MX-1, GC3, LNCaP, HCT-116, and PC-3).[15]

The SAR described here will be primarily focused on five regions: the epoxide of fragment A and the prodrugs of its corresponding chlorohydrin, the substitution of the phenyl group of fragment

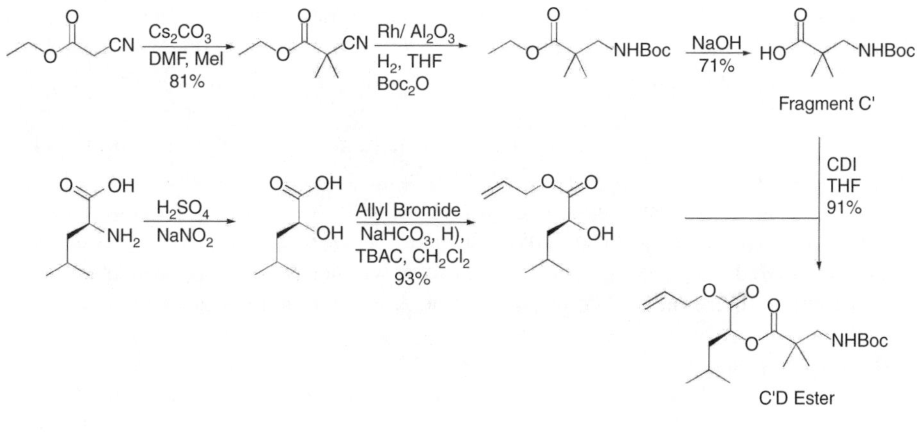

FIGURE 9.8 Preparation of Fragment C′D.

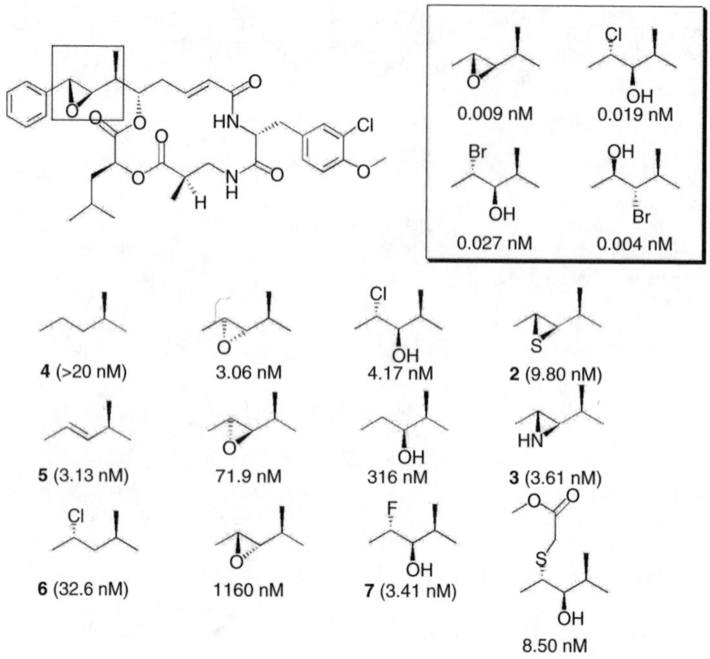

FIGURE 9.9 Structure–activity relationship of β-epoxide region of Fragment A.

A, the modification of the D-chloromethyltyrosine unit of fragment B, the C/D ester bond, and the C6, C7 positions of fragment C.

A. SAR OF THE β-EPOXIDE REGION OF FRAGMENT A

Early SAR efforts quickly established the importance of the absolute stereochemistry of the four consecutive stereogenic centers of this fragment. One very important structure–activity feature of the cryptophycins is centered on the configuration of the styrylepoxide portion of the molecule where only the β-epoxide or its masked equivalents (such as the corresponding β-chlorohydrin, β-bromohydrin and the reversed β-bromohydrin) possessed potent cytotoxicity (IC_{50} = 0.004–0.027 nM) in cell culture assays (Figure 9.9). Epoxides with other configurations (*R/S*, *S/R*, and *S/S*) were much less active. Replacement of the oxygen atom of the epoxide with sulfur (episulfide, **2**) or nitrogen (aziridine, **3**) led to major loss in activity.[16] Removal of the oxygen functionality at C18/C19 also led to inactive phenethyl (**4**), styryl (**5**), and chlorobenzyl (**6**) derivatives of cryptophycin. The corresponding diol and the fluorohydrin (**7**) derivatives of β-epoxide were also much less active in the *in vitro* assays.

Pharmacokinetic data (*vide infra*) clearly suggested that the β-chlorohydrin was rapidly converting to the corresponding β-epoxide *in vivo* in rodents and could account for the majority of the antitumor activity seen in murine tumors and human tumor xenografts. The drawback of the chlorohydrins was their lack of long-term stability in a suitable formulation, which prompted a study to stabilize these derivatives and improve their aqueous solubility by preparing several ester prodrugs. Although several of these derivatives showed weaker *in vitro* activity (CCRF-CEM human leukemia) as compared with cryptophycins 52 and 55, it was necessary to evaluate these derivatives *in vivo* to determine their potential conversion to the active epoxide species (Figure 9.10).[17]

B. SAR OF THE PHENYL GROUP OF FRAGMENT A

From this SAR it was concluded that the activity of the cryptophycins can be modulated by introducing electron-donating or electron-withdrawing groups on the phenyl ring of the styrene

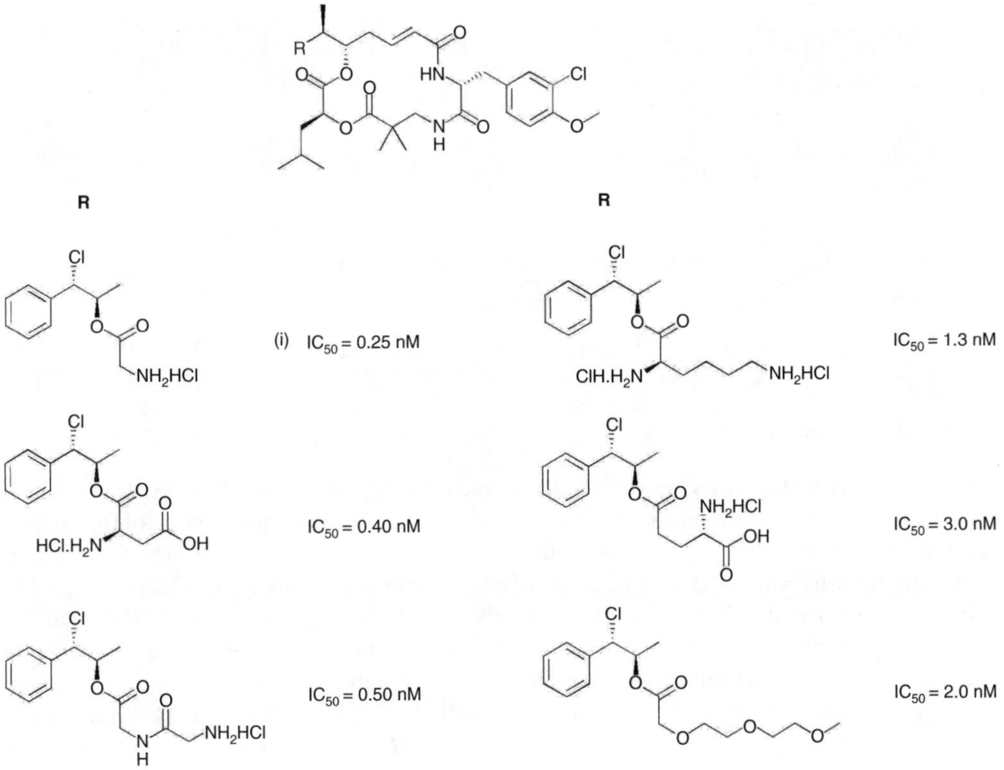

FIGURE 9.10 IC_{50s} of prodrugs in CCRF-CEM cells.

oxide region.[18] Introduction of a group larger than a fluorine in the ortho position led to loss of activity, whereas the meta and para positions tolerated many large functional groups. Electron-donating groups such as methyl, hydroxymethyl, and methoxy usually led to highly potent cryptophycin derivatives *in vitro* (*m*-methyl, IC_{50} = 0.15 nM; *p*-methoxy, IC_{50} = 0.28 nM; *p*-hydroxymethyl, IC_{50} = 0.004 nM; compared with IC_{50} = 0.022 nM for cryptophycin 52, measured in CCRF-CEM cells). Introduction of *p*-methyl ester also resulted in an active derivative (IC_{50} = 0.037 nM; cryptophycin 55, IC_{50} = 0.05 nM). Along the same trend, introduction of fluorine at either the *ortho*-, *meta*-, or *para*- positions relative to the oxirane ring of the styrene oxide also led to derivatives with potencies ranging between 0.01 and 0.06 nM.

Taking advantage of the potential substitution at the para position to improve the aqueous solubility of the molecule, amines **9–11** were prepared, starting with benzyl chloride **8** (Figure 9.11).[18] Their potency ranged between 0.001 and 0.092 nM, compared with 0.05 nM for cryptophycin 55, which proved that not only are large groups tolerated but these might provide more potent analogs than cryptophycin 52 or 55.

Replacement of the phenyl group of the styrene oxide with other heteroatom-containing aromatics such as pyridine, thiophene, furan, or thiazole, however, all led to a major loss of cytotoxicity in cell culture. Replacement of the phenyl group with simple alkyl groups (methyl, ethyl, *n*-propyl, methoxy, formyl, and *t*-butyl ester) also led to loss of activity in cell culture studies and to completely inactive analogs in *in vivo* animal studies.

C. SAR OF THE D-CHLOROTYROSINE OF FRAGMENT B

SAR studies around Fragment B indicated that this is probably the region that is least amenable to structural modification. Various structural modifications of this region of the molecule all led to

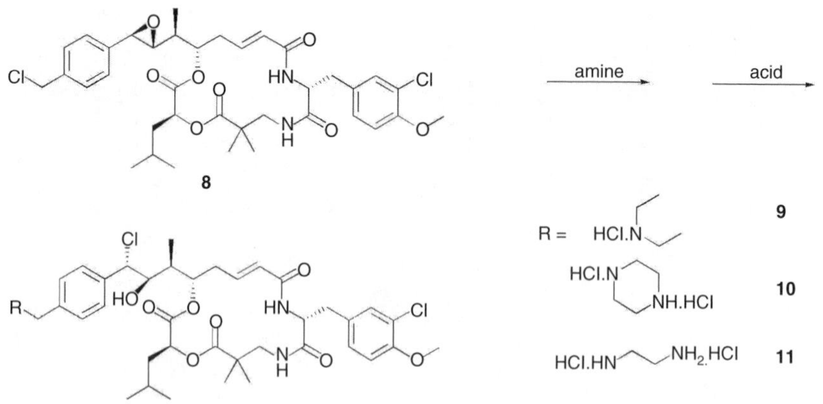

FIGURE 9.11 Preparation of amines 9–11.

significant loss of cytotoxicity *in vitro*.[19] These included replacing the chlorine atom with the fluorine, introduction of additional halogen atoms onto the phenyl ring, modification of the methoxy ether, replacing the chloromethoxyphenyl with other aromatic or nonaromatic groups, and replacing chloromethyltyrosine with other simple natural and unnatural amino acids (such as glycine, valine, proline, *p*-aminophenylalanine, *p*-hydroxyphenylglycine, phenylethylglycine, etc.). Methylation of the asymmetric methine carbon at C10 also gave inactive analogs. The only tolerable modification in fragment B was the replacement of the chlorine with a bromine or a methyl group on the phenyl ring of the tyrosine. This led to the corresponding methyl D-tyrosine derivative, which had only a twofold decrease of cytotoxicity *in vitro* (IC_{50} = 0.04 n*M*; cryptophycin 52, IC_{50} = 0.019 n*M*, measured in CCRF-CEM cells).

D. SAR OF THE C/D ESTER BOND AND THE C6, C7 POSITIONS

Some of the early studies of cryptophycin 1 indicated that the ester bond between fragments C and D of the macrocyclic ring may be metabolically labile to nucleophilic attack. This is supported by the observation that arenastatin (a closely related natural product isolated from Japanese sponge *Dysidea arenaria*) is completely devoid of activity *in vivo*, although it has potent activity in cell culture.[20] Two approaches were then investigated for the stabilization of the C–D ester bond: (a) geminal substitution — to the carbonyl group and (b) isosteric replacement of the ester bond with other functionalities such as an amide. Both approaches produced very active derivatives — in particular, introduction of a gem-dimethyl group at the C6 position gave an analog (cryptophycin 52, LY355703) with a very much improved *in vivo* antitumor efficacy profile compared with cryptophycin 1. Although geminal disubstitution was effective in protecting the carbonyl group from nucleophilic attack and could be successfully extended to the spirocyclopropyl group (12), however, it was found that potency was rapidly diminished when larger groups (gem-diethyl, gem-di-n-propyl, spirocyclopentyl, and spirocyclohexyl) were introduced into the C6 position (Figure 9.12).[21]

The amide isosteric replacement of the C–D ester bond also gave very active analogs 13 and 14, regardless of whether there was a monomethyl or a geminal dimethyl group at the C6 position.[22] Both agents exhibited potent *in vitro* activity (IC_{50} = 0.015 n*M*); however, it was found that the *in vivo* potency for the gem-dimethyl amide analog was significantly lower than that of cryptophycin 52 when compared head-to-head in the Panc 03 model. Replacement of the C–D ester with the corresponding ether linkage (15) causes a major loss of activity in cell culture (240-fold).[22] As part of the SAR studies of Fragment C, it was also found that moving the substituent ($R = CH_3$, 16) from the C6 to the C7 position retains the activity *in vitro*. However, similar to what was observed for arenastatin, absence of substituents at the C6 position caused the compound to be completely inactive *in vivo* presumably because of the metabolic instability issue with the C–D ester bond.

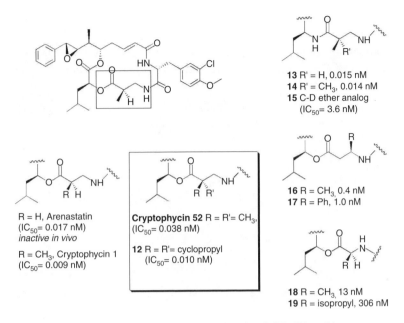

13 R' = H, 0.015 nM
14 R' = CH$_3$, 0.014 nM
15 C-D ether analog
(IC$_{50}$= 3.6 nM)

R = H, Arenastatin
(IC$_{50}$= 0.017 nM)
inactive in vivo

R = CH$_3$, Cryptophycin 1
(IC$_{50}$= 0.009 nM)

Cryptophycin 52 R = R'= CH$_3$,
(IC$_{50}$= 0.038 nM)

12 R = R'= cyclopropyl
(IC$_{50}$= 0.010 nM)

16 R = CH$_3$ 0.4 nM
17 R = Ph, 1.0 nM

18 R = CH$_3$, 13 nM
19 R = isopropyl, 306 nM

FIGURE 9.12 Structure–activity relationship of CD ester bond and C6, C7 positions.

Larger substituents at the C7 position (*R* = phenyl, **17**) did not improve the activity of the compound either *in vitro* or *in vivo*. Excising the C6 carbon atom of the peptolide ring was also attempted, and the result was disappointing. The smaller 15-membered macrocycles **18** and **19** were significantly less active in cell culture (1,400-fold) than the natural cryptophycins.[23]

The key findings from the SAR studies conducted around the various regions of the molecule are summarized in Figure 9.13. Overall, it can be concluded, first, that the absolute stereochemistry of all seven asymmetric centers is essential for eliciting the biological activity of the cryptophycins, and the opposite configuration at these centers led to major loss of activity. Second, the β-epoxides (*R/R*) at the **18** and **19** positions (Fragment A) are consistently much more active than the other isomers (*R/S*, *S/R*, and *S/S*). Third, substitution of the amide nitrogen of the macrocyclic ring led to inactive analogs; x-ray and nuclear magnetic resonance studies indicate that an intrapeptolide

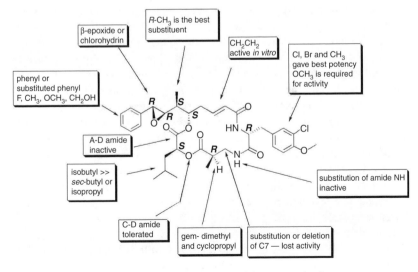

FIGURE 9.13 Key structure–activity relationship findings of the cryptophycins.

TABLE 9.1

Comparison of the *In Vitro* Cytotoxicity and the *In Vivo* Antitumor Activity of Cryptophycins 52, 55 and 55-glycinate in Human Tumor Xenografts

	Crypto 52 (355703)	Crypto 55 (355702)	Crypto 55-glycinate (368422)
Cytotoxicity (IC_{50}/nM/GC3 cells)	0.2–0.7	0.2–0.75	1.0
In Vivo Potency	10–12 mg/kg (CDI/nu/nu)	40–48 mg/kg (CD1/nu/nu)	48–64 mg/kg (CD1/nu/nu)
In vivo Antitumor Activity[a]			
GC3 Colon	+	+	+++
LNCap Prostate	++++	++++	++++
MX-1 Mammary	++	+++	NT
H116 Colon	+++	++++	NT

[a] All *in vivo* antitumor activity are expressed in log cell kill units: Log Kill = T-C/3.2 × Td.
+: 0.7–1.2: ++: 1.3–1.9: +++: 2.0–2.8: ++++: >2.8.

hydrogen bond may exist between the NH(8) and CO(12) groups for the macrocyclic ring to adopt certain conformations. Fourth, amide replacement of the ester linkage may be tolerated between the fragment C/D but not between the A/D fragments. Fifth, substituents on the phenyl ring of fragment A are well tolerated; however, substitutions and modifications of the chloromethyl tyrosine unit of Fragment B are limited to a few options.

VI. COMPARISON OF CRYPTOPHYCINS 52, 55, AND 55-GLYCINATE

A comparison of the *in vitro* cytotoxicity and the *in vivo* antitumor activity of cryptophycin 52 (epoxide, LY355703), cryptophycin 55 (chlorohydrin derivative of cryptophycin 52, 355702), and the glycinate derivative of chlorohydrin 55 (analog [i] in Figure 9.10, 368422) are summarized in Table 9.1. All three compounds were found to exert potent cytotoxic activities (200 p*M* ~ 1 n*M*) against a number of human tumor cell lines (CCRF-CEM, GC3, HeLa), with the epoxide and chlorohydrin being slightly more potent than the corresponding glycinate derivative. However, it was found that the epoxide compound exhibited the most potent effect *in vivo* when compared on a milligram-by-milligram basis. To achieve similar levels of antitumor activities *in vivo* (maximum tolerated doses), it was usually necessary to dose four- to five-fold more of the chlorohydrin or glycinate as compared to the epoxide. This is primarily because both chlorohydrin and glycinate were found to act as prodrugs (*vide infra*) of the epoxide species *in vivo*. All three cryptophycins (355703, 355702, and 368422) were found to be active against a broad panel of human tumor xenografts including colon, mammary, prostate, pancreatic, and lung. The level of antitumor activity exhibited by the cryptophycins is superior to most of the commonly used anticancer agents such as the vincas, 5-fluorouracil, cisplatin, and doxorubicin. The only other class of compound that demonstrated comparable levels of *in vivo* activity as the cryptophycins in these tumors were the taxanes (such as paclitaxel).

Various studies demonstrated that the chlorohydrin derivative of cryptophycin 52 (355702) can be converted to the more potent epoxide under *in vitro* as well as *in vivo* conditions. It was found that the chlorohydrin derivatives could quite rapidly be transformed into the corresponding epoxide in the formulation (contains 5% ethanol and 5% cremophor) under various storage conditions. This

TABLE 9.2
**Antitumor Activity (Log Kill*) of LY355703 and Other
Oncolytics Against Human Tumor Xenografts**

Agent	HCT-116 Colon	H125 Lung	LNCaP Prostate	TSU Prostate	PC-3 Prostate
Cryptophycin 52	+++	++	++++	++	+
Paclitaxel	+++	+	++++	++++	++
Doxorubicin	NT	—	—	—	—
Etoposide	—	+	+	—	—
VLB/VCR	+	+++	++++	++++	++
Cytoxan	—	—	++++	++++	—
Cisplatin	—	—	—	—	—
5-FU	—	—	—	—	—

* All *in vivo* antitumor activity are expressed in log cell kill units: Log Kill = T-C/3.2 × Td. +: 0.7–1.2; ++: 1.3–1.9; +++: 2.0–2.8; ++++: >2.8.

conversion was also found to take place in plasma (*in vitro*) and in animal studies. For example, in mice dosed intravenously with 80 mg/kg of cryptophycin 55, the plasma levels of cryptophycin 52 (converted directly from cryptophycin 55) were almost identical to those found after administration of 20 mg/kg of cryptophycin 52 itself (a dose relationship that was consistent with the findings of the efficacy studies). This observation indicated that the chlorohydrin derivative cryptophycin 55 may act as a prodrug of cryptophycin 52. For the glycinate derivative to convert to the active component epoxide, it requires an enzymatic cleavage step of the ester bond, via the esterases, to generate chlorohydrin 55, which subsequently forms cryptophycin 52. Although the ester hydrolysis of the glycinate (368422) seems quite efficient in mouse plasma, it was discovered that the ester hydrolysis of the glycinate to give the epoxide was significantly reduced in dog and human plasma.

VII. CRYPTOPHYCIN 52 (LY355703)

A. *In Vivo* Antitumor Activity of Cryptophycin 52 (LY355703)

The antitumor activity of cryptophycin 52 *in vivo*, however, is summarized here to illustrate the broad spectrum of activity of this novel agent against a variety of syngeneic murine solid tumors and human tumor xenografts. Because of the lipophilic nature of the cryptophycins, cryptophycin 52 was formulated in a mixture of 5% Cremophor and 5% ethanol in 90% saline, a formulation that is used clinically for paclitaxel. The treatment schedule varied but was generally consistent, with at least five repeated intermittent intravenous injections, usually every other day (q2d), and the doses used in the xenograft studies are usually in the range of 6–10 mg/kg for cryptophycin 52. The maximum tolerated dose for cryptophycin 52 in nude mice is 10 mg/kg per injection, with a total dose of ca. 50–60 mg/kg. Table 9.2 summarizes the antitumor activity (expressed in Log Kill) of cryptophycin 52 against five human tumor xenografts (HCT-116 colon, H125 lung, LNCaP prostate, TSU prostate, and PC-3 prostate carcinomas). In comparison, the antitumor activity of a group of commonly used oncolytic agents (paclitaxel, doxorubicin, etoposide, vincristine/vinblastine, cytoxan, cisplatin, and 5-fluorouracil) is also included.[24] It is clear from these studies that cryptophycin 52 is a very active anticancer agent with superior antitumor activity compared to many of the commonly used oncolytic drugs. In addition, cryptophycin 52 also exhibited excellent activity against tumors that are resistant to adriamycin (Mamm16C/Adr, Mamm17/Adr, both are syngeneic murine tumors) and vinblastine (UCLA-P3), whereas paclitaxel, doxorubicin, etoposide, and vinblastine are completely inactive in these multidrug-resistant tumors.

B. CLINICAL EVALUATION OF CRYPTOPHYCIN 52 (LY355703)

Cryptophycin 52, with its broad antitumor activity and potency, was chosen as the first-generation cryptophycin to undergo clinical evaluation, with 25 patients receiving the drug as a 2-h intravenous infusion on days 1 and 8 every 3 weeks. The doses were escalated from 0.1 to 2.22 mg/m^2 where neurological toxicities at or over 1.84 mg/m^2 were found to be dose limiting. An alternative twice-weekly schedule was also investigated with 11 patients in an attempt to reduce the neurotoxicity. Unfortunately, doses over 0.75 mg/m^2 on a days 1, 4, 8, and 11 were also not tolerated as a result of nausea and constipation.[25] As a result of phase I studies showing that toxicity was not schedule dependent but most likely associated with cumulative dose, the dose for phase II was set at 1.5 mg/m^2 for a day 1 and 8 schedule.

In Phase II clinical trials, LY355703 was used as second-line therapy for metastatic colorectal cancer, for example, or in non–small cell lung cancer patients who had received previous platinum-based chemotherapy. Although partial responses were observed in two patients with platinum-resistant advanced ovarian cancer, no significant activity or stable disease was reported in patients with other tumor types. Toxicities such as arthralgia, constipation, myalgia, and neuropathy were also observed; therefore, these results, along with the lack of efficacy, led to the termination of the clinical trials.

VIII. CONCLUSIONS

The discovery of the cryptophycins as a new class of antimitotic agents has attracted the attention of many synthetic groups to find better routes for their preparation and to investigate the activity of new related analogs.[26] In addition, investigators specializing in antimicrotubule agents have extensively studied their mode of action.[27] The research described in this chapter was the collaborative effort of Eli Lilly, University of Hawaii, and Wayne State University scientists resulting in the identification of cryptophycin 52 as the first generation cryptophycin to undergo clinical evaluation.[28] Extensive SAR studies were conducted and facilitated by the support provided by Lilly's process chemistry group. A convergent and highly efficient synthetic route to cryptophycin 52 was developed and used to both support the SAR and provide the clinical material.

The SAR studies led to a large number of cryptophycins with a broad array of structural diversity and activity. They also revealed some critically important areas for structure activity optimization. These included the styrene β-oxide region, the C/D ester bond region, and the phenyl region of fragment A. From the assessment of the collective data of *in vitro/in vivo* pharmacology, biopharmaceutical properties, ADME/pharmacokinetic evaluation, and toxicological findings, cryptophycin 52 (LY355703) was selected as the first-generation cryptophycin compound for clinical evaluation. The greatest advantage identified during the preclinical investigation of this series was the effectiveness of many of the cryptophycins against tumors that overexpressed the multidrug resistance phenotype. This new class of macrocyclic depsipeptides has proven to be quite novel in its mode of action and in its preclinical antitumor activity.

ACKNOWLEDGMENTS

We wish to thank all our colleagues and collaborators who contributed to this work: Andrew H. Fray, Michael J. Martinelli, Eric D. Moher, Bryan H. Norman, Vinod F. Patel, Richard M. Schultz, John E. Toth, David L. Varie, Daniel C. Williams, John F. Worzalla, Tony Y. Zhang, Thomas H. Corbett (Wayne State University), Marc Tius (University of Hawaii), Richard E. Moore (University of Hawaii), Leslie Wilson, and Mary Ann Jordan (University of California at Santa Barbara). Special thanks to Jenniffer Phillips and Susan Koppleman for their help with the references and formatting of the document. We also wish to thank Drs. Gordon Cragg, David Newman (National Cancer Institute), and David Kingston (Virginia Tech University) for giving us the opportunity to summarize the work of the cryptophycin team and for their critical evaluation of this chapter.

REFERENCES

1. Schwartz, R. E. et al., Pharmaceuticals from cultured algae. *J. Ind. Microbiol.*, 5, 118–126, 1990.
2. Patterson, G. M. et al., Antineoplastic activity of cultured blue-green-algae (Cyanophyta). *J. Phycol.*, 27, 530–536, 1991.
3. (a) Corbett, T. H. et al., Cytoxic anticancer drugs: models and concepts for drug discovery and development; Valeriote, F. A., Corbett, T. H., Baker, L. H., Eds., Kluwer Academic Publishers, 35–87, 1992. (b) Valeriote, F. A. et al., Discovery and development of anticancer agents, Valeriote, F. A., Corbett, T. H., Baker, L. H., Eds., Kluwer Academic Publishers, 67–93, 1997.
4. (a) Golakoti, T. et al., Total structures of cryptophycins, potent antitumor depsipeptides from the blue-green alga Nostoc sp. Strain GSV 224. *J. Am. Chem. Soc.*, 116, 4729–4737, 1994. (b) Golakoti, T. et al., Structure determination, conformational analysis, chemical stability studies, and antitumor evaluation of the cryptophycins. Isolation of 18 new analogs from Nostoc sp. strain GSV 224, *J. Am. Chem. Soc.*, 117, 12030–12049, 1995.
5. (a) Corbett, T. H. et al., *Anticancer Drug Development Guide*. Teicher, B. A., Ed., Humana Press, 75–99, 1997. (b) Corbett, T. H. et al., Discovery of cryptophycin-1 and BCN-183577: examples of strategies and problems in the detection of antitumor activity in mice. *Invest. New Drugs,* 15, 207–218, 1997.
6. Barrow, R. A. et al., Total synthesis of cryptophycins. Revision of the structures of cryptophycins A and C. *J. Am. Chem. Soc.*, 117, 2479–2490, 1995.
7. (a) Kerksiek, K. et al., Interaction of cryptophycin 1 with tubulin and microtubules. *FEBS Lett.*, 377, 59–61, 1995. (b) Smith, C. D., Zhang, X. Mechanism of action of cryptophycin. Interaction with the Vinca alkaloid domain of tubulin. *J. Biol. Chem.*, 271, 6192–6198, 1996. (c) Bai, R. et al., Characterization of the interaction of cryptophycin 1 with tubulin: binding in the Vinca domain, competitive inhibition of dolastatin 10 binding, and an unusual aggregation reaction. *Cancer Res.*, 56, 4398–4406, 1996. (d) Mooberry, S. L., Taoko, C. R., Busquets, L. Cryptophycin 1 binds to tubulin at a site distinct from the colchicine binding site and at a site that may overlap the vinca binding site. *Cancer Lett.*, 107, 53–57, 1996. (e) Panda, D. et al., Mechanism of action of the unusually potent microtubule inhibitor cryptophycin 1. *Biochemistry*, 36, 12948–12953, 1997.
8. Mooberry, S. L., Busquets, L., Tien, G. Induction of apoptosis by cryptophycin 1, a new antimicrotubule agent. *Int. J. Cancer*, 73, 440–448, 1997.
9. (a) Smith, C. D. et al., Cryptophycin: a new antimicrotubule agent active against drug-resistant cells. *Cancer Res*, 54, 3779–3784, 1994. (b) Wagner, M. M. et al., *In vitro* pharmacology of cryptophycin 52 (LY355703) in human tumor cell lines. *Cancer Chemother. Pharmacol.,* 43, 115–125, 1999.
10. Panda, D. et al., Antiproliferative mechanism of action of cryptophycin-52: kinetic stabilization of microtubule dynamics by high-affinity binding to microtubule ends. *Proc. Nat. Acad. Sci. USA*, 95, 9313–9318, 1998.
11. (a) Farrell, K. W., Jordan, M. A. A kinetic analysis of assembly-disassembly at opposite microtubule ends. *J. Biol. Chem.*, 257, 3131–3138, 1982. (b) Farrell, K. W. et al., Phase dynamics at microtubule ends: the coexistence of microtubule length changes and treadmilling. *J. Cell Biol.*, 104, 1035–1046, 1987. (c) Wilson, L., Jordan, M. A. Pharmacological probes of microtubule function. *Modern Cell Biol.*, 13, 59–83, 1994. (d) Jordan, M. A., Wilson, L. Microtubules and actin filaments: dynamic targets for cancer chemotherapy. *Curr. Opin. Cell Biol.*, 10, 123–130, 1998. (e) Yvon, A-M. C., Wadsworth, P., Jordan, M. A. Taxol suppresses dynamics of individual microtubules in living human tumor cells. *Mol. Biol. Cell*, 10, 947–959, 1999. (f) Wilson, L., Panda, D., Jordan, M. A. Modulation of microtubule dynamics by drugs: a paradigm for the actions of cellular regulators. *Cell Struct. Funct.*, 24, 329–335, 1999. (g) Jordan, M. A. Mechanism of action of antitumor drugs that interact with microtubules and tubulin. *Curr. Med. Chem: Anti-cancer Agents*, 2, 1–17, 2002. (h) Jordan, M. A., Wilson, L. Microtubules as a target for anticancer drugs. *Nat. Rev. Cancer*, 4, 253–265, 2004.

12. Fray. A. H., Intramolecular aminolysis of trichloroethyl esters: a mild macrocyclization protocol for the preparation of cryptophycin derivatives. *Tetrahedron Asymmetry*, 9, 2777–2781, 1998.
13. (a) Martinelli, M. J. et al., Dibutyltin oxide catalyzed selective sulfonylation of β-chelatable primary alcohols. *Org. Lett.*, 1, 447–450, 1999. (b) Martinelli, M. J., Vaidyanathan, R., Khau, V. V., Selective monosulfonylation of internal 1,2-diols catalyzed by di-*n*-butyltin oxide. *Tetrahedron Lett.*, 41, 3773–3776, 2000.
14. For other approaches to Fragment A see: (a) Varie, D. L. et al., Bioreduction of (R)-carvone and regioselective Baeyer-Villiger oxidations: application to the asymmetric synthesis of cryptophycin fragment A. *Tet. Lett.*, 39, 8405–8408, 1998. (b) Dhokte, U. P. et al., A novel approach for total synthesis of cryptophycins via asymmetric crotylboration protocol. *Tet. Lett.*, 39, 8771–8774, 1998. (c) Furuyama, M., Shimizu, I. A short enantioselective synthesis of a component of cryptophycin A and arenastatin A. *Tetrahedron Asymmetry*, 9, 1351–1357, 1998. (d) Liang, J. et al., Synthesis of unit A of cryptophycin via a [2,3]-Wittig rearrangement. *J. Org. Chem.*, 64, 1459–1463, 1999. (e) Liang, J. et al., Synthesis of cryptophycin 52 using the Sharpless asymmetric dihydroxylation: diol to epoxide transformation optimized for a base-sensitive substrate. *J. Org. Chem.*, 65, 3143–3147, 2000. (f) Pousset, C; Haddad, M., Larcheveque, M. Diastereocontrolled synthesis of unit A of cryptophycin. *Tetrahedron*, 57, 7163–7167, 2001. (g) Raghavan, S., Tony, K. A. Sulfinyl moiety as an internal nucleophile. 1. Efficient stereoselective synthesis of fragment A of cryptophycin 3. *J. Org. Chem.*, 68, 5002–5005, 2003. (h) Phukan, P., Sasmal, S., Maier, M. E. Flexible routes to the 5-hydroxy acid fragment of the cryptophycins. *Eur. J. Org. Chem.*, 9, 1733–1740, 2003.
15. Moore, R. E. et al., The search for new antitumor drugs from blue-green algae. *Current Pharmaceutical Design*, 2, 317–330, 1996.
16. Moore, R. E. Cyclic peptides and depsipeptides from cyanobacteria: a review. *J. Ind. Microbiol.*, 16, 134–143, 1996.
17. Al-awar, R. S. et al., Preparation of novel cryptophycin pharmaceuticals. PCT Int. Appl., 293 pp. WO 9808505 A1 19980305, 1998.
18. Al-awar, R. S. et al., A convergent approach to cryptophycin 52 analogues: synthesis and biological evaluation of a novel series of fragment a epoxides and chlorohydrins. *J. Med. Chem.*, 46, 2985–3007, 2003.
19. Patel, V. F. et al., Novel cryptophycin antitumor agents: synthesis and cytotoxicity of fragment "B" analogues. *J. Med. Chem.*, 42, 2588–2603, 1999.
20. (a) Kobayashi, M. et al., Arenastatin A, a potent cytotoxic depsipeptide from the okinawan marine sponge *Dysidea arenaria*. *Tetrahedron Lett.*, 35, 7969–7972, 1994. (b) Kobayashi, M., et al., The absolute stereostructure of arenastatin A, a potent cytotoxic depsipeptide from the Okinawan marine sponge *Dysidea arenaria*. *Chem. Pharm. Bull.*, 42, 2196–2198, 1994.
21. Varie, E. et al., Synthesis and biological evaluation of cryptophycin analogs with substitution at C-6 (fragment C region). *Bioorg. Med. Chem. Lett.*, 9, 369–374, 1999.
22. Norman, B. H. et al., Total synthesis of cryptophycin analogs. Isosteric replacement of the C-D ester. *J. Org. Chem.*, 63, 5288–5294, 1998.
23. Shih, C. et al., Synthesis and biological evaluation of novel cryptophycin analogs with modification in the β-alanine region. *Bioorg. Med. Chem. Lett.*, 9, 69–74, 1999.
24. (a) Worzalla, J. F. et al., LY355702 and LY355703, new cryptophycin analogues with antitumor activity against human tumor xenografts. *Proc. Am. Assoc. Cancer Res.*, 38, 225, 1997, (Abs. 1516). (b) Polin, L. et al., Preclinical antitumor activity of cryptophycin-52/55 (C-52; C-55) against human tumors in SCID mice. *Proc. Am. Assoc. Cancer Res.*, 38, 225, 1997, (Abs. 1514). (c) Polin, L. et al., Treatment of human prostate tumors PC-3 and TSU-PR1 with standard and investigational agents in SCID mice. *Invest. New Drugs*, 15, 99–108, 1997.
25. (a) Sessa, C. et al., Phase I and pharmacological studies of the cryptophycin analogue LY355703 administered on a single intermittent or weekly schedule. *Eur. J. Cancer*, 38, 2388–2396, 2002. (b) Stevenson, J. P. et al., Phase I trial of the cryptophycin analogue LY355703 administered as an intravenous infusion on a day 1 and 8 schedule every 21 days. *Clin. Cancer Res.*, 8, 2524–2529, 2002.
26. (a) Salamonczyk, G. M. et al., Total synthesis of cryptophycins via a chemoenzymatic approach. *J. Org. Chem.*, 61, 6893–6900, 1996. (b) de Muys, J-M. et al., Synthesis and *in vitro* cytotoxicity of cryptophycins and related analogs. *Bioorg. Med. Chem. Lett.*, 6, 1111–1116, 1996. (c) Rej, R. et al., Total synthesis of cryptophycins and their 16-(3-phenylacryloyl) derivatives. *J. Org. Chem.*, 61,

6289–6295, 1996. (d) Gardinier, K. M., Leahy, J. W. Enantiospecific total synthesis of the potent antitumor macrolides cryptophycins 1 and 8. *J. Org. Chem.,* 62, 7098–7099, 1997. (e) Ali, S. M., Georg, G. I. Formal syntheses of cryptophycin A and arenastatin A. *Tetrahedron Lett.,* 38, 1703–1706, 1997. (f) Georg, G. I., Ali, S. M., Stella, V. J. Halohydrin analogues of cryptophycin 1: synthesis and biological activity. *Bioorg. Med. Chem. Lett.,* 8, 1959–1962, 1998. (g) Eggen, M., Georg, G. I. Enantioselective synthesis of a 3β-dephenylcryptophycin synthon. *Bioorg. Med. Chem. Lett.,* 8, 3177–3180, 1998. (h) White, J. D., Hong, J., Robarge, L. A. Total synthesis of cryptophycins-1, -3, -4, -24 (arenastatin A), and -29, cytotoxic depsipeptides from cyanobacteria of the Nostocaceae. *J. Org. Chem.,* 64, 6206–6216, 1999. (i) Chakraborty, T. K., Das, S. Diastereoselective opening of trisubstituted epoxy alcohols: application in the studies directed toward the synthesis of octenoic acid moiety of cryptophycins. *J. Indian Chem. Soc.,* 76, 611–616, 1999. (j) Christopher, J. A. et al., A synthesis of cryptophycin 4 using a planar chiral molybdenum cationic complex. *Synlett.,* 4, 463–466, 2000. (k) Ghosh, A. K., Bischoff, A. A convergent synthesis of (+)-cryptophycin B, a potent antitumor macrolide from *Nostoc* sp. Cyanobacteria. *Org. Lett.,* 2, 1573–1575, 2000. (l) Barrow, R. A. et al., Synthesis of 1-aza-cryptophycin 1, an unstable cryptophycin. An unusual skeletal rearrangement. *Tetrahedron,* 56, 3339–3351, 2000. (m) Eggen, M. et al., Total synthesis of cryptophycin-24 (Arenastatin A) amenable to structural modifications in the C16 side chain. *J. Org. Chem.,* 65, 7792–7799, 2000. (n) Eggen, MJ., Nair, S. K., Georg, G. I. Rapid entry into the cryptophycin core via an acyl-β-lactam macrolactonization: total synthesis of cryptophycin-24. *Org. Lett.,* 3, 1813–1815, 2001. (o) Smith, A. B. et al., First generation design, synthesis, and evaluation of azepine-based cryptophycin analogues. *Org. Lett.,* 3, 4063–4066, 2001. (p) Martinelli, M. J. et al., Reaction of cryptophycin 52 with thiols. *Tetrahedron Lett.,* 43, 3365–3367, 2002. (q) Li, L-H., Tius, M. A. Stereospecific synthesis of cryptophycin 1. *Org. Lett.,* 4, 1637–1640, 2002. (r) Hoard, D. W. et al., Synthesis of cryptophycin 52 using the Shi epoxidation. *Org. Lett.,* 4, 1813–1815, 2002. (s) Smith, A. B. et al., Design, synthesis, and evaluation of azepine-based cryptophycin mimetics. *Tetrahedron,* 59, 6991–7009, 2003. (t) Vidya, R. et al., Synthesis of cryptophycins via an N-acyl-β-lactam macrolactonization. *J. Org. Chem.,* 68, 9687–9693, 2003. (u) Ghosh, A. K., Swanson, L. Enantioselective synthesis of (+)-cryptophycin 52 (LY355703), a potent antimitotic antitumor agent. *J. Org. Chem.,* 68, 9823–9826, 2003.

27. (a) Smith, C. D., Zhang, X. Mechanism of action of cryptophycin. Interaction with the Vinca alkaloid domain of tubulin. *J. Biol. Chem.,* 271, 6192–6198, 1996. (b) Kessel, D., Luo, Y. Cells in cryptophycin-induced cell-cycle arrest are susceptible to apoptosis. *Cancer Lett.,* 151, 25–29, 2000. (c) Barbier, P. et al., *In vitro* effect of cryptophycin 52 on microtubule assembly and tubulin: molecular modeling of the mechanism of action of a new antimitotic drug. *Biochemistry,* 40, 13510–13519, 2001. (d) Watts, N. R. et al., The cryptophycin-tubulin ring structure indicates two points of curvature in the tubulin dimer. *Biochemistry,* 41, 12662–12669, 2002. (e) Drew, L et al., The novel antimicrotubule agent cryptophycin 52 (LY355703) induces apoptosis via multiple pathways in human prostate cancer cells. *Clinical Cancer Res.,* 8, 3922–3932, 2002.

28. For other reviews on the cryptophycins see: (a) Shih, C., Teicher, B. A. Cryptophycins: a novel class of potent antimitotic antitumor depsipeptides. *Curr. Pharm. Design,* 7, 1259–1276, 2001. (b) Shih, C. et al., Synthesis and structure-activity relationship studies of cryptophycins: a novel class of potent antimitotic antitumor depsipeptides. *ACS Symposium Series,* 796 (Anticancer Agents), 171–189, 2001. (c) Eggen, MJ., Georg, G. I. The cryptophycins: their synthesis and anticancer activity. *Med. Res. Rev.,* 22, 85–101, 2002. (d) Tius, M. A. Synthesis of the cryptophycins. *Tetrahedron,* 58, 4343–4367, 2002. (e) Li, T., Shih, C. Structure activity relationships of cryptophycins, a novel class of antitumor antimitotic agents. *Frontiers Biotech. Pharm.,* 3, 172–192, 2002. (f) Hong, J., Zhang, Li. Novel approaches to the synthesis of cryptophycin and related depsipeptides. *Frontiers Biotech. Pharm.,* 3, 193–213, 2002. (g) Tius, M. A. Cryptophycin synthesis. *Handbook Environ. Chem.,* 3 (Part P), 265–305, 2003.

10 Chemistry and Biology of the Discodermolides, Potent Mitotic Spindle Poisons

Sarath P. Gunasekera and Amy E. Wright

CONTENTS

I. Introduction 171
II. Natural Sources 172
III. Chemistry 174
IV. Synthesis...174
    A. Synthesis of Discodermolide..174
    B. Synthesis of Analogs 175
V. Biological Activity 175
    A. Immunosuppressive Properties..175
    B. Antitumor Properties 179
VI. Structure–Activity Studies 183
VII. Clinical Investigation 187
References 187

I. INTRODUCTION

(+)-Discodermolide (**1**) is a highly functionalized, marine sponge–derived polyketide δ-lactone presumably formed by a combination of propionate and acetate units. **1** was first isolated and characterized by Gunasekera et al.[1] at the Harbor Branch Oceanographic Institution (HBOI) from the marine sponge *Discodermia dissoluta*. Crude extracts of the sponge showed strong potency against the murine P388 lymphocytic leukemia cell line, and bioassay-guided fractionation using this assay resulted in the purification of **1**. The structure was defined by spectroscopic methods and was confirmed by X-ray diffraction analysis. The initial reports of (+)-discodermolide's strong *in vitro* antiproliferative and *in vitro* and *in vivo* immunosuppressive activities[2,3] created a wide interest among the synthetic organic chemistry community. The absolute stereochemistry was subsequently defined by total synthesis conducted by the Schreiber group.[4]

(+)-Discodermolide is a member of the group of antimitotic agents known to act by microtubule stabilization, other members of which include Taxol[®5] (**2**), the epothilones A[6] (**3**), B[6] (**4**), and D[7] (**5**), eleutherobin[8] (**6**), laulimalide[9] (**7**), and recently, dictyostatin 1[10] (**8**). Because of its strong microtubule stabilizing properties and its excellent activity against multiple-drug-resistant tumors, HBOI licensed (+)-discodermolide (**1**) to Novartis Pharmaceutical Corporation in early 1998 for development as an anticancer drug.

The discovery that discodermolide's mechanism of action is similar to that of Taxol (**2**)[5] (Taxol is a registered trademark of Bristol-Myers Squibb Company, NY), coupled with its limited supply from the natural source, prompted the scientific community to prepare synthetic **1** and its analogs via a number of routes. The unique chemical structure with 13 stereogenic centers posed a major synthetic challenge for the preparation of sufficient material to allow for the preclinical and clinical studies. Over 15 publications describe this work. A major milestone in the development of **1** was the progression of a large-scale synthesis of (+)-discodermolide (**1**) by the Smith group.[11] This, coupled with other synthetic schemes, led to a refined process in which over 60 g of synthetic (+)-discodermolide was produced. At present, synthetic (+)-discodermolide is undergoing phase I clinical trials for solid tumor malignancies at the Cancer Therapy and Research Center in San Antonio, Texas.

II. NATURAL SOURCES

The genus *Discodermia* du Bocage belongs to the family Theonellidae. This genus is characterized by the presence of discotriaene spicules in addition to the presence of desmas, which are characteristic of all lithistid sponges. The microsclere compliment of the genus *Discodermia* consists of microxeas and microrhabds.[12] *Discodermia* is represented by at least three valid species, *D. dissoluta* Schmidt, *D. polydiscus* du Bocage, and *D. verrucosa*, in the Caribbean region of the central Atlantic Ocean. Additional as-yet-undescribed species have been collected by HBOI and are in their collection.

(+)-Discodermolide (**1**) was first isolated[1] from a sponge sample collected on March 27, 1987, off Lucaya, Grand Bahamas Island, at a depth of 30 m. Subsequently, sufficient quantity of **1** was isolated from an earlier 1985 collection to allow for the completion of the structure elucidation and preliminary biological evaluations. The 1987 sponge sample was slightly dark in appearance, and the compound responsible for this color coeluted in trace amounts with the first isolate of (+)-discodermolide. This trace impurity gave a pink coloration to the solution of (**1**), and further purification attenuated the yield of **1**. The freshly thawed 434 g of sponge sample from the 1987 collection yielded 7 mg of pure (+)-discodermolide. The samples collected in 1985 and most of the subsequent collections did not contain the pink impurity, and the average yield of **1** was 10 mg/kg of the wet sponge sample.

Harbor Branch chemists noted the existence of two chemically distinguishable forms of *D. dissoluta*. Both sponges contained **1** as a minor secondary metabolite and either the tetramic acid–derived alkaloid discodermide (**9**)[13] or a mixture of depsipeptides trivially named polydisca-mides as the major metabolites. The major metabolites discodermide (**9**) and polydiscamide A (**10**)[14] occur in gram quantities in 1 kg of the wet sponge sample, but **9** and **10** have not been observed to co-occur in the same sponge sample. The need for larger amounts of discodermolide for preclinical investigation prompted the collection of 20 kg of *Discodermia* spp. in November 1998. From this material, four naturally occurring analogs have been isolated. One sponge sample (wet weight of 1931 g) was collected by submersible on November 23, 1998, at a depth of 147 m, from the Bell Channel Buoy, Grand Bahama Island, and yielded 12.0 mg of **1** and 0.3 mg of 2-*epi*-discodermolide[15] (**11**, yield, 0.00001% of wet weight). The morphology of this sponge is described as a ranging from a large cup shape to irregular cups forming branches. A second batch of 3570 g of the *Discodermia* sp. collected at a depth of 157 m off Lucaya, Grand Bahama Island, furnished **1** (35 mg), 2-*des*-methyldiscodermolide[15] (**12**, 1.0 mg, yield, 0.00003% wet weight) and 5-hydroxy methyldiscodermolate[15] (**13**, 1.1 mg, yield, 0.00003% wet weight). This sponge sample was tan in color, and the morphology was a cluster of fingers. A third form of *Discodermia* sp. (2480 g) was described as thin, irregular anastomosing branches that are cream in color. It was collected from the same location as the latter sample and yielded **1** (25 mg) and 19-*des*-aminocarbonyldiscodermolide[15] (**14**, 1.1 mg, yield, 0.00006% wet weight). A cyclic analog of **1**, trivially named 9-13-cyclodiscodermolide[15] (**15**), was isolated from a collection of *Discodermia* sp. made in May 1993 off Tartar Bank, south of Cat Island, Bahamas, at a depth of 183–198 m. This specimen was cup shaped and white to yellow in color, and the 2500 g sample furnished only 1.2 mg (yield, 0.00005% wet weight) of 9-13-cyclodiscodermolide (**15**).

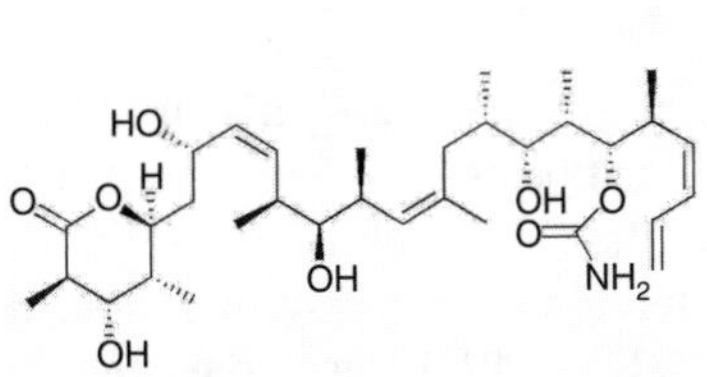

1. (+)-Discodermolide

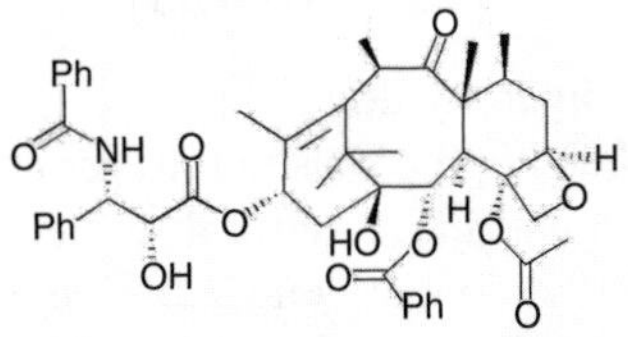

2. Taxol

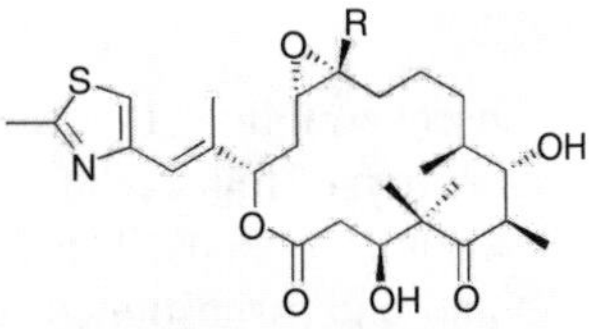

3. Epothilone A, R = H
4. Epothilone B, R = Me

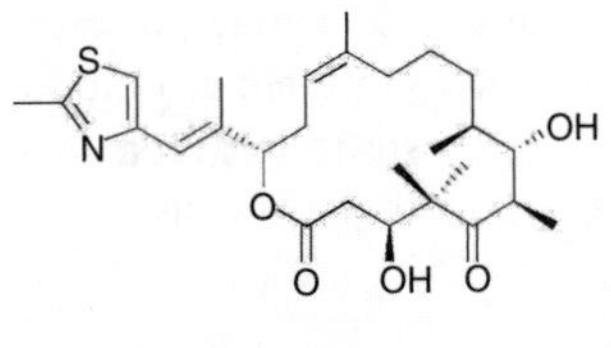

5. Epothilone D

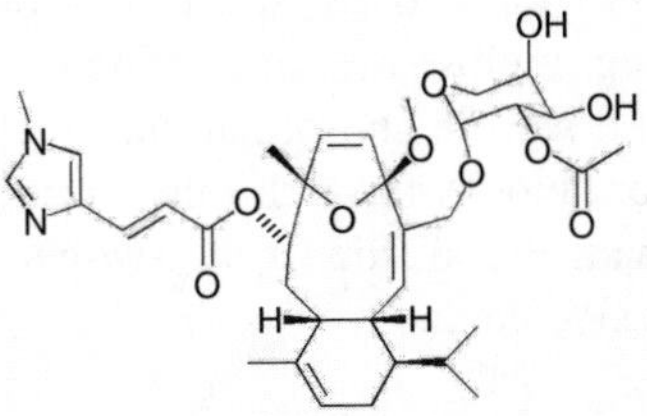

6. Eleutherobin

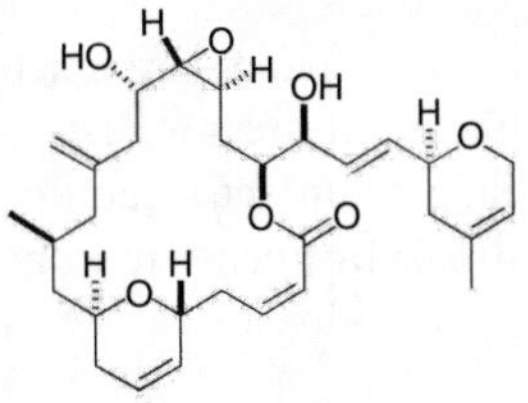

7. Laulimalide

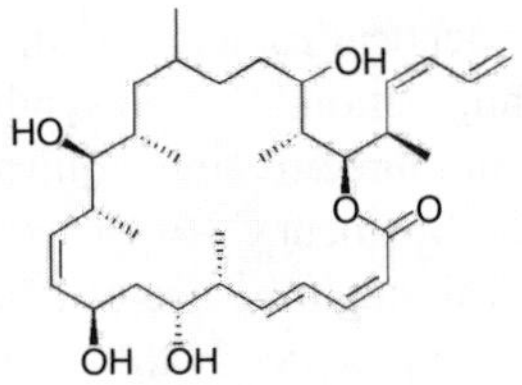

8. Dictyostatin-1

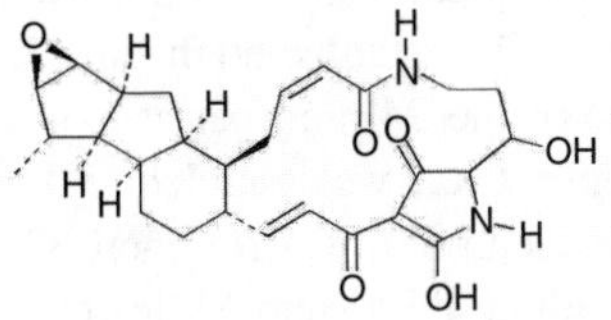

9. Discodermide

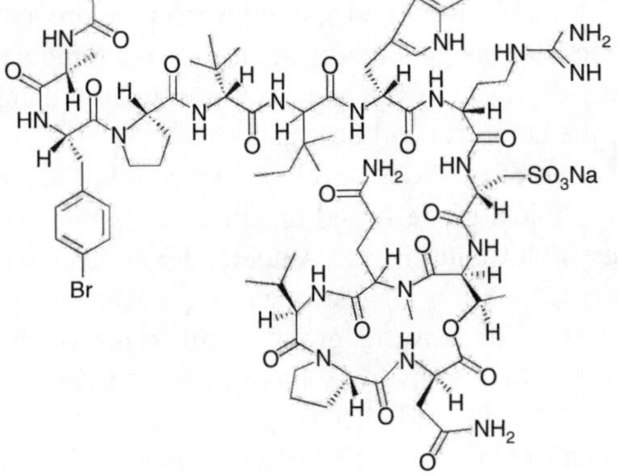

10. Polydiscamide A

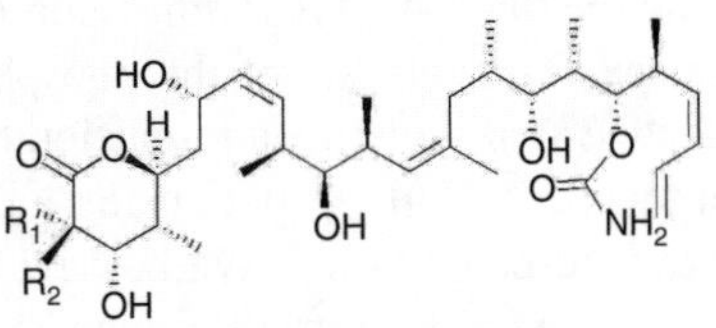

11. 2-*Epi*-descidermolide, R_1 = Me, R_2 = H
12. 2-*Des*-methyldiscodermolide, R_1 = H, R_2 = H

STRUCTURES 1–12

III. CHEMISTRY

(+)-Discodermolide (**1**) is a polyhydroxy dodecaketide δ-lactone presumably biosynthesized by a combination of four acetate and eight propionate units. Structurally, it bears 13 stereo centers with 15 rotatable σ-bonds. The δ-lactone ring is tetrasubstituted, and the side chain contains one di-(C8) and one trisubstituted (C13) Z-double bonds, as well as a terminal Z-diene system. Four of the five free secondary hydroxyl groups are located at C3, C7, C11, and C17 positions, and the fifth hydroxyl group is functionalized to a pendant carbamate group. Both in the solid state (as evident from the crystal structure[1]) and in the solution state (determined by NOE studies in conjunction with Monte Carlo conformational studies[16]), the structure of **1** exists in a helical conformation, arranging the C1–C19 region into a U-shaped conformation that brings the δ-lactone and the C19 pendant carbamate group in close proximity. The tetrasubstituted δ-lactone preferred a boatlike conformation. This adopted hairpin conformation, in which the two internal (Z)-olefins in the side chain act as conformational locks through minimization of (1,3) strain between their respective substituents in concert with the avoidance of *syn*-pentane interactions, results in the restricted mobility of the 19-carbon side chain.

IV. SYNTHESIS

A. Synthesis of Discodermolide

The first total synthesis of discodermolide was achieved by the Schreiber group at Harvard University, and their synthesis resulted in the preparation of (–)-discodermolide[17] based on the stereochemistry reported in the first publication.[1] Subsequently, this group, using the identical route to synthesize the natural antipode (+)-discodermolide (**1**), established the absolute configuration.[4,18] Their synthetic plan was adapted on a disconnection of the 24-member carbon chain into three roughly equivalent segments. The δ-lactone-end fragment C1–C7 was coupled to the middle fragment C8–C15, and the carbon skeleton was completed by attachment of a fragment C15–C24 through the use of enolate alkylation chemistry. Since then, Smith et al.[11,19] and Myles et al.[20,21] have prepared both antipodes of discodermolide. Marshall and Jones[22] and Paterson et al.[23] have synthesized (+)-discodermolide. All these groups used similar triply convergent approaches with different coupling and reaction conditions to achieve the total synthesis. A review by Kalesse[24] described the various approaches used by these groups to synthesize small quantities of discodermolide. In 1999, the Smith group improved on their previously published method[19] and synthesized **1** in 6% overall yield, a procedure that was amenable to gram-scale production.[11,25] They improved the triply convergent approach by synthesizing the three subunits from a common precursor, using a modified Negishi coupling using alkyl zinc species, synthesis of the phosphonium salt via ultrahigh pressure, and finally, using a chemoselective Wittig olefination reaction to give the complete carbon chain.

Recently, the Novartis group, using a combination of methods developed by Smith, Paterson, and their own research group, synthesized 60 g of clinical grade **1,** and this material is now being used in the phase I clinical trials at the Cancer Research Center in San Antonio, Texas. This large-scale synthesis of clinical grade **1** involved 39 steps, with the longest linear sequence having 26 steps and 17 large-scale chromatographic purifications. The Novartis group modified the synthesis of the Smith group Weinreb amide[25] that contains the methyl-hydroxy-methyl stereo triad that is repeated in three discodermolide fragments C1–C6, C9–C14, and C15–C21 for pilot plant production.[26] The total synthesis was recently described in a five-part series.[26–30] The final paper describes the linkage of the methylketone-containing C1–C6 fragment and the C7–C24 aldehyde fragment using a reagent-controlled stereoselective boron enolate aldol coupling procedure developed by Paterson's group.[31] The total synthesis by the Novartis group of 60 g of crystalline (+)-discodermolide using good manufacturing processes represents a significant milestone in both natural product drug development and organic synthesis.

B. Synthesis of Analogs

The Schreiber group was the first to prepare discodermolide analogs for structure–activity relationship (SAR) studies.[4,18] They prepared C16 (**16**) and C17 (**17**) epimers, 16-normethyl (**18**), epimers of C1 thiophenyl (**19, 20**), and C1-β thiophenyl 16-normethyldiscodermolide (**21**). Other analogs prepared by the Schreiber group are the C17 tritium-labeled discodermolide enantiomers (**22, 23**), the Fragments A (**24**) and B (**25**), C17-keto (**26**), pentyl-NHFmoc ester (**27**), and C24-hydroxylbutylester (**28**). The last two analogs and several similar analogs were prepared to study the binding properties of (+)-discodermolide.[18] The Paterson group[23] synthesized 5-*epi*-, 7-*epi*-, and 5,7-bis-*epi*-discodermolides (**29–31**) during their aldol-based total synthesis of (+)-discodermolide. They selectively accessed both C7 epimers in the final C6–C7 aldol coupling, which allowed the synthesis of the above epimers by stereocontrolled reduction of the C5 ketone and subsequent deprotection. The Schreiber group[4] observed that changes in the C17–C20 region of the discodermolide backbone resulted in loss of activity, and therefore Smith's group[32] focused their modifications on the C1–C14 segment of **1** and prepared four analogs including 3-deoxy-discodermolide-2-ene (**32**), 3,7-dideoxy-discodermolide-2-ene (**33**), 8,9-*E*-discodermolide (**34**), and 14-normethyldiscodermolide (**35**) for SAR studies.

Using a small quantity of natural (+)-discodermolide isolated from sponge samples, Gunasekera et al.[33] prepared eight acetylated discodermolide analogs (**36–43**) to study the importance of hydrogen bonding by the hydroxyl groups in the molecule on the cytotoxicity and microtubule stabilizing activity. These analogs were prepared by treating discodermolide with acetic anhydride in pyridine at a controlled temperature and by separating the resulting acetylated mixture by high-performance liquid chromatography methods. Subsequently, the same group has prepared 3-deoxy-discodermolide-2-ene (**32**) and four acetylated analogs of 3-deoxy-discodermolide-2-ene (**44–47**), using acetylated analogs of **1** as the starting material.[34] Temperature- and time-controlled catalytic hydrogenation of **1** resulted in three hydrogenated discodermolide analogs (**48–50**) and one 7-deoxy-hexahydro analog (**51**), which on acetylation, yielded the 7-deoxy-hexahydro-triacetate (**52**). These compounds were used to study the importance of the double bonds in contributing to the activity. In addition, one 6,7-*seco*-analog (**53**) has been generated during the attempted allylic oxidation of the C7 hydroxyl group in **1**, using MnO_2 as an oxidizing agent. This compound **53** on stirring with titanium(IV) isopropoxide in CH_2Cl_2 at room temperature for 2 weeks furnished the 19-hydroxy-6,7-*seco*-analog (**54**). These compounds were used to ascertain the contribution of the C1–C7 unit on the activity.[34]

During the final stages of the 60-g synthesis of **1** conducted by the Novartis group,[30] the three *t*-butyldimethylsilanyl protecting groups in (**55**) were removed by a one-step acid catalyzed procedure. This hydrolysis procedure with concomitant lactonization gave crude **1** in 85% purity and quite a number of various combinations of silyl ether cleavage side products and several cyclized side-products in trace quantities. These minor cyclized side-products (**56–61**) have been purified, and their structures were determined by nuclear magnetic resonance studies. The formation of five-cyclopentyl (**56–60**) and one-pyrano (**61**) analogs indicated that the hairpin confirmation of **1** in solution brings the two isolated double bonds closer together in correct geometry for the proton catalyzed cyclization.[35]

V. BIOLOGICAL ACTIVITY

A. Immunosuppressive Properties

(+)-Discodermolide (**1**) could be considered a unique compound that exhibits both strong immunosuppressive activity and strong antiproliferative activity.[1] Initial testing of **1** indicated potent immunosuppressive activity in the bi-directional murine mixed lymphocyte assay (MLR) and antiproliferative activity toward P388 murine leukemia cells. Further studies by Longley et al.

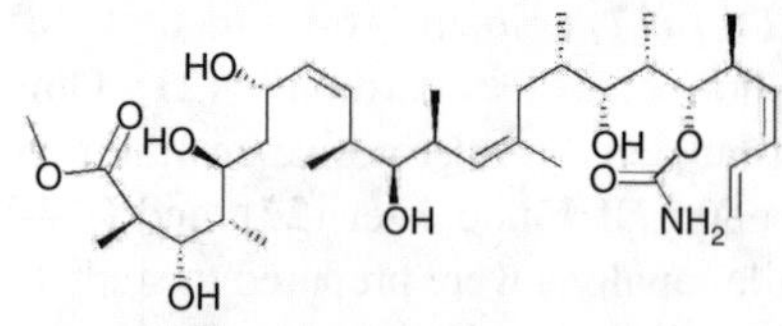

13. 5-Hydroxymethyldiscodermolate

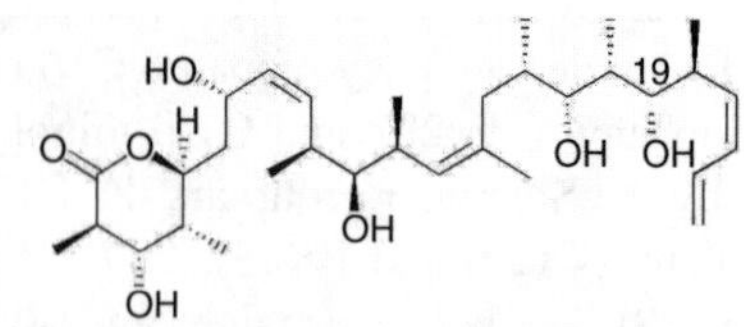

14. 19-*Des*-aminocarbonyldiscodermolide

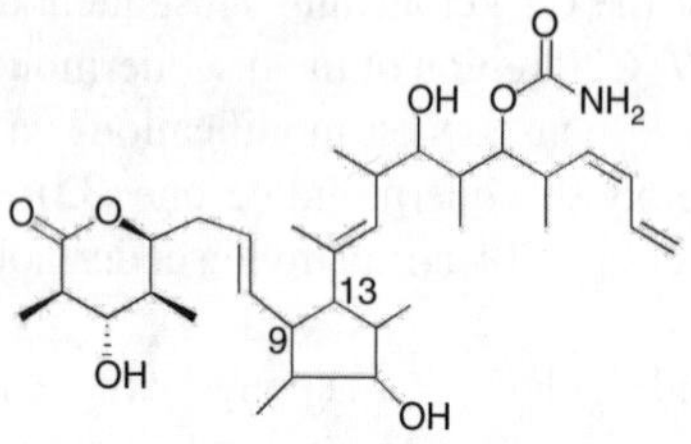

15. 9-13-Cyclodiscodermolide

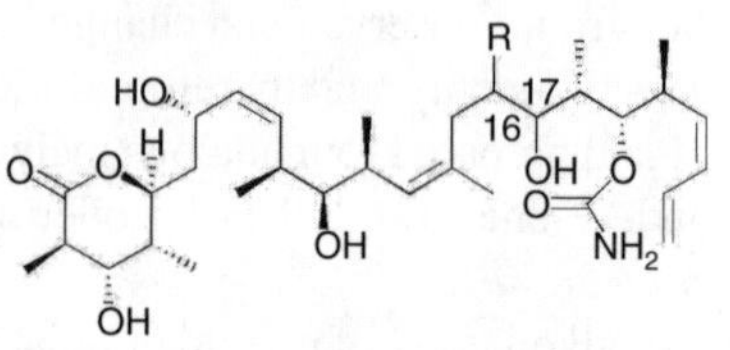

16. R = Me, (C16*R*, C17*R*)
17. R = Me, (C16*S*, C17*S*)
18. R = H, (C17*R*)

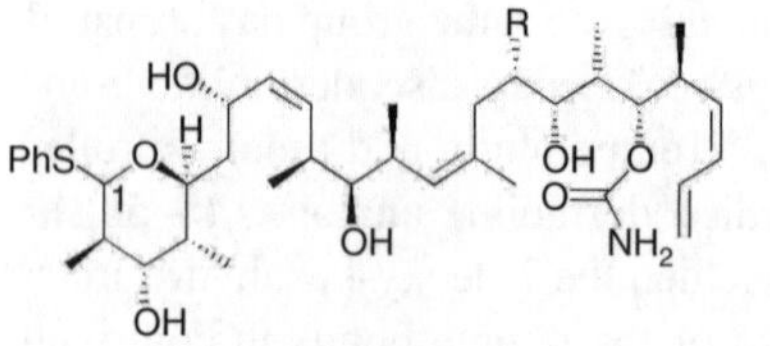

19. β-PhS, R = Me
20. α-PhS, R = Me
21. β-PhS, R = H

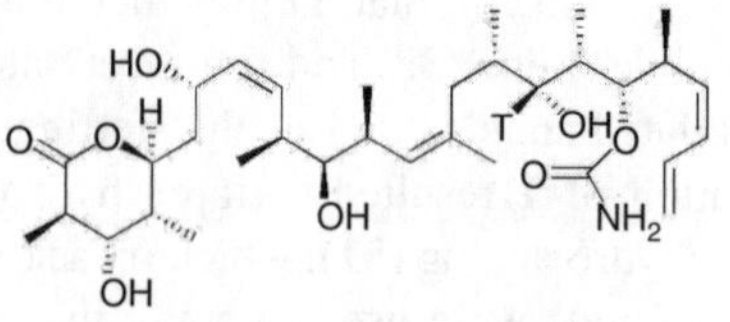

22. ³H-(+)-Discodermolide

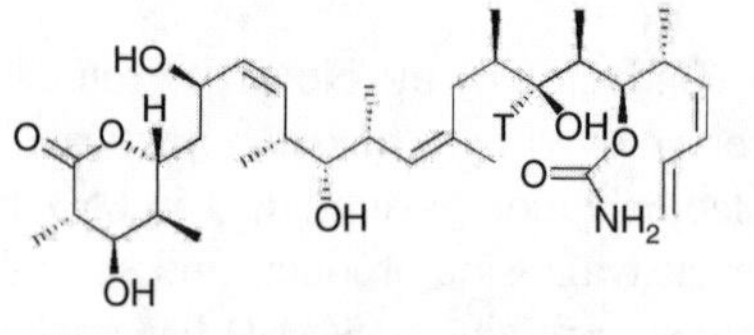

23. ³H-(−)-Discodermolide

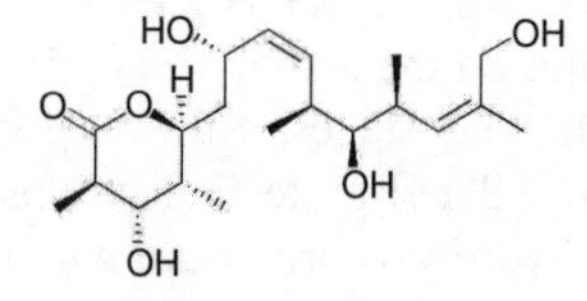

24. Fragment A

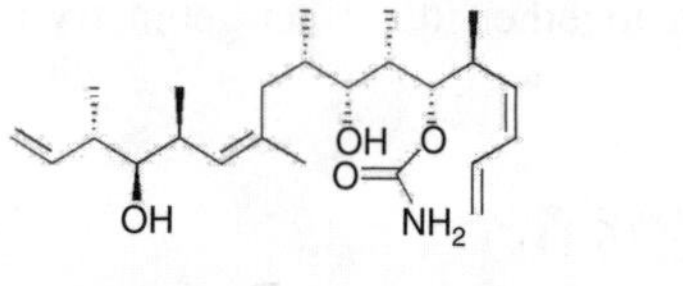

25. Fragment B

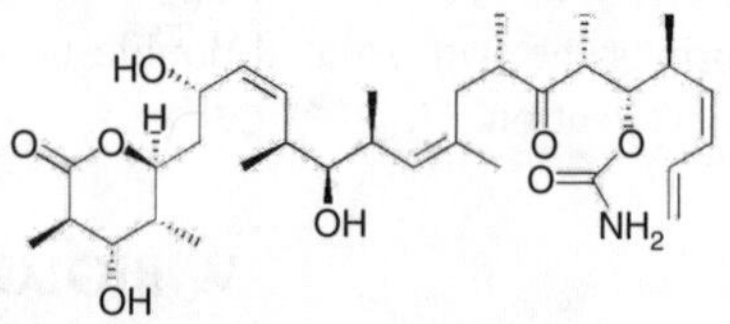

26. (+)-17-*Oxo*-discodermolide

STRUCTURES 13–26

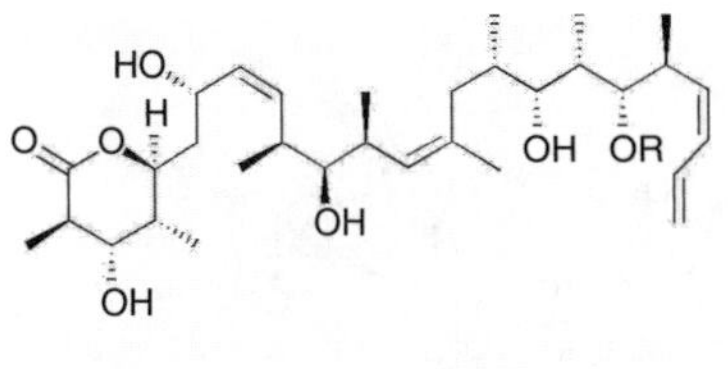

27. R = CO(CH₂)₅NHFmoc

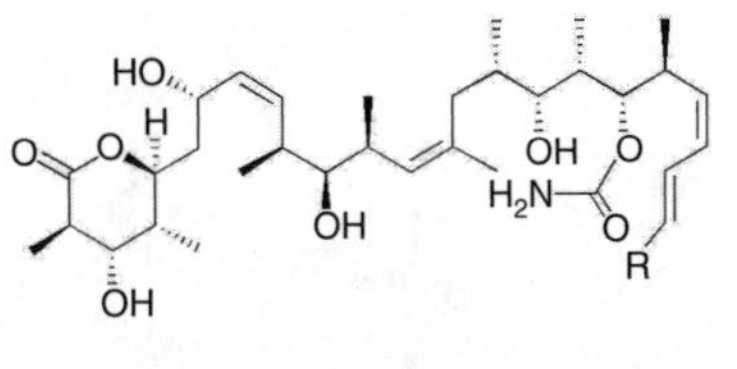

28. R = –(CH2)₄OPv

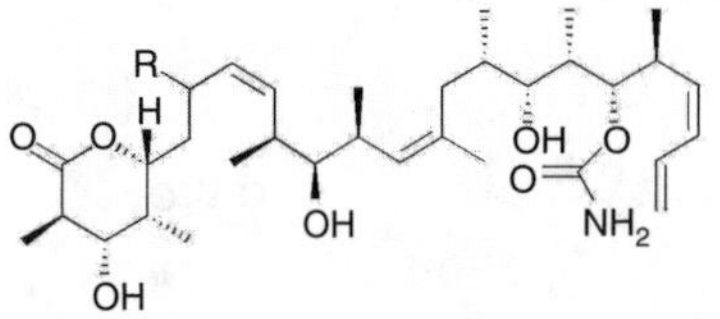

29. 5-*Epi*-discodermolide, R = αOH
31. 5,7-Bis-*epi*-discodermolide, R = βOH

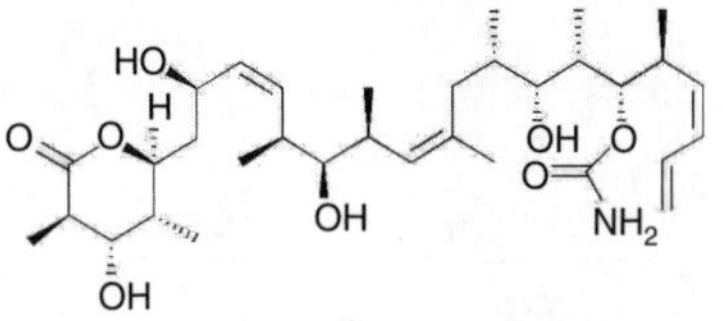

30. 7-*Epi*-discodermolide

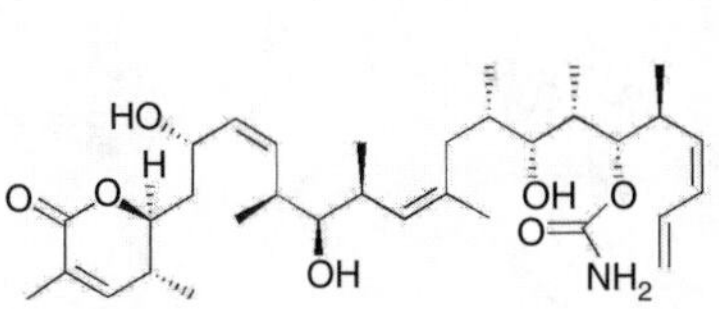

32. 3-Deoxy-discodermolide-2-ene

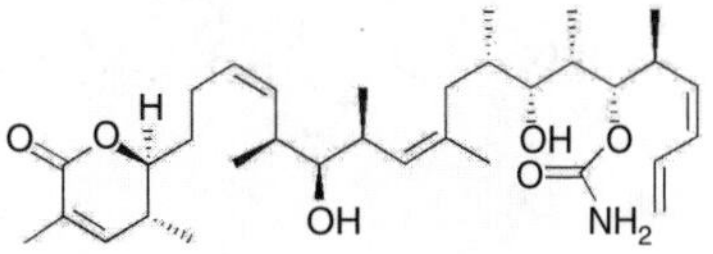

33. 3,7-Dideoxy-discodermolide-2-ene

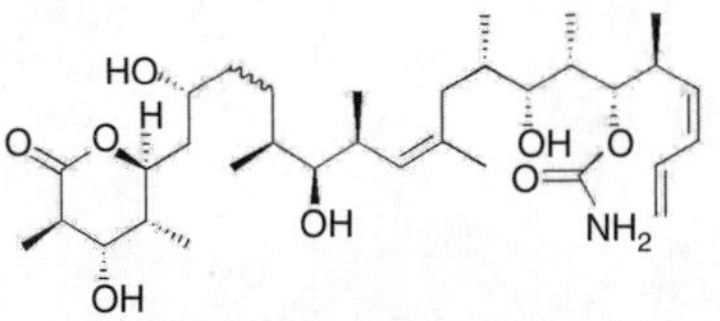

34. 8,9-*E*-Discodermolide

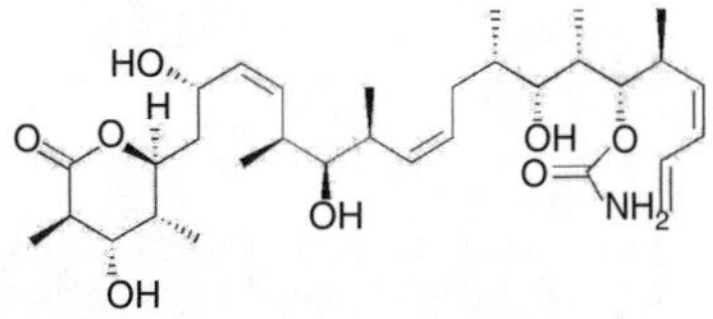

35. 14-Normethyldiscodermolide

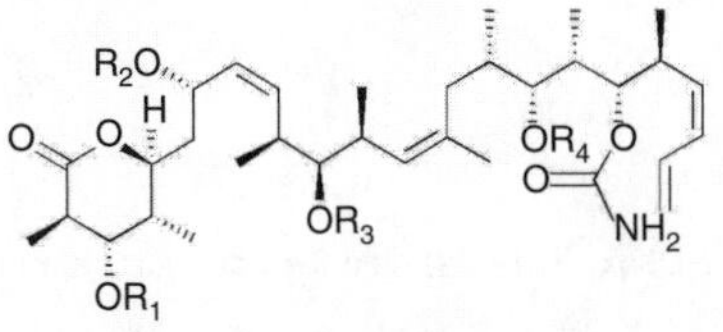

36. Discodermolide-3-acetate, R₁ = COCH₃, R₂ = R₃ = R₄ = H
37. Discodermolide-7-acetate, R₂ = COCH₃, R₁ = R₃ = R₄ = H
38. Discodermolide-3,7-diacetate, R₁ = R₂ = COCH₃, R₃ = R₄ = H
39. Discodermolide-3,11-diacetate, R₁ = R₃ = COCH₃, R₂ = R₄ = H
40. Discodermolide-3,17-diacetate, R₁ = R₄ = COCH₃, R₂ = R₃ = H
41. Discodermolide-3,7,11-triacetate, R₁ = R₂ = R₃ = COCH₃, R₄ = H
42. Discodermolide-3,7,17-triacetate, R₁ = R₂ = R₄ = COCH₃, R₃ = H
43. Discodermolide-3,7,11,17-tetraacetate, R₁ = R₂ = R₃ = R₄ = COCH₃

STRUCTURES 27–43

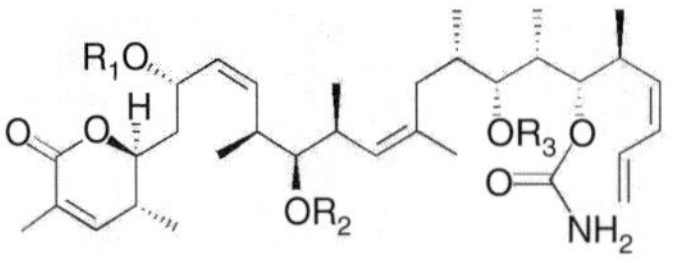

44. 3-Deoxy-discodermolide-2-en-11-acetate, $R_1 = R_3 = H$, $R_2 = COCH_3$
45. 3-Deoxy-discodermolide-2-en-17-acetate, $R_1 = R_2 = H$, $R_3 = COCH_3$
46. 3-Deoxy-discodermolide-2-en-11,17-diacetate, $R_1 = H$, $R_2 = R_3 = COCH_3$
47. 3-Deoxy-discodermolide-2-en-7,11,17-triacetate, $R_1 = R_2 = R_3 = COCH_3$

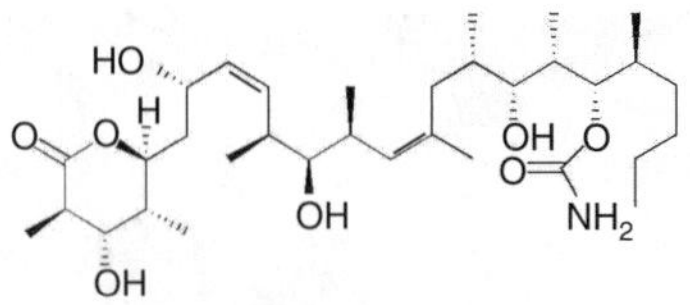

48. 21,23-Tetrahydrodiscodermolide

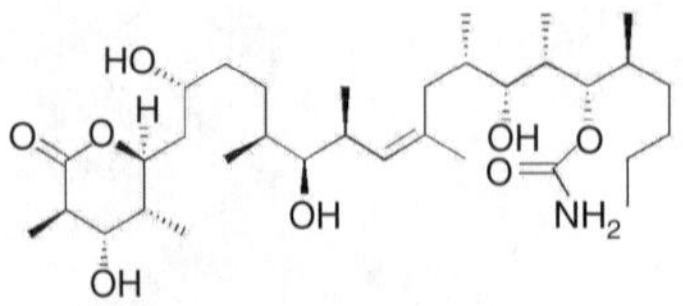

49. 8,21,23-Hexahydrodiscodermolide

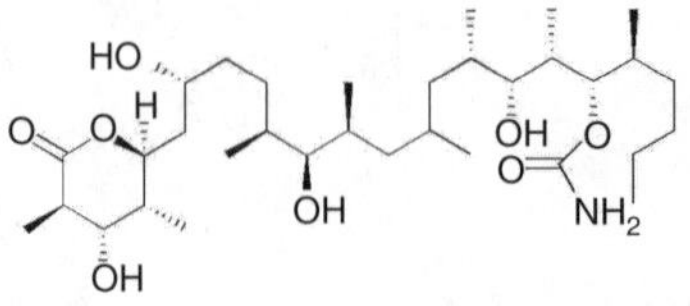

50. 8,13,21,23-Octahydrodiscodermolide

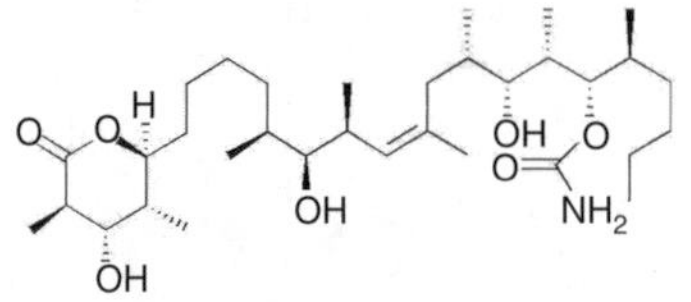

51. 7-Deoxy-8,21,23-hexahydrodiscodermolide

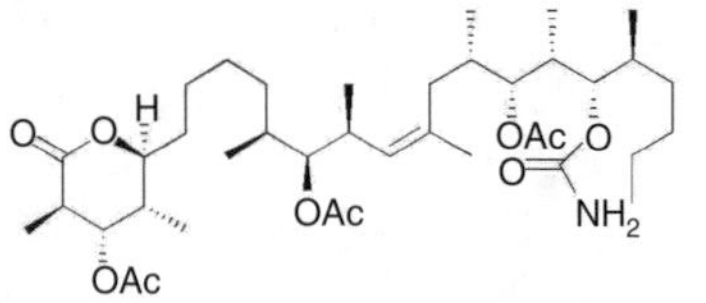

52. 7-Deoxy-8,21,34-hexhydrodiscodermolide-3,11,17-triacetate

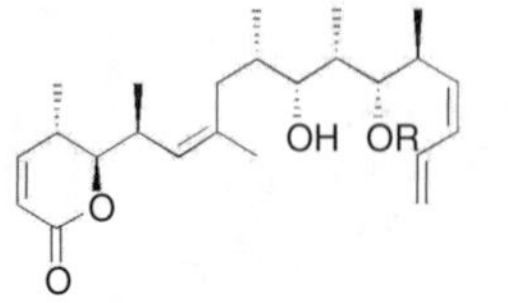

53. 6,7-*Seco*-discodermolide, R = $CONH_2$
54. 19-*Des*-aminocarbonyl-6,7-*seco*-discodermolide, R = H

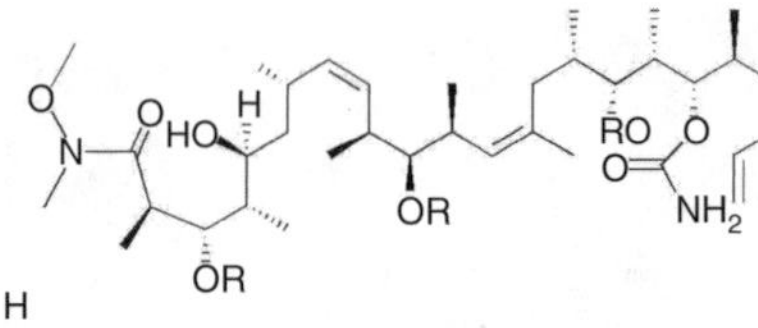

55. R = $Si(CH_3)_2C(CH_3)_3$

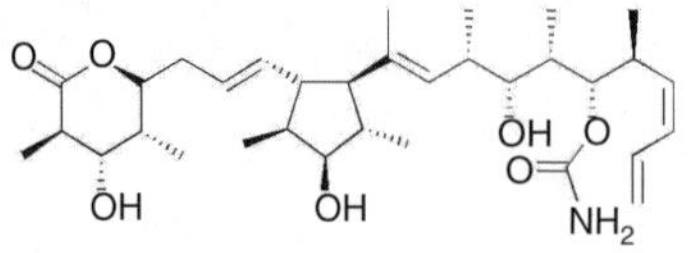

56. 7-Deoxy-7, 14-dien-9-13-cyclodiscodermolide

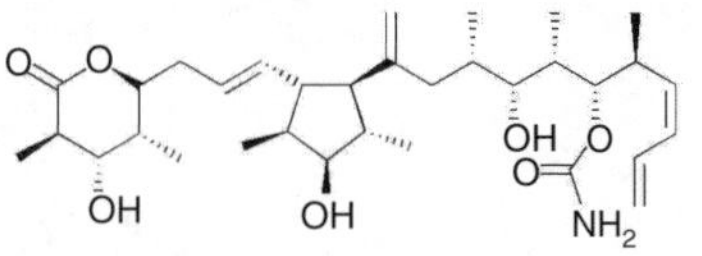

57. 7-Deoxy-7, 14(29)-dien-9-13-cyclodiscodermolide

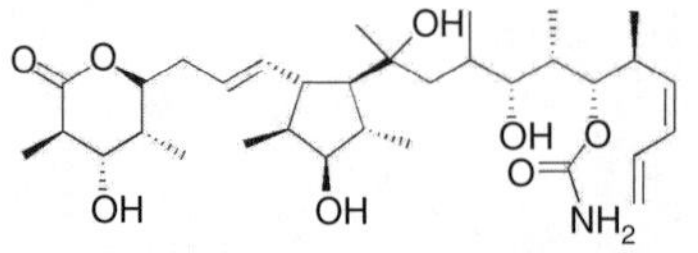

58. 14(*R*)-hydroxy-16(*S*)-9-13-cyclodiscodermolide
59. 14(*S*)-Hydroxy-16(*S*)-9-13-cyclodiscodermolide
60. 14(*S*)-Hydroxy-16(*R*)-9-13-cyclodiscodermolide

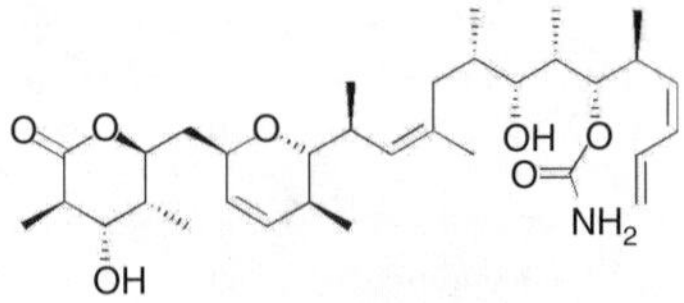

61. 7-(11)-Oxycyclodiscodermolide

STRUCTURES 44–61

demonstrated immunosuppressive activity in a number of in *vitro* assays,[2,36] and in an *in vivo* experimental model of graft-versus-host reaction,[3] the Simonsen splenomegaly assay.[37] These studies indicated that **1** suppresses the proliferative responses of splenocytes in the murine two-way mixed lymphocyte reaction, and concavalin A (Con A) stimulated cultures, with IC_{50} values of 0.24 μ*M* and 0.19 μ*M*, respectively, with no detection of cytotoxicity for murine splenocytes at concentrations of **1** as high as 1.26 μ*M*. In further studies, **1** suppressed the proliferative responses of human peripheral blood leukocytes (PBL) in the two-way mixed lymphocyte reaction, Con-A and phytohemagglutinin mitogenesis assays, with IC_{50} values of 5.65 μ*M*, 28.02 μ*M*, and 30.12 μ*M*, respectively, with no evidence of cytotoxicity toward human PBL at (+)-discodermolide concentrations as high as 80.64 μ*M*.[3]

(+)-Discodermolide was more potent than the clinically used cyclosporin A, a cyclic decapeptide (CsA) isolated from a fungus *Trichoderma inflatum*,[38] in its ability to suppress the PMA –ino-mycin-induced proliferation of purified murine T-cells. The IC_{50} values observed in this assay were 9.0 and 14.0 μ*M* for **1** and CsA, respectively. **1** had no effect on interleukin 2 (IL-2) production or IL-2 mRNA expression; however, the expression of the low-affinity IL-2 receptor in PMA–ion-omycin-stimulated, purified, murine T-lymphocytes, and that of PHA-stimulated human PBL blasts, was affected. *In vivo* studies by Longley et al.[3] reported that **1** is effective in the graft-versus-host splenomegaly response of BALB/c → CB6F$_1$ (BALB/c × C57Bl/6J)F$_1$ grafted mice. Mice that were administered daily, intraperitoneal injections with dose levels of 1.25, 0.625, and 0.313 mg/kg remained healthy after 7 days on the regimen and continued to demonstrate suppression of sple-nomegaly with 106%, 72%, and 76% suppression, respectively.

Splenocytes obtained from (+)-discodermolide-treated allogeneic grafted mice were suppressed in their ability to respond *in vitro* to optimal mitogenic concentrations of Con A, and natural killer cell activity directed against YAC-1 tumor cells, compared with vehicle-treated, allogenic grafted control mice. Lower dosages of **1**, however, did not affect the subsequent ability of splenocytes obtained from these mice to produce IL-2 following *in vitro* stimulation with Con A. The unusual *in vivo* immunosuppressive properties of **1** as compared to CsA indicated that **1** has a specific *in vivo* mechanism of action. Using the murine DO11.10 T-hybridoma cell line, Longley et al.[36] reported that the antiproliferative activity of **1** was a result of its ability to block cell proliferation in the G$_2$/M phase of the cell cycle. Control cultures of DO11.10 cells demonstrated a characteristic pattern of cell cycling. Approximately 68% of the control cells make up the G$_1$ phase of the cell cycle, and approximately 31% of the cells make up the S phase. Only 1% of these cells are demonstrable in the G$_2$/M phase in asynchronous cultures at any one time, mainly because of the high proliferative rate of DO11.10 cells. At 24 h post–culture initiation, after treatment with 1.0 μg/mL **1**, the percentage of cells in G$_1$ phase decreased to 25%, and those in S to 16%. However, the percentage of cells in G$_2$/M increased to 58%. The results from this study concluded that **1** acts by blocking cells in the G$_2$/M phase of the cell cycle.[36]

B. ANTITUMOR PROPERTIES

(+)-Discodermolide (**1**) was originally isolated by Gunasekera et al.[1] from the marine sponge *Disco-dermia dissoluta,* following a cytotoxicity-based bioassay-guided approach. The crude ethanol extracts of the source sponge initially revealed strong antiproliferative activity toward P388 murine leukemia cells in addition to potent immunosuppressive activity in the bidirectional murine mixed lymphocyte assay. (+)-Discodermolide was evaluated for its antiproliferative effects in the National Cancer Insti-tute's (NCI's) 60–human tumor cell-line screen.[39] The results of the NCI analyses were reported as growth inhibition 50 (GI_{50}) values, in which the growth of 50% of the cells is inhibited; total growth inhibition (TGI) values, in which the growth of all cells is inhibited (i.e., cytostasis); and lethal concentration 50 (LC_{50}), in which 50% of the cells are killed. **1** demonstrated selective cytotoxicity for 32 cell lines for GI_{50} and 18 cell lines for TGI. According to LC_{50} measurements for the NCI cell lines, **1** was particularly effective against human non–small cell lung cancer (NCI-H23), human colon cancers (COLO 205 and HCC-2998), human melanoma (M14), two of six central nervous system cancer cell lines (SF 295 and SF 539), and two of eight breast cancer cell lines (MDA-MB-435 and MDA-N).

A statistical analysis of this selective toxicity pattern (the COMPARE algorithm)[40] indicated that the cytotoxicity pattern of **1** matched that of several previously tested microtubule interactive agents, including the pattern generated by Taxol® (**2**). Computer-assisted structure–activity analysis by Ernest Hamel at NCI predicted antimitotic properties.[41] This finding led to the collaboration of the HBOI group with Billy Day of the University of Pittsburgh and Ernest Hamel of the NCI to begin a study on the antimitotic properties of (+)-discodermolide (**1**). The group found[41] that **1** was equally effective as Taxol (**2**) in inhibiting the growth of two human breast carcinoma cell lines, estrogen receptor–positive MCF-7 cells (IC_{50} = 2.4 nM for **1** and 2.1 nM for **2**), and estrogen receptor–negative MDA-MB-231 cells (the IC_{50} values are not reported). Complete inhibition of growth occurred with 10 nM or greater of each drug and was not reversed by removal. (+)-Discodermolide (**1**) treated cells exhibited highly fragmented condensed nuclei. Flow cytometric studies of cells treated with **1** or **2** at 10 nM showed that both compounds caused cell cycle perturbation and induction of a hypodiploid cell population. **1** caused these effects more extensively and at an earlier time point. Immunofluorescence patterns of cells treated with **1** revealed remarkable rearrangement of the microtubule network, demonstrating promotion of microtubule assembly and indicating a Taxol-like effect (Figure 10.1).

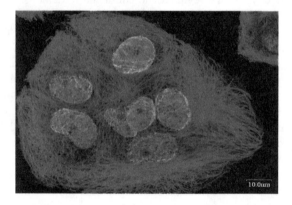

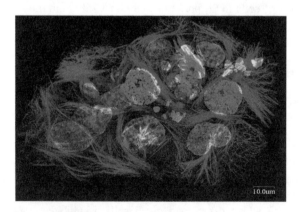

FIGURE 10.1 (See color insert following page 304.) Immunofluorescence images of A549 cells stained with anti-α-tubulin (green) and propidium iodide (red) and observed by confocal microscopy. Cells were exposed to (A) 0.05% ethanol (vehicle control), or (B) (+)-discodermolide at a concentration of 100 nM.

1 was also analyzed for its interactions with purified tubulin. It was found that **1** induced the polymerization of tubulin to form microtubules in both the presence and absence of microtubule-associated proteins and guanosine 5′-triphosphate (GTP), and the resulting microtubules induced polymerization at lower temperatures than **2** and were more stable to the cold (4°C) than those formed by **2**. (+)-Discodermolide-induced polymer differed from Taxol-induced polymer in that it was completely stable at 0°C in the presence of high concentrations of Ca^{++} ions. In a quantitative assay designed to select for compounds more potent than Taxol (**2**) in its ability to polymerize tubulin, **1** showed an EC_{50} value of 3.2 μ*M*, versus 23 μ*M* for **2**.[41] Hung et al.[4,42] in the Schreiber group reported similar results using synthetic (+)-discodermolide. Their studies demonstrate similar arrest of MG63 human osteosarcoma cells in the M-phase of the cell cycle, induction of polymerization of purified tubulin, and "bundling" of the microtubule network in Swiss 3T3 cells. They discovered that the binding of microtubules by **1** and **2** were mutually exclusive in competitive binding assays (i.e., **1** was capable of displacing **2** and vice versa in binding to microtubules). **1** was also found to bind to microtubules with a much higher affinity than **2**.

Longley et al.[39] at HBOI continued further studies on the effect of **1** on cell cycle progression and apoptosis of A549 human lung and GI-101A human breast cells in comparison to Taxol (**2**). The GI-101A cell line used by the Longley group was provided by Dr. Josephine Hurst, Goodwin Cancer Center, Plantation, Florida, and is an adherent breast cancer cell line derived from the original parent mammary tumor xenograft cell line, GI-101.[43] The line was originally derived from a local first recurrence of an infiltrating ductal adenocarcinoma (Stage IIIa, T3N2MX) in a 57-yr-old female who had previously not received any chemotherapy or radiation therapy. Tumors transplanted into murine hosts are positive for normal human breast tumor markers and for the p53 antigen but are negative for c-erbB-2 oncogene. The A549 human lung and GI-101 human breast cells were incubated at 37°C in 5% CO_2 in the presence or absence of 10 n*M* **1** or **2** for 24 h. Cells were harvested, fixed in ethanol, and stained with 0.5 mg/ml of propidium iodide together with 0.1 mg/ml of RNAse A. Stained preparations were analyzed on a Coulter EPICS ELITE flow cytometer with 488 n*M* excitation, and the resulting DNA histograms were collected from at least 10,000 propidium iodide (P.I.) stained cells at an emission wavelength of 690 n*M*. Raw histogram data were further analyzed using a cell cycle analysis program. A549 and GI-101 cells without the drug exhibited a typical pattern of cell cycling, with a large percentage of the cell population comprising the G_1 population, with lesser percentages comprising the S and G_2/M phases of the cell cycle, as shown in Figure 10.2. Cells treated with 10 n*M* **1** or **2** for 24 h exhibited decreased percentages of cells comprising the G_1 population, with corresponding increased percentages in both S and G_2/M phases of the cell cycle, indicating the ability of **1** and **2** to induce G_2/M block, as shown in Figure 10.3. Cells undergoing apoptosis are also evident as a peak immediately to the left of the G_1 peak.

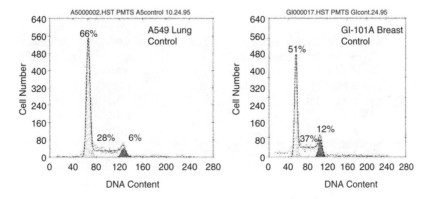

FIGURE 10.2 Cell cycle analysis of A549 and GI-101 control cells.

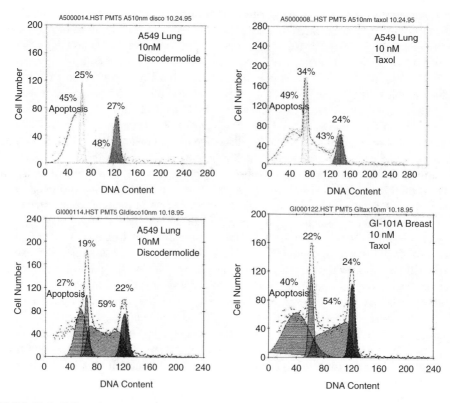

FIGURE 10.3 Cell cycle analysis of A549 and GI-101A treated with (+)-discodermolide (**1**) and Taxol (**2**).

(+)-Discodermolide (**1**) indicated strong cytotoxicity in all cancer cell lines tested *in vitro*, and the activity was comparable to that of Taxol. The results are shown in Table 10.1. However, it was interesting to note that **1** was less cytotoxic to the normal kidney epithelial cell line of the African green monkey. These results indicated that **1** selectively targets dividing cancer cells, and thus suggested microtubule stabilization is probably the only mechanism of action of (+)-discodermolide.

TABLE 10.1
In Vitro Cytotoxicity of (+)-Discodermolide (1) in Selected Cancer Cell Lines

Cell Line	Activity Data (IC$_{50}$ in µM)
Murine P388 leukemia cells	0.033
Human A549 lung cancer cell line	0.015
Human MCF-7 breast cancer cell line	0.0024*
Human MDA-MB-231 breast cancer cell line	0.0024*
Human NCI/ADR	0.017*
PANC-1	0.049
VERO	30
For Taxol*	0.0021

A549 (human lung adenocarcinoma cell line); P388 (cultured murine leukemia cell line); MCF-7 (human breast cancer cell line); NCI/ADR [adriamycin drug resistant cell line]; PANC-1 (human pancreatic cancer cell line); VERO (normal kidney epithelial cell line of African green monkey).

Longley's group at HBOI continued studies on the characterization of the molecular basis of apoptosis in (+)-discodermolide-treated tumor cells.[39] This study was designed to detect and characterize the (+)-discodermolide-induced molecular pathways of apoptosis and to evaluate the differences with Taxol-induced apoptosis, by probing Bcl-2, Bcl-Xl, and Raf-1 proteins in JURKAT and MCF-7 cell lines by Western blot analysis, and following the phosphorylation patterns in response to **1** and **2** treatments. The HBOI group detected that both **1** and **2** phosphorylate the Bcl-2 family proteins, and this appears to be accompanied by a loss of Bcl-2 function. They also measured the apoptotic responses by detecting and quantitating DNA fragmentation in various cell populations (MCF-7, DO11.10, and JURKAT) under the treatment of drugs, such as Fas-antigen, dexamethasone, Taxol (**2**), and (+)-discodermolide (**1**), using a time course (6, 12, 24 h) and concentration-dependent (10, 100, 1000 nM of drug) experiments. The DNA fragmentation analysis, correlated by flow cytometric studies and by microscopy, indicated that **1** induces apoptosis in a dose-dependent manner and appears to do so in the same way as **2**. Also, it appears that the dose range for inducing these apoptotic events is more or less the same depending on the cell line used. In comparative studies of **1**, with the epothilones (**3**, **4**) and eleutherobin (**6**) against a Taxol-dependent human lung carcinoma cell line A549-T12, Martello et al.[32] found that **1** was unable to act as a substitute for **2**, whereas **3**, **4**, and **6** were able to maintain the viability of the cell line. The presence of low concentrations of **2** amplified the cytotoxicity of **1** 20-fold against this cell line and thus suggested that **1** and **2** may constitute a promising chemotherapeutic combination.[44]

VI. STRUCTURE–ACTIVITY STUDIES

The Schreiber group was the first to synthesize the unnatural antipode, (–)-discodermolide and a number of structural analogs, thus laying the ground work for SAR analysis.[17,18] Both enantiomers exhibited antiproliferative activity, but at different stages of the cell cycle. The natural (+)-discodermolide (**1**) blocks the cells in the G2/M phase, whereas the (–)-discodermolide inhibits the cell cycle in the S-phase. The two antipodes also showed significantly different potency in cytotoxicity-based assays. In a [³H]-thymidine incorporation assay using NIH3T3 cells, synthetic (**1**) had an IC_{50} value of 7 nM, compared to135 nM for the ()-discodermolide, or a 19-fold loss of activity.[4,18] Removal of the C16S methyl group (e.g., 16-normethyldiscodermolide [**18**]) did not affect the potency. However, epimerization at C16 (e.g., C16R [**16**]) or at C17 (e.g., C16S, C17S [**17**]) showed a significant reduction in activity (IC_{50} = 300 and >300 nM, respectively). Interestingly, the thiophenyl derivatives (**19**, **20**), both α and β anomers, as well as the 16-normethyl analog (**18**) were equally active as **1**, indicating that the carbonyl functionality and C16Me are not essential for the cytotoxicity of **1**. The other analogs prepared by the Schreiber group,[18] the left and right fragments of **1** (**24**, **25**), 17-oxo-discodermolide (**26**) and pentyl-NHFmoc ester (**27**) and 24-hydroxybutyl ester (**28**), retain a low level of activity of IC_{50} = 70 nM in the [³H]thymidine incorporation assay using MG63 cells. The three epimeric analogs 5,7-bis-epi-, 5-epi-, and 7-epi-discodermolides (**29–31**) prepared by the Paterson group[23] are structurally interesting; however, the biological activities of these compounds were not presented in the synthetic publication.

The Smith group[32] prepared four interesting discodermolide analogs, 3-deoxy-discodermolide-2-ene (**32**), 3,7-dideoxy-discodermolide-2-ene (**33**), 8,9-E-discodermolide (**34**), and 14-normethyl-discodermolide (**35**) — during their total synthesis of **1**, which used a highly efficient, triply convergent approach. This was the first report that indicated that microtubule stabilization by **1** is extremely sensitive to simple modifications in the molecular structure. The researchers' in vitro studies demonstrated that **1** and the structural analogs induced tubulin assembly in the absence of GTP and that the microtubules formed were stable under Ca++-induced depolymerizing conditions. The initial rates of polymerization reflect the nucleation process. It was observed that the rate of tubulin nucleation of the four analogs (**32–35**) was reduced as compared to that of **1**. Analogs **32** and **35** showed similar levels of polymerization to that of **1**, and analogs **33** and **34** demonstrated reduced polymerization, attaining approximately 50% of the level of polymerization observed with

TABLE 10.2
Biological Data of Taxol (2), (+)-Discodermolide (1) and Discodermolide Acetates (36–43)

Compound	P388 IC$_{50}$ nM	A549 IC$_{50}$ nM	G2/M Block 100 nM	Microtubule Bundling (nM)
Taxol (**2**)	7.7	—	+	Yes (100)
(+)-Discodermolide (**1**)	35	34	+	Yes (10)
3-Acetate (**36**)	9	9.4	+	Yes (10)
7-Acetate (**37**)	2.5	0.5	+	Yes (10)
3,7-Diacetate (**38**)	0.7	3.7	+	Yes (10)
3,11-Diacetate (**39**)	103	295	—	No (100)
3,17-Diacetate (**40**)	1149	736	—	No (100)
3,7,11-Triacetate (**41**)	166	554	—	No (100)
3,7,17-Triacetate (**42**)	>6825	>1307	—	No (100)
3,7,11,17-Tetraacetate (**43**)	837	1307	—	No (100)

(+)-discodermolide (**1**). Electron microscopy examination of the microtubules formed in the in vitro assays confirmed that normal microtubules were formed in the presence of the compounds. The microtubules that assembled in the presence of **1** were approximately fivefold shorter than those formed by analogs (**32**) and (**35**); 12-fold shorter than those formed by analogs (**33**) and (**34**), and fourfold shorter than those formed by Taxol (**1**). The same group showed that in a drug-binding competition assay, **1** and analogs **32**, **33**, and **35** are competitive inhibitors of [³H]Taxol binding to preformed microtubules. At 1 μM concentration, Taxol (**2**), (+)-discodermolide (**1**), and analogs (**32**, **33**, **35**) exhibited very similar inhibition, whereas analog **34** was essentially inactive in displacing [³H]Taxol binding. At 100 μM, **1** displayed 95% inhibition, **2** displayed 80% inhibition, and analog (**32**) exhibited >90% inhibition of [³H]Taxol binding to preformed microtubules. Analogs (**33–35**) demonstrated a range of inhibition between 45 and 65% at 100 μM.

Using natural (+)-discodermolide (**1**) isolated from the sponge *Discodermia*, Gunasekera et al.[34] in the Harbor Branch group prepared eight acetylated analogs (**36–43**) by acetylation of the hydroxyl groups at C3, C7, C11, or C17. These analogs were assayed for biological activity in cultured murine P388 leukemia and human A549 lung adenocarcinoma tumor cell lines to determine the structural requirements for tubulin interaction and cytotoxic effects. The biological data are presented in Table 10.2. The P388 and A549 assay results indicated that acetylation of the hydroxyl groups in the left-hand side of the molecule at C3 and C7 confers a greater cytotoxicity to the discodermolide structure, as seen with analogs (**36–38**). Interestingly, the discodermolide acetates (**39, 41**) with acetyl groups at C11 showed reduced cytotoxicity as compared to the parent molecule, whereas compounds (**40, 42, 43**) that include acetylation at C17 caused a dramatic reduction in cytotoxicity. These results confirmed the previous finding by the Schreiber group[18] that acetylation at C17 reduced the activity by tenfold in the [³H]-thymidine incorporation assay, using MG63 cells. Cell cycle analysis by flow cytometry using A549 cells revealed that the highly cytotoxic analogs caused the accumulation of cells in the G2/M phase. The effect on microtubule bundling follows the cytotoxicity as seen by analogs (**33–38**).

The HBOI group recently isolated and reported three natural discodermolide analogs (**11–13**) that showed modification in the lactone ring system of the molecule. In addition the group prepared through semisynthesis 3-deoxy-discodermolide-2-ene (**32**) and four 3-deoxy-2-ene acetylated analogs (**44–47**) for SAR studies. The activity data reported in Table 10.3 indicate that the ring-opened methyl ester (**13**) resulted in two- and fivefold reduction in activity in the P388 and A549 assays, respectively. Epimerization at C2 (**11**) results in a fivefold reduction in cytotoxicity and a reduction in the level of tubulin polymerization. Dehydration of the 3-hydroxyl functionality to

TABLE 10.3
Biological Data of Natural Analogs (11–13) and 3-Deoxy-Discodermolide-2-ene (32) and Analogs (44–47)

Compound	P388, IC$_{50}$ nM	A549, IC$_{50}$ nM	G2/M Block, 100 nM	Microtubule Bundling, 10 nM
2-*Epi*-discodermolide (**11**)	134	67	+Yes	+
2-Des-methyldiscodermolide (**12**)	172	120	+Yes	+++
5-Hydroxymethyldiscodermolate (**13**)	65	74	+Yes	+
3-Deoxy-discodermolide-2-ene (**32**)	34	21	+Yes	+++
3-Deoxy-discodermolide-2-en-11-acetate (**44**)	519	519	No	—
3-Deoxy-discodermolide-2-en-17-acetate (**45**)	8103	8103	No	—
3-Deoxy-discodermolide-2-en-11,17-diacetate (**46**)	7587	7587	No	—
3-Deoxy-discodermolide-2-en-7-11-17-triacetate (**47**)	7076	7133	No	—

yield 3-deoxy-discodermolide-2-ene (**32**) results in no loss of biological activity compared to that of the parent (+)-discodermolide (**1**). Interestingly, the 2-des-methyldiscodermolide (**12**) showed greater than fivefold reduction in cytotoxicity, yet the microtubule bundling properties were comparable to **1** and **32**. The analogs (**11–13, 32**) showed cell cycle arrest at the G2/M checkpoint and induction of microtubule bundling. These SAR studies confirmed that minor changes in the functional groups on the lactone ring do not significantly affect the antimitotic properties of (+)-discodermolide (**1**). The acetylated analogs of 3-deoxy-discodermolide-2-ene (**44–47**) showed a significant reduction in cytotoxicity and the complete loss of microtubule-stabilizing properties on acetylation of the C11 or C17 hydroxyl groups, indicating the importance of these two hydroxyl groups for the microtubule stabilizing properties of (+)-discodermolide (**1**).

As shown in Table 10.4, both the 21,23-tetrahydrodiscodermolide (**48**) in which the terminal conjugated double bond has been hydrogenated and 8,21,23-hexahydrodiscodermolide (**49**) have similar activity to the parent (+)-discodermolide (**1**), indicating that the C8 double bond and the terminal diene are not essential for the cytotoxicity and tubulin polymerizing action of (+)-disco-dermolide. The 8,13,21,23-octahydrodiscodermolide (**50**) shows significantly less cytotoxicity (8.2 μM) and no effect on tubulin polymerization. The retention of the microtubule bundling properties in 21,23-tetrahydrodiscodermolide (**48**) and 19-des-aminocarbonyl-discodermolide (**14**) indicated that the terminal conjugated double bond and the C19-carbamate groups are not essential for the microtubule bundling properties of (+)-discodermolide. Further saturation of **48** to **49** showed a

TABLE 10.4
Biological Data of Hydrogenated Analogs (48–52) and 19-Des-Aminocarbonyldiscodermolide (14)

Compound	P388, IC$_{50}$ nM	A549, IC$_{50}$ nM	G2/M Block, 100 nM	Microtubule Bundling, 10 nM
21,23-Tetrahydrodiscodermolide (**48**)	10	8	Yes	++
8,21,23-Hexahydrodiscodermolide (**49**)	33	68	Yes	++
8,13,21,23-Octahydrodiscodermolide (**50**)	8292	8292	No	—
7-Deoxy-8,21,23-hexahydro-discodermolide (**51**)	309	377	No	++
7-Deoxy-8,21,23-hexahydro-discodermolide-3,11,17- triacetate(**52**)	7052	NT	No	—
19-Des-aminocarbonyldiscodermolide (**14**)	128	74	Yes	++

TABLE 10.5

Biological Data of 6,7-Seco-Discodermolide Analogs (53, 54) and Cyclodiscodermolide Analogs (56–61)

Compound	P388, IC$_{50}$ nM	A549, IC$_{50}$ nM	G2/M Block, 100 nM	Microtubule Bundling, 10 nM
6,7-*seco*-Discodermolide (**53**)	3686	6774	No	—
19-*des*-Aminocarbonyl-6,7-*seco*-discodermolide (**54**)	12787	12787	No	—
(7*E*,9*S*,13*S*,14*E*)-7-deoxy-7,14-dien-9-(13)-cyclodiscodermolide (**56**)	2130	6640	No	NT
(7E,9*S*,13*S*)-7-deoxy-7,14(29)-dien-9-(13)-cyclodiscodermolide (**57**)	190	100	NT	NT
(7*E*,9*S*,13*S*,14*R*)-7-deoxy-14-hydroxy-7-en-9-(13)-cyclodiscodermolide (**58**)	2420	840	NT	NT
(7*E*,9*S*,*13S*,14*S*, 16S)-7-deoxy-14-hydroxy-7-en-9-(13)-cyclodiscodermolide (**59**)	2090	2640	NT	NT
(7*E*,9*S*,13*S*,14*S*,16*R*)-7-deoxy-14-hydroxy-7-en-9-(13)-cyclodiscodermolide (**60**)	7650	>5000	NT	NT
(7*R*,11*S*)-7-(11)-Oxycyclodiscodermolide (**61**)	1000	1460	NT	NT

fivefold reduction in cytotoxicity in the A549 assay, but comparable activity in the P388 cytotoxicity assay and in the formation of microtubule bundles. These data indicated that the C8 Z-double bond is not essential for the biological activity of **1**. Comparison of this finding with the biological activity reported earlier by the Smith group for the C8 E-analog (**34**) indicated that a C8 Z-double bond or C8 σ-bond type flexibility is required to maintain the high activity of (+)-discodermolide. The 7-deoxy-8, 21,23-hexahydrodiscodermolide (**51**) showed considerable reduction in cytotoxicity but, interestingly, retained the microtubule stabilizing properties of (+)-discodermolide. A remarkable loss of activity in the P388 and A549 assays on saturation of the C13 Z-double bond, as seen in the 8,13,21,23-octahydrodiscodermolide (**50**), indicated that the C13 Z-double bond, which partly controls the folding of the molecule, is essential for the cytotoxicity and microtubule stabilizing properties of (+)-discodermolide.

As shown in Table 10.5, the reduction of biological activity in 6,7-seco-analogs (**53, 54**), C9-C13 cyclo-analogs (**56–59**) except **57**, and C7-C11-oxycyclo analog (**61**) indicate that the combination of the C7 and C11 hydroxy groups are contributing to the optimum biological activity of (+)-discodermolide (**1**). The C16(R)-analog **60** that has the opposite chirality to that of (+)-discodermolide showed the least cytotoxicity among all assayed cyclo-analogs. This observation is significant because of the earlier finding by the Schreiber group[18] that C16(S) is essential for biological activity of (+)-discodermolide (**1**). The significant loss of biological activity in five of the six cyclo-analogs as shown in Table 10.5 that have the rigid ring structure from C6 through C13 could be attributed to the loss of the hairpin folding observed in (+)-discodermolide (**1**), and therefore, these observations confirmed the importance of the hairpin conformation for biological activity in (+)-discodermolide (**1**). These SAR studies confirmed that the C7, C11, and C17 hydroxy groups; the C16(S) stereochemistry; and the C13 Z-double bond that folds the molecule in to the hairpin conformation are essential for the strong microtubule stabilizing properties of (+)-discodermolide (**1**).

A number of models have been proposed regarding the binding of (+)-discodermolide (1) to β-tubulin. Inhibition of binding of [³H]-Taxol to microtubules in the presence of **1** indicates that the binding sites are similar or overlap. The core structures (pharmacophores) of Taxol (**2**), (+)-discodermolide (**1**), epothilones (**3–5**), and eleutherobin (**6**) are all proposed to bind in a pocket formed by Gly368, Thr274, His227, and Asp224 in β-tubulin, and their side chains fit into a pocket formed by His 227 and Asp 224.[45] Both the x-ray and solution structures of (+)-discodermolide (**1**) have demonstrated that the molecule arranges the C1-C19 region into a hairpin or U conformation, which brings the δ-lactone and C19 side chain into close proximity. Modeling studies indicate that the backbone of discodermolide mimics the northern region of Taxol (**2**), and in a preferred model for (+)-discodermolide (**1**) binding to the Taxol binding site, the δ-lactone and C19 side chain correspond to the C13 and C2 chains of Taxol, respectively.[32] None of the models fully accounts for the differences in activity, such as temperature dependence and stability of polymerization or increased nucleation rates, observed between (+)-discodermolide (**1**) and Taxol (**2**). Other factors, such as the lack of cross resistance for (+)-discodermolide in the A549-T12 Taxol-resistant cell line and the lack of cross resistance for epothilone-resistant cell lines with class I β-tubulin mutations, indicate that (+)-discodermolide may bind an overlapping site on β-tubulin. Factors such as these, as well as the increased affinity of (+)-discodermolide for β-tubulin, may indicate the presence of additional discrete contact sites with the β-tubulin structure. For example, none of the models proposed to date addresses the near-complete loss of activity when the C17 hydroxyl is functionalized, indicating that this hydroxyl may be an additional contact or binding point with tubulin.

VII. CLINICAL INVESTIGATION

Based on preclinical studies in which (+)-discodermolide (**1,** XAA296A) stabilized microtubule polymers more potently than **2** and demonstrated prominent activity against Taxol-refractory xenograft models, **1** was taken into phase I trial in September 2002 at the Cancer Therapy and Research Center in San Antonio, Texas. In the trial, **1** is administered by intravenous infusion once every 22 d to adult patients with advanced solid malignancies. The study is ongoing, but at the time of writing, 21 patients (median age 59.4 yr; 12 males, 9 females) have been treated with the following dose levels (mg/m^2): 0.6 (three patients), 1.2 (three patients), 2.4 (four patients), 4.8 (four patients), 9.6 (four patients), and 14.4 (three patients). Karnofsky performance status was as follows 70% (seven patients), 80% (eight patients), 90% (five patients), and 100% (one patient). The drug is well tolerated with no peripheral neuropathy or neutropenia. No objective responses have been observed; however, four of 21 patients have exhibited stable disease as their best response. Pharmacokinetics were evaluated in 18 patients receiving discodermolide over a dose range of 0.6–9.6 mg/m^2. Following a short infusion, (+)-discodermolide blood concentrations had decreased to <10% of C_{max} (29–460 ng/mL). Nonlinear kinetics, characterized by a second broad peak, were observed at the terminal phase of blood concentration–time profiles. In the majority of patients, the second rising curve started at 24 h postdose, with peak concentrations of 101%–208% of C_{24h} at 48–72 h. Drug concentrations then gradually declined to 56%–173% of C_{24h} at 168 h. The average (mean ± standard deviation) partial mean residence time (last measurable concentration) was 81.77 ± 43.02 h. Additional evaluation of patients at 14.4 mg/m^2 dose level is in progress. Accrual will continue to define the recommended phase II dose.[46]

REFERENCES

1. Gunasekera, S.P. et al., Discodermolide: a new bioactive polyhydroxylated lactone from the marine sponge *Discodermia dissoluta*, *J. Org. Chem.*, 55, 4912, 1990 (correction: 56.1346, 1991).
2. Longley, R.E. et al., Discodermolide: a new marine-derived immunosuppressive compound. I. *In vitro* studies, *Transplantation*, 52, 650, 1991.
3. Longley, R.E. et al., Discodermolide: a new marine-derived immunosuppressive compound. II. *In vivo* studies, *Transplantation*, 52, 656, 1991.

4. Hung, D. T., Nerenberg, J.B., and Schreiber, S.L., Distinct binding and cellular properties of synthetic (+)- and (-)-discodermolides, *Chem. Biol.*, 1, 67, 1994.

5. Parness, J. and Horwitz, S.B., Taxol binding to polymerized tubulin *in vitro, J. Cell Biol.*, 91, 479, 1981.

6. Bollag, P.A. et al., Epothilones, a new class of microtubule-stabilizing agents with a Taxol-like mechanism of action, *Cancer Res.*, 55, 2325, 1995.

7. Hardt, I.H. et al., New natural epothilones from *Sorangium cellulosum*, strains Soce90/B2 and So ce90/D13: isolation, structure elucidation and SAR studies, *J. Nat. Prod.*, 64, 847, 2001.

8. Lindel, T. et al., Eleutherobin, a new cytotoxin that mimics Paclitaxel (Taxol) by stabilizing micro-tubules, *J. Am. Chem. Soc.,* 119, 8744, 1997.

9. Mooberry, S.L. et al., Laulimalide and isolaulimalide, a new paclitaxel-like microtubule-stabilizing agent, *Cancer Res.*, 59, 653, 1999.

10. Isbrucker, R.A. et al., Tubulin polymerizing activity of dictyostatin-1, a polyketide of marine sponge origin, *Biochem. Pharmacol.*, 66, 75, 2003.

11. Smith, III, A.B. et al., Gram-scale synthesis of (+)-discodermolide, *Org. Lett.*, 1, 1823, 1999.

12. Kelly-Borges, M. et al., Species differentiation in the marine sponge genus *Discodermia* (Demospon-giae: Lithistida): the utility of ethanol extract profiles as species-specific chemotaxonomic markers, *Biochem. Syst. Ecol.*, 22, 353, 1994.

13. Gunasekera, S.P., Gunasekera, M., and McCarthy, P., Discodermide: a new bioactive macrocyclic lactam from the marine sponge *Discodermia dissoluta, J. Org. Chem.*, 56, 4830, 1991.

14. Gulavita, N.K. et al., Polydiscamide A: a new bioactive depsipeptide from the marine sponge *Disco-dermia* sp., *J. Org. Chem.*, 57, 1767, 1992.

15. Gunasekera, S.P. et al., Five new discodermolide analogs from the marine sponge *Discodermia* sp., *J. Nat. Prod.*, 65, 1643, 2002.

16. Smith, III, A.B., LaMarche, M.J., and Falcone-Hindley, M., Solution structure of (+)-discodermolide, *Org. Lett.*, 3, 695, 2001.

17. Nerenberg, J.B. et al., Total synthesis of the immunosuppressive agent (-)-discodermolide, *J. Am. Chem. Soc.*, 115, 12621, 1993.

18. Hung, D. T., Nerenberg, J.B., and Schreiber, S.L., Synthesis of discodermolides useful for investigating microtubule binding and stabilization, *J. Am. Chem. Soc.*, 118, 11054, 1996.

19. Smith, III, A.B. et al., Total synthesis of (-)-discodermolide, *J. Am. Chem. Soc.*, 117, 12011, 1995.

20. Harried, S.S. et al., Total synthesis of (-)-discodermolide: an application of a chelation-controlled alkylation reaction, *J. Org. Chem.*, 62, 6098, 1997.

21. Harried, S.S. et al., Total synthesis of (+)-discodermolide, a microtubule-stabilizing agent from the sponge *Discodermia dissoluta, J. Org. Chem.*, 68, 6646, 2003.

22. Marshall, J.A. and Jones, B.A., Total synthesis of (+)-discodermolide, *J. Org. Chem.*, 63, 7885, 1998.

23. Paterson, I. et al., A practical synthesis of (+)-discodermolide and analogs: fragment union by complex aldol reactions, *J. Am. Chem. Soc.*, 123, 9535, 2001.

24. Kalesse, M., The chemistry and biology of discodermolide, *Chem. BioChem.*, 1, 171, 2000.

25. Smith, III, A.B. et al., Evolution of a gram-scale synthesis of (+)-discodermolide, *J. Am. Chem. Soc.*, 122, 8654,.

26. Mickel, S.J. et al., Large-scale synthesis of the anti-cancer marine natural product (+)-discodermolide. Part 1: synthetic strategy and preparation of a common precursor, *Org. Process Res. Dev.,* 8, 92, 2004.

27. Mickel, S. J. et al., Large-scale synthesis of the anti-cancer marine natural product (+)-discodermolide. Part 2: synthesis of fragments C_{1-6} and C_{9-14}, *Org. Process Res. Dev.,* 8, 101, 2004.

28. Mickel, S.J. et al., Large-scale synthesis of the anti-cancer marine natural product (+)-discodermolide. Part 3: synthesis of fragment C_{15-21}, *Org. Process Res. Dev.,* 8, 107, 2004.

29. Mickel, S.J. et al., Large-scale synthesis of the anti-cancer marine natural product (+)-discodermolide. Part 4: preparation of fragment C_{7-24}, *Org. Process Res. Dev.,* 8, 113, 2004.

30. Mickel, S.J. et al., Large-scale synthesis of the anti-cancer marine natural product (+)-discodermolide. Part 5: linkage of fragments C_{1-6} and C_{7-24} and finale, *Org. Process Res. Dev.,* 8, 122, 2004.

31. Paterson, I. et al., Total synthesis of the antimicrotubule agent (+)-discodermolide using boron-mediated aldol reactions of chiral ketones, *Angew. Chem. Int. Ed.*, 39, 377, 2000.

32. Martello, L.A. et al., The relationship between Taxol and (+)-discodermolide: synthetic analogs and modeling studies, *Chem. Biol.*, 8, 843, 2001.

33. Gunasekera, S P., Longley, R.E., and Isbrucker, R.A., Acetylated analogs of the microtubule-stabilizing agent discodermolide: preparation and biological activity, *J. Nat. Prod.*, 64, 171, 2001.

34. Gunasekera, S.P., Longley, R.E., and Isbrucker, R.A., Semisynthetic analogs of the microtubule-stabilizing agent discodermolide: preparation and biological activity, *J. Nat. Prod.*, 65, 1830, 2002.

35. Gunasekera, S.P. et al., Synthetic analogs of the microtubule-stabilizing agent (+)-discodermolide: preparation and biological activity, *J. Nat. Prod.*, 67, 749, 2004.

36. Longley, R.E. et al., Immunosuppression by discodermolide, *Ann. N. Y. Acad. Sci.*, 696, 94, 1993.

37. Simonsen, M., Graft versus host reactions: their natural history and applicability as tools of research, *Prog. Allergy*, 6, 349, 1962.

38. Borel, J.F. et al., Biological effects of cyclosporin A: a new antilymphocytic agent, *Agent Action,* 6, 468, 1977.

39. Longley, R.E., Gunasekera, S.P., and Pomponi, S.A., Discodermolide compounds. U. S. Patent 5,840,750 issued November 24, 1998.

40. Boyd, M.R., The NCI in *vitro* anticancer drug discovery screen, *Cancer Lett.*, 15, 6, 1989.

41. ter Harr, E. et al., Discodermolide, a cytotoxic marine agent that stabilizes microtubules more potently than taxol, *Biochemistry*, 35, 243, 1996.

42. Hung, D.T., Chen, J., and Schreiber, S.L., (+)-Discodermolide binds to microtubules in stoichiometric ratio to tubulin dimers, blocks taxol binding and results in mitotic arrest, *Chem. Biol.*, 3, 287, 1996.

43. Hurst, J. et al., A novel model of a metastatic breast tumour xenograft line, *Br. J. Cancer*, 68, 274, 1993.

44. Giannakakou, P. et al., A common pharmacophore for epothilone and taxanes: molecular basis for drug resistance conferred by tubulin mutations in human cancer cells, *Proc. Natl. Acad. Sci. USA*, 97, 2904, 2000.

45. He, L. et al., A common pharmacophore for Taxol® and the epothilones based on the biological activity of a taxane molecule lacking a C-13 side chain, *Biochemistry,* 39, 3972, 2000.

46. Mita, C. et al., Phase 1 and pharmacokinetic study of XAA296A (discodermolide) administered once every three weeks in adult patients with advanced solid malignancies, Abstract A252 2003, AACR-NCI-EORTC Meeting, Molecular Target and Cancer therapeutics, Boston, MA, November 17–21, 2003.

11 The Dolastatins: Novel Antitumor Agents from *Dolabella auricularia*

Erik Flahive and Jayaram Srirangam

CONTENTS

I. Introduction and History 191
II. Origin, Isolation, and Structure of the Dolastatins...191
III. Structure and Synthesis..195
IV. Bioactivity and Mechanisms of Action 198
V. Structural Modifications and SAR..202
VI. Clinical Update 206
References...208

I. INTRODUCTION AND HISTORY

Within the field of natural products, marine organisms represent a relatively untapped resource for novel compounds of medicinal interest. The systematic evaluation of biological constituents from marine organisms as potential drug candidates did not begin in earnest until the late 1960s, when the U.S. National Cancer Institute (NCI) initiated a massive screening program directed at the discovery of new anticancer drug leads.[1,2] Over the next 30 yr, evaluation of isolates from the marine shell-less mollusk *Dolabella auricularia* (Aplyidae family) has led to the discovery of an impressive number of novel compounds possessing antineoplastic activity. Among the number of substances characterized to date, dolastatins 10 (**1**) and 15 (**2**) and the synthetic compounds derived from them show the most promising antitumor activity and are currently in human clinical trials in the United States, Europe, and Japan. This article will provide highlights of this work and will review recent progress in the discovery and development of the dolastatins as potential cancer chemotherapeutic drugs, with particular emphasis on dolastatins 10 and 15.

II. ORIGIN, ISOLATION, AND STRUCTURE OF THE DOLASTATINS

D. auricularia (Figure 11.1) and other gastropods of the order Anaspidea (Aplysiomorpha), known as "sea hares" because of the resemblance of the auriculate tentacula of these mollusks to the ears of a hare,[3] typically lack the protection of an external shell. A remarkable feature of *D. auricularia* and other sea hares is their demonstrated ability to sequester and store secondary metabolites from dietary sources in their digestive glands for long periods.[4–6] These accumulated substances, which typically occur in vanishingly small quantities, have long been thought to serve as a chemical defense mechanism against predators, though there is some debate as to their true role.[6–8] Although

FIGURE 11.1 (See color insert following page 304.) *Dollabella auricularia.*

the dynamics of sequestration and differentiation processes for this apparent chemical defense mechanism are still unclear, the adaptive strategies of *D. auricularia* and other shell-less mollusks are nonetheless truly remarkable.

Some understanding of both the pharmacological utility as well as the potential dangers of extracts of *D. auricularia* has been known since ancient times.[9–13] However, the potential of *D. auricularia* with respect to modern medical problems was not recognized until the initial report from Pettit and coworkers at Arizona State University concerning the potent antineoplastic activity of crude extracts of the Indian Ocean variety of this inconspicuous marine animal.[14] Bioassay-guided (murine P388 leukemia cells [PS system]) isolation of the active principles of the crude extracts of *D. auricularia* collected off the coast of the Island of Mauritius in the Western Indian Ocean by Pettit's group eventually led to a considerable number of pure compounds exhibiting an impressive spectrum of antineoplastic and cytostatic activity.

Dolastatins 3–15 (Figure 11.2) were among the first group of compounds discovered and were first described in the chemical literature from 1982 to 1989.[15] Structurally, the dolastatins are a series of novel linear and cyclic peptides that exhibit remarkable potency against neoplastic cells *in vitro* (see Table 11.1). These unique peptide constituents possess many interesting structural features including hitherto unknown, modified amino acid residues and were isolated in very low yield (from 10^6% to 10^8% yield based on crude wet weight), requiring several recollections of the marine animal. The isolation and structure elucidation[16,17] of dolastatins 3–15, which required total synthesis in most cases to verify stereochemistry, has been reviewed.[15,18] Dolastatin 10 (**1**), with an ED_{50} in the subnanomolar range against a number of cancer cell lines, was the most potent antineoplastic substance known at the time of its discovery.[19]

The last of the bioactive peptides isolated by Pettit to date were dolastatins 16–18.[20–22] These compounds, isolated from *D. auricularia* specimens collected in Papua New Guinea, possess unique structural elements typical of peptides in this series including the novel alkynyl -β-amino acid of dolastatin 17 (**9**) and β-branched amino acid dolaphenvaline of dolastatin 16 (**8**). Among these, dolastatin 16 was the most potent inhibitor of cell growth against a minipanel of the NCI's human cancer cell lines (mean GI_{50} 2.5×10^4 μg/mL, ~18-fold less potent than dolastatin 10). Isolation of a closely related but less bioactive compound, homodolastatin 16, has recently been

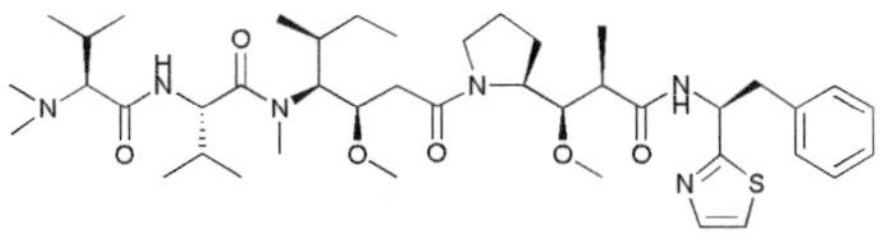

Dolastatin 10 (**1**)

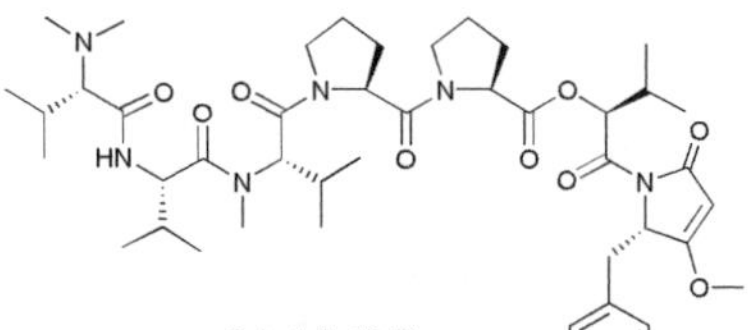

Dolastatin 15 (**2**)

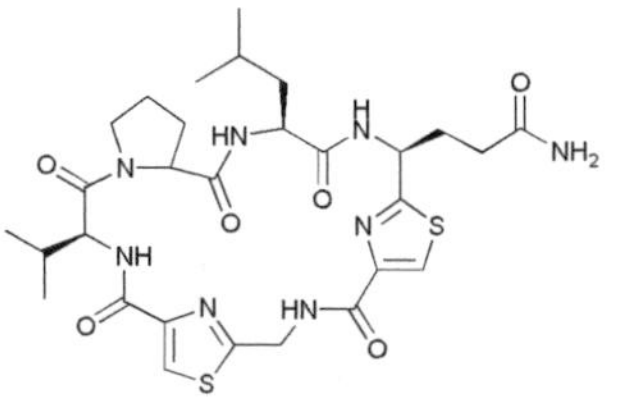

Dolastatin 3 (**3**)

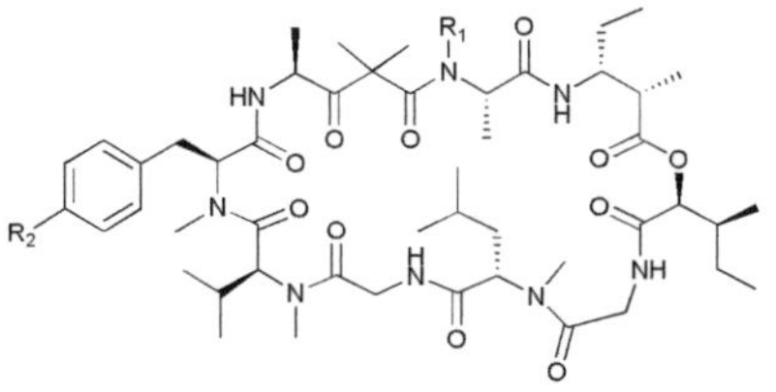

Dolastatin 11 (**4**, R_1 = H, R_2 = OCH_3)
Dolastatin 12 (**5**, R_1 = CH_3, R_2 = H)

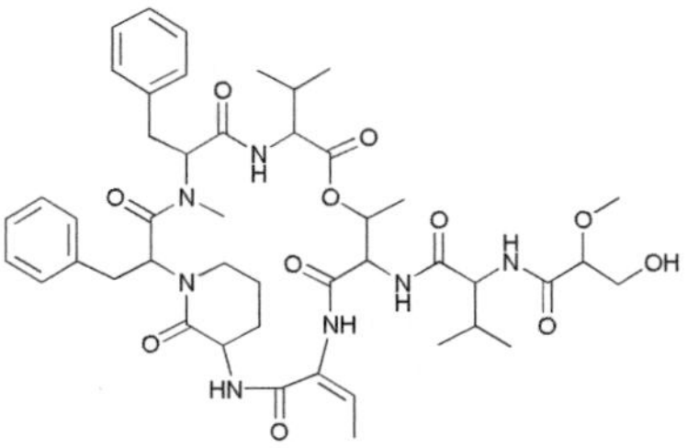

Dolastatin 13 (**6**)

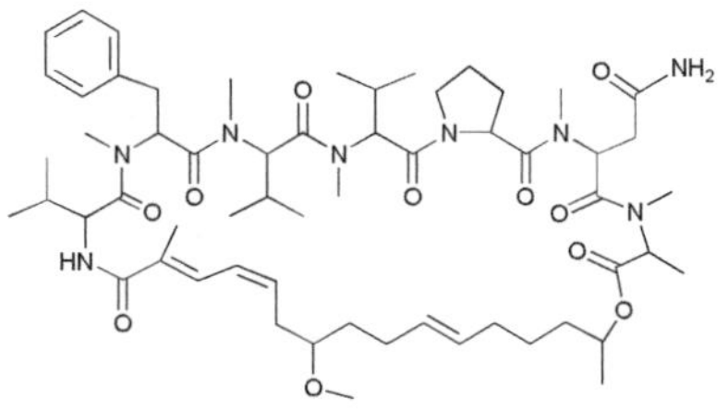

Dolastatin 14 (**7**)

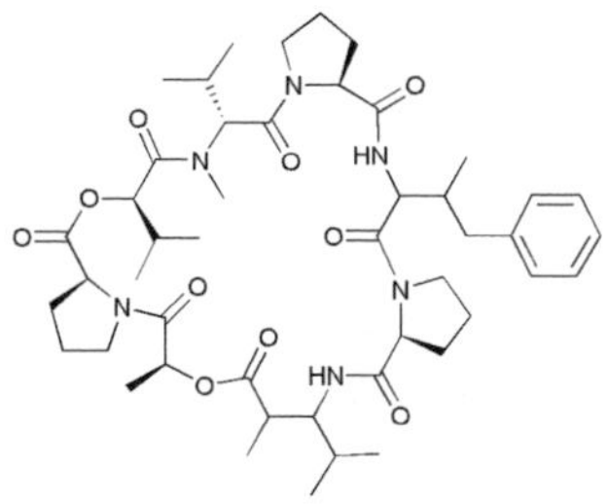

Dolastatin 16 (**8**)

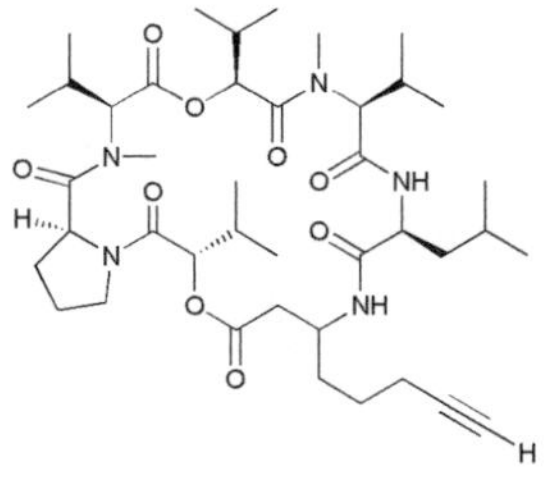

Dolastatin 17 (**9**)

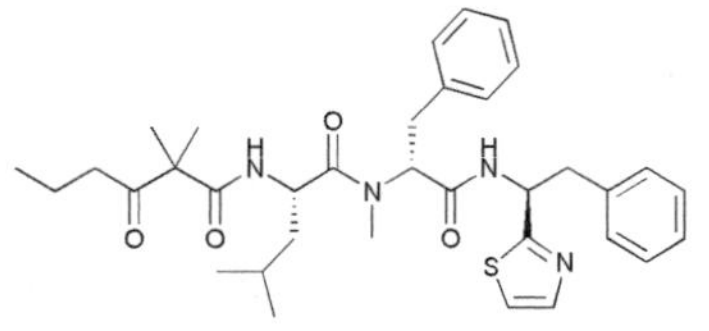

Dolastatin 18 (**10**)

FIGURE 11.2 Dolastatins 1–18.

reported.[23] Yamada and coworkers of Nagoya University initiated a large-scale collection of *D. auricularia* from Japanese waters (Shima peninsula) in the mid-1980s that led to the discovery of yet another series of unique compounds with antitumor activity (Figure 11.3).[24] The majority of these sea hare constituents (Dolastatins C-E,[25–28] dolastatin G/nordolastatin G,[29,30] dolastatin H/isodolastatin H,[31] dolastatin I,[32] aurilide,[33,34] dolabellin,[35] and doliculide[36]) are linear or cyclic pseudopeptides and exhibit potent *in vitro* antitumor activity in some cases. Doliculide (**21**), containing a novel *ortho*-iodo-N-methyl-D-tyrosine residue, and dolastatin H (**16**)/isodolastatin H (**17**), which bear close structural resemblance to dolastatin 10, are the most potent inhibitors of antineoplastic growth (see Table 11.1). In addition to the pseudopeptide constituents described by Yamada, a number of sea hare–derived macrolides with antineoplastic activity were reported by these same researchers (dolabellides A and B[24] and doliculols A and B[37]). Despite the structural similarity between the dolastatins discovered by Pettit's and Yamada's groups, it is interesting to note that neither group has yet to isolate an identical compound from *D. auricularia*.

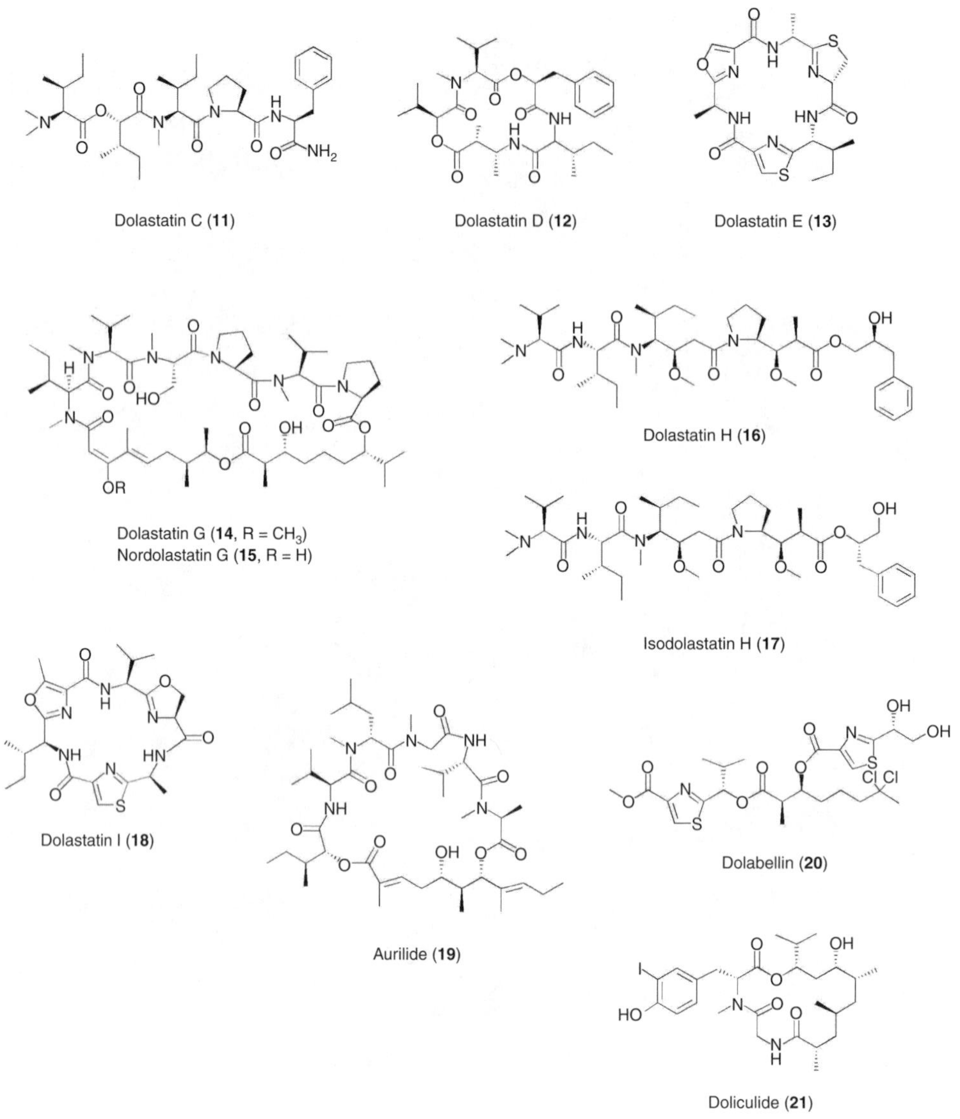

Dolastatin C (**11**)   Dolastatin D (**12**)   Dolastatin E (**13**)

Dolastatin G (**14**, R = CH₃)
Nordolastatin G (**15**, R = H)

Dolastatin H (**16**)

Isodolastatin H (**17**)

Dolastatin I (**18**)

Aurilide (**19**)

Dolabellin (**20**)

Doliculide (**21**)

FIGURE 11.3 Dolastatins isolated from Shima Peninsula collection of *Dollabella auricularia*.[42]

The origin of this impressive array of bioactive secondary metabolites found within sea hares has been the subject of much interest. Although direct biosynthesis or biosynthesis by symbiotic microorganisms has been considered, these theories have slowly given way to strong evidence for exogenous procurement of such metabolites from dietary sources.[38] It is now generally accepted that cyanobacteria consumed by sea hares are the true biosynthetic source of the rich structural diversity represented by the dolastatins. This idea is supported by a number of observations. Collections of the same species from different locations have led to the isolation of similar, but structurally distinct, dolastatins. Furthermore, the incorporation of structurally unique amino acids that is typical of the dolastatins seems to favor a diet-derived origin based on the prevalence of such novel, nonproteinogenic amino acids among marine algae.[39] Moreover, a number of these compounds (dolastatins 3,[40] 10,[41] 12–14[42,43] and 16[44]) have now been isolated directly from varieties of the blue-green algae *Lyngbya majuscule* and *Symploca hydnoides*, along with new bioactive

TABLE 11.1
Summary of Antitumor Agents from the Marine
Mollusk *Dollabella auricularia*[79]

Compound	Collection site	*In vitro* activity (μg/mL)
Dolastatin 3 (**3**)	Mauritius	1.6×10^{-1a}
Dolastatin 10 (**1**)	Mauritius	3.4×10^{-6b}
Dolastatin 11 (**4**)	Mauritius	1.9×10^{-4b}
Dolastatin 12 (**5**)	Mauritius	8.8×10^{-2b}
Dolastatin 13 (**6**)	Mauritius	3.5^{b}
Dolastatin 14 (**7**)	Mauritius	2.7×10^{-3b}
Dolastatin 15 (**2**)	Mauritius	2.8×10^{-3b}
Dolastatin 16 (**8**)	Mauritius	9.6×10^{-4b}
Dolastatin 17 (**9**)	Mauritius	4.5×10^{-1b}
Dolastatin 18 (**10**)	Papua New Guinea	3.9×10^{-1b}
Dolastatin C (**11**)	Papua New Guinea	17^{c}
Dolastatin D (**12**)	Papua New Guinea	2.2^{c}
Dolastatin E (**13**)	Shima Peninsula (Japan)	$22\text{-}40^{c}$
Dolastatin G (**14**)	Shima Peninsula (Japan)	1^{c}
Nordolastatin G (**15**)	Shima Peninsula (Japan)	5.3^{c}
Dolastatin H (**16**)	Shima Peninsula (Japan)	2.2×10^{-2c}
Isodolastatin H (**17**)	Shima Peninsula (Japan)	1.6×10^{-2c}
Dolastatin I (**18**)	Shima Peninsula (Japan)	12^{c}
Doliculide (**21**)	Shima Peninsula (Japan)	1.1×10^{-2c}
Dolabellin (**20**)	Shima Peninsula (Japan)	6.1^{c}
Doliculide (**21**)	Shima Peninsula (Japan)	1×10^{-3c}

[a] ED_{50}, P388 (lymphocytic leukemia cells).
[b] GI_{50}, NCI-H460 (non–small cell lung carcinoma).
[c] IC_{50}, HeLa-S_3 (cervical carcinoma cells).

compounds with closely related structures such as the recently described symplostatins 1–3[45–47] and somamides A and B.[43] This link between the dolastatins and cyanobacteria has important economic and ecological relevance for the future discovery and development of new anticancer drug leads.[48]

III. STRUCTURE AND SYNTHESIS

Though structurally complex in many cases, total synthesis of most of the dolastatins has been achieved to date. The importance of these synthetic endeavors should be underscored, as these efforts have allowed further study of this interesting family of compounds in both *in vivo* bioassays and human clinical trials. In this section, an overview of progress made in the synthesis of the various dolastatins and related compounds is presented, with particular attention to the methods devised for preparation of dolastatins 10 and 15 and important derivatives thereof.

Structure determination for dolastatin 3 (**3**), the first in the series to be described, required detailed spectral (nuclear magnetic resonance [NMR] and tandem mass spectrometry [MS] sequencing) and hydrolytic studies, as well as total synthesis.[15] The structurally interesting thiazole-containing amino acids of dolastatin 3 are related to similar cysteine-derived[49] thiazole amino acids found in other biologically active natural products, such as the antineoplastic marine sponge constituents patellamides A–C[50], the antineoplastic lower plant metabolite bleomycin,[51] and the broad-spectrum antibiotic thiostrepton,[52] among others. Several total syntheses of dolastatin 3 have been reported,[49,53–56] highlighted by the synthesis of the thiazole amino acids (Gly)Thz and (Gln)Thz.

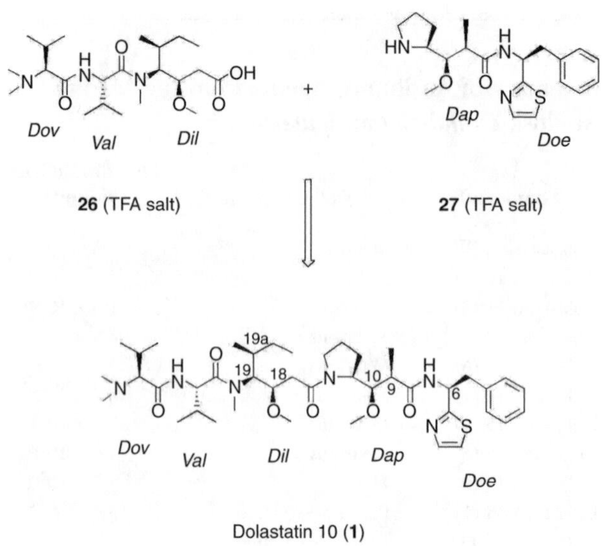

Dolastatin 10 (1)

FIGURE 11.4 Dolastatin 10 structural units.

The synthesis of the most bioactive of the dolastatins, dolastatin 10 (**1**), was an extremely challenging undertaking because of the total of nine asymmetric centers of unknown configuration that make up the structure. The most notable structural features of dolastatin 10 are the two novel γ-amino-β-methoxy-acid residues, dolaproine (*Dap*, **22**) and dolaisoleucine (*Dil*, **23**), each containing three chiral centers (Figure 11.4). Similar structural units are also found in the marine tunicate antineoplastic components didemnins A–C[57] and the lower plant metabolite bleomycin.[51] Also present is a novel thiazole-containing C-terminal unit dolaphenine (*Doe*, **24**), derived from phenylalanine, that is also found in dolastatin 18 (**10**). Several stereoselective syntheses[58] of the unique *Dap, Dil,* and *Doe* subunits have appeared in the chemical literature.

The absolute configuration of dolastatin 10 was confirmed by total synthesis,[59] first reported in 1989 by Pettit's group. Because of the exceptional promise of dolastatin 10 as an anticancer drug, several partial and total syntheses[58] of this peptide have been subsequently reported. Assembly of the amino acid constituents in both stepwise (by Shiori,[60,61] Koga,[62] and Poncet[63]) and convergent fashion has been successfully demonstrated. The general approach followed by Pettit's group for the preparation of initial quantities of dolastatin 10 for preclinical studies was convergent, as shown in Figure 11.4. The absence of a chiral center adjacent to the carboxylic acid of tripeptide *Dov-Val-Dil* (**26**) and the ease of its coupling with a proline derivative made this approach strategically attractive and free of epimerization concerns. In addition, this strategy was particularly amenable to the synthesis of shortened peptide segments and synthetic analogs that were used for elucidation of structure–activity relationships (SARs), as well as to provide insights into the mechanism of action for this drug, as discussed in more detail in section V of this chapter. Yamada and coworkers also found this general [3 + 2] coupling approach useful for the preparation of the structurally similar compounds dolastatin H (**16**) and isodolastatin H (**17**).[31]

Although the crude structures of dolastatins 11 (**4**) and 12 (**5**) were reported in 1989,[64] the absolute stereochemistry of these cyclopeptides remained undetermined until a total synthesis was reported by Bates and coworkers in 1997.[65,66] Dolastatins 11 and 12 are closely related cyclodepsipeptides bearing a novel β-amino acid residue and are structurally very similar to the antifungal natural products majusculamide C[67] and 57-normajusculamide C[68] — components of the marine blue-green alga *Lyngbya majuscula*. This synthetic achievement has allowed for the unique biochemical effects of dolastatin 11 on tubulin and actin to be studied in more detail (see section IV). Dolastatin 13 (**6**)[69] is a structurally complex cyclodepsipeptide containing a novel piperidine unit

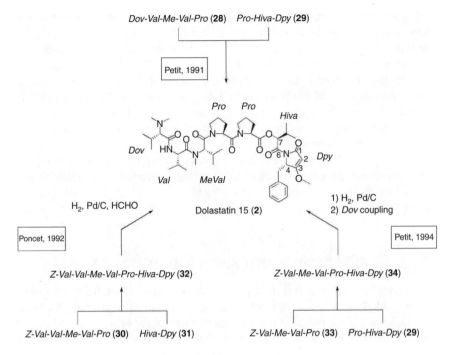

FIGURE 11.5 Synthetic approaches to dolastatin 15.

and a dehydroamino acid residue. Although no total synthesis has appeared to date, it is presumed that the recent discovery of the structurally analogous cyanobacterial metabolites symplostatin 2[46] and somamides A and B[43] will spark interest in construction of this peptide core. Total synthesis of the moderately active dolastatin 14 (**7**), a cyclodepsipeptide containing a polyene hydrocarbon fragment, has not been achieved to date. Several groups have published complementary approaches to the four possible isomers of the polyene subunit (dolatrienoic acid).[70–72] The absolute configurations of both dolastatins 13 and 14 are presently unknown, and their biological effects have not been thoroughly evaluated because of a lack of sufficient material.

The second most potent *D. auricularia* metabolite discovered to date is the linear depsipeptide dolastatin 15 (**2**). This unique peptide component bears many structural similarities to dolastatin 10 (**1**), including the presence of an N,N-dimethylvaline (*Dov*) N-terminus (Figure 11.5). The most interesting structural feature of dolastatin 15 is the novel C-terminal dolapyrrolidone unit (*Dpy*), which is presumably derived from biosynthetic modification of N-acyl phenylalanine. A total synthesis[73] of **2** was achieved by Pettit's group in 1991, and the absolute configuration was assigned accordingly. A novel synthesis from Poncet's group appeared in 1992[74] and featured reductive bis-methylation of the N-terminus as a final step. A second submission from Pettit in 1994 outlined some subtle improvements to the original approach,[75] and recently, a solid-phase synthesis of dolastatin 15 was also reported.[76] All of the reported solution–phase syntheses of dolastatin 15 are convergent and take advantage of the reduced tendency of peptide segments coupled at proline to epimerize (see Figure 11.5). More recent approaches incorporate the *Dov* unit in the final step to circumvent purification difficulties associated with early introduction of the N,N-dimethylated N terminus. As mentioned above for dolastatin 10, these complementary, convergent approaches are well suited for rapid preparation of synthetic derivatives useful for uncovering important SARs. The ample supply of synthetic dolastatin 15 has allowed for the biological activity of this unique peptide to be studied in great detail.[77]

Dolastatin 16 (**8**) is a unique cyclodepsipeptide containing novel β-amino acid (dolamethylleucine) and β-branched valine (dolaphenvaline) units reported by Pettit and coworkers in 1997. The structure and amino acid sequence was determined using high-field NMR and tandem mass spectral interpretations, and the stereochemistry of most of the subunits was determined via chiral high-performance liquid chromatography analysis of the acidic hydrolysate of the natural peptide. This compound exhibits marked antiproliferative effects on cancer cells *in vitro*, but the lack of available synthetic material has prevented further study of its biochemical effects.[20] The structure and partial stereochemistry of dolastatin 17 (**9**) were elucidated in a manner similar to that of dolastatin 16 above. The absolute stereochemistry of both dolastatin 16 and 17 has not been reported to date. The absolute stereochemistry of the C-terminal dolaphenine unit of dolastatin 18 (**10**) was recently determined via total synthesis and confirmed by x-ray crystal structure.[78]

Among the group of dolastatins (and related compounds) described by Yamada (Figure 11.3),[24] the absolute configuration for each is known, and a synthesis has been described for nearly every compound. This represents a tremendous achievement given the structural complexity of these substances. The various published syntheses of pseudopeptides **11–21** have been reviewed in detail by Poncet.[79]

IV. BIOACTIVITY AND MECHANISMS OF ACTION

The dolastatins are all strong inhibitors of cancer cell growth, and indeed, the discovery of all these substances was guided by bioassay of extracts of the crude animal matter against actual cancer cells *in vitro*. Dolastatin 10 is the most potent antineoplastic agent of the dolastatins, followed by dolastatin 15 (on average, dolastatin 15 appears to be about sevenfold less active than dolastatin 10 in most preclinical models[80]). Dolastatin 10 has been studied most intensively because of its exceptionally potent antiproliferative effect on cancer cells and initial promise as an anticancer drug. Both dolastatins 10 and 15 are currently being pursued as clinical antitumor drug candidates and have reached phase II clinical trials in the United States, Europe, and Japan (see section VI). Dolastatin 10 also exhibits therapeutic potential against other disease indications, as evidenced by its reported antiviral,[81] antibacterial,[82,83] antifungal,[84] and antimalarial[83,84] activities. A comparison of the growth inhibitory effects of dolastatins 10–16 against a selection of human cancer cell lines is presented in Table 11.2.[15]

TABLE 11.2
Comparative ACTIVITIES (GI$_{50}$ µg/mL) of Dolastatins 10–16 Against a Human Tumor Cell Line Minipanel from the National Cancer Institute Primary Screen

Dolastatin	OVCAR-3[a]	SF-295[b]	A498[c]	NCI-H460[d]	KM20L2[e]	SK-MEL-5[f]
10 (**1**)	9.5×10^{-7}	7.61×10^{-6}	2.6×10^{-5}	3.4×10^{-6}	4.7×10^{-6}	7.4×10^{-6}
11 (**4**)	0.04	0.031	0.023	1.9×10^{-4}	0.037	0.034
12 (**5**)	0.14	0.37	0.78	0.088	0.29	0.32
13 (**6**)	2.5	3.9	4.5	3.5	2.4	2.7
14 (**7**)	4.1×10^{-3}	2.0×10^{-3}	0.038	2.7×10^{-3}	0.020	6.4×10^{-5}
15 (**2**)	1.3×10^{-4}	4.3×10^{-4}	2.7×10^{-4}	2.8×10^{-4}	1.1×10^{-4}	1.7×10^{-4}
16 (**8**)	—	5.2×10^{-3}	—	9.6×10^{-4}	1.2×10^{-3}	3.3×10^{-3}

[a] Ovarian adenocarcinoma.
[b] Brain (glioma).
[c] Renal carcinoma.
[d] Non–small cell lung carcinoma.
[e] Colon carcinoma.
[f] Melanoma.

TABLE 11.3
Dolastatin 10 and 15 Activity versus Human Leukemia Cell Lines

Cell Line	IC$_{50}$ (nM)	
	Dolastatin 10 (1)	Dolastatin 15 (2)
BV-173[a]	4.7×10^{-2}	5.9×10^{-1}
KM-3[a]	8.7×10^{-2}	8.9×10^{-1}
K-562[b]	1.2×10^{-1}	1
U-937[c]	1.8×10^{-2}	1.2×10^{-2}
HL-60[d]	5.5×10^{-2}	2.5×10^{-2}
KARPAS-299[e]	9.8×10^{-2}	2.8×10^{-2}
B-Cells	1.0	>2.0
T-Cells	3.7×10^{-1}	>2.0

[a] Pre-B-cell leukemia.
[b] Erythroleukemia.
[c] Monocytic leukemia.
[d] Myeloid leukemia.
[e] T-cell lymphoma.

Both dolastatin 10 and 15 are potent antimitotic agents and, as such, interfere with cell division via disruption of microtubule function.[85] Accumulation of cells trapped in metaphase arrest is observed in cancer cells treated with either of these agents, and at higher concentrations, intracellular microtubules disappear completely.[86] The cytostatic effects of dolastatins 10 and 15 on cells *in vitro* are derived from the binding interaction between these peptides and the β-subunit of the dimeric microtubule protein tubulin.[87] The role that this biochemical interaction plays in the cytotoxic and antitumor effects of these compounds is discussed below. The impressive performance of dolastatins 10 and 15 against cancer cells *in vitro* was followed directly by evaluation of these compounds in several murine tumor models. Both compounds were found to be effective against several human tumor xenografts such as the OVCAR-3 human ovary and LOX human melanoma *in vivo* systems.[15] A number of studies concerning the activity of both compounds against human leukemia[88–92] and lymphoma[93–95] cell lines as well as its efficacy for the treatment of prostate cancer[96] have been reported. An interesting result to come out of these studies is that both agents are potently cytostatic (log 3 to log 4 more effective than the clinically established drug vincristine) but not cytotoxic at drug concentrations required to achieve maximum growth inhibition. Importantly, the two dolastatin compounds showed low toxicities to resting or proliferating nonleukemic cells. Furthermore, the observed growth inhibition was reversed once treatment was terminated.[90] A summary of results for dolastatins 10 and 15 against human leukemia and lymphoma cells appears in Table 11.3.

The antiproliferative effects of dolastatin 10 on purified tubulin and cellular microtubule assembly generally parallel those of the well-known vinca alkaloids, but its inhibition of cancer cell growth *in vitro* is much more potent. Both dolastatin 10 and the structurally unrelated mycotoxin phomopsin A are strong noncompetitive inhibitors of the binding of radiolabeled vinca alkaloids. In addition, dolastatin 10 is a potent inhibitor of nucleotide exchange and tubulin-dependent guanine 5-triphosphate (GTP) hydrolysis (Table 11.4). The hydrolysis of GTP bound to the exchangeable nucleotide site on the β-subunit of tubulin is believed to play an important regulatory role in the polymerization of tubulin. Results of extensive competition experiments with dolastatin 10 and other vinca domain drugs[87] have led to a proposed model of the binding sites of this class of antimitotics. Hamel and coworkers have proposed that dolastatin 10 binds to a "peptide site" near both the vinca site and the exchangeable nucleotide site.[97] According to the model, when dolastatin 10 is bound to its critical active site, the C-terminal end of the peptide effectively blocks access to

TABLE 11.4
Relative Inhibitory Effects of Vinca Domain Drugs on [³H]Vincristine Binding, on the Polymerization of Purified Tubulin,[a] on Guanine 5-Triphosphate (GTP) Nucleotide Exchange, and on Tubulin-Dependent GTP Hydrolysis

Drug	Inhibition [³H]Vincristine Binding (type)	IC_{50}[b] Tubulin Polymerization (μM)	IC_{50} GTP Exchange (μM)	IC_{50} GTP Hydrolysis (μM)
Vinblastine	Competitive	1.5	>80	6
Maytansine	Competitive	3.4	9	4
Rhizoxin	Competitive	6.9	29	13
Phomopsin A	Noncompetitive	1.4	5	3
Dolastatin 10	Noncompetitive	1.2	8	3
Dolastatin 15	NI[c]	23	NI[c]	33
Halichondrin B	Noncompetitive[d]	6.1	17	12
Spongistatin 1	Noncompetitive[d]	5.3	87	—

[a] Standard tubulin assay: preincubation (15 min) of a drug–tubulin mixture (10 μM tubulin) in 1 M monosodium glutamate + 1 mM MgCl$_2$ at 37°C was followed by the addition of GTP, and the polymerization reaction at 37°C was monitored, using a spectrophotometer. The tubulin assay used in the current structure–activity study was a modified method as discussed in section 10.3 of reference 87.

[b] IC_{50} refers to the drug dosage required for 50% inhibition.

[c] Noninhibitory.

[d] Competition experiments with halichondrin B and spongistatin 1 were carried out using [³H]vinblastine rather than [³H]vincristine.

the vinca alkaloid binding site and the exchangeable nucleotide site. This idea is supported by several structure–activity observations[98] and, in particular, by the fact that a shortened tripeptide segment not containing the latter two C-terminal units (26; Figure 11.4) retains tubulin antiproliferative activity but exhibits reduced inhibition of vinca alkaloid binding and nucleotide exchange.[86]

The effects of dolastatin 15 on the proliferation of cellular microtubules and purified tubulin in vitro have also been studied.[97] Similar to dolastatin 10, the structurally similar dolastatin 15 appears to derive its potent antineoplastic effects by interfering with the assembly of microtubules and causing mitotic arrest, though in the case of the latter peptide, the connection is less clear. Dolastatin 15 is roughly 1/7 as active as dolastatin 10 but is nearly seven times more active than vinblastine with respect to cell growth inhibition of L1210 murine leukemia cells.[99] Despite its potent antitumor activity, the activity of dolastatin 15 in the tubulin assembly assay is much weaker than for any of the other drugs that bind in the vinca domain (Table 11.4). In addition, dolastatin 15 does not inhibit the binding of the vinca alkaloids to tubulin and does not affect nucleotide exchange to any degree. Tubulin-dependent GTP hydrolysis is inhibited by dolastatin 15, but to a lesser degree than encountered with dolastatin 10. Hamel et al. have interpreted these results as an indication of a relatively weak drug–tubulin interaction for dolastatin 15. More recently, Jordan and coworkers have shown evidence that the dolastatin 15 derivative cemadotin (38; Figure 11.6) may exert its antitumor activity via suppression of dynamic instability of microtubules assembled to steady state.[100] These authors suggest that cemadotin may bind at a novel site in tubulin, distinct from the vinca site. Recent competition studies with radiolabeled dolastatin 15 seem to indicate that the binding sites for antimitotic peptides in the dolastatin family may consist of a series of overlapping domains.[101] Dolastatin 10 and 15 are nevertheless presumed to interact with a similar specific binding site in the vinca domain based on structural homology and GTP hydrolysis considerations,[97] although other targets cannot be ruled out.

With regard to the demonstrated capacity of dolastatin 10 and 15 to disrupt microtubule organization, it is important to note that the exact mechanism by which antimitotic drugs cause

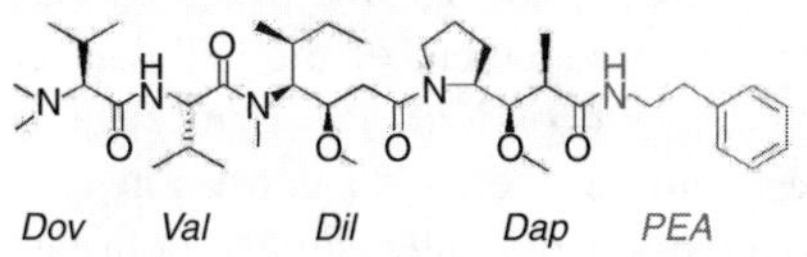

Soblidotin (**37**)

(TZ-1027, Auristatin PE)

$GI_{50} = 3.0 \times 10^{-6}$ mg/mL

NCI-H460 (Lung-NSC)

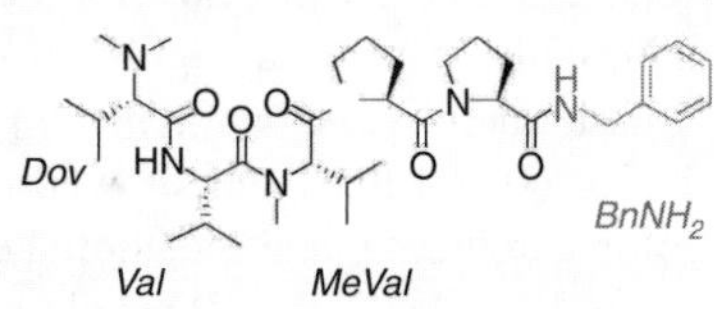

Cemadotin (**38**)

(LU103793)

$IC_{50} = 7.0 \times 10^{-4}$ mg/mL

HeLa-S_3 (cervical carcinoma)

FIGURE 11.6 (See color insert following page 304.) Synthetic derivatives of dolastatin 10 (soblidotin) and dolastatin 15 (cematodin) in phase II clinical trials.

cell death remains unclear. Some researchers have suggested that apoptosis induced by phosphorylation of the oncoprotein bcl-2 (an inhibitor of programmed cell death) may play an important role in the mechanism of action of some antitumor agents that are also microtubule poisons,[102,103] and some evidence for this explanation has been shown for dolastatins 10 and 15.[94,104,105]

Some biology of newly discovered, natural analogs of dolastatin 10 was also recently divulged. Cyanobacterial metabolite symplostatin 1 (**35**; Figure 11.7) reportedly exhibited potent microtubule inhibition and was furthermore found to be effective *in vivo* against two drug-insensitive, murine tumor models.[41] A recent study of the molecular pharmacology of symplostatin 1[106] revealed that this closely related substance shares a very similar mechanism of action with dolastatin 10. Symplostatin 3 (**36**; Figure 11.7) is about 10–100-fold less cytotoxic than dolastatin 10 *in vitro*.[47] Soblidotin (**37**, TZT-1027; Figure 11.6), a synthetic derivative of dolastatin now in clinical trials, inhibits microtubule polymerization and interferes with microtubule assembly/disassembly equilibria through interaction with tubulin, as expected from similar work with the parent peptide. Apoptosis by a direct mechanism (cytotoxicity) versus the known antimitotic effect has been observed with soblidotin at higher doses *in vitro*[107] and *in vivo* with murine B16 melanoma xenografts.[108] Of considerable interest is the observation made by Japanese researchers that treatment of an advanced murine tumor model (colon 26 adenocarcinoma) with soblidotin induces tumoral vascular collapse and tumor cell death, necrosis, and tumor regression.[109] This indicates that soblidotin may exert its antitumor activity by both conventional microtubule disruption and by a unique, antitumoral vascular mechanism of action that could prove to be a powerful combination for cancer chemotherapy.

Among other dolastatins whose biology has been studied in detail, dolastatin 11 is particularly interesting because of its unique biochemical effects on tubulin and actin proteins *in vitro*.[110] A

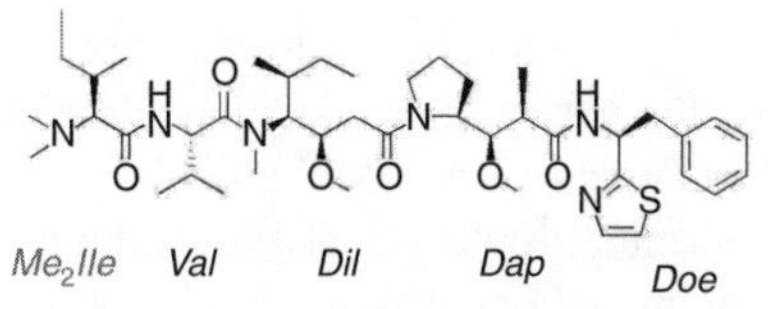

Symplostatin 1 (**35**)

$IC_{50} = 3.0 \times 10^{-4}$ mg/mL

KB cells (epidemoid carcinoma)

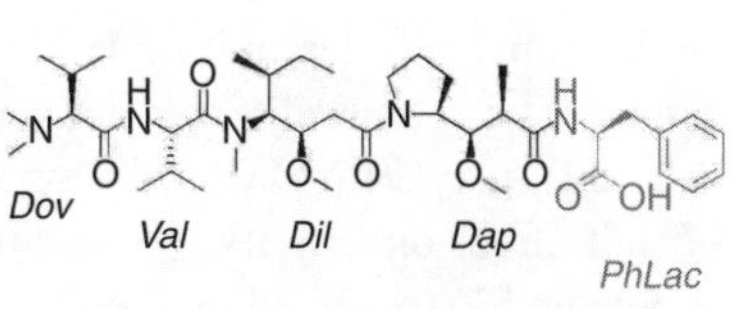

Symplostatin 3 (**36**)

$IC_{50} = 3.9 \times 10^{-3}$ mg/mL

KB cells (epidemoid carcinoma)

FIGURE 11.7 (See color insert following page 304.) Naturally occurring derivatives of dolastatin 10.

recent study concerning the mechanism of action of this peptide showed that dolastatin 11 arrests cells at cytokinesis by causing a rapid and massive rearrangement of the cellular actin filament network.[110] Dolastatin 16 has shown a mean panel GI_{50} value of 2.2×10^1 µg/mL against the NCI's complete panel of 60 cell lines and showed marked inhibitory effects against non–small cell lung carcinoma, colon carcinoma, and melanoma cells in particular.[20] Interestingly, dolastatin 16 shows a relatively low GI_{50}-COMPARE[111] correlation with dolastatins 10 and 15, which may indicate a different mechanism of action for this cyclic peptide relative to the others.

V. STRUCTURAL MODIFICATIONS AND SAR

The initial discovery of dolastatin 10 and its potent inhibitory effect on cancer cell growth at unprecedentedly low drug concentrations was soon followed by structure–activity studies on this structurally unique peptide. These efforts were aimed at elucidation of key structural requirements for activity and at reducing structural complexity where possible. The first structural modifications of dolastatin 10 to be prepared and evaluated against tumor cells *in vitro* were 19 of the possible 128 stereoisomers of the natural peptide, which were synthesized in connection with absolute configuration determinations for the non-proteinogenic *Dil, Dap,* and *Doe* residues.[98,112] Among the various stereoisomers tested, only two retained antitumor activity comparable to the natural peptide. Change of chirality at most stereocenters resulted in moderate to substantial decrease in activity against the murine P388 cell line, with the exception of the (6R)-isomer [isodolastatin-(6R), **42**; Table 11.5] and (19aR)-isomer, where reversal of stereochemistry at the *Doe* α-carbon (position *6*) and *Dil* side chain (position *19a*) was attended by no loss of antitumor activity (Figure 11.8). Moreover, a clear correlation between inhibition of tubulin polymerization and inhibition of L1210 leukemia cell growth was demonstrated for all stereoisomers that retained some antitumor activity. Interestingly, the shortened tripeptide segment *Dov-Val-Dil-OMe* ester is completely inactive against tumor cell growth and yet retains substantial inhibitory effects on tubulin polymerization. Hamel and coworkers have speculated that the *Dil* subunit may be essential for the interaction of dolastatin 10 with tubulin, although the exact nature of the link between affinity for tubulin-binding and antineoplastic activity is not completely clear.[86] This early work set the stage for more fundamental changes to the peptide backbone of dolastatin 10, carried out mainly by the research groups of Pettit,[86,112–114] Poncet,[115–117] and Miyazaki.[118,119]

Most of the attention given to the preparation of dolastatin 10 analogs has centered on modifications of the C terminus (e.g., *Doe* unit). Modifications at this position allow for structure simplification (e.g., thiazole replacement) without diminution of antineoplastic activity and may furthermore permit fine-tuning of selectivity in some cases. A survey of interesting results from some of these SAR studies is given in Table 11.5. In particular, the phenethylamine core of the C terminus was found to be essential for antitumor activity, and the corresponding derivative, described by both Pettit[113,114] and Miyazaki,[118,119] has quickly advanced as a drug candidate (soblidotin, **37**; Figure 11.6), currently in PII clinical trials. The length of the alkyl chain between the amide and phenyl group does appear to be important for antitumor activity (derivatives **43**, **44**, **37**, and **48**), whereas the effect of aryl substitution is less obvious (derivatives **45–47**). Several amino acid methyl esters are also adequate replacement groups for the usual C terminus, although the nature of the amino acid employed [e.g., aryl (Phe-OMe, **52**), alkyl (Ile-OMe, **55**), or sulfur-containing (Met-OMe, **54**)] does not appear to have a significant effect on activity. Interestingly, C-terminal replacement groups containing basic functionality (**50** and **51**) exhibited markedly reduced antineoplastic activity and furthermore showed very little correlation to the activity profile of the parent peptide. Derivatives in which the *Dap* subunit was either omitted (**58**) or replaced with proline (**57**) were ineffective. Modification to certain positions of the other subunits also produced derivatives that showed antitumor activity comparable to that of dolastatin 10. However, none of the modifications described resulted in any meaningful simplification of the core structure. It is interesting to note that one of the active *Dov*-modified, synthetic analogs (Me$_2$Ile N terminus) described by Pettit et al.[114] has since been discovered as a naturally occurring

TABLE 11.5
Survey of Dolastatin 10 Modifications and Effect on Inhibition of Tumor Cell Proliferation and Tubulin Polymerization.

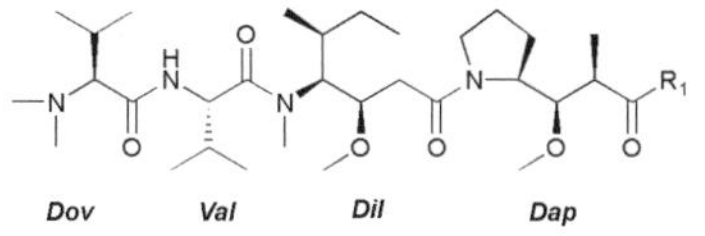

No.	Compound	Mean GI_{50} ($\times 10^8$ M)[a]	Correlation Coefficient[b]	L1210 IC_{50} (nM)[c]	Tubulin IC_{50} ($\mu g/mL$)[d]
1	Dolastatin 10	0.012	1.00	0.5	1.3 ± 0.2
42	Isodolastatin-(6R)	0.1[e]	—	—	—

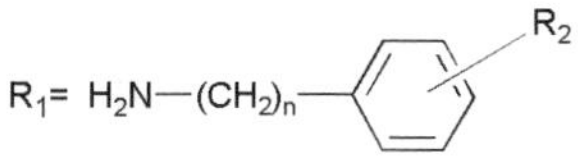

$R_1 = H_2N-(CH_2)_n-$ (phenyl with R_2)

43	n = 0, R_2 = H	1.0	0.828	60	1.8 ± 0.2
44	n = 1, R_2 = H	0.174	0.851	80	1.6 ± 0.2
37	n = 2, R_2 = H	0.049	0.901	0.6	1.3 ± 0.2
45	n = 2, R_2 = *ortho*-Cl	—	—	3	1.2 ± 0.2
46	n = 2, R_2 = *meta*-Cl	—	—	2	1.6 ± 0.2
47	n = 2, R_2 = *para*-Cl	0.076	0.835	0.4	1.5 ± 0.2
48	n = 3, R_2 = H	0.182	0.805	20	1.3 ± 0.05
49	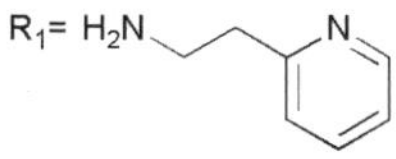 $R_1 = H_2N$ (ethyl-pyridine)	0.085	0.847	5	1.2 ± 0.2
50	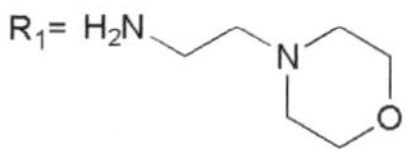 $R_1 = H_2N$ (ethyl-morpholine)	3.8	0.667	500	2.6 ± 0.5
51	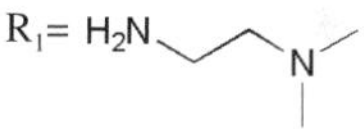 $R_1 = H_2N$ (ethyl-dimethylamine)	8.2	<0.6	>1000	2.1 ± 0.3
52	R_1 = Phe-OMe	0.020	0.903	5	1.6 ± 0.04
53	R_1 = Phe-NH_2	0.282	0.852	10	1.0 ± 0.2
54	R_1 = Met-OMe	0.040	0.904	6	1.4 ± 0.2
55	R_1 = Ile-OMe	0.016	0.899	7	1.7 ± 0.2
56	R_1 = OMe	—	—	30	1.3 ± 0.1
57	*Dap-Doe* → *Pro-PEA*[f]	3.8	0.667	800	2.9 ± 0.3
58	*Dov-Val-Dil-Doe* (*Dap* omitted)	19.5	0.711	200	2.1 ± 0.1

[a] Mean GI_{50}s obtained from quadruplicate screenings of compounds indicated against the National Cancer Institute's full 60–cancer cell line panel.

[b] Correlation coefficients from COMPARE algorithm analyses of averaged Total Growth Inhibition (TGI) mean graph obtained from quadruplicate screening results against the full 60–cancer cell line panel, using the mean value for dolastatin 10 as the seed for all correlations.

[c] Murine L1210 (lymphocytic leukemia cells).

[d] Inhibition of the polymerization of purified, bovine brain tubulin.

[e] ED_{50} value for isodolastatin-(6R) against murine P388 leukemia cells. This compounds were not tested on other assays listed above.

[f] *Pro-PEA* corresponds to the phenethylamide of L-proline.

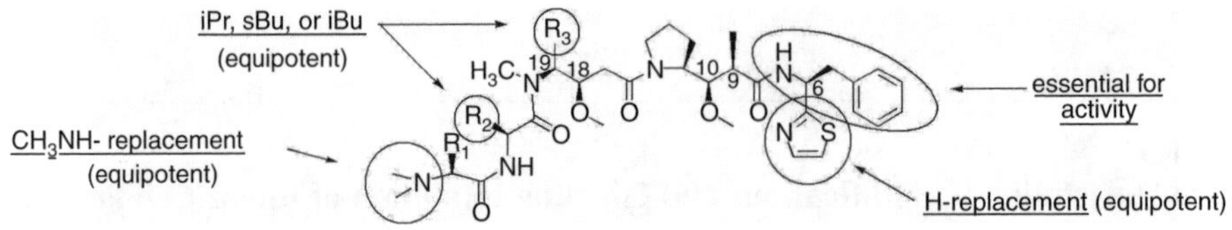

FIGURE 11.8 Effects of structural modifications on dolastatin 10 activity.

cyanobacterial metabolite (symplostatin 1, **35**; Figure 11.7).[45] A summary of some SARs for dolastatin 10 is given in Figure 11.8.

Preparation of a single crystal of the (6R)-isomer of dolastatin 10 [isodolastatin-(6R), **42**] afforded the first opportunity for study of the three-dimensional conformation of this peptide, and a molecular modeling study was undertaken by Pettit and Srirangam. Computer-assisted modeling studies for this isomer of dolastatin 10, supported by data from an x-ray crystal structure, revealed a ~21 kcal/mol energy barrier between *cis* and *trans* conformers, visible by NMR at room temperature. Interestingly, the crystalline structure of isodolastatin-(6R) contains all *trans*-amide bonds except for the *Dil–Dap* bond, which proved to be *cis*.[112] Following this work on stereoisomer **42**, Poncet and coworkers reported a detailed molecular modeling study on the natural stereoisomer of both dolastatin 10 and dolastatin 15, aimed at elucidation of the structural requirements for antitumor activity. The preferred, low-energy conformations of these peptides in DMSO-d_6 solution were determined via two-dimensional NMR measurements, followed by molecular dynamics simulations.[115] These calculations confirmed that the two conformers observed for dolastatin 10 in solution are a result of *cis-trans* isomerization of the *Dil–Dap* amide bond, with an energy barrier of 8 kcal/mol in favor of the *cis* conformer.

The unique, *cis-trans* isomerization of dolastatin 10 could potentially provide insights into the active conformation of the peptide in association with its presumed intracellular target tubulin. Poncet synthesized a cyclic derivative of the peptide (**39**), joined through a threonine-like ester linkage between the "*Dov*" and "*Doe*" residues, to probe the effect of locking the backbone in a folded conformation (Figure 11.9).[117] Although the resulting compound was completely inactive, it is possible that *cis-trans* isomerization of dolastatin 10 plays a central role in its interaction with tubulin, a mechanism that will likely be more well-defined after further studies. Among the D10/D15 hybrid compounds **40** and **41** prepared by Poncet (Figure 11.9), both peptides retained partial

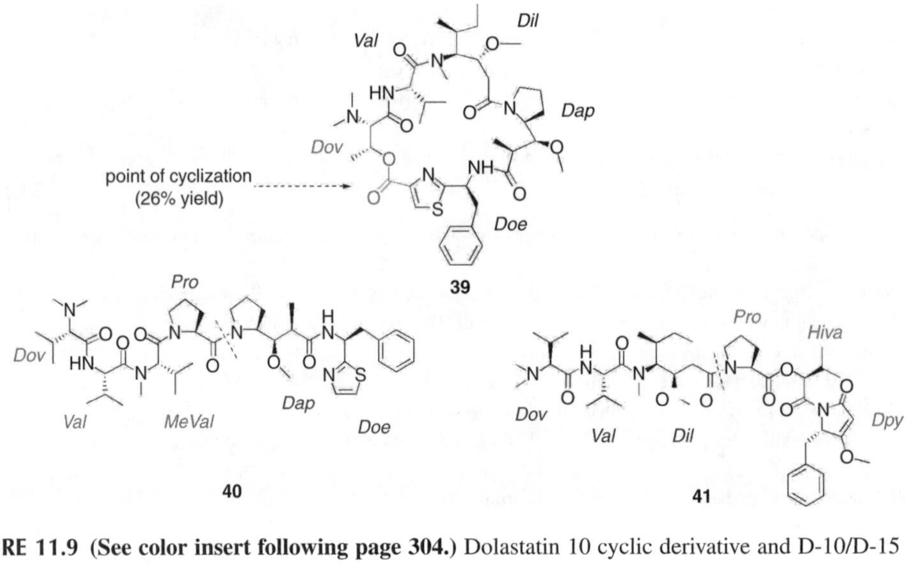

FIGURE 11.9 (See color insert following page 304.) Dolastatin 10 cyclic derivative and D-10/D-15 hybrid compounds.

TABLE 11.6
Survey of Dolastatin 15 N-Terminus Modifications and Effect on Inhibition of Tumor Cell Proliferation and Tubulin Polymerization

No.	N-Terminus	L1210 IC_{50} (nM)[a]	P388 ED_{50} (ng/mL)[b]	NCI-H460 ED_{50} (ng/mL)[c]	Tubulin IC_{50} (μg/mL)[d]
1	(7S)-Hiva-(4S)-Dpy	0.15	2.4	0.28	5.2
59	(7R)-Hiva-(4S)-Dpy	0.30	—	—	—
60	(7S)-Hiva-(4R)-Dpy	0.31	—	—	—
61	(7S)-Hiva-(4S)-Dpy	1.78	—	—	—
62	(7S)-Hiva-anilide	3	0.52	0.032	26
63	(7S)-Hiva-benzamide	1	0.42	0.25	20
64	(7S)-Hiva-phenethylamide	2	0.26	0.044	15
65	(7S)-Hiva-phenpropylamide	—	0.38	1.1	—
66	(7S)-Lac-phenethylamide	—	0.46	0.37	—
67	L-Val-phenethylamide	—	0.59	2.5	—
68	L-MeVal-phenethylamide	—	>1000	3600	—

[a] Murine L1210 (lymphocytic leukemia cells).
[b] Murine P388 (lymphocytic leukemia cells).
[c] Human non–small cell lung carcinoma.
[d] Inhibition of the polymerization of purified, bovine brain tubulin.

antitumor activity, whereas only peptide **41** (*Dov-Val-Dil-Pro-Hiva-Dpy*) retained capacity to inhibit tubulin polymerization *in vitro*.[117] This result appears to support Hamel's assertion that the *Dil* residue is critical for molecular interactions between the peptide and tubulin.[86]

Because dolastatins 11–14, dolastatins 16–18, and Yamada's series of *D. auricularia* metabolites have not undergone extensive biochemical evaluation to date, less is presently known about SARs for these compounds. Recently, Russo and coworkers in Italy began preliminary studies aimed at elucidation of SAR for dolastatins 11 and 12 using molecular mechanics and dynamics calculations in aqueous solution.[120] Now that Bates and coworkers have devised a total synthesis of dolastatins 11 and 12,[66] the door remains open for further elucidation of its unique biochemical effects on actin through preparation of synthetic derivatives. These and other advances should help provide more insight into the specific mechanisms of action for these as-yet largely unstudied peptides.

Similar to dolastatin 10, efforts toward understanding SARs for dolastatin 15 are further advanced than for other members of this series, and a number of research groups have prepared derivatives and published studies. In contrast to dolastatin 10, dolastatin 15 was found to be surprisingly insensitive to structural and stereochemical modifications at the C terminus (Table 11.6; Figure 11.10). Change in the configuration of the stereogenic centers that comprise the *Hiva* and *Dpy* unit was found to minimally affect *in vitro* antineoplastic activity (compounds **59–61**).[121] As with dolastatin 10, a slight trend favoring the phenethylamine C terminus over other arylalkyl chain lengths was observed, but the effect was much more subtle in the case of dolastatin 15 (compounds **62–65**).[122] Results of SAR studies for dolastatin 15 clearly indicate that the structurally complex *Dpy* C terminus plays a limited role in the cell growth inhibitory effects of this peptide.

The interaction of dolastatin 15 with purified tubulin *in vitro* is considerably weaker than that of dolastatin 10 (about 20-fold) and other antimitotic drugs that bind in the vinca domain of tubulin,[99] despite its potent antiproliferative effect on cellular microtubules (one-seventh as active as dolastatin 10, but 10 times more potent than the clinically useful anticancer drug vinblastine). Thus far, results from SAR studies on dolastatin 15 have not established a clear connection between cancer cell growth inhibition and a tubulin-based mechanism of action. Although all structural modifications of the C terminus of dolastatin 15 exhibited potent antineoplastic activity, most showed modest reduction of

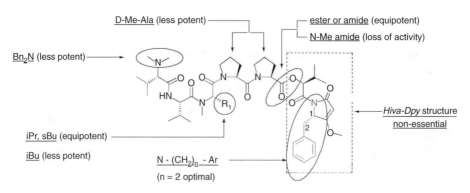

FIGURE 11.10 (See color insert following page 304.) Effects of structural modifications on dolastatin 15 activity.

inhibitory activity against tubulin polymerization. The opposite has been observed for dolastatin 10 derivatives, where potent inhibitory effect on tubulin polymerization ($1–4$-μM range) has been retained in spite of significant loss of antitumor activity (>1000-fold in some cases).[122] It has been suggested that the potent cytotoxicity and inhibitory effects on microtubule assembly exhibited by dolastatin 15 are not a result of the parent molecule but of a more active metabolite derived from it.[99] In this context, the reduced activity of dolastatin 15 in the purified tubulin assay could be explained by the absence of proteolytic enzymes normally present in cells. Furthermore, the fact that dolastatin 15 is active against prokaryotic cells that lack tubulin indicates a second cellular target. Derivatives **66–68** were prepared by Pettit's group to probe whether intracellular hydrolysis of the ester bond of this depsipeptide might be a necessary condition for *in vitro* activity.[123] Interestingly, replacement of the ester with an N-methyl amide function (compound **68**) led to complete loss of antitumor activity. Although the N-methyl amide modification would be expected to impart greater stability of the peptide bond toward enzymatic cleavage, conformational changes in the peptide backbone induced by N-methylation cannot be ruled out as the cause for the lack of potency in this compound.

Among the most important discoveries made in the dolastatin series is that made by researchers at BASF, who found that the *Hiva-Dpy* units of dolastatin 15 may be completely replaced with a simple benzamide group without loss of activity (cemadotin, LU103793, **38**; Figure 11.6).[124] The progress of this water-soluble D15 derivative through clinical trials in humans is discussed in the next section. Hu and coworkers have described a series of derivatives based on the structure of cemadotin, in which modifications of the *Dov* (Bn$_2$Val) and Pro (D-MeAla) residues as well as aromatic ring substitutions were evaluated.[125] Although aromatic ring substituents altered the cytostatic properties only slightly, other modifications described led to more pronounced loss of activity in this series. Kessler and coworkers have prepared a peptidomimetic compound based on the structure of dolastatin 15, using N-substituted glycine monomers, but the resulting peptoid was devoid of cytostatic properties.[126] A summary of dolastatin 15 structural requirements for antitumor activity is shown in Figure 11.10.

The minimum-energy, solution-phase conformation of dolastatin 15 has been predicted by Poncet et al., using two-dimensional NMR techniques and molecular dynamics simulations.[115] Despite the known tendency for proline residues to form peptide turn structures because of a decreased energy barrier between *cis* and *trans* amide bond conformations, dolastatin 15, which contains two contiguous proline units, appears to adopt a less-folded, linear shape (all *trans* amide bonds) in solution.[121]

VI. CLINICAL UPDATE

Of the naturally occurring dolastatins, only dolastatin 10 has progressed to human clinical trials. Although dolastatin 15 has not been studied directly, synthetic derivatives of both dolastatins 10 (soblidotin, **37**) and 15 (cemadotin, **38**) are currently under evaluation in phase II clinical trials. The progress of these studies and the future potential of the dolastatins and its congeners are discussed in this section.

Among the naturally occurring peptides of *D. auricularia*, dolastatin 10 was selected over dolastatin 15 for further evaluation in clinical trials because of its inhibitory effect on cancer cell growth at extremely low concentrations. Preclinical studies on both peptides indicated that myelotoxicity would be a potentially limiting factor, as both compounds were found to inhibit bone marrow cell colony formation at drug concentrations ~100-fold lower than that required for antitumor activity in human tumor cell lines *in vitro*.[127,128] Preclinical animal toxicity evaluations in mice, rats, and dogs showed that myelotoxicity was dose-limiting in all three species, with mice being the least sensitive, and a phase I clinical trial starting dose of 65 μg/M^2 was recommended.[127]

Among phase I patients with advanced solid tumors, a maximum tolerated dose was reached at 300 μg/M^2, and granulocytopenia (myelosuppression) was the dose-limiting toxicity, observed in 33% of patients at this dose.[129] Other mild-to-moderate toxicities were also noted such as fatigue, nausea, and peripheral neuropathy (40% of patients).[128] Clinical pharmacokinetic studies with five patients detected a single dolastatin 10 metabolite, corresponding to a more water-soluble, N-desmethyl derivative.[129,130] Dolastatin 10, administered as an intravenous bolus injection, showed a plasma elimination half-life of 5–7 hours.[80] Phase II clinical trials for dolastatin 10 in a number of cancer indications (metastatic prostate adenocarcinoma,[131] metastatic melanoma,[132] metastatic colorectal cancer,[133] advanced non–small cell lung carcinoma,[134] and platinum-sensitive ovarian carcinoma[135]) have been completed. The MTD has varied from patient to patient from 300 to 400 μg/M^2. No significant antitumor activity has been observed to date for dolastatin 10, administered as a single agent to patients with advanced solid tumors. Dolastatin 10 does act synergistically with other anticancer substances (e.g., vincristine, bryostatin 1),[95] providing some hope that this highly potent peptide may find use in drug combination therapies.

Structure/activity work by Miyazaki[118,119] and Pettit[136] led to the discovery of the dolastatin 10 synthetic derivative soblidotin (**37**) that is currently in phase II clinical trials. Although less information concerning the results of preclinical work and phase I clinical trials is presently available for soblidotin, a pattern similar to that of dolastatin 10 would be expected because of their close structural analogy. Some preliminary results from a few phase I studies have been recently summarized.[137,138] Soblidotin (single dose, intravenous feed at 1.35–2.4 mg/M^2) was generally well tolerated among patients with advanced solid tumors, and induced antitumor effects were observed.[139] A more recent study conducted with non–small cell lung carcinoma patients has generated some promising results. The drug (single dose, intravenous infusion) was reportedly well tolerated up to 4.8 mg/M^2, and complete or partial responses were seen in 4 of 44 patients.[140,141] These results are a positive sign for how this promising drug may fare in phase II clinical trials, although no further data are presently available.

The third and final drug in the dolastatin family to reach phase II clinical trials in humans as of this writing is the dolastatin 15 derivative cemadotin (**38**). As with dolastatin 10, myelotoxicity (neutropenia) was found to be the principal dose-limiting toxicity, and a maximum tolerated dose of 12.5 mg/M^2 was established for phase II patients.[142] In an earlier phase I clinical study, cardiovascular toxicity and hypertension were the dose-limiting toxicities, but these were later easily controlled via modification of the treatment schedule.[143] Other nonhematologic toxicities were nausea and fatigue and were moderate. Interestingly, the primary metabolite detected for cemadotin is not due to modification of the N terminus (as with dolastatin 10) but, rather, from loss of the C-terminal benzamide group. This may constitute further evidence that the bioactive species in the case of dolastatin 15 is actually the peptide metabolite resulting from enzymolysis of the *Pro-Hiva* bond (for dolastatin 15) or *Pro*-benzamide bond (for cemadotin), as discussed in section V. Phase II results have been reported for trials involving patients with non–small cell lung carcinoma and metastatic breast cancer, but no positive responses have been noted with the current dose and schedule.[144,145] Researchers at BASF have also pursued the use of cemadotin in drug combination therapies (e.g., with taxanes).[146] Although the dolastatins have yet to demonstrate clinical efficacy against cancer, it should be noted that these compounds have been studies in biological domains for a relatively short period of time, and in this regard, much of what the dolastatins may ultimately offer for the treatment of human diseases such as cancer remains to be determined.

REFERENCES

1. *Chemistry of marine natural products,* Scheuer, P. J. Ed., Academic Press, New York, NY, Vols. I and II, 1973; Vol. III 1980; Vol. IV, 1981; Vol. V. 1983.
2. Faulkner, D. J., Marine natural products, *Nat. Prod. Rep.,* 12, 223, 1995.
3. Roberts, M., in *A popular history of the mollusca,* Reeve & Benham, London, 1851.
4. Paul, V. J. and Pennings, S. C., Diet-derived chemical defenses in the sea hare *Stylocheilus longicauda* (Quoy and Gaimard 1824), *J. Expl. Mar. Biol. Ecol.,* 151, 227, 1991.
5. Pennings, S. C., Nadeau, M. T., and Paul, V. J., Selectivity and growth of the generalist herbivore *Dolabella auricularia* feeding upon complementary resources, *Ecology,* 74, 879, 1993.
6. Pennings, S. C. and Paul, V. J., Sequestration of dietary secondary metabolites by three species of sea hares: location, specificity and dynamics, *Mar. Biol.,* 117, 535, 1993.
7. De Nys, R. et al., Quantitative variation of secondary metabolites in the sea hare *Aplysia parvula* and its host plant *Delisea pulchra, Mar. Ecol. Prog. Ser.,* 130, 135, 1996.
8. Faulkner, D. J., Chemical defenses of marine molluscs, in *Ecological roles of marine natural products,* Paul V. J., Ed., Comstock Publishing Associates, Ithaca, NY, 1992, 119.
9. Pliny, in *Historica naturalis,* 60 AD, Lib. 9, Lib. 32.
10. Donati, G. and Porfirio, B., Marine pharmacology and toxicology. The dolastatins, *Conchiglia,* 16,1984.
11. Eales, N. B., Typical british marine plants and animals, aplysia, in *L.M.B.C. Memoirs* Hardmann, W. A. and Johnstone, J., Eds, Liverpool, England, 1921. Vol. 24.
12. Halstead, B. W., Poisonous sea hares, in *Poisonous and venemous marine animals of the world,* Government Printing Office, Washington, DC, 1965. Vol. 1, 709
13. Sorokin, M., Human poisoning by ingestion of a sea hare (*Dolabella auricularia*), *Toxicon,* 26, 1095, 1998.
14. Pettit, G. R. et al., Antineoplastic components of marine animals, *Nature,* 227, 962, 1970.
15. Pettit, G. R., The dolastatins, *Fortschr. Chem. Org. Naturst.,* 70, 1, 1997.
16. Pettit, G. R. et al., Anti-neoplastic agents. 69. The isolation of loliolide from an Indian Ocean opisthobranch mollusk, *J. Nat. Prod.,* 43, 752, 1980.
17. Pettit, G. R. et al., Anti-neoplastic agents. 72. Marine animal biosynthetic constituents for cancer chemotherapy, *J. Nat. Prod.,* 44, 482, 1981.
18. Pettit, G. R. et al., Isolation of dolastatins 10-15 from the marine mollusc *Dolabella auricularia, Tetrahedron,* 49, 9151, 1993.
19. Pettit, G. R., The isolation and structure of a remarkable marine animal antineoplastic constituent: dolastatin 10, *J. Am. Chem. Soc.,* 109, 6883, 1987.
20. Pettit, G. R. et al., Isolation and structure of the human cancer cell growth inhibitory cyclodepsipeptide dolastatin 16, *J. Nat. Prod.,* 60, 752, 1997.
21. Pettit, G. R. et al., Antineoplastic agents. 369. Isolation and structure of dolastatin 17, *Heterocycles,* 47, 491, 1998.
22. Pettit, G. R. et al., Antineoplastic agents. 370. Isolation and structure of dolastatin 18, *Bioorg. Med. Chem. Lett,* 7, 827, 1997.
23. Davies-Coleman, M. T. et al., Isolation of homodolastatin 16, a new cyclic depsipeptide from a Kenyan collection of *Lyngbya majuscula, J. Nat. Prod.,* 66, 712, 2003.
24. Yamada, K. and Kigoshi, H., Bioactive compounds from the sea hares of two genera: *Aplysia* and *Dolabella, Bull. Chem. Soc. Japan,* 70, 1479, 1997.
25. Sone, H. et al., Isolation, structure, and synthesis of dolastatin C, a new depsipeptide from the sea hare *Dolabella auricularia, Tetrahedron Lett.,* 34, 8445, 1993.
26. Sone, H. et al., Isolation, structure, and synthesis of dolastatin D, a cytotoxic cyclic depsipeptide from the sea hare *Dolabella auricularia, Tetrahedron Lett.,* 34, 8449, 1993.

27. Ojika, M. et al., Dolastatin E, a new cyclic hexapeptide isolated from the sea hare *Dolabella auricularia*, *Tetrahedron Lett.*, 36, 5057, 1995.

28. Nakamura, M. et al., Stereochemistry and total synthesis of dolastatin E, *Tetrahedron Lett.*, 36, 5059, 1995.

29. Mutou, T. et al., Isolation and stereostructures of dolastatin G and nordolastatin G, cytotoxic 35-membered cyclodepsipeptides from the Japanese sea hare *Dolabella auricularia*, *J. Org. Chem.*, 61, 6340, 1996.

30. Mutou, T. et al., Synthesis of dolastatin G and nordolastatin G, cytotoxic 35-membered cyclodepsipeptides of marine origin, *Tetrahedron Lett.*, 37, 7299, 1996.

31. Sone, H. et al., Dolastatin H and isodolastatin H, potent cytotoxic peptides from the sea hare *Dolabella auricularia*: isolation, stereo structures, and synthesis, *J. Am. Chem. Soc.*, 118, 1874, 1996.

32. Sone, H., Kigoshi, H., and Yamada, K., Isolation and stereostructure of dolastatin I, a cytotoxic cyclic hexapeptide from the Japanese sea hare *Dolabella auricularia*, *Tetrahedron*, 53, 8149, 1997.

33. Suenaga, K. et al., Isolation and stereostructure of aurilide, a novel cyclodepsipeptide from the Japanese sea hare I, *Tetrahedron Lett.*, 37, 6771, 1996.

34. Mutou, T. et al., Enantioselective synthesis of aurilide, a cytotoxic 26-membered cyclodepsipeptide of marine origin, *Synlett*, 199, 1997.

35. Sone, H. et al., Dolabellin, a cytotoxic bisthiazole metabolite from the sea hare *Dolabella auricularia*: structural determination and synthesis, *J. Org. Chem.*, 60, 4774, 1995.

36. Ishiwata, H. et al., Isolation and stereostructure of doliculide, a cytotoxic cyclodepsipeptide from the Japanese sea hare *Dolabella auricularia*, *J. Org. Chem.*, 59, 4710, 1994.

37. Ojika, M., Nemoto, T., and Yamada, K., Doliculol-A and doliculol-B, the nonhalogenated C-15 acetogenins with cyclic ether from the sea hare *Dolabella-auricularia*, *Tetrahedron Lett.*, 34, 3461, 1993.

38. Faulkner, D. J., Marine natural-products — metabolites of marine-algae and herbivorous marine mollusks, *Nat. Prod. Rep.*, 1, 251, 1984.

39. Fattorusso, E. and Piattelli, M., Amino acids from marine algae, in *Marine Natural Products, Chemical and Biological Perspectives,* Scheuer, P. J., Ed., Academic Press, New York, 1980, Vol. III, 95.

40. Mitchell, S. S. et al., Dolastatin 3 and two novel cyclic peptides from a Palauan collection of *Lyngbya majuscula*, *J. Nat. Prod.*, 63, 279, 2000.

41. Luesch, H. et al., Isolation of dolastatin 10 from the marine cyanobacterium *Symploca* species VP642 and total stereochemistry and biological evaluation of its analogue symplostatin 1, *J. Nat. Prod.*, 64, 907, 2001.

42. Harrigan, G. G. et al., Isolation, structure determination, and biological activity of dolastatin 12 and lyngbyastatin I from *Lyngbya majuscula Schizothrix calcicola* cyanobacterial assemblages, *J. Nat. Prod.*, 61, 1221, 1998.

43. Nogle, L. M., Williamson, R. T., and Gerwick, W. H., Somamides A and B, two new depsipeptide analogues of dolastatin 13 from a Fijian cyanobacterial assemblage of *Lyngbya majuscula* and *Schizothrix* species, *J. Nat. Prod.*, 64, 716, 2001.

44. Nogle, L. M. and Gerwick, W. H., Isolation of four new cyclic depsipeptides, antanapeptins A-D, and dolastatin 16 from a Madagascan collection of *Lyngbya majuscula*, *J. Nat. Prod.*, 65, 21, 2002.

45. Harrigan, G. G. et al., Symplostatin 1: a dolastatin 10 analogue from the marine cyanobacterium *Symploca hydnoides*, *J. Nat. Prod.*, 61, 1075, 1998.

46. Harrigan, G. G. et al., Symplostatin 2: a dolastatin 13 analogue from the marine cyanobacterium *Symploca hydnoides*, *J. Nat. Prod.*, 62, 655, 1999.

47. Luesch, H. et al., Symplostatin 3, a new dolastatin 10 analogue from the marine cyanobacterium *Symploca* sp VP452, *J. Nat. Prod.*, 65, 16, 2002.

48. Luesch, H. et al., The cyanobacterial origin of potent anticancer agents originally isolated from sea hares, *Curr. Med. Chem.*, 9, 1791, 2002.

49. Schmidt, U. et al., Amino acids and peptides; 60. Synthesis of biologically active cyclopeptides; 10. Synthesis of 16 structural isomers of dolastatin 3; ii: synthesis of the linear educts and the cyclopeptides, *Synthesis*, 236, 1987.

50. Hamada, Y., Kato, S., and Shioiri, T., New methods and reagents in organic-synthesis 51. A synthesis of ascidiacyclamide, a cyto-toxic cyclic peptide from ascidian — determination of its absolute-configuration, *Tetrahedron Lett.*, 26, 3223, 1985.

51. Arai, H. et al., Synthesis of the pyrimidine moieties of bleomycin and epibleomycin, *J. Am. Chem. Soc.*, 102, 6631, 1980.

52. Bodanszky, M., Scozzie, J. A., and Muramatsu, I., Dehydroalanine residues in thiostrepton, *J. Antibiot.*, 23, 9, 1970.

53. Pettit, G. R. et al., Antineoplastic agents. 150. The structure and synthesis of dolastatin-3, *J. Am. Chem. Soc.*, 109, 7581, 1987.

54. Schmidt, U. and Utz, R., Synthetic studies on the elucidation of the structure and configuration of dolastatin 3, *Angew. Chem. Int. Ed. Eng.*, 23, 725, 1984.

55. Hamada, Y., Kohda, K., and Shioiri, T., Proposed structure of the cyclic peptide dolastatin-3, a powerful cell-growth inhibitor, should be revised, *Tetrahedron Lett.*, 25, 5303, 1984.

56. Holzapfel, C. W. and Van Zyl, W. J., The synthesis and conformation in solution of cyclo[l-pro-l-leu-(gln)thz-(gly)thz-l-val] (dolastatin 3), *Tetrahedron*, 46, 649, 1990.

57. Rinehart, Jr., K. L. et al., Biologically-active peptides and their mass-spectra, *Pure Appl. Chem.*, 54, 2409, 1982.

58. Pettit, G. R., Dolastatin 10, *Fortschr. Chem. Org. Naturst.*, 70, 14, 1997.

59. Pettit, G. R. et al., Antineoplastic agents. 189. The absolute-configuration and synthesis of natural (-)-dolastatin-10, *J. Am. Chem. Soc.*, 111, 5463, 1989.

60. Hamada, Y., Hayashi, K., and Shioiri, T., Efficient stereoselective synthesis of dolastatin-10, an antineoplastic peptide from a sea hare, *Tetrahedron Lett.*, 32, 931, 1991.

61. Shioiri, T., Hayashi, K., and Hamada, Y., Stereoselective synthesis of dolastatin 10 and its congeners, *Tetrahedron*, 49, 1913, 1993.

62. Tomioka, K., Kamai, M., and Koga, K., An expeditious synthesis of dolastatin-10, *Tetrahedron Lett.*, 32, 2395, 1991.

63. Roux, F. et al., Synthesis of dolastatin-10 and [r-doe]-dolastatin-10, *Tetrahedron*, 50, 5345, 1994.

64. Pettit, G. R. et al., Antineoplastic agents. 173. Isolation and structure of the cell growth inhibitory depsipeptides dolastatins 11 and 12, *Heterocycles*, 28, 553, 1989.

65. Bates, R. B. and Gangwar, S., Asymmetric syntheses of 3-amino-2-methylpentanoic acids. Configurations of the beta-amino acid in majusculamide C, 57-normajusculamide C and dolastatins 11 and 12, *Tetrahedron Asymm.*, 4, 69, 1993.

66. Bates, R. B. et al., Dolastatins. 26. Synthesis and stereochemistry of dolastatin 11, *J. Am. Chem Soc.*, 119, 2111, 1997.

67. Carter, D. C. et al., Structure of majusculamide-C, a cyclic depsipeptide from lyngbya-majuscula, *J. Org. Chem.*, 49, 236, 1984.

68. Mynderse, J. S., Hunt, A. H., and Moore, R. E., 57-normajusculamide-C, a minor cyclic depsipeptide isolated from *Lyngbya-majuscula*, *J. Nat. Prod.*, 51, 1299, 1988.

69. Pettit, G. R. et al., Antineoplastic agents. 174. Isolation and structure of the cytostatic depsipeptide dolastatin 13 from the sea hare *Dolabella auricularia*, *J. Am. Chem. Soc.*, 111, 5015, 1989.

70. Sih, C. J. et al., Biocatalytic methods for enantioselective synthesis, *Stereocontrolled Organic Synthesis*, 399, 1994.

71. Moune, S. et al., Total synthesis of dolatrienoic acid: A subunit of dolastatin 14, *J. Org. Chem.*, 62, 3332, 1997.

72. Duffield, J. J. and Pettit, G. R., Synthesis of (7s,15s)- and (7r,15s)-dolatrienoic acid, *J. Nat. Prod.*, 64, 472 2001.

73. Pettit, G. R. et al., Antineoplastic agents. 220. Synthesis of natural (-)-dolastatin 15, *J. Am. Chem. Soc.*, 113, 6692, 1991.

74. Patino, N. et al., Total synthesis of the proposed structure of dolastatin-15, *Tetrahedron*, 48, 4115, 1992.

75. Pettit, G. R. et al., The dolastatins-20 — a convenient synthetic route to dolastatin-15, *Tetrahedron*, 50, 12097, 1994.

76. Akaji, K. et al., Convergent synthesis of dolastatin 15 by solid phase coupling of n-methylamino acid, *Peptide Science*, 35, 9, 1999.

77. Pettit, G. R., Structural modifications of dolastatin 10, *Fortschr. Chem. Org. Naturst.*, 70, 42, 1997.

78. Pettit, G. R., Hogan, F., and Herald, D. L., Synthesis and x-ray crystal structure of the *Dolabella auricularia* peptide dolastatin 18, *J. Org Chem.*, 69, 4019, 2004.

79. Poncet, J., The dolastatins, a family of promising antineoplastic agents, *Current Pharm Design*, 5, 139, 1999.

80. Aherne, G. W. et al., Antitumour evaluation of dolastatins 10 and 15 and their measurement in plasma by radioimmunoassay, *Cancer Chemother. Pharmacol.*, 38, 225, 1996.

81. Pietra, F., A secret world, in *The natural products of marine life*, Basel, Switzerland, 1990, 39.

82. Yamazaki, M., Iijima, R., and Kosuna, K., Antibacterial and antifungal peptide from *Dolabella auricularia*, Japanese Patent JP10251297, 22 SEP 1998.

83. Donia, M. and Hamann, M. T., Marine natural products and their potential applications as anti-infective agents, *Lancet Infect. Dis.*, 3, 338, 2003.

84. Fennell, B. J. et al., Effects of the antimitotic natural product dolastatin 10, and related peptides, on the human malarial parasite *Plasmodium falciparum*, *J. Antimicrobial Chemother.*, 51, 833, 2003.

85. Hamel, E., Interaction of tubulin with small ligands, in *Microtubule proteins*, Avila, J., Ed., CRC Press, Boca Raton, FL, 1990, 89.

86. Bai, R., Pettit, G. R., and Hamel, E., Structure activity studies with chiral isomers and with segments of the antimitotic marine peptide dolastatin-10, *Biochem. Pharmacol.*, 40, 1859, 1990.

87. Bai, R., Pettit, G. R., and Hamel, E., Binding of dolastatin-10 to tubulin at a distinct site for peptide antimitotic agents near the exchangeable nucleotide and vinca alkaloid sites, *J. Biol. Chem.*, 265, 17141, 1990.

88. Quentmeier, H. et al., Cytostatic effects of dolastatin 10 and dolastatin 15 on human leukemia cancer cell lines, *Leukemia Lymphoma*, 6, 245, 1992.

89. Steube, K. G. et al., Inhibition of cellular proliferation by the natural peptides dolastatin 10 and dolastatin 15, *Mol. Biol. Haematapoiesis*, 2, 567, 1992.

90. Steube, K. G. et al., Dolastatin 10 and dolastatin 15: Effect of two natural peptides on growth and differentiation of leukemia cells, *Leukemia*, 6, 1048, 1992.

91. Hu, Z.-B. et al., Effects of dolastatins on human B-lymphocytic leukemia cell lines, *Leukemia Res.*, 17, 333, 1993.

92. Mohammad, R. M. et al., Successful treatment of human chronic lymphocytic leukemia xenografts with combination biological agents auristatin PE and bryostatin 1, *Clin. Cancer Res.*, 4, 1337, 1998.

93. Beckwith, M., Urba, W. J., and Longo, D. L., Growth inhibition of human lymphoma cell lines by the marine natural products dolastatins 10 and 15, *J. Natl. Cancer Inst.*, 85, 483, 1993.

94. Maki, A. et al., Effect of dolastatin 10 on human non-Hodgkin's lymphoma cell lines, *Anti-Cancer Drugs*, 7, 344, 1996.

95. Mohammad, R. M. et al., Synergistic interaction of selected marine animal anticancer drugs against human diffuse large cell lymphoma, *Anti-Cancer Drugs*, 9, 149, 1998.

96. Turner, T. et al., Treatment of human prostate cancer cells with dolastatin 10, a peptide isolated from a marine shell-less mollusc, *Prostate*, 34, 175, 1998.

97. Hamel, E., Natural products which interact with tubulin in the vinca domain: Maytansine, rhizoxin, phomopsin a, dolastatins 10 and 15 and halichondrin B, *Pharmacol. Ther.*, 55, 31, 1992.

98. Pettit, G. R. et al., Antineoplastic agents. 205. Chiral modifications of dolastatin-10 — the potent cytostatic peptide (19ar)-isodolastatin-10, *J. Med. Chem.*, 33, 3132, 1990.

99. Bai, R. et al., Dolastatin-15, a potent antimitotic depsipeptide derived from *Dolabella-auricularia* — interaction with tubulin and effects on cellular microtubules, *Biochem Pharmacol.*, 43, 2637, 1992.

100. Jordan, M. A. et al., Suppression of microtubule dynamics by binding of cemadotin to tubulin: possible mechanism for its antitumor action, *Biochemistry*, 37, 17571, 1998.

101. Cruz-Monserrate, Z. et al., Dolastatin 15 binds in the vinca domain of tubulin as demonstrated by Hummel-Dreyer chromatography, *Eur. J. Biochem.*, 270, 3822, 2003.

102. Haldar, S., Basu, A., and Croce, C. M., Bcl2 is the guardian of microtubule integrity, *Cancer Res.*, 57, 229, 1997.

103. Haldar, S., Basu, A., and Croce, C. M., Serine-70 is one of the critical sites for drug-induced bcl2 phosphorylation in cancer cells, *Cancer Res.*, 58, 1609, 1998.

104. Maki, A. et al., The bcl-2 and p53 oncoproteins can be modulated by bryostatin-1 and dolastatins in human diffuse large-cell lymphoma, *Anti-Cancer Drugs*, 6, 392, 1995.

105. Ali, M. A. et al., Dolastatin 15 induces apoptosis and bcl-2 phosphorylation in small cell lung cancer cell lines, *Anticancer Res.*, 18, 1021, 1998.

106. Mooberry, S. L. et al., The molecular pharmacology of symplostatin 1: a new antimitotic dolastatin 10 analog, *International J. Cancer*, 104, 512, 2003.

107. Watanabe, J. et al., Induction of apoptosis in human cancer cells by TZT-1027, an antimicrotubule agent, *Apoptosis*, 5, 345, 2000.

108. Ikeda, R. et al., Induction of apoptosis in mice with B16 melanoma by treatment with the antimicrotubule agent TZT-1027, *Acta Histochem. Cytochem.*, 33, 341, 2000.

109. Otani, M. et al., TZT-1027, an antimicrotubule agent, attacks tumor vasculature and induces tumor cell death, *Jpn. J. Cancer Res*, 91, 837, 2000.

110. Bai, R. et al., Dolastatin 11, a marine depsipeptide, arrests cells at cytokinesis and induces hyperpolymerization of purified actin, *Mol. Pharmacol.*, 59, 462, 2001.

111. Boyd, M. R. and Paull, K., Some practical considerations and applications of the National-Cancer-Institute *in-vitro* anticancer drug discovery screen, *Drug Dev. Res.*, 34, 91, 1995.

112. Pettit, G. R. et al., The dolastatins. 21. Synthesis, x-ray crystal-structure, and molecular modeling of (6r)-isodolastatin-10, *J. Org. Chem.*, 59, 6127, 1994.

113. Pettit, G. R. et al., Antineoplastic agents. 337. Synthesis of dolastatin 10 structural modifications, *Anti-Cancer Drug Design*, 10, 529, 1995.

114. Pettit, G. R. et al., Antineoplastic agents 365. Dolastatin 10 SAR probes, *Anti-Cancer Drug Design*, 13, 243, 1998.

115. Alattia, T. et al., Conformational study of dolastatin-10, *Tetrahedron*, 51, 2593, 1995.

116. Poncet, J. et al., Synthesis and antiproliferative activity of a cyclic analog of dolastatin 10, *Bioorg. Med. Chem.*, 8, 2855, 1998.

117. Poncet, J. et al., Synthesis and biological activity of chimeric structures derived from the cytotoxic natural compounds dolastatin 10 and dolastatin 15, *J. Med. Chem.*, 41, 1524, 1998.

118. Miyazaki, K. et al., Synthesis and antitumor-activity of novel dolastatin-10 analogs, *Chem. Pharm. Bull.*, 43, 1706, 1995.

119. Kobayashi, M. et al., Antitumor activity of TZT-1027, a novel dolastatin 10 derivative, *Jpn. J. Cancer Res.*, 88, 316, 1997.

120. Alcaro, S. et al., Theoretical comparison between structural and dynamical features of dolastatins 11 and 12 antineoplastic depsipeptides, *SAR QSAR Environ. Res.*, 14, 475, 2003.

121. Roux, F. et al., Synthesis and in-vitro cytotoxicity of diastereoisomerically modified dolastatin-15 analogs, *Bioorg. Med. Chem. Lett.*, 4, 1947, 1994.

122. Pettit, G. R. et al., Antineoplastic agents 360. Synthesis and cancer cell growth inhibitory studies of dolastatin 15 structural modifications, *Anti-Cancer Drug Design*, 13, 47, 1998.

123. Flahive, E. J., The dolastatins: synthesis and structural modification of dolastatin 15, Ph.D. Dissertation, Arizona State University, Tempe, AZ, 1996.

124. De Arruda, M. et al., Lu103793 (NSC D-669356): a synthetic peptide that interacts with microtubules and inhibits mitosis, *Cancer Res.*, 55, 3085, 1995.

125. Hu, M. K. and Huang, W.-S., Synthesis and cytostatic properties of structure simplified analogues of dolastatin 15, *J. Peptide Res.*, 54, 460, 1999.

126. Schmitt, J. et al., Synthesis of dolastatin 15 mimetic peptoids, *Bioorg. Med. Chem. Lett.*, 8, 385, 1998.

127. Mirsalis, J. C. et al., Toxicity of dolastatin 10 in mice, rats and dogs and its clinical relevance, *Cancer Chemother. Pharmacol.*, 44, 395, 1999.

128. Schwartsmann, G. et al., Marine-derived anticancer agents in clinical trials, *Exp. Opin. Investig. Drugs*, 12, 1367, 2003.

129. Madden, T. et al., Novel marine-derived anticancer agents: a phase I clinical, pharmacological and pharmacodynamic study of dolastatin 10, *Clinical Cancer Res.*, 6, 1293, 2000.

130. Garteiz, D. A. et al., Quantitation of dolastatin 10 using HPLC/electrospray ionization mass spectrometry: application in a phase I clinical trial, *Cancer Chemother. Pharmacol.*, 41, 299, 1998.

131. Vaishampayan, U. et al., Phase II study of dolastatin 10 in patients with hormone-refractory metastatic prostrate adenocarcinoma, *Clinical Cancer Res.*, 6, 4205, 2000.

132. Margolin, K. et al., Dolastatin 10 in metastatic melanoma: a phase II and pharmacokinetic trial of the California Cancer Consortium, *Investig. New Drugs*, 19, 335, 2001.

133. Aguayo, A. et al., Phase II study of dolastatin 10 administered intravenously every 21 days to patients with metastatic colorectal cancer, *Proc. Am. Soc. Clin. Oncol.*, 1125 (Abs), 2000.

134. Krug, L. M., Phase II study of dolastatin 10 in patients with non-small-cell lung carcinoma, *Ann. Oncol.*, 11, 227, 2000.

135. Hoffman, M. A., Blessing, J. A., and Lentz, S. S., A phase II trial of dolastatin 10 in recurrent platinum-sensitive ovarian carcinoma: A gynecologic oncology group study, *Gynecol. Oncol.*, 89, 95, 2003.

136. Pettit, G. R., Williams, M. D., and Srirangam, J. K., Preparation of dolastatin analog pentapeptide methyl esters as anticancer agents, Eur. Pat. Appl. EP 95-305130, 1995.

137. Bayés, M., Rabasseda, X., and Prous, J. R., Gateways to clinical trials, *Methods Find. Exp. Clin. Pharmacol.*, 25, 387, 2003.

138. Bayés, M., Rabasseda, X., and Prous, J. R., Gateways to clinical trials, *Methods Find. Exp. Clin. Pharmacol.*, 26, 53, 2004.

139. Verweij, J. et al., Phase I and pharmacologic study of TZT-1027, a novel dolastatin 10 derivative, administered in a day 1 and 8 iv schedule every 3 weeks in patients (pts) with advanced solid tumors, *Clin. Cancer Res.*, 16 (Suppl.), A255 (Abs.), 2003.

140. Horti, J., Juhasz, E., and Bodrogi, I., Preliminary results of a phase I trial of TZT-1027, an inhibitor of tubulin polymerization, in patients with advanced non-small cell lung carcinoma, *Proc. Am. Assoc. Cancer Res.*, 43, 2744 (Abs.), 2002.

141. Horti, J. et al., A phase I trial of TZT-1027, an inhibitor of tubulin polymerization, in patients with advanced non-small cell lung cancer (NSCLC), *Clin. Cancer Res.*, 16 (Suppl.), A256 (Abs.), 2003.

142. Supko, J. G. et al., A phase I clinical and pharmacokinetic study of the dolastatin analogue cemadotin administered as a 5-day continuous intravenous infusion, *Cancer Chemother. Pharmacol.*, 46, 319, 2000.

143. Mross, K. et al., Phase I clinical and pharmacokinetic study of lu103793 (cemadotin hydrochloride) as an intravenous bolus injection in patients with metastatic solid tumors, *Onkologie*, 19, 405, 1996.

144. Marks, R. S. et al., A phase II study of the dolastatin 15 analogue lu103793 in the treatment of advanced non-small cell lung cancer, *Am. J. Clin. Oncol.*, 26, 336, 2003.

145. Kerbrat, P. et al., Phase II study of LU103793 (dolastatin analogue) in patients with metastatic breast cancer, *Eur. J. Cancer*, 39, 317, 2003.

146. Barlozzari, T. and Haupt, A., Dolastatin-15 derivatives in combination with taxanes, U.S. Patent 6,103, 618, 15 AUG 2000.

12 Ecteinascidin 743 (ET-743; Yondelis™), Aplidin, and Kahalalide F

Rubén Henríquez, Glynn Faircloth, and Carmen Cuevas

CONTENTS

I. Introduction 215
II. Ecteinascidin 743 (ET-743; Yondelis) 216
   A. Background 216
   B. Mechanism of Action 217
   C. Preclinical Drug Development 218
      1. *In Vitro* and *In Vivo* Testing...218
      2. Preclinical Toxicology 218
   D. Chemical Synthesis...220
      1. Synthetic Routes to Ecteinascidin 743 (Yondelis)....................221
   E. Clinical Studies...221
      1. Yondelis (ET-743) in STS (Soft Tissue Sarcoma)....................221
      2. Yondelis in Ovarian Cancer...223
      3. Safety Profile 223
III. Aplidin...223
   A. Background 223
   B. Mechanism of Action 224
   C. Preclinical Drug Development 227
      1. *In Vitro* and *In Vivo* Testing...227
      2. Preclinical Toxicology 227
   D. Chemical Synthesis...229
   E. Clinical Studies...229
IV. Kahalalide F 230
   A. Background 230
   B. Mechanism of Action 231
   C. Drug Development...233
      1. *In Vitro* and *In Vivo* Testing...233
      2. Preclinical Toxicology 233
   D. Synthesis of Kahalalide F..234
   E. Clinical Trials 234
References 236

I. INTRODUCTION

The natural world, both terrestrial and marine, is a varied and bountiful source for new drugs. Yet until recently only natural products from terrestrial sources have been developed into, or have

FIGURE 12.1 (**See color insert following page 304.**) Structure of Ecteinascidin 743 (Yondelis™) and source organism, *Ecteinascidia turbinata*.

formed the basis for, modern medicines. The advent of more sophisticated and efficient means to collect marine organisms over the last few decades has allowed marine biologists, natural products chemists, and medicinal scientists from academia and the pharmaceutical industry to collaborate in unique and enterprising ways to sample the great variety and bounty of intertidal, shallow, and deep-water sea life in the world. Recently these efforts have resulted in the development of a first generation of drugs from the sea into clinical trials that includes the PharmaMar compounds, Yondelis™, Aplidin®, Kahalalide, and ES 285.

II. ECTEINASCIDIN 743 (ET-743; YONDELIS)

A. Background

Ecteinascidia turbinata is a tunicate species from the Caribbean and the Mediterranean that belongs to the class Ascidiacea within the subphylum Tunicata (also called Urochordata). Ascidians, or sea squirts, are small, bottom-dwelling soft-bodied marine animals that form colonies comprising many individuals, called zooids. The name tunicate is derived from their characteristic protective covering, or tunic, which functions to a certain extent as an external skeleton and consists of some cells, blood vessels, and a secretion of a variety of proteins and carbohydrates, including cellulose — an unusual finding in animals. Within the tunic is the muscular body wall, which controls the opening of the siphons used for feeding. Colonies are formed by asexual reproduction through budding; that is, outgrowths on the parent break off as new individuals. Sexual reproduction, in contrast, leads to a fertilized egg that develops into a free-swimming tadpole larva. The larval stage is brief and is used to find an appropriate place for the adult to live. In 1969, early assays with crude aqueous ethanol extracts of *Ecteinacidia turbinata* showed impressive *in vitro* cytotoxicity against a variety of cancer cell lines. Many unsuccessful attempts at isolation of the active compounds were followed by concerted efforts that led to the characterization of six alkaloids called ecteinas-cidins 729, 743, 745, 729A, 759B, and 770 by Kenneth Rinehart and his coworkers from the University of Illinois, Urbana-Champaign, of which ET 743 was the most abundant (0.0001% yield) (Figure 12.1).[1]

Ecteinascidin-743 (ET-743; molecular formula: $C_{39}H_{43}N_3O_{11}S$; relative molecular mass: 761.8) has the international nonproprietary name of trabectedin, is being developed under the trademark name of Yondelis, and the full chemical name of [6R-(6α, 6aβ, 7β, 13β, 14β, 16α, 20R*)]-5-(acetyloxy)-3,4,6,6a,7,13,14,16,-octahydro-6,8,14-trihydroxy-7,9-dimethoxy-4,10,23-trimethyl-spiro[6,16(epithiopropanoxymethano)-7,13-imino-12H-1,3-dioxolo[7,8]-isoquino[3,2-b][3]benza-zocine-20, 1 (2H)-isoquinolin]-19-one. It is a white to pale yellow amorphous powder, is hydrophobic, and is very soluble in methanol, ethanol, chloroform, ethyl acetate, and acetone.

The novel and unique chemical structure of ecteinascidins is formed by a monobridged pentacyclic skeleton composed of two fused tetrahydroisoquinoline rings linked to a 10-member lactone bridge through a benzylic sulfide linkage. Most ecteinascidins have an additional tetrahydroisoquinoline ring attached to the rest of the structure through a spiro ring. This is one of the features distinguishing these molecules from saframycins, safracins, and renieramycins,[2] compounds isolated from bacteria and sponges.

B. Mechanism of Action

Ecteinascidin 743 (ET743; trabectedin) is the first of a new class of DNA binding agents with a complex, transcription-targeted mechanism of action. Ecteinascidin 743 inhibits tumor cell proliferation via a promoter-specific mechanism and transcription-dependent nucleotide excision repair. Cell cycle studies on tumor cells *in vitro* reveal that ET-743 decreases the rate of progression of the cells through S phase toward G2 and causes a prolonged blockade in G2/M at biologically relevant concentrations (20–80 n*M*).[3,4] The cell cycle block is p53 independent and eventually leads to a strong apoptotic response. The major findings that shed light on the primary mechanisms of action are detailed below.

Ecteinascidin 743 binds to DNA through the minor groove and, through its C21 carbon, alkylates the N2 position of the guanine residue, preferentially flanked by guanine or cytosine on the 3 side and a pyrimidine on the 5 side.[5–7] Remarkably, the binding of ET 743 to DNA is reversible under nondenaturing conditions, and the rate of reversion is greatest at nonpreferred sequences, allowing it to "walk" from its less favored target sequences toward its most favored target sequences.[8]

Ecteinascidin 743 strongly inhibits the activation of the transcription of certain genes without affecting basal levels of transcription.[9,10] It appears to selectively target a set of cellular genes in a promoter-specific manner, of which notable examples are p21(WAF1/CIP1), a fundamental regulator of cell division that is very frequently found mutated in tumors; c-Fos and c-Jun, which are oncogenes functioning as transcription factors central to cell regulation; and MDR1, a gene critically involved in resistance to many chemotherapeutic agents.[11] The expression of genes targeted by ET-743 is normally under the control of transcription factor Sp1, and current studies implicate this drug as a transcription factor–specific, CCAAT box–dependent inhibitor of transcription activation.[12] ET-743 has also been shown to counteract the activation of the steroid and xenobiotic receptor (SXR).[13] The orphan nuclear SXR controls MDR1 and the liver enzyme P450 3A4, thereby providing a possible mechanism for modulating degradation and efflux of xenobiotic drugs in normal liver cells and cells lining the intestine, respectively. Finally, ET-743 also blocks transcription activation of reporter genes. It has been shown that ET-743 blocked Trichostatin A-induced activation of the Sp1-regulated p21 gene at concentrations that had a minimal effect on uninduced expression.[18]

Through a mechanism not yet fully understood, the transcription-dependent nucleotide excision repair system is also involved in mediating some of the actions of this drug.[14–17] Unexpectedly, mutations rendering defects in the cellular transcription–coupled DNA repair machinery were reported to result in resistance to ET-743, and reduced cytotoxic activity of Yondelis against cells that are deficient in the nucleotide repair machinery was observed, indicating that some component of this machinery is involved in mediating tumor cell killing by ET 743. A model has been proposed explaining the role of the TC-NER process in the production of abortive intermediates of an incomplete repair process, leading to cell death. Other anticancer drugs such as cisplatin, in clinical use, and Irofulven, in trials, exhibit the opposite behavior with regard to TC-NER condition. These compounds are more active against cells with a faulty TC-NER, as is expected for traditional DNA alkylating agents.

Taking all these data into account, it is very tempting to propose a mechanistic link between ET-743 and the transcription and repair protein complexes associated with RNA pol II while plowing through coding regions of activated genes. TC-NER or other transcriptional targets may be important in the drug's specific cellular toxicity profile.

C. PRECLINICAL DRUG DEVELOPMENT

1. *In Vitro* and *In Vivo* Testing

Preclinical data generated during the development of ET-743 have provided important insight for the selection and design of the clinical trials. Early *in vitro* studies carried out by PharmaMar and the National Cancer Institute (NCI) identified potent (1 pM to 10 nM) activity of ET-743 against cell line subpanels containing murine leukemias and human solid tumors, such as melanoma, breast, non–small cell lung (NSCL), and ovarian cancer. ET-743 was subsequently evaluated in a human tumor colony forming unit (TCFU) assay derived from surgically ablated primary tumor specimens.[18] Over a range of concentrations (1 nM to 10 μM), ET-743 activity was seen against cancers of the head and neck, ovarian, mesothelioma, breast, kidney, peritoneum, NSCL, melanoma, sarcomas, and primary tumor specimens. The NCI COMPARE analysis of ET-743 (NSC648766) and the Standard Anticancer Agent Database showed significant correlation coefficients (0.962–0.876) to that of the DNA intercalators such as the anthracyclines and the morpholino compounds (NCI data).

ET-743 has been tested in a great variety and number of models against tumors of murine origin (P388 leukemia and B16 melanoma) and human xenografts (melanoma, MEXF 989; NSLC, LXFL 529; breast, MX-1 early and advanced; and ovarian, HOC 22 and HOC 18) (Table 12.1).[19–22]

Drug-resistant lines have also been tested that include melanoma (MEXF 514), NSLC (LXFL 629), and ovarian (HOC 18) tumor lines. These first efforts showed that ET-743 has a broad spectrum of antineoplastic activity, with several tumor types showing selectivity; namely, melanoma, NSCL, and ovarian carcinomas. As a further example of strong activity and long-lasting antitumor effects, the action of ET-743 on human endometrial carcinoma xenografts (HEC-1-B) results in complete regression lasting for more than 125 days. These *in vivo* findings correlate strongly with the results from *in vitro* cross-resistance studies in which ET-743 was very active against many resistant tumors and only moderately cross-resistant to several of the standard agents.[18]

The determination that soft tissue sarcomas (STSs) are more sensitive to ET-743 than other solid tumors was not predicted during the preclinical development of the drug. The finding was serendipitous and came from the prevalence of responding or stable STS patients in the clinical trials. As a result, there has been a considerable effort to confirm this finding to supplement the extensive nonclinical profile already characterized for the antineoplastic effect of ET-743 against other solid tumors. In fact, a variety of sarcomas are differentially sensitive to ET-743, showing IC$_{50}$ potencies in the picomolar and subpicomolar range compared to the nanomolar concentrations established against non-STS solid tumors (Table 12.2).[23,24]

During its clinical development, several *in vivo* studies of ET-743 were performed in representative sarcoma tissues.[19–31] ET-743 shows a strong activity against several sarcoma types, especially chondrosarcomas, where tumor growth can be inhibited as much as 90% at relatively low doses (about 20% Maximum Tolerated Dose [MTD] in mice) and correlates with the *in vitro* findings.[24] In contrast, the *in vivo* effects of ET-743 against some soft-tissue sarcoma models, such as SW634 and UW2237 fibrosarcomas, appear to be less marked than in chondrosarcomas or osteosarcomas.[25]

2. Preclinical Toxicology

Early preclinical toxicology studies were independently carried out both by the NCI (Bethesda, MD)[32,33] and by PharmaMar in collaboration with the New Drug Development Office (Amsterdam). More specific nonclinical studies are currently being performed in collaboration with Johnson &

TABLE 12.1
In vivo Antitumor Activity of ET-743 in Murine and Human Tumors

Cancer Type (Cell Line)	Regimen	Optimal Dose (µg/kg per d)	Antitumor Effect	Tumor-free Survivors
Leukemia i.p. P388	i.p. qd×5	27	48% ILS	None
Melanoma				
i.p.B16	i.p. qd×9	13.4	78% ILS	1/10 (d60)
s.c. MEXF 989[a]	i.v. q4d×3	100	0.2% T/C	6/6 (d35)
	i.v. (qd×3) × 2	50	1.8% T/C	none
s.c. MEXF 514[a]	i.v. q4d × 3	100	37% T/C	none
DTIC resistant	i.v. (qd×3) × 2	50	76% T/C (inactive)	none
NSCL				
s.c. LXFL 529[a]	i.v. q4d×3	150	0.1% T/C	3/4 (d63)
	i.v. (qd×3)×2	50	0.1% T/C	2/4 (d40)
s.c. LXFA 629[b]	i.v. q4d×3	200	70% T/C (inactive)	none
Ifos resistant	i.v. qd×3	200	90% T/(inactive)	none
Ovarian				
s.c. HOC 22	i.v. q4d×3	200	<1.0% T/C	5/6 (d102)
	i.v. (qd×3) × 3	100	2.0% T/C	3/5 (d102)
s.c. HOC 18	i.v. q4d×3	200	5.8% T/C	4/8 (d100)
CDDP resistant	i.v. (qd×3) × 2		11.8% T/C	2/6 (d100)
Breast				
s.c. MX-1 early	i.p. qd×9	9	no activity	none
	i.p. qd×5	18	no activity	none
	i.v. qd×5	40	>160% growth delay	7/10
	i.v. qd×5	60	0.1% T/C	10/10 (d47)
	i.v. q4d×3	75	0.1% T/C	9/10 (d23)
s.c. MX-1 advanced[c]	i.v. qd×1 (d9)	250	203% growth delay	4/10 (d58)
	i.v. qd×5	50	224% growth delay	3/10 (d58)
	i.v. q3d×4	41.7	161% growth delay	3/10 (d58)
s.c. G101metastatic[d]	i.v. qd×1	250	50% (d70) delay	0/10 (d30)
Endometrial s.c. HEC-1-B[e]	i.v. q4d×3	200	< 1% T/C	3/6 (d125)
Prostate s.c. PC-3	i.v. q4d×3	200	30% T/C	0/6 (d30)
Soft Tissue Sarcomas				
Chondrosarcoma (CHSA)	i.v. qd×2	100	24% T/C	ND
Fibrosarcoma (UV2237)	i.v. qd×1	200	37% T/C	ND
Osteosarcoma (OSA-FH)	i.v. qd×5	40	33% T/C	ND
Rhabdomyosarc (TE-671)	i.v. qd×1	300	[toxic]	ND
Reticulocell of ovary (M5076)	i.v. q7d×2	150	36% T/C	ND

Note: d = study day, i.p. = intraperitoneal, ILS = % increase in life span, i.v. = intravenous, s.c. = subcutaneous, T/C = % of control versus treated tumor volumes.

[a] Fiebig, Freiburg, Germany.

[b] Langdon, Edinburgh, Scotland.

[c] Jacqueline Plowman, National Cancer Institute, Frederick, Maryland.

[d] Josephine Hurst, Goodwin Institute for Cancer Research, Plantation, Florida (unpublished data).

[e] Raffaella Giavazzi, Mario Negri Institute, Milan, Italy.

TABLE 12.2
Comparative *In Vitro* Activity of ET-743 in Human Cancer Cell Lines[23]

Tumor Type	Cell Lines	IC50 Potency
Malignant fibrous histiocytoma	HS-90M/M-8805/M-9005	< 0.1 pM
Fibrosarcoma	HT-1080	< 0.1 pM
Mesenchymal chondrosarcoma	HS-16	4 pM–100 pM
Liposarcoma	HS-18	4 pM–100 pM
Hemangiopericytoma	HS-30	4 pM–100 pM
Mesenchymoma	HS-42	4 pM–100 pM
Colon	HCT-8	10 nM
Colon	HT-29	3 nM
Colon	HCT-16	3 nM
Breast	MCF-7	20 nM

Johnson Pharmaceutical Research & Development in anticipation of drug registration in the United States and Europe. Acute toxicity studies in mice (CD1, MF1), rats (WISTAR, Fisher 344), dogs (beagles), and monkeys (cynomolgus) by intravenous (i.v.) administration, either as a single-bolus dose or using an intermittent five times daily (d×5) schedule, provided the safe starting dose for phase I clinical trials as 50 µg/m^2 for a single dose and 10 µg/m^2 per day for fractionated dose (d×5) schedules. Furthermore, chronic toxicity studies (3 cycles) using several infusion schedules in rats (Sprague–Dawley) and monkeys (Cynomolgus) have been performed as a part of the integrated development plan.

ET-743 toxicity, regardless of the animal species tested, is characterized by a common pattern that includes: hepatotoxicity; bone marrow toxicity, both *in vitro*[32,33] and *in vivo*; and damage at the injection site. Qualitative and quantitative differences of this common pattern can be noted depending on the animal species tested.

In rats, gender differences were detectable (although not in other species), with female animals displaying more severe toxicity. Gender differences in rats have also been detected in metabolic behavior.[42] Liver toxicity in rats was characterized by proliferative/inflammatory and epithelial changes in the bile ducts. Cholangitis is mainly found in rats, but not in humans. Bone marrow toxicity has also been detected in rats, and this toxicity shows a tendency to recover after 3 weeks.

In contrast to rats, monkeys express metabolic profile similarities to humans[42] and are thus considered to be a more relevant and predictive animal model of hepatoxicity than are rats, and particularly than female rats. The pattern of hepatotoxicity observed in monkey was also similar to that observed in patients (noncumulative, reversible, predominance of transaminase increase). Bone marrow toxicity was also detected, and it was characterized by myelosuppression (lymphoid depletion in the spleen and thymus, hypocellularity in the bone marrow), with recovery after 3 weeks.

D. CHEMICAL SYNTHESIS

Ecteinascidia turbinata has been successfully grown and harvested in aquaculture facilities located along the Mediterranean coast. The purification of the active ingredient is then accomplished on an industrial scale, using chromatographic procedures that represent a more practical and environmentally sound practice than harvesting the creature from the wild. Nevertheless, in recent years several synthetic schemes have been developed for industrial production of ET-743 in the quantities and quality required for a drug product (Yondelis) that will be used in clinical studies worldwide and manufacturing for commercialization.

1. Synthetic Routes to Ecteinascidin 743 (Yondelis)

To date, two distinct total synthetic routes to Yondelis (ET-743) have been reported. The seminal work of E. J. Corey and coworkers[34] provided for the first time a total synthesis of this complex molecule. This breakthrough scheme resolved one of the main roadblocks to the synthesis — the cyclization to obtain the 10-member lactone bridge through a benzylic sulfide linkage. More recently, Fukuyama et al.[35] were able to successfully direct previous work on synthetic routes to the saframycins to converge with the general approach established by Corey for the later stages of the synthesis.

The procedures outlined above represent some of the most outstanding work in recent organic chemistry. However, the long and involved procedures for total synthesis of the molecule represent a tremendous barrier to industrial manufacture of the drug, which is particularly challenging in the face of regulatory requirements for pharmaceuticals. This problem was finally solved with the development of a semisynthetic procedure[36,37] representing the first industrially feasible route to the manufacture of the drug on a large scale. This semisynthetic approach provides access not only to Yondelis (ET-743) but also to other members of the ecteinascidins and analogous compounds. The procedure uses safracin B, an antibiotic available through fermentation of the bacteria *Pseudomonas fluorescens,* as the starting point. Optimization of the fermentation process, followed by its transformation according to Scheme 12.1, provided a robust, easily scaled-up procedure for manufacturing the drug.

Cyanosafracin B was converted into **2** via a five-step sequence. The amino and phenol groups were protected as the BOC and MOM derivatives, respectively. The hydrolysis of the methoxy-p-quinone and subsequent reduction of the p-quinone afforded the unstable hydroquinone, which was treated with bromochloromethane to give the methylendioxy ring. Alkylation of the remaining phenol gave compound **2**. Removal of the BOC and MOM groups from **2** was followed by amide cleavage via an Edman degradation protocol providing **3** in 68% yield. Protection of the phenol allowed for the diazotization of the primary amine for conversion to alcohol **4**. Next, esterification with the protected cysteine and protection of the phenol with the MEMCl produces compound **5**. The synthesis of ET-743 was completed using the chemistry from Corey et al. on similar substrates. A five-step sequence was used to form **9**. Deprotection of the allyl group and oxidation of the phenol, followed by dehydration and cyclization under Swern conditions, gave the intermediate **7**. Removal of the MEM and BOC protecting groups was followed by ketone formation. Finally, Pictet-Spengler reaction and carbinolamine formation produces ET-743.

E. Clinical Studies

Yondelis (ET-743) is currently in the advanced stages of clinical development. Worldwide, more than 2000 patients have already been treated with this innovative drug. ET-743 is currently being studied in phase II clinical trials for ovarian, STS, breast, endometrial, prostate, NSCLC, and pediatric indications. Yondelis was granted Orphan Drug designation by the European Commission for the indications of STS and ovarian cancer.

1. Yondelis (ET-743) in STS (Soft Tissue Sarcoma)

STS is an aggressive type of cancer with unfavorable prognosis and limited treatment options. The current survival rate after 5 years of chemotherapy treatment in metastatic disease is 8%, highlighting the urgent need for improved therapeutic options.

Results of an exploratory, pooled analysis that assessed the efficacy of Yondelis and the tumor growth rate in patients with advanced STS participating in three pivotal phase II trials have been released.[38] These trials used 1500 µg/m^2 of Yondelis in a 24-h infusion regimen every 3 weeks. Of the 183 patients with progressive disease evaluated, 7.7% (n = 14) achieved objective response (>50% tumor shrinkage), a further 7.7% (n = 14) achieved minor response (25% to 50% tumor

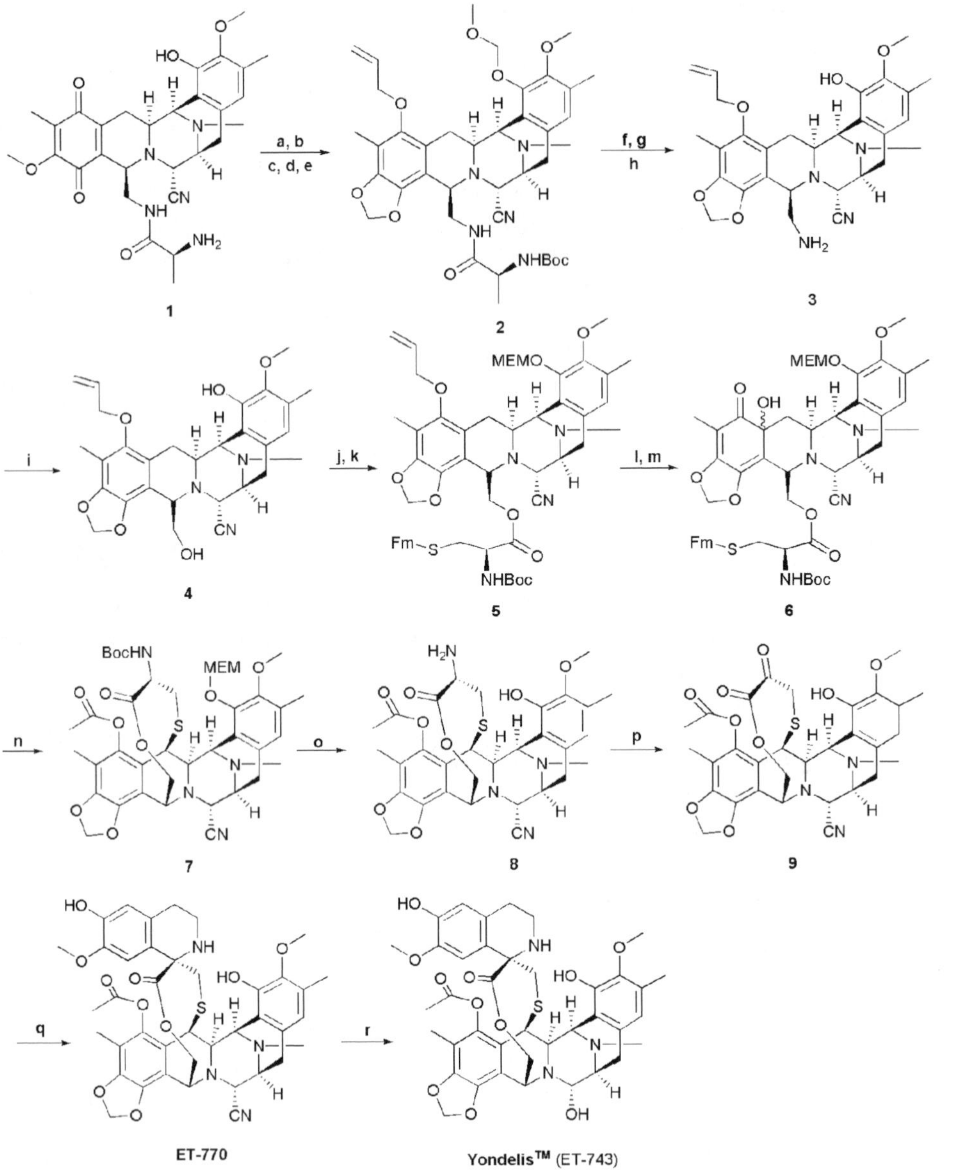

Reagents: (a) Boc₂O, EtOH, 23 °C, 23 h, 81%; (b) MOMBr, *i*-Pr₂NEt, DMAP, CH₃CN, 40 °C, 6 h, 83%; (c) NaOH 1M, MeOH, 20 °C, 2.5 h, 68%; (d) H₂, 10% Pd/C, 23 °C, 2 h; ClBrCH₂, Cs₂CO₃, DMF, 110 °C, 2.5 h; (e) Allyl bromide, Cs₂CO₃, DMF, 23 °C, 3 h, 56% for two steps; (f) TFA, CH₂Cl₂, 23 °C, 4 h, 95%; (g) Phenyl isothiocyanate, CH₂Cl₂, 23 °C, 3 h, 87%; (h) HCl/Dioxane 4.3M, 23 °C, 1 h, 82%; (i) NaNO₂, AcOH, THF, H₂O, 0 °C, 3 h, 50%; (j) (*S*)-N-[(trichloroethoxy)carbonyl]-S-(9-fluorenyl-methyl)cysteine, EDC·HCl, DMAP, CH₂Cl₂, 23 °C, 2 h, 95%; (k) MEMCl, *i*-Pr₂NEt, DMAP, CH₃CN, 23 °C, 6 h, 88%; (l) Bu₃SnH, (PPh₃)₂PdCl₂, AcOH, CH₂Cl₂, 23 °C, 15 min, 90%; (m) (PhSeO)₂O, CH₂Cl₂, -10 °C, 15 min, 91%; (n) DMSO, Tf₂O, CH₂Cl₂, -40 °C, 35 min; *i*-Pr₂NEt, 0 °C, 45 min; *t*-BuOH, 0 °C, 5 min; (CH₃)₂C=N-*t*-Bu, 23 °C, 40 min; Ac₂O, 23 °C, 1 h, 58%; (o) *p*-TsOH, CHCl₃, 23 °C, 1 h, 71%; (p) [*N*-methylpyridinium-4-carboxaldehyde]⁺I⁻, DBU, (CO₂H)₂, 23 °C, 4 h, 57%; (q) 3-Hydroxy-4-methoxyphenylethylamine, silica gel, EtOH, 23 °C, 12 h, 90%; (r) AgNO₃, CH₃CN, H₂O, 23 °C, 16 h, 90%.

SCHEME 12.1 Semisynthesis of ET-743 from Saframycin.

shrinkage), and 36.1% (n = 66) had stabilization of the disease with a median duration of 9 months. Therefore, in total, 51.4% (n = 94) of patients showed clinical benefit in terms of tumor growth control from treatment with Yondelis. The 6-month progression-free survival (PFS) rate, an endpoint that assesses the percentage of patients whose disease has not progressed by a defined time point, was 19.8%, which is favorable according to results of an European Organization for Research and Treatment of Cancer study that considers drugs with a 6-month PFS rate above 14% as efficacious, and those below 8% as inactive.[39] Following Yondelis treatment, 47.5% and 29.3% of patients with advanced, metastatic disease were alive after 1 and 2 years, respectively. Median overall survival time was 10.3 months, with 29% of patients surviving at 2 years or beyond. Yondelis proved particularly effective in cases of advanced sarcoma that had relapsed or that were resistant to conventional therapy. Phase I and II trials showed that a shorter treatment schedule (3-h infusions at 1300 µg/m^2 every 3 weeks) was feasible.

2. Yondelis in Ovarian Cancer

Ovarian cancer is one of the deadliest gynecological cancers. Unfortunately, detection of ovarian cancer is difficult, and the disease is often diagnosed too late for successful treatment. The overall 5-year survival rate of patients with advanced disease (stage III and IV) is approximately 20%, thus, there is an urgent need for improved therapy. In a study of Yondelis administered using a 3-h i.v. infusion regimen in patients with ovarian cancer who had previously been treated with platinum-taxane combination therapy, 10 of 29 patients (34.5%) achieved an objective response (tumor shrinkage), with a recurrence of 6 months or longer after the platinum regimen.[40] Those with recurrence before 6 months had a confirmed objective response rate of 6.7% (2 of 30 patients).[40]

First results of a phase I study of a 3-h i.v. infusion of Yondelis administered in combination with cisplatin every 3 weeks in patients with a variety of cancers revealed no evidence of severe hematological and nonhematological toxicities, and the pharmacokinetic data demonstrated that Yondelis does not influence the plasma disposition of cisplatin, indicating that the combination regime should be feasible for clinical use.[41] This study of Yondelis in combination with cisplatin, as well as eight other studies of Yondelis in combination with doxorubicin, liposomal doxorubicin, docetaxel, carboplatin, paclitaxel, capecitabine, gemcitabine and oxaliplatin, are currently ongoing.

Yondelis has also been tested in advanced breast cancer in patients failing to respond to standard chemotherapy (anthracyclines and taxanes). This heavily pretreated population has a very poor prognosis, with few responses to available agents. Yondelis has been tested in this setting with a response rate of 13.6% (3 out of 22 patients, 9 patients had stable disease and 8 had a greater than 50% decrease of tumor marker CA 15.3 levels in blood).[103]

3. Safety Profile

In all clinical studies, Yondelis has shown a good safety and tolerability profile. In STS patients treated with Yondelis, no mucositis, alopecia, neurotoxicity, cardiotoxicity, or cumulative toxicities have been observed. The most frequent adverse event appears to be neutropenia, which is reversible. Transaminase elevations were also reported but were transient.

III. APLIDIN

A. Background

Aplidin (dehydrodidemnin B) is a potent depsipeptide that was isolated from a Mediterranean tunicate, *Aplidium albicans*.[43] This compound is structurally related to the Didemnins, agents that were previously isolated from the Caribbean marine tunicate *Trididemnum solidum*.[44,45] The structures of the didemnins were reported to contain a 23-membered macrocycle with an attached side chain. The macrocycle is made up of six subunits, (S)-Leu, (S)-Pro, (1S, 2R)-Thr, (S)-N(Me)-O(Me)-Tyr,

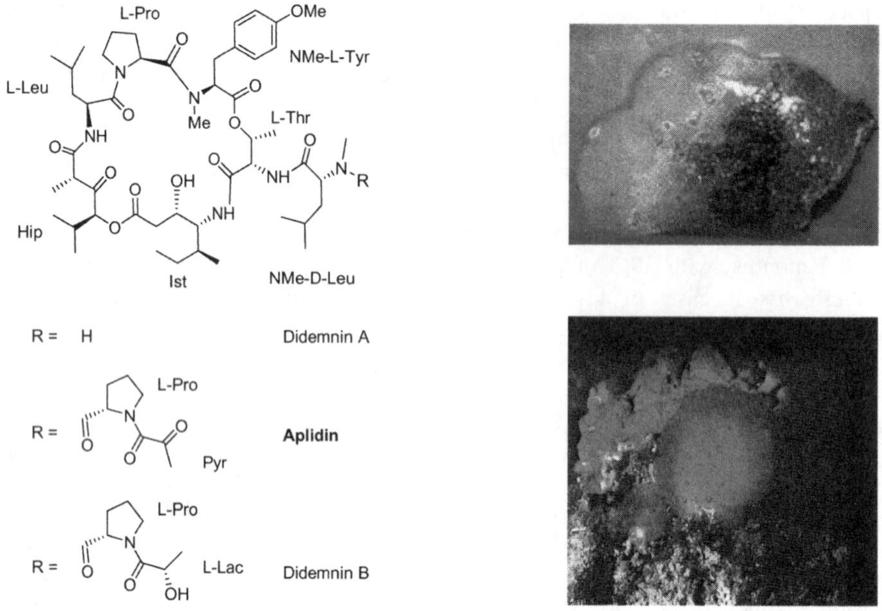

FIGURE 12.2 (See color insert following page 304.) Structure and source organism of Aplidin.

(3S,4R,5S)-isostatin (Ist), and (2S,4S)-3-oxo-4-hydroxy-2,5-dimethylhexanoic acid, also known as α-(α-hydroxyisovaleryl)propionic acid. The side chain, whose first amino acid is always (R)-N(Me)-Leu, is joined to the Thr of the macrocycle, and its structure differentiates the various didemnins. Aplidin contains pyruvil-L-Pro as the side chain (Figure 12.2).

Aplidin (molecular formula: $C_{57}H_{87}N_7O_{15}$; relative molecular mass: 1110.3) has the registered name Aplidin®, the common name dehydrodidemnin B, and the chemical name (2S)-N-{(1R)-1-[({(3S,6R,7S,10R,11S,15S,17S,20S,25aS)-11-hydroxy-20-isobutyl-15-isopropyl-3-(4-methoxyben-zyl)-2,6,17-trimethyl-10-[(1S)-1-methylpropyl]-1,4,8,13,16,18,21-heptaoxodocosahydro-15H-pyr-rolo[2,1-f][1,15,4,7,10,20]dioxatetraazacyclotricosin-7-yl}amino)carbonyl]-3-methylbutyl}-N-methyl-1-pyruvoyl-2-pyrrolidinecarboxamide. Several investigators have reported their results of various biological activities for a wide variety of didemnins.[46–48] Among them Aplidin exhibited enhanced cytotoxic activity *in vitro* against several solid tumor and leukemic cell lines with IC_{50} values that were about 10-fold more potent than those for didemnin B. This enhanced level of *in vitro* potency focused attention on further use of the Aplidin congener. The compound was then chosen for development as an antineoplastic agent because of its antitumor activity, and in early 1999, Aplidin entered phase I clinical trials in Europe and Canada.

B. MECHANISM OF ACTION

The primary mechanism of action of Aplidin is still unknown and is the subject of active investiga-tion.[49–62] Aplidin induces a G_1 block or delay in the S phase and a G_2 blockade in nonsynchronized cells.[49] New cells are blocked in G_1 or become apoptotic. S phase–synchronized cells treated with Aplidin proceed to G_1 and then either die (apoptosis) or stop growing. This effect is clear from 1–100 n*M* at between 1 and 24 h of exposure. Proportionately stronger apoptosis occurs with higher concen-trations or length of exposure to Aplidin. An NCI COMPARE analysis between Aplidin (NSC:638719) and the Standard Anticancer Agent Database showed moderate (0.7–0.8) but statistically significant correlation coefficients to pancratistatin, bactobolin, Didemnin B, bisantrene HCl, and anguidine.

Early studies on the effects of Didemnins identified at least three effects on cellular biochemistry. First, inhibition of protein synthesis by modulating the GTP-dependent elongation factor 1-α.[50]

Second, inhibition of palmitoyl protein thioesterase, which modulates G-protein-dependent signal transduction pathways for cellular proliferation.[51] Aplidin inhibits palmitoyl protein thioesterase in an uncompetitive manner through a selective complex formation with substrate-bound palmitoyl protein thioesterase similar to Didemnins A and B and with similar inhibition IC_{50} values. Third, Aplidin reduces cellular ornithine decarboxylase activity. However, the effect is not directly on enzyme activity but some degree of inhibition of the level of transcriptional expression of the gene, as observed in the DU145 human prostate carcinoma cell line.[52,53] In these experiments, one course of Aplidin significantly (twofold or more) down-regulated the expression of 81 (<1%) genes encoding proteins involved in lipid metabolism, receptors and other membrane proteins, and DCMTase. The latter finding may be relevant because DCMTase is strongly implicated in carcinogenesis, and a link between DCMTase and ODC has been postulated.[54] Twelve genes were up-regulated by a single course of Aplidin. Two courses of Aplidin significantly down-regulated 34 genes, including the angiotensin receptor, while transglutaminase, a marker of cellular differentiation, was significantly up-regulated. In a separate microchip array assay using MOLT-4 (p53+) leukemic cells exposed for 1 h to the drug, Aplidin decreased cyclin E levels by 24 h coincident to a G1 block of the cell cycle. There is no effect on the expression of CDK2, CDC2, CDK4, CYCB, CYCD1, p21, or p53.[55]

Several new mechanistic proposals for Aplidin are emerging to explain its strong cytotoxicity and apoptogenic activity to human leukemic and solid tumor cell lines.[57–62] In human T-cell leukemia (MOLT-4) cells, Aplidin inhibits the secretion of vascular endothelial growth factor (VEGF) by an unexplained mechanism in which liberation to the extracellular media rather than VEGF synthesis is blocked[58] (Figure 12.3).

At the same time, Aplidin downregulates the VEGF receptor (flt-1), thereby blocking an essential autocrine loop for cell proliferation. Ectopic expression or addition to the culture media of VEGF partially palliates cytotoxicity of Aplidin in this system, further supporting the relevance of the VEGF autocrine loop. Aplidin also seems to exert a direct antiangiogenic activity (unpublished data). *In vitro*, Aplidin inhibited endothelial cell proliferation. *In vivo*, Aplidin dose-dependently inhibited angiogenesis in the chick embryo chorioallantoic membrane assay (up to 80% inhibition at 10 n*M*). Moreover, Aplidin was highly active in the process of angiogenesis induced in the chick embryo chorioallantoic membrane by VEGF. These data indicate that, in addition to its direct antitumor activity, Aplidin also affects tumor angiogenesis by preventing the production of angiogenic factors by tumor cells and by directly inhibiting the angiogenic process. In fact, in studies on solid tumor cell types Aplidin was found to significantly reduce VEGF production in the human ovarian tumor cell line (IGROV-1) and inhibited the intraperitoneal growth of human ovarian carcinoma cells xenografted into nude mice (unpublished data).

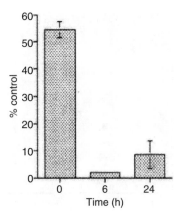

FIGURE 12.3 Vascular endothelial growth factor concentration in the medium of ALL-PO cells treated for 1 h with 20 n*M* Aplidin and evaluated at 0.6 and 24 h after drug washout. The values express the percentage of VEGF concentration in the medium of treated cells with regard to control cells.[57]

In promyelocytic leukemia (HL-60 and primary) and breast cancer (MCF-7) cells, Aplidin cytotoxicity occurs by a selective apoptotic action along the signal pathway mediated through Fas/CD95 (reversed by use of an antagonistic anti-Fas antibody) and early mitochondrial release of cytochrome c. Overexpressed bcl-2 blocked the onset of these actions by Aplidin,[61] and in breast (MDA-MB-231) and renal (A-498 and ACHN) cancer cells, Aplidin induces apoptosis via glutathione depletion and sustained activation of the EGFR, Src, JNK, and p38 MAP kinases along the stress response pathway.[62,63] Apoptosis was not induced by Aplidin in breast cancer cells (MCF-7) that do not express caspase-3, known for its proapoptotic action. However, Aplidin did induce apoptosis in cell lines normally expressing caspase-3, and its action was inhibited when caspase-3 activity was blocked.

Identification of new antileukemic agents is essential for improving the survival of acute lymphoblastic leukemia (ALL) patients with high-risk or refractory leukemia. Aplidin shows potent *in vitro* and *in vivo* activity against both human hematological and solid tumor cells.[46,47,58,59] The antileukemia activity of Aplidin was investigated on a number of continuously growing ALL cell lines, ALL cell lines resistant to standard therapy, and leukemia cells freshly isolated from bone marrow of children with ALL at diagnosis (nine cases) or at the time of relapse (five cases).[58] A dose-dependent growth inhibitory effect was observed at nanomolar concentrations in all ALL cell lines (IC_{50} ranging from 5 to 20 nM). Aplidin acts by triggering apoptosis in the ALL cell lines as assessed by TdT-dUTP flow cytometric analysis. Taken together, these data indicate that Aplidin is a potent antileukemic drug to be evaluated *in vitro* in refractory and relapsing ALL patients. Analyzed within ALL only, Aplidin did not show a significant cross-resistance to any of eight cytotoxic drugs commonly used in leukemia.

In a separate line of study, investigators have been measuring the *in vitro* sensitivity of childhood leukemias to Aplidin.[59] The dose-limiting toxicity of Aplidin in phase I/II trials has been muscular toxicity, with a remarkable lack of severe myelosuppression. Studies of the cytotoxicity of Aplidin on bone marrow and peripheral blood lymphocytes samples of healthy children and on samples from children with different types of leukemia have been performed to assess the potential use of Aplidin as a cytotoxic agent in pediatric leukemia. When the sensitivity to Aplidin of a number of leukemic cells from patients suffering from acute lymphoblastic or myelogenous leukemia was compared with that of normal bone marrow and peripheral blood lymphocytes from healthy children, a significant difference in the level of sensitivity was observed — an indication of the potential for this drug in the clinical treatment of these severe hematological diseases (Figure 12.4).

In the context of hematological tumors, Aplidin has also been reported to have potent *in vitro* activity in multiple myeloma tumor cell lines.[60] In these studies, Aplidin is active at pharmacologically

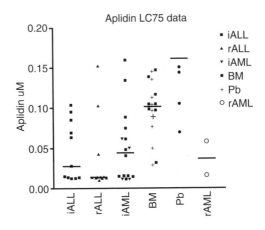

FIGURE 12.4 Aplidin LC_{75} values and median in different subgroups of leukemia and normal bone marrow and peripheral blood (PB) samples. [59]

relevant concentrations (<10 nM) against most of the cell lines in a multiple myeloma panel that includes cells resistant to conventional or novel agents used in multiple myeloma therapies.

Taken together, these separate lines of investigation indicate the great potential relevance to the ongoing clinical studies, particularly to determine the effects of Aplidin on VEGF levels in phase II clinical trials.

C. PRECLINICAL DRUG DEVELOPMENT

1. *In Vitro* and *In Vivo* Testing

The preclinical data including *in vitro*, *in vivo*, and toxicology results were important for the selection and design of the phase I and II programs. The selection of tumor types and design of the studies have been directly related to the observation that Aplidin has consistent cytotoxic activity against human leukemia, multiple myeloma, and lymphoma cell lines, as well as selected human solid tumors.

In early drug development, Aplidin was shown to be broadly cytotoxic in the range of IC_{50} concentrations between 0.16 and 0.51 nM to murine L1210 lymphocytic leukemia cells (0.29 nM), murine P388 lymphoid neoplasm (0.16 nM), murine B16 melanoma (0.23 nM), and human HeLa cervix epithelioid carcinoma (0.51 nM). Moreover, there was noted a selectivity toward human A549 NSCLC (0.08 nM) and less selectivity, but adequate potency, to human KB oral epidermoid carcinoma (5.3 nM).[46–48,64,65] A further evaluation in the human *in vitro* screening panel of tumor cell lines at the NCI confirmed that Aplidin was strongly cytotoxic against a broad spectrum of tumor types with selectivity for melanomas, NSCL, prostate, ovarian, and colorectal cancer cell lines, with IC_{50}s in the range of 0.18 to 0.45 nM (NCI data). Corollary studies have extended the profile by evaluating *in vitro* selectivity against prostate tumors showing species-independent specificity for this tumor type: rat MATLyLu (3.40 nM, IC_{50}), human LNCaP (0.32 nM, IC_{50}), PC-3 (27.0 nM, IC_{50}), and DU 145 (10.23 nM, IC_{50}).[66]

Aplidin demonstrated significant schedule-dependent antitumor activity in a modified *in vitro* human tumor colony-forming assay, with potent activity noted against gastric carcinoma and lymphomas.[67,68] These studies indicate that Aplidin is a phase-specific agent, a conclusion subsequently supported by data obtained with a human LoVo colon carcinoma cell line showing a cell cycle block at G1, G2/M.[69] Moreover, cell cycle perturbation studies on LoVo (0.1 pM, IC_{50}) and the adriamycin-resistant MDR overexpressing subline LoVo/DX (1 μM, IC_{50}) indicate that Aplidin is a Pgp substrate affected by the classical multidrug-resistant phenotype.

Aplidin has been tested in several tumors of murine origin, transplantable human tumor xenografts, and human cancer cell lines in hollow fibers.[45–47,64,65,70,71] Activities were seen against leukemias and solid tumors, with dose–response relationships evident in most models. The optimal modes of administration are i.v. infusion and the intraperitoneal route. From the earliest antitumor studies, significant increases in the life span were observed in mice implanted intraperitoneally with murine P388 leukemia and B16 melanoma, followed by intraperitoneal administration of the drug on days 1–9. Significant antitumor activity was also seen using the same dosing regimen against subcutaneously implanted Lewis lung tumors, with complete remissions recorded at the highest dose. The full spectrum of *in vivo* antitumor activity from the early studies of Aplidin is summarized below (Table 12.3).

During this time it was found that i.v. bolus activity was not a route to observe optimal antitumor activity (not published). However, follow-up studies demonstrated that i.v. infusions could restore the activity of the drug and, hence, form the basis for proceeding to clinical trials.

2. Preclinical Toxicology

Preliminary toxicology studies were carried out in mice (CD1) and rats (WISTAR). The most significant observations from acute i.v. toxicity studies of Aplidin in mice and rats are bone marrow and reticuloendothelial system toxicities when given either as a single injection on an intermittent (d×5) schedule. In both species, local injection site effects were also seen (tail lesions and discoloration). To varying

TABLE 12.3

In vivo **Antitumor Activity against Murine and Human Tumors**

Tumor	Type	Regimen	Opt. Dose	Antitumor Effect	Score
Leukemia	Intraperitoneal P388	Intraperitoneally, qd×5	MTD[a]	88% ILS	++++
Melanoma	Intraperitoneal B16	Intraperitoneally, qd×9	MTD[a]	98% ILS	++++
Ovary	Intraperitoneal M5076	Intraperitoneally, q4d×5	MTD[a]	48% ILS	+++
NSC Lung	Subcutaneous Lewis	Intraperitoneally, qd×9	MTD[a]	0% T/C	++++
Breast	Subcutaneous MX-1	Intraperitoneally, q4d×3	MTD[b]	37% T/C	+
Prostate	Subcutaneous PC-3	Intraperitoneally, q4d×3	MTD[a]	34% T/C	+
			1/2 MTD[a]	38% T/C	+
		Intraperitoneally, qd×9	MTD[c]	25% T/C	++
			1/2 MTD[c]	30% T/C	++
Burkitts	Subcutaneous P3HR1	Intraperitoneally, q4d×3	MTD[c]	32% T/C	+
			1/2 MTD[c]	48% T/C	+/-
		Subcutaneously, 7days	MTD[d]	36% T/C	+
Gastric	Subcutaneous MRIH254	Intraperitoneally, Q d×9	MTD[b]	19% T/C	+++
			1/2 MTD[b]	17% T/C	+++
		Intraperitoneally, Q4 d×3	MTD[c]	18% T/C	+++
			1/2 MTD[c]	38% T/C	+

[a] MTD = 2.10 mg/kg.
[b] MTD = 0.80 mg/kg.
[c] MTD = 1.25 mg/kg (drug product).
[d] 1.80 mg/kg, and ILS = increases in the life span.

degrees by species and schedule, there was evidence of leukopenia and gastrointestinal tract inflammation and transient signs of hepatotoxicity. The depletion of circulating lymphocytes is more persistent in rats. Effects on the stomach were seen in both species — crypt cell necrosis in the rat and hyperplasia of the squamous epithelium in mice. In all cases, recovery was apparent toward the end of the studies. A summary table of the early preclinical MTD levels is given below (Table 12.4). The MTD for formulated product in mice is 3300–3750 $\mu g/m^2$ and 3420–4200 $\mu g/m^2$ in rats.

The *in vitro* toxicity of Aplidin on cultured human bone marrow hematopoietic progenitors was evaluated comparatively to doxorubicin on each of three pre–stem cell lineages: myeloid, erythroid, and megakaryocytic cells.[72] Regardless of the origin of the hematopoietic progenitors, the toxicity of Aplidin was lower than that observed in tumor cell lines (IC_{50}: 0.0002–0.027). Significantly, plasma concentrations in humans receiving maximum tolerated doses in phase I clinical trials were below the IC_{50} for hematopoietic progenitors (maximum 0.0034 μM), and no myelotoxicity has been observed to date, indicating that Aplidin is less toxic than doxorubicin in all hematopoietic progenitors studies.

TABLE 12.4

MTD Determinations in Mice and Rats for Aplidin

Regimen	Species	MTD Weight ($\mu g/kg$)	MTD Area ($\mu g/m^2$)	Safe Clinical Starting Dose ($\mu g/m^2$)[a]
Intravenous, d×1	Mouse	1250	3750	342–375
	Rat	570	3420	350

[a] 1/10 of mean MTDs ($\mu g/m^2$).

D. Chemical Synthesis

Aplidin belongs to the didemnin family,[73] and as a result of the biological activity of these compounds, several research laboratories have made them synthetic targets. Rinehart et al. reported the first total synthesis of the didemnins in 1987. Subsequently, different research groups published several syntheses, differing mainly in the selection of the two amino acids to achieve macrocyclization (Figure 12.5).

To date, the highest-yielding cyclization was reported by Schmidt et al.[74–76] and by Jou et al.[77] The first group synthesized didemnin A, B, and C, and Jou et al. described a process to obtain Didemnin A and Aplidin (dehydrodidemnin B); the uniqueness of this synthesis is the use of new coupling reagents based on uranium and phosphonium salts. With this process, it is now possible to prepare a large number of analogs.

The synthesis published by Jou et al. involved the synthesis of the Boc-protected didemnin macrocycle, followed by its coupling with Boc-(R)-N(Me)-Leu-OH to give Boc-didemnin A, which was coupled with Pyr-Pro-OH to afford Aplidin (Scheme 12.2).

E. Clinical Studies

Phase I trials with Aplidin focused on the evaluation of its safety and pharmacokinetic profile and on the optimization of the infusion schedule.[78–82] Results of phase I clinical trials presented at the 2001 European Conference on Clinical Oncology showed promising antitumor activity of Aplidin against pretreated colorectal and renal carcinoma, melanoma, and different neuroendocrine tumors (specifically in medullary thyroid carcinoma and carcinoid), using doses below the MTD.[83] Updated results on a pooled analysis on medullary thyroid carcinoma have been presented.[84] In further phase I studies, Aplidin has also demonstrated activity against bronchial carcinoid and NSCLC and has been shown to cause tumor shrinkage and improve symptoms. From these phase I studies, an i.v. infusion of 5 mg/m^2 Aplidin every 2 weeks has emerged as the preferred dosage schedule, and this schedule is being used in most of the ongoing phase II studies.[85]

Aplidin is currently in phase II trials for melanoma, pancreatic cancer, head and neck tumors, non-Hodgkin lymphoma, NSCLC, and SCLC. Phase II trials in gastric, bladder, and prostate tumors,

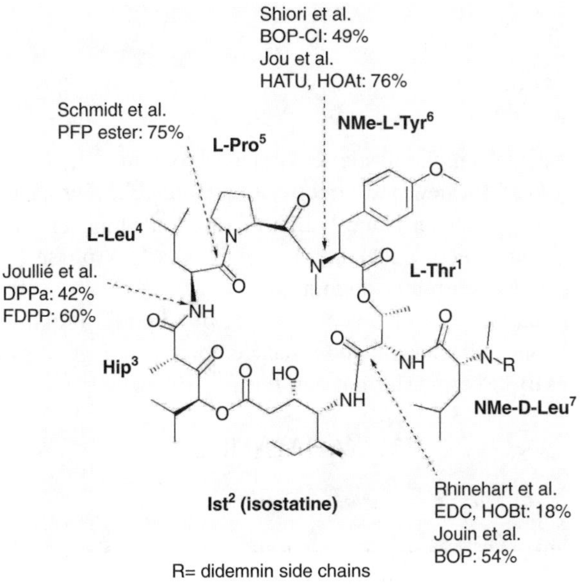

FIGURE 12.5 Amino acids to achieve the cyclization of Aplidin.

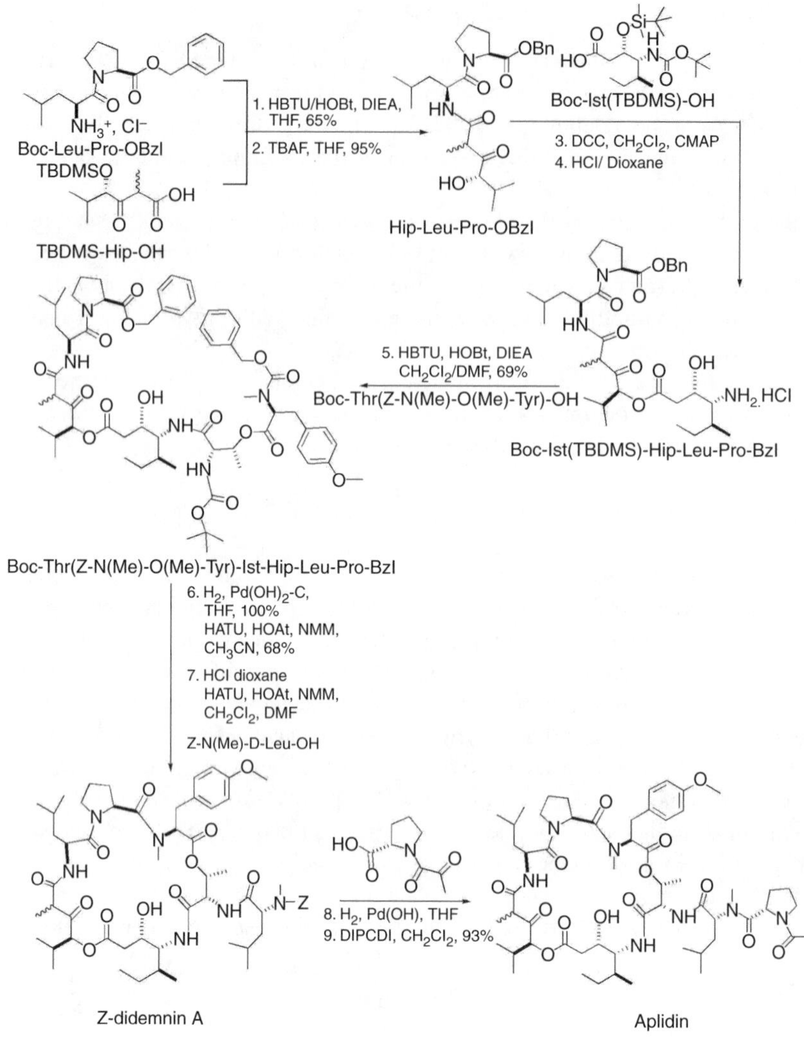

SCHEME 12.2 Synthesis of Aplidin.

multiple myeloma, and acute lymphoblastic leukemia are ongoing. In 2003, based on preclinical evidence and the urgent need for new and effective treatments, the European Commission granted Aplidin Orphan Drug designation for acute lymphoblastic leukemia. This type of leukemia is the leading cause of death from cancer under the age of 35 years. A phase I trial in pediatric solid tumors and acute leukemias is currently ongoing.

To date, more than 325 patients have been treated with Aplidin in nine clinical studies across seven trial centers in Europe and Canada, and it has demonstrated a good safety profile and broad-spectrum antitumor activity (PharmaMar data on file).

IV. KAHALALIDE F

A. BACKGROUND

Kahalalide F is a partially cyclic depsipeptide with the unusual feature of having a short chain fatty acid amide at its amino terminal residue (Figure 12.6).[86,87] A number of other natural Kahalalides have also been described.[88] This compound, together with a number of other natural Kahalalides, was

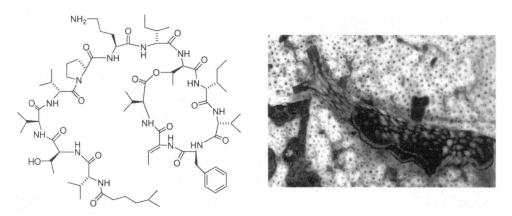

FIGURE 12.6 (See color insert following page 304.) Structure and source organism of Kahalalide F.

discovered in the mollusc *Elysia rufescens* (a sea slug; a marine gastropod of the ophistobranchia subclass) collected at Kahala (Muanalua Bay) in O'ahu (Hawai'i). Several species of animals of this class are known to acquire, process, and accumulate chemicals produced by the algae on which they feed. Surprisingly, some species are able to retain chloroplasts from these algae that remain active in photosynthesis within the animal for prolonged periods of time.[89] In the case of Kahalalide F, the compound is found in the algae *(Bryopsis pennata)* on which the *Elysia* molluscs feed, albeit in much reduced concentration.[90] In this natural setting a biological role has been proposed for Kahalalides as a deterrent to the feeding behavior of predators of the *Elysia* nudibranchs.

Kahalalide F (molecular formula: $C_{75}H_{124}N_{14}O_{16}$; relative molecular mass: 1477.8) has the chemical name, (2R)-N-{(1S)4-amino-1-[({(1R,2R)-1-[({(3S,6Z,9S,12R,15R,18R,19S)-9-benzyl-6-eth-ylidene-3,12-diisopropyl-19-methyl-15-[(1R)-1-methylpropyl]-2,5,8,11,14,17-hexaoxo-1-oxa-4,7,10,13,16-pentaazacyclononadecan-18-yl}amino)carbonyl]-2-methylbutyl}amino) carbonyl] butyl}-1-{(2R,5S,8S,11R)-8-[(1R)-1-hydroxyethyl]-2,5,11-triisopropyl-17-methyl-4,7,10,13-tetraoxo-3,6,9,12-tetraazaoctadec-1-anoyl}-2-pyrrolidinecarboxamide.

B. MECHANISM OF ACTION

The primary mechanism of action of Kahalalide F has not been identified; however, early experimental results have provided some insight into the physiological events that correlate with tumor cell killing by this compound. No significant correlations to any other standard chemotherapeutic agents were discovered when the actions of Kahalalide F on the NCI tumor cell line panel were analyzed through the COMPARE algorithm, indicating that the compound may possess a unique way to achieve its biological effectiveness. In these *in vitro* studies, selectivity was noted to colon, central nervous system, melanoma, prostate, and breast tumor cell lines, where PC-3 and DU-145 prostate tumor cells are the most sensitive at LC_{50} concentrations below 1 μM (NCI data).

Neither protein nor nucleic acid synthesis are found to be inhibited in cultured cells by sublethal concentrations of Kahalalide F. Topoisomerase enzymatic activities (I or II) are also not affected, and no damage to DNA has been specifically correlated with exposure to Kahalalide F. A cell cycle block in G_0/G_1 has been identified in a variety of tumor cell lines that include prostate (DU145), cervical (HeLa), colon (HT29), head and neck (HN30), and NSCL (HOP62), all with IC_{50}s in the 1-μM range.[91]

Kahalalide F is strongly cytotoxic to both wild-type p53 and mutated p53 tumor cells in the NCI panel. A number of cell lines overexpressing MDR (PC-3 prostate, CACO-2 colon, UO-31 renal, and MCF7 breast), as well as cell lines resistant to topoisomerase II inhibitors, are sensitive to Kahalalide F, indicating that it acts independent of the respective resistance mechanisms. Cultured cells exposed to biologically relevant concentrations of Kahalalide F detach from their substrate and become markedly swollen, which is associated with the formation of large intracellular vacuoles.[92]

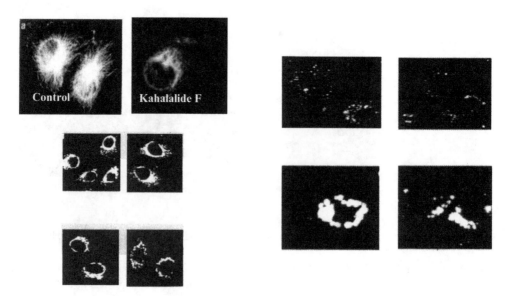

FIGURE 12.7 (See color insert following page 304.) Effect on the microtubule network and membrane-bound organelles.[92]

Within minutes, these engorged vesicles move from the periphery of the cell to a perinuclear location, as observed by confocal laser scanning microscopy. Confocal fluorescence microscopy studies have identified major and immediate effects on lysosomes (Figure 12.7). Moreover, there is an increase in lysosomal pH. The morphologies of the endoplasmic reticulum and the Golgi apparatus appear to be unaffected by the action of Kahalalide F. The effect on cell adhesion has not been explained by changes in organization of cytoskeletal structures, and in particular the microtubule network, which appear intact.

It has been suggested that the subcellular effect might be explained if Kahalalide F is inserted as an ionophore in membranes favoring an increase in cation permeability, thus causing a passive water influx and resulting in cisternal dilation. The hydrophobic nature of the compound would not be incompatible with a model in which membrane-associated events trigger cell death. These effects would be similar to those of compounds like the carboxylic ionophore, monensin.[93]

The compound has also been suggested to block the EGF receptor and inhibit TGF-β gene expression.[94] *In vitro* studies have shown that Kahalalide F is selectively cytotoxic to neu+ cells overexpressing Her2, indicating that it may interfere with erbB2 transmembrane tyrosine kinase activity. However, it does not inhibit autophosphorylation of the receptors or MEK kinase activity (unpublished data). More recently, Kahalalide F has been shown to cause rapid and potent cytotoxic effects in ErbB2 (HER2/neu) overexpressing breast cancer cell lines (SKBR3, BT474), which was associated with the induction of a hypodiploid cell population, dramatic cell swelling, and permeabilization of the plasma and lysosomal membranes; however, overexpression of HER2 does not affect the sensitivity of MCF7 breast cancer cells to Kahalalide F.[95] Several markers of caspase-dependent apoptosis were negative after Kahalalide F exposure, including the externalization of phosphatidyl serine, release of cytochrome c out of mitochondria, and cleavage of caspase-3 and PARP. Moreover, molecular or chemical inhibition of caspases by ectopic overexpression of Bcl-2 or a pan-caspase inhibitor (zVAD-fmk), respectively, failed to protect against Kahalalide F cytotoxicity. Specific inhibitors of cathepsin B (CA-074 Me, zFA-fmk) or D (pepstatin A) also failed to protect against Kahalalide F–induced cell death. Together, these results indicate that Kahalalide F–induced cytotoxicity is predominantly a result of a process of a necrotic cell death involving oncosis rather than apoptosis. This effect has also been reported in hepatoma cell lines.[96]

The sensitivity to Kahalalide F in a panel of human tumor cell lines derived from breast (SKBR3, BT474, MCF7), vulval (A431), NSCLC (H460, A549, SW1573, H292), and hepatic carcinoma

(SKHep1, HepG2, Hep3B), significantly correlated with protein expression levels of ErbB3 (HER3), but not other ErbB receptors. Downregulation of ErbB3 expression was observed in Kahalalide F–sensitive cell lines within 4 h exposure to Kahalalide F, as well as inhibition of the PI3K-Akt/PKB signaling pathway, which is directly linked to ErbB3.[97] Moreover, ectopic expression of a constitutively active Akt mutant had a protective effect against Kahalalide F cytotoxicity. This suggests that ErbB3 and the Akt pathway are major determinants of Kahalalide F action on these cell lines.

C. Drug Development

1. *In Vitro* and *In Vivo* Testing

Early preclinical data identified Kahalalide F as a potent new chemical entity showing significant cytotoxic activity below 10 μM (IC_{50}) against solid tumor cell lines. Further evaluation demonstrated that this activity is selective for, but not restricted to, prostate tumor cells. Subsequent studies have identified tumor cells that overexpress the Her2/neu and Her3/neu oncogenes as potentially sensitive targets for Kahalalide F. Moreover, Kahalalide F cytotoxicity is not schedule dependent (not published), and it is not a strong MDR substrate, as it is effective against many multidrug-resistant tumor cell lines (NCI data).

Preliminary *in vitro* screening studies identified micromolar activity of Kahalalide F against murine leukemia (P388) and two human solid tumors: NSCL (A549) and colon (HT29). Gastric (HS746T, 0.01 μM) and prostatic (PC-3, 0.08 μM) tumors were shown to be very sensitive as well. *In vitro* studies in cell lines of human origin evaluated by the NCI confirmed these results and identified selective activity against colon, NSCL, CNS, melanoma, prostate, and breast cancer cells, with potencies ranging from 200 nM (prostate) to 10 μM (leukemia). Extended *in vitro* selectivity studies reveals that Kahalalide F is active against neu+ (Her2 overexpressing) human breast tumor cells[95] and some primary sarcoma lines, but not against hormone-sensitive LNCAP tumor cells (Table 12.5).

In vivo antitumor activity was observed against breast, colon, prostate, and NSCL human tumor cells xenografted into athymic mice. Interestingly, DU-145 (hormone refractory prostate) tumors, resistant to most agents, responded to Kahalalide F at the MTD and 1/2 MTD levels. Similar studies were completed with PC-3 xenografted tumors, producing the same effect of tumor growth inhibition (between 31% and 48% T/C; unpublished data).

2. Preclinical Toxicology

The pharmacokinetic behavior of Kahalalide F was characterized in mice and confirmed in rats, in conjunction with *in vitro* and *in vivo* antitumor activity studies, to discern the pattern of systemic exposure to the drug associated with efficacious dosing regimens.[98–100]

TABLE 12.5
***In Vitro* Cytotoxic Activity of Kahalalide F against Solid Tumors**

Tumor	Line	IC_{50} (Molar)
Chondrosarcoma	CHSA	1.58 μM
Osteosarcoma	OSA-FH	1.65 μM
Prostate	PC-3	1.02 μM
Prostate	DU-145	1.78 μM
Prostate	LNCAP	Not active
Breast	SK-BR-3	2.50 μM
Breast	BT-474	2.00 μM

A dose of 278 µg/kg given as a rapid i.v. injection afforded an initial plasma concentration of 1.55 µM. The plasma concentration–time profile was distinctly biexponential, with half-lives of 15.8 min and 4.4 h for the initial and terminal disposition phases, respectively. The apparent volumes of distribution of the drug were very large, more than 100 times the body weight, indicating that the compound distributes extensively into peripheral tissues. The total body clearance, 14.5 mL/min per kilogram, was only 23% of hepatic blood flow. The C_{max} of the MTD i.v. dose is comparable to the *in vitro* IC_{50} values of the most sensitive human tumor cell lines.

No drug was found in mouse plasma 24 h after i.v. injection. There was no accumulation on repeated i.v. injection at an interval of 24 h. Surprisingly, although a slightly greater than MTD dose is too toxic when given as an i.v. single bolus, daily serial i.v. injections of the MTD dose are not toxic. These preliminary data indicate that Kahalalide F is rapidly eliminated from plasma with limited binding to extravascular tissues. This suggests tolerance to repeat dosing without acute cumulative adverse effects, which would be favorable to clinical development of the drug when confirmed in the human setting.

Conditions in which a number of tumor cell lines were exposed *in vitro* to Kahalalide F for 24 h exhibited IC_{50} values ranging from 0.01 to 2.51 µM. Bone marrow hemopoietic progenitors show no significant inhibition by Kahalalide F after 24 h of exposure at this range of concentrations; in fact, myeloid and erythroid bone marrow hemopoietic progenitors show no significant inhibition by Kahalalide F after 24 h of exposure at higher (10 µM) concentrations.[99] The results indicate sensitivity of tumor cells to doses of Kahalalide F that are not toxic to normal hematopoietic cells, indicating that neutropenias and thrombocytopenias as a consequence of Kahalalide F treatment would not be expected.

A dose of 250 µg/kg of a single bolus injection was established to be the maximal tolerated dose in rodents, although there are gender differences. The dose-limiting toxicity is predominantly renal toxicity, reversible by the day 29 necropsy, with minimal histologic evidence of nephrotoxicity at the 1/2 MTD. Signs of liver function alteration were seen on day 4 at the MTD dose, with recovery by day 29. If the maximum tolerated dose is exceeded there are signs of neurotoxicity, responsible for the lethalities. In intermittent dose studies in beagle dogs, slight nonregenerative anemia was seen in males and females at all dose levels. The effect was typically most severe on day 8, with general recovery by days 15 and 22. No other hematological changes were observed. No correlative histologic changes, for example, in the bone marrow, spleen, or liver, were seen. No other indices of toxicity were seen, and drug-related clinical signs or histologic lesions were not apparent.

D. Synthesis of Kahalalide F

Only a single synthetic approach has been published to date to produce Kahalalide F.[87] This procedure makes use of solid-phase chemistry to prepare a linear chain using the Fmoc/tBu strategy on 2-chlorotrityl chloride-resin, allowing the cleavage of the peptide under mild acid conditions. The amino acids, D-allo-Thr, and the Thr precursor of the Z-Dhb, were both introduced without protection of the hydroxyl function. HATU/DIEA was used to form the amide functional groups. With the linear peptide in hand, cyclization in solution phase and final deprotection allows the preparation of the natural compound in a straightforward way. Before detaching the peptide from the resin, the alloc group is removed under standard conditions. The cyclization reaction is performed with PyBOP/DIEA, using DMF as solvent, and finally the deprotection of the Boc group yields the natural compound. This solid-phase methodology is easily scaled up and applicable to the generation of a wide variety of new analogs (Scheme 12.3).

E. Clinical Trials

Results from a dose-finding phase I study in patients with advanced androgen-resistant prostate cancer were presented at the 2002 annual meeting of the American Society for Clinical

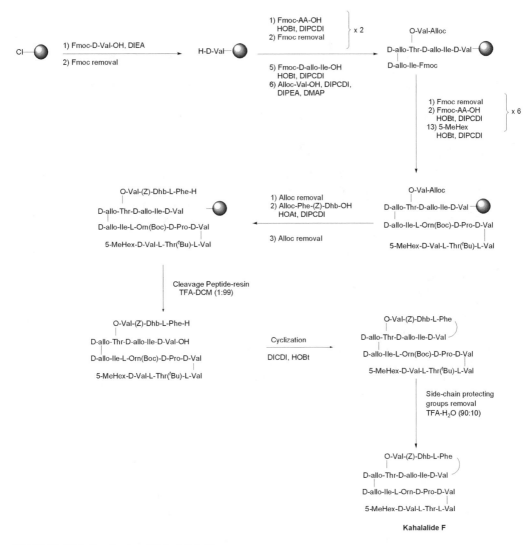

SCHEME 12.3 Synthesis of Kahalalide F.

Oncology.[101] In this study, Kahalalide F was found to have a rapid plasma clearance, which could be beneficial for serial dosing and could further enhance its efficacy in patients. In addition, the dose could be escalated up to 930 µg/m² per day. Kahalalide F also demonstrated a very favorable safety profile, and treatment-related side effects were noncumulative and rapidly reversible.

Data from a phase I study in patients with advanced solid tumors that had failed to respond to previous chemotherapies were presented at the 2002 joint meeting of the European Organization for Research and Treatment of Cancer, the National Cancer Institute, and the American Association for Cancer Research.[102] In this study, Kahalalide F was administered as weekly 1-h i.v. infusions, and the dose could be escalated up to 1200 µg/m² per week. Signs of activity in a variety of cancers were observed at 400–1200 µg/m² per week. The data support the favorable safety profile of Kahalalide F, with no bone marrow or renal toxicities, mucositis, alopecia, or cumulative toxicity reported.

REFERENCES

1. Rinehart, K.L., et al., Ecteinascidins 729, 743, 745, 759A, 759B, and 770: potent antitumor agents from the Caribbean tunicate *Ecteinascidia turbinata* [Erratum to document cited in CA113(9):75189d], *J. Org. Chem.*, 55, 4512–4515, 1990.
2. He, H., Faulkner, D.J., Renieramycins E and F from the sponge *Reniera sp.*: reassignment of the stereochemistry of the renieramycins, *J. Org. Chem.*, 54, 5822–5824, 1989.
3. Erba, E., et al., Mode of action of Ecteinascidin 743, a natural marine compound with antitumor activity, *Ann. Oncol.*, 9, 535, 1998.
4. Erba, E., et al., Ecteinascidin 743, a natural marine compound with a unique mechanism of action, *Eur. J. Cancer*, 37, 97–105, 2001.
5. Moore, B.M., et al., Mechanism for the catalytic activation of ecteinascidin 743 and its subsequent alkylation of guanine N2, *J. Am. Chem. Soc.*, 120, 2490–2491, 1998.
6. Seaman, F.C., Hurley, L.H., Molecular basis for the DNA sequence selectivity of Ecteinascidin 736 and 743: evidence for the dominant role of direct readout via hydrogen bonding, *J. Am. Chem. Soc.*, 120, 13028–13041, 1998.
7. Hurley, L.H., DNA and its associated processes as targets for cancer therapy, *Nat. Rev. Cancer*, 2, 188–200, 2002.
8. Zewail-Foote, M., Hurley, L.H., Differential rates of reversibility of Ecteinascidin 743-DNA covalent adducts from different sequences lead to migration to favoured bonding sites, *J. Am. Chem. Soc.*, 123, 6485–6495, 2001.
9. Sheng, K., et al., Ecteinascidin 743, a transcription-targeted chemotherapeutic that inhibits MDR1 activation, *Proc. N.Y. Acad. Sci.*, 97, 6775–6779, 2000.
10. Minuzzo, M., et al., Interference of transcriptional activation by the antineoplastic drug Ecteinascidin 743, *Proc. N.Y. Acad. Sci.*, 97, 6780–6784, 2000.
11. Jin, S., Hu, Z., Scotto, K., The antitumor agent Ecteinascidin 743 inhibits transcriptional activation of the MDR1 gene by multiple inducers, *Clin. Cancer Res.*, 5, 302, 1999.
12. Mantovani, R., et al., Effect on ET-743 on the interaction between transcription factors and DNA, *Ann. Oncol.*, 9, 534, 1998.
13. Synold, T.W., Dussault, I., Forman, B.M., The orphan nuclear receptor SXR coordinately regulates drug metabolism and efflux, *Nat. Med.*, 7, 584–590, 2001
14. Takebayashi, Y., et al., Antiproliferative activity of ecteinascidin 743 is dependent upon transcription-coupled nucleotide-excision repair, *Nat. Med.*, 7, 961–966, 2001.
15. Zewail-Foote, M., et al., The inefficiency of incisions of Ecteinascidin 743-DNA adducts by the UvrABC nuclease and the unique structural feature of the DNA adducts can be used to explain the repair-dependent toxicity of this antitumor agent. *Chem. Biol.*, 8, 1033–1049, 2001.
16. Bonfanti, M., et al., Effect of Ecteinascidin 743 on the interaction between DNA binding proteins and DNA, *Anticancer Drug Des.*, 14, 179–186, 1999.
17. Friedman, D., et al., Ecteinascidin 743 inhibits activated by not constitutive transcription. *Cancer Res.*, 62, 3377–3381, 2002.
18. Izbicka, E., et al., Incomplete cross-resistance between ET-743 and standard chemotherapeutic agents against primary tumors in human tumor cloning assay, *Ann. Oncol.*, 9, 130, 1998.
19. Valoti, G., et al., Ecteinascidin-743, a new marine natural product with potent antitumor activity on human ovarian carcinoma xenografts, *Clin. Cancer Res.*, 4, 1977–1983, 1998.
20. Hendriks, H.R., et al., High antitumor activity of ET-743 against human tumor xenografts from melanoma, non-small cell lung and ovarian cancer, *Ann. Oncol.*, 10, 1233–1240, 1999.
21. Riccardi, R., et al., Effective combinations of ET-743 and doxorubicin for tumor growth inhibitions against murine and human sarcomas in athymic mice, Proc. Am. Assoc. Cancer Res., 42, 1132, 2001.
22. D'Incalci, M., et al., The combination of Yondelis and cisplatin is synergistic against human tumor xenografts, *Eur. J. Cancer*, 39, 1920–1926, 2003.
23. Li, W.W., et al., Sensitivity of soft tissue sarcoma cell lines to chemotherapeutic agents: identification of ecteinascidin 743 as a potent cytotoxic agent, *Clin. Cancer Res.*, 7, 2908–2911, 2001.
24. Takahashi, N., et al., Sequence-dependent enhancement of cytotoxicity produced by ecteinascidin 743 with doxorubicin or paclitaxel in soft tissue sarcoma cells, *Clin. Cancer Res.*, 7, 3251–3257, 2001.

25. Faircloth, G.T., In vivo study using tumor containing hollow fibers in the nude mouse, PharmaMar USA report no. 02E01HF, January 15, 2001.

26. Hornicek, F., Weissbach, L., Invasiveness of chondrosarcoma cells in vitro is altered by ET743, *Proc. Am. Assoc. Cancer Res.*, 43, 539, 2002.

27. Faircloth, G.T., Grant, W., Hornicek, F., In vivo combinations of chemotherapeutic agents with ET743 against solid tumors, *Clin. Cancer Res.*, 7, 387, 2001.

28. Faircloth, G.T., et al., Dexamethasone potentiates the activity of Ecteinascidin 743 in preclinical melanoma and osteosarcoma models, *Proc. Am. Assoc. Cancer Res.*, 43, 73, #379, 2002.

29. Morioka, H., et al., Antiangiogenesis treatment combined with chemotherapy produces chondrosarcoma necrosis, *Clin. Cancer Res.,* 9, 1211–1217, 2003.

30. Morioka, H., et al., Antiangiogenesis treatment combined with chemotherapy produces chondrosarcoma necrosis, *Clin. Cancer Res.*, 8, 1211–1217, 2003.

31. Faircloth, G.T., Grant, W., Jimeno, J., Enhancing the preclinical in vivo antitumor activity of Ecteinascidin 743, a marine natural product currently in phase II clinical trials, *Clin. Cancer Res.*, 5, 306, 1999.

32. Ghielmini, M., et al., *In vitro* schedule dependancy of myelotoxicity and cytotoxicity of Ecteinascidin 743, *Ann. Oncol.*, 9, 989–993, 1998.

33. Albella, B., et al., *In vitro* toxicity of ET743 and Apldine, two marine-derived antineoplastics, on human bone marrow haematopoietic progenitors: comparison with the clinical results, *Eur. J. Cancer*, 38, 1395–1404, 2002.

34. Corey, E.J., Gin, D.Y., Kania, R.S., Enantioselective total synthesis of Ecteinascidin 743, *J. Am. Chem. Soc.*, 118, 9202, 1996.

35. Endo, A., et al., Total synthesis of Ecteinascidin 743, *J. Am. Chem. Soc.*, 124, 6552, 2002.

36. Cuevas, C., et al., Synthesis of Ecteinascidin ET-743 and Phthalascidin Pt-650 from Cyanosafracin B, *Org. Lett.*, 2, 2545, 2000.

37. Menchaca, R., et al., Synthesis of natural Ecteinascidins (ET-729, ET-745, ET-759B, ET-736, ET-637, ET-594) from Cyanosafracin B. *J. Org. Chem.*, 68, 8859, 2003.

38. Lopez-Martin, A.J., et al., An exploratory analysis of tumor growth rate (TGR) variations induced by trabectedin (ecteinascidin-743, ET-743) in patients (pts) with pretreated advanced soft tissue sarcoma (PASTS), *Proc. Am. Soc. Clin. Oncol.*, 22, 819, abst. 3293, 2003.

39. Van Glabbeke, M., et al., On behalf of the EORTC Soft Tissue and Bone Sarcoma Group. Progression-free rate as the principal end-point for phase II trials in soft-tissue sarcomas, *Eur. J. Cancer*, 38, 543–549, 2002.

40. Colombo, N., et al., Phase II and pharmacokinetics study of 3-hr infusion ET-743 in ovarian cancer patients failing platinum-taxanes, *Proc. Am. Soc. Clin. Oncol.*, 21, 221a, abst. 880, 2002.

41. Grasselli, G., et al., Phase I and pharmacokinetic (PK) study of ecteinascidin-743 (ET, Trabectedin) and cisplatin (P) combination in pre-treated patients (pts) with selected advanced solid tumors, *Proc. Am. Soc. Clin. Oncol.*, 22, 135, abst. 542, 2003.

42. Vermier, M., et al., An Interspecies Comparison of the Metabolism of the Anticancer Agent Yondelis™ (Trabectedin, ET-743). Cytochrome P450, Biochemistry, Biophysics and Drug Metabolism. 13th International Conference on Cytochromes P450, Prague, Czech Republic, June 29–July 3, 2003.

43. Rinehart, K.L., Lithgow-Berelloni, A.M., Novel Antiviral and Cytotoxic Agent, PCT Int. Pat. Appl. WO 91.04985, Apr. 18, 1991; GB Appl. 89/22,026, Sept. 29, 1989; Chem. Abstr., 1991, 115, 248086q.

44. Rinehart, K.L., et al., Structures of the didemnins, antiviral and cytotoxic depsipeptides from a caribbean tunicate, *J. Am. Chem. Soc.*, 103, 1857–1859, 1981.

45. Rinehart, K.L., et al., Didemnins and tunichlorin: novel natural products from the marine tunicate *Trididemnum solidum*, *J. Nat. Prod.*, 51, 1–21, 1988.

46. Urdiales, J.L., et al., Antiproliferative effect of dehydrodidemnin B (DDB), a depsipeptide isolated from Mediterranean tunicates, *Cancer Lett.*, 102, 31–37, 1996.

47. Faircloth, G.T., et al., Dehydrodidemnin B, a new marine derived antitumor agent with activity against experimental tumor models, *Ann. Oncol.*, 7, 34, 1996.

48. Sakai, R., et al., Structure activity relationships of didemnins, *J. Med. Chem.*, 39, 2819–2834, 1996.

49. Erba, E., et al., Mechanism of antileukemic activity of Aplidin, *Proc. Amer. Assoc. Cancer Res.*, 40, 3, 1999.

50. Crews, C.M., et al., GTP-dependent binding of the antiproliferative agent didemnin to elongation factor 1-alpha, *J. Biol. Chem.*, 269, 15411–15414, 1994.

51. Crews, C.M., Lane, W.S., Schreiber, S.L., Didemnin binds to the palmitoyl protein thioesterase responsible for infantile neuronal ceroid lipofuscinosis, *Proc. N.Y. Acad. Sci.*, 93, 4316–4319, 1996.

52. Erba, E., et al., Is Aplidin acting as an ornithine decarboxylase (ODC) inhibitor? AACR-EORTC-NCI Conference, 312, 1999.

53. Izbicka, E., et al., Evaluation of molecular targets for Aplidin, a novel anticancer agent, 11th NCI-EORTC-AACR Symposium on New Drugs in Cancer Therapy, 213, 2000.

54. Duranton, B., et al., Concomitant changes in polyamine pools and DNA methylation during growth inhibition of human colonic cancer cells, *Exp. Cell Res.*, 243, 319–325, 1998.

55. Erba, E., et al., Mechanism of antileukemic activity of Aplidin, *Proc. Amer. Assoc. Cancer Res.*, 40, 3, 1999.

56. Broggini, M., et al., Changes in gene expression in tumor cells exposed to the two marine compounds ET-743 and Aplidin by using cDNA microarrays, 10th NCI-EORTC-AACR Symposium on New Drugs in Cancer Therapy, 310, 1998.

57. Broggini, M., et al., Aplidin-induced apoptosis in Molt-4 cells is mediated by its ability to block VEGF secretion, 12th NCI-EORTC-AACR Symposium on New Drugs in Cancer Therapy, 384, 2002.

58. Erba, E., et al., Effect of Aplidin in acute lymphoblastic leukaemia cells, *Br. J. Cancer*, 89, 763–773, 2003.

59. Bresters, D., et al., In vitro cytotoxicity of Aplidin and crossresistance with other cytotoxic drugs in childhood leukemias and normal bone marrow and blood samples: a rational basis for clinical development, *Leukemia*, 17, 1338–1343, 2003.

60. Mitsiades, C.S., et al., Preclinical studies of the clinical development of Aplidin® for the treatment of multiple myeloma, American Society of Hematology, 45th Ann. Meeting and Exposition, abst. 250, 2003.

61. Gajate, C., An, F., Mollinedo, F., Rapid and selective apoptosis in human leukemia cells induced by Aplidin through a Fas/CD95- and mitochondrial-mediated mechanism, *Clin. Cancer Res.*, 9, 1535–1545, 2003.

62. Cuadrado, A., et al., Aplidin™ induces apoptosis in human cancer cells via glutathione depletion and sustained activation of the epidermal growth factor receptor, Src, JNK, and p38 MAPK, *J. Biol. Chem.*, 278, 241–250, 2003.

63. Garcia-Fernandez, L.F., et al., Aplidin™ induce the mitochondrial apoptotic pathway via oxidative stress-mediated JNK and p38 and protein kinase C, *Oncogene*, 21, 7533–7544, 2002.

64. Lobo, C., et al., Effect of Dehydrodidemnin B on human colon carcinoma cell lines, *Anticancer Research*, 17, 333–336, 1997.

65. Faircloth, G.T., et al., Schedule-dependency of Aplidin, a marine depsipeptide with antitumor activity, 10th NCI-EORTC-AACR Symposium on New Drugs in Cancer Therapy, 394, 1998.

66. Mastbergen, S.C., et al., Cytotoxicity and neurocytotoxicity of Aplidin, a new marine anticancer agent evaluated using in vitro assays, 10th NCI-EORTC-AACR Symposium on New Drugs in Cancer Therapy, 131, 1998.

67. Faircloth, G.T., et al., Preclinical characterization of Aplidin, a new marine anticancer depsipeptide, *Proc. Am. Assoc. Cancer Res.*, 38, 692, 1997.

68. Depenbrock, H., et al., In vitro characterization of Aplidin, a new marine derived anticancer compound, on freshly explanted clonogenic human tumor cells and haematopoietic precursor cells, *Br. J. Cancer*, 78, 739–744, 1998.

69. D'Incalci, M., et al., New drugs of marine origin, *Ann. Oncol.*, 7, 19, 1996.

70. Faircloth, G.T., et al., Preclinical development of Aplidin, a novel marine-derived agent with potent antitumor activity, 10th NCI-EORTC-AACR Symposium on New Drugs in Cancer Therapy, 129, 1998.

71. Faircloth, G.T., et al., Aplidin is a novel, marine derived depsipeptide with in vivo antitumor activity, *Proc. Am. Assoc. Cancer Res.*, 39, 227, 1986.

72. Gomez, S.G., et al., In vitro hematotoxicity of Aplidin on human bone marrow and cord blood progenitor cells, *Toxicology In Vitro*, 15, 347–350, 2001.

73. Vera, M.D., Joullié, M.M., Natural products as probes of cell biology: 20 years of didemnin research, *Med. Res. Rev.*, 22, 102–145, 2002.

74. Schmidt, U., Kroner, M., Griesser, H., Total synthesis of the didemnins — I. Synthesis of the peptolide ring, *Tet. Lett.*, 29, 3057, 1988.

75. Schmidt, U., Kroner, M., Griesser, H., Total synthesis of the didemnins — II. Synthesis of didemnins A, B, C and prolyldidemnin A, *Tet. Lett.*, 29, 4407, 1988.

76. Schmidt, U., Kroner, M., Griesser, H., Total synthesis of the didemnins — III. Synthesis of proptected (2R, 3S)-allo-isoleucine and (3S, 4R, 5S)-isostatine derivatives-amino acids from hydroxyl acids, *Synthesis*, 832–835, 1989.

77. Jou, G., et al., Total synthesis of dehydrodidemnin B. Use of uranium and phosphonium salt coupling reagents in peptide synthesis in solution, *J. Org. Chem.*, 62, 354, 1997.

78. Anthoney, A., Paz-Ares, L., Twelves, C., Phase I And Pharmacokinetic (PK) Study of Aplidin (APL) using A 24-hour, weekly schedule, *Proc. Am. Soc. Clin. Oncol.*, abst. 734, 2000.

79. Ciruelos, E.M., et al., Phase I clinical and pharmacokinetic study of the marine compound Aplidin (APL) administered as a 3 hour infusion every 2 weeks, *Proc. Am. Soc. Clin. Oncol.*, abst. 422, 2002.

80. Bowman, A., et al., Phase I clinical and pharmacokinetic (PK) study of the marine compound Aplidin (APL), administered as A 1 hour weekly infusion, *Proc. Am. Soc. Clin. Oncol.*, abst. 476, 2001.

81. Armand, J.P., et al., Phase I and pharmacokinetic study of Aplidin (APL) given as a 24-hour continuous infusion every other week (Q2w) in patients (Pts) with solid tumor (ST) and lymphoma (NHL), *Proc. Am. Soc. Clin. Oncol.*, abst. 477, 2001.

82. Maroun, J.A., et al., Phase I study of Aplidin in a 5 day bolus Q 3 weeks in patients with solid tumors and lymphomas, *Proc. Am. Soc. Clin. Oncol.*, abst. 2082, 2001.

83. Raymond, E., et al., Phase I (PI) trials with aplidin (APL), a new marine derived anticancer compound, *Eur. J. Cancer*, 37, S32, 2001.

84. Raymond, E., et al., Activity of APL, a new marine compound, against medullary thyroid carcinoma (MTC): phase I trials as screening tools for rare tumors, *Ann. Oncol.*, 13, 22, 2002.

85. Jimeno, J., et al., Progress in the clinical development of new marine-derived anticancer compounds, *Anticancer Drugs*, 15, 321–329. 2004.

86. Hamann, M.T., Scheuer, P.J., Kahalalide F., A bioactive depsipeptide from the sacoglossan mollusk *Elysia rufescens* and the green alga *Bryopsis sp.*, *J. Am. Chem. Soc.*, 115, 5825–5826, 1993.

87. López-Macia, A., et al., Synthesis and structural determination of Kahalalide F, *J. Am. Chem. Soc.*, 123, 11398–11401, 2001.

88. Hamann, M.T., et al., Kahalalides, bioactive depsipeptides from the sacoblassan mollusk *Elysia rufescens* and the green alga *Bryopsis sp.*, *J. Org. Chem.*, 61, 6594–6600, 1996.

89. Green, B.J., et al., Mollusc-algal chloroplast endosymbiosis. Photosynthesis, thylakoid protein maintenance, and chloroplast gene expression continue for many months in the absence of the algal nucleus, *Plant Physiol.*, 124, 331–342, 2000.

90. Bercero, M.A., et al., Chemical defences of the sacoglossan mollusk *Elysia rufescens* and its host alga *Briopsis sp.*, *J. Chem. Ecology*, 27, 2287–2299, 2001.

91. Córdoba, S., et al., *In vitro* chemosensitivity, cell cycle redistribution and apoptosis induction by Kahalalide F, a new marine compound in a panel of human tumoral cells, *Eur. J. of Cancer Suppl.*, 1, S173, 2003.

92. García-Rocha, M., Bonay, P., Avila, J., The antitumoral compound Kahalalide F acts on cell lysosomes. *Cancer Lett.*, 99, 43–50, 1996.

93. Tartakoff A.M., Perturbations of vesicular traffic with the carboxylic ionophore monensin, *Cell*, 32, 1026–1028, 1983.

94. Wosikowski, K., Schuurhuis, D., Johnson, K., Identification of epidermal growth factor receptor and c-erbB2 pathway inhibitors by correlation with gene expression patterns, *J. Nat. Cancer Inst.*, 89, 1505–1515, 1997.

95. Suarez, J., et al., A new marine-derived compound induces oncosis in human prostate and breast cancer cells, *Mol. Cancer Ther.*, 2, 863–872, 2003.

96. Sewell, J.M., et al., Kahalalide F appears to promote necrotic cell death in hepatoma cell lines, *Proc. Am. Assoc. Cancer Res.*, abst. 1509, 2004.

97. Janmaat, M.L., et al., Kahalalide F induces caspase-independent cytotoxicity that correlates with HER2 and/or HER3 expression levels and is accompanied by down-regulation of AKT signaling, *Proc. Am. Assoc. Cancer Res.*, abst. 5328, 2004.

98. Nuijen, B., et al., Development of a lyophilized, parenteral pharmaceutical formulation of the investigational polypeptide marine anticancer agent Kahalalide F, *Drug Dev. Ind. Pharm.*, 27, 767–780, 2001.

99. Gómez, S.G., et al., In vitro toxicity of three new antitumoral drugs (trabectedin, aplidin and Kahalalide F) on hematopoietic progenitors and stem cells, *Exp. Hemat.*, 31, 1104–1111, 2003.

100. Brown, A.P., et al., Preclinical toxicity of Kahalalide F, a new anticancer agent: single and multiple dosing regimens in the rat, *Cancer Chemother. Pharmacol.*, 50, 333–340, 2002.

101. Schellens, J.H., et al., Phase I and pharmacokinetic study of Kahalalide F in patients with advanced androgen refractory prostate cancer, *Proc. Am. Soc. Clin. Oncol.*, 451, 2002.

102. Circuelos, C., et al., A phase I clinical and pharmacokinetic study with Kahalalide F in patients with advanced solid tumors with a continuous weekly 1-hour iv infusion schedule, *Eur. J. Cancer*, 38, 33, 2002.

103. Zelek, L., et al., Preliminary results of Phase II study of ecteinascidin 743 (ET-743) with the 24-hour continuous infusion q3 week schedule in pretreated advanced/metastatic breast cancer patients [Abstract], Proc. 11th NCI-EORTC-AACR Symposium, 6, 45085, 2000.

13 Discovery of E7389, a Fully Synthetic Macrocyclic Ketone Analog of Halichondrin B

Melvin J. Yu, Yoshito Kishi, and Bruce A. Littlefield

CONTENTS

I. Introduction 241
II. Background 242
III. Halichondrin Structure–Activity Relationship 243
IV. Total Synthesis and Compound Supply...244
V. Structure–Activity Relationship of Structurally Simplified Analogs............................248
VI. Synthesis of HB Analogs..257
VII. *In Vitro* Characterization of E7389..257
VIII. *In Vivo* Characterization of E7389 258
IX. Conclusion..259
Acknowledgments..262
References 262

I. INTRODUCTION

E7389, a fully synthetic analog of the structurally complex marine natural product halichondrin B (HB), exhibits unique effects on tubulin dynamics, along with striking anticancer activity in preclinical models of human cancer. The nature of E7389's antitumor activity in such models, combined with its novel tubulin-based effects and superiority over paclitaxel in several animal models, indicates that this agent may exhibit a unique spectrum of antitumor activities in the clinic that will differentiate it from existing anticancer drugs.

Limited supply of natural halichondrins from the wild, a serious obstacle common to other structurally complex marine natural products isolated in low yield, represented a critical issue that all but precluded development of these compounds as drug candidates. Three issues in particular affected the halichondrins. These included securing the required absolute amount of material with a consistent impurity profile to support drug development activities, separating the desired halichondrin from other members of the family closely related in structure, and removing highly toxic coisolated metabolites such as okadaic acid (a known contaminant in extracts from *Halichondria okadai*). The limited supply of halichondrins and strong interest on the part of the U.S. National Cancer Institute (NCI) to develop HB as a new anticancer drug created the need to identify alternative sources of this important class of marine natural products.

Despite the chemical complexity of the halichondrin family, total synthesis represented the most attractive option to not only identify an optimized analog for drug development but also provide the means for a reliable and renewable source of material. With the ability to address problems at the molecular level, to fine-tune pharmacological properties, and to adjust physico-chemical parameters, E7389 was identified as the most promising analog, with biological characteristics matching or surpassing those of the natural product that inspired it. Although the chemical structure of E7389 is significantly less complex than HB, it nevertheless remains a challenging synthetic target. However, contemporary synthetic organic chemistry has both the capacity and the potential to meet this challenge. With preclinical development work completed, clinical evaluation of this novel agent is now underway to establish the therapeutic value of halichondrin-based drugs as anticancer therapeutics.

II. BACKGROUND

HB is a polyether macrolide that exhibits remarkable anticancer activity in preclinical animal models of human cancer. First isolated in the mid-1980s from the western Pacific marine sponge *Halichondria okadai* Kadota,[1,2] members of the halichondrin family were subsequently found in *Axinella* sp.,[3] *Phakellia carteri*,[4] and most recently, *Lissodendoryx* sp.[5] Trace quantities of halichondrins were also detected in an extract of the black New Zealand shallow-water sponge *Raspalia agminata*.[6] Members of the halichondrin family include halichondrins B and C, norhalichondrins A, B, and C, homohalichondrins A, B, and C, halistatins 1 and 2, isohomohalichondrin B, neonorhalichondrin B, neohomohalichondrin B, and 55-methoxyisohomohalichondrin B (Figure 13.1). Differences between members of this family lie primarily in the level of oxygenation at C.10, C.12, and C.13 and in the length of the carbon backbone. Of these compounds the most potent include HB (**1a**) and homohalichondrin B (**4b**). The name halipyran was proposed by Pettit and coworkers for the conserved unsubstituted C.1–C.38 carbon skeleton common to the halichondrin family of marine natural products.[4]

Mechanistically, the halichondrins make up a subtype of tubulin-interactive antimitotic agent with subtle differences in mechanism of action from those of other antimitotics such as the vinca alkaloids.[7] Although classified as a tubulin depolymerizer, the pattern of tubulin interaction by the halichondrins is highly specific.[8] Testing against the 60 human tumor cell lines of the NCI's drug screening program demonstrated that HB exhibited subnanomolar growth-inhibitory activity against a wide variety of cancer cell types *in vitro*, consistent with a tubulin-based antimitotic mechanism.[9] These conclusions were further supported by findings that low micromolar concentrations of the drug inhibited various aspects of tubulin function *in vitro*, including tubulin polymerization, microtubule assembly, β^s-tubulin crosslinking, GTP and vinblastine binding to tubulin, and tubulin-dependent GTP hydrolysis.[10,11] Although such *in vitro* findings are consistent with a vinca alkaloid–like antitubulin mechanism, they must still be viewed in light of the $\sim 10^4$-fold discrepancy between the micromolar concentrations required for antitubulin effects *in vitro* and the subnanomolar concentrations that inhibit cell growth. This discrepancy is shared by other tubulin-active drugs and may reflect direct or indirect effects on microtubule dynamics at low drug concentrations rather than net changes in microtubule content seen at higher drug concentrations.[12,13] Other members of the halichondrin family reported to disrupt microtubule networks by preventing microtubule assembly include isohomohalichondrin B.[14]

Recent studies by Jordan and coworkers with E7389, a synthetic analog of HB, helped shed further light on the unique mechanism of microtubule disruption by the halichondrin-class molecules.[15] Although many tubulin-based antimitotic agents exert their biological effects through direct interactions with microtubules, the halichondrins represented by E7389 suppressed microtubule dynamics by inducing nonproductive tubulin aggregates, leading to suppression of spindle microtubule growth events and reduced microtubule polymer mass. This mechanism differs significantly from that of other tubulin-interactive drugs such as vinblastine or paclitaxel and is consistent with the highly specific pattern of tubulin interaction by HB reported by Hamel and coworkers. The novel mechanism of E7389 relative to existing tubulin-based drugs further supports the halichondrins as

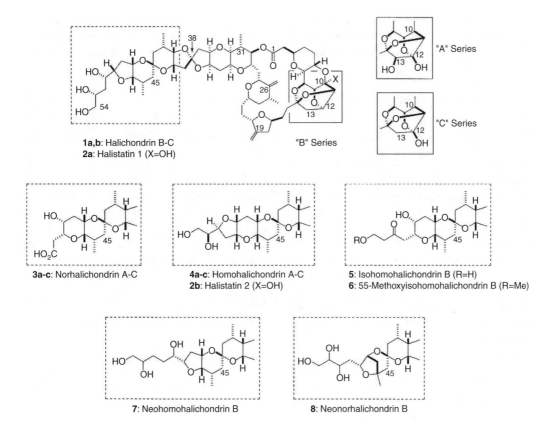

FIGURE 13.1 Halichondrin family (X = H unless otherwise noted).

a unique subtype of tubulin-based antimitotics and raises the possibility that compounds from this class may exhibit unique characteristics in the clinic as cancer chemotherapeutic agents.

The high *in vitro* potency of HB is mirrored by its extraordinary anticancer activity in a variety of preclinical animal models. HB was reported by Hirata and Uemura in 1986 to increase mean survival times of mice bearing murine B16 melanoma and P388 and L1210 leukemias.[2] Efficacy in these animal models was also reported in the patent literature by Fujisawa scientists that same year.[16] In 1996, Fodstad and coworkers described the remarkable activity of halichondrins isolated from natural sources in a number of human tumor xenograft models including melanoma, osteogenic sarcoma, lung cancers, and colon cancers.[7] Synthetic HB prepared by Kishi and coworkers at Harvard University similarly exhibited impressive activity in human tumor xenograft models.[17] For example, dose-dependent inhibition of subcutaneous LOX human melanoma xenografts was observed in nude mice following intraperitoneal administration (5–25 µg/kg). At 50 µg/kg, tumor growth was completely suppressed, with no visible signs of toxicity.[17]

The extraordinary *in vitro* potency, *in vivo* efficacy, and distinct antitumor profile of the halichondrins in human tumor models led the NCI to select HB for preclinical development as an anticancer chemotherapeutic agent in 1992.

III. HALICHONDRIN STRUCTURE–ACTIVITY RELATIONSHIP

Based on limited cytotoxicity data of naturally occurring halichondrins A–C, norhalichondrin A, and homohalichondrin B in B16 melanoma cells *in vitro*, Hirata and Uemura[2] proposed that the 2,6,9-trioxatricyclo[3.3.2.0^{3,7}]decane system should be relatively lipophilic (HB-type) and that the terminal moiety should contain two or three hydroxyl groups, but not a carboxylate, for maximum expression of antitumor activity. Because the length of the halichondrin molecules was estimated to be 30–35 Å, corresponding to half the distance of a lipid biomembrane, Hirata and Uemura speculated that the 2,6,9-trioxatricyclo[3.3.2.0^{3,7}]decane portion of the molecule may be important for insertion into the lipid bilayer.

Activity in the NCI 60–human tumor cell line panel of several naturally occurring and semi-synthetic halichondrin derivatives was summarized in a review by Munro.[6] Hydrolysis of the macrolactone ring to the corresponding seco acid and reduction of the exo-olefin groups at C.19 and C.26 were both detrimental to *in vitro* biological activity. Acid-induced rearrangement of homohalichondrin B similarly provided products with diminished biological activity. Treatment with excess trifluoroacetic acid in CH_2Cl_2 at 20°C, for example, led to spiroacetal epimerization at C.38 or rearrangement of the 2,6,9-trioxatricyclo[3.3.2.0^{3,7}]decane ring system to a furan derivative.[18] Of the two acid-induced rearrangement pathways, epimerization at C.38 was reportedly more detrimental to biological activity, affording compounds with a two-log shift in GI_{50} values in the NCI 60–human tumor cell line panel, along with correspondingly major changes in molecular conformation. Extrapolating these results to other naturally occurring members of the halichondrin family, Munro speculated that epimerization at C.38 would, in general, lead to halichondrin derivatives with diminished biological activity. For example, the C.38 epimer of isohomohalichondrin was less active against murine leukemia P388 cells *in vitro* compared with natural material.[5] However, minor modifications to the terminal region of the halichondrins were well tolerated, with minimal impact on *in vitro* potency.

Using HB as the seed compound, correlation coefficients from a COMPARE analysis using a GI_{50}-centered mean graph profile as a qualitative index of antimitotic ability were plotted against GI_{50} values for CCRF-CEM cells as an index of cytotoxicity. The resulting scatter did not allow Munro and coworkers to draw meaningful conclusions regarding structure–activity relationships for these compounds based on subtle differences in the antimitotic mean graph profile relative to HB. However, significant differences in the average solution conformation in $CDCl_3$ were noted for the less cytotoxic compounds despite similar COMPARE correlation coefficients. Thus, Munro concluded that conformational changes resulting from structural modifications diminished biological activity without altering the antimitotic profile characteristic of HB.

IV. TOTAL SYNTHESIS AND COMPOUND SUPPLY

Because of the extraordinary *in vivo* potency of the halichondrins, the amount of material required to support preclinical development and eventual clinical investigation was estimated to be approximately 10 g.[6] Assuming success in clinical trials, supply of the halichondrins to meet commercial needs was similarly estimated to be relatively small, at 1–5 kg/yr. However, securing even these modest amounts through collection from the wild was determined to be unlikely because of the extremely low yield of halichondrins isolated from natural sources, the low abundance of known HB-producing sponges in the world, and the environmental effect of subjecting deep sea sponges to large-scale harvesting, with risk of extinction. As a result, compound supply represented a significant obstacle for both development and possible future commercialization of HB as an anticancer drug. Although initial aquaculture feasibility trials on *Lissodendoryx* sp. 1 demonstrated the potential for halichondrins to be isolated from a cultured sponge,[6] the ability of scale-up operations to meet commercial demands of bulk drug substance that satisfy purity specifications in a reliable, economic, and consistent manner remains unknown.

The difficulties associated with securing the needed absolute amount of material for development and possible commercialization are not unique to the halichondrins[19] and represent only one aspect of the supply issue surrounding natural products isolated in extremely low yield from the producing organism. For example, separation of the desired halichondrin (e.g., HB) from other closely related family members, and the need to remove structurally unrelated but highly potent coisolated bioactive metabolites, added purification concerns to the list of compound supply challenges associated with collection from the wild or through aquaculture techniques. These three issues made development of the halichondrins difficult; unfortunately, NCI's collaborative efforts to identify more widespread sources of the natural product met with only limited success, placing the program at risk of stalling at an early stage of preclinical development and leaving compound supply a serious concern for the NCI Drug Development Group (http://dtp. nci.nih.gov/docs/ddg/ddg_descript.html). Hirata and Uemura proposed that symbiotic microorganisms rather than the sponge itself may be responsible for producing the halichondrins as secondary metabolites.[2] Although the isolation of halichondrins from several marine sponges supported this proposal, no reports regarding the isolation, tissue culture, or fermentation of the putative producing organism or organisms have appeared in the literature, making this approach an unlikely near-term solution.

With compound supply an obstacle for preclinical development of this promising class of anticancer agents, alternative sources of the natural product were critically needed. The extraordinary biological activity of the halichondrins coupled with the extremely challenging and unique structural architecture of these compounds attracted the attention of chemistry research groups throughout the world. Although a number of laboratories reported synthetic studies on the halichondrins,[20–23] the only successful total synthesis to date was reported by Kishi and coworkers in 1992 for the total synthesis of HB and norhalichondrin B.[20] Kishi's approach was highly convergent, bringing together four major fragments corresponding to the C.1–C.13, C.14–C.26, C.27–C.38, and C.39–C.54 regions of the molecule for final assembly (Figure 13.2). The convergent nature of this approach allowed the chemistry of individual fragments to be further optimized independent of one another, with minimal effect on late-stage coupling or protecting group strategies.[20] In addition, Kishi's synthesis was extremely flexible, allowing structure modifications to be made in a modular fashion, thereby facilitating use of key stockpiled intermediates to prepare different members of the halichondrin family (e.g., synthesis of HB or norhalichondrin B from common "right half" intermediate **11**).

The availability of pure natural product via total synthesis was an important factor in the NCI Decision Network Committee's decision to select HB for preclinical development in March 1992.[24] Total synthesis provided not only a reliable source of material free from potentially toxic coisolated contaminants such as okadaic acid but also the tools for directed structure modification and the opportunity to identify novel analogs with optimized biopharmaceutical and physicochemical properties not accessible through semisynthesis or derivatization of the natural product itself. The ability to address problems at the molecular level, to fine-tune pharmacological properties as well as metabolic and chemical stability, aqueous solubility, and compound purity, were all essential factors for identifying and bringing the best possible analog forward for clinical investigation.

Although a number of synthetic intermediates and model systems were generated in connection with the synthetic efforts launched by research groups worldwide, biological data for only a handful of compounds were reported in the literature. In 1992, synthetic HB and several synthetic intermediates from the Kishi group at Harvard University were kindly provided to Eisai Research Institute for *in vitro* and *in vivo* biological evaluation as anticancer agents. As expected, potent cell growth-inhibitory activity was observed for synthetic HB. Surprisingly, however, synthetic intermediate **11** corresponding to the C.1–C.38 "right half" fragment of the natural product exhibited *in vitro* activity within an order of magnitude of the natural product in a 3–4-d growth-inhibition assay, using DLD-1 human colon cancer cells.[20,25,26] This compound represented the first example of a structurally simplified synthetic analog of HB exhibiting potent antiproliferative activity, thereby providing experimental evidence that biological activity was associated with the macrocyclic moiety. Both HB and **11** blocked cells at the G2/M phase of the cell cycle, both disrupted normal mitotic spindle architecture through microtubule destabilization with few if any effects on other phases of the cell cycle, and both exhibited a similar profile in the NCI 60–cell line screen. The exciting discovery of a structurally simplified HB fragment that retains the potent

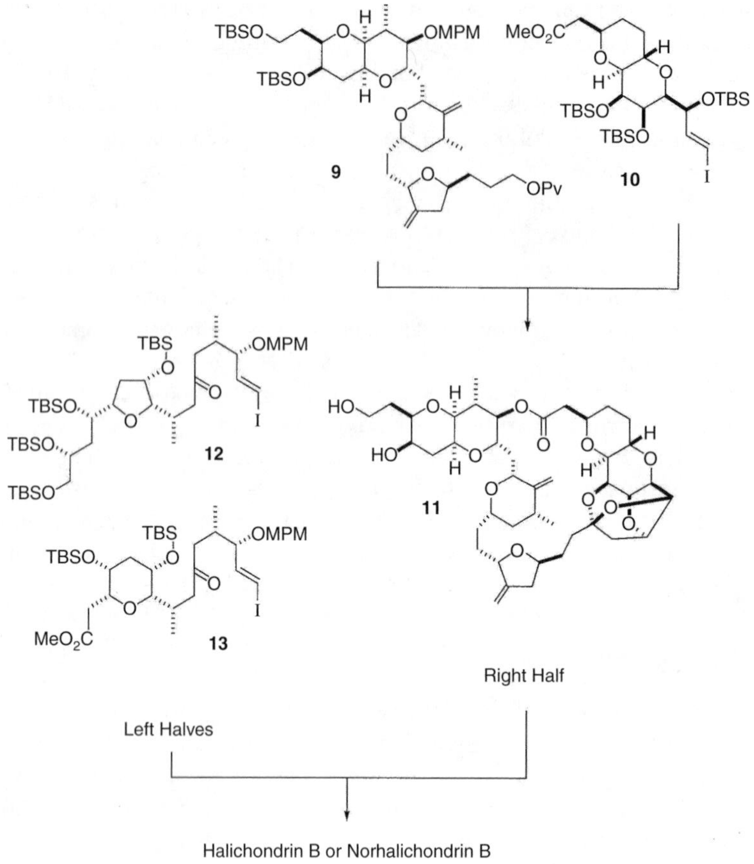

FIGURE 13.2 Kishi total synthesis of halichondrin B and norhalichondrin B.

cell growth-inhibitory activity of the natural product formed the basis for a research program at Eisai Research Institute to identify novel halichondrin-based anticancer agents using chemistry technology developed in the Kishi laboratory at Harvard University.

Results obtained with biotinylated macrolactone analog **14** (Figure 13.3) provided the first direct proof of any halichondrin class molecule binding directly to tubulin itself.[27] Under the affinity chromatography conditions, it was not possible to identify the tubulin subtype targeted by the halichondrin macrolactone, as binding to either the α- or the β-tubulin subtype present in α/β-tubulin heterodimers would result in retention of both. The pattern of cell growth inhibition, cell cycle effects, mitotic spindle disruption, and affinity chromatography results taken together indicated that the structurally simplified macrolactone analogs share a similar or identical mechanism to parental HB. Thus, the structural elements necessary for anticancer biological activity reside in the macrocyclic sector of the molecule, providing a clear and compelling starting point for a structure–activity relationship investigation to identify the minimum pharmacophoric substructure for this family of marine natural products.

The discovery that a structurally simplified and synthetically accessible fragment of HB retained potent anticancer activity created the possibility that supply problems associated with the natural product itself could be circumvented. Specifically, if readily accessible halichondrin analogs that retain the full spectrum of remarkable biological activity associated with this family of compounds could be identified through total synthesis, then clinical evaluation of the halichondrin class of

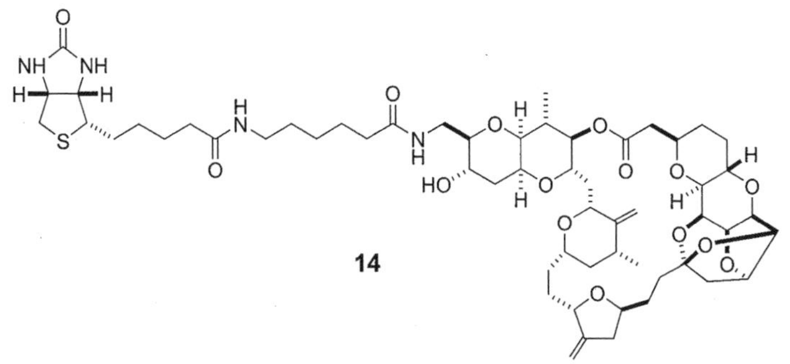

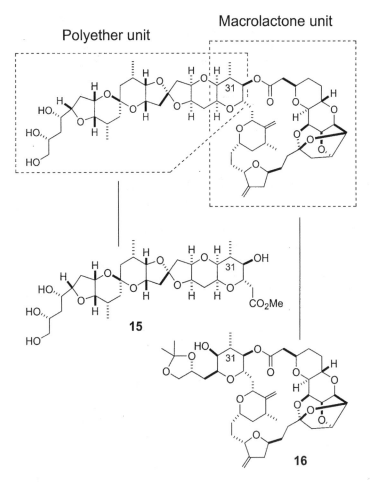

FIGURE 13.3 Biotinylated derivative of compound **11**.

marine natural products could proceed with an optimized derivative to determine their value as chemotherapeutic agents in humans.

Horita and Yonemitsu in 1997 described a similar approach from their laboratory that bifurcated HB into partially overlapping polyether and macrolactone units (Figure 13.4).[28] Although the synthesis for each (compounds **15** and **16**, respectively) was reported in the literature, biological activity was presented only for early precursors rather than for the target compounds themselves

Polyether unit

Macrolactone unit

FIGURE 13.4 Yonemitsu synthetic study.

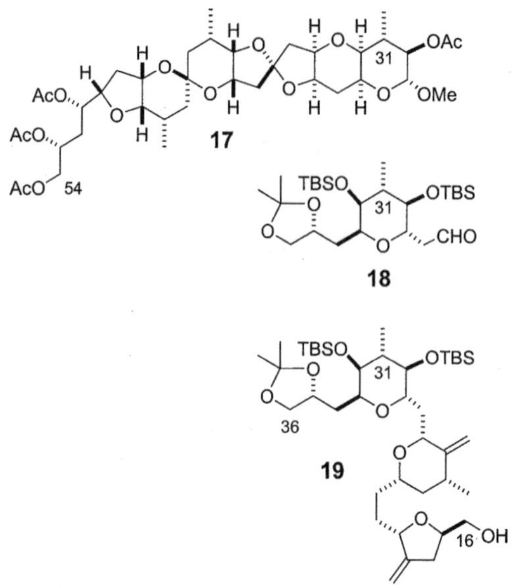

FIGURE 13.5 Representative synthetic intermediates reported by Yonemitsu et al.[28]

(Figure 13.5 and Table 13.1). Interestingly, activity was noted for intermediates associated with both sections, although those corresponding to the polyether unit exhibited a higher level of cytotoxicity against KB and L1210 cells relative to those corresponding to the macrolactone region. Polyether analogs corresponding to the C.37–C.54 region of HB were described in the patent literature as antitumor agents with micromolar activity against a panel of 39 human tumor cell lines[29] and were shown in a university publication to have modest activity against MCF-7 breast and KE-4 esophagus cancer cells.[30] Because no details regarding mechanism of action for these compounds were provided, it is unclear whether any of the fragments inhibited cell growth through a tubulin-based antimitotic mechanism associated with the natural product.

V. STRUCTURE–ACTIVITY RELATIONSHIP OF STRUCTURALLY SIMPLIFIED ANALOGS

Despite the *in vitro* similarities, and in contrast to HB, compound **11** was not active in the *in vivo* LOX human melanoma xenograft model. Cell growth-inhibitory activity for both HB and **11** were similar under continuous exposure conditions in the 3–4-d cell growth assay. Flow cytometric analysis of U937 human histiocytic lymphoma cells, however, revealed a clear difference in ability of the two compounds to maintain a complete mitotic block (CMB) 10 h after drug washout.[31]

TABLE 13.1
In Vitro Activity of Halichondrin B Synthetic
Intermediates Reported by Yonemitsu et al.[28]

Compound	IC$_{50}$ (nM)	
	KB	L1210
17	8.3	0.64
18	3.9	3.1
19	>100	28

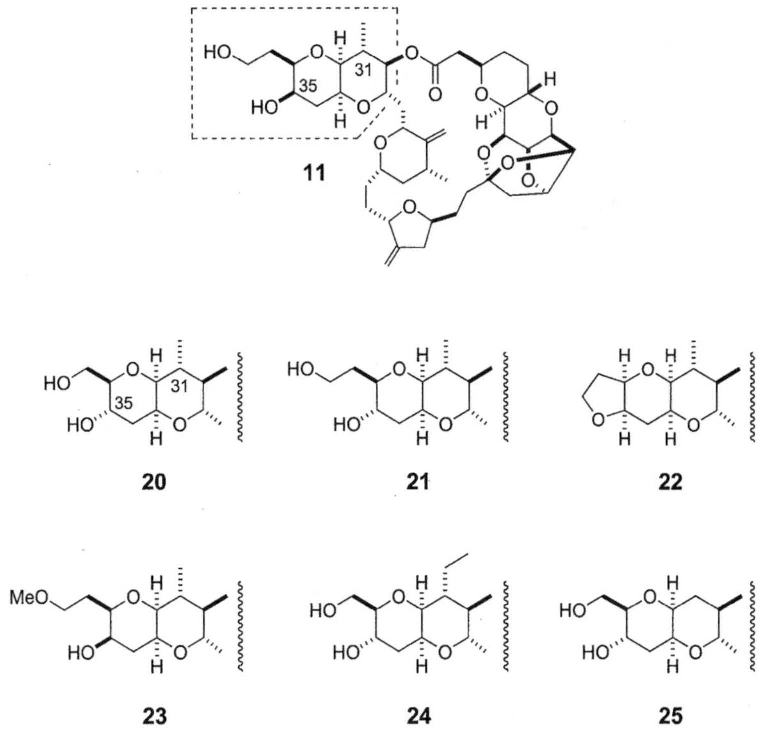

FIGURE 13.6 Representative octahydropyrano[3,2-b]pyran macrolactone analogs.

Although a 10-nM concentration of HB was sufficient to induce and maintain a CMB, concentrations of **11** as high as 1 µM (highest tested concentration) were ineffective against this challenge. Thus, in contrast to the natural product, the antimitotic effects of **11** were reversible on washout. Cell viability dose–response curves 5 d postwashout tracked 10-h postwashout rather than 0-h prewashout IC$_{99}$ values, indicating that long-term cell viability was dependent on continued maintenance of a CMB, rather than its initial induction. The reversible nature of **11** indicated that cell growth inhibition would only occur under continuous exposure conditions and that tumor inhibitory activity *in vivo* would cease when drug concentrations in blood fell below a certain critical threshold. On the basis of this working hypothesis, a proprietary cell-based assay was established at Eisai Research Institute that allowed measurement of a compound's ability to maintain a CMB following drug washout as an index of potential antitumor efficacy in *in vivo* settings, using standard intermittent dosing schedules.[32]

Macrolactone analogs modified at the C.30–C.38 region of the molecule were subsequently screened for both intrinsic cell growth-inhibitory activity under continuous exposure conditions as well as the ability to induce an irreversible CMB (Figure 13.6 and Table 13.2).[33] Although intrinsic potency was relatively insensitive to variations in this region of the molecule, the ability to induce an irreversible CMB was structure specific. For example, compound **20**, prepared by the Kishi group at Harvard University in connection with an improved synthesis of the macrolactone fragment of HB, was the first structurally simplified macrolactone analog found to induce an irreversible CMB (postwashout IC$_{99}$ = 220 nM). With a shortened side chain at C.36 and an epimeric hydroxyl group at C.35, this compound represented two structural changes from lead compound **11**, which exhibited a postwashout IC$_{99}$ value greater than 1000 nM. Inverting the C.35 hydroxyl group of **11** alone (cf. **11** and **21**) was not sufficient to improve activity in the mitotic block reversibility assay. Consistent with the idea that induction of an irreversible CMB would permit manifestation of *in vivo* activity, compound **20** exhibited good antitumor activity in the LOX human melanoma

TABLE 13.2
In Vitro **Biological Activity of Octahydropyrano[3,2-b]Pyran Macrolactone Analogs**

Compound	Intrinsic Potency, DLD-1 IC$_{50}$ (nM)[a]	Mitotic Block Reversibility		
		U937 IC$_{99}$ 0 h (nM)[b]	U937 IC$_{99}$ 10 h (nM)[c]	Reversibility Ratio
11	4.6	30	>880	>29
20	3.4	9	220	24
21	3.6	10	>1000	>100
22	2.6	20	>1000	>50
23	1.2	7	>650	>93
24	6.0	10	1000	100
25	610.0	—	—	—
26	9.1	30	>1000	>33
27	59	300	>1000	>3
28	650	—	—	—

[a] Growth-inhibitory potency against DLD-1 cells after 3–4 d continuous drug exposure
[b] Induction of CMB in U937 cells after 12 h drug exposure (0 h of washout)
[c] Maintenance of CMB 10 h after drug washout

xenograft model.[34] In contrast, analog **22**, a direct substructure of the natural product, exhibited a nearly identical *in vitro* profile to diol **11**, as did methoxyl derivative **23**. Increasing the size of the C.31 methyl substituent to an ethyl group (e.g., **24**) led to a modest loss of potency, but removing this group altogether to give compound **25** reduced activity by over two orders of magnitude.

In addition to improving the *in vitro* profile of macrolactone **11**, structural simplicity and compound accessibility were also major factors driving the structure–activity relationship investigation. For example, if the C.19, C.25, or C.26 substituents of compound **11** could be removed without significantly affecting biological activity (Figure 13.7), then the synthesis of the compounds could be simplified with enhanced access to these compounds. Analogs **26** and **27** indicated that complete removal of the C.19 and C.26 exo-olefins was tolerated, but removal of the C.25 methyl

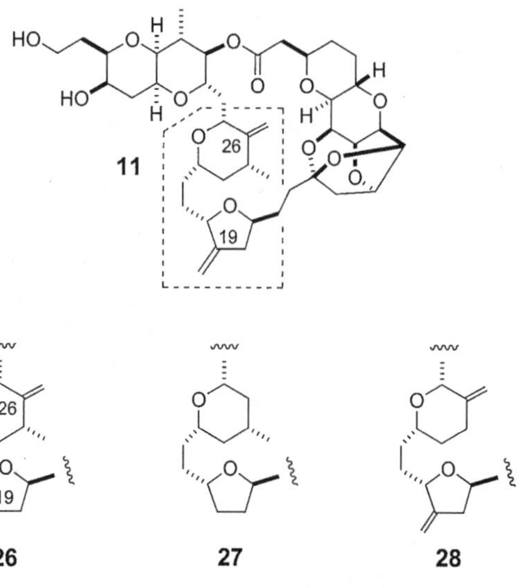

FIGURE 13.7 Representative deletion macrolactone analogs.

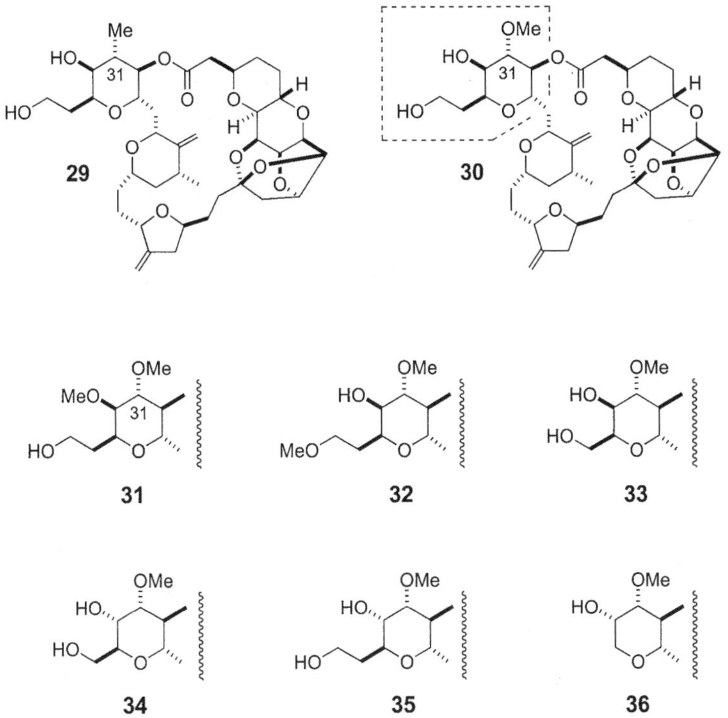

FIGURE 13.8 Representative tetrahydropyran macrolactone analogs.

group to give macrolactone **28** reduced activity 100-fold (Table 13.2). Because modifying centers contained directly within the macrocyclic core uniformly led to analogs with equivalent or reduced activity (these results apply only to the particular macrolactone analogs synthesized and tested. The possibility exists that other modifications [e.g., substitution patterns or conversion of the macrolactone ring to the corresponding macrolactam, macrocyclic ketone, carbocyclic ring, etc.] may lead to analogs where the same changes may not negatively affect activity), it was hypothesized that the macrocyclic ring may represent the minimum pharmacophoric substructure of the halichondrins for cell growth-inhibitory activity.

As a result, the macrolactone ring was fixed and held constant as modifications to the fused C.29–C.32 pyran sector were explored. In support of this hypothesis, compound **29** (Figure 13.8) with a single tetrahydropyran ring in place of the C.29–C.36 octahydropyrano[3,2-b]pyran ring system exhibited *in vitro* biological activity comparable to lead compound **20** in terms of both intrinsic potency and ability to induce an irreversible CMB (Table 13.3). Modifications to the tetrahydropyran ring substituents, especially with regard to the C.32 hydroxyl group, however, had little effect on intrinsic potency, but a significant effect on the compound's reversibility characteristics.

The earlier structure–activity relationship investigation around the original octahydropyrano[3,2-b]pyran series of analogs demonstrated the importance of the C.31 substituent. Because small changes in substituent size (e.g., methyl to ethyl) were tolerated, extrapolation to the tetrahydropyran series suggested that replacing the C.31 methyl group with a methoxyl substituent would allow more ready access from commercially available carbohydrate precursors without compromising the biology activity of the final compound. The C.31 methoxyl derivative **30** was found to be equipotent with the corresponding C.31 methyl analog **29**, supporting the interchangeability of these two groups at this center of the molecule.

As observed with previous macrolactone analogs, modifications to the tetrahydropyran ring had no effect on intrinsic cell growth-inhibitory potency. Individually converting each of the hydroxyl groups of diol **30** to methyl ethers **31** and **32**, in contrast, provided highly reversible compounds,

TABLE 13.3

In Vitro **Biological Activity of Tetrahydropyran and Tetrahydrofuran Macrolactone Analogs**

Compound	Intrinsic Potency, DLD-1 IC$_{50}$ (nM)[a]	Mitotic Block Reversibility		
		U937 IC$_{99}$ 0 h (nM)[b]	U937 IC$_{99}$ 10 h (nM)[c]	Reversibility Ratio
Tetrahydropyran Series				
29	2.5	10	170	17
30	1.8	10	300	30
31	2.1	10	>1000	>100
32	1.8	10	>1000	>100
33	2.6	10	300	30
34	2.0	3	65	22
35	1.8	2	100	50
36	>1000	—	—	—
Tetrahydrofuran Series				
37	1.0	3	100	33
38	>1000	—	—	—
39	0.97	7	100	14
40	6.4	10	100	10
41	1.4	3	30	10
42	>1000	—	—	—
43	220	—	—	—
44	0.67	1.5	15	10

[a] Growth-inhibitory potency against DLD-1 cells after 3–4 d continuous drug exposure.
[b] Induction of CMB in U937 cells after 12 h drug exposure (0 h of washout).
[c] Maintenance of CMB 10 h after drug washout.

as measured by the U937 mitotic block reversibility assay. Shortening the C.33 side chain to a one-carbon-lower homolog or inverting the hydroxyl stereochemistry at C.32 had little effect (e.g., **33–35**), but removing the C.33 side chain altogether (i.e., **36**) led to complete loss of activity.

Because incremental changes to the C.29–C.33 tetrahydropyran ring system did not provide a clear direction for improvement, an alternative strategy was pursued based on a working hypothesis that the C.31 substituent played a critical role in stabilizing the bioactive conformation of the macrolactone. As noted earlier, presence of the C.31 substituent was critical for good *in vitro* biological activity, albeit with a degree of steric tolerance. Thus, if the bioactive conformation could be reinforced by an alternative ring system, then the apparent potency limitations associated with the tetrahydropyran analogs might be circumvented.

The x-ray crystal structure of norhalichondrin A *p*-bromophenacyl ester reported in 1985 provided an empirically derived starting point for conformational analysis of the C.29–C.33 tetrahydropyran ring.[1] Although the solid-state conformation may not necessarily represent the bioactive conformation of the halichondrins, it does represent a reasonable low-energy structure for examining the octahydropyrano[3,2-b]pyran ring conformation. ^{1}H-nmr studies reported by Munro indicate that the average solution conformation of HB in CDCl$_3$ may be similar to the solid-state conformation of norhalichondrin A *p*-bromophenacyl ester.[6] In the x-ray structure, the C.29–C.33 tetrahydropyran ring adopts a twist boat conformation that places the C.31 methyl group in a pseudoequatorial position.[1,35] Although a tetrahydropyran ring can adopt a similar twist boat conformation, the tetrahydrofuran ring would be expected to do so with less conformational freedom (Figure 13.9). By stabilizing the pseudoequatorial orientation of the C.31 methyl group relative to

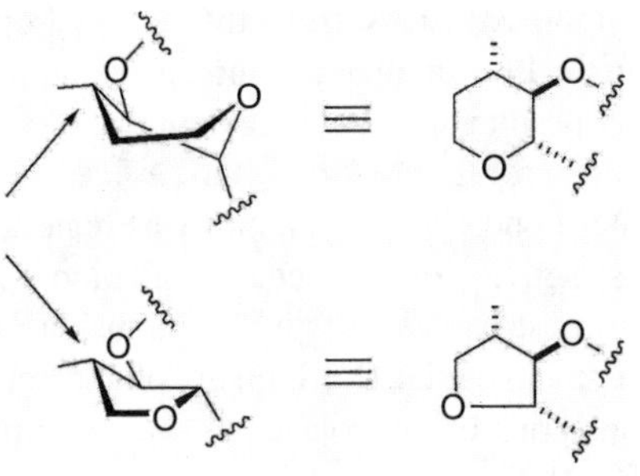

FIGURE 13.9 Conformational analysis of tetrahydropyran and tetrahydrofuran analogs.

the macrocyclic ring, the tetrahydrofuran analogs might be expected to exhibit superior biological activity relative to the tetrahydropyran derivatives based on favorable entropic factors.

Consistent with this working hypothesis, the *in vitro* biological activity of tetrahydrofuran **37** (Figure 13.10) was virtually indistinguishable from that of tetrahydropyran **34**. Inverting the stereocenter at C.29 to give compound **38**, however, provided a compound that was inactive at the highest concentration tested, consistent with the need to maintain a particular low-energy macrolactone ring conformation (these results apply only to the particular macrolactone analogs synthesized and tested). Extending the C.32 hydroxymethyl side chain by either one carbon (e.g., **39**) or to a vicinal propanediol (e.g., **40** and **41**) had little effect, albeit showing a modest difference between the C.34 *R*- and *S*-isomers in the DLD-1 cell growth assay. As expected, based on conformational arguments, activity was critically dependent on the configuration at C.30 and C.31 (e.g., **42** and **43**, respectively; these results apply only to the particular macrolactone analogs synthesized and tested). Further structural and stereochemical optimization ultimately led to tetrahydrofuran **44**, whose intrinsic potency and ability to maintain a CMB exceeded those of the tetrahydropyran derivatives.

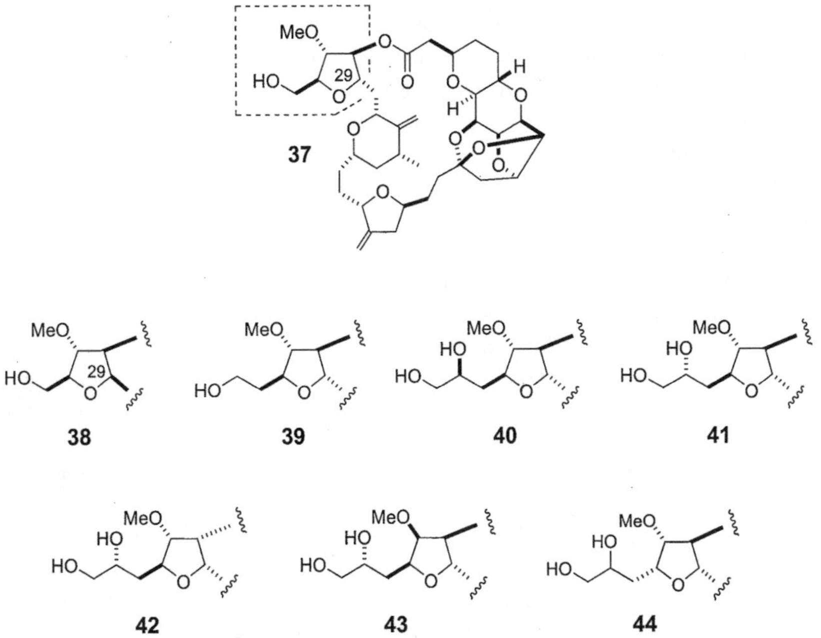

FIGURE 13.10 Representative tetrahydrofuran macrolactone analogs.

Compounds **30** and **44**, representatives from the tetrahydropyran and tetrahydrofuran series, respectively, were evaluated in the LOX human melanoma xenograft model. Unexpectedly, both failed to demonstrate *in vivo* efficacy despite an equal or superior *in vitro* profile relative to lead compound **20**. Although several explanations for this lack of *in vivo* efficacy in the mouse xenograft model were possible, instability of the ester bond in mouse serum was regarded as the most likely explanation. In support of this hypothesis, activity of the octahydropyrano[3,2-b]pyran series of macrolactone analogs (e.g., **11**) was for the most part unaffected by inclusion of up to 1% (v/v) mouse serum in the 3–4-d growth-inhibition assay using DLD-1 human colon carcinoma cells. However, activities of the simplified tetrahydropyran series of macrolactones were completely abrogated by mouse serum concentrations as low as 0.01%–0.05% in the cell culture medium.[36] Elimination of biological activity did not occur with either bovine or human sera, regardless of heat inactivation status, consistent with high levels of nonspecific esterases present in mouse serum (18 U/mg protein) compared to bovine (0.3 U/mg) or human sera (0.6 U/mg). Because the seco acid of macrolactone **11** was not active *in vitro*, lack of *in vivo* efficacy for compounds **30** and **44** in the mouse xenograft model was attributed to sensitivity of the macrolactone ring to nonspecific esterases present in mouse serum.

The 3–4-d DLD-1 human colon carcinoma cell growth assay was adapted to provide a functional measure of compound stability in the presence of mouse serum. At a fixed drug concentration of 100 nM, analogs were examined for cell growth-inhibitory activity in the presence of varying concentrations of mouse serum. The "mouse serum IC$_{50}$ value" from this assay corresponded to the percentage of mouse serum needed to suppress 100-nM drug-mediated cell growth inhibition by 50%. Both the tetrahydropyran and tetrahydrofuran series of compounds were found to be highly sensitive to mouse serum inactivation, with compounds **35** and **37** exhibiting mouse serum IC$_{50}$ values of just 0.002%. However, the reasons underlying the stability of **11** to nonspecific esterases present in mouse serum relative to the sensitivity seen in structurally simplified tetrahydropyran or tetrahydrofuran analogs were not well understood. Although conformational factors may play a role, molecular modeling failed to provide a clear explanation for the striking loss of cell growth-inhibitory activity in the presence of mouse serum esterases, although speculative, protein-binding calculations indicate that differential low-affinity binding of these compounds to serum proteins with nonspecific esterase activity may account at least in part for their observed behavior.

Attempts to further improve the *in vitro* biological profile of the macrolactones met with limited success. As summarized in Figure 13.11, most changes were detrimental to biological activity, with greater latitude observed for those centers external to the C.1–C.30 macrolactone ring (e.g., C.32–C.38). Because minimal opportunities existed to modify the 2,6,9-trioxatricyclo[3.3.2.0^{3,7}]decane ring system without significantly altering the global conformation of the molecule, and as members of the halichondrin family with oxygenation at C.10, C.12, or C.13 (e.g., halichondrin A, halistatins 1–2) were less potent relative to their nonhydroxylated counterparts, the C.1–C.13 fragment of HB was held constant as structure–activity relationships were explored around the remainder of the molecule. Despite extensive investigation, instability of the macrolactone ring remained problematic.

Lacking a predictive model with sufficient resolution to guide structure modification efforts, nonhydrolyzable ester bioisosteres were evaluated to identify derivatives that retain activity in the presence of mouse serum. As expected, the C.1 amide, hydroxyl, ether, and ketone analogs exhibited improved stability against nonspecific esterases. However, of the surrogate groups examined, only the ketone bioisostere provided analogs with sufficient *in vitro* potency to warrant further investigation.[32]

From these studies, compound **45** (ER-076349, NSC 707390) emerged as the prototypical macrocyclic ketone for *in vivo* evaluation (Figure 13.12). This analog exhibited antitumor effects in a variety of human tumor xenograft models including MDA-MB-435 breast carcinoma, COLO-205 colon carcinoma, LOX melanoma, and NIH:OVCAR-3 ovarian carcinoma, with no evidence of toxicity based on body weight or water consumption at the doses tested.[27]

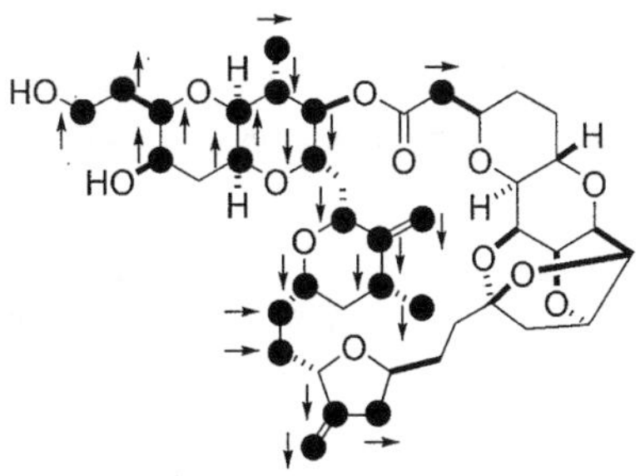

FIGURE 13.11 Closed circles indicate carbon centers that were modified to explore structure–activity relationships. The arrows indicate whether changes ultimately led to an increase, decrease or no effect on *in vitro* potency.

Consistent with the bifurcation model of HB, in which the macrocyclic region represented the minimum pharmacophoric substructure for biological activity, structure modifications to regions of the molecule distant from the macrocyclic ring (e.g., the C.32 side chain) had little effect on *in vitro* biological activity. For example, treatment of **45** with p-toluenesulfonyl chloride and several nucleophiles afforded a variety of derivatives differing at C.35 but exhibiting a nearly identical *in vitro* biological profile (Table 13.4).[37] Inverting the C.34 stereocenter (e.g., **46**) similarly had little effect on potency, as did cleavage of the C.34–C.35 vicinal diol to furnish alcohol **50**.

Treatment of **45** with p-toluenesulfonyl chloride followed by ammonia, however, provided an amine (compound **51**; E7389, previously ER-086526, also NSC-707389) that for the first time exhibited a reversibility ratio of one, indicating that even a 10-h drug washout had no effect on the compound's ability to maintain a CMB (Figure 13.13). Several other amine derivatives were subsequently prepared and also found to exhibit a low reversibility ratio (e.g., **52**). In addition to conferring superior reversibility characteristics to this series, the amino group also provided a convenient handle for late-stage derivatization (e.g., acylation). Although simple amides of **51** were

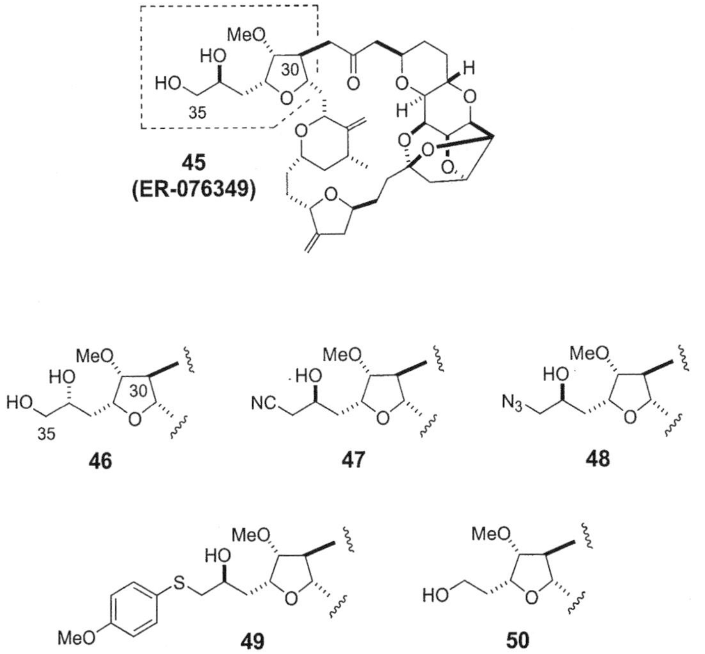

FIGURE 13.12 Representative macrocyclic ketone analogs.

TABLE 13.4
In Vitro Biological Activity of Macrocyclic Ketone Analogs

		Mitotic Block Reversibility		
Compound	Intrinsic Potency, DLD-1 IC_{50} (n*M*)[a]	U937 IC_{99} 0 h (n*M*)[b]	U937 IC_{99} 10 h (n*M*)[c]	Reversibility Ratio
45	1.0	3	38	13
46	1.9	3	100	33
47	0.49	2.0	30	15
48	0.53	3	30	10
49	0.71	3	30	10
50	0.66	3	100	33
51 (E7389)	20	11	11	1
52	3.3	10	10	1
53	1.2	1	10	10
54	0.47	1	10	10
55	0.43	3	10	3.3
56	1.3	3	30	10
57	0.69	3	20	6.7
58	0.72	2	10	5
59	0.87	3	10	3.3
60	1.8	3	30	10

[a] Growth-inhibitory potency against DLD-1 cells after 3–4 d continuous drug exposure.
[b] Induction of CMB in U937 cells after 12 h drug exposure (0 h of washout).
[c] Maintenance of CMB 10 h after drug washout.

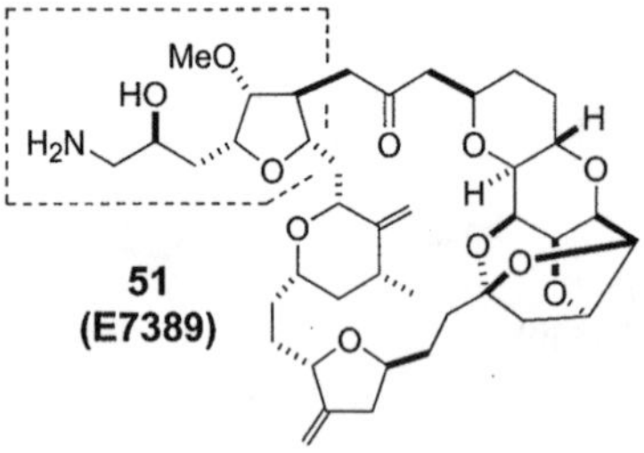

51
(E7389)

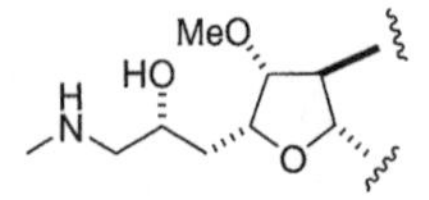

52

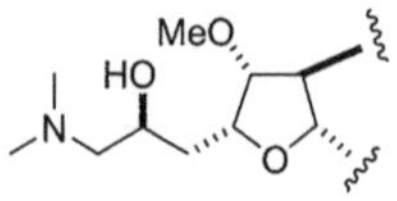

53

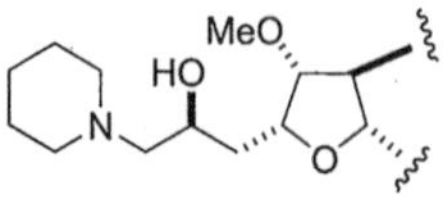

54

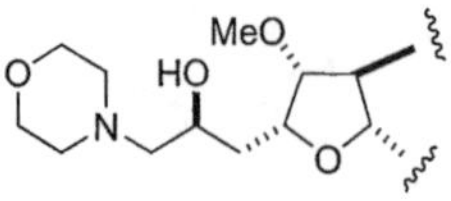

55

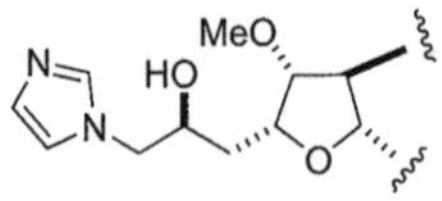

56

FIGURE 13.13 Representative C.35 amine analogs.

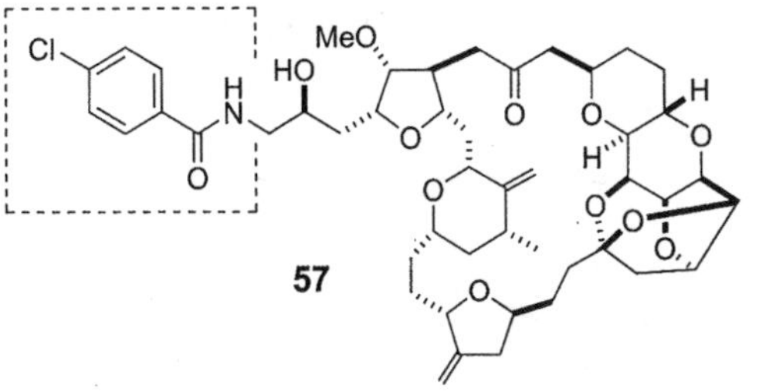

FIGURE 13.14 Representative C.35 amide analogs.

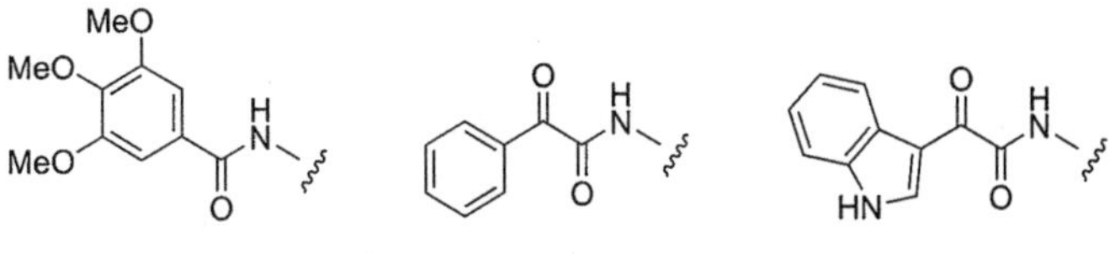

highly potent *in vitro* (Figure 13.14), they exhibited diminished *in vivo* activity in human tumor xenograft models (data not shown). From the analogs prepared, compound **51** (E7389) emerged as the most promising candidate and was therefore selected for preclinical development by Eisai Co., Ltd., as a new anticancer chemotherapeutic agent.

VI. SYNTHESIS OF HB ANALOGS

The macrocyclic ketone analogs were prepared in a highly convergent manner, taking advantage of key building blocks, protecting group strategies, and carbon–carbon bond disconnections developed by Kishi and coworkers for the total synthesis of HB and norhalichondrin B (Figure 13.15). Having established the minimum pharmacophore for cytotoxic activity, the C.1–C.13 and C.14–C.26 synthetic intermediates were fixed and stockpiled, allowing final coupling of the three key fragments to proceed under a standard set of conditions. In this manner, the synthesis of new analogs was reduced to preparation of various C.27–C.35 tetrahydrofuran derivatives for conversion to final product. For example, aldehyde **61** was coupled with vinyl iodide **62** under Nozaki–Hiyama–Kishi conditions[38,39] and cyclized with base. Removal of the MPM protecting group and introduction of the sulfone moiety afforded intermediate **63**. Coupling with C.1–C.13 aldehyde **64** and functional group modification around the C.1 ketone provided an intermediate that underwent a smooth intramolecular Nozaki–Hiyama–Kishi reaction to close the macrocyclic ring between C.13 and C.14. Finally, treatment with tetrabutylammonium fluoride and pyridinium p-toluenesulfonate generated the 2,6,9-trioxatricyclo[3.3.2.0^{3,7}]decane system to furnish **45** in good yield. Modification of the C.32 side chain subsequently afforded the desired macrocyclic ketone analogs.

VII. *IN VITRO* CHARACTERIZATION OF E7389

The halichondrins exhibited a unique pattern of interaction with tubulin compared with other antimitotic drugs,[10] indicating that E7389 may similarly demonstrate unique interactions with tubulin and microtubules. Both E7389 and HB shared low- to subnanomolar cell growth-inhibitory activity against multiple human tumor cell lines *in vitro*, both performed similarly in the NCI 60-cell line screen, both induced G$_2$/M cell cycle arrest with mitotic spindle disruption consistent with a tubulin-based antimitotic mechanism, and both directly inhibited tubulin polymerization with nearly identical IC$_{50}$ values.

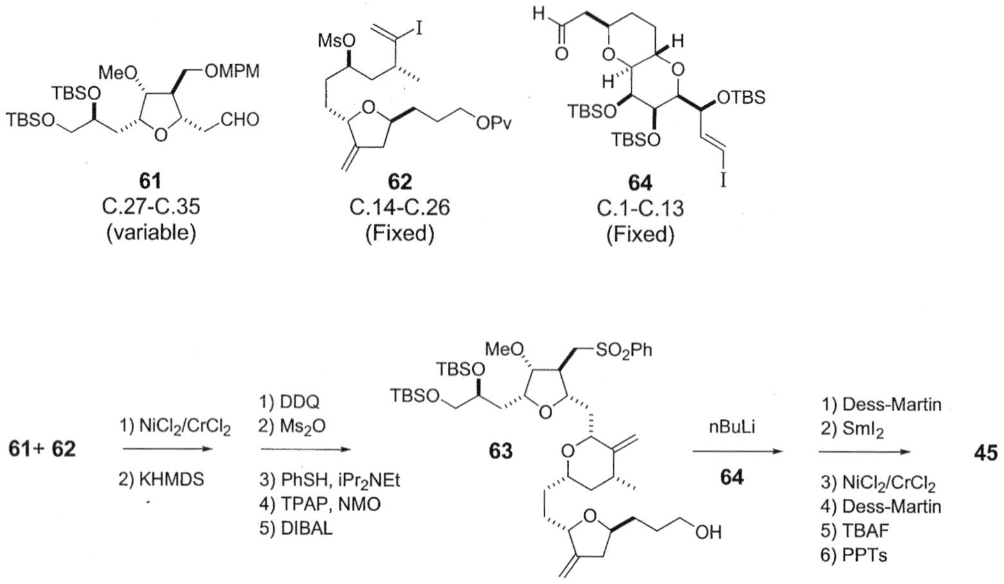

FIGURE 13.15 Synthesis of macrocyclic ketone analogs.

Recent work by Jordan and coworkers revealed that E7389 suppressed microtubule dynamics through a unique mechanism that differed significantly from those of other tubulin-active drugs such as paclitaxel and vinblastine.[15] For example, unlike other antimitotic drugs, this compound induced the formation of nonproductive tubulin aggregates from soluble α/β-tubulin heterodimers that effectively reduced the concentration of soluble tubulin available for productive microtubule growth. Interestingly, there was no effect on treadmilling of microtubules *in vitro*, and in living cells E7389 did not alter catastrophe and rescue frequencies or the shortening events of microtubule dynamic instability at the concentrations tested. The unique characteristics of E7389 in these studies further emphasized the mechanistic differences that set the halichondrin class of compounds apart from other tubulin-interactive agents. Because the role of microtubule-associated proteins and microtubule-regulatory proteins in tubulin aggregate formation induced by E7389 remains unclear at this time, further studies will be needed to elucidate what effects, if any, this agent may have on the interactions between such proteins and microtubules in the context of drug-induced alterations in microtubule function and dynamics. In addition, studies with radio-labeled E7389 may help clarify whether it binds directly to microtubules or only to soluble tubulin heterodimers.

In vitro metabolism studies using a series of human enzyme preparations including liver microsomal preparations, S9 fractions, recombinant drug-metabolizing cytochrome P450 and phase 2 conjugative enzymes, and primary human hepatocytes indicate that E7389 is mainly metabolized by CYP3A4 at pharmacologically relevant concentrations. Suppression of CYP3A4 activity was modest, with an apparent K_i of 3–10 μM, as determined by form-specific nifedipine dehydration, testosterone 6ß-hydroxylation, and R-warfarin 10-hydroxylation. No induction of CYP1A and CYP3A enzymes was observed immunochemically and enzymatically using primary human hepatocytes. Thus, based on a low anticipated therapeutic dose in humans, clinical drug–drug interactions with E7389 are anticipated to be unlikely.[40]

VIII. *IN VIVO* CHARACTERIZATION OF E7389

The subnanomolar growth-inhibitory activity of E7389 *in vitro* was reflected in the marked antitumor activity of this agent *in vivo* at 0.05–1 mg/kg against several human cancer xenografts, including MDA-MB-435 breast cancer, COLO 205 colon cancer, LOX melanoma, and NIH:OVCAR-3 ovarian cancer (Figure 13.16).[27]

In the MDA-MB-435 breast cancer model (Figure 13.16A), treatment with E7389 led to regression of measurable tumors that were present at the start of dosing. No evidence of toxicity was observed at any dose based on body weight losses or decreased water consumption. Notably, all doses of E7389 over a fourfold range were either equally efficacious or superior to paclitaxel, which was dosed at its empirically determined MTD. In contrast, paclitaxel exhibited complete tumor suppression only over a narrow 1.7-fold dose range in this model (data not shown). The ability of E7389 to completely suppress tumor growth at several doses below the maximum tolerated dose (MTD) is unusual for a cytotoxic agent and may be a consequence of its unique mechanism of action.

In the COLO 205 colon cancer xenograft model (Figure 13.16B), E7389 also showed frank regression and long-term suppression of tumor regrowth superior to that seen with paclitaxel at its empirically determined MTD. As in the MDA-MB-435 model, no evidence of toxicity based on body weight losses or decreased water consumption was observed. In the LOX melanoma model (Figure 13.16C), treatment with E7389 led to complete tumor suppression over a fivefold dose range, again representing an unusually wide therapeutic window for a cytotoxic agent. In this model, tumor regrowth rates in the 0.5 mg/kg E7389 group were delayed significantly beyond that seen with 12.5 mg/kg paclitaxel. Finally, in the NIH:OVCAR-3 ovarian cancer xenograft model (Figure 13.16D), E7389 led to complete tumor suppression equivalent to paclitaxel, both during dosing and throughout the regrowth phase.

The ability of E7389 to induce complete tumor suppression over a relatively wide dose range below the MTD is in sharp contrast to the behavior of standard cytotoxic drugs, irrespective of mechanistic class. Expression of *in vivo* anticancer efficacy by paclitaxel, doxorubicin, or the vinca alkaloids, for example, required dosing at or near the MTD, reflecting a delicate balance between efficacy and toxicity. Although speculative, the exceptionally wide therapeutic window for E7389 in preclinical xenograft models may be related to its novel mode of interaction with tubulin and raises the possibility that this agent may exhibit unique characteristics in the clinic that will distinguish it from existing tubulin-based chemotherapeutic agents.

IX. CONCLUSION

In the 18 years since Hirata and Uemura first described the halichondrins, each of the challenges associated with bringing this class of structurally complex marine natural products through the discovery and preclinical development pipeline was successfully met with dramatic results. The first — and to date only — total synthesis of HB reported by Kishi and coworkers provided the basis for directed structure modifications by Eisai scientists to transform the original polyether macrolactone to a structurally simplified macrocyclic ketone through a rational step-wise series of modifications that did not compromise the natural product's remarkable biological profile. Starting from synthetic intermediate **11** prepared in connection with Kishi's total synthesis, the minimum pharmacophoric substructure of the halichondrins was determined to be the C.1–C.34 macrolactone fragment. Optimization of the C.29–C.38 ring system ultimately led to structurally simplified macrolactones that retained the natural product's extraordinary biological activity. Stabilization of the macrolactone ring against serum inactivation and final structure optimization afforded a series of highly potent analogs from which E7389 emerged as the most promising, with outstanding pharmacological, toxicological, and physicochemical characteristics.

With the identification of an optimized fully synthetic drug candidate, the obstacles commonly encountered during the development of structurally complex marine natural products were circumvented.[19] For example, limited compound supply was a critical issue that all but precluded development of HB as an anticancer agent in the early 1990s. Access to sufficient quantity of the producing organism to secure the needed absolute amount of material in a reliable, consistent, and

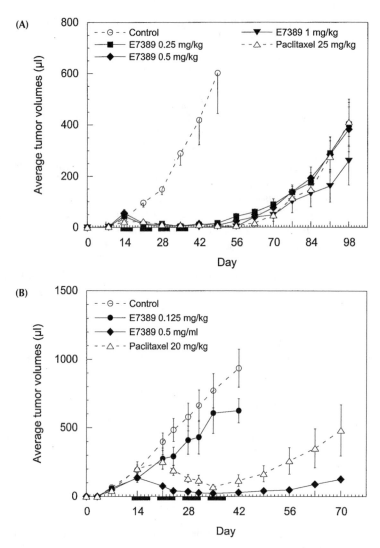

FIGURE 13.16 Inhibition of human tumor xenograft growth *in vivo* by E7389 and paclitaxel. Nude mice bearing MDA-MB-435 breast cancer (A), COLO 205 colon cancer (B), LOX melanoma (C), and NIH:OVCAR-3 ovarian cancer (D). Periods of dosing are indicated by gray bars underlying the *x* axes. COLO 205 and MDA-MB-435 models: E7389 was administered intravenously (MDA-MB-435) or intraperitoneally (COLO 205), using a q2d×3 schedule for four weekly cycles, beginning on day 13. LOX model: E7389 was administered by daily intraperitoneal injection on days 3–7 and 10–14 inclusive. NIH:OVCAR-3 model: E7389 was administered intravenously beginning on day 40 and continuing on a q2d×3 schedule for three weekly cycles. Control groups in all of the studies received appropriate vehicle injections. Paclitaxel was administered at its empirically determined MTD for each treatment regimen in all of the experiments. Plotted mean tumor volumes occurring after onset of any complete remissions represent averages of only those animals continuing to bear measurable tumors. (Modified from Towle, et al., *Cancer Res.*, 61, 1013, 2001. With permission.)

(Continued)

economic manner; separation of the desired drug substance from other structurally related members of the halichondrin family; and removing trace amounts of highly toxic coisolated contaminants to ensure safety were contributing factors that made collection from the wild or through aquaculture techniques impractical, difficult, or unknown. Total synthesis, however, allowed ready access to halichondrin analogs with a defined and consistent impurity profile in sufficient quantity to support

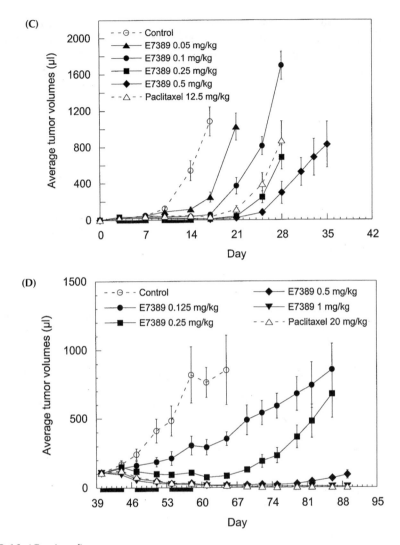

FIGURE 13.16 (*Continued*)

all preclinical research activities, satisfy quality criteria, and prevent compound supply from becoming a critical obstacle in the path for clinical evaluation.

Despite significant simplification relative to the natural product that inspired it, the chemical structure of E7389 remains a synthetically challenging target that exceeds the structural complexity of any other totally synthetic drug on the market today. Whereas success in clinical trials will unequivocally establish the potential of E7389 to become an exciting and important drug in the war on cancer, it will be the development of an economically feasible route for possible commercial production that will determine the long-term success of this program. Although hurdles remain, contemporary synthetic organic chemistry has both the capacity and the potential to meet this challenge.[20]

Tubulin-based antimitotics such as paclitaxel represent a class of agents that is clinically proven. In several human tumor xenograft animal models, E7389 exhibited superior efficacy to paclitaxel, with antitumor activity observed at dose levels well below its MTD. The exceptionally wide therapeutic window of E7389 in preclinical animal models, combined with its novel mechanism of tubulin heterodimer aggregation and superior efficacy to paclitaxel, raises the possibility that E7389 may exhibit a unique profile in the clinic that will differentiate it from other anticancer drugs.

With preclinical development completed and work well underway to address future compound supply needs, the stage is finally set to establish the therapeutic value of halichondrin-based agents in humans. E7389 is currently undergoing phase I clinical trials as an anticancer chemotherapeutic agent with first-in-man studies initiated collaboratively between Eisai Co., Ltd., and NCI.[41] If results in the clinic parallel the outstanding preclinical activity observed in animal models, then E7389 will offer new hope for cancer patients and their families and will help bring the dream of getting drugs from the sea another step closer to reality.

ACKNOWLEDGMENTS

The discovery and development of E7389 was possible only through the dedication, vision, and spirit of innovation shared by the many Eisai scientists who made this human healthcare program a reality. In addition, we express our appreciation for valuable assistance of the NCI through a collaborative research and development agreement to help bring the halichondrin class of marine natural products from laboratory concept to clinical investigative drug.

REFERENCES

1. Uemura, D., et al., Norhalichondrin A: an antitumor polyether macrolide from a marine sponge, *J. Am. Chem. Soc.*, 107, 4796, 1985.
2. Hirata, Y. and Uemura, D., Halichondrins — antitumor polyether macrolides from a marine sponge, *Pure Appl. Chem.*, 58, 701, 1986.
3. Pettit, G. R., et al., Isolation and structure of the cell growth inhibitory constituents from the western pacific marine sponge *Axinella* sp., *J. Med. Chem.*, 34, 3339, 1991.
4. Pettit, G. R., et al. Isolation and structure of halistatin 1 from the eastern Indian Ocean marine sponge *Phakellia carteri*, *J. Org. Chem.*, 58, 2538, 1993.
5. (a) Litaudon, M., et al., Antitumor polyether macrolides: new and hemisynthetic halichondrins from the New Zealand deep-water sponge *Lissodendoryx* sp., *J. Org. Chem.*, 62, 1868, 1997; (b) Gravelos, D.G., et al., Halichondrins: cytotoxic polyether macrolides, European Patent Application, EP 0 572 109, 1993.
6. Lill, J.B., et al., The halichondrins: chemistry, biology, supply and delivery, in *Drugs from the Sea*, Fusetani, N., Ed., Karger, New York, 2000, 134.
7. Fodstad, Ø., et al., Comparative antitumor activities of halichondrins and vinblastine against human tumor xenografts, *J. Exp. Ther. Oncol.*, 1, 119, 1996.
8. Hamel, E., Natural products which interact with tubulin in the vinca domain: maytansine, rhizoxin, phomopsin A, dolastatins 10 and 15 and halichondrin B, *Pharmacol. Ther.*, 55, 31, 1992.
9. Bai, R.L., et al., Halichondrin B and homohalichondrin B, marine natural products binding in the vinca domain of tubulin. Discovery of tubulin-based mechanism of action by analysis of differential cytotoxicity data, *J. Biol. Chem.*, 266, 15882, 1991.
10. Ludueña, R.F., et al., Interaction of halichondrin B and homohalichondrin B with bovine brain tubulin, *Biochem. Pharmacol.*, 45, 421, 1993.
11. Bai, R., et al., Spongistatin 1, a highly cytotoxic, sponge-derived, marine natural product that inhibits mitosis, microtubule assembly, and the binding of vinblastine to tubulin, *Mol. Pharmacol.*, 44, 757, 1993.
12. Jordan, M.A., Thrower, D., and Wilson, L. Mechanism of inhibition of cell proliferation by vinca alkaloids, *Cancer Res.*, 51, 2212, 1991.
13. Jordan, M.A., et al., Mechanism of mitotic block and inhibition of cell proliferation by taxol at low concentrations, *Proc. Natl. Acad. Sci. USA*, 90, 9552, 1993.
14. García-Rocha, M., García-Gravalos, M.D., and Avila, J., Characterization of antimitotic products from marine organisms that disorganize the microtubule network: ecteinascidin 743, isohomohalichondrin-B and LL-15, *Br. J. Cancer*, 73, 875, 1996.
15. Kamath, K., et al., E7389, a synthetic analog of halichondrin B, suppresses microtubule dynamics in living MCF7 cells by a novel mechanism, presented at the annual meeting of the American Association for Cancer Research, Washington, DC, July 11–14, 2003, LB-58.

16. Tsukitani, Y., et al., Japanese Patent Application JP 61191687, 1986; Japanese Patent 92001751, 1992.

17. Towle, M.J., et al., *In vitro* and *in vivo* anticancer properties and cell cycle effects of synthetic halichondrin B, presented at the annual meeting of the American Association for Cancer Research, Toronto, Ontario, March 18–22, 1995, 2342.

18. Hart, J.B., Blunt, J.W., and Munro, M.H.G., Acid-catalyzed reactions of homohalichondrin B, a marine sponge-derived antitumor polyether macrolide, *J. Org. Chem.*, 61, 2888, 1996.

19. Fusetani, N., Introduction, in *Drugs from the Sea*, Fusetani, N., Ed., Karger, New York, 2000, 1.

20. (a) Aicher, T.D. and Kishi, Y., Synthetic studies towards halichondrins, *Tetrahedron Lett.*, 28, 3463, 1987; (b) Aicher, T.D., et al., Total synthesis of halichondrin B and norhalichondrin B. *J. Am. Chem. Soc.*, 114, 3162, 1992; (c) Aicher, T.D., et al., Synthetic studies towards halichondrins: synthesis of the C.27-C.38 segment, *Tetrahedron Lett.*, 33, 1549, 1992; (d) Buszek, K.R., Synthetic studies towards halichondrins: synthesis of the left half of halichondrins, *Tetrahedron Lett.*, 33, 1553, 1992; (e) Fang, F.G., et al., Synthetic studies towards halichondrins: synthesis of the left halves of norhalichondrins and homohalichondrins, *Tetrahedron Lett.*, 33, 1557, 1992; (f) Duan, J.J. and Kishi, Y., Synthetic studies on halichondrins: a new practical synthesis of the C-1–C-12 segment, *Tetrahedron Lett.*, 34, 7541, 1993; (g) Chen, C., Tagami, K., and Kishi, Y., Ni(II)/Cr(II)-mediated coupling reaction: an asymmetric process, *J. Org. Chem.*, 60, 5386, 1995; (h) Stamos, D.P. and Kishi, Y., Synthetic studies on halichondrins: a practical synthesis of the C.1-C.13 segment, *Tetrahedron Lett.*, 37, 8643, 1996; (i) Stamos, D.P. and Kishi, Y., Synthetic studies on halichondrins: a practical synthesis of the C.1-C.13 segment, *Tetrahedron Lett.*, 37, 8643, 1996; (j) Stamos, D.P., Taylor, A.G., and Kishi, Y., A mild preparation of vinyliodides from vinylsilanes, *Tetrahedron Lett.*, 37, 8647, 1996; (k) Stamos, D.P., Chen, S.S., and Kishi, Y., New synthetic route to the C.14 – C.38 segment of halichondrins, *J. Org. Chem.*, 62, 7552, 1997; (l) Stamos, D.P., et al., Ni(II)/Cr(II)-mediated coupling reaction: beneficial effects of 4-tert-butylpyridine as an additive and development of new and improved workup procedures, *Tetrahedron Lett.*, 38, 6355, 1997; (m) Stamos, D.P., Chen, S.S., and Kishi, Y., New synthetic route to the C.14-C.38 segment of halichondrins, *J. Org. Chem.*, 62, 7552, 1997; (n) Xie, C., Nowak, P., and Kishi, Y., Synthesis of the C20-C26 building block of halichondrins via a regiospecific and stereoselective SN2' reaction, *Org. Lett.*, 4, 4427, 2002; (o) Wan, Z.-K., et al., Asymmetric Ni(II)/Cr(II)-mediated coupling reaction: stoichiometric process, *Org. Lett.*, 4, 4431, 2002; (k) Choi, H., et al., Asymmetric Ni(II)/Cr(II)-mediated coupling reaction: catalytic process, *Org. Lett.*, 4, 4435, 2002; (p) Choi, H., et al., Synthetic studies on the marine natural product halichondrins, *Pure Appl. Chem.*, 75, 1, 2003.

21. (a) Kim, S. and Salomon, R.G., Total synthesis of halichondrins: highly stereoselective construction of a homochiral pentasubstituted H-ring pyran intermediate from -D-glucose, *Tetrahedron Lett.*, 30, 6279, 1989; (b) Cooper, A.J. and Salomon, R.G., Total synthesis of halichondrins: enantioselective construction of a homochiral pentacyclic C1-C15 intermediate from D-ribose, *Tetrahedron Lett.*, 31, 3813, 1990; (c) DiFranco, E., Ravikumar, V.T., and Salomon, R.G., Total synthesis of halichondrins: enantioselective construction of a homochiral tetracyclic KLMN-ring intermediate from D-mannitol, *Tetrahedron Lett.*, 34, 3247, 1993; (d) Cooper, A.J., Pan, W., and Salomon, R.G., Total synthesis of halichondrin B from common sugars: an F-ring intermediate from D-glucose and efficient construction of the C1 to C21 segment, *Tetrahedron Lett.*, 34, 8193, 1993.

22. (a) Burke, S.D., Buchanan, J.L., and Rovin, J.D., Synthesis of a C(22) C(34) halichondrin precursor via a double dioxanone-to-dihydropyran rearrangement, *Tetrahedron Lett.*, 32, 3961, 1991; (b) Burke, S.D., et al., An expeditious synthesis of the C(1)-C(14) subunit of halichondrin B, *Tetrahedron Lett.*, 35, 703, 1994; (c) Burke, S.D., Zhang, G., and Buchanan, J.L, Enantioselective synthesis of a halichondrin B C(20)–C(36) precursor, *Tetrahedron Lett.*, 36, 7023, 1995; (d) Burke, S.D., et al., Synthetic studies toward complex polyether macrolides of marine origin, *Spec. Publ. R. Soc. Chem.*, 198 (Anti-Infectives), 73, 1997; (e) Burke, S.D., Austad, B.C., and Hart, A.C., An expeditious synthesis of the C(38)-C(54) halichondrin B subunit, *J. Org. Chem.*, 63, 6770, 1998; (f) Burke, S.D., Quinn, K.J., and Chen, V.J., Synthesis of a C(22)-C(34) halichondrin B precursor via ring opening-double ring closing metathesis, *J. Org. Chem.*, 63, 8626, 1998; (g) Burke, S.D., et al., Halichondrin B: synthesis of the C(1)-C(15) subunit, *J. Org. Chem.*, 65, 4070, 2000; (h) Austad, B.C., Hart, A.C., and Burke, S.D., Halichondrin B: synthesis of the C(37)-C(54) subunit, *Tetrahedron*, 58, 2011, 2002; (i) Jiang, L. and Burke, S.D., A novel route to the F-ring of halichondrin B. Diastereoselection in Pd(0)-mediated meso and C2 diol desymmetrization, *Org. Lett.*, 4, 3411, 2002; (j) Jiang, L., Martinelli, J.R.,

and Burke, S.D., A practical synthesis of the F-ring of halichondrin B via ozonolytic desymmetrization of a C2-symmetric dihydroxycyclohexene. *J. Org. Chem.*, 68, 1150, 2003; (k) Lambert, W.T. and Burke, S.D., Halichondrin B: synthesis of a C1-C14 model via desymmetrization of (+)-conduritol E, *Org. Lett.*, 5, 515, 2003.

23. (a) Horita, K., et al., Chiral synthesis of polyketide-derived natural products. 46. Synthetic studies of halichondrin B, an antitumor polyether macrolide isolated from a marine sponge. 1. Stereoselective synthesis of the C1-C13 fragment via construction of the B and A rings by kinetically and thermo-dynamically controlled intramolecular Michael reactions, *Synlett*, 38, 1994; (b) Horita, K., et al., Chiral synthesis of polyketide-derived natural products. 47. Synthetic studies of halichondrin B, an antitumor polyether macrolide isolated from a marine sponge. 2. Efficient synthesis of C16-C26 fragments via construction of the D ring by a highly stereocontrolled iodoetherification, *Synlett*, 40, 1994; (c) Horita, K., Chiral synthesis of polyketide-derived natural products. 49. Synthetic studies of halichondrin B, an antitumor polyether macrolide isolated from a marine sponge. 4. Synthesis of the C37-C54 subunit via stereoselective construction of three consecutive J, K, and L rings, *Synlett*, 46, 1994; (d) Horita, K., et al., Chiral synthesis of polyketide-derived natural products. 48. Synthetic studies of halichondrin B, an antitumor polyether macrolide isolated from a marine sponge. 3. Synthesis of C27-C36 subunit via completely stereoselective C-glycosylation to the F ring, *Synlett*, 43, 1994; (e) Horita, K., et al., Synthetic studies of halichondrin B, an antitumor polyether macrolide isolated from a marine sponge. 5. A highly concise and efficient synthesis of the C37-C54 tricyclic JKL-ring part, *Heterocycles*, 42, 99, 1996; (f) Horita, K., et al., Synthetic studies of halichondrin B, an antitumor polyether macrolide isolated from a marine sponge. 6. Synthesis of the C1-C15 unit via stereoselective construction of the B and A rings by kinetically and thermodynamically controlled Michael reactions with the aid of computational search for dominant conformers, *Chem. Pharm. Bull.*, 45, 1265, 1997; (g) Horita, K., et al., Synthetic studies of halichondrin B, an antitumor polyether macrolide isolated from a marine sponge. 7. Synthesis of two C27-C36 units via construction of the F ring and completely stereoselective C-glycosylation using mixed Lewis acids, *Chem. Pharm. Bull.*, 45, 1558, 1997; (h) Horita, K., et al., Synthetic studies of halichondrin B, an antitumor polyether macrolide isolated from a marine sponge. 8. Synthesis of the lactone part (C1-C36) via Horner-Emmons coupling between C1-C15 and C16-C36 fragments and Yamaguchi lactonization, *Tetrahedron Lett.*, 38, 8965, 1997; (i) Horita, K., et al., Synthetic studies on halichondrin B, an antitumor polyether macrolide isolated from a marine sponge. 9. Synthesis of the C16-C36 unit via stereoselective construction of the D and E rings, *Chem. Pharm. Bull.*, 46, 1199, 1998; (j) Horita, K., Nishibe, S., and Yonemitsu, O., Research on antitumor active site of marine source natural product, halichondrin B. *International Congress Series*, 1157 (Towards Natural Medicine Research in the 21st Century), 327, 1998; (k) Yonemitsu, O., Yamazaki, T., and Uenishi, J.-I., On the stereoselective construction of the B and A rings of halichondrin B. A PM3 study, *Heterocycles*, 49, 89, 1998; (l) Horita, K., Nishibe, S., and Yonemitsu, O., Synthetic study of a highly antitumorigenic marine phytochemical, halichondrin B, in *Phytochemicals and Phytopharmaceuticals*, Shahidi, F. and Ho, C.-T., Eds., AOCS Press, Champaign, IL, 2000, chap. 35.

24. Michael R. Boyd, Personal communication, 1992.

25. Towle, M. J., Kishi, Y., and Littlefield, B. A., Unpublished observations, 1992.

26. Yoon, S.K., et al., U.S. Patent Application 849769, 1992; U.S. Patent 5338865, 1994; U.S. Patent 5436238, 1995.

27. Towle, M.J., et al., *In vitro* and *in vivo* anticancer activities of synthetic macrocyclic ketone analogs of halichondrin B, *Cancer Res.*, 61, 1013, 2001.

28. Horita, K., Nishibe, S., and Yonemitsu, O., Research on anti-tumor active site of marine source natural product, halichondrin B, International Congress Series, 1157 (Towards Natural Medicine Research in the 21st Century), 327, 1998.

29. Horita, K., Takahashi, N., and Nishizawa, M., Japanese Patent JP 2003261447, 2003.

30. Horita, K., et al., Research on anti-tumor active site of marine source compound, halichondrin B. Inhibitory activity of C37-C54 part derivatives toward the breast and the esophagus cancer cell lines, *Bull. Marine Biomed. Inst. Sapporo Med. Univ.*, 5, 35, 2002.

31. Towle, M.J., et al., *In vivo* anticancer activity of synthetic halichondrin B macrocyclic ketone analogs ER-076349 and ER-086526 correlates with ability to induce irreversible mitotic blocks, presented at

the annual meeting of the American Association for Cancer Research, New Orleans, LA, March 24–28, 2001. Abstract 1976.

32. (a) Littlefield, B.A., et al., U.S. Patent 6214865, 2001; (b) Littlefield, B.A., et al., U.S. Patent 6365759, 2002; Littlefield, B.A., et al., U.S. Patent 6469182, 2002; (c) Littlefield, B.A. and Towle, M.J., U.S. Patent 6653341, 2003.

33. Wang, Y., et al., Structure-activity relationships of halichondrin B analogs: modifications at C.30-C38, *Bioorg. Med. Chem. Lett.*, 10, 1029 (2000).

34. Eisai Research Institute, 1995.

35. Supplementary crystallographic data for norhalichondrin A *p*-bromophenacyl ester (CCDC 231897). http://www.ccdc.cam.ac.uk/data_request/cif.

36. Towle, M.J., et al., Halichondrin B macrocyclic ketone analog E7389: medicinal chemistry repair of lactone ester instability generated during structural simplification to clinical candidate, presented at the annual meeting of the American Association for Cancer Research, San Francisco, CA, April 6–10, 2002, 5721.

37. Zheng, W., et al., Synthetic macrocyclic ketone analogs of halichondrin B: structure-activity relationships, presented at the annual meeting of the American Association for Cancer Research, San Francisco, CA, April 1–5, 2000, 1915.

38. (a) Jin, H, et al., Catalytic effect of nickel(II) chloride and palladium(II) acetate on chromium(II)-mediated coupling reaction of iodo olefins with aldehydes, *J. Am. Chem. Soc.*, 108, 5644, 1986; (b) Takai, K., et al., Reactions of alkenylchromium reagents prepared from alkenyl trifluoromethane-sulfonates (triflates) with chromium(II) chloride under nickel catalysis, *J. Am. Chem. Soc.*, 108, 6048, 1986.

39. (a) Nozaki, H. and Takai, K., Nucleophilic addition of organochromium reagents to carbonyl compounds, *Proc. Jpn. Acad.*, 76, 123, 2000; (b) Fürstner, A., Carbon-carbon bond formations involving organochromium(III) reagents, *Chem. Rev.*, 99, 991, 1999; (c) Wessjohann, L.A. and Scheid, G., Recent advances in chromium(II)- and chromium(III)-mediated organic synthesis, *Synthesis* 1, 1999; (d) Avalos, M., et al., Synthetic variations based on low-valent chromium: new developments, *Chem. Soc. Rev.*, 28, 169, 1999; (e) Saccomano, N.A., Organochromium reagents, in *Comprehensive Organic Synthesis*, Trost, B. M., Fleming, I., Eds., Pergamon Press, Oxford, UK, 1991; vol. 1, chap. 1.6.

40. Zhang, Z.-Y., et al., Characterization of *in vitro* metabolism of an anticancer agent E7389: prediction of the potential risk of clinical drug-drug interactions, presented at the 12th North American meeting of the International Society for the Study of Xenobiotics, Providence, RI, October 12–16, 2003, 367.

41. Synold, T.W., et al., Human pharmacokinetics of E7389 (halichondrin B analog), a novel anti-microtubule agent undergoing phase I investigation in the California Cancer Consortium (CCC), presented at the American Society of Clinical Oncology, Chicago, IL, May 31–June 3, 2003, 575.

14 HTI-286, A Synthetic Analog of the Antimitotic Natural Product Hemiasterlin

Raymond J. Andersen and Michel Roberge

CONTENTS

I.   Introduction 267
II. Natural Product Lead Structures..267
III. Biological Activity of the Natural Products...269
IV. Synthesis of Hemiasterlin 271
V.   Synthesis of Analogs and SAR..275
VI. Biological Activity of HTI-286 277
VII. Preclinical and Clinical Development of HTI-286 277
VIII. Summary 278
Acknowledgment...279
References 279

I. INTRODUCTION

HTI-286 (**1**) is a synthetic experimental anticancer drug currently in phase II clinical trials for the treatment of non–small cell lung cancer.[1] It shows activity in a wide variety of tumor xenograft models, including several multidrug-resistant tumors. The lead structure for the development of HTI-286 was the sponge tripeptide hemiasterlin (**2**), a microtubule depolymerizing agent that kills cells by causing mitotic arrest, leading to apoptosis. The sequence of discoveries that led to the development of HTI-286 and a profile of its biological activities are described here.

II. NATURAL PRODUCT LEAD STRUCTURES

The hemiasterlins, criamides, and milnamides are a small family of tri- and tetrapeptide cytotoxins isolated from a variety of marine sponges. Hemiasterlin (**2**) was first isolated by Kashman and coworkers in 1994 from the sponge *Hemiasterella minor* (Kirkpatrick) collected in Sodwana Bay, South Africa.[2] At virtually the same time, Crews and coworkers reported the isolation of milnamide A (**3**) from the sponge *Auletta* cf. *constricta* collected in Milne Bay, Papua New Guinea.[3] Both source sponges also contained cyclic depsipeptide cytotoxins. *H. minor* yielded jaspamide (jasplak-inolide) (**4**) and geodiamolide TA (**5**) in addition to hemiasterlin (**2**), and *A. constricta* contained jaspamide (**4**) and milnamide A (**3**).

Hemiasterlin (**2**) is a tripeptide composed of the rare amino acid *tert*-leucine and the two novel amino acids 4-amino-2,5-dimethylhex-2-enoic acid and N,N,β,β-tetramethyltryptophan. *Tert*-Leu had been previously reported only from the discodermins,[4] isolated from the marine sponge *Discodermia kiiensis*, and the bottromycin peptides, produced by cultured actinomycetes.[5]

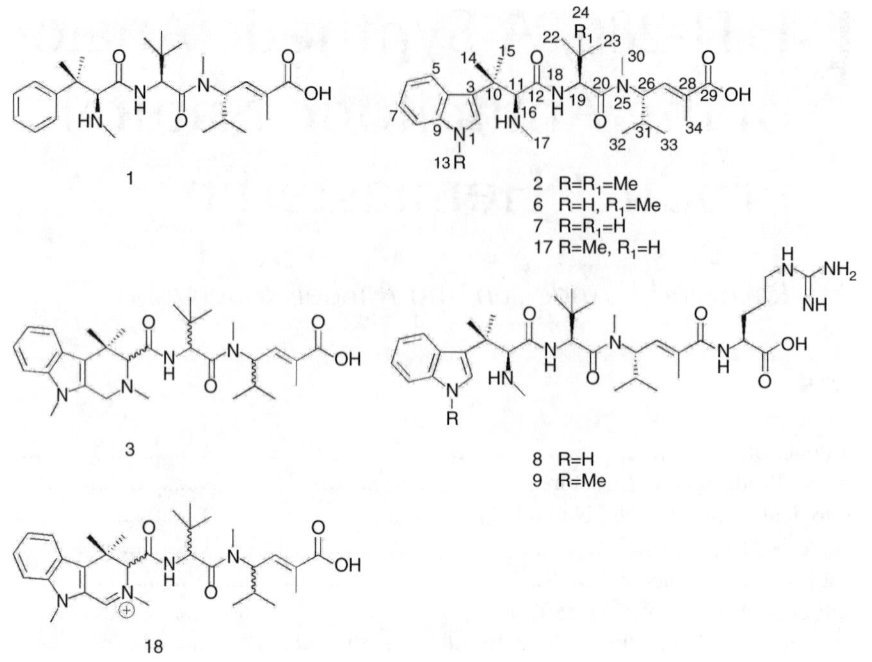

STRUCTURES 1–3, 6–9, 17, 18.

Milnamide A (**3**) has a β-carboline, formally derived from N,N,β,β-tetramethyltryptophan via a Pictet Spengler condensation with formaldehyde, in place of the N,N,β,β-tetramethyltryptophan residue in hemiasterlin (**2**). The 4-amino-2,5-dimethylhex-2-enoic acid residue in **2** and **3** appears to have a biogenetic origin involving N-methylvaline and propionic acid building blocks. There were no relative or absolute configurations assigned to the amino acids in hemiasterlin (**2**) or milnamide A (**3**) in the original reports of their structures. Kashman noted there was an NOE observed between the olefinic methyl (H-34) and tryptophan N-methyl (H-17) proton resonances in hemiasterlin (**2**), indicating a cyclic conformation of the C11 to C29 peptide backbone, probably stabilized by an electrostatic attraction between the spatially close C-terminal carboxylate anion and N-terminal N-methyl ammonium cation of the zwitterionic linear peptide.[2]

In 1995, Andersen and coworkers reported the isolation of hemiasterlin (**2**), hemiasterlins A (**6**) and B (**7**), criamides A (**8**) and B (**9**), and geodiamolides A (**10**) to G (**16**) from the sponge *Cymbastela* sp. collected on reefs off of Madang, Papua New Guinea.[6] All three amino acids in hemiasterlin were shown to have the L configuration by a combination of Marfey's analysis on the natural product and single-crystal x-ray diffraction analysis of hemiasterlin methyl ester.[7] It was also shown that the arginine residue in the criamides had the L configuration. The solid-state conformation of hemiasterlin methyl ester has an extended linear peptide backbone, which supports Kashman's suggestion that ionic interactions play an important role in the formation of a cyclic peptide backbone conformation for hemiasterlin in solution.

Boyd and coworkers reported finding hemiasterlins in two sponges also collected in Papua New Guinea.[8] An *Auletta* sp. yielded the known compounds hemiasterlin (**2**), hemiasterlin A (**6**), and several geodiamolides, along with the new tripeptide hemiasterlin C (**17**). Two *Siphonochalina* spp. were also found to contain the same hemiasterlin tripeptides **2**, **6**, and **17**, along with geodiamolides. No evidence for the presence of milnamide A (**3**) or hemiasterlin B (**7**) was found in any of the extracts studied. Ireland's group reported isolating jaspamide, hemiasterlin (**2**), milnamide A (**3**), and the new compound milnamide D (**18**) from specimens of *Cymbastela* sp. collected in Milne Bay, Papua New Guinea.[9]

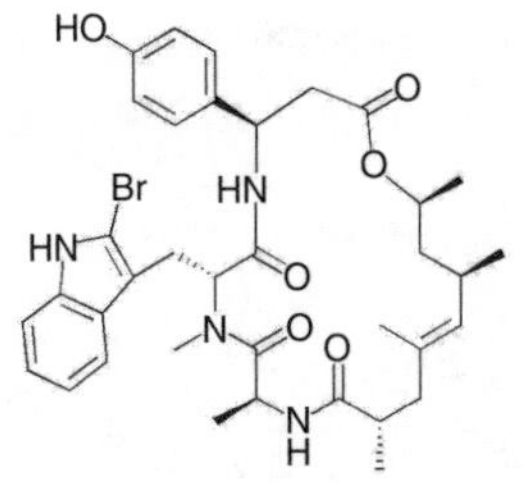

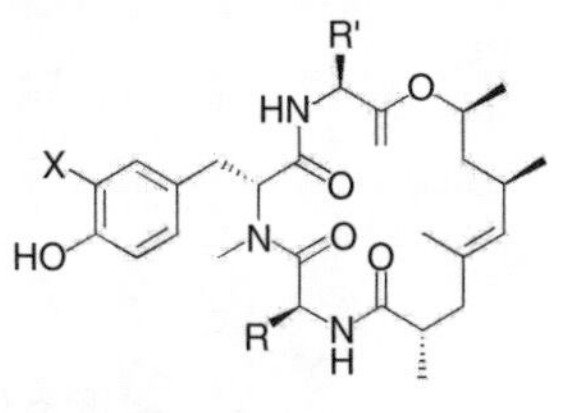

5   R' = isopropyl, R = methyl, X=I
10 R'=R=Me, X=I
11 R'=R=Me, X=Br
12 R'=R=Me, X=Cl
13 R'=Me, R=H, X=I
14 R'=Me, R=H, X=Br
15 R'=Me, R=H, X=Cl

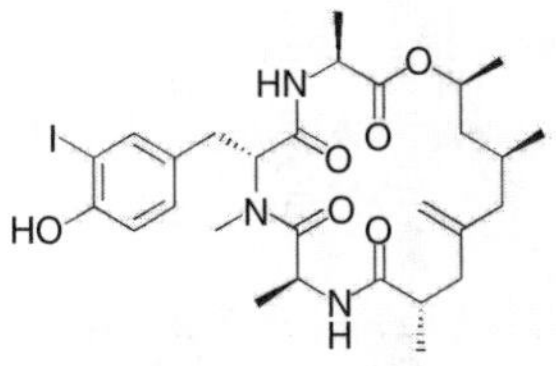

STRUCTURES 4, 5, 10–16, 19.

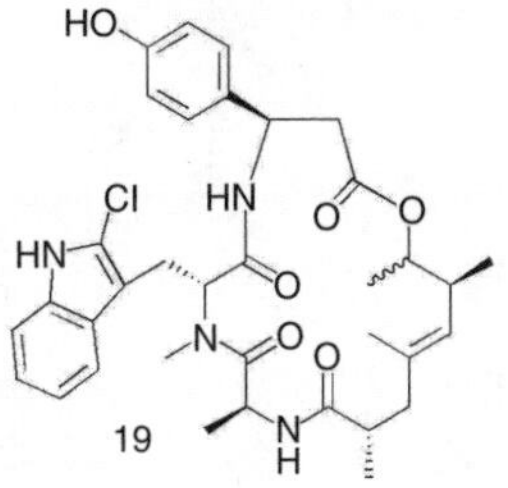

Table 14.1 lists the taxonomic classification of the sponges that have been the source of hemiasterlins and milnamides.[10] The source sponges in the genera *Auletta* and *Cymbastela* belong to the same family *Axinellidae* and are, therefore, closely related taxonomically. In contrast, the sponges *Hemiasterella minor* and *Siphonochalina sp.* belong to different orders, and each is in a different order than *Auletta* and *Cymbastela*. Therefore, the genera *Hemiasterella*, *Siphonochalina*, and *Cymbastela/Auletta* are taxonomically very distant, which indicates a microbial rather than sponge cell origin for the hemiasterlins/milnamides. To date, all occurrences of hemiasterlins/milnamides have been accompanied by cyclic depsipeptides in the jaspamide/geodiamolide family. There is a report of cultured myxobacteria making chondramides (e.g., chondramide A (**19**)), which are close analogs of jaspamide.[11] This strongly supports a microbial origin for the jaspamide/geodiamolide cyclic depsipeptides. The absolute fidelity in the cooccurrence of the hemiasterlin/milnamide and jaspamide/geodiamolide classes of peptides perhaps indicates that there is a single microorganism in each sponge that makes both families. The variation in metabolite content from one source sponge to another might then result either from microbial strain or growth condition differences at each site.

III. BIOLOGICAL ACTIVITY OF THE NATURAL PRODUCTS

Kashman reported that hemiasterlin (**2**) showed *in vitro* cytotoxicity against murine leukemia P388 cells at a concentration of about 19 nM.[2] However, Kashman also cautioned that his sample might have contained impurities — namely the cytotoxic depsipeptide geodiamolide TA (**5**) — so the assays needed to be repeated when additional hemiasterlin became available. Andersen and coworkers isolated sufficient quantities of hemiasterlin from *Cymbastela* to allow the first determination of accurate IC$_{50}$s for its *in vitro* cytotoxicity.[6] The researchers reported an IC$_{50}$ of 87 pM against P388, which is over two orders of magnitude lower than Kashman's value. Potent activity was also observed against a small panel of human cancer cell lines (breast cancer MCF-7 IC$_{50}$, 169 nM; glioblastoma/astrocytoma U373 IC$_{50}$, 23 nM; ovarian carcinoma HEY IC$_{50}$, 3 nM). The tetrapeptide criamides (**8** and **9**) were found to be roughly 100-fold less potent than hemiasterlin. Crews reported

TABLE 14.1
Source Sponges Taxonomy

Research Group	Collection Site	Genus/species	Class	Order	Family	Compounds
Kashman (1994)	Sodwana Bay, South Africa	Hemiasterella minor	Demospongiae	Hadromerida	Hemiasterellidae	Hemiasterlin (2) Geodiamolide TA (5)
Crews (1994)	Milne Bay, Papua New Guinea	Auletta cf. constricta	Demospongiae	Halichondrida	Axinellidae	Milnamide A (3) Jaspamide (4)
Andersen (1995)	Madang, Papua New Guinea	Cymbastela sp.	Demospongiae	Halichondrida	Axinellidae	Hemiasterlins (2,6,7) Criamides (8, 9) Geodiamolides (10-16)
Boyd (1999)	Papua New Guinea	Auletta sp.	Demospongiae	Halichondrida	Axinellidae	Hemiasterlins (2,6,17) Geodiamolides
Boyd (1999)	Papua New Guinea	Siphonochalina sp.	Demospongiae	Haplosclerida	Callyspongiidae	Hemiasterlins (2,6,17) Geodiamolides
Ireland (2003)	Milne Bay, Papua New Guinea	Cymbastela sp.	Demospongiae	Halichondrida	Axinellidae	Hemiasterlin (2) Milnamide D (18) Jaspamide (4)

that milnamide A (**3**) also had *in vitro* cytotoxicity against P388 (IC_{50}, 1.4 μM) and a small panel of human tumor cell lines (IC_{50}s, 5.4–7.9 μM), but its potency was modest compared with hemiasterlin.[3] Ireland found that milnamide D (**18**) (IC_{50}, 67 nM) was more potent than milnamide A (**3**) (IC_{50}, 1600 nM) against human colon carcinoma HCT-116 cells *in vitro*, but that both were less potent than hemiasterlin (**2**) (IC_{50}, 6.8 nM), confirming that the presence of the β-carboline N terminus in **3** and **18** decreased activity relative to **2**.[9]

Andersen and Kashman isolated the hemiasterlins as cytotoxic agents with potent subnanomolar to nanomolar activity against P388 mouse leukemia and other cell lines.[2,6] However, the structure of these compounds and their potent cytotoxic activity gave no clue to their mechanism of action. Roberge and coworkers examined the effects of hemiasterlins on cell cycle progression in human mammary carcinoma MCF-7 cells and found that hemiasterlin, hemiasterlin A, and hemiasterlin B all block cells in mitosis at the same concentrations that showed cytotoxicity.[12] Examination of the hemiasterlin mitotic arrest phenotype using immunofluorescence microscopy showed that the cells arrested at a metaphase-like stage, with effects on the morphology of the mitotic spindle that were similar to those caused by tubulin-targeting agents such as vinblastine and nocodazole. High concentrations of hemiasterlin caused complete microtubule depolymerization. At low hemiasterlin concentrations, cells displayed bipolar spindles with long astral microtubules or multiple asters. Chromosomes failed to align at metaphase and were unable to proceed through later stages of mitosis because of activation of the spindle checkpoint. This study concluded that hemiasterlins exert their cytotoxic effects by inhibiting spindle microtubule dynamics at mitosis.

The study by Anderson[12] implied, but did not demonstrate, that hemiasterlin targets microtubules directly. Lassota, at Wyeth, subsequently showed that hemiasterlin inhibits the *in vitro* assembly of tubulin into microtubules and noncompetitively inhibits vincristine binding to tubulin.[13] Researchers at the National Cancer Institute (NCI) then examined in detail the interaction of hemiasterlin with tubulin *in vitro*. Gamble et al. showed that hemiasterlin, hemiasterlin A, and hemiasterlin C potently inhibited tubulin polymerization *in vitro* (IC_{50}, 1μM).[8] Bai et al.[14] confirmed that hemiasterlin inhibits noncompetitively the binding of vinblastine to tubulin and inhibits competitively the binding of dolastatin 10, indicating that hemiasterlin binding causes steric interference with the binding of vinca alkaloids to tubulin.[15] The researchers further demonstrated that hemiasterlin inhibits nucleotide exchange on β-tubulin and that it induces the formation of tubulin aggregates with a ring-like structure and a diameter of about 40 nm. Overall, these effects resembled, but were not identical to, those of dolastatin and cryptophycin, which bind to the same site in tubulin.

The tubulin aggregates induced by hemiasterlin *in vitro* were recently shown to be closed, single-walled, ring-like nanostructures containing 14 α,β–tubulin dimers.[16] These rings form at high concentrations and are unstable — they dissociate even at the relatively high hemiasterlin concentration of 100 nM. This is unlike the rings induced by cryptophycin, which are very stable, and those induced by dolastatin, which have an intermediate stability. Because the tubulin rings are induced only at hemiasterlin concentrations about 100-fold above the concentration necessary to induce mitotic arrest and to inhibit cell proliferation, they likely do not contribute to hemiasterlin's cytotoxicity.[16]

The first *in vivo* studies of hemiasterlin were carried out by Theresa Allen at the University of Alberta. As reported by Ireland et al.,[17] Allen found high antileukemic activity in an *in vivo* P388 murine leukemia model (%T/C 308 with five doses at 0.45 μg/mouse). Hemiasterlin also showed activity in an *in vivo* Gzhi (metastatic murine breast cancer) model: 60% of the mice were long-term survivors with five doses at 1.0 μg/mouse.

IV. SYNTHESIS OF HEMIASTERLIN

The potent *in vitro* and *in vivo* cytotoxic activity of hemiasterlin, its ability to depolymerize microtubules, a validated anticancer target, and its relatively simple but novel tripeptide structure

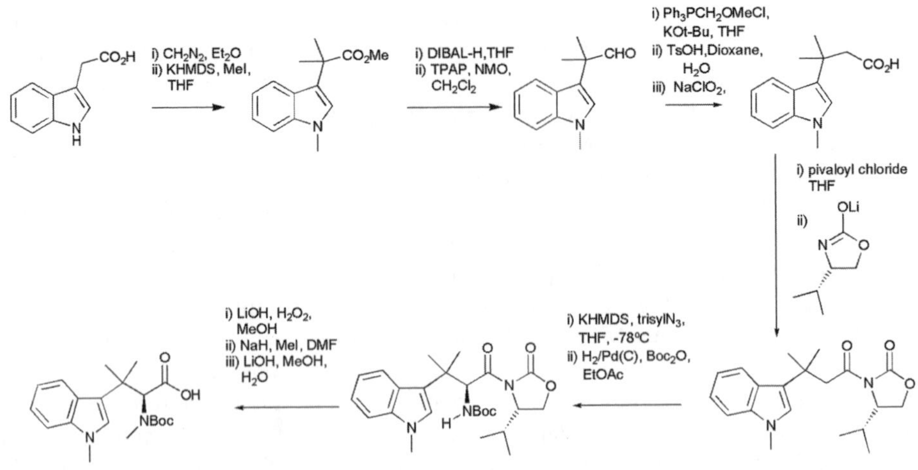

SCHEME 14.1 Reagents for Scheme 14.1:

i) CH$_2$N$_2$, Et$_2$O
ii) KHMDS, MeI, THF

i) DIBAL-H, THF
ii) TPAP, NMO, CH$_2$Cl$_2$

i) Ph$_3$PCH$_2$OMeCl, KOt-Bu, THF
ii) TsOH, Dioxane, H$_2$O
iii) NaClO$_2$,

i) pivaloyl chloride THF
ii)

i) KHMDS, trisylN$_3$, THF, -78°C
ii) H$_2$/Pd(C), Boc$_2$O, EtOAc

i) LiOH, H$_2$O$_2$, MeOH
ii) NaH, MeI, DMF
iii) LiOH, MeOH, H$_2$O

SCHEME 14.1 Synthesis of Boc-tetramethyltryptophan.

made it an attractive target for total synthesis and structure–activity relationship (SAR)-driven modification. A reasonably practical synthesis of hemiasterlin and structural analogs was also required to provide sufficient quantities of compounds for preclinical evaluation and, ultimately, clinical trials. Andersen and Piers reported the first total synthesis of hemiasterlin in 1997.[18,19] The synthesis employed a convergent approach that involved independent preparation of appropriately protected forms of the novel N-terminal tetramethyltryptophan and C-terminal 4-amino-2,5-dimethylhex-2-enoic acid residues, followed by peptide coupling with the commercially available L-*tert*-leucine.

The synthesis of Boc-protected tetramethyltryptophan, starting from indole-3-acetic acid, is outlined in Scheme 14.1. Methylation, first with diazomethane and then with potassium hexamethylsilazide and methyl iodide, introduced the side chain gem-dimethyl and indole N-methyl substituents present in the desired tetramethyltryptophan residue. Homologation of the permethylated indole acetic acid was accomplished in a straightforward fashion to give an intermediate indole-3-propionic acid derivative that contained the carbon skeleton of the final amino acid. Introduction of the α-amino functionality with the correct L configuration was carried out using Evan's methodology. The Evan's chiral auxiliary was introduced by first making a mixed anhydride with 2,2-dimethylpropionyl chloride, which was then treated with the lithiated isopropyl oxazolidone to give an amide. Sequential treatment of the amide with base and 2,4,6-triisopropylbenzenesulfonyl azide produced an α-azide, which was reduced *in situ* to an amine, and then reacted with di-*tert*-butyl dicarbonate to give the Boc-protected α-amine functionality with the correct configuration. The chiral auxiliary was removed with lithium hydroperoxide, and the resulting acid was bismethylated. Hydrolysis of the ester formed in the methylation reaction gave Boc-protected (L)- N,N,β,β-tetramethyltryptophan, ready for peptide coupling.

Scheme 14.2 shows the synthesis of Boc-protected 4-amino-2,5-dimethylhex-2-enoic acid ethyl ester. Commercially available (L)-N-Boc-N-methylvaline was first converted into the Weinreb amide, followed by reduction with LiAlH$_4$, to give the corresponding aldehyde. A Wittig reaction on the aldehyde gave the fully protected C-terminal amino acid residue with high stereoselectivity.

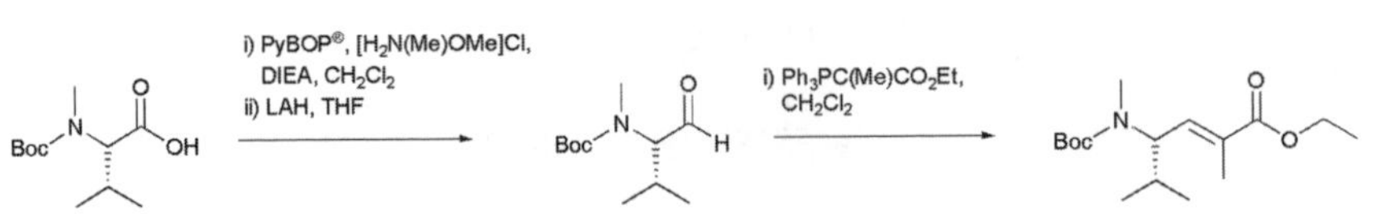

i) PyBOP®, [H$_2$N(Me)OMe]Cl, DIEA, CH$_2$Cl$_2$
ii) LAH, THF

i) Ph$_3$PC(Me)CO$_2$Et, CH$_2$Cl$_2$

SCHEME 14.2 Synthesis of Boc-4-amino-2,5-dimethylhex-2-enoic acid ethyl ester.

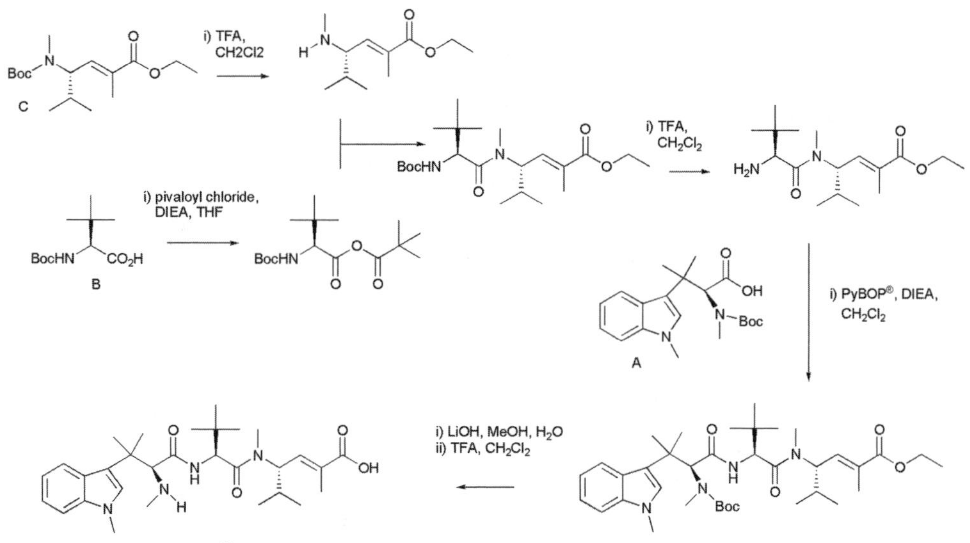

SCHEME 14.3 Andersen synthesis of hemiasterlin.[18,19]

The reactions involved in coupling the N-Boc-amino acids and deprotection to form the desired tripeptide hemiasterlin (**2**) are shown in Scheme 14.3. In the discussions that follow, the N-terminal N,N,β,β-tetramethyltryptophan will be designated as the "A" residue, *tert*-leucine as "B," and 4-amino-2,5-dimethylhex-2-enoic acid as "C." Formation of the N-Boc-B-C-OEt dipeptide was accomplished by removal of the N-Boc group of the fully protected C residue with TFA, followed by coupling with N-Boc-amino acid B as either the mixed anhydride or in the presence of the uronium coupling agent HATU (not shown[19]). Fully protected hemiasterlin resulted from treatment of the TFA salt of the dipeptide B-C-OEt with the N-Boc protected amino acid A, using PyBOP® as the coupling agent. Deprotected hemiasterlin (**2**) was obtained by saponification of the ethyl ester, followed by removal of the Boc group with TFA.

A second, shorter, and more efficient synthesis of hemiasterlin (**2**) that takes advantage of N-benzothiazol-2-sulfonyl (Bts)–protected amino acid derivatives was subsequently reported by Vedejs in 2001.[20] The Vedejs approach to hemiasterlin (**2**), outlined in Scheme 14.4, parallels the Andersen synthesis, passing through many of the same intermediates. Commercially available (S)-valinol is the starting point for the Vedejs synthesis of the ethyl ester of 4-amino-2,5-dimethylhex-2-enoic acid. Bts-protected (S)-valinol was readily N-methylated under very mild conditions by treatment with iodomethane and potassium carbonate in DMF. Oxidation using either Dess-Martin periodinane or Swern conditions gave the corresponding aldehyde that was subjected to the same ester-stabilized Wittig reaction employed by Andersen. Removal of the Bts protecting group by treatment with PhSH/K_2CO_3 afforded the C residue ethyl ester ready for peptide coupling.

L-*tert*-Leucine was protected as its Bts amide, and the free carboxylic acid was converted to an acid chloride in preparation for peptide coupling. Biphasic conditions (NaHCO$_3$–Na$_2$CO$_3$, CH$_2$Cl$_2$–H$_2$O) were used to couple the activated L-*tert*-leucine with the C residue amine. This coupling reaction, involving a N-methyl amine, proceeded in higher yield (86%) than the mixed anhydride coupling (69%) employed by Andersen to make the same peptide bond. Removal of the Bts protecting group by treatment with PhSH/K_2CO_3 gave the BC dipeptide ethyl ester, ready for coupling to the A residue.

Synthesis of a protected N,N,β,β-tetramethyltryptophan A residue started with the trimethylated indole-3-acetaldehyde intermediate from the Andersen synthesis. An asymmetric Strecker sequence

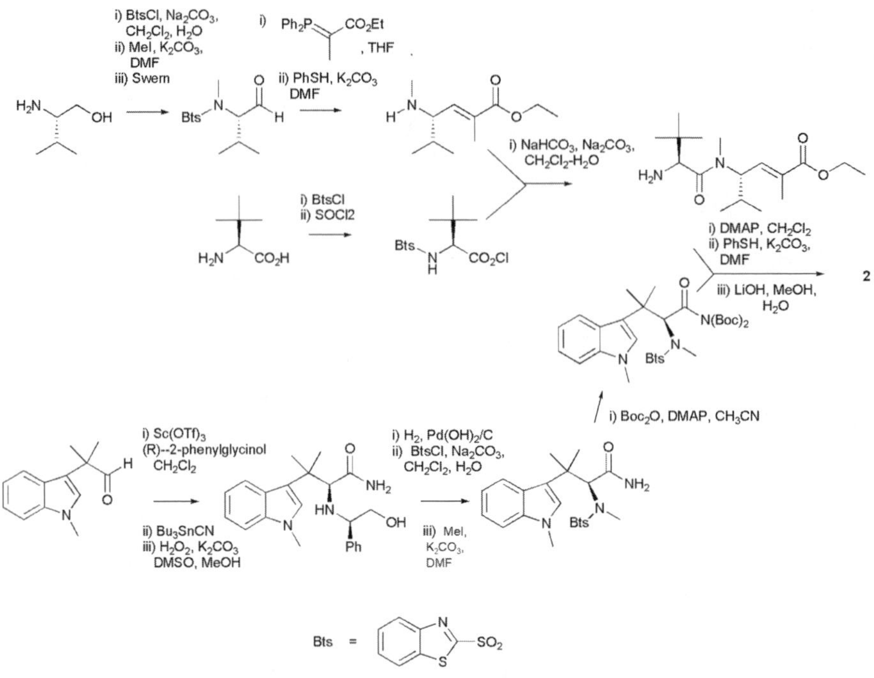

SCHEME 14.4 Vedejs synthesis of hemiasterlin.[20]

using (R)-2-phenylglycinol as the chiral auxiliary and Bu$_3$SnCN as a source of cyanide was employed to introduce the α-amino group with the required stereochemistry. The resulting nitrile was hydrolyzed to a primary amide with H$_2$O$_2$/K$_2$CO$_3$·1.5H$_2$O in 10:1 MeOH:DMSO, and the chiral auxiliary was removed by hydrogenolysis to give a primary α-amine. Treatment of the amine with BtsCl followed by methylation with MeI/K$_2$CO$_3$ gave the N-Bts-N-methyl amide, which was converted to the bis-Boc derivative by reaction with Boc$_2$O/DMAP in CH$_3$CN. The bis-Boc amide was reactive enough that with DMAP catalysis it directly formed a peptide bond with the amino group of the BC dipeptide ethyl ester (97% yield). Straightforward removal of the Bts (PhSH/K$_2$CO$_3$) and ethyl ester (LiOH) protecting groups gave hemiasterlin (**2**).

Durst's group has reported two approaches to a convergent asymmetric synthesis of the N,N,β,β-tetramethyltryptophan A residue in hemiasterlin (**2**) (Scheme 14.5).[21] The key sequence of reactions in the first approach are a SnCl$_4$-mediated ring opening of a cyanoepoxide in the

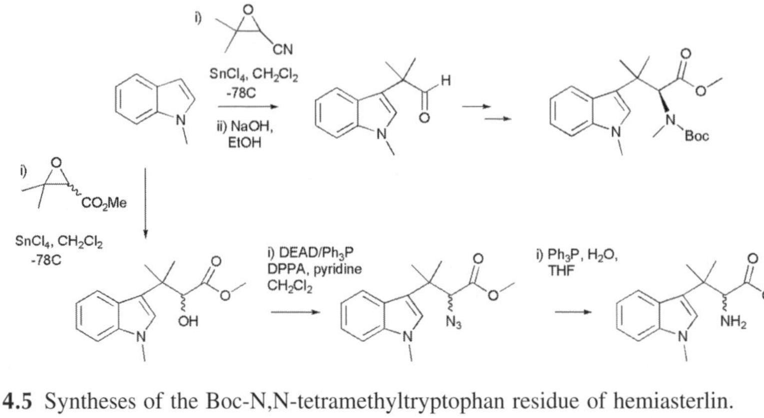

SCHEME 14.5 Syntheses of the Boc-N,N-tetramethyltryptophan residue of hemiasterlin.

presence of N-methylindole to give a cyanohydrin, followed by hydrolysis with NaOH to give the trimethylated indole-3-acetaldehyde intermediate from the Andersen and Vedejs syntheses. This intermediate was then converted into the N-Boc methyl ester derivative of N,N,β,β-tetramethyltryptophan by a chiral Strecker synthesis similar to that employed by Vedejs. Durst's second approach involved the use of a glycidic ester in place of the cyanoepoxide to directly generate an α-hydroxy methyl ester (Scheme 14.5). Mitsunobu azidization of the α-hydroxy ester followed by reduction with triphenyphosphine in wet THF gave the methyl ester of N,N,β,β-tetramethyltryptophan. Reaction of optically active glycidic ester with N-methyl indole gave the α-hydroxy methyl ester with the same optical purity as the glycidic ester. Durst proposed that further transformation of this compound via Mitsunobu azidization and reduction would lead to enantiomerically enriched N,N,β,β-tetramethyltryptophan methyl ester, although this was not verified experimentally.

V. SYNTHESIS OF ANALOGS AND SAR

The convergent nature of the Andersen synthesis[18] of hemiasterlin made it ideally suited for preparing analogs to probe the structural requirements for potent antimitotic activity. Using this synthesis, each of the A, B, and C amino acid residues was altered in turn to prepare a small library of roughly 30 tripeptides that were assessed for biological activity.[19] The synthetic modifications focused on the unique structural features of the natural product hemiasterlin. These included the extensive methylation of the N-terminal tryptophan residue — and in particular the β,β-dimethylation — the presence of the 4-amino-2,5-dimethylhex-2-enoic acid residue at the C terminus, and occurrence of the rare *tert*-leucine residue in the center of the tripeptide. At the outset, it was felt that these structural features might play important roles in protecting hemiasterlin from proteolysis and in stabilizing a biologically active conformation. A major objective of the analog synthesis program at the University of British Columbia (UBC) was to identify a compound that was more potent and easier to synthesize than the natural product hemiasterlin (**2**).

The development of a convenient cell-based assay for the quantitative determination of cells arrested in mitosis[22] enabled determination of the antimitotic activity of the synthetic hemiasterlin analogs prepared at UBC. A comparison of the IC_{50} of these compounds for antimitotic activity with their IC_{50} for cytotoxicity showed a remarkably linear relation ($r = 0.958$) over a concentration range of almost six orders of magnitude (Figure 14.1).[19] This quantitative relation between antimitotic activity and cytotoxic activity provided strong evidence that the sole mechanism of cytotoxicity of this class of compounds is antimitotic activity (i.e., that the hemiasterlins probably have no secondary cellular targets of relevance to cell proliferation or survival).

The synthetic analog study identified structural elements that are critical for potent activity and regions in which single structural changes could be made without loss of potent activity (Figure 14.2). It was found that the tryptophan β-methyl substituents, the N-methyl substituent on the tryptophan α-amino group, and the C26 isopropyl substituent were all extremely important for potent biological activity. In addition, L configurations for the N,N,β,β-tetramethyltryptophan and 4-amino-2,5-dimethylhex-2-enoic acid residues, as well as the presence of the $\Delta^{27,28}$ alkene, were required for maximal activity. Conversely, the C-terminal carboxylic acid could be converted to a methyl ester, the N-25 methyl substituent could be removed, or the *tert*-leucine residue could be replaced with valine or butyrine as single structural changes without any significant loss in potency. Replacing the N-methylindole fragment of the tetramethyltryptophan residue with either methyl or phenyl groups was also well tolerated.

Only one of the synthetic tripeptide analogs tested was more potent than hemiasterlin against MCF-7 mp53 cells. This compound, designated SPA110 (synthetic peptide analog 110), was the phenyl alanine analog of hemiasterlin.[19,23] The synthesis of SPA110 was considerably shorter and more efficient than the synthesis of hemiasterlin, primarily because the carbon skeleton of the trimethylphenylalanine N-terminal residue could be prepared in a single, high-yield step from

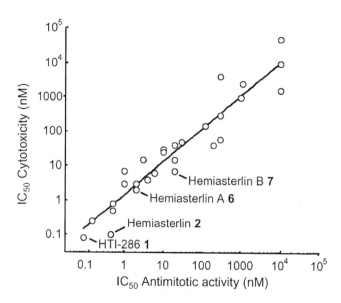

FIGURE 14.1 Plot of *in vitro* cytotoxicity versus antimitotic activity against human breast cancer MCF-7 cells expressing mutant p53, for natural hemiasterlins, HTI-286, and other synthetic analogues (unlabeled circles) generated by Nieman et al.[19]

benzene and 3-methyl-2-butenoic acid, as shown in Scheme 14.6. SPA110 was roughly threefold more potent than hemiasterlin in Roberge's cell-based antimitotic assay (SPA110 IC_{50}, 0.08 n*M*; hemiasterlin IC_{50}, 0.3 n*M*). On the basis of its relative ease of synthesis and enhanced potency, Wyeth Pharmaceuticals licensed SPA110 from UBC, and they gave it the code number HTI-286 (Hemiasterlin Tubulin Inhibitor 286).

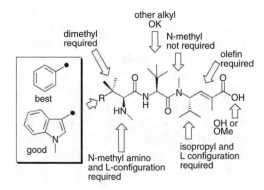

FIGURE 14.2 Structural requirements for optimal cytotoxicity ($IC_{50} < 1$ n*M*) and antimitotic activity ($IC_{50} < 1$ n*M*) of hemiasterlin analogs.

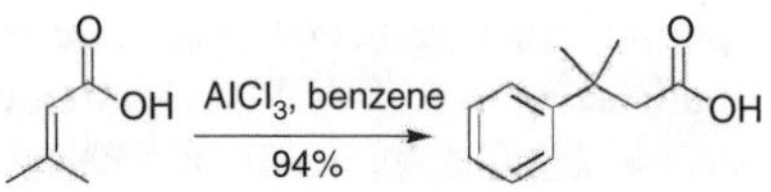

SCHEME 14.6 Synthesis of trimethylphenylalanine N-terminal residue of HTI-286.

VI. BIOLOGICAL ACTIVITY OF HTI-286

Loganzo et al. recently published an in-depth study of the tubulin-binding and antiproliferative properties of HTI-286.[24] HTI-286 potently inhibited tubulin polymerization *in vitro* (about 50% at 0.1 μM and totally at 1 μM), about tenfold more potently than reported by others for hemiasterlin ($IC_{50} = 0.98$ μM).[8] The effects of HTI-286 on the morphology of mitotic cells were similar to those described for hemiasterlin by Anderson.[12] Flow cytometry showed that HTI-286 caused a pronounced accumulation of cells in the G_2 or M phases of the cell cycle after 24 h, consistent with mitotic arrest as the mechanism of action. After 48 h there was an increase in apoptotic cells.

The antiproliferative activity of HTI-286 was evaluated against a panel of 18 tumor cell lines.[24] The compound showed broad antiproliferative activity, independent of tumor origin, with an average IC_{50} of 2.5 ± 2.1 nM. HTI-286 was more potent than paclitaxel ($IC_{50} = 128 \pm 369$ nM) for all cell lines. Importantly, cells expressing P-glycoprotein, and consequently resistant to many drugs including paclitaxel, retained nearly complete sensitivity to HTI-286. In addition, the KB-8-5 cell line that has been selected for expression of drug efflux pumps by chronic drug exposure was resistant to paclitaxel, docetaxel, vinblastine, vinorelbine, dolastatin-10, colchicine, and doxorubicin but retained sensitivity to HTI-286. The KB-V1 cell line that expresses very high levels of P-glycoprotein showed resistance to HTI-286 (81-fold), which was much less than the resistance shown to paclitaxel (1400–1800-fold). HTI-286 was selected for further preclinical development over other hemiasterlin analogs in large part because of this ability to inhibit the growth of cell lines expressing moderate to high levels of P-glycoprotein.

Researchers at Wyeth[25] found that tritium-labeled benzophenone photoaffinity analogs of HTI-286 specifically and solely photolabel α-tubulin. The binding site maps to sheet 8- helix 10 region of α-tubulin, a region of lateral and longitudinal contacts with other tubulin subunits. In contrast, the binding sites for colchicine are at the α–β-tubulin interface, and the binding site for paclitaxel is within β-tubulin. This is the first evidence that a tubulin-binding peptide interacts with α-tubulin and is consistent with the observation that mutations in α-tubulin have been reported in HTI-286-resistant cells.

VII. PRECLINICAL AND CLINICAL DEVELOPMENT OF HTI-286

The initial encouraging *in vivo* efficacy of hemiasterlin in the mouse leukemia P388 model reported by Coleman et al.,[6] combined with the availability of HTI-286, a potent, water-soluble, and more readily synthesized analog, prompted Wyeth to examine the ability of HTI-286 to inhibit tumor growth in a variety of mouse models.[24] HTI-286 administered intravenously has *in vivo* efficacy against a variety of xenograft tumor models, including LOX-IMV melanoma, KB-3-1 epidermoid carcinoma, PC3-MM2 prostate carcinoma, SW620 colon carcinoma, U87 glioma, MCF-7 breast carcinoma, and LOVO colon carcinoma. Moreover, HTI-286 was able to inhibit the growth of even large established tumors derived from LOX-IMV, KB-3-1, LOVO, and PC3MM2. HTI-286 administered for 4 days inhibited the growth of 2.5 g LOX-IMV tumors, and subsequent dosing caused 93% tumor regression.

In keeping with results observed with cell lines in culture, HTI-286 also displayed efficacy against tumors derived from cell lines with inherent or acquired multidrug resistance; KB-8-5 tumors, MX-1W human breast carcinoma, DLD-1 human colon carcinoma, and HCT-15 human colon carcinoma all overexpressed P-glycoprotein and showed resistance to paclitaxel or vincristine but were sensitive to HTI-286.[24] Interestingly, ovarian carcinoma cells resistant to paclitaxel or epothilone, and an A549 lung carcinoma line selected for epothilone resistance because of mutations in the α-tubulin gene, also retained sensitivity to HTI-286.[24]

Wyeth also examined the metabolism of HTI-286 *in vitro* in liver microsomes from a variety of animal species, and *in vivo* in rats and dogs. HTI-286 was metabolized to a major biologically inactive N-demethylated species.[26] Wyeth's preclinical studies also showed that HTI-286 has efficacy when administered orally.[24] These favorable preclinical parameters, combined with the relative ease of synthesis and water solubility of HTI-286, indicated that HTI-286 may have clinical utility in patients, particularly those with intrinsic or acquired resistance to paclitaxel.

A phase I clinical trial was carried out in 2002 with patients with metastatic or advanced-stage malignant solid tumors to assess the safety, tolerability, and pharmacokinetics of HTI-286. HTI-286 dissolved in saline was administered intravenously over 30 min once every 21 d at doses ranging from 0.06 to 2.0 mg/m^2. This investigation has been published in abstract form by Ratain et al.[27] Pain, hypertension, and neutropenia were observed as dose-limiting toxicities. Preliminary pharmacokinetic data indicated considerable interindividual variability in clearance, steady-state volume of distribution, and half-life. Serum concentrations above 0.5 ng/mL were associated with neutropenia. The study recommended a phase II dose of 1.5 mg/m^2 and additional studies to determine the basis for the marked interindividual variability in clearance.

An additional phase I dose escalation study of HTI-286 administered in combination with carboplatin is underway. A phase II open-label study of HTI-286 as a single agent for the treatment of non–small cell lung cancer for disease reoccurrence following platinum-based therapy is also currently underway.[1]

VIII. SUMMARY

Natural products continue to be an important source of lead structures for the development of new experimental drugs for the treatment of cancer.[28] The development path leading to HTI-286 illustrates some of the challenges and opportunities encountered in bringing a marine natural product–derived drug to clinical trials.[29]

Marine sponges, which are the most primitive metazoans in the oceans, are the richest source of new marine natural product structures.[29,30] Secondary metabolites isolated from sponge tissues can be either true metabolites of the sponge's own enzymes or metabolites of microorganisms living in association with the sponge. Large-scale wild harvest of the source sponge is environmentally unacceptable, and the problems associated with sponge aquaculture and with laboratory culture of sponge-associated microorganisms make it difficult to routinely use nature's biosynthetic machinery to make sponge-derived compounds in bulk. Therefore, the major practical source of sponge secondary metabolites for drug development at the moment is via laboratory synthesis. As a consequence, the current goal of examining sponge-derived natural products as part of drug discovery programs is primarily to find "chemical inspiration" about new pharmacophores. Research aimed at overcoming the limitations of culturing sponges and their associated microorganisms, partly involving genetic approaches to cloning entire biosynthetic pathways, is ongoing, so this situation might well change in the near future. As suggested above, hemiasterlin, which was the chemical inspiration for the HTI-286 antimitotic pharmacophore, is presumably a microbial secondary metabolite. Several attempts by the Andersen group at UBC to isolate and culture a hemiasterlin-producing microorganism from *Cymbastela* tissues have all failed.

A number of properties of hemiasterlin made it stand out as an extremely exciting natural product lead structure for an anticancer drug development program. The first things that attracted attention were its exceptionally potent *in vitro* and *in vivo* cytotoxicities against P388 and its relatively simple tripeptide structure. Identification of tubulin as its molecular target elevated the interest in the compound and really provided the impetus for initiating the synthetic program. It is interesting to speculate whether a tripeptide antimitotic agent such as HTI-286 would have ever emerged from a combinatorial chemistry screening program. Some of the first combinatorial libraries comprised small peptides made from protein amino acids, and a Trp-Val-Val tripeptide would have been a logical compound to make as part of a small peptide library. However, the SAR studies on hemiasterlin showed that such an unmodified tripeptide would not have shown significant activity in a cytotoxin or cell-based antimitotic screen. Nature revealed through the natural product hemiasterlin the additional requirements of extensive β- and N-methylation, along with modification of the C-terminal valine residue, that are necessary to create potent antimitotic activity and likely proteolytic stability starting from this simple tripeptide template. These combined modifications are unlikely to have been part of a combinatorial synthesis program before the chemical inspiration provided by the discovery of the natural product hemiasterlin.

Lack of a reliable industrial-scale source of the natural compound or bioactive synthetic analogs is often a major impediment to developing a natural product lead into a drug. The relatively easy synthetic accessibility to hemiasterlin and analogs played an important role in moving the development of this class of antimitotic peptides forward. Synthesis provided hemiasterlin and selected analogs in sufficient quantities for more detailed preclinical *in vitro* and *in vivo* evaluation and guaranteed an unlimited supply of compound for human trials and eventual clinical use. It is important to recognize that the natural product hemiasterlin already had very potent *in vitro* and *in vivo* activity against human tumor cell lines, so nature had done most of the work of optimizing the SAR for this pharmacophore. Analog synthesis did identify HTI-286, which shows an incremental improvement in activity profile, but all other analogs that have been made and tested at UBC showed a loss of activity relative to the natural product.[19]

Two important milestones in the preclinical decision process to move ahead with HTI-286 were its superior *in vitro* cytotoxicity profile versus multidrug-resistant human cancer cells lines relative to the natural product hemiasterlin, and the fact that it retained activity against paclitaxel and vinca alkaloid-resistant cell lines in mouse tumor xenograft models. Other desirable attributes were its chemical stability and solubility in saline, making formulation straightforward, and its oral activity in mouse models. The discovery that it interacts with α-tubulin distinguished it mechanistically from other tubulin depolymerization agents and raised its development profile.

HTI-286 appears to be a very promising experimental anticancer drug. It represents a new "natural product–inspired" chemical structural class of antimitotic agents that target tubulin and inhibit its polymerization into microtubules. HTI-286 shows excellent intravenous and oral activity against drug-sensitive and multidrug-resistant human tumors in mouse xenograft models, both preventing the growth of newly implanted tumors and also regressing well-established tumors. The compound has successfully completed a phase I clinical trial, which established a maximum tolerated dose in humans and identified the dose limiting toxicities. Phase II clinical trials with HTI-286 are ongoing, and as with all experimental drugs, there is the hope that the high level of efficacy observed in preclinical animal models will translate into real benefit for human cancer patients.

ACKNOWLEDGMENT

Financial support for the research at UBC was provided by the National Cancer Institute of Canada (MR, RJA), the U.S. National Institutes of Health (RJA), and the Natural Sciences and Engineering Research Council of Canada (RJA, EP). We thank Theresa Allen, Ed Piers, Hilary Anderson, John Coleman, Mike LeBlanc, Dilip de Silva, Debra Wallace, Lynette Lim, and Jim Nieman, who all played important roles in the research leading to HTI-286 (SPA-110) at UBC. Wyeth Pharmaceuticals has been a wonderful partner in taking HTI-286 from a basic laboratory discovery to clinical trials. Many individuals at Wyeth, too numerous to mention, made major contributions to the basic science of HTI-286 and its preclinical development. Several that deserve special mention are Phil Frost, Lee Greenberger, Frank Loganzo, Carolyn Discafani, Arie Zask, and Semiramis Ayral-Kaloustian.

REFERENCES

1. Margaret and Charles Juravinski Cancer Centre. Cancer Services. Available at: http://www. hrcc.on.ca/ActiveTrials.asp.
2. Talpir, R., et al., Hemiasterlin and geodiamolide TA: two new cytotoxic peptides from the marine sponge *Hemiasterella minor* (Kirkpatrick), *Tetrahedron Lett.*, 35, 4453, 1994.

3. Crews, P., et al., A highly methylated cytotoxic tripeptide from the marine sponge *Auletta cf constricta*, *J. Org. Chem.*, 59, 2932, 1994.

4. Matsunaga, S., Fusetani, N., and Konosu, S., Bioactive metabolites VI. Structure elucidation of discodermin A, an antimicrobial peptide from the marine sponge *Discodermia kiiensis*, *Tetrahedron Lett.*, 25, 5165, 1984.

5. Kaneda, M., Studies on the Bottromycins. I. proton and carbon-13 NMR assignments of bottromycin A2, the main component of the complex, *J. Antibiot.*, 45, 792, 1992.

6. Coleman, J. E., et al., Cytotoxic peptides from the marine sponge *Cymbastela sp*, *Tetrahedron*, 51, 10653, 1995.

7. Coleman, J. E., et al., Hemiasterlin methyl ester, *Acta Crystal.*, 52, 1525, 1996.

8. Gamble, W. R., et al., Cytotoxic and tubulin-interactive hemiasterlins from *Auletta* sp. and *Siphonochalina* spp. sponges, *Bioorg. Med. Chem.*, 7, 1611, 1999.

9. Chevallier, C., et al., A new cytotoxic and tubulin-interactive milnamide derivative from a marine sponge *Cymbastela* sp., *Org. Lett.*, 5, 3737, 2003.

10. Hooper, J. N. A. and van Soest, R. W. M., *Systema Porifera, a Guide to the Classification of Sponges*, Kluwer Academic/Plenum, New York, 2002.

11. Kunze, B., et al., Chondramines A-D, new antifungal and cytostatic depsipeptides from *Chondromyces crocatus* (Myxobacteria). Production, physico-chemical and biological properties, *J. Antibiot.*, 48, 1262, 1995.

12. Anderson, H. J., et al., Cytotoxic peptides hemiasterlin, hemiasterlin A and hemiasterlin B induce mitotic arrest and abnormal spindle formation, *Cancer Chemother. Pharmacol.*, 39, 223, 1997.

13. Personal communication from Peter Lassota.

14. Bai, R., et al., Interactions of the sponge-derived antimitotic tripeptide heniasterlin with tubulin: comparison with dolastatin 10 and cryptophycin 1, *Biochemistry*, 38, 14302, 1999.

15. Hamel, E., Interactions of antimitotic peptides and depsipeptides with tubulin, *Biopolymers*, 66, 142, 2002.

16. Boukari, H., Nossal, R., and Sackett, D. L., Stability of drug-induced tubulin rings by fluorescence correlation spectroscopy, *Biochemistry*, 42, 1292, 2003.

17. Ireland, C. M., et al., Anticancer agents from unique natural products sources, *Pharmaceutical Biol.*, in press, 2004.

18. Andersen, R. J., et al., Total synthesis of (-)-hemiasterlin, a structurally novel tripeptide that exhibits potent cytotoxic activity, *Tetrahedron Lett.*, 38, 317, 1997.

19. Nieman, J., et al., Synthesis and antimitotic/cytotoxic activity of hemiasterlin analogs, *J. Nat. Prod.*, 66, 183, 2003.

20. Vedejs, E. and Kongkittingam, C., A total synthesis of (-)-hemiasterlin using N-Bts methodology, *J. Org. Chem.*, 66, 7355, 2001.

21. Reddy, R., et al., Asymmetric synthesis of the highly methylated tryptophan portion of the hemiasterlin tripeptides, *Org. Lett.*, 4, 695, 2002.

22. Roberge, M., et al., Cell-based screen for antimitotic agents and identification of analogues of rhizoxin, eleutherobin and paclitaxel in natural extracts, *Cancer Res.*, 60, 5052, 2000.

23. Andersen, R. J., et al. WO Patent 99/32509, 1999.

24. Loganzo, F., et al., HTI-286, a synthetic analog of the tripeptide hemiasterlin, is a potent antimicrotubule agent that circumvents P-glycoprotein-mediated resistance *in vitro* and *in vivo*, *Cancer Res.*, 63, 1838, 2003.

25. Nunes, M., et al., Two photoaffinity analogs of HTI-286, a synthetic analog of hemiasterlin, interact with alpha-tubulin, *Eur. J. Cancer*, 38, S119, 2002.

26. Wang, C. P., et al., *In vitro* and *in vivo* metabolism of HTI-286, an antimicrotubule anticancer agent, *Drug Metabol. Rev.*, 35 Suppl. 2, 42, 2003.

27. Ratain, M. J., et al., Phase 1 and pharmacological study of HTI-286, a novel antimicrotubule agent: correlation of neutropenia with time above a threshold serum concentration, *Proc. Am. Soc. Clin. Oncol.*, 22, 129, 2003.

28. Newman, D. J., Cragg, G. M., and Snader, K. M., Natural products as sources of new drugs over the period 1981–2002, *J. Nat. Prod.*, 66, 1022, 2003.

29. Andersen, R. J. and Williams, D. E., in *Chemistry in the Marine Environment*, Vol. 13, R. E. Hester and R. M. Harrison, Eds., The Royal Society of Chemistry, Cambridge, 2000, 55.

30. Blunt, J., et al., Marine natural products, *Nat. Prod. Rep.*, 20, 1, 2003.

15 The Actinomycins

Anthony B. Mauger and Helmut Lackner

CONTENTS

I. Historical Introduction 281
II. Separation and Nomenclature 282
III. Structures 283
IV. Conformation 283
V. Mechanism of Biological Action 284
VI. *In Vitro* Antitumor Activity 285
VII. *In Vivo* Antitumor Activity and Toxicity 286
VIII. Synthesis 286
IX. Analogs 287
    A. Directed Biosynthesis 287
    B. Partial Synthesis 287
    C. Total Synthesis 288
X. Structure–Activity Relationships 289
    A. Introduction 289
    B. Natural Actinomycin Variants 290
    C. Chromophoric Analogs 291
    D. Peptidic Analogs 291
XI. Clinical Applications of Actinomycins 292
References 292

I. HISTORICAL INTRODUCTION

The actinomycins, a family of structurally related chromopeptide antibiotics with a common phenoxazinone chromophore attached to two pentapeptide lactone moieties (Figure 15.1), vary in their amino acid content (Table 15.1). They emerged from the pioneering work of Selman Waksman on soil microorganisms, and in particular from his work on the *Streptomyces*. The first actinomycin, isolated in 1940 from *Streptomyces antibioticus*,[1,2] was the first of several antibiotics discovered by Waksman, the first crystalline antibiotic, and the first to display antitumor activity. Many other actinomycins have since been isolated from other *Streptomyces* species, and it was also found in an unrelated genus, *Micromonospora*.[3]

The first actinomycin was described as a red crystalline substance active against Gram-positive microorganisms,[2] but its toxicity[4] precluded its clinical use as an antibiotic. Interest in the actinomycins revived in 1952 when the antitumor activity of another actinomycin (C) in mouse and rat tumors was found by Hackmann[5] and when Schulte reported the first clinical studies.[6] The discovery that actinomycin inhibits DNA-primed RNA synthesis[7,8] resulted in its widespread use in studies of macromolecular biosynthesis and virus replication.[9]

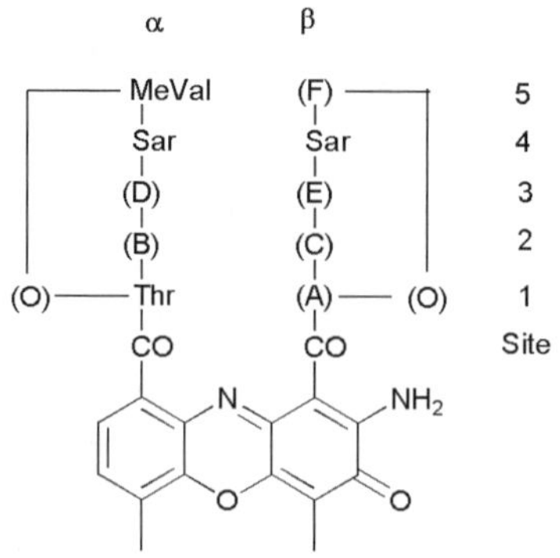

FIGURE 15.1 Structures of 18 naturally occurring actinomycins (see Table 15.1).

II. SEPARATION AND NOMENCLATURE

With the first actinomycin preparation designated A, subsequent isolates from different species or strains of *Streptomyces* were termed B,[10] C,[11] D,[12] I,[13] X,[14] Z,[15] and so forth. That they were mixtures was first shown when actinomycin C separated into components C_1, C_2, and C_3 in countercurrent distribution.[16,17] Paper partition chromatographic studies confirmed that most other actinomycin preparations were also multicomponent complexes.[16,18–21] Techniques for their separation have been reviewed.[22]

TABLE 15.1
Structures of Natural Actinomycins (See Figure 15.1)

Actinomycin	A	B	C	D	E	F
I=$X_{0\beta}$	Thr	D-Val	D-Val	Pro	Hyp	MeVal
$X_{0\delta}$	Thr	D-Val	D-Val	Pro	aHyp	MeVal
II=F_8	Thr	D-Val	D-Val	Sar	Sar	MeVal
III*=F_9	Thr	D-Val	D-Val	Pro	Sar	MeVal
IV=X_1=C_1=D	Thr	D-Val	D-Val	Pro	Pro	MeVal
V=X_2	Thr	D-Val	D-Val	Pro	4-oxo-Pro	MeVal
X_{1a}	Thr	D-Val	D-Val	Sar	4-oxo-Pro	MeVal
$X_{0\gamma}$	Thr	D-Val	D-Val	Sar	Pro	MeVal
VI=C_2	Thr	D-Val	D-aIle	Pro	Pro	MeVal
i-C_2	Thr	D-aIle	D-Val	Pro	Pro	MeVal
VII=C_3	Thr	D-aIle	D-aIle	Pro	Pro	MeVal
Z_1	4-OH-Thr	D-Val	D-Val	3-OH-5-MePro	4-oxo-5-MePro	MeAla
Z_2	Thr	D-Val	D-Val	3-OH-5-MePro	4-oxo-5-MePro	MeAla
Z_3	4-Cl-Thr	D-Val	D-Val	3-OH-5-MePro	4-oxo-5-MePro	MeAla
Z_4	Thr	D-Val	D-Val	5-MePro	4-oxo-5-MePro	MeAla
Z_5	4-Cl-Thr	D-Val	D-Val	5-MePro	4-oxo-5-MePro	MeAla
ZP	Thr	D-Val	D-Val	5-MePro	5-MePro	MeVal
G†	4-OH-Thr	D-Val	D-Val	Pro	3-OH-5-MePro	MeAla

Note: Nonstandard amino acid abbreviations: MeVal = N-methylvaline; MeAla = N-methylalanine; 5-MePro = 5-methylproline; all amino acids are L except where prefixed by "D".

* Actinomycin III is a mixture of two isomers with Pro and Sar interchangeable at sites D and E.
† Actinomycin G refers to the product from *Streptomyces HKI-0155*.

This situation was simplified when it was found that different complexes had components in common. Thus, actinomycins A, B, and X contained the same five compounds, albeit in different proportions, and a system of nomenclature was proposed[23] that named them actinomycins I through V in the order of increasing lipophilicity. The actinomycin C components were named IV, VI, and VII, of which only IV was also present in the A, B, and X complexes. Actinomycin D was found to consist solely of IV. The five components of the actinomycin Z complex lie outside of this system and are termed Z_1 through Z_5.

III. STRUCTURES

The structures of the best-characterized actinomycins are shown in Figure 15.1 and Table 15.1. The elucidation of the structure of actinomycin C_3 (VII) in 1956 by Brockmann et al. was a notable achievement in the era before nuclear magnetic resonance and mass spectrometry were available.[24] The structure comprises a chromophoric 2-aminophenoxazin-3–one ring system bearing two methyl groups and two identical pentapeptide lactone moieties. The chromophore, including the two carboxyls, is termed actinocin, so an actinomycin can be described as an actinocinyl-*bis*(pentapeptide lactone).

The structure of actinomycin D (IV), which has been abbreviated "AMD," was found to be identical to that of C_3, except that both D-*allo*-Ile residues are replaced by D-Val.[25] As distinct from *iso*-actinomycins such as D (IV) and C_3 (VII), the *aniso*-actinomycin C_2 (VI)[26] has a minor regioisomer i-C_2 (originally termed C_{2a}).[27] The two isomers were distinguished via oxidative degradation of the chromophore, permitting separation of the two peptides.[28]

The structures of actinomycins I and V emerged from studies on the reduction of actinomycin X_2 in various ways to generate $X_{0\beta}$ (I), $X_{0\delta}$, and X_1 (IV).[29] The oxygenated proline residues were located in the β-peptide by oxidative degradation of the chromophore. These and other experiments indicated that X_2 contains 4-oxoproline, which is usually destroyed during hydrolysis, and this was confirmed[30] by its isolation. Minor components of the X complex termed $X_{0\gamma}$ and X_{1a} were also described[31] (Table 15.1).

Actinomycins II and III are trace components of the A, B, and X complexes and were characterized after addition of sarcosine to the culture medium for *S. antibioticus* markedly increased their relative concentrations.[32] The added sarcosine acted as a biosynthetic precursor, replacing one or both proline residues.[33] That this involved no change in amino acid sequence was also confirmed.[34] Chromatography revealed that the *aniso*-actinomycin III was a mixture of major and minor regioisomers.[35]

The actinomycin Z complex contains at least five components comprising a far greater diversity of amino acid content than actinomycins I–VII. Early studies[15] revealed that they differ from the other actinomycins in that they all contain N-methylalanine, but no proline. Instead, there are several unusual proline congeners found in these actinomycins, including *cis*-5-MePro,[36,37] 4-oxo-5-MePro,[38,39] and *trans*-3-hydroxy-*cis*-5-MePro.[40,41] Also, 4-hydroxy-Thr[42] was found in Z_1. More recently, two-dimensional nuclear magnetic resonance and mass spectrometry techniques permitted the elucidation of the structures of actinomycins Z_1 through Z_5 and unexpectedly revealed that Z_3 and Z_5 contained 4-chloro-Thr.[43] In addition, actinomycin G from *S. HKI-0155* has been found to have an α-peptide, which is similar to that of AMD, and a Z-type β-peptide.[44] Another actinomycin named ZP has 5-MePro in both peptides but otherwise is identical to AMD.[45]

Numerous other actinomycins have been described, but they are less well characterized structurally than those included in Table 15.1. Information on them is available in several reviews.[46–49]

IV. CONFORMATION

The first information on the conformation of an actinomycin (D) emerged from an x-ray crystallographic study of its complex with deoxyguanosine (Figure 15.2).[50] Later, x-ray studies of uncomplexed

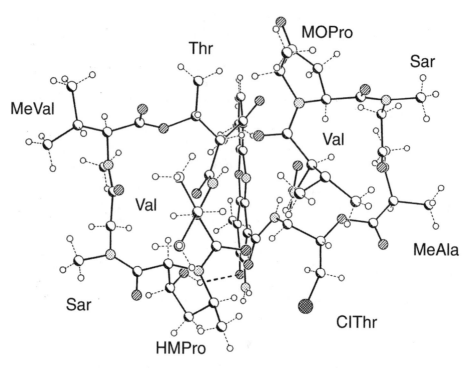

FIGURE 15.2 High-resolution x-ray crystallographic structure of actinomycin Z_3, shown from the chromophoric side of the molecule with the β-peptide ring on the right.[52] Abbreviations: MOPro = *cis*-5-methyl-4-oxo-L-proline; HMPro = *trans*-3-hydroxy-*cis*-5-methyl-L-proline (Table 15.1).

actinomycins D,[51,52] X$_2$,[53] and Z$_3$ (Figure 15.3)[52] indicated that their conformations were similar except for an orientation change in the ester linkages. The molecule possesses pseudo-C2 symmetry, and the two peptide moieties are held in unique juxtaposition by two antiparallel hydrogen bonds between the D-Val NH in one peptide and the D-Val C=O in the other, and vice versa. The D-Val-Pro and Pro-Sar peptide bonds are *cis*. Nuclear magnetic resonance studies of actinomycins[54] and of actinomycin-DNA complexes[55] indicate an essentially similar conformation in solution that is remarkably stable in solvents of different polarity.

V. MECHANISM OF BIOLOGICAL ACTION

The biological activity of the actinomycins is exerted by inhibition of DNA-dependent RNA synthesis,[7,8] which thereby inhibits protein synthesis.[8] This effect is explained by the strong binding of actinomycin to double-helical DNA.[56,57] This complexation has usually been detected by absorption spectroscopy (bathochromic shift of about 20 nm in the visible maximum of an actinomycin normally at 440 nm)[7,58,59] or by elevation in the thermal denaturation temperature (*Tm*) of the DNA.[60,61] It is required that the DNA be helical[62,63] and that it contains guanine residues.[62,64,65] An intercalative model for the actinomycin-DNA complex was proposed[66] in which the actinomycin chromophore is inserted adjacent to a G–C pair, and this hypothesis was supported by several solution studies involving equilibrium, kinetic and hydrodynamic techniques, and sedimentation coefficients.[67]

Interaction of deoxyguanosine with AMD in solution produces spectral shifts similar to those with DNA.[68] It was therefore of much interest when the 2:1 complex was obtained in crystalline form for x-ray studies.[50] The structure that emerged (Figure 15.2) has approximate two-fold symmetry, with the AMD chromophore "sandwiched" between the two guanine ring systems. The complex is held together by π-complex interactions between the three ring systems, by

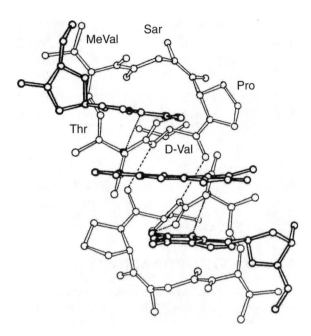

FIGURE 15.3 X-ray crystallographic structure of the 1:2 complex of actinomycin D with deoxyguanosine.[50] The actinomycin chromophore and the deoxyguanosine molecules are shown in bold. Hydrogen bonds are shown as dotted lines. Reprinted from *Journal of Molecular Biology*, Vol 61, Jain and Sobell, pp.1, 1972, with permission from Elsevier.

hydrogen bonds between the guanine 2-amino groups and the threonine carbonyl oxygens, and by hydrophobic interactions. This structure provided a model for the geometry of the AMD–DNA interaction, and this model is supported by x-ray structures of DNA complexes of AMD with d(GC)[69] and d(ATGCAT).[70]

The hydrophobic interactions that enhance the binding of actinomycins in the narrow groove of DNA depend on the hydrophobic inner surface of the peptide units, and the nature of the outer surfaces, when the binding occurs near the pause or rho-dependent termination sites of the DNA, terminates transcription by RNA polymerase.[71] Also, actinomycins belong to a class of antineoplastic agents that inhibit both topoisomerase I and topoisomerase II. It forms a ternary complex of topoisomerase II, inducing DNA strand breaks,[72] and it also stimulates topoisomerase I–mediated DNA cleavage.[73] Studies have also been reported on the binding of actinomycin D to single-stranded DNA.[74] Recently, evidence has emerged of the capability of actinomycin to down-regulate the effect of the transcription factors TBP and Sp1 by blocking access to their specific binding sites.[75]

VI. *IN VITRO* ANTITUMOR ACTIVITY

Early studies (1958) showed that AMD displays equal growth inhibitory activity against both normal and malignant cells.[76,77] The cytotoxic effects of several actinomycins were compared in HeLa cells.[78] More recently, AMD and several other actinomycins have been evaluated in the 60 human tumor cell line screening system[79] of the National Cancer Institute (NCI). For AMD, the mean molar log GI_{50} (50% growth inhibition), averaged over many experiments, was –8.73. The corresponding numbers for the various panels of cell lines (mean values from several lines) were leukemia, –9.2; non–small cell lung, –8.6; small cell lung, –8.8; colon, –8.7; CNS, –9.0; melanoma, –8.9; ovarian, –8.4; renal, –8.5; prostate, –8.6; and breast, –8.7. Thus, the most sensitive cell lines were those of leukemia, CNS cancers, and melanoma. In the COMPARE program,[80] AMD most closely resembles other topoisomerase II inhibitors.

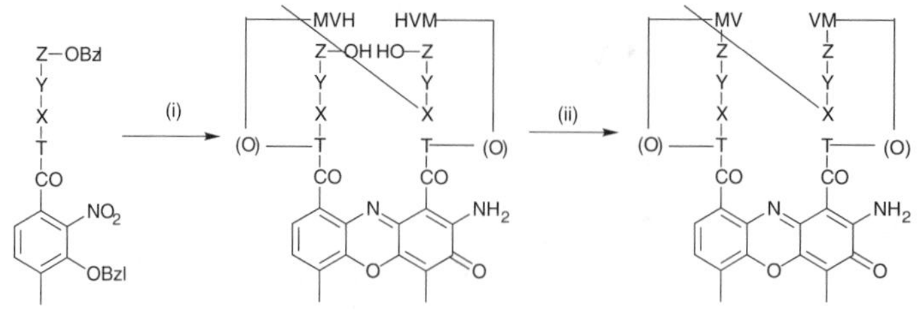

T = Threonine
X = D-*allo*-Isoleucine
Y = Proline
Z = Sarcosine
MV = Methylvaline

FIGURE 15.4 First total synthesis of an actinomycin (C_3 = VII).[89] Reagents: (i) Cbz-MeVal-OH/carbonyldi-imidazole; H_2/Pd, then $K_3Fe(CN)_6$; (ii) ClCOOEt/Bu_3N.

VII. *IN VIVO* ANTITUMOR ACTIVITY AND TOXICITY

The first observations of antitumor activity of an actinomycin (C) involved the suppression of the Ehrlich carcinoma in mice and the Wilms tumor in rats.[5,81] These were quickly followed by numerous studies in various laboratories; the early work has been reviewed.[82] Among the best responders were the Ridgway osteogenic carcinoma in mice[83] and the Wilms tumor in rats,[84] especially when combined with radiation.[85] Other experimental tumors susceptible to actinomycins include the P388 and L1210 leukemias, B16 melanoma, and adenocarcinoma 755.[86] Actinomycins C_1 (AMD), C_2, and C_3 were equally efficacious.

Actinomycins are highly toxic[4]; the intravenous LD_{50}s of AMD in mice and rats are 1.2 and 0.6 mg/kg per day, respectively, and the corresponding subcutaneous figures are 1.4 and 0.80.[87] In common with many other antitumor agents, the toxic effects manifest primarily in rapidly prolif-erating tissues such as the bone marrow, intestinal mucosa, and lymphoid organs.

VIII. SYNTHESIS

The first synthesis of an actinomycin (C_3) involved the oxidative coupling of two molecules of a 3-hydroxy-4-methylanthraniloyl O-pentapeptide, followed by cyclization of both peptide moieties at the Sar-MeVal peptide bonds[88,89] (Figure 15.4). Interestingly, this route failed to distinguish between the accepted structure for actinomycins and an alternate version that had been proposed,[90,91] in which the C terminus of one peptide is lactonized by the Thr OH of the other, and vice versa, forming a cyclode-capepetide dilactone. Subsequently, several unambiguous syntheses of actinomycins have been described involving a pentapeptide lactone derivative as the key intermediate. In the first example[92] (Figure 15.5), the 3-benzyloxy-4-methylanthraniloyl-pentapeptides were lactonized in syntheses of actinomycins C_1, C_2, i-C_2, and C_3. Analogous peptide lactones were subsequently synthesized in higher yields via cycliza-tion at the Pro-Sar[93] or D-Val-Pro[94] peptide bonds. This intermediate can also be constructed from the

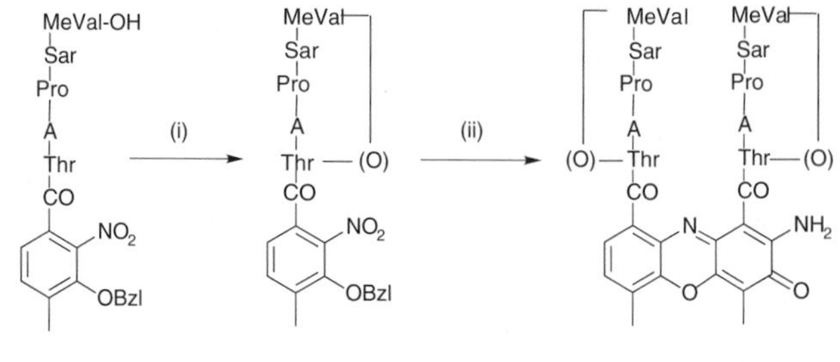

FIGURE 15.5 Synthesis of actinomycins C_1, C_2, i-C_2 and C_3.[92] A = D-Val or D-*a*Ile. Reagents: (i) AcCl/imi-dazole; (ii) H_2/Pd, then $K_3Fe(CN)_6$.

cyclic pentapeptide by acid-catalyzed N,O-acyl shift followed by acylation of the Thr amino group.[95] Regioselective syntheses of actinomycins C_2 and i-C_2 have been reported,[96,97] as well as that of $X_{0\beta}$.[53]

IX. ANALOGS

A. DIRECTED BIOSYNTHESIS

The amino acid content of actinomycins can be manipulated by the addition of certain amino acids to the medium in which the producing organism is cultured, a method also known as "controlled biosynthesis."[98] Addition of isoleucine to cultures of *Streptomyces chrysomallus* generated new actinomycins, termed E_1 and E_2,[99] containing one and two N-Me-*allo*-Ile residues (respectively) in place of N-MeVal.[100] Similar studies with *Streptomyces antibioticus* and *Streptomyces parvulus* uncovered a more complex situation, in which D-Ile and D-*allo*-Ile were both able to replace D-Val.[101] Addition of sarcosine to *S. chrysomallus* cultures produced a mixture in which one or both prolines of C_2 and C_3 were replaced by sarcosine, and the products were termed (respectively) F_3 and F_1 (analogs of C_2) and F_4 and F_2 (analogs of C_3).[99,102]

One or both of the proline residues in actinomycins can be replaced by several Pro analogs using directed biosynthesis. They include the *cis* and *trans* isomers of 4-methyl-,[103] 4-chloro-, and 4-bromo-prolines[35]; thiazolidine-4-carboxylic acid[104]; azetidine-2-carboxylic acid; and pipecolic acid.[105] When one Pro is replaced, the product is a mixture of two difficultly separable regioisomers. In the case of the 4-substituted prolines, only the four analogs formed using *cis*- and *trans*-4-MePro were characterized.[103] Following the early observation[105] that new actinomycins were formed by *S. antibioticus* in the presence of azetidine-2-carboxylic acid, two analogs, named azetomycins I and II, were isolated,[106] with one or both prolines (respectively) replaced by azetidine-2-carboxylic acid. When pipecolic acid is added to *S. antibioticus* cultures, not only does it replace Pro but actinomycins containing 4-hydroxy- and 4-oxo-pipecolic acid are generated as well[107] (Table 15.2). This is analogous to the presence of Hyp and 4-oxo-Pro (respectively) in the natural actinomycins I and V.

B. PARTIAL SYNTHESIS

Mild acid hydrolysis of actinomycins replaces the chromophoric NH_2 by OH, and subsequent treatment with thionyl chloride produces 2-chloro-2-deaminoactinomycin.[108] Treatment of this compound with a variety of primary and secondary amines provided 34 2-N-substituted actinomycins.[109–113] In addition to its hydrogenation to 2-deaminoactinomycin,[108] the 2-chloro derivative is also an intermediate in the preparation of analogs substituted with Cl and Br in the 7-position.[114] Other groups substituted in the 7-position include nitro,[109,115] amino,[110,115] hydroxy,[110] and methoxy.[116]

TABLE 15.2

Biosynthetic Actinomycin Analogs[a] from *Streptomyces antibioticus* with Pipecolic Acid[105]

Actinomycin	Proline	Pipecolic Acid	4-OH-Pipecolic Acid	4-Oxo-Pipecolic Acid
Pip 1α	0	1	0	1
Pip 1β	1	1	0	0
Pip 1γ	0	1	1	0
Pip 1δ	1	0	0	1
Pip 1ε	1	0	1	0
Pip 2	0	2	0	0

[a] The number of residues of each amino acid at the two 3-sites are shown; the remainder of each structure is similar to that of actinomycin D.

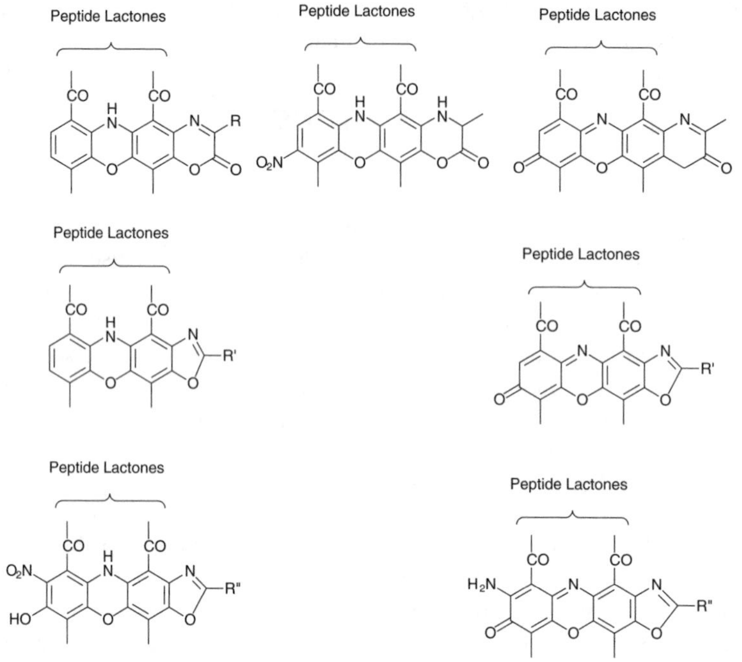

FIGURE 15.6 Tetracyclic analogs of actinomycin D. R = H, CH_3, C_2H_5, CH_2F, $n\text{-}C_6H_{13}$, CH_2COOH, CH_2CH_2COOH, C_6H_5, C_6F_5, $o\text{-}ClC_6H_4$, $m\text{-}ClC_6H_4$, $p\text{-}ClC_6H_4$, $2,4\text{-}Cl_2C_6H_3$, 2-naphthyl; R' = CH_3, C_6H_5, C_6F_5; R″ = CH_3, C_6H_5, CH_2COOH, $CH_2CONH(CH_2)_4NH_2$.

N-Acetyl, N-pivaloyl, and N-stearoyl derivatives of 7-amino-AMD have been described,[67] as well as nine N-alkyl and N-aryl derivatives.[117] 7-Hydroxy-8-amino-AMD[118] and nine O-alkyl and four O-acyl derivatives of 7-hydroxy-AMD have also been reported.[113]

Hydrogenation of actinomycin produces dihydroactinomycin, which, unlike actinomycin, can be acetylated. Subsequent reoxidation furnishes N-acetyl-actinomycin.[119] Dihydroactinomycin reacts with α-ketoacids to generate tetracyclic analogs with a fused oxazine ring.[109,120,121] Reaction of AMD with aldehydes leads to tetracyclic analogs with a fused oxazole ring, and further manipulations produced related compounds substituted in the 7 and 8 positions.[118,121,122] Some of these tetracyclic analogs are shown in Figure 15.6.

The lactone rings of actinomycins were opened with alkali to produce actinomycinic acids[123] from which dimethyl esters and di-O-acetyl derivatives have also been prepared.[119] One or both lactone rings can also be opened by microbial degradation by *Actinoplanes* species.[124] The peptide lactones of actinomycin C_3 have been cleaved at the Sar-MeVal peptide bond with cold, concentrated HCl to produce the *bis-seco*-actinomycin.[125]

The OH groups of Hyp and *allo*-Hyp in actinomycins $X_{0\beta}$ and $X_{0\delta}$ have been converted to their O-acetyl and O-hexadecanoyl derivatives.[29,126] Likewise, an O-acetate and di-O-acetate have been prepared from actinomycin Z_1.[39] The 4-oxo-5-MePro residue in this actinomycin has been reduced to two diastereoisomers of 4-OH-5-MePro, and a triacetate of the resulting actinomycins was prepared.[127] The 4-oxo-Pro in actinomycin $X_{1\alpha}$ has been reduced to Hyp, of which the O-acetyl derivative has also been produced.[31] Actinocinyl-gramicidin S has been described[128]; the two couplings were effected at the ornithine δ-NH_2 groups.

C. TOTAL SYNTHESIS

The two chromophoric methyl groups of AMD have been replaced by H, Br, OMe, Et, and *t*-Bu.[129] Also, the 6-Me and both 4- and 6-Me groups have been replaced by CF_3.[130] An analog termed *pseudo*-actinomycin C_1 has been synthesized in which the chromophoric methyls and the peptide moieties are exchanged.[131] In another analog, the actinocinyl chromophore is replaced by 4-methylphenazine-1,9-dicarboxylic acid.[132]

A number of peptidic analogs of actinomycin have been synthesized by methods similar to those used for the natural actinomycins (Table 15.3); they include enantio-AMD.[133]

X. STRUCTURE–ACTIVITY RELATIONSHIPS

A. INTRODUCTION

For biological activity mimicking the natural actinomycins, the chromophore and both intact cyclopeptide moieties are required. Even the scission of one peptide lactone abolishes biological activity.[96,124] Some actinocinyl derivatives bind to DNA but are inactive. In the case of actinomine (actinocinyl-*bis*[diethylaminoethylamide]),[67] this phenomenon was shown to derive from the rapid rate of dissociation from its complex with DNA, which was 1000 times faster than that of AMD. Thermodynamic and other studies[62,70] indicated an important hydrophobic interaction between the peptide lactones of actinomycins and DNA, and their biological activity depends on the slow dissociation of the complex.

The number of actinomycin analogs that have been described is very large (>100), and the review that follows is not exhaustive. A more detailed review of the work reported through 1977 has appeared.[48] Natural actinomycins and their analogs have been evaluated in a variety of ways, including binding to DNA, inhibition of RNA synthesis, and antimicrobial and antitumor activities. Because antitumor activity is of primary interest, for those compounds for which such data are available, other activities are usually omitted. For other compounds, antimicrobial data, albeit a poor predictor of antitumor activity, is provided. Some analogs described above, which are biologically inactive, are not discussed below. In many cases, analogs were produced in insufficient quantities for *in vivo* evaluation, and none reached clinical trial, although some partially synthetic analogs surpassed AMD in *in vivo* antitumor activity.

TABLE 15.3
Amino Acid Content of Synthetic Peptidic Actinomycin Analogs

Site 1	Site 2	Site 3	Site 4	Site 5	References
Thr	D-Val	α-Hyp, β-Pro	Sar	MeVal	53
Thr	D-Val	Hyp	Sar	MeVal	53
Ser	D-Val	Pro	Sar	MeVal	134
Dpr	D-Val	Pro	Sar	MeVal	135
Dbu	D-Val	Pro	Sar	MeVal	136
Thr	D-Ala	Pro	Sar	MeVal	134
Thr	D-Leu	Pro	Sar	MeVal	134
Thr	D-Thr	Pro	Sar	MeVal	137
Thr	D-Val	Meg	Sar	MeVal	137
Thr	D-Val	Pro	Gly	Val	138
Thr	D-Val	Pro	Sar	MeAla	139
Thr	D-Val	Pro	Sar	MeLeu	137
D-Thr	Val	D-Pro	Sar	D-MeVal	133

Note: Nonstandard amino acid abbreviations: Dpr = L-2,3-diaminopropionic acid; Dbu = L-2,3-diamino-*n*-butyric acid; Meg = N-[2-(methoxycarbonyl)ethyl]-glycine.

TABLE 15.4
Biological Activities of Actinomycins I–VII[a]

Actinomycin	Toxicity	Antitumor Efficacy[b]	Streptomyces aureus[c]	Streptomyces subtilis[c]
I = $X_{0\beta}$	10	+	25	5–15
II	10	++	45	35
III	50	++	35	25–33
IV = AMD = X_1 = C_1	100	++	100	100
V = X_2	800	±	200	200
VI = C_2	100	++	70	90
VII = C_3	70	++	95	100

[a] Numerals indicate activity relative to AMD = 100.
[b] Evaluated by increased life span in mice implanted with leukemia P388, leukemia L1210, or B16 melanoma.
[c] Antimicrobial activities.

B. NATURAL ACTINOMYCIN VARIANTS

Comparative biological data for actinomycins I–VII are shown in Table 15.4.[140,141] In addition, comparison of II, III, and IV in several mouse ascitic tumors revealed that in the most sensitive tumor, the Gardner lymphosarcoma, these actinomycins had comparable efficacy.[142] These data indicate that replacement of the β-Pro by Hyp or by sarcosine reduces toxicity, antitumor potency, and antimicrobial activity, whereas antitumor efficacy remains about the same. In contrast, replacement by 4-oxo-Pro increases toxicity and antimicrobial activity but reduces antitumor efficacy. More recent work[43] comparing actinomycins Z_1, Z_3, and Z_5 with AMD is shown in Table 15.5.

TABLE 15.5
Cytotoxicities (Micromolar) in Three Human Tumor Cell Lines and Minimal Inhibitory Concentrations (MICs) (μg/mL) against *Bacillus subtilis* for Actinomycins D, Z_1, Z_3, and Z_5

Actinomycin and Level	HMO2 (Stomach)	HEP G2 (Liver)	MCF 7 (Breast)	Antimicrobial MIC
D				
GI50	0.2	1.0	0.5	0.78
TGI	0.8	4.0	2.2	
LC50	>50	>50	>50	
Z_1				
GI50	0.75	0.95	<0.5	12.5
TGI	5.8	5.5	>50	
LC50	>50	>50	>50	
Z_3				
GI50	<0.1	<0.1	<0.1	0.20
TGI	<0.1	1.4	<0.1	
LC50	0.28	>50	0.5	
Z_5				
GI50	<0.1	1.5	<0.1	0.75
TGI	<0.1	10	0.12	
LC50	0.50	>50	0.5	

Actinomycin Z_3 is several times as potent as AMD against all three human tumor cell lines studied, as well as against *Bacillus subtilis*. Z_1 is considerably less active, and Z_5 is roughly comparable with AMD. In earlier studies,[143] Z_1 and Z_5 were less effective than AMD in inhibiting RNA synthesis in *B. subtilis* and in HeLa cells. The lower activity of Z_1 probably results from the presence of 4-hydroxythreonine in a location where its OH group is close to the chromophoric amino group.[52]

C. Chromophoric Analogs

The importance of the chromophoric 2-NH$_2$ group is illustrated by the lack of biological activity when it is replaced by H, OH, or Cl[108] or is acylated.[119] Most 2-N-alkylated actinomycins were only tested against *B. subtilis* and were at least 10-fold less active that the natural variants.[109,111] However, 2-N-γ-hydroxypropyl-AMD, although possessing only 10% of the antimicrobial activity of AMD, was equal in antitumor activity.[112] Also, in P388 leukemia (mice), 2-N-3′-amino-*n*-propyl-AMD surpassed AMD in terms of increased life span (ILS).[113] Replacement of the chromophoric Me groups by H or OMe weakened binding of the analog to DNA and reduced antimicrobial activity to 1%–5% of that of AMD, whereas their replacement by Et reduced antimicrobial activity to 30%–35%, and the di-*t*-Bu analog lacked activity, presumably because these bulky groups obstruct DNA intercalation.[129] Despite the nearly equal size of Me and CF$_3$ groups, antibacterial and cytotoxic activities of the 6-CF$_3$ and 4, 6-di-CF$_3$ analogs were reduced to 25% and almost 0% (respectively) of the parent actinomycin, probably because of electronic effects.[130]

Substitution in the chromophoric 7-position produces a variety of effects. 7-Chloroactinomycin C_3[114] and 7-nitroactinomycin C_2[110] have about half the antimicrobial activity of the parent compounds, whereas that of 7-bromoactinomycin C_3[114] is 50% higher. 7-Nitro- and 7-amino-AMD were roughly comparable with AMD in *in vivo* antitumor activity.[144] *In vivo* antitumor studies (mice, P388) of nine N-substituted 7-amino-AMDs gave ILS values superior to those of AMD for 7-amino-AMD and 7-N-(3′,4′-dichlorophenyl)amino-AMD.[117] In similar studies of 7-hydroxy-AMD and its O-alkyl and O-acyl derivatives, superior ILS values were seen in 7-hydroxy-AMD and 7-adamantoyloxy-AMD.[113] No rationale was proposed for the enhanced efficacy resulting from these bulky substitutions in the chromophoric 7-position.

Chromophoric analogs of AMD bearing a fourth oxazinone ring hydrolyze to AMD above pH7. In P388 murine leukemia, the compounds with R = CH$_2$F and C$_6$H$_5$ (Figure 15.6) had ILS values superior to that of AMD.[120] These analogs presumably act as AMD prodrugs, but their enhanced efficacy has not been explained. Dihydro-AMD analogs bearing a fourth oxazole ring were inactive unless the A ring was oxidized to the iminoquinone, which is metabolized to 7-hydroxy-AMD.[121] The 8-amino-iminoquinones had antitumor activity superior to that of AMD when R″ = C$_6$H$_5$ or CH$_2$CONH(CH$_2$)$_4$NH$_2$ (Figure 15.6), presumably resulting from their metabolism to 7-hydroxy-8-amino-AMD.[122]

D. Peptidic Analogs

Synthetic variations at the Thr sites of actinomycin involving replacement by Ser, Dpr, or Dbu (Table 15.3) gave antimicrobial activities of 20%, 10%, and 10% (respectively) of that of AMD.[134–136] Antitumor potency of the Dbu analog was 10–20 times lower than that of AMD. At the D-Val sites, replacement by D-Ala or D-Leu reduces antimicrobial activities to 1% and 10% (respectively) of that of AMD,[134] and D-Thr produces a reduction to 10%–20% of that of AMD.[137]

Replacement of one Pro by pipecolic acid (Pip 1β) has little effect on antimicrobial activity, whereas replacement of both (Pip 2) reduces this activity 5- to 15-fold, and replacement by one 4-ketopipecolic acid (Pip 1δ) reduces it further.[145] The *in vivo* antitumor efficacy of Pip 1 is comparable with that of AMD at an eightfold higher dosage, whereas Pip 2 is less active.[86] The *in vivo* antitumor activity of the compounds in which one or both prolines were replaced by azetidine-2-carboxylic acid, termed azetomycins I and II, respectively, have been extensively evaluated; azetomycin I, the more active of the two, was found equivalent to AMD in some experiments[146] and somewhat superior in others.[147] Actinomycins in which one or both prolines were replaced by *cis-* or *trans*-4-MePro (termed K_{1c}, K_{2c}, K_{1t}, and K_{2t} respectively) had lower antimicrobial activity than AMD in the sequence AMD > K_{1t} > K_{1c} > K_{2t} > K_{2c}.[103]

Replacement of both MeVal residues by MeLeu[137] produced much stronger binding to DNA than AMD, and this analog was at least 10-fold more potent than AMD in the NCI *in vitro* screen comprising 60 human tumor cell lines.[79] (This analog was initially reported[137] to be approximately 100 times as potent as AMD in the NCI screen, but evaluation of data from several experiments has since required a revision of that ratio.) Presumably the MeLeu side-chains enhance hydrophobic interaction with DNA. In contrast, the AMD analog with both MeVal residues replaced by MeAla binds to DNA less strongly than AMD and has approximately 10-fold lower antimicrobial activity,[139] indicating a poorer hydrophobic interaction with DNA. This effect does not prevent actinomycins Z_3 and Z_5, which both have one MeVal replaced by MeAla, from being at least as active as AMD, but there are other structural differences, including additional methyl groups in the 5-position of both prolines.[43]

XI. CLINICAL APPLICATIONS OF ACTINOMYCINS

Many early clinical studies with actinomycins C and D in a variety of cancers demonstrated that their effectiveness was most prominent in trophoblastic tumors in females[148] (including methotrexate-resistant choriocarcinoma[149]), in metastatic carcinoma of the testis,[148,150,151] and in Wilms' tumor (nephroblastoma) in children.[152] In rhabdomyosarcoma, efficacy was enhanced by combination with vincristine.[153] Cures were obtained in Wilms' tumor by a combination of surgery, radiotherapy, and actinomycin therapy.[154,155] The most effective route of administration is intravenous.

Actinomycin D, also known as Dactinomycin and Cosmegen, soon became the only form of actinomycin in clinical use, as it is produced by *S. parvulus* as a single component, whereas actinomycin C is a mixture, and they are equally effective. Actinomycin D is still in use in therapy of the above-mentioned tumors, as well as in treatment of Ewing's sarcoma. Side-effects include bone marrow depression, oral and intestinal lesions, skin reactions, alopecia, and immunosuppression. As a consequence, in common with other classical anticancer drugs, its effectiveness is constrained by dose-limiting toxicity. Nonetheless, many complete remissions have occurred.

REFERENCES

1. Waksman, S.A. and Woodruff, H.B., Bacteriostatic and bacteriocidal substances produced by a soil actinomyces, *Proc. Soc. Exptl. Biol. Med.*, 45, 609, 1940.
2. Waksman, S.A. and Woodruff, H.B., *Actinomyces antibioticus*, a new soil organism antagonistic to pathogenic and nonpathogenic bacteria, *J. Bacteriol.*, 42, 231, 1941.
3. Fisher, W.P., Charney, J. and Bolhofer, W.A., An actinomycin from a species of the genus *Micromonospora, Antibiot. Chemother.*, 1, 571, 1951.
4. Waksman, S.A. et al., Toxicity of actinomycin, *Proc. Soc. Exptl. Biol. Med.*, 47, 261, 1941.
5. Hackmann, C., Experimentelle untersuchungen über die wirkung von actinomycin C, *Z. Krebsforsch.*, 58, 607, 1952.
6. Schulte, G., Erfahrungen mit neuen zytostatischen mitteln bei hämoblastosen und carcinomen und die abgrenzung ihrer wirkungen gegen roentgentherapie, *Z. Krebsforsch*, 58, 500, 1952.
7. Kirk, J.M., The mode of action of actinomycin D, *Biochim. Biophys. Acta*, 42, 167, 1960.
8. Slotnick, I.J., Mechanism of action of actinomycin D in microbiological systems, *Ann. N.Y. Acad. Sci.*, 89, 342, 1960.
9. Shatkin, A.J., The effects of actinomycin on virus replication, in *Actinomycin*, Waksman, S.A., ed., Interscience Publishers, New York, 1968, 69.

10. Lehr, K. and Burger, J., The isolation of a crystalline actinomycin-like antibiotic, *J. Arch. Biochem.*, 23, 503, 1949.

11. Brockmann, H. and Grubhofer, N., Actinomycin C, *Naturwiss.*, 36, 376, 1949.

12. Manaker, R.A. et al., Actinomycin. III. The production and properties of a new actinomycin, *Antibiot. Annu.*, 853, 1954/55.

13. Brockmann, H. and Gröne, H., Reine actinomycine, *Naturwiss.*, 41, 65, 1954.

14. Brockmann, H., Linge, H. and Gröne, H., Zur kenntnis des actinomycins X, *Naturwiss.*, 40, 224, 1953.

15. Bossi, R. et al., Stoffwechselprodukte von actinomyceten, 14. Mitt. actinomycin Z, *Helv. Chim. Acta*, 41, 1645, 1958.

16. Brockmann, H. et al., Über das actinomycin C (antibiotica aus actinomyceten, V. Mitt.), *Chem. Ber.*, 84, 260, 1951.

17. Brockmann, H. and Pfennig, N., Auftrennung von actinomycin C durch gegenstromverteilung, *Naturwiss.*, 39, 429, 1952.

18. Vining, L.C. and Waksman, S.A., Paper chromatographic identification of the actinomycins, *Science*, 120, 389, 1954.

19. Gregory, F.J., Vining, L.C. and Waksman, S.A., Actinomycin. IV. Classification of the actinomycins by paper chromatography, *Antibiot. Chemother.*, 5, 409, 1955.

20. Vining, L.C., Gregory, F.J. and Waksman, S.A., Actinomycin. V. Chromatographic separation of the actinomycin complexes, *Antibiot. Chemother.*, 5, 417, 1955.

21. Roussos, G.G. and Vining, L.C., Isolation and properties of pure actinomycins, *J. Chem. Soc.*, 2469, 1956.

22. Mauger, A.B. and Katz, E., Actinomycins, in *Antibiotics: isolation, separation and purification, J. Chromatogr. Library*, Wagman, G.H. and Weinstein, M.J., Eds., Elsevier, New York, 1978, 15, 1.

23. Waksman, S.A., Katz, E. and Vining, L.C., Nomenclature of actinomycins, *Proc. Natl. Acad. Sci. U.S.A.*, 44, 602, 1958.

24. Brockmann, H. et al., Zur konstitution der actinomycine, *Angew. Chem.*, 68, 70, 1956.

25. Bullock, E. and Johnson, A.W., The structure of actinomycin D, *J. Chem. Soc*, 3280, 1957.

26. Brockmann, H., Boldt, P. and Petras, H.-S., Die aminosäuresequenz von actinomycin C_1 und actinomycin C_2, *Naturwiss.*, 47, 62, 1960.

27. Brockmann, H. and Franck, B., Actinomycin C_{2a}, ein isomeres des actinomycins C_2, *Naturwiss.*, 47, 15, 1960.

28. Brockmann, H. and Boldt, P., Oxidative abspaltung des β-pentapeptidlactonringes aus actinomycin C_2 und C_3; ein beweis für die bis-pentapeptidlacton-struktur der actinomycine, *Chem. Ber.*, 101, 1940, 1968.

29. Brockmann, H. and Manegold, J.H., Ueberfuehrung von actinomycin X_2 in die actinomycine C_1, $X_{0β}$ und $X_{0β}$; actinomycine XXIII, *Chem. Ber.*, 93, 2971, 1960.

30. Diegelman, R., Mauger, A. and Katz, E., Isolation and identification of oxoproline in actinomycin X_2 (V) hydrolysates, *J. Antibiot.*, 22, 85, 1969.

31. Brockmann, H and Manegold, J.H., Actinomycin X_{1a} and $X_{0β}$, *Chem. Ber.*, 95, 1081, 1962.

32. Waksman, S.A. et al., Directed formation of the actinomycins — chemical structure and biological activities, *Science*, 129, 1290, 1959.

33. Johnson, A.W. and Mauger, A.B., The isolation and structure of actinomycins II and III, *Biochem. J.*, 73, 535, 1959.

34. Mauger, A.B., Degradation of peptides to diketopiperazines: application of pyrolysis-gas chromatography to sequence determination in actinomycins, *J. Chem. Soc. Chem. Commun.*, 39, 1971.

35. Katz, E. and Mauger, A.B., unpublished data, 1977.

36. Brockmann, H. and Stähler, E.A., Zur konstitution des actinomycins Z_5, *Tetrahedron Lett.*, 2567, 1973.

37. Katz, E., Mason, K.T. and Mauger, A.B., Identification of *cis*-5-methylproline in hydrolysates of actinomycin Z_5, *Biochem. Biophys. Res. Commun.*, 52, 819, 1973.

38. Brockmann, H. and Stähler, E.A., 4-Oxo-5-methylprolin, ein baustein des actinomycins Z_1, *Naturwiss.*, 52, 391, 1965.

39. Brockmann, H. and Manegold, J.H., Actinomycine, XXV; antibiotica aus actinomyceten, LIV. Actinomycin Z_1; zur quantitativen aminosäureanalyse der actinomycine, *Hoppe-Seyler's Z. Physiol. Chem.*, 343, 86, 1965.

40. Katz, E., Mason, K.T. and Mauger, A.B., 3-Hydroxy-5-methylproline, a new amino acid identified as a component of actinomycin Z_1, *Biochem. Biophys. Res. Commun.*, 63, 502, 1975.
41. Mauger, A.B. et al., Synthesis and stereochemistry of 3-hydroxy-5-methylproline, a new naturally occurring imino acid, *J. Org. Chem.*, 42, 1000, 1977.
42. Katz, E., Mason, K.T. and Mauger, A.B., The presence of α-amino-β,γ-dihydroxybutyric acid in hydrolysates of actinomycin Z_1, *J. Antibiot.* 27, 952, 1974.
43. Lackner, H. et al., Structures of five components of the actinomycin Z complex from *Streptomyces fradiae*, two of which contain 4-chlorothreonine, *J. Nat. Prod.*, 63, 352, 2000.
44. Lackner, H. et al., A new actinomycin-type chromopeptide from *Streptomyces* Sp. HKI-0155, *J. Antibiot.*, 53, 84, 2000.
45. Hanada, M. et al., Protactin, a new antibiotic metabolite and a possible precursor of the actinomycins, *J. Antibiot.*, 45, 20, 1992.
46. Brockmann, H., Die actinomycine, *Fortschr. Chem. Org. Naturst.*, 18, 1, 1960.
47. Umezawa, H., Index of antibiotics from Actinomycetes, *Japan Scientific Societies Press*, State College Pennsylvania, 1967 (Vol.I) and Univ. Park Press, Baltimore, MD, 1978 (Vol. II).
48. Meienhofer, J. and Atherton, E., Structure-activity relationships in the actinomycins, in *Structure-activity relationships among the semisynthetic antibiotics*, Perlman, D., ed., John Wiley & Sons, New York, 1977, 427.
49. Mauger, A.B., The actinomycins, in *Topics in antibiotic chemistry*, Vol. 5, Sammes, P.G., ed., Ellis Horwood, Chichester, 1980, 223.
50. Jain, S.C. and Sobell, H.M., Stereochemistry of actinomycin binding to DNA. I. Refinement and further structural details of the actinomycin-deoxyguanosine crystalline complex, *J. Mol. Biol.*, 68, 1, 1972.
51. Ginell, S., Lessinger, L. and Berman, H.M., The crystal and molecular structure of the anticancer drug actinomycin D — some explanations for its unusual properties, *Biopolymers*, 27, 843, 1988.
52. Schäfer, M., Sheldrick, G.M. et al., Crystal structures of actinomycin D and actinomycin Z_3, *Angew. Chem. Internat. Ed.*, 37, 2381, 1998.
53. Lifferth, A. et al., Synthese und struktur von prolinring-modifizierten actinomycinen des X-typs, *Z. Naturforsch.*, 54b, 681, 1999.
54. Lackner, H., Three-dimensional structure of the actinomycins, *Angew. Chem. Int. Ed.*, 14, 375, 1975.
55. Liu, X., Chen, H. and Patel, D.J., Solution structure of actinomycin-DNA complexes: drug intercalation at isolated G-C sites, *J. Biomol. NMR*, 1, 323, 1991.
56. Kawamata, J. and Imanishi, M., Interactions of actinomycin with deoxyribonucleic acid, *Nature*, 187, 1112, 1960.
57. Kersten, H., Kersten, W. and Rauen, H.M., Action of nucleic acids on the inhibition of growth by actinomycin of *Neurospora crassa*, *Nature*, 187, 60, 1960.
58. Rauen, H.M., Kersten, H. and Kersten, W., Zur wirkungsweise von actinomycinen, *Hoppe-Seyler's Z. Physiol. Chem.*, 321, 139, 1960.
59. Kawamata, J. and Imanishi, M., Mechanism of action of actinomycin, with special reference to its interaction with deoxyribonucleic acid, *Biken J.*, 4, 13, 1961.
60. Reich, E. et al., Actinomycin: Effect of actinomycin D on cellular nucleic acid synthesis and virus production, *Science*, 134, 556, 1961.
61. Haselkorn, R., Actinomycin D as a probe for nucleic acid secondary structure, *Science*, 143, 682, 1964.
62. Reich, E. and Goldberg, I.H., Actinomycin and nucleic acid function, *Prog. Nucleic Acid Res. Mol. Biol.*, 3, 183, 1964.
63. Gellert, M. et al., Actinomycin binding to DNA: mechanism and specificity, *J. Mol. Biol.*, 11, 445, 1965.
64. Goldberg, I.H., Rabinowitz, M. and Reich, E., Basis of actinomycin action. I. DNA binding and inhibition of RNA-polymerase synthetic reactions by actinomycin, *Proc. Natl. Acad. Sci. U.S.A.*, 48, 2094, 1962.
65. Cerami, A. et al., The interaction of actinomycin with DNA: requirement for the 2-amino group of purines, *Proc. Natl. Acad. Sci. U.S.A.*, 57, 1036, 1967.
66. Kahan, E., Kahan, F.M. and Hurwitz, J., The role of deoxyribonucleic acid in ribonucleic acid synthesis. VI. Specificity of action of actinomycin D, *J. Biol. Chem.*, 238, 2491, 1963.
67. Müller, W. and Crothers, D.M., Studies on the binding of actinomycin and related compounds to DNA, *J. Mol. Biol.*, 35, 251, 1968.

68. Kersten, W., Interaction of actinomycin C with constituents of nucleic acids, *Biochim. Biophys. Acta*, 47, 610, 1961.

69. Takusagawa, F. et al., The structure of a pseudo intercalated complex between actinomycin and the DNA binding sequence d(GpC), *Nature*, 296, 466, 1982.

70. Takusagawa, F. et al., Crystallization and preliminary X-ray study of a complex between d(ATGCAT) and actinomycin D, *J. Biol. Chem.*, 259, 4714, 1984.

71. Takusagawa, F., The role of the cyclic depsipeptide rings in antibiotics, *J. Antibiot.*, 38, 1596, 1985.

72. Ross, W.E., Glaubiger, D. and Kohn, K.W., Qualitative and quantitative aspects of intercalator-induced DNA strand breaks, *Biochim. Biophys. Acta*, 562, 41, 1979.

73. Wasserman, K. et al., Effects of morpholinyl doxorubicins, doxorubicin and actinomycin D on mammalian DNA topoisomerases I and II, *Mol. Pharmacol.*, 38, 38, 1990.

74. Wadkins, R.M. et al., Actinomycin D binding to single-stranded DNA: sequence specificity and hemi-intercalation model from fluorescence and 1H NMR spectroscopy, *J. Mol. Biol.*, 262, 53, 1996.

75. Gniazdowski, M. et al., Transcription factors as targets for DNA-interacting drugs, *Curr. Med. Chem.* 10, 909, 2003.

76. Foley, G.E. and Eagle, H., The cytotoxicity of anti-tumor agents for normal human and animal cells in first tissue culture passage, *Cancer Res.*, 18, 1012, 1958.

77. Eagle, H. and Foley, G.E., Cytotoxicity in human cell cultures as a primary screen for the detection of anti-tumor agents, *Cancer Res.*, 18, 1017, 1958.

78. Reich, E., Goldberg, I.H. and Rabinowitz, M., Structure-activity correlations of actinomycins and their derivatives, *Nature*, 196, 743, 1962.

79. Monks, A. et al., Feasibility of a high-flux anticancer drug screen utilizing a diverse panel of human tumor cell lines in culture, *J. Natl. Cancer Inst.*, 83, 757, 1991.

80. Paul, K.D. et al., Display and analysis of patterns of differential activity of drugs against human tumor cell lines: development of mean graph and COMPARE algorithm, *J. Natl. Cancer Inst.*, 81, 1088, 1989.

81. Hackmann, C., HBF 386 (Actinomycin C), ein cytostatisch wirksamer naturstoff, *Strahlentherapie*, 90, 296, 1953.

82. Stock, J.A., Antitumor antibiotics, *Exptl. Chemother.*, 4, 239, 1966.

83. Sugiura, K. and Schmidt, M.S., Effects of antibiotics on the growth of a variety of mouse and rat tumors, *Proc. Am. Assoc. Cancer Res.*, 2, 151, 1956.

84. Hirono, I. et al., Antitumor effect of actinomycin D on transplantable Wilm's tumor in rats, *Gann.*, 59, 473, 1968.

85. Priestly, J.B., The laboratory basis for the clinical treatment of Wilms tumor, *J. Urol*, 107, 696, 1972.

86. Goldin, A. and Johnson, R., Evaluation of actinomycins in experimental systems, *Cancer Chemother. Reps.*, Part 1, 58, 63, 1974.

87. Philips, F.S. et al., The toxicity of actinomycin D, *Ann. N.Y. Acad. Sci.*, 89, 348, 1960.

88. Brockmann, H. and Lackner, H., Totalsynthese von actinomycin C_3, *Naturwiss.*, 47, 230, 1960.

89. Brockmann, H. and Lackner, H., Totalsynthese von actinomycin C_3 über bis-seco-actinomycin C_3, *Chem. Ber.*, 100, 353, 1967.

90. Bachmann, H.G. and Müller, W., X-ray diffraction of actinomycin C_3, *Nature*, 201, 261, 1964.

91. Perutz, M.F., X-ray diffraction of actinomycin C_3, *Nature*, 201, 814, 1964.

92. Brockmann, H. and Lackner, H., Totalsynthese der actinomycine C_1, C_2, i-C_2 und C_3 über N-(2-amino-3-hydroxy-4-methylbenzoyl)-pentapeptidlactone; ein weiterer beweis für die bispentapeptidstruktur der actinomycine, *Chem. Ber.*, 101, 2231, 1968.

93. Meienhofer, J., Synthesis of actinomycin and analogs. III. A total synthesis of actinomycin D (C_1) *via* peptide cyclization between proline and sarcosine, *J. Am. Chem. Soc.*, 92, 3771, 1970.

94. Vlasov, G.P., Lashkov, V.N. and Glibin, E.N., Synthesis of actinomycin D and its peptide analogs. IV. Synthesis of de(di-3'-proline)actinomycin D, *Zh. Org. Khim.* S.S.S.R., 15, 983, 1979.

95. Mauger, A.B. and Stuart, O.A., Synthesis of an actinomycin-related peptide lactone from the corresponding cyclic peptide by N,O-acyl shift, *Int. J. Peptide Protein Res.*, 34, 196, 1989.

96. Lackner, H., Totalsynthese von aniso-actinocinyl-peptiden und aniso-actinomycinen über deuterium-markierte vorstufen, *Chem. Ber.*, 103, 2476, 1970.

97. Lackner, H., Synthesen von selectiv deuterierten pentapeptidlactonen und stellungsspezifisch im α-oder β-peptidring markierten actinomycinen, *Chem. Ber.*, 104, 3653, 1971.

98. Katz, E., Controlled biosynthesis of actinomycins, *Cancer Chemother. Rep.*, Part 1, 58, 83, 1974.
99. Schmidt-Kastner, G., Actinomycin E und actinomycin F, zwei neue biosynthetische actinomycingemische, *Naturwiss.*, 43, 131, 1956.
100. Katz, E., Kawai, Y. and Shoji, J., Configuration of the N-methylisoleucine in the actinomycins, *J. Biochem. Biophys. Res. Commun.*, 43, 1035, 1971.
101. Yajima, T., Grigg, M.A. and Katz, E., Biosynthesis of antibiotic peptides with isoleucine isomers, *Arch. Biochem. Biophys.*, 151, 565, 1972.
102. Schmidt-Kastner, G., The production of actinomycins by controlled biosynthesis: the F actinomycins, *Ann. N.Y. Acad. Sci.*, 89, 299, 1960.
103. Katz, E. et al., Novel actinomycins formed by biosynthetic incorporation of *cis*- and *trans*-4-methylproline, *Antimicrob. Agents Chemother.*, 11, 1056, 1977.
104. Nishimura, J.S. and Bowers, W.F., Evidence for the incorporation of L-thiazolidine-4-carboxylate into actinomycins by *Streptomyces antibioticus*, *Biochem. Biophys. Res. Commun.*, 28, 665, 1967.
105. Katz, E., Biogenesis of the actinomycins, *Ann. N.Y. Acad. Sci.*, 89, 304, 1960.
106. Formica, J.V. and Apple, M.A., Production, isolation and properties of azetomycins, *Antimicrob. Agents Chemother.*, 9, 214, 1976.
107. Formica, J.V. and Katz, E., Isolation, purification, and characterization of pipecolic acid-containing actinomycins, Pip 2, Pip 1α and Pip 1β, *J. Biol. Chem.*, 248, 2066, 1973.
108. Brockmann, H., Gröne, H. and Pampus, G., Chloractinomycine, *Chem. Ber.*, 91, 1916, 1958.
109. Brockmann, H., Pampus, G. and Mecke, R., Actinomycine, XXII; antibiotica aus actinomyceten, XLIV: N-Alkyl-actinomycine, *Chem. Ber.*, 92, 3082, 1959.
110. Brockmann, H., Müller, W. and Peterssen-Borstel, H., Actinomycin-derivate aus dihydroactinomycinen, *Tetrahedron Lett.*, 3531, 1966.
111. Brockmann, H., Hocks, P. and Müller, W., N-2-Substituierte actinomycine, *Chem. Ber.*, 100, 1051, 1967.
112. Moore, S. et al., Synthesis and antitumor activity of 2-deamino- and N²-(γ-hydroxypropyl)actinomycin D, *J. Med. Chem.*, 18, 1098, 1975.
113. Sengupta, S.K. et al., N²- and C-7-substituted actinomycin D analogues: Synthesis, DNA-binding affinity, and biochemical and biological properties. Structure-activity relationship, *J. Med. Chem.*, 24, 1052, 1981.
114. Brockmann, H., Ammann, J. and Müller, W., 7-Chlor- und 7-brom-actinomycine, *Tetrahedron Lett.*, 3595, 1966.
115. Sengupta, S.K. et al., Synthesis and DNA-binding properties of 7-amino-actinomycin D — a selective fluorescent binding agent, *Fed. Proc., Fed. Am. Soc. Exptl. Biol.*, 30, 342, 1971.
116. Müller, W., Bindung von actinomycin und actinomycin-derivaten an desoxyribonucleinsäuren, *Naturwiss.*, 49, 156, 1962.
117. Madhavarao, M.S., Chaykovsky, M. and Sengupta, S.K., N⁷-Substituted 7-aminoactinomycin D analogs. Synthesis and biological properties, *J. Med. Chem.*, 21, 958, 1978.
118. Sengupta, S.K., Kelly, C. and Sehgal, R.K., "Reverse" and "Symmetrical" analogs of actinomycin D: metabolic activation and *in vitro* and *in vivo* tumor growth inhibitory activities, *J. Med. Chem.*, 28, 620, 1985.
119. Brockmann, H. and Franck, B., Hydrierende acetylierung von actinomycin, *Angew. Chem.*, 68, 68, 1956.
120. Sengupta, S.K. et al., Actinomycin D oxazinones as improved antitumor agents, *J. Med. Chem.*, 22, 797, 1979.
121. Sengupta, S.K. et al., Tetracyclic chromophoric analogs of actinomycin D: Synthesis, structure elucidation and interconvertability from one form to another, antitumor activity, and structure-activity relationships, *J. Med. Chem.*, 26, 1631, 1983.
122. Sengupta, S.K., et al. New actinomycin D analogs as superior chemotherapeutic agents against primary and advanced colon tumors and colon xenografts in nude mice, *J. Med. Chem.*, 31, 768, 1988.
123. Brockmann, H. and Franck, B., Aufspaltung der actinomycine zu actinomycinsäuren, *Angew. Chem.*, 68, 68, 1956.
124. Perlman, D., Mauger, A.B. and Weissbach, H., Microbial transformations of peptide antibiotics. I. Degradation of actinomycins by *Actinoplanes* species, *Antimicrob. Agents Chemother.*, 581, 1966.
125. Brockmann, H. and Sunderkötter, W., Abbau von actinomycin C₃ mit konzentrierter salzsäure, *Naturwiss.*, 47, 229, 1960.

126. Brockmann, H., Pampus, G. and Manegold, J.H., Actinomycin $X_{0\beta}$; zur systematik und nomenklatur der actinomycine, *Chem. Ber.*, 92, 1294, 1959.

127. Brockmann, H. and Stähler, E.A., 3-Hydroxy-4-oxo-5-methylprolin als baustein von actinomycin Z_1, *Tetrahedron Lett.*, 3585, 1973.

128. Mauger, A.B. and Wade, R., The synthesis of actinomycin analogs. Part II. Actinocyl-gramicidin S, *J. Chem. Soc.*, 1406, 1966.

129. Brockmann, H. and Seela, F., 4,6-Didesmethyl-actinomycin C_1 und dessen 4,6-dimethoxy-, 4,6-diäthyl- und 4,6-di-*tert*-butyl-derivat, *Chem. Ber.*, 104, 2751, 1971.

130. Giencke, A. and Lackner, H., Desmethyl(trifluormethyl)-actinomycine, *Liebigs Ann. Chem.*, 569, 1990.

131. Brockmann, H. and Schulze, E., Synthese von pseudo-actinomycin C_1, *Tetrahedron Lett.*, 1489, 1971.

132. Mosher, C.W. et al., A phenazine analog of actinomycin D, *J. Med. Chem.*, 22, 918, 1979.

133. Brockmann, H. and Schramm, W., Synthese von actinomycin-(thr-Val-pro-sar-meval), dem antipoden von actinomycin C_1, *Tetrahedron Lett.*, 2331, 1966.

134. Brockmann, H. and Lackner, H., Aufbau von actinomycinen über actinomycinsäuren; ein zweiter syntheseweg, *Chem. Ber.*, 101, 1312, 1968.

135. Moore, S. et al., Synthesis and some properties and antitumor effects of the actinomycin lactam analog [di(1′-L-α,β-diaminopropionic acid)]actinomycin D, *J. Med. Chem.*, 19, 766, 1976.

136. Atherton, E. et al., Actinomycin D lactam, [1′1′-bis(*threo*-α,β-diaminobutyric acid)]actinomycin D, *J. Med. Chem.*, 16, 355, 1973.

137. Mauger, A.B. and Stuart, O.A., Synthesis and properties of some peptide analogs of actinomycin D, *J. Med. Chem.*, 34, 1297, 1991.

138. Mosher, C.W. and Goodman, M., Synthesis of [4′,4′-bis(glycine), 5′,5′-bis(valine)]actinomycin D, a tetra-N-demethylactinomycin, *J. Org. Chem.*, 37, 2928, 1972.

139. Mauger, A.B. et al., Conformation and dimerization of actinomycin-related peptide lactones in solution and in the solid state, *J. Am. Chem. Soc.*, 107, 7154, 1985.

140. Pugh, L.H., Katz, E. and Waksman, S.A., Antibiotic and cytostatic properties of the actinomycins, *J. Bacteriol*, 72, 660, 1956.

141. Katz, E. and Pugh, L.H., Variations on a theme. The actinomycins, *Appl. Microbiol.*, 9, 263, 1961.

142. Pugh, L.H. and Solotorovsky, M., Comparative effects of actinomycins II, III and IV on several ascitic tumors in mice, *Ann. N.Y. Acad. Sci.*, 89, 373, 1960.

143. Mason, K., Katz, E. and Mauger, A.B., Studies on the biological activities of actinomycins Z_1 and Z_5, *Arch. Biochem. Biophys.*, 160, 402, 1974.

144. Modest, E.J. and Sengupta, S.K., 7-Substituted actinomycin D (NSC-3053) analogs as fluorescent DNA-binding and experimental antitumor agents, *Cancer Chemother. Reps.* Part I, 58, 35, 1974.

145. Formica, J.V., Shatkin, A.J. and Katz, E., Actinomycin analogs containing pipecolic acid: relationship of structure to biological activity, *J. Bacteriol.*, 95, 2139, 1968.

146. Johnson, R.K., private communication.

147. Rose, W.C. et al., Comparative antitumor activity of actinomycin analogs in mice bearing Ridgeway osteogenic sarcoma or P388 leukemia, *Cancer Treatment Reps.*, 62, 779, 1978.

148. Li, M.C., Management of choriocarcinoma and related tumors of uterus and testis, *Med. Clin. N. Am.*, 45, 661, 1961.

149. Ross, G.T., Stohlbach, L.L. and Hertz, R., Actinomycin in the treatment of methotrexate-resistant trophoblastic disease in women, *Cancer Res.*, 22, 1015, 1962.

150. Li, M.C. et al., Effects of combined drug therapy of metastatic cancer of the testis, *J. Am. Med. Assoc.*, 174, 1291, 1960.

151. MacKenzie, A.R., Chemotherapy of metastatic testis cancer — results in 154 patients, *Cancer*, 19, 1369, 1966.

152. Tan, C.T.C., Dargeon, H.W. and Burchenal, J.H., The effect of actinomycin D in cancer in childhood, *Pediatrics*, 24, 544, 1959.

153. James, D.H. et al., Childhood malignant tumors. Concurrent chemotherapy with dactinomycin and vincristine sulfate, *J. Am. Med. Assoc.*, 19, 1043, 1966.

154. Farber, S., Chemotherapy in the treatment of leukemia and Wilms' tumor, *J. Am. Med. Assoc.*, 198, 826, 1966.

155. Fernbach, D.J. and Martyn, D.T., Role of dactinomycin in the improved survival of children with Wilms' tumor, *J. Am. Med. Assoc.*, 195, 1005, 1966.

16 Anthracyclines

Federico Maria Arcamone

CONTENTS

I.   Introduction 299
II. Development of the Anthracycline Drugs 301
   A.   Daunorubicin..301
   B.   Doxorubicin 301
   C.   Second-Generation Antitumor Anthracyclines...301
III. Mechanism of Action...302
   A.   DNA Complex and Topoisomerase II Poisoning..302
   B.   Natural and Acquired Resistance to Anthracyclines.....................................303
   C.   Cardiotoxicity 304
IV. Synthesis of the Drugs...304
   A.   Doxorubicin 304
   B.   Epirubicin...304
   C.   Idarubicin 304
V.   Medicinal Chemistry of Antitumor Anthracyclines 305
   A.   Analogs Modified in the Aglycone Moiety 305
      1.   Deoxyderivatives...305
      2.   8-Fluoroderivatives 306
      3.   14-α-Fluorohydrins..307
   B.   Analogs Modified in the Sugar Moiety 308
      1.   Monosaccharide Analogs...308
      2.   Disaccharide Analogs 309
   C.   Anthracycline–Triplex Helix-Forming Oligonucleotide Conjugates............312
VI. Clinical Activity of Anthracyclines 313
   A.   Daunorubicin..313
   B.   Doxorubicin 313
   C.   Epirubicin...314
   D.   Idarubicin 314
   E.   Aclacinomycin 315
   F.   Sabarubicin 315
   G.   Liposomal Anthracyclines 315
References 316

I. INTRODUCTION

Anthracycline aminoglycosides represent a wide class of antibiotics obtained by submerged aerobic fermentation of different microorganisms belonging to the genus *Streptomyces* and are called anthracyclines because the aglycone moiety is a tetracyclic system bearing an anthraquinone chromophore.[1] The anthracyclines are microbial metabolites belonging, from the biogenetic standpoint,

299

to the large family of the polyketide natural products, comprising a wide range of biologically active structural groups: the antibacterial antibiotics, tetracyclines, and the macrolides of the erythromycin type; immunosuppressants such as FK506; antiparasitic compounds such as monensin; and the avermectins. These molecules are synthesized by multifunctional polyketide synthase enzymes, which catalyze repeated condensation cycles between simple acylthioesters with the formation of a growing carbon chain containing β keto groups. These keto groups may undergo a number of intramolecular reactions, leading to aromatic derivatives, such as the aglycone of the anthracyclines, the anthracyclinones, and a series of reductive steps. The final products are characterized by a considerable molecular diversity.[2]

The antitumor activity of a biosynthetic anthracycline was first reported in 1959.[3] In fact, different samples of a mixture of red pigments isolated from the cultures of a *Streptomyces* strain at Farmitalia Research Laboratories were found to be active against murine Ehrlich carcinoma and sarcoma 180 test systems.[3] The presence of compounds possessing the hydroxyanthraquinone chromophore was deduced on the basis of the typical electronic spectrum — red in acid and blue in alkaline solutions. The pigments were isolated by countercurrent distribution, and their relationship with the anthracyclines — previously studied and described by different authors, and more precisely with the rhodomycin complex[1] — was established. Subsequently, useful pharmacological properties were found that were associated with the compounds produced by a novel species, *Streptomyces peucetius,* and related strains. Doxorubicin (**Ia**), the best known component of this group, is the 14-hydroxylated derivative of the main fermentation product daunorubicin (**Ib**), and since its registration in the early 1970s, doxorubicin has been one of the most widely used drugs in cancer chemotherapy.[4] (The trademark of doxorubicin formulations in the United States is Adriamycin®.)

The successful therapeutic application of **Ia** stimulated a considerable effort in the chemical and biological study of the antitumor anthracyclines aimed at the development of better analogs by chemical modification of the parent drugs or by total synthesis of new structurally related compounds. Although **Ia** remains one of the most effective agents in the medical treatment of a range of solid tumors, different new members of this chemical group, among which epirubicin (**II**) and idarubicin (**III**) are the most important, are presently used in the medical practice. Such new agents were already described in previous books[5,6] and are currently known as second-generation anthracyclines.

In 1992 it was observed that an analog truly superior to doxorubicin had not been identified. In fact, after the synthesis and biological testing of hundreds of new anthracycline glycosides in different laboratories, neither an analog with a really improved therapeutic index nor one with a substantially different spectrum of activity had been found.[7] It should be noted that analogs with considerable structural variations were submitted to testing procedures that were mostly limited to murine models of transplantable leukemias, which is possibly not the best way to take an appropriate selection of new interesting compounds. The present tumor type–oriented pharmacological evaluation, based on the use of a large number of human tumor cell lines and on human tumor heterotransplants in nude mice, coupled with rational drug design and a closer attention to aspects such as drug bioavailability, tissue distribution, and pharmacokinetics, might increase the probability of obtaining results closer to the desired goal.[8]

Antitumor activity also has been associated with other biosynthetic anthracyclines generally differing from the daunorubicin–doxorubicin group in the chemical structure of both the tetracyclic aglycone and the sugar moiety, with the latter often present as an oligosaccharide fragment. One such compound, aclacinomycin, has been introduced in clinical use.[9] The biosynthetic anthracyclines have been reviewed, together with the products of microbial transformations of individual anthracycline antibiotics, by Oki.[10] More recently, an overview of the different biosynthetic anthracyclines, describing the producing microorganisms, the chemical structure, and when available, the pharmacological data of over 250 compounds has been published.[11] A series of outstanding review articles on the fascinating molecular biology of anthracycline biosynthesis is also available.[12-14]

II. DEVELOPMENT OF THE ANTHRACYCLINE DRUGS

A. DAUNORUBICIN

In December 1958 and in February 1959, two similar crude preparations, obtained from a different *Streptomyces* species isolated from a soil sample collected in Apulia at Caste del Monte close to the ancient castle of King Frederick II, were shown to be endowed with significantly superior antitumor properties in terms of tumor size reduction and time of survival of the treated animals versus untreated ones in the above-mentioned mouse tests. The new microorganism and the main compound obtained from the active samples were named *S. peucetius* and daunomycin, respectively, because of the names of the ancient populations living in the region in early historical times, the Peucetii and the Daunii. The significant antitumor activity of daunomycin was confirmed in a wider spectrum of experimental systems.[15] These results stimulated a detailed investigation of the chemical structure of daunomycin to determine the differences between it and the classical anthracyclines — differences that were responsible for the markedly improved pharmacological properties. The products of acid hydrolysis, namely, aglycone, daunomycinone, and the sugar moiety, daunosamine, were studied by means of chemical degradation reactions and of physicochemical analyses. In brief, both fragments were new compounds[16,17] indicating that the microorganism produced a novel family of anthracycline glycosides. The complete structure and stereochemistry of daunomycin were finally established as represented in **Ib**,[18–20] and they were eventually confirmed by the x-ray diffraction analysis of the corresponding N-bromoacetylderivative.[21] In the meantime, daunomycin was identified with rubidomycin, isolated by French researchers from *Streptomyces coeruleorubidus*.[22] The World Health Organization suggested "daunorubicin" for the compound's name in recognition of the independent contributions from Farmitalia and Rhone-Poulenc laboratories. Clinical studies, showing its high activity in acute leukemias, were presented in a meeting held at St. Louis Hospital in Paris in 1967.[23]

B. DOXORUBICIN

Doxorubicin (**Ia**, originally named Adriamycin), a compound with marked activity in experimental systems and a wide spectrum of antitumor efficacy, especially in solid tumors, was isolated as a pigment coproduced with daunorubicin in the culture medium of *S. peucetius*–derived strains.[24] The structure of doxorubicin was determined, *inter alia*, by nuclear magnetic resonance analysis of the aglycone, adriamycinone, and by direct semisynthesis from daunorubicin.[25] Because of the very small quantities of doxorubicin present in the fermentation broths of the wild strains, the semisynthetic process was developed at a pilot plant scale and eventually was scaled up to the industrial scale, starting in the early 1970s. The publications from the Istituto Nazionale Tumori, Milan, concerning laboratory[26] and clinical[27] results aroused high interest, and the efficacy of doxorubicin in a spectrum of solid tumors was highlighted in an International Symposium held in Milan in September 1971. Early experience with doxorubicin in the treatment of human cancer diseases was reviewed by S.K. Carter.[28]

C. SECOND-GENERATION ANTITUMOR ANTHRACYCLINES

Daunosamine, 3-amino-2,6 dideoxy-L-*lyxo*-hexose,[17] is the natural aminosugar present in all anthracycline glycosides produced by the strains belonging to the species *S. peucetius* — the original producer of the daunorubicin group of antibiotics — and related strains.[11] Daunosamine has never been identified as a constituent of other anthracyclines or other natural products. An early approach to new analogs consisted of the synthesis of related aminosugars, differing from daunosamine in stereochemistry or in structure, and of the coupling of the same with the aglycone moiety, with the aim of modifying the pharmacokinetics or metabolism of the natural glycosides. This approach afforded a number of compounds with a wide variation in bioactivity, out of which the analog of

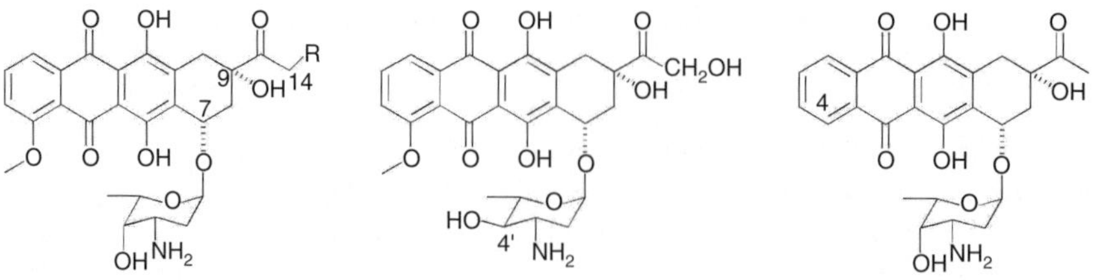

Ia: R=OH
Ib: R=H

II

III

STRUCTURES I–III

structure **II**, containing the aminosugar acosamine, namely, 3-amino-2,6-dideoxy-L-*arabino*-hexose, was selected for further development because of a significantly higher chemotherapeutic index when compared with doxorubicin.[4] The compound, known by the generic name of epirubicin, exhibited superior performance when compared with doxorubicin in experimental models and in clinical studies and was registered in most countries in the 1980s as Farmorubicin® or Pharmorubicin®. However, registration in the United States was obtained only more recently, and the compound is currently marketed in this country as Ellence®.

Daunomycinone, the aglycone of daunorubicin, is characterized by the tetracyclic trihydroxyanthraquinone chromophore typical of natural anthracyclinones, albeit containing specific structural features that identify the daunomycin group of metabolites; namely, the ketone function in the C9 side chain, the absence of a substituent at C10, and the presence of a methoxyl group instead of a hydroxyl at C4. In the closely related carminomycin, the C4 hydroxyl is not methylated. As methods for the total synthesis of daunomycinone and related compounds were made available, it is reasonable that analogs showing no substitution at C.4 were prepared and tested. In fact, their synthesis was easier in comparison with that of the natural compounds because of the absence of regioisomers in the construction of the anthraquinone system. The first compound synthesized in this series, 4-demethoxydaunorubicin (**III**) (idarubicin), has been developed to the clinical stage because of its powerful activity in experimental leukemia models and reduced cardiotoxicity. Idarubicin exhibits features rendering this compound unique among clinically available anthracyclines. The results of preclinical studies indicate that the higher lipophilicity of **III** leads to faster accumulation in the nuclei, superior DNA-binding capacity, and consequently greater cytotoxicity compared to **Ib**. A major advantage over the 4-methoxylated parent drug is its ability to partially overcome multidrug resistance. Its major metabolite, idarubicinol, is as active as the parent compound.[29] Idarubicin is marketed with the trademark Zavedos®. As seen below, a 4-demethoxy aglycone has become a favorable structural feature in more recently designed analogs.

III. MECHANISM OF ACTION

Anthracyclines, as with several other classes of antitumor agents, work by DNA intercalation, a specific mode of binding in which the drug chromophore is inserted between adjacent DNA base pairs.[30] However, the mechanism of action cannot be described in such a simple way, and the picture appears to be rather more complex.

A. DNA Complex and Topoisomerase II Poisoning

The binding of anthracyclines to DNA has been extensively studied by a variety of biochemical and biophysical techniques.[31,32] The drug intercalates at a pyrimidine–purine step with a preference for the cytidine-guanosine (CG) sequence and acts as a groove binder, as the sugar moiety interacts with the minor groove.[33–35] DNA binding is a necessary but not sufficient condition for drug activity.

In other words, there is no simple relationship between DNA binding affinity and cytotoxicity, and a number of different mechanisms have been proposed. Briefly, poisoning of topoisomerases is likely to be the mechanism most closely associated with their cytoxicity, although other molecular inter-actions may play an important role as well. Topoisomerases are enzymes that are involved in the topological interconversions of DNA during DNA transcription and replication; their functions are numerous and important because they are critical for DNA functions and cell survival. It was found that the DNA breaks induced by doxorubicin were associated with the enzyme topoisomerase II.[36] In a study including different anthracyclines, it was found that for all bioactive anthracyclines in the entire SV40 DNA a single region of prominent cleavage sites lies between nucleotides 4237 and 4294. A good correlation was recorded between cytotoxicity and intensity of topoisomerase II–medi-ated DNA breakage.[37] All of the doxorubicin-stabilized sites had an adenosine at the 3 terminus of at least one member of each pair of strand breaks that would constitute a topoisomerase II double-strand scission. Conversely, none of the enzyme-only sites had an adenosine simultaneously at the corresponding positions on opposite strands. The requirement of a 3-adenosine for doxorubicin-stabilized cleavage is therefore incompatible with enzyme-only cleavage and explains the mutual exclusivity of the two classes of sites.[38] The strength of DNA binding does not correlate with the stimulatory effect of anthracyclines on topoisomerase II–mediated DNA cleavage and with the cytotoxic potency. However, drug intercalation is still required for optimal drug activity, indicating that the specific mode of DNA interaction, rather than the strength of binding, is important in determining the cytotoxic potency.[39] Again, on the basis of a study including different semisynthetic doxorubicin analogs, it was concluded that cytotoxic potency of anthracyclines might be the result of an interplay of different factors, namely, the level, the persistence, and the genomic localization of topoisomerase II–mediated DNA cleavage.[40] In conclusion, the main steps of the mode of action of antitumor anthracyclines would be the formation of a drug-DNA-enzyme ternary complex, in which the enzyme is covalently linked to the broken DNA strand, and the consequent DNA damage consisting of protein-associated double-strand breaks that trigger apoptotic cell death.[41,42]

B. NATURAL AND ACQUIRED RESISTANCE TO ANTHRACYCLINES

Drug resistance is a major problem in cancer chemotherapy, and it is also clearly apparent in the case of antitumor anthracyclines. A number of important tumors of great clinical importance do not respond to currently available anthracycline drugs. These tumors include colon cancers, lung cancers, pancreatic and renal cancers, and malignant melanoma, to cite only some examples of "naturally resistant" tumors. Other diseases, such as gastric and small cell lung cancers and advanced ovary and breast tumors, are only partially responsive, and often the benefit of drug treatment is marginal. So far, only the classic multidrug resistance phenotype, which is a result of the presence of P-glycoprotein in the plasma membrane (a "pump" that can extrude a wide range of anticancer drugs), has been shown to contribute to resistance in clinical conditions. The other possible mechanisms of resistance are a non-P-glycoprotein-mediated multidrug resistance, a phenotype characterized by expression of other proteins in the plasma membrane that are also able to extrude anticancer drugs; changes in the intracellular distribution of the drug; glutathione transferases and detoxification mechanisms; low levels of expression or alterations in topoisomerase II; and finally, an increased DNA repair. However, evidence for these mechanisms of resistance in cancer patients is still missing.[43]

A relationship was found between doxorubicin resistance and glutathione transferase activity, but not between multidrug resistance–associated protein and P-glycoprotein, in human colon ade-nocarcinoma cell lines,[44] and enzyme-catalyzed formation of the adduct, 7-deoxy-7-S-glutathionyl-adriamycinone, may be relevant to tumor cell resistance to Adriamycin.[45] Introduction of HER-2 oncoprotein gene *in vitro* induces resistance to several anticancer drugs, including taxanes, cisplatin, and 5-fluorouracil, in breast cancer cells. Interestingly, in clinical studies, patients with HER-2 oncoprotein overexpression, an important prognostic factor associated with a poor prognosis in breast cancer, responded better to an anthracycline-based regimen than did patients with low HER-2 expression, and their overall survival was also superior. In addition, treatment with trastuzumab, an anti-HER-2 antibody, increased drug sensitivity to anthracyclines.[46]

C. Cardiotoxicity

Together with the onset of resistance, a major dose-limiting factor in the repeated treatment with doxorubicin is the development of cardiotoxicity at cumulative dosages higher than 400 mg/m^2. In the mouse, the most prominent and persistent myocardial change induced by repeated doxorubicin administrations was an increased reactivity of the sarcoplasmic reticulum with the zinc iodide–osmium tetroxide reagent that appeared to be related with the observed cardiotoxic effects.[47] The zinc iodide–osmium tetroxide reagent allows detection of disulfide bonds in polypeptides, cystine, and oxidized glutathione, substances that react more promptly than other potential substrates such as unsaturated lipids and catecholamines. This observation indicates the presence of an oxidative stress that, as already demonstrated by different authors,[48,49] may be caused by a higher then normal oxygen tension in the heart tissue. Excess oxygen could be the result of the markedly long-lasting inhibition of nucleic acid synthesis found in the heart tissue of laboratory animals treated with the drug,[50] as well as a consequent reduction of metabolic consumption of oxygen. Also, inhibition by doxorubicin of mitochondrial DNA transcription leads to a diminished ATP production and consequently to cell damage in high-energy-demanding tissues.[51] In any case, current evidence appears to be consistent with a fundamental role for free radicals, whatever their origin, in the cardiotoxic actions of the anthracyclines.[41]

IV. SYNTHESIS OF THE DRUGS

A. Doxorubicin

Doxorubicin **Ia** is prepared by semisynthesis starting from biosynthetic daunorubicin **Ib**, currently obtained in bulk quantities by fermentation of different *Streptomyces* strains, mostly derived from *S. peucetius*. The reaction sequence comprises an electrophilic bromination of **Ib** followed by an acid treatment to give the 14-bromoketone that is converted to **Ia** via the intermediate 14-O-formate.[4] As with all commercial anthracyclines, the compound is crystallized to higher than 98% purity as the hydrochloride and is formulated in freeze-dried form with addition of lactose. Doxorubicin ready-to-use solutions, both with and without stabilizers, have been patented.[52,53]

B. Epirubicin

Epirubicin (**II**) is prepared by a similar reaction sequence starting from 4-epidaunorubicin, originally prepared by the glycosidation of daunorubicin aglycone, daunomycinone, with N,O-di-trifluoroacetyl-L-acosaminyl chloride and silver triflate, followed by removal of the protecting trifluoroacetyl groups.[4] A direct conversion of biosynthetic daunorubicin to the 4-epi derivative **VII** has also been carried out.[54] It involves oxidation at C4 of N-trifluoroacetyl derivative **IV** to ketone **V** and selective reduction of the latter with sodium borohydride to **VI**, followed by deprotection with base (Scheme 16.1). This procedure is the basis of current industrial production of **II**.

C. Idarubicin

Idarubicin (**III**) was originally prepared by glycosidation of 4-demethoxydaunomycinone (idarubicinone), which was obtained by total synthesis using a modification of the Wong procedure.[4] A stereoselective synthesis of Wong's tetralin has been published very recently.[55] In contrast, a semisynthetic process useful for the industrial preparation of idarubicinone starting from daunomycinone **VIII**, easily prepared on acid hydrolysis of biosynthetic daunorubicin, has been carried

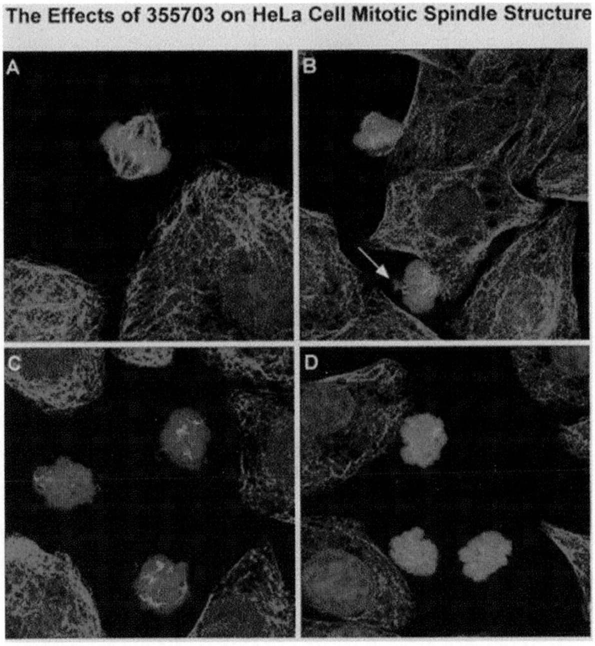

The Effects of 355703 on HeLa Cell Mitotic Spindle Structure

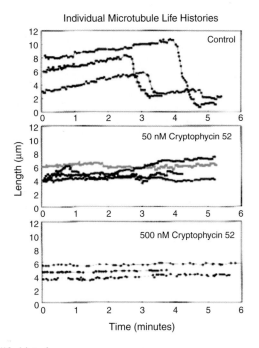

FIGURE 9.2 Fluorescence confocal micrographs of HeLa cells. Panel A shows a metaphase mitotic spindle in an untreated control cell. The metaphase spindle is bipolar, with the chromosomes located in a compact metaphase plate at the midpoint between the spindle poles. Panel B shows cells treated with 30 pM of LY355703 (approx. IC_{50}) for 8–10 h. The mitotic spindles look relatively normal. The most obvious abnormality involves the displacement of chromosomes from the compact metaphase plate. In the lower spindle, a chromosome (red) can be seen near the spindle pole (arrow). Panel C (8–10 h; 100 pM of LY355703) and D (8–10 h; 300 pM of LY355703) show increased fragmentation of the spindle microtubules and disorganization of the chromosome mass into ball-like clusters.

FIGURE 9.3 Microtubule life histories.

(A)

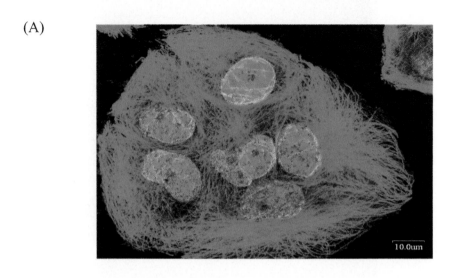

(B)

FIGURE 10.1 Immunofluorescence images of A549 cells stained with anti-α-tubulin (green) and propidium iodide (red) and observed by confocal microscopy. Cells were exposed to (A) 0.05% ethanol (vehicle control), or (B) (+)-discodermolide at a concentration of 100 n*M*.

FIGURE 11.1 *Dollabella auricularia.*

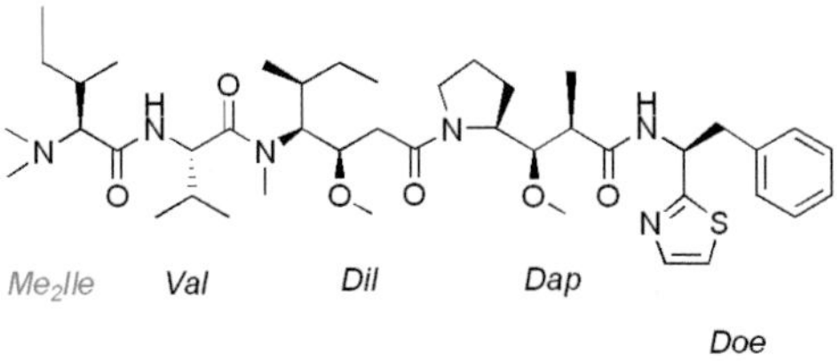

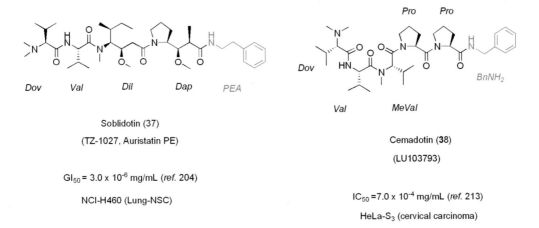

Symplostatin 1 (35)

$IC_{50} = 3.0 \times 10^{-4}$ mg/mL

KB cells (epidermoid carcinoma)

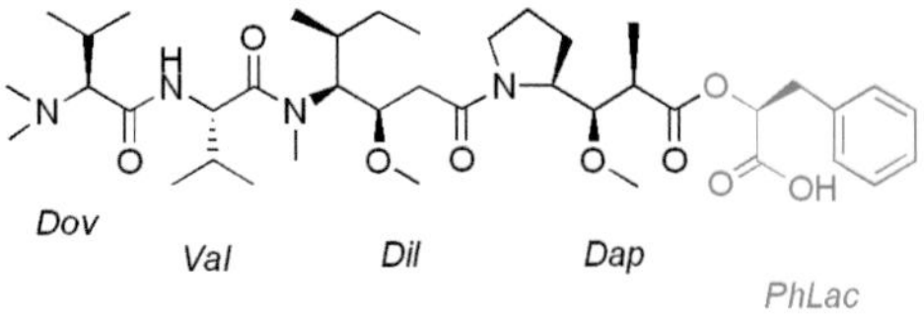

Symplostatin 3 (36)

$IC_{50} = 3.9 \times 10^{-3}$ mg/mL

KB cells (epidermoid carcinoma)

FIGURE 11.6 Synthetic derivatives of dolastatin 10 (soblidotin) and dolastatin 15 (cematodin) in phase II clinical trials.

Soblidotin (37)

(TZ-1027, Auristatin PE)

$GI_{50} = 3.0 \times 10^{-6}$ mg/mL (*ref. 204*)

NCI-H460 (Lung-NSC)

Cemadotin (38)

(LU103793)

$IC_{50} = 7.0 \times 10^{-4}$ mg/mL (*ref. 213*)

HeLa-S$_3$ (cervical carcinoma)

FIGURE 11.7 Naturally occurring derivatives of dolastatin 10.

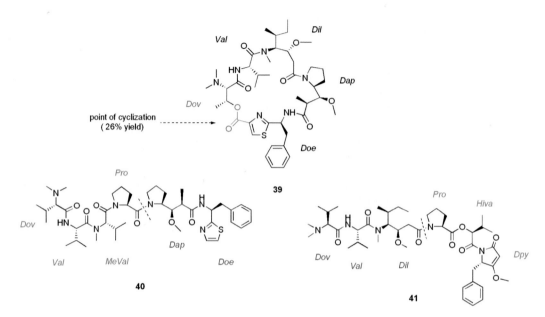

FIGURE 11.9 Dolastatin 10 cyclic derivative and D-10/D-15 hybrid compounds.

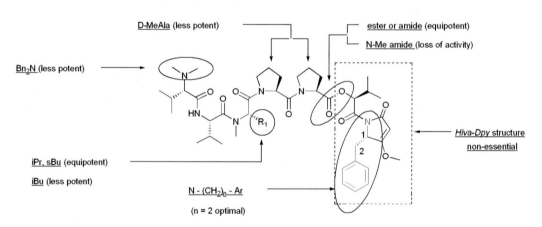

FIGURE 11.10 Effects of structural modifications on dolastatin 15 activity.

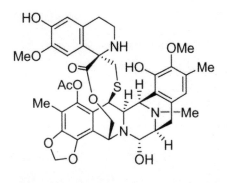

FIGURE 12.1 Structure and source organism of Yondelis.

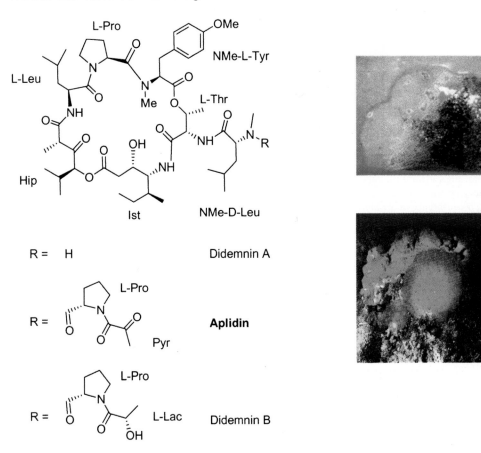

L-Pro

OMe

NMe-L-Tyr

L-Leu

Me

L-Thr

NH

OH

N—R

Hip

NH

Ist

NMe-D-Leu

R =   H   Didemnin A

L-Pro

R =   **Aplidin**

Pyr

L-Pro

R =   L-Lac   Didemnin B

OH

FIGURE 12.2 Structure and source organism of Aplidin.

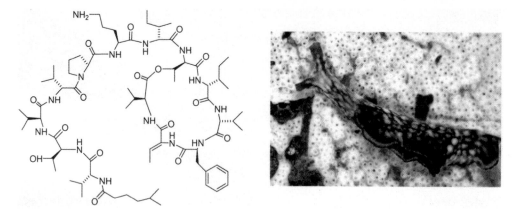

FIGURE 12.6 Structure and source organism of Kahalalide F.

Trypsinized COS-1 cells incubated in the absence or presence of 2 µM Kahalalide F. A reduction of cytoskeletal space is found, but the microtubule network is otherwise intact.

Effect on ER and Golgi Morphology

Control and treated cells with 2 µM Kahalalide F. Stained for ER with anti-PDI and for Golgi apparatus with anti-Golgi alpha mannosidase and showing no difference in organelle morphologies.

Control

ER NRK Cells

Kahalalide F

Golgi

NRK cells   BHK cells

Control

Kahalalide F

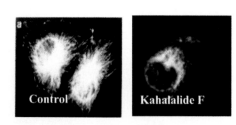

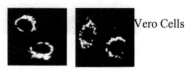 Vero Cells

Effect on Lyosomal Membranes

Control and treated cells with 2 µM Kahalalide F. Labeled with the fluorescent acidophilic probe LysoTracker Green and observed under the laser scanning microscope showing an increase in lysosomal volume of treated cells.

FIGURE 12.7 Effect on the microtubule network and membrane-bound organelles.[92]

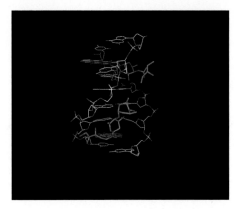

FIGURE 16.1 Representation of the sabarubicin intercalation complex with d(CGATCG)2 according to a x-ray crystal diffraction study. The drawing was kindly provided by Dr. Giovanni Ughetto, Laboratorio del CNR di Montelibretti (Roma).

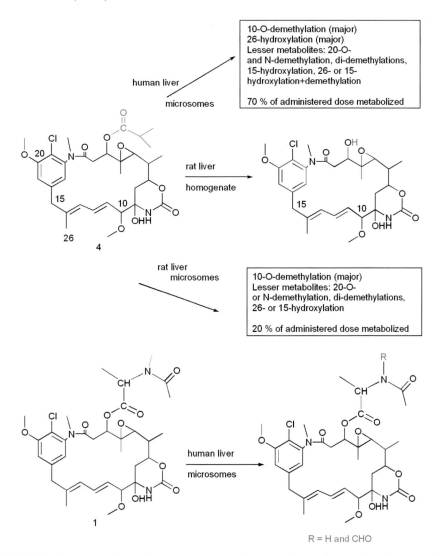

FIGURE 17.6 Preliminary results on the mammalian metabolism of ansamitocin P-3 and maytansine.[67]

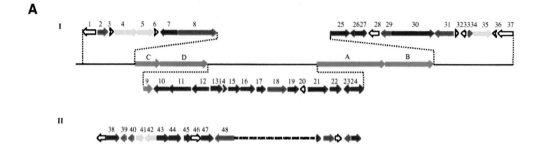

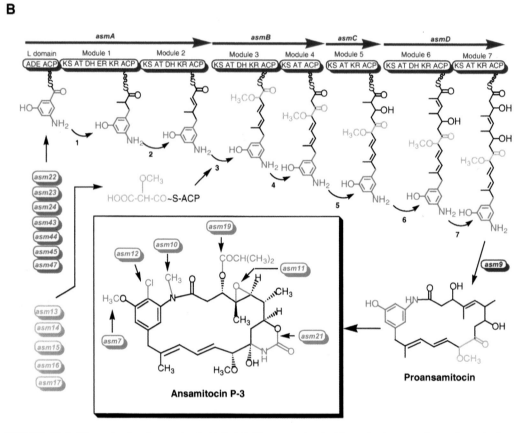

FIGURE 17.8 Biosynthesis of ansamitocin P-3: (A) The ansamitocin biosynthetic gene cluster from *Actinosynnema pretiosum*, (B) assembly of proansamitocin on a type I polyketide synthase and post-PKS modification to ansamitocin.[102]

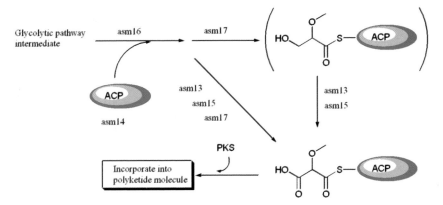

FIGURE 17.9 Formation of 2-methoxymalonyl-ACP, the substrate for the incorporation of the "glycolate" chain extension unit into ansamitocin.[105]

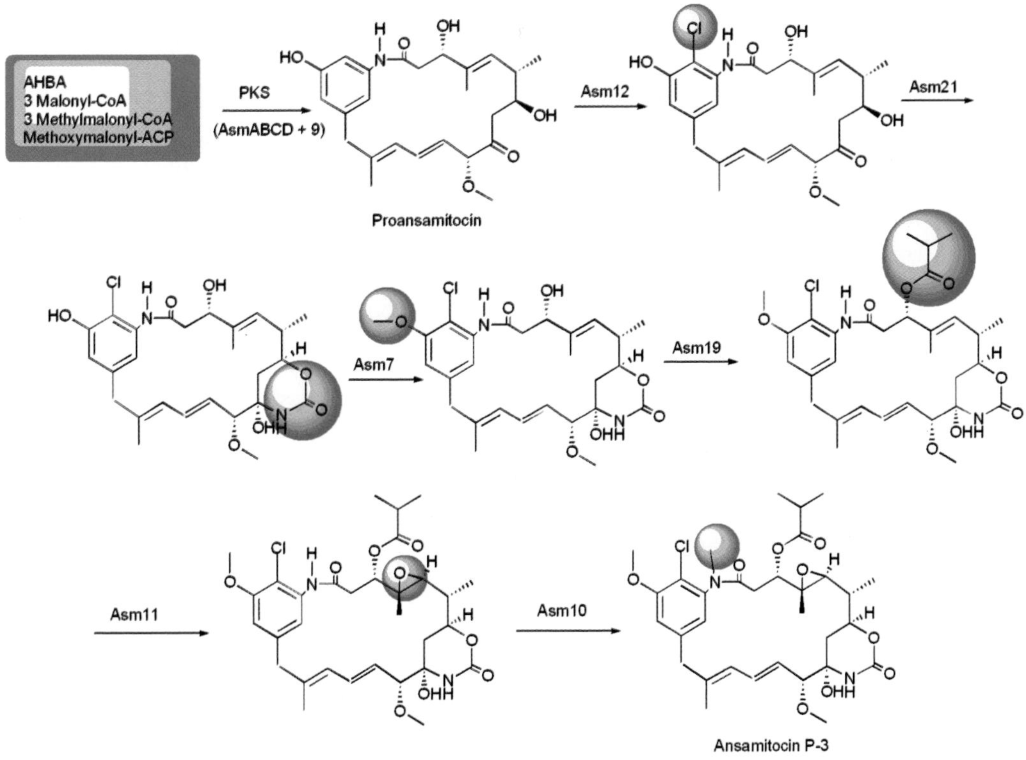

FIGURE 17.10 The sequence of post-PKS modification reactions converting proansamitocin into ansamitocin P-3.[108]

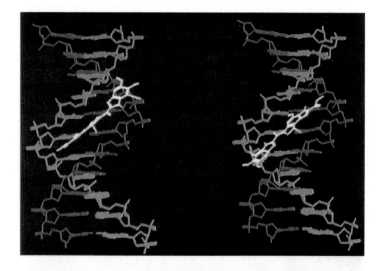

ImPyLDu86; **76**

77 X = -CO(CH₂)₄CO⁻

STRUCTURES 76–77

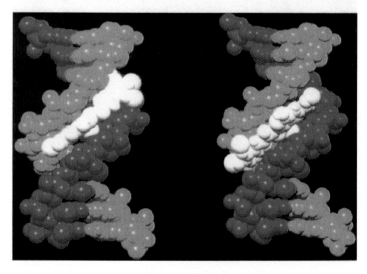

FIGURE 20.2 Comparison stick and space-filling models of the (+)-duocarmycin SA (left) and ent-()-duocarmycin SA (right) alkylation at the same site within w794DNA: duplex 5-d(GACTAATTTTT). The natural enantiomer binding extends in the 3 → 5 direction from the adenine N3 alkylation site 5-CTAA. The unnatural enantiomer binding extends in the 5 → 3 direction across the site 5-AATT.

(a)

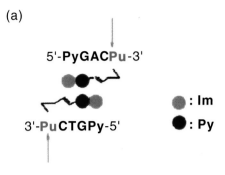

5'-**PyGACPu**-3'

● : Im

● : Py

3'-**PuCTGPy**-5'

(b)   5'-**PyGGCAGCCPu**-3'

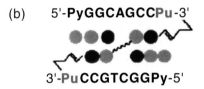

3'-**PuCCGTCGGPy**-5'

FIGURE 20.5 (A) The chemical structure of ImPyLDu86 76 and a schematic representation of the recognition of the 5-PyGACPu-3 sequence by the homodimer of 76. The arrows indicate the site where alkylation takes place. (B) A putative binding mode of the 1:2 complex of covalent dimer of 76, 77, with ImImPy to 5-PyGGCAGCCPu-3 sequence.

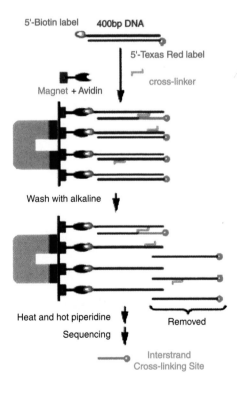

FIGURE 20.6 Determination of interstrand cross-linking sites.

(a)

(b)

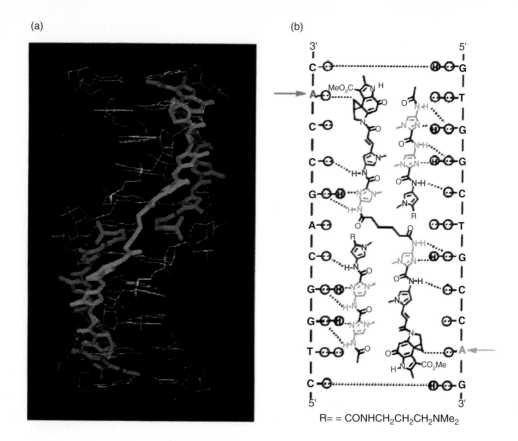

R= = CONHCH₂CH₂CH₂NMe₂

FIGURE 20.7 (A) Energy-minimized structure of 77-ImImPy₂-d(CTGGCTGCCAC)/d(GTGGCAGCCAG) interstrand cross-linked complex. DNA is drawn in white. Py and Im residues of 77 and ImImPy are drawn in blue and red, respectively. (B) A DNA interstrand cross-linking mode of the 1:2 complex of 77 with ImImPy.

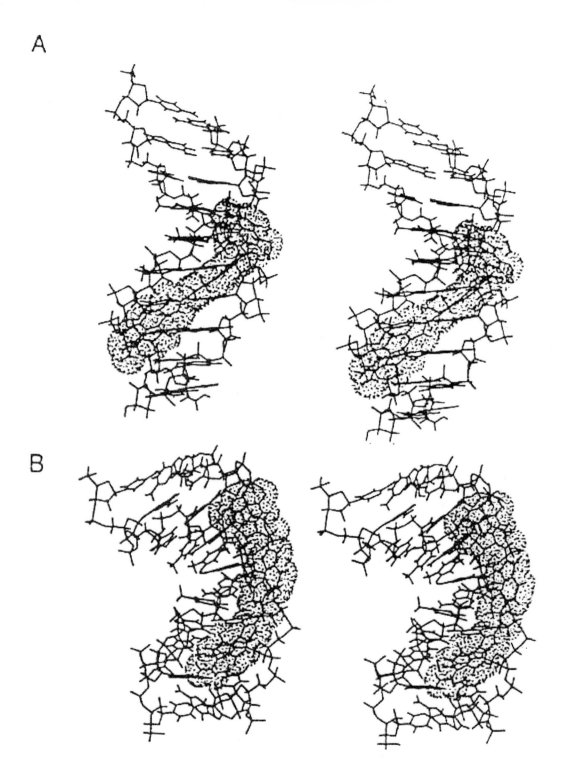

FIGURE 20.9 Stereoviews of (A) the initial model involving docking of the CPI-CDPI$_2$ (dotted van der Waals surface is shown) to a regular B-DNA decamer and (B) NOESY distance restrained, MORASS hybrid matrix/molecular dynamics final refined structure of the decamer-CPI-CDPI$_2$ complex starting from the initial model. The A strand C1 is depicted in the upper left of the structures, with the minor groove at A17 facing away or to the right of the viewer. Note the sharp bend in the duplex near the site of alkylation in B.

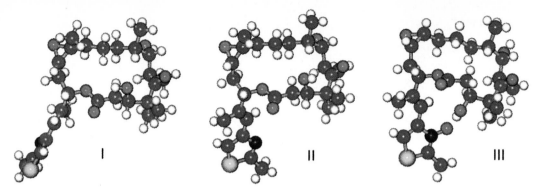

FIGURE 21.9 Crystal structures of epothilone B (1b) from dichloromethane/petroleum ether (I), methanol/water (II), and epothilone A-*N*-oxide (III).

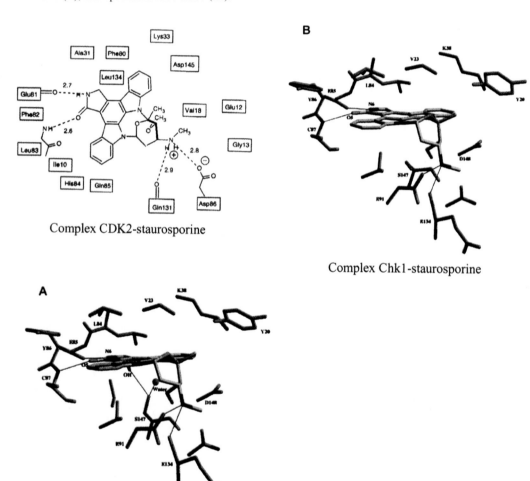

Complex CDK2-staurosporine

Complex Chk1-staurosporine

Complex Chk1-UCN-01

FIGURE 24.2 Crystal structures of staurosporine and UCN-01 complexes with kinases.

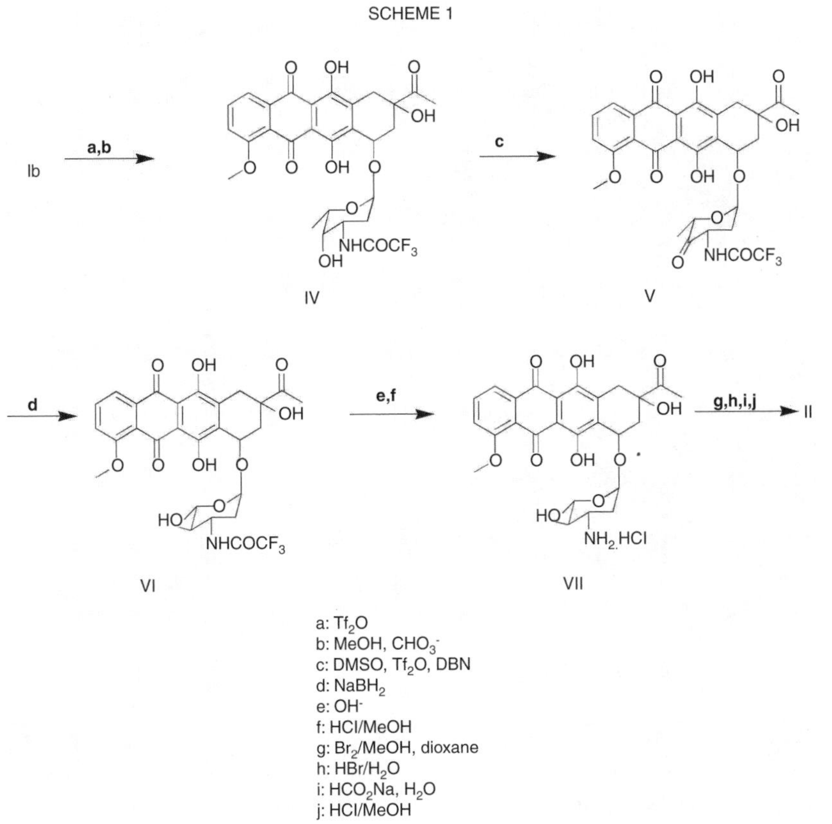

SCHEME 1

a: Tf$_2$O
b: MeOH, CHO$_3^-$
c: DMSO, Tf$_2$O, DBN
d: NaBH$_2$
e: OH$^-$
f: HCl/MeOH
g: Br$_2$/MeOH, dioxane
h: HBr/H$_2$O
i: HCO$_2$Na, H$_2$O
j: HCl/MeOH

SCHEME 16.1 Synthesis of epirubicin.

out.[56] As shown in Scheme 16.2, this process involves demethylation of **VIII** to carminomycinone **IX** and the conversion of the latter to the 13-ethylidene ketal **X** and then to the 4-O-triflate **XI**. The subsequent step, leading to 4-deoxy derivative **XII,** is based on the deoxygenation reaction of phenols discovered by Cacchi.[57] Removal of the protecting group provides idarubicinone **XIII,** which is glycosylated to **III**. The method is particularly interesting because daunorubicin is also a source of the aminosugar daunosamine — the other starting material for the synthesis of **III**.

V. MEDICINAL CHEMISTRY OF ANTITUMOR ANTHRACYCLINES

A. ANALOGS MODIFIED IN THE AGLYCONE MOIETY

1. Deoxyderivatives

The substitution of a hydroxyl group with a hydrogen atom has been extended to positions 6 and 11, as specific interest in the C6 and C11 deoxyderivatives stems from the hypothesis that antitumor activity or cardiotoxicity of anthracyclines might be related with the redox behavior or with the ability to form metal complexes. Synthesis of the corresponding daunorubicin and doxorubicin analogs led to the conclusion that such modifications, although bearing noticeable consequences on the polarographic behavior or on the chemical outcome of anaerobic quinone reduction, did not result in a marked change of bioactivity.[59] A significant structural modification in terms of structure–activity relationships was instead the substitution of the C9 hydroxyl group on ring A with a hydrogen atom. In fact 9-deoxy-daunorubicin, although showing the same affinity for DNA and similar electrochemical behavior as the parent daunorubicin, was two orders of magnitude less cytotoxic on cultured HeLa cells. Therefore, it

SCHEME 2

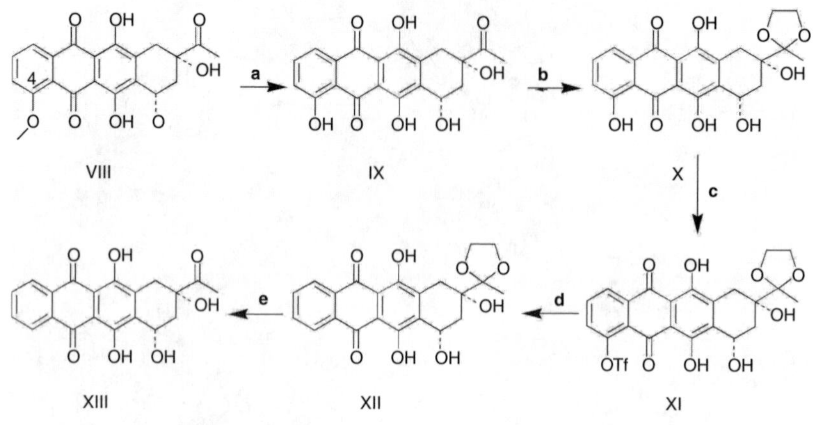

a: AlCl$_3$, CH$_2$Cl$_2$, reflux
b: (CH$_2$OH)$_2$, cat. PTSA, toluene/i-PrOH, reflux
c: iPr$_2$EtN, DMAP, (CF$_3$SO$_2$)$_2$O, Py
d: Pd(OAc)$_4$, 1,1'-bis(diphenylphosphine)ferrocene, HCO$_2$H, Et$_3$N, DMF, 40°
e: CF$_3$CO$_2$H, 0°

SCHEME 16.2 Synthesis of idarubicinone.

was deduced that the 9-hydroxyl group, which according to the well-known molecular model of the daunorubicin–DNA complex protrudes outside the double helix, might be involved in a specific inter-action relevant to the bioactivity of the drug in terms of topoisomerase II blockade in the ternary complex.[60]

2. 8-Fluoroderivatives

According to the established importance of the 9-hydroxyl group mentioned above, a study aimed at the improvement of doxorubicin anticancer activity via introduction of a fluorine atom at the α position with respect to the 9 position has been carried out. The derivatives substituted at C8 and at C14 (the latter in the doxorubicin series) resulted in the most interesting compounds. The substitution of a hydrogen atom with a fluorine at position 8 gave rise to compounds with markedly different chemical and biological properties depending on the stereochemistry at the new chiral center. 8(S)-Fluoro aglycones **XIVa** and **XIVb** were obtained from 8,9-oxirane intermediates. These where either prepared by total synthesis when R = H or, in the case of **XIVb**, in enantiomeric (8S) form by semisynthesis from "pink compound" **XV** via the corresponding 7,8-epoxide, **XVIa**, and finally **XVIb**. Glycosidation of **XIVa** and **XIVb**, followed by diastereomeric separation when R = H, afforded glycosides **XVIIa** and **XVIIb**, respectively. Compound **XVIIb** was eventually converted to 8(S)-fluorodoxorubicin, albeit in low yields because of an inertness toward electrophilic bromination at C14 (possibly a sign of the strong electrostatic effect of the fluorine substitution). A significantly lower cytotoxic potency exhibited by **XVIIa** compared to unsubstituted idarubicin was related to the β conformation of ring A, predominating in the solution of **XVIIa** over the α conformation typical of idarubicin and the other antitumor anthracyclines. To obtain the corresponding epimeric *cis* 8R,9S-fluorohydrins, Diels-Alder adduct **XVIII** was treated with N-bromosuccinimide and tetrabutylam-monium dihydrogentrifluoride to give, together with a 8-bromo-9-fluoro derivative, the desired 8-fluoro-9-bromo isomer **XIX** that was converted to the 9,13-epoxide with base. Opening of the oxirane ring with acid and oxidation of the side chain alcohol gave a ketone that was functionalized at C7 via radical bromination and hydrolysis to give the desired 8,9-*cis*-fluorohydrin **XIVc**.

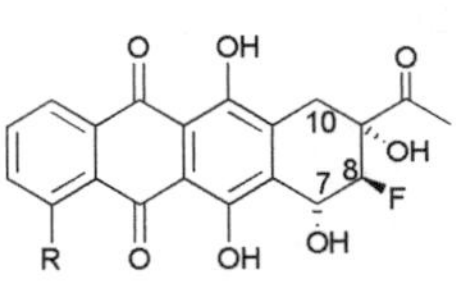

XIVa: R=H; XIVb: R=OCH₃;
XIVc: R=H, *epi* at C-8

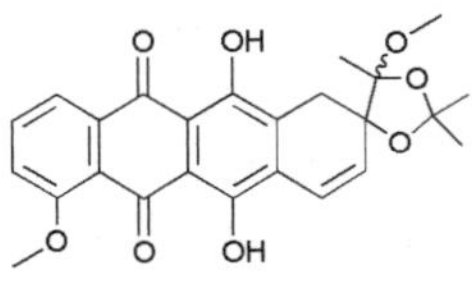

XV

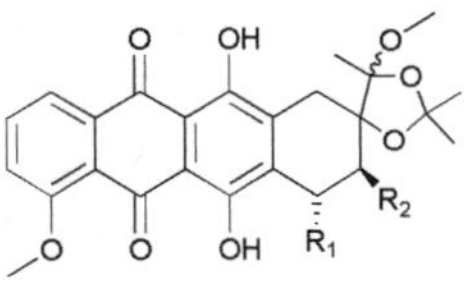

XVIa: R₁=OCH₃, R₂=OH;
XVIb: R₁=OCH₃, R₂=F

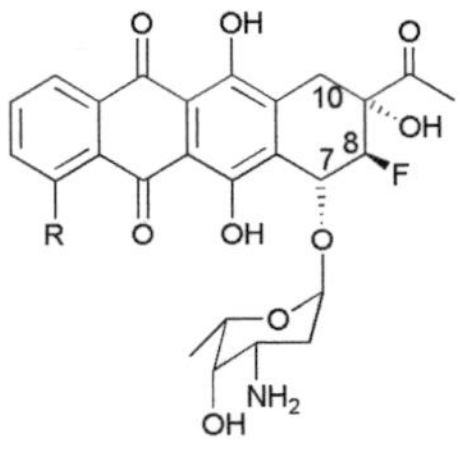

XVIIa: R=H; XVIIb: R=OCH₃;
XVIIc: R=H, *epi* at C-8

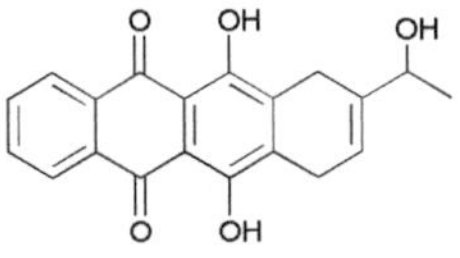

XVIII

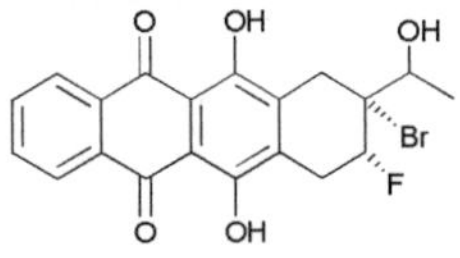

XIX

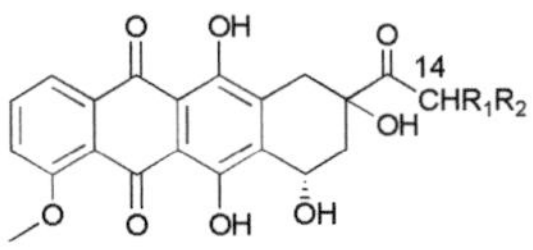

XXa: R₁=H, R₂=Br
XXb: R₁=H, R₂=F
XXc: R₁=F, R₂=Br
XXd: R₁=F, R₂=OH

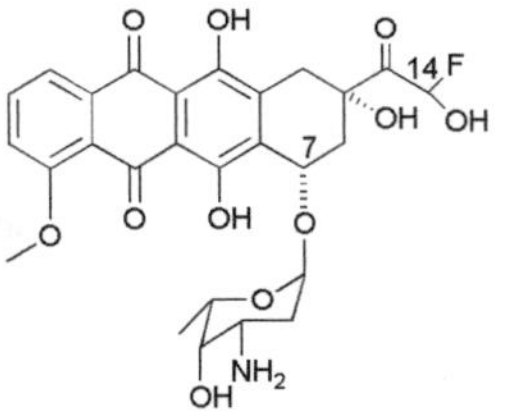

XXI

STRUCTURES XIV–XXI

Glycosidation of the latter, followed by separation of diastereomeric glycosides and deprotection, gave 8(R)-fluoroidarubicin XVIIc, which was found to be as potent as idarubicin in the inhibition of the growth of tumor cell lines *in vitro*. Interestingly, XVIIc displayed marked cytotoxic activity against doxorubicin-resistant cell lines (Table 16.1). When delivered intravenously at a dosage of 1.5 mg/kg to immunodepressed mice bearing A2780 human ovarian carcinoma xenografts, the new compound inhibited tumor growth by 93%, whereas in the same system, doxorubicin at 7 mg/kg inhibited by 83% and idarubicin at 1.2 mg/kg inhibited by only 36%. Values of "log cell kill" (difference in mean time required for tumors of treated and control mice to reach 2 g divided by the product of mean doubling time of the tumor and by 3.32) were equal to 3 for the 8R-fluoro derivative, and 1.8 for doxorubicin.[61,62]

3.   14-α-Fluorohydrins

Synthesis of 14-fluorodoxorubicin **XXI** did not appear to be an easy task. However, this α-fluorohydrin, a compound with an unusual atomic arrangement, has been obtained starting from daunomycinone **VIII**, brominated to known **XXa**, whose treatment with AgBF₄ and dimethylsulfoxide gave, interestingly, fluoroderivative **XXb**. Bromination and reaction of compound **XXc** with AgBF₄ afforded **XXd**. Desired **XXI** was eventually obtained through glycosylation of **XXc** with 1-p-nitrobenzoyl-3,4-*N,O*-diallyloxycarbonyl daunosamine and treatment of the resulting protected glycoside with the AgBF₄-DMSO reagent, followed by removal of the protecting groups. Bioactivity of **XXI** is currently under investigation.[63]

TABLE 16.1
Inhibition of Growth of Human Tumor Cell Lines by Doxorubicin (Ia), Idarubicin (III), and Epimeric 8(S) and 8(R)-Fluoroidarubicins (XVIIa and XVIIc, Respectively)

Cell line	I	III	XVIIa	XVIIc
A2780[a]	13	2.8	14	5
A2780/DX	250	10	60	9
LOVO[b]	41	3	12	3
LOVO/DX	989	20	147	11
MCF-7[c]	13	2	6	3
MCF-7/DX	2460	27	291	16
POVD[d]	37	4	6	9
POVD/DX	350	48	70	9
POGB[d]	16	2	20	2
POGB/DX	390	24	140	4

Note: Results are expressed as IC_{50} values (ng/ml) (adjusted from Ref. 62).

[a] Ovarian tumor.
[b] Colon tumor.
[c] Breast tumor.
[d] Small cell lung cancer.

B. ANALOGS MODIFIED IN THE SUGAR MOIETY

1. Monosaccharide Analogs

4-Deoxydaunosamine was obtained from daunosamine and used for the preparation of 4-deoxy-doxorubicin (esorubicin) — a compound showing activity comparable to that of the parent, doxorubicin — and lower cardiotoxicity in experimental studies[5] (albeit not confirmed in clinical phase II studies[64]). Similarly, 4-iodo-4-deoxydoxorubicin, showing promising results in preclinical studies,[65] was also developed up to clinical phase II but did not exhibit therapeutic properties suggesting improved anticancer activity.[66] Instead, the compound was found to be the lead for a novel class of drugs for amyloid-related diseases.[67,68] A doxorubicin derivative named pirarubicin, namely, 4-O-dihydropyranyl-doxorubicin, was studied in Japan and has entered into clinical use as a doxorubicin analog endowed with lower side effects.[9] Annamycin **XXII**, a lipophilic iodinated analog, devoid of the sugar amino group and of the methoxy group at C4, was shown, especially when entrapped in liposomes, to be partially effective in cell cultures and in mouse tests against

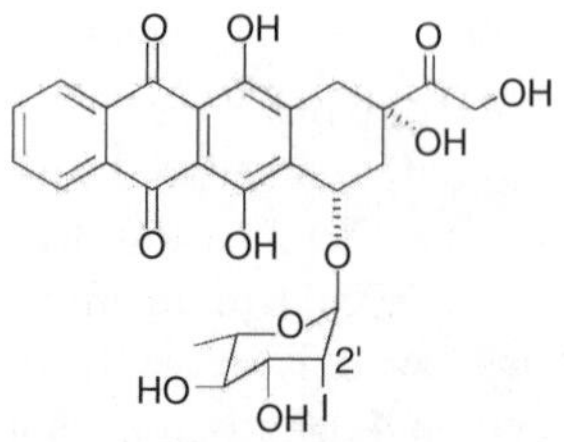

XXII

STRUCTURE XXII

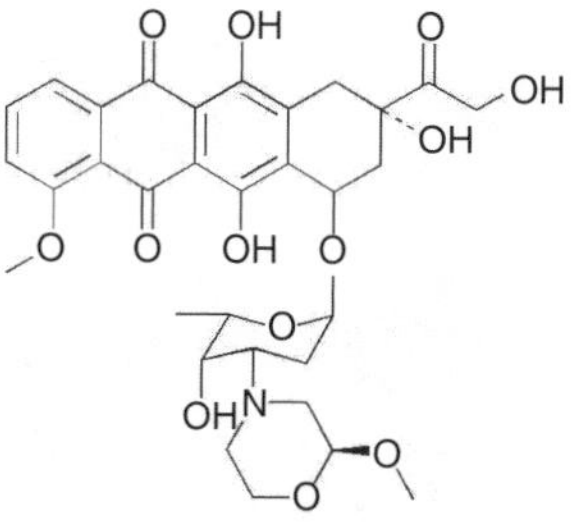

XXIII

STRUCTURE XXIII

doxorubicin-resistant tumors.[69] However, no efficacy was found in breast cancer patients not responding to doxorubicin.[70] An interesting group of analogs is represented by the morpholinyl anthracyclines. These new derivatives were discovered by Acton et al. and found to be about two orders of magnitude more cytotoxic than the known antitumor anthracyclines — very likely owing their high potency to an alkylating mechanism of action.[71] Methoxymorpholinyl derivative **XXIII** was developed up to the clinical stage, but a low response rate was observed in patients with tumors endowed with intrinsic resistance to chemotherapy.[72] A different, not DNA-related, biochemical mechanism of action has been suggested for the very lipophilic doxorubicin derivative, N-benzyldoxorubicin 14-O-valerate.[73]

2. Disaccharide Analogs

More recently, attention has been given to analogs containing two sugar residues. Anthracycline disaccharides have been obtained both as biosynthetic products or by semisynthesis. The disaccharide 4-O-daunosaminyldaunorubicin was present in the medium of the daunorubicin fermentation, and the same, as well as the corresponding analogs in which the second sugar was acosamine or 2-deoxy-L-rhamnose, was also obtained by semisynthesis and found to be cytotoxic, albeit less potent against murine transplantable leukemias than daunorubicin.[4] Other related glycosides of daunorubicin have been synthesized[74,75]; however, the first bearing a nonaminated sugar directly linked to the aglycone, namely, 7-O-(3-O-α,L-daunosaminyl-2-deoxy- α,L-rhamnosyl)-daunomycinone, was biologically inactive, apparently because of the different position of the second glycosidic linkage.[76] In fact, as suggested by the DNA binding geometry and by the conclusions concerning the topoisomerase-related mechanism of action of antitumor anthracyclines, a rational design of disaccharide analogs might be based on the hypothesis that it would be beneficial to increase the molecular recognition properties through a functional enrichment of the portion of the anthracycline molecule involved in the stabilization of the "cleavable ternary complex." Structure activity relationships pointed to the importance of ring A as a "scaffold" determining the orientation of the C7 and C9 substituents, in particular that of the sugar moiety, whose structure and stereochemistry are critical for the stabilization of the ternary complex.[77,78]

Synthesis of disaccharide derivatives of the daunorubicin/doxorubicin and of the 4-demethoxy (idarubicin) series in which the first sugar moiety linked to the aglycon was a nonaminated sugar, namely, 2-deoxy-L-fucose or 2-deoxy-L-rhamnose, and the second moiety was either daunosamine or acosamine or a 3-hydroxy or 4-deoxy analog of these aminosugars, was performed starting from the corresponding aglycon that was glycosylated with the appropriately protected and activated disaccharide.[79] The desired final products were obtained by sequential deprotections and purification. Disaccharide glycosides containing 2-deoxy-L-fucose **XXIVa,b** exhibited superior pharmacological properties in respect to the stereoisomers **XXVa,b** containing 2-deoxy-L-rhamnose.[80] 7-O-(α-L-Daunosaminyl-α(1-4)-2-deoxy-L-fucosyl)-4-demethoxyadriamycinone (**XXIVc**) was synthesized as shown in Scheme 16.3 and was developed to the clinical stage because of its favorable pharmacological

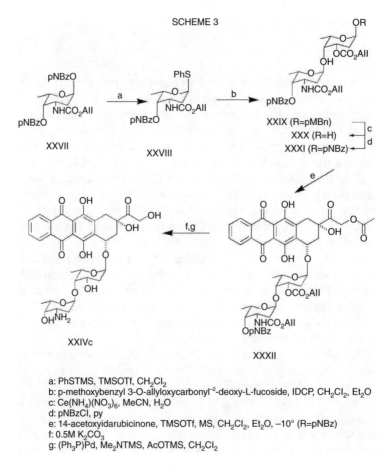

STRUCTURES XXIV–XXV

properties. The high activity of **XXIVc** was related to the activation of p53-independent apoptosis.[81] The new compound exhibited, especially at selected DNA sites, a more marked topoisomerase II–mediated cleavage that was accompanied by a superior antitumor efficacy in the experimental systems when compared with doxorubicin. Very significant activity was found against a spectrum of human tumors such as breast, ovary, and lung cancer xenografts in athymic nude mice, where the

SCHEME 16.3 Synthesis of 7-O-(-L-daunosaminyl-(1-4)-2-deoxy-L-fucosyl)-4-demethoxyadriamycinone.

compound appeared markedly superior to doxorubicin in inhibiting tumor growth and in terms of an increased number of disease-free survivors among treated animals.[82,.83] The generic name selected for **XXIVc** by World Health Organization is "sabarubicin."

Spectrophotometric and fluorescence titrations with calf thymus DNA,[85] and with the d(CGATCG) hexamer of **XXIVc**, have revealed spectral patterns very similar to those obtained with doxorubicin, implying that the binding mechanism and the stability of the resulting complexes are nearly the same. X-ray diffraction data of the orange–red crystals of the sabarubicin–hexamer complex[86] allowed the definition of the asymmetric unit containing two DNA strands forming one duplex, two bound drugs, and 35 solvent molecules. Similar to the well-known complexes of other antitumor anthracyclines with the same DNA hexamer, intercalation occurs at each end of the DNA duplex in the CpG steps, with the drug chromophoric system perpendicular to the long axis of stacked base pairs. However, the two drug molecules appear to have different conformations, and the two binding sites show large differences. In one site, both sugar rings lie in the minor groove, whereas in the other site, the arrangement of the drug is such that the first sugar resides in the minor groove and, although rotated by about 45° around the O7- C1 bond with respect to the corresponding doxorubicin complex, is in the chair conformation with the 3-hydroxyl group still in van der Waals contact with the guanine residue. The second sugar protrudes out in the solvent region where the amino and hydroxyl groups interact with two water molecules (Figure 16.1).

Interestingly, a tight interaction takes place between the amino group of the disaccharide moiety and a guanine residue of a second DNA molecule, different from the one where the drug is intercalated. As a consequence, in the crystals, the end-over-end stacked duplexes do not run parallel, as in other crystallized anthracycline-DNA complexes, but are crossing each other at right angles along the tetragonal c axis, leading to the formation of two layers of stacked duplexes mutually perpendicular to one another. The packing contacts include the amino group on the second sugar

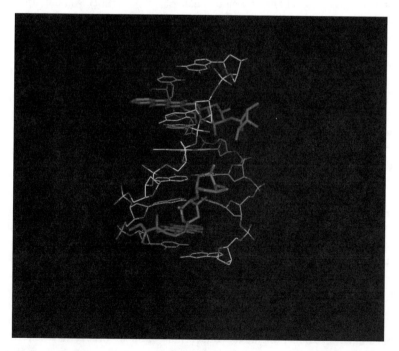

FIGURE 16.1 (See color insert following page 304.) Representation of the sabarubicin intercalation complex with d(CGATCG)2 according to a x-ray crystal diffraction study. The drawing was kindly provided by Dr. Giovanni Ughetto, Laboratorio del CNR di Montelibretti (Roma).

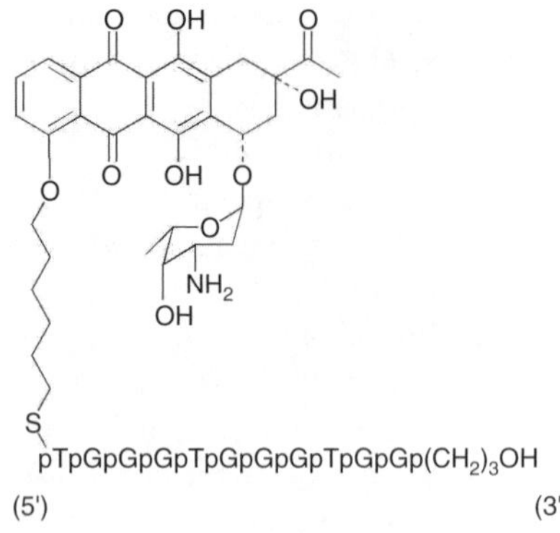

pTpGpGpGpTpGpGpGpTpGpGp(CH$_2$)$_3$OH

(5')   (3')

XXVI

STRUCTURE XXVI

that is directly hydrogen bonded to N7 and, through a bridging water, to O6 of a guanine, providing the first example of an anthracycline/DNA complex in which a cross-link with a second DNA helix has been observed. It seems, therefore, that the second sugar may exert interactions with other cellular targets in close proximity with DNA.

C. ANTHRACYCLINE–TRIPLEX HELIX-FORMING OLIGONUCLEOTIDE CONJUGATES

Triplex helix-forming oligonucleotides (TFOs) that bind to homopurine:homopyrimidine sequences in DNA in a sequence-specific manner can be used to selectively inhibit gene expression and provide the opportunity for rational design of novel DNA sequence–specific cancer therapeutics. Chemical modifications of the TFO that would enhance triplex stability without affecting sequence specificity, such as conjugation with an anthracycline molecule, further increase the potential efficacy of this approach.[87] The resulting conjugates may correctly position the TFO moiety in the major groove of the duplex and allow it to form Hoogsteen hydrogen bonds with the complementary bases of the purine-rich strand of the target DNA while allowing ring A of the intercalated anthracycline, together with the sugar moiety, to protrude into the minor groove, as in the well-known model of the daunomycin–DNA complex. When the aminosugar is lacking (i.e., the TFO is conjugated with a nonglycosylated anthracyclinone), the stabilization of the triple helix is still obtained, albeit one order of magnitude lower.[88]

In a recent study, DNA binding and antigene activity of anthracycline-conjugated TFOs targeted to a sequence near the major P2 promoter in the *c-myc* gene were synthesized. GT-rich 11-mer TFOs were covalently attached through a hexamethylene bridge at the hydroxyl on position 4 of carminomycin. As shown by footprinting and electrophoretic experiments, as well as by spectroscopic measurements, conjugate **XXVI** had high binding affinity and formed a much more stable triplex than the nonconjugated TFO. Binding of **XXVI** was also highly sequence specific, as indicated by the fact that it protected a single region corresponding to the 11-basepair target sequence in the *c-myc* promoter. Conjugates with different TFOs, which were partially complementary to the *c-myc* target sequence, bound with much lower affinity than **XXVI**. The latter inhibited transcription from the P2 promoter in *in vitro* transcription assays and also inhibited transcription in luciferase reporter assays when cotransfected into prostate cancer cells along with a pGL-myc promoter reporter plasmid, but it did not affect the activity of the control pRL-SV40 plasmid. These data indicate that attachment of the anthracycline to the TFO increased its binding affinity and allowed formation of a very stable complex with duplex DNA. Moreover, **XXVI** was able to selectively inhibit transcription of the *c-myc* gene both *in vitro* and in cells. These results encourage further testing of this approach for the development of novel antigene cancer therapeutics.[89]

VI. CLINICAL ACTIVITY OF ANTHRACYCLINES

Since the first clinical trials, daunorubicin and doxorubicin have gained rapid acceptance as major therapeutic agents in the medical treatment of cancer. Both drugs, but especially doxorubicin, have been widely used as part of combination regimens for the effective treatment of acute nonlymphocytic leukemia, Hodgkin and non-Hodgkin lymphomas, breast cancer, and sarcomas.[90] Epirubicin, idarubicin, and liposomal doxorubicin offer less toxic and, in some instances, more effective alternatives to older anthracyclines for leukemia, breast cancer, ovarian cancer, and other diseases.[91]

A. DAUNORUBICIN

Daunorubicin (**Ib**), the first anthracycline introduced into therapeutic use, is endowed with a potent antitumor activity whose spectrum is narrower than that of doxorubicin and is accompanied by a significant hematological toxicity. The compound exhibits delayed cardiotoxicity. Main indications are acute leukemias, but significant activity was recorded against neuroblastomas and rhabdomyosarcomas. Induction of remission in over 70% of patients with acute lymphoblastic leukemia is obtained on treatment with daunorubicin, vincristine, and prednisone, with disease-free survival at 5 yr in up to 45% of patients receiving a consolidation/intensification therapy after remission.[92] Intensive induction chemotherapy using 7-d courses of cytarabine and daunorubicin or amsacrine produces remission in 60%–85% of patients with acute myelogenous leukemia, a condition that was invariably fatal 30 yr ago. Median remission duration is 9–16 mo. In some series, 20%–40% of patients are in continuous remission for 2 yr or more; many of these patients remain in remission for 5 yr or longer, and some may be cured.[93]

B. DOXORUBICIN

A notable feature of the antitumor properties of doxorubicin (**Ia**) is the wide spectrum of activity including (as particularly responsive) acute leukemias, Hodgkin and non-Hodgkin lymphomas, lung microcytomas, soft-tissues sarcomas, Ewing's tumor, thyroid carcinoma, ovarian cancer, breast cancer, rhinopharyngeal tumors, Wilms' tumor, and other pediatric malignancies. It was recognized early on that in these diseases, doxorubicin shows efficacy when administered alone as a slow intravenous injection every 3 wk at a dosage of 65 mg/m^3, with its most usual administration being, however, in combination with other active agents.[94] Out of 113 drug combinations containing doxorubicin (Adriamycin) already used in clinical studies worldwide at the end of the first decade after the original demonstration of therapeutic efficacy of the drug, cyclophosphamide, an alkylating agent, was the most frequent component (54 combinations), followed by the spindle poison vincristine or other vinca alkaloids such as vinblastine (42 combinations), the metabolic inhibitor 5-fluorouracil (24 combinations), bleomycin and cis-platinum (22 combinations each), and methotrexate (12 combinations), with other less frequent components being mitomycin, etoposide, or nitrosoureas.[95] Doxorubicin was frequently accompanied by cyclophosphamide and vincristine or by 5-fluorouracyl in the treatment of breast cancer (the acronyms used for the main combinations are FAC and VAC). Cyclophosphamide and vincristine or etoposide were the most frequent components in the treatment of small cell lung tumors, whereas doxorubicin was accompanied by cyclophosphamide and methotrexate or either cis-platinum or mitomycin or a nitrosourea in the case of the less responsive non–small cell lung cancer. In the low respondent gastrointestinal tract malignancies, doxorubicin was combined with 5-fluorouracyl and mitomycin or a nitrosourea and with cyclophosphamide, a vinca alkaloid, bleomycin, and prednisone in the generally more responsive lymphomas. Cis-platinum and cyclophosphamide were mainly associated with doxorubicin in the treatment of genital tract tumors.

Doxorubicin is characterized by a rapid distribution phase and a slow elimination phase. The successive half-lives of doxorubicin in plasma are approximately 5 min, 1 h, and 30 h. Total plasma clearance is about 30 L/h/m^2, and its total volume of distribution is approximately 15 L/kg. As the other anthracyclines, the drug is excreted mostly through the bile, and special care is taken in patients with hepatic dysfunction.[96] Doxorubicinol, the product of the reduction of the 13-keto group to a secondary alcohol function, is the major human metabolite of doxorubicin, albeit present in plasma at concentrations well below those of the parent drug.[97]

C. Epirubicin

Phase II studies of epirubicin (**II**) indicated that the drug produces a pattern of acute toxicity qualitatively similar to but quantitatively lower than that of doxorubicin at identical doses, the most frequently employed dose schedules being 75 or 90 mg/m² intravenously every 3 wk. Epirubicin showed activity in a variety of tumors such as breast carcinoma, soft tissue sarcomas, lymphomas, leukemias, ovarian cancer, and gastric cancer, and preliminary evidence of activity was found in melanoma, rectal cancer, and pancreatic cancer, indicating a broad spectrum of activity. As to chronic cardiac toxicity, preliminary findings from a randomized comparison of the two drugs in breast cancer indicated that epirubicin might have a lower cumulative cardiotoxicity than doxorubicin. According to the first clinical studies, pharmacokinetics of epirubicin in human patients showed a multiexponential decrease of plasma levels, with the same pattern being observed for the metabolite epidoxorubicinol, together with a high plasma clearance (0.9–1.4L/min), a terminal half-life of about 30–40 hr, and a large volume of distribution.[98] According to a subsequent review,[99] to achieve hematological equitoxicity of the two drugs, the dose of epirubicin should be approximately 20% higher than that of doxorubicin, giving rise to a higher dosage level for epirubicin. In contrast, epirubicin is significantly less cardiotoxic than doxorubicin. Thus, the maximum cumulative dose, which for doxorubicin is 500 mg/m², for epirubicin is close to 1000 mg/m². More recent conclusions indicate that epirubicin, used alone or in combination with other cytotoxic agents in the treatment of a variety of malignancies, is better tolerated than, and achieves equivalent objective response rates and overall median survival to, similar doxorubicin-containing regimens in the treatment of advanced and early breast cancer, non–small cell lung cancer, small cell lung cancer, non-Hodgkin lymphoma, ovarian cancer, gastric cancer, and nonresectable primary hepatocellular carcinoma.

Dose-intensive epirubicin-containing regimens, which have been designed because of the drug's lower myelosuppression and cardiotoxicity and the definite dose–response relationship of anthracyclines, have produced high response rates in early breast cancer, a potentially curable malignancy, as well as advanced breast and lung cancers. Furthermore, such improved response rates have improved quality of life in some clinical settings, but whether this leads to prolonged survival is yet to be determined.[100] The metabolism of epirubicin is characterized by a biotransformation to relatively or totally inactive metabolites, including a 13-dihydro derivative, epirubicinol, two glucuronides, and four aglycones. Quantitatively, the glucuronides of epirubicin and epirubicinol are very important, and this pathway, which is unique to epirubicin metabolism in humans, might explain the better tolerability of this drug compared with doxorubicin. Epirubicin pharmacokinetics may be described by a three-compartment model, with median half-life values of 3.2 min, 1.2 h, and 32 h for each phase. Total plasma clearance is 46 L/h/m², and the volume of distribution at steady state is 1000 L/m². The pharmacokinetics of epirubicin appear to be linear for doses up to 150–180 mg/m², which is the maximal tolerated dose.[101]

D. Idarubicin

Idarubicin (**III**) is currently used in the medical treatment of acute nonlymphocytic leukemia in adults. Idarubicin is at least as effective as daunorubicin in this disease, and two studies indicate greater activity and longer survival with an idarubicin–cytarabine regimen than with a daunorubicin–cytarabine regimen. The drug is active in acute lymphocytic leukemia, lymphomas, and breast cancer. Reportedly, idarubicin is more potent and less cardiotoxic than daunorubicin or doxorubicin. However, a maximum cumulative dose, to prevent chronic cardiomyopathy, has not yet been defined. Unlike the other clinically useful anthracyclines, idarubicin has significant oral bioavailability (average 28%), and an oral dosage form has been developed.[102] The oral formulation has demonstrated efficacy in advanced breast cancer, non-Hodgkin lymphoma, and myelodysplastic syndromes as well as in first-line induction therapy of acute myelogenous leukemia. It also has potential for ameliorating blast crisis of chronic myelogenous leukemia and in multiple myeloma. The most frequent acute adverse effects associated with oral idarubicin are those commonly found with anthracyclines, but there appears to be minimal significant cardiotoxicity. Available data concerning oral idarubicin in hematological malignancies and advanced breast cancer are sufficiently encouraging to warrant further research.[103]

E. ACLACINOMYCIN

Aclacinomycin (ACM) has been extensively evaluated in patients with relapsed leukemia and advanced malignant lymphoma. Analysis of results compiled from Europe, Japan, and the United States shows that ACM is probably equivalent to doxorubicin for remission induction of patients with relapsed acute nonlymphoblastic leukemia. The antitumor activity of ACM in patients with acute lymphoblastic leukemia or malignant lymphoma who have previously received doxorubicin or daunorubicin is low, and the issue of whether ACM lacks clinical cross-resistance to other anthracyclines is unresolved. Acute cardiac arrhythmias have been observed following administration of ACM, but congestive cardiomyopathy has been uncommon. It has been reported that ACM has fulfilled its early expectations of antileukemic activity and reduced toxicity.[104]

F. SABARUBICIN

Clinical tolerability of weekly dosed sabarubicin (**XXIVc**) has been determined by researchers at EORTC, who reported the results of a phase I study and recommended a intravenous dosages of 30 and 40 mg/m^2 for three consecutive weeks followed by 1 wk rest in, respectively, pretreated and naïve patients.[105] Pharmacokinetics of sabarubicin were studied in patients from two different phase I dose schedules, one in which the drug was administered once every 3 wk (the current schedule of anthracyclines), and the other in which it was administered once every week for 3 wk, followed by 1 wk rest. The plasma peak concentration levels and AUC after each dosage showed linear relationship with the administered dose. Plasma concentrations were particularly high, in the ranges of 0.474–0.612 µg/mL for the lowest (4 mg/m^2) and 9.538–21.587 µg/mL for the highest (110 mg/m^2) dose used. There was no accumulation of the drug before the next infusion in the weekly regimen. The mean elimination half-life was 20.7 h, which is significantly shorter than that for doxorubicin or epirubicin. The volume of distribution, much smaller than that of the said anthracyclines, was 95.6 L/m^2 and showed a standard deviation of 43.4 L/m^2. In fact, a particularly high interpatient variation of pharmacokinetic parameters was found.[106]

A dosage of 80–90 mg/m^2 every 3 wk was used in poor-prognosis patients with advanced or metastatic ovarian cancer who had received prior chemotherapy. Out of 14 evaluable patients, one partial response and eight stabilizations of the disease were recorded.[107] The activity of sabarubicin has been demonstrated acting as a single agent in a poor-prognosis population of patients with lung cancers. The drug was well tolerated when used at an intravenous dose of 80–90 mg/m^2 every 3 wk in patients with non–small cell lung cancer, as second line therapy, and with small cell lung cancer as first line therapy, with dose-limiting toxicity being hematological. One partial response and five stable diseases were observed in 16 evaluable patients with non–small cell lung cancer, and one complete response, nine partial responses, and three stable diseases were observed in 21 evaluable, pretreated small cell lung cancer patients.[108,109]

G. LIPOSOMAL ANTHRACYCLINES

Liposomes are closed vesicular lipidic structures that envelop water-soluble molecules. Liposomal anthracyclines have been extensively studied in humans with a variety of cancer types, including one containing daunorubicin and two preparations containing doxorubicin. Results have indicated that liposomal encapsulation may lead to toxicity attenuation while retaining or even enhancing the efficacy of the parent anthracyclines. Approved indications have been achieved for liposomal daunorubicin against AIDS-related Kaposi sarcoma and for liposomal doxorubicin against AIDS-related Kaposi sarcoma and ovarian cancers.[110] In particular, pegylated liposomal doxorubicin shows a pharmacokinetic profile characterized by an extended circulation time and a reduced volume of distribution. Clinical studies in humans have included patients with AIDS-related Kaposi sarcoma and with a variety of solid tumors, including ovarian, breast, and prostate carcinomas. The pharmacokinetic profile in humans at doses between 10 and 80 mg/m^2 is biphasic, exhibiting an initial phase with a half-life of 1–3 h and a second phase with a half-life of 30–90 h. The AUC after a dose of 50 mg/m^2 is reported to be approximately 300-fold greater than that with free drug. Clearance and volume of distribution are drastically reduced (at least 250-fold and 60-fold, respectively). The toxicity profile is characterized by dose-limiting mucosal and cutaneous toxicities, mild myelosuppression, and decreased cardiotoxicity compared with free doxorubicin and minimal alopecia. Interestingly, pegylated liposomal doxorubicin represents a new class of chemotherapy delivery system that may significantly improve the therapeutic index of doxorubicin.[111] A retrospective study on 40 patients indicated that pegylated liposomal doxorubicin at a dosage of 40–45 mg/m^2 every 4 wk is clinically active and well tolerated in women with metastatic breast cancer.[112]

REFERENCES

1. Brockmann, H., Anthracyclinone und anthracycline (Rhodomycinone, Pyrromycinone und Ihre Glycoside), in *Forschritte der Chemie Organische Naturstoffe*, Zechmeister, L, Ed., Springer, Vienna, 21, 121-182, 1963.
2. O'Hagan, D., *The polyketide metabolites*, Ellis Horwood, Chichester, 1991.
3. Arcamone, F. et al., Isolamento ed attività antitumorale di un antibiotico da Streptomyces sp., *Giorn. Microbiol.*, 9, 83, 1961.
4. Arcamone, F., *Doxorubicin anticancer antibiotics*, De Stevens, G. Ed., Academic Press, New York, 1981.
5. Arcamone, F., The development of new antitumor anthracyclines, in *Anticancer agents based on natural products models*. Cassady, J. and Douros, J. Eds., Academic Press, New York, 1980, chapter 1.
6. Arcamone, F. and Penco, S., Synthesis of new doxorubicin analogs, in *Anthracycline and anthracene-dione-based anticancer agents*, Lown, J.W. Ed., Elsevier, Amsterdam, 1988, chapter 1.
7. Weiss, R.B., The anthracyclines: will we ever find a better doxorubicin? *Semin. Oncol.* 19, 670, 1992.
8. Arcamone, F. et al., New developments in antitumor anthracyclines, *Pharmacol. Ther.* 76, 117, 1997.
9. Takeuchi, T. Antitumor antibiotics discovered and studied at the Institute of Microbial Chemistry. *J. Cancer Res. Clin. Oncol.*, 121, 505, 1995.
10. Oki, T., Antitumor anthracycline antibiotics from microbial origin, in ref. (5), p.103.
11. Arcamone, F. and Cassinelli, G., Biosynthetic anthracyclines, *Curr. Med. Chem.* 5, 391, 1998.
12. Fujii, I. and Ebizuka, Y., Anthracycline biosynthesis in *Streptomyces galilaeus*, *Chem. Rev.*, 97, 2511, 1997.
13. Hutchinson, C.R., Biosynthetic studies of daunorubicin and tetracenomycin C, *Chem. Rev.*, 97, 2525, 1997.
14. Hopwood, D.A., Genetic contributions to understanding polyketide synthase, *Chem. Rev.*, 97, 2465, 1997.
15. Di Marco, A. et al., Studi sperimentali sull'attività antineoplastica del nuovo antibiotico daunomicina, *Tumori*, 49, 203, 1963.
16. Arcamone, F. et al., Daunomycin. I. The structure of daunomycinone, *J. Am. Chem. Soc.*, 86, 5334, 1964.
17. Arcamone, F. et al., Daunomycin. II. The structure and stereochemistry of daunosamine, J. *Am. Chem. Soc.*, 86, 5335, 1964.
18. Arcamone, F. et al., The structure of daunomycin, *Tetrahedron Lett.*, 30, 3349, 1968.

19. Arcamone, F. et al., The total absolute configuration of daunomycin, *Tetrahedron Lett.*, 30, 3353, 1968-

20. Arcamone, F. et al., Struttura e stereochimica della daunomicina, *Gazz. Chim. Ital.*, 100, 848, 1970.

21. Angiuli, R. et al., Structure of daunomycin; x-ray analysis of N-Br-acetyl-daunomycin solvate, *Nat. New Biol.*, 234, 78, 1971.

22. Dubost, M. et al., Rubidomycin, a new antibiotic with cytostatic properties, *Cancer Chemother. Rep.*, 41, 35, 1964.

23. Bernard, J. et al., *Recent results in cancer research: rubidomycin*, Springer, Berlin, 1969.

24. Arcamone, F. et al., Adriamycin, 14-hydroxydaunomycin, a new antitumor antibiotic from *S. peucetius var. caesius. Biotechnol. Bioeng.*, 11, 1101, 1969.

25. Arcamone, F. et al., Adriamycin(14-hydroxydaunomycin), a novel antitumor antibiotic, *Tetrahedron Lett.*, 13, 1007, 1969.

26. Di Marco, A., Gaetani, M. and Scarpinato, B.M., A new antibiotic with antitumor activity, *Cancer Chemother. Rep.*, 53, 33, 1969.

27. Bonadonna, G. et al., Clinical evaluation of Adriamycin a new antitumor antibiotic, *Brit. Med. J.*, 3, 503, 1969.

28. Carter, S.K., Adriamycin — a review, *J. Nat. Cancer Inst.*, 55, 1265, 1975.

29. Borchmann, P. et al., Idarubicin: a brief overview on pharmacology and clinical use, *Int. J. Clin. Pharmacol. Ther.*, 3, 80, 1997.

30. Wang, A.H.J., Intercalative drug binding to DNA, *Curr. Opin. Struct. Biol.*, 2, 361, 1992.

31. Stutter, E., Schuetz, H. and Berg, H., Quantitative determination of cooperative anthracycline-DNA binding, *ref. 5*, p. 245.

32. Ughetto, G., X.-Ray diffraction analysis of anthracycline-oligonucleotide complexes, *ref. 5*, p. 296.

33. Pullman, B., Binding affinities and sequence selectivity in the interaction of antitumor anthracyclines and anthracenediones with double stranded polynucleotides and DNA, *ref. 5*, p. 371.

34. Chaires, J.B., Herrera, J.E., and Waring, J.M., Preferential binding of daunomycin to 5' at CG and 5' at GC sequences revealed by footprinting titration experiments, *Biochemistry*, 29, 6145, 1990.

35. Frederick, C.A. et al., The structural comparison of anticancer drug-DNA complexes: Adriamycin and daunomycin, *Biochemistry*, 29, 2538, 1990.

36. Tewey, K.M. et al, Adriamycin induced DNA damage mediated by mammalian DNA topoisomerase II, *Science*, 226, 466, 1984.

37. Capranico, G. et al., Sequenze selective topoisomerase Ii inhibition by anthracycline derivatives in SV 40 DNA: relationship with DNA binding affinity and cytotoxicity, *Biochemistry*, 29, 562, *1990*.

38. Capranico, G., Kohn, K.W. and Pommier, Y., Local sequence requirements for DNA cleavage by mammalian topoisomerase II in the presence of doxorubicin, *Nucleic Acids Res.*, 22, 6611, 1990.

39. Zunino, F. and Capranico G., DNA topoisomerase II as the primary target of anti-tumor anthracyclines, *Anticancer Drug Design*, 4, 307, 1990.

40. Binaschi, M. et al., Relationship between lethal effects and topoisomerase II-mediated double-stranded DNA breaks produced by anthracyclines with different sequence specificity, *Mol. Pharmacol.*, 6, 1053, 1997.

41. Gerwitz, D.A., A critical evaluation of the mechanisms of action proposed for the antitumor effects of the anthracyclines antibiotics Adriamycin and daunorubicin, *Biochem. Pharmacol.*, 57, 727, 1992.

42. Pommier, Y., DNA topoisomerase II inhibitors, in *Cancer therapeutics, experimental and clinical agents*, Techter B.A. Ed., Humana Press, Totowa, NJ, 153, 1997.

43. Nielsen, D., Maare, C. and Skovsgaard, T., Cellular resistance to anthracyclines, *Gen. Pharmacol.*, 27, 255, 1996.

44. Beaumont, P.O. et al., Role of glutathione S-transferases in the resistance of human colon cancer cell lines to doxorubicin, *Cancer Res.*, 58, 947, 1998.

45. Gaudiano, G., Resing, K. and Koch, T.H., Reaction of reduced anthracyclines with glutathione — formation of aglycone conjugates, *J. Am. Chem. Soc.*, 116, **6537**, 1994.

46. Kim, R., The role of HER-2 oncoprotein in drug-sensitivity in breast cancer, *Oncol. Rep.*, 9, 3, 2002.

47. Bellini, O. and Solcia, E., Early and late sarcoplasmic reticulum changes in doxorubicin cardiomyopathy, *Virchows Arch. (Cell Pathol.)*, 49, 137, 1985.

48. Ishikawa, T. and Sies, H., Cardiac transport of glutathione disulfide and S-conjugates. Studies with isolated perfused rat heart during hydroperoxide metabolism, *J. Biol. Chem.*, 239, 3838, 1984.

49. Olson, R.D. et al., Regulatory role of glutathione and soluble sulfidryl groups in the toxicity of Adriamycin, *J. Pharmacol. Exp. Ther.,* 215, 450, 1980.

50. Formelli, F. et al., Effect of Adriamycin on DNA synthesis in mouse and rat heart, *Cancer Res.,* 38, 3286, 1978.

51. Ellis, C.N., Ellis, M.B. and Blakemore, W.S., Effect of Adriamycin on heart mitochondrial DNA, *Biochem. J.,* 245, 309, 1987.

52. Gatti, G. et al., Injectable stabilized solutions containing an antitumor anthracycline glucoside, *BP 2178311 (Feb. 11, 1987), Chem. Abs.* 107, 223275, 1997.

53. Gatti, G. et al., Ready to use injections containing anthracycline glycosides in acidic solution, *Germ. Offen. 3741037 (Jun. 9, 1988), Chem. Abs.* 110, 141560, 2000.

54. Suarato, A. et al., Anthracycline chemistry: direct conversion of daunorubicin into the L-arabino, L-ribo, and L-lyxo analogues, and selecrive deoxygenation at C-4, *Carbohydr. Res.,* 98, C1-C3, 1981.

55. Badalassi, F. et al., Efficient enantioselective synthesis of (R)-2-hydroxy-5,8-dimethoxy-1,2,3,4-tetrahydronaphthalene, the key intermediate in the synthesis of anthracycline antibiotics, *Tetrahedron Asymmetry,* 12, 3155, 2001.

56. Cabri, W. et al., Process for preparing anthracyclinones, U.S. Patent 5,180,758 (Jan. 19, 1993).

57. Cacchi, S. et al., Palladium-catalyzed triethylammonium formate reduction of aryl triflates, a selective method for the deoxygenation of phenols, *Tetrahedron Lett.,* 27, 5541, 1986.

58. Arcamone, F. and Penco, S., Chemical derivatives of anticancer antibiotics with different DNA binding properties, in *Molecular mechanisms of carcinogenic and antitumor activity,* Chagas, C. and Pullman B. Eds. Pontificiae Academiae Scientiarum Scripta Varia, Vatican City, 70, 225, 1987.

59. Arcamone, F. and Penco, S., Relationship of structure to anticancer activity and toxicity in anthracyclines, *Gann Monogr. Cancer Res.,* 36, 81, 1989.

60. Capranico, G. et al., Role of DNA breakage to cytotoxicity of doxorubicin, 9-deoxydoxorubicin, and 4-demethyl-6-deoxydoxorubicin in murine leukemia P388 cells, *Cancer Res.,* 49, 2022, 1989.

61. Pasqui, F. et al., Synthesis of ring A fluorinated anthracyclines, *Tetrahedron,* 52, 185, 1996.

62. Animati, F. et al., Biochemical and pharmacological activity of novel 8-fluoroanthracyclines: influence of stereochemistry and conformation, *Mol. Pharmacol.,* 50, 603, 1996.

63. Berettoni, M. et al., Synthesis of 14-fluorodoxorubicin, *Tetrahedron Lett.,* 43, 2867, 2002.

64. Ringenberg, Q.S. et al., Clinical cardiotoxicity of esorubicin (4-deoxydoxorubicin, DxDx): prospective studies with serial gated heart scans and reports of selected cases. A Cancer and Leukemia Group B report, *Invest. New Drugs,* 8, 221, 1990.

65. Barbieri, B. et al., Chemical and biological characterization of 4-iodo-4-deoxydoxorubicin, *Cancer Res.,* 47, 4001, 1987.

66. Sessa, C. et al., Phase II studies of 4-iodo-4-deoxydoxorubicin in advanced non-small cell lung, colon and breast cancers, *Ann. Oncol.,* 2, 727, 1991.

67. Merlini, G. et al., Interaction of the anthracycline 4-iodo-4-deoxydoxorubicin with amyloid fibrils: inhibition of amyloidogenesis, *Proc. Natl. Acad. Sci. U.S.A.,* 92, 2959, 1995.

68. Gertz, M.A. et al., A multicenter phase II trial of 4-iodo-4deoxydoxorubicin (IDOX) in primary amyloidosis (AL), *Amyloid,* 9, 24, 2002.

69. Priebe, W. et al., Non-cross-resistant anthracyclines with reduced basicity and increased stability of the glycosidic bond, in *ACS symposium series 574: Anthracycline antibiotics, new analogues, methods of delivery, and mechanism of action,* Priebe, W. Ed., American Chemical Society, Washington, DC, 1995.

70. Booser, D.J. et al., Phase II study of liposomal annamycin in the treatment of doxorubicin-resistant breast cancer, *Cancer Chemother. Pharmacol.,* 50, **6,** 2002.

71. Acton, M., Wasserman, K. and Newman, R.A., Morpholinyl anthracyclines, *ref. 6,* chapter 2.

72. Bakker, M. et al., Broad phase II and pharmacokinetic study of methoxy-morpholino doxorubicin (FCE 23762-MMRDX) in non small cell lung cancer, renal cancer and other solid tumor patients, *Br. J. Cancer,* 77, 139, 1998.

73. Roaten, J.B., Interaction of the novel anthracycline antitumor agent N-benzyladriamycin-14-valerate with the C1-regulatory domain of protein kinase C: structural requirements, isoform specificity, and correlation with drug cytotoxicity. *Mol. Cancer Therapeutics,* 1, 483, 2002.

74. Boivin J., Monneret, C. and Pais, M., Approche de la synthèse d'anthracyclines oligosaccharidiques. Hemisynthèse de la 4-O-(2.deoxy-L-fucosyl)-daunorubicine, *Tetrahedron,* 37, 4219, 1981.

75. El Khadem, H.S. and Matsura, D., Synthesis of 7-(3-amino-2,3,6-trideoxy-4-*O*-(2,6 dideoxy-α-L-*lyxo*-hexopyranosyl)-α-L-*lyxo*-hexopyranosyl)-daunomycinone, *Carbohydr. Res.*, 101, C1, 1982.

76. Horton, D. et al., Synthesis and antitumor activity of anthracycline disaccharide glycosides containing daunosamine, *J. Antibiotics*, 46, 1720, 1993.

77. Capranico, G. et al., Influence of structural modifications at the 3 and 4 positions of doxorubicin on the drug ability to trap topoisomerase II and to overcome drug resistance, *Mol. Pharmacol.*, 45, 908, 1994.

78. Capranico, G., Butelli, E. and Zunino, F., Change of sequence specificity of daunorubicin-stimulated topoisomerase II DNA cleavage by epimerisation of the amino group of the sugar moiety, *Cancer Res.*, 55, 312, 1995.

79. Animati, F. et al., New anthracycline disaccharides. Synthesis of L-daunosaminyl-α(1-4)-2-deoxy-L-rhamnosyl and of L-daunosaminyl-α(1-4)-2-deoxy-L-fucosyl daunorubicin analogues, *J. Chem. Soc. Perkin Trans. I*, 1327, 1996.

80. Arcamone, F. et al., Configurational requirements of the sugar moiety for the pharmacological activity of anthracycline disaccharides, *Biochem. Pharmacol.*, 57, 1133, 1999.

81. Arcamone, F. et al., A doxorubicin disaccharide analog: apoptosis related improvement of efficacy *in vivo*, *J. Natl. Cancer Inst.*, 89, 1217, 1997.

82. Pratesi, G. et al., Improved efficacy and enlarged spectrum of activity of a novel anthracycline disaccharide analogue of doxorubicin against human tumor xenografts, *Clin. Cancer Res.*, 4, 2833, 1998.

83. Pratesi, G. et al., A comparative study of cellular and molecular pharmacology of doxorubicin and MEN 10755, a disaccharide analogue, *Biochem. Pharmacol.*, 62, 63, 2001.

84. Cirillo, R. et al., Comparison of doxorubicin and MEN 10755-induced long term progressive cardiotoxicity in the rat, *J. Cardiovasc. Pharmacol.*, 35, 100, 2000.

85. Messori, L. et al., Solution chemistry and DNA binding properties of MEN 10755, a novel disaccharide analogue of doxorubicin, *Bioorg. Med. Chem.*, 9, 938, 2001.

86. Temperini, C. et al., The crystal structure of the complex between a disaccharide anthracycline and the DNA hexamer d(CGATCG) reveals two different binding sites involving two DNA duplexes, *Nucleic Acids Res.*, 31, 1464, 2003.

87. Garbesi, A. et al., Synthesis and binding properties of conjugates between oligodeoxynucleotides and daunorubicin derivatives, *Nucleic Acids Res.*, 25, 2121, 1997.

88. Capobianco, M.L. et al., New TFO conjugates containing a carminomycinone derived chromophore, *J. Bioconjugate Chem.*, 12, 523, 2001.

89. Catapano C. et al., Synthesis of daunomycin conjugated triple-forming oligonucleotides that act as selective repressors of c-myc gene expression, in *Proc. 92nd Annual Meeting Am. Ass. Cancer Res.*, Abstract 4563, New Orleans LA, 2001.

90. Young, R..C., Ozols, R.F., and Myers, C.E., Medical progress. The anthracycline antineoplastic drugs, *New Engl. J. Med.*, 305, 149, 1981.

91. Geffen, D.B. and Man, S., New drugs for the treatment of cancer, 1990-2001, *Isr. Med. Assoc. J.*, 4, 1124, 2002.

92. Hoelzer, D. and Gale, R.P., Acute lymphoblastic leukemia in adults: recent progress, future directions, *Semin. Hematol.*, 24, 27, 1987.

93. Gale, R.P. and Foon, K.A., Therapy of acute myelogenous leukaemia, *Semin. Hematol.*, 24, 40, 1987.

94. Bonadonna, G., Principi di chemioterapia antitumorale, in *Manuale di Oncologia Medica*, Bonadonna, G. and Robustelli della Cuna, G. Eds., Masson Italia, Milan, 1981, chap. 18.

95. Beretta, G., *Cancer chemotherapy regimens*, Farmitalia-Carlo Erba SpA, Milan, 1983.

96. Robert, J. and Gianni, L., Pharmacokinetics and metabolism of anthracyclines, *Cancer Surv.*, 17, 219, 1993.

97. Gill, P. et al., Time dependency of Adriamycin and adriamycinol kinetics, *Cancer Chemother. Pharmacol.*, 10, 120, 1983.

98. Ganzina, F., 4'-Epidoxorubicin, a new analogue of doxorubicin: a preliminary overview of preclinical and clinical data, *Cancer Treat Rev.*, 10, 1, 1983.

99. Mouridsen, H.T. et al., Current status of epirubicin (Farmorubicin) in the treatment of solid tumours, *Acta Oncol.*, 29, 257, 1990.

100. Plosker, G.L. and Fauld, D., Epirubicin. A review of its pharmacodynamic and pharmacokinetic properties, and therapeutic use in cancer chemotherapy, *Drugs,* 45, 788, 1993.

101. Robert, J., Clinical pharmacokinetics of epirubicin, *Clin Pharmacokinet.,* 26, 428, 1994.

102. Cersosimo, R.J., Idarubicin: an anthracycline antineoplastic agent, *Clin. Pharm.,* 11, 152, 1992.

103. Buckley, M.M. and Lamb, H.M., Oral idarubicin. A review of its pharmacological properties and clinical efficacy in the treatment of haematological malignancies and advanced breast cancer, *Drugs Aging,* 11, 61, 1997.

104. Warrell Jr., R.P., Aclacinomycin A: clinical development of a novel anthracycline antibiotic in the haematological cancers, *Drugs Exp. Clin. Res.,* 12, 275, 1986.

105. Schrivers, D. et al., Phase I study of MEN-10755, a new anthracycline in patients with solid tumors: a report from the European Organization for Research and Treatment of Cancer, Early Clinical Studies Group, *Ann. Oncol.,* 13, 385, 2002.

106. Bos, A.M.E. et al., Pharmacokinetics of MEN-10755, a novel anthracycline disaccharide analogue, in two phase I studies in adults with advanced solid tumours, *Cancer Chemother. Pharmacol.,* 48, 361, 2001.

107. Caponigro, F. et al., A phase II study of MEN 10755 in patients with advanced or metastatic ovarian cancer. EORTC-NCI-AACR Meeting, (Abstr. 42), *Eur. J. Cancer,* 38 (Suppl. 7) S18, 2002.

108. Tjan-Heijnen, V.C.G. et al., A phase II study with MEN 10755 as second line therapy in advanced non small cell lung cancer (NSCLC), 27th ESMO Congress, Nice, France, Oct. 2002. Abstr. 542P, *Ann. Oncol.,* 13, 55, 2002.

109. Comandini, A. et al., MEN 10755 in lung cancers: a survey of phase II studies in patients with advanced or metastatic NSCLC and SCLC-ED. 24th Winter Meeting of the EORTC-PAMM Group, Florence, Italy, Feb. 5-8, 2003.

110. Muggia, F.M., Liposomal encapsulated anthracyclines: new therapeutic horizons, *Curr Oncol Rep.,* 3, 156, 2001.

111. Gabizon, A., Shmeeda, H. and Barenholz, Y., Pharmacokinetics of pegylated liposomal Doxorubicin: review of animal and human studies, *Clin Pharmacokinet.,* 42, 419, 2003.

112. Perez, A.T et al., Pegylated liposomal doxorubicin (Doxil) for metastatic breast cancer: the Cancer Research Network, Inc. experience, *Cancer Invest.,* 20 (Suppl. 2), 22, 2002.

17 Ansamitocins (Maytansinoids)

Tin-Wein Yu and Heinz G. Floss

CONTENTS

I.   Introduction 321
II. Naturally Occurring Maytansinoids 321
III. Biological Activity and Mechanism of Action 323
IV. Total Synthesis 324
V.   Medicinal Chemistry: Structure Modification, Structure–Activity
     Relationships, and Metabolism 324
VI. Preclinical and Clinical Developments 328
VII. Biosynthesis 330
References 333

I. INTRODUCTION

The maytansinoids represent a family of antitumor agents of extraordinary potency. Following the discovery of the parent compound, maytansine, in 1972,[1] the chemistry and biology of the maytansinoids were the subject of intense studies in the 1970s and early 1980s, culminating in clinical trials of maytansine against a variety of human cancers. After the disappointing outcome and eventual termination of the phase II clinical trials, interest in these compounds subsided considerably. The last decade has brought a certain resurgence of interest in the maytanisinoids, which were shown to have considerable potential as "warheads" in immunoconjugates with tumor-specific antibodies — an application that is currently under intense investigation. Other studies are addressing the biosynthesis of the maytansinoids, its genetic control, and the role of plants versus microorganisms in their formation.

The earlier work, from the initial discovery to about 1984, has been summarized in several authoritative reviews.[2–5] This chapter only briefly covers some of this earlier work and focuses primarily on the more recent developments in this field, which have occurred since these first reviews were published.

II. NATURALLY OCCURRING MAYTANSINOIDS

In 1972, Kupchan and colleagues, working under the auspices of the National Cancer Institute (NCI), reported the isolation and structure elucidation of maytansine (1) (Figure 17.1).[1] They followed up on the observation of activity *in vitro* against cells of the human epidermoid carcinoma of the nasopharynx (KB) and *in vivo* against mouse L1210 and P388 leukemia, mouse sarcoma 180, and Lewis lung carcinoma and rat Walker 256 carcinosarcoma of extracts of *Maytenus serrata* (Celastraceae) collected in Ethiopia in 1961/1962. Bioassay-guided fractionation led to the isolation of 1 as the principal active constituent in a yield of 0.2 mg/kg of dried plant material. The targeted isolation of such a trace compound was only possible because of its potent cytotoxic activity (ED_{50} 10^4–10^5 µg/mL against KB cells). The structure of 1 was determined primarily by x-ray crystallography. Subsequently, other

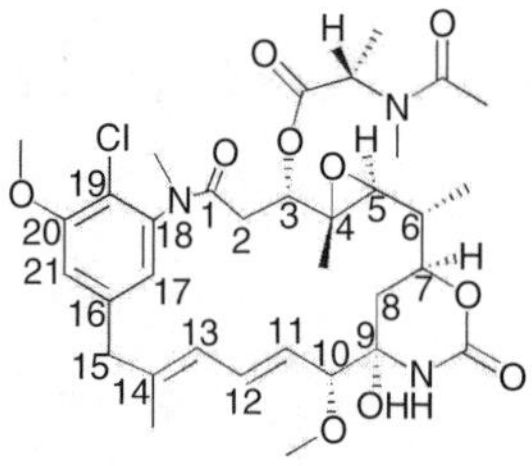

Maytansine (1)

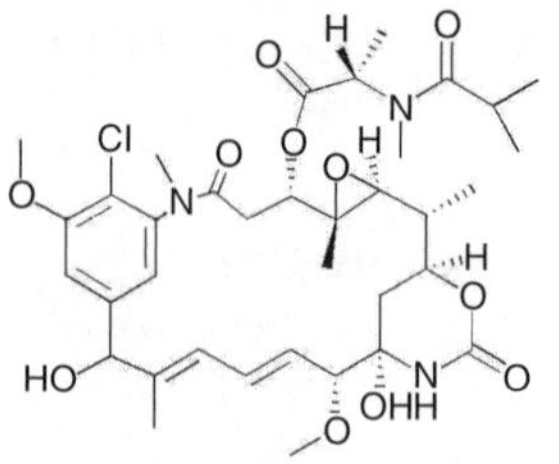

Colubrinol (2)

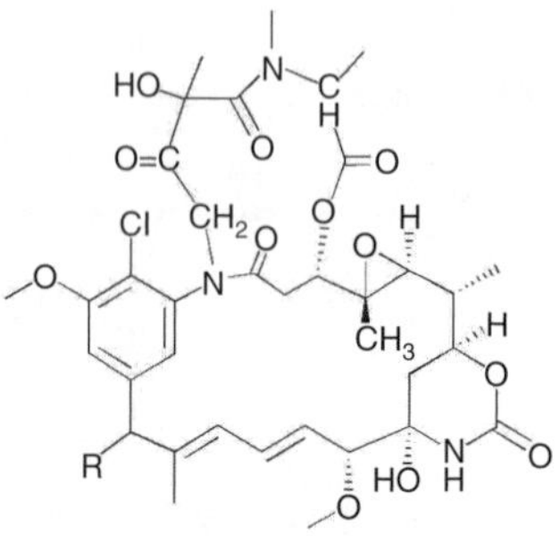

R=OCH₃, N-Methyltrenudone (3)
R=H, Maytanbicyclinol (8)

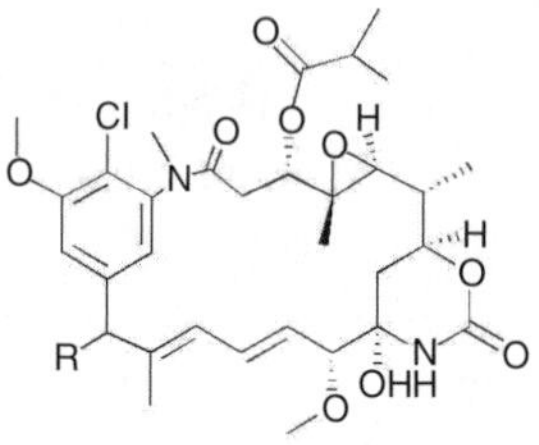

R=H, Ansamitocin P-3 (4)
R=OCH₃, 15-Methoxyansamitocin P-3 (5)

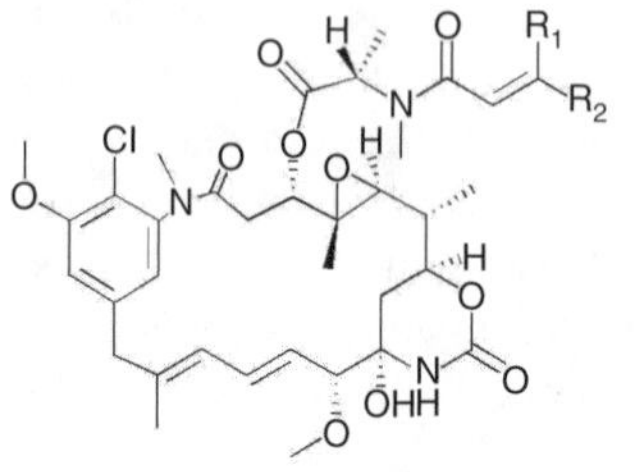

R₁=CH₂CH₃, R₂=CH₃, Mallotusine (6)
R₁=CH₃, R₂=CH₂CH₃, Isomallotusine (7)

FIGURE 17.1 Structures of naturally occurring maytanisinoids.

Maytenus species (e.g., *Maytenus buchananii*), as well as another Celastraceae, *Putterlickia verrucosa*, were established as better sources of **1**, containing seven to eight times higher levels of the compound.[3]

Following the discovery of maytansine, substantial numbers of analogs were also isolated from the same or related plant sources. Most of these contain different C3 ester side-chains, lack the ester side-chain or the *N*-methyl group, or carry a hydroxy/methoxy substituent at C15. Members of two other plant families, *Colubrina texensis* (Rhamnaceae) and *Trewia nudiflora* (Euphorbiaceae), also yielded maytanisinoids — the former some 15-oxygenated compounds, such as colubrinol (**2**),[6] and the latter structures with cyclized ester moieties, such as *N*-methyltrenudone (**3**).[7]

The structures of the maytansinoids are unusual for higher plant products but closely resemble microbial metabolites of the ansamycin family.[8] This raised questions as to the true biosynthetic origin of these compounds and led to a search for microbial producers of maytanisinoids. This search culminated in a report by scientists at Takeda Chemical Industries[9] on the isolation of a microorganism originally designated a *Nocardia* species but later reclassified as *Actinosynnema pretiosum* (Actinomycetes),[10] which produces a series of compounds — the ansamitocins (e.g., ansamitocin P-3, **4**) — with structures closely resembling and in some cases identical to those of plant-derived maytansinoids. Again, these compounds, to the extent that they carry ester side chains at C3, such as **4**, are very potent cytotoxic agents.

To date, 49 fully characterized, naturally occurring maytansinoids have been reported in the literature.[11] Most of these were isolated during the 1970s and early 1980s, and their structures are listed in the earlier reviews.[3–5] Since then, only five new compounds have been discovered, and these represent minor variations of known structures. Two groups reported the isolation of **4** from several species of mosses,[12,13] together with three related maytansinoids including one new compound: 15-methoxyansamitocin P-3 (**5**).[12] Feng et al.[14] reported two new compounds, mallotusine (**6**) and isomallotusine (**7**), from *Mallotus anomalus*, and Sneden and coworkers[15] isolated 2-*N*-demethylmaytanbutine (normaytanbutine) and, notably, maytanbicyclinol (**8**), from *M. buchananii*. There are also allusions to additional microbial maytansinoids in the literature (e.g., ansamitocin S-3 carrying a carbamoylated pentose moiety in place of the *N*-methyl group of **4**),[16] but details of their characterization have apparently not been published.

III. BIOLOGICAL ACTIVITY AND MECHANISM OF ACTION

The antitumor activity of maytansine and the ansamitocins has been discussed extensively in previous reviews.[2–5] **1** showed significant inhibitory activity against solid tumors, Lewis lung carcinoma, and B16 melanocarcinoma and antileukemic activity against P388 lymphocytic leukemia over a large (50–100-fold) dose range.[17,18] In the 60–human cancer cell line screen of the NCI,[19] **1** showed an activity pattern indicative of tubulin-interactive agents.[20] Related compounds carrying the C3 ester function (e.g., maytanprine, maytanbutine, and normaytansine), exhibited a similar activity spectrum and potency as **1**.[21] The 15-oxygenated maytansinoids colubrinol (**2**) and colubrinol acetate isolated from *C. texensis* also showed potencies comparable to those of **1** against KB cells *in vitro* and P388 leukemia *in vivo*.[6] Similarly, the bicyclic maytansinoids isolated from *T. nudiflora* and *M. buchananii* retain activity against P388 leukemia and KB cells.[7,15] The ansamitocins P-3 and P-4 exhibit a very similar activity spectrum and dose range as maytansine (e.g., antitumor activity against P388 leukemia *in vivo* at daily doses as low as 0.8 µg/kg), and they are also active against B16 melanoma, sarcoma 180, Ehrlich carcinoma, and P815 mastocytoma, though not very active against L1210 leukemia.[9]

It was established early on that maytansine[22–24] and ansamitocin P-3[25] inhibit cell division by binding to tubulin, blocking the assembly of microtubules. The cells are arrested in metaphase; the cytological changes resemble those induced by vincristine. Maytansine competitively inhibits vincristine binding to tubulin, and vincristine — but not colchicine — competitively inhibits tubulin binding of **1**.[26] Thus, maytansinoids apparently bind to tubulin at a site overlapping the vinca alkaloids binding site and different from the colchicine binding site. The binding site of the vinca alkaloids has been mapped to the β-subunit of tubulin,[27] and maytansine also binds to this subunit.[28] Several studies have explored features of the β-tubulin structure that are important for maytansinoid binding/activity (e.g., a conserved asparagine residue[29,30] and key sulfhydryl groups[31]). Fluorescent and photoaffinity probes based on ansamitocin P-3 have been described as reagents for probing the tubulin structure.[32] Because tubulin binding of maytansinoids correlates with their activity against P388 leukemia[33,34] and no other primary effects were apparent, the interaction with tubulin is considered the principal mechanism of antitumor action of the maytansinoids.[3,22]

In addition to their antitumor activity, the maytansinoids also show inhibitory activity against other eukaryotic systems, such as fungi and yeasts, protozoa, insects, and plants, but they have little if any antibacterial activity.[35] Among a large number of fungi and yeasts tested, the yeast *Filobasidium uniguttulatum* (IFO 0699) was selected as a particularly sensitive test organism for the bioassay of maytansinoids. An inhibition zone of 20 mm was obtained with 45 ng of ansamitocin P-4.[36] The antitubulinic activity of maytansinoids can be tested using protozoa, such as *Tetrahymena pyriformis*, which have cilia containing microtubule bundles. Deciliated *Tetrahymena* naturally recover their cilia, and hence motility, within 60 min, and this process is inhibited by maytansinoids (e.g., 100% by 1 μM **4**).[37,38] A similar assay observes axopod retraction in *Actinophris sol* or *Actinosphaerium eichhorni*.[39] Insecticidal activity against members of the Lepidoptera and Coleoptera and against the chicken body louse was reported for extracts and isolated maytansinoids from *T. nudiflora* seeds.[40]

IV. TOTAL SYNTHESIS

The decade following the original report on the isolation of **1** saw intense efforts aimed at the structure modification of the naturally occurring maytansinoids and at the total synthesis of this class of compounds. The work published up to 1983 has been summarized,[3,5] and in particular detail in the review by Reider and Roland.[4] Two groups, those of Meyers and of Corey, reported the total synthesis of (±) and (–)-*N*-methylmaysenine (not a naturally occurring maytansinoid).[41–43] The first synthesis of a natural maytansinoid, (±)-maysine, was reported by Meyers et al. in 1979,[44] followed in short order by syntheses of (±)-maytansinol (Meyers)[45] and (–)-maytansine (Corey)[46] in 1980. A third group, that of Isobe and Goto, also completed a total synthesis of (±)-maytansinol in 1982.[47] These approaches all converged on an aminoaldehyde or aminoester intermediate containing the complete functionalized structural framework except carbons 1 and 2, which were introduced and ring-closed to the macrocyclic lactam in different ways.[4] In a stereocontrolled synthesis of (–)-maytansinol reported by Isobe et al. in 1984, such an intermediate was prepared stereospecifically from D-mannose.[48] Other approaches to the assembly of the macrolactam structure of the maytansinoids were explored (e.g., by the groups of Ho,[4] Barton,[4] Fried,[49] Ganem,[4] Confalone,[50] Goodwin,[51] and Hodgson et al.[52]) but did not lead to completed syntheses. Subsequent to the early synthetic work, two more total syntheses were reported. Gao's group in Shanghai completed a synthesis of maytansine in 1988 (Figure 17.2),[53] and Bénéchie and Khuong-Huu, building on the earlier efforts of the group at Gif-sur Yvette, reported a new synthesis of (–)-maytansinol in 1996 (Figure 17.3).[54] However, all these syntheses, although challenging and creative, are lengthy and low yielding and neither solved the supply issue nor contributed significantly to the generation of structural analogs for structure–activity studies.

V. MEDICINAL CHEMISTRY: STRUCTURE MODIFICATION, STRUCTURE–ACTIVITY RELATIONSHIPS, AND METABOLISM

Despite the small amounts of material available from the plant sources, the Kupchan group explored the chemistry of maytansine to generate some analogs for preliminary structure–activity relationship studies. The 9-OH group was converted into various ethers, all of which showed no significant cytotoxic activity. Attempts to cleave the C3 ester group by hydrolysis led instead to elimination to give the olefin, maysine. The ester was cleaved successfully by $LiAlH_4$ reduction at low temperature to give the alcohol, maytansinol. $LiAlH_4$ reduction under more vigorous conditions led to the corresponding deschloro compound. None of the compounds lacking the ester side chain had significant activity, whereas numerous naturally occurring analogs with different ester groups all showed potent activity. All this chemistry has been described in detail in the previous reviews.[3–5]

With access to more starting material — compound **4** from the microbial fermentation — scientists at Takeda Chemical Industries explored the structure–antitumor activity of the maytansinoids in more detail. Through chemical and microbial transformations, the researchers prepared a large number of analogs of **4** and tested their antitumor activity, as reported in a series of publications[5,55–60] and a large number of patents.[11] Microbial transformations of **4** led to *N*-demethylation,[55] 20-*O*-demethylation,[56] 15-hydroxylation with either R or S stereochemistry,[57] and 3-*O*-deacylation.[55] Except for the deacylated compounds, all of these transformation products

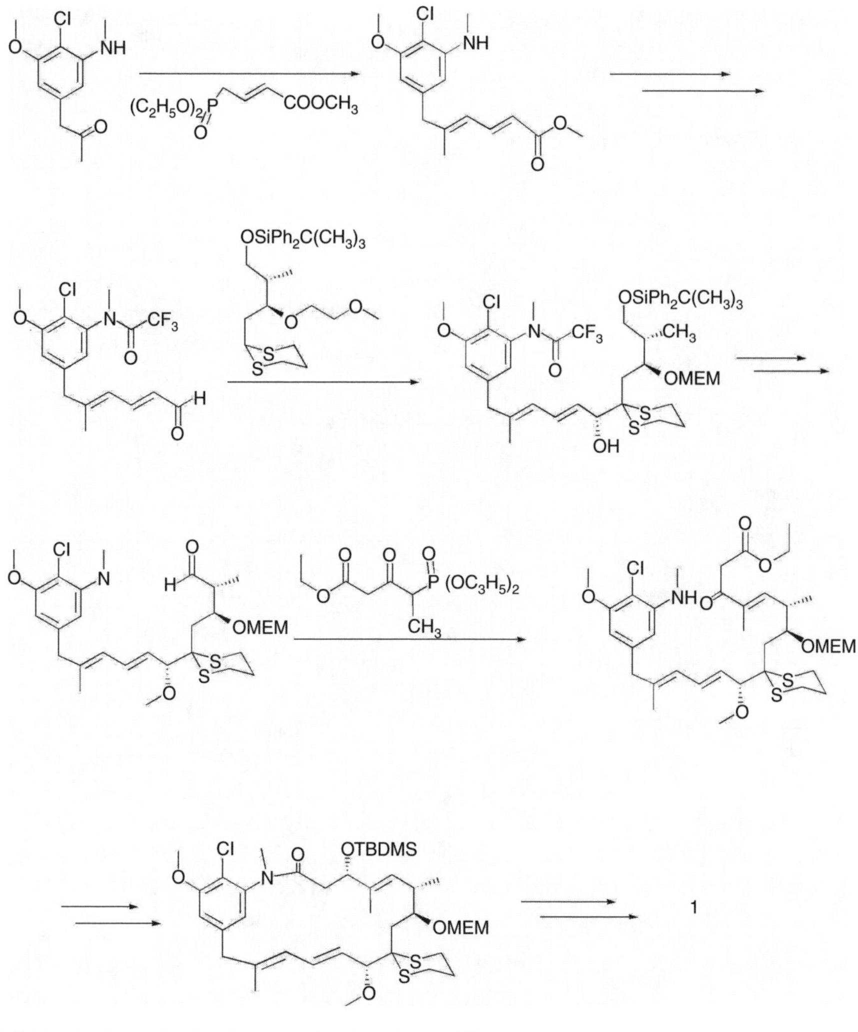

FIGURE 17.2 Total synthesis of maytansine by Gu et al.[53]

retained significant antitumor activity. The 3-, 15-, and 20-hydroxy compounds served as starting materials for the synthesis of large numbers of alkyl and acyl derivatives at these positions, most of them showing varying degrees of biological activity.[60] Interestingly, modifications of the 3-*O*-ester moiety modulated the activity of the molecule, with some compounds of low *in vitro* activity having high *in vivo* activity. One compound, the 3-*O*-phenylglycinate ester, showed a two to four times better than that of **1**.[60] Removal of the epoxide function to give an *E*-4,5-double bond did not abolish the activity,[58] but inversion of the stereochemistry at C3 by oxidation of maytansinol to the 3-ketone, reduction with NaBH$_4$, and reesterification gave 3-*epi*-ansamitocin P-3, which was devoid of cytotoxic activity.[59] Deschloro compounds were also prepared by reduction of **4** with excess LiAlH$_4$ followed by reesterification, with little effect on the biological activity.[61]

More recently, Sneden and coworkers explored further the reductive chemistry of maytansinoids (Figure 17.4).[62,63] Reduction of maytanbutine, the analog of **1** with an *N*-isobutyryl instead of the *N*-acetyl group, with lithium borohydride/lithium triethylborohydride (10:1) cleaved the carbinolamide function to give the 7-carbamoyloxy-9-hydroxy derivative with a 1000-fold decrease in cytotoxic activity.[62] In contrast, hydrogenation of **1** over Pd/C generated 11,12,13,14-tetrahydromaytansine, with little effect on the biological activity.[63] The structure–antitumor activity relationships of the

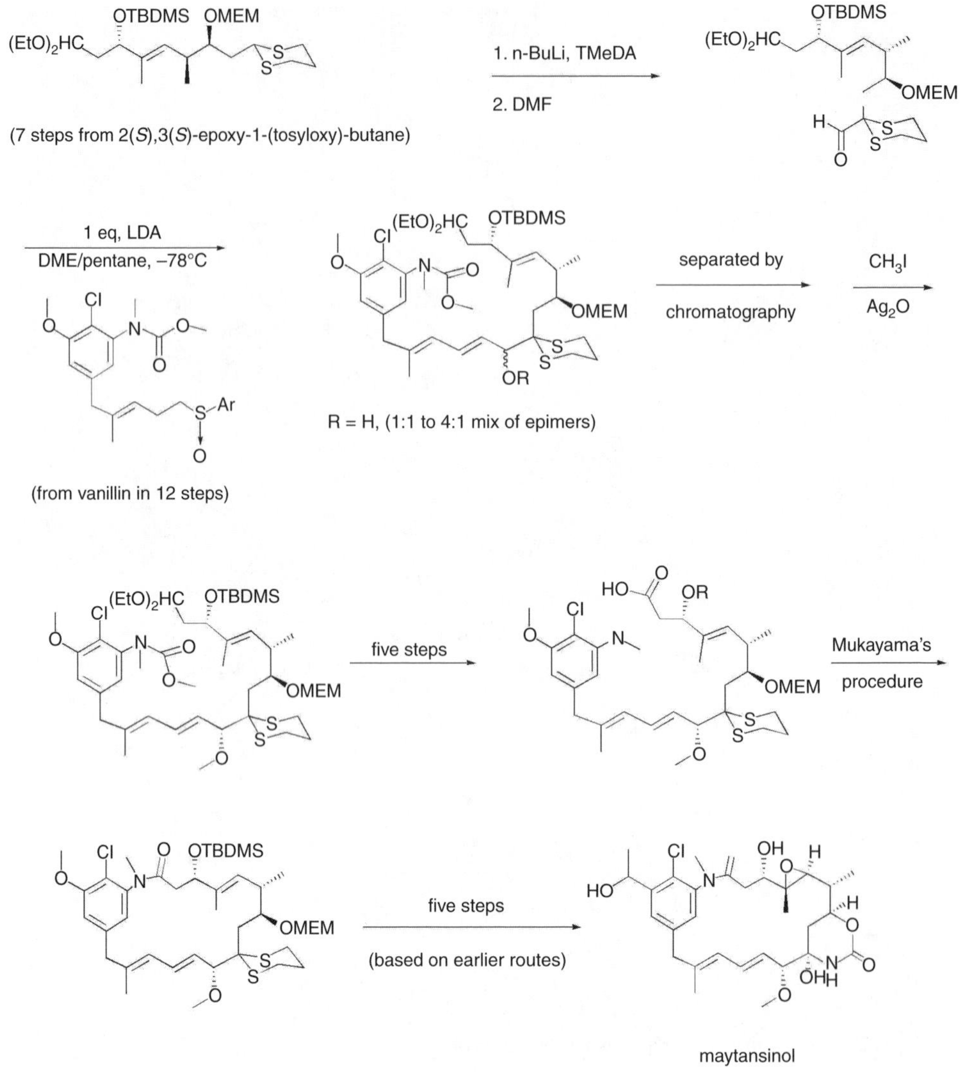

FIGURE 17.3 Total synthesis of (–)-maytansinol by Bénéchie and Khuong-Huu.[54]

maytansinoids revealed by all these studies can be summarized, as shown in Figure 17.5. An ester group at the 3-position and an unblocked carbinolamide function at C9 are absolute requirements for potent antitumor activity; most other functional group modifications have either no or only a modulating effect on the biological activity.

Suchocki and Sneden also examined the long-term stability of **1**, showing that in outdated clinical samples 40% of the maytansine had decomposed, mostly by elimination of the C3 side chain to give maysine.[64] One report[65] and a patent[66] deal with improved methods for the preparation of maytansinol (Figure 17.3) — an important starting material for the synthesis of immunoconjugates of maytansinoids (*vide infra*).[67]

Although maytansine progressed all the way to phase II clinical trials, there are no data in the literature on the mammalian metabolism of the compound. The reasons probably are that the amounts of material available at the time were extremely small and that the analytical techniques then available were not sensitive enough to follow the metabolism of a compound that is active at such low concentrations. Chan and colleagues have recently begun to address this problem.[11,67] Ansamitocin

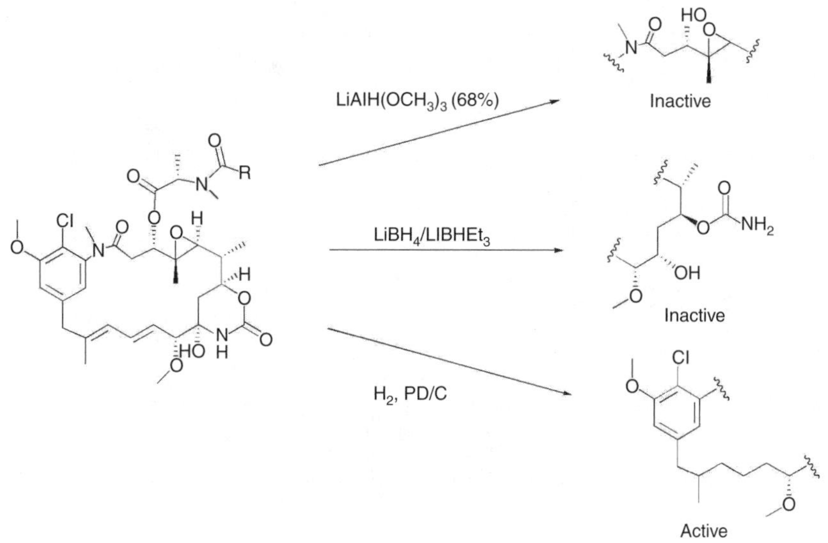

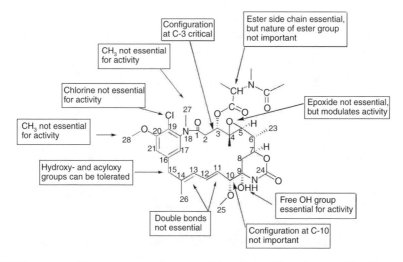

FIGURE 17.4 Reductive modifications of maytanisinoids.[62,63,65]

P-3 was incubated with rat and human liver microsomes, with a rat liver homogenate, and administered to a whole rat. The metabolites were analyzed by electrospray tandem mass spectrometry, identifying relevant compounds by the presence of the chlorine isotopes and determining their likely structures by comparing their fragmentation patterns to those of the parent compound. Early results indicated that the most prominent route of metabolism in microsomes was demethylation of the C10 methoxy group, with several other oxidative metabolites detectable. Human liver microsomes metabolized 70% of the administered dose, compared to 20% for rat liver microsomes. In contrast, the most prominent metabolite in rat liver homogenate was maytansinol, indicating the presence of an esterase in the soluble enzyme fraction. In a similar study, the metabolism of maytansine in human liver microsomes was found to involve most prominently the oxidation and removal of the *N*-methyl group in the ester side chain. These preliminary results are summarized in Figure 17.6.

FIGURE 17.5 Summary of the structure–antitumor activity relationships of maytanisinoids.

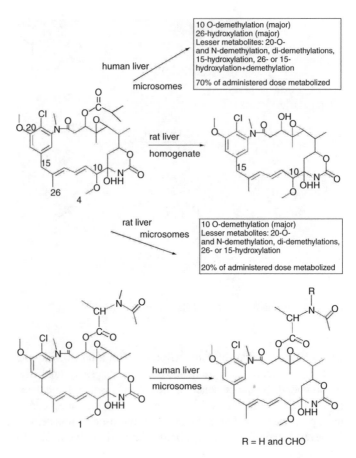

FIGURE 17.6 (See color insert following page 304.) Preliminary results on the mammalian metabolism of ansamitocin P-3 and maytansine.[67]

VI. PRECLINICAL AND CLINICAL DEVELOPMENTS

Given the high potency of maytansine against various tumor types, including some solid tumors, the development of the compound for clinical trials was a high priority of the NCI. Following the isolation of larger, although still limited, quantities of **1** and satisfactory preclinical pharmacological and toxicological evaluation, the compound was entered into phase I clinical trials in patients with advanced disease refractory to conventional therapy in 1975.[2,3] The tolerated dose was established as 1–2 mg/m²; observed side effects were primarily gastrointestinal toxicity, neurotoxicity, nausea, vomiting, and diarrhea; and a few patients showed promising partial responses. This led to the scheduling of a large number of phase II trials of **1** alone and in combination with other agents. These gave rather disappointing results, showing no substantial clinical benefits.[68–84] As summarized in a 1984 report by the NCI to the Food and Drug Administration on the Investigational New Drug Application (INDA) for maytansine,[85] the evaluation of the compound as a single agent against 36 different types of tumors in 819 patients showed only one complete response and 20 partial responses while at the same time revealing substantial gastrointestinal and neurological toxicity. At this point the NCI terminated the clinical trials of **1**.

The 1990s have seen a certain resurgence of clinical interest in the maytansinoids, specifically as cytotoxic agents ("warheads") for antibody-directed targeted delivery to tumor cells. The

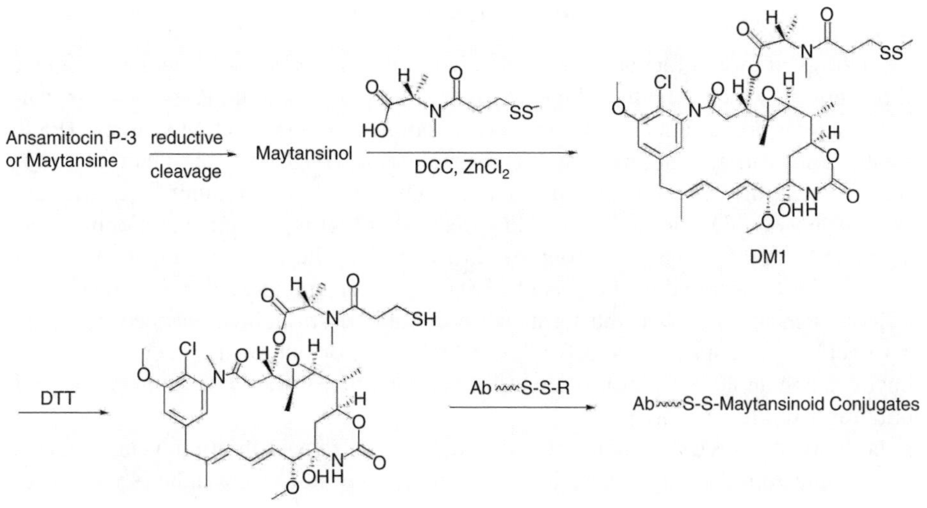

FIGURE 17.7 Structural modification of maytansinoids for conjugation to antibodies and receptor-specific carriers.[86–89]

antibody-targeted delivery approach, although conceptually appealing, has had limited practical success because the amount of drug that can be delivered to the target by an antibody is too small for most clinically used anticancer drugs to be effective. To overcome this problem, the warhead should be a highly potent drug, such as a maytansinoid. At the same time, the target-specific delivery holds the promise of minimizing the general toxic effects of the drug. Two approaches have been explored.

Iwasa and coworkers at Takeda[16] developed a bispecific monoclonal antibody that recognizes both ansamitocin P-3 and the human transferrin receptor, which is highly expressed on various human tumor cells. The **4** targeted to human A431 epidermoid carcinoma xenografts in nude mice was more effective than unconjugated **4** and eventually eradicated this tumor while at the same time showing reduced side effects.

An alternative approach was pursued by Chari et al.[86] at ImmunoGen, Inc., who modified maytansine for conjugation to antibodies or receptor agonists through disulfide linkages to the ester side chain. As shown in Figure 17.7, maytansinol generated either from **1** or **4** was converted by procedures detailed in the patent literature[87,88] into a compound, designated DM1, with a methyl disulfide function in the ester side chain. DM1, on reduction with dithiothreitol gives the corresponding thiol, which can be conjugated to an antibody through either thioether or, preferably, disulfide linkages. The effectiveness of this approach was demonstrated with conjugates to the TA.1, 3E9, and A7 antibodies directed against the *c-erb-2* oncogene protein, the human transferrin receptor, and human colon cancer cell lines, respectively. Multiple DM1 molecules could be conjugated to the TA.1 antibody (one to six molecules per antibody) and gave cytotoxicities against SB-BR-3 breast cancer cells up to IC_{50} 1.2×10^{11} M, which were antagonized by the unconjugated antibody and dependent on expression of the target epitope.[86] Subsequently, encouraging preclinical results were presented with a DM1 conjugate to the monoclonal antibody C242, which recognizes a glycoprotein expressed in human colorectal cancers.[89] The C242-DM1 conjugate cured mice carrying xenografts of the COLO 205 human colon tumor — even very large tumors — at doses that showed very little toxicity. Effectiveness against two other colon tumor types was also demonstrated. The ImmunoGen technology was considered clinically promising and has been licensed to a number of major pharmaceutical companies.[90] Several different DM1 conjugates are now undergoing clinical trials.[91–94] The ImmunoGen group also reported a DM1 conjugate to a folate derivative designed to target cancer cells that overexpress the folate receptor.[95]

VII. BIOSYNTHESIS

The biosynthesis of ansamitocins has been studied in *A. pretiosum* by feeding experiments with labeled precursors, which established the building blocks from which the molecule is generated. The maytansinoids are ansamacrolides that are assembled on a polyketide synthase (PKS) from an aromatic starter unit, 3-amino-5-hydroxybenzoic acid (AHBA),[96] by chain extension with three acetate units, three propionate units, and one "glycolate" extender unit.[97] Functional groups added to the polyketide backbone include three methyl groups from methionine,[97] a cyclic carbinolamide from the carbamoyl group of citrulline[97] and the ester side chain at C3, which is derived from the corresponding fatty acid or its precursor amino acid.[98] The spectrum of the maytansinol esters produced in the fermentation could be strongly influenced by feeding the respective acid (e.g., isobutyric acid for ansamitocin P-3) or one of its precursors (e.g., isobutyraldehyde, isobutanol, or valine) to direct the fermentation towards predominant (up to 90%) production of a single ansamitocin.[98]

On the basis of knowledge about the biosynthesis of AHBA in the rifamycin producer, *Amycolatopsis mediterranei*,[99–101] the gene cluster encoding ansamitocin biosynthesis was cloned from *A. pretiosum*, using the AHBA synthase gene as a probe.[102] Analysis of the cloned DNA revealed the presence of the relevant genes in two clusters, I and II, each carrying an AHBA synthase homologue (Figure 17.8A). The clusters are separated by a 30-kilobase, nonessential DNA segment. Cluster I contains the majority of the biosynthetic genes, including four large type I polyketide synthase genes (*asmA–D*), which together with the cognate downloading enzyme, encoded by *asm9*, are predicted to catalyze the assembly of the required linear octaketide and its release and cyclization to proansamitocin (Figure 17.8B). Also present are some, but not all, of the AHBA synthesis genes (*asm22–24*) — the genes required to synthesize the substrate for the third chain extension step incorporating the "glycolate" extender unit (*asm13–17*), genes for the post-PKS modifications of proansamitocin into **4** (*asm7,10–12,19,21*), and potential regulatory (*asm2,8,18,29,31,34*) and transport/resistance genes (*asm4,5,35*). Cluster II only contains the remaining genes essential for AHBA biosynthesis (*asm43–45,47*); potential regulatory (*asm39,40,48*) and transport/resistance genes (*asm41,42*) present can only affect the AHBA supply, as deletion of cluster II gave a nonproducing mutant in which full **4** production could be restored by supplementation with AHBA.[102]

On the basis of analogy to the other chain extension substrates, malonyl-CoA for acetate and methylmalonyl-CoA for propionate units, one would predict the substrate for incorporation of "glycolate" units to be 2-hydroxy- or 2-methoxymalonyl-CoA. This, however, is not the case, as the corresponding *N*-acetylcysteamine thioesters were not incorporated into **4**.[103] By comparison with the gene cluster for FK520,[104] which also contains "glycolate" extender units, the products encoded by *asm13–17* were recognized as probably involved in forming the substrate for this unusual chain extension step. *Asm14* encodes a unique acyl carrier protein, which could substitute for the coenzyme A moiety as a carrier of the hydroxy- or methoxymalonate unit. Inactivation of each of these genes and phenotypic analysis of the mutants confirmed this deduction, showing loss of **4** production in each mutant, but formation of 10-demethoxy-**4** in a low yield in the *asm13*, *asm15*, *asm16*, and *asm17* mutants. 10-Demethoxy-**4** was not formed in the *asm14* mutant, indicating that the unique *asm14*-encoded ACP is necessary for the aberrant incorporation of malonyl-CoA in the absence of the "glycolate" substrate.[11,103] *Asm13–17* are not only indispensable but also sufficient to generate this "glycolate" substrate, as shown in *Streptomyces lividans* by coexpression of an *asm13–17* cassette[105] with a cassette of the erythromycin PKS, modified by replacement of the methylmalonate-specific AT6 with the presumably hydroxymalonate-specific AT8 domain of the FK520 PKS.[106] The cotransformant produced a new compound, 2-methoxy-2-desmethyl-6-deoxyerythronolide B, resulting from the incorporation of a "glycolate" in place of the terminal propionate unit. Deletion of the methyltransferase gene *asm17* from the *asm13–17* cassette gave no new product, indicating that the hydroxymalonate moiety must be methylated before incorporation.[105] Hence the substrate for the third chain extension

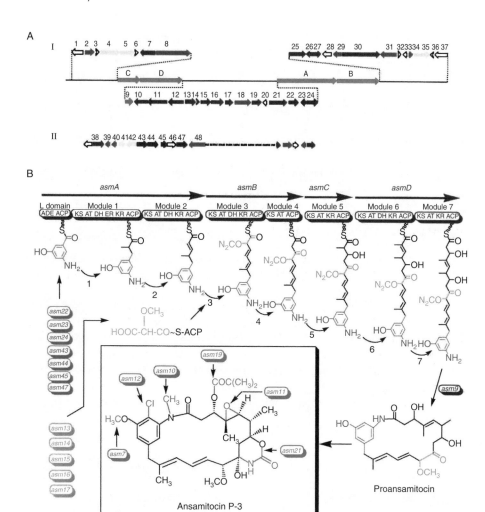

FIGURE 17.8 (See color insert following page 304.) Biosynthesis of ansamitocin P-3: (A) The ansamitocin biosynthetic gene cluster from *Actinosynnema pretiosum*, (B) assembly of proansamitocin on a type I polyketide synthase and post-PKS modification to ansamitocin.[102]

step must be 2-methoxymalonyl-ACP, which is formed from an as-yet-unidentified intermediate of the glycolytic pathway, as shown in Figure 17.9.

The enzymes catalyzing the post-PKS modification steps from proansamitocin to **4** and the order in which they function were examined by inactivation of each corresponding gene and, in some cases, by heterologous expression in *Escherichia coli* and characterization of the enzymatic activity.[107,108] Halogenation and carbamoylation, catalyzed by Asm12 and Asm21, respectively, appear to be the first and second step in the sequence. As predicted by the results of Hatano et al.,[98] the acyltransferase Asm19 catalyzing the addition of the ester side chain is rather promiscuous regarding its acyl substrate. Surprisingly, this reaction is not the terminal biosynthetic step, as expected from the natural occurrence of maytansinol, but occurs before the epoxidation by Asm11 and *N*-methylation by Asm10.[107] The predominant pathway from proansamitocin to **4** is depicted in Figure 17.10. However, many of these post-PKS modifying enzymes are not highly substrate specific; hence, alternate minor pathways to **4** also operate, resulting in a metabolic grid.[108]

The results delineated above provide the tools for the genetic manipulation of the biosynthetic pathway to maytansinoids to generate structural analogs for biologic evaluation, which would be

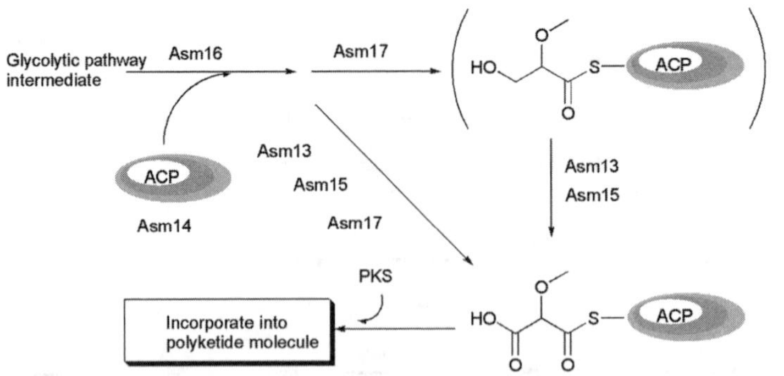

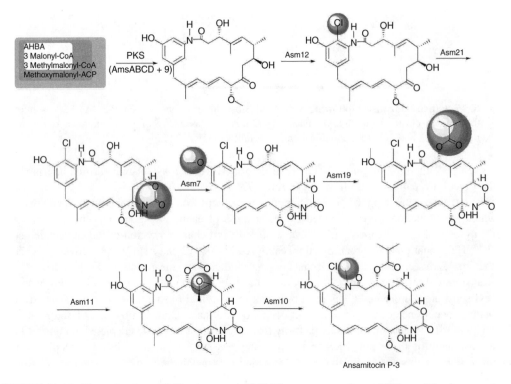

FIGURE 17.9 (See color insert following page 304.) Formation of 2-methoxymalonyl-ACP, the substrate for the incorporation of the "glycolate" chain extension unit into ansamitocin.[105]

difficult to synthesize chemically. Such genetic manipulations could increase the structural diversity by altering the polyketide backbone and its functionality, or the downstream modifications of the macrolide.

The question of the biosynthetic source of the maytansinoids isolated from higher plants has been investigated by Leistner and colleagues in *Putterlickia* species.[39] Several lines of indirect evidence indicate that at least the polyketide framework of **1** is not biosynthesized by the plant, although the plants may have a role in determining the final structures elaborated: greenhouse-grown plants do not contain maytansine[39]; cell cultures of *Maytenus*[109,110] and *Putterlickia*[39] species do not produce maytansinoids; the **1**-content of individual *Putterlickia* plants (three species) varied dramatically — a phenomenon apparently also encountered by the U.S. Department of Agriculture workers who recollected *Maytenus* species for large-scale isolation of **1**[39]; and a

FIGURE 17.10 (See color insert following page 304.) The sequence of post-PKS modification reactions converting proansamitocin into ansamitocin P-3.[108]

polymerase chain reaction–based search for AHBA synthase-homologous genes in the plant genome gave negative results.[39]

The focus is now on the isolation and screening of microorganisms found in, on, or near 1-producing plants. None of a substantial number of fungal endosymbionts isolated from three *Putterlickia* species carried an AHBA synthase homologue in its genome. Among the bacterial community of *P. verrucosa* and *P. retrospinosa*, one promising candidate was isolated from the rhizosphere and identified as *Kitasatospora putterlickiae* sp. nov.[111] Several ansamitocin-specific gene homologues were found in its genome, but the ability to produce maytansinoids has not yet been demonstrated for this organism.[11]

REFERENCES

1. Kupchan, S.M. et al., Maytansine, a novel antileukemic ansa macrolide from *Maytenus ovatus*, *J. Am. Chem. Soc.*, 94, 1354-1356, 1972.
2. Issell, B.F. and Crooke, S.T., Maytansine, *Cancer Treat. Rev.*, 5, 199-207, 1978.
3. Komoda, Y. and Kishi, T., Maytansinoids, in *Anticancer Agents Based on Natural Product Models*, Douros, J. and Cassady, J.M., Eds., Academic Press, New York, 1980, pp. 353-389.
4. Reider, P.J. and Roland, D.M., Maytansinoids, in *The Alkaloids*, Brossi, A., Ed., Academic Press, New York, 1984, Vol. 23, pp. 71-156.
5. Smith Jr., C.R. and Powell, R.G., Chemistry and pharmacology of maytansinoid alkaloids, in *Alkaloids*, Pelletier, S.W., Ed., John Wiley and Sons, New York, 1984, Vol. 2, pp. 149-204.
6. Wani, M.C., Taylor, H.L., and Wall, M.E., Plant antitumor agents: Colubrinol acetate and colubrinol, antileukaemic ansa macrolides from *Colubrina texensis, J. Chem. Soc. Chem. Commun.*, 390, 1973.
7. Powell, R.G. et al., Treflorine, trenudine, and *N*-methyltrenudone: Novel maytansinoid tumor inhibitors containing two fused macrocyclic rings, *J. Am. Chem. Soc.*, 104, 4929-4934, 1982.
8. Rinehart Jr., K.L. and Shield, L.S., Chemistry of the ansamycin antibiotics, *Fortschr. Chem. Org. Nat.*, 33, 231-307, 1976.
9. Higashide, E. et al., Ansamitocins, a group of novel maytansinoid antibiotics with antitumor properties from *Nocardia*, *Nature*, 270, 721-722, 1977.
10. Hasegawa, T. et al., Motile actinomycetes: *Actinosynnema pretiosum* subsp. *pretiosum* sp. nov., subsp. nov., and *Actinosynnema pretiosum* subsp. *auranticum* subsp. nov., *Int. J. Syst. Bacteriol.*, 33, 314-320, 1983.
11. Cassady, J.M. et al., Recent developments in the maytansinoid antitumor agents, *Chem. Pharm. Bull.*, 52, 1-26, 2004.
12. Sakai, K. et al., Antitumor principles in mosses: The first isolation and identification of maytansinoids, including a novel 15-methoxyansamitocin P-3, *J. Nat. Prod.*, 51, 845-850, 1988.
13. Suwanborirux, K. et al., Ansamitocin P-3, a maytansinoid, from *Claopodium crispifolium* and *Anomodon attenuatus* or associated actinomycetes, *Experientia*, 46, 117-120, 1990.
14. Feng, S.C. et al., Studies on the chemical constituents of *Mallotus anomalus*. II. Structures of the antitumor components from *Mallotus anomalus*, *Chinese Chem. Lett.*, 5, 743-746, 1994.
15. Larson G.M., Schaneberg B.T., and Sneden A.T., Two new maytansinoids from *Maytenus buchananii*, *J. Nat. Prod.*, 62, 361-363, 1999.
16. Okamoto, K. et al., Therapeutic effect of ansamitocin targeted to tumor by a bispecific monoclonal antibody, *Jpn. J. Cancer Res.*, 83, 761-768, 1992.
17. Kupchan, S.M. et al., Novel maytansinoids. Structural interrelations and requirements for antileukemic activity, *J. Am. Chem. Soc.*, 96, 3706-3708, 1974.
18. Kupchan, S.M. et al., Maytanprine and maytanbutine, new antileukemic ansa macrolides from *Maytenus buchananii*, *J. C. S. Chem. Comm.*, 1065, 1972.
19. Boyd, M.R. and Paull, K.D., Some practical considerations and applications of the National Cancer Institute *in vitro* anticancer drug discovery screen, *Drug Dev. Res.*, 34, 91-109, 1995.
20. Cragg, G.M., personal communication, 2002.
21. Kupchan, S.M. et al., The maytansinoids. Isolation, structural elucidation, and chemical interrelation of novel ansa macrolides, *J. Org. Chem.*, 42, 2349-2357, 1977.

22. Wolpert-DeFilippes, M.K. et al., Initial studies on the cytotoxic action of maytansine, a novel ansa macrolide, *Biochem. Pharmacol.*, 24, 751-754, 1975.

23. Remillard, S. et al., Antimitotic activity of the potent tumor inhibitor maytansine, *Science*, 189, 1002-1005, 1975.

24. Wolpert-DeFilippes, M.K. et al., Initial studies on maytansine-induced metaphase arrest in L1210 murine leukemia cells, *Biochem. Pharmacol.*, 24, 1735-1738, 1975.

25. Ootsu, K. et al., Effects of new antimitotic antibiotics, ansamitocins, on the growth of murine tumors *in vivo* and on the assembly of microtubules *in vitro*, *Cancer Res.*, 40, 1707-1717, 1980.

26. Mandelbaum-Shavit, F., Wolpert-DeFilippes, M.K., and Johns, D.G., Binding of maytansine to rat brain tubulin, *Biochem. Biophys. Res. Commun.*, 72, 47-54, 1976.

27. Rai, S.S. and Wolff, J., Localization of the vinblastine-binding site on beta-tubulin, *J. Biol. Chem.*, 271, 14707-14711, 1996.

28. Hamel E., Natural products which interact with tubulin in the vinca domain: Maytansine, rhizoxin, phomopsin A, dolastatin 10 and 15 and halichondrin B, *Pharmacol. Therap.*, 55, 31-51, 1992.

29. Takahashi, M. et al., Molecular basis for determining the sensitivity of eucaryotes to the antimitotic drug rhizoxin, *Mol. Gen. Genet.*, 222, 169-175, 1990.

30. Li, Y. et al., Binding selectivity of rhizoxin, phomopsin A, vinblastine, and ansamitocin P-3 to fungal tubulins: Differential interactions of these antimitotic agents with brain and fungal tubulins, *Biochem. Biophys. Res. Comm.*, 187, 722-729, 1992.

31. Luduena, R.F. and Roach, M.C., Tubulin sulfhydryl groups as probes and targets for antimitotic and antimicrotubule agents, *Pharmacol. Ther.*, 49, 133-152, 1991.

32. Sawada, T. et al., A fluorescent probe and a photoaffinity labeling reagent to study the binding site of maytansine and rhioxin on tubulin, *Bioconjugate Chem.*, 4, 284-289, 1993.

33. Ikeyama, S. and Takeuchi, M., Antitubulin activities of ansamitocins and maytansinoids, *Biochem. Pharmacol.*, 30, 2421-2425, 1981.

34. York, J. et al., Binding of maytansinoids to tubulin, *Biochem. Pharmacol.*, 30, 3239-3243, 1981.

35. Tanida, S. et al., Ansamitocins, maytansinoid antitumor antibiotics. Producing organism, fermentation, and antimicrobial activities, *J. Antibiot.*, 33, 192-198, 1980.

36. Hatano, K., Higashide, E., and Yoneda, M., Bioassay of ansamitocin P-3, an antitumor antibiotic, *Agric. Biol. Chem.*, 48, 1889-1890, 1984.

37. Tanida, S., Higashide, E., and Yoneda, M., Inhibition of cilia regeneration of *Tetrahymena* by ansamitocins, new antitumor antibiotics, *Antimicrob. Agents Chemother.*, 16, 101-103, 1979.

38. Tanida, S., Hasegawa, T., and Higashide, E., A simple method for determining antitubulinic activities of ansamitocins and related compounds using a cilia regeneration system with deciliated *Tetrahymena*, *Agric. Biol. Chem.*, 44, 1847-1853, 1980.

39. Pullen, C.B. et al., Occurrence and non-detectability of maytansinoids in individual plants of the genera *Maytenus* and *Putterlickia*, *Phytochemistry*, 62, 377-387, 2003.

40. Freedman, B. et al., Biological activities of *Trewia nudiflora* extracts against certain economically important insect pests, *J. Chem. Ecol.*, 8, 409-418, 1982.

41. Corey, E.J. et al., The total synthesis of (±)-*N*-methylmaysenine, *J. Am. Chem. Soc.*, 100, 2916-2918, 1978.

42. Meyers, A.I. et al., Progress toward the total synthesis of maytansinoids. Synthesis of (±)-4,5-deoxymaysine (*N*-methylmaysenine), *J. Am. Chem. Soc.*, 101, 4732-4734, 1979.

43. Corey, E.J. et al., Total synthesis of ()-*N*-methylmaysenine, *J. Am. Chem. Soc.*, 102, 1439-1441, 1980.

44. Meyers, A.I. et al., Total synthesis of (±)-maysine, *J. Am. Chem. Soc.*, 101, 7104-7105, 1979.

45. Meyers, A.I., Reider, P.J., and Campbell, A.L., Total synthesis of (±)-maytansinol. The common precursor to the maytansinoids, *J. Am. Chem. Soc.*, 102, 6597-6598, 1980.

46. Corey, E.J. et al., Total synthesis of maytansine, *J. Am. Chem. Soc.*, 102, 6613-6615, 1980.

47. Isobe, M., Kitamura, M., and Goto, T., Stereocontrolled total synthesis of (±)-maytansinol, *J. Am. Chem. Soc.*, 104, 4997-4999, 1982.

48. Kitamura, M. et al., Stereocontrolled total synthesis of (-)-maytansinol, *J. Am. Chem. Soc.*, 106, 3252-3257, 1984.

49. Petrakis, K.S. and Fried, J., A novel synthesis of the cyclic carbinolamide moiety of the maytansinoids, *Tetrahedron Lett.*, 24, 3065-3066, 1983, and references therein.

50. Confalone, P.N., The use of heterocyclic chemistry in the synthesis of natural and unnatural products, *J. Heterocyclic Chem.*, 27, 31-46, 1990, and references therein.

51. Goodwin T.E. et al., Synthesis of two new maytansinoid model compounds from carbohydrate precursors, *J. Carbohydr. Chem.*, 17, 323-339, 1998, and references therein.

52. Hodgson, D.M., Parsons, P.J., and Stones, P.A., A short and efficient synthesis of the C-3 to C-10 portion of the maytansinoids, *Tetrahedron*, 47, 4133-4142, 1991.

53. Gu, X. et al., Studies on the total synthesis of maytansinoids (IV) — synthesis of maytansine, *Sci. Sinica, Ser. B.*, 31, 1342-1351, 1988, and references therein.

54. Bénéchie, M. and Khuong-Huu, F., Total synthesis of ()-maytansinol, *J. Org. Chem.*, 61, 7133-7138, 1996.

55. Nakahama, K. et al., Microbial conversion of ansamitocin, *J. Antibiot.*, 34, 1581-1586, 1981.

56. Izawa, M. et al., Demethylation of ansamitocins and related compounds, *J. Antibiot.*, 34, 1587-1590, 1981.

57. Izawa, M. et al., Hydroxylation of ansamitocin P-3, *J. Antibiot.*, 34, 1591-1595, 1981.

58. Kawai, A. et al., Chemical modification of ansamitocins. I. Synthesis and properties of 4,4-deoxy-maytansinoids, *Chem. Pharm. Bull.*, 32, 2194-2199, 1984.

59. Akimoto, H. et al., Chemical modification of ansamitocins. II. Synthesis of 3-epimaytansinoids *via* 3-maytansinones, *Chem. Pharm. Bull.*, 32, 2565-2570, 1984.

60. Kawai, A. et al., Chemical modification of ansamitocins. III. Synthesis and biological effects of 3-acyl esters of maytansinol, *Chem. Pharm. Bull.*, 32, 3341-3351, 1984.

61. Miyashita, O. and Akimoto, H., Dechloromaytansinoids, Takeda Chemical Industries, Ltd., Japan, European Patent EP 11277 19800528, 1980.

62. Suchocki, J.A. and Sneden, A.T., New maytansinoids: Reduction products of the C(9)-carbinolamide, *J. Org. Chem.*, 53, 4116-4118, 1988.

63. Schaneberg, B.T. and Sneden, A.T., Abstracts, 218th ACS Natl. Meeting, New Orleans, Aug. 22-26, 1999.

64. Suchocki, J.A. and Sneden, A.T., Characterization of decomposition products of maytansine, *J. Pharm. Sci.*, 76, 738-743, 1987.

65. Sneden, A.T., Synthetic modifications of biologically interesting natural products, *Synlett*, 5, 313-322, 1993.

66. Terfloth, G., SmithKline Beecham Corp., USA, Process for preparation and purification of maytansinol, Int. Appl. WO 02/074775 A1, 2002.

67. Liu, Z., Floss, H.G., Cassady, J.M., and Chan, K.K., unpublished work, 2003.

68. O'Connell, M.J. et al., Phase II trial of maytansine in patients with advanced colorectal carcinoma, *Cancer Treat. Rep.*, 62, 1237-1238, 1978.

69. Cabanillas, F. et al., Results of a phase II study of maytansine in patients with breast carcinoma and melanoma, *Cancer Treat. Rep.*, 63, 507-509, 1979.

70. Creagen, E.T. et al., Phase II evaluation of maytansine in patients with advanced head and neck cancer, *Cancer Treat. Rep.*, 63, 2061-2062, 1979.

71. Neidhart, J.A. et al., Minimal single-agent activity of maytansine in refractory breast cancer: A Southwest Oncology Group study, *Cancer Treat. Rep.*, 64, 675-677, 1980.

72. Ahmann, D.L. et al., Phase II study of maytansine and chlorozotocin in patients with disseminated malignant melanoma, *Cancer Treat. Rep.*, 64, 721-723, 1980.

73. Franklin, R. et al., A phase I-II study of maytansine utilizing a weekly schedule, *Cancer*, 46, 1104-1108, 1980.

74. Rosenthal, S. et al., Phase II study of maytansine in patients with advanced lymphomas: An Eastern Cooperative Oncology Group pilot study, *Cancer Treat. Rep.*, 64, 1115-1117, 1980.

75. Edmonson, J.H. et al., Phase II study of maytansine in advanced breast cancer, *Cancer Treat. Rep.*, 65, 536-537, 1981.

76. Borden, E.C. et al., Phase II evaluation of dibromodulcitol, ICRF-159, and maytansine for sarcomas, *Am. J. Clin. Oncol.*, 5, 417-420, 1982.

77. Creech, R.H. et al., Phase II study of cisplatin, maytansine, and chlorozotocin in small cell lung carcinoma (EST 2578), *Cancer Treat. Rep.*, 66, 1417-1419, 1982.

78. Ratanatharathorn, V. et al., Phase II evaluation of maytansine in refractory non-Hodgkin's lymphoma: A Southwest Oncology Group study, *Cancer Treat. Rep.*, 66, 1587-1588, 1982.

79. Thigpen, J.T. et al., Phase II study of maytansine in the treatment of advanced or recurrent adenocarcinoma of the ovary: A Gynecology Oncology Group study, *Am. J. Clin. Oncol.*, 6, 273-275, 1983.

80. Edmonson, J.H. et al., Phase II study of maytansine in advanced sarcomas, *Cancer Treat. Rep.*, 67, 401-402, 1983.

81. Thigpen, J.T. et al., Phase II study of maytansine in the treatment of advanced or recurrent squamous cell carcinoma of the cervix, *Am. J. Clin. Oncol.*, 6, 427-430, 1983.

82. Sabio, H. et al., Maytansine in refractory childhood acute lymphocytic leukemia: A Pediatric Oncology Group study, *Cancer Treat. Rep.*, 67, 1045, 1983.

83. Ravry, M.J., Omura, G.A., and Birch, R., Phase II evaluation of maytansine (NSC 153858) in advanced cancer: A Southwestern Cancer Study Group trial, *Am. J. Clin. Oncol.*, 8, 148-150, 1985.

84. Kalser, M.H. et al., Phase II trials of maytansine, low-dose chlorozotocin, and high-dose chlorozotocin as single agents against advanced measurable adenocarcinoma of the pancreas, *Cancer Treatment Rep.* 69, 417-420, 1985.

85. Anon., Annual Report to the FDA by DCT, NCI, on Maytansine, NSC 153858, IND 11857, February 1984.

86. Chari, R.J. et al., Immunoconjugates containing novel maytansinoids: Promising anticancer drugs, *Cancer Res.*, 52, 127-131, 1992.

87. Chari, R.J. et al., Cytotoxic agents comprising maytansinoids and their therapeutic use, ImmunoGen, Inc., USA, U.S. Patent 5,208,020 (1993).

88. Chari, R.J. and Widdison, W.C., Process for the preparation and purification of thiol-containing maytansinoids, ImmunoGen, Inc., USA, U.S. Patent 6,333,410 (2001).

89. Liu, C. et al., Eradication of large colon tumor xenografts by targeted delivery of maytansinoids, *Proc. Natl. Acad. Sci. USA*, 93, 8618-8623, 1996.

90. ImmunoGen Web site. Available at: http://www.immunogen.com.

91. Liu, C. and Chari, R.J., The development of antibody delivery systems to target cancer with highly potent maytansinoids, *Expert Opin. Investig. Drugs*, 6, 169-171, 1997.

92. Johnson, R., Progress in targeted anticancer therapy using immunotoxins, *Pharmaceut. J.*, 265, 490, 2000.

93. Erickson, S. and Schwall, R., Humanized anti-ErbB2 antibody-maytansinoid conjugates and uses thereof in cancer therapy, Genentech, Inc., WO 2000, US 17229, 2001.

94. Ross, S. et al., Prostate stem cell antigen as therapy target: Tissue expression and *in vivo* efficacy of an immunoconjugate, *Cancer Res.*, 62, 2546-2553, 2002.

95. Ladino, C.A. et al., Folate-maytansinoids: Target-selective drugs of low molecular weight, *Int. J. Cancer*, 73, 859-864, 1997.

96. Hatano, K. et al., Biosynthetic origin of aminobenzenoid nucleus (C7N-unit) of ansamitocin, a group of novel maytansinoid antibiotics, *J. Antibiot.*, 35, 1415-1417, 1982.

97. Hatano, K. et al., Biosynthetic origin of the *ansa*-structure of ansamitocin P-3, *Agric. Biol. Chem.*, 49, 327-333, 1985.

98. Hatano, K. et al., Selective accumulation of ansamitocins P-2, P-3 and P-4, and biosynthetic origins of their acyl moieties, *Agric. Biol. Chem.*, 48, 1721-1729, 1984.

99. Yu, T.-W. et al., Mutational analysis and reconstituted expression of the biosynthetic genes involved in the formation of 3-amino-5-hydroxybenzoic acid, the starter unit of rifamycin biosynthesis in *Amycolatopsis mediterranei* S699, *J. Biol. Chem.*, 276, 12546-12555, 2001, and references therein.

100. Guo, J. and Frost, J.W., Kanosamine biosynthesis: A likely source of the aminoshikimate pathway's nitrogen atom, *J. Am. Chem. Soc.*, 124, 10642-10643, 2002.

101. Arakawa, K. et al., Characterization of the early stage aminoshikimate pathway in the formation of 3-amino-5-hydroxybenzoic acid: The RifN protein specifically converts kanosamine into kanosamine 6-phosphate, *J. Am. Chem. Soc.*, 124, 10644-10645, 2002.

102. Yu, T.-W. et al., The biosynthetic gene cluster of the maytansinoid antitumor agent ansamitocin from *Actinosynnema pretiosum*, *Proc. Natl. Acad. Sci. USA*, 99, 7968-7973, 2002.

103. Carroll, B.J. et al., Identification of a set of genes involved in the formation of the substrate for the incorporation of the unusual "glycolate" chain extension unit in ansamitocin biosynthesis, *J. Am. Chem. Soc.*, 124, 4179-4177, 2002.

104. Wu, K. et al., The FK520 gene cluster of *Streptomyces hygroscopicus* var. *ascomyceticus* (ATTC 14891) contains genes for biosynthesis of unusual polyketide extender units, *Gene*, 251, 81-90, 2000.

105. Kato, Y. et al., Functional expression of genes involved in the biosynthesis of the novel polyketide chain extension unit, methoxymalonyl-acyl carrier protein, and engineered biosynthesis of 2-desme-thyl-2-methoxy-6-deoxyerythronolide B, *J. Am. Chem. Soc.*, 124, 5268-5269, 2002.
106. Reeves, C.D. et al., A new substrate specificity for acyl transferase domains of the ascomycin polyketide synthase in *Streptomyces hygroscopicus*, *J. Biol. Chem.*, 277, 9155-9159, 2002.
107. Moss, S.J. et al., Identification of Asm19 as an acyltransferase attaching the biologically essential ester side chain of ansamitocins using *N*-desmethyl-4,5-desepoxymaytansinol, not maytansinol, as its substrate, *J. Am. Chem. Soc.*, 124, 6544-6545, 2002.
108. Spiteller, P. et al., The post-polyketide synthase modification steps in the biosynthesis of the antitumor agent ansamitocin by *Actinosynnema pretiosum*, *J. Am. Chem. Soc.*, 125, 14236-14237, 2003.
109. Kutney, J.P. et al., Isolation and characterization of natural products from plant tissue cultures of *Maytenus buchananii*, *Phytochemistry*, 20, 653-657, 1981.
110. Dymowski, W. and Furmanowa, M., Searching for cytostatic substances in plant tissue of *Maytenus molina* by *in vitro* culture. III. Release of substances from active biological fractions from the callus extract of *Maytenus wallichiana* R. and B., *Acta Pol. Pharm.*, 47, 51-54, 1990.
111. Groth, I. et al., *Kitasatospora putterlickiae* sp. *nov.*, isolated from rhizosphere soil, transfer of *Streptomyces kifunensis* to the genus *Kitasatospora* as *Kitasatospora kifunensis* comb. nov., and amended description of *Streptomyces aureofaciens* Duggar 1948, *Int. J. Syst. Evol. Microbiol.*, 53, 2033-2040, 2003.

18 Benzoquinone Ansamycins

Kenneth M. Snader

CONTENTS

I. Introduction 339
II. Mechanism of Action..340
III. Synthesis of the Drug 344
IV. Medicinal Chemistry of the Drug...350
V. Clinical Applications...353
References 355

I. INTRODUCTION

Ansamycins have been defined as cyclic molecules with long aliphatic chains that connect two opposite points of an aromatic ring, like a handle or ansa (hence the name).[1] Most commonly the ansa ring is a macrocyclic lactam whose biosynthesis resembles that of other macrolides by being acetate derived. The aromatic rings are usually divided into two groups: those containing naphthalene rings, whose members include rifamycins, streptovaricins, tolypomycin, and naphthomycin, and those containing benzene rings. Furthermore, the most common, and usually the most biologically active, aromatic rings are quinones or hydroquinones. This discussion will be limited to the biologically active benzoquinone ansamycins, whose members include macbecins, herbimycins, and geldanamycin, with particular emphasis on geldanamycin and its analogs. Geldanamycin derivatives especially are enjoying a resurgence of interest as a result of the discovery of their novel mechanism of cytotoxicity and potential utility as anticancer agents.

The first of the benzoquinoid ansamycins to be discovered was geldanamycin (GA). It was isolated from a strain of *Streptomyces hygroscopicus* by the group at Upjohn,[2] and the structure was published the same year by the Rinehart group.[3] GA showed modest activity against Gram-positive organisms and some fungi and parasites, but most interesting was the cytotoxicity that was also demonstrated against various cancer cell lines *in vitro*. Indeed, GA has shown a broad range of strong activity against the National Cancer Institute (NCI) panel of 60 human tumor cell lines.[4] The structure was determined,[3] and the absolute configuration has been confirmed by x-ray crystallographic determination[5] to be 6(S)-, 7(S)-, 10(S)-, 11(R)-, 12(S)-, 14(R)-, as shown in Figure 18.1. Its strong *in vivo* activity in several models prompted its initial clinical trial, but that was discontinued because of hepatotoxicity, which was further elaborated on in a pharmacological evaluation at the NCI.[6]

Although the positions of the substituents listed in Figure 18.1 are reported in a consistent manner in the papers published on the benzoquinone ansamycins, the designation of the relative stereochemistry of some of the substituents, particularly those at C14 and C15, varies between papers. These structural variations are shown in Scheme 18.1 (note that the references cited in Scheme 18.1 are not in numerical sequence) and are also reflected in the discussions of the synthetic approaches presented in this review.

Not long after the initial discovery, two other benzoquinoid ansamycins were discovered. In 1979, herbimycin A was discovered in another strain of *Streptomyces hygroscopicus* var. *geldanus*,[7]

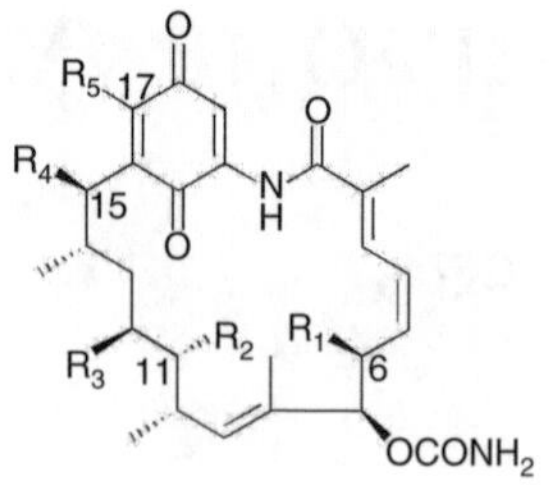

	R1 (C6)	R2 (C11)	R3 (C12)	R4 (C15)	R5 (C17)
Geldanamycin	-OMe	-OH	-OMe	-H	-OMe
Herbimycin A (TAN 420F)	-OMe	-OMe	-OMe	-OMe	-H
Herbimycin B	-OMe	-OH	-OMe	-H	-H
Herbimycin C (TAN 420D)	-OMe	-OH	-OMe	-OMe	-H
Macbecin I	-Me	-OMe	-OMe	-OMe	-H
Macbecin II (hydroquinone)	-Me	-OMe	-OMe	-OMe	-H
TAN 420A (hydroquinone)	-OMe	-OH	-OH	-OMe	-H
TAN 420B	-OMe	-OH	-OH	-OMe	-H
TAN 420C (hydroquinone)	-OMe	-OH	-OMe	-OMe	-H
TAN 420E (hydroquinone)	-OMe	-OMe	-OMe	-OMe	-H

FIGURE 18.1 Geldanamycin and related compounds.

and x-ray crystallographic studies[8] proved it to be quite similar to GA, with the same macrolide ring system but differing in the substituents at C11, C15, and C17. Not unsurprising was the retention of chirality between the two compounds in all shared chiral centers. Two other isomers, herbimycin B and herbimycin C, were also isolated and differed from the parent by variations at C11 and C15, as shown in Figure 18.1. Herbimycin B is also known as 17-demethoxygeldanamycin. Although somewhat less potent in *in vitro* studies, the herbimycins still share many of the same biological activities with GA.[9]

A third benzoquinoid ansamycin to be discovered was macbecin I, together with its hydroquinone form, macbecin II. This compound was produced by a *Nocardia sp.* and showed a biological activity profile similar to GA.[10,11] Once again the macrolide ring system and stereochemistry were the same as GA, with differences only at C6, C11, C15, and C17, as shown in Figure 18.1. It is not clear whether the hydroquinone is deliberately produced by the organism or is an artifact of the fermentation and isolation procedure. Because the quinones are more stable chemical entities when exposed to air, they have been most commonly studied, but the hydroquinones can be isolated under the appropriate conditions.

In 1986 another series of ansamycins were isolated from a *Streptomyces hygroscopicus* strain and reported as the TAN 420 series.[12] Isomers A, C, and E were isolated as the hydroquinones, and their corresponding quinone forms were reported as B, D, and F. TAN 420D was shown to be identical to herbimycin C, and TAN 420F was shown to be identical to herbimycin A. Only TAN 420A and its quinone, TAN 420B, were new isomers, differing from the reported herbimycin structures by having an hydroxy group at C12 instead of the more usual methoxy group.

II. MECHANISM OF ACTION

The original discovery of the antitumor activity of GA prompted a study of its mechanism of action. Initial reports were that the benzoquinone ansamycins were tyrosine kinase inhibitors[4,13–15] although they also demonstrated several other activities, including virucidal,[16] ischemia protection,[17] herbicidal,[7] antiprotozoal,[2] and antimalarial.[18] The presence of a quinone ring as part of the GA structure

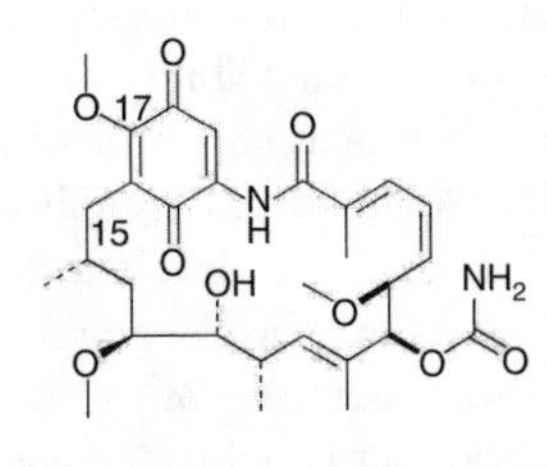

Geldanamycin after Andrus et al.[39]
DNP[56] has 6 & 7 as Herbimycin B

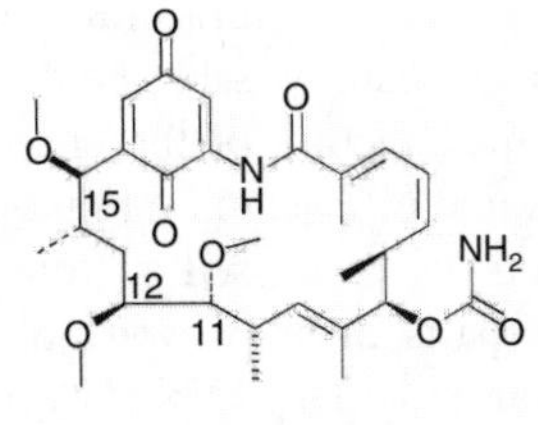

Macbecin I
after Panek et al.[41]

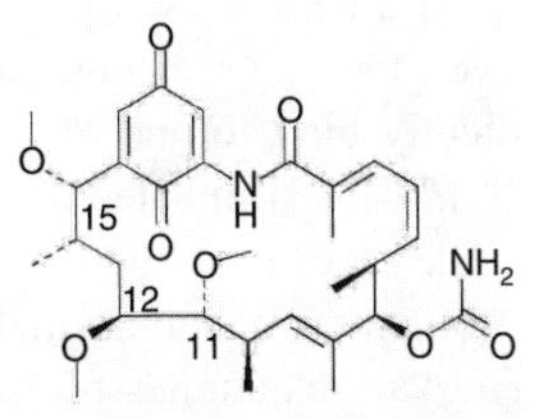

Macbecin I, Antibiotic C14919E1
after Baker et al.[33] & DNP
Geldanamycin numeration

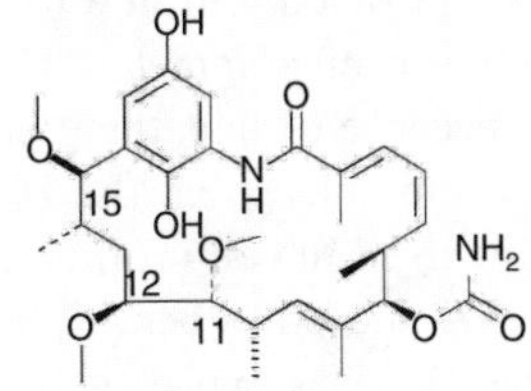

Macbecin II
after Panek et al.[41]

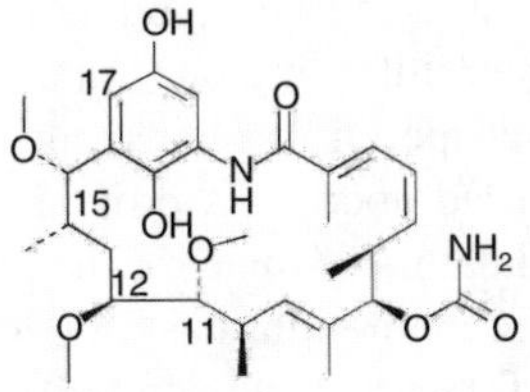

Macbecin II, Antibiotic C14919E2
after Baker et al.[33] & DNP
Geldanamycin numeration

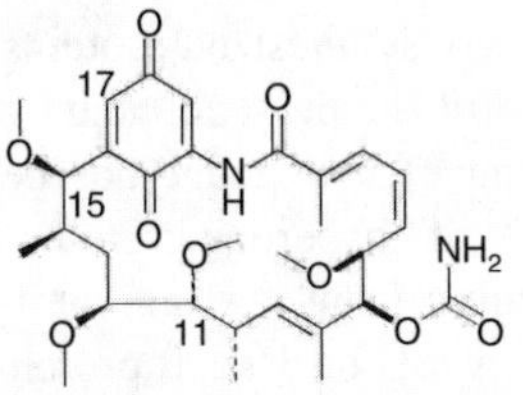

Herbimycin A, Antibiotic TAN 420F
after Martin et al.[57]
DNP has 6 & 7 as Herbimycin B

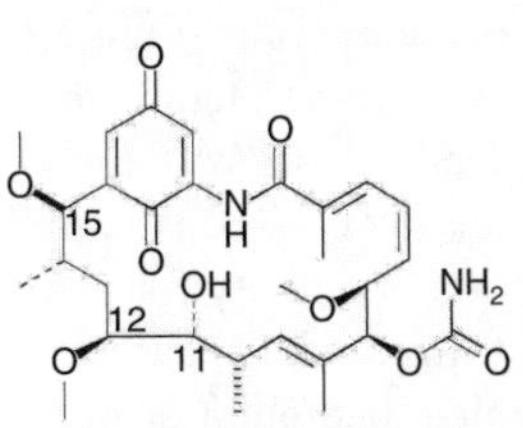

Herbimycin A
after Panek et al.[41]

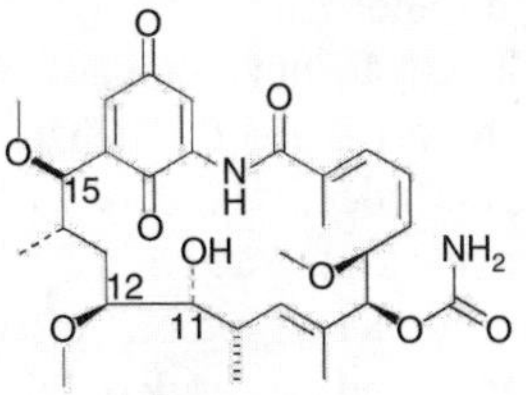

Herbimycin B
after Panek et al.[41]

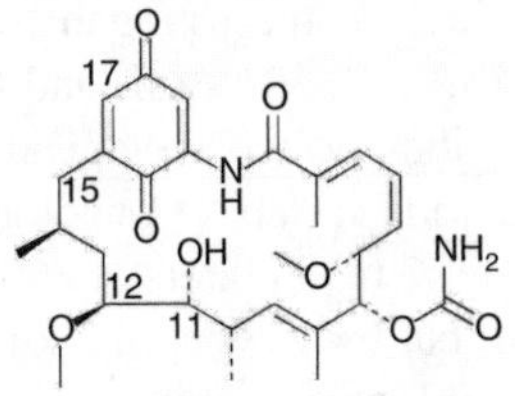

Herbimycin B
DNP[56]

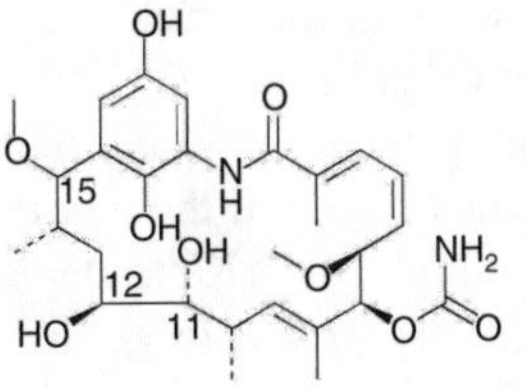

Antibiotic TAN 420A
DNP[56]

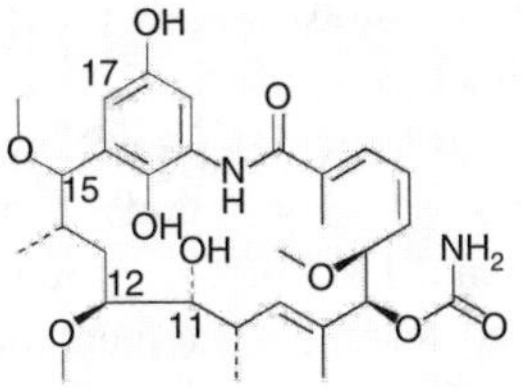

Antibiotic TAN 420C
DNP[56]

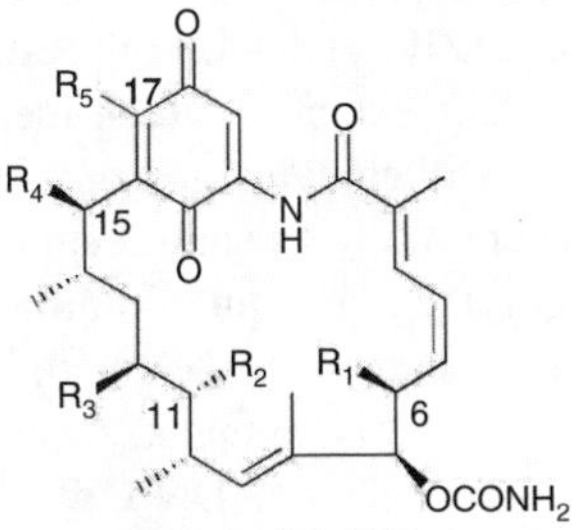

Antibiotic TAN 420B
DNP[56]

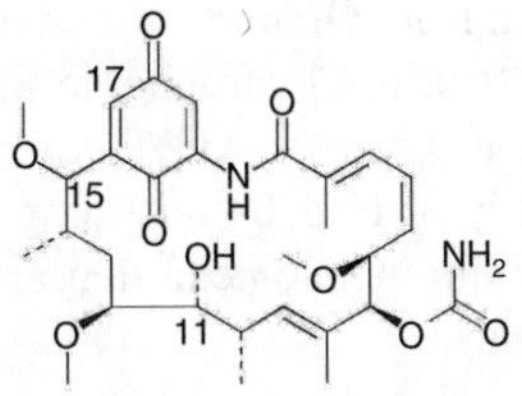

Herbimycin C, Antibiotic TAN 420D
DNP[56]

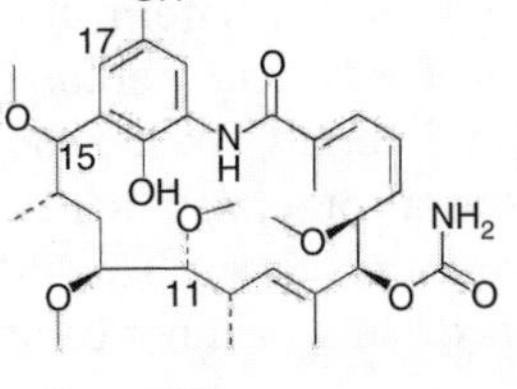

Antibiotic TAN 420E
DNP[56]

SCHEME 18.1 Benzoquinone ansamycins.

prompted a detailed study on the generation of intracellular free radicals through redox cycling.[19] However, the major change in emphasis occurred with the discovery of the unique ability of GA to specifically bind to and antagonize the function of the chaperone protein, heat shock protein 90 (Hsp90).[4] This stimulated a large volume of research into the role of these mediators in replicating cells. The fact that Hsp90 is constitutively expressed at from 2 to 10 times higher levels in tumor cells, compared to their normal counterparts, makes it even more important as an anticancer target.[20,21]

The various studies on the inhibition of Hsp90 by GA and the other members of the benzo-quinoid ansamycins have been thoroughly reviewed,[20,22] but a few important conclusions and the current hypothetical model can be restated here. Chaperone proteins are regulatory molecules. They control the correct assembly of other polypeptides into active molecules without becoming incorporated into the final structure. They are generally thought of as being involved with the folding of newly synthesized proteins or in their unfolding once their function has been completed. Hsp90 is one of the most abundant of these chaperone proteins, and perhaps because of this, the greatest amount of knowledge has been acquired about this protein. It has become clear that the Hsp90 family of chaperone proteins are modulators of many of the proteins involved in cell proliferation, including many tyrosine and serine/threonine kinases. The action of GA and other benzoquinoid ansamycins on Hsp90 proteins would then also show an effect on these kinases, which was their originally observed action.

In their reviews, both Neckers and Ochel note that Hsp90 also regulates several other "client proteins" that are intimately involved with the proliferation and survival of cancer cells (Figure 18.2). These include the estrogen receptor, the androgen receptor, mutant p53, the Src family kinases, p185erbB2, Raf-1/v-Raf, and CDK4/CDK6. It thus becomes more understandable why the earlier publications on the biological activity reported for the benzoquinoid ansamycins could implicate such a wide variety of oncologic markers. Indeed, the very earliest reports of activity for GA were that they were inhibitors of p60v-src tyrosine kinase activity.[23,24] It now fits the model not as a primary activity but by viewing the kinase inhibition as an indirect effect, in which disruption of the p60v-src-Hsp90 complex mimics Hsp90 depletion and causes destabilization and loss of the kinase protein.[4] It also helps to explain the very high potency of these compounds, which are frequently active at nanograms per milliliter in *in vitro* cell systems such as rapidly proliferating leukemia cell lines like P388, L1210, and KB cells. Affecting numerous targets provides an amplification factor in GA's action. The extent to which the heat shock proteins regulate all the proteins in the cell replication cycle is still evolving and will undoubtedly provide research opportunities for some years to come.

There has been significant effort in attempting to understand the exact nature of the interaction between the benzoquinoid ansamycins and Hsp90. The group at Memorial Sloan Kettering[25] was able to cocrystallize GA with an amino terminal fragment of human Hsp90 and to publish the structure of this complex. At approximately the same time, another group[26] described a structural portion of bacterial DNA gyrase that contained the ATP binding sites and has an amino acid sequence that is nearly identical to that contained in the Hsp90 sequence, which binds ansamycins. In addition, the ATP binding region is also fully contained within the benzoquinoid ansamycin binding region. This information, together with the binding study data from Grenert's laboratory,[27] has led to the current working model in which ATP binds to Hsp90 in the hydrophobic pocket studied by Stebbins, and the conformation adopted by the model indicates that the ATP/GA binding site acts as a conformational switch region that regulates the assembly of Hsp90-containing multichaperone complexes. This model also implies that any Hsp90 binding phenomena that are not ATP dependent would not be affected by the benzoquinoid ansamycins.

FIGURE 18.2 Mechanistic role of Hsp90.

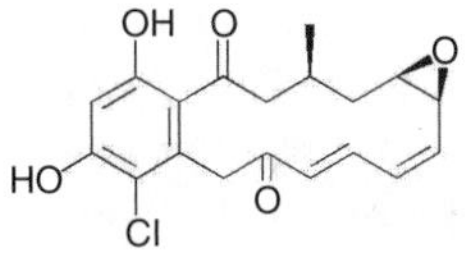

FIGURE 18.3 Radicicol.

One other observation that has the potential to provide substantial medicinal chemistry development is the discovery[28-30] of a non-ansamycin compound, radicicol (Figure 18.3), a fungal product that also binds to Hsp90 and competes with GA for binding to the amino-terminal binding pocket of Hsp90. This entry into the molecular modeling pool indicates that substantial latitude is available for more synthetic modifications and possibly some less complex structures — an ideal model for combi-chem approaches. Further encouragement can be derived from the recent report that a small purine, PU-3 (Figure 18.4), competes with GA in binding assays to Hsp90, although the relative affinity has only reached the 15–20 μM level so far.[31]

A full discussion of the structure of heat shock proteins is not within the scope of this review but can be found in one of the two excellent reviews already published.[20,22] Some specifics relating to individual proteins identified with cancer cells are appropriate. Hsp90 exists in two forms, α and β. These forms are dimeric in the cytosol of unstressed cells but accumulate in the nucleus when the cell is stressed. Hsp90 interacts with 14 proteins that have been identified as important in cancer because of their involvement in the proliferation or survival of cancer cells.

The original reference to the interaction of benzoquinoid ansamycins with Hsp90 was from the National Institutes of Health.[4] They began with identifying a high correlation between biological effects and the ability to bind to Hsp90. Three groups independently identified the Hsp90 binding site for benzoquinoid ansamycins. Serendipitously, the pooled results of these studies showed that the hydrophobic binding pocket for GA-Hsp90 also fully contains the ATP-binding sites and thus offered an explanation for the effect of nucleotide binding (ATP) to Hsp90. The key to benzoquinoid ansamycin sensitivity is target protein destabilization, which is generally mediated by proteosomes.

One of the major proteins in defining cancer cells is p53 protein. The current working hypothesis suggests that normal (wild-type) p53 has a tumor suppressant effect. A mutated p53 protein reacts with Hsp90/p23 complexes to achieve their correct conformation. GA interferes with this process and thus destabilizes mutant p53 to help restore the normal transcriptional activity of wild-type p53.

The transforming protein of Rous sarcoma virus (p60v-src) is a tyrosine kinase. It is complexed with Hsp90 and another protein and is inactive as a kinase in this form to allow transport to the plasma membrane, where it then becomes active. Presence of GA disrupts this complex, causing destabilization and loss of the kinase protein. This has been offered as a partial explanation of the original observations that GA was a tyrosine kinase inhibitor, although it may be an insufficient reason. It has been shown[32] that a hybrid of the phosphoinositol-3 kinase (PI3K) inhibitor LY294002 with GA is a direct inhibitor of DNA-dependent protein kinase (DNA-PK) and is over two orders of magnitude more active against DNA-PK than PI3K (see Section IV).

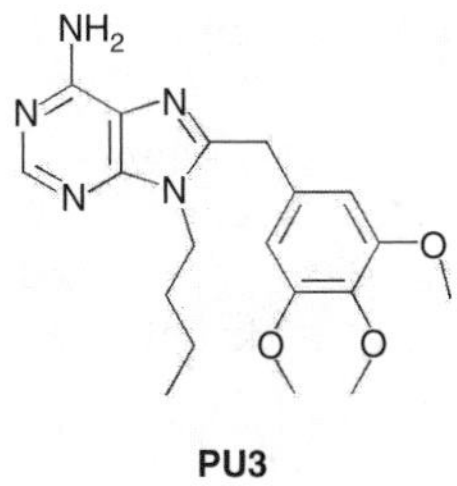

PU3

FIGURE 18.4 PU-3.

In prostate and breast cancers, p185erbB2 (a receptor tyrosine kinase) is overexpressed, and GAs were shown to stimulate its rapid degradation. They cause rapid polyubiquitination of this kinase, followed by proteosome-dependent degradation. The p185erbB2 protein associates with another protein, Grp 94 (a Hsp90 homolog), and GAs bind to the Grp94. The full effect of Hsp90 in cancer cells is not yet fully elaborated; however, several important points can be made: Hsp90 is up-regulated in many tumor cells both *in vivo* and *in vitro*; Hsp90 has been correlated with poor survival in breast cancer; Hsp90 is essential for normal eukaryotic cell survival, and cancer cells seem to be especially sensitive to interference; and Hsp90 interference may also be of value in those instances in which steroid hormone antagonism is indicated, such as hormone-dependent prostate cancer and breast cancer.

Ochel[22] notes that there are at least three functions for Hsp90 proteins: they prevent aggregation and promote refolding of proteins after various stresses, they suppress phenotypic traits that only become expressed after certain stresses, and they stabilize certain key regulatory proteins such as progesterone and gluco-corticosteroid receptors. He also notes that the role of p53 in cancer cells is to act as a tumor suppressor but is mutated or absent in about one-half of all human cancers. It has been clearly demonstrated that treatment with GA results in destabilization of mutated p53, although it has no effect on wild-type p53. Thus, the benzoquinone ansamycins are able to selectively destabilize mutated p53.

III. SYNTHESIS OF THE DRUG

The synthesis of all three of the benzoquinoid ansamycins has been accomplished. The first total synthesis of (+)-macbecin I was published by Baker in 1990[33] and was rapidly followed by Evans[34] and Panek.[35] In addition, a formal total synthesis was also published by Martin.[36] Similarly, herbimycin was also prepared by Tatsuta et al.[37,38] However, it took until 2002 for the synthesis of GA to be accomplished.[39]

With seven stereogenic sites on the macrocyclic lactam, an isolated trisubstituted double bond, and a (Z,E) conjugated diene lactam, these molecules are a synthetic challenge that bring out the most creative solutions. There are at least three different methods for introducing the chiral centers, and a brief outline of how these centers were introduced is presented here. None of the syntheses were intended to replace the fermentation process for production of the parent antibiotic, but all of them offer the opportunity to examine the individual elements of the molecule for their relevance in the molecular binding to Hsp90 and subsequent enhancement of activity. Presented here is a drastically shortened description of the synthetic strategies of three of these syntheses intended to illustrate the different chiral reagents used for introduction of the stereochemical centers.

For the construction of macbecin I, the Evans group[34,40] used chiral boron enolate methodology to introduce the chiral centers. The molecule was broken down into three subunits, each containing appropriate stereochemistry, as shown in Figure 18.5. The first synthon was prepared from the benzaldehyde **4** plus the chiral oxazolidinone **5** as the boron enolate made with di-n-butyl boronate triflate (n-Bu$_2$BOTf). The chirality derived from the selection of 4R,5S-norephedrine as starting material for the oxazolidinone established the centers at C15 and C14 of the ansa ring. Further elaboration of this intermediate led to the first synthon, **1** (Figure 18.6).

Using the same oxazolidinone and boron enolate condensation with trans-cinnamaldehyde, the chirality around C11 and C10 was established as shown. The chain was extended with appropriate phosphonate chemistry to lead to an intermediate aldehyde, **6**, which was again condensed with a third use of the oxazolidinone to fix the stereochemistry at C7 and C6. Cleavage of the double bond in **7** originally present in the cinnamaldehyde, followed by conversion to an aldehyde produced the second synthon **2** (Figure 18.7).

Coupling the two synthons, **1** and **2**, with TiCl$_4$, followed by Dess Martin oxidation, gave the precursor **8** for macbecin with all the stereochemistry from C15 through C5. All that remained was to extend the chain by building the cis,trans diene (C1–C4)–conjugated ester using appropriate

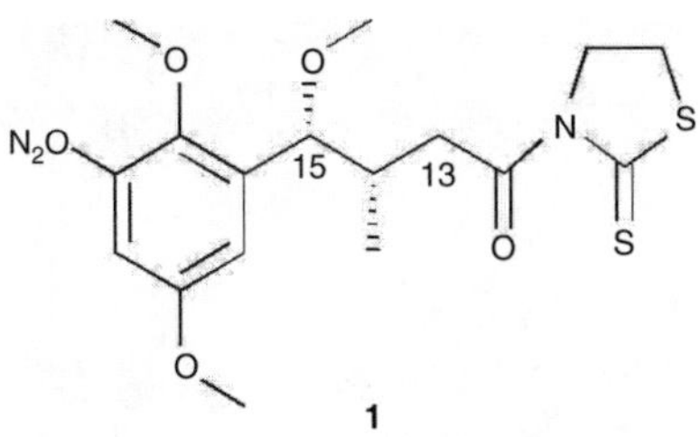

FIGURE 18.5 Evans synthesis synthons.

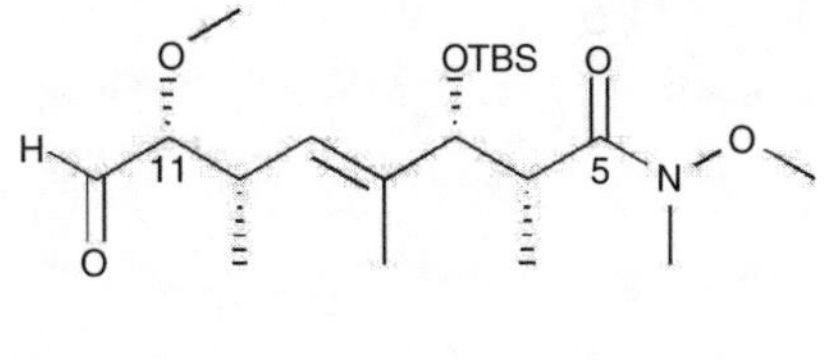

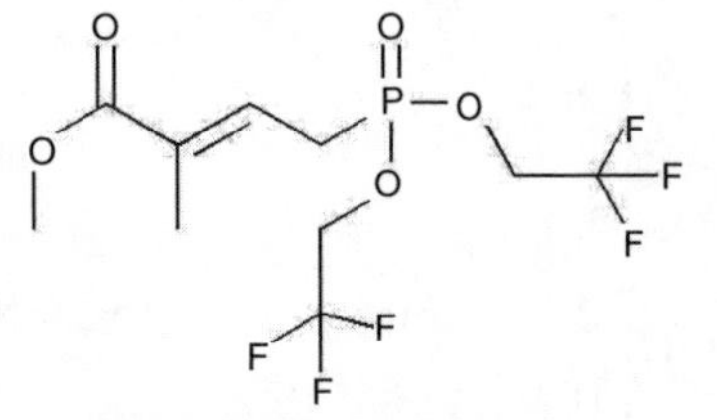

vinlogous phosphorus-based reagents to produce the open-chain intermediate **9**. Closure was accomplished by reduction of the nitro to amine, followed by ring closure using BOPCl and Hunig's base, as previously used by Baker,[33] to produce the des-carbamoyl macbecin I (**10**) (Figure 18.8).

Although the first method used a partial convergent approach, Panek's group[35,41] assembled the entire chain sequentially using three crotyl silation steps. Starting with the aromatic aldehyde dimethylacetal (**11**) and the (E)-crotylsilane reagent **12**, the stereochemistry at C15 and C14 was established. Hydroboration of the olefin led to the correct hydroxylation at the C12 position in a 10:1 ratio, and the chain was shortened by converting the terminal ester to an alcohol followed by cleavage of the diol to the aldehyde **13** (synthon 1) (Figure 18.9).

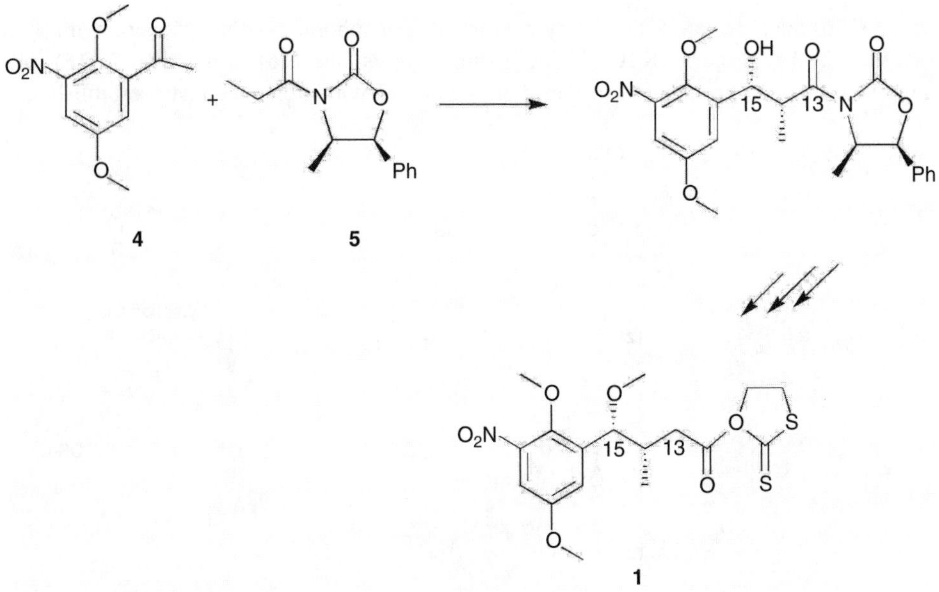

FIGURE 18.6 Evans synthesis synthon 1.

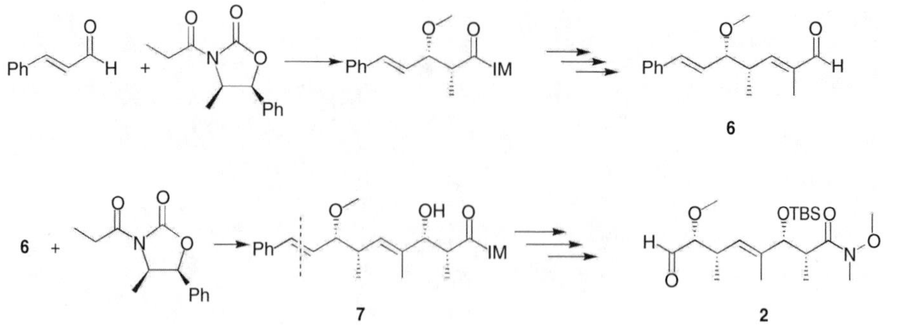

FIGURE 18.7 Evans synthesis synthon 2.

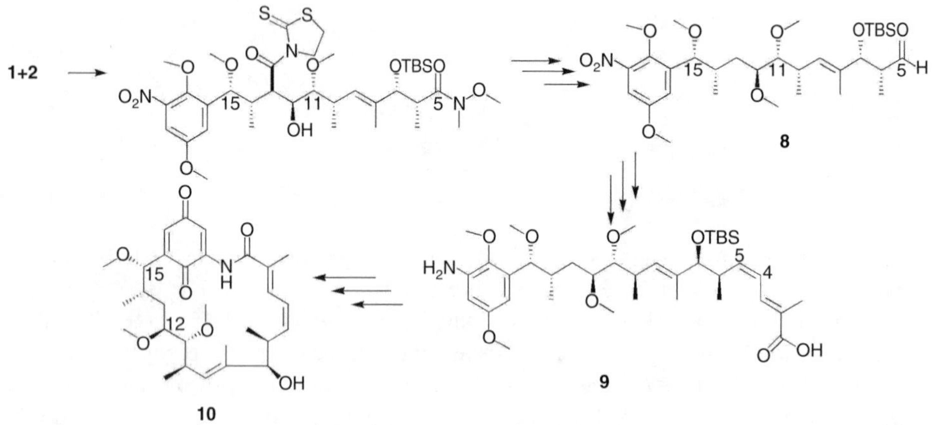

FIGURE 18.8 Evans synthesis: descarbamoyl-macbecin I.

Condensation of synthon 1 (**13**) with a second crotylsilane (**14**) gave the intermediate with stereochemistry established at C12 and C11 in a 12:1 diastereomeric ratio. Cleavage of the olefin by ozonolysis to the aldehyde followed by appropriate phosphorane olefin extension with pyruvate gave synthon 2 (**15**) (Figure 18.10). Using a third crotylsilane (**16**) on synthon 2 (**15**), plus an exchange of protective groups, led to the intermediate **17** with stereochemistry established at C7

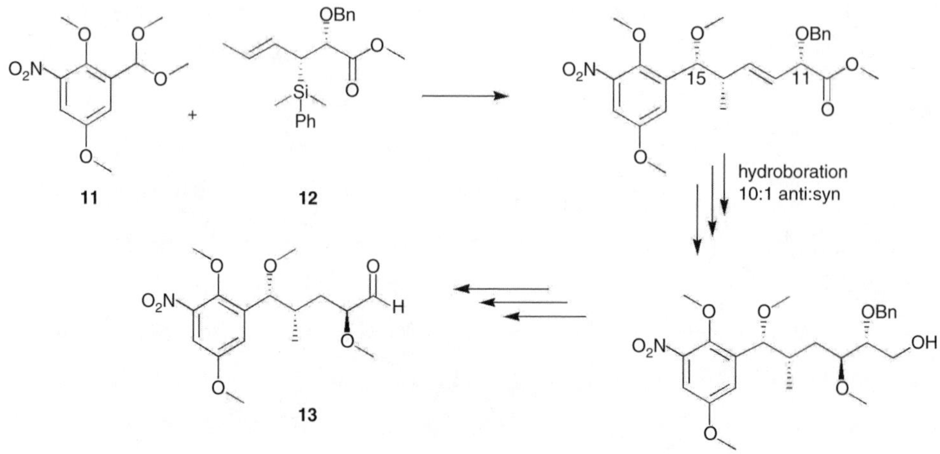

FIGURE 18.9 Panek synthesis synthon 1.

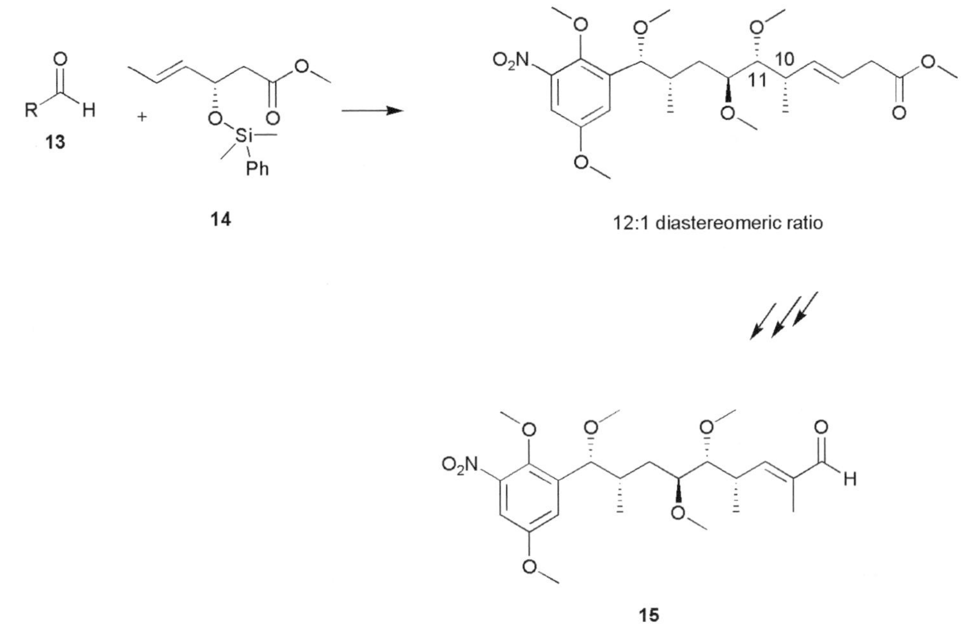

14

12:1 diastereomeric ratio

15

FIGURE 18.10 Panek synthesis synthon 2.

and C6. Cleavage of the olefin once again led to the aldehyde synthon 3 (**18**) (Figure 18.11). Once again, introduction of the olefins was done with appropriate phosphorane chemistry to give the open-chain nitro ester. Following the procedures that have precedence for final macrocyclic ring closure, developed for the maytansine ansamitocins, then completed the synthesis of descarbamoylmacbecin (**19**) (Figure 18.12).

As a final example, the recent preparation of GA [39,42,43] represents an even more difficult challenge. The chiral synthesis of geldanmycin is more complicated than that of macbecin and herbimycin by a further two additional features. First, the lack of an oxygen-bearing functionality at C15 discourages the use of an aldol reaction to form the C15–C14 bond. Second, the functionality at C11 is an hydroxyl rather than a methoxyl group, which implies that the synthesis must take care to protect and distinguish this hydroxyl from the other oxygenated function at C12. Finally, the unexpected behavior of the trimethoxy benzene precursor made the final oxidation to quinone much more complicated than the syntheses of the other benzoquinoid ansamysins. Nevertheless,

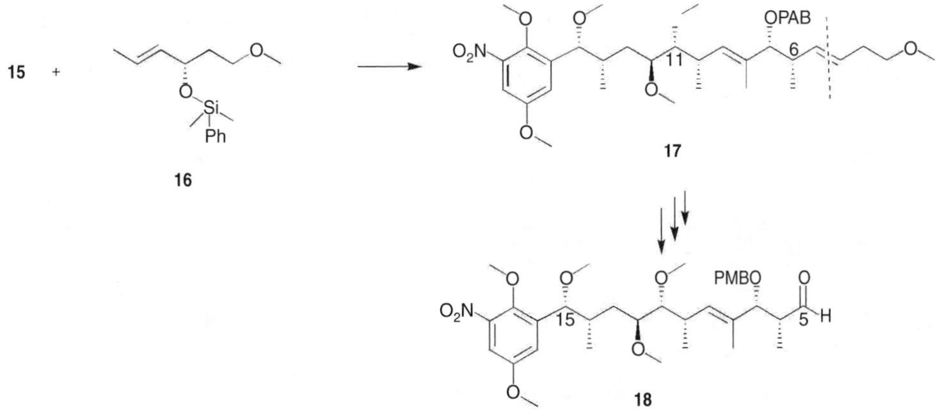

16

17

18

FIGURE 18.11 Panek synthesis synthon 3.

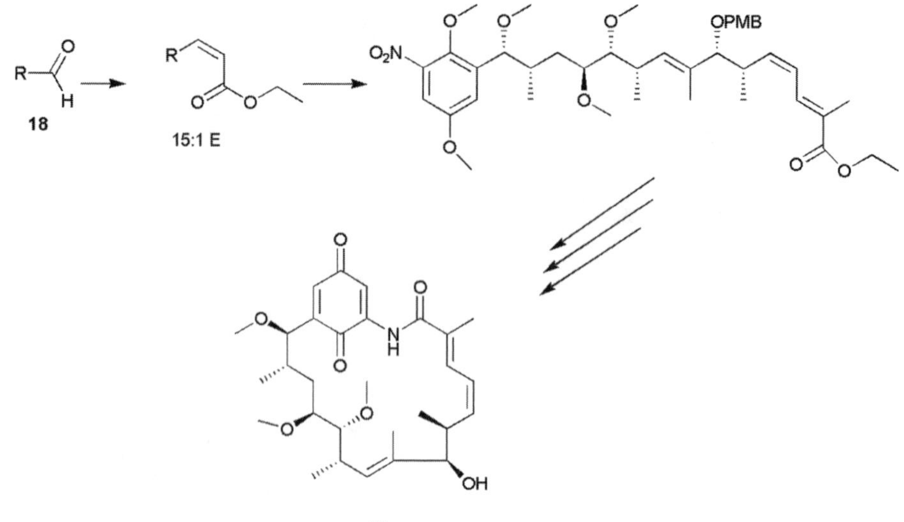

FIGURE 18.12 Panek synthesis; descarbamoyl-macbecin I.

the synthesis was successfully completed in 41 steps and produced several new compounds whose pharmacological properties should contribute significantly to the developing structure–activity studies between GA and Hsp90.

Once again, the synthesis was a sequential construct beginning with the trimethoxy aromatic **20**. The first chiral center was then introduced through the use of asymmetric Evans alkylation. The resulting oxazolidinone was subsequently converted to the chiral aldehyde, synthon 1 (**21**) (Figure 18.13). The next chiral center was introduced by aldol condensation with the dioxanone **22**. After rearranging protecting groups, methylation of the hydroxyl at C12 and reduction, the aldehyde synthon 2 (**23**) was obtained (Figure 18.14). The aldehyde was then homologated once again with an antiglycolate aldol condensation and then converted to the terminal methylene **24**. Careful hydroboration, followed by deprotection, gave the S configuration at the C10 position in **25**. Oxidation led to the aldehyde which, after condensation with pyruvate, provided the aldehyde synthon 3 (**26**) (Figure 18.15).

The aldehyde was then condensed once again to complete the chain, using a new (–)-norephedrine-derived glycolate, which was subjected to boron–enolate aldol condensation to give the syn adduct **27** in a ratio better than 20:1. Introduction of the Z and E olefin constructs followed routes similar to those of earlier investigators in macbecin and herbimycin syntheses and led to a carboxylate ester. The carboxylate residue was then elaborated into the protected allyl ester (synthon 4) **28** (Figure 18.16), which became the precursor for asymmetric hydroboration to the diol with the S configuration at C10.

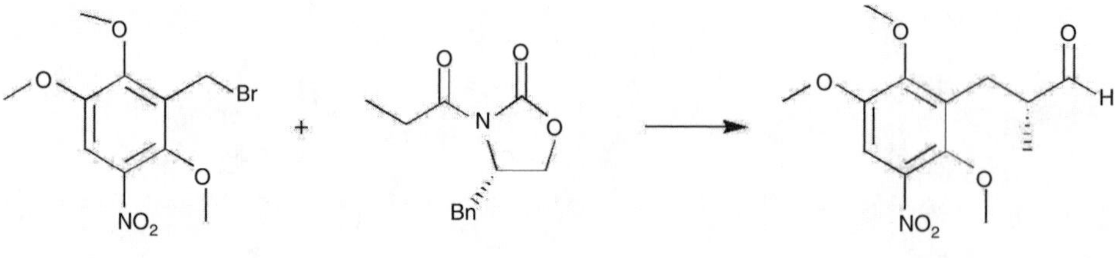

FIGURE 18.13 Andrus synthesis synthon 1.

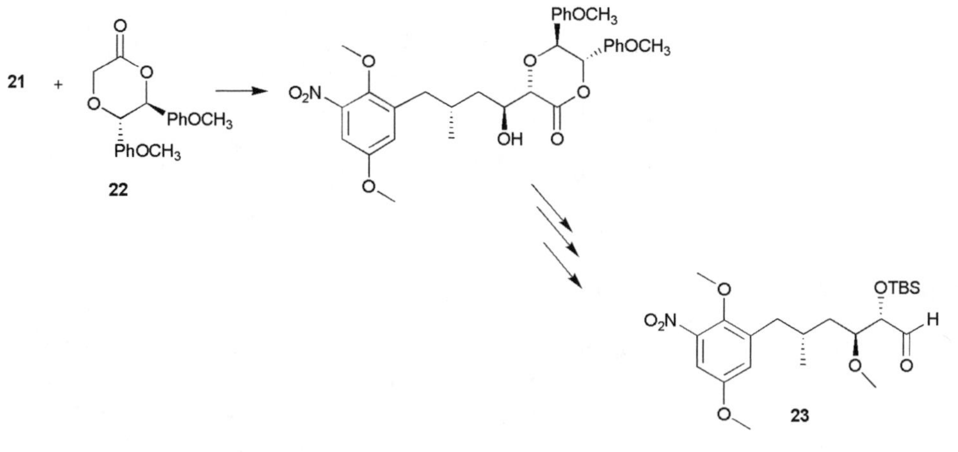

FIGURE 18.14 Andrus synthesis synthon 2.

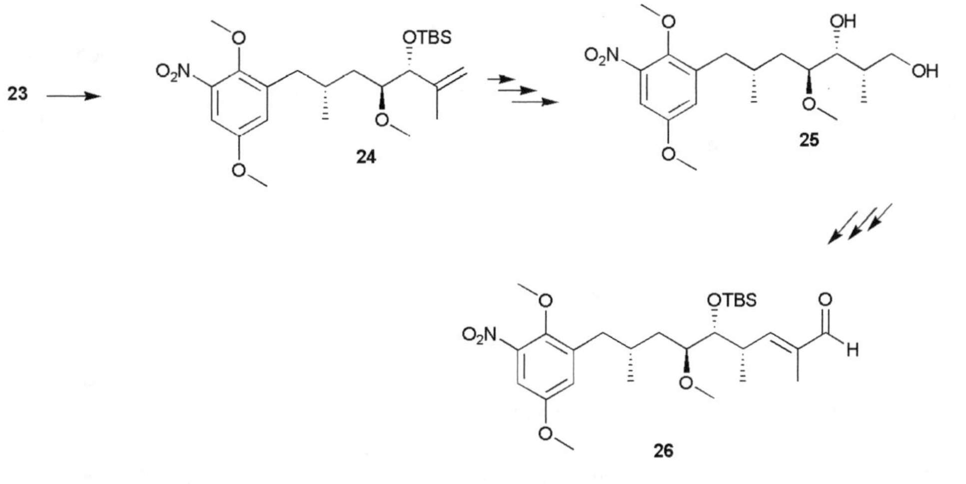

FIGURE 18.15 Andrus synthesis synthon 3.

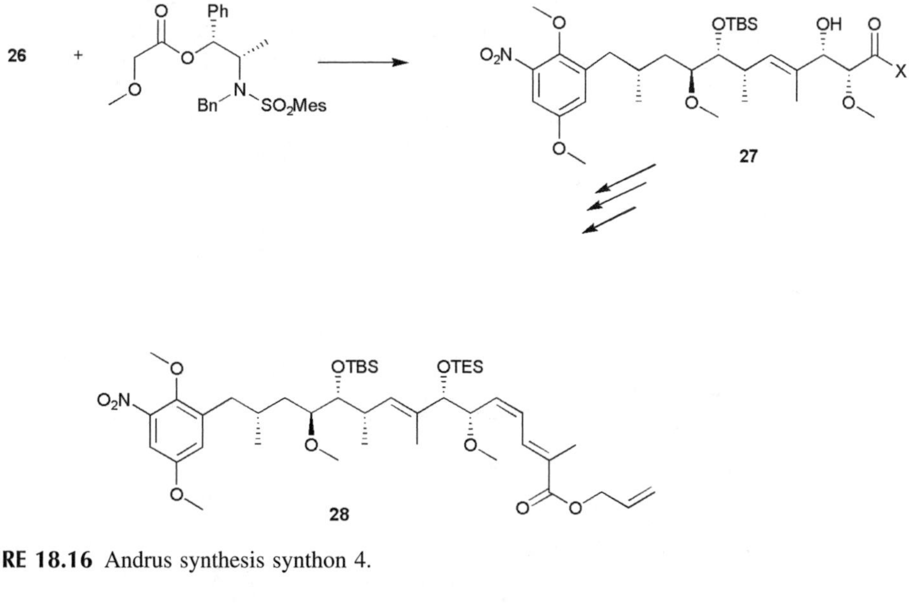

FIGURE 18.16 Andrus synthesis synthon 4.

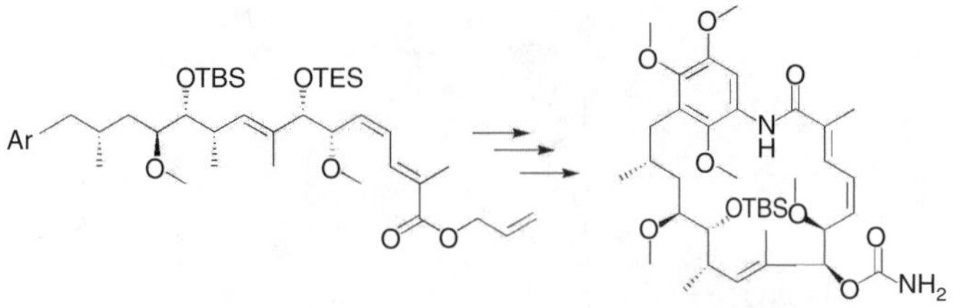

FIGURE 18.17 Andrus synthesis synthon 5.

Ring closure using the now-familiar macrocyclic lactamization procedures smoothly gave the trimethoxy macrocycle. The carbamoyl was attached using trichloroacetylisocyanate to avoid premature deprotection to give synthon 5 (**29**) (Figure 18.17). All that remained was the required oxidation to quinone. Conversion of the trimethoxy benzene to the p-quinone proved to be much more complex than the other benzoansamycins. Most oxidation techniques led to either azaquinone **30** or the orthoquinone **31** (Figure 18.18).

Ultimately, GA was obtained through a very brief exposure (~1 min) to nitric acid, followed by rapid quenching of the reaction in $NaHCO_3$. A careful examination of the electronic conformation of the hydroquinone system offered some explanation for this bias but did not lead to a good oxidation system. The side reactions led to at least two new ansamycin analogs for biological studies, although all of the other analogs were less cytotoxic in a cell-based assay known to be dependent on Hsp90 kinase signaling. The azageldanamycin showed significant activity and may be important in future design studies.

IV. MEDICINAL CHEMISTRY OF THE DRUG

Early studies on GA attempted to improve solubility of this hydrophobic molecule. A variety of prodrug esters and quinone derivatives were prepared[44] in an attempt to improve solubility and drug delivery. Unfortunately, none of these early modifications approached the potency of the parent and so were abandoned.

One of the first published, systematic evaluations of the structure–activity relationships of GA came from the group at Pfizer.[45] Using both GA and 4,5-dihydrogeldanamycin, which are readily prepared by fermentation, the group exploited the opportunity offered by the facile Michael addition that is possible at the C17 location on the quinone ring. They noted that it was possible to treat both GA and dihydrogeldanamycin with various amines to produce the 17-amino-17-desmethoxy analogs (**32**) (Figure 18.19).

It was only some time later that it was shown that the binding with Hsp90 is not affected by substitutions in this location of the molecule.[25] The group also prepared a series of derivatives at

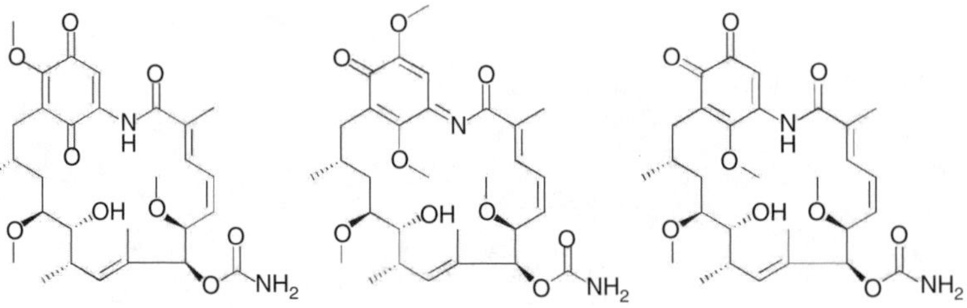

Geldanamycin	Azaquinone	*ortho*-Geldanamycin
	30	**31**

FIGURE 18.18 Andrus synthesis: geldanamycin and analogs.

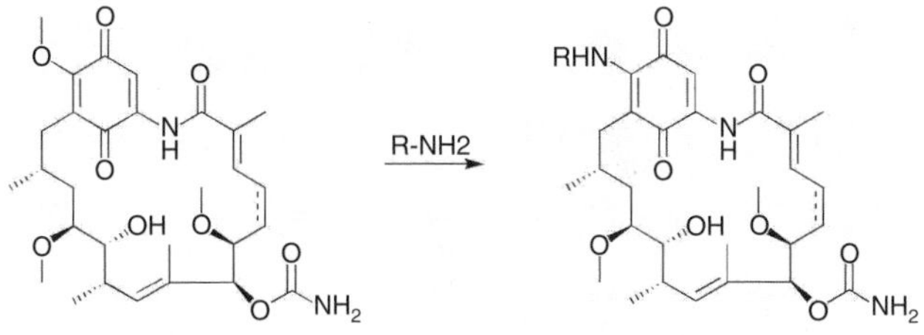

FIGURE 18.19 Pfizer aminogeldanamycin analogs.

both the C17 and C19 position by condensation with diamines. Activity was evaluated *in vitro* by measuring depletion of erB2 in SKBR cells. With this model they found that C17 substituents were best tolerated and so prepared a series of more than 80 derivatives. *In vitro* activity was followed by *in vivo* evaluation, and although the researchers noted that the two activities did not correlate well, 17-allylamino-17-desmethoxylgeldanamycin was one of the best candidates for further study. Indeed this analog is currently in phase I clinical trial in spite of difficulty in intravenous admin-istration because of the poor solubility characteristics. The group also studied a series of fused ring compounds similar to geldanoxazine (**33**) and geldanazine (**34**) that were originally reported by Rinehart (Figure 18.20).[44] In summary, the researchers found that these fused-ring compounds were not significantly active in their *in vivo* systems.

In a second discussion,[46] the authors expanded their description of the 17-amino derivatives including 17-AAG. Careful evaluation of the N-22 acylated analogs was also explored but was abandoned when it was found that they were readily cleaved *in vitro*. The researchers also explored the decarbamoylated products at C7, the 11-O-acyl derivatives, the C7 ketone, and the open ansa ring compound. From their studies, several conclusions were made: simple alkylation of the N22 substituent generally reduced activity, except for the phenacyl derivative, where the x-ray supported a hydrogen-bonded structure that preserved the ansa ring configuration; decarbamoylation at C7 generally reduced the *in vitro* activity; acylation of the C11 hydroxyl generally reduced the *in vitro* activity; opening the ansa ring always destroyed any biological activity; the ketone formed at C7 was inactive, but a ketone at C11 was quite potent *in vitro* — almost as good as the corresponding hydroxy analog; and herbimycins A and C (containing a methoxy group at C15 and a proton at C17) are active *in vitro* but are about fourfold less potent than GA.

Unfortunately, the Pfizer group did not have the benefit of x-ray crystal studies, which later demonstrated the binding of their analogs to a highly conserved pocket in the molecular chaperon Hsp90.[25,47,48] Among other observations, these studies showed that substitution at C17 of the GA structure was well tolerated by the Hsp90 binding. The group at Sloan-Kettering[49] exploited this tolerance by creating several dimers with molecules that were selected to improve their selective

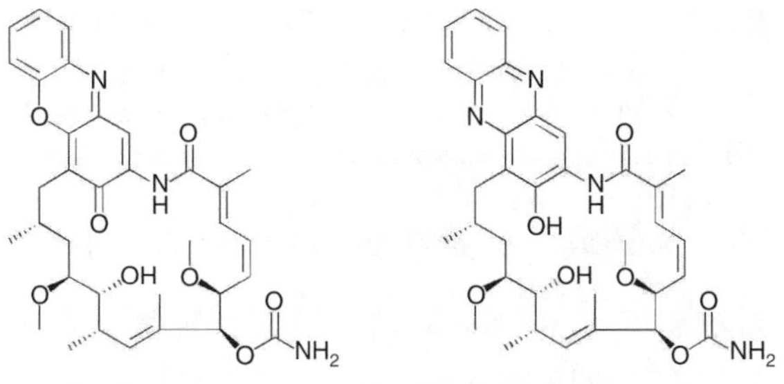

Geldanoxazine
33

Geldanazine
34

FIGURE 18.20 Geldanazine and derivatives.

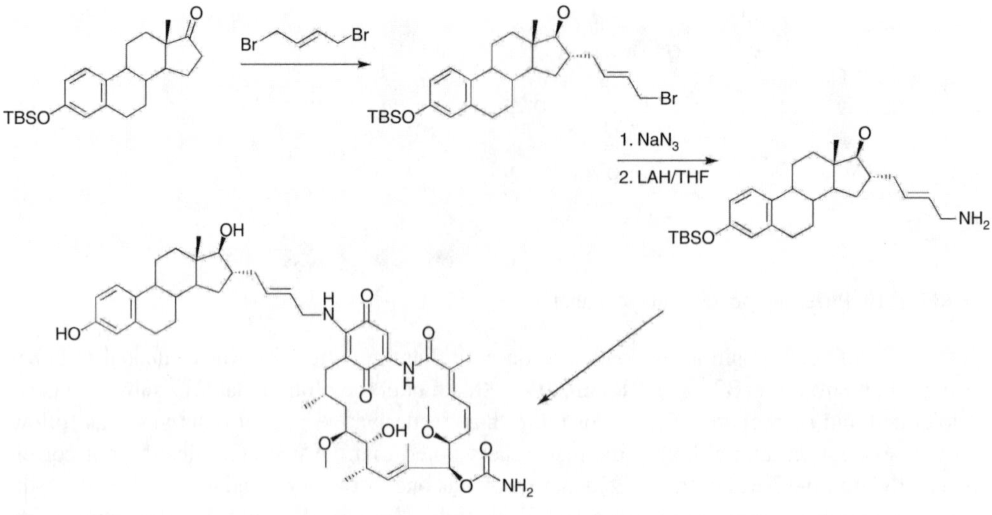

FIGURE 18.21 Memorial Sloan-Kettering estradiol heterodimers.

adsorption by specific cell types. The first attempt involved the construction of heterodimers linking GA to estradiol through an aminoalkyl group, which was prepared as shown in Figure 18.21.

A series of linkage chains were prepared and tested for their ability to affect the steady-state levels of HER2, ER (estrogen receptor), and Raf in MCF-7 breast cancer cells. HER2 is a transmembrane kinase that is amplified and overexpressed in a significant number of breast cancers. The researchers' studies showed that all of the heterodimers were less potent than the parent GA. The nature of the linker was important for selective activity. The best compound incorporated the 1,4-(2-butenyl) linker (**35**), which showed quite reasonable activity against the three marker proteins but was essentially inactive against IGF1R marker protein and had only marginal activity against Raf-1, both of which are not associated with the estrogen receptor. Also supporting the researchers' hypothesis was the demonstration that the analog was essentially inactive against a prostate cancer cell line, whereas GA induced the loss of an androgen receptor. With this success in hand, the group[50] prepared a series of heterodimers with a testosterone attachment, as shown in Figure 18.22.

In this case, activity was evaluated *in vitro* against LNCaP prostate cells. Once again, the nature of the linker was important for activity, and it was also important that the attachment of the testosterone molecule was linked to the C17α position of the steroid. The best analog of this series was (**36**; n = 4), which contained a six-carbon chain. It was substantially less active than GA against AR-independent cells (up to 7.5-fold) but was more potent that GA against the wild-type prostate cells (LAPC4), which strongly expresses the androgen receptor.

A third study[31] prepared a series of hybrid dimers between the phosphoinositol kinase (PI3K) inhibitors LY 294002 or LY292223 (Figure 18.23) and GA, with the intention that it would then affect the complex formed between the LY-binding protein and Hsp90 and would produce a better inhibitor of PI3Kinase. PI3Ks are critical in signal transduction, regulating p53 and certain DNA damage-repair mechanisms. The test system for evaluating biological effect involved direct inhibition of PI3K, which was purified from MCF-7 breast cancer cells. A series of dimers were prepared according to the scheme shown in Figure 18.24.

On testing the various analogs, the dimers (**37**) showed inhibition of MCF-7-derived PI3K, which was weaker than their parents, while GA did not inhibit the kinases at all. Some of the dimers were actually more potent than the parent LY 294002. Addition of Hsp90 to the system followed by measurement of the inhibition showed dramatic selectivity of the dimer against the DNA-PK with virtually no activity against the ubiquitous PI3K, whereas the parent was equally

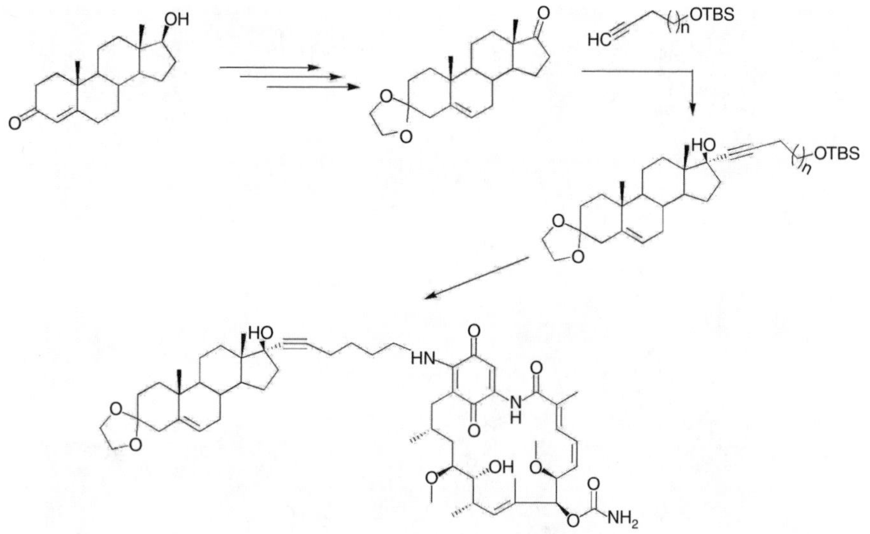

FIGURE 18.22 Memorial Sloan-Kettering testosterone heterodimers.

active against both systems demonstrating the desired selectivity. As further evidence for selectivity, the authors showed direct binding of the dimers to Hsp90, indicating that hybridization did not interfere with this association.

A similar study with GA dimers[51] was made in an attempt to circumvent the broad toxicity of the parent GA. The objective was to improve the selectivity of GA against HER-kinases that may require association with Hsp90 chaperones and thus reduce overall toxicity. Analogs were prepared by the route shown in Figure 18.25.

The best analog in this case was the four-carbon link (**38**; n = 4), which was almost as potent against HER2 as GA but was much less effective against Raf-1. The researchers also prepared several unsymmetrical derivatives containing only a single GA residue, but these were generally no more selective than the parent GA. Once again, the opening of the ansa ring destroyed any activity. Further mechanism studies indicated that the dimer, similar to GA, affects HER2 expression by inducing protein degradation. Further tests on cell lines for which the HER-kinase family are not expressed showed that GA was still a potent inhibitor at 3 nM but the dimer was not.

Further evaluation of all of these dimers in *in vivo* systems will have to done, but clearly the possibility of tailoring the molecular structure to build enzyme specificity is a viable possibility and should produce new therapeutics.

V. CLINICAL APPLICATIONS

Many of the various 17-aminogeldanamycin analogs were evaluated at the NCI using their 60–human tumor cell line panel and *in vivo* models. A combination of potency, reduced toxicity,

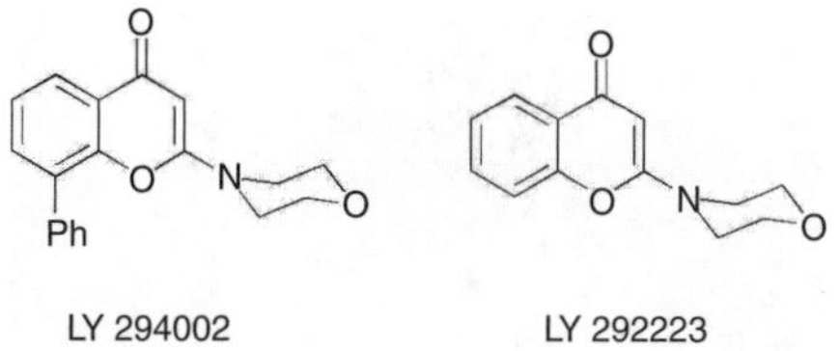

LY 294002   LY 292223

FIGURE 18.23 LY 294002 and LY 29223.

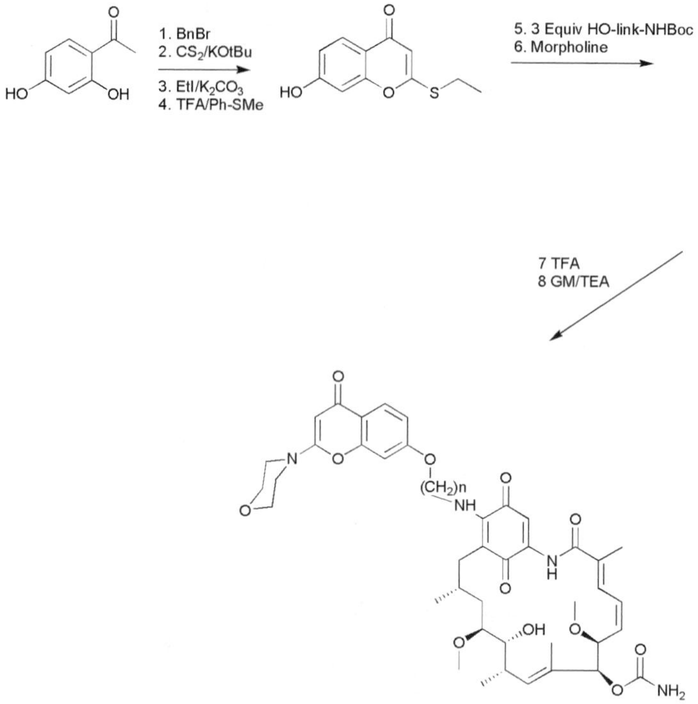

FIGURE 18.24 Memorial Sloan-Kettering LY heterodimers.

and the discovery of a novel mechanism of activity (Hsp90 binding) prompted the selection of 17-allylamino-17-desmethoxygeldanamycin (17-AAG) for clinical studies,[52] and it is currently in six phase I clinical trials.[53] These trials are directed toward solid tumors, epithelial cancer, malignant lymphoma, sarcoma, prostate, and chronic myelogenous leukemia.

One of the greatest obstacles to the development of 17-AAG is its very poor solubility, which prompted the development of a microdispersion preparation for drug delivery.[54] This pharmacokinetic problem with the GA analogs has prompted other groups to pursue improvements in their physical properties.[55] Subsequent examination of other analogs suggested the selection of another, more soluble analog — 17-dimethylaminoethylamino-17-desmethoxygeldanamycin (17-DMAG) — for clinical development. This analog is currently undergoing preclinical evaluation at the NCI.

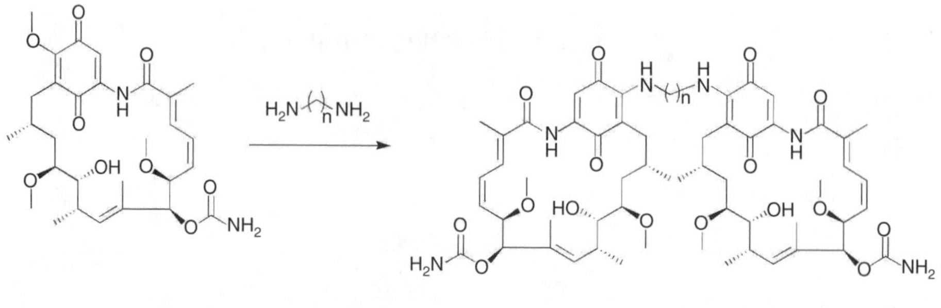

FIGURE 18.25 Memorial Sloan-Kettering GA dimers.

REFERENCES

1. Lancini, G. and Parenti, F., *Antibiotics: An Integrated View*, Springer, New York, 1982, 10.
2. DeBoer, C. et al., Geldanamycin, a new antibiotic, *J. Antibiot.*, **23**, 442, 1970.
3. Rinehart, K.L. et al., Geldanamycin I. Structure assignment, *J. Am. Chem. Soc.*, 92, 7591, 1970.
4. Whitesell L. et al., Inhibition of heat shock protein HSP90-pp60v-src heteroprotein complex formation by benzoquinone ansamycins: essential role for stress proteins in oncogenic transformation, *Proc. Natl. Acad. Sci. USA*, 91, 8324, 1994.
5. Schnur, R.C. and Corman, M.L., Tandem [3,3]-sigmatropic rearrangements in an ansamycin: Stereospecific conversion of an (S)-allylic alcohol to an (S)-allylic amine derivative, *J. Org. Chem.*, 59, 2581, 1994.
6. Supko, J.G. et al., Preclinical pharmacologic evaluation of geldanamycin as an antitumor agent, *Cancer Chemother. Pharmacol.*, 36, 305, 1995.
7. Omura, S. et al., Herbimycin, a new antibiotic produced by a strain of *Streptomyces*, *J. Antibiot.*, 32, 255, 1979.
8. Furusaki, A. et al., Herbimycin A: an ansamycin antibiotic; x-ray crystal structure, *J. Antibiot.*, 33, 781, 1980.
9. Fukazawa, H. et al., Specific inhibition of cytoplasmic protein tyrokinases by herbimycin A *in vitro*, *Biochem. Pharmacol.*, 42, 1661, 1991.
10. Tanida, S., Hasegawa, T., and Higashida, E. Macbecins I and II, new antitumor antibiotics. I., *J. Antibiot.*, 33, 199, 1980.
11. Muroi, M. et al., Macbecins I and II, new antitumor antibiotics. II., *J. Antibiot.*, 33, 205, 1980.
12. Shibata, K. et al., TAN420 Antibiotics, *J. Antib.*, 39, 1630, 1986.
13. Miller, P. et al., Depletion of the erbB-2 gene product p185 by benzoquinoid ansamycins, *Cancer Res.* 54, 2724, 1994.
14. Miller, P. et al., Binding of benzoquinoid ansamycins to p100 correlates with their ability to deplete the erbB2 gene product p185, *Biochem. Biophys. Res. Comm.*, 210, 1313, 1994.
15. Sepp-Lorenzino, L. et al., Herbimycin A induces the 20S proteasome and ubiquitin dependent degradation of receptor tyrosine kinases, *J. Biol. Chem.*, 270, 16580, 1995.
16. Li, L.H. et al., Effects of geldanamycin and its derivatives on RNA-directed DNA polymerase and infectivity of Rausher leukemia virus, *Cancer Treat. Rep.*, 61, 815, 1977.
17. Conde, A.G., et al., Induction of heat shock proteins by tryosine kinase inhibitors in rat cardiomylocytes and myogenic cells confers protection against simulated ischemia, *J. Mol. Cell Cardiol.*, 29, 1927, 1997.
18. Kumar, R., Musiyenko, A., and Barik, S., The heat shock protein 90 of *Plasmodium falciparum* and antimalarial activity of its inhibitor, geldanamycin, *Malaria J.*, 2, 1, 2003.
19. Benchekroun, N.M., Myers, C.E., and Sinha, B.K., Free radical formation by ansamycin benzoquinone in human breast tumor cells; implications for cytotoxicity and resistance, *Free Radic. Biol. Med.*, 17, 191, 1994.
20. Neckers, L., Schulte, T.W., and Mimnaugh, E., Geldanamycin as a potential anti-cancer agent: Its molecular target and biochemical activity, *Investig. New Drugs*, 17, 361, 1999.
21. Ferarini, M. et al., Unusual expression and localization of heat-shock proteins in human tumor cells, *Int. J. Cancer*, 51, 613, 1992.
22. Ochel, H-J., Eichhorn, K., and Gademann, G., Geldanamycin: the prototype of a class of antitumor drugs targeting the heat shock protein 90 family of molecular chaperones, *Cell Stress Chaperones*, 6, 105, 2001.
23. Uehara, Y., et al., Phenotypic change from transformed to normal induced by benzoquinoid ansamycins accompanies inactivation of p60src in rat kidney cells infected with Rous sarcoma virus, *Mol. Cell Biol.*, 6, 2198, 1986.
24. Uehara, Y. et al., Effects of herbimycin derivatives on src oncogene function in relation to antitumor activity, *J. Antibiot. (Tokyo)*, 41, 831, 1988.
25. Stebbins, C.E. et al., Crystal structure of an Hsp90-geldanamycin complex: targeting of a protein chaperone by an antitumor agent, *Cell*, 89, 239, 1997.
26. Bergerat, A. et al., An atypical topoisomerase II from Archaea with implications for meiotic recombination, *Nature*, 386, 414, 1997.

27. Grenert et al., The amino-terminal domain of heat shock protein 90 (hsp90) that binds geldanamycin is an ATP/ADP switch domain that regulates hsp90 conformation, *J. Biol. Chem.* 272, 23843, 1997.
28. Schulte, T.W. et al., Antibiotic radicicol binds to the N-terminal domain of Hsp90 and shares important biologic activities with geldanamycin, *Cell Stress Chaperones*, 3, 100, 1998.
29. Sharma, S.V., Agatsuma, T., and Nakano, H., Targeting of the protein Hsp90 by the transformation suppressing agent radicicol, *Oncogene*, 16, 2639, 1998.
30. Roe, S.M. et al., Structural basis for inhibition of the Hsp90 molecular chaperone by the antitumor antibiotics radicicol and geldanamycin, *J. Med. Chem.*, 42, 260, 1999.
31. Chiosis, G. et al., A small molecule designed to bind to the adenine nucleotide pocket of Hsp90 causes Her2 degradation and the growth arrest and differentiation of breast cancer cells, *Chem. Biol.*, 8, 289, 2001.
32. Chiosis, G., Rosen, N., and Sepp-Kirenzino, L., LY294002-geldanamycin heterodimers as selective inhibitors of the PI3K and PI3K-related family, *Bioorg. Med. Chem. Lett.*, 11, 909, 2001.
33. Baker, R and Castro, L., Total synthesis of (+)-macbecin I, *J. Chem. Soc., Perkin Trans. I.* 47, 1990.
34. Evans, D.A. et al., Asymmetric synthesis of macbecin I, *J. Org. Chem.*, 57, 1067, 1992.
35. Panek, J.S. and Xu, F., Total synthesis of (+)-macbecin I, *J. Am. Chem. Soc.*, 117, 10587, 1995.
36. Martin, S.F. et al., A formal total synthesis of (+)-macbecin I, *J. Org. Chem.*, 57, 1070, 1992.
37. Nakata, M. et al., Total synthesis of herbimycin A, *Tetrahedron Lett.*, 32, 6015, 1991.
38. Nakata, M. et al., The total synthesis of herbimycin-a, *Bull. Chem. Soc. Jpn.*, 65, 2974, 1992.
39. Andrus, M.B. et al., Total synthesis of (+)-geldanamycin and (-)-o-quinogeldanamycin with use of asymmetric anti- and syn-glycolate aldol reactions, *Org. Lett.*, 4, 3549, 2002.
40. Evans, D.A., Miller, S.J. and Ennis, M.D., Asymmetric synthesis of the benzoquinoid ansamycin antitumor antibiotics: Total synthesis of (+)-macbecin, *J. Org. Chem.*, 58, 471, 1993.
41. Panek, J.S., Xu, F., and Rondon, A.C., Chiral crotylsilane-based approach to benzoquinoid ansamycins: Total synthesis of (+)-Mabecin I, *J. Am. Chem. Soc.*, 120, 4113, 1998.
42. Andrus, M.B. et al., Total synthesis of (+)-geldanamycin and (-)-quinogeldanamycin: asymmetric glycolate aldol reactions and biological evaluation, *J. Org. Chem.*, 68, 8162, 2003.
43. Andrus, M.B., Meredith, E.L., and Soma Sekhar, B.B.V., Synthesis of the left-hand portion of geldanamycin using an anti-glycolate aldol reaction, *Org. Lett.*, 3, 259, 2001.
44. Rinehart, K.L. et al., Synthesis of phenazine and phenoxazinone derivatives of geldanamycin as potential polymerase inhibitors, *Bioorg. Chem.*, 6, 353, 1977.
45. Schnur, R.C. et al., Inhibition of the oncogene product p185[erbB-2] *in vitro* and *in vivo* by geldanamycin and dihydrogeldanamycin derivatives, *J. Med. Chem.*, 38, 3806, 1995.
46. Schnur, R.C. et al., *erbB-2* Oncogene inhibition by geldanamycin derivatives: synthesis, mechanism of action and structure-activity relationships, *J. Med. Chem.*, 38, 3813, 1995.
47. Prodromou, C. et al., Identification and structural characterization of the ATP/ADP binding site in the Hsp90 molecular chaperone, *Cell*, 90, 65, 1997.
48. Prodromou, C. et al., A molecular clamp in the crystal structure of the N-terminal domain of the yeast Hsp90 chaperone, *Nat. Struct. Biol*, 4, 477, 1997.
49. Kudkuk, S.D. et al., Synthesis and evaluation of geldanamycin-estradiol hybrids, *Bioorg. Med. Chem. Lett.*, 9, 1233, 1999.
50. Kudkuk, S.D. et al., Synthesis and evaluation of geldanamycin-testosterone hybrids, *Bioorg. Med. Chem. Lett.*, 10, 1303, 2000.
51. Zheng, F.F., et al., Identification of a geldanamycin dimer that induces the selective degradation of HER-family tyrosine kinases, *Cancer Res.*, 60, 2090, 2000.
52. Sausville, E.A., Tomaszewski, J.E., and Ivy, P., Clinical development of 17-allylamino-17-demethoxygeldanamycin, *Curr. Cancer Drug Targets*, 3, 377, 2003.
53. A Phase I and Pharmacologic Study of 17-(Allylamino)-17-demethoxygeldanamycin (AAG, NSC 330507) in Adult Patients with Solid Tumors, NIH Clinical Research Studies, Protocol Number: 99-C-0054. Available at: http://clinicaltrials.gov.
54. Tabibi, S.E. et al., Water-insoluble drug delivery system, U.S. Patent, 6,682,758 B1, 2004.
55. Kasuya, Y. et al., Improved synthesis and evaluation of 17-substituted aminoalkylgeldanamycin derivatives applicable to drug delivery systems, *Bioorg. Med. Chem. Lett.*, 11, 2089, 2001.
56. Dictionary of Natural Products on CD-ROM, Version 12:2. Chapman & Hall, Boca Raton, FL, 2004.
57. Martin, S.F. et al., Novel approach to the ansamycin antibiotics Macbecin I and Herbimycin A. A formal total synthesis of (+)-Macbecin I, *Tetrahedron*, 52, 3229, 1996.

19 Bleomycin Group Antitumor Agents

Sidney M. Hecht

CONTENTS

I. Introduction 357
II. Mechanism of Action..358
    A.   DNA Degradation 358
    B.   RNA Degradation 362
          1. Transfer RNAs and tRNA Precursor Transcripts.....................................362
          2. Other RNA Substrates for BLM 362
          3. Chemistry of RNA Cleavage..363
          4. Nonoxidative Cleavage of RNA 363
    C.   Protein Synthesis Inhibition 363
III. Development of Novel Antitumor Agents 365
IV. Synthesis of Bleomycin 368
V. Medicinal Chemistry..370
    A.   Study of Individual "Amino Acid" Constituents 370
          1. Bithiazole Moiety and C Substituent...370
          2. Linker Domain...371
          3. Metal-Binding Domain 372
    B.   A BLM Combinatorial Library 372
VI. Clinical Applications...373
Acknowledgment...374
References 374

I. INTRODUCTION

The bleomycins are structurally related antitumor antibiotics first isolated from *Streptomyces verticillus* as Cu(II) chelates by Umezawa and his colleagues.[1,2] As illustrated in Figure 19.1, the bleomycins may be considered to be glycosylated oligopeptide-derived species, albeit composed of some rather unusual amino acid constituents. Some of these may actually be derived biosynthetically from proteinogenic amino acids, such as the bithiazole moiety, which is apparently formed by dehydrative cyclization and oxidation of β-alanylcysteinylcysteine.[3] However, recent studies have revealed that the biosynthesis also depends in part on transformations used to assemble polyketide antibiotics.[4]

As discussed in some detail below, the bleomycins exhibit characteristic sequence-selective DNA binding and degradation properties. DNA affinity resides to a significant extent in the "tripeptide S"[5,6] portion of bleomycin, which is composed of the contiguous threonine, bithiazole, and C-terminal moieties. The N terminus of bleomycin consists of the pyrimidoblamic acid and glycosylated β-hydroxyhistidine moieties, which are responsible for metal chelation and oxygen

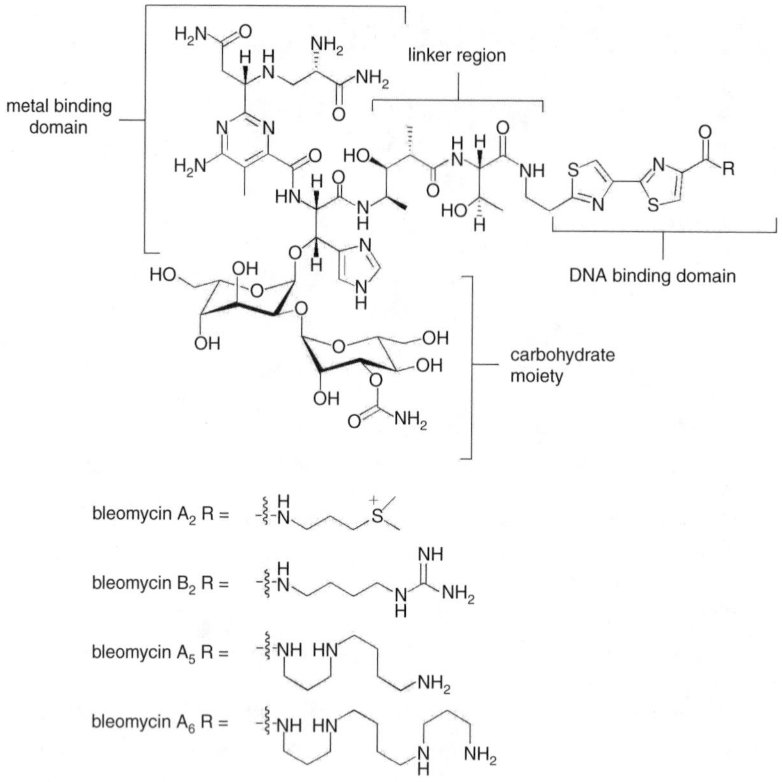

FIGURE 19.1 Structures of some naturally occurring bleomycins.

activation. These domains also bind to DNA and constitute the primary determinant of the sequence selectivity of DNA cleavage by bleomycin.[7-10]

The N- and C-terminal domains of bleomycin are connected by a linker, (2S, 3S, 4R)-4-amino-3-hydroxy-2-methylvaleric acid. Although possible functions of this domain were unclear for some time, it has now been shown convincingly that this linker assumes a compact, folded structure that is essential for efficient sequence-selective DNA cleavage by bleomycin.[11,12]

Although the bleomycins have been used clinically for a number of years, the complex structures of agents in this class have thus far precluded the identification of synthetic congeners worthy of clinical development on the basis of improved properties. This chapter focuses in part on the unrealized potential of the bleomycins and the newly implemented synthetic strategies that may permit bleomycin analogs having improved properties to be prepared and characterized with facility.

II. MECHANISM OF ACTION

Early mechanistic studies of bleomycin explored the effects of this class of compounds on DNA and RNA polymerases, DNA and RNA nucleases, and DNA ligase.[13-16] However, the finding that bleomycin could mediate damage to DNA *in vitro*[17-20] and *in vivo*[21-23] focused efforts on a detailed characterization of this process.

A. DNA DEGRADATION

Bleomycin-mediated DNA degradation requires a metal ion, which forms a complex with the N-terminal domain of bleomycin (cf. Figure 19.1). The first metallobleomycin studied was Fe(II)•BLM,[24-26] but Cu(I)•BLM,[27,28] Mn(II)•BLM,[29,30] and VO(IV)•BLM[31] have now been shown

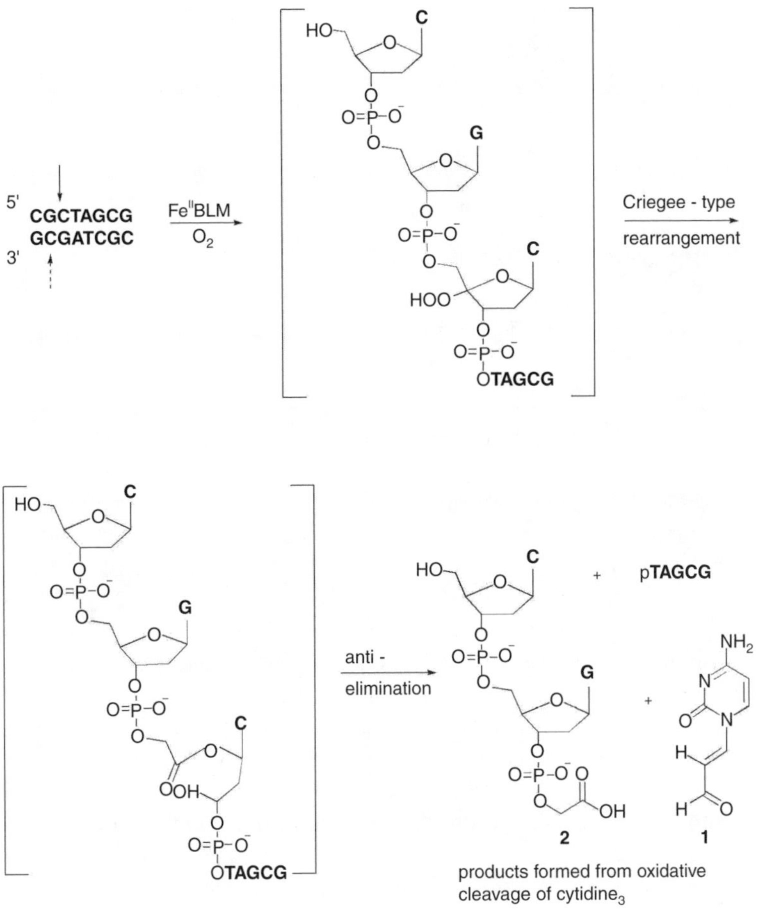

FIGURE 19.2 Pathway leading to frank strand scission of DNA by Fe•BLM.

to mediate DNA degradation as well. To the extent that information is available, these species appear to effect DNA degradation in the same fashion as Fe(II)•BLM. DNA degradation by Fe(II)•BLM also depends on the availability of O_2, which is used to form an activated complex with the metallobleomycin. The nature of the activated Fe•BLM has been the subject of considerable debate as regards metal ion ligation sites, stoichiometry of the complex, and oxidation state of the metal ion.[32] One useful insight regarding the nature of activated Fe•BLM derives from the observation that activated Fe•BLM mediates chemical transformations of small molecules in a fashion most similar to that of cytochrome P450 and its analogs.[28,33,34] This includes the oxidation and oxygenation of compounds such as stilbene and styrene and the NIH shift of deuterium in *p*-deuterioanisole.[34] In further analogy with cytochrome P450 model compounds, Fe(III)•BLM can be activated with H_2O_2[35,36] and with oxygen surrogates such as periodate and iodosylbenzene.[34]

The actual chemistry of DNA degradation by activated Fe•BLM has been studied using both DNA and DNA oligonucleotides. Because the cleavage of DNA by Fe•BLM is sequence selective, the use of DNA oligonucleotides containing a limited number of cleavage sites has permitted a detailed analysis of the nature of DNA degradation.[17–19] Two pathways for Fe•BLM-mediated DNA degradation have been defined and are outlined in Figures 19.2 and 19.3. Shown in Figure 19.2 is a pathway that leads to frank DNA strand scission.[37] The self-complementary octanucleotide d(CGCTAGCG) was employed as a substrate for Fe•BLM based on the earlier finding that the cleavage of DNA is sequence selective and involves primarily a subset of 5-GT-3 and 5-GC-3

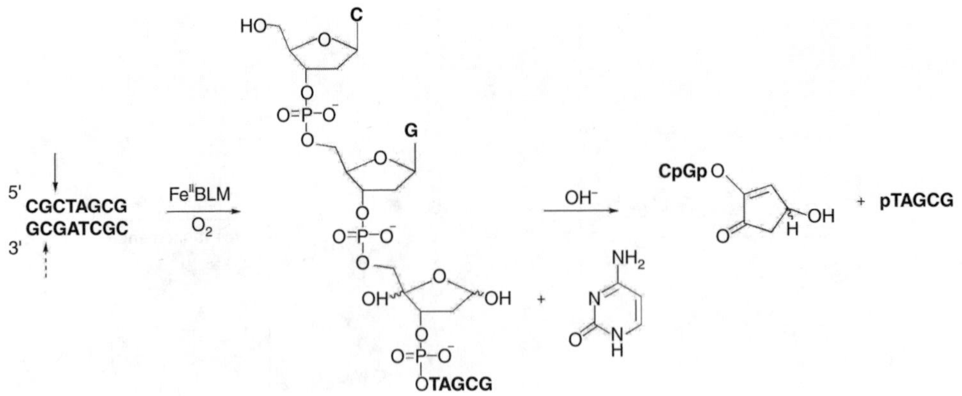

FIGURE 19.3 BLM-mediated formation and base-induced decomposition of the alkali–labile lesion.

sequences. In fact, the substrate shown in Figure 19.2 was cleaved quite efficiently by activated Fe•BLM and predominantly at the two 5-GC-3 sequences. The observed products, verified by independent chemical synthesis, were *trans*-3-(cytosin-1-yl)propenal (**1**), 5-GMP and a CpG derivative terminating at its 3-end with a 3-(phosphoro-2-*O*-glycolate) moiety (**2**).[7,31,38] The yield of base propenal was approximately equal to the sum of the yields of 5-GMP and **2**.[7,36] Although the relative amounts of 5-GMP and **2** formed could be altered by changing the BLM congener employed, in all cases the amount of **1** formed closely approximated the combined yields of 5-GMP and **2**. The mechanistic scheme shown in Figure 19.2 rationalizes the nature and the amounts of products obtained. Initial abstraction of a H atom would afford a C4 deoxyribose radical that could combine with O_2 to form a C4 hydroperoxy radical. Reduction and protonation of this intermediate, or recombination with a H atom, would afford a C4 hydroperoxide, the latter of which would be expected to undergo a Criegee-type rearrangement, as shown, with scission of the C3-C4 deoxyribose bond. Conversion of the resulting hemiacetal to the aldehyde, and elimination of elements of glycolic acid, would then afford the observed products. It may be noted that in the example shown in Figure 19.2, $CpGp_{CH_2 COOH}$ (**2**) would result only from cleavage at cytidine$_3$, while 5-GMP would result only from cleavage at cytidine$_7$. Two other oligomeric products would be expected to form (one oligomeric 5-phosphate and one oligomeric 3-phosphoroglycolate) but were not conveniently analyzed. It may be noted that a reasonable alternative for the scheme shown in Figure 19.2 has been proposed.[39]

In addition to the products shown in Figure 19.2, another set of products was also formed.[40–43] These included free cytosine and a DNA lesion denoted the alkali labile lesion, as subsequent treatment with alkali resulted in strand scission. Strand scission was found to occur at the same sites as those outlined in Figure 19.2. In fact, the proportion of frank strand scission products and alkali labile lesions was a function of oxygen tension; higher oxygen tension favored frank strand scission products.[43–45] The structure determined for the initially formed alkali labile lesion was that of a C4-OH apurinic acid, which presumably arose from depurination of 4-OH deoxycytidine (Figure 19.3). Because the ratio of products formed via the two pathways was influenced by the amount of oxygen available, the two pathways likely have a common intermediate.[44,45] It seems likely that in the absence of sufficient O_2, the initially formed C4 deoxyribose radical (Figure 19.2) undergoes a further one-electron oxidation to afford the carbocation.[46–48] The latter could then react with water, affording the putative 4-OH deoxycytidine intermediate. Independent support for the intermediacy of a C4-deoxyribose radical has been obtained by the use of a C4-deuterated DNA oligonucleotide as a substrate for Fe(II)•BLM.[49] A primary kinetic isotope effect was observed at individual thymidine residues, indicating that abstraction of the C4-H is actually rate limiting for Fe•BLM-mediated DNA degradation.

It is interesting to note that although activated Fe•BLM is capable of effecting both the oxidation and oxygenation of small molecules,[50,51] the degradation of DNA according to the pathways shown in Figures 19.2 and 19.3 involves only DNA substrate oxidation.[48] Dioxygen is required both for Fe(II)•BLM activation and for reaction with the putative C4 deoxyribose radical but is apparently not transferred to DNA from the oxygenated metalloBLM.

The catalytic cycle for Fe(II)•BLM was studied further using both electrochemical activation at a glassy carbon electrode[52] and ^{17}O nuclear magnetic resonance spectroscopy.[53] Electrochemical activation permitted correlation of the number of electrons consumed and DNA oligonucleotide products produced under conditions in which each Fe•BLM molecule was employed up to about 13 times in producing DNA lesions. In the electrochemical system, it was also clear that Fe(II)•BLM could not be activated reductively once it had bound to DNA. Thus, the obligatory order of events must have been Fe(II)•BLM binding by O_2, reduction to afford activated Fe•BLM, and then DNA binding and cleavage.[52] Fe(II)•BLM activation was also studied by ^{17}O nuclear magnetic resonance spectroscopy in the absence and presence of a DNA substrate.[53] These studies verified that in the absence of any DNA substrate, Fe(II)•BLM catalyzes a 4 e- reduction of O_2 to H_2O. The production of $H_2^{17}O$ in the presence of DNA was entirely consistent with the mechanisms outlined in Figures 19.2 and 19.3.

An additional noteworthy feature of DNA cleavage by Fe(II)•BLM is the formation of double-strand DNA breaks. The proportion of double-strand cleavage is a function both the specific BLM congener[54] and the DNA substrate employed[55-58]; under favorable circumstances, such breaks can constitute up to 20% of all DNA lesions. It has been demonstrated convincingly that double-strand cleavage results from a specific set of mechanistic events at a DNA sequence conducive to such cleavage[55-58] and not from the random accumulation of single-strand breaks. For some DNA sequences, it has been argued that a significant rearrangement of structure must occur after the first break to orient the activated BLM in a fashion conducive to the cleavage of a second strand.[58,59] However, at least for one substrate, this would seem to be unnecessary. The oligonucleotide shown in Figure 19.4 underwent 43% double-strand cleavage from a bound conformation in which bleomycin was found to be oriented close to the cleavage sites on both strands.[60]

Although the individual metallobleomycins noted above all cleave DNA in a superficially similar fashion, for most metallobleomycins other than Fe•BLM there is a paucity of information on which to base mechanistic comparisons. Those species that have been studied to some extent include Cu(I)•BLM, which has been shown to form a metal complex with BLM quite different than that formed by Fe(II),[27,61] to undergo aerobic activation much less quickly than Fe(II)•BLM,[28] and to produce sequence specific DNA lesions in different ratios.[28] Co(III)•BLM has a significantly higher DNA affinity than other metalloBLMs,[62] requires light but not oxygen for DNA cleavage,[63-67] binds and cleaves DNA at some sites distinct from those targeted by Fe(II)•BLM,[68] and cleaves DNA with minimal production of the compounds formed in Figure 19.2.[67] This metalloBLM has also been postulated to employ a reactive species different than activated Fe•BLM.[68] Accordingly, mechanistic generalizations between metalloBLMs should be made only with great caution.

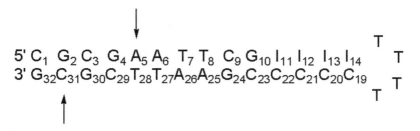

FIGURE 19.4 Oligonucleotide substrate that underwent efficient double strand cleavage by Fe•BLM.

B. RNA Degradation

1. Transfer RNAs and tRNA Precursor Transcripts

Following an initial report by Magliozzo et al.[69] in which a high (300 μM) concentration of activated Fe•BLM was shown to effect limited degradation of yeast tRNA[Phe] and a few other tRNA isoacceptors, the Hecht laboratory carried out a systematic study of RNA degradation by Fe(II)•BLM.[70–74]

For transfer RNAs and tRNA precursor transcripts, most of the species studied were refractory to activated Fe•BLM at all tested concentrations. A few tRNAs were cleaved with modest efficiency at intermediate concentrations, notably a *Schizosaccharomyces pombe* amber suppressor tRNA[Ser] construct and mature *Escherichia coli* tRNA$_1$[His].[73] By far the most efficient cleavage was achieved with a *Bacillus subtilis* tRNA[His] precursor (Figure 19.5), which was cleaved with surprising efficiency at U$_{35}$ and weakly at several other positions.[70,73] Although examples of cleavage occurred in many positions of the cloverleaf structure of tRNA, and at sequences not well represented in DNA cleavage studies, in the aggregate the majority of cleavage sites involved 5-G•pyr-3 sequences and were located at what are believed to be the junctions between single- and double-stranded regions of the tRNAs.[10,70–73]

Two tRNA substrates that were cleaved with reasonable efficiency by Fe•BLM were particularly helpful in defining the RNA microenvironments conducive to cleavage. These were yeast tRNA[Phe] and cytoplasmic yeast tRNA[Asp],[75] the structures of which have been defined crystallographically.[76–80] By systematic variation of these two RNA structures, Giegé, Florentz, and coworkers have defined the identity elements for these two RNAs required for recognition by their cognate aminoacyl–tRNA synthetases.[81–85] The use of these hybrid tRNA structures as substrates for Fe•BLM revealed a surprisingly complex set of cleavage patterns.[75] Perhaps more than any other study, this report underscored the ability of Fe•BLM to recognize subtle changes in RNA structure.

2. Other RNA Substrates for BLM

Although most of the studies of RNA cleavage by bleomycin have involved tRNAs, several other RNAs have been tested as substrates. Perhaps the most successful of these has been yeast 5S ribosomal RNA, which was found to be cleaved quite efficiently at three sites by Fe(II)•BLM.[73] Interestingly, each of the three cleavage sites involved the uridine moiety in a 5-GUA-3 sequence, and each had a one-base bulge on the cleaved strand one or two nucleosides to the 3-side of the cleavage site. However, it has not been shown that any rRNA within an intact ribosome is cleaved by Fe(II)•BLM, and studies of the effects of Fe(II)•BLM on a eukaryotic protein synthesizing system indicated that ribosomes are not a primary target for BLM (*vide infra*).[86]

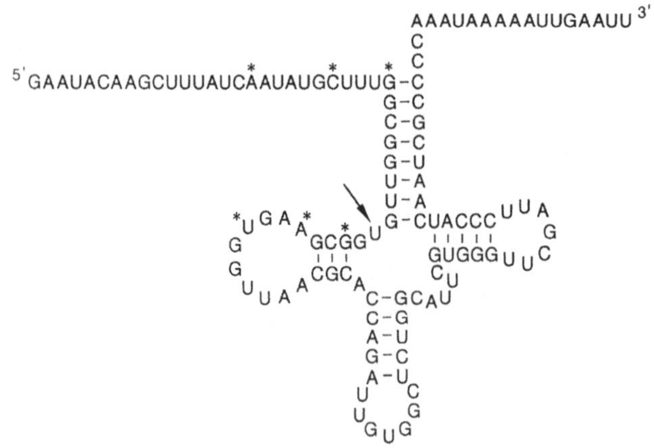

FIGURE 19.5 A *Bacillus subtilis* tRNA[His] precursor transcript that undergoes efficient cleavage by Fe•BLM.

Several messenger RNAs have been studied as substrates for Fe(II)•BLM. RNAs found to undergo cleavage have included HIV-1 reverse transcriptase mRNA,[70] the iron regulatory element of bullfrog ferritin mRNA,[87] firefly luciferase mRNA,[86] and *E. coli* dihydrofolate reductase mRNA.[88] Also studied has been the cleavage of an RNA–DNA heteroduplex prepared by reverse transcription of *E. coli* 5S rRNA.[74] Both the RNA and DNA strands were cleaved under comparable conditions and at a limited number of sites. Significantly, cleavage of the RNA strand of the heteroduplex involved cleavage sites different than those observed for the original rRNA, indicating that cleavage must actually have involved a heteroduplex structure.

3. Chemistry of RNA Cleavage

The chemistry of RNA cleavage by Fe(II)•BLM has been studied less intensively than that of DNA, but a few useful observations have been made. By the use of yeast 5S rRNA ^{32}P-labeled alternatively at the 5 and 3 ends, it was shown that the mobility of individual bands was consistent with that expected if the chemistry of RNA cleavage was analogous to that of DNA.[73] This conclusion was also supported by experiments using the self-complementary oligonucleotide CGCTAGCC in which deoxycytidine$_3$ was replaced by ribocytidine.[70] However, the latter experiment indicated the presence of greater amounts of free cytosine than would have been anticipated on the basis of products resulting from initial abstraction of C4 H (cf. Figures 19.2 and 19.3). In the belief that an altered conformation at position 3 in the minor groove might lead to H atom abstraction from C1, the initially formed products were treated with diaminobenzene in the hope of capturing the putative intermediate (**i**) resulting from Criegee-type rearrangement of a C1 hydroperoxynucleotide (Figure 19.6). In fact, a product having chromatographic properties identical with synthetic quinoxaline **3** was observed. Assuming that this product arose by the pathway shown in Figure 19.6, about 10% of the products resulting from admixture of C$_3$-*ribo* CGCTAGCG and Fe(II)•BLM were formed by initial abstraction of C1 H from ribocytidine$_3$.[89,90] Repetition of this experiment using C$_3$-*ara* CGCTAGCG gave an amount of **3** corresponding to 58% C1 H abstraction.[89] This is consistent with the known conformational alteration of the latter species.[91]

Not yet resolved experimentally is the issue of the formation of an alkali labile lesion from RNA analogous to the intermediate formed from DNA (Figure 19.3). Although the intrinsic lability of the RNA internucleotide linkage to alkali precludes the use of this reagent to unmask the alkali labile lesion selectively, it has been shown by admixture of hydrazine to Fe•BLM-treated *B. subtilis* tRNA^{His} precursor transcript that some currently uncharacterized lesion is present.[90]

4. Nonoxidative Cleavage of RNA

One particularly surprising finding has been reported by Keck and Hecht.[92] In the absence of any added metal ion, BLM A$_2$ was found to mediate RNA strand scission at a number of sites. Cleavage was found to be suppressed by admixture of metal ions that bind to BLM. As illustrated in Figure 19.7A, cleavage was observed at all unmodified pyrimidine–purine sites in the tRNA. Analysis of the nature of the formed cleavage products indicated that each contained a 2,3-cyclic phosphate at the 3 end. Logically, these might be thought to have resulted from a phosphoryl transfer reaction in which the internucleotide phosphate linkage at certain positions underwent nucleophilic attack by the 2-OH group of ribose. Oligonucleotide 2,3-cyclic phosphate moieties could then form concomitant with RNA strand scission (Figure 19.7B).

Although the possible relevance of this transformation to the mechanism of action of BLM is unclear, it may be mentioned that the facility of this transformation is entirely comparable to that involving oxidative RNA cleavage by Fe(II)•BLM (*vide infra*).

C. Protein Synthesis Inhibition

The ability of Fe(II)•BLM to cleave some tRNAs, mRNAs, and rRNAs suggested that this metalloleomycin might be able to inhibit the synthesis of proteins. This was studied in a DNA-independent

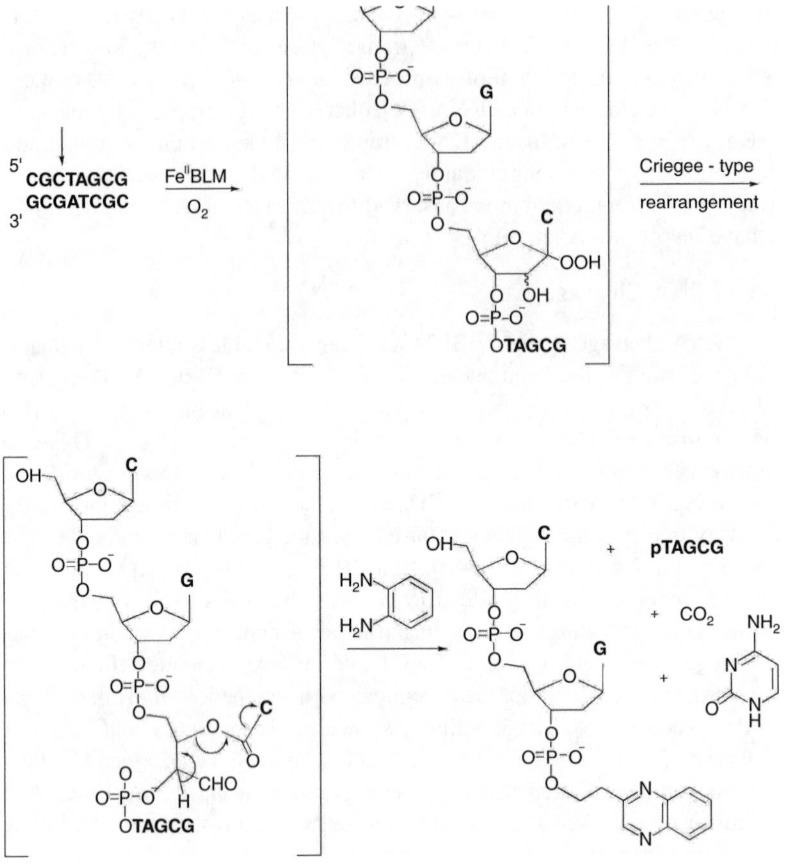

FIGURE 19.6 Products potentially accessible from RNA by abstraction of the C1 H of deoxyribose by Fe•BLM. *C* represents aracytidine or ribocytidine.

cell-free protein biosynthesizing system containing rabbit reticulocyte lysate and programmed with luciferase mRNA.[86] As shown in Table 19.1, treatment with 100 µM BLM A$_5$ alone reduced luciferase synthesis by ~26%, either because of adventitious metal ions or because of nonoxidative RNA cleavage, as described above. Treatment with 50 µM Fe(II)•BLM A$_5$ reduced protein synthesis by about 68%, and 100 µM Fe(II)•BLM A$_5$ caused ~86% reduction. To determine which RNAs had been affected by Fe(II)•BLM A$_5$, the reaction mixtures were treated with additional ribosomes, luciferase mRNA, or a mixture of tRNA isoacceptors following Fe(II)•BLM treatment. As indicated in the table, only admixture of tRNAs had any effect in retaining luciferase synthesis. Production of this protein was completely restored for the treatment involving 50 µM Fe(II)•BLM and was partially restored for the 100-µM treatment. Thus, one or more transfer RNAs may represent a critical cellular target for BLM.

In this context, it is interesting to note that the antitumor ribonuclease onconase[93–97] has been employed in an experiment analogous to the one summarized in Table 19.1, and virtually identical results were obtained.[98] Because onconase and BLM have been shown to have remarkably similar cytotoxicity profiles toward a panel of cultured tumor cells,[86] the specific tRNAs cleaved by both agents were compared. The two antitumor agents were found to effect cleavage of several tRNAs in common, although a comparison of cleavage patterns with one isoacceptor revealed cleavage at quite different sites in that tRNA.

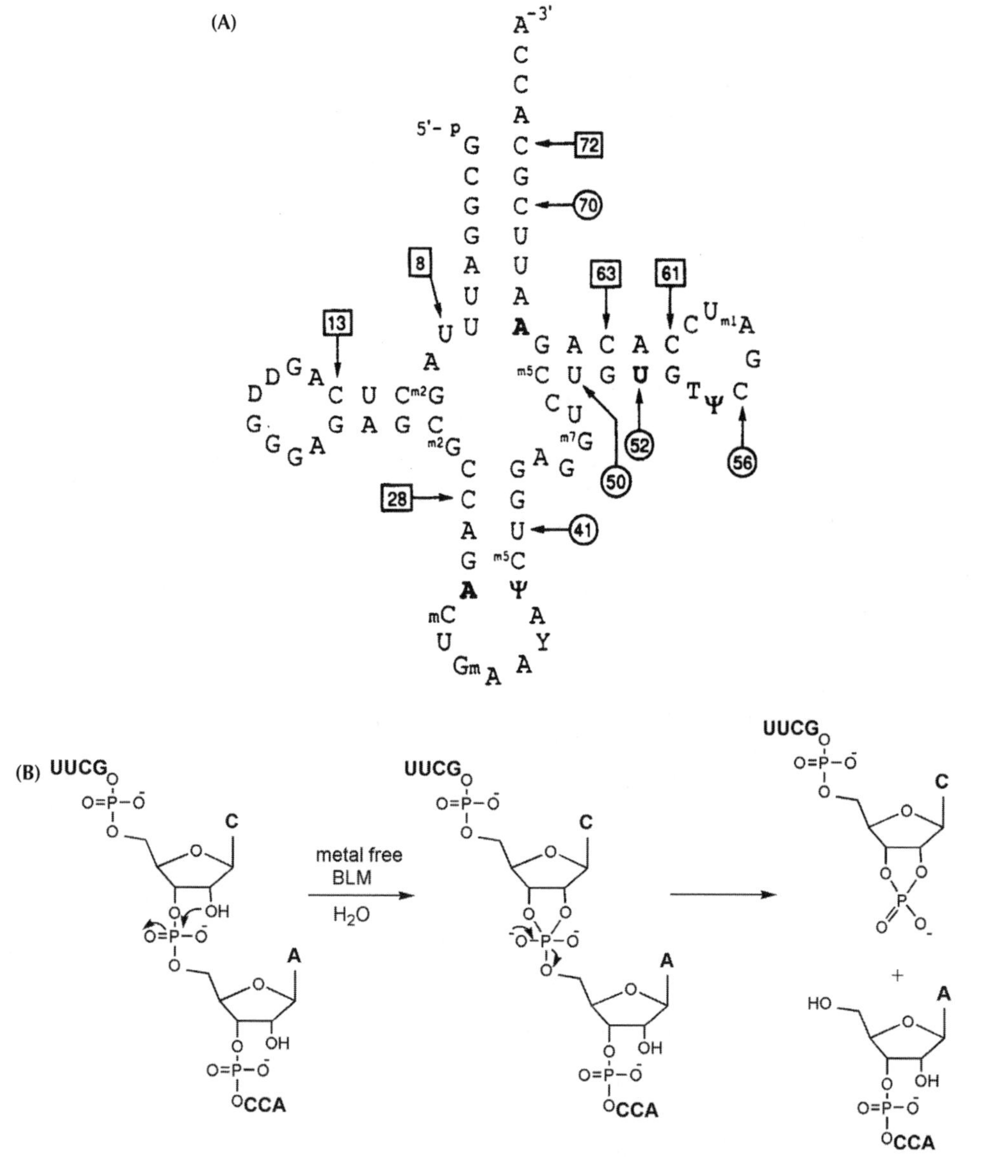

FIGURE 19.7 (A) Sites of cleavage of tRNA^{Phe} by metal-free BLM. Major cleavage sites are in boxes; minor cleavage sites in circles. (B) Chemistry of tRNA degradation by metal-free BLM. Cleavage between C_{72} and A_{73} is exemplified.

Although the specific tRNA(s) exhibiting maximal sensitivity to these agents have yet to be reported, it is interesting to note that following treatment of uninfected or HIV-1 infected H9 lymphocytes with onconase, tRNA$_3^{Lys}$ was found to be cleaved by onconase.[99] Although the role of RNA cleavage in the cancer chemotherapeutic effects of BLM remains to be established, the foregoing experiments support the possible involvement of one or more tRNA isoacceptors.

III. DEVELOPMENT OF NOVEL ANTITUMOR AGENTS

The earliest members of the bleomycin family to be identified were the phleomycins, which contain a thiazolinylthiazole moiety rather than the bithiazole present in bleomycin (cf. Figures 19.1 and 19.8).

TABLE 19.1
**Effect of tRNAs on Luciferase Synthesis Following
Treatment with Fe(II)•BLM A$_5$**

Addition	Luciferase Synthesis (%)
—	100
100 µM Fe^{2+}	113
100 µM BLM A$_5$	74
50 µM Fe(II)•BLM A$_5$	32
50 µM Fe(II)•BLM A$_5$, then 40 µg/mL tRNAs	120
100 µM Fe(II)•BLM A$_2$	14
100 µM Fe(II)•BLM A$_2$, then 40 µg/mL tRNAs	53

The phleomycins and bleomycins each contain several different C substituents, as illustrated in Figures 19.1 and 19.8. Although the members of both series each cleave DNA with similar characteristics with regard to sequence selectivity, individual congeners do display different biochemical properties, even at the level of cell culture.[100]

Although discovered before the bleomycins, the phleomycins exhibited unacceptable toxicity when tested as a mixture of congeners. Further, the thiazoline moiety is susceptible to hydrolytic ring opening and possibly also to racemization. For these reasons, attention was focused on the bleomycins as clinical candidates.

The complexity of the bleomycin structure precluded total chemical synthesis at the time of their initial isolation. Accordingly, most of the congeners studied were naturally occurring variants that differed exclusively at the C terminus (Figure 19.1). It was also found that additional congeners

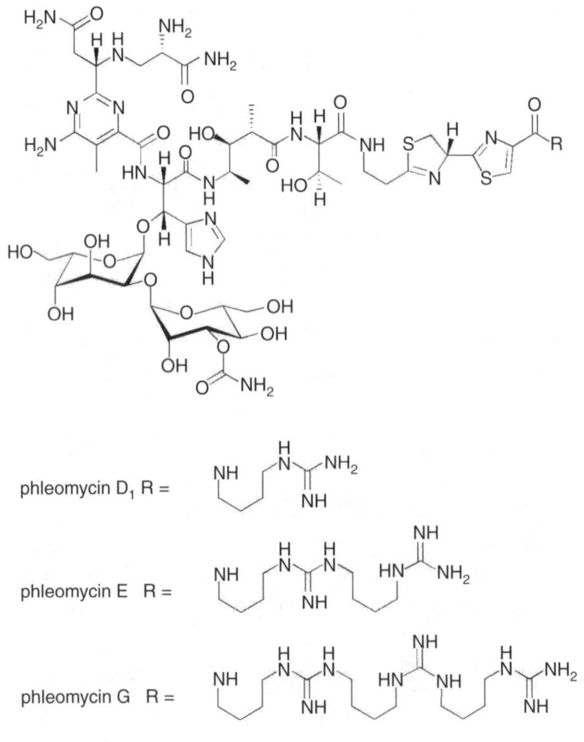

FIGURE 19.8 Structures of some naturally occurring phleomycins.

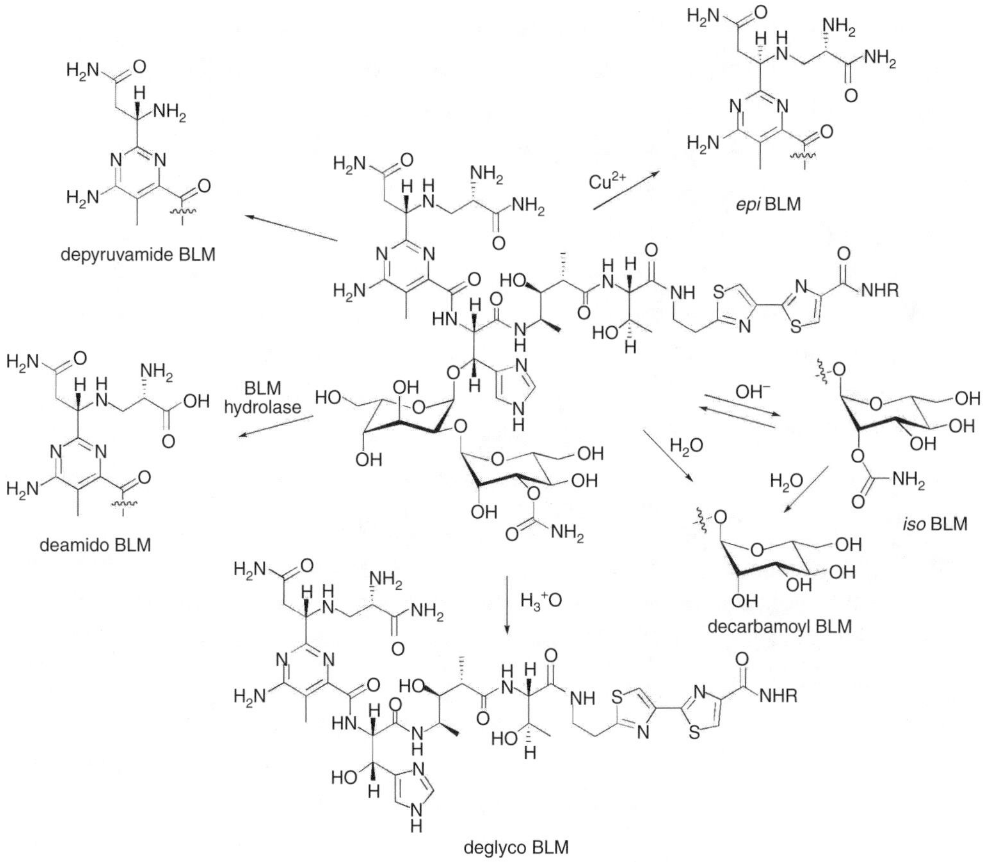

FIGURE 19.9 Analogs of bleomycin accessible by semisynthesis from BLM itself.

differing at the C terminus could be prepared via fermentation by including in the incubation medium a high concentration of the amine intended for incorporation into the C terminus of bleomycin.[101] Semisynthetic bleomycins differing at the C terminus were also accessible following initial chemical removal of the C substituent of bleomycin A_2; the latter was realized by means of a series of chemical transformations to afford bleomycinic acid.[102–104] Treatment of bleomycin B_2 with an acylagmatine hydrolase also provided access to bleomycinic acid.[105] Condensation of bleomycinic acid with different polyamines, for example, then afforded new analogs of bleomycin.

Although most of the early structural modifications of the bleomycins were limited to alterations at the C terminus, several other types of transformations of the natural product were reported. As summarized in Figure 19.9, this included the preparation of *epi*-BLM, *iso* and decarbamoyl BLMs, and deamido and depyruvamide BLMs.[106] These were found to differ somewhat in their animal organ distribution profiles, as well as in their intrinsic biological properties.[106] One of the most remarkable analogs was deglycoBLM, the simplified analog lacking the carbohydrate moiety of BLM.[107,108] This species was found to mediate DNA cleavage with almost the same sequence selectivity as BLM itself and to produce the same chemical products.[108,109]

Other naturally occurring variants of the BLMs were also identified. Perhaps the best known are the tallysomycins, which have also been used in clinical trials. As shown in Figure 19.10, the tallysomycins have a third carbohydrate moiety, in the form of a talose residue, which is attached to the peptide backbone of tallysomycin via a structurally novel glycosylcarbinolamide linkage. The tallysomycins cleave DNA predominantly at 5-GC-3 and 5-GT-3 sequences, as do the bleomycins, but the patterns of preferred sites of cleavage are somewhat altered.[110]

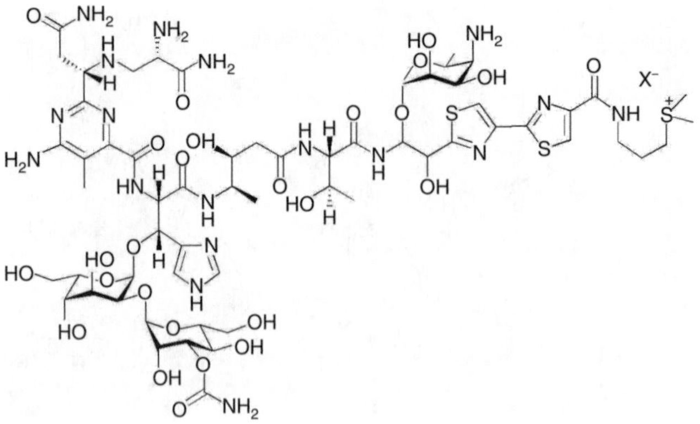

FIGURE 19.10 Structure of tallysomycin S$_2$B.

In the United States, Europe, and Japan, bleomycin is used clinically as a mixture of congeners that differ at the C terminus. The major constituents are bleomycin A$_2$ (~60%–65%) and bleomycin B$_2$ (~30%). In animal tumor models, this mixture was reported to exhibit efficacy better than single congeners.[1] It has been reported that BLM congeners having free amino groups within the C substituent exhibited enhanced pulmonary toxicity — an observation that led to the preparation and evaluation of analogs such as liblomycin[111] and BAPP.[104,106] Interestingly, BLM A$_5$ (which has free amino groups within the C substituent in the form of a spermidine) is used clinically in Russia and China.

IV. SYNTHESIS OF BLEOMYCIN

Three laboratories have reported the synthesis of naturally occurring bleomycins. The first two syntheses were reported in 1982 by the Umezawa[112] and Hecht[113] laboratories. The development of methodology relevant to the synthesis of bleomycin has continued to this time. As a result, two additional syntheses of bleomycin have been described recently by the Boger[114] and Hecht[115] laboratories, as have the syntheses of closely related species.[54,115–117] Of special interest in this context has been the recent report of the total synthesis of deamidobleomycin, the sole reported catabolite formed following therapeutic administration of bleomycin.[117]

An important advance in the synthesis of bleomycin congeners was realized in 2000 with the first report of the preparation of a deglycobleomycin by solid-phase synthesis.[118] The synthesis involved the use of Tentagel resin functionalized with a trityl linker. The "amino acid" constituents of bleomycin were condensed stepwise onto the resin-bound linker from the C terminus. Following condensation of Boc pyrimidoblamic acid,[119] the formed deglycoBLM A$_5$ was deblocked and cleaved from the resin by treatment with trifluoroacetic acid/triisopropylsilane/water/dimethylsulfide.[118]

Although the use of a linker incorporating a trityl group did provide synthetic access to deglycoBLM, there were some limitations to this strategy. One critical limitation was the obligatory cleavage of deglycoBLM from the resin concomitant with removal of the Boc protecting group. While reducing the number of steps in the overall procedure, this simultaneous deblocking/cleavage precluded the elaboration of fully deblocked deglycoBLM still attached to the synthesis resin. Because it had previously been shown that natural bleomycins conjugated to solid supports via their C substituents retained the ability to effect sequence-selective DNA cleavage,[120] it seemed desirable to be able to test the properties of newly synthesized deglycoBLMs before their removal from the synthesis resin.

The use of a dimedone-based resin linker, cleavable by treatment with hydrazine, proved successful in permitting the elaboration of fully protected deglycoBLMs attached to a Tentagel

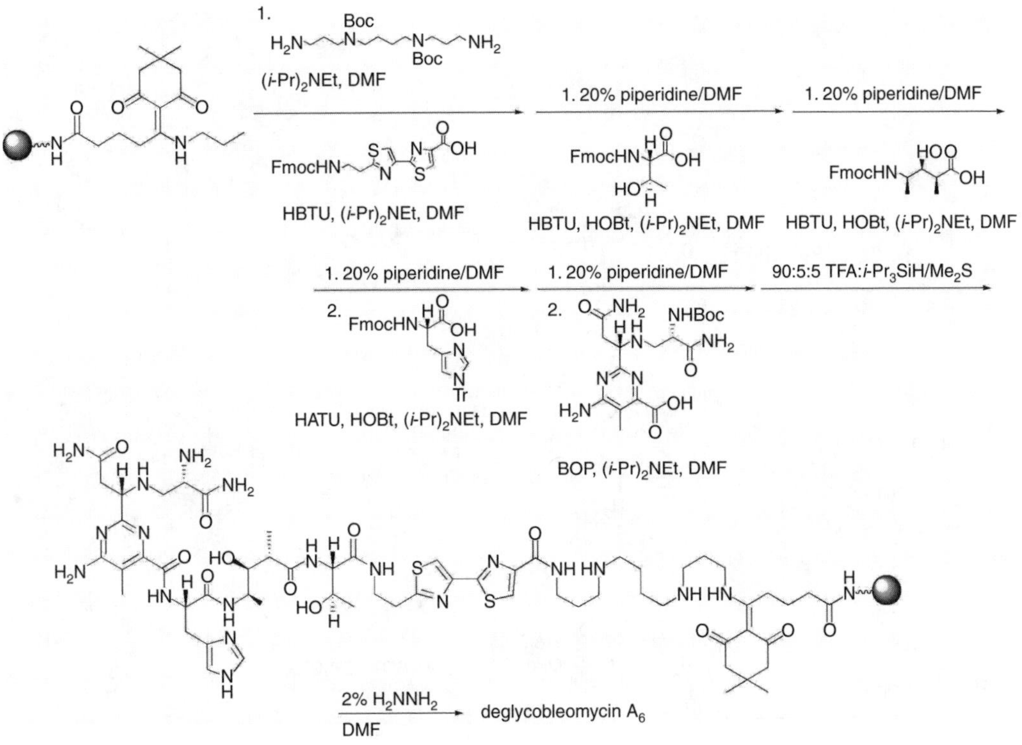

FIGURE 19.11 Scheme employed for the solid-phase synthesis of deglycoBLM.

support.[121] The synthetic scheme is outlined in Figure 19.11. As shown, after attachment of each Fmoc-protected fragment to the resin-bound linker, subsequent treatment with piperidine effected Fmoc cleavage and formation of a dibenzofulvene-piperidine adduct, which has been characterized spectroscopically.[122] This permitted stepwise coupling yields to be determined. The synthesis of deglycoBLM A_{2-a} was realized in an overall yield of 68% using this procedure.[121] In a parallel study, optimization of the C substituent permitted determination of the C substituent optimal to permit on-resin assay of the synthesized deglycoBLM.[123]

A further study addressed the issue of preparing by solid-phase synthesis BLMs containing the disaccharide moiety of BLM itself, or other carbohydrate analogs. As outlined in Figure 19.12, this was realized by preparing a suitably protected β-hydroxyhistidine derivative containing the disaccharide moiety present in BLM. The synthesis was based on earlier syntheses described by Ohno[124] and Boger[125] but employed protecting groups compatible with the solid-phase synthesis outlined in Figure 19.11. In this fashion, bleomycin A_5 was synthesized in reasonable yield,[126] and with much greater facility than the respective solution phase synthesis.[115] As described in a section below, this strategy has also been employed successfully to prepare carbohydrate analogs of bleomycin. It may be mentioned that the stepwise linear construction realized by solid-phase synthesis was different than the coupling scheme employed in any of the solution phase syntheses.

As regards the preparation of other bleomycin group antibiotics, the absolute stereochemistry of the methine H in the thiazoline moiety of PLM has been described,[127] potentially enabling its total synthesis. However, the synthesis of phleomyin itself has not been reported as yet. In contrast, Sznaidman and Hecht recently reported a strategy for elaborating the glycosylcarbinolamide moiety of tallysomycin[128] in spite of the fact that the absolute stereochemistry has not yet been defined at some of the stereocenters.

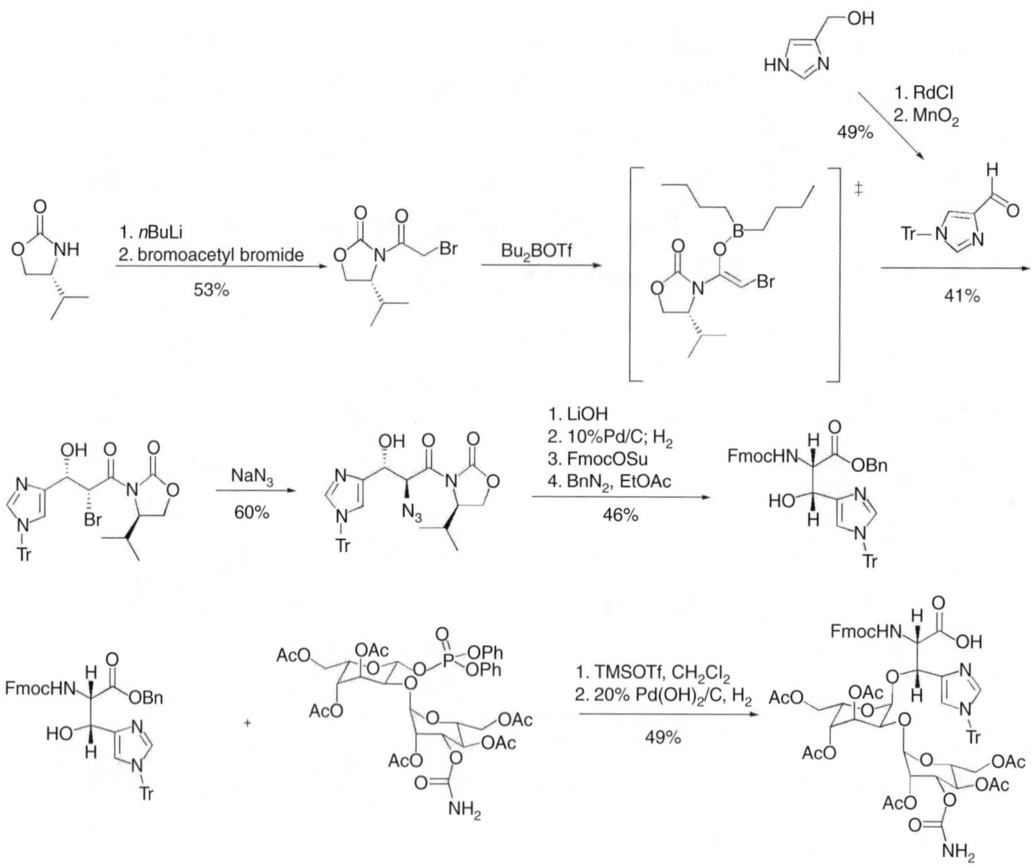

FIGURE 19.12 Synthesis of the carbohydrate moiety of BLM protected in a fashion suitable for use in the solid phase synthesis of the natural product.

V. MEDICINAL CHEMISTRY

A. STUDY OF INDIVIDUAL "AMINO ACID" CONSTITUENTS

1. Bithiazole Moiety and C Substituent

In contrast to efforts on the total synthesis of bleomycin, which have been pursued for 30 years, systematic evaluation of those structural elements in bleomycin essential for its nucleic acid degradation properties has been addressed only more recently. Alteration of the bithiazole moiety in a systematic fashion has yet to be reported, although several studies have contributed to our understanding of the way this moiety functions in DNA binding. These include photochemical transformations of the bithiazole moiety[129,130] and replacement with mono and bishalogenated bithiazoles.[131] Although neither of these types of alterations effected obvious changes in the DNA-binding properties of the derived (deglyco)BLMs under dark conditions, irradiation of the deglycoBLMs containing halogenated thiazoles did permit confirmation that these structural elements can reside in the minor groove of DNA.[131]

The ability of the tripeptide S moiety of BLM (consisting of the threonine, bithiazole, and C-terminal substituent) to bind DNA with essentially the same affinity as BLM[5,6] argues that much of the DNA binding affinity of BLM resides within this domain. Fully consistent with this interpretation were the findings that deglycoBLM analogs containing monothiazole constituents cleaved DNA inefficiently and without demonstrable sequence selectivity.[132] As might have been expected, the replacement of the bithiazole moiety with a pyrrole-based heterocyclic system resulted in a change in sequence selectivity to afford cleavage at sequences reminiscent of the sequence-selective binding of distamycin and netropsin.[133,134]

On the basis of the foregoing observations, it might have been reasonable to conclude that the sequence selectivity of DNA cleavage by the (deglyco)BLMs is determined by the bithiazole moiety, as actually suggested in an early study.[135] However, there are now a few lines of experimentation that make it clear that the actual situation is more complex. These include the findings that photolysis of halogenated bithiazoles in the presence of DNA resulted in sequence-selective DNA damage, albeit not with a sequence selectivity characteristic of BLM,[136,137] whereas an analog of the pyrim- idoblamic acid moiety effected sequence selective cleavage of DNA in the same fashion as BLM.[9] Also, BLM congeners having the same bithiazole + C terminus, but differing in their metal binding domains, exhibited altered strand selectivity of DNA cleavage.[7]

Perhaps the most compelling evidence was obtained for a series of analogs of deglycoBLM in which the threonine moiety was replaced by glycine or oligoglycine, thus systematically altering the distance between the metal binding and bithiazole domains.[8] On the assumption that increasing the number of glycines between the two binding domains constrained them on average to be at greater distances from each other when bound to a DNA substrate, the pattern of DNA cleavage should then reflect which binding partner (i.e., the metal binding domain or bithiazole) is the predominant structural element in defining the sequence selectivity of DNA binding. It was shown experimentally that all of the analogs were cleaved at the same sites, arguing that the metal binding domain is the primary determinant of the sequence selectivity of DNA binding by BLM.[8]

The selectivity for 5-GC-3 and 5-GT-3 sequences may well reflect the reality that the metal binding domain of BLM is able to bind to DNA the most effectively at the widest, shallowest part of the minor groove.[10] This suggestion is supported by modeling studies[138,139] and by the observation that BLM cleaves DNA and RNA duplexes most efficiently at sites having a widened minor groove. Examples might be thought to include DNA[140] and RNA bulges,[73] sites of DNA platination,[141] the 5-junction of a DNA triplex structure with its DNA duplex target,[142] and the putative junctions between single- and double-stranded structures in RNAs.[70,73]

Although not yet established experimentally, it seems likely that the most efficient cleavage of DNA by BLM occurs at those sites that permit efficient binding and H atom abstraction by the metal binding domain of BLM while also allowing the bithiazole to bind to its own preferred sites.

Naturally occurring BLMs have several different C substituents, most of which are polyamines, or polyamines containing guanidine moieties. Virtually all of the naturally occurring BLMs have C substituents that are charged at physiological pH, and the charged C substituent is clearly important for productive binding to the (negatively charged) nucleic acid substrates for BLM. BLM congeners lacking positively charged C substituents, notably (deglyco)BLM demethyl A_2, have been demonstrated to bind to and cleave DNA less efficiently.

Although not studied in a thorough fashion, it appears that two positively charged substituents within the C substituent[123] that are appropriately positioned[143] are sufficient to support DNA cleavage by that (deglyco)BLM congener in an optimal fashion. Although the nature of the C substituent has often been ignored in studies of (deglyco)BLM, cell culture studies make it clear that this substituent is of considerable importance from the perspective of medicinal chemistry.[100]

2. Linker Domain

Several lines of evidence have suggested that (deglyco)BLM assumes a folded conformation when it is bound to a DNA duplex substrate.[11,12,139,144,145] The structural elements that control this folded conformation have now been defined convincingly by studies from the Boger and Stubbe labora- tories.[11,12] Alterations of the threonine moiety did not affect the sequence selectivity of DNA cleavage but did affect efficiency of cleavage, as shown earlier by the Hecht laboratory.[8] The ratio of single- to double-strand cleavage was also affected by changes in the threonine moiety.[11] Careful analysis of the effect of substituents within the methylvalerate moiety indicated that substituents in the 2- and 4-positions are critical in facilitating folding into a stable conformation optimal for DNA cleavage.[11,12] Recently, an effort has been made to exploit the putative folded conformation of (deglyco)BLM by the preparation of conformationally rigid analogs containing structurally altered methylvalerate moieties.[146,147] Although the prepared analogs were capable of binding Fe^{2+} and activating oxygen for transfer to low–molecular weight substrates, none mediated sequence-selective DNA cleavage in the same fashion as BLM. The most likely interpretation of these findings is that none of the analogs was a sufficient mimic of the preferred folded conformation of BLM to permit efficient DNA cleavage to proceed. It may be noted, however, that it is possible that productive DNA binding and cleavage by BLM may actually require some flexibility within the ligand for mechanistic reasons. It seems possible that this structural domain of BLM may be of use in differentiating the binding of different classes of substrates (e.g., DNA and RNA) by BLM.

3. Metal-Binding Domain

The metal-binding domain of BLM is composed of pyrimidoblamic acid and (glycosylated) β-hydroxyhistidine moieties. Although many of the early analogs of BLM, prepared by semisynthesis from naturally occurring BLMs, were altered within this domain, relatively little has been done to systematically vary this domain.

Changes made within the β-hydroxyhistidine domain of (deglyco)BLM have mainly involved those reported by Boger and coworkers. These have been important in defining the role of the π-nitrogen of the imidazole moiety in sequence selective DNA cleavage,[148] consistent with the role of this N atom as a metal ion ligand, and the lack of importance of the OH group of β-hydroxyhistidine to the function of deglycoBLM as a DNA cleaving agent.[149]

Recently, some reports have appeared that are pertinent to the roles of the disaccharide moiety of BLM. These have included two studies in which monosaccharide derivatives have been prepared that explored the role of the L-gulose moiety. Both studies found that monosaccharide analogs of BLM containing L-gulose attached to β-hydroxyhistidine were the most effective in maintaining potency of DNA cleavage.[116,126] This monosaccharide analog was also found to be more efficient in cleaving a good RNA substrate for BLM than congeners containing D-mannose or L-rhamnose.[126] It seems clear that the carbohydrate moiety of BLM is exceptionally important in determining the efficiency of RNA cleavage by individual BLM analogs.

B. A BLM COMBINATORIAL LIBRARY

Although the studies described above have provided a good starting point for a program of systematic modification of BLM structure to identify analogs with improved properties, they provide little insight into the potential effects of preparing analogs that differ from BLM in multiple positions. Given the complexity of the structure of BLM, multiple alterations represent a particularly attractive opportunity to derive congeners having properties substantially different than those of the parent antitumor agent.

To address this opportunity, the Hecht laboratory used its methodology for solid-phase BLM synthesis (cf. Figures 19.11 and 19.12) to prepare a library of 108 deglycoBLM analogs through parallel solid-phase synthesis.[150] The library, prepared with the spermine C substituent of BLM A_6, involved the use of three analogs of the bithiazole moiety, three analogs of threonine, three analogs of the methylvalerate moiety, and four analogs of β-hydroxyhistidine (Figure 19.13). Each of the analogs was purified by preparative reversed-phase high performance liquid chromatography, characterized by ^{1}H nuclear magnetic resonance spectroscopy and mass spectrometry, and then used to assess DNA cleavage.

A few important lessons were learned from this initial library, including the observations that a number of the analogs retained the ability to effect sequence-selective DNA cleavage, and two

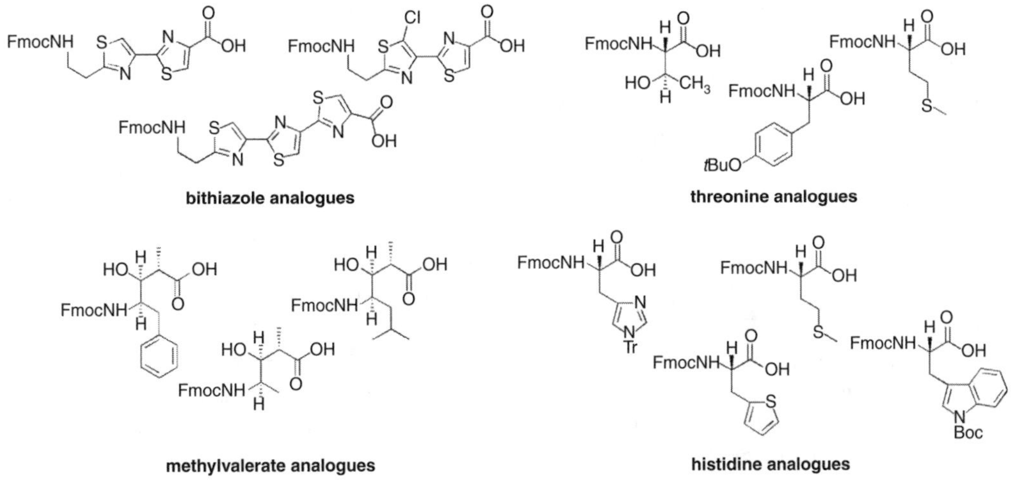

FIGURE 19.13 Building blocks used for the preparation of a deglycoBLM library.

of the analogs were more potent than the parent molecule from which they were derived in mediating relaxation of a supercoiled plasmid DNA. Critically, for any amino acid constituent whose inclusion rendered the parent molecule poorly functional in DNA cleavage, no combination of changes elsewhere in the molecule were capable of restoring function.

The ability to prepare parallel libraries of BLMs on beads should be readily extensible to mix-and-split libraries [i.e., libraries containing mixtures of beads, each of which contains multiple copies of a single (deglyco)BLM congener]. In concert with the use of DNA molecular beacons already described as substrates for BLM,[151] in principle it should be possible to identify BLM congeners with improved properties by a process of selection rather than design. Clearly, the realization of this strategy could dramatically accelerate the development of our understanding of the medicinal chemistry of BLM.

VI. CLINICAL APPLICATIONS

Bleomycin has been employed in the clinic as an antitumor agent for more than 25 years; the early studies and major findings have been summarized in two books.[152,153] Bleomycin was found to have clinically apparent activity as a single agent when used to treat squamous cell carcinomas, including those of the skin and head and neck.[154] Although initial studies involved limited numbers of patients, good response rates were also observed in the treatment of primary brain tumors and superficial bladder tumors.

A more usual application for bleomycin at this time is in combination with other antitumor agents. Bleomycin is effective as part of a regimen for the treatment of Hodgkin's lymphomas, testicular cancers, germ cell ovarian cancers, and some non-Hodgkin lymphomas.[155,156] One feature that has promoted the study of bleomycin as an agent in combination therapy is its lack of myelosuppressive activity. Pulmonary fibrosis, observable in about 10% of patients, is generally the dose-limiting toxicity for bleomycin in the clinic.[157,158]

Bleomycin is well established in the clinic as an antitumor agent in certain regimens, where it has been shown to contribute to favorable clinical outcomes. A good example is its activity in the treatment of germ cell carcinomas.[159] In spite of its long history of clinical trials and applications, new uses continue to be sought for this agent. Numerous clinical trials have been reported within the last several years for a variety of indications. Recent trials with potentially promising outcomes include those involving BLM treatment of symptomatic hemangiomas[160] in children and craniopharyngiomas.[161–163] Bleomycin has also been used recently with cisplatin and vinblastine in a comparative study for the treatment of epithelial ovarian cancer[164] and with etoposide and cisplatin for the treatment of advanced disseminated germ cell tumors.[165,166] Other recent studies include the use of bleomycin in combination with radiotherapy for the treatment of patients with inoperable head and neck cancer.[167] In a somewhat different vein, BLM has also been used to treat resistant warts by intralesional injection.[168,169]

Following reports that the toxicity of BLM in cultured cells was increased dramatically by electroporation,[170] this technique has also been incorporated into clinical trials.[171,172] Not unexpectedly, a significant increase in toxicity has been observed in some cases. The uptake of bleomycins into cells, and the factors that control cellular and nuclear uptake, is an important issue whose resolution could provide direct clinical benefit.

One striking feature of the clinical studies with BLM is that all have involved naturally derived BLMs, and most of these have differed only in the nature of the C-terminal substituent. Some clinical studies were conducted with the natural product tallysomycin,[173,174] which differs structurally from the other BLM group antibiotics (cf. Figures 19.1 and 19.8) and produces a somewhat different spectrum of DNA damage.[110,175]

The absence of clinical trials of any synthetic or semisynthetic bleomycin no doubt reflects the difficulty of obtaining such materials in quantities sufficient for preclinical and clinical evaluation. Hopefully, that situation may now have changed with the development of more efficient synthetic methodology and the use of solid-phase synthesis. Another key element that needs to be addressed is the development of assays that permit the facile characterization and understanding of the therapeutic importance of key parameters of BLM analog function, including tissue distribution and cell uptake, catabolism by BLM hydrolase and possibly other mechanisms, and the actual cellular locus of action of the drug (i.e., DNA, RNA, or both).

ACKNOWLEDGMENT

Work in our laboratory on bleomycin is supported by National Institutes of Health research grants CA76297 and CA77284, awarded by the National Cancer Institute.

REFERENCES

1. Umezawa, H., Advances in bleomycin studies, in *Bleomycin: Chemical, Biochemical and Biological Aspects*, Hecht, S. M., Ed., Springer, New York, 1979, pp 24-36.
2. Crooke, S. T., Bleomycin: a brief review, in Carter, S. K., Crooke, S. T., and Umezawa, H., Eds. *Bleomycin: Current Status and New Developments*, Academic, New York, 1978, pp 1-8.
3. Fujii, A., Biogenetic aspects of bleomycin-phleomycin group antibiotics, in *Bleomycin: Chemical, Biochemical and Biological Aspects*, Hecht, S. M., Ed., Springer, New York, 1979, pp 75-91.
4. Shen, B. et al., Cloning and characterization of the bleomycin biosynthetic gene cluster from *Streptomyces verticillus* ATCC15003, *J. Nat. Prod.*, 65, 422, 2002.
5. Chien, M., Grollman, A. P., and Horwitz, S. B., Bleomycin-DNA interactions: fluorescence and proton magnetic resonance studies, *Biochemistry*, 16, 3641, 1977.
6. Boger, D. L. et al., Total synthesis of bleomycin A$_2$ and related agents. 1. Synthesis and DNA binding properties of the extended C-terminus: tripeptide S, tetrapeptide S, pentapeptides S, and related agents, *J. Am. Chem. Soc.*, 116, 5607, 1994.
7. Sugiyama, H. et al., DNA strand scission by bleomycin: catalytic cleavage and strand selectivity, *J. Am. Chem. Soc.*, 108, 3852, 1986.
8. Carter, B. J. et al., A role for the metal binding domain of bleomycin in determining the DNA sequence selectivity of Fe-bleomycin, *J. Biol. Chem.*, 265, 4193, 1990.
9. Guajardo, R. J. et al., [FePMA]^{n+}n = 1,2: good models of Fe-bleomycins and examples of mononuclear non-heme iron complexes with significant O$_2$-activation capabilities, *J. Am. Chem. Soc.*, 115, 7971, 1993.

10. Hecht, S. M., RNA degradation by bleomycin, a naturally occurring bioconjugate, *Bioconjugate Chem.*, 5, 513, 1994.

11. Boger, D. L. et al., A systematic evaluation of the bleomycin A_2 L-threonine side chain: its role in preorganization of a compact conformation implicated in sequence-selective DNA cleavage, *J. Am. Chem. Soc.*, 120, 9139, 1998.

12. Boger, D. L et al., Definition of the effect and role of the bleomycin A_2 valerate substituents: preorganization of a rigid, compact conformation implicated in sequence-selective DNA cleavage, *J. Am. Chem. Soc.*, 120, 9149, 1998.

13. Tanaka, N., Yamaguchi, H., and Umezawa, H., Mechanism of action of phleomycin. I. Selective inhibition of the DNA synthesis in *E. coli* and HeLa cells, *J. Antibiot.*, 16A, 86, 1963.

14. Falaschi, A. and Kornberg, A., Phleomycin, an inhibitor of DNA polymerase, *Fed. Proc.*, 23, 940, 1964.

15. Mueller, W. E. and Zahn, R. K., Effect of bleomycin on DNA, RNA, protein, chromatin and on cell transformation by oncogenic RNA viruses, *Prog. Biochem. Pharmacol.*, 11, 28, 1976.

16. Ohno, T. et al., Actions of bleomycin on DNA ligase and polymerases, *Prog. Biochem. Pharmacol.*, 11, 48, 1976.

17. Hecht, S. M., The chemistry of activated bleomycin, *Acc. Chem. Res.*, 19, 383, 1986.

18. Stubbe, J. and Kozarich, J. W., Mechanisms of bleomycin-induced DNA degradation, *Chem. Rev.*, 87, 1107, 1987.

19. Natrajan, A. and Hecht, S. M., Bleomycin: mechanism of polynucleotide recognition and oxidative degradation, in *Molecular Aspects of Anticancer Drug–DNA Interactions*, Neidle, S. and Waring, M., Eds., MacMillan, London, 1994, pp 197-242.

20. Hecht, S. M., Bleomycin: new perspectives on the mechanism of action, *J. Nat. Prod.*, 63, 158, 2000.

21. Barranco, S. C. and Humphrey, R. M., Effects of bleomycin on survival and cell progression in Chinese hamster cells, *Cancer Res.*, 31, 1218, 1971.

22. Hittelman, W. N. and Rao, P. N., Bleomycin-induced damage in prematurely condensed chromosomes and its relation to cell cycle progression in CHO cells, *Cancer Res.*, 34, 3433, 1974.

23. Barlogie, B. et al., Pulse cytophotomeric analysis of cell cycle perturbation with bleomycin *in vitro*, *Cancer Res.*, 36, 1182, 1976.

24. Ishida, R. and Takahashi, T., Increased DNA chain breakage by combined action of bleomycin and superoxide radical, *Biochem. Biophys. Res. Commun.*, 66, 1432, 1975.

25. Sausville, E. A., Peisach, J., and Horwitz, S. B., Effect of chelating agents and metal ions on the degradation of DNA by bleomycin, *Biochemistry*, 17, 2740, 1978.

26. Sausville, E. A. et al., Properties and products of the degradation of DNA by bleomycin and iron II, *Biochemistry*, 17, 2746, 1978.

27. Ehrenfeld, G. M. et al., Copper I-bleomycin: structurally unique complex that mediates oxidative DNA strand scission, *Biochemistry*, 24, 81, 1985.

28. Ehrenfeld, G. M et al., Copper dependent cleavage of DNA by bleomycin, *Biochemistry*, 26, 931, 1987.

29. Ehrenfeld, G. M., Murugesan, N., and Hecht, S. M., Activation of oxygen and mediation of DNA degradation by manganese-bleomycin, *Inorg. Chem.*, 23, 1496, 1984.

30. Burger, R. M. et al., DNA degradation by manganese II-bleomycin plus peroxide, *Inorg. Chem.*, 23, 2215, 1984.

31. Kuwahara, J., Suzuki, T., and Sugiura, Y., Effective DNA cleavage by bleomycin-vanadium IV. complex plus hydrogen peroxide, *Biochem. Biophys. Res. Commun*, 129, 368, 1985.

32. Claussen, C. A. and Long, E. C., Nucleic acid recognition by metal complexes of bleomycin, *Chem. Rev.*, 99, 2797, 1999.

33. Murugesan, N., Ehrenfeld, G. M., and Hecht, S. M., Oxygen transfer from bleomycin-metal complexes, *J. Biol. Chem.*, 257, 8600, 1982.

34. Murugesan, N. and Hecht, S. M., Bleomycin as an oxene transferase. Catalytic oxygen transfer to olefins, *J. Am. Chem. Soc.*, 107, 493, 1985.

35. Burger, R. M., Peisach, J., and Horwitz, S. B., Activated bleomycin. A transient complex of drug, iron, and oxygen that degrades DNA, *J. Biol. Chem.*, 256, 11636, 1981.

36. Natrajan, A. et al., A study of O_2 vs H_2O_2-supported activation of Fe•bleomycin, *J. Am. Chem. Soc.*, 112, 3997, 1990.

37. Sugiyama, H. et al., An efficient, site-specific DNA target for bleomycin, *J. Am. Chem. Soc.*, 107, 7765, 1985.
38. Sugiyama, H. et al., DNA strand scission by bleomycin group antibiotics, *J. Nat. Prod.*, 48, 869, 1985.
39. McGall, G. H. et al., New insight into the mechanism of base propenal formation during bleomycin-mediated DNA degradation, *J. Am. Chem. Soc.*, 114, 4958, 1992.
40. Sugiyama, H. et al., Structure of the alkali-labile product formed during FeII•bleomycin-mediated DNA strand scission, *J. Am. Chem. Soc.*, 107, 4104, 1985.
41. Rabow, L. E. et al., Identification of the alkali-labile product accompanying cytosine release during bleomycin-mediated degradation of dCGCGCG., *J. Am. Chem. Soc.*, 108, 7130, 1986.
42. Sugiyama, H. et al., Chemistry of the alkali-labile lesion formed from ironII-bleomycin and dCGCTT-TAAAGCG., *Biochemistry*, 27, 58, 1988.
43. Burger, R. M., Peisach, J., and Horwitz, S. B., Stoichiometry of DNA strand scission and aldehyde formation by bleomycin, *J. Biol. Chem.*, 257, 8612, 1982.
44. Wu, J. C., Kozarich, J. W., and Stubbe, J., The mechanism of free base formation from DNA by bleomycin. A proposal based on site specific tritium release from polydA.dU, *J. Biol. Chem.*, 258, 4694, 1983.
45. Wu, J. C., Stubbe, J., and Kozarich, J. W., Mechanism of bleomycin: evidence for 4-ketone formation in polydA-dU associated exclusively with free base release, *Biochemistry*, 24, 7569, 1985.
46. Rabow, L. E., Stubbe, J., and Kozarich, J. W., Identification and quantitation of lesion accompanying base release in bleomycin-mediated DNA degradation, *J. Am. Chem. Soc.*, 112, 3196, 1990.
47. Rabow, L. E. et al., Identification of the source of oxygen in the alkaline-labile product accompanying cytosine release during bleomycin-mediated oxidative degradation of dCGCGCG., *J. Am. Chem. Soc.*, 112, 3203, 1990.
48. Kane, S. A. and Hecht, S. M., Polynucleotide recognition and degradation by bleomycin, *Prog. Nucleic Acid Res. Mol. Biol.*, 49, 313, 1994.
49. Kozarich, J. W. et al., Sequence-specific isotope effects on the cleavage of DNA by bleomycin, *Science*, 245, 1396, 1989.
50. Heimbrook, D. C., Mulholland, R. L., and Hecht, S. M, Multiple pathways in the oxidation of *cis*-stilbene by Fe•bleomycin, *J. Am. Chem. Soc.*, 108, 7839, 1986.
51. Heimbrook, D. C. et al., On the mechanism of oxygenation of *cis*-stilbene by Fe•bleomycin, *Inorg. Chem.*, 26, 3835, 1987.
52. Van Atta, R. B. et al., Electrochemical activation of oxygenated Fe•bleomycin, *J. Am. Chem. Soc.*, 111, 2722, 1989.
53. Barr, J. R. et al., FeII•BLM-mediated reduction of O_2 to H_2O: an ^{17}O-NMR study, *J. Am. Chem. Soc.*, 112, 4058, 1990.
54. Boger, D. L. et al., Total synthesis of bleomycin A_2 and related agents. 3. Synthesis and comparative evaluation of deglycobleomycin A_2, epideglycobleomycin A_2, deglycobleomycin A_1, and desacetamido-, descarboxamido-, desmethyl-, and desimidazolyldeglycobleomycin A_2, *J. Am. Chem. Soc.*, 116, 5631, 1994.
55. Povirk, L. F., Han, Y.-H., and Steighner, R. J., Structure of bleomycin-induced DNA double-strand breaks: predominance of blunt ends and single-base 5 extensions, *Biochemistry*, 28, 5808, 1989.
56. Steighner, R. J. and Povirk, L. F., Bleomycin-induced DNA lesions at mutational hot spots: implications for the mechanism of double-strand cleavage, *Proc. Natl. Acad. Sci. U.S.A.*, 87, 8350, 1990.
57. Absalon, M. J., Kozarich, J. W., and Stubbe, J., Sequence specific double-strand cleavage of DNA by Fe-bleomycin. 1. The detection of sequence-specific double-strand breaks using hairpin oligonucleotides, *Biochemistry*, 34, 2065, 1995.
58. Absalon, M. J. et al., Sequence-specific double-strand cleavage of DNA by Fe-bleomycin. 2. Mechanism and dynamics, *Biochemistry*, 34, 2076, 1995.
59. Hoehn, S. T. et al., Solution structure of CoIII-bleomycin-OOH bound to a phosphoglycolate lesion containing oligonucleotide: implications for bleomycin-induced double-strand DNA cleavage, *Biochemistry*, 40, 5894, 2001.
60. Keck, M. V., Manderville, R. A., and Hecht, S. M., Chemical and structural characterization of the interaction of bleomycin A_2 with dCGCGAATTCGCG.$_2$. Efficient, double-strand DNA cleavage accessible without structural reorganization, *J. Am. Chem. Soc.*, 123, 8690, 2001.

61. Oppenheimer, N. J. et al., Copper I•bleomycin. A structurally unique redox-active complex, *J. Biol. Chem.*, 256, 1514, 1981.

62. Wu, W. et al., Studies of Co•bleomyicn A$_2$ green: its detailed structural characterization by NMR and molecular modeling and its sequence-specific interaction with DNA oligonucleotides, *J. Am. Chem. Soc.*, 118, 1268, 1996.

63. Chang, C.-H. and Meares, C. F., Light-induced nicking of deoxyribonucleic acid by cobaltIII•bleomycins, *Biochemistry*, 21, 6332, 1982.

64. Chang, C.-H. and Meares, C. F., Cobalt-bleomycins and deoxyribonucleic acid: sequence-dependent interactions, action spectrum for nicking, and indifference to oxygen, *Biochemistry*, 23, 2268, 1984.

65. Wensel, T. G., Chang, C.-H., and Meares, C. F., Diffusion-enhanced lanthanide energy-transfer study of DNA-bound cobaltIII. bleomycins: comparisons of accessibility and electrostatic potential with DNA complexes of ethidium and acridine orange, *Biochemistry*, 24, 3060, 1985.

66. Subramanian, R. and Meares, C. F., Photosensitization of cobalt bleomycin, *J. Am. Chem. Soc.*, 108, 6427, 1986.

67. Saito, I. et al., Photoinduced DNA strand scission by cobalt bleomycin green complex, *J. Am. Chem. Soc.*, 111, 2307, 1989.

68. McLean, M. J., Dar, A., and Waring, M. J., Differences between sites of binding to DNA and strand cleavage for complexes of bleomycin with iron or cobalt, *J. Mol. Recog.*, 1, 184, 1989.

69. Magliozzo, R. S., Peisach, J., and Ciriolo, M. R., Transfer RNA is cleaved by activated bleomycin, *Mol. Pharmacol.*, 35, 428, 1989.

70. Carter, B. J. et al., Site-specific cleavage of RNA by FeII•bleomycin, *Proc. Natl. Acad. Sci. U.S.A.*, 87, 9373, 1990.

71. Carter, B. J., Reddy, K. S., and Hecht, S. M., Polynucleotide recognition and strand scission by Fe•bleomycin, *Tetrahedron*, 47, 2463, 1991.

72. Carter, B. J. et al., Metal-catalyzed polynucleotide strand scission, *Nucleosides Nucleotides*, 10, 215, 1991.

73. Holmes, C. E., Carter, B. J., and Hecht, S. M., Characterization of ironII.-bleomycin-mediated RNA strand scission, *Biochemistry*, 32, 4293, 1993.

74. Morgan, M. and Hecht, S. M., IronII•bleomycin-mediated degradation of a DNA-RNA heteroduplex, *Biochemistry*, 33, 10286, 1994.

75. Holmes, C. E. et al., Fe•bleomycin as a probe of RNA conformation, *Nucleic Acids Res.*, 24, 3399, 1996.

76. Kim, S. H. et al., Three-dimensional tertiary structure of yeast phenylalanine transfer RNA, *Science*, 185, 435, 1974.

77. Robertus, J. D. et al., Structure of yeast phenylalanine tRNA at 3Å resolution, *Nature*, 250, 546, 1974.

78. Stout, C. D. et al., Crystal and molecular structure of yeast phenylalanyl transfer RNA. Structure determination, difference Fourier refinement, molecular conformation, metal and solvent binding, *Acta Crystallogr.*, 354, 1529, 1978.

79. Moras, D. et al., Crystal structure of yeast tRNA[Asp], *Nature*, 288, 699, 1980.

80. Westhof, E., Dumas, P., and Moras, D., Crystallographic refinement of yeast aspartic acid transfer RNA, *J. Mol. Biol.*, 184, 119, 1985.

81. Giegé, R. et al., Exploring the aminoacylation function of transfer RNA by macromolecular engineering approaches-involvement of conformational features in the charging process of yeast transfer RNA[Asp], *Biochimie*, 72, 453, 1990.

82. Perret, V. et al., Conformation in solution of yeast transfer RNA[Asp] transcripts deprived of modified nucleotides, *Biochimie*, 72, 735, 1990.

83. Pütz, J. et al., Identity elements for specific aminoacylation of yeast transfer RNA[Asp] by cognate aspartyl-transfer RNA synthetase, *Science*, 252, 1696, 1991.

84. Perret, V. et al., Effect of conformational features on the aminoacylation of transfer RNAs and consequences on the permutation of transfer RNA specificities, *J. Mol. Biol.*, 226, 323, 1992.

85. Rudinger, J. et al., Determinant nucleotides of yeast tRNA[Asp] interact directly with aspartyl-tRNA synthetase, *Proc. Natl. Acad. Sci. U.S.A.*, 89, 5882, 1992.

86. Abraham, A. T. et al., RNA cleavage and inhibition of protein synthesis by bleomycin, *Chem. Biol.*, 10, 45, 2003.

87. Dix, D. J. et al., The influence of the base-paired flanking region on structure and function of the ferritin mRNA iron regulatory element, *J. Mol. Biol.*, 231, 230, 1993.

88. Snow, A., Ph.D. Thesis, University of Virginia, 1995.

89. Duff, R. J. et al., Evidence for C-1' hydrogen abstraction from modified oligonucleotides by Fe•bleomycin, *J. Am. Chem. Soc.*, 115, 3350, 1993.

90. Holmes, C. E. et al., On the chemistry of RNA degradation by FeII•BLM, *Bioorg. Med. Chem.*, 5, 1235, 1997.

91. Pieters, J. M. L. et al., Hairpin structures in DNA containing arabinofuranosylcytosine. A combination of nuclear magnetic resonance and molecular dynamics, *Biochemistry*, 29, 788, 1990.

92. Keck, M. V. and Hecht, S. M., Sequence-specific hydrolysis of yeast tRNA[Phe] mediated by metal free bleomycin, *Biochemistry*, 34, 12029, 1995.

93. Ardelt, W., Mikulski, S. M., and Shogen, K., Amino-acid-sequence of an antitumor protein from rana-pipiens oocytes and early embryos-homology to pancreatic ribonucleases, *J. Biol. Chem.*, 266, 245, 1991.

94. Darzynkiewicz, Z. et al., Cystostatic and cytotoxic effects of pannon P-30. A novel anticancer agent, *Cell Tissue Kinet.*, 21, 169, 1988.

95. Mikulski, S. M. et al., Tamoxifen and trifluoroperazine (stelazine) potentiate cystostatic cytotoxic effects of P-30 protein. A novel protein possessing antitumor activity, *Cell Tissue Kinet.*, 23, 237, 1990.

96. Mikulski, S. M. et al., Striking increase of survival of mice bearing M109 madison carcinoma treated with a novel protein from amphibian embryos, *J. Natl. Cancer Inst.*, 82, 151, 1990.

97. Mikulski, S. M. et al., Phase-1 human clinical trial of onconase R (P-30 protein) administered intravenously on a weekly schedule in cancer patients with solid tumors, *Int. J. Oncol.*, 3, 57, 1993.

98. Lin, J.-J. et al., Characterization of the mechanism of cellular and cell free protein synthesis inhibition by an anti-tumor ribonuclease, *Biochem. Biophys. Res. Commun.*, 204, 156, 1994.

99. Saxena, S. K. et al., Entry into cells and selective degradation of tRNAs by a cytotoxic member of the RNase A family, *J. Biol. Chem.*, 277, 15142, 2002.

100. Berry, D. E., Chang, L.-H., and Hecht, S. M., DNA damage and growth inhibition in cultured human cells by bleomycin congeners, *Biochemistry*, 24, 3207, 1985.

101. Hecht, S. M., Bleomycin group antitumor agents, in *Cancer Chemotherapeutic Agents*, Foye, W. O., Ed., American Chemical Society, Washington, DC, 1995, pp 369-388.

102. Takita, T. et al., Chemical cleavage of bleomycin to bleomycinic acid and synthesis of new bleomycins, *J. Antibiot.*, 26, 252, 1973.

103. Tanaka, W. and Takita, T., Pepleomycin, the second generation bleomycin chemically derived from bleomycin A$_2$, *Heterocycles*, 13, 469, 1979.

104. Umezawa, H., Recent chemical studies of bioactive microbial products: genetics, active structures, development of effective agents with potential usefulness, *Heterocycles*, 13, 23, 1979.

105. Umezawa, H. et al., Preparation of bleomycinic acid: hydrolysis of bleomycin B$_2$ by a Fusarium acylagmatine amidohydrolase, *J. Antibiot.*, 26, 117, 1973.

106. Tanaka, W., Development of new bleomycins with potential clinical utility, *J. Antibiot.*, 30, S-41, 1977.

107. Saito, S. et al., A new synthesis of deglyco-bleomycin A$_2$ aiming at the total synthesis of bleomycin, *Tetrahedron Lett.*, 23, 529, 1982.

108. Aoyagi, Y. et al., Deglycobleomycin: total synthesis and oxygen transfer properties of an active bleomycin analog, *J. Am. Chem. Soc.*, 104, 5237, 1982.

109. Oppenheimer, N. J. et al., Deglyco-bleomycin. Degradation of DNA and formation of a structurally unique FeII•CO complex, *J. Biol. Chem.*, 257, 1606, 1982.

110. Kross, J. et al., Specificity of deoxyribonucleic acid cleavage by bleomycin, phleomycin and tallysomycin, *Biochemistry*, 21, 4310, 1982.

111. Takahashi, K. et al., Liblomycin, a new analog of bleomycin, *Cancer Treat. Rev.*, 14, 169, 1987.

112. Takita, T. et al., Total synthesis of bleomycin A$_2$, *Tetrahedron Lett.*, 23, 521, 1982.

113. Aoyagi, Y. et al., Total synthesis of bleomycin, *J. Am. Chem. Soc.*, 104, 5537, 1982.

114. Boger, D. L. and Honda, T., Total synthesis of bleomycin A$_2$ and related agents. 4. Synthesis of the disaccharide subunit 2-O-3-O-carbamoyl-α-D-mannopyranosyl-L-gulopyranose and completion of the total synthesis of bleomycin A$_2$, *J. Am. Chem. Soc.*, 116, 5647, 1994.

115. Katano, K. et al., Total synthesis of bleomycin group antibiotics. Total synthesis of bleomycin demethyl A$_2$, bleomycin A$_2$, and decarbamoyl bleomycin demethyl A$_2$, *J. Am. Chem. Soc.*, 120, 11285, 1998.

116. Boger, D. L., Teramoto, S., and Zhou, J., Key synthetic analogs of bleomycin A_2 that directly address the effect and role of the disaccharide: demannosylbleomycin A_2 and α-D-mannopyranosyldeglycobleomycin A_2, *J. Am. Chem. Soc.*, 117, 7344, 1995.

117. Zou, Y. et al., Total synthesis of deamido bleomycin A_2, the major catabolite of the antitumor agent bleomycin, *J. Am. Chem. Soc.*, 124, 9476, 2002.

118. Leitheiser, C. J. et al., Solid phase synthesis of bleomycin group antibiotics. Elaboration of deglycobleomycin A_5, *Org. Lett.*, 2, 3397, 2000.

119. Aoyagi, Y. et al., Synthesis of pyrimidoblamic acid and epipyrimidoblamic acid, *J. Org. Chem.*, 55, 6291, 1990.

120. Abraham, A. T., Zhou, X., and Hecht, S. M., DNA cleavage by FeII•bleomycin conjugated to a solid support, *J. Am. Chem. Soc.*, 121, 1982, 1999.

121. Smith, K. L. et al., Deglycobleomycin: solid-phase synthesis and DNA cleavage by the resin-bound ligand, *Org. Lett.*, 4, 1079, 2002.

122. Gordeev, M. F. et al., Combinatorial chemistry of natural products: solid phase synthesis of D-and L-cycloserine derivatives, *Tetrahedron*, 54, 15879, 1998.

123. Tao, Z.-F. et al., Solid phase synthesis of deglycobleomycins: a C-terminal tetramine linker that permits direct evaluation of resin-bound bleomycins, *Bioconjugate Chem.*, 13, 426, 2002.

124. Owa, T., Otsuka, M., and Ohno, M., An efficient synthesis of *erythro*-β-hydroxy-L-histidine, the pivotal amino acid of bleomycin-FeII•O_2 complex, *Chem. Lett.*, 1873, 1988.

125. Boger, D. L. and Menezes, R. F., Synthesis of tri- and tetrapeptide S: the extended C-terminus of bleomycin A_2, *J. Org. Chem.*, 57, 4331, 1992.

126. Thomas, C. J. et al., Solid phase synthesis of bleomycin A_5 and three monosaccharide analogs. Exploring the role of the carbohydrate moiety in RNA cleavage, *J. Am. Chem. Soc.*, 124, 12926, 2002.

127. Hamamichi, N. and Hecht, S. M., Determination of the absolute configuration of the thiazolinylthiazole moiety of phleomycin, *J. Am. Chem. Soc.*, 115, 12605, 1993.

128. Sznaidman, M. L. and Hecht, S. M., Studies on the total synthesis of tallysomycin. Synthesis of the threonylbithiazole moiety containing a structurally unique glycosylcarbinolamide, *Org. Lett.*, 3, 2811, 2001.

129. Morii, T. et al., Phototransformed bleomycin antibiotics. Structure and DNA cleavage activity, *J. Am. Chem. Soc.*, 108, 7089, 1986.

130. Morii, T. et al., New lumibleomycin-containing thiazolylisothiazole ring, *J. Am. Chem. Soc.*, 109, 938, 1987.

131. Zuber, G., Quada Jr., J. C., and Hecht, S. M., Sequence selective cleavage of a DNA octanucleotide by chlorinated bithiazoles and bleomycins, *J. Am. Chem. Soc.*, 120, 9368, 1998.

132. Hamamichi, N., Natrajan, A., and Hecht, S. M., On the role of individual bleomycin thiazoles in oxygen activation and DNA cleavage, *J. Am. Chem. Soc.*, 114, 6278, 1992.

133. Otsuka, M. et al., Man-designed bleomycin with altered sequence specificity in DNA cleavage, *J. Am. Chem. Soc.*, 112, 838, 1990.

134. Owa, T. et al., Man-designed bleomycins: significance of the binding sites as enzyme models and of the stereochemistry of the linker moiety, *Tetrahedron*, 48, 1193, 1992.

135. Kuwahara, J. and Sugiura, Y., Sequence-specific recognition and cleavage of DNA by metallobleomycin: minor groove binding and possible interaction mode, *Proc. Natl. Acad. Sci. U.S.A.*, 85, 2459, 1988.

136. Quada Jr., J. C., Zuber, G. F., and Hecht, S. M., Interaction of bleomycin with DNA, *Pure Appl. Chem.*, 70, 307, 1998.

137. Quada Jr., J. C., Boturyn, D., and Hecht, S. M., Photoactivated DNA cleavage by compounds structurally related to the bithiazole moiety of bleomycin, *Bioorg. Med. Chem.*, 9, 2303, 2001.

138. Manderville, R. A., Ellena, J. F., and Hecht, S. M., Interaction of ZnII•bleomycin with dCGCTAGCG$_2$. A binding model based on NMR experiments and restrained molecular dynamics calculations, *J. Am. Chem. Soc.*, 117, 7891, 1995.

139. Sucheck, S. J., Ellena, J. F., and Hecht, S. M., Characterization of ZnII•deglycobleomycin A_2 and interaction with dCGCTAGCG$_2$. Direct evidence for minor groove binding of the bithiazole moiety, *J. Am. Chem. Soc.*, 120, 7450, 1998.

140. Williams, L. D. and Goldberg, I. H., Selective strand scission by intercalating drugs at DNA bulges, *Biochemistry*, 27, 3004, 1988.

141. Gold, B. et al., Alteration of bleomycin cleavage specificity in a platinated DNA oligomer of defined structure, *J. Am. Chem. Soc.*, 110, 2347, 1988.
142. Kane, S. A. et al., Specific cleavage of a DNA triple helix by Fe^{II}•bleomycin, *Biochemistry*, 34, 16715, 1995.
143. Kross, J. et al., Structural basis for the DNA affinity of bleomycins, *Biochemistry*, 21, 3711, 1982.
144. Lui, S. M. et al., Structural characterization of Co•bleomycin A$_2$ brown: free and bound to dCCAG-GCCTGG., *J. Am. Chem. Soc.*, 199, 9603, 1997.
145. Wu, W. et al., NMR studies of Co•deglycobleomycin A$_2$ green and its complex with dCCAGGC-CTGG., *J. Am. Chem. Soc.*, 120, 2239, 1998.
146. Rishel, M. J. et al., Conformationally constrained analogs of bleomycin A$_5$, *J. Am. Chem. Soc.*, 125, 10194, 2003.
147. Cagir, A. et al., Solid-phase synthesis and biochemical evaluation of conformationally constrained analogs of deglycobleomycin A$_5$, *Bioorg. Med. Chem.*, 11, 5179, 2003.
148. Boger, D. L., Ramsey, T. M., and Cai, H., Synthesis and evaluation of potential N$^\pi$ and N$^\sigma$ metal chelation sites with the beta-hydroxy-L-histidine subunit of bleomyin A$_2$: functional characterization of imidazole N$^\pi$ metal complexation, *Bioorg. Med. Chem.*, 4, 195, 1996.
149. Boger, D. L., Teramoto, S., and Cai, H., Synthesis and evaluation of deglycobleomycin A$_2$ analogs containing a tertiary *N*-methyl amide and simple ester replacement for the L-histidine secondary amide: direct functional characterization of the requirement for secondary amide metal complexation, *Bioorg. Med. Chem.*, 4, 179, 1996.
150. Leitheiser, C. J. et al., Solid phase synthesis of bleomycin group antibiotics. Construction of a 108-member deglycobleomycin library, *J. Am. Chem. Soc.*, 125, 8218, 2003.
151. Hashimoto, S., Wang, B., and Hecht, S. M., Kinetics of DNA cleavage by FeII•bleomycins, *J. Am. Chem. Soc.*, 123, 7437, 2001.
152. *Bleomycin: Current Status and New Developments*, Carter, S. K., Crooke, S. T., and Umezawa, H., Eds., Academic Press, Orlando, FL, 1978.
153. *Bleomycin Chemotherapy*, Sikic, B. I., Rozencweig, M., and Carter, S. K., Eds., Academic Press, Orlando, FL, 1985.
154. Bennett, J. M. and Reich, S. D., Bleomycin, *Ann. Intern. Med.*, 90, 945, 1979.
155. Einhorn, L. H. and Donohue, J., Cis-diamminedichloroplatinum, vinblastine, and bleomycin combination chemotherapy in disseminated testicular cancer, *Ann. Intern. Med.*, 87, 293, 1977.
156. Carlson, R. W. et al., Combination cisplatin, vinblastine, and bleomycin chemotherapy (PVB) for malignant germ-cell tumors of the ovary, *J. Clin. Oncol.*, 1, 645, 1983.
157. Comis, R. L. et al., Role of single-breath carbon monoxide-diffusing capacity in monitoring the pulmonary effects of bleomycin in germ cell tumor patients, *Cancer Res.*, 39, 5076, 1979.
158. Muggia, F. M., Louie, A. C., and Sikic, B. I., Pulmonary toxicity of antitumor agents, *Cancer Treat. Rev.*, 10, 221, 1982.
159. Levi, J. A. et al., The importance of bleomycin in combination chemotherapy for good-prognosis germ cell carcinoma, *J. Clin. Oncol.*, 11, 1300, 1993.
160. Kullendorff, C. M., Efficacy of bleomycin treatment for symptomatic hemangiomas in children, *Ped. Surg. Int.*, 12, 526, 1997.
161. Mottolese, C. et al., Intracystic chemotherapy with bleomycin in the treatment of craniophrryngiomas, *Childs Nervous System*, 17, 724, 2001.
162. Alen, J. F. et al., Intratumoural bleomycin as a treatment for recurrent cystic craniopharyngioma. Case report and review of the literature, *Neurocirugia*, 13, 479, 2002.
163. Jiang, R., Liu, Z., and Zhu, C., Preliminary exploration of the clinical effect of bleomycin on craniopharyngiomas, *Stereotactic Funct. Neurosurg.*, 78, 84, 2002.
164. Tokuhashi, Y. et al., A randomized trial of cisplatin, vinblastine, and bleomycin versus cyclophospha-mide, aclacinomycin, and cisplatin in eipthelial ovarian cancer, *Oncology*, 54, 281, 1997.
165. Nichols, C. R et al., Randomized comparison of cisplatin and etoposide and either bleomycin or ifosfamide in treatment of advanced disseminated germ cell tumors: an Eastern Cooperative Oncology Group, Southwest Oncology Group, and Cancer and Leukemia Group B study, *J. Clin. Oncol.*, 16, 1287, 1998.
166. Christian, J. A. et al., Intensive induction chemotherapy with CBOP/BEP in patients with poor prognosis germ cell tumors, *J. Clin. Oncol.*, 21, 871, 2003.

167. Minatel, E. et al., Combined radiotherapy and bleomycin in patients with inoperable head and neck cancer with unfavourable prognostic factors and severe symptoms, *Oral Oncol.*, 34, 119, 1998.

168. Munn, S. E. et al., A new method of intralesional bleomycin therapy in the treatment of recalcitrant warts, *Br. J. Dermatol.*, 135, 969, 1996.

169. Pollock, B. and Sheehan-Dare, R., Pulsed dye laser and intralesional bleomycin for treatment of resistant viol hand warts, *Lasers Surg. Med.*, 30, 135, 2002.

170. Poddevin, B. et al., Very high cytotoxicity of bleomycin introduced into the cytosol of cells in culture, *Biochem. Pharmacol.*, 42, 567, 1991.

171. Horiuchi, A. et al., Enhancement of antitumor effect of bleomycin by low-voltage in vivo electroporation: a study of human uterine leiomyosarcomas in nude mice, *Int. J. Cancer*, 88, 640, 2000.

172. Nanda, G. S. et al., Electroporation enhances therapeutic efficacy of anticancer drugs: treatment of human pancreatic tumor in animal models, *Anticancer Res.*, 18, 1361, 1998.

173. Nicaise, C. et al., Phase II study of tallysomycin S10b in patients with advanced colorectal cancer, *Cancer Chemother. Pharmacol.*, 26, 221, 1990.

174. Nicaise, C. et al., Phase II study of tallysomyin S10b in patients with advanced head and neck cancer, *Invest. New Drugs*, 8, 325, 1990.

175. Mirabelli, C. K., Huang, C. H., and Crooke, S. T., Comparison of DNA damage and single- and double-strand breakage activities on PM-2 DNA by tallysomycin and bleomycin analogs, *Cancer Res.*, 40, 4173, 1980.

20 Biochemical and Biological Evaluation of (+)-CC-1065 Analogs and Conjugates with Polyamides

Rohtash Kumar and J. William Lown

CONTENTS

I. Introduction 383
II. Drug–DNA Interactions 385
III. Modification in the Pharmacophore Unit 386
    A. CPI Analogs 386
    B. Enantiomers of CPI Analogs 387
    C. CC-1065 Analogs 389
IV. Interaction of CC-1065 with Chromatin 392
V. Polyamide Conjugates and Related Sequence-Directed Structures 394
VI. Bisalkylators (Dimers) 400
VII. Structural Characterization of Drug DNA Complexes 402
VIII. Cellular and Pharmacological Studies 403
    A. *In Vitro* Studies 403
    B. *In Vivo* Studies 406
IX. Clinical Studies 407
X. Conclusions and Prospects 407
References 408

I. INTRODUCTION

CC-1065 (**1**) is an extremely potent antitumor antibiotic isolated from *Streptomyces zelensis* by the Upjohn company.[1,2] It is one of the most potent anticancer compounds having a wide spectrum of activities against tumor cells *in vitro* and *in vivo* as well as against microbial organisms.[3,4] Structurally, the CC-1065 molecule consists of three repeating pyrroloindole subunits, one of which contains the DNA reactive cyclopropylpyrroloindole (CPI) moiety (**2**) while the other two subunits mediate noncovalent binding interactions with DNA. CC-1065 alkylates the N3 position of adenine with its reactive cyclopropyl group in the minor groove of double helical DNA in a sequence specific manner with a consensus sequence of 5-(T/A)(T/A) A* (where the asterisk indicates the site of alkylation). The preferred site of alkylation by CC-1065 was determined to be the 5-TTA* sequence.[4-10] Despite the highly potent activity of CC-1065 *in vitro*[3,11] and *in vivo*,[2] clinical development was precluded by an unusual hepatotoxicity, which led to delayed death in mice at therapeutic doses.[12] Subsequently, an aggressive and highly successful synthetic program at the

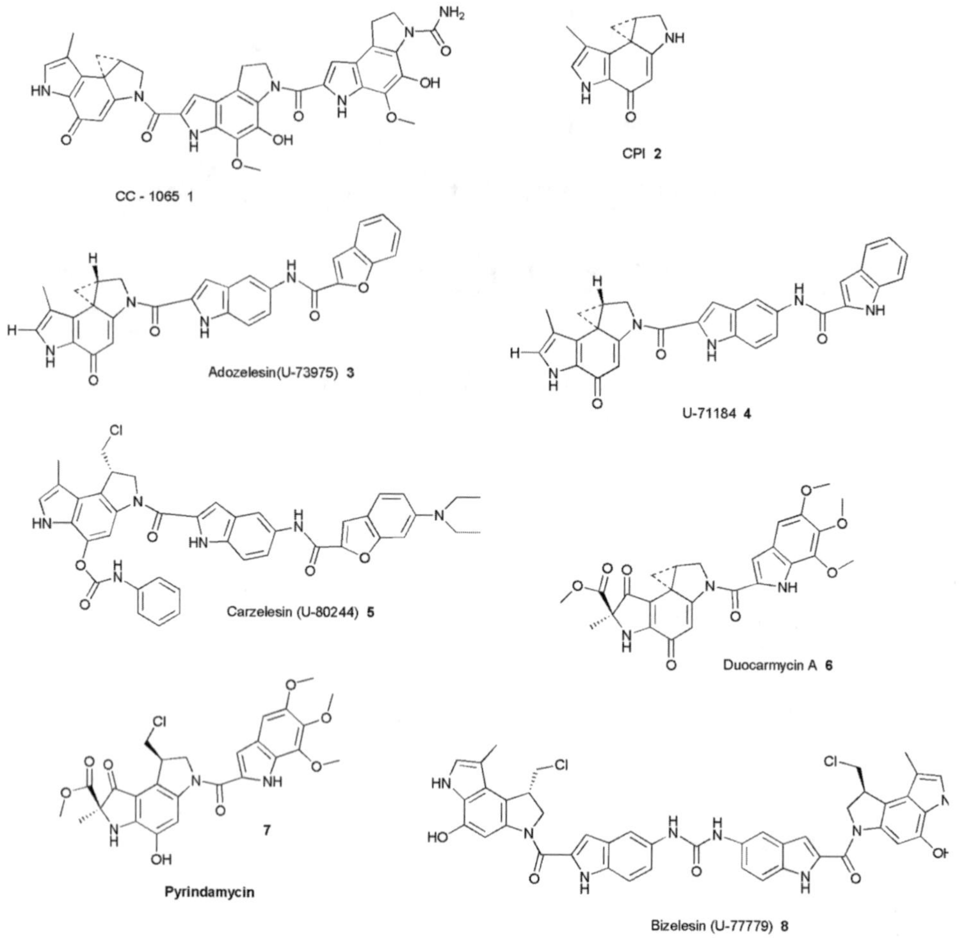

CC - 1065 1

CPI 2

Adozelesin(U-73975) 3

U-71184 4

Carzelesin (U-80244) 5

Duocarmycin A 6

7

Pyrindamycin

Bizelesin (U-77779) 8

STRUCTURES 1–8

Upjohn company provided a series of three new compounds that appeared to be more suitable for clinical development. These agents, adozelesin (**3**), U-71184 (**4**), and U-80224 (**5**), have been shown to have improved antitumor efficacy over CC-1065 and do not show delayed lethality.[13,14]

Duocarmycin (**6**) and pyrindamycin (**7**) are structurally related natural products, with (**6**) possessing the CC-1065-DNA reactive cyclopropane ring.[15,16] The structural element of CC-1065 that produces delayed toxicity has been elucidated, but its biochemical mechanism is still unknown. A number of CC-1065 analogs, which maintain potent antitumor activity without exhibiting delayed death, were subsequently synthesized, and the unique feature of all these synthetic analogs is the cyclopropyl indole moiety (CPI; **2**). Bizelesin (**8**) is a bifunctional analog, shows good antitumor efficacy both *in vitro* and *in vivo*, and is generally 20–30-fold more potent than adozelesin (a monofunctional analog) when tested against human carcinoma cells.[17] Adozelesin **3** is currently in phase II clinical trials, and bizelesin (**8**) is currently in phase I clinical trials.

Certain synthetic analogs of CC-1065 and duocarmycin, including dimeric structures, show promise and are also currently undergoing clinical trials,[17] indicating that synthetic efforts in this area can be useful. A new class of agents bearing the 1,2,9,9a-tetrahydrocyclopropa[c] benz [e] indole-4-one (**9**) pharmacophore (CBI) has been synthesized.[18]

In this chapter, in addition to delineating the cytotoxic and antitumor activities of CC-1065, we also examine the most recent analogs of CPI and CBI and their cytotoxicities. Certain CPI–polyamide conjugates have shown good cytotoxic potency[19,20]; therefore, we also describe various new CPI and

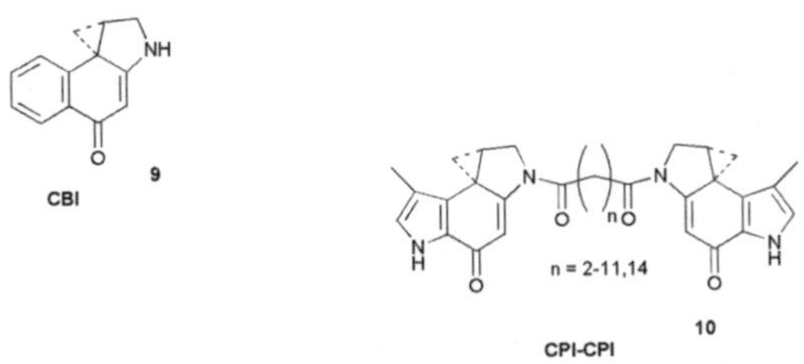

STRUCTURES 9–10

CBI polyamide conjugates. Since CPI–CPI dimers were very active against tumor cells, a large number of CBI–CBI dimers with varying lengths of the aliphatic chain linkers and with specific polyamides were synthesized in our group, and their cytotoxic properties against tumor cells are summarized below.

II. DRUG–DNA INTERACTIONS

The mechanisms of cytotoxic action of CC-1065 and its analogs have been shown to involve binding to the minor groove of DNA and alkylation at the N-3 position of adenine with concomitant opening of the reactive CPI unit (Figure 20.1). Thermal treatment of this covalently modified DNA results in depurination of the alkylated bases followed by strand cleavage, producing a 5-phosphate and 3-modified deoxyribose.[4,5,21]

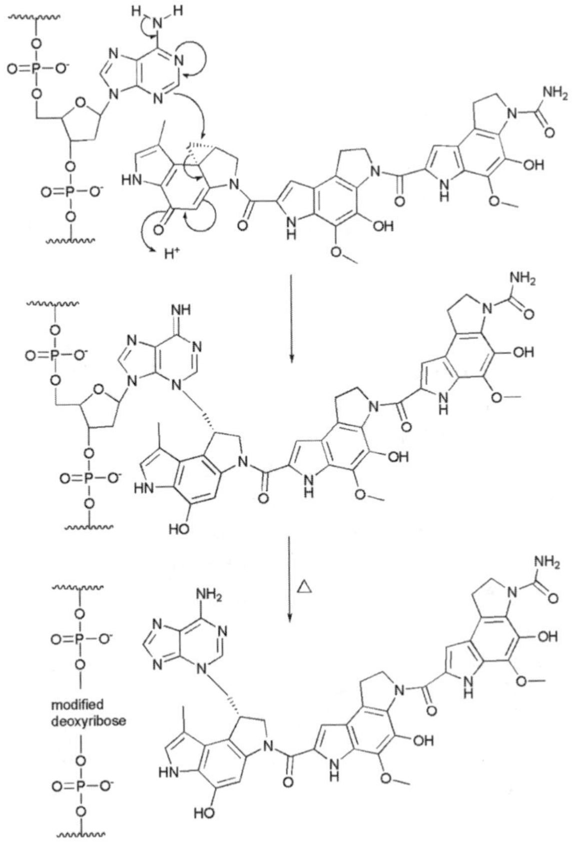

FIGURE 20.1 The mechanisms of cytotoxic action of CC-1065 and its analogs.

Both enantiomers primarily alkylate N-3 of adenine, but the orientation of the drug is different. For the natural (+) enantiomer, the B and C subunits lie on the 5 site of alkylation and on the 3 site for the (–) enantiomer, as shown by DNA footprinting. They also show different sequence selectivity, although both prefer A•T rich sites. However, (+) and (–) analogs that contain only the A subunit do not show different sequence selectivity, demonstrating the importance of the B and C subunits in determining the site of alkylation. The unnatural enantiomer appears to be less reactive and provides less stabilization of the DNA helix as judged by the change in melting temperature of the alkylated DNA. A slight increase in the amount of alkylation on guanine was observed for (–)-CC-1065. Only the natural enantiomer appears to cause delayed death.[7,12,22,23]

Recent studies have shown that many analogs of CC-1065 reversibly alkylate DNA. However, CC-1065 itself appears to alkylate DNA irreversibly under the conditions tested. This difference is attributed to the strong noncovalent binding of CC-1065, which either prevents the reverse reaction after alkylation has occurred or keeps the drug bound so that it realkylates the same site. Boger and Johnson have suggested that irreversibility of the alkylation may be related to the delayed toxicity of some of these compounds.[24]

Several studies have shown that the unnatural enantiomers or epimers of this class of compounds also alkylate DNA and possess cytotoxic activity. Terashima's group synthesized (+)-duocarmycin A and its three possible stereoisomers and compared their *in vitro* cytotoxicities against P388 murine leukemia, demonstrating that both compounds bearing natural configurations at the cyclopropane moiety were more potent than those having unnatural configurations. This result indicated the importance of chirality in the cyclopropyl ring in duocarmycin A.[25,26]

Boger and Johnson[24] synthesized both enantiomers of CC-1065 and duocarmycin A, duocarmycin SA, and epimers of duocarmycin A and compared cytotoxic potencies against L1210 and relative DNA alkylation efficiencies. In the case of duocarmycin A, both (–) forms, which had opposite chirality in the cyclopropyl ring to that of natural (+)-duocarmycin A, had almost no effect on the DNA alkylation and cytotoxic potency. In the case of duocarmycin SA, the natural enantiomer was found to alkylate DNA at concentrations approximately 10 times that required for the nonnatural products.[27]

III. MODIFICATION IN THE PHARMACOPHORE UNIT

A. CPI Analogs

Compounds derived from the pharmacophore of CC-1065 but bearing different electron-withdrawing constituents in the benzene ring and different leaving groups on the C3 methylene have been synthesized and evaluated.[28] These synthetic agents were less reactive at the common preferred sites than CC-1065, and higher concentrations are required to produce detectable cleavage. On the basis of alkylation intensities and densitometric data, the overall binding intensity is in the order of CC-1065 > 12 > 13 > 11. From those experiments, it appears that there are two factors that are critical for the formation of the cyclopropane ring by cyclization: The first is the electron density of the benzoyl moiety where an electron-withdrawing group retards the cyclization, whereas an electron-donating group promotes the cyclization, and the second factor is the leaving ability of the group on the C3 methylene carbon. A better leaving group will favor the cyclization.

The evidence from gel electrophoresis showed that when supercoiled PM2 DNA is incubated with 11, 12, 13, and CC-1065 at 80 μ*M* for 20h, the CCC DNA form was completely converted to OC DNA. This indicated that, like CC-1065, 11, 12, and 13 also exert their cytotoxicity through covalent binding to DNA. However, additional binding at the G residues by these synthetic analogs may represent an alternative or additional mechanism for this recognition of DNA.[29,30] In this connection, the DNA sequence specificity of the pyrrolo [1,4] benzodiazepine antitumor antibiotics, which bind through N2 of guanine, has been supported by theoretical calculations.[31]

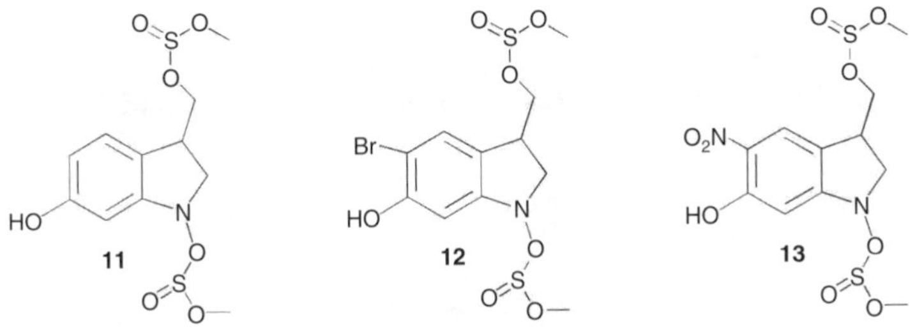

STRUCTURES 11–13

A series of N-substituted CPI-analogs (**14–40**) has been synthesized, and certain structural features have been identified that affect reactivity toward nucleophiles and those influencing non-covalent interactions with DNA. Those features have then been related to biological activity and potency.[13] Table 20.1 lists the biological activities of CC-1065 and the racemic CPI analogs. Solvolytic reactivity is indicated by the pseudo-first-order rate constant for acid-catalyzed opening of the cyclopropyl ring in methanolic buffer. When CC-1065 is incubated with calf thymus DNA for 24 h at 25°C (Table 20.1), the observed induced circular dichroism (ICD) is entirely caused by the binding. Although the racemic analogs display no circular dichroism on their own, in the presence of DNA some of them show significant ICD. From the ICD it appears that two structural features of the CPI moiety strongly influence DNA binding, the size of the ring attached to the pyrrolidine nitrogen of CPI, and the length of the aromatic amide chain. An ICD is almost exclusively associated with structures having a five-membered heteroaryl ring attached to CPI. Methyl substituents on the five-membered ring significantly diminish the ICD, as is also the case when the nitrogen atom is replaced with oxygen or sulfur. Analogs with six-membered rings attached to CPI fail to show an appreciable ICD. Table 20.1 shows that cytotoxic potency is qualitatively correlated to DNA binding.

The DNA alkylation selectivity of the analogs of (+) CC-1065 has been studied.[32,33] The sites of alkylation of double-stranded DNA were examined for simple derivatives of 7-methyl-1, 2,8,8a-tetrahydrocycloprop[1,2-C] pyrrolo [2,2-e] indole-4 (5H)-one (CPI; **2**). (+) CC-1065 (**1**) and 1,2,7,7a- tetrahydrocycloprop [1,2-C] indole-4-one (CI; **41**) left-hand subunits. It was found that the CI subunit of the agents is a much more reactive alkylating agent than the natural CPI alkylation subunit of CC-1065. The simple derivatives **42, 43, 44,** and **45** of CI were found to alkylate double-stranded DNA under milder conditions than did simple derivatives of CPI, and the marked similarities in CI and CPI DNA alkylation profiles indicate that CI represents the minimum pharmacophore of CPI. By use of the Fe (III)-CDPI$_3$ EDTA (FeIII-**42, -43**), they demonstrated that the noncovalent binding is a prerequisite for DNA alkylations. The relative cleavage efficiency observed with Fe(III)-**46** was found to parallel the relative alkylation efficiency of (+)- or net (–)-CC-1065.

B. ENANTIOMERS OF CPI ANALOGS

Boger's group[34] made a detailed evaluation of the DNA-alkylation selectivity of (+)-CC-1065 and its enantiomer (–) CC-1065 and a series of analogs (**47–51**) possessing the CPI alkylation subunit. The natural enantiomers bind in the minor groove in the 3 → 5 direction starting from the adenine N3 alkylation site across a 2-base (N-BocCPI **47**; i.e., 5-AA), 3.5-base (CPI-CDPI$_1$ **48**/CPI-PDEI$_1$ **51**; i.e., 5-AAA), 5-base (CC-1065/CPI-CDPI$_2$ **49**; i.e., 5-AAAAA), or 6.5-base (CPI-CDPI$_3$ **50**; i.e., 5-AAAAAA) AT-rich site. In contrast, the unnatural enantiomers bind in the reverse 5 → 3 direction in the minor groove, and the binding site starts at the first 5 base preceding the adenine N3 alkylation site and extends across the alkylation site to the adjacent 3 base covering an AT-rich site of 2 bases (NBoc-CPI **47**; i.e., 5AA), 5 bases (CC-1065/CPI-CDPI$_2$; i.e., 5-AAAAA), or 6.5 bases (CPI-CDPI$_3$; i.e., 5-AAAAAA). Three-dimensional models of natural and unnatural enantiomer alkylations clearly

TABLE 20.1
Physicochemical and Biological Properties of Racemic CC-1065 and CPI Analogs

No.	R	Formula	R_2[a]	$K \times 10^5$[a]	$ICD \times 10^3$[c]	ID_{50}[d]	P388 In Vivo[e] OD[f]	%ILS[g]
14	(PDE-I)$_2$	C$_{37}$H$_{33}$N$_7$O$_8$	0.39		280	0.07	0.1	62
15	SO$_2$Me	C$_{13}$H$_{14}$N$_2$O$_3$S	0.82	0.3	0	100	25	60
16	SO$_2$Ph	C$_{18}$H$_{16}$N$_2$O$_3$S	0.58			900		
17	H	C$_{12}$H$_{12}$N$_2$O	0.87	Stable	0	900	10(HDT)	NA
18	CO$_2$C(CH$_3$)$_3$	C$_{17}$H$_{20}$N$_2$O$_3$	0.35	0.5	0	300		
19	COCH$_3$	C$_{14}$H$_{14}$N$_2$O$_2$	0.90		0	20	3	45
20	CO(CH$_2$)$_8$CH$_3$	C$_{22}$H$_{30}$N$_2$O$_2$	0.07		0	100	40(HDT)	NA
21	COPh	C$_{19}$H$_{16}$N$_2$O$_2$	0.63	0.9	<1	3	12	71
22	CO-4-(NHCOPh)C$_6$H$_4$	C$_{26}$H$_{21}$N$_3$O$_3$	0.42		11	0.5	0.5	83
23	CO-2-pyridyl	C$_{18}$H$_{15}$N$_3$O$_2$	0.76	1.2	0	10	15	69
24	CO-2-naphthyl	C$_{23}$H$_{18}$N$_2$O$_2$	0.46		0	7	6	73
25	CO-2-quinolyl	C$_{22}$H$_{17}$N$_3$O$_2$	0.50	1.2	1	10	6	142
26	CO-6-(NHCO(CH$_2$)$_4$CH$_3$)-2-quinolyl	C$_{28}$H$_{28}$N$_4$O$_3$	0.15		h	100	21	127
27	CO-2-pyrrolyl	C$_{17}$H$_{15}$N$_3$O$_2$	0.70		10	2	1	71
28	CO-2-indolyl	C$_{21}$H$_{17}$N$_3$O$_2$	0.45	1.0	36	0.2	0.4	90
29	CO-1-Me-2-indolyl	C$_{22}$H$_{19}$N$_3$O$_2$	0.43	1.0	3	10	8	44
30	CO-2-benzofuranyl	C$_{21}$H$_{16}$N$_2$O$_3$	0.47	1.1	6	0.04	0.25	110
31	CO-2-benzothiophenyl	C$_{21}$H$_{16}$N$_2$O$_2$S	0.41	1.1	7	0.3	0.80	100
32	CO-3-Me-2-indenyl	C$_{23}$H$_{20}$N$_2$O$_2$	0.39		4	30	17 (HDT)	NA
33	5-OMe	C$_{22}$H$_{19}$N$_3$O$_3$	0.47	1.0	22	0.01	0.05	95
34	6-OH, 7-OMe	C$_{22}$H$_{19}$N$_3$O$_4$	0.63	1.0	37	0.1	0.05	55
35	5-NHCONH$_2$	C$_{22}$H$_{19}$N$_5$O$_3$	0.89	1.0	47	0.6	0.06	65
36	5-NHCOPh	C$_{28}$H$_{22}$N$_4$O$_3$	0.43		36	0.07	0.05	82
37	5-NHCO-2-indolyl	C$_{30}$H$_{23}$N$_5$O$_3$	0.30	i	70	0.01	0.05	159(4)
38	5-NHCO-2-benzofuranyl	C$_{30}$H$_{22}$N$_4$O$_4$	0.30	1.0	50	0.02	0.05	170(1)
39	5-NHCO-5-(NHCONH$_2$)-2-indolyl	C$_{31}$H$_{25}$N$_7$O$_4$			75	0.3	0.1	121(2)
40	5-NHCO-5-(NHCOPh)-2-indolyl	C$_{37}$H$_{28}$N$_6$O$_4$	0.30		33	0.01	0.1	77

Note: HDT, highest dose tested, toxicity not reached; NA, no activity.

[a] Chromatographic measure of lipophilicity determined by liquid–liquid chromatography on C$_{18}$-bonded silica gel.

[b] Pseudo-first-order rate constant, in sec¹, for solvolytic ring opening at pH3, followed spectrophotometrically.

[c] Molar ellipticity, in the presence of calf thymus DNA.

[d] ID_{50}, the nanomolar concentration of drug required to inhibit, by 50%, the growth of murine L1210 leukemia cells in a 3-day assay.

[e] Drug given by interperitoneal injection to mice implanted interperitoneally with 10^6 P388 leukemia cells.

[f] Optimum dose in milligrams per kilogram per day, given on days 1, 5, and 9.

[g] Percentage increase in life span of treated animals over that of control mice bearing tumor, when treated at the optimal dose. Numbers in parentheses are the numbers of mice that survived for 30 days.

[h] Extreme insolubility of the drug gave anomalous and irreproducible results.

[i] The drug was not soluble in this medium in the concentration range used.

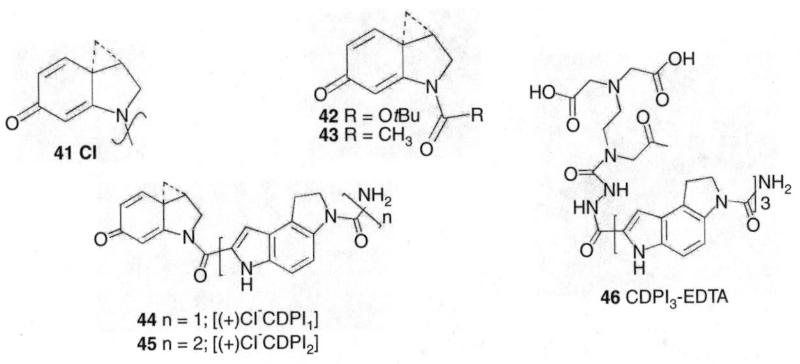

STRUCTURES 41–46

illustrate the offset binding sites. A simple model for the CC-1065 DNA alkylation reaction that accommodates the behavior of both enantiomers is provided, in which the sequence selectivity is derived from the noncovalent binding selectivity of the agents in the narrower sterically more accessible AT-rich minor groove. The inherent steric accessibility to the adenine N3-alkylation site accompanies deep penetration of the agent into the minor groove within the AT-rich site (Figure 20.2).

In the DNA binding and cytotoxic activity studies of two enantiomers of bis-indole analogs of CC-1065, the isomer with the same stereochemical configuration as natural (+)-CC-1065 was a potent cytotoxic agent, but its enantiomer was essentially inactive.[35] Both enantiomers showed significant binding to DNA, but the biologically less active isomer showed less overall binding. In all cases the agents preferred AT-rich DNA, and all bound to similar regions in DNA, as evidenced by the positions of drug-initiated thermal breaks in single end-labeled fragments of Phi X174 RF DNA. The overall similarity in site specificity for binding of the structurally diverse agents indicates that much of the specificity observed in the binding of the agent to DNA lies in the DNA itself.

C. CC-1065 ANALOGS

A simple method to detect CPI binding sites on double-stranded DNA has been devised.[8] The technique uses a modified form of bacteriophage T7 polymerase to synthesize a radiolabeled nascent strand from double-stranded DNA that has been allowed to react *in vitro* with the CC-1065 analog U-73,975 (adozelesin) **3**. The reaction products were electrophoresed on sequencing gels containing 8 *M* urea and visualized by autoradiography. The transit of this DNA polymerase is inhibited at the sites where CPIs are bound to the template strand. The positions of polymerase inhibition can be determined by comparison of CPI-treated and untreated DNA reactions. This modified dideoxynucleotide technique has been used to establish the sequence selectivity of U-73,975.

Bizelesin (U-77,779) **8** is a symmetrical dimer of the spirocyclopropyl alkylating subunit of (+)-CC-1065 in which the linker consists of two indole subunits separated by a ureido group. A comparison

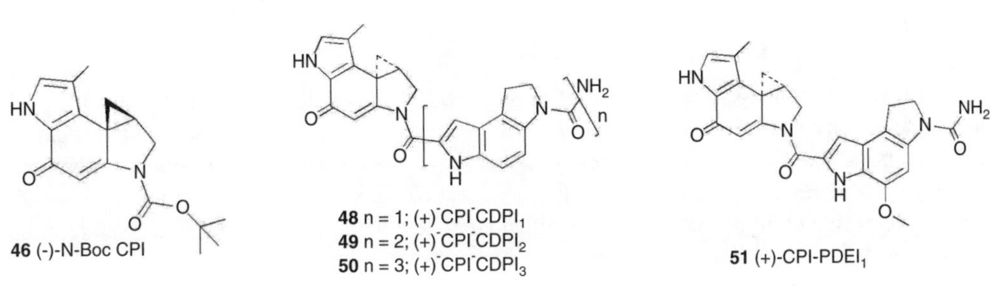

STRUCTURES 47–51

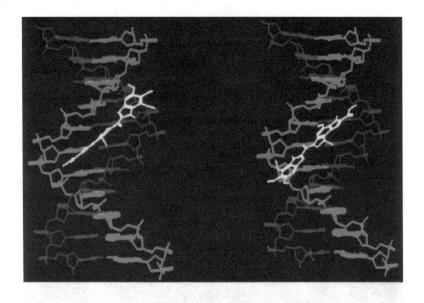

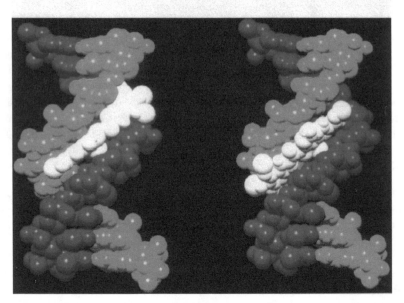

FIGURE 20.2 (See color insert following page 304.) Comparison stick and space-filling models of the (+)-duocarmycin SA (left) and ent-()-duocarmycin SA (right) alkylation at the same site within w794DNA: duplex 5-d(GACTAATTTTT). The natural enantiomer binding extends in the 3 → 5 direction from the adenine N3 alkylation site 5-CTAA. The unnatural enantiomer binding extends in the 5 → 3 direction across the site 5-AATT.

of bizelesin with a monoalkylating analog of (+)-CC-1065[36] shows that it appears to have increased sequence selectivity such that monoalkylating compounds like (+)-CC-1065 react at more than one site. In contrast, however, bizelesin reacts only at sites where there are two suitably positioned alkylation sites. Bizelesin generally forms interstrand cross-links with adenine primarily spaced 6 base pairs apart (including the covalently modified adenines) while occupying the intervening minor groove.

Bizelesin alkylates DNA through guanine in restriction fragments in which there is a suitably positioned adenine contained in a highly reactive monoalkylation sequence 5-TTTTT N*, and in which N was G, C, or T. Such fragments were prepared to evaluate the cross-linking potential of bizelesin at nonadenine bases. Kinetic analysis of monoalkylation and cross-linking events demonstrates that it is the reaction at N (guanine or cytosine) that results in the cross-link that is the slow step. On the basis of this analysis and the normal unreactivity of guanine and cytosine to alkylation by the CPI alkylating moiety of (+)-CC-1065, the molecular mechanism for this type of cross-linking reaction most probably involves a covalent immobilization of the second alkylating arm in a proximity-driven reaction.

The sequence selectivity for DNA interstrand cross-links (ISCs) by bizelesin was determined by sequencing gel analysis[37] and showed that bizelesin induced two distinct forms of DNA-ISC, one of which spanned six nucleotides and linked two adenine N3- positions within an AT-rich sequence, and the second spanned seven nucleotides, also linking two adenine N3- positions, with a preference for contiguous runs of adenines. Three major six-nucleotide DNA ISCs were identified and found to occur within 5-TAATTA-3, 5-TAAATA-3, and 5-TAAAAA-3 sequences. The major seven-nucleotide DNA-ISC was found to occur within 5-TAAAAAA-3 sequences, and within this sequence, the formation of the seven-nucleotide DNA-ISC was preferred over the six-nucleotide DNA-ISC by a ratio of approximately 2:1. DNA ISC formation within adenine tracts eliminated the inherent DNA bending associated with such sequences. Further, chemical probing of each isolated DNA ISC with diethyl pyrocarbonate (A-specific) and potassium permanganate (T-specific) showed that the major DNA conformational changes, such as helical distortion, were localized within the cross-linked sequence.

These results indicate that a significant degree of DNA distortion may also be demonstrated[38] and that covalent DNA adducts induced by bizelesin at the adenine N3- position undergo two subsequent competing reactions. One reaction causes DNA strand cleavage via depurination, and one reaction proceeds through loss of the DNA adduct (adduct reversal with restoration of DNA integrity). The results were obtained by studying the chemical stability of synthetic oligonucleotides, which contained either a distinct monofunctional adduct or DNA interstrand cross-links. Quantification of adduct reversal was performed on the basis that drug-modified DNA, on exposure to heat followed by hot piperidine treatment, was resistant to strand cleavage at the site of alkylation. The rate of adduct reversal was found to increase with increasing temperature and was found to be maximum at 70°–80°C. The rate of adduct reversal was also found to increase with increasing pH and ionic strength. Adduct reversal was favored in DNA containing cross-links, whereas rapid depurination occurred preferentially within monofunctionally alkylated DNA.

The ligand-mediated polymerase chain reaction technique has been used to map drug-induced DNA alkylation sites in single-copy genes at the nucleotide level in human cells.[39–41] This technique has been used to map the alkylation sites induced by adozelesin and bizelesin within DNA in human cells.[42] The pattern of the drug alkylation sites is mapped within the human PGK1 and P53 genes, with the monofunctional alkylating agent, adozelisin, alkylating genomic DNA predominantly within 5-(A/T)(A/T) A* (A* = alkylated adenine) sequences, and additional sites of alkylation were observed within 5-(A/T)(G/C)(A/T) A* sequences; however, these were considered to represent sites of medium to low preference. In contrast, bizelesin, a bifunctional analog capable of both DNA monofunctional alkylation and DNA interstrand cross-link formation, was also found to alkylate 5-(A/T) (A/T) A* sequences. Putative bizelesin DNA interstrand cross-link sites indicated that AT-rich sequences are preferred in the intervening sequence between the two cross-linked adenines. Both six- and seven-nucleotide regions were identified as putative sites of DNA interstrand cross-link formation, with 5-TTTTTTA*, 5- TTTATCA*, and 5-GTACTAA* sequences being preferred. Nonadenine bases were not observed as potential intracellular sites of either DNA interstrand cross-linking formation or monofunctional alkylation.

The DNA base pair preferences of the antitumor antibiotic CC-1065 and two analogs of CC-1065 were studied by following the rate of covalent bond formation (N3 adenine adduct) with DNA oligomers containing the 5-NNTTA* and 5-NNAAA* sequences (N = nucleotide).[43] The rate of adduct formation of CC-1065 is greatly affected by DNA base changes at the fourth and fifth positions of the binding sites for the 5-NNAAA sequences, but not the 5-NNTTA sequences. However, an analog of the CC-1065 ring system, with additional methylene and oxygen substituents, shows similar rates of adduct formation for all sequences. A second analog of CC-1065, containing a fused three-ring system, but not the methylene and oxygen substituents of CC-1065, shows similar rates of adduct formation, with the same sequence dependence as CC-1065, but does not distinguish between the sequence to the degree shown by CC-1065. Adduct formation of CC-1065, but not the analogs, competes with a reversibly bound species. Thymine bases to the 3 side of a potentially reactive adenine or cytosine base at the fifth position from the bonding adenine create reversible binding sites that decrease the rate of adduct formation of CC-1065. The sequence 5-GCGAATT binds CC-1065 only reversibly. This sequence can compete for CC-1065 with covalent binding sequences if the sites are located in different oligomers, or if the sites are located (overlapped or not overlapped) in the same oligomer. The results of these competitive binding experiments indicate that the transfer of CC-1065 from the reversible binding site to the covalent bonding site with both sites located on a single DNA duplex, not overlapped, occurs through an equilibrium of CC-1065 in solution, not by migration of CC-1065 in the minor groove.

A number of indirect observations that indicate a conformational acceleration in the action of CC-1065 and its analogs have been summarized by Boger et al.[44,45] The DNA alkylation reaction is derived in part from a DNA binding–induced conformational change in the agents, which substantially increases their inherent reactivity. The ground-state destabilization results from a binding-induced twist in the linkage N2 amide and requires a rigid extended N2-amide substituent, which disrupts the vinylogous amide stabilization and activates the agents for DNA alkylation.

The calf thymus DNA (CT-DNA) and poly (dI-dC). poly (dI-dC) binding properties of the natural antitumor antibiotic and selected analogs of CC-1065, U-66,664, U-68,819, U-66,694, and U-65,415 were studied by circular dichroism (CD) and absorbance methods.[46] The results indicate that the intense long-wavelength DNA–induced CD band of these molecules originates from a chiral electronic transition, which is delocalized over the whole molecule. Both the covalently bound species exhibit the characteristic spectral behavior of an inherently disymmetric chromophore when these agents bind within the minor groove of B-form DNA. This mechanism of optical activity accounts for why CC-1065 shows a weak CD in buffer but a very intense induced CD at long wavelength when bound to DNA, why the intensity of the induced CD of CC-1065 analogs depends on how many fused ring systems the analog contains, and why covalently bound analogs, having the mirror image configuration of the natural configuration also exhibit an intense, positive, induced CD at long wavelength.

IV. INTERACTION OF CC-1065 WITH CHROMATIN

Most of the CC-1065 studies have used natural DNA or synthetic oligonucleotides for binding studies. However, nuclear DNA does not exist as naked DNA but is packaged as chromatin. Chromatin consists of repeating nucleosome subunits composed of an interior core of histones, around which are wrapped approximately 140 base pairs of the B-form DNA. Moy et al.[47] reported on the interactions of CC-1065 and adozelesin with DNA, chromatin, and histones. CD binding studies show that CC-1065 and adozelesin bind to P388 chromatin and that induced CD spectra are similar to those obtained when CC-1065 binds to P388 DNA and many synthetic polynucleotides. The intensity of the induced CD spectra and the rate of conversion of reversibly bound drug to irreversibly bound drug are less for chromatin than for naked DNA. Because the histones themselves do not bind CC-1065, this indicates that histones act indirectly to restrict the binding of CC-1065 to chromatin. This could occur because fewer binding sites are available on chromatin as compared to free DNA or because of a less dissymmetric environment of chromatin. It was assumed that a good fit in the minor groove leads to a more chiral environment for the drug chromophore and, hence, a more intense induced CD.

A new type of CC-1065 analog, **52–58**, where the cyclopropylpyrroloindole (CPI) moiety is replaced by 1,2-dihydro-1-(chloromethyl)-5-hydroxy-8-methyl-3H-furano [3,2-c] indole (CFI), has

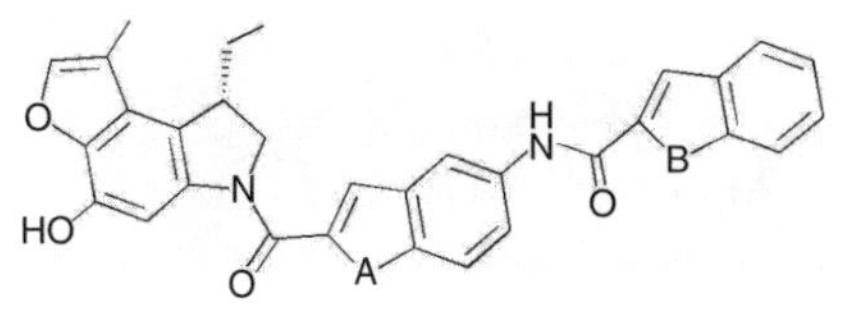

52 A = B = NH   **55** A = S, B = NH
53 A = NH, B = O   **56** A = S, B = O
54 A = NH, B = S   **57** A = B = S

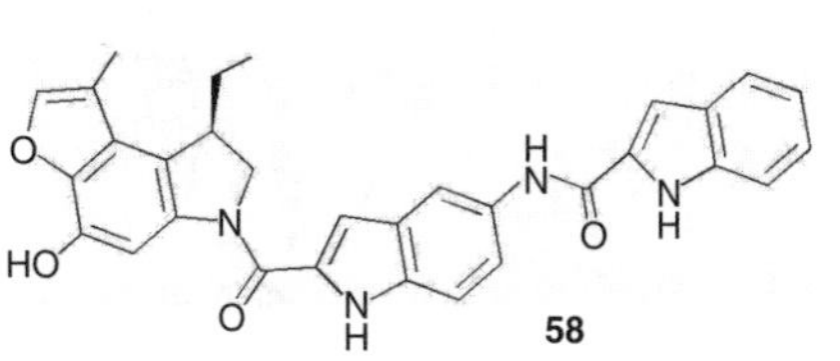

STRUCTURES 52–58

been synthesized.[48] The major site of alkylation with W794 duplex DNA for **55** and (+)-CC-1065 proved to be identical. 5-d(AATTA) 3- alkylation occurs at the 3 adenine and agent binding is in the 3′→5′ direction. The profile of DNA alkylation selectivity for **52** also proved to be identical to that of adozelesin (**3**). However, compound **58** exhibited a different and less selective profile of DNA alkylation. The high-affinity site for **58** with W794 duplex DNA was 5′-(AATTT)-3′, with alkylation at the 5 adenine and agent binding in 5′→3′ direction. The profile of DNA alkylation exhibited by **58** proved essentially identical to that of the unnatural enantiomers (–)-CC-1065 and (–)-CPI-CDPI3.[49] A more significant distinction between **52** and **58** is the relative efficiency or intensity of DNA alkylation. Consistent with the relative *in vitro* cytotoxic potency of **52,** it proved to be comparable or slightly less effective (one to five times) at alkylating the duplex DNA than (+)-CC-1065 and much more effective than compound **58**.

A new CBI (1,2,9,9a-tetrahydrocyclopropa[c] benz [e] indole-4-one; **9**) analog of CC-1065, CBI-PDEI$_2$ (**59**) has been synthesized and evaluated.[18] Compound **59** was shown to have similar DNA sequence selectivity and structural effects on DNA as (+)-CC-1065. The effect of CBI-PDE-I dimer was also compared with (+)-CC-1065 in the inhibition of duplex unwinding by helicase II and nick sealing by T4 ligase and found to be quantitatively similar. The *in vitro* and *in vivo* potencies of CBI compounds correspond very closely to the corresponding CPI derivatives.

A number of 1,2,9,9a-tetrahydro cyclopropa[c]benz[e]indole-4-one (CBI; **9**)–based and 9a-chloromethyl 1,2,9,9a-tetrahydro cyclopropa[c]benz[e]indole-4-one (C2BI; **60**)–based analogs of CC-1065 (**50**) and **61** have been synthesized and evaluated by Boger et al.[50,51] It was found that CBI analogs are four times more stable and potent than the corresponding CPI analogs. Similarly, the CBI-based agents alkylate DNA with unaltered sequence selectivity at an enhanced rate with greater efficiency than the corresponding CPI analogs. In comparison with the CBI-based agents, the C$_2$BI-based agents proved to be approximately 100–10,000 times less effective at DNA alkylation and 100–10,000 times less potent in cytotoxicity assays. These effects are suggested to be the consequence of a significant steric deceleration of the adenine N3-alkylation reaction attributed to the additional 9a-chloromethyl substituent. Consistent with this interpretation, the noncovalent binding constant of C$_2$BI-CDPI$_2$ (**61**) for poly [dA]-poly [DT] proved nearly identical to CDPI$_3$ under kinetic binding conditions, and prolonged incubation of C$_2$BI-CDPI$_2$ (**61**) with poly [dA]-poly [dT] (72 h, 25°C) provided covalent complexes with a helix stabilization comparable to that observed with (+) or (–)-CPI-CDPI$_2$, indicating that the size of the C$_2$BI subunit inhibits, but does not preclude, productive DNA alkylation.

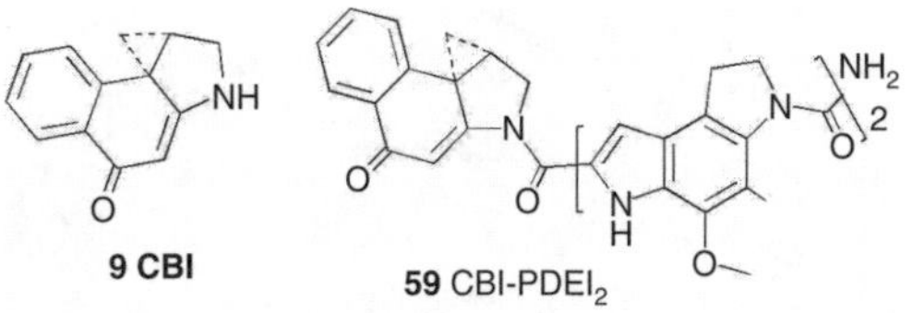

9 CBI   **59** CBI-PDEI$_2$

STRUCTURES 9, 59

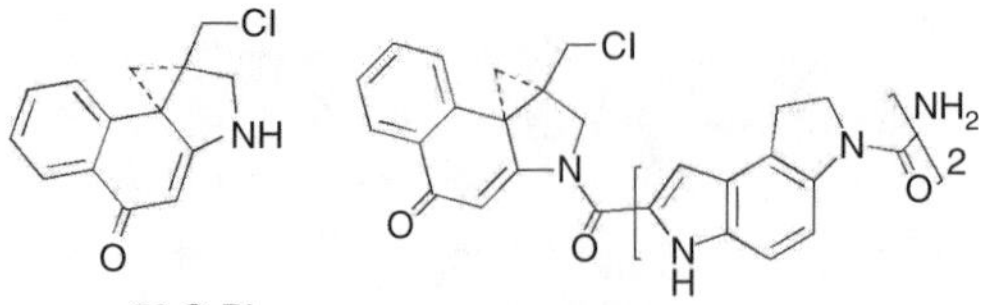

STRUCTURES 60–61

A library of 132 CBI (1,2,9,9a-tetrahydrocyclopropa[c]benzo[e]indole-4-one) analogs of CC-1065 and the duocarmycins was prepared using the solution-phase technology of acid–base liquid–liquid extraction for the isolation and purification steps (Figure 20.3).[52] The 132 analogs constituted a systematic study of the DNA binding domain with the incorporation of dimers composed of monocyclic, bicyclic, and tricyclic (hetero) aromatic subunits. From their examination, clear trends in cytotoxic potency and DNA alkylation efficiency emerged, highlighting the importance of the first-attached DNA-binding subunit (A subunit): tricyclic is more active than bicyclic, which in turn is far more active than monocyclic (hetero) aromatic subunits. Notably, the trends observed in the cytotoxic potencies paralleled those observed in the relative efficiencies of DNA alkylation. Their interpretation of these results is that the trends represent the partitioning of the role of the DNA-binding subunits into two distinct contributions, the first of which is derived from an increase in DNA-binding selectivity and affinity, which leads to property enhancements of 10–100-fold and is embodied in the monocyclic series. The second, which is additionally embodied in the bicyclic and tricyclic heteroaromatic subunits, is a contribution to the catalysis of the DNA alkylation reaction that provides additional enhancements of 100–1000-fold. The total enhancements thus exceed 25,000-fold. Aside from the significance of these observations in the design of future CC-1065/duocarmycin analogs, their importance to the design of hybrid structures containing the CC-1065/duocarmycin alkylation subunit should not be underestimated. Those that lack an attached bicyclic or tricyclic A subunit, that is, duocarmycin/distamycin hybrids, can be expected to be intrinsically poor or slow DNA alkylating agents.

V. POLYAMIDE CONJUGATES AND RELATED SEQUENCE-DIRECTED STRUCTURES

In a novel approach to explore the properties of cyclopropylindole (CPI), Lown et al. envisaged certain CC-1065 analogs, **62-71**, in which different numbers of N-methylpyrrole amide (a DNA sequence recognizing moiety) groups are attached with different linkers to the CPI moiety.[19] They postulated that the positively charged functionality of the protonated dimethylamino group in the

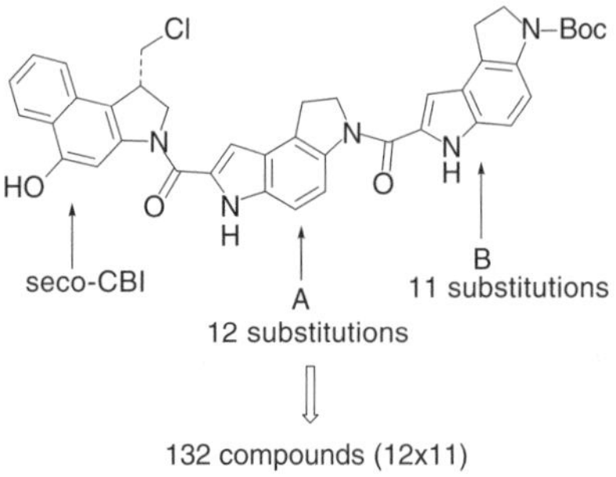

FIGURE 20.3 Solution-phase synthesis of 132 CC-1065 analogs based on CBI. All the compounds were purified by liquid–liquid acid–base extraction.

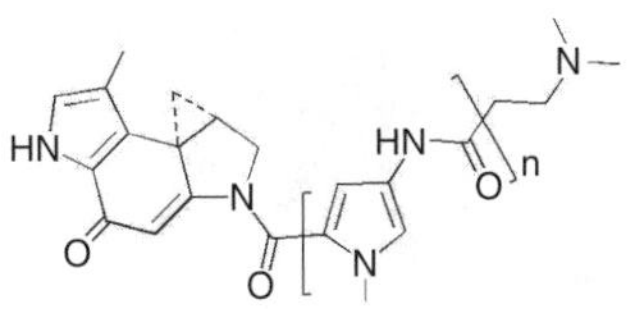

62 n = 1
63 n = 2
64 n = 3

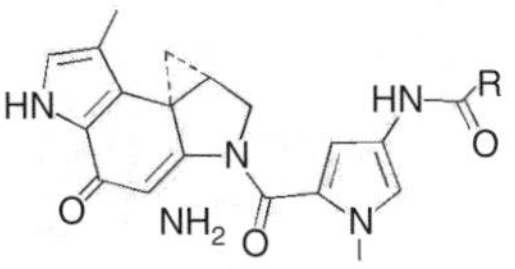

65 R = CH$_3$
66 R = CH$_2$CH$_3$
67 R = CH$_2$CH$_2$CH$_3$

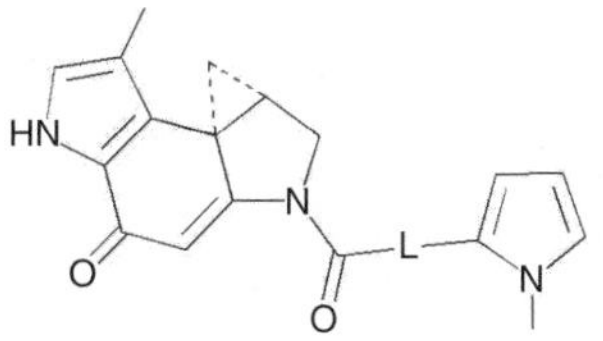

68 L = no linker
69 L = CH$_2$
70 L = CH$_2$CH$_2$
71 L = trans (CH=CH)$_2$

STRUCTURES 62–71

oligopeptide might contribute to reduced cytotoxic potency by reducing intracellular accessibility and, therefore, should be replaced by an uncharged moiety. It was also considered likely that the linker between the CPI unit and the oligopeptide would be important to the binding potency of the conjugate molecules to DNA to maintain proper contact over the whole of the molecular recognition region.[53] The extent and the relative rates of DNA cleavage following alkylation and thermal treatment of these conjugates were determined by agarose gel mobility shift assay and high-resolution polyacrylamide gel electrophoresis and contrasted with that of CC-1065. The CPI-N methyl pyrrole agents avoid the major alkylation sites of CC-1065, but all alkylate the minor CC-1065 site of 5ATAA and exhibit a consensus sequence of 5-NA/TA/TA. The cytotoxicities of these compounds were determined against KB human tumor cells *in vitro*. Compound **67** bearing a 4-butyramide group in the N-methyl pyrrole is 100 times more potent than compound **68**, which lacks an amide group, whereas compound **71**, which bears a rigid trans double-bond linker is 1000 times more potent than its flexible ethyl-linked counterpart **70**.

Sugiyama's group synthesized the novel hybrid molecules **72–73**, and they alkylate the 3 end of A in AT-rich sequences, as does the parent duocarmycin. More significantly, these hybrids alkylate G residues of predetermined DNA sequences effectively and with high specificity by formation of a heterodimer with distamycin.[54] They also synthesized compounds **74** and **75**,[55] which were designed to alkylate the target sequences (A/T) G (A/T)$_2$N(A/G) and (A/T) G (A/T) CN (A/G), respectively, according to Dervan's ring-pairing rule. High-resolution denaturing gel electrophoresis indicated that compound **74** exclusively alkylated the 5-TGTAAAA-3 within a 400–base pair DNA fragment. Similarly, alkylation of **75** occurred exclusively at the G of the 5AGTCAGA-3 sequence with efficiency at the nanomolar concentration (Figure 20.4)

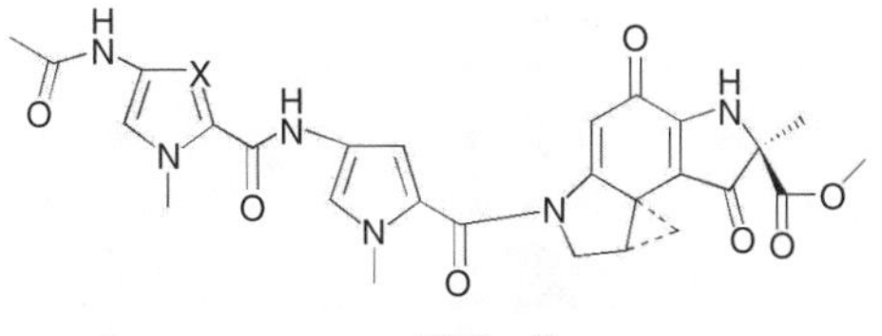

72 X = N
73 X = CH

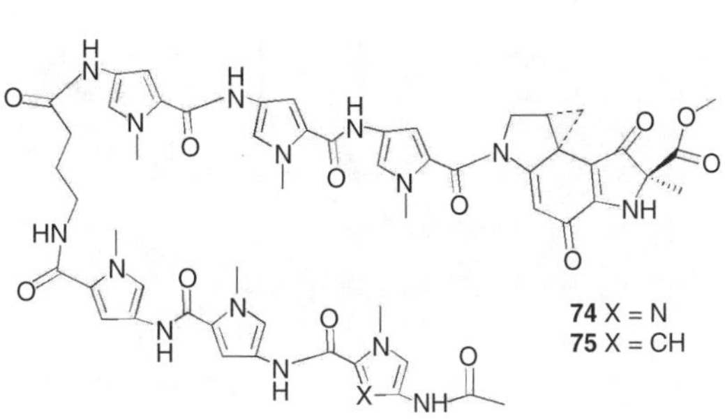

74 X = N
75 X = CH

STRUCTURES 72–75

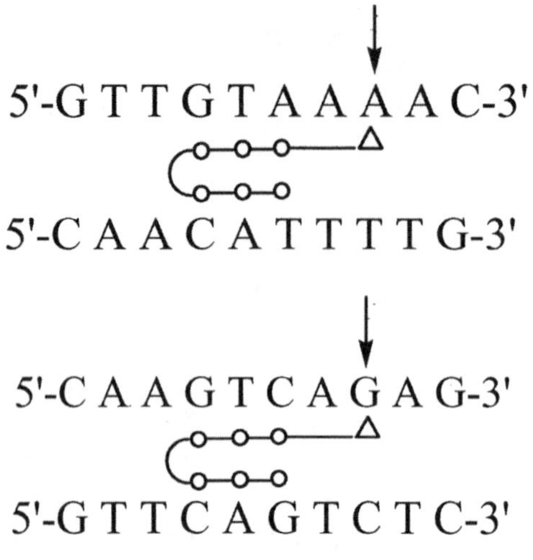

5'-G T T G T A A A A C-3'

5'-C A A C A T T T T G-3'

5'-C A A G T C A G A G-3'

5'-G T T C A G T C T C-3'

FIGURE 20.4 DNA alkylation sites of **74** and **75**.

Recently, Bando et al.[56] developed a novel type of DNA interstrand cross-linking agent by synthesizing dimers of a pyrrole (Py)/imidazole (Im)-diamide-CPI conjugate, ImPyLDu86 (**76**), using seven different linkers for connection (Figure 20.5). The tetramethylene linker compound, **77**, efficiently produces DNA interstrand cross-links at the 9–base pair sequence, 5-PyGGC(T/A)GCCPu-3, but only in the presence of a partner triamide, ImImPy. For efficient cross-linking by **77** with ImImPy, one AT base pair between two recognition sites was required to accommodate the linker region. Elimination of the AT base pair and insertion of an additional AT base pair, and substitution with the GC base pair, significantly reduced the degree of cross-linking. The sequence specificity of the interstrand cross-linking by **77** was also examined in the presence of various triamides. The presence of ImImIm slightly reduced the formation of a cross-linked product compared with ImImPy. The mismatch partners ImPyPy and PyImPy did not produce an interstrand cross-link product with **77**, whereas ImPyPy and PyImPy induced efficient alkylation at their matching site with **77**.

The interstrand cross-linking abilities of **77** were further examined using denaturing polyacrylamide gel electrophoresis with 5-Texas Red–labeled 400-and 67–base pair DNA fragments. The sequencing gel analysis of the 400–base pair DNA fragment with ImImPy demonstrated that **77** alkylates several sites on the top and bottom strands, including one interstrand cross-linkage on longer DNA fragments. A simple method using biotin-labeled complementary strands was developed (Figure 20.6) and produced a band corresponding to the interstrand cross-linked site on both top and bottom strands. Densitometric analysis indicated that the interstrand cross-links in the observed alkylation bands was approximately 40% of the total. This compound efficiently cross-linked both strands at the target sequence. This system consisted of a 1:2 complex of the alkylating agent and its partner ImImPy and caused an interstrand cross-linking in a sequence-specific fashion according to the base pair recognition rule of Py-Im polyamides (Figure 20.7).

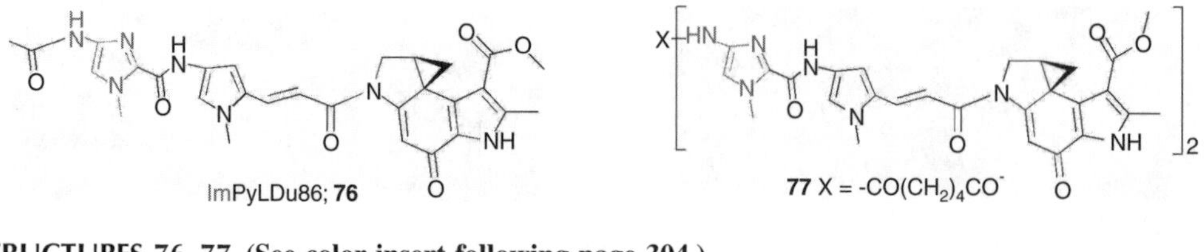

ImPyLDu86; **76**

77 X = -CO(CH$_2$)$_4$CO$^-$

STRUCTURES 76–77 (See color insert following page 304.)

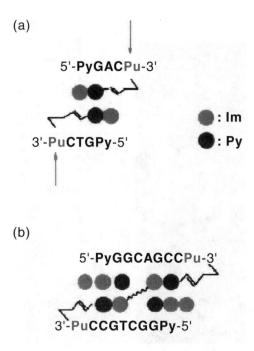

FIGURE 20.5 (See color insert following page 304.) (A) The chemical structure of ImPyLDu86 **76** and a schematic representation of the recognition of the 5-PyGACPu-3 sequence by the homodimer of **76**. The arrows indicate the site where alkylation takes place. (B) A putative binding mode of the 1:2 complex of covalent dimer of **76**, **77**, with ImImPy to 5-PyGGCAGCCPu-3 sequence.

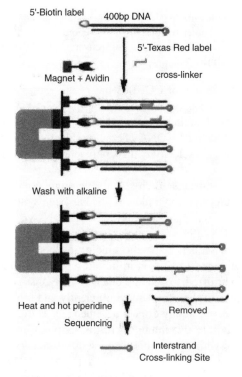

FIGURE 20.6 (See color insert following page 304.) Determination of interstrand cross-linking sites.

(a)   (b)

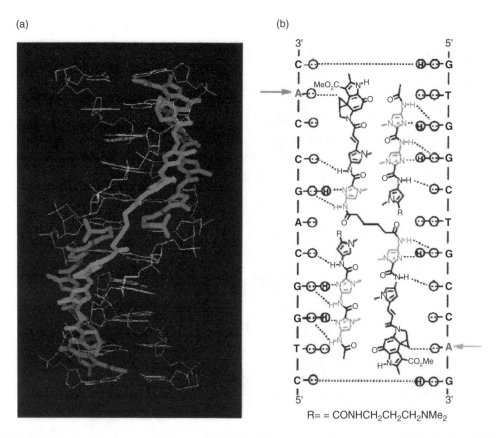

R= = CONHCH₂CH₂CH₂NMe₂

FIGURE 20.7 (See color insert following page 304.) (A) Energy-minimized structure of **77-ImImPy₂**-d(CTG-GCTGCCAC)/d(GTGGCAGCCAG) interstrand cross-linked complex. DNA is drawn in white. Py and Im residues of **77** and **ImImPy** are drawn in blue and red, respectively. (B) A DNA interstrand cross-linking mode of the 1:2 complex of **77** with **ImImPy**.

A series of hybrid agents, **78–84,** of CC-1065 has been synthesized and evaluated and incorporate the CBI analog of the DNA-alkylation subunit of the CC-1065 linked to the C-terminus di- and tripeptide S-DNA binding domain of bleomycin A2.[57] The attachment of the DNA binding domain of bleomycin A2 did not alter the DNA-alkylation selectivity of the CBI alkylation's subunit (5A**A**> 5T**A**), nor did it enhance the selectivity that approaches the 5 base pairs' AT-rich alkylation's selectivity of CC-1065.

A new class of hybridization-triggered cross-linkable oligodeoxyribonucleotides (ODNs) that are conjugated to the reactive cyclopropylindole (CPI) has been synthesized and studied for their properties.[58] Here, the authors conjugated racemic CPI to ODNs via a terminal phosphorothioate at either the 3 or 5 end of ODNs (**85**). This replaces the minor groove binding B and C subunits of CC-1065 with many more sequence-specific binding agents (ODN). These conjugates were stable in aqueous solution at neutral pH even in the presence of strong nucleophiles. When a 3-CPI-ODN conjugate was hybridized to a complementary DNA strand at 37°C, the CPI moiety alkylated nearby adenine bases of the complement efficiently and rapidly, with a half-life of a few minutes. CPI-ODN conjugates are highly effective sequence-specific inhibitors of single-stranded viral DNA replication or gene-selective inhibitors of transcription initiation.

Lown and coworkers[59–61] synthesized a bis functionalized precursor of CBI, **86,** and with this precursor synthesized bis-polyamide CBI conjugates of symmetrical **87** and unsymmetrical **88** types by incorporating pyrrole rings on both sides for AT specificity and pyrrole on one side and imidazole on the other side of CBI for mixed sequence recognition. The authors also explored[62,63] solid-phase

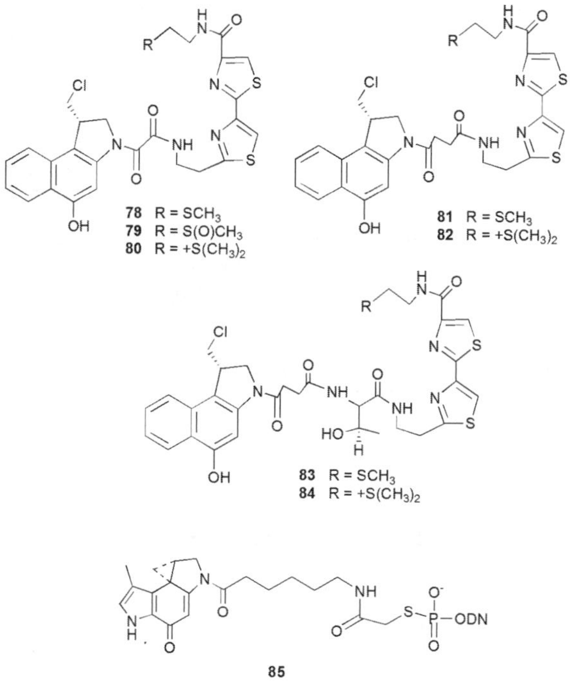

STRUCTURES 78–85

techniques for the synthesis of seco-CBI **90** and conjugates, which bears two polyamide moieties on either side of the pharmacophore **89** through an intermediate **91**. Lown et al.[19,20] have demonstrated that incorporation of a vinyl linker between CPI or CBI and the carrier dramatically enhances the efficiency of DNA alkylation as well as cytotoxicity. In confirmation of this significant finding, a diamide CPI conjugate possessing a vinyl linker **92** was also synthesized by Tao et al.[64] Molecular modeling indicated that the insertion of a vinyl linker (L) between polyamide and CPI adjusts the location of the cyclopropane ring of the conjugates and in this way improves the alkylation efficacy

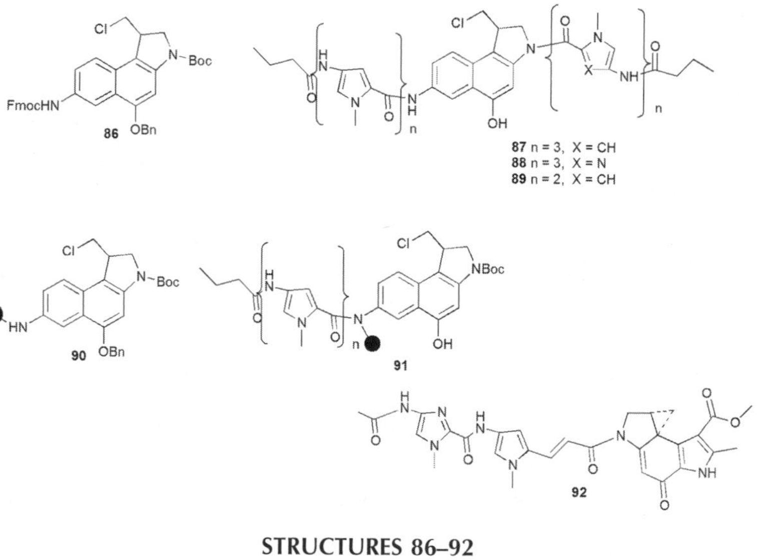

STRUCTURES 86–92

of the conjugates. Sequence-selective alkylation of double-stranded DNA by **92** was investigated by high-resolution denaturing gel electrophoresis, using 400-bp DNA fragments. Highly efficient alkylation predominantly occurs simultaneously at the purines of the 5-PyG(A/T) Cpu-3 site on both strands at nanomolar concentrations of **92**. These results indicate that the homodimer of conjugate **92** dialkylates both strands according to Dervan's pairing rule,[65] together with a new mode of recognition in which the Im-vinyl linker (L) pair target G/C base pairs. In addition to the major dialkylation sites, a minor alkylation site was also observed at 5-GT(A/T) GC-3. This alkylation can be explained by an analogous slipped homodimer recognition mode in which the L–L pair recognizes the A/T base pair. High performance liquid chromatography analysis revealed that the conjugate **92** simultaneously alkylates GN3/AN3 of the target sequences on both strands of ODNs.

Eight-ring hairpin polyamides conjugated with *seco*-CBI were prepared by Chang et al.[66] Alkylation yields and specificity were determined on a restriction fragment containing a 6–base pair match and mismatch series. Alkylation was observed at a single adenine flanking the polyamide binding site, and strand-selective cleavage could be achieved on the basis of the enantiomer of seco-CBI conjugate; a near quantitative cleavage was observed after 12 h.

VI. BISALKYLATORS (DIMERS)

The CPI bisalkylator, U-77,779 (bizelesin), in which the two alkylating moieties are connected with a rigid linker 1,3-bis (2-carbonyl-1H-indole-5-yl) urea group, showed promising antitumor activity,[17,67] and this compound is in clinical trials. Boger et al.[68] synthesized a series of four dimers, **93–96**, derived from head-to-tail coupling of the two enantiomers of duocarmycin SA alkylation subunits. All the agents proved to be potent in cytotoxicity assays and displayed a two- to threefold higher activity than duocarmycin SA. CPI dimers have shown good cytotoxic and antitumor activities, and now it is well documented that the activity of the dimeric drugs is strongly related to the length and the position of the linker. To investigate the structure–activity relationship systematically, Lown and coworkers[61,62,69–71] have designed three types of seco-CBI dimers, viz. C7-C7, N3-N3, and N3-C7 **97-102**, which contain two racemic CBI moieties linked from two different positions by a flexible methylene chain of variable length and with *n* numbers of pyrrole, imidazole, or thiazole moieties with the polyamides. Dimers, which are connected by a flexible methylene chain, are generally more active against human cancer cell lines as compared to the corresponding monomer, connected to a polyamide. Recently, Lown and coworkers developed the glycosylated water-soluble version of seco-CBI polyamide conjugates (unpublished work), **103–104**, which are more active against the human cancer cell lines when compared with their non-water-soluble version.

A set of 10 compounds, each combining the seco-1, 2,9,9a-tetrahydrocyclo-propa[c]benz[e]indol-4-one (seco-CBI) and pyrrolo[2,1-c]benzodiazepine pharmacophores (**105**), was designed and prepared.[72] These compounds were anticipated to cross-link between N3 of adenine and N2 of guanine in the minor groove of DNA. The compounds, which differ in the chain length separating the two alkylation subunits, and the configuration of the CBI portion, showed great variation in cellular toxicity (over four orders of magnitude in a cell line panel), with the most potent example exhibiting IC_{50}s in the picomolar range. Cytotoxicity correlated with the ability of the compounds to cross-link naked DNA. Cross-linking was also observed in living cells, at much lower concentrations than for a related symmetrical pyrrolo[2,1-c]benzodiazepine dimer. A thermal cleavage assay was used to assess sequence selectivity, demonstrating that the CBI portion controlled the alkylation sites, whereas the pyrrolo[2,1-c]benzodiazepine substituent increased the overall efficiency of alkylation. Several compounds were tested for *in vivo* activity using a tumor growth delay assay against WiDr human colon carcinoma xenografts, with one compound (the most cytotoxic and most efficient cross-linker) showing a statistically significant increase in survival time following a single intravenous dose.

STRUCTURES 93–102

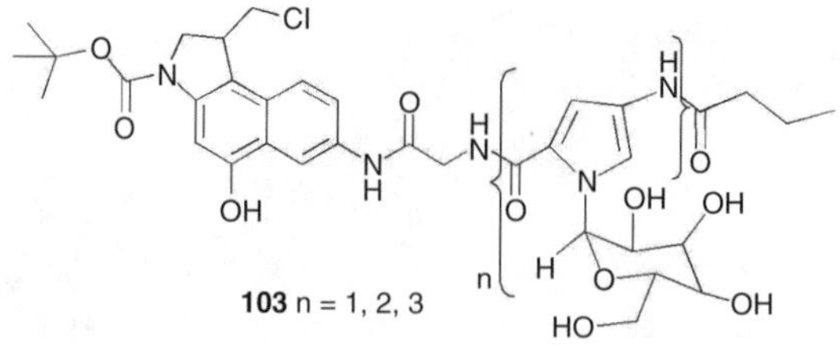

103 n = 1, 2, 3

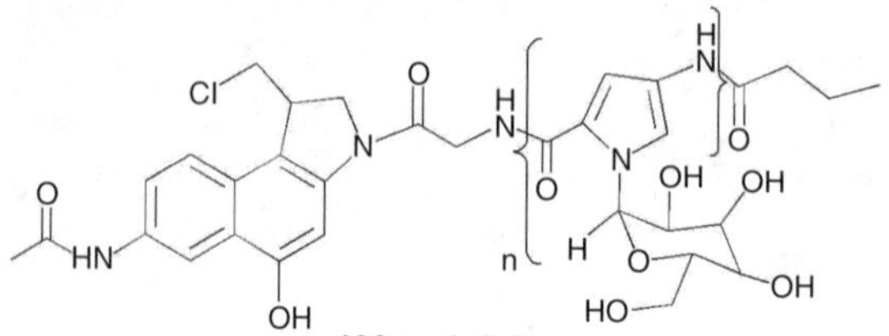

104 n = 1, 2, 3

STRUCTURES 103–104

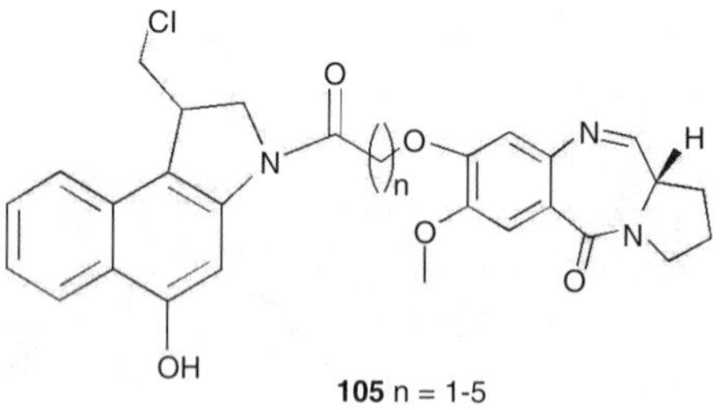

105 n = 1-5

STRUCTURE 105

VII. STRUCTURAL CHARACTERIZATION OF DRUG DNA COMPLEXES

Bizelesin forms an adduct with d-(CGTAATTACG)$_2$ and ^{1}H nuclear magnetic resonance (NMR) analysis of this adduct indicated that adenines 6 base pairs apart on opposite DNA strands are cross-linked, yielding two major adducts differing in the central duplex region (5AATT3).[73] In the major product, both adenines are *syn* oriented and Hoogsteen base paired to thymines (5HG model), whereas the other adduct contains an *anti*-oriented AT- step adenine that shows no evidence of hydrogen bonding with pairing thymines (50P model). Bizelesin's size, rigidity, and cross-linking properties restrict the DNA adduct's range of motion and freeze out DNA conformations adopted during the cross-linking process. In several sequences, bizelesin cross-links and shows overwhelming preference for a 7–base pair sequence over a possible 6–base pair sequence[74] (Figure 20.8).

The role of the central GC base pairs in the formation of the 7–base pair cross link was investigated using two-dimensional ^{1}H NMR studies that concentrated on the 7–base pair cross-link formed with the sequence 5TTAGTTA-3. ^{1}H NMR analysis coupled with restrained molecular

$$5'C^1\text{-}G^2\text{-}T^3\text{-}T^4\text{-}A^5\text{-}G^6\text{-}T^7\text{-}T^8\text{-}A^9\text{-}C^{10}\text{-}G^{11}$$

$$G_{22}\text{-}C_{21}\text{-}A_{20}\text{-}A_{19}\text{-}T_{18}\text{-}C_{17}\text{-}A_{16}\text{-}A_{15}\text{-}T_{14}\text{-}G_{13}\text{-}C^{5'}_{12}$$

FIGURE 20.8 Crosslinking of adenines 6 base pairs apart on opposite DNA strands.

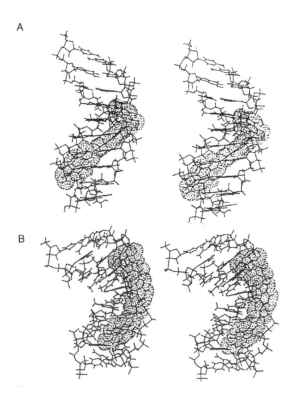

FIGURE 20.9 (**See color insert following page 304.**) Stereoviews of (A) the initial model involving docking of the CPI-CDPI$_2$ (dotted van der Waals surface is shown) to a regular B-DNA decamer and (B) NOESY distance restrained, MORASS hybrid matrix/molecular dynamics final refined structure of the decamer-CPI-CDPI$_2$ complex starting from the initial model. The A strand C1 is depicted in the upper left of the structures, with the minor groove at A17 facing away or to the right of the viewer. Note the sharp bend in the duplex near the site of alkylation in B.

dynamics provided evidence for distortion around the covalently modified adenines. The modified bases are twisted toward the center of the duplex adduct because of distortion, and this effectively reduces the cross-linked distance. The restrained molecular dynamics study also showed that a hydrogen bond is formed between the exocyclic amine of the central guanine and the carbonyl of the ureylene linker. It is possible to speculate on the role of the central GC bases in this sequence preference and to propose a mechanism by which bizelesin forms a 7–base pair rather than a 6–base pair cross-link with 5-TTAGTTA-3 on the basis of the observation of the distortion in the duplex and the hydrogen bonding between the drug and DNA.

Multidimensional NMR studies of the CPI-CDPI2.d-(CGCTTAAGCG)$_2$ complex indicate that CPI-CDPI$_2$ binds to the decamer in the same manner as CC-1065[75] (Figure 20.9). The effect of CPI-CDPI$_2$ on the ^{1}H and ^{31}P spectra of the decamer was consistent with a minor groove-binding motif, with the drug alkylating at A17 with CDPI rings oriented toward the 5-end of the alkylated strand. NMR data indicate formation of one major and one minor adduct. The minor adduct formation could represent drug alkylation of the DNA at a secondary site or alternative orientation of the rings.

VIII. CELLULAR AND PHARMACOLOGICAL STUDIES

A. *In Vitro* Studies

Adozelesin is an extremely potent cytotoxic agent that causes 90% lethality after 2 h exposure *in vitro* of Chinese hamster ovary (CHO) and lung (V79), mouse melanoma (B16), and human carcinoma (A2780) cells at 0.33, 0.19, 0.2, and 0.25 ng/mL, respectively. The relative drug sensitivity of the cell lines (A2780 > V79, B16, CHO) was correlated to the relative amounts of [³H] adozelesin that alkylated DNA. The greater sensitivity of A2780 resulted in greater cell kill at comparable DNA alkylation levels. Phase-specific toxicity studies show that adozelesin was least lethal to CHO cell in mitosis and very early G1. Lethality increased as the cell progressed through G1 and was maximum in late G1 and early S. Adozelesin had three different effects on progression of CHO, V79, B16, and A2780 through the cell cycle: it slowed progression through S, which resulted in significantly increasing the percentage of S-phase cells (though this effect was transient); cell progression was blocked in G2 for a long time period; and the response of the cell lines to G2 block differed. CHO and V79 cells escaped G2 block by dividing and became tetraploid. In contrast, B16 and A2780 cells remained blocked in G2 and did not become tetraploid. Subsequent G2 blockade was followed by a succession of events (polyploidy and unbalanced growth) that resulted in cell death.

The cell cycle kinetics have been studied by flow cytometry[76] and were performed on five human gynecological cancer cell lines, AN3, AE7, BG1, HEC1A, and SKUT1B, with exposure to adozelesin at near confluency at 0, 0.1, 0.2, 0.5, 1.0, and 5X, with X = 10 pg/mL as a reference concentration for 90 min. Cell samples were taken by trypsinization at 0, 24, 48, 72, 96, and 168 h for flow cytometry. The results revealed that there was a spectrum of cell-cycle perturbations that included biphasic S and G2 blocks, reverse dose-dependent G2 blocks, and a sequential relationship of S and G2 blocks. This study demonstrated that the cell kinetics response to adozelesin depended on several variables such as cell lines, drug sensitivity, concentrations, and sampling time.

The effects of CC-1065 and its analogs U-66,664, U-66,819, U-66,694, and U-71,184 have been studied on inhibition of CHO cell survival, cell progression, and their phase-specific toxicity.[77,78] Lethality of these compounds after 2 h drug exposure was in the following order (50% lethal dose in nanomolars in parentheses), CC-1065 (0.06) > U-71,184 (1.3) > U-66,694 (3.2) > U-68,819 (171) > U- 66,664 (greater than 1200). The phase-specific cytotoxicity of U-71,184 and U-66,694 was different from that of CC-1065. CC-1065 was most cytotoxic to cells in M and early G1, and toxicity decreased as cells entered late G1 and S. In contrast, U-66,694 and U-71,184 were most toxic to cells in late G1. The biochemical and cellular effects of U-71,184 were then studied in detail; after 2 h exposure to 3 ng/mL, 90% cell kill or growth inhibition was observed, whereas 100 ng/mL was needed for similar inhibition of DNA and RNA synthesis. This discrepency between the doses indicated that inhibition of nucleic acid synthesis may not be causally related to lethality. Further studies showed that when the drug was removed after 2 h exposure, DNA synthesis continued to be inhibited, whereas RNA and protein synthesis reached levels higher than the control. Therefore, it is likely that the stimulation of RNA combined with protein synthesis leads to unbalanced growth and cell death.

The adozelesin cytotoxic potency was compared with doxorubicin, cis-platin, 5-fluorouracil, and cytoxan by the adenosine triphosphate chemosensitivity assay in 10 gynecologic-cancer cell lines.[79] By comparing the mean drug concentrations required to reduce the adenosine triphosphate concentrations by 50%, adozelesin was found to be approximately 100 times more potent than doxorubicin, cisplatin, 5-fluorouracil, or cytoxan. Similarly, experiments designed to investigate the cytotoxicity of adozelesin on fresh cervical and ovarian carcinoma samples demonstrated that at least a 4000-fold lower level of this agent was required compared to cisplatin in the same experiment,[80] with similar results being found when the adenosine triphosphate assay referred to earlier was used with 10 gynecologic cancer cell lines.[81]

The cytotoxicity against CHO cells of several agents combined with adozelesin has been reported.[82] The additive, synergistic, or antagonistic nature of the combined drug effect was determined for most combinations using the median effect principle. In experiments with alkylating agents, combinations of adozelesin with melphalan or cisplatin were usually additive or slightly synergistic. Adozelesin–tetraplatin combinations were synergistic at several different ratios of exposure to drug. By using methylxanthines, adozelesin combined synergistically with noncytotoxic doses of caffeine and pentoxifylline and resulted in several logs increase in adozelesin cytotoxicity. In experiments with hypomethylating agents, adozelesin combined synergistically with 5-azacytidine (5-aza-CR) and 5-aza-2-deoxy cytidine (5-aza-2-CdR), and combinations of adozelesin with tetraplatin or 5-aza-2-CdR were also tested against B16 melanoma cell *in vitro* and were found to be additive and synergistic, respectively.

Hematopoietic clonal assays were used to evaluate the effects of bizelesin on granulocyte-macrophage (CFU-gm) colony formation.[83] Marrow cells were exposed *in vitro* to bizelesin (0.001–1000 nM) for 1 or 8 h and were then assayed for colony formation. There was a 3-log difference in drug concentration at which 100% colony inhibition occurred (1 or 8 h) for murine CFU-gm versus human or canine CFU-gm. The IC_{70} value after an 8 h bizelesin exposure for human CFU-gm (0.006 ± 0.002 nM) was 2220 times lower than for murine CFU-gm (13.32 ± 8.31 nM). At any given concentration, an 8-h drug exposure resulted in greater colony inhibition than a 1-h exposure for all species ($P < 0.05$). Increasing exposure time from 1 to 8 h increased toxicity to human and canine CFU-gm much more than to murine CFU-gm. The results of these *in vitro* clonal assays were quantitatively consistent with those seen in whole animal studies, indicating that bizelesin could be a potent myelosuppressive agent in the clinic. Because the dose-limiting toxicity in preclinical models is myelosuppression, and because the *in vitro* sensitivity of human and canine CFU-gm is similar, the canine maximum tolerated dose is better than the murine maximum tolerated dose to determine a safe starting dose for phase I clinical trials.

The DNA damage and differential cytotoxicity produced in human carcinoma cells by adozelesin and bizelesin have been studied.[84] The concentration of adozelesin required to produce a 1-log cell kill in six human tumor cell lines varied from 20 to 60 pM, whereas with bizelesin, the comparable concentrations ranged from 1 to 20 pM. The cytotoxicity of adozelesin and bizelesin was found to be independent of the guanine O6-alkyl transferase phenotype. The sensitivities of the BE and HT-29 human colon carcinoma cells were increased when the time of drug exposure was varied from 2 to 6 h, with DNA interstrand cross-links, as measured by the technique of alkaline elution, only detected when HT-29 or BE cells were exposed to extremely high concentrations of bizelesin for 6 h.

The cytotoxic actions of adozelesin against the human colon (HT-29, DLD-1) and lung (SK) carcinoma cell lines have been studied.[85] The concentrations of adozelesin that produced 50% cell kill for 4- and 24-h exposures were in the range of 0.001–0.02 ng/mL for both colon and lung carcinoma cells, indicating that this analog is a very potent cytotoxic agent. Because most clinical regimens for tumor therapy consist of several drugs, the antineoplastic action of adozelesin in combination with 5-aza-2-CdR, a potent inhibitor of DNA methylation, or cytosine arabinoside (Ara-C), a potent inhibitor of DNA synthesis, was examined. The adozelesin plus 5-Aza-CdR combination showed a synergistic effect on cytotoxicity against DLD-1 colon carcinoma cells for both a 6- and 24-h exposure. However, the combination of adozelesin and Ara-C for a 6-h exposure showed an antagonistic effect, whereas 24-h exposure showed a synergistic effect. These preclinical results provided some preliminary data on possible drugs that might be useful in combination with adozelesin in clinical trials.

Bis-indolyl-(seco)-1,2,9a-tetrahydrocyclopropa[c] benz [e] indole-4-one compounds are synthetic analogs of CC-1065 that are highly cytotoxic toward a broad spectrum of tumor cell lines. One of these compounds, DC1, was conjugated to antibodies via novel cleavable disulfide linkers. Conjugates of DC1 with the murine mAbs anti-Bu and N901 directed against tumor-associated antigens CD19 and CD56, respectively, proved to be extremely potent and antigen selective in killing target cells in culture.[86] DC1 conjugates with humanized versions of anti-B4 and N901 antibodies were also examined and shown to be as cytotoxic and selective as the respective murine antibody conjugates. The anti-B4-DC1 conjugate showed antitumor efficacy in an aggressive metastatic human B-cell lymphoma survival model in SCID mice and completely cured animals bearing large tumors. Anti-B4-DC1 was considerably more effective in this tumor model than doxorubicin, cyclophosphamide, etoposide, or vincristine at their maximum tolerated doses.

The mutagenicity for V-79 cell (6-thioguanine resistance and salmonella histidine auxotropy or aza guanine resistance) of selected CPI analogs was compared with DNA-binding activity, and the structure–activity relationship was determined. The compounds CC-1065, U-62,736, U-66,866, U-66,694, U-67,786, and U-68,415 all have an A segment with an intact cyclopropyl group and different B segments,[87] whereas the cyclopropyl group is absent from U-66,226 and U-63,360. This absence of the cyclopropyl ring diminished the cytotoxic and mutagenic potency of the compounds such that U-63,360 was nearly three orders of magnitude less potent than CC-1065 in V-79 cells. For the compounds with an intact cyclopropyl group, the order of cytotoxic and mutagenic potency in V79 cells generally correlated with binding to calf thymus DNA and increased with the length of the B segment; the order of cytotoxicity being CC-1065> U-68,415> U-66,694 > U-66,866> U-62,736.

These results show that an electrophilic carbon afforded by an intact cyclopropyl group of this type is necessary but not sufficient to account for the high cytotoxic and mutagenic potency of CC-1065 and U-68,415. The size and characteristics of the B segment also affect the potency. The authors speculate that the more cytotoxic analogs are less mutagenic because they may have greater structure directed binding to less mutable DNA sites in the minor groove.

B. *In Vivo* Studies

The analog of CPI or its kinetically equivalent ring opened seco precursor, carzelesin (5), was evaluated against a series of human xenografts growing at the subcutaneous sites.[88] The model consisted of seven colon adenocarcinomas and six pediatric rhabdomyosarcomas. Carzelesin was given a single intravenous injection, and the tumor volumes were determined at 7-day intervals. At the highest dose (0.5 mg/kg), the dose producing 10% lethality (LD10), carzelesin significantly inhibited growth in four of the colon tumor lines; in contrast, on lowering the dose to 0.25 mg/kg, only two of the seven colon adenocarcinomas exhibited significant growth inhibition. However, with the rhabdomyosarcoma lines, all were significantly arrested in growth. There was no apparent cross-resistance to carzelesin in two rhabdomyosarcomas selected for vincristin resistance (RH 12/VCR, Rh 18/VCR) or in Rh 28/LPAM xenografts selected for primary resistance to the bifunctional alkylating agent, melphalan. Carzelesin maintained full activity against Rh 18/Topo tumors selected *in situ* for resistance to topotecan.

Adozelesin (3), which was selected from a series of analogs of CC-1065 for its superior *in vivo* and antitumor activity, was highly active when administered intravenously against intraperitonel or subcutaneous implanted murine tumors, including L1210 leukemia, B16 melanoma, MS076 sarcoma, and colon 38 carcinoma, and produced long-term survivors in mice bearing intravenous inoculated L1210 and Lewis lung carcinomas.[14] In addition, modest activity was shown against the highly drug-resistant pancreas 02 carcinoma. Adozelesin was also highly effective against human tumor xenografts, implanted subcutaneously in athymic (nude) mice, including colon CX-1 adenocarcinoma, lung LX-1 tumor, clear cell caki-1 carcinoma, and ovarian 2780 carcinoma. Its broad spectrum of *in vivo* activity compared favorably with the three widely used antitumor drugs: cisplatin, cyclophosphamide, and doxorubicin. Adozelesin appeared to be more effective than these drugs in the treatment of various resistant tumors such as subcutaneously implanted mouse B16 melanoma, pancreatic 02 carcinoma, human colon CX-1, and human lung LX-1 tumor xenografts, and based on its high potency and high efficacy, it was chosen for clinical investigation and development.

The mutagenic potential of the antitumor compounds CC-1065 and adozelesin were assessed[89] to track the *in vivo* fate of their unique modifications at the nucleotide level. Mice were inoculated with a single therapeutic dose of these agents and sacrificed at 18 h, 3 d, or 15 d for extraction and analysis of liver DNA. Although undetectable at 18 h posttreatment, by 72 h a threefold increase in mutant frequency was observed in drug-treated animals such that sequence analysis of drug-induced mutations could be performed and a direct comparison made between *in vitro* and *in vivo*

DNA alkylation. Base substitution involving guanine or cytosine accounted for 64% of the 41 mutations that occurred at a cyclopropylindole alkylation site, whereas 23 mutations occurred 1–4 nucleotides from a potentially alkylated adenine.

Following intravenous administration of bizelesin (15 μg/kg) to male CD2F1 mice, the plasma elimination of cytotoxic activity resulting from bizelesin in the L1210 cell bioassay was described by a two-compartment open model[90]; the A-phase ($t_{1/2}$ A) and B-phase ($t_{1/2}$ B) half lives, steady-state volume distribution, and total body clearance were 3.5 min, 7.3 h, 7.641 mL/kg, and 16.3 mL/min per kilogram, respectively. The drug level following intraperitoneal administration was at least 10 times lower than that resulting from intravenous infusion. Following intravenous or intraperitoneal administration, the recovery of material in urine was <0.1% of the delivered dose. The low urine recovery indicated an extensive interaction of the parent drug and other reactive species with macromolecules or further metabolism or degradation to inactive species.

IX. CLINICAL STUDIES

Adozelesin is a potent synthetic analog that was chosen for clinical development because it had a preclinical antitumor spectrum that was similar to CC-1065 at therapeutic doses but did not produce deaths. Phase I evaluations using a variety of adozelesin treatment schedules have been reported.[91,92] Adozelesin was given as a 10-min intravenous infusion for five consecutive days every 21 d on recovery from toxicity. The dose range evaluated was 6–30 μg/m² per day. All patients had refractory solid tumors and had received prior cytotoxic drug treatment. Thirty-three patients (22 men, 11 women) were entered in the study, and 87 courses of treatment were initiated. Dose-limiting toxicity was cumulative myelosuppression (leukopenia, thrombocytopenia). The maximum tolerated dose was 30 μg/m² per day. The only other significant toxicity was an anaphylactoid syndrome that occurred in two patients. A partial response was observed in a patient with refractory soft tissue sarcoma. The recommended phase II starting dose of adozelesin using 10-min intravenous infusion for five consecutive days was 25 μg/m² per day, to be repeated every 4–6 weeks to allow recovery from myelotoxicity.

As of late September 2004, the only agent that is still listed as being in clinical trials is carzelesin, which is quoted in the Prous Integrity™ database as being in phase II and under "active development." The corresponding National Cancer Institute (NCI) database does not list the compound as being in any current trial. The differences could be a result of the NCI database not yet containing all clinical trial data from pharmaceutical companies.

X. CONCLUSIONS AND PROSPECTS

The sequence selectivity inherent in even the simplest minor groove ligands has enabled them to be used as probes of DNA structure and dynamics. From these data has arisen the concept of devising sequence-specific agents such as polyamides capable of single-gene recognition, and thus of being targeted to single deleterious genes. An extensive series of studies has led to the development of synthetic approaches to the natural products CC-1065 and the duocarmycins. The present understanding of the origin of the properties of the natural products emerged from the design and evaluation of synthetic substrates and agents containing systematic structural modifications, in addition to extensive studies conducted on the natural product themselves. As a consequence of these efforts, and despite expectations of the unique behavior of the natural products, potent and efficacious antitumor activity has not only been observed with such analogs but both their potency and efficacy may even exceed those of the natural products. This illustrates that their useful properties may be enhanced by well-founded and well-designed structural modifications. The active units, when coupled with certain appropriate linkers as dimers, have proven to be even more potent and are in clinical trials. Conjugating the pharmacophore unit of CC-1065 with the DNA-binding polyamides indicates a promising additional approach for developing new types of sequence-specific

DNA alkylating agents. The evident advantages of single-gene agents for molecular medicine will determine much of the future emphasis of this area, provided the formidable problems of converting them into pharmacologically acceptable drugs can be overcome.

REFERENCES

1. Hanka, L. J. et al., CC-1065 (NSC-298223), a new antitumor antibiotic: Production, *in vitro* biological activity, microbiological assays and taxonomy of the producing microorganisms, *J. Antibiot.*, 31, 1211, 1978.
2. Martin, D. G. et al., CC-1065 (NSC 298223), a potent new antitumor agent: Improved production and isolation, characterization and antitumor activity, *J. Antibiot.*, 34, 1119, 1981.
3. Bhuyan, B. K. et al., CC-1065 (NSC-298223), a most potent antitumor agent, kinetics of inhibition of growth DNA synthesis, and cell survival, *Cancer Res.*, 42, 3532, 1982.
4. Reynolds, V. L. et al., Reaction of the antitumor antibiotic CC-1065 with DNA location of the site of thermally induced strand breakage analysis of DNA sequence specificity, *Biochemistry*, 24, 6228, 1985.
5. Hurley, L. H. et al., Reaction of the antitumor antibiotic CC-1065 with DNA: Structure of a DNA adduct with DNA sequence specificity, *Science*, 226, 843, 1984.
6. Hurley, L. H. et al., Molecular basis for sequence-specific DNA alkylation by CC-1065, *Biochemistry*, 27, 3886, 1988.
7. Hurley, L. H. et al., Sequence specificity of DNA alkylation by the unnatural enantiomer of CC-1065 and its synthetic analogs, *J. Am. Chem. Soc.*, 112, 4633, 1990.
8. Weiland, K. L. and Dooley, J. P., *In vitro* and *in vivo* DNA binding by the CC-1065 analogue U-73975, *Biochemistry*, 30, 7559, 1991.
9. Boger, D. L., Munk, S. A., and Zarrinmayeh, H., (+)-CC-1065 DNA alkylation: Key studies demonstrating a no covalent binding selectivity contribution to the alkylation selectivity, *J. Am. Chem. Soc.*, 113, 3980, 1991.
10. Boger, D. L. et al., An alternative and convenient strategy for generation of substantial quantities of singly5-^{32}P-end labeled double stranded DNA for binding studies. Development of a protocol for examination of functional features of (+)-CC-1065 and ducaromycins that contribute to their sequence selective DNA alkylation properties, *Tetrahedron*, 47, 2661, 1991.
11. Li, H. et al., CC-1065 (NSC 298223), a novel antitumor agent that interacts strongly with double-stranded DNA, *Cancer Res.*, 42, 999, 1982.
12. McGovren, J. P. et al., Preliminary toxicity studies with the DNA-binding antibiotic, CC-1065, *J. Antibiot.*, 37, 63, 1984.
13. Warpehoski, M. A. et al., Stereoelectronic factors influencing the biological activity and DNA interaction of synthetic antitumor agents modeled on CC-1065, *J. Med. Chem.*, 31, 590, 1988.
14. Li, L. H. et al., Adozelesin, a selected lead among cyclopropyl pyrroleindole analogs of the DNA binding antibiotic, CC-1065, *Invest. New Drugs*, 9, 137, 1991.
15. Ichimura, M. et al., Duocarmycin SA, a new antitumor antibiotic from *Streptomyces* sp., *J. Antibiot.*, 43, 1037, 1990.
16. Sugiyama, H. et al., Covalent alkylation of DNA with duocarmycin A. Identification of a basic site structure, *Tetrahedron Lett.*, 31, 7197, 1990.
17. Mitchell, M. A. et al., Synthesis and DNA cross-linking by a rigid CPI dimer, *J. Am. Chem. Soc.*, 113, 8994, 1991.
18. Aristoff, P. A. et al., Synthesis and biochemical evaluation of the CBI-PDE-I dimer, a benzannelated analog of (+)-CC-1065 that also produces delayed toxicity in mice, *J. Med. Chem.*, 36, 1956, 1993.
19. Wang, Y. et al., Design, synthesis, cytotoxic properties and preliminary DNA sequencing evaluation of cpi-n-methyl pyrrole hybrids. Enhancing effect of a trans double bond linker and role of the terminal amide functionality on cytotoxic potency, *Anticancer Drug Des.*, 11, 15, 1996.
20. Iida, H. and Lown, J. W., Design and synthesis of cyclopropylpyrroloindole (CPI)-lexitropsin conjugates and related structure, *Synth. Org. Chem.*, 1, 1998.
21. Reynolds, V. L., McGovren, J. P., and Hurley, L. H., The chemistry, mechanism of action and biological properties of CC-1065, a potent antitumor antibiotic, *J. Antibiot.*, 39, 1986.

22. Boger, D. L. et al., Synthesis and evaluation of aborted and extended CC-1065 functional analogs: (+)- and ()-cpi-pde-i1, (+)- and ()-cpi-cdpi1, and (+-.)-, (+)-, and ()-CPI-CDPI3. Preparation of key partial structures and definition of an additional functional role of the CC-1065 central and right-hand subunits, *J. Am. Chem. Soc.*, 112, 4623, 1990.

23. Boger, D. L., Johnson, D. S., and Yun, W., (+)- and ent-(–)-duocarmycin SA and (+)- and ent-(–)-n-boc-dsa DNA alkylation properties. Alkylation site models that accommodate the offset at-rich adenine N3 alkylation selectivity of the enantiomeric agents, *J. Am. Chem. Soc.*, 116, 1635, 1994.

24. Boger, D. L. and Johnson, D. S., CC-1065 and the duocarmycins: Understanding their biological function through the mechanistic studies, *Angew. Chem. Int. Ed. Engl.*, 35, 1439, 1996.

25. Fukuda, Y. et al., Synthetic studies on duocarmycin. 1. Total synthesis of dl-duocarmycin A and its 2-epimer, *Tetrahedron*, 50, 2793, 1994.

26. Fukuda, Y., Nakatani, K., and Tershima, S., Synthetic studies on duocarmycin. 2. Synthesis and cytotoxicity of natural (+)-duocarmycin A and its three possible stereoisomers, *Tetrahedron*, 50, 2809, 1994.

27. Boger, D. L. and Mesini, P., DNA alkylation properties of CC-1065 and duocarmycin analogs incorporating the 2,3,10,10a-tetrahydrocyclopropa[d]benzo[f]quinol-5-one alkylation subunit: Identification of subtle structural features that contribute to the regioselectivity of the adenine N3 alkylation reaction, *J. Am. Chem. Soc.*, 117, 11647, 1995.

28. Wang, Y. et al., CC-1065 functional analogues possessing different electron withdrawing substituents and leaving groups: Synthesis, kinetics and sequence specificity of reaction with DNA and biological evaluation, *J. Med. Chem.*, 36, 4172, 1993.

29. Lown, J. W. et al., Molecular recognition between oligopeptides and nucleic acids: Novel imidazole containing oligopeptides related to netropsin that exhibited altered DNA sequence specificity, *Biochemistry*, 27, 3886, 1983.

30. Sugiyama, H. et al., A novel guanine N3 alkylation by antitumor antibiotic duocarmycin a, *Tetrahedron Lett.*, 34, 2179, 1993.

31. Dickerson, R. E., Kopka, M. L., and Pjura, P. E., The major and minor grooves of the DNA helix as conduits for information transfer, in *DNA ligand interaction, from drugs to proteins*, Guschlbauer, W. and Saenger, W., Eds., Plenum Press, New York, 1987, 45.

32. Boger, D. L. et al., Demonstration of a pronounced effect of nonvovalent binding selectivity on the (+) CC-1065 DNA alkylation and identification of the pharmacophore of the alkylation subunit, *Proc. Natl. Acad. Sci. USA*, 88, 1431, 1991.

33. Boger, D. L., Zhou, J., and Cai, H., Demonstration and defination of the noncovalent binding selectivity of agents related to CC-1065 by an affinity cleavage agent: Non covalent binding coincidental with alkylation, *Bioorg. Med. Chem.*, 4, 859, 1996.

34. Boger, D. L. et al., Molecular basis for sequence selective DNA alkylation by (+)- and enant-(–)-CC-1065 and related agents: Alkylation site models that accommodate the offest at-rich adenine n3 alkylation selectivity, *Bioorg. Med. Chem.*, 2, 115, 1994.

35. Swenson, D. H. et al., Evaluation of DNA binding characteristics of some CC-1065 analogues, *Chem. Biol. Interactions*, 67, 199, 1985.

36. Ding, Z. M. and Hurley, L. H., DNA interstrand cross-linking, DNA sequence specificity, and induced conformational changes produced by dimeric analogue of (+)-CC-1065, *Anticancer Drug Design*, 6, 427, 1991.

37. Lee, C. S. and Gibson, N. W., Nucleotide preferences for DNA interstrand crosslinking induced by the cyclopropyl pyrrole indole analogue U-77, 779, *Biochemistry*, 32, 2592, 1993.

38. Lee, C. S. and Gibson, N. W., DNA interstrand cross-links induced by the cyclopropyl pyrroloindole antitumor agent bizelisin are reversible upon exposure to alkali, *Biochemistry*, 32, 9108, 1993.

39. Muller, P. R. and Wold, B., *In vivo* footprinting of a muscle specific enhancer by ligation mediated PCR, *Science*, 246, 780, 1989.

40. Pfeifer, G. P. et al., Genomic sequencing and methylation analysis by ligation mediated PCR, *Science*, 246, 810, 1989.

41. Garrity, P. A. and Wold, B. J., Effects of different DNA polymerases in ligation mediated PCR: Enhanced genomic sequencing and *in vivo* footprinting, *Proc. Natl. Acad. Sci. USA*, 89, 1021, 1992.

42. Lee, C. S., Pfeifer, G. P., and Gibson, N. W., Mapping of DNA alkylation sites induced by adozelesin and bizelesin in human cells by ligation-mediated polymerase chain reaction, *Biochemistry*, 33, 6024, 1994.

43. Krueger, W. C. et al., DNA sequence recognition by the antitumor antibiotic CC-1065 and analogues of CC-1065, *Chem. Biol. Interactions*, 79, 265, 1991.

44. Boger, D. L. and Johnson, D. S., CC-1065 and duocarmycins: Unraveling the keys to a new class of naturally derived DNA alkylating agents, *Proc. Natl. Acad. Sci. USA*, 92, 3642, 1995.

45. Boger, D. L. and Garbaccio, R. M., Catalysis of the CC-1065 and duocarmycin DNA alkylation reaction: DNA binding induced conformational change in the agent results inactivation, *Bioorg. Med. Chem.*, 5, 263, 1997.

46. Krueger, W. C. and Prairie, M. D., The origin of the DNA induced circular dichroism of CC-1065 and analogues, *Chem. Biol. Interact.*, 79, 137, 1991.

47. Moy, B. C. et al., Interaction of CC-1065 and its analogues with mouse DNA and chromatin, *Cancer Res.*, 49, 1983, 1989.

48. Mohamdi, F. et al., Total synthesis and biological properties of novel antineoplastic (chloromethyl)furanoindolines, an asymmetric hydroboration mediated synthesis of the alkylation subunit, *J. Med. Chem.*, 37, 232, 1994.

49. Coleman, R. S. et al., Synthesis and evaluation of aborted and extended CC-1065 functional analogues: (+) and (−) CPI-PDEI3 preparation of key partial structures and definition of an additional functional role of CC-1065 central and right hand subunits, *J. Am. Chem. Soc.*, 112, 4623, 1990.

50. Boger, D. L. et al., Evaluation of functional analogues of CC-1065 and the duocarmycins incorporating the cross-linking 9a-chloromethyl-1, 2,9,9a-tetrahydrocyclopropa[c] benz [e] indole-4-one (C2BI) alkylation subunit, *Bioorg. Med. Chem*, 1, 27, 1993.

51. Boger, D. L., Yun, W., and Han, N., 1,2,9,9a-tetrahydro cyclopropa[c] benz [e] indole-4-one (CBI) analogues of CC-1065 and the duocarmycins: Synthesis and evaluation, *Bioorg. Med. Chem.*, 3, 1429, 1995.

52. Boger, D. L. et al., Parallel synthesis and evaluation of 132 (+)-1,2,9,9a-tetrahydrocyclopropa[c]benz[e]indol-4-one (CBI) analogues of CC-1065 and the duocarmycins defining the contribution of the DNA-binding domain, *J. Org. Chem.*, 66, 6654, 2001.

53. Reddy, B. S. P., Sharma, S. K., and Lown, J. W., Recent developments in sequence selective minor groove DNA effectors, *Curr. Med. Chem.*, 8, 475, 2001.

54. Tao, Z. F. et al., Sequence-specific DNA alkylation by hybrid molecules between segment A of duocarmycin A and pyrrole/imidazole diamide, *Angew. Chem. Int. Ed. Engl.*, 38, 650, 1999.

55. Tao, Z. F. et al., Rational design of sequence-specific DNA alkylating agents based on duocarmycin A and pyrrole-imidazole hairpin polyamides, *J. Am. Chem. Soc.*, 121, 4961, 1999.

56. Bando, T. et al., Highly efficient sequence-specific DNA interstrand cross-linking by pyrrole/imidazole CPI conjugates, *J. Am. Chem. Soc.*, 125, 3471, 2003.

57. Boger, D. L. and Han, N., CC-1065/duocarmycin and bleomycin A_2 hybrid agents. Lack of enhancement of DNA alkylation by attachment to non-complementary DNA binding subunits, *Bioorg. Med. Chem.*, 5, 233, 1997.

58. Lukhatanov, E. A. et al., Rapid and efficient hybridization-triggered cross linking within a DNA duplex by an oligodeoxyribonucleotide bearing a conjugated cyclopropyl pyrrole indole, *Nucleic Acids Res.*, 24, 683, 1996.

59. Jia, G., Iida, H., and Lown, J. W., Synthesis of bis-lexitropsin-1,2,9,9a-tetrahydrocyclopropa[c]benz[e]indole-4-one (CBI) conjugates, *Heterocyclic Comm.*, 4, 1998.

60. Jia, G., Iida, H., and Lown, J. W., Synthesis of an unsymmetrical bis-lexitropsin-1,2,9,9a-tetrahydrocyclo-propa[c]benz[e]indole-4-one (CBI) conjugates, *Chem. Comm.*, 119, 1999.

61. Jia, G., Iida, H., and Lown, J. W., Design and synthesis of 1,2,9,9a-tetrahydrocyclopropa[c]benz[e]indole-4-one (seco-CBI) dimers, *Heterocyclic Comm.*, 5, 497, 1999.

62. Jia, G., Iida, H., and Lown, J. W., Solid-phase synthesis of 1-chloromethyl-1,2-dihydro-3h-benz[e]indole (seco-CBI) and polyamide conjugate, *Synlett.*, 603, 2000.

63. Jia, G. and Lown, J. W., Design, synthesis and cytotoxicity evaluation of 1-chloromethyl-5-hydroxy-1,2-dihydro-3h-benz[e]indole (seco-CBI) dimers, *Bioorg. Med. Chem.*, 8, 1607, 2000.

64. Tao, Z. F., Saito, I., and Sugiyama, H., Highly cooperative DNA dialkylation by the homodimer of imidazole-pyrrole diamide-CPI conjugate with vinyl linker, *J. Am. Chem. Soc.*, 122, 1602, 2000.

65. Wemmer, D. E. and Dervan, P. B., Targeting the minor groove of DNA, *Curr. Opin. Struct. Biol.*, 7, 355, 1997.

66. Chang, A. Y. and Dervan, P. B., Strand selective cleavage of DNA by diastereomers of hairpin polyamide-seco-CBI conjugates, *J. Am. Chem. Soc.*, 122, 4856, 2000.

67. Mitchell, M. A. et al., Interstrand DNA cross-linking with dimers of the spirocyclopropyl alkylating moiety of CC-1065, *J. Am. Chem. Soc.*, 111, 6428, 1989.

68. Boger, D. L. et al., Bifunctional alkylating agents derived from duocarmycin SA: Potent antitumor activity with altered sequence selectivity, *Bioorg. Med. Chem. Lett.*, 10, 495, 2000.

69. Kumar, R. and Lown, J. W., Design and synthesis of bis 1-chloromethyl-5-hydroxy-1, 2-dihydro-3h-benz [e] indole (seco-CBI)-pyrrole polyamide conjugates, *Org. Lett.*, 4, 1851, 2002.

70. Kumar, R. et al., Synthesis of novel thiazole-contaning bis 1-chloromethyl-5-hydroxy-1, 2-dihydro-3h-benz [e] indole (seco-CBI)-polyamide conjugates, as anticancer agents, *Heterocyclic Commun.*, 8, 521, 2003.

71. Kumar, R. and Lown, J. W., Synthesis and cytotoxicity evaluation of novel c7-c7, c7-n3 & n3-n3 dimers of 1-chloromethyl-5- hydroxy-1,2-dihydro-3h-benz [e]indole (seco-CBI) with pyrrole and imidazole polyamide conjugates, *Org. Biomol. Chem.*, 1, 2630, 2003.

72. Tercel, M. et al., Unsymmetrical DNA cross-linking agents: Combination of the CBI and PBD pharmacophores, *J. Med. Chem.*, 46, 2132, 2003.

73. Seaman, F. C. and Hurley, L. H., Interstrand cross-linking by bizelesin produces a Watson-Crick to Hoogsteen base pairing transition region in d(CGTAATTACG)$_2$, *Biochemistry*, 32, 12577, 1993.

74. Thompson, A. S. et al., Determination of structural role of the internal guanine-cytosine base pair in recognition of a seven base pair sequence cross-linked by bizelesin, *Biochemistry*, 34, 11005, 1995.

75. Powers, R. and Gorenstein, D. G., Two-dimensional ^{1}H and ^{31}P NMR spectra restrained molecular dynamics structure of covalent CPI-CDPI2-oligodeoxy ribonucleotide decamer complex, *Biochemistry*, 29, 9994, 1990.

76. Nguyen, H. N. et al., Spectrum of cell-cycle kinetics of alkylating agent adozelesin in gynecological cancer cell lines: Correlation with drug induced cytotoxicity, *J. Cancer Res. Clin. Oncol.*, 118, 515, 1992.

77. Adams, E. G., Bardiner, G. J., and Bhuyan, B. K., Effects of U-71, 184 and several other CC-1065 analogues on cell survival and cell cycle of Chinese hamster ovary cells, *Cancer Res.*, 48, 109, 1988.

78. Zsido, T. J. et al., Resistance of CHO cells expressing p-glycoprotein to cyclopropylpyrroloindole (CPI) alkylating agents, *Biochem. Pharmacol.*, 43, 1817, 1992.

79. Nguyen, H. N. et al., *In vitro* evaluation of novel chemotherapeutic agent, adozelesin, in gynecologic-cancer cell lines, *Cancer Chemother. Pharmacol.*, 30, 37, 1992.

80. Hightower, R. D. et al., Comparison of U-73,975 and cisplatin cytotoxicity in fresh cervical and ovarian carcinoma specimens with ATP-chemosensitivity assay, *Gynecol. Oncol.*, 47, 186, 1992.

81. Hightower, R. D. et al., *In vitro* evaluation of the novel chemotherapeutic agents U-73, 975, U-77, 779, and U-80, 244 in gynecologic cancer cell lines, *Cancer Invest.*, 11, 276, 1993.

82. Smith, K. S. et al., Synergistic and additive combinations of several antitumor drugs and other agents with the potent alkylating agent adozelesin, *Cancer Chemother. Pharmacol.*, 35, 471, 1995.

83. Volpe, D. A. et al., Myelotoxic effects of the bifunctional alkylating agent bizelesin on human, canine and murine myeloid progenitor cells, *Cancer Chemother. Pharmacol.*, 39, 143, 1996.

84. Lee, C. S. and Gibson, N. W., DNA damage and differential cytotoxicity produced in human carcinoma cells by CC-1065 analogues, U-73, 975 and U-77, 779, *Cancer. Res.*, 51, 6586, 1991.

85. Cote, S. and Momparler, R. L., Evaluation of the antineoplastic activity of adozelesin alone and in combination with 5-aza-2-deoxycytidine and cytosine arabinoside on DLD-1 human colon carcinoma cells, *Anticancer Drugs*, 4, 327, 1993.

86. Chari, R. V. et al., Enhancement of the selectivity and antitumor efficacy of a CC-1065 analogue through immuno conjugate formation, *Cancer Res.*, 55, 4079, 1995.

87. Harbach, P. R. et al., Mutagenicity of the antitumor antibiotic CC-1065 and its analogues in mammalian (V79) cells and bacteria, *Cancer Res.*, 48, 32, 1988.

88. Houghton, P. J. et al., Therapeutic efficacy of the cyclopropylpyrroloindole, carzelesin, against xenografts derived from adult and childhood solid tumors, *Cancer Chemother. Pharmacol.*, 36, 45, 1995.

89. Monroe, T. J. and Mitchell, M. A., *In vivo* mutagenesis induced by CC-1065 and adozelesin DNA alkylation in a transgenic mouse model, *Cancer Res.*, 53, 5690, 1993.

90. Walker, D. L., Reid, J. M., and Ames, M. M., Preclinical pharmacology of bizelesin, a potent bifunctional analogue of the DNA-binding antibiotic CC-1065, *Cancer Chemother. Pharmacol.*, 34, 317, 1994.
91. Shamdas, G. J. et al., Phase I study of adozelesin (U-73, 975) in patients with solid tumors, *Anticancer Drug Des*, 5, 10, 1994.
92. Foster, B. J. et al., Phase I trial of adozelesin using the treatment schedule of daily x 5 every three weeks, *Invest. New Drugs*, 13, 321, 1996.

21 Epothilone, a Myxobacterial Metabolite with Promising Antitumor Activity

Gerhard Höfle and Hans Reichenbach

CONTENTS

I.   Introduction 413
II.   Natural Epothilones..415
III. Mechanism of Action..420
IV.   Total Synthesis and Semisynthesis...429
      A.   Total Synthesis..429
      B.   Semisynthesis..430
V.   Structure–Activity Relationships 435
VI.   Preclinical Studies..438
VII. Clinical Applications 440
Note Added in Proof...441
Acknowledgments...442
References 442

I. INTRODUCTION

Epothilone is a secondary metabolite of the myxobacterium *Sorangium cellulosum*. It is a macrolactone of a novel structural type characterized by an epoxy and a ketogroup in the lactone ring and a side chain with a thiazole ring (Figure 21.1) — functional groups from which the name epo-thi-lone was coined.[1] Two main variants, epothilone A and epothilone B, are normally isolated from the bacterial culture.[2,3] The B variant is a homolog with an extra methyl group at the epoxide, which makes it by a factor of 10 more active. Similarly, minor and trace components are always observed as pairs of homologs, with the B-series being invariably more active.[4] Epothilone, originally spelled epothilon, was discovered by us in 1986 because of its antifungal activity. Its structure was first disclosed in 1991 in the Annual Scientific Report of our institute, Gesellschaft für Biotechnologische Forschung (GBF).[1] The structure was elucidated mainly by spectroscopic methods and corroborated by x-ray crystallography, which also provided the absolute configuration.[3] Besides antifungal activity, strong cytotoxicity against mammalian cells was also reported. A German patent was filed in November 1991 (granted in May 1994), and an international one November 1992,[5] though it was withdrawn in April 1994 because the impressive antitumor activity of the epothilones was not, as yet, known, and there was limited interest in purely cytotoxic compounds. The patent strain, *Sorangium cellulosum* So ce90, was deposited at the German Culture Collection (Deutsche Sammlung von Mikroorganismen und Zellkulturen, DSMZ) in Braunschweig in October 1991 under number DSM 6773 and has been available to the public since then.

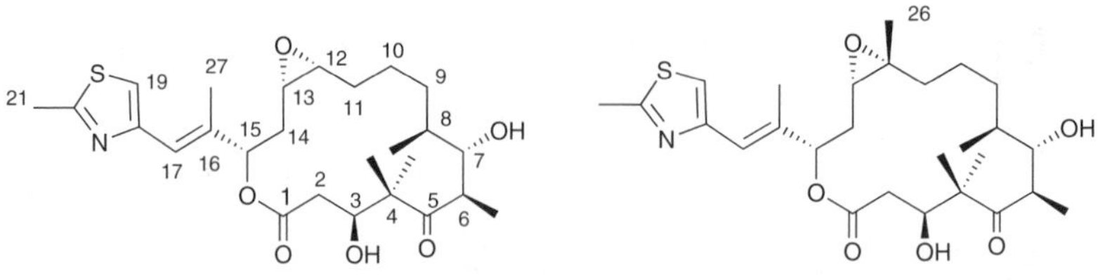

Epothilone A   **1a**   Epothilone B   **1b**

FIGURE 21.1 Structures of the major epothilones.

The discovery of epothilone was the result of a decade of basic research at the GBF with the aim of identifying new producers of diverse secondary metabolites among bacteria. This has indeed been achieved, with the discovery that myxobacteria synthesize a wealth of compounds, most of which possess diverse and novel structures and unusual and interesting mechanisms of action. Thus far, we have isolated and characterized more than 600 different compounds from those organisms, representing around 120 structural types,[6] and it is quite obvious that our many publications have stimulated more extensive investigation of myxobacteria for secondary metabolites.

In the 1980s and 1990s Taxol® (paclitaxel) was developed into a very successful antitumor drug with a new mechanism of action, viz, stimulation of tubulin polymerization and stabilization of microtubuli. As Taxol initially had to be produced from the bark of slow-growing Pacific yew trees, and there was little chance of development of an economically viable total synthesis, several pharmaceutical companies started screening for Taxol mimics from diverse natural sources, at first without any success in spite of the enormous numbers of compounds and extracts tested.[7] Colleagues at Merck, Sharp, and Dohme screened extracts from myxobacteria, having obtained access to the collection of *Sorangium* strains of Dr. John E. Peterson, a retired botanist and professor at Emporia State University in Kansas, who had worked on those organisms 30 years ago. They soon found the desired activity, and with the small quantities of substance they could recover from plate cultures they elucidated the structure, only to find out that they had rediscovered epothilone. They retained our name, just adding the "e," and cited our patent. The publication of their results, including the mechanism of action,[8] which we had failed to determine, started an avalanche of patents and publications on total synthesis, biosynthesis, genetics, molecular biology, biochemistry, and preclinical and clinical work, which still is increasing in volume at a rapid pace,[9] including numerous reviews.[10]

It soon turned out that epothilone has several properties that could make it superior to Taxol. It acts at very low doses (in the nano- to picomolar range), is active against multidrug-resistant (MDR), including Taxol-resistant, cancer cells, and is more water soluble, so that certain derivatives can even be administered orally. Further, it can be produced by fermentation without difficulty, and as a smaller and simpler molecule, is also accessible by total synthesis, enabling the production of many variants (so far about 1000).

The epothilone story following publication of the 1995 Merck article[8] is a good example of the often-overlooked fact that basic research and its reduction to practice is far from a straightforward process. When the Merck group discovered our patents, they terminated the project without contacting us regarding possible collaboration. Later, we ourselves tried to interest four major German, one Swiss, and three U.S. pharmaceutical companies in the possibility of collaboration in epothilone development, supplying test material to each company. After over 18 months of consideration, we finally came to an agreement with Bristol-Myers Squibb. The Swiss (Novartis) and one of the German companies (Schering) soon started independent epothilone projects, Schering using total synthesis, and Novartis with material produced by fermentation, using our patent strain, So ce90. By 1997 we had already developed fermentation and separation processes with mutant strains, allowing us to produce 10 g amounts of pure epothilone A and B.[10c] The yields could later be increased using traditional methods to several hundred milligrams per liter, so that, contrary to

1985	Dr. Reichenbach at GBF isolates *Sorangium cellulosum* strain So ce90 from a soil sample
1987	The strain is screened positive for antifungal activity, epothilone A and B were isolated, and their structures elucidated by Dr. Gerth and Dr. Bedorf
1991	Immunosuppressive activity studied by Ciba Geigy
	First publication of structure in Sci. Ann. Rep. of GBF
1992	Application tests and field trials for plant protection by Ciba-Geigy
1994	Work on epothilones terminated at GBF, international patent application abandoned
	Good activity in NCI 60 cell line antitumor screening
1995	Rediscovery of epothilone A and B as taxol mimics at MSD
	Disclosure of the absolute configuration as personal communications by Dr. Höfle
1996	First total syntheses by the Danishefsky, Nicolaou, and Schinzer groups
1997	Large scale production at GBF by fermentation
	Joint semisynthesis program started at BMS and GBF
1999	Clinical trials with natural, synthetic, and semisynthetic epothilones commence
2000	Biosynthesis genes cloned by Novartis and Kosan
2003	Phase II/III clinical trials with epothilone B-lactam (Ixabepilone)

FIGURE 21.2 Timeline for epothilone discovery and development.

statements in the literature, it was possible to produce all the material needed for development and future applications using *Sorangium cellulosum*, which can be cultivated without problem on cheap media based on soy meal in industrial bioreactors on a 60–100 m³ scale. The time course of epothilone discovery and development is shown in Figure 21.2.

II. NATURAL EPOTHILONES

The epothilones were discovered in cultures of the myxobacterium *Sorangium cellulosum*. Myxobacteria are Gram-negative bacteria that move by gliding along surfaces and are notable for their highly developed intercellular communication systems and their ability to reproduce, under unfavorable living conditions, in a cooperative action involving hundreds of thousands of cell fruiting bodies of a remarkably sophisticated shape and structure.[11,12] The producing organism (Figure 21.3), strain So ce90, is a cellulose degrader and was isolated at the GBF in 1985 from a soil sample collected on the banks of the Zambesi river in southern Africa. In addition to epothilones A and B, the strain synthesized in large scale a family of novel spiroketal polyene polyketides, named spirangiens,[1,13] which also showed antifungal traits and cytotoxicity. The epothilones aroused immediate interest because of their strong immunosuppressive effect and good activity against oomycetes — important parasites of agricultural plants.[5] Both effects lacked practical applications, however, because of the high toxicity of the compounds. The need for gram amounts of epothilones for relevant tests had already led at that time to the establishment of a production and isolation process at the GBF; however, because of the presence of an excess of spirangiens, the isolation of pure compounds was cumbersome and only achieved by multistep chromatography.

The discovery in 1995 that *S. cellulosum* strain SMP44 also synthesizes epothilones made a second producer strain available.[8] This strain is used for practically all studies in the Unites States, with the exception of the BMS/GBF project. Epothilone production with SMP44 was performed in large (150 mm) Petri dishes on agar for 10 d at 28°C. From 80 plates, 2.7 mg of pure epothilone A and 0.9 mg of epothilone B could be recovered, which was a sufficient amount to elucidate the structures by nuclear magnetic resonance and mass spectroscopy.[8] Similarly, the initial yields of epothilones from the GBF strain were rather low. As a first step in yield improvement, strain So ce90 was mutated to eliminate spirangien synthesis, which competed for precursors (the first spirangien negative mutant was obtained in October 1996).[14] This was possible only after the strain

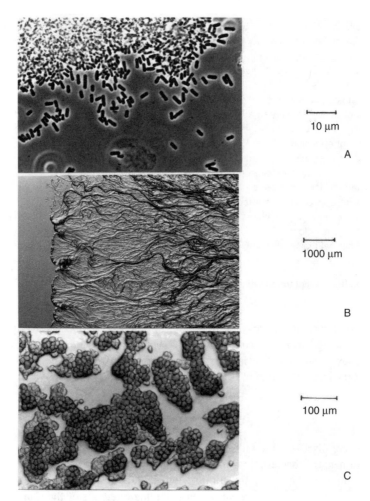

FIGURE 21.3 The myxobacterium, *Sorangium cellulosum*, producer of epothilone. (A) Vegetative cells; (B) vegetative, spreading swarm colony, (C) fruiting bodies, consisting of tiny sporangioles.

had been adapted to grow in liquid media in single cell suspension and to develop colonies from individual cells after plating, which are both time-consuming procedures in myxobacteria. A first optimization of the fermentation medium was then undertaken.

Cellulose-degrading *S. cellulosum* is much more versatile metabolically than most myxobacteria, notably *Myxococcus xanthus*, and grows on complex as well as on very simple media. Our first production medium consisted of defatted soybean meal, potato starch, glucose, yeast extract, and mineral salts.[2] As epothilones are excreted into the broth, a hydrophobic adsorber resin, Amberlite XAD-16, was added during fermentation. The resin increased yields by a factor of five by avoiding feedback inhibition by the end product, and by prevention of its degradation.[9,15] The medium was later improved, among others, by adding skim milk powder.[16] Production was during log and early stationary phase, with yields of 22 mg/L epothilone A and 11 mg/L epothilone B.[2] The generation time was around 16 h, and cell densities of up to 2×10^9 were achieved. After 7 d of fermentation, the resin was harvested by sieving, and the epothilones were extracted with methanol and processed further. Later, the yields could be increased substantially by further optimizing the medium, and especially by a mutation program during which more than 24,000 mutant clones were generated by treatment either with ultraviolet light or nitroso guanidine (NTG)

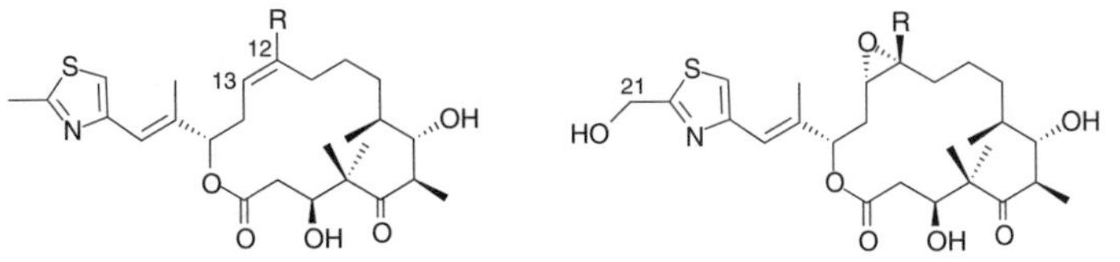

Epothilone C **2a** R = H
Epothilone D **2b** R = CH₃

Epothilone E **3a** R = H
Epothilone F **3b** R = CH₃

FIGURE 21.4 Structures of the minor epothilones.

and were characterized by high performance liquid chromatography analysis.[17] Thereby, not only high producers but also clones with defective biosynthetic pathways were obtained,[15–17] among them strains that produced epothilone A only, but none that produced pure epothilone B. The same was observed with about 40 further epothilone producers discovered among the 1700 *S. cellulosum* strains of the GBF collection. The positive strains came from soil samples collected on four continents and represented between 1% and 2.5% of all strains, depending on the area of origin.[9] The various strains differed not only in the epothilone pattern but also in the physiology of production. Thus, for example, epothilone synthesis is stimulated by glucose in So ce90 but inhibited in strain So ce1198.[9] So far, epothilones have not been found in any other myxobacterial species, nor in any other organism.

From high-producing *S. cellulosum* strains, small amounts of epothilones C (**2a**), D (**2b**), E (**3a**), and F (**3b**) (Figure 21.4) were isolated, in addition to epothilones A and B.[4,18] Whereas in C and D the epoxide is replaced by a cis double bond, an additional hydroxyl group is present at the thiazole methyl group in the variants E and F. Later, both desoxyepothilones **2** and hydroxyepothilones **3** became available by fermentation of a P450 defective mutant and by biotransformation with *S. cellulosum*, respectively. This greatly facilitates their isolation, but separation of the homologous pairs still requires careful reversed-phase chromatography.

From a large-scale fermentation (700 L) of So ce90 mutant B2, 35 variants could be isolated in very small amounts in addition to epothilones A–F.[4] Of those, only epothilones C_7 and C_9 may be regarded as regular biosynthetic products carrying extra hydroxyl groups at C14β and C27. The majority of the other variants result from aberrant biosynthesis. They lack methyl groups, carry extra methyl groups, have an oxazole instead of a thiazole ring and extra double bonds, and have reduced (by two C-atoms) or expanded ring sizes (epothilones K and I). Several small open-chain molecules are obviously derived from biosynthetic intermediates that had escaped from the enzyme complex.[4]

The epothilones A–F are colorless solids, readily soluble in polar organic solvents and stable at ambient temperature in the pH range of 5–9. Epothilones A, B, and D are crystalline, with melting points in the range of 76°–128°C, depending on the solvents of crystallization (the correct melting point of epothilone A is 76°–78°C [ethyl acetate/toluene], and 85°–87°C [methanol/water].[3,19–21] Several crystal structures are available for crystals from lipophilic and polar solvents exhibiting different conformations (Figure 21.7; Section V).[3,20] Aqueous solubility depends crucially on the presence of epoxide and C12 methyl groups and ranges from 700 mg/L for epothilone A and 200 mg/L for B to 16 mg/L for epothilone D.[3,21] Thus, epothilone D is about as poorly water soluble as paclitaxel (6–30 mg/L),[22] posing similar problems with its administration at therapeutically useful doses. Nevertheless, because of its promising therapeutic window, epothilone D was identified very early as a candidate for clinical studies by the Danishefsky group.[23] Total synthesis was upscaled and production by fermentation optimized. Under favorable conditions, our So ce90 P450 defective mutant produces 70 mg/L of epothilone D.

At Kosan Biosciences, the genes for epothilone synthesis were cloned from strain SMP44, first into *Streptomyces coelicolor*[24] and then into another myxobacterium, *M. xanthus*.[25] Both strains, however, produced less than 0.2 mg/L of epothilones. By eliminating the P450 epoxidase gene,

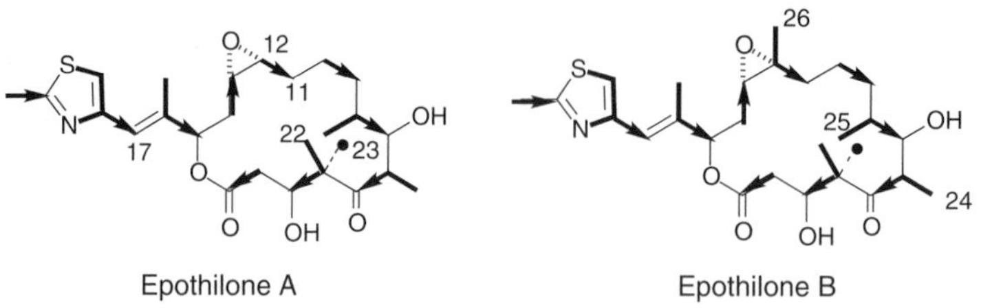

Epothilone A   Epothilone B

FIGURE 21.5 Incorporation of acetate, propionate, methionine, and cysteine in epothilones A and B.

clones that produced only epothilones C and D (0.45 µg/mL) were obtained.[26] Later, the yields could be increased to 23 mg/L and to 90 mg/L[26,27] by optimizing the fermentation process. This process requires the addition of methyl oleinate, which complicates work-up; thus, only 63 mg/L of epothilone D could be recovered.

Biogenesis of epothilones has been studied in *S. cellulosum* on three levels: actual biosynthesis using feeding experiments and blocked mutants; analysis of the genes coding for the relevant enzymes; the genes were cloned and sequenced from both producer strains, So ce90 and SMP44; and *in vitro* synthesis using part of the enzymatic apparatus cloned and expressed in *Escherichia coli*.

By feeding ^{13}C and radioactively labeled precursors to strain So ce90, it was quickly established that the carbon chain is synthesized from acetate, propionate, and cysteine (Figure 21.5; the correct stereochemical assignment of the 22-Me and 23-Me groups is given in references 184 and 185).[16] Methyl C21 comes from acetate, methyl C22 from S-adenosyl methionine, and all other methyls arise from propionate. This is also the case with methyl C26 at the epoxide site. The incorporation of propionate at this site in place of acetate is a result of a low specificity of the enzyme in this region.

So far it has not been possible to isolate strains or mutants that synthesize exclusively the more active epothilone B, although strains restricted to epothilone A are available.[16] Synthesis starts by joining cysteine and acetate to yield methyl thiazole. This requires participation of a nonribosomal peptide synthetase (NRPS). The rest of the molecule is essentially provided by polyketide synthase (PKS) modules.[16] As was corroborated by the genetic approach, the whole biosynthetic machinery consists of a huge hybrid multienzyme complex composed of one NRPS and eight PKS modules. The oxygen of the epoxy group is introduced by a post-PKS monooxygenase using molecular oxygen. The PKS module responsible for this section of the molecule has, however, no dehydratase sequence that could produce the double bond required as the substrate of the monooxygenase. As we could isolate two epothilone variants, one with a hydroxyl on C13 and the other with a C12/C13 double bond, which is perhaps produced by a dehydratase (DH) in another module, the substrate of the monooxygenase may arise in this way.[17]

The dehydration must happen at an early stage because the hydroxy precursors of epothilones C and D could not be detected.[17] Increasing amounts of epothilones A or B in the medium inhibit *de novo* synthesis of those molecules. At the same time, epothilones C and D, present only in traces in the control, rise to very high levels.[17] Obviously, the monooxygenase responsible for the introduction of the epoxy group is feedback inhibited by epothilones A and B, and epothilones C and D are the end products of the PKS. When epothilones C or D are fed to the nonproducing So ce90 C2 mutant, the compounds are bound immediately to the cells and transformed into epothilones A/B. The monooxygenase is found only in epothilone-producing strains and was never seen in other *Sorangium* strains.[17] It appears that epothilone C has a higher affinity for the monooxygenase than epothilone D, as it is preferentially epoxidized, and even slight and remote changes in the molecule may prevent epoxidation. The block mutant D48 is defective in its monooxydase and is thus a producer of epothilones C and D.[17]

In cultures of So ce90 without XAD resin, yields of epothilones A/B go down dramatically, and a small proportion of the material becomes hydroxylated on C21 to give epothilones E/F.[15] Degradation starts with the opening of the lactone ring by an esterase, which can be demonstrated

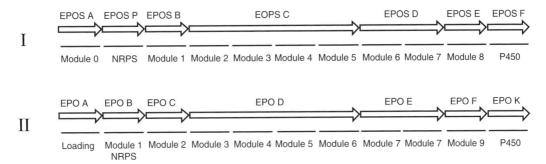

SCHEME 21.1 Analysis and assignment of epothilone biosynthesis genes by the Novartis (I) and Kosan groups (II).

in the culture supernatant. The hydroxylation is performed by another epothilone A/B–induced monooxygenase, which appears to be a P450 enzyme, for it is inhibited when iconazole is added to the culture.[15] A screening of 95 wild strains of *S. cellulosum* showed that this monooxygenase is widely distributed among nonproducing *Sorangium* strains.

Comparable biosynthetic studies are not available for strain SMP44 and the heterologous host, *M. xanthus*. For those strains, it was shown that the composition of the medium may have an influence not only on epothilone yields but also on the pattern of variants produced.[28,29] In *M. xanthus*, addition of serine resulted in a substantial increase of epothilones that have an oxazole instead of the thiazole ring. Oxazole epothilones are normally found only in trace amounts, also in So ce90 fermentations.[4] Their synthesis is obviously a result of a lack of absolute specificity of the NRPS module taking occasionally a serine instead of a cysteine precursor. The epothilone genes have been cloned and sequenced from both producer strains, So ce90[30] and SMP44.[31] Both sequences are patented.[32,33] As mentioned, the SMP44 genes have been modified and used to produce epothilones heterologously.

Good summaries of the organization of the epothilone gene cluster are available[30,31,34] (Scheme 21.1), and because of space limitations, details can not be discussed here. Briefly, in SMP44, all seven genes required for epothilone synthesis, *epoA* to *epoF* and *epoK*, are clustered together in a 56–kilobasepair stretch of DNA.[31] *EpoA* is the loading domain, *EpoB* the NRPS module, and *EpoC* to *EpoF* are PKS modules. Overall, there are nine modules plus the loading domain and *EpoK*, a P450 monooxygenase responsible for the introduction of the epoxy group. All genes are transcribed in one direction and probably in one operon. A methyl transferase domain in module 8 (EpoE) may be responsible for the introduction of one of the methyl groups on C4. When EpoK was expressed in *E. coli*, the isolated protein could be used to epoxidize epothilone D into epothilone B.[31] There is an anomaly in the sequence of PKS module 4: the DH required for the introduction of the C-12/13 double bond of the PKS end product is lacking. The hypothesis that the DH of the neighboring module 5 may be responsible for creating that double bond has recently been proven by destroying that domain.[35] Although the exact mechanism of the enzymatic reaction is not yet elucidated, a forward–backward shuffling of the growing chain bweeen modules 4 and 5 is suggested. Another anomaly is found in module 9 (*Epo F*), where the DH and ER are inactive[31] ([3-^{13}C, 3-^{18}O] labeled epothilones were isolated after feeding of [1-^{13}C, 1-^{18}O] propionate[21]).

The group sequencing the So ce90 epothilone genes used a slightly different nomenclature for the same general organization of the gene cluster[30] (Scheme 21.1). In this case also, transcription is unidirectional. It is suggested that all those genes form one operon of more than 56 kilobasepair. The finished polyketide, epothilone C/D, appears to be released from the PKS by a thioesterase; a sequence showing conserved motifs of that enzyme has been identified next to module 8 in *epoE*.[30] A methyltransferase domain in *epoD* is required to produce one of the gem methyl groups at C4. The sequence of *epoF* indicates a cytochrome P450 oxygenase. A short distance in front and behind the epothilone genes, two ORFs, *orf3* and *orf14*, seem to code for transport proteins that may be involved in export of epothilone from the producing cell.[30] In summary, the genetic apparatus for epothilone synthesis is remarkably similar, in fact, almost identical, in two producing strains of *S. cellulosum*, although these are of completely different origin. Among 56,000 basepairs, only 407 (= 0.7%) differ in the two gene clusters, which is an astonishing and very unusual correspondence not yet seen with any other *Sorangium* compounds (Rolf Müller, personal communication).

The epothilone gene cluster from SMP44 was cloned and expressed in heterologous hosts that actually produced epothilones.[24,25] The *M. xanthus* system was also used to make structural variants by manipulating the gene cluster. Thus, by inactivating the monooxygenase gene, strains were obtained that synthesized preferentially epothilone D.[25-27] Using site-specific mutation, epoD strains were further modified by inactivating the enoyl reductase domain in module 5 to make 10,11-didehydro-epoD (=Epo490),[36] and by inactivating the ketoreductase domain in module 6 to synthesize 9-keto-epothilone D.[37a] In addition, 4-desmethyl epothilones, molecules with an additional ether ring, and two fragmentary epothilones were obtained.[37a] Recently, it was demonstrated that an *E. coli* strain engineered to express the last half of the epothilone biosynthetic pathway is able to process an advanced precursor to epothilone C.[37b] In the future, this approach possibly may allow for the exclusive production of epothilone D and its analogs from simple synthetic precursors, thus avoiding tedious separation of C/D homologs. By biotransformation using the oxygen-introducing bacterium, *Amycolata autotrophica* (formerly *Nocardia autotrophica*), hydroxylated variants were produced from epothilone D and 10,11-didehydro-epothilone D.[38] Several of the analogs created by those different methods are also known from So ce90 fermentations.[4] The genes of the first three modules of the cluster, *EpoA, EpoB,* and *EpoC*, were expressed in *E. coli*, the resulting proteins purified, and their enzymatic reactions studied *in vitro*.[34,39-43] Products were usually obtained in minute quantities but could be identified by comparison with synthetic material. The system is of particular interest because it allows us to study PKS/NRPS/PKS interfaces in this hybrid multienzyme complex. It turned out that the individual enzymes of the epothilone cluster show considerable flexibility with respect to their substrate specificity, which could be useful for combinatorial biosynthesis. However, before application, the problem of producing reasonable yields has to be solved.

In a recent study, it was demonstrated that the base sequence of the KS domain of *epoB* in So ce90 (corresponds to *epoC* in SMP44) differs in several respects from the KS domains of all other epothilone modules.[44] Although those, in a phylogenetic analysis, cluster closely together, as is the rule in PKS genes, the *epoB* KS goes with PKS modules of other organisms. Also, the *epoB* KS deviates significantly in its GC composition. This was taken as a hint that the *epoB* KS sequence may have been introduced into a rudimentary epothilone gene sequence by horizontal gene transfer. Also recently, the exact position of epothilone bound to a protein could be analyzed for the first time.[45] The crystal structure of the P450epoK monooxygenase was analyzed in the free form and again with the substrate and product bound. The enzyme has a very large binding cavity, but otherwise it resembles other P450 enzymes, particularly P450eryF from the erythromycin cluster (30% identity, 48% similarity).

III. MECHANISM OF ACTION

The biological and biochemical effects of the epothilones have been studied in many laboratories, applying diverse experimental methods, in the United States and in Europe, and results are not always strictly comparable. Clearly, individual cell lines and the various epothilone analogs behave quite differently. Nevertheless, the overall picture is fairly consistent, and as a rule, safe generalizations can be made. In the ensuing discussion, frequent reference is made to synthetic and structure-activity studies discussed in detail in Sections IV and V, respectively.

Epothilone does not act on bacteria, and among the filamentous fungi and yeasts tested, only *Mucor hiemalis* (Zygomycetes) responded (minimal inhibitory concentration 20 μg/mL).[2]

In greenhouse experiments, various oomycetes were also inhibited. The cytotoxicity of the epothilones in animal cell cultures was noted immediately after their discovery,[1,2,5] but the effects they produce on the tubulin cytoskeleton were only realized much later.[8] In the beginning, only the two main fermentation products, epothilones A and B, were available for study. Although epothilone B differs only in an additional methyl group on C13, its efficacy is consistently higher by a factor of 5–10.[4,8,10e,m,46–51] This difference also remains with analogs of the two (e.g., epothilones C and D[4,17,46,49,52] or the oxazole derivatives[4,53,54]) and is even seen in binding and polymerization experiments with tubulin.[47] The epothilones act on a wide variety of cell types, including many human tumor cell lines. The concentration range for cytotoxicity is IC_{50} = 0.3–4 ng/mL[10n] or 1.3–39 nM[8,10m,46–48] for epothilone A, and 0.1–1.5 ng/mL[4] or 0.06–40 nM for epothilone B.[8,10m,46–48] The figures for Taxol are 0.2–95 nM.[8,46,47] (The molecular weights are Taxol, 854; epothilone A, 494; epothilone B, 508; i.e., Taxol is 1.73 and 1.68 times larger than epothilone A and B, respectively). Efficacy of Taxol equals that of epothilone A and is clearly less than that of epothilone B.[8,46,47] For instance, with 0.05–0.06 vs. 0.22–0.6 nM, epothilone B is 6–10-fold more active than Taxol against human prostrate cancer cells.[52] It came as a big surprise and is of considerable practical relevance that the epothilones still act on Taxol-resistant and MDR cells in general.[8,47] Although the resistant cell lines tolerated 100- to several thousand–fold higher concentrations of the selective agent, such as Taxol, vinblastine, or adriamycin, resistance to epothilones rose only by a factor of 1.5–10, and rarely higher. For epothilones C and D, it was even found to decrease in some instances.[46] It was reported that epothilone B–lactam and epothilone C–lactam lose activity on drug-resistant cell lines,[48,55,56] but this was not corroborated by another study, using several drug-resistant human cancer cell lines.[57] *In vivo* studies also did not support that statement. Epothilone B–lactam indeed shows about the same activity as epothilone A. Resistance to epothilone itself will be discussed below.

The epothilones promote tubulin polymerization and stabilize microtubules (MTs), as does Taxol. The morphological changes produced in treated cells resemble those effected by Taxol. In interphase cells, massive bundles of MTs arise in the cytoplasm, apparently randomly and independent from centrosomes, originating from multiple nucleation centers.[8,47,58,59] This transformation can already be seen 2 h after application, and after 24 h, virtually all MTs are in bundles (epothilone B).[47] When cells are kept at 4°C, their MTs decay completely within 1 h, but in the presence of 25 μM epothilone A or B or Taxol, they remain stable.[8] Already, 2 h after the addition of epothilones, mitotic defects can be observed, reaching a maximum after 6 h.[47]

At concentrations of 10 μM, normal spindles are no longer seen.[47] The organization of MTs in mitotic cells differs from that in interphase cells, so that a considerable turnover of tubulin still must occur in spite of the MT-stabilizing effect of the drug.[8,47] An explanation could be that before mitosis, MT-destabilizing proteins, such as katanin or stathmin/op18, are activated, or that MTs are stabilized less rigorously, perhaps because MT-associated proteins (MAPs) have dissociated after having become phosphorylated.[60] Multipolar spindles arise in a concentration range between 5 nM and 1 μM.[8] The spindles produced in the presence of epothilone A, epothilone B, and Taxol seem to differ somewhat in their morphology.[47] The result of those mitotic aberrations is an arrest of the cell cycle at the G2–M stage (e.g., with 100 nM epothilone A or B after 72 h).[8,47,52,57] The DNA content is then 4n.[52] The mitotic blockade is complete after 8 h (at the ID_{50}, 7.5 nM of epothilone B–lactam,[57] or in another system, after 20 h at 3.5 nM epothilone B[60]). Mitotic arrest and cytotoxicity are seen at much lower doses than aberrations in interphase MTs (e.g., at 3–30 nM vs. 6–40 nM vs. more than 1 μM, respectively).[8] A certain percentage of cells seem to pass through aberrant mitosis and become arrested in the following G1 phase as multimininucleate cells.[8] With epothilone B at its IC_{50}, 38% of cells become multimininucleate, and with epothilone A and Taxol, 10%–15% do so. Delayed mitosis takes place in those cells 24–30 h after the addition of the drug, and although they die, too, they do so after a longer time. The phenomenon is preceded by an aberrant chromosome alignment in metaphase and is probably related to the apoptotic fragmentation of nuclei discussed below. In MCT7 cell cultures, 2.5% of control cells are in mitosis.[60] When epothilone B is added, many cells enter mitosis but do not progress beyond metaphase. The ratio of anaphase to metaphase cells in the control was 0.18; with 1 n*M* epothilone B, 0.03; and at 3 n*M*, 0. Many of the cells that were not blocked in mitosis became multinucleate, probably because of an abortive mitosis; the behavior of cells exposed to epothilone B at a dose too low to lead to mitotic arrest is of considerable interest. In this case, multipolar spindles are induced, resulting in aneuploid cells which also are not viable.[61]

The epothilones lead to apoptosis, as indicated by DNA ladders. The DNA is nicked specifically in G2–M blocked cells.[8] Further, fragmentation of the cell nuclei is seen. This can be monitored automatically by an image analyzer.[62] With L926 mouse fibroblasts, fragmentation of nuclei can be seen already after 1 d and reaches its maximum after 3 d (100 ng/mL epothilone B). Fragmentation increases linearly with the logarithm of concentration between 1 and 1000 ng/mL epothilone A.[62] Treated cells appear to remain vital for some time, at least the reduction of tetrazolium salts continues in mouse macrophages unrestricted for 48 h (epothilone B, 0.02–20 ng/mL).[59]

A comparison of epothilones B and D showed that total uptake of epothilone B is higher by a factor of 4.[63] In addition, the intracellular distribution of the two compounds differs dramatically. Although nearly all of epothilone D is found in the protein fraction of the cytosol, almost 50% of the epothilone B resides in the nucleus.[63] This different behavior was corroborated by a more detailed study using tritiated 6-propyl-epothilones B and D (pEpts B and D), which show almost the same inhibitory effects as the natural 6-methyl analogs.[64] The study compared two cell lines, A431 (human epidermoid) and NCI/Adr (human mammary carcinoma, Taxol resistant up to more than 1μM). Epothilone D and pEpt D acted much more slowly than epothilone B and pEpt B, which produced almost maximal and lasting growth inhibition after only 4 h exposure in A431 cells. Drug uptake and intracellular pools turned out to be dose dependent. Efflux of pEpt D was faster (almost complete after 60 min) than that of pEpt B, 60%–70% of which was still inside the cells after 4 h when the loading dose was high (70 n*M*). When it was low (3.5 n*M*), NCI/Adr, in contrast to A431 cells, lost pEpt B quickly. This demonstrates that the two cell lines have different transport mechanisms for the two analogs. Apparently, the export pump of NCI/Adr cells becomes blocked when pEpt B is present in high concentrations.

In both cell lines there seems to exist a MDR1-independent, yet verapamil-sensitive, export. Verapamil, a Ca antagonist and inhibitor of the MDR1 P-glycoprotein (Pgp) shuttle, stimulates uptake of both drugs at low drug concentrations, but at high concentrations stimulates uptake only of pEpt D. Intracellular pools may reach up to 100 pmol/10⁶ cells. Drug distribution was the same in both cell lines: pEpt D (and Taxol) was mainly in the soluble part of the cytosol (80%), and only little was in the nucleus (5%–15%), whereas half of the pEpt B was in the nucleus (40%–50% vs. 40%–60% in the cytosol); there it was located on nuclear proteins, presumably tubulin, and nothing was on the DNA. This distribution may contribute to the differences in drug transport and efficacy of the two analogs. The influence of epothilone B on MT dynamics was directly measured on time-lapse photographs of MTC7 cells that were stably transfected with green-fluorescent-protein α-tubulin.[60] Cells were analyzed 6 h after the addition of the drug, when the intracellular equilibrium was reached.

The growth rate of MTs in control cells was 13.3 μm/min, the shortening rate 21 μm/min, and dynamicity (length of growth + length of shortening during total observation period per minute) 10.4 μm/min. With 2 n*M* epothilone B, those figures were reduced to 8.3, 19.3, and 6.3 μm/min, and with 3.6 n*M*, to 7.5, 14.9, and 3.9 μm/min. Thus, while under all conditions MTs shortened twice as fast as they grew, epothilone B clearly reduced MT dynamics. Also, at 3.5 n*M* most MTs were completely stabile, although their ends showed minute (below 5 μm/min) changes in length (they "chatted"). The few MTs that still were dynamic grew and shortened more slowly and for shorter lengths of time; also, pause times, at which no changes occurred, became longer. The concentration and timescales of the MT effects correlate closely with those of mitotic arrest, and it was concluded that reduction of MT dynamics is the direct cause of mitotic arrest.

Spindle MTs may be particularly sensitive to interference with their dynamics because they have to grow for a longer time and to greater length than interphase MTs. Essentially the same reactions as just described for epothilone B were seen with Taxol, only at somewhat higher concentrations. However similar the reactions of cells to Taxol and epothilone are, they still are not identical. An example is the behavior of mouse macrophages (*in vitro*).[59] At very high doses (0.1–10 µg/mL) Taxol stimulates the macrophages to produce endotoxin-like effects; for example, to synthesize nitric oxide (*Salmonella typhimurium* lipopolysaccharide does so at 1–10 ng/mL). Epothilone B (up to 10 µg/ml) neither induces macrophages to release nitric oxide nor prevents it when they are treated with lipopolysaccharide, yet epothilone B acts normally on MTs inside the cell (at 12.5 ng/mL). Also, the degrees of resistance of normal cells to epothilone and Taxol do not always match.[65]

Studies on the effects of epothilones on tubulin *in vitro* showed that they make tubulin polymerize under conditions not applicable in the absence of the drug; for instance, in the absence of GTP[8,47] or MAPs or both, and in the presence of Ca^{++} (3 mM)[8] and at low temperatures (0°C).[47] They stabilize existing MTs at cold temperatures (4°C for 30 min,[8] 0° for at least 1 h[47]), with epothilone B doing so even after the addition of 5 mM $CaCl_2$ (not, however, epothilone A or Taxol).[47] Although initial studies suggested that epothilone A is more effective than epothilone B,[8] this was later corrected when purified tubulin was used.[47] The observed effects resemble those produced by Taxol, and indeed it was shown that epothilones displace [^{3}H]-Taxol from polymerized tubulin[8,47] (apparent K_i values, depending on the method of data analysis: epothilone A, 1.4/0.6 µM; epothilone B, 0.7/0.4 µM; docetaxel, 1.2 µM).[47] This indicates both that the epothilones dock to the binding site of Taxol (i.e., β-tubulin), although initially allosteric effects or overlapping binding sites could not be excluded, and that epothilone B binds more tightly than epothilone A and Taxol. More thorough studies showed that epothilone A is close to Taxol in efficacy while there are still subtle differences.[47] Thus, for example, in experiments without GTP but with MAPs, epothilone A causes better polymerization than Taxol (both epothilones allow assembly at 10°C, maximum at 25°C, whereas Taxol produces no MTs below 25°C, maximum at 37°C), or with GTP but without MAPs at 25°C, Taxol is more active than epothilone A (epothilone B is already near its maximum) and reaches its maximum at 37°C, when it is twice as active as epothilone A. In the absence of both MAPs and GTP, significant polymerization takes place with all three drugs, only at higher concentrations (i.e., more than 10 µM per 10 µM tubulin, which was the ratio in the former experiments). Polymerization in the absence of GTP is somewhat enigmatic.

There are two GTPs bound to the tubulin dimer: one on α-tubulin at the α, β-interface, which is neither exchangeable nor hydrolyzable, and one near the carboxy end of β-tubulin that is exchangeable and becomes hydrolyzed to GDP soon after the addition of the heterodimer to the plus end of the growing MT (which is capped with GTP–tubulin). It appears that the GTP hydrolysis changes the shape of the dimer from straight to (in the free form) bean shaped.[66] This in turn may introduce some stress in the MT, produce the helical arrangement of the protofilaments, and may be a prerequisite for decay of the MT. Epothilone also reduces the number of protofilaments in MTs formed *in vitro* from 55% with 14 (control) to 67% with 13 (20 µM epothilone B, 10 µM tubulin).[66] The idea that the epothilones stimulate nucleation of tubulin polymerization is supported by two further experiments.[47] The critical concentration of tubulin, $[T]_{cr}$ (i.e., the lowest concentration at which polymerization begins under standard conditions), is regarded to be proportional to nucleation. The variable $[T]_{cr}$ was, without drug, 3 µM and in the presence of any one of the three drugs, less than 1 µM. In the presence of only MAPs or only GTP, $[T]_{cr}$ with Taxol or epothilone A, was 1.4–1.7/4.3–4.5 µM, and with epothilone B, 0.8/2 µM. Without both MAPs and GTP, $[T]_{cr}$ with Taxol, was 22 µM, with epothilone A was 19 µM, and with epothilone B was 7.4 µM.

In addition, the average length of the MTs formed in the presence of the three drugs was measured in electronmicroscopic pictures. With MAPs and GTP but without drugs, the MTs were 4.3 µm long, with epothilones A or B, they were 2.8 and 1.1 µm, and with Taxol, they were 1.9 µm (i.e., shorter than with epothilone A). When GTP was lacking, the MTs became longer, and without MAPs even more so, yet with epothilone B, they were always half as long as with the other drugs. The morphology of the MTs was the same with or without drug, viz, tubules with occasional ribbon-like open sections.

In summary, the experiments just described demonstrate that the epothilones bind to, or close to, the Taxol site on β-tubulin, that they (also) bind to polymerized tubulin, that they seem to stimulate nucleation, and that MAPs are not required for drug binding. There is an obvious discrepancy between the concentrations effective in producing cytotoxicity and mitotic arrest (lower nanomole range) and those interfering with MT turnover *in vitro* (micromole range). This may be explained by a several hundred–fold accumulation of epothilones inside the cells.[63,67] When HeLa cells, containing about 25 μM tubulin, were exposed to 10 nM epothilone, the drug reached intracellular saturation levels after 2 h. The cells then contained 4.22 μM epothilone A, or 2.55 μM epothilone B, respectively. Exposure to 100 nM drug reversed the relative internal levels to 17 μM epothilone A or 26 μM epothilone B.[67] Thus, even at relatively low doses, both drugs reached intracellular concentrations sufficient to disturb MT dynamicity, which may lead to a lasting mitotic arrest after a short exposure of the cells to the drug.

In the last few years a much better understanding of the molecular interaction between Taxol/epothilone and β-tubulin has been gained. The yeast *Saccharomyces cerevisiae* is not inhibited by epothilone B (up to 150 μM), which is likely a result of a lack of uptake of the drug,[68,69] as purified yeast tubulin is induced by epothilone B to polymerize *in vitro* (EC$_{50}$ = 1.3 μM in the presence of 5 μM tubulin and 50 μM GTP; with bovine brain tubulin, the EC$_{50}$ is 1.2 μM).[69] Interestingly, Taxol has no such effect. Epothilone B reduces the dynamicity of yeast tubulin by a factor of nine.[69]

In the presence of epothilone B, yeast tubulin also assembles without GTP. Yeast contains only one β-tubulin gene, and its β-tubulin differs in 124 amino acids from mammalian brain tubulin. With 75% similarity, it is close enough to coassemble into MTs. By analyzing available data on Taxol binding to bovine brain β-tubulin, five amino acids that differed in yeast tubulin were assumed to be responsible for preventing Taxol binding. Those amino acids were exchanged by site-specific mutation, viz, Ala19 to Lys19, Thr23 to Val23, Gly26 to Asp26, Asn227 to His227, and Tyr270 to Phe270.[68] The mutated yeast tubulin now polymerized with Taxol (EC$_{50}$ 1.55 μM, at 5 μM tubulin under conditions that do not allow spontaneous assembly) even more efficiently than bovine brain tubulin, which is perhaps a consequence of the 5%–10% lower critical concentration of yeast tubulin. The EC$_{50}$ of epothilone B remained the same (1.45 μM) with yeast wild type and mutant tubulin. The molar binding ratio of Taxol:tubulin for wild-type yeast tubulin was 0.13, and for mutant tubulin it was 1.01, which was identical to that for bovine brain tubulin.[68] Radioactive Taxol bound to mutant MTs could be replaced almost completely by a 10-fold excess of epothilone B.[68] Clearly, the five inserted amino acids are not required for epothilone binding but are essential for interaction of Taxol with β-tubulin. This shows that the binding site of epothilone need not be exactly identical with that of Taxol. The unaltered EC$_{50}$ of epothilone B for the mutant tubulin seems to indicate that epothilone indeed binds at a different location to Taxol.

Efforts have been made to identify the binding site of epothilone by mapping point mutations conferring resistance. The studies concentrated on the highly variable carboxy end of the otherwise conserved 450 amino acids of α- and β-tubulin, because this is the location of the nucleotide and MAP binding sites. Also, this part of the molecule is important for MT stability, because it is responsible for lateral contact of the protofilaments. In addition, most residues involved in Taxol binding have been located in this area. The mutants showed an increase in epothilone tolerance by a factor of 20 to 470.[70–73] Several of them had at the same time become hypersensitive to MT-destabilizing drugs (colchicine, vinblastine) or become outright epothilone dependent.[71,73] This was explained by a destabilizing effect of the mutation that is counteracted by the stabilizing activity of epothilone (or Taxol).[74]

The drug-resistant cell lines grow more slowly than the parent lines.[71] Expression of the MDR1 (Pgp) and the MPR (multidrug-resistance associated protein, another membrane protein) export systems are not increased[71,73] or even absent.[72] Also, there is no reduction of intracellular [³H]-Taxol accumulation.[73] All this means that resistance is most likely caused by reduced epothilone

TABLE 21.1
Amino Acids in β-Tubulin Suspected to Be Involved in Epothilone Binding

Method	Responsible Amino Acids	Isotype	Reference
Resistant mutants	Thr274 to Ile, Arg282 to Gln		70
Resistant mutants	Pro173 to Ala, Gln292 to Lys, Tyr422 to Tyr/Lys (heterozyg.)	βI	71
Resistant mutants	Pro173 to Pro/Ala, Tyr422 to Tyr/Lys	βI	75
Resistant mutants	Ala231 to Thr, Gln292 to Glu	βI	73
Predicted from epothilone binding to P450epoK	Leu217, His229, Thr276, Arg284		45
Not required in yeast tubulin (in contrast to Taxol®)	Lys19, Val23, Asp26, His227, Phe270		68

binding. This has been shown to be the case by determining the percentage of tubulin polymerized; the tubulin decreased with increasing resistance from 28% (parental) to 6.9%.[73] In one cell line selected against 300 n*M* epothilone D, the mutated tubulin did not bind epothilone to any extent and also could not be stimulated to polymerize by the drug.[73] As a rule, epothilone resistance goes parallel to Taxol resistance,[73] but sometimes considerable discrepancies were observed — and even more so with respect to docetaxel.[70,71,73]

Usually, only one point mutation was discovered in resistant cell lines, but occasionally two could arise.[73] Most mutations were expressed on the protein level.[73] The identified amino acid exchanges were usually close to the M loop and on helices H7 and H9. Thus, the mutated residues were indeed near the Taxol binding site. Thr274 appears to be particularly critical for epothilone binding; it lies exactly at the site of interaction between β-tubulin and the C7–OH of Taxol.[72] However, the mapped mutations concerned a considerable variety of residues (see Table 21.1). This may be because epothilone resistance is a more complex phenomenon than it may seem at first look. For one thing, resistance may vary with different cell types, among others, because they vary in their β-tubulin isotype composition, including posttranslational modifications. Only recently it became possible to determine the exact isotype profile of an experimental cell line by liquid chromatography combined with electrospray ionization mass spectrometry.[75] This permits assurance that the isotypes present in the genome are really expressed under the conditions of the experiment, which is not necessarily the case.[76] The isotypes are encoded by different genes, and the various genes are preferentially expressed in specific cell types. They are more than 80% homologous and vary mainly at their carboxy ends. The isotypes differ in conformation, assembly, dynamics, and ligand binding.[63] Among the seven known isotypes of β-tubulin, I appears to be the preferred target of epothilone,[71,73,75] and II that of Taxol.[63] In HeLa cells, the prominent isotypes of β-tubulin were I and IVb. Two epothilone-resistant sublines contained the same isotypes with a separate heterozygous mutation, each in I. It could be shown that the mutant alleles were expressed, and apparently at a higher level than the wild type.[75]

Altered expression of isotypes[74] and changes in posttranslational modification of β-tubulin,[75] which could influence binding of MAPs, have to be considered as possibilities in producing resistance, although this has not yet been shown for epothilone. Nor can one be sure that tubulin mutations that lead to drug resistance are located in the β-isomer. In a Taxol-resistant, Taxol-dependent A549 cell line, the β1-tubulin was unaltered, and Taxol still bound to it, but there was a Ser379 to Arg exchange in the Kα1 isotype of β-tubulin.[74] Interestingly, there was an increased expression of βIII-tubulin, which also seemed to contribute to Taxol resistance, apparently by changing MT dynamicity.[74,75]

A major contribution to resistance may come from altered binding of MAPs to tubulin. MAPs may stabilize or destabilize MTs, and mutations in tubulin could influence the interaction between the two proteins, but of course, there could also be mutations in the MAPs themselves; in the enzymes that activate or deactivate them, usually kinases and phosphatases; or in the systems that control the expression of MAPs. There is experimental evidence that those assumptions are not mere theory. In the Taxol-resistant (and Taxol-dependent) cell lines, A549-T12 and A549-T24, a rise in MT dynamicity is seen in the absence of Taxol.[74,75] In this case, the responsible mutation is in the α-tubulin. The point mutation is near the carboxy end and presumably at a site of MAP4 and stathmin interaction. In response to rising Taxol concentrations (2 nM is required for survival), the status of the two regulatory proteins is also changed. Whereas the active, hyperphosphorylated form of stathmin (which destabilizes MTs) was barely detectable, the intracellular level of the nonphosphorylated, inactive form increased twofold, and the phosphorylated, and therefore inactive, form of MAP4 (which is a MT stabilizer) had clearly risen.[74] In another study it was shown that tubulin-interactive drugs, like epothilone A, lead to phosphorylation of MAP4, probably by a mitotic kinase.[76] The phosphorylated MAP4 dissociates from the MTs, which then become disorganized. Drug-resistant cells do not phosphorylate MAP4, and their MTs remain intact with MAP4 associated.

At very high doses, however, the same events as in sensitive cells take place, which shows that the interaction between tubulin and MAP4 is undisturbed. Also, expression of MAP4 is not influenced by the drugs in sensitive and resistant cells. Thus, it appears that MAP4 takes part in the drug-induced killing process, and resistance is somehow connected with the failure to phosphorylate the molecule. Although the exact mechanism of this sequence of events is not yet understood, it is obvious that the binding of epothilone to tubulin activates downstream events that contribute to cytotoxicity.

In a more general approach, genes were identified by analyzing cDNA microarrays, which were overexpressed in epothilone A–resistant cell lines.[65,77,78] The STAT1 gene, interferon-inducible genes, and a MT-associated GTPase were found to be upregulated. The latter contributes perhaps to resistance by destabilizing MTs in the resistant cells. A recent study, also using cDNA microarrays, demonstrated that 41 genes had an altered expression after epothilone B (10 nM) treatment of cells, and that many of them were connected with the TNF (tumor necrosis factor) stress response pathway.[77]

Another important determinant in epothilone resistance could be the regulation of transport systems (i.e., of the membrane proteins involved in uptake and thus intracellular accumulation of the drug) and of the MDR1 and MPR machinery responsible for export. The latter are of particular interest in connection with epothilone, because they appear to be much less active on epothilone than on other cytotoxic drugs, including Taxol.

Studies on epothilone resistance repeatedly demonstrated that intracellular accumulation of tritium-labeled Taxol was unimpaired.[73] Thus, in those cases, resistance could not be attributed to impeded uptake of the drug. Similar experiments have been done with tritium-labeled epothilone analogs.[63,64] [^{3}H]-6-Propyl epothilones B and D were used to study uptake by nonresistant A431 (human epidermoid) and doxorubicin (Adriamycin)-resistant NCI/Adr tumor cells (which also are resistant to epothilones).[63] The resulting picture turned out to be rather complex, as pointed out already. In several epothilone B–resistant A549 and HeLa cell lines, the MDR1 gene was not expressed at all, and although the MRP system was expressed, its level of expression did not increase after the cells became epothilone resistant.[71] Also, in epothilone D–resistant leukemia cell lines, no rise in the levels of the MDR1 and MRP systems was seen, which could have explained resistance.[73] In fact, it already had been discovered that epothilones are not substrates for Pgp.[8,47] A human neuroblastoma cell line that was constitutive for MDR1 still responded to epothilone A, but not to Taxol.[58] The level of Pgp even rose during drug treatment (i.e., MDR1 was induced by epothilone A). In contrast to epothilone, calcein was exported from the cells ever more efficiently the higher the Pgp levels became, which proved that the transport system was fully functional. Thus, it seems that the cellular export systems do not play a major role in epothilone resistance, nor is there evidence that impaired uptake may be a decisive factor.

How does epothilone kill cells? All studies dealing with the killing mechanism of epothilone agree that the drug binds to MTs within cells, which leads to mitotic arrest by disorganization of the normal turnover of tubulin, to an accumulation of the cell population in the G2–M phase of the cell cycle, to a decay of the cell nuclei, and to apoptosis. However, there still remains to be explained how exactly apoptosis is achieved, and whether apoptosis is the only way by which epothilones become cytotoxic. The answer is of more than academic interest, as it may have implications for the treatment of cancer. Thus far, no fully consistent picture can be presented. Among other reasons, this may be because investigators dealing with that problem used different cell types and various epothilone analogs.

It appears that the tumor suppressor protein, p53, does not play a role in the killing of cells by epothilone, as human neuroblastoma cell lines, which constitutively sequester p53 in the cytoplasm and thus prevent its effect as an intranuclear transcription factor, still were fully sensitive to epothilone A (and Taxol).[58] In contrast, it was shown that, although cells with mutated p53 normally become less sensitive to cytostatic compounds, their sensitivity to epothilone B (and Taxol) rises, perhaps because a higher proportion of cells with mutant p53 is in mitosis — the target of epothilone.[79]

Apoptosis is a rather complex phenomenon induced by a receptor-mediated extrinsic or a mitochondrial pathway, and it is modulated by a large number of proteins. Very briefly (for more details see, e.g., refs. 80 and 81), the extrinsic pathway starts with death receptors (e.g., DR4 and DR5) in the cell membrane. They are activated by binding a death ligand (e.g., Apo2-2L/TRAIL=TNF-related apoptosis-inducing ligand) and then, after several more steps, form a death-initiating signaling complex, which results in autocatalytic activation of initiator caspase 8, and the following effector caspase cascade finally triggers apoptosis.

The mitochondrial pathway begins with release of cytochrome c from mitochondria into the cytosol. This, which again involves several steps, leads to the formation of apoptosomes (consisting of oligomers of APAF-1 [apoptotic protease-activating factor 1]), which activate initiator caspase 9 and, in turn, executioner caspase 3 (and caspase 7). Those caspases cleave a number of cellular proteins and finally lead to apoptosis. The sequence is modulated by the Bcl-2 and IAP protein families. One study applying (water-soluble) epothilone B–amine (BMS 310705) on ovarian carcinoma cells taken from a patient who no longer responded to treatment with platinum and Taxol demonstrated that the drug killed the cells via the mitochondrial pathway of apoptosis.[80] The cells were exposed for 1 h to 10–500 nM of the drug and then incubated in drug-free medium. After 24 h, more than 25% of cells showed apoptosis, and after 96 h, 90% were dead. An increase in activity of initiator caspase 9 and effector caspase 3, but not of initiator caspase 8 (typical for the extrinsic pathway) was seen. Also, 12 h after treatment, cytochrome c was detected in the cytosol. Another study using epothilone B–lactam on cell cycle–synchronized human ovarian carcinoma cells found indicators of the mitochondrial pathway (leakage of cytochrome c into the cytosol, activation of caspase 3, and downregulation of the protective proteins survivin, cIAP1, and XIAP, which inhibit caspase 3, caspase 7, and caspase 9), but also a rise of the death receptors DR4 and DR5 in the cytoplasmic membrane, which sensitized the cells to the death ligand, Apo2-2L/TRAIL.[81] Thus, in this case the extrinsic pathway seemed also to be in operation. The study also showed that epothilone B–lactam is active on Taxol-resistant cells.

Still another study showed a role of the Bcl-2 family of apoptosis-modulating proteins in the killing process.[82] Again, epothilone B–lactam was used. The drug induced a conformational change in the proapoptotic Bax protein, which resulted in its transfer from the cytosol into mitochondria and a concomitant release of cytochrome c from mitochondria into the cytosol, with the consequences pointed out above. Apoptosis could be mitigated by overexpression of Bcl-2 and enhanced by the Bcl-2 antagonist, BAK-BH3 peptide. The conclusion was that epothilone B–lactam induces apoptosis by a Bcl-2-suppressible pathway, and that treatment of breast cancer can perhaps be improved by a combined therapy with epothilone B–lactam and a Bcl-2 antagonist. Involvement of the Bcl-2 family was also seen when human ovarian carcinoma cells were treated with epothilone A.[76] The epothilone A–tubulin interaction, which must be unimpaired, results in inactivation of the prosurvival proteins Bcl-2 and Bcl-x_L by phosphorylation, and in lowering of the Mcl-1 level by increased turnover. As Mcl-1 inactivates the proapoptotic Bax protein by forming a complex with it, reduced Mcl-1 and a concomitant overexpression of Bax clearly would support apoptosis. Typically, in resistant cells Mcl-1 remains high in the presence of epothilone A, Bax is lowered, and Bcl-x_L is not phosphorylated.

In contrast to the studies just discussed, the authors of an investigation of the effects of epothilone B on non–small cell lung cancer cells concluded that epothilone kills the cells by an as-yet-unknown, caspase-independent mechanism.[83] At low doses, cytotoxic effects (disruption of the MT cytoskeleton, arrest of cells in the G2–M phase, aberrant mitosis, multinucleated cells) are seen rather soon, before 24 h after the addition of the drug, yet symptoms of apoptosis [nuclear condensation, apoptotic bodies, cytochrome c release, cleavage of poly (ADP-ribose) polymerase, activation of caspase 3 and caspase 9] appear only much later, and therefore may be just side effects. Furthermore, the cells would still die when the mitochondrial or the extrinsic pathway of apoptosis were neutralized by overexpression of the antiapoptotic proteins, Bcl-2 or Bcl-x_L, or of the dominant negative FAS-associated death domain. Even the caspases could be blocked by natural (response modifier A) or artificial (Z-Val-Ala-Asp-fluoromethylketone) antagonists without interfering with cell death. If this hypothesis is correct, drugs may be found that induce that mechanism more efficiently and, thus, may be even more beneficial in cancer therapy.

Several studies report synergistic effects between epothilone and other drugs, which may become of interest in cancer therapy. Many solid tumors show interstitial hypertension, which impedes transfer of drugs from capillaries to tumor tissue. This high interstitial fluid pressure appears to be mediated by the activity of platelet-derived growth factor, the receptors of which are expressed in many solid tumors. The tyrosine kinase inhibitor, STI571 (Novartis Pharma), blocks the kinase activity of those receptors, which results in lowered interstitial fluid pressure and improves uptake of drugs by the tumor. When severe combined immunodeficiency (SCID) mice with human anaplastic thyroid carcinoma were treated with a combination of epothilone B and STI571, the tumors became more than 40% smaller than after treatment with epothilone B alone.[84] The level of epothilone B in the tumor increased threefold but remained unchanged in liver, kidney, and the intestinal tract. However, STI571 (100–150 mg/kg) had to be administered on three consecutive days at a minimum, and its effects faded away 2 d after the last application.

The synthetic flavonoid, flavopiridol, has aroused considerable interest in recent years on account of its antitumor activity. It arrests cells in the G1–S (or G2–M) phase of the cell cycle and induces apoptosis. The mechanism appears to be inhibition of mRNA synthesis, perhaps by blocking transcriptional elongation or by binding to duplex DNA. In consequence of this, cyclins and antiapoptotic members of the IAP and Bcl-2 families of proteins are downregulated.[85] Further, flavopiridol blocks cyclin-dependent kinases and brings about conformational changes of Bax, with the sequels outlined above. In human breast cancer cells, flavopiridol (100–500 nM) acted synergistically with epothilone B (20 nM) in inducing apoptosis, but only when applied after, and not in advance of, that drug.[85] In a Bcl-2 overexpressing cell line, epothilone B and flavopiridol, alone or in sequence, induced only minor apoptosis, which demonstrates again that Bcl-2 acts as an epothilone antagonist.

Specific farnesyl transferase inhibitors are potent antitumor drugs. The explanation is that they prevent posttranslational farnesylation of certain proteins (e.g., of the p21ras family), which is a prerequisite of their insertion into the cell membrane. With human breast cancer cells, it was shown that those inhibitors (e.g., L-744,822; Merck) increase sensitivity to antimitotic drugs, such as Taxol and epothilone.[86] The synergistic effect was additive. It is speculated that this happens because a farnesylated protein controls the checkpoint of mitosis. Also, human prostate cancer cells, including a line that was rather resistant to L-744,822, became considerably more sensitive to the drug when they were pretreated with epothilones.[52] This was regarded as a promising strategy for therapy of advanced prostate cancer. A broad synergistic effect was seen in colon cancer cells treated with a combination of epothilone D and 5-fluorouracil.[87]

Histone-deacetylase inhibitors, like LAQ824 (Novartis), lead to altered gene expression, with changes in the levels of several important regulatory proteins. As a result they enhance epothilone B–induced apoptosis in human breast cancer cells.[88]

Epothilone B–lactam sensitizes human lung cancer cells to radiation, *in vitro* as well as *in vivo* (athymic nude mice).[89] *In vitro*, the induction of apoptosis by drug and radiation was additive, and a higher dose of the drug for a shorter exposure time was more efficient than a lower dose for a longer time. The *in vivo* effect is considered to be a result of a combination of mitotic arrest and reoxygenation of the tumor, so that the timing of the irradiation after drug exposure becomes a critical factor.

IV. TOTAL SYNTHESIS AND SEMISYNTHESIS

A. TOTAL SYNTHESIS

The discovery of the epothilones as the first Taxol mimics, and reports of their promising pharmacological properties, triggered enormous worldwide synthetic activities in 1995. With seven stereogenic centers in a 16-membered macrocycle, their total synthesis appeared, though challenging, to be far less difficult than that of Taxol. Thus, even smaller groups entered the field and made epothilone probably the most often synthesized natural product in recent years. Apart from preliminary synthetic studies, more focused work started in late 1995, when the absolute configuration became available from our lab (before publication in July 1996, the absolute configuration of epothilone A and B had been communicated to interested scientists from October 1995 on). First total syntheses of epothilone A and, later, of epothilone B were reported by the groups of Danishefsky, Nicolaou, and Schinzer in 1996 and 1997.[10a] Key bond-forming steps in these early syntheses were Yamaguchi macrolactonization, a highly diastereoselective C6–C7 aldol reaction, and olefin metathesis at C12–C13. Soon after, macroaldolization forming the C2–C3 bond and a novel B-alkyl Suzuki coupling of C11–C12 were introduced by the Danishefsky group. Over the years, these basic approaches have been developed further by the original authors and adopted by many other groups for their own synthetic strategies. Many more alternatives were developed, omitting only a few carbon–carbon bonds not used in the assembly of the macrocycle and the attachment of the thiazole side-chain (Figure 21.6). Thanks to the enormous efforts of many groups, the early synthetic strategies mentioned above proved their value and were adapted to plant-scale syntheses. It is certainly beyond the scope of this review to discuss in detail the more than 20 direct and formal total syntheses published and reviewed already[10a,b,d,f,j–p]; however, a few points on the bond-forming strategies should be made.

In most syntheses, the first strategic goals are, as in biosynthesis, the 12,13-olefins epothilone C and D (occasionally named also dEpoA and dEpoB, **2a**, **2b**), which are epoxidized in a more or less stereoselective manner to epothilone A and B.[125,126] Alternatively, the sensitive epoxide is

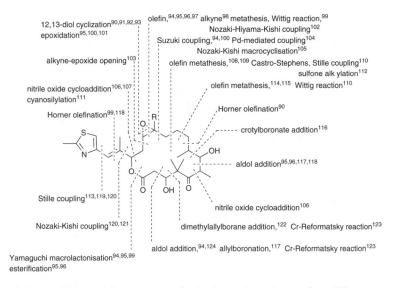

12,13-diol cyclization[90,91,92,93] olefin,[94,95,96,97] alkyne[98] metathesis, Wittig reaction,[99]
epoxidation[95,100,101] Nozaki-Hiyama-Kishi coupling[102]
Suzuki coupling,[94,100] Pd-mediated coupling[104]
Nozaki-Kishi macrocyclisation[105]
alkyne-epoxide opening[103] olefin metathesis,[108,109] Castro-Stephens, Stille coupling[110]
sulfone alkylation[112]
nitrile oxide cycloaddition[106,107] olefin metathesis,[114,115] Wittig reaction[110]
cyanosilylation[111]
Horner olefination[99,118] Horner olefination[90]
crotylboronate addition[116]
aldol addition[95,96,117,118]
Stille coupling[113,119,120]
nitrile oxide cycloaddition[106]
Nozaki-Kishi coupling[120,121] dimethylallylborane addition,[122] Cr-Reformatsky reaction[123]
aldol addition,[94,124] allylboronation,[117] Cr-Reformatsky reaction[123]
Yamaguchi macrolactonisation[94,95,99]
esterification[95,96]

FIGURE 21.6 C,C- and C,O-bond-forming strategies in the total syntheses of epothilones.

introduced at an early stage with high stereoselectivity, and with the necessary precautions, carried through the synthesis by the Mulzer group.[90] Soon after completion of the synthesis of the four major epothilones, A–D, the groups of Danishefsky and Nicolaou and others started to explore structure–activity relationships for cytotoxicity and tubulin binding. Following their own philosophies, either a limited number of selected analogs or large libraries were synthesized via combinatorial chemistry. In this work, all stereocenters were probed for their relevance for biological activity, the alkyl groups on C8[127] and C16[128,129] were removed, or new ones were added at C10[130] and C14.[130,131] Further, the C6[64] and C12 methyl groups were replaced by larger alkyl and modified alkyl groups.[49,115,119,128,132–136] C17–C18 Stille coupling allowed the introduction of a great variety of five- and six-membered heterocyclic side-chains, optionally carrying additional substituents.[113,119,129] Another obvious modification, the ring closure of C27 with the thiazole ring, was first realized by the Novartis group only in 2000 with the synthesis of benzthiazole, benzoxazole, and quinoline analogs.[137] The 12,13-*cis* epoxide was changed to *trans* and, moreover, replaced by a cyclopropane,[120,121,138–141] cyclobutane,[121,140] or aziridin ring.[142]

Whereas 12,13-trans olefins and the epoxides derived from them were regularly unwanted byproducts in the ring-closing olefin metathesis, Altmann et al. synthesized them in a stereoselective manner to prove their unexpected biological activity,[143] which had been observed earlier by Nicolaou and coworkers.[144] To rigidify the macrocycle and enforce certain conformations, 12,13-benzo,[145] 10,11,12-benzo,[64,146] and 10,11-cyclopropano[108] analogs, and a C23, C24 ethanol bridged analog,[91] were synthesized. Similarly, the C4 geminal dimethyl group was integrated in a cyclopropane ring,[53,54] 11-hydroxy and 11-fluoro substituents were introduced,[105] and oxygen was introduced as isosteric replacement of C10.[147] Epothilones with extra double bonds at C9, C10[101,115b,148,149] and C10, C11,[108] or with a C9, C10 alkyne bond,[150] became available *en route* to epothilone B and D.

Occasionally, natural product building blocks were introduced, such as nerol[123,151] and carene[152] in the northern ring segment. The use of pantolactone[64,153] for the C1–C5 ring segment and malic acid[64] for the C13–C16 fragment provided the C3 and C15 stereocenters. In one and the same synthesis, both northwestern and southeastern ring segments were constructed from glucose by extensive functional group manipulations performed by Potier et al.[154] Chemo-enzymatic methods were applied to establish the desired configuration at the C3, C7, and C15 hydroxyl groups. Thus, resolution was achieved by commercial esterases and lipases,[155,156] by a genetically engineered lipase,[157] or in case of C15, by a catalytic aldolase antibody.[158–160] An aldolase-catalyzed asymmetric synthesis was developed by Wong et al.,[161] providing building blocks with the C7 and C15 stereocenters in enantiomerically pure form. In a polymer-supported synthesis of an epothilone C library, Nicolaou et al. used an olefinic linker to the polymer support, which is cleaved in the final step by participation in an olefin metathesis cyclization.[144,162] In this way, only successfully cyclized molecules are released from the polymer to give immediately pure products. In an alternative approach, Ley et al.[163] synthesized epothilone B, using polymer-supported reagents and scavengers while the growing product was kept in solution. Thus, tedious extraction and purification steps were avoided, making this approach amenable for large-scale synthesis. However, the advantages of the overall process may be doubtful in terms of cost effectiveness when costs for regeneration of the polymer reagents are included.

In summary, during the past eight years of epothilone synthesis, around 160 papers have been published and more than 100 patents filed and, in part, granted. Apart from academic syntheses of natural epothilones and the development of new synthetic methods, around 1000 epothilone analogs may have been synthesized for structure–activity relationships studies. In two cases, epothilone D and epothilone C6–modified epothilone B analogs, large-scale processes were developed, taken to plant-level production by the Danishefsky and Schering groups, and introduced in clinical trials.

B. SEMISYNTHESIS

In the course of strain and production optimization at GBF, epothilone A and B became available in larger amounts in 1996 and 1997 and were used in a semisynthesis program in collaboration with

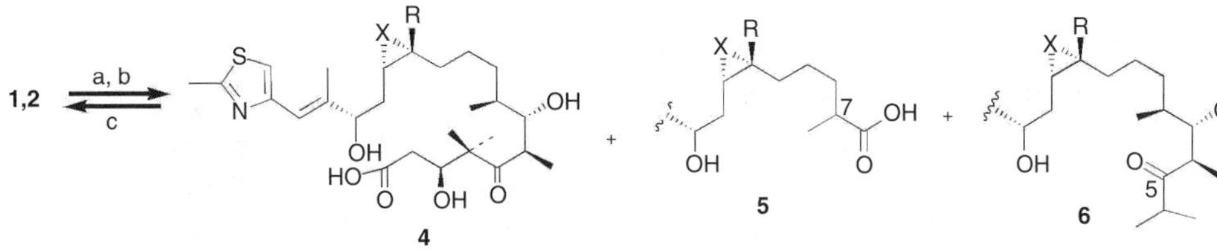

a. Pig liver esterase, DMSO/buffer, 37°C
b. LiOH, CH_3OH/H_2O, RT
c. 2,4,6-trichlorobenzoyl chloride, $N(C_2H_5)_3$, DMAP, Toluene

R = H or CH_3
X = O

SCHEME 21.2

Bristol-Myers Squibb. Later, a block-mutant strain was added, supplying epothilones C and D.[17] Early one-step transformations were performed with epothilone A, including mono- and diacylation of the 3,7-dihydroxy groups, Swern oxidation to 3,5- and 5,7-diketones, and borohydride reduction of the C5 ketone to a 1:1 mixture of epimeric alcohols.[164] Similarly, solvolysis of the epoxide produced mixtures of stereoisomeric diols (cat. H_2SO_4), chlorohydrins (HCl), or rearranged 14-membered lactones (TFA, BF_3OEt_2). Dihydroxylation of epothilone C with OsO_4/NMO gave a 2:1 mixture of epimeric *cis* diols, dimethyl dioxirane, and MCPBA mixtures of natural epothilone A — its β-epoxy isomer — along with other oxidation products.[165,166] 12,13-Dihydro-epothilone C was formed selectively with diimine, whereas catalytic reduction produced complex mixtures because of hydrogenolysis of the allylic ester moiety.[165] Disappointingly, all these derivatives were devoid of significant biological activity.

Hydrolysis of the lactone in epothilones A–D was achieved by LiOH in $MeOH/H_2O$ to give *seco*-acids **4a–d**,[21,167] and depending on reaction time, by retroaldol reactions, the C7 aldehyde and C5 isopropyl-ketone fragments (Scheme 21.2).[21] A very clean cleavage to *seco*-acids was achieved with pig liver esterase.[21,51,167] Interestingly, epothilone A was restored by Yamaguchi macrolactonization of 3,7-nonprotected *seco*-acid **4a** in 25% yield.[21] Although C3 and C7 hydroxyl groups occupy β-positions to carbonyl groups, they are not eliminated under these and other more rigorous conditions. Only the 3,7-diformyl epothilones **7** on treatment with DBU gave in good yield trans-2,3-olefins **8**, which act as Michael acceptors (e.g., for cyanide to give a 1:1 mixture of 3α- and 3β-cyano epothilones **9a** and **9b**; Scheme 21.3).[168]

Another ideal target for structural modification was the 12,13-epoxide. Thus, epothilone A was selectively opened to bromohydrin **11** with $MgBr_2$ in ether, oxidation to the bromoketone, α addition of cyanide followed by ring closure gave the *trans*-like cyanoepoxide **12**.[168] Its structure was proven by x-ray crystallography. Alternatively, the bromohydrin was converted to the azido alcohol, and Mitsunobu inversion and *O*-mesylation set the stage for phosphine reduction and cyclization to aziridin **10a** with the natural α-configuration.[142] Subsequently, the nitrogen was alkylated and acylated with a broad range of residues to **10b** (Scheme 21.4).

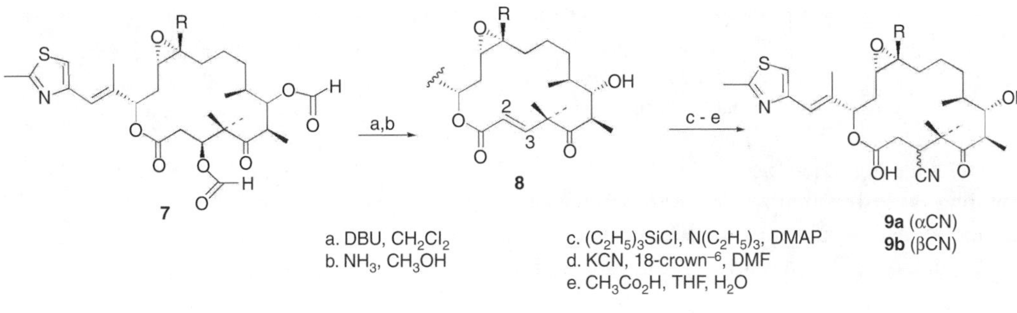

a. DBU, CH_2Cl_2
b. NH_3, CH_3OH

c. $(C_2H_5)_3SiCl$, $N(C_2H_5)_3$, DMAP
d. KCN, 18-crown-6, DMF
e. CH_3Co_2H, THF, H_2O

9a (αCN)
9b (βCN)

SCHEME 21.3

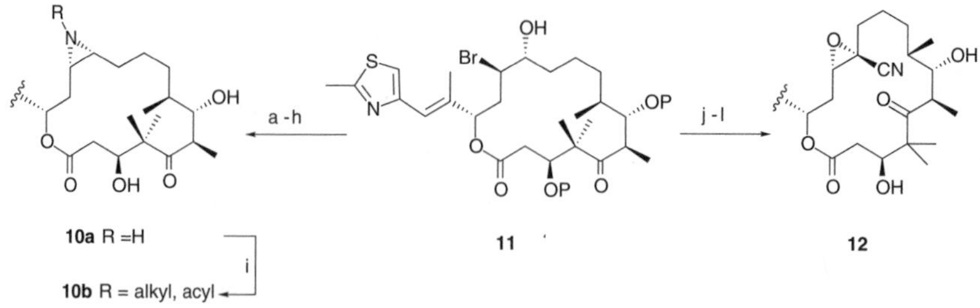

a. (C₂H₅)₃SiCl, N(C₂H₅)₃, DMAP, DMF
b. MgBr₂, (C₂H₅)₂O/CH₂Cl₂ -20 to -5 °C
c. NaN₃, DMF, 42 °C
d. p-NBA, DEAD, Ph₃P
e. NH₃, CH₃OH
f. MsCl, N(C₂H₅)₃, CH₂Cl₂
g. P(CH₃)₃, THF/H₂O
h. TFA/CH₂Cl₂, -10 °C
i. e.g. (CH₃)₂SO₄, proton sponge, THF
   or RCOCl/Hunig's base

j. PCC, Py, CH₂Cl₂
k. KCN, 18-crown-6, THF
l. TFA/CH₂Cl₂, -10 °C

SCHEME 21.4

Whereas epoxidation of the C12, C13 double bond is the final step in epothilone A and B biosynthesis and in many total syntheses, it was desirable to use the rich supply of epothilone A and B from fermentation for the reverse reaction. This was achieved by the BMS group in good yield and retention of (Z)-configuration by reduction of unprotected **1a** and **1b** with TiCp₂Cl₂/Mg and WCl₆/n-butyl lithium to epothilone C and D, respectively.[139] Cyclopropanation of the C12, C13 double bond of epothilones C and D turned out to be a tough problem but was eventually achieved in low yield with dibromo- and dichlorocarbene generated under phase transfer conditions to give exclusively α-cyclopropane **13a**.[139] Subsequently, the halogens were removed by tinhydride reduction to **13b** (Scheme 21.5).

Cleavage of the C12, C13 double bond of epothilone C and D by either ozonation or with OsO₄/Pb(OAc)₄ or, alternatively, diol formation and periodate cleavage of epothilone A and B gave valuable C12 aldehyde and ketone building blocks ready for, for example, Wittig condensation with synthetic ring segments.[169,170] Inspired by the ring-closing olefin metathesis in the total synthesis of epothilones, the reverse reaction was investigated in presence of an excess of ethylene and second-generation Grubbs's catalysts. The expected *seco*-diene **14** was obtained in good yield and processed further by replacement of the C13–C15 ring segment with a synthetic analog to give **15**. Ring-closing olefin metathesis afforded, for example, the alkyne analog **16** of epothilone C along with the expected stereoisomer at C12/C13 (Scheme 21.6). Subsequently, **16** was epoxidized to the alkyne analog of epothilone A.[171] More direct modifications of the linker group C16,C17 were catalytic hydrogenation, epoxidation followed by reduction to a 16-hydroxyl group,[164] and its elimination to a C16 exomethylene derivative.[21]

Several epothilones, including the 16-desmethyl analog of epothilone A, could be isomerized to 6:4 mixtures with their Z-isomers **17** by benzophenone-sensitized irradiation (Scheme 21.7).[171] C16-ketones **18** were obtained from epothilone A and B by ozonolysis and used as synthetic building blocks.[164] Apart from standard carbonyl derivatives of the acetyl group, new side-chains were introduced by aldol condensation to, for example, **19**, whereas Wittig-like olefinations only worked well with the boron analog (C₂H₄O₂B)₂CHLi yielding vinyl boronic acid **20** (Scheme 21.8). Suzuki coupling of **20** required reactive iodo compounds, such as iodo benzene; otherwise, it was transformed by N-iodosuccinimide (NIS) to the corresponding iodovinyl epothilone and introduced in Stille couplings,[166] as described by Nicolaou et al.[119,120]

Initially, modification of the thiazole ring other than by replacement seemed hardly possible in the presence of the sensitive functional groups in the macrocycle. However, with *sec*-butyl and

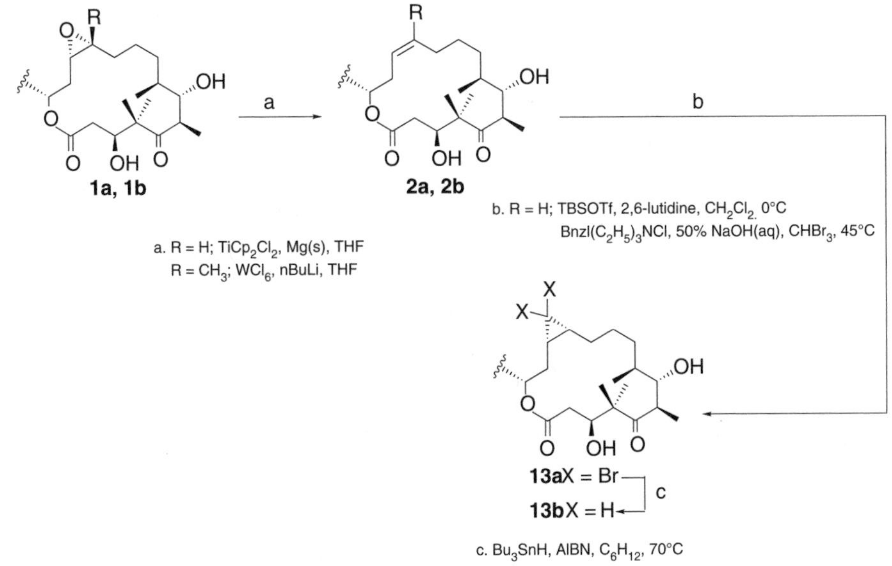
1a, 1b
2a, 2b
a. R = H; TiCp2Cl2, Mg(s), THF
R = CH3; WCl6, nBuLi, THF
b. R = H; TBSOTf, 2,6-lutidine, CH2Cl2, 0°C
BnzI(C2H5)3NCl, 50% NaOH(aq), CHBr3, 45°C
13aX = Br
13bX = H
c. Bu3SnH, AIBN, C6H12, 70°C

SCHEME 21.5

t-butyl lithium at low temperatures, deprotonation of C19 and, to a lesser extent, C21 methyl was possible, and after quenching with deuterium chloride or carbon and heteroatom electrophiles, the corresponding derivatives were obtained in moderate to low yields.[172] An unexpected and selective access to C21 derivatives was opened by thiazole *N*-oxidation using *m*-chloroperbenzoic acid (MCPBA). Expectedly, on treatment of the *N*-oxide **21** with acetic anhydride, a Polonovsky-like

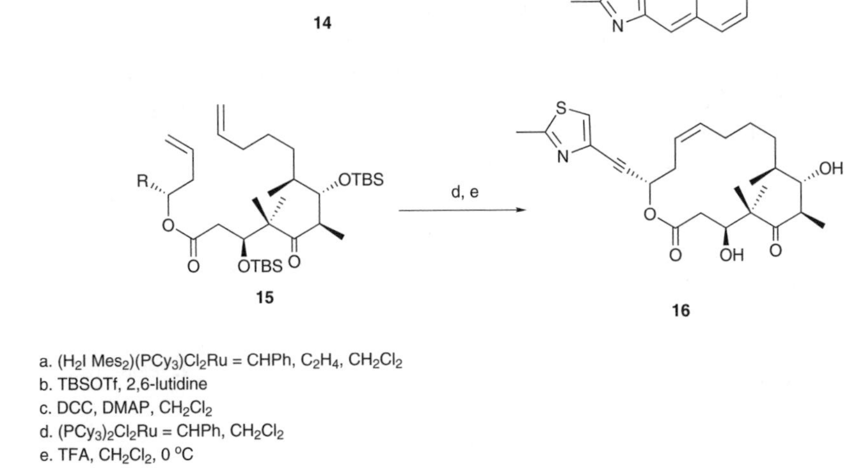
2a
14
15
16
a. (H2I Mes2)(PCy3)Cl2Ru = CHPh, C2H4, CH2Cl2
b. TBSOTf, 2,6-lutidine
c. DCC, DMAP, CH2Cl2
d. (PCy3)2Cl2Ru = CHPh, CH2Cl2
e. TFA, CH2Cl2, 0 °C

SCHEME 21.6

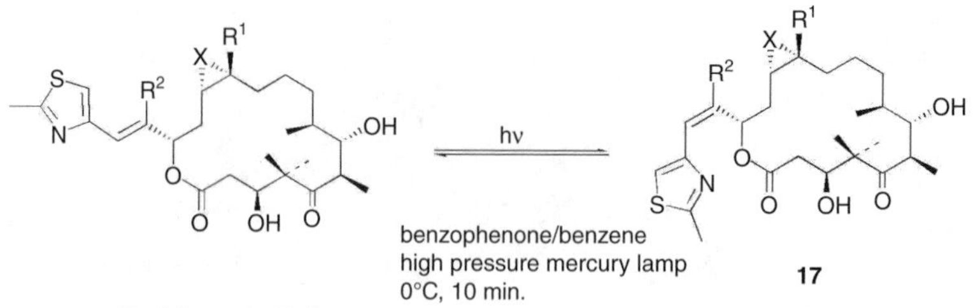

Epothilones A - D, A_8

17

x = O or Δ

SCHEME 21.7

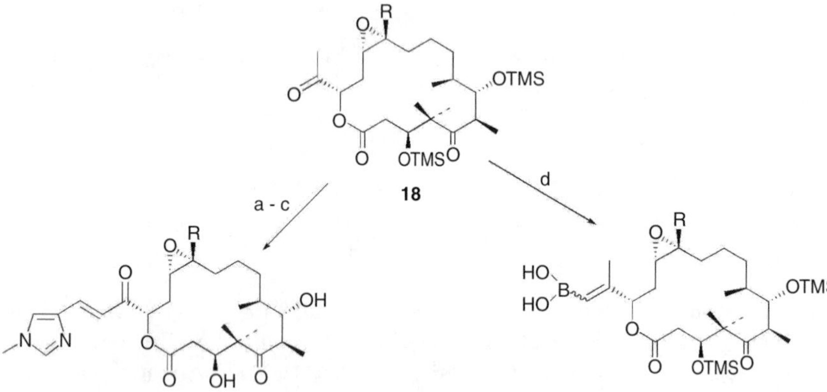

18

19

20 (E:Z = 7:3)

a. LiTMP, THF, –70°C
b. 1-methyl-3-imidazocarbaldehyde
c. citric acid

d. $(C_2H_4O_2B)_2CHLi$, CH_2Cl_2/THF 1:1

SCHEME 21.8

rearrangement transposed the oxygen to the C21 methyl, and on mild hydrolysis, epothilones E and F were obtained.[50] It should be mentioned that all these steps were carried out without 3,7-OH protection (Scheme 21.9). As benzylic alcohols, epothilones E and F were easily oxidized to the corresponding aldehydes, and both alcohols and aldehydes were further modified by the introduction of halogen and a great variety of N, O, S, and carbon substituents.[50,173] One of these, 21-amino epothilone B **22**, was introduced into clinical trials by the BMS. An alternative to the *N*-oxide route, C21 hydroxylation, was achieved by biotransformation using *Amycolata* or *Actinomyces* strains.[174] In addition to C21, *Am. autotrophica* also hydroxylates C9, C14, and the C12 methyl.[38]

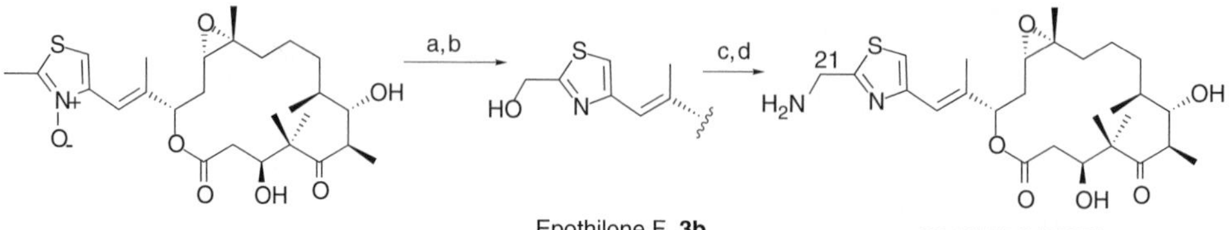

21

Epothilone F **3b**

22 (BMS-310705)

a. TFA anhydride
b. 1%NaOH in CH_3OH/H_2O

c. DPPA, DBU, DMF
d. H_2/Pt, C_2H_5OH

SCHEME 21.9

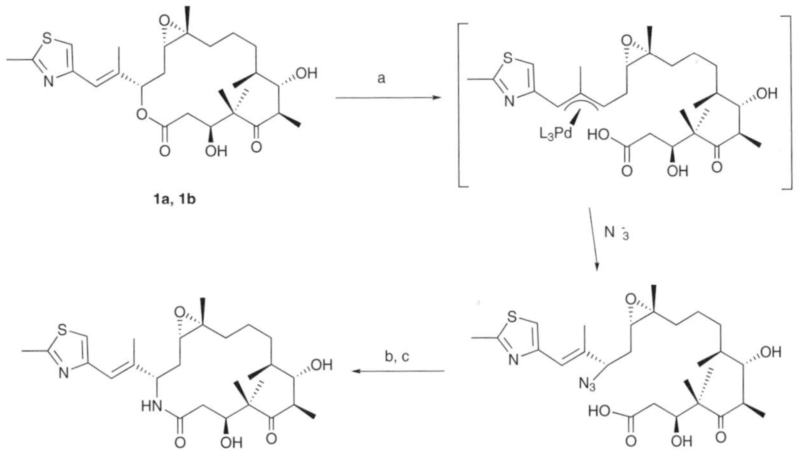

23 (BMS-247550)

a. Pd(PPh₃)₄, NaN₃, THF/H₂O, 45 °C
b. P(CH₃)₃, THF; then NH₃/H₂O, 45 °C
c. EDC, HOBT, CH₃CN/DMF

24

SCHEME 21.10

Although obvious, isosteric replacements of the carboxyl group were not covered in the earlier total syntheses but were first realized by semisynthesis. In a very short and elegant reaction sequence, Vite and coworkers at BMS transformed epothilone B lactone to the lactam. Using palladium chemistry, the allylic lactone was opened and the C15 oxygen replaced by azide with overall retention of configuration. Reduction of the azide to an amine, and cyclization by standard peptide chemistry, produced the lactam in good yield (Scheme 21.10).[51] Remarkably, the whole sequence did not require 3,7-OH protection and could be performed as a one-pot reaction. Because of its favorable cytotoxic and pharmacological profile, epothilone B–lactam **23** (BMS-247550, Ixabepilone) was introduced into clinical trials in 2000.[10q]

V. STRUCTURE–ACTIVITY RELATIONSHIPS

Investigation of structure–activity relationships started very early at GBF, with more than 39 natural epothilones isolated from large-scale fermentations and simple one-step semisynthetic derivatives. Shortly after, the Danishefsky and Nicolaou groups published biological data for a great variety of analogs, and industrial groups at Novartis and Schering followed. Unfortunately, the data generated for cytotoxicity/growth inhibition by different groups are hardly comparable on a quantitative basis because of the broad range of sensitivity of the various cell lines used. *In vitro* tubulin polymerization depends even more on the experimental conditions used.[138,175] In this review, we classify activity as high (>50% of the parent compound), moderate (5%–50%), low (1%–5%), and inactive. It is generally accepted that a 16-membered macrocycle and the stereochemical assignment of substituents within the box (Figure 21.7) are a prerequisite for biological activity.[113] However, good to moderate activity was observed with certain 17- and 18-membered epothilones.[113,176] The substituents in the box may be modified to a certain extent.

Epothilones with an (*E*)-2,3-double bond from elimination of the 3-hydroxyl are still moderately active, whereas 3β-cyano analogs are highly active.[168] This means that the 3-hydroxyl is not a hydrogen bridge donor but, rather, an acceptor in the tubulin bound state of epothilone. Both 4-desmethylepothilones A₁ and A₂[4] and 16-desmethylepothilone[129] are highly active, whereas replacement of 8-Me by hydrogen drastically reduces activity.[127] Extension of the C6 methyl to ethyl increases activity fourfold compared with epothilone B, and allyl and propyl derivatives are only slightly less active than the parent compound.[64] An extra methyl in the 14α position reduces activity

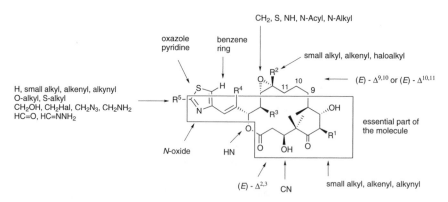

FIGURE 21.7 Structure–activity relationships of epothilone.

slightly, whereas a 14β methyl causes refolding of the corresponding ring segment and loss of activity.[131] The vinyl spacer C16/C15 in the side-chain and the presence and position of the sp^2 nitrogen are of crucial importance for biological activity (Figure 21.8).[49] Shorter spacers or those extended by one atom,[166] (Z)-configuration of the double bond,[171] an alkyne spacer,[171] epoxides, or conformationally flexible derivatives with a C16/C17 single bond[164] are inactive. The thiazole may be replaced by oxazole or α-pyridyl (**25**, X = O, **26**), without loss of activity whereas β- and γ-pyridyl analogs are only moderately active or inactive.[4,53,54,64,119]

Likewise, thiazoles linked by their C2 or C5 position are inactive.[113] Integration of the C16, C17 spacer in a benzene ring to give benzothiazole, benzoxazole (**28**), and quinoline (**27**) analogs confers high biological activity.[64,137] In contrast, introduction of any substituent at the thiazole C19 (**29**) or alkylation of the nitrogen (**30**) abolishes activity.[172] The C20 methyl on the thiazole may be replaced by hydrogen, small alkyl, alkenyl, or alkynyl groups[173] or heteroatom substituents with retention of high activity.[141,178] In general, three-atom (except hydrogen) residues are still highly to moderately active, and bigger ones are weakly active or inactive.[173] The methyl group C21 may be modified by a variety of halogen, oxygen, nitrogen, and sulfur substituents or oxidized to the aldehyde and further derivatized without loss of activity, provided the space demand of the entire residue does not exceed three to four atoms.[173] The thiazole-2-carboxylic acid derivative is inactive.[173]

Another part of great structural flexibility is the C12, C13 epoxide. Cyclopropane,[120,139,140] cyclobutane,[140] episulfide,[21,177] and aziridin analogs including N-alkyl or N-acyl modifications[142] are highly active, whereas the corresponding olefins, epothilone C and D, are moderately active.[97,113,179]

FIGURE 21.8 Side-chain modified epothilones.

FIGURE 21.9 (See color insert following page 304.) Crystal structures of epothilone B (**1b**) from dichloromethane/petroleum ether (I), methanol/water (II), and epothilone A-*N*-oxide (III).

Remarkably, the latter are almost as active as their epoxides in the *in vitro* tubulin assay,[162,179] a fact that rules out an early hypothesis of covalent binding to tubulin protein via epoxide opening. Whereas β-epoxides and their analogs are entirely inactive, inversion of the configuration only at C12, corresponding to a C12, C13-*trans* configuration, retains activity.[162] The same applies for the introduction of *trans* double bonds in the C9, C10,[115,148,149] and C10, C11 position.[4,108,109] The C12 methyl is extremely important for high activity of the epothilone B series. However, it may be replaced with retention of activity by small alkyl, alkenyl, and alkynyl groups or modified by introduction of halogen and oxygen substituents.[49,132,136] Interestingly, the fluorescent epothilone D analog with a large *m*-dimethylaminobenzoyl residue is still about as active as Taxol.[180] Last but not least, replacement of the lactone group of epothilone B, D, and F by a lactam reduces cytotoxic activity and tubulin binding slightly, while other properties such as the therapeutic window or p.o. efficacy are improved.[51,57] Remarkably, lactam analogs obtained by total synthesis rather than semisynthesis seem to be less active.[48,181]

Studies on the solution conformation of epothilone A started even before the absolute configuration was published. Thus, Georg's group predicted[182] the relative configuration and conformation of epothilone A on the basis of nuclear magnetic resonance measurements and molecular dynamics calculations. One of the proposed structures showed the correct relative configuration for the independent western and eastern ring segments. We found NOEs and coupling constants of epothilone B in DMSO/D$_2$O solution in good agreement with a conformation similar to that in the crystal.[3] Only the thiazole side-chain is apparently rotating around both its single bonds. In a detailed conformational analysis, Taylor and Zajicek[183] added a second minor conformer, designated B, with the 3-OH rotated from an axial to an equatorial position, accompanied by rotations in the C6/C7 segment. Interestingly, an equatorial 3-OH is also found in the crystal structure of epothilone B-*N*-oxide, where it is stabilized by a hydrogen bridge to the *N*-oxide, while the conformation around C6-C8 is not influenced.[50] Apparently, a weak intramolecular hydrogen bond of the 3-OH to the thiazole nitrogen is also able to stabilize this 3-OH$_{eq}$ conformation in solution, as we found in (21*R,S*)-diastereomeric C21-substituted epothilone A and B derivatives.[173] Thus, in the 20-epoxyethyl derivative **31** (Figure 21.8), the 3-OH nuclear magnetic resonance signal appears duplicated, and a strong NOE is observed between 17H and 3H. The latter indicates a C16–C19 *syn*-periplanar conformation as it is observed in the crystal structure of epothilone B from an aqueous solvent (Figure 21.9).

Most remarkably, the same conformational changes were recently proposed for the tubulin-bound state of epothilone A from sophisticated nuclear magnetic resonance measurements (transferred NOEs and transferred cross-correlation rates) by the Griesinger group.[184,185] However, the unusual experimental conditions employed and the lack of appropriate controls cast some doubts on the relevance of these findings.[186] In fact, changes may have been induced by the aqueous solvent of high ionic strength rather than specific binding to tubulin. Earlier results on the bioactive conformation of epothilones based on computational methods were published in a series of papers in 1999 and 2000. Inspired by some intriguing structural similarities of epothilone and Taxol and by the fact that Taxol is replaced from its binding site by epothilone, common pharmacophore models were postulated and corroborated by mapping of cross-resistance to certain amino acid replacements in β-tubulin.[187–189] Recently, a tubulin point mutation outside the Taxol binding site was also observed that confers high resistance to both Taxol and epothilone.[73] Even though the binding models proposed are contradictory, there are some similarities. Thus, after discovering that the C13 side-chain of Taxol is not absolutely necessary for tubulin binding, Kingston and Horwitz[189] proposed that the C2 benzoyl residue of Taxol and the thiazole side-chain of epothilone occupy the same region of the protein. The same applies for the (less-favored) binding mode **I** proposed by Giannakakou et al.[70] Surprisingly, little use is made of conformationally restrained epothilone analogs and the highly active C12,C13 *trans* isomers in these modeling studies. From replacement of the thiazole by a pyridine and modifications of the C16–C17 linker we know that the nitrogen has to be exactly positioned in the binding site serving presumably as hydrogen bond acceptor. The bioactive conformation of the thiazole ring is clearly C16–C19 *syn*-periplaner from the fact that no substituents are tolerated at C19[166,172] and from the high activity of the conformationally *syn*-locked benzothiazole and quinoline analogs.[64,137] Interestingly, the opposite, a C16–C19 anti-periplaner conformation of epothilone D and B, is adopted in the crystal structure of the complex with cytochrome P450epok,[45] the biosynthetic enzyme that introduces the epoxide. Clearly, major advances of our knowledge of the epothilone–tubulin binding have to come from crystallographic and photoaffinity labeling studies, as was the case with the taxoids.

VI. PRECLINICAL STUDIES

Epothilone effects, as far as they are published, have been determined mostly in athymic nude mice but occasionally also in severe combined immunodeficiency (SCID) mice, nude rats, and beagle dogs, in most cases with xenografts of human cancer cells. The data are even more heterogeneous than those with cell cultures because of differences of the method and schedule of application and because of the kinds of cancer cells, epothilone analogs, and experimental animals used. The studies established, nevertheless, that epothilones act *in vivo*; that they inhibit human cancers, often resulting in cures of the animals; and that they are superior to Taxol and other cytostatic drugs in many cases, especially with MDR tumors. Severe combined immunodeficiency mice tolerate epothilone B well at a dose of 0.3 mg/kg; at 1 mg/kg, weight loss is observed; and at 3 mg/kg the animals show severe toxic effects (once per week, subcutaneously, vehicle 30% PEG + 70% 0.9% NaCl solution).[84]

In a preliminary study on nude mice, a dose of 0.6 mg/kg epothilone B, daily for 4 days, intraperitoneally, killed all animals, whereas all survived treatment with 25 mg/kg, daily for 5 days, of epothilone D.[46] It was suggested that the epoxy group at C12/C13 may be responsible for high *in vivo* toxicity.[55] Apart from leukopenia (43% decrease of lymphocytes), no hematological or other pathological changes in various organs were seen with either drug.[46] Toxicity of epothilone B could be mitigated by an altered application schedule: at 0.6 mg/kg, 5 times every second day, intraperitoneally, in DMSO, only three of seven mice died, and with the same dosage, but intravenously, all animals survived. Although epothilone D treatment resulted in appreciable regression and even cures of human mammary and ovarian carcinoma xenografts, epothilone B had too narrow a therapeutic window for good antitumor effects. The mode of application was later optimized intravenously by infusion over 6 h rather than by injection, using the vehicle cremophore-ethanol.[23a,56]

The benefits of epothilones are most impressive on MDR tumors. Thus, in nude mice epothilone D was >35,000-fold more active than Taxol on MDR human adenocarcinoma. It also was fully curative for human T-cell leukemia.[23a] A comparison of several epothilones, Taxol, and other cytostatic drugs showed superior performance of epothilone D and its C21–OH analog (d Epothione F) in many cases.[23a,b] Epothilone D (30 mg/kg, intravenously) was better than Taxol in controlling xenografts in nude mice of human solid lung, mammary, and one of two ovarian adenocarcinomas, and in particular of MDR sublines. It was inferior to Taxol, however, with colon, colon adeno, prostate adeno, and one of the ovarian adenocarcinoma lines. Complete remission was only achieved with MX-1 mammary carcinoma (as well by Taxol). Epothilone D was superior to Taxol, with complete remission, in four human leukemias, including MDR lines, and was inferior to Taxol only in the HL-60 line. Epothilone B and epothilone B–lactam were much less successful because of their narrow therapeutic windows (but see below). The same study showed that resistance to epothilone D developed only very slowly and to low levels (twofold in 21 months).[56] In mouse plasma, epothilone D was quickly inactivated ($t_{1/2}$ = 20 min), probably because of esterase activity, but was much more stable in human plasma ($t_{1/2}$ = >3 h).[56] Beagle dogs fully tolerated a single dose of 2–6 mg/kg, intravenously (40–120 mg/m^2); they developed diarrhea and leukopenia and lost weight at 12 mg/kg (240 mg/m^2); and 20 mg/kg (400 mg/m^2) produced severe toxic effects (e.g., necrosis of the intestinal mucosa, cell decay in the bone marrow) and was lethal.[56] At 6 mg/kg, $t_{1/2}$ was >5 h, and plasma concentrations after 24 h were 0.045 µg/mL (0.092 µM), which is 10 times the IC$_{50}$ (0.0095 µM) for leukemia cells.[56]

Recently, more effective epothilone analogs have been tested.[136] Introduction of a C9, C10 double bond in epothilone D makes the compound 10 times more active on MX-1 mammary carcinoma xenografts in nude mice, but also more toxic (with a maximum tolerated dose [MTD] of 3 mg/kg, versus that of epothilone D [30 mg/kg]). Also, although the compound may lead to a complete remission of tumors, they often later come back. Performance could be much improved by further replacing the CH$_3$ on C12 by CF$_3$, which bestows higher stability ($t_{1/2}$ = 212 min in mouse plasma, 10.5 h in human plasma; epothilone D, 46 min) and bioavailability (solubility in water: 20 µg/mL; epothilone D: 9.4 µg/mL). Thus, for example, with xenografts of human slow-growing lung carcinoma, A549, 25 mg/kg, 6 times every second day twice 8 days apart, resulted in 99.5% remission and, after two more doses, in complete elimination (epothilone D could not eradicate the tumor). Growth of a Taxol-resistant subline was completely stopped, and tumor size reduced by 24% at 20 mg/kg, 7 times every second day.[136]

Another novel compound, semisynthetic C-21 thiomethyl epothilone B (ABJ879), has just been proposed by Novartis for a phase I clinical study.[190] It shows about the same tubulin-polymerizing activity as Taxol but is clearly more potent in blocking cell proliferation in cultures (average IC$_{50}$ for seven human carcinomas: 0.09 nM vs. 4.7 nM of Taxol). The drug is fully active on MDR1 cells and lines with resistance resulting from β-tubulin mutations. In nude mice, single-dose treatment (2–3 mg/kg, intravenously, 6–7 d after implantation of the tumor) produces transient but long-lasting remission (25%–95%). Positive effects were seen with slow-growing lung adenocarcinoma H-596, colon tumor HT-29, and large-cell lung tumor H-460. A preliminary pharmacokinetics–pharmacodynamics drug-response model has also been presented.[191] Doses of 1.5 and 1.8 mg/kg, every second week, were acceptable, but 1.5 and 2 mg/kg, once a week, caused intolerable weight losses. The drug penetrates tissues, where its level becomes 10 times higher than in plasma, and is retained there for a long time, so that plasma concentrations are not representative of tissue exposure — a fact that may be true also for other epothilone analogs.

The controversy about the efficacy of epothilone B–lactam (BMS-247550) has been fueled by decidedly positive results reported by BMS scientists.[57] In addition to various established tumor cell lines, three primary xenografts from biopsies were tested. A human ovarian carcinoma, Pat-7, from a patient who no longer responded to Taxol and other cytostatic drugs was highly responsive to epothilone B–lactam in nude mice at the optimal dose of 4.8–6.3 mg/kg, intravenously, 5 times every second day (the optimal dose of Taxol is 24–36 mg/kg). The log cell kill (LCK) was 2.9 (Taxol, 0.8). The same tumor in nude rats at the optimal dose of 3 mg/kg, intravenously, two times eight days apart, could be eradicated in four of six animals (LCK > 5; Taxol, LCK = 2.2, no cures [0/6]). Good LCK values and growth delay were also seen in nude mice with Taxol-resistant human ovarian and colon carcinomas. Also, in several other instances of Taxol-sensitive tumors in nude mice, the animals could be cured, and high LCK values (>6.3) were seen, even when the compound was not better than Taxol. Furthermore, epothilone B–lactam gave excellent results when administered orally. At the MTD (60–80 mg/kg, 5 times every second day), epothilone B–lactam was highly active on the Pat-7 xenograft (LCK = 3.1, growth delay of 32.8 d; in comparison, Taxol, intravenously, LCK = 1.3, growth delay of 9.8 d, orally inactive). HCT116 colon carcinoma was eradicated in all eight mice treated with 90 mg/kg, 5 times every second day (equivalent to the best intravenous regimen). In contrast to epothilone A and B, epothilone B–lactam is completely stable in rodent and human plasma and in liver microsomal preparations and shows clearly reduced toxicity.[57]

Epothilone D, too, was reported to be active in rats and beagle dogs after oral administration.[192,193] The drug was well tolerated, and >50% (dogs) and 10%–20% (rats) was found in the bloodstream, with half-lives of 9–11 h (dogs) and 6 h (rats), respectively. A study by the Sloan-Kettering Institute for Cancer Research yielded far fewer positive results with respect to epothilone β lactam in comparison to their own epothilone D.[55] At 6 mg/kg, 4 times every day, intravenous infusion, MX-1 mammary tumor xenografts in nude mice were somewhat inhibited in growth, but no remission was seen, and the effect was not much improved by treatment with doses near the MTD (9 mg/kg). Similar results were obtained with K562 leukemia, which could be cured, however, when the animals subsequently were treated with C-21-OH epothilone D (30 mg/kg).

A study, to date only *in vitro*, indicates that the dosing regime could have a massive influence on tumor angiogenesis and metastasis.[194] Human vascular endothelial cell lines were more efficiently killed by a "metronomic" administration of cytostatic drugs, including epothilone B; that is, by extended (144 h) application of low (10–100 pM) concentrations rather than high doses for a short time (24 h). IC$_{50}$ values were thus reduced to 25–140 pM, although those for tumor cells and fibroblasts remained high (550 pM to >1 nM), so that this strategy appears to selectively affect endothelial cells.

Not much has been published about the metabolic fate of epothilones in animals. Degradation of epothilone B (10 mg/kg, intravenously) in the liver of nude mice was studied by capillary high performance liquid chromatography combined with mass spectrometry.[195] One hour after administration, three metabolites could be recovered and identified guided by a key fragment at m/z166, an acylium ion derived from the very stable thiazole side-chain. The following degradation pathway was reconstructed: opening of the epoxide, opening of the macrolactone ring to the hydroxycarboxylic acid, reduction of the C5-keto group, and reaction of the C5-hydroxy group with the carboxylic acid group to give a six-membered lactone ring or, after conjugation with taurine, to yield a six-membered lactam ring at the end of the chain.

VII. CLINICAL APPLICATIONS

Five epothilone analogs have been introduced into clinical studies (Figure 21.10). Epothilone B (**1b**, EPO906, Patupilone, Novartis) started phase I, beginning early in 1999, followed by phase II.[10n,196] Epothilone B lactam (**23**, BMS-247550, Bristol-Myers Squibb) started phase I beginning October 1999 and completed September 2000; phase I in children started in April 2002; phase II trials have been operative since February 2001, and in 2002, there were 15 National Cancer Institute–sponsored phase II studies; it is presently in phase III.[10q,197] Epothilone B C21-amine (**22**, BMS-310705) is in phase I (2002),[10s,198] and epothilone D (**2b**, KOS-862, Kosan) is in phase I.[136,192,193] Also in phase I is a synthetic epothilone of undisclosed structure, ZK-EPO (**32**, Schering). Epothilone B C2-thiomethyl (**33**, ABJ879, Novartis) is a recent candidate for a phase I study.[190,191] Although most data are only published as abstracts, several excellent reviews summarize the results,[10n,q,s,196,199] so that details need not be repeated here. The epothilone B lactam studies are by far the best documented ones.

The epothilones have been tested in patients with many different carcinomas, mostly solid tumors, including MDR cases and others that are difficult to treat (e.g., cervical adenocarcinomas of the squamous type).[200] The responses have varied between complete and partial remission, tumor stabilization, reduction of CA 125 levels (cancer antigen found at elevated levels in the blood of many cancer patients, especially those with ovarian cancer), and continued growth. The individual epothilone analogs appear to differ in the MTD, the dose-limiting toxicity (DLT), the side effects, half-life in the patient, and efficacy. It may still be too early to assess the quality of the drugs conclusively.

For epothilone B lactam, the MTD was estimated to be 6 mg/m^2, intravenously, over 1 h, daily for 5 days every 3 wk.[201] At 8 mg/m^2 dose-limiting neutropenia was seen, but many patients tolerated higher doses when the drug was applied in combination with filgrastim (granulocyte colony-stimulating

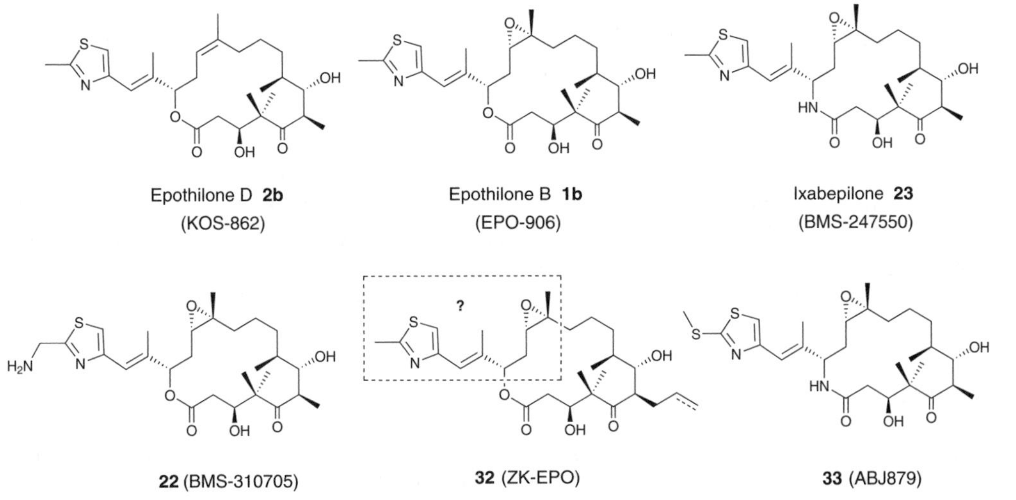

FIGURE 21.10 Epothilones in clinical trials.

factor, which mitigates myelosuppression and neutropenia). Other side effects were peripheral neuropathy, fatigue, nausea, stomatitis, myalgia, and arthralgia, usually in mild forms.[10q,201,202] DLT was neutropenia. Two patients died of sepsis, probably as a consequence of neutropenia. The acceptable weekly dose may be 20–30 mg/m². The drug was quickly absorbed by the tissue, where its half-life was 24–48 h, whereas $t_{1/2}$ in the plasma was only 16.8 h. This was also supported by a study that analyzed effects in peripheral blood mononuclear cells in patients.[197] It showed that MTs formed bundles 1 h after application (at the already toxic dose of 50 mg/m², intravenously, over 1 h, once a week every 3 wk, in 70% of cells; at 40 mg/m² in 63%), but after 24 h, the effect was much reduced (to 25% and 16%–23%, respectively). Cytotoxic effects followed with a time delay on MT bundling but were expressed within 24 h: Cleavage of poly(ADP-ribose)polymerase, indicating cell death (apoptosis and necrosis), was at base level after 1 h (3%) but rose to 28% of cells after 24 h. For treatment of progressive metastatic prostate cancer, a dose of 35 mg/m², intravenously, over 3 h, once every 3 weeks (five cycles) in combination with estramustine phosphate (an estradiol analog without affinity to the estradiol receptor) is recommended.[202] Positive responses and decline in prostate-specific antigen levels were seen in 40%–60% of patients. It appears that in aqueous solution the epoxide ring of epothilone B lactam opens to give the inactive diol.[198]

For epothilone B,[199] the MTD is 6 mg/m², intravenously, once every 3 weeks, or 2.5 mg/m², intravenously, once every week. The DLT was diarrhea. Other toxic effects (neuropathy, nausea) were mild to moderate, and significant myelosuppression, mucositis, and alopecia were not registered. Blood distribution was multiphasic. An initial half-life of 7 min was followed by a terminal half-life of 4–5 d, which indicates fast absorption by the tissue. There was no excretion of unaltered compound via the kidneys (<0.1%).

For epothilone D, reports on clinical performance still are very preliminary.[193] Five patients were treated with doses of 9 and 18 mg/m² (intravenously, once every 3 weeks), at which no DLT was seen (mild to moderate emesis and anemia). The elimination half-life of the drug was 5–10 h, and plasma concentrations declined rapidly after infusion in a biphasic pattern.

NOTE ADDED IN PROOF

Since the completion of this article in July 2004 a considerable number of papers on biology and chemistry of epothilones were published. Among the most important ones is that by Nettles et al. on the structure of the epothilone A/tubulin complex.[203] As with taxol, the structure was solved from two-dimensional crystals by electron crystallography and molecular modelling. Expectedly, epothilone occupies part of the taxol binding site, however, its conformation and contacts to the protein are profoundly different from previous proposals.[204]

ACKNOWLEDGMENTS

We thank all former group members at the GBF for their contributions and the colleagues at Bristol-Meyers Squibb for a fruitful collaboration in the epothilone project. We further thank Mrs. Monica Kirchner for taking care for literature references and typing the manuscript. Financial support from Bristol-Myers Squibb and the Fonds der Chemischen Industrie is gratefully acknowledged.

REFERENCES

1. Höfle, G., Biologically active secondary metabolites from myxobacteria — isolation and structure elucidation, in *Scientific Annual Report of the GBF*, Walsdorff, H.-J., Ed., Gesellschaft für Biotechnologische Forschung, Braunschweig, 1991, 65-68.
2. Gerth, K. et al., Epothilones A and B: Antifungal and cytotoxic compounds from *Sorangium cellulosum* (Myxobacteria), *J. Antibiot.*, 49, 560, 1996.
3. Höfle, G. et al., Epothilone A and B — Novel 16-membered macrolides with cytotoxic activity: isolation, crystal structure, and conformation in solution, *Angew. Chem. Int. Ed.*, 35, 1567, 1996.
4. Hardt, I.H. et al., New natural epothilones from *Sorangium cellulosum*, strains So ce90/B2 and So ce90/D13: isolation, structure elucidation, and SAR studies, *J. Nat. Prod.*, 64, 847, 2001.
5. Höfle, G. et al., German Patent 4138042, filed Nov. 11, 1991, granted Oct. 14, 1993; World Patent 9310121 filed Nov. 19, 1992.
6. See, e.g., Reichenbach, H. and Höfle, G., Biologically active secondary metabolites from myxobacteria, *Biotech. Adv.*, 11, 219, 1993; Höfle, G. and Reichenbach, H., The biosynthetic potential of the myxobacteria, in *Sekundärmetabolismus bei Mikroorganismen*, Kuhn, W. and Fiedler, H.-P., Eds., Attempto, Tübingen, 1995, 61-78; Reichenbach, H. and Höfle, G., Myxobacteria as producers of secondary metabolites, in *Drug Discovery from Nature*, Grabley, S. and Thiericke, R., Eds., Springer, Berlin, 1999, 149-179.
7. Lavelle, F., What's new about new tubulin/microtubules binding agents? *Exp. Opin. Invest. Drugs*, 4, 771, 1995.
8. Bollag, D.M. et al., Epothilones, a new class of microtubule-stabilizing agents with a Taxol-like mechanism of action, *Cancer Res.*, 55, 2325, 1995.
9. Gerth, K. et al., Myxobacteria, proficient producers of novel natural products with various biological activities past and future biotechnological aspects with the focus on the genus *Sorangium*, *J. Biotechnol.*, 106, 233, 2003.
10. a) Nicolaou, K.C., Roschangar, F., and Vourloumis, D., Chemical biology of epothilones, *Angew. Chem. Int. Ed. Engl.* 37, 2014, 1998; b) Mulzer, J., Martin, H.J., and Berger, M., Progress in the synthesis of chiral heterocyclic natural products: epothilone B and tartrolon B, *J. Heterocyclic Chem.*, 36, 1421, 1999; c) G. Höfle, Epothilone — a natural product on the road to becoming a medicine, in *Scientific Annual Report of the GBF*, Jonas, R., Ed., 1999/2000, 21; d) Altmann, K.-H. et al., Epothilones and their analogs — potential new weapons in the fight against cancer, *Chimia*, 54, 612, 2000; e) Altmann, K.-H., Wartmann, M., and O'Reilly, T., Epothilones and related structures — a new class of microtubule inhibitors with potent *in vivo* antitumor activity, *Biochim. Biophys. Acta*, 1470, M79, 2000; f) Mulzer, J., Epothilone B and its derivatives as novel antitumor drugs: total and partial synthesis and biological evaluation, *Monatshefte für Chemie*, 131, 205, 2000; g) Winssinger, N. and Nicolaou, K.C., Epothilones and sarcodictyns: from combinatorial libraries to designed analogs, ACS Symposium Series 796, *Anticancer Agents, Frontiers in Cancer Therapy*, Ojima, I., et al., Eds., American Chemical Society, 2001, 148; h) Altmann, K.-H., Microtubule-stabilizing agents: a growing class of important anticancer drugs, *Curr. Opin. Chem. Biol.*, 5, 424, 2001; i) Stachel, S.J., Biswas, K., and Danishefsky, S.J., The epothilones, eleutherobins, and related types of molecules, *Curr. Pharm. Design* 7, 1277, 2001; j) Nicolaou, K.C., Ritzén, A., and Namoto, K., Recent development in the

chemistry, biology and medicine of the epothilones, *Chem. Commun.*, 1523, 2001; k) Flörsheimer, A. and Altmann, K.-H., Epothilones and their analogs — a new class of promising microtubule inhibitors, *Expert Opin. Ther. Patents*, 11, 951, 2001; m) Wartmann, M. and Altmann, K.-H., The biology and medicinal chemistry of epothilones, *Curr. Med. Chem. AntiCancer Agents,* 2, 123, 2002; n) Borzilleri, R.M. and Vite, G.D., Epothilones: new tubulin polymerization agents in preclinical and clinical development, *Drugs Future*, 27, 1149, 2002; o) Altmann, K.-H., Epothilone B and its analogs — a new family of anticancer agents, *Med. Chem.*, 3, 149, 2003; p) Mulzer, J. and Martin, H.J., *Chem. Rec.*, 3, 258, 2004; q) Lin, N., Brakora, K., and Seiden, M., BMS-247550 Bristol-Myers Squibb/GBF, *Curr. Opin. Investig. Drugs*, 4, 746, 2003; r) Rivkin, A. et al., Application of ring-closing metathesis reactions in the synthesis of epothilones, *J. Nat. Prod.*, 67, 139, 2004; s) Goodin, S., Kane, M.P., and Rubin, E.H., Epothilones: mechanism of action and biological activity, *J. Clin, Oncol.,* 10, 2015, 2004.

11. Reichenbach, H., Order Myxococcales, in *Bergey's Manual of Systematic Bacteriology*, 2nd edition, Garrity, G., Ed., Springer, New York, 2004.

12. *Myxobacteria II*, Dworkin, M. and Kaiser, D., Eds., American Society for Microbiology, Washington, DC, 1993, 1-404

13. Höfle, G. et al., (GBF) German Patent 4211056, 1993.

14. K. Gerth, personal communication.

15. Gerth, K. et al., Studies on the biosynthesis of epothilones: hydroxylation of epo A and B to epothilones E and F, *J. Antibiot.*, 55, 41, 2002.

16. Gerth, K. et al., Studies on the biosynthesis of epothilones: The biosynthetic origin of the carbon skeleton, *J. Antibiot.*, 53, 1373, 2000.

17. Gerth, K. et al., Studies on the biosynthesis of epothilones: the PKS and epothilone C/D monooxygenase, *J. Antibiot.*, 54, 144, 2001.

18. Reichenbach, H. et al., (GBF) German Patent 9647580 (1996).

19. Hofmann, H. et al., (Novartis) World Patent 9942602 (1999).

20. Hecht, H.-J. and Höfle, G., unpublished results; Crystallographic data of the structure have been deposited with the Cambridge Crystallographic Data Centre as supplementary publication no. CCDC-241333 and CCDC-241334 — copies of the data can be obtained free of charge on application to CCDC, 12 Union Road, Cambridge CB2 1EZ, UK [fax: (+44)1223-336-033; e-mail: deposit@ccdc.cam.ac.uk].

21. Höfle, G. et al., unpublished results.

22. Straubing, R.M., Biopharmaceuticals of paclitaxel, in *Taxol®, Science Application*, Suffness, M., Ed., CRC Press, Boca Raton, FL, 1995, 238.

23. a) Chou, T.-C. et al., Desoxyepothilone B is curative against human tumor xenografts that are refractory to paclitaxel, *Proc. Natl. Acad. Sci. USA*, 95, 15798, 1998; b) Lee, C.B. et al., Total synthesis and antitumor activity of 12,13-desoxyepothilone F: An unexpected solvolysis problem at C15, mediated by remote substitution at C21, *J. Org. Chem.,* 65, 6525, 2000.

24. Tang, L. et al., Cloning and heterologous expression of the epothilone gene cluster, *Science*, 287, 640, 2000.

25. Julien, B. and Shah, S., Heterologous expression of epothilone biosynthetic genes in *Myxococcus xanthus, Antimicrob. Agents Chemother.*, 46, 2772, 2002.

26. Lau, J. et al., Optimizing the heterologous production of epothilone D in *Myxococcus xanthus, Biotechnol. Bioeng.*, 78, 280, 2002.

27. Arslanian, R.L. et al., Large-scale isolation and crystallization of epothilone D from *Myxococcus xanthus* cultures, *J. Nat. Prod.*, 65, 570, 2002.

28. Regentin, R. et al., Nutrient regulation of epothilone biosynthesis in heterologous and native production strains, *Appl. Microbiol. Biotechnol.*, 61, 451, 2003.

29. Frykman, S. et al., Modulation of epothilone analog production through media design, *J. Ind. Microbiol. Biotechnol.*, 28, 17, 2002.

30. Molnar, I. et al., The biosynthetic gene cluster for the microtubule-stabilizing agents epothilones A and B from *Sorangium cellulosum* So ce90, *Chem. Biol.*, 7, 97, 2000.

31. Julien, B. et al., Isolation and characterization of the epothilone biosynthetic gene cluster from *Sorangium cellulosum, Gene*, 249, 153, 2000.

32. Schupp, T. et al., World Patent 9966028-A (Dec. 23, 1999), (Novartis AG).

33. Julien, B. et al., World Patent 200031247-A (June 2, 2000), (Kosan Bioscience).

34. Walsh, C.T. and O'Connor, S.E., Polyketide-nonribosomal peptide epothilone antitumor agents: the EpoA, B, C subunits, *J. Ind. Microbiol. Biotechnol.*, 30, 448, 2003.

35. Tang, L. et al., Elucidating the mechanism of *cis* double bond formation in epothilone biosynthesis, *J. Am. Chem. Soc.*, 126, 46, 2004.

36. Arslanian, R.L. et al., A new cytotoxic epothilone from modified polyketide synthases heterologously expressed in *Myxococcus xanthus*, *J. Nat. Prod.*, 65, 1061, 2002.

37. a) Starks, C.M. et al., Isolation and characterization of new epothilone analogs from recombinant *Myxococcus xanthus* fermentations, *J. Nat. Prod.*, 66, 1313, 2003; b) Boddy, C.N. et al., Precursor-directed biosynthesis of epothilone in *Escherichia coli*, *J. Am. Chem. Soc.*, 126, 7436, 2004.

38. Tang, L. et al., Generation of novel epothilone analogs with cytotoxic activity by biotransformation, *J. Antibiot.*, 56, 16, 2003.

39. O'Connor, S.E., Walsh, C.T., and Liu, F., Biosynthesis of epothilone intermediates with alternate starter units: engineering polyketide-nonribosomal interfaces, *Angew. Chem. Int. Ed.*, 42, 3917, 2003.

40. Schneider, T.L., Walsh, C.T., and O'Connor, S.E., Utilization of alternate substrates by the first three modules of the epothilone synthetase assembly line, *J. Am. Chem. Soc.*, 124, 11272, 2002.

41. Chen, H. et al., Epothilone biosynthesis: assembly of the methylthiazolylcarboxy starter unit on the EpoB subunit, *Chem. Biol.*, 8, 899, 2001.

42. O'Connor, S. E., Chen, H., and Walsh, C. T., Enzymatic assembly of epothilones: the epoC subunit and reconstitution of the epoA-ACP/B/C polyketide and nonribosomal peptide interfaces, *Biochemistry*, 41, 5685, 2002.

43. Schneider, T.L., Shen, B., and Walsh, C.T., Oxidase domains in epothilone and bleomycin biosynthesis: thiazoline to thiazole oxidation during chain elongation, *Biochemistry*, 42, 9722, 2003.

44. Lopez, J.V., Naturally mosaic operons for secondary metabolite biosynthesis: variability and putative horizontal transfer of discrete catalytic domains of the epothilone polyketide synthase locus, *Mol. Gen. Genomics*, 270, 420, 2003.

45. Nagano, S. et al., Crystal structures of epothilone D-bound, epothilone B-bound, and substrate-free forms of cytochrome P450epoK, *J. Biol. Chem.*, 278, 44886, 2003.

46. Chou, T.-C. et al., Desoxyepothilone B: An efficacious microtubule-targeted antitumor agent with a promising *in vivo* profile relative to epothilone B, *Proc. Natl. Acad. Sci. USA*, 95, 9642, 1998.

47. Kowalski, R.J., Giannakakou, P., and Hamel, E., Desoxyepothilone B is curative against human tumor xenografts that are refractory to paclitaxel, *J. Biol. Chem.*, 272, 2534, 1997.

48. Schinzer, D. et al., Synthesis and biological evaluation of Aza-epothilones, *Chem. Bio. Chem.*, 1, 67, 2000.

49. Su, D.-S. et al., Structure–activity relationships of the epothilones and the first *in vivo* comparison with paclitaxel, *Angew. Chem. Int. Ed.*, 36, 2093, 1997.

50. Höfle, G. et al., N-Oxidation of epothilone A–C and *O*-Acyl rearrangement to C-19- and C21-substituted epothilones, *Angew. Chem. Int. Ed.*, 38, 1971, 1999.

51. Borzilleri, R.M. et al., A novel application of a Pd(0)-catalyzed nucleophilic substitution reaction to the regio- and stereoselective synthesis of lactam analogs of the epothilone natural products, *J. Am. Chem. Soc.*, 122, 8890, 2000.

52. Sepp-Lorenzino, L. et al., The microtubule-stabilizing agents epothilones A and B and their desoxy-derivatives induce mitotic arrest and apoptosis in human prostate cancer cells, *Prostate Cancer Prost. Dis.*, 2, 41, 1999.

53. Nicolaou, K.C. et al., Total synthesis of oxazole- and cyclopropane-containing epothilone A analogs by the olefin metathesis approach, *Chem. Eur. J.* 3, 1957, 1997.

54. Nicolaou, K.C. et al., Total synthesis of oxazole- and cyclopropane-containing epothilone B analogs by the macrolactonization approach, *Chem. Eur. J.*, 3, 1971, 1997.

55. Stachel, S.J. et al., On the interactivity of complex synthesis and tumor pharmacology in the drug discovery process: total synthesis and comparative *in vivo* evaluations of the 15-Aza epothilones, *J. Org. Chem.*, 66, 4369, 2001.

56. Chou, T.-C. et al., The synthesis, discovery, and development of a highly promising class of micro-tubule stabilisation agents: curative effects of desoxyepothilones B and F against human tumor xenografts in nude mice, *Proc. Natl. Acad. Sci. USA*, 98, 8113, 2001.

57. Lee, F.Y.F. et al., BMS-247550: a novel epothilone analog with a mode of action similar to paclitaxel but possessing superior antitumor efficacy, *Clin. Cancer Res.*, 7, 1429, 2001.

58. Wolff, A., Technau, A., and Brandner, G., Epothilone A induces apoptosis in neuroblastoma cells with multiple mechanisms of drug resistance, *Int. J. Oncol.*, 11, 123, 1997.

59. Mühlradt, P.F. and Sasse, F., Epothilone B stabilizes microtubuli of macrophages like Taxol without showing Taxol-like endotoxin activity, *Cancer Res.*, 57, 3344, 1997.

60. Kamath, K. and Jordan, M.A., Suppression of microtubule dynamics by epothilone B is associated with mitotic arrest, *Cancer Res.*, 63, 6026, 2003.

61. Chen, J.G. and Horwitz, S.B., Differential mitotic responses to microtubule-stabilizing and -destabilizing drugs, *Cancer Res.*, 62, 1935, 2002.

62. Iffert, R. and Sasse, F., Analyse des Apoptischen Zellkernzerfalls durch Bildverarbeitung, *BIOforum*, 22, 631, 1999.

63. Lichtner, R.B. et al., Subcellular distribution of epothilones in human tumor cells, *Proc. Natl. Acad. Sci. USA*, 98, 11743, 2001.

64. Klar, U. et al., Synthesis and biological activity of epothilones, ACS Symposium Series 796, *Anticancer Agents, Frontiers in Cancer Therapy*, Ojima, Iwao et al., Eds., American Chemical Society, 2001, 131–146.

65. Atadja, P., Isolation and characterisation of epothilone A resistant cells, *Proc. Am. Assoc. Cancer Res.*, 41, 803, 2000.

66. Meurer-Grob, P., Kasparian, J., and Wade, R.H., Microtubule structure at improved resolution, *Biochemistry*, 40, 8000, 2001.

67. Wartmann, M. et al., Epothilones A and B accumulate several hundred fold inside cells, *Proc. Am. Assoc. Cancer Res.* 41, 213, 2000, Abstr. No. 1362.

68. Gupta, M.L. et al., Understanding tubulin-Taxol interactions: mutations that impart Taxol binding to yeast tubulin, *Proc. Natl. Acad. Sci. USA*, 100, 6394, 2003.

69. Bode, C.J. et al., Epothilone and paclitaxel: Unexpected differences in promoting the assembly and stabilisation of yeast microtubules, *Biochemistry*, 41, 3870, 2002.

70. Giannakakou, P. et al., A common pharmacophore for epothilone and taxanes: molecular basis for drug resistance conferred by tubulin mutations in human cancer cells, *Proc. Natl. Acad. Sci. USA*, 97, 2904, 2000.

71. He, L., Huang Yang, C.-P., and Horwitz, S.B., Mutations in β-tubulin map to domains involved in regulation of microtubule stability in epothilone-resistant cell lines, *Mol. Cancer Ther.*, 1, 3, 2001.

72. Giannakakou, P. et al., A β-tubulin mutation confers epothilone resistance in human cancer cells, *Proc. Am. Assoc. Cancer Res.*, 40, 284, 1999, Abstr. No. 1885.

73. Verrills, N.M. et al., Microtubule alterations and mutations induced by desoxyepothilone B: implications for drug-target interactions, *Chem. Biol.*, 10, 597, 2003.

74. Martello, L.A. et al., Elevated levels of microtubule destabilizing factors in a Taxol-resistant/dependent A549 cell line with an β-tubulin mutation, *Cancer Res.*, 63, 1207, 2003.

75. Verdier-Pinard, P. et al., Direct analysis of tubulin expression in cancer cell lines by electrospray ionization mass spectrometry, *Biochemistry*, 42, 12019, 2003.

76. Poruchynsky, M.S. et al., Accompanying protein alterations in malignant cells with a microtubule-polymerizing drug-resistance phenotype and a primary resistance mechanism, *Biochem. Pharmacol.*, 62, 1469, 2001.

77. Khabele, D. et al., Tumor necrosis factor-α related gene response to epothilone B in ovarian cancer, *Gynecol. Oncol.*, 93, 19, 2004.

78. Atadja, P. et al., Gene expression profiling of epothilone A–resistant cells, in *Mechanisms of Drug Resistance in Epilepsy: Lessons from Oncology*, Novartis Foundation Symposium, 2002, pp. 119-136.

79. Ioffe, M. et al., Epothilone induced cytotoxicity is dependent on p53 in prostate cells, *Proc. Am. Assoc. Cancer Res.*, 44, Abstr. No. 829, 2003.

80. Uyar, D. et al., Apoptotic pathways of epothilone BMS 310705, *Gynecol. Oncol.*, 91, 173, 2003.

81. Griffin, D. et al., Molecular determinants of epothilone B derivative (BMS 247550) and Apo-2L/TRAIL-induced apoptosis of human ovarian cancer cells, *Gynecol. Oncol.*, 89, 37, 2003.

82. Yamaguchi, H. et al., Epothilone B analog (BMS-247550)-mediated cytotoxicity through induction of bax conformational change in human breast cancer cells, *Cancer Res.*, 62, 466, 2002.

83. Bröker, L.E. et al., Late activation of apoptotic pathways plays a negligible role in mediating the cytotoxic effects of discodermolide and epothilone B in non-small cell lung cancer cells, *Cancer Res.*, 62, 4081, 2002.

84. Pietras, K. et al., STI571 enhances the therapeutic index of epothilone B by a tumor-selective increase of drug uptake, *Clin. Cancer Res.*, 9, 3779, 2003.

85. Wittmann, S. et al., Flavopiridol down-regulates antiapoptotic proteins and sensitizes human breast cancer cells to epothilone B-induced apoptosis, *Cancer Res.*, 63, 93, 2003.

86. Moasser, M.M. et al., Farnesyl transferase inhibitors cause enhanced mitotic sensitivity to Taxol and epothilones, *Proc. Natl. Acad. Sci. USA*, 95, 1369, 1998.

87. Zhou, Y. et al., KDS-862 (epothilone D) with 5-fluorouracil in human colon cancer cell lines: synergistic antitumor effects, *Proc. Am. Assoc. Cancer Res.*, 44, Abstr. No. 2746, 2003.

88. Fuino, I. et al., Histone deacetylase inhibitor IAQ824 down-regulates Her-2 and sensitizes human breast cancer cells to trastuzumab, taxotere, gemcitabine, and epothilone B, *Mol. Cancer Ther.*, 2, 971, 2003.

89. Kim, J.-C. et al., Potential radiation-sensitizing effect of semisynthetic epothilone B in human lung cancer cells, *Radiother. Oncol.*, 68, 305, 2003.

90. Martin, H.J., Drescher, M., and Mulzer, J., How stable are epoxides? A novel synthesis of epothilone B, *Angew. Chem. Int. Ed.*, 39, 581, 2000.

91. Martin, H.J. et al., The 12,13-diol cyclization approach for a truly stereocontrolled total synthesis of epothilone B and the synthesis of a conformationally restrained analog, *Chem. Eur. J.*, 7, 2261, 2001.

92. Mulzer, J., Karig, G., and Pojarliev, P., A novel highly stereoselective total synthesis of epothilone B and of its (12*R*, 13*R*) acetonide, *Tetrahedron Lett.*, 41, 7635, 2000.

93. Meng, D. et al., Studies toward a synthesis of epothilone A: use of hydropyran templates for the management of acyclic stereochemical relationships, *J. Org. Chem.*, 61, 7998, 1996.

94. Meng, D. et al., Total syntheses of epothilones A and B, *J. Am. Chem. Soc.*, 119, 10073, 1997.

95. Schinzer, D. et al., Total synthesis of (–)-epothilone A, *Angew. Chem. Int. Ed.*, 36, 523, 1997.

96. Nicolaou, K.C. et al., The olefin metathesis approach to epothilone A and its analogues, *J. Am. Chem. Soc.*, 119, 7960, 1997.

97. Meng, D. et al., Remote effects in macrolide formation through ring-forming olefin metathesis: an application to the synthesis of fully active epothilone congeners, *J. Am. Chem. Soc.*, 119, 2733, 1997.

98. Fürstner, A., Mathes, C., and Grela, K., Concise total syntheses of epothilone A and C based on alkyne metathesis, *Chem. Commun.*, 1057, 2001.

99. Nicolaou, K.C. et al., Total syntheses of epothilones A and B via a macrolactonization-based strategy, *J. Am. Chem. Soc.*, 119, 7974, 1997.

100. Balog, A. et al., Total synthesis of (–)-epothilone A, *Angew. Chem. Int. Ed.*, 35, 2801, 1996.

101. Yang, Z. et al., Total synthesis of epothilone A: the olefin metathesis approach, *Angew. Chem. Int. Ed.*, 36, 166, 1997.

102. Taylor, R.E. and Chen, Y. Total synthesis of epothilones B and D, *Org. Lett.*, 3, 2221, 2001.

103. Liu, Z.-Y., Yu, C.-Z., Yang, J.-D., Chiral synthesis of the C_{3-13} segment of epothilone A, *Synlett*, 12, 1383, 1997.

104. Schinzer, D., Bauer, A., and Schieber, J., Synthesis of epothilones: stereoselective routes to epothilone B, *Synlett*, 8, 861, 1998.

105. Njardson, J.T., Biswas, K., and Danishefsky, S.J., Application of hitherto unexplored macrocyclization strategies in the epothilone series: novel epothilone analogs by total synthesis, *Chem. Commun.*, 2759, 2002.

106. Bode, J.W. and Carreira, E.M., Stereoselective syntheses of epothilones A and B in nitrile oxide cycloadditions and related studies, *J. Org. Chem.*, 66, 6410, 2001.

107. Bode, J. W. and Carreira, E.M., Epothilone B, *J. Am. Chem. Soc.*, 123, 3611, 2001.

108. Biswas, K. et al., Highly concise routes to epothilones: the total synthesis and evaluation of epothilone 490, *J. Am. Chem. Soc.*, 124, 9825, 2002.

109. Yamamoto, K. et al., Effects of temperature and concentration in some ring closing metathesis reactions, *Tetrahedron Lett.*, 44, 3297, 2003.

110. White, J.D. et al., Total synthesis of epothilone B, epothilone D, and *cis*- and *trans*-9,10-dehydroepothilone D, *J. Am. Chem. Soc.*, 123, 5407, 2001.

111. Sawada, D., Kanai, M., and Shibasaki, M., Enantioselective total synthesis of epothilones A and B using multifunctional asymmetric catalysis, *J. Am. Chem. Soc.*, 122, 10521, 2000.

112. Mulzer, J., Mantoulidis, A., and Öhler, E., Easy access to the epothilone family — synthesis of epothilone B, *Tetrahedron Lett.*, 39, 8633, 1998.

113. Nicolaou, K.C. et al., Total synthesis of epothilone E and analogs with modified side chains through the Stille Coupling Reaction, *Angew. Chem. Int. Ed.*, 37, 84, 1998.

114. Bertinato, P. et al., Studies toward a synthesis of epothilone A: stereocontrolled assembly of the acyl region and models for macrocyclization, *J. Org. Chem.*, 61, 8000, 1996.

115. a) Rivkin, A. et al., Complex target-oriented total synthesis in the drug discovery process: the discovery of a highly promising family of second generation epothilones, *J. Am. Chem. Soc.*, 125, 2899, 2003; b) Sun, J. and Sinha S.C., Stereoselective total synthesis of epothilones by the metathesis approach involving C9-C10 bond formation, *Angew. Chem. Int. Ed.*, 41, 1381, 2002.

116. May, S.A. and Grieco, P.A., Total synthesis of ()-epothilone B, *Chem. Commun.*, 1597, 1998.

117. Mulzer, J. and Mantoulidis, A., Synthesis of the C(1)-C(9) segment of the cytotoxic macrolides epothilon A and B, *Tetrahedron Lett.*, 37, 9179, 1996.

118. Schinzer, D., Limberg, A., and Böhm, O.M., Studies towards the total synthesis of epothilones: asymmetric synthesis of the key fragments, *Chem. Eur. J.*, 2, 1477, 1996.

119. a) Nicolaou, K.C. et al., Total synthesis of epothilone E and related side-chain modified analogs via a Stille Coupling based strategy, *Bioorg. Med. Chem.*, 7, 665, 1999; b) Nicolaou, K.C. et al., Chemical synthesis and biological properties of pyridine epothilones. *Chem. Biol.*, 7, 593, 2000.

120. Nicolaou, K.C. et al., Chemical synthesis and biological evaluation of novel epothilone B and *trans*-12,13-cyclopropyl epothilone B analogs, *Tetrahedron*, 58, 6413, 2002.

121. Nicolaou, K.C. et al., Synthesis and biological evaluation of 12,13-cyclopropyl and 12,13-cyclobutyl epothilones, *Chem. Bio. Chem.*, 1, 69, 2001.

122. Ramachandran, P.V. et al., Preparative-scale synthesis of both antipodes of B-γ,γ-dimethylallyldiisopinocampheylborane: application for the synthesis of C_1–C_6 subunit of epothilone, *Tetrahedron Lett.*, 45, 1011, 2004.

123. Gabriel, T. and Wessjohann, L., The chromium-reformatsky reaction: asymmetric synthesis of the aldol fragment of the cytotoxic epothilons from 3-(2-bromoacyl)-2-oxazolidinones, *Tetrahedron Lett.*, 38, 1363, 1997.

124. Panicker, B., Karle, J.M., and Avery, M.A., An unusual reversal of stereoselectivity in a boron mediated aldol reaction: enenantioselective synthesis of the C_1–C_6 segment of the epothilones, *Tetrahedron*, 56, 7859, 2000.

125. Stachel, S.J. and Danishefsky, S.J., Chemo- and stereoselective epoxidation of 12,13-desoxyepothilone B using 2,2-dimethyldioxirane, *Tetrahedron Lett.*, 42, 6785, 2001.

126. Liu, Z.-Y. et al., Total synthesis of epothilone A through stereospecific epoxidation of the *p*-methoxybenzyl ether of epothilone C, *Chem. Eur. J.*, 8, 3747, 2002.

127. Balog, A. et al., Stereoselective syntheses and evaluation of compounds in the 8-desmethylepothilone A. Series: some surprising observations regarding their chemical and biological properties, *Tetrahedron Lett.*, 38, 4529, 1997.

128. Nicolaou, K.C. et al., Synthesis of 16-desmethylepothilone B: improved methodology for the rapid, highly selective and convergent construction of epothilone B and analogs, *Chem. Commun.*, 519, 1999.

129. Nicolaou, K.C. et al., Total synthesis of 16-desmethylepothilone B, epothilone B_{10}, epothilone F, and related side chain modified epothilone B analogs, *Chem. Eur. J.*, 6, 2783, 2000.

130. Taylor, R.E. et al, Conformation–activity relationships in polyketide natural products. Towards the biologically active conformation of epothilone, *Org. Biomol. Chem.*, 2, 127, 2004.

131. Taylor, R.E., Chen, Y., and Beatty, A., Conformation–activity relationships in polyketie natural products: a new perspective on the rational design of epothilone analogs, *J. Am. Chem. Soc.*, 125, 26, 2003.

132. Nicolaou, K.C. et al, Total synthesis of 26-hydroxyepothilone B and related analogs, *Chem. Commun.*, 2343, 1997.

133. Nicolaou, K.C. et al., Total synthesis of 26-hydroxy-epothilone B and related analogs *via* a macrolactonization based strategy, *Tetrahedron*, 54, 7127, 1998.

134. Chappell, M.D. et al., Probing the SAR of dEpoB via chemical synthesis: a total synthesis evaluation of C26-(1,3-dioxolanyl)-12,13-desoxyepothilone B, *J. Org. Chem.*, 67, 7730, 2002.

135. Koch, G., Loiseleur, O., and Altmann, K.-H., Total synthesis of 26-fluoro-epothilone B, *Synlett*, 4, 693, 2004.

136. Chou, T.-C. et al., Design and total synthesis of a superior family of epothilone analogs, which eliminate xenograft tumors to a nonrelapsable state, *Angew. Chem. Int. Ed.*, 42, 4762, 2003.

137. Altmann, K.-H. et al., Synthesis and biological evaluation of highly potent analogs of epothilones B and D, *Bioorg. Med. Chem. Lett.*, 10, 2765, 2000.

138. Buey, R.M. et al., Interaction of epothilone analogs with the paclitaxel binding site: relationship between binding affinity, microtubule stabilisation, and cytotoxicity, *Chem. Biol.*, 11, 225, 2004.

139. Johnson, J. et al., Synthesis, structure proof, and biological activity of epothilone cyclopropanes, *Org. Lett.*, 2, 1537, 2000.

140. Nicolaou, K.C. et al., Chemical synthesis and biological evaluation of *cis*- and *trans*-12,13-cyclopropyl and 12,13-cyclobutyl epothilones and related pyridine side chain analogs, *J. Am. Chem. Soc.*, 123, 9313, 2001.

141. Nicolaou, K.C. et al., Design, synthesis, and biological properties of highly potent epothilone B analogs, *Angew. Chem. Int. Ed.*, 42, 3515, 2003.

142. Regueiro-Ren, A. et al., Synthesis and biological activity of novel epothilone aziridines, *Org. Lett.*, 3, 2693, 2001.

143. Altmann, K.-H. et al., The total synthesis and biological assessment of *trans*-epothilone A, *Helvetica Chim. Acta*, 85, 4086, 2002.

144. Nicolaou, K.C. et al., Synthesis of epothilone A and B in solid and solution phase, *Nature*, 387, 268, 1997.

145. Glunz, P.W. et al., The synthesis and evaluation of 12,13-benzodesoxyepothilone B: a highly convergent route, *Tetrahedron Lett.*, 40, 6895, 1999.

146. End, N. et al., Synthetic epothilone analogs with modifications in the northern hemispere and the heterocyclic side-chain-synthesis and biological evaluation, Fourth International Electronic Conference on Synthetic Organic Chemistry (ECSOC-4), September 1–30, 2000.

147. Quintard, D. et al., Enantioselective synthesis of 2,3-dehydro-3-desoxy-10-oxa epothilone D, *Synlett*, 13, 2033, 203.

148. Rivkin, A. et al., On the introduction of a trifluoromethyl substituent in the epothilone setting: chemical issues related to ring forming olefin metathesis and earliest biological findings, *Org. Lett.*, 4, 4081, 2002.

149. Yoshimura, F. et al., Synthesis and conformational analysis of (*E*)-9,10-dehydroepothilone B: a suggestive link between the chemistry and biology of epothilones, *Angew. Chem. Int. Ed.*, 42, 2518, 2003.

150. White, J.D. and Sundermann, K.F., Synthesis, conformational analysis, and bioassay of 9,10-didehydroepothilone D, *Org. Lett.*, 4, 995, 2002.

151. Zhang, H.S., Zhong, C.F., and Bao, X.P., Studies of the total synthesis of epothilone B and D: a facile synthesis of C7-C14 and C15-C21 fragments, *Chinese Chem. Lett.*, 14, 115, 2003.

152. Akbutina, F.A. et al., Approaches to epothilone carboanalogs starting from Δ^3-carene, *Russ. J. Org. Chem.*, 39, 75, 2003.

153. Akbutina, F.A. et al., Chiral synthetic block based on (*R*)-pantolactone, *Mendeleev Commun.*, 151, 2003.

154. Ermolenko, M.S. and Potier, P., Synthesis of epothilones B and D from D-glucose, *Tetrahedron Lett.*, 43, 2895, 2002.

155. Machajewski, T.D. and Wong, C.-H., Chemoenzymatic synthesis of Key epothilone fragments, *Synthesis*, 1469, 1999.

156. Shioji, K. et al., Synthesis of C1-C6 fragment for epothilone A via lipase-catalyzed optical resolution, *Synth. Comm.*, 31, 3569, 2001.

157. Bornscheuer, U.T., Altenbuchner, J., and Meyer, H.H., Directed evolution of an esterase for the stereoselective resolution of a key intermediate in the synthesis of epothilones, *Biotechnol. Bioeng.*, 58, 554, 1998.

158. Sinha, S.C. et al., Synthesis of epothilone analogs by antibody-catalyzed resolution of thiazole aldol synthons on a multigram scale. Biological consequences of C-13 alkylation of epothilones, *Chem. Bio. Chem*, 2, 656, 2001.

159. Sinha, S.C. at al., Sets of aldolase antibodies with antipodal reactivities. Formal synthesis of epothilone E by large-scale antibody-catalyzed resolution of thiazole aldol, *Org. Lett.*, 1, 1623, 1999.

160. Sinha, S.C., Barbas III, C.F., and Lerner, R.A., The antibody catalysis route to the total synthesis of epothilones, *Proc. Natl. Acad. Sci. USA*, 95, 14603, 1998.

161. Liu, J. and Wong, C.-H., Aldolase-catalyzed asymmetric synthesis of novel pyranose synthons as a new entry to heterocycles and epothilones, *Angew. Chem. Int. Ed.*, 41, 1404, 2002.

162. Nicolaou, K.C. et al., Designed epothilones: combinatorial synthesis, tubulin assembly properties, and cytotoxic action against Taxol-resistant tumor cells, *Angew. Chem. Int. Ed.*, 36, 2097, 1997.

163. Storer, R.J. et al., A total synthesis of epothilones using solid-supported reagents and scavengers, *Angew. Chem. Int. Ed.*, 42, 2521, 2003; Storer, R.I. et al., Multi-step application of immobilized reagents and scavengers: a total synthesis of epothilone C, *Chem. Eur. J.*, 10, 2529, 2004.

164. Sefkow, M. et al., Oxidative and reductive transformations of epothilone A, *Bioorg. Med. Chem. Lett.*, 8, 3025, 1998.

165. Sefkow, M., Kiffe, M., and Höfle, G., Derivatization of the C12-C13 functional groups of epothilones A, B and C, *Bioorg. Med. Chem. Lett.*, 8, 3031, 1998.

166. Höfle, G. et al., Epothilone A–D and their thiazole-modified analogs as novel anticancer agents, *Pure Appl. Chem.*, 71, 2019, 1999.

167. Boddy, C.N. et al., Epothilone C macrolactonization and hydrolysis are catalyzed by the isolated thioesterase domain of epothilone polyketide synthase, *J. Am. Chem. Soc.*, 125, 3428, 2003.

168. Regueiro-Ren, A. et al., SAR and pH stability of cyano-substituted epothilones, *Org. Lett.*, 4, 3815, 2002.

169. Vite, G.D. et al., (BMS), US Patent, 6399638 B1 (June, 4, 2002).

170. Dong, S.D. et al., Rapid access to epothilone analogs via semisynthetic degradation and reconstruction of epothilone D, *Tetrahedron Lett.*, 45, 1945, 2004.

171. Karama, U. and Höfle, G., Synthesis of epothilone 16,17-alkyne by replacement of the C13-C15(O)-ring segment of natural epothilone C, *Eur. J. Org. Chem.*, 1042, 2003.

172. Sefkow, M. and Höfle, G., Substitutions at the thiazole moiety of epothilone, *Heterocyles*, 48, 2485, 1998.

173. Glaser, N., Doctoral Thesis, Technical University of Braunschweig, 2001.

174. Matson, J.A. et al., (BMS), International Patent Application, WO 00/39276 (July 6, 2000).

175. Hamel, E., Evaluation of antimitotic agents by quantitative comparisons of their effects on the polymerization of purified tubulin, *Cell Biochem. Biophys.*, 38, 1, 2003.

176. a) Nicolaou, K.C. et al., Variation der Ringgröße von Epothilonen — Totalsynthese von [14]-, [15], [17]- und [18]Epothilon A, *Angew. Chem.*, 110, 85, 1998; b) Rivkin, A. et al., Total syntheses of [17]- and [18]dehydrodesoxyepothilones B via a concise ring-closing metathesis-based strategy: correlation of ring size with biological activity in the epothilone series, *J. Org. Chem.*, 67, 7737, 2002.

177. U.S. Patent Application Serial No. 09/280,210, U.S. Patent No. 6,498,257 B1, 2002.

178. Nicolaou, K.C. et al., Synthesis and biological properties of C12,13-cyclopropyl-epothilone A and related epothilones, *Chem. Biol.*, 5, 365, 1998.

179. Su, D.-S. et al., Total synthesis of (-)-epothilone B: an extension of the Suzuki coupling method and insights into structure-activity relationships of the epothilones, *Angew. Chem. Int. Ed.*, 36, 757, 1997.

180. Ganesh, T. et al., Synthesis and biological evaluation of fluorescently labeled epothilone analogs for tubulin binding studies, *Tetrahedron*, 59, 9979, 2003.

181. Stachel, S. J. et al., On the total synthesis and preliminary biological evaluations of 15(R) and 15(S) Aza-dEpoB: a Mitsunobu inversion at C15 in pre-epothilone fragments, *Org. Lett.*, 2, 1637, 2000.

182. Victory, S.F. et al., Relative stereochemistry and solution conformation of the novel paclitaxel-like antimitotic agent epothilone A, *Bioorg. Med. Chem. Lett.*, 6, 893, 1996.

183. Taylor, R.E. and Zajicek, J., Conformational properties of epothilone, *J. Org. Chem.*, 64, 7224, 1999; ibid. 65, 5449, 2000.

184. Carlomagno, T. et al., The high-resolution solution structure of epothilone A bound to tubulin: an understanding of the structure-activity relationships for a powerful class of antitumor agents, *Angew. Chem. Int. Ed.*, 42, 2511, 2003.

185. Carlomagno, T. et al., Derivation of dihedral angles from CH-CH dipolar-dipolar cross-correlated relaxation rates: a C-C torsion involving a quaternary carbon atom in epothilone A bound to tubulin, *Angew. Chem. Int. Ed.*, 42, 2515, 2003.

186. The spectral data for the supposed epothilone/tubulin complex in aqueous buffer are compared with that of epothilone in dichloromethane, and the tubulin is stated not to polymerise even in the presence of a 100 fold molar excess of epothilone.

187. Ojima, I. et al., A common pharmacophore for cytotoxic natural products that stabilize microtubules, *Proc. Natl. Acad. Sci. USA*, 96, 4256, 1999.

188. Wang, M. et al., A unified and quantitative receptor model for the microtubule binding of paclitaxel and epothilone, *Org. Lett.*, 1, 43, 1999.

189. He, L. et al., A common pharmacophore for Taxol and the epothilones based on the biological activity of a taxane molecule lacking a C13 side chain, *Biochemistry*, 39, 3972, 2000.

190. Wartmann, M., Preclinical pharmacological profile of ABJ879, a novel epothilone B analog with potent and protracted anti-tumor activity, AACR Ann. Meet. Orlando, March 27–31, 2004, Abstr. No. 5440.

191. DiLea, C. et al., A PK-PD dose optimization strategy for the microtubule stabilizing agent ABJ879, AACR Ann. Meet. Orlando, March 27–31, 2004, Abstr. No. 5132.

192. Johnson et al., Oral bioavailability of KOS-862 (epothilone D), a potent stabilizer of microtubulin polymerization in rats and Beagle dogs. *Proc. Am. Assoc. Cancer Res.*, 44, Abstr. No. 5339, 2003.

193. Rosen, P.J. et al., KOS-862 (epothilone D): results of a phase I dose-escalating trial in patients with advanced malignancies, *Proc. Am. Soc. Clin. Oncol.* 21: Abstr. No. 413, 2002.

194. Bocci, G., Nicolaou, K.C., and Kerbel, R.S., Protracted low-dose effects on human endothelial cell proliferation and survival *in vitro* reveal a selective antiangiogenic window for various chemotherapeutic drugs, *Cancer Res.*, 62, 6938, 2002.

195. Blum, W. et al., *In vivo* metabolism of epothilone B in tumor-bearing nude mice: identification of three new epothilone B metabolites by capillary high-pressure liquid chromatography/mass spectrometry/tandem mass spectrometry, *Rapid Commun. Mass Spectrom.*, 15, 41, 2001.

196. Altaha, R. et al., Epothilones: a novel class of non-taxane microtubule-stabilizing agents, *Curr. Pharm. Design*, 8, 1707, 2002.

197. McDaid, H.M. et al., Validation of the pharmacodynamics of BMS-247550, an analog of epothilone B, during a phase I clinical study, *Clin. Cancer Res.*, 8, 2035, 2002.

198. Borman, S., First disclosures of clinical candidates, *Chem. Eng. News*, 80, 35, 2002.

199. Rothermel, J. et al., EPO906 (epothilone B): A promising novel microtubule stabilizer, *Sem. Oncol.*, 30, Suppl. 6, 51, 2003.

200. Agrawal, M. et al., Treatment of recurrent cervical adenocarcinoma with BMS-247550, an epothilone B analog, *Gynecol. Oncol.*, 90, 96, 2003.

201. Abraham, J. et al., Phase I trial and pharmacokinetic study of BMS-247550, an epothilone B analog, administered intravenously on a daily schedule for five days, *J. Clin. Oncol.*, 21, 1866, 2003.

202. Smaletz, O. et al., Pilot study of epothilone B analog (BMS-247550) and estramustine phosphate in patients with progressive metastatic prostate cancer following castration, *Ann. Oncol.*, 14, 1518, 2003.

203. Nettles, J.H. et al., The Binding Mode of Epothilone A on α,β-Tubulin by Electron Crystallography, *Science*, 305, 866, 2004.

204. Heinz, D.H. et al., Much Anticipated — The Bioactive Conformation of Epothilone and Its Binding to Tubulin, *Angew. Chem. Int. Ed.*, 44, 1298, 2005.

22 Enediynes

*Philip R. Hamann, Janis Upeslacis, and
Donald B. Borders*

CONTENTS

      I.   Introduction 451
     II.   History...451
    III.   Structural Classes and Producing Organisms...454
     IV.   Biosynthesis 455
      V.   Molecular Mechanisms of Action...456
     VI.   Cellular Mechanisms of Action 458
    VII.   Preclinical Studies...459
   VIII.   Clinical Experience 463
     IX.   Summary 464
References 465

I. INTRODUCTION

The enediyne antitumor antibiotics represent some of the most potent biologically active natural products ever discovered. These compounds possess both antibacterial and antitumor activity and are related by containing enediyne functionality constrained within 9- or 10-member rings, with varying degrees of chemical complexity in the remainder of the molecules. Most of these antitumor antibiotics are produced by microorganisms classified as the actinomycetes. The challenging chemical structures of these enediynes, their mechanisms of action, and their biological activity have attracted considerable scientific attention. Two of these compounds, neocarzinostatin and esperamicin A1, were advanced into anticancer clinical trials as single agents. Esperamicin A1 was eventually abandoned from further development because of toxicity issues. However, neocarzinostatin has been approved for human use in Japan for a number of cancer indications, both as the natural product itself as well as a part of a polymer-based conjugate. A monoclonal antibody conjugate of calicheamicin was successful in clinical trials and has been commercialized for the treatment of acute myeloid leukemia in adults.

Some of the previous reviews covering the enediynes have included a comprehensive summary in book form[1] as well as a number of focused reviews detailing structure determinations, mechanism of action, synthetic approaches, biosynthesis, and biological activity.[2–8] A book devoted to neocarzinostatin has also been published.[9] This review is limited to a selective coverage of highlights, with an emphasis on recent developments that have produced, or have the potential to produce, marketed products based on the enediynes.

II. HISTORY

The structures of the major components of the calicheamicins and esperamicins were published as back-to-back articles in a 1987 issue of the *Journal of the American Chemical Society* based on

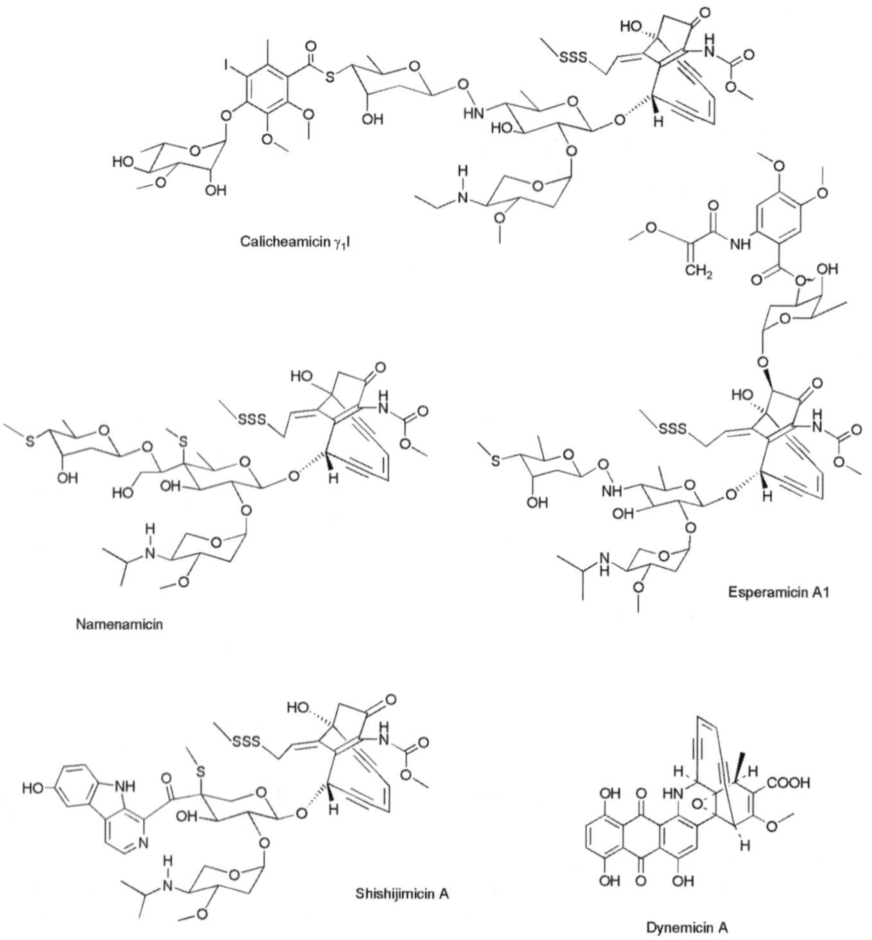

FIGURE 22.1 Structures of representative 10-membered ring enediyne antibiotics.

research at two different pharmaceutical companies.[10–13] It is fortunate that these publications appeared together because, taken individually, the structures with 10-membered ring enediynes were difficult to believe. These remarkable compounds were structurally quite different from one another, but surprisingly, both had the same novel 10-member ring enediyne system, an allylic methyl trisulfide group, a hydroxyamino-linked disaccharide unit, and unique amino- and thio-containing carbohydrates (Figure 22.1). In addition, the calicheamicins possessed a highly substituted halobenzene unit. Both classes of compounds showed extremely potent biological activity with a mechanism of action involving free-radical damage to DNA at selective sites. In the case of the calicheamicins, a high percentage of the damage involved double-strand cleavage of DNA.[14] In time, the structures of additional members of both families of compounds were elucidated.[15,16]

A key factor in the discovery of the calicheamicin complex in fermentation extracts was a newly developed *in vitro* test termed the biochemical induction assay that detected DNA damaging agents.[17] The assay was extremely sensitive for detecting the very low concentrations of the calicheamicin complex present in the early fermentation. Even very crude preparations of the complex were significantly more potent in this assay than the reference compounds, and this provided the first suggestion that the compounds were novel. Pure calicheamicin γ_1^I was extremely active in the biochemical induction assay and against Gram-positive bacteria in the picogram per milliliter range.[18,19] It was also very active against Gram-negative bacteria.

Once the structures of calicheamicin and esperamicin were firmly established, the very elegant and earlier structural work with the more remarkable nine-membered ring enediyne chromophoric component and associated apoprotein of the antitumor antibiotic neocarzinostatin could be fully appreciated.[20] This compound could only be isolated along with the apoprotein, which provided stability to the highly reactive 9-member ring enediyne chromophoric component. It became apparent that calicheamicin and esperamicin represented 10-member ring enediyne antitumor antibiotics that were stable without an associated protein, whereas neocarzinostatin represented another type, the 9-member chromoprotein enediynes, requiring a protein for stabilization.

A number of the enediyne antibiotics listed in Table 22.1 have various synonyms. The numeric designations were given by the pharmaceutical companies at the initial stages of their investigations and then were changed to more conventional names if the compounds showed promise for further development. Thus, the esperamicin complex was originally designated BMY 28175 by Bristol-Myers. Two other companies working on components of the same complex at approximately the same time as Bristol-Myers designated their compounds as FR900405 and FR900406 (Fujisawa) or veractamycins A and B (Parke-Davis). The BMY 28175 designation was subsequently changed to esperamicin and eventually shown to be identical to the products isolated by the other two companies.[21] Lederle Laboratories originally named the LL-E33288 complex "calichemicin," and that name appeared in the initial structure papers. However, a conflict caused by the resemblance of the name of "calichemicin" to that of a chemical company required changing the name to "calicheamicin."

TABLE 22.1
Summary of Enediyne Antibiotic Complexes Produced by Various Organisms

Name	Synonyms	Producing Organism	Structure Type[a]
Calicheamicin	LL-E33288 calichemicin	*Micromonospora echinospora,* ssp. *calichensis*	10-mr enediyne with allylic MeSSS
Esperamicin	BMY 28175 FR900405 FR900406 Veractamycins A and B	*Actinomadura verrucosospora*	10-mr enediyne with allylic MeSSS
Namenamicin		*Polysyncraton lithostrotum* (ascidian)	10-mr enediyne with allylic MeSSS
Shishijimicin		*Didemnum proliferum* (ascidian)	10-mr enediyne with allylic MeSSS
Dynemicin	BU 3420T	*Micromonospora chersina Micromonospora globosa*	10-mr enediyne with anthraquinone
Endynamicin		*Micromonospora globosa*	10-mr enediyne with anthraquinone
Actinoxanthin		*Streptomyces globisporus*	Presumed 9-mr enediyne with apoprotein
Auromomycin	Macromomycin I	*Streptomyces macromyceticus*	Presumed 9-mr enediyne with apoprotein
Kedarcidin		*Streptoalloteichus* sp.	9-mr enediyne with apoprotein
Lidamycin	C-1027	*Streptomyces globisporus*	9-mr enediyne with apoprotein
Maduropeptin	BBM 1644	*Actinomadura madurae*	9-mr enediyne with apoprotein
Neocarzinostatin	Zinostatin NSC 69856	*Streptomyces carzinostaticus*	9-mr enediyne with apoprotein
Antibiotic N1999A2		*Streptoverticillium* sp.	9-mr enediyne

[a] 9- and 10-membered rings are abbreviated as 9-mr and 10-mr, respectively.

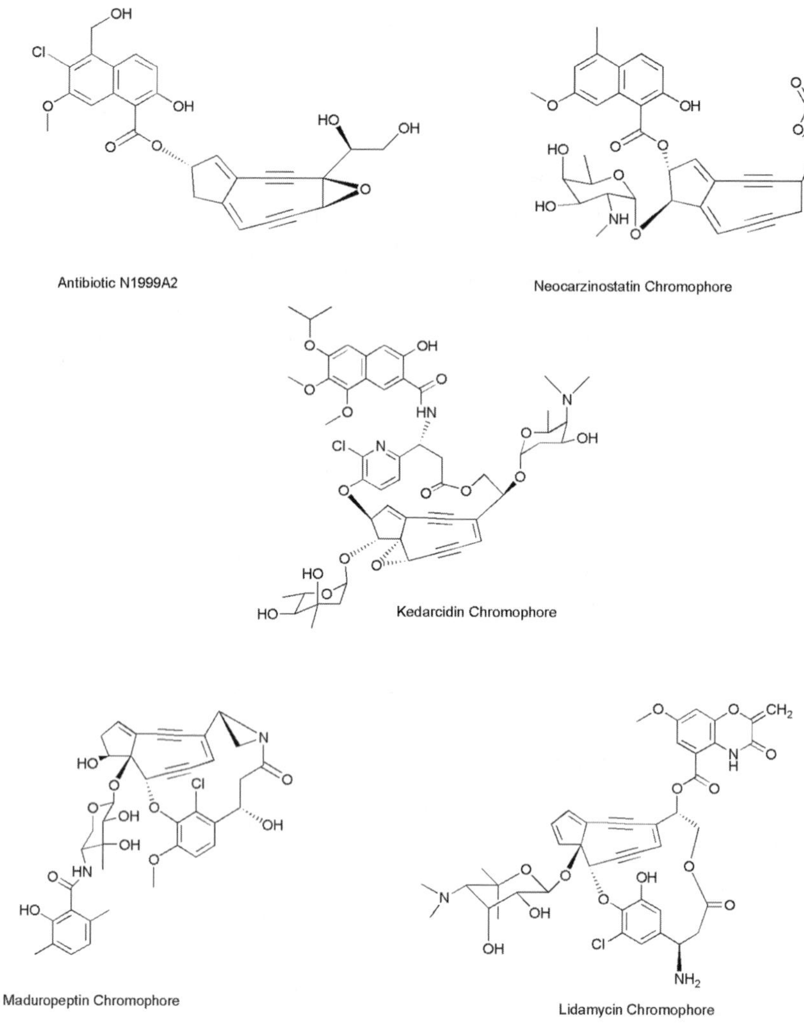

FIGURE 22.2 Structures of representative nine-membered ring enediyne antibiotics.

III. STRUCTURAL CLASSES AND PRODUCING ORGANISMS

The enediyne antitumor antibiotics fall into two general classes. Representative examples wherein the conjugated enediyne system is a part of a 10-member macrocycle are shown in Figure 22.1, and examples of the nine-member enediynes identified so far are shown in Figure 22.2. Most of the 10-member ring enediynes are related to calicheamicin and esperamicin and have an allylic methyl trisulfide group that can act as a trigger to initiate formation of a diradical through cycloaromatization of the core ring. These diradicals are responsible for the resultant DNA damage. Although dynemicin and endynamicin, differing from each other by virtue of stereochemistry, also possess a 10-member ring enediyne, they contain no methyl trisulfide group to initiate radical formation. Instead, these compounds contain an epoxide and bridging anthraquinone unit that is involved in redox activation, putting them into a different subclass than the other 10-member ring enediynes.

With the exception of antibiotic N1999A2,[22] the 9-member ring enediynes are molecules that are associated noncovalently with coproduced apoproteins (Figure 22.2) that stabilize the chromophores by binding them in a hydrophobic cavity.[23] Although the apoproteins examined so far exhibit significant homology, they differ for each individual antibiotic. For example, the neocarzinostatin, kedarcidin, and lidamycin apoproteins contain 113, 114, and 110 amino acids, respectively.[24–26] The x-ray crystal structure of the neocarzinostatin apoprotein was determined to a resolution of 0.15 nm. The structure of this apoprotein was then compared to crystal structure data for the apoproteins of actinoxanthin and auromomycin to show a three-dimensional similarity to these homologous proteins.[27] The apoproteins stabilize the enediynes in a cellular environment by excluding glutathione from the reaction center through electronic repulsions.[28] Supra C-1027, a significantly stabilized analog of the lidamycin chromoprotein, was prepared by incorporation of deuterated glycine into the apoprotein.[29] This was accomplished by expressing the dideutero-Gly apoprotein in *Escherichia coli* containing a vector with the lidamycin apoprotein coding sequence. The dideutero-Gly apoprotein was then complexed with the chromophore to obtain the chromoprotein. Peptidase activity has also been attributed to the apoproteins, which may help expose the DNA to the enediyne chromophores.[30,31]

The structures of the enediynes have been determined by spectral methods, mainly two-dimensional nuclear magnetic resonance and mass spectrometry techniques, chemical degradations, and in some cases, x-ray crystallography. Because of the considerable interest in this class of compounds and the challenge of synthesizing these unusual structures, a number of the enediyne antibiotics have been prepared by total syntheses. This has either confirmed the structures, particularly individual points of asymmetry, or resulted in modifications of the proposed structures. Calicheamicin γ_1^I has one of the most complex and elaborate structures, yet interestingly it was the first enediyne antibiotic prepared by total synthesis.[32–34] The structure of the kedarcidin chromophore was originally deduced mainly from nuclear magnetic resonance and mass spectrometry data but was recently revised to the structure shown in Figure 22.2 as a result of total synthesis studies.[35–37]

Total syntheses have been achieved for several additional enediynes after their structures were elucidated, including the neocarzinostatin chromophore,[20,38] antibiotic N1999A2,[22,39,40] and dynemicin.[41–43] Enediyne antibiotics whose structures have been solved but have only led to partial synthesis studies are lidamycin chromophore[44,45] and maduropeptin chromophore.[46] Thus, the structure of maduropeptin shown in Figure 22.2 is one of several possibilities. Structures for namenamicin and the shishijimicin complex (Figure 22.1) have been determined[47,48] but no synthetic efforts have been reported. Only limited structural information has been published for actinoxanthin and auromomycin chromophores.[27,49–51] The apoprotein for auromomycin is macromomycin.

Most of the enediyne antitumor antibiotics are produced by microorganisms classified as actinomycetes. Within this general class, however, there is considerable diversity of genus and species for the organisms that produce the natural products (Table 22.1). Recently, macroscopic organisms have been implicated in producing 10-member ring enediynes related to calicheamicin and esperamicin. Marine ascidians were credited for producing namenamicin and shishijimicin.[47,48] However, the ascidian that produced namenamicin was later found to coexist with several *Micromonospora* species in a symbiotic relationship, which led to speculation that one of the *Micromonospora* sp. may have actually produced the enediynes rather than the ascidian.

IV. BIOSYNTHESIS

Most of the enediyne antibiotics are generated as complexes of related components by the producing organisms. The various components of the calicheamicin complex — at last count about 20 — differ not only by the halogen on the aromatic ring but also by the presence or absence of the terminal sugars. In addition, the amine of the terminal amino sugar is elaborated with a methyl, ethyl, or isopropyl group. The natural components of the calicheamicin complex contained a bromine atom, which is an unusual halogen for a terrestrial organism. Attempts to improve fermentation yields by adding potassium iodide to the media resulted in a significant improvement and gave new components containing iodine instead of bromine.[52] Surprisingly, addition of iodide to the fermentation media of esperamicin and dynemicin also resulted in increased fermentation yields even though these products do not contain a halogen.[21] This indicates that iodide or one of the other halides may be involved in intermediate steps in the biosynthesis of the enediyne antibiotics.

Biosynthetic incorporation studies using labeled acetate and other potential polyketide precursors revealed the pathway for assembly of the esperamicin, neocarzinostatin chromophore, and dynemicin biosynthetic precursors.[53–56] Cloning, sequencing, and characterization of the complete calicheamicin and lidamycin chromophore biosynthesis gene clusters have been reported.[8,57,58] The polyketide synthases for these enediynes are quite unique from previously reported polyketide synthases. A comparison of the lidamycin chromophore locus with that of calicheamicin revealed that the enediyne polyketide synthase is highly conserved in spite of producing 9- and 10-member ring enediynes, respectively. This has led to a proposed general polyketide pathway for the biosynthesis of both classes of enediyne antibiotics. In addition, a phylogenetic tree has been suggested that relates the synthesis of both 9- and 10-member enediynes and yields a genomic method to screen for other enediynes.[59]

Various unprecedented enzymes have been uncovered in investigating the biosynthesis of the enediynes. For example, exploration of the biosynthesis of the unique structural elements in calicheamicin has led to various unique enzymes involved in those pathways.[60] The gene cluster responsible for the biosynthesis of lidamycin also was found to contain a gene that encodes for a new type of aminomutase that converts L-tyrosine into (S)-β-tyrosine.[61] Further biochemical processing of the (S)-β-tyrosine converts it to the subunit of lidamycin. In addition, the genes that encode the production of the lidamycin chromophore in *Streptomyces globisporus* are clustered with the cagA gene that encodes the lidamycin apoprotein.[62] This suggests that some type of genetic coordination exists between production of the chromophore and the apoprotein in this organism.

It has been speculated that the lidamycin chromophore is sequestered by binding to a preapoprotein to form a complex that is transported out of the cell by an efflux pump and processed by removing a leader peptide sequence to yield the chromoprotein.[62] This may serve to protect the producing organism from the toxic effects of the lidamycin. Another mechanism has been discovered for calicheamicin that may prove to be a general phenomenon. A protein, CalC, has been discovered that is coproduced with calicheamicin. It binds to the triggered natural product before the Bergman cyclization better than it binds to the intact form, thereby scavenging the triggered form through "self-sacrifice" before it can damage the DNA.[63]

Previous studies with kedarcidin showed that exposure of the producing organism to the chromophore resulted in significantly upregulated apoprotein production by as much as 10- to 20-fold.[51] However, decoupling of the joint biosynthesis of apoprotein and chromophore has also been achieved. When the neocarzinostatin-producing organism was fermented in a synthetic medium containing $MgSO_4$, the free chromophore was observed in the culture filtrate.[64]

V. MOLECULAR MECHANISMS OF ACTION

As a class, the enediynes cause single- and double-strand cleavage of DNA by rearrangement of their core to an aromatic diradical through a Bergman (10-mr) or Bergman-like (9-mr) cyclization.[65–67] How the molecules interact with DNA differs somewhat. Calicheamicin is the first known DNA binder that uses a carbohydrate sequence to recognize DNA and binds to the minor groove in a nonintercalative manner. In addition, the carbohydrate portion of calicheamicin has a novel pyrimidine selectivity as a result of a shape-dependent induced-fit interaction.[68–72] Esperamicin binds through a combination of carbohydrate-guided minor groove fit as well as base-pair intercalation, and dynemicin and antibiotic N1999A2 bind through purely intercalative mechanisms.[22,39,73]

Triggering of the methyl trisulfide-containing enediynes within cells requires a reductive cleavage of the trisulfide by glutathione and subsequent rearrangements to form the diradical. This chemistry is illustrated for calicheamicin γ_1^I in Figure 22.3. The reduction does not necessarily have to occur with the enediyne prebound to DNA. For example, the pre-Berman cyclization intermediate for calicheamicin γ_1^I has a half-life of about 4 sec.[74] However, how the enediyne and product diradical

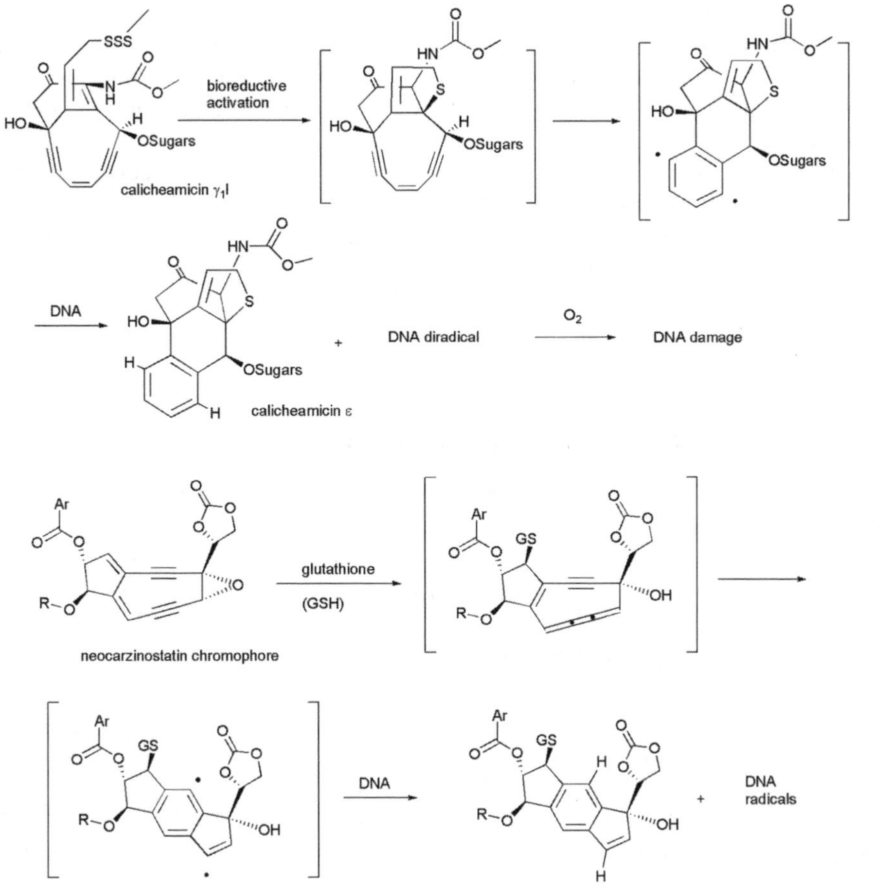

FIGURE 22.3 Diradical formation from calicheamicin γ_1^I/neocarzinostatin and their interaction with DNA.

are positioned in the DNA determines the type of damage that occurs. A diradical that spans both strands produces double-strand breaks, whereas a diradical with access to only one strand leads to single-strand breaks.[14,74–76] The structure of a calicheamicin γ_1^I-DNA duplex complex was studied in aqueous solution using two-dimensional nuclear magnetic resonance techniques and molecular dynamics calculations. A predominate species of the hairpin duplex complex with DNA containing (TCCT) (AGGA) was observed and gave further insight into the binding and sequence specificity.[77]

The Bergman cyclization of the dynemicin-like enediynes is the same as illustrated for calicheamicin, but the trigger in this case is reduction of the anthraquinone, which allows the electrons on nitrogen to open the epoxide vinylogously through the aromatic ring.[78,79]

Triggering of the 9-member enediynes involves glutathione, just like most of the 10-member ring enediynes, but in a conjugate addition to the poly-unsaturated system. They all owe their existence to an epoxide functionality (e.g., neocarzinostatin and kedarcidin chromophores) or a bridging chain across the enediyne (e.g., maduropeptin chromophore), or both, that keep the diyne portion of the core separated. The triggering of neocarzinostatin is illustrated in Figure 22.3. As is also true for the 10-mr enediynes, the triggering is achieved by structural changes, such as opening of the epoxide, disruption of the bridging structure, or binding to DNA, that permit the reacting ends of an enediyne to approach within bonding distance of 2.9 to 3.4 Å.[67]

A review of the interaction of enediynes with DNA has recently been published.[80] Lidamycin causes sequence-specific double-strand DNA cleavage.[81] However, when oxygen is depleted from the reaction of lidamycin with DNA, sequence-specific covalent DNA drug adducts and DNA interstrand cross links are formed with the drug.[82] These types of reactions may be important in the central regions of large tumors where relatively anaerobic conditions prevail. Lidamycin cleaved DNA more efficiently in cells than in a cell-free environment. Mixed single- and double-strand breaks were observed in intracellular episomal, mitochondrial, and genomic DNA at low nanomolar concentrations.[83]

Calicheamicin γ_1^I cleavage of plasmid DNA was not inhibited by excess tRNA or protein, showing that this enediyne specifically targets DNA.[84] The ratio of DNA double-strand breaks to single-strand breaks caused by calicheamicin γ_1^I in cellular DNA of human fibroblasts was 1:3, close to the 1:2 ratio observed with purified plasmid DNA.[84] In addition, calicheamicin γ_1^I induced a strong double-strand break response in human fibroblasts at concentrations that would deliver fewer than 1000 molecules per cell. In these and other studies, however, the cells were fully capable of repairing these breaks.[85]

VI. CELLULAR MECHANISMS OF ACTION

Enediynes exert their biological effects through free-radical damage to DNA, and as a result they have been called radiomimetics. It is clear, however, that the cellular response to each of the various enediynes is unique and differs significantly from the effects of radiation. Although the traditional responses to DNA damage such as apoptosis (programmed cell death) have been observed, other types of responses occur as well. The specific cellular response is in part guided by the type of DNA damage induced, but it is also clearly dependent on the inherent make-up of the cells and the resultant tendencies to follow certain biological pathways.

Lidamycin appears to cause normal apoptosis in HL-60 leukemia cells[86] but abnormal apoptosis at higher concentrations in BEL-7402 hepatoma cells.[87] Although laddering of DNA and caspase involvement were seen, the nuclear membrane remained intact, and no apoptotic bodies were observed. This was ascribed to direct DNA shredding by lidamycin preceding the onset of the actual apoptotic signals. Consistent with this, a separate study indicated that the DNA laddering appears to occur before caspase involvement.[88]

Many of the responses that were ataxia–telangiectasia mutation–dependent for ionizing radiation and bleomycin, such as p53 phosphorylation, were shown to occur by ataxia–telangiectasia mutation–independent mechanisms with lidamycin.[89] However, Mylotarg® (Wyeth), an antibody conjugate of calicheamicin, has been shown to cause G2 arrest associated with the ataxia–telangiectasia mutation/Chk1 and Chk2 phosphorylation pathway and caspase-3 mediated apoptosis or G2 arrest without apoptosis.[90] Lidamycin, as well as neocarzinostatin, has been reported to inhibit DNA replication by an indirect mechanism involving the hyperphosphorylation of replication protein A.[91]

Dynemicin A induces apoptosis.[92] Curiously though, a series of simplified synthetic analogs of this enediyne appear to induce apoptosis by a non-DNA-dependent mechanism,[92] although it is possible that initial DNA damage triggers a secondary mechanism. Other related analogs not capable of generating free radicals were also shown to block apoptosis induced by a variety of unrelated agents, such as the anthracyclines, camptothecin, and Ara-C.

Calicheamicin γ_1^I, as well as other members of this family, show different effects on different cell lines. SH-SY5Y N-type neuroblastoma cells undergo apoptosis at subnanomolar concentrations of the compound, whereas SH-EP1 S-type neuroblastoma cells proceed to differentiation.[93] The same differential effects were seen for esperamicin A1, dynemicin, and neocarzinostatin. Work in the same labs with this latter agent has demonstrated the involvement of the Bcl-2/caspase-3 pathway in apoptosis of PC12 pheochromocytoma cells.[94]

The synthetic enediyne calicheamicin θ differs from γ_1^I only by having a thioester replacement for the methyl trisulfide. Thus, hydrolysis activates this analog as opposed to the reductive mechanism required for the natural calicheamicins. This compound causes apoptosis in Molt-4 leukemia cells, as indicated by chromatin condensation and DNA degradation.[95] However, when examined over a broader concentration range in LS neuroblastoma cells, it showed a variety of effects.[96] At femtomolar concentrations, cytostasis with upregulation of p53 occurred. At picomolar concentrations, it showed signs of typical apoptosis, and at nanomolar concentrations the derivative produced rapid cell lysis. Other studies have indicated that calicheamicin θ–induced apoptosis is p53 independent and Bax dependent, involving the usual cytochrome C release and caspase-3 and caspase-9 involvement.[97]

Upregulation of p53 that correlated with DNA damage has also been reported for calicheamicin γ_1^I.[98] Studies with Mylotarg®, an antibody conjugate of calicheamicin, on bone marrow samples from acute myeloid leukemia (AML) patients indicated a general lack of intrinsic apoptotic potential in these cells and a low rate (25%) of caspase involvement after drug exposure.[99] This may indicate that nonapoptotic mechanisms predominate in this case, or that the inherently low apoptotic potential in many of the samples may be responsible for a lack of response in some patients.

One similarity has been noted between radiation damage and the cellular response to esperamicin and neocarzinostatin. Cytotoxicity for all these treatments shows a dependence on oxygen, making hypoxic cells that occur in the center of solid tumors more resistant.[100] Radiosensitive cells are also more responsive to calicheamicin and neocarzinostatin as well as to bleomycin, although no mechanistic details have been investigated.[101] Although poly(ADP-ribose) polymerase has been reported to be involved in double-strand break repair caused by radiation, it does not appear to be involved in repairing double-strand breaks caused by neocarzinostatin.[102]

Similar to radiation sensitivity, the cytotoxicity of one enediyne, calicheamicin in the form of its antibody conjugate Mylotarg, has been shown to be cell-cycle dependent.[103] Causing cells to leave G_0 phase and cycle by treatment with INF or GM-CSF resulted in more rapid cell death and increased sensitivity.

In marked contrast to radiation sensitivity, neocarzinostatin is more cytotoxic to neuroblastoma cells when Bcl-2 is upregulated, even though Bcl-2 normally renders cells resistant to apoptosis.[104] This has been correlated to an increase in glutathione in these cells, which is responsible for activating neocarzinostatin.[105] Such an increase in sensitivity does not occur for a synthetic enediyne that is activated by a nonreductive pathway. This may not be general for all enediynes or all cell lines. In related work, 6-mercaptodopamine, a thiol taken up by the dopamine pathway, increases the efficacy of neocarzinostatin in neuroblastoma cell lines both *in vitro* and *in vivo*.[106]

VII. PRECLINICAL STUDIES

The enediynes are among the most cytotoxic compounds known, but they tend to have rather narrow therapeutic windows. As a class they also exhibit poor pharmacokinetic properties because of their lipophilicity. There has only been limited preclinical pharmacology published for most of the unmodified natural products. Availability of the enediynes themselves has been a deterrent in this regard because fermentation production levels, even after considerable refinement, have remained low compared with most natural products. Studies with semisynthetic derivatives have also been hampered by the same constraints, and even though totally synthetic routes have been developed for many of the natural enediynes, these have been undertaken for the sheer intellectual challenge of assembling these molecules from available starting materials rather than with any sourcing goals in mind. Two notable exceptions exist to these generalities. The first is calicheamicin θ, a totally synthetic derivative further discussed below. The second is dynemicin A, structurally one of the simplest members of the natural enediynes. Even though the total syntheses reported were well over 20 steps each and were still not a practical source for these compounds, they led to a variety of studies designed to identify the minimal structural units necessary for high biological activity. A relatively recent review summarizes both the chemistry and biology of these efforts,[107] so only derivatives designed for tumor targeting are summarized here. It has been such attempts at targeting the enediynes, particularly by attaching them to antibodies, that have given this class of natural products significant medicinal promise.

Lidamycin has been attached to an RGD-containing nonapeptide. This targeted construct inhibited pulmonary metastases in a PG giant-cell lung carcinoma model.[108] Antibody conjugates have also been reported and were made with a monosulfide bond to native antibody (3G11) or a Fab or scFv antibody fragment against type-IV collagenase.[109–111] These constructs inhibited KB epidermoid cells at 2 fM and decreased the growth of hepatoma 22 xenograft tumors by 90%. An antigastric cancer antibody, 3H11, has also been used for targeting this enediyne.[112] The conjugate had an IC_{50} of 0.02 fM with 400-fold selectivity versus a nontargeting control conjugate and was active against BGC823 gastric cell xenografts.

Dynemicin A has been examined by a different strategy. ADEPT and GDEPT are approaches that use a less cytotoxic prodrug that is activated at the tumor site by an enzyme that is either pretargeted by an antibody or expressed by an introduced or induced gene, respectively. A simplified version of dynemicin missing the anthraquinone and containing an alternate carbamate activating group has been reported.[113] The nitrobenzyl carbamate is activated by nitroreductases.[114] This derivative showed a 90-fold increase in potency in the presence of the enzyme, and it was also 21- to 135-fold more cytotoxic to cell lines that express a nitroreductase.[115] Related simplified versions of dynemicin have been attached to lexitropsins and were cytotoxic to a panel of cancer cell lines in the 20–50 μM range.[116] Nuclear uptake was indicated by confocal microscopy.

Calicheamicin and esperamicin have been used to make numerous targeted constructs. Esperamicin has been conjugated to alpha-fetoprotein to make a conjugate that showed activity both *in vitro* and *in vivo*, where subcutaneous P388 tumors were completely eliminated in over 90% of the animals.[117] Calicheamicin has also been conjugated to the same protein.[118] This conjugate was cytotoxic to the QOS T-lymphoma cell line at 0.15 nM and was more potent than calicheamicin itself. The structure of the actual alpha-fetoprotein-calicheamicin conjugate is not evident from these reports, however. The conjugation methodology and structures reported are not entirely compatible with what is known about the chemistry of calicheamicin, though a disulfide-linked construct attached through lysines can be assumed. An antisense oligodeoxynucleotide conjugate of calicheamicin has also been prepared,[119] but no further details are available.

Nearly all of the antibody conjugates prepared from calicheamicin analogs have been based on semisynthetic modification of the natural products themselves. An N-acetyl derivative of calicheamicin γ_1^I has been used most extensively for preparing such constructs, even though this derivative is about one order of magnitude less potent than the corresponding parent molecule.[120] An unusual displacement of the methyl trisulfide functionality by thiols has been used to introduce linkers for eventual antibody conjugation.[121] This transformation has had three beneficial consequences. First, conversion to a disulfide stabilizes the molecule. Second, a wide variety of thiols is useful for this displacement, allowing further tailoring of how readily the reductive activation step can occur, modulating the stability of the construct in serum, and allowing introduction of a functional group for antibody conjugation.[121] And third, the cleavage of the disulfide releases calicheamicin γ_1^I in "triggered" form, significantly limiting the potential half-life of the resultant triggered intermediates.[74]

For conjugates of N-acetyl calicheamicin γ_1^I possessing stabilized ("α,α-dimethyl") disulfides, three different methods of making conjugates have been reported. The first, direct conjugation to aldehyde groups generated by oxidation of carbohydrate residues on the antibodies, gives hydrazone linkages that are capable of releasing drug by mild acid hydrolysis ("carbohydrate" conjugates). The second method is the conjugation of activated esters to lysines on the antibodies to produce constructs with only the disulfide available as a point of drug release ("amide" conjugates). The third method also attaches drug to the lysines, but these conjugates contain an added hydrolytic site of release by virtue of a hydrazone linkage in the center of the linker ("hybrid" conjugates). The carbohydrate-based conjugates were the first to be reported and were capable of curing a number of xenograft tumors.[122]

Two specific examples of the calicheamicin amide conjugates have been published. One was made from an anticanine lymphoma antibody designated 231.[123] This conjugate was curative of canine lymphoma xenografts in mice with no apparent lethality, and all control groups (antibody, free calicheamicin derivative, and admixtures of the two) were essentially inactive. The second was a conjugate of the anti-PEM (MUC1) antibody CT-M-01. This conjugate, CMB-401, was curative of various solid tumor xenografts.[124] In addition, CMB-401 was superior to the corresponding carbohydrate-based conjugate against multidrug resistant (MDR)-resistant cell lines.[125]

Three examples of "hybrid" calicheamicin conjugates have been reported. One is Mylotarg (formerly CMA-676), the first Food and Drug Administration–approved antibody-targeted chemotherapeutic agent that recognizes the CD33 antigen expressed on differentiated myeloid cells and is used to treat AML.[126,127] N-acetyl-calicheamicin γ_1^I was converted to the disulfide derivative of α,α-dimethylpropionyl hydrazide, and this in turn was reacted with 4-(4-acetylphenoxy)-butyric acid (AcBut). Activation of the carboxylic acid as a hydroxysuccinimide ester followed by conjugation to the antibody hP67.6 produced Mylotarg. The product, containing the AcBut "hybrid" linker, is relatively stable in circulation but releases the calicheamicin derivative in the acidic environment of the lysosomes.[128]

In vitro, Mylotarg was superior to the corresponding amide conjugate versus HL-60 promyelocytic leukemia cells[125] and was also an improvement over the corresponding carbohydrate conjugate.[128] It showed an IC_{50} in cell culture against HL-60 cells of 0.46 pg of calicheamicin equivalents per milliliter, was 78,000-fold selective versus control cells that were equally sensitive to N-acetyl calicheamicin γ_1^I, and was curative of HL-60 xenografts in mice over a wide dose range. Clinical support studies have shown that sensitivity to Mylotarg is cell cycle dependent and that some internalization can occur through nonspecific endocytosis.[129] Such studies have also largely confirmed the intended mechanism of CD33 targeting and internalization.[130]

The other two "hybrid" calicheamicin conjugates differ primarily in the antibody used to produce the constructs. CMC-544 targets the CD22 lymphoid antigen, makes use of the G544 antibody, and is to B-cell lymphomas what Mylotarg is to leukemia.[131] It had an IC_{50} of 10–500 pM against a panel of lymphoma cell lines and was 20-fold more potent than a control conjugate made from a nonbinding antibody. It produced long-term, tumor-free survivors in various xenograft models and was active in a disseminated model of B-cell lymphoma. The other conjugate targets the Lewis^Y antigen and uses the antibody designated Hu3S193.[132] This conjugate was more efficacious in several different solid-tumor xenograft models (gastric, colon, and prostate) than a control conjugate, producing cures in the gastric model.

Four conjugates of calicheamicin θ have been reported that target three different antigens. This enediyne was also linked to antibodies through a disulfide-containing spacer, but attachment was through the amino sugar residue, which was first modified with N-hydroxysuccinimido-3-(2-pyridyldithio)propionate. This activated intermediate was then reacted with antibodies that had been modified with iminothiolane to produce the conjugates.[133] Two of these conjugates targeted ganglioside GD_2 expressed on neuroblastomas.[134] One was produced using the chimeric CH 14.18 antibody, where IC_{50} values against three different cell lines ranged between 0.1 and 6 nM. The second, more thoroughly studied, conjugate was prepared from antibody 14G2a.[133] This conjugate exhibited an IC_{50} against NXS2 cells of around 0.1 pM, or about 100-fold less cytotoxic than calicheamicin θ itself, and had some activity in a model of neuroblastoma metastatic to the liver. No data were reported for nontargeting conjugates or control cell lines.

An anti-CD19 conjugate of calicheamicin θ has also been described.[135] The IC_{50} was between 1 and 100 pM and tripled the life span of mice implanted with the BCR/Abl-positive pre-B cell leukemia Nalm-6. Last, an antirenal gamma-glutamyl transferase conjugate made using antibody 138H11 had an IC_{50} of 50 pM against Caki-1 cells, with 40-fold selectivity over a control conjugate and a control cell line.[136] Suppression of tumor growth for at least 3 week was seen in a xenograft model of Caki-1 cells at the maximum tolerated dose, where some weight loss was also seen. At a half-maximal dose, a growth delay of 2 wk was seen, whereas no significant activity was observed with a nonbinding control conjugate. Mild reduction of the same antibody cleaved some of the surface-accessible disulfide bonds to generate thiols. A calicheamicin θ conjugate has also been constructed through these thiols, showing an IC_{50} of 16 nM against Caki-2 cells.[137]

Numerous publications describe neocarzinostatin conjugates, all prepared with the chromophore inside the apoprotein, using the antiadenocarcinoma antibody A7. These have included not only the whole antibody as the targeting unit but also smaller antigen-recognizing fragments such as F(ab)$_2$s and Fabs, as well. All such constructs have been prepared using traditional N-hydroxysuccinimido-3-(2-pyridyldithio)propionate (SPDP)–based cross-linking strategies. Whole-antibody conjugates of A7 and neocarzinostatin were active in a xenograft model of pancreatic cancer when injected either intravenously or intratumorally.[138] The same conjugate also showed activity against gastric cell lines.[139] *In vitro*, the conjugate was 10-fold more active against antigen positive MKN45 cells than antigen negative MKN1 control cells with an IC$_{50}$ of 70 ng/mL in neocarzinostatin equivalents. Tumor nodule formation was inhibited by 90% in a disseminated intraperitoneal model. However, the effect was not significantly different than that seen with neocarzinostatin alone, although the diminished toxicity data indicated a wider therapeutic window for the conjugate. This was confirmed in a subsequent report where the MTD was improved by about fivefold, with the conjugate being less toxic, even though the significantly better pharmacokinetic properties would be expected to increase the total exposure to neocarzinostatin.[140] More than twice as much antibody A7 and its conjugate localized to xenografts compared to a control antibody, yet both localized to an antigen negative tumor in equal amounts.

A neocarzinostatin conjugate prepared from Fab fragments of A7 inhibited colony formation of antigen positive colon[141] and pancreatic[142] cells in an antigen-specific manner, but the specificity was only about twofold. Modest antigen-specific activity was also seen in xenografts of colorectal cancer,[143] but complete, antigen-dependent tumor suppression was seen in a pancreatic xenograft model[144] and was greater than the response seen with a whole-antibody conjugate.[145]

The tumor-targeting ability of the Fab construct has been studied relative to whole antibody and a F(ab)$_2$ construct using ^{125}I radiolabeling.[146–148] The Fab conjugate targeted the tumor faster, and more radiolabel was associated with the implant at the earlier time points. Of the three, whole antibody showed the least amount of tumor targeting at 2 h,[148] so even though Fab fragments are usually considered to be cleared too rapidly for drug targeting, this may be appropriate for neocarzinostatin, which itself has a relatively short half-life in circulation. When this Fab conjugate (chFabA7-neocarzinostatin) was administered to cancer patients, kidney toxicity was observed,[141,143] probably because of the rapid clearance in general of Fab fragments through the kidneys.

In an approach involving targeting through an avidin–biotin system, both neocarzinostatin and an anti-CEA antibody were first biotinylated. Cells were treated with the modified antibody, followed by avidin, and finally the modified neocarzinostatin.[149,150] The multivalent avidin protein binds tightly to both biotinylated molecules stepwise in sandwich fashion. When colon adenocarcinoma cells were treated for 72 h, no difference was seen with or without the biotinylated antibody. However, when the cells were washed 7 min after the final exposure, a fivefold increase was seen in the cytotoxicity with the biotinylated antibody compared to controls.[151]

TES-23 is an antibody raised against rat tumor–derived endothelial cells.[152] As such, it targets tumor vasculature in human xenografts about 20 times more effectively than a control antibody. A neocarzinostatin conjugate was active in suppressing KMT-17 fibrosarcoma tumors in rats (the same tumor used for making the antibody) at 17 µg/kg, injected three times.[152] Tumor hemorrhagic necrosis was observed. Less toxicity was seen with the conjugate, even though the activity was not significantly different from neocarzinostatin itself. Antigen-specific distribution of the conjugate to KMT-17 rat and Sarcoma-180 murine tumors was demonstrated,[153] as well as to Meth-A and Colon 26 murine tumors.[154] The conjugate was active against Meth-A tumors implanted subcutaneously in mice with three out of four survivors in a group treated with 50 µg/kg of neocarzinostatin equivalents.[154]

Four other antibodies have been conjugated to neocarzinostatin.[155] These were prepared using thiol–maleimide coupling to give a monosulfide linkage instead of the disulfides produced by SPDP–based couplings. The antibodies used were 791T/36 (anti-gp72, targeting osteosarcoma and colorectal carcinoma), BW 431/26 (CEA, targeting colon and stomach cancer), SWA11 (anti-SCLC-Ag cluster w4/gp45, targeting small cell lung cancer), and OV-TL3 (anti-OA3/p85, targeting ovarian carcinoma). The *in vitro* selectivity of the first three conjugates relative to antigen-negative cells was low, whereas the last one was about eight-fold selective over background. The potency of these conjugates ranged from 16 to 5100 ng/mL.

Neocarzinostatin has also been conjugated to transferrin, in this case, through a monosulfide between dithiolane and a maleimide.[156] An increase in cytotoxicity was seen compared to neocarzinostatin alone; however, additional experiments with modified incubation conditions and with excess transferrin indicated that the effects were not caused by internalization of the conjugate but by neocarzinostatin that had disassociated from the conjugate before cell entry.

VIII. CLINICAL EXPERIENCE

Esperamicin A1 has been examined in phase I and II clinical trials for the treatment of advanced adenocarcinoma of the pancreas.[157–159] No clearly demonstrable cases of response were seen in the 14 patients treated. Nausea/vomiting, malaise/fatigue, thrombocytopenia, and hepatic toxicity were the most common side effects.

Neither calicheamicin γ_1^1 itself nor other members of this family have been examined in the clinic. Instead, all the clinical experience regarding this enediyne has been based on antibody-targeted forms. Clinical data are available for two such conjugates. CMB-401, which targets the PEM (MUC-1) antigen with the hCT-M-01 antibody, has been evaluated through phase II clinical trials. An imaging study was conducted first with ^{111}In labeled antibody to confirm tumor targeting in ovarian cancer.[160] A 7.6-fold increase of radioactivity in tumor versus normal tissues was seen at 0.1 mg/kg protein dose, and a 14-fold increase was seen at 1 mg/kg. In a phase I trial of CMB-401, 34 patients were pretreated with 35 mg/m^2 of unmodified hCT-M-01 antibody to clear circulating antigen, followed by doses of the calicheamicin conjugate ranging from 2 to 16 mg/m^2.[161] The maximum tolerated dose was defined by malaise, hematological toxicities, and gastrointestinal hemorrhage. Four patients had >50% decrease in CA125 levels, and three patients showed reduction in tumor bulk. In this trial, the half-life of the conjugate was estimated to be 36 h, and up to four cycles of treatment led to no allergic reactions with only three increases in antibody titer.[162]

Ovarian cancer patients who had relapsed after standard chemotherapy were treated with 35 mg/m^2 of native antibody followed by 16 mg/m^2 of CMB-401 in a phase II trial.[163] Of the 19 evaluable patients, only four had >50% decreases in CA125 levels, and there were no objective responses. Two reasons cited for possible failure were complications arising from the attempts to clear circulating antigen and from use of an amide-based linking system for calicheamicin, relying on only the reductive mechanism of drug release. There are no data given to support either of these hypotheses, and numerous other possibilities certainly exist.

Mylotarg (gemtuzumab ozogamicin) is the first antibody-targeted chemotherapeutic agent registered for humans. It was approved by the Food and Drug Administration in May 2000 for use in relapsed AML in patients over the age of 60 yr who no longer respond to conventional chemotherapy.[126,127] Approval was based on phase II data, in which about 30% of patients achieved a complete remission. Myelosuppression was the dose-limiting toxicity, along with transient, reversible liver effects. A common infusional syndrome seen with most antibody therapies was also observed. Mylotarg is significantly less toxic than traditional chemotherapy, which has resulted in a savings regarding hospitalization time and the associated costs.[164,165] However, a low incidence of veno-occlusive disease, an unusual and sometimes fatal toxicity, has been seen with Mylotarg. This may be CD33 antigen related.[166] The occurrence of veno-occlusive disease appears to depend on the history of the patient as well as other coadministered therapies.[167-169] Reviews of Mylotarg efficacy and toxicities have recently been published.[170,171]

The lack of a universal response to Mylotarg in AML may have various causes. Some possibilities include P-glycoprotein expression,[172–174] other efflux pumps,[175] variations in CD33 levels,[176–178] and in apoptosis.[179]

Although AML most commonly affects the elderly, Mylotarg is also being examined for pediatric AML as well as acute lymphoblastic leukemia.[180,181] The drug has been used before hematopoietic stem cell transplant[182] and appears to be useful in relapse after hematopoietic stem cell transplant.[183,184] However, the occurrence of veno-occlusive disease may be increased if Mylotarg is used pre–hematopoietic stem cell transplant [185] or if sufficient time has not elapsed since the stem cell transplant.[186] Mylotarg has also been evaluated as a treatment for myelodysplastic syndrome, but with less than encouraging results so far.[187] However, there is an indication of benefit when it is used in CD33-positive acute lymphoblastic leukemia patients.[188]

Mylotarg has been evaluated in combinations with various high-dose regimens involving traditional anticancer agents. These include cytarabine,[189,190] all-trans retinoic acid,[191] IL-11,[192,193] topotecan/Ara-C,[194,195] fludarabine/Ara-C/G-CSF/idarubicin,[196] troxatyl,[197] daunorubicin/Ara-C with or without thioguanidine,[198] Ara-C/mitoxantrone/amifostine,[199] daunorubicin/Ara-C/thioguanidine/mitoxantrone/etoposide,[169] mitoxantrone/Ara-C/etoposide,[200] imatinib,[201] fludarabine/Ara-C/cyclosporin,[202] Ara-C/idarubicin,[203,204] and daunorubicin/Ara-C/cyclosporin.[205] Although most of the data on these combinations are preliminary, these studies have the potential to increase the usefulness of Mylotarg.

Neocarzinostatin has been examined in the clinic both by itself and as several forms of polymer-based conjugates. The natural product is approved for a number of cancers in Japan, but not elsewhere. These cancers include stomach, pancreas, liver, bladder, and brain, as well as leukemia. The main side effect with neocarzinostatin is myelosuppression. More recently it failed to outperform intravesical Bacille Calmette-Guerin therapy for recurrent superficial bladder cancer in combination with cytarabine.[206] However, it has shown some promise in refractory or recurrent AML in combination with enocitabine and aclarubicin, with 57% of patients achieving complete remission.[207]

Conjugates of neocarzinostatin and the anticolon antibody A7 have been in clinical trials. Patients given between 15 and 90 mg of whole-antibody conjugate showed no adverse effects; three out of eight patients with liver metastases secondary to colon cancer showed a decrease in tumor size, and three experienced pain relief, but this treatment had no effect on other metastases.[208] However, most of these patients mounted a human anti-mouse response against the murine antibody. Experiments with sera from these patients showed that the human anti-mouse response blocked conjugate cytotoxicity.[209]

Styrene maleic acid copolymer neocarzinostatin conjugate is approved in Japan for patients with liver cancer or liver metastases and is being explored for cancers of the kidney, lung, and pancreas. It is coadministered as a mixture with Lipiodol, a heavily iodinated poppy seed oil. Recent reviews covering the discovery, development, and clinical use of this therapy are available.[210–213] A somewhat related approach that has not progressed to clinical trials yet is the use of chitin derivatives to deliver neocarzinostatin.[214]

IX. SUMMARY

The isolation and preliminary biological activity of neocarzinostatin as an anticancer agent was first reported in 1965.[215] During subsequent years, the characterization and x-ray crystal structure of this natural product were reported, and the mechanism of DNA cleavage was elucidated. A variety of conjugates were prepared, including styrene maleic acid copolymer neocarzinostatin conjugate as well as various antibody-based constructs, in an effort to overcome the short biological half-life of this compound. Clinical trials were initiated with neocarzinostatin, in Japan as well as in the United States, yet these studies were all done with the conviction that neocarzinostatin was, in fact, an anticancer protein. It was not until 1979 that the protein was shown to be associated with a much smaller chromophoric subunit that was responsible for the observed biological activity,[216] and it took another 6 yr before an enediyne macrocyclic structure was assigned to the chromophore, though without much of the stereochemistry.[20] The last chapter relating to the structure of the molecule was closed in 1998, with the report of the total synthesis of the neocarzinostatin chromophore.[38]

In the meantime, similarities between the neocarzinostatin apoprotein and several other "anticancer proteins" soon led to work that showed that some of these sequestered enediynes as well. However, these early enediynes were soon upstaged by the more stable 10-member ring analogs: isolation and biological activity for esperamicin was published in 1985[217] and its structure in 1987,[11] along with that of calicheamicin γ_1^I the same year.[13] The complex and extremely novel structures of these compounds captured the attention and resources of many major academic total synthesis groups in a race to be first to construct these molecules. Biochemists and biologists helped elucidate the activity and the mechanisms of action for these compounds. As more members of this family were isolated and their structures were assigned, the mechanisms and pathways by which microorganisms assembled these molecules became an important and ongoing challenge.

What remained, however, was to make the enediynes into useful drugs. In spite of extreme potency, their physical characteristics, their short half-lives, and their small therapeutic windows precluded using most of these natural products as drugs themselves. Neocarzinostatin and calicheamicin γ_1^I were eventually converted to anticancer agents useful in humans via conjugation to a polymer and an antibody, respectively. We have tried to demonstrate in this review that the research into enediynes continues in an effort to understand the biology of these molecules better and thereby expand their clinical applications.

REFERENCES

1. *Enediyne Antibiotics as Antitumor Agents*, Borders, D.B. and Doyle, T.W., Eds., Marcel Dekker, New York, 1995.
2. Lee, M.D., Ellestad, G.A., and Borders, D.B., Calicheamicins: Discovery, structure, chemistry, and interaction with DNA, *Acc. Chem. Res.*, 24, 235, 1991.
3. Nicolaou, K.C. and Dai, W.M., Chemistry and biology of the enediyne anticancer antibiotics, *Angew. Chem., Int. Ed. Engl.*, 30, 1387, 1991.
4. Smith, A.L. and Nicolaou, K.C., The enediyne antibiotics, *J. Med. Chem.*, 39, 2103, 1996.
5. Nicolaou, K.C., Sorensen, E.J., and Winssinger, N., The art and science of organic and natural products synthesis, *J. Chem. Educ.*, 75, 1225, 1998.
6. Xi, Z. and Goldberg, I.H., DNA-damaging enediyne compounds, in *Comprehensive Natural Products Chemistry*, Barton, D.S., Nakanishi, K., and Meth-Cohn, O., Eds., Pergamon Press, Oxford, UK, 1999, 7, 553.
7. Jones, G.B. and Fouad, F.S., Designed enediyne antitumor agents, *Curr. Pharmaceut. Design*, 8, 2415, 2002.
8. Shen, B., Liu, W., and Nonaka, K., Enediyne natural products: Biosynthesis and prospect towards engineering novel antitumor agents, *Curr. Med. Chem.*, 10, 2317, 2003.
9. *Neocarzinostatin: The Past, Present and Future of an Anticancer Drug*, Maeda, H., Edo, K., and Ishida, N., Eds., Springer, New York, 1997.
10. Golik, J. et al., Esperamicins, a novel class of potent antitumor antibiotics. 2. Structure of esperamicin X., *J. Am. Chem. Soc.*, 109, 3461, 1987.
11. Golik, J. et al., Esperamicins, a novel class of potent antitumor antibiotics. 3. Structures of esperamicins A1, A2, and A1b, *J. Am. Chem. Soc.*, 109, 3462, 1987.
12. Lee, M.D. et al., Calichemicins, a novel family of antitumor antibiotics. 1. Chemistry and partial structure of calichemicin γ_1^I, *J. Am. Chem. Soc.*, 109, 3464, 1987.
13. Lee, M.D. et al., Calichemicin, a novel family of antitumor antibiotics. 2. Chemistry and structure of calichemicin γ_1^I, *J. Am. Chem. Soc.*, 109, 3466, 1987.
14. Zein, N. et al., Calichemicin γ_1^I: An antitumor antibiotic that cleaves double-stranded DNA site specifically, *Science*, 240, 1198, 1988.
15. Lee, M.D. et al., Calicheamicins, a novel family of antitumor antibiotics. 4. Structure elucidation of calicheamicins β_1^{Br}, γ_1^{Br}, α_2^I, α_3^I, β_1^I, γ_1^I, and δ_1^I, *J. Am. Chem. Soc.*, 114, 985, 1992.

16. Golik, J.A. et al., *Antitumor Antibiotics*, U.S. patent 5,086,045, 1992.

17. White, R.J., Maiese, W.M., and Greenstein, M., in *Manual of Industrial Microbiology and Biotechnology*, Demain, A.L. and Solomon, N.A., Eds., American Society of Microbiology, Washington, DC, 1986, 24.

18. Lee, M.D. et al., Calichemicins, a novel family of antitumor antibiotics. 3. Isolation, purification and characterization of calicheamicins β_1^{Br}, γ_1^{Br}, α_2^I, α_3^I, β_1^I, γ_1^I, and δ_1^I, *J. Antibiot.*, 42, 1070, 1989.

19. Maiese, W.M. et al., Calichemicins, a novel family of antitumor antibiotics: Taxonomy, fermentation and biological properties, *J. Antibiot.*, 42, 558, 1989.

20. Edo, K. et al., The structure of neocarzinostatin chromophore possessing a novel bicyclo-7.3.0-dodecadiyne system, *Tetrahedron Lett.*, 26, 331, 1985.

21. Lam, K.S. and Forenza, S., Fermentation and isolation of esperamicins, in *Enediyne Antibiotics as Antitumor Agents*, Borders, D.B. and Doyle, T.W., Eds., Marcel Dekker, New York, 1995, chap. 10.

22. Ando, T. et al., A new non-protein enediyne antibiotic N1999A2: Unique enediyne chromophore similar to neocarzinostatin and DNA cleavage feature, *Tetrahedron Lett.*, 39, 6495, 1998.

23. Takashima, H., Amiya, S., and Kobayashi, Y., Neocarzinostatin: Interaction between the antitumor-active chromophore and the carrier protein, *J. Biochem.*, 109, 807, 1991.

24. Hofstead, S.J. et al., Kedarcidin, a new chromoprotein antitumor antibiotic. II. Isolation, purification and physico-chemical properties, *J. Antibiot.*, 45, 1250, 1992.

25. Kim, K.H. et al., Crystal structure of neocarzinostatin, an antitumor protein-chromophore complex, *Science*, 262, 1042, 1993.

26. Okuno, Y., Otsuka, M., and Sugiura, Y., Computer modeling analysis for enediyne chromophore-apoprotein complex of macromolecular antitumor antibiotic C-1027, *J. Med. Chem.*, 37, 2266, 1994.

27. Teplyakov, A. et al., Crystal structure of apo-neocarzinostatin at 0.15-nm resolution, *Eur. J. Biochem.*, 213, 737, 1993.

28. Chin, D.H., Rejection by neocarzinostatin protein through charges rather than sizes, *Chem.-Eur. J.*, 5, 1084, 1999.

29. Usuki, T. et al., Rational design of supra C-1027: Kinetically stabilized analog of the antitumor enediyne chromoprotein, *J. Am. Chem. Soc.*, 126, 3022, 2004.

30. Zein, N. et al., Selective proteolytic activity of the antitumor agent kedarcidin, *Proc. Natl. Acad. Sci. U.S.A.*, 90, 8009, 1993.

31. Sakata, N. et al., Aminopeptidase activity of an antitumor antibiotic, C-1027, *J. Antibiot.*, 45, 113, 1992.

32. Nicolaou, K.C. et al., Total synthesis of calicheamicin γ_1^I: 3. The final stages, *J. Am. Chem. Soc.*, 115, 7625, 1993.

33. Hitchcock, S.A. et al., A convergent total synthesis of calicheamicin γ_1^I, *Angew. Chem., Int. Ed. Engl.*, 33, 858, 1994.

34. Nicolaou, K.C. et al., Total synthesis of calicheamicin γ_1^I, *J. Am. Chem. Soc.*, 114, 10082, 1992.

35. Leet, J.E. et al., Kedarcidin, a new chromoprotein antitumor antibiotic: Structure elucidation of kedarcidin chromophore, *J. Am. Chem. Soc.*, 114, 7946, 1992.

36. Leet, J.E. et al., Kedarcidin chromophore: Structure elucidation of the amino sugar kedarosamine, *Tetrahedron Lett.*, 33, 6107, 1992.

37. Kawata, S. and Hirama, M., Synthetic study of the kedarcidin chromophore: Efficient construction of the aryl alkyl ether linkage, *Tetrahedron Lett.*, 39, 8707, 1998.

38. Myers, A.G. et al., Total synthesis of (+)-neocarzinostatin chromophore, *J. Am. Chem. Soc.*, 120, 5319, 1998.

39. Kobayashi, S. et al., The first total synthesis of N1999-A2: Absolute stereochemistry and stereochemical implications into DNA cleavage, *J. Am. Chem. Soc.*, 123, 11294, 2001.

40. Kobayashi, S. et al., Investigation of the total synthesis of N1999-A2: Implication of stereochemistry, *J. Am. Chem. Soc.*, 123, 2887, 2001.

41. Konishi, M. et al., Crystal and molecular structure of dynemicin A: A novel 1,5-diyn-3-ene antitumor antibiotic, *J. Am. Chem. Soc.*, 112, 3715, 1990.

42. Myers, A.G. et al., A convergent synthetic route to (+)-dynemicin A and analogs of wide structural variability, *J. Am. Chem. Soc.*, 119, 6072, 1997.

43. Shair, M.D. et al., The total synthesis of dynemicin A leading to development of a fully contained bioreductively activated enediyne prodrug, *J. Am. Chem. Soc.*, 118, 9509, 1996.

44. Yoshida, K.-I. et al., Structure and cycloaromatization of a novel enediyne, C-1027 chromophore, *Tetrahedron Lett.*, 34, 2637, 1993.
45. Iida, K.-I. et al., Synthesis and absolute stereochemistry of the aminosugar moiety of antibiotic C-1027 chromophore, *Tetrahedron Lett.*, 34, 4079, 1993.
46. Schroeder, D.R. et al., Isolation, structure determination and proposed mechanism of action for artifacts of maduropeptin chromophore, *J. Am. Chem. Soc.*, 116, 9351, 1994.
47. McDonald, L.A. et al., Namenamicin, a new enediyne antitumor antibiotic from the marine ascidian *Polysyncraton lithostrotum*, *J. Am. Chem. Soc.*, 118, 10898, 1996.
48. Oku, N., Matsunaga, S., and Fusetani, N., Shishijimicins A-C, novel enediyne antitumor antibiotics from the ascidian *Didemnum proliferum*, *J. Am. Chem. Soc.*, 125, 2044, 2003.
49. Samy, T.S. et al., Primary structure of macromomycin, an antitumor antibiotic protein, *J. Biol. Chem.*, 258, 183, 1983.
50. Suzuki, H. et al., Biological activities of non-protein chromophores of antitumor protein antibiotics: Auromomycin and neocarzinostatin, *Biochem. Biophys. Res. Commun.*, 94, 255, 1980.
51. Doyle, T.W. and Borders, D.B., Enediyne antitumor antibiotics, in *Enediyne Antibiotics as Antitumor Agents*, Borders, D.B. and Doyle, T.W., Eds., Marcel Dekker, New York, 1995, chap. 1.
52. Fantini, A.A. and Testa, R.T., Taxonomy, fermentation and yield improvement, in *Enediyne Antibiotics as Antitumor Agents*, Borders, D.B. and Doyle, T.W., Eds., Marcel Dekker, New York, 1995, chap. 3.
53. Hensens, O.D., Giner, J.-L., and Goldberg, I.H., Biosynthesis of NCS Chrom A, the chromophore of the antitumor antibiotic neocarzinostatin, *J. Am. Chem. Soc.*, 111, 3295, 1989.
54. Lam, K.S. et al., Biosynthesis of esperamicin A1, an enediyne antitumor antibiotic, *J. Am. Chem. Soc.*, 115, 12340, 1993.
55. Lam, K.S. et al., Inhibition of esperamicin A1 synthesis by cerulenin, *Abstr. Annu. Meet. Am. Soc. Microbiol.*, 90, 266, 1990.
56. Tokiwa, Y. et al., Biosynthesis of dynemicin A, a 3-ene-1,5-diyne antitumor antibiotic, *J. Am. Chem. Soc.*, 114, 4107, 1992.
57. Ahlert, J. et al., The calicheamicin gene cluster and its iterative type I enediyne PKS, *Science*, 297, 1173, 2002.
58. Liu, W. et al., Biosynthesis of the enediyne antitumor antibiotic C-1027, *Science*, 297, 1170, 2002.
59. Wen, L. et al., Rapid PCR amplification of minimal enediyne polyketide synthase cassettes leads to a predictive familial classification model, *Proc. Natl. Acad. Sci. U. S. A.*, 100, 11959, 2003.
60. Thorson, J.S. et al., Understanding and exploiting nature's chemical arsenal: The past, present and future of calicheamicin research, *Curr. Pharmaceut. Design*, 6, 1841, 2000.
61. Christenson, S.D. et al., A novel 4-methylideneimidazole-5-one-containing tyrosine aminomutase in enediyne antitumor antibiotic C-1027 biosynthesis, *J. Am. Chem. Soc.*, 125, 6062, 2003.
62. Liu, W. and Shen, B., Genes for production of the enediyne antitumor antibiotic C-1027 in *Streptomyces lobisporus* are clustered with the cagA gene that encodes the C-1027 apoprotein, *Antimicrob. Agents Chemother.*, 44, 382, 2000.
63. Biggins, J.B., Onwueme, K., and Thorson, J.S., Resistance to enediyne antitumor antibiotics by CalC self-sacrifice, *Science*, 301, 1537, 2003.
64. Kudo, K. et al., Production of a free chromophore component of neocarzinostatin (NCS) in the culture filtrate of *Streptomyces carzinostaticus* var. F-41, *J. Antibiot.*, 35, 1111, 1982.
65. Bergman, R.G., Reactive 1,4-dehydroaromatics, *Acc. Chem. Res.*, 6, 25, 1973.
66. Chen, W.C., Chang, N.Y., and Yu, C.H., Density functional study of Bergman cyclization of enediynes, *J. Phys. Chem. A*, 102, 2584, 1998.
67. Schreiner, P.R., Cyclic enediynes — relationship between ring size, alkyne carbon distance, and cyclization barrier, *Chem. Comm.*, 483, 1998.
68. Biswas, K. et al., The molecular basis for pyrimidine-selective DNA binding: Analysis of calicheamicin oligosaccharide derivatives by capillary electrophoresis, *J. Am. Chem. Soc.*, 122, 8413, 2000.
69. Walker, S.L., Andreotti, A.H., and Kahne, D.E., NMR characterization of calicheamicin bound to DNA, *Tetrahedron*, 50, 1351, 1994.
70. Walker, S. et al., Analysis of hydroxylamine glycosidic linkages: Structural consequences of the NO bond in calicheamicin, *J. Am. Chem. Soc.*, 116, 3197, 1994.
71. Paloma, L.G. et al., Interaction of calicheamicin with duplex DNA: Role of the oligosaccharide domain and identification of multiple binding modes, *J. Am. Chem. Soc.*, 116, 3697, 1994.

72. Ikemoto, N. et al., Calicheamicin-DNA complexes: Warhead alignment and saccharide recognition of the minor groove, *Proc. Natl. Acad. Sci. U.S.A.*, 92, 10506, 1995.

73. Miyagawa, N. et al., DNA cleavage characteristics of non-protein enediyne antibiotic N1999-2A, *Biochem. Biophys. Res. Commun.*, 306, 87, 2003.

74. De Voss, J.J., Hangeland, J.J., and Townsend, C.A., Characterization of the in-vitro cyclization chemistry of calicheamicin and its relation to DNA cleavage, *J. Am. Chem. Soc.*, 112, 4554, 1990.

75. Zein, N. et al., Exclusive abstraction of nonexchangeable hydrogens from DNA by calicheamicin γ_1^I, *J. Am. Chem. Soc.*, 111, 6888, 1989.

76. De Voss, J.J. et al., Site-specific atom transfer from DNA to a bound ligand defines the geometry of a DNA-calicheamicin γ_1^I complex, *J. Am. Chem. Soc.*, 112, 9669, 1990.

77. Kumar, R.A., Ikemoto, N., and Patel, D.J., Solution structure of the calicheamicin γ_1^I-DNA complex, *J. Mol. Biol.*, 265, 187, 1997.

78. Sugiura, Y. et al., Reductive and nucleophilic activation products of dynemicin A with methyl thioglycolate. A rational mechanism for DNA cleavage of the thiol-activated dynemicin A, *Biochemistry*, 30, 2989, 1991.

79. Nicolaou, K.C. et al., Molecular design, chemical synthesis, kinetic studies, calculations, and biological studies of novel enediynes equipped with triggering, detection, and deactivating devices: Model dynemicin A epoxide and cis-diol systems, *J. Am. Chem. Soc.*, 115, 7944, 1993.

80. Cosgrove, J.P., and Dedon, P.C., Binding and reaction of calicheamicin and other enediyne antibiotics with DNA, in *Small Molecule DNA and RNA Binders*, Demeunynck, M., Bailly, C., and Wilson, W.D., Eds., Wiley-VCH, Weinheim, Germany, 2003, 2, 609.

81. Xu, Y.-J., Zhen, Y.-S., and Goldberg, I.H., C1027 chromophore, a potent new enediyne antitumor antibiotics, induces sequence-specific double-strand DNA cleavage, *Biochemistry*, 33, 5947, 1994.

82. Xu, Y.J., Zhen, Y.S., and Goldberg, I.H., Enediyne C1027 induces the formation of novel covalent DNA interstrand cross-links and monoadducts, *J. Am. Chem. Soc.*, 119, 1133, 1997.

83. Cobuzzi, R.J., Jr. et al., Effects of the enediyne C-1027 on intracellular DNA targets, *Biochemistry*, 34, 583, 1995.

84. Elmroth, K. et al., Cleavage of cellular DNA by calicheamicin γ_1^I, *DNA Repair*, 2, 363, 2003.

85. Susanne, M. et al., Activation of the DNA-dependent protein kinase by drug-induced and radiation-induced DNA strand breaks, *Radiation Research*, 160, 291, 2003.

86. Jiang, B., Li, D.D., and Zhen, Y.S., Induction of apoptosis by enediyne antitumor antibiotic C1027 in HL-60 human promyelocytic leukemia cells, *Biochem. Biophys. Res. Commun.*, 208, 238, 1995.

87. He, Q. and Li, D., Interaction of direct breaking DNA strands and initiating of apoptotic pathway contributed to high potent cytotoxicities of enediyne antibiotic lidamycin (C1027) to human tumor cells, *Proc. Am. Assoc. Cancer Res.*, 42, 643, 2001.

88. Wang, Z. et al., Non-caspase-mediated apoptosis contributes to the potent cytotoxicity of the enediyne antibiotic lidamycin toward human tumor cells, *Biochem. Pharmacol.*, 65, 1767, 2003.

89. Dziegielewski, J. and Beerman, T.A., C-1027-induced activation of DNA-damage responses is ATM-independent, *Proc. Am. Assoc. Cancer Res.*, 42, 199, 2001.

90. Amico, D. et al., Differential response of human acute myeloid leukemia cells to gemtuzumab ozogamicin *in vitro*: Role of Chk1 and Chk2 phosphorylation and caspase 3, *Blood*, 101, 4589, 2003.

91. McHugh, M.M. et al., The cellular response to DNA damage induced by the enediynes C-1027 and neocarzinostatin includes hyperphosphorylation and increased nuclear retention of replication protein A (RPA) and trans inhibition of DNA replication, *Biochemistry*, 40, 4792, 2001.

92. Hiatt, A. et al., Regulation of apoptosis in leukemic cells by analogs of dynemicin A, *Bioorg. Med. Chem.*, 2, 315, 1994.

93. Hartsell, T.L. et al., Determinants of the response of neuroblastoma cells to DNA damage: The roles of pre-treatment cell morphology and chemical nature of the damage, *J. Pharmacol. Exp. Ther.*, 277, 1158, 1996.

94. Liang, Y. et al., Early events in Bcl-2-enhanced apoptosis, *Apoptosis*, 8, 609, 2003.

95. Nicolaou, K.C. et al., Calicheamicin θ_1^I: A rationally designed molecule with extremely potent and selective DNA cleaving properties and apoptosis inducing activity, *Angew. Chem., Int. Ed. Engl.*, 33, 183, 1994.

96. Wrasidlo, W. et al., Modulation of p53 expression in neuroblastoma cell lines by the DNA cleaving agent calicheamicin theta, *Proc. Am. Assoc. Cancer Res.*, 38, 610, 1997.

97. Prokop, A. et al., Induction of apoptosis by enediyne antibiotic calicheamicin θ proceeds through a caspase-mediated mitochondrial amplification loop in an entirely Bax-dependent manner, *Oncogene*, 22, 9107, 2003.

98. Siegel, J. et al., Enhanced p53 activity and accumulation in response to DNA damage upon DNA transfection, *Oncogene*, 11, 1363, 1995.

99. Andreeff, M. et al., Caspase independent cell death in AML: An *in vivo* study in patients undergoing chemotherapy, *Blood*, 98, 208b, 2001.

100. Batchelder, R.M. et al., Oxygen dependence of the cytotoxicity of the enediyne anti-tumour antibiotic esperamicin A-1, *Br. J. Cancer*, 74, S52, 1996.

101. Van Duijn-Goedhart, A. et al., Differential responses of Chinese hamster mutagen sensitive cell lines to low and high concentrations of calicheamicin and neocarzinostatin, *Mutat. Res.*, 471, 95, 2000.

102. Noel, G. et al., Poly(ADP-ribose) polymerase (PARP-1) is not involved in DNA double-strand break recovery, *BMC Cell Biology*, http://www.biomedcentral.com/content/pdf/1471-2121-4-7.pdf, 2003.

103. Jedema, I. et al., Sensitivity of acute myeloid leukemia cells to anti-CD33-calicheamicin (gemtuzumab ozogamicin) is cell cycle-dependent and can be increased by interferon gamma, *Blood*, 98, 120a (abst. 503), 2001.

104. Cortazzo, M. and Schor, N.F., Potentiation of enediyne-induced apoptosis and differentiation by Bcl-2, *Cancer Res.*, 56, 1199, 1996.

105. Kappen, L.S. and Goldberg, I.H., Activation and inactivation of neocarzinostatin-induced cleavage of DNA, *Nucleic Acids Res.*, 5, 2959, 1978.

106. Schor, N.F. et al., The use of 6-mercaptodopamine to target neuroblastoma cells for enediyne-induced apoptosis and antineoplastic effect, *Proc. Am. Assoc. Cancer Res.*, 35, 316, 1994.

107. Maier, M.E., Bosse, F., and Niestroj, A.J., Design and synthesis of dynemicin analogs, *Eur. J. Org. Chem.*, 1, 1999.

108. Zhen, Y.S. et al., Lidamycin and its RGD-containing peptide conjugate inhibit angiogenesis and metastasis, *Proc. Am. Assoc. Cancer Res.*, 41, 645 (abst. 4098), 2000.

109. Liu, X.Y. et al., Antitumor effects of novel immunoconjugates with downsized-molecule prepared by linking lidamycin to Fab′ and scFv antibody, *Proc. Am. Assoc. Cancer Res.*, 41, 290 (abst. 1848), 2000.

110. Wang, F., Shang, B., and Zhen, Y., Antitumor effects of the molecule-downsized immunoconjugate composed of lidamycin and Fab′ fragment of monoclonal antibody directed against type IV collagenase, *Sci. China, Ser. C Life Sci.*, 47, 66, 2004.

111. Feng-Qiang, W., Bo-Yang, S., and Yong-Su, Z., Antitumor effects of the immunoconjugate composed of lidamycin and monoclonal antibody 3G11, *Acta Pharmaceutica Sinica*, 38, 515, 2003.

112. Zhen, Y.S. et al., Conjugate of antibiotic C1027 and monoclonal antibody 3H11 shows potent antitumor effect against gastric cancer, *Proc. Am. Assoc. Cancer Res.*, 35, 511 (abst. 3047), 1994.

113. Wrasidlo, W. et al., *In vivo* efficacy of novel synthetic enediynes, *Acta Oncol.*, 34, 157, 1995.

114. Hay, M.P., Wilson, W.R., and Denny, W.A., A novel enediyne prodrug for antibody-directed enzyme prodrug therapy (ADEPT) using *E. coli* B nitroreductase, *Bioorg. Med. Chem. Lett.*, 5, 2829, 1995.

115. Hay, M.P., Wilson, W.R., and Denny, W.A., Nitrobenzyl carbamate prodrugs of enediynes for nitroreductase gene-directed enzyme prodrug therapy (GDEPT), *Bioorg. Med. Chem. Lett.*, 9, 3417, 1999.

116. Xie, Y. et al., Enediyne-lexitropsin DNA-targeted anticancer agents. Physicochemical and cytotoxic properties in human neoplastic cells *in vitro*, and intracellular distribution, *Anti-Cancer Drug Des.*, 12, 169, 1997.

117. Severin, S.E. et al., Targeted delivery of esperamicin A1 by using oncofetal protein alpha-fetoprotein, *Eur. J. Cancer*, 33, S89, 1997.

118. Severin, S.E. et al., Alpha-fetoprotein-mediated targeting of anti-cancer drugs to tumor cells *in vitro*, *Biochem. Mol. Biol. Int.*, 37, 385, 1995.

119. Capson, T. and Ruffner, D., Preparation of calicheamicin-oligodeoxynucleotide conjugates for use as antisense therapeutics, *Pharm. Res. (New York)*, 13, S153, 1996.

120. Durr, F.E. et al., Biological activities of calicheamicin, in *Enediyne Antibiotics as Antitumor Agents*, Borders, D.B. and Doyle, T.W., Eds., Marcel Dekker, 1995, chap. 8.

121. Ellestad, G.A. et al., Reactions of the trisulfide moiety in calicheamicin, *Tetrahedron Lett.*, 30, 3033, 1989.

122. Hinman, L.M. et al., Preparation and characterization of monoclonal antibody conjugates of the calicheamicins: A novel and potent family of antitumor antibiotics, *Cancer Res.*, 53, 3336, 1993.

123. Jeglum, K.A. et al., Antitumor effects of a calicheamicin immunoconjugate with anti-canine lymphoma monoclonal antibody 231, *Proc. Am. Assoc. Cancer Res.*, 35, 506 (abstr. 3016), 1994.
124. Hinman, L.M. et al., Lysine-linked conjugates of calicheamicin γ_1^I with a fully humanized anti-polymorphic epithelial mucin antibody show potent antitumor effects in ovarian tumor xenografts, *Proc. Am. Assoc. Cancer Res.*, 34, 479 (Abs 2860), 1993.
125. Hamann, P.R. et al., An anti-CD33 antibody-calicheamicin conjugate for treatment of acute myeloid leukemia. Choice of linker, *Bioconj. Chem.*, 13, 40, 2002.
126. Bross, P.F. et al., Approval summary: Gemtuzumab ozogamicin in relapsed acute myeloid leukemia, *Clin. Cancer Res.*, 7, 1490, 2001.
127. Bross, P.F. et al., Approval summary: Gemtuzumab ozogamicin in relapsed acute myeloid leukemia (correction to previous reference), *Clin. Cancer Res.*, 8, 300, 2002.
128. Hamann, P.R. et al., Gemtuzumab ozogamicin, a potent and selective anti-CD33 antibody-calicheamicin conjugate for treatment of acute myeloid leukemia, *Bioconj. Chem.*, 13, 47, 2002.
129. Jedema, I. et al., Internalization and cell cycle-dependent killing of leukemic cells by gemtuzumab ozogamicin: Rationale for efficacy in CD33-negative malignancies with endocytic capacity, *Leukemia*, 18, 316, 2004.
130. van der Velden, V.H.J., Berger, M.S., and Van Dongen, J.J.M., Mylotarg therapy in acute myeloid leukemia: Mechanism of action and implications for future treatment protocols, *Haematol. Blood Transfusion*, 41, 169, 2003.
131. DiJoseph, J.F. et al., Antibody-targeted chemotherapy with CMC-544: A CD22-targeted immunoconjugate of calicheamicin for the treatment of B-lymphoid malignancies, *Blood*, 103, 1807, 2004.
132. Boghaert, E.R. et al., Antibody-targeted chemotherapy with the calicheamicin conjugate hu3S193-N-acetyl gamma calicheamicin dimethyl hydrazide targets Lewis^y and eliminates Lewis^y-positive human carcinoma cells and xenografts, *Clinical Cancer Research*, 10, 4538, 2004.
133. Lode, H.N. et al., Targeted therapy with a novel enediyne antibiotic calicheamicin θ_1^I effectively suppresses growth and dissemination of liver metastases in a syngeneic model of murine neuroblastoma, *Cancer Res.*, 58, 2925, 1998.
134. Burchardt, C.A. et al., Cytotoxic effects of new immunoconjugates in different neuroblastoma cell-lines, *J. Cancer Res. Clin. Oncol.*, 124, 137, 1998.
135. Bernt, K.M. et al., CD19 targeted calicheamicin theta is effective against high risk ALL, *Blood*, 96, 467a, 2000.
136. Knoll, K. et al., Targeted therapy of experimental renal cell carcinoma with a novel conjugate of monoclonal antibody 138H11 and calicheamicin θ_1^I, *Cancer Res.*, 60, 6089, 2000.
137. Schmidt, C.S. et al., Chemoimmunoconjugates with the monoclonal antibody 138H11 for targeting the cytotoxic prodrug calicheamicin θ to renal cell carcinomas, *Tumor Target.*, 4, 271, 1999.
138. Otsuji, E. et al., The effect of intravenous and intra-tumoural chemotherapy using a monoclonal antibody-drug conjugate in a xenograft model of pancreatic cancer, *Eur. J. Surg. Oncol.*, 21, 61, 1995.
139. Okamoto, K. et al., Targeted chemotherapy in mice with peritoneally disseminated gastric cancer using monoclonal antibody-drug conjugate, *Cancer Lett.*, 122, 231, 1998.
140. Kitamura, K. et al., The role of monoclonal antibody A7 as a drug modifier in cancer therapy, *Cancer Immunol. Immunother.*, 36, 177, 1993.
141. Yamaguchi, T. et al., Production, binding and cytotoxicity of human/mouse chimeric monoclonal antibody-neocarzinostatin conjugate, *Jpn. J. Cancer Res.*, 84, 1190, 1993.
142. Otsuji, E. et al., Antitumor effect of neocarzinostatin conjugated to human/mouse chimeric Fab fragments of the monoclonal antibody A7 on human pancreatic carcinoma, *J. Surg. Oncol.*, 57, 230, 1994.
143. Yamaguchi, T. et al., *In vivo* efficacy of neocarzinostatin coupled with Fab human/mouse chimeric monoclonal antibody A7 against human colorectal cancer, *Jpn. J. Cancer Res.*, 85, 167, 1994.
144. Otsuji, E. et al., Effects of neocarzinostatin-chimeric Fab conjugates on the growth of human pancreatic carcinoma xenografts, *Br. J. Cancer*, 73, 1178, 1996.
145. Matsumura, H. et al., Increased effects of neocarzinostatin-chimeric Fab conjugates on the growth of human pancreatic carcinoma xenografts, *Cancer Biother. Radiopharm.*, 13, 60, 1998.
146. Otsuji, E. et al., Biodistribution of neocarzinostatin conjugated to chimeric Fab fragments of the monoclonal antibody A7 in nude mice bearing human pancreatic cancer xenografts, *Jpn. J. Cancer Res.*, 85, 530, 1994.

147. Yamaguchi, T. et al., Production of human/mouse chimeric monoclonal antibody-neocarzinostatin conjugate and distribution in tumor bearing athymic mice, *Proc. Controlled Release Soc.*, 22, 544, 1995.

148. Otsuji, E. et al., Applicability of monoclonal antibody Fab fragments as a carrier of neocarzinostatin in targeting chemotherapy, *J. Surg. Oncol.*, 61, 149, 1996.

149. Otsuji, E. et al., Distribution of neocarzinostatin conjugated to biotinylated chimeric monoclonal antibody Fab fragments after administration of avidin, *Br. J. Cancer*, 74, 597, 1996.

150. Otsuji, E. et al., Decreased renal accumulation of biotinylated chimeric monoclonal antibody-neocarzinostatin conjugate after administration of avidin, *Jpn. J. Cancer Res.*, 88, 205, 1997.

151. Nakaki, M., Takikawa, H., and Yamanaka, M., Targeting immunotherapy using the avidin-biotin system for a human colon adenocarcinoma *in vitro*, *J. Int. Med. Res.*, 25, 14, 1997.

152. Tsunoda, S. et al., Specific binding of TES-23 antibody to tumour vascular endothelium in mice, rats and human cancer tissue: A novel drug carrier for cancer targeting therapy, *Br. J. Cancer*, 81, 1155, 1999.

153. Makimoto, H. et al., Tumor vascular targeting using a tumor-tissue endothelium-specific monoclonal antibody as an effective strategy for cancer chemotherapy, *Biochem. Biophys. Res. Commun.*, 260, 346, 1999.

154. Wakai, Y. et al., Effective cancer targeting using an anti-tumor tissue vascular endothelium-specific monoclonal antibody (TES-23), *Jpn. J. Cancer Res.*, 91, 1319, 2000.

155. Maibucher, A. et al., Neocarzinostatin immunoconjugates: *In vitro* evaluation of therapeutic potential, *Pharm. Sci. Commun.*, 4, 253, 1994.

156. Schonlau, F. et al., Mechanism of free and conjugated neocarzinostatin activity: Studies on chromophore and protein uptake using a transferrin-neocarzinostatin conjugate, *Z. Naturforsch., C: Biosci.*, 52, 245, 1997.

157. Brown, T. et al., A phase II trial of BMY-28175 in advanced pancreatic carcinoma, *Proc. Am. Soc. Clin. Oncol.*, 13, 219, 1994.

158. Melink, T. et al., A phase I trial of BMY-28175 on a single dose schedule, *Proc. Am. Soc. Clin. Oncol.*, 10, 120, 1991.

159. Sessa, C. et al., Phase I study of BMY-28175 (esperamicin A1) on a single intermittent schedule, *Proc. Am. Soc. Clin. Oncol.*, 9, 66, 1990.

160. Van Hof, A.C. et al., Biodistribution of [111]indium-labeled engineered human antibody CTM01 in ovarian cancer patients: Influence of protein dose, *Cancer Res.*, 56, 5179, 1996.

161. Gillespie, A.M. et al., Phase I open study of the effects of ascending doses of the cytotoxic immunoconjugate CMB-401 (hCTM01-calicheamicin) in patients with epithelial ovarian cancer, *Ann. Oncol.*, 11, 735, 2000.

162. Broadhead, T.J. et al., CMB-401 — a new agent for immunotherapy of epithelial ovarian cancer: Immune response and pharmacokinetics, *Br. J. Cancer*, 78, 44, 1998.

163. Chan, S.Y. et al., A phase 2 study of the cytotoxic immunoconjugate CMB-401 (hCTM01-calicheamicin) in patients with platinum-sensitive recurrent epithelial ovarian carcinoma, *Cancer Immunol. Immunother.*, 52, 243, 2003.

164. Mallick, R., Sievers, E., and Berger, M., A decision analysis of expected cost savings with outpatient administration of gemtuzumab ozogamicin (Mylotarg™, CMA-676) in patients in first relapse of acute myeloid leukemia (AML), *Blood*, 96, 439a, 2000.

165. Lang, K. et al., Outcomes in patients treated with gemtuzumab ozogamicin for relapsed acute myelogenous leukemia, *Am. J. Health-Syst. Pharm.*, 59, 941, 2002.

166. Rajvanshi, P. et al., Hepatic sinusoidal obstruction after gemtuzumab ozogamicin (Mylotarg) therapy, *Blood*, 99, 2310, 2002.

167. Erba, H.P. et al., Final results of a multivariate logistic regression analysis to determine factors contributing to the risk of developing hepatic veno-occlusive disease (VOD) following treatment with gemtuzumab ozogamicin, *Blood*, 100, 339a (abstr. 1313), 2002.

168. Erba, H. et al., Results of a multivariate logistic regression analysis to determine factors contributing to the risk of developing hepatic veno-occlusive disease (VOD) following treatment with gemtuzumab ozogamicin, *Proc. Am. Soc. Clin. Oncol.*, 21, 270a, 2002.

169. Kell, W.J. et al., A feasibility study of Mylotarg (gemtuzumab ozogamicin, GO) in combination with standard induced therapy in acute myeloid leukaemia (AML), *Br. J. Haematol.*, 117, 19, 2002.

170. Stadtmauer, E.A., Trials with gemtuzumab ozogamicin (Mylotarg®) combined with chemotherapy regimens in acute myeloid leukemia, *Clinical Lymphoma*, 2, S24, 2002.
171. Sievers, E.L., Antibody-targeted chemotherapy of acute myeloid leukemia using gemtuzumab ozogamicin (Mylotarg), *Blood Cells, Molecules, & Diseases*, 31, 7, 2003.
172. Naito, K. et al., Calicheamicin-conjugated humanized anti-CD33 monoclonal antibody (gemtuzumab ozogamicin, CMA-676) shows cytocidal effect on CD33-positive leukemia cell lines, but is inactive on p-glycoprotein-expressing sublines, *Leukemia*, 14, 1436, 2000.
173. Linenberger, M.L. et al., Multidrug-resistance phenotype and clinical responses to gemtuzumab ozogamicin, *Blood*, 98, 988, 2001.
174. Matsui, H. et al., Reduced effect of gemtuzumab ozogamicin (CMA-676) on p-glycoprotein and/or CD34-positive leukemia cells and its restoration by multidrug resistance modifiers, *Leukemia*, 16, 813, 2002.
175. Walter, R.B. et al., Multidrug resistance protein (MRP) attenuates gemtuzumab ozogamicin (GO)-induced cytotoxicity in acute myeloid leukemia (AML), *Blood*, 100, 224b (abstr. 4415), 2002.
176. Jilani, I. et al., Differences in CD33 intensity between various myeloid neoplasms, *Am. J. Clin. Pathol.*, 118, 560, 2002.
177. Jilani, I. et al., Quantitative differences in CD33 intensity between various myeloid neoplasms, *Blood*, 98, 586a (abstr. 2458), 2001.
178. Becker, M.W., Qian, D., and Clarke, M.F., CD33 is variably expressed on leukemic stem cells in acute myelogenous leukemia, *Blood*, 98, 587a (abstr. 2459), 2001.
179. Introna, M. et al., Differential response of acute myeloid leukaemia cells to gemtuzumab ozogamicin (Mylotarg®) *in vitro*. Role of Chk1 and 2 phosphorylation and caspase 3, *Blood*, 100, 319a (abstr. 1237), 2002.
180. Sievers, E. et al., Preliminary report of an ascending dose study of gemtuzumab ozogamicin (Mylotarg™, CMA-676) in pediatric patients with acute myeloid leukemia, *Blood*, 96, 217b, 2000.
181. Zwaan, Ch. M. et al., Gemtuzumab ozogamicin in pediatric CD33-positive acute lymphoblastic leukemia: First clinical experiences and relation with cellular sensitivity to single agent calicheamicin, *Leukemia*, 17, 468, 2003.
182. Sievers, E. et al., Final report of prolonged disease-free survival in patients with acute myeloid leukemia in first relapse treated with gemtuzumab ozogamicin followed by hematopoietic stem cell transplantation, *Blood*, 100, 89a (abstr. 327), 2002.
183. Cohen, A.D. et al., Gemtuzumab ozogamicin (Mylotarg) monotherapy for relapsed AML after hematopoietic stem cell transplant: Efficacy and incidence of hepatic veno-occlusive disease, *Bone Marrow Transplant.*, 30, 23, 2002.
184. Sievers, E.L. et al., Gemtuzumab ozogamicin (Mylotarg) as a single agent to evaluate safety and determine maximum tolerated dose (MTD) in post hematopoietic stem cell transplant patients with relapsed, CD33+ acute myeloid leukemia (AML), *Blood*, 100, 336a (abstr. 1304), 2002.
185. Goldberg, S.L. et al., Gemtuzumab ozogamicin (Mylotarg) prior to allogeneic hematopoietic stem cell transplantation increases the risk of hepatic veno-occlusive disease, *Blood*, 100, 415a (abstr. 1611), 2002.
186. Stadtmauer, E. et al., Analysis of predisposing factors for hepatic veno-occlusive disease after treatment with gemtuzumab ozogamicin (Mylotarg®, CMA-676), *Blood*, 98, 124a (abstr. 520), 2001.
187. Raza, A. et al., Preliminary analysis of a randomized phase 2 study of the efficacy of 1 vs. 2 doses of gemtuzumab ozogamicin (Mylotarg®) in patients with high risk myelodysplastic syndrome, *Blood*, 100, 795a (abstr. 3140), 2002.
188. Zwaan, Ch. M. et al., Gemtuzumab ozogamicin in pediatric CD33-positive acute lymphoblastic leukemia: First clinical experiences and relation with cellular sensitivity to single agent calicheamicin, *Leukemia*, 17, 468, 2003.
189. Leopold, L.H. et al., Comparative efficacy and safety of gemtuzumab ozogamicin monotherapy and high-dose cytarabine combination therapy in the treatment of patients with acute myeloid leukemia in first relapse, *Blood*, 96, 504a, 2000.
190. Baccarani, M. et al., Preliminary report of a phase 2 study of the safety and efficacy of gemtuzumab ozogamicin (Mylotarg®) given in combination with cytarabine in patients with acute myeloid leukemia, *Blood*, 100, 341a (abstr. 1322), 2002.

191. Estey, E.H. et al., Experience with gemtuzumab ozogamicin ("Mylotarg") and all-trans retinoic acid in untreated acute promyelocytic leukemia, *Blood*, 99, 4222, 2002.

192. Estey, E. et al., Mylotarg +/- IL-11 in patients age 65 and above with newly-diagnosed AML/MDS: Comparison with idarubicin + Ara-C, *Blood*, 98, 720a (abstr. 3007), 2001.

193. Estey, E.H. et al., Gemtuzumab ozogamicin with or without interleukin 11 in patients 65 years of age or older with untreated acute myeloid leukemia and high-risk myelodysplastic syndrome: Comparison with idarubicin plus continuous-infusion, high-dose cytosine arabinoside, *Blood*, 99, 4343, 2002.

194. Alvarez, R.H. et al., Pilot trial of Mylotarg, topotecan and Ara-C in relapsed/refractory acute myeloid leukemia (AML) and myelodysplastic syndrome (MDS), *Blood*, 100, 340a (abstr. 1321), 2002.

195. Cortes, J. et al., Mylotarg combined with topotecan and cytarabine in patients with refractory acute myelogenous leukemia, *Cancer Chemother. Pharmacol.*, 50, 497, 2002.

196. Raj, K. et al., Gemtuzumab ozogamicin is not associated with an increased risk of veno-occlusive disease in relapsed or refractory AML when used alone or in combination with flag chemotherapy, *Br. J. Haematol.*, 117 (Suppl. 1), 20, 2002.

197. Giles, F. et al., Fatal hepatic veno-occlusive disease in a phase I study of Mylotarg and troxatyl in patients with refractory acute myeloid leukemia or myelodysplastic syndrome, *Acta Haematol. (Basel)*, 108, 164, 2002.

198. Kell, J.W. et al., Mylotarg (gemtuzumab ozogamicin: GO) given simultaneously with intensive and/or consolidation therapy for AML is feasible and may improve the response rate, *Blood*, 100, 199a (abstr. 746), 2002.

199. Venugopal, P. et al., Phase II study of gemtuzumab ozogamicin (Mylotarg) combined with intensive induction chemotherapy using high dose Ara-C and mitoxantrone followed by amifostine in poor prognosis acute myeloid leukemia: Preliminary results, *Blood*, 100, 341a (abstr. 1323), 2002.

200. Amadori, S. et al., Sequential administration of gemtuzumab ozogamicin (GO) and intensive chemotherapy for remission induction in previously untreated patients with AML over the age of 60: Interim results of the EORTC leukemia group AML-15A phase II trial, *Blood*, 98, 587a (abstr. 2465), 2001.

201. Sallah, A.S. et al., Combination imatinib and gemtuzumab ozogamicin in patients with chronic myeloid leukemia in myeloid blast crisis, *Blood*, 100, 319b (abstr. 4824), 2002.

202. Tsimberidou, A.M. et al., Mylotarg, fludarabine, cytarabine (Ara-C) and cyclosporine (MFAC) combination regimen in patients with CD33-positive primary resistant or relapsed acute myeloid leukemia (AML), *Blood*, 100, 339a (abstr. 1317), 2002.

203. Haran, V., Hiemenz, J., and Ballester, O.F., A phase I-II study of intensified induction and consolidation therapy with gemtuzumab ozogamicin for patients with high-risk acute myeloid leukemia: The AIM protocol, *Blood*, 100, 269b (abstr. 4608), 2002.

204. Alvarado, Y. et al., Pilot study of Mylotarg, idarubicin and cytarabine combination regimen in patients with primary resistant or relapsed acute myeloid leukemia, *Cancer Chemother. Pharmacol.*, 51, 87, 2003.

205. Apostolidou, E. et al., Mylotarg, liposomal daunorubicin, Ara-C and cyclosporine (MDAC) regimen in patients with relapsed/refractory acute myelogenous leukemia (AML), *Blood*, 100, 268b (abstr. 4600), 2002.

206. Tanaka, M. et al., Prophylactic intravesical instillation of BCG in recurrent superficial bladder cancer: Is BCG treatment superior to other chemotherapeutic agents? *J. Urol.*, 151, Suppl., 1994.

207. Takeshita, T., Salvage combination chemotherapy with enocitabine, aclarubicin and neocarzinostatin for recurrent or refractory acute myelogenous leukemia, *Med. J. Kagoshima Univ.*, 52, 33, 2000.

208. Takahashi, T. et al., Clinical application of monoclonal antibody-drug conjugates for immunotargeting chemotherapy of colorectal carcinoma, *Cancer*, 61, 881, 1988.

209. Otsuji, E. et al., Effects of idiotypic human anti-mouse antibody against *in vitro* binding and antitumor activity of a monoclonal antibody-drug conjugate, *Hepato-Gastroenterology*, 50, 380, 2003.

210. Fang, J., Sawa, T., and Maeda, H., Factors and mechanism of "EPR" effect and the enhanced antitumor effects of macromolecular drugs including SMANCS, *Adv. Exp. Med. Biol.*, 519, 29, 2003.

211. Maeda, H., Sawa, T., and Konno, T., Mechanism of tumor-targeted delivery of macromolecular drugs, including the EPR effect in solid tumor and clinical overview of the prototype polymeric drug SMANCS, *J. Controlled Release*, 74, 47, 2001.

212. Maeda, H., SMANCS and polymer-conjugated macromolecular drugs: Advantages in cancer chemotherapy, *Adv. Drug Delivery Rev.*, 46, 169, 2001.

213. Greish, K. et al., Macromolecular therapeutics: Advantages and prospects with special emphasis on solid tumor targeting, *Clin. Pharmacokinetics*, 42, 1089, 2003.

214. Tokura, S. et al., Induction of drug-specific antibody and the controlled release of drug by 6-O-carboxymethyl-chitin, *J. Controlled Release*, 28, 235, 1994.

215. Nishikawa, T. et al., Cytological study on the effect of neocarzinostatin on sarcoma 180 *in vivo*, *J. Antibiot.*, 18, 223, 1965.

216. Napier, M.A. et al., Neocarzinostatin: Spectral characterization and separation of a non-protein chromophore, *Biochem. Biophys. Res. Commun.*, 89, 635, 1979.

217. Konishi, M. et al., Esperamicins, a novel class of potent antitumor antibiotics. I. Physico-chemical data and partial structure, *J. Antibiot.*, 38, 1605, 1985.

23 The Mitomycins

William A. Remers

CONTENTS

  I. Introduction 475
 II. Mechanism of Action..479
III. Analogs and Derivatives 484
      A. Structural Analogs and Derivatives...484
      B. Conjugated Derivatives..485
 IV. Structure–Activity Relationships 487
  V. Synthesis..489
 VI. Clinical Applications..489
References 491

I. INTRODUCTION

In 1956 Hata and colleagues at the Kitasato Institute isolated mitomycins A and B by chromatography of a mixture compounds from cultures of *Streptomyces caespitosis*.[1] A subsequent investigation on the same organism by Wakaki's group at Kyowa Fermentation Industry yielded mitomycin C, which proved to be the most active anticancer agent among these, and subsequent, natural mitomycins.[2] Mitomycin C (Chart 23.1) was introduced as an anticancer drug by Kyowa, and it caught on rapidly.[3] It was given to an estimated 60% of patients receiving chemotherapy in Japan during the 1960s. In contrast, the first clinical trials in the United States were very unsatisfactory, with 11 patients dying from the effects of myelosuppression.[4] The reason for this serious toxicity was that the mitomycin C was given on the schedule used for 5-fluorouracil — daily injection for five consecutive days — but it has cumulative and delayed toxicity to bone marrow. Bristol-Myers finally received approval in 1974 for clinical use of mitomycin C based on a revised dosage schedule.[5]

The earliest mitomycins all had the azirino[2,3:3,4]pyrrolo[1,2-*a*]indole ring system as their basic skeleton.[6] They were divided into three structural types based on the nature and stereochemistry of their side chains, as shown in Chart 23.1. Thus, mitomycin A and B types differ in the stereochemistry of the side chain,[7] whereas mitomycin G types have only a methylene group.[8] Two mitomycin-like structures (mitomycinoids) with somewhat different skeletons, isomytomycin A and albomitomycin A, were discovered by Kono and colleagues in 1987,[9] and the somewhat related compound FR-900482 was isolated in 1987 (Chart 23.1).[9]

The complicated systematic chemical names of the mitomycins led investigators to develop a distinctive nomenclature for them. Compounds with the complete skeleton including the aziridine ring, saturation at carbons 9 and 9a, and the quinone oxidation state are called mitosanes,[6] and the corresponding hydroquinones are called leucomitosanes.[10] If the structures are the same, except that there is a 9,9a-double bond, the compounds are called aziridinomitosenes and leucoaziridinomitosenes. If these compounds lack the aziridine ring, they are mitosenes (Figure 23.1).[6]

The structures of mitomycins A, B, and porfiromycin (Chart 23.1) were elucidated by Webb and colleagues at American Cyanamid in 1962 and confirmed by Tulinsky's x-ray diffraction analysis.[11]

CHART 23.1
Natural Mitomycins

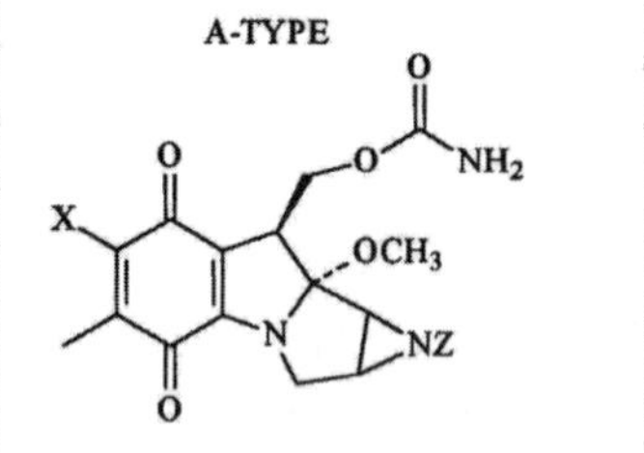

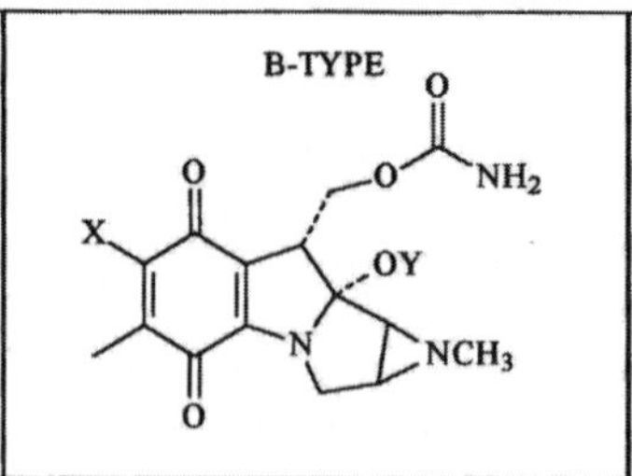

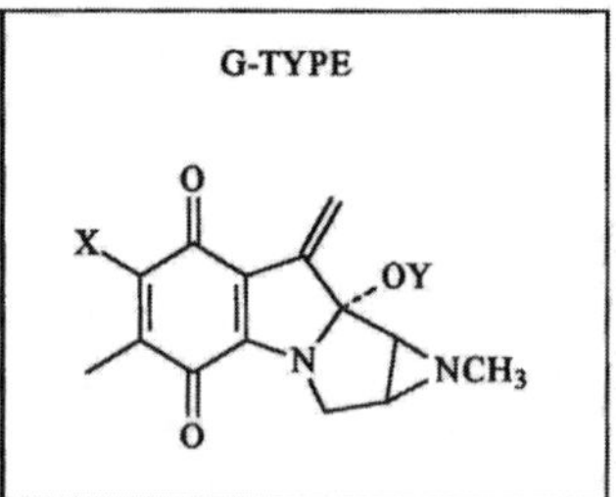

Mitomycin	X	Z
A	CH_3O	H
C	H_2N	H
F	CH_3O	CH_3
Porfiromycin	H_2N	CH_3 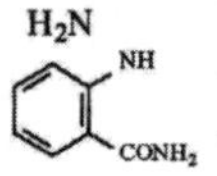
M	CH_3NH	H

Mitomycin	X	Y
B	CH_3O	H
D	H_2N	H
F	H_2N	CH_3
I	(structure)	H
J	CH_3O	CH_3
L	CH_3NH	H

Mitomycin	X	Y
G	H_2N	CH_3
H	CH_3O	H
K	CH_3O	CH_3

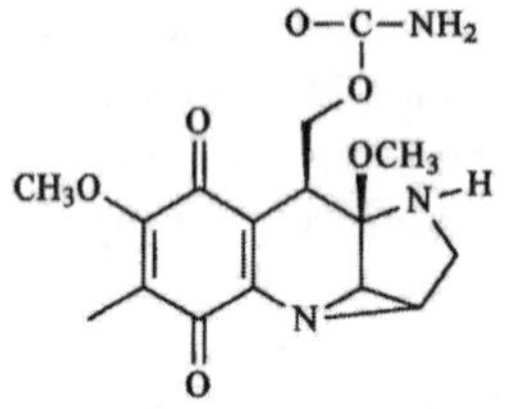

Isomitomycin A

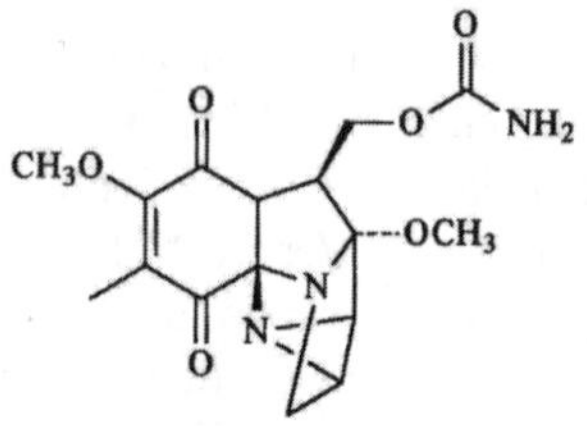

Albomitomycin A

One noteworthy finding in the course of degradative studies on mitomycins was the formation of 1-hydroxy-2-aminomitosenes when the mitomycins were treated with acid (Scheme 23.1).[6] The stereochemistry of these "apomitomycins" and other acid hydrolysis products were determined by Taylor and Remers and by Shaber and Rebek.[12,13] Other chemical transformations of mitomycins that became important in later mode of action studies included selective hydrolysis of the 7-amino group of mitomycin C to a hydroxyl group[14] and selective hydrolysis of the carbamate group to a hydroxyl group (Scheme 23.1).[15] Acylation of the aziridinyl nitrogen of certain mitomycins was used in preparing many conjugates, as described below.

Catalytic reduction of the mitomycin quinone ring produces a leucoaziridinomitosene, which can be reoxidized to an aziridinomitosene.[10] Other two-electron-reduction processes typically afford 1,2-disubstituted mitosenes with the 1-substituent and the stereochemistry dependent on the nucleophile.[16] Reduction of mitomycins in the presence of reactive thiols resulted in replacement of the carbamate group in the side chain, as well as opening of the aziridine ring.[17] Hornemann found that sodium dithionite reduction of mitomycin C afforded a mitosane in which the 9a-methoxy group was replaced by sodium bisulfite (Scheme 23.2).[18] When mitomycin C was reduced at lower pH (5.0), it formed a 2,7-diaminomitosene by trapping a proton from the solvent. Kohn and coworkers reported that proton capture at C(1) occurs with high stereoselectivity: the proton is

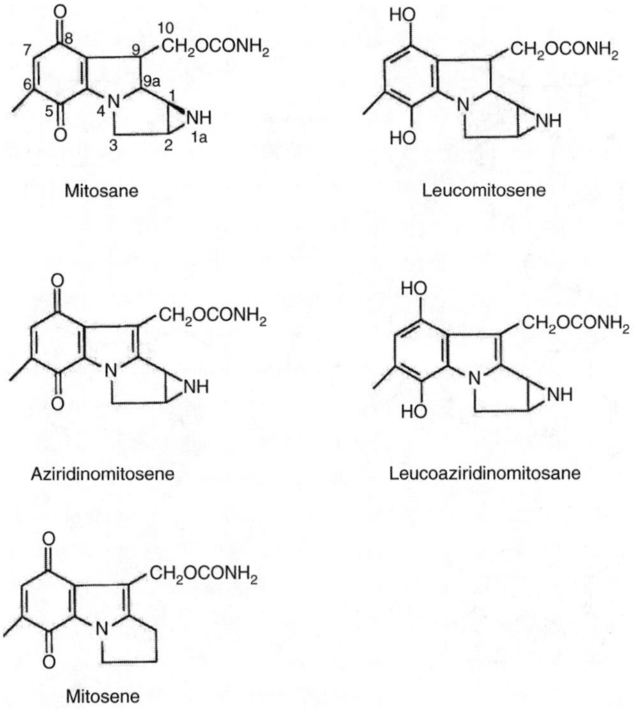

FIGURE 23.1 Nomenclature for mitomycins and derivatives.

added from the opposite side of the ring to the protonated 2-amino group. They suggested further that a quinone methide intermediate participates in this process (Scheme 23.3).[19]

Another interesting discovery by Kohn's group was that mitomycin C analogs containing a second amino group in the C(7) substituent undergo cyclization to give a C(8) imine, provided that there is an appropriate spacer between the two nitrogens (Scheme 23.4).[20] When the two nitrogens were separated by a three-methylene chain in which the central atom was dimethylated, the product was an iminoquinone with an albomitomycin structure (Scheme 23.4).[21]

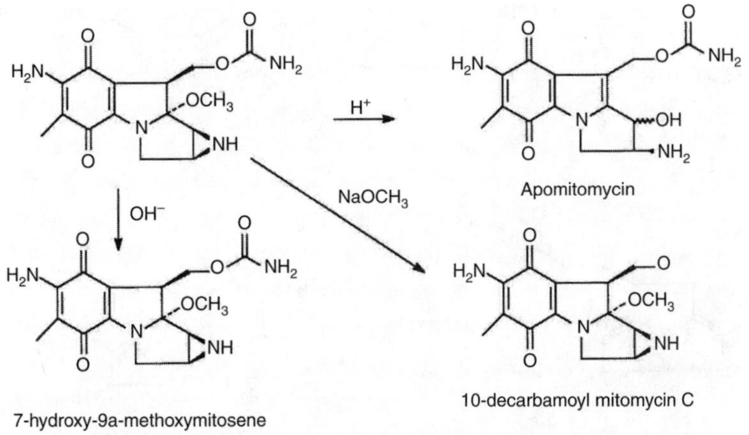

SCHEME 23.1 Chemical transformations of mitomycin C.

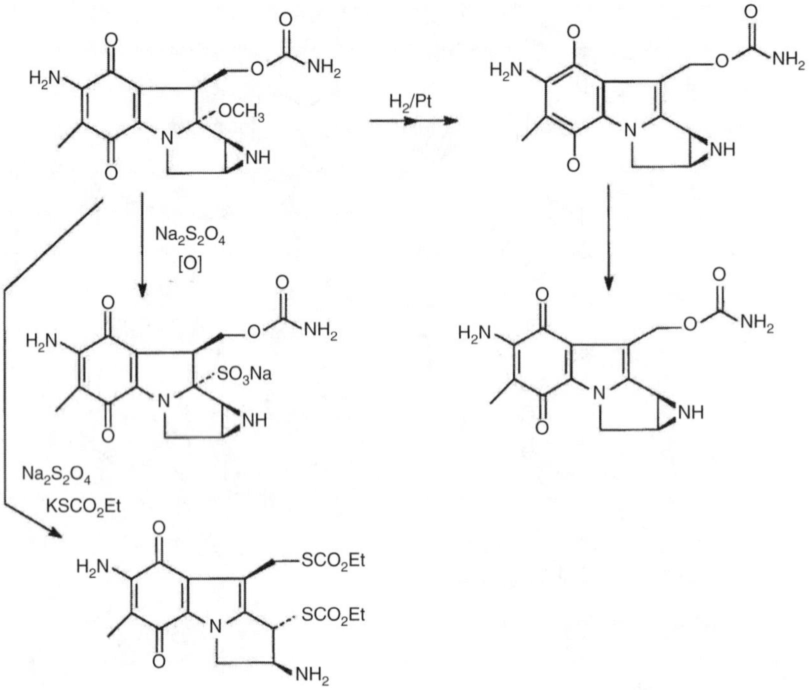

SCHEME 23.2 Reductive transformations of mitomycin C.

Mitomycin C reduction by biological agents such as xanthine oxidase or NADPH -cytochrome C P450 reductase involves an initial one-electron transfer to give a semiquinone radical, which can be detected by electron paramagnetic resonance.[22] Electochemical or catalytic reduction of mitomycin C in inert solvents also produced semiquinone radicals.[23] In biological systems, these species undergo disproportionation to the hydroquinones and quinones.[24]

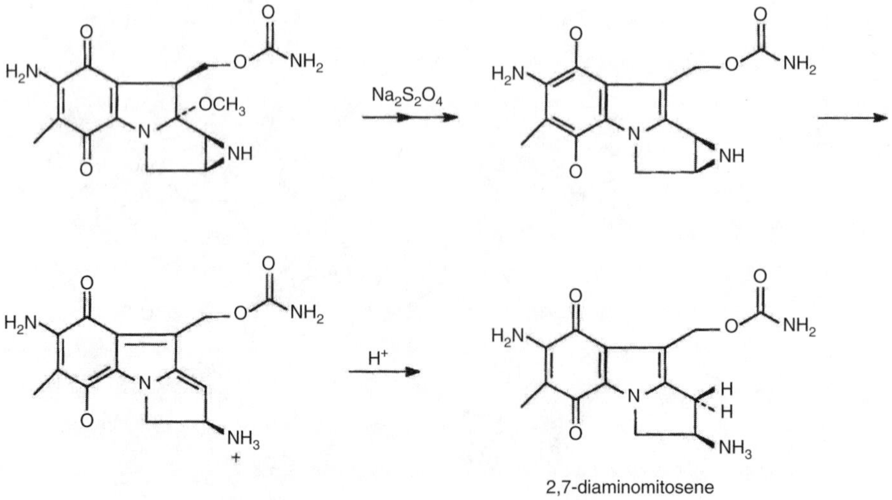

SCHEME 23.3 Formation of 2,7-diaminomitosene.

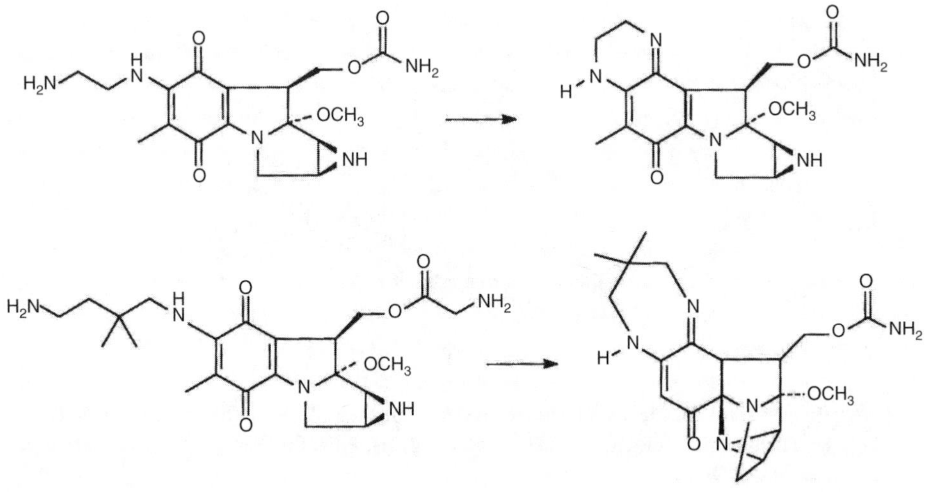

SCHEME 23.4 Cyclization reactions of mitosanes with amino groups on the c(7)-substituent.

Early studies on the biosynthesis of mitomycins showed that the O-methyl and N-methyl groups, but not the C-methyl group, are derived from L-methionine.[25] Hornemann and coworkers then demonstrated that D-glucosamine is incorporated intact into mitomycins, and that L-citrulline donates the intact $CONH_2$ group.[26] More recently, Sherman and colleagues determined that the aminomethylbenzoquinone unit of mitomycin C comes from 3-amino-5-hydroxybenzoic acid (Figure 23.2).[27] Floss and coworkers had previously shown that 3-amino-5-hydroxybenzoic acid is formed by a modified shikimate pathway involving 3,4-dideoxy-4-amino-D-arabinoheptulosonic acid-7-phosphate.[28]

In 1990 Islam and Skibo prepared pyrrolobenzimidazole analogs of mitosenes.[29] The compound with 6-aziridino and 1-carbamoyloxy substituents (Scheme 23.5) was highly potent against human solid-tumor cell lines. It resembled doxorubicin in its DNA-strand cleaving ability and cardiotoxicity.[30] The S-enantiomer was selectively toxic over the R-enantiomer for ovarian cancer cells.[31]

II. MECHANISM OF ACTION

The antimicrobial and antitumor activities of mitomycins result in part from alkylation of DNA. Approximately 10% of their covalent interactions result in cross-linking of double-helical DNA, and the remaining linkages are monofunctional.[32,33] Although alkylation of DNA occurred readily in living cells, this process could not be demonstrated *in vitro* until Iyer and Szybalski discovered that NADPH-dependent cell extracts or chemical reducing agents were required.[34] Under these conditions, only marginal amounts of mitomycin C were incorporated, and there was no cross-linking. An ingenious solution to this problem in mitomycin activation was provided by Tomasz, who postulated

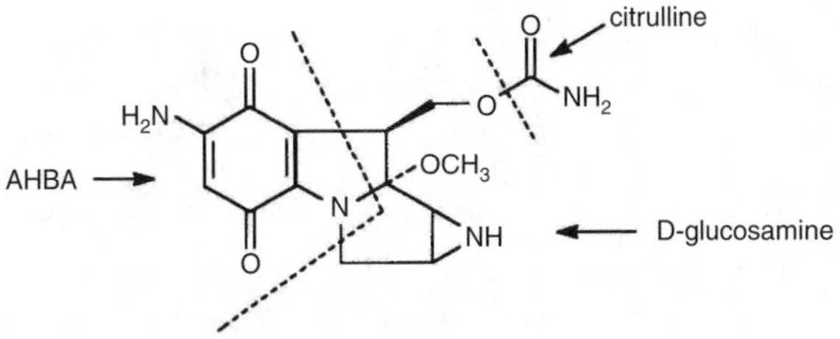

FIGURE 23.2 Biosynthetic precursors of mitomycin C.

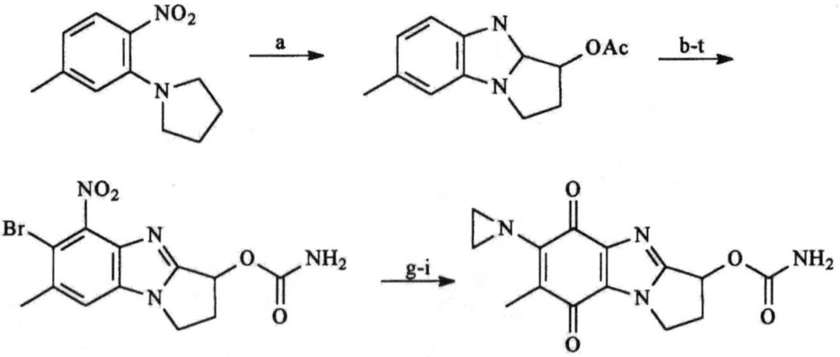

Reagents: (a) Zn Cl₂, Ac₂O; (b) Br₂, HOAc; (c) HNO₃, H₂SO₄; (d) NaOCH₃, MeOH;
(e) phenyl chloroformate; (f) NH₃, MeOH; (g) H₂, Pd/c; (h) Fremy's salt; (i) ▷NH,
aerobic MeOH.

SCHEME 23.5 Preparation of 6-aziridino-1-carbamoyloxy-2,3-dihydro-7-methyl-[1,2-a]benzimidazole-5,8-dione.

that the initial binding of mitomycins involved semiquinone radicals rather than hydroquinones and devised a method of maximizing the concentration of semiquinone radicals by adding the reducing agent, sodium dithionite, in portions and maintaining an excess of the mitomycin.[35] These conditions produced a high binding ratio of mitomycins, and cross-links were obtained.

The effects of different oxidoreductase enzymes on the activation of mitomycin C were studied. Gustafson and Pritsos found that xanthine dehydrogenase activated it under both aerobic and anaerobic conditions, and they proposed that mitomycin reduction occurs at the NAD⁺ site on this enzyme.[36] Enhancement of mitomycin C toxicity by compounds that induce D-T diaphorase was found to occur in tumor cells by Begleiter and coworkers.[37] In contrast, D-T diaphorase down regulation in MKN28 gastric cancer cells produced resistance to mitomycin C.[38] Sartorelli and colleagues found that in Chinese hamster ovary cells that overexpressed human NADPH:cytochrome C reductase, mitomycin C was highly active under both oxygenated and hypoxic conditions, whereas porfiromycin was markedly more cytotoxic under hypoxic than aerobic conditions.[39]

In 1964 Iyer and Szybalski proposed a mechanism of action for mitomycins based on bifunctional alkylation and cross-linking of DNA. It involved reduction of the quinone ring to a hydroquinone followed by loss of the elements of methanol to give a mitosene hydroquinone.

In this hydroquinone, the aziridine ring opens to provide a carbonium ion that can alkylate DNA. Cross-linking occurs by alkylation of DNA by C(10) of the mitomycin following elimination of the carbamoyloxy group.[34] Moore modified this mechanism to include the intermediacy of quinone methides (Scheme 23.6).[40] Kohn's group found that mitomycin D and mitomycin-9a-sulfonate gave binding profiles comparable to that of mitomycin C under reductive conditions. These results indicated that neither the leaving group at C(9a) nor the stereochemistry at C(9) and C(9a) influence the sites of DNA binding and supported the concept of quinone methide formation before bonding at C(1).[41]

Boryah and Skibo determined the pKa values of mitomycin C in its quinone and hydroquinone forms. The quinone form has pKa values as follows: 1.2 for the protonated quinone ring, 2.7 for the protonated indole nitrogen, and 7.6 for the protonated aziridine nitrogen. In comparison, the hydroquinone form has pKa values of 1.2 for the protonated 7-amino group, 5.1–6.1 for the protonated indole nitrogen, and 9.1 for the protonated aziridine nitrogen. Protonation of the indoline nitrogen in the quinone form prevents the elimination of methanol; however, there is a 10⁵-fold increase in the rate constant for methanol elimination in the hydroquinone form. The carbocation resulting from methanol elimination (pKa = 7.1) dissociates to a quinone methide. At pH values below 7.1 the carbocation traps water, and at pH values above 7.1 it traps both water and a proton.[42]

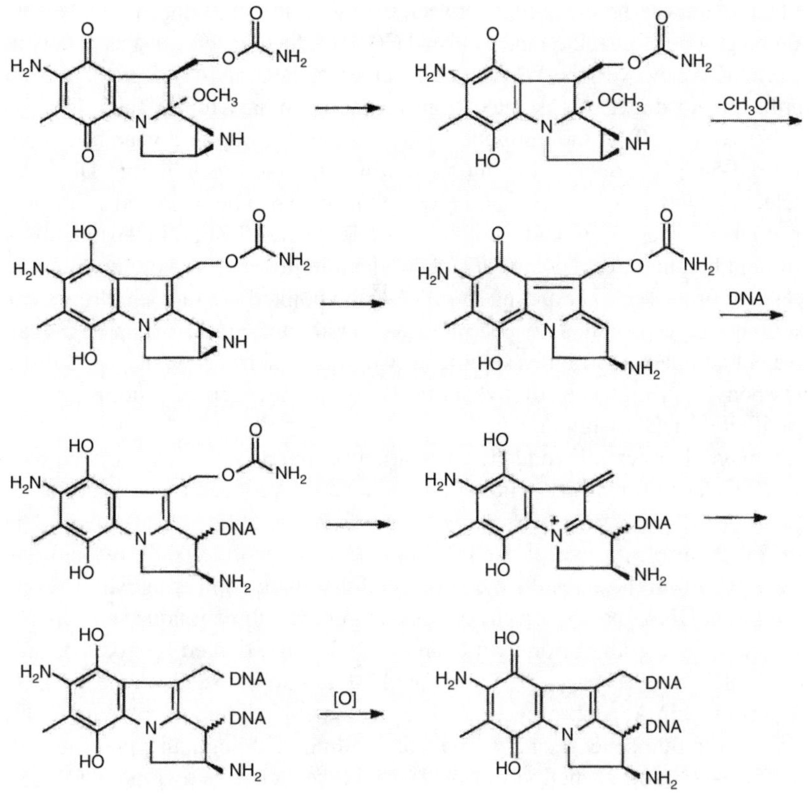

SCHEME 23.6 Proposed mechanism of bifunctional alkylation of DNA by mitomycin C.

The specific atoms involved in the alkylation of DNA by mitomycin C were determined by Tomasz, Nakanishi, and coworkers. They hydrolyzed the mitomycin-bound DNA with enzymes including snake venom diesterase and alkaline phosphatase to obtain the mitomycin-bound nucleosides and then elucidated their structures by a variety of analytical methods.[43–48] Monoalkylation occurs preferentially at C(1) of the mitomycin, with the covalent bond formed with the 2-amino group of a guanine residue. Among the three different products isolated in the initial experiments, the main product had the amino groups attached to C(1) and C(2) of the mitosene moiety in a *cis* orientation. The minor products were the C(1) epimer and the 10-decarbamoyl derivative.[49] Dialkylation with cross-linking involves alkylation of the 2-amino group of guanine in one strand of the DNA double helix by C(1) of mitomycin C and alkylation of the 2-amino group of guanine in the opposite strand by C(10) of the same mitomycin C molecule (Figure 23.3).[50,51] Tomasz and coworkers isolated a

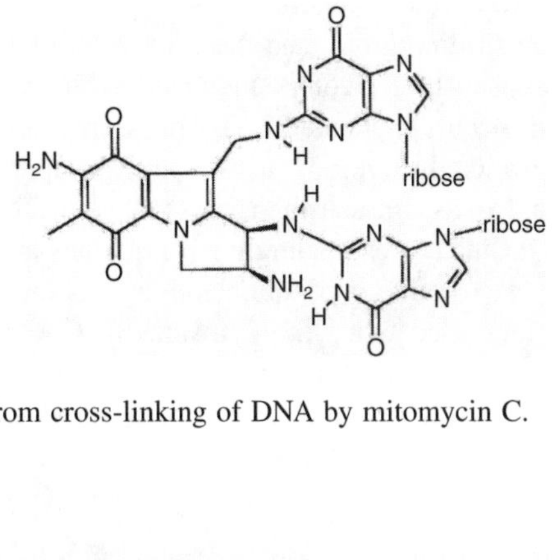

FIGURE 23.3 Main product from cross-linking of DNA by mitomycin C.

product that had an intrastrand cross-link between the DNA and mitosene. The covalent bonds were between 2-amino groups of guanines and C(1) and C(10) of the mitosene, just as in the intermolecular cross-linked products; however, the DNA double helix was bent at 14.6 degrees near the cross-link site.[50,52] A further monoadduct was isolated from mouse mammary tumor cells. It had the covalent bond formed between C(10) of the mitosene and the 2-amino group of a guanine residue.[53]

The preferred 5-CpG sequence for mitomycin binding and cross-linking DNA was first suggested by molecular modeling studies,[54] and it was confirmed by chemical and spectroscopic studies in a number of laboratories.[51,55,56] The laboratories of Tomasz and Kohn independently showed that the initial covalent bonding event occurs at C(1) of the mitomycin in the preferred 5-CpG sequence and is independent of the second bonding step.[57–59] Kohn pointed out that a hydrogen bond between the C(10) carbonyl oxygen of an activated mitomycin and the 2-amino group on a guanine residue in the precovalent complex contributes significantly to the sequence specificity.[59] Substitutions such as methoxycarbonyl and methanesulfonyl on the aziridine nitrogen of mitomycin C changed the sequence specificity for DNA binding.[60]

The mechanism of enzymatic reductive activation of mitomycin C was revised by Tomasz and colleagues in 1997. They found that when mitomycin C was reduced by a variety of chemical and enzymic agents in the presence of *Micrococcus lysodeicticus*, the overall yield of adducts was higher at pH 7.4 than at 5.8, except for DT-diaphorase reduction. The cross-links were always greater at lower pH. From these results they proposed that three competing pathways for activation afford three different DNA-reactive electrophiles that generate three unique sets of DNA adducts.[61]

DNA alkylation by certain mitomycin C analogs was investigated by Tomasz and coworkers, who reported that in the absence of reduction 2,7-diaminomitosene bound to DNA, whereas mitomycin C, mitomycin A, porfiromycin, and certain analogs did not. Protonation of the 2-amino group of 2,7-diaminomitosene strongly affected binding, although the presence of functional groups including 1-OH and N-methyl did not.[62] Under reductive activation, C(10) of 2,7-diaminomitosene alkylated DNA on N(7) of a guanine residue.[63] Decarbamoyl mitomycin C was surprisingly more cytotoxic than mitomycin C to hypoxic EMT6 mouse mammary tumor cells and CHO cells. The frequency of total adducts was 20–30-fold higher with decarbamoyl mitomycin C, although the frequency of cross-links was equal. These results are in contrast to the alkylation of calf thymus DNA *in vitro*, wherein mitomycin C is more reactive.[64] The unnatural enantiomer of mitomycin C formed both a mono and a bis adduct on reduction in the presence of DNA. These adducts were stereoisomers of the main adducts formed by natural mitomycin C.[65] Reduction of racemic 1,10-bisacetoxy-7-methoxymitosene in the presence of DNA gave five products. In the two major adducts, the stereochemistry at C(1) was the same as that found for natural mitomycin C adducts, but the C(2) chirality was reversed.[66]

Alkylation sites on reduced mitosenes were assessed by Ouyang and Skibo, using ^{13}C nuclear magnetic resonance. They found that reduction by dithionite gave sulfite esters as well as sulfonates, both the 6-amino group of adenine and the 2-amino group of guanine are alkylated, both nitrogen and oxygen centers in purines residues are alkylated, and the main fate of the iminium ions formed after reduction is head-to-tail polymerization (Scheme 23.7).[67]

Zhu and Skibo found that when indoloquinones with 5-aziridino substituents were reduced to hydroquinones, the fate of this aziridino group and the cytotoxicity were strongly influenced by the adjacent 6-substituent. In the 6-unsubstituted compound the aziridine ring may assume a conformation in which a 1,5-sigmatropic shift occurs (Scheme 23.8). This shift results in loss of the aziridine ring and drastically reduces cytotoxicity. In contrast, a 6-methyl group provides steric hindrance that prevents the sigmatropic shift and allows the aziridine ring to alkylate DNA, thus retaining cytotoxicity. Similar results were observed for the corresponding benzoquinones and benzimidazolequinones.[68]

The binding of mitomycin C to modified oligonucleotides was investigated. Millard and Beachy found that cross-linking was increased when the cytosines in CpG-CpG were methylated. They suggested that this enhancement resulted from an electronic effect.[69] Oligonucleotides in which 2,6-diaminopurine was substituted for guanine were readily alkylated by reduced mitomycin C.

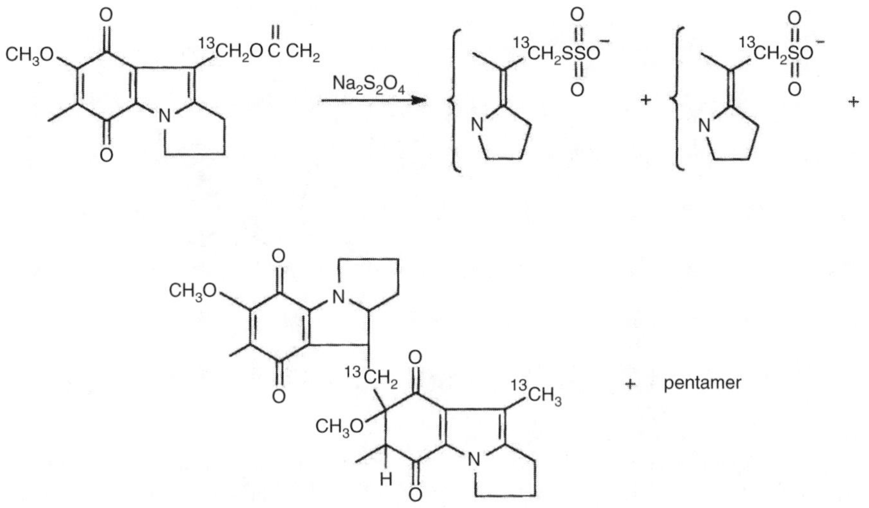

SCHEME 23.7 Products from dithionite reduction of a ^{13}C-labeled mitosene.

The sequence specificity was TpD (analogous to CpG), and cross-links occurred in TpD-TpD and TpG.CpD regions.[70]

Glutathione does not reduce mitomycin C or form conjugates with it; however, in the presence of chemical or enzymic reducing systems, three monoconjugates and two bisconjugates are formed. Similar results were found with mercaptoethanol or N-acetylcysteine. Glutathione accelerated the reduction of mitomycin C by "slow" agents such as cytochrome C reductases. Glutathione was considered by Tomasz to be involved in further reduction of the initially formed semiquinone radical.[71] A ternary glutathione–mitomycin–DNA adduct was isolated from reduction of mitomycin C in the presence of DNA or synthetic duplex oligonucleotides. The mitosene residue was linked to the nucleotide by its C(1) atom, and the glutathione residue was linked to C(10). Overall, glutahione has an inhibitory effect on mitomycin C alkylation of DNA, which may be a factor in mitomycin resistance by tumor cells with high glutathione levels.[72] In support of this idea, Xu and coworkers found that the cytotoxicity of mitomycin C increased significantly in cells pretreated with ethacrynic acid, which inhibits glutathione transferase and depletes glutathione levels.[73]

The effects of DNA alkylation by mitomycin C on other biological processes were examined by Tomasz and collaborators. The action of DNA polymerases was terminated nearly quantitatively at the nucleotide on the 3 side of each mitomycin monoadduct site on the DNA. These monoadducts do not distort the DNA, but the resulting increase in thermodynamic stability may prevent replication.[74] In selected tissue culture cell lines, mitomycin C and decarbamoylmitomycin C, but not 2,7-diaminomitosene, induced p53 protein levels and increased levels of p21am and Gadd45 mRNA. They also

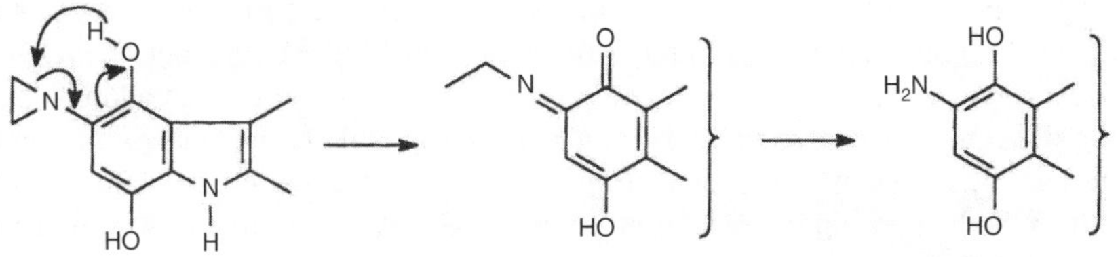

SCHEME 23.8 1,5-Sigmatropic shift in a 5-aziridinoindolohydroquinone.

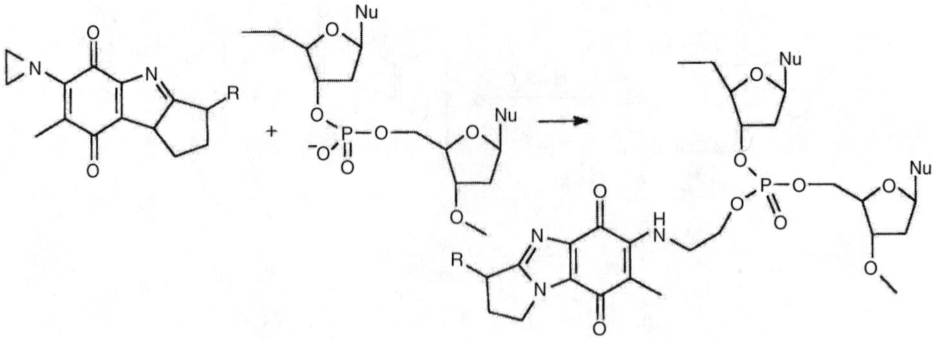

SCHEME 23.9 Alkylation of a phosphate group in DNA by a 6-aziridinopyrrolo[1,2-a]imidazole.

induced apoptosis in ML-1 cells. This marked the first observation that adducts of a multiadduct type of DNA-damaging agent were differentially recognized by DNA damage sensor pathways.[75]

Benzimidazole analogs with 6-aziridino substituents (Scheme 23.9) cleave DNA under reducing conditions, specifically at G + A bases.[76] Skibo and Schultz showed that after reduction of the quinone ring, the aziridino group of these compounds alkylate phosphate groups in the DNA backbone to form phosphotriesters that are labile to hydrolysis. They also alkylate N(7) of guanine and adenine residues. The observed base specificity was attributed to Hoogsteen-type hydrogen bonding of the hydroquinone in the major groove of the DNA double helix.[77]

A second important mechanism of action for mitomycins is the generation under aerobic conditions of superoxide and hydroxyl radicals that can cause DNA strand cleavage.[78-82] This process is promoted by metal ions such as Cu(II) and inhibited by enzymes such as catalase and superoxide dismutase, as well as by free-radical scavengers and metal-ion sequestering agents.[83] Pritsos and Sartorelli showed that the free radicals generated by quinone antibiotics, including mitomycins, cause cytotoxicity that depends on the extent of DNA damage and the ability of cells to repair the damage.[84] This damage is easier to repair than that caused by mitomycin cross-linking of DNA.

III. ANALOGS AND DERIVATIVES

A. STRUCTURAL ANALOGS AND DERIVATIVES

The rather broad antitumor spectrum of mitomycin C, together with its serious toxicity, spurred the quest for analogs that might have better therapeutic indices. Hundreds of mitosanes, aziridinomitosenes, and mitosenes, as well as indoloquinones, were synthesized and screened.[85] Despite the development of numerous compounds with very impressive activity in animal screens, there have been no new clinical agents approved to date.

Among the mitosanes, the position most readily substituted is C(7) in the quinone ring. In particular, the replacement of the 7-methoxy group, as found in mitomycin A, with amines and alkoxides has been much investigated (Scheme 23.10). The earliest studies were by Kinoshita and colleagues.[86-88] Mitomycin A is not produced in quantity by fermentation, but it is readily obtained by hydrolysis of mitomycin C to the 7-hydroxy derivative (Scheme 23.1), followed by O-methylation using diazomethane (Scheme 23.10).[14] The corresponding compounds with a methyl group on the aziridine nitrogen can be prepared from porfiromycin. An important feature of the 7-amino analogs is that both the quinone-reduction potential and the lipophilicity can be controlled by varying the substituents on the nitrogen. A large variety of substituted amines, including ethylamines, secondary amines, arylamines, and amidines were prepared and tested by Remers and coworkers and by chemists at Bristol-Myers Squibb.[89-98] As discussed in the following sections, the arylamines and amidines were highly potent, and ethylamines bearing disulfide groups were

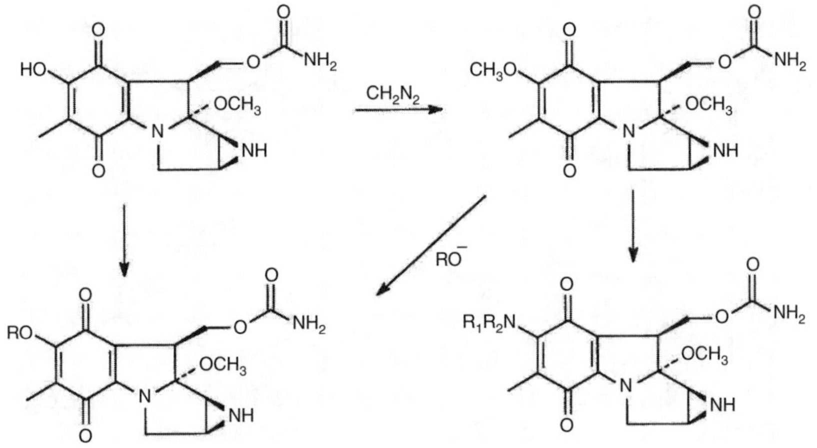

SCHEME 23.10 Preparation of 7-substituted mitosanes.

efficacious. New 7-alkoxymitosanes were prepared by alkoxide exchange on mitomycin A or treatment of 7-hydroxymitosane with 3-substituted 1-phenyltriazenes (Scheme 23.10). Many of them showed high potency.[99]

Mitomycin C analogs with structural changes at other positions included acyl derivatives on the aziridine nitrogen,[86] the potent 10-decarbamoyl derivative,[15] which was converted into 10-chloro and 10-bromoderivatives,[100] and the 6-carboxaldehyde.[101]

Aziridinomitosenes corresponding to mitomycin A and its N-methyl derivative were prepared from the parent compounds by catalytic reduction and air oxidation. The methoxy group was then replaced by a variety of amines. As described below, potency depended on the ease of quinone reduction.[102] Mitosenes without the aziridine ring, but with substituents at C(1) or at C(1) and C(2), were prepared by total synthesis by Remers and coworkers.[103,104] 1-Acetoxy-mitosenes with a variety of substituents in the quinone ring, and 7-methoxymitosenes with new substituents at C(1) and C(10), were also investigated.[105,106] None of these mitosenes were comparable in activity to the mitosanes.

The earliest work on synthesis and evaluation of mitomycin analogs was conducted by Allen and coworkers on indoloquinones and directed toward the development of new antibacterial agents. It provided important information on the relative importance of various functional groups in mitomycin-related structures.[107–110]

B. CONJUGATED DERIVATIVES

A variety of conjugates has been prepared from mitomycin C in attempts to improve its therapeutic properties. The earliest one, which was prepared by Iwata and Goldberg, involved coupling con-conavilin A with N(1a)-succinylated mitomycin C in the presence of a carbodiimide. An improved antitumor effect was claimed.[111] Conjugation with bovine serum albumin required prior acetylation to prevent protein polymerization in the presence of carbodiimides. The conjugate of mitomycin C with glutarated albumin had high mitomycin content, and it was stable.[112]

Many conjugates were based on the delivery of mitomycin C to selected targets using monoclonal antibodies. In general, these conjugates were prepared by substituting the aziridino nitrogen with acyl groups having a free carboxyl group at the other end of the chain and then coupling with an antibody (Scheme 23.11). Sometimes the antibody is linked to an intermediary, which provides many binding sites for the mitomycin. Acylated aziridines are less stable than ordinary amides, and they undergo gradual hydrolysis to liberate mitomycin C at the target. The following conjugates were formed in this manner: monoclonal anti-MM46 immunoglobulin M antibody, with or without

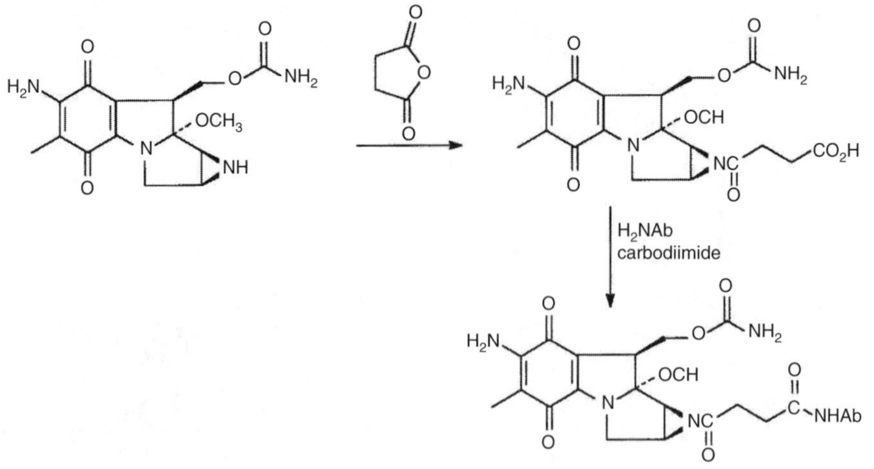

SCHEME 23.11 Synthesis of monoclonal antibody conjugates of mitomycin C.

serum albumin as intermediary[113]; anti-alpha-fetoprotein antibody[114]; monoclonal antibody MGb2 with dextran T-40 as intermediary[115]; and anti–neural cell adhesion molecule monoclonal antibody.[116] None of these conjugates has become an established therapeutic agent to date, although the MGb2 monoclonal antibody conjugate was reported to have enhanced activity against human gastric cancer cells in nude mice.[115]

Two types of conjugates were directed toward binding with DNA. Kohn's group treated decarbamoylporfiromycin with carbonyldiimidazole followed by H$_2$N(CH$_2$)$_6$P=S(OH)R, wherein R is the base sequence for human A-raf-1-gene or human FGFR1 gene.[117] Tomasz and colleagues prepared conjugates in which 1, 2, or 3 N-methylpyrrolecarboxamide units were tethered by five methylene groups to the 7-amino group of mitomycin C. These N-methylpyrrole units are found in distamycin, and they direct molecules to AT-rich regions of DNA.[118]

Some derivatives of mitomycin C were prepared to modify its physical properties for improved absorption and distribution. Thus, alkyl and alkenylazacycloalkananone derivatives exhibit improved penetration through rat skin.[119] Another lipophilic substituent was the docosahexenoyl group.[120] Derivatives with hydrophilic properties were prepared by attaching a linear (1-3)-β-D-glucan to the aziridine nitrogen,[121] or a glucopyranosyl group linked through a phenolic group to N(7).[122] A conjugate with dextran, prepared by using CO(CH$_2$)$_4$CH$_2$NHC(N=H) as the linker, showed strong affinity to lymph cells.[123] Other hydrophilic mitomycin C derivatives were formed by condensing poly[N5-(2-hydroxyethyl)-L-glutamine] through a tetrapeptide spacer[124] and by coupling a N(1a)succinyl mitomycin C with a chitosan. The latter derivative had high drug content, pH-dependent drug release, and long duration in plasma.[125]

A novel prodrug containing a 1,2-diacyl lipid linked to mitomycin C by a thiolytically cleavable dithiobenzyl carbamate (Figure 23.4) was formulated in STEALTH liposomes. It was reported to be well retained in the liposomes and to show increased tumor inhibition over mitomycin C.[126]

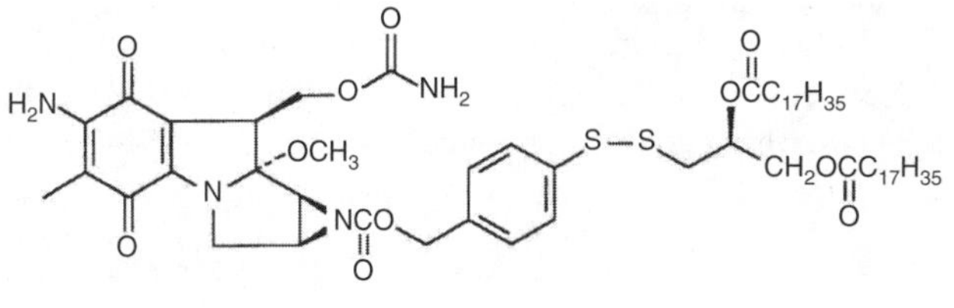

FIGURE 23.4 Thiolytically cleavable lipophilic prodrug of mitomycin C.

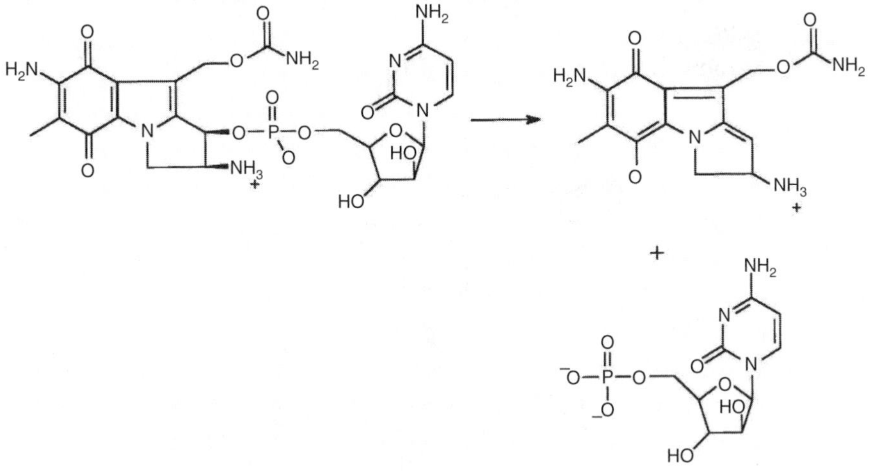

SCHEME 23.12 Prodrug for mitomycin and 5-ara-CMP.

Conjugates of 2,7-diaminomitosenes with cytidine 5-monophosphate and 5-*ara*-CMP were prepared as prodrugs that would liberate both the nucleotides and an alkylating intermediate on reduction. This process actually worked, as shown by liberation of the nucleotides, and the 5-*ara*-CMP conjugate prepared from mitomycin C was more potent than either parent compound (Scheme 23.12).[127,128]

IV. STRUCTURE–ACTIVITY RELATIONSHIPS

One key to understanding structure–activity relationships among mitomycins and their analogs is that the functional groups are so distributed on the pyrrolo[1,2-a]indole nucleus that they interact with each other in a chain of events that produces potent alkylating functionalities. Another key to understanding is that at least three factors determine antitumor potency. They include the ability to enter cells, which is determined in part by their partition coefficient; the ease of their reduction to reactive intermediates, which is determined by the quinone reduction potential; and the presence of optimal alkylating groups such as the aziridine ring and the carbamoyloxymethyl substituent.[86,87,95]

Mitomycins are relatively stable in their quinone state; however, as described in Section II, reduction of the quinone ring strongly activates the molecule. The nitrogen atom common to two rings becomes more electron rich and is able to participate in the loss of the 9a-substituent to give a leukoaziridinomitosene, which can undergo opening of the aziridine ring and loss of the carbamoyl group (Scheme 23.6). The ease of reduction of the quinone ring is obviously important in initiating the action of mitomycins, and this can influence antitumor potency. Thus mitomycin A, which has a 7-methoxy substituent and half-wave reduction potential ($E_{1/2}$) of –0.21 V, is more readily reduced and a more potent cytotoxic agent than mitomycin C, which has a 7-amino group and $E_{1/2}$ of –0.45 V.[91] Kunz and coworkers subjected a set of 30 mitomycin C and mitomycin A analogs to a QSAR study using multiple linear regression analysis based on cytotoxic potency in a panel of human tumor cells in culture (log 1/C), $E_{1/2}$, and the logarithm of the partition coefficient (log P). A statistically significant correlation with an equation log (1/C) = 10.1 + 6.59 $E_{1/2}$ + 0.35 log P was obtained. The parameter $E_{1/2}$ accounted for 48% of the variance, and log P accounted for 21%.[95] When the analysis was limited to mitomycin C analogs, $E_{1/2}$ was the only significant variable, whereas when the analysis was limited to mitomycin A analogs, log P was the only significant variable. These results were explained by the consideration that all of the mitomycin A analogs were 7-alkoxymitosanes with identical low $E_{1/2}$ values, meaning that cell uptake is the controlling

TABLE 23.1
Mitomycins Receiving Clinical Trials

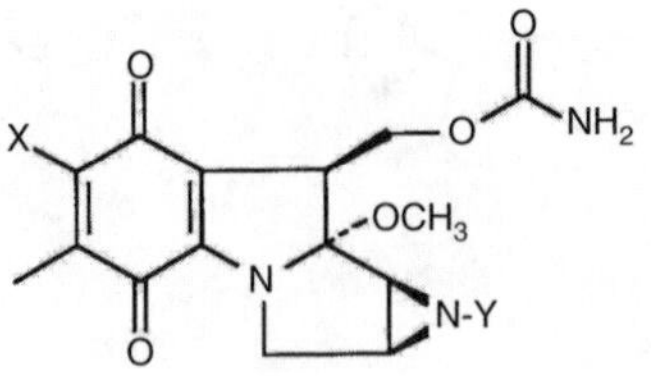

Identification	X	Y
Mitomycin C	H_2N	H
Porfiromycin	H_2N	CH_3
M83	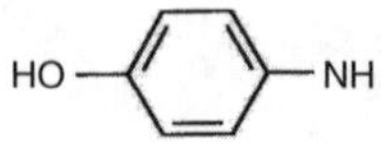 HO—⟨benzene⟩—NH	H
BMY-25282	$(CH_3)_2N\text{–}C\text{=}N\text{–}$	H
BMY-25067	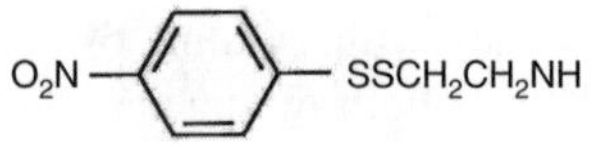 O_2N—⟨benzene⟩—$SSCH_2CH_2NH$	H
KT 6149	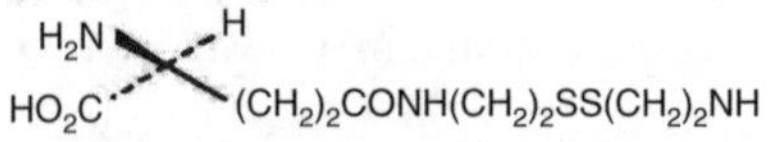 H_2N, H / HO_2C⤬$(CH_2)_2CONH(CH_2)_2SS(CH_2)_2NH$	H

factor in their potency, whereas the mitomycin C analogs, being harder to reduce in biological systems, are limited by reduction potential rather than by cell uptake. The nature of the assay system may be crucial in determining the relative potencies of mitomycins, and presumably other types of drugs. Thus, in the murine leukemia assay, the most potent mitomycin C and mitomycin A analogs were the most hydrophilic ones, whereas in the cell culture study described above, the most lipophilic analogs were the most potent ones.[129,130] This result indicates that the more lipophilic compounds may be concentrating in fatty tissues in the mice. It also indicates that optimizing a set of analogs in an *in vitro* assay might not be valuable for subsequent *in vivo* assays.

An ethyleneimino substituent at C(7) in the quinone ring gave especially potent antitumor activity against P-388 leukemia in mice.[91] Skibo and coworkers subsequently showed that this substituent alkylates phosphate groups in the DNA backbone, giving enhanced covalent binding energy.[76] Quinones bearing an ethyleneimine group are reduced more readily than those with alkylamines.[92] Another C(7) substituent that confers increased antitumor potency to mitomycins is the amidine group, as found in BMY-25282 (Table 23.1).[98,131]

Elimination of the elements of water or methanol from C(9) and C(9a) of mitomycins produces aziridinomitosenes. They have all of the appropriate functionality for efficient DNA cross-linking, but they are less potent than the corresponding mitosanes because their indoloquinone chromophores are more difficult to reduce than the corresponding benzoquinone chromophores of mitosanes.[132] Mitosenes without the aziridine ring are also less potent than mitosanes, although the presence of two alkylating functionalities at C(1) and C(10) permits bisalkylation with cross-linking of DNA. For example, the 1,10-diacetoxymitosene prepared by Orlemans *et al.* possessed antitumor activity, although the spectrum was more limited than that of mitomycin C.[133] Even indoloquinones with good leaving groups at C(2), which is equivalent to C(1) in mitosenes, have potent antibacterial activity in culture.[134]

Mitosanes lacking the carbamoyl substituent are surprisingly active antitumor agents.[64] Their potency may be explained by facile loss of hydroxide from an activated intermediate to give the

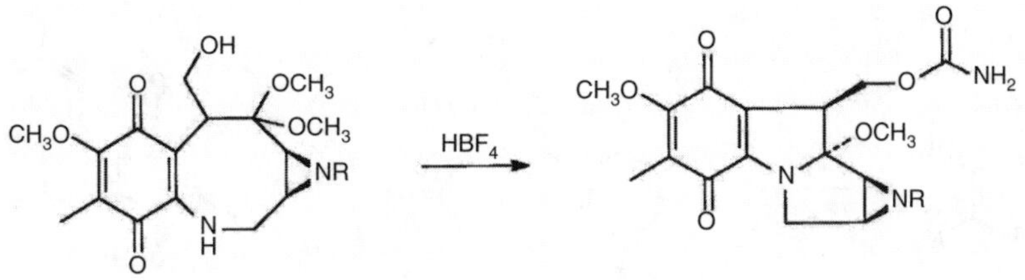

SCHEME 23.13 Formation of the mitosane nucleus in the Kishi synthesis.

same quinonemethide that mitomycins provide. The antitumor activity of mitomycin G, which has only a methylene group for C(10),[8] can be rationalized in the same way.

V. SYNTHESIS

The total synthesis of mitomycins has presented a formidable challenge to chemists because of the highly interactive functionality arranged on a compact skeleton. Allen and colleagues at American Cyanamid made early syntheses of indoloquinines and mitosenes related to the mitomycins, with the goal of producing new antibacterial agents.[135–137] This research was extended by Remers and coworkers to 1-substituted and 1,2-disubstituted mitosenes.[103,104,138] Groups led by Hirata and by Franck showed how to form the aziriding ring on the pyrrolo[1,2-a]indole nucleus.[139,140] Kametani and coworkers introduced the 9a-methoxy group into the pyrrolo[1,2-a]indole system.[141]

The real breakthrough in mitomycin synthesis was the landmark total synthesis of mitomycin C published by Kishi's group in 1977.[142,143] Although this synthesis was quite long, it showed how to solve most of the vexing problems in handling the reactive functionalities of mitomycins. A key step was the transannular closure of the two five-membered rings of the pyrrolo[1,2-a] indole system from an eight-membered ring (Scheme 23.13). In 1989, Fukuyama and Lang published an elegant and practical synthesis of mitomycin C by a route that went through isomitomycin A (Scheme 23.14).[144] Danishefsky and colleagues have reported a number of ingenious, but unsuccessful, attempts to synthesize mitomycin C.[145] They did, however, succeed in the total syntheses of mitomycin K and the mitomycin-related compound FR-900482.[146,147]

VI. CLINICAL APPLICATIONS

Mitomycin C was very widely used in cancer chemotherapy in Japan during the 1960s, and it subsequently received some important applications in the United States and other countries. It showed a broad spectrum of antitumor activity, although it was seldom the most effective agent against specific tumors. More recently, its use has declined because of its high toxicity and the introduction of newer agents of other chemical types. Nevertheless, it is still effective in cancer chemotherapy, especially in combinations with other anticancer drugs.

In 1968 Mortel and colleagues found that mitomycin C produced objective responses in patients with tumors of the colon, pancreas, stomach, liver, and gallbladder.[148] It also showed activity against non–small cell lung cancer, some head and neck cancers, breast cancer, and chronic myelogenous leukemia.[149–151] Early and coworkers showed that mitomycin C is active in topical administration to superficial bladder tumors.[152] Furthermore, systemic activity was found for mitomycin C in advanced biliary, ovarian, and cervical squamous cell carcinoma.[153]

Combination regimens including mitomycin C have been used against a variety of cancers. One of the most common combinations is the FAM regimen, which contains 5-fluorouracil, Adriamycin, and mitomycin C and is used in treating gastric and pancreatic carcinomas.[154] Mitomycin C and bleomycin in combination produced response rates up to 80% in cancers of the uterine cervix.[149] The MOB regimen includes mitomycin C, Oncovin, and bleomycin.

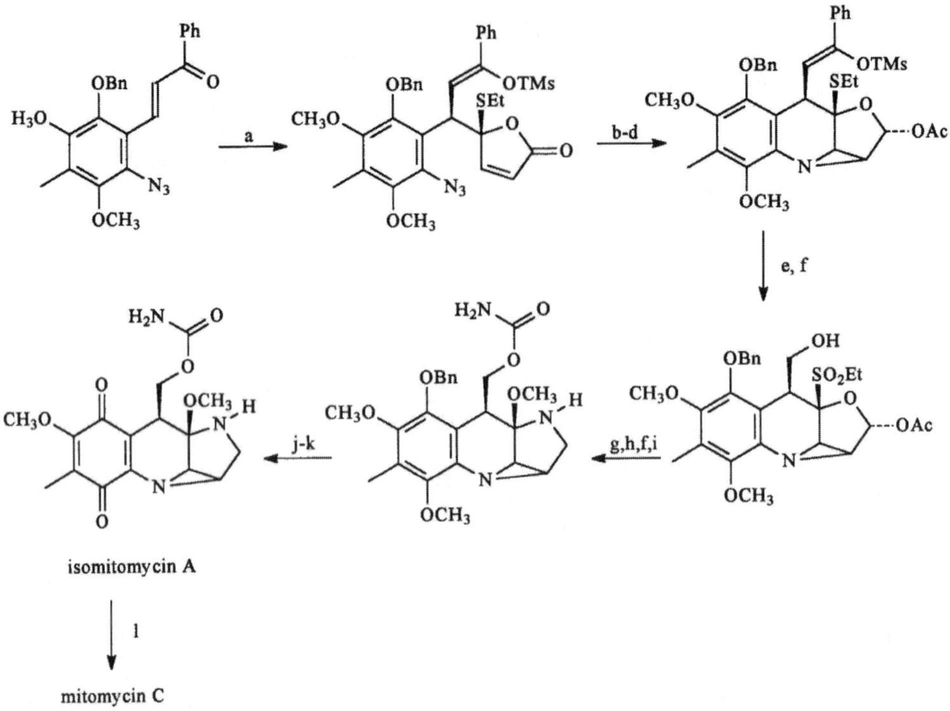

Reagents: (a) 5-Ethylthio-2-trimethylsiloxyfuran, $SnCl_4$, CH_2Cl_2, -78°C; (b) toluene 110°C; (c) DIBAL, THF, -78°C; (d) Ac_2O, Py; (e) RuO_2, $NaIO_4$; (f) $NaBH_4$/MeOH; (g) $Cl_3CCONCO$; (h) NH_3/MeOH; (i) $ClSO_3H$/MeOH; (j) H_2, Pd-C; (k) DDQ, acetone-H_2O, -78°C; (l) NH_3, MeOH

SCHEME 23.14 Total synthesis of mitomycin C.

Mitomycin C (Mutamycin®) is supplied as lyophilized powder also containing mannitol in 5- and 20-mg vials. It is diluted with sterile water and administered by slow intravenous push or by continuous intravenous infusion. Extravasation produces severe local necrosis. It is cleared rapidly from the vascular compartment, and the highest level appears in the kidneys, with lower detectable levels found in muscle, intestine, stomach, and eye.[155] The most serious toxicity of mitomycin C is bone marrow suppression, which reduces blood levels of platelets, leukocytes, and erythrocytes.[151] It is delayed and cumulative, which requires careful minimization of lifetime doses to under 50 to 60 mg/m^2. Gastrointestinal disturbances, renal toxicity, and alopecia also occur, and fatigue and drowsiness have been observed.[151,153,155]

The troublesome toxicity of mitomycin C has prompted the synthesis of hundreds of analogs. Among them, four compounds with novel substituents at C(7) have been introduced into clinical trials (Table 23.1). The *p*-hydroxyanilino derivative known as M83 was introduced by Kyowa Hakko Kogyo, and it showed excellent activities in animal models;[90] however, it was not superior to mitomycin C in the clinic. Bristol-Myers Squibb developed the highly potent dimethylamidino derivative, BMY–25282,[156–157] but it was cardiotoxic. Following interesting preclinical studies on the supposed mercaptoethylamino derivative, which was shown to actually be the corresponding disulfide,[98,158,159] a variety of disulfides cleavable to mercaptoethylamino derivatives was prepared.[160] Bristol-Myers Squibb conducted clinical trials on the [2-(4-nitrophenyl)dithio]ethylamino derivative, BMY-25067,[161] but eventually decided not to pursue it further. The 2-[2-(L-glutamylamino)ethyl]dithioethylamino derivative, KT 6149,[96] was introduced into the clinic by Kyowa Fermentation Industries and is still under investigation.

Porfiromycin was discovered at Upjohn,[162] and they gave it a brief clinical trial. It has much the same pattern of activity and toxicity as mitomycin C, but it is less potent.

REFERENCES

1. Hata, T. et al., Mitomycin, a new antibiotic from streptomyces, *J. Antibiot. (Jpn.), Ser. A*, 9, 141, 1956.
2. Wakaki, S. et al., Isolation of new fractions of antitumor mitomycins, *Antibiot. Chemother.* 8, 288, 1958.
3. Frank, W. and Osterberg, A.E., Mitomycin C (NSC-26988) — an evaluation of the Japanese reports, *Cancer Chemother. Rep.*, 9, 114, 1960.
4. Jones, Jr., R., Mitomycin C. A preliminary report of studies of human pharmacology and initial therapeutic trial, *Cancer Chemother. Rep.* 2, 3, 1959.
5. Sutow, W.W. et al., Evaluation of dosage schedules of mitomycin C (NSC-26988) in children, *Cancer Chemother. Rep., Part 1*, 55, 285, 1971.
6. Webb, J.S. et al., The structures of mitomycins A, B, C, and porfiromycins — parts I and II, *J. Am. Chem. Soc.*, 84, 3185, 3186, 1962.
7. Yahashi, R. and Matsubara, I., The molecular structure of 7-demethoxy-7-*p*-bromoanilino mitomycin B, *J. Antibiot* (Tokyo), 29, 10, 1976.
8. Urakawa, C. et al., Mitomycin derivatives, *Jpn. Kokai Tokkyo Koho*, 1980.
9. Kono, M. et al., Albomitomycin A and isomitomycin A: products of novel intramolecular rearrangement of mitomycin A, *J. Am. Chem. Soc.*, 109, 7224, 1987.
10. Feigelson, G.B. et al, On the reaction of leucomitosenes with osmium tetroxide. A route to novel mitomycins, *J. Org. Chem.*, 1988, 53, 3390.
11. Tulinsky, A. and van den Hende, J.H., The crystal and molecular structure of n-brosylmitomycin A, *J. Am. Chem. Soc.*, 89, 2905, 1967.
12. Taylor, W.G. and Remers, W.A., Structures and stereochemistry of mitomycin hydrolysis products, *Tetrahedron Lett.*, 2483, 1974.
13. Rebek, J. et al., Total synthesis of a mitosene, *J. Org. Chem.*, 49, 5164, 1984.
14. Garrett, E.R., The physical chemical characterization of the products, equilibria, and kinetics of the complex transformations of the antibiotic porfiromycin, *J. Med. Chem.*, 6, 488, 1963.
15. Kinoshita, S. et al., Chemical transformations of mitomycins: structure-activity relation of mitomycin derivatives, *Prog. Antimicrob. Anticancer Chemother.* II, 112, 1970.
16. Underberg, W.J.M. and Lingeman, H., Aspects of chemical stability of mitomycin and porfiromycin in solution, *J. Pharm. Sci.*, 72, 549, 1974.
17. Schiltz, P. and Kohn, H., Studies on the reactivity of reductively activated mitomycin C, *J. Am. Chem. Soc.*, 115, 10510, 1993.
18. Hornemann, Ho, Y.-K., Mackey, Jr., J.K., and Srivastava, S.C., Studies on the mode of action of mitomycin antibiotics. Reversible conversion of mitomycin C into sodium 7-aminomitosane-9a-sulfonate, *J. Am. Chem. Soc.*, 98, 7069, 1976.
19. Han, I., Russell, D.J., and Kohn, H., Studies on the mechanism of mitomycin C(1) electrophilic transformations: structure-activity relationships, *J. Org. Chem.*, 57, 1799, 1992.
20. Wang, S. and Kohn, H., Studies on the mechanism of activation of C(7) ethylenediamine substituted mitomycins. Relevance of the proposed mode of action of BMY-25067 and KW-2149, *Tetrahedron Lett.*, 37, 2337, 1996.
21. Wang, S. and Kohn, H., Mitomycin betaines: synthesis, structure and solvolytic reactivity, *J. Org. Chem.*, 61, 9202, 1996.
22. Pan, S.-S. et al., Reductive activation of mitomycin C and mitomycin C metabolites catalyzed by NADPH-cytochrome P-450 reductase and xanthine oxidase, *J. Biol. Chem.*, 259, 959, 1984.
23. Danishefsky, S.J. and Egbertson, M., The characterization of intermediates in the mitomycin activation cascade: a practical synthesis of an aziridinomitosene, *J. Am. Chem. Soc.*, 108, 4648, 1986.
24. Hoey, B.M., Butler, J., and Swallow, A.J., Reductive activation of mitomycin C, *Biochemistry*, 27, 2608, 1988.
25. Kirsch, E.J. and Korshalla, J.D., Influence of biological methylation on the biosynthesis of mitomycin A, *J. Bacteriol.*, 87, 247, 1964.
26. Hornemann, U. et al., D-Glucosamine and L-citrulline, precursors in mitomycin biosynthesis by *Streptomyces verticillatus*, *J. Am. Chem. Soc.*, 96, 320, 1974.
27. Mao, Y., Varoglu, M., and Sherman, D.H., Mitomycin resistance in *Streptomyces lavendulae* includes a novel drug-binding-protein-dependent export system, *J. Bact.*, 181, 2199, 1999.

28. Kim, C.G. et al., Formation of 3-amino-5-hydroxybenzoic acid, the precursor of mC7 units in ansamycin antibiotics, by a new variant of the shikimate pathway, *J. Am. Chem. Soc.*, 111, 4941, 1992.

29. Islam, I. and Skibo, E.B., Synthesis and physical studies of azamitosene and iminoazamitosene reductive alkylating agents. Iminoquinone hydrolytic stability, syn/anti isomerization, and electrochemistry, *J. Org. Chem.*, 55, 3195, 1990.

30. Islam, I. et al, Structure-activity studies of antitumor agents based on pyrrolo[1,2-a]benzimidazoles: new reductive alkylating DNA cleaving agents, *J. Med. Chem.*, 34, 2994, 1991.

31. Huang, X., Suleman, A., and Skibo, E.B., Rational design of pyrrolo[1,2-a]benzimidazole based antitumor agents targeting the DNA major groove, *Bioorg. Chem.*, 28, 324, 2000.

32. Mercado, C.M. and Tomasz, M., Inhibitory effects of mitomycin-related compounds lacking the C1-C2 aziridine ring, *Antimicrob. Agents Chemother.*, 1, 73, 1972.

33. Szybalski, W. and Iyer, V.N., The mitomycins and porfiromycin, in *Antibiotics I, Mechanism of Action*, Gottlieb, D. and Shaw, P.D., Eds., Springer, New York, 1967, pp. 221-245.

34. Iyer, V.N. and Szybalski, W., Mode of interaction of mitomycin C with deoxyribonucleic acid and other polynucleotides *in vitro*, *Science*, 145, 55, 1964.

35. Tomasz, M. et al., Mode of interaction of mitomycin C with deoxyribonucleic acid and other polynucleotides *in vitro*, *Biochemistry*, 13, 4878, 1974.

36. Gustafson, D.L. and Pritsos, C.A., Kinetics and mechanism of mitomycin C bioactivation by xanthine dehydrogenase under aerobic and hypoxic conditions, *Cancer Res.*, 53, 5470, 1993.

37. Wang, X. et al., Enhanced cytotoxicity of mitomycin C in human tumor cells with inducers of DT-diaphorase, *Br. J. Cancer*, 80, 1223, 1999.

38. Sagara, N. and Katoh, M., Mitomycin C resistance induced by TCF-3 overexpression in gastric cancer cell line MKN28 is associated with DT-diaphorase down regulation, *Cancer Res.*, 60, 5959, 2000.

39. Belcourt, M.F. et al., Differential toxicity of mitomycin C and porfiromycin to aerobic and hypoxic Chinese hamster ovary cells overexpressing human NADPH:cytotcrome (P-540) reductase, *Proc. Natl. Acad. Sci. USA*, 93, 456, 1996.

40. Moore, H.W., Bioactivation as a model for drug design bioreductive alkylation, *Science*, 197, 527, 1977.

41. Li, V.-S. et al., Structural requirements for mitomycin C DNA binding, *Biochemistry*, 34, 7120, 1995.

42. Boruah, R.C. and Skibo, E.B., Mitomycin C redox couple pKa values studied by spectrophotometric titration, pH rate profiles, and Nernst-Clark plots, *J. Org. Chem.*, 60, 2232, 1995.

43. Tomasz, M. et al., Isolation and characterization of a major adduct between mitomycin C and DNA, *J. Am. Chem. Soc.*, 110, 5892, 1988.

44. Tomasz, M., Chawla, A.K., and Lipman, R., Mechanism of monofunctional and bifunctional alkylation of DNA by mitomycin C, *Biochemistry*, 27, 3182, 1988.

45. Tomasz, M. et al., Full structure of a mitomycin C dinucleotide adduct. Use of differential FT-IR spectroscopy in microscale structural studies, *J. Am. Chem. Soc.*, 105, 2059, 1983.

46. Verdine, G.L. and Nakanishi, K., Use of differential second-derivative UV and FTIR spectroscopy in structural studies with multichromophoric compounds, *J. Am. Chem. Soc.*, 107, 6118, 1985.

47. Tomasz, M. et al., Reassignment of the guanine-binding mode of reduced mitomycin C. *Biochemistry*, 25, 4337, 1986.

48. Tomasz, M. et al., Reaction of DNA with chemically or enzymatically activated mitomycin C: isolation and structure of the major covalent adduct, *Proc. Natl. Acad. Sci. USA*, 83, 670, 1986.

49. Borowy-Borowski, H. et al., Duplex oligodeoxyribonucleotides crosslinked by mitomycin C at a single site: synthesis, properties, and crosslink, *Biochemistry*, 19, 2992, 1990.

50. Norman, B. et al., NMR and computational characterization of mitomycin cross-linked to adjacent deoxyguanosines in the minor groove of the d(TACGA).d(TACGA) duplex, *Biochemistry*, 29, 2861, 1990.

51. Weidner, M.F., Millard, J.T., and Hopkins, P.B., Determination at single-nucleotide resolution of the sequence specificity of DNA interstrand cross-linking agents in DNA fragments, *J. Am. Chem. Soc.*, 111, 9720, 1989.

52. Rao, S.N., Singh, U.C., and Kollman, P.A., Conformations of the noncovalent and covalent complexes between mitomycins A and C and d(GCGCGCGCGC)₂, *J. Am. Chem. Soc.*, 108, 2058, 1986.

53. McGuiness, B.F. et al., Reductive alkylation of DNA by mitomycin A, a mitomycin with high redox potential, *Biochemistry*, 30, 6444, 1991.

54. Remers, W.A. et al., Conformations of complexes between mitomycins and decanucleotides. 3. Sequence specificity, binding at C-10, and mitomycin analogs, *J. Med. Chem.*, 31, 1612, 1988.

55. Teng, S.P., Woodson, S.A., and Crothers, D.M., DNA sequence specificity for mitomycin crosslinking, *Biochemistry*, 28, 3901, 1989.

56. Borowy-Borowski, H., Lipman, R., and Tomasz, M., Recognition between mitomycin C and specific double-strand DNA sequences for cross-link formation, *Biochemistry*, 29, 2999, 1990.

57. Kumar, S., Johnson, W.S., and Tomasz, M., Orientation isomers of the mitomycin C interstrand cross-links in non-self-complementary DNA. Differential effect of two isomers on restriction endonuclease cleavage at a nearby site, *Biochemistry*, 32, 1364, 1993.

58. Kumar, S., Lipman, R., and Tomasz, M., Recognition of specific DNA sequences by mitomycin C for alkylation, *Biochemistry*, 31, 1399, 1992.

59. Li, V.-S. et al., Role of the C-10 subsituent in mitomycin C-1-DNA binding, *J. Am. Chem. Soc.*, 118, 2326, 1996.

60. Kohn. H. et al., On the origins of the DNA sequence selectivity of mitomycin monoalkylation transformations, *J. Am. Chem. Soc.*, 114, 9218, 1992.

61. Kumar, S.G. et al., Mitomycin C-DNA adducts generated by DT-diaphorase. Revised mechanism of the enzymatic reductive activation of mitomycin C, *Biochemistry*, 36, 14128, 1997.

62. Kumar, G.S. et al., Binding of 2,7-diaminomitosene to DNA: Model for recognition of DNA by activated DNA, *Biochemistry*, 34, 2662, 1995.

63. Kumar, G.S. et al., 2,7-Diaminomitosene, a monofunctional mitomycin C derivative, alkylates DNA in the major groove. Structure and base-sequence specificity of the DNA adduct and mechanism of alkylation, *J. Am. Chem. Soc.*, 118, 9209, 1996.

64. Palom, Y. et al., Relative toxicities of DNA cross-links and monoadducts: new insights from studies of decarbamoyl mitomycin C and mitomycin C, *Chem. Res. Tox.*, 15, 1398, 2002.

65. Gargulio, D. et al., Alkylation and crosslinking of DNA by the unnatural enantiomer of mitomycin C: mechanism of the DNA-sequence specificity of mitomycins, *J. Am. Chem. Soc.*, 117, 9388, 1995.

66. Maliepaard, M. et al., Mitosene-DNA adducts. Characterization of two major DNA adducts formed by 1,10-bis(acetoxy)-7-methoxymitosene upon reductive activation., *Biochemistry*, 36, 9211, 1997.

67. Ouyang, A. and Skibo, E., The iminium ion chemistry of mitosene DNA alkylating agents. Enriched ^{13}C-NMR studies, *Biochemistry*, 39, 5718, 2000.

68. Xing, C. and Skibo, E.B., Sigmatropic reactions of the aziridinyl semiquinone species. Why aziridinyl benzoquinones are metabolically more stable than aziridinyl indoloquinones, *Biochemistry*, 39, 10770, 2000.

69. Millard, J.T. and Beachy, T.M., Selective recognition of m5Cpg dinucleotide sequence in DNA by mitomycin C for alkylation and crosslinking, *Biochemistry* 32, 12850, 1993.

70. Tomasz, M. et al., The purine 2-amino group as the critical recognition element for sequence-specific alkylation and crosslinking of DNA by mitomycin C, *J. Am. Chem. Soc.*, 120, 11581, 1998.

71. Sharma, M. and Tomasz, M., Conjugation of glutathione and other thiols with bioreductively activated mitomycin C. Effect of thiols on reductive activation rate, *Chem. Res. Toxicol.*, 7, 390, 1994.

72. Sharma, M., He, Q.-Y., and Tomasz, M., Effects of glutathione on alkylation and crosslinking of DNA by mitomycin C. Isolation of a ternary glutathione-mitomycin-DNA adduct, *Chem. Res. Toxicol.*, 7, 401, 1994.

73. Xu, B.H., Gupta, V., and Singh, S.V., Mitomycin sensitivity in human bladder cancer cells: possible role of glutathione and glutathione transferase in resistance, *Arch. Biochem. Biophys.*, 308, 164, 1994.

74. Basu, A.K. et al., Effect of site specifically located mitomycin C-DNA monoadducts on *in vitro* DNA synthesis by DNA polymerases, *Biochemistry*, 32, 4708, 1993.

75. Abbas, T. et al., Differential activation of p53 by the various adducts of mitomycin C, *J. Biol. Chem.*, 277, 40513, 2002.

76. Skibo, E.B. and Schultz, W.G., Pyrrolo[1,2-a]benzimidazole based aziridinyl quinones, *J. Med. Chem.*, 36, 3050, 1993.

77. Schultz, W.G., Nieman, R.A., and Skibo, E.B., Evidence for DNA phosphate backbone alkylation and cleavage by pyrrolo[1,2-a]benzimidazoles, small molecules capable of causing sequence specific phosphodiester bond hydrolysis, *Proc. Natl. Aced. Sci. USA*, 92, 11854, 1995.

78. Bachur, N.R., Gordon, S.L., and Gee, M.V., A general mechanism for microsomal activation of quinone anticancer agents to free radicals, *Cancer Res.*, 38, 1745, 1978.
79. Tomasz, M., Hydrogen peroxide generation during the redox cycle of mitomycin C and DNA-bound mitomycin C, *Chem. Biol. Interact.*, 13, 89, 1976.
80. Lown, J.W., The molecular mechanism of antitumor action of the mitomycins, in *Mitomycin C, Current Status and New Developments*, Carter, S.T. and Crooke, S.T., Eds., Academic Press, Orlando, FL, 1979, pp. 5-26.
81. Ueda, K. et al., Inactivation of bacteriophage NX174 by mitomycin C in the presence of sodium hydrosulfite and cupric ions, *Chem. Biol. Interact.*, 29, 145, 1980.
82. Pan, S.-S., Andrews, P.A., and Bachur, N.R., Reduction of mitomycin C by NADPH cytochrome p450 reductase and xanthine oxidase, *Fed. Proc.*, 41, 1541, 1982.
83. Hashimoto, Y., Shudo, K., and Okamoto, T., Structures of modified nucleotides isolated from calf thymus DNA alkylated with reductively activated mitomycin C, *Tetrahedron Lett.*, 23, 677, 1982.
84. Pritsos, C.A. and Sartorelli, A.C, Generation of reactive oxygen radicals through bioactivation of mitomycin antibiotics, *Cancer Res.*, 46, 3528, 1986.
85. Remers, W.A. and Dorr, R.T. in *The Alkaloids: Chemical and Biological Perspectives*, Pelletier, S.W., Ed., Wiley, New York, 1988, Vol. 6.
86. Kinoshita, S. et al., Mitomycin derivatives. 1. Preparation of mitosane and mitosene compounds and their biological activities, *J. Med. Chem.*, 14, 103, 1971.
87. Kinoshita, S. et al., Mitomycin derivatives. 2. Derivatives of decarbamoylmitosane and decarbamoylmitosene, *J. Med. Chem.*, 14, 109, 1971.
88. Kojima, R. et al., Some structure-activity relationships for mitomycin C (NSC-26980) derivatives in the treatment of l1210 leukemia, *Cancer Chemother. Rep.*, 3, 121, 1972.
89. Iyengar, B.S. et al., Mitomycin C and porfiromycin analogs with substituted ethylamines at position 7 *J. Med. Chem.*, 26, 16, 1983.
90. Imai, R. et al., Antitumor activity of 7-N-phenyl derivatives of mitomycin C in the leukemia p388 system, *Gann*, 71, 560, 1980.
91. Iyengar, B.S. et al., Development of new mitomycin C and porfiromycin analogs. *J. Med. Chem.*, 24, 975, 1981.
92. Iyengar, B.S. et al., Mitomycin C analogs with secondary amines at position 7, *J. Med. Chem.*, 26, 1453, 1983.
93. Sami, S.M. et al., Mitomycin C analogs with aryl substituents on the 7-amino group, *J. Med. Chem.*, 27, 701, 1984.
94. Iyengar, B.S. et al., Metal complexes of mitomycins, *J. Med. Chem.*, 29, 144, 1986.
95. Kunz, K.R. et al., Structure-activity relationships for mitomycin C and mitomycin A analogs, *J. Med. Chem.*, 34, 2281, 1991.
96. Ishioka, C. et al., Comparative studies on the action of 7-N-[2-[[2-(-L-glutamylamino)ethyl]dithio]ethyl]mitomycin C and of mitomycin C on cultured NL60-cells and isolated phage and plasmid DNA, *Cancer Chemother. Pharmacol.*, 26, 117, 1990.
97. Bradner, W.T. et al, Antitumor activity and toxicity in animals of N-7[2-(4-nitrophenyldithio)ethyl]mitomycin C (BMY-25067), *Invest. New Drugs*, 8, 51, 1990.
98. Bradner, W.T. et al., Antitumor activity and toxicity in animals of RR-150 (7-cysteaminomitosane), a new mitomycin derivative, *Cancer Res.*, 44, 5619, 1984.
99. Sami, S.M. et al., Preparation and antitumor activity of new mitomycin A analogs, *J. Med. Chem.*, 30, 168, 1987.
100. Li, V.-S. et al., The role of the C-10 substituent in mitomycin C-1-DNA binding, *J. Am. Chem. Soc.*, 118, 2326, 1966.
101. Takanagi, H. et al., One step preparation of 6-demethyl-6-formylmitomycin C and cytotoxicity of its derivatives, *J. Antibiot.*, 45, 1815, 1992.
102. Iyengar, B.S., Remers, W.A., and Bradner, W.T., Preparation and antitumor activities of 7-substituted-1,2-aziridinomitosenes, *J. Med. Chem.*, 29, 1864, 1986.
103. Fost, D.L., Ekwuribe, N.N., and Remers, W.A., Synthesis of 1-substituted-7-methoxymitosenes, *Tetrahedron Lett.*, 131, 1973.
104. Taylor, W.G. et al., Mitomycin antibiotics. Synthesis and antineoplastic activity of mitosene analogs of the mitomycins, *J. Med. Chem*, 24, 1184, 1981.

105. Casner, M.L., Remers, W.A., and Bradner, W.T., Synthesis and biological activity of 6-substituted analogs of the mitomycins, *J. Med. Chem*, 28, 921, 1985.

106. Hodges, J.C., Remers, W.A., and Bradner, W.T., Synthesis and antineoplastic activity of mitosene analogs of the mitomycins, *J. Med. Chem*, 24, 1184, 1981.

107. Allen, Jr., G.R., Poletto, J.F., and Weiss, M.J., The mitomycin antibiotics. Synthetic studies. II. Synthesis of 7-methoxymitosene, an antibacterial agent, *J. Am. Chem. Soc.*, 86, 3878, 1964.

108. Allen, Jr., G.R., Poletto, J.F., and Weiss, M.J., The mitomycin antibiotics. Synthetic studies. XV. The preparation of a related indoloquinone, *J. Med. Chem.*, 10, 1, 1967.

109. Allen, Jr., G.R. and Weiss, M.J., The mitomycin antibiotics. Synthetic studies XVII. Preparation of 1-(β-substituted ethyl)indoloquinone analogs, *J. Med. Chem.*, 10, 23, 1967.

110. Remers, W.A., Roth, R.H., and Weiss, M.J., The mitomycin antibiotics. Synthetic studies. VIII. Indoloquinone analogs with variations at C-5, *J. Org. Chem.*, 31, 1012, 1966.

111. Iwata, H. and Goldberg, E.P., Tissue-binding macromolecular antitumor drugs for localized therapy: mitomycin C-concanavalin A conjugates, *Polymer Reprints*, 24, 73, 1983.

112. Kaneko, Y., Tanaka, T., and Iguchi, S., Preparation and properties of a mitomycin C-albumin conjugate, *Chem. Pharm. Bull.*, 38, 2614, 1990.

113. Umemoto, N. et al., Conjugates of mitomycin C with the immunoglobulin M monomer fragment of a monoclonal anti-MM46 immunoglobulin M antibody with or without serum albumin as intermediary, *J. Appl. Biochem.*, 6, 297, 1984.

114. Tsukada, Y., Characterization of human AFP producing stomach cancer xenotransplanted in nude mice and the effect of mitomycin C and antibody to human AFP of this tumor, *Nippon Shokakibyo Gakki Zasshi*, 82, 18, 1985.

115. Li, S. et al., Preparation of antigastric cancer monoclonal antibody MGb2-mitomycin C conjugate with improved antitumor activity, *Bioconjugate Chem.*, 1, 245, 1990.

116. Tanaka, J.-I. et al., Preparation of a conjugate of mitomycin C and anti-neural cell adhesion molecule monoclonal antibody for specific chemotherapy against biliary tract carcinoma, *Surg. Today*, 28, 1217, 1998.

117. Huh, N. et al., Design, synthesis, and evaluation of mitomycin-tethered phosphorothioate oligodeoxynucleotides, *Bioconjugate Chem.*, 7, 659, 1996.

118. Paz, M.M, Das, A., and Tomasz, M., Mitomycin C linked to minor groove binding agents: synthesis, reductive activation, DNA binding and cross-linking properties and *in vitro* antitumor activity, *Bioorg. Med. Chem.*, 7, 2713, 1999.

119. Okamoto, H. et al., Effect of 1-alkyl or 1-alkenyl-azocycloalkanone derivatives on penetration of mitomycin C through rat skin, *Chem. Pharm. Bull.*, 35, 4605, 1987.

120. Shikano, M. et al., 1a-Docosahexaenoyl mitomycin C: a novel inhibitor of protein tyrosine kinase, *Biochem. Biophys. Res. Commun.*, 248, 858, 1998.

121. Nagai, K. et al., Synthesis and antitumor activities of mitomycin C (1-3)-beta-D-glucan conjugate, *Chem. Pharm. Bull.*, 40, 986, 1992.

122. Ghiorghis, A., Talebaum, A., and Clarke, R., *In vitro* antineoplastic activity of C-7-substituted mitomycin C analogs MC-77 and MC-62 against human breast-cancer cell lines, *Cancer Chemother. Pharmacol.*, 29, 290, 1992.

123. Yan, Z. et al., Synthesis of the anticancer drug mitomycin C-dextran conjugate with affinity to lymph and studies on its biological properties, *Zhongguo Yaowu Huaxue Zashi*, 9, 240, 1999.

124. Hoste, K., Schacht, and E., Rihova, B., Synthesis and biological evaluation of PEG-substituted macromolecular prodrugs from mitomycin C, *J. Bioactive Compatible Polymers*, 17, 123, 2002.

125. Kato, Y., Onishi, H., and Machida, Y., A novel water-soluble N-succinyl-chitosan-mitomycin C conjugate prepared by direct carbodiimide coupling: physicochemical properties, antitumor characteristics, and systemic retention, *S.T.P. Pharma Sci.*, 10, 133, 2000.

126. Zalipsly, S. et al., New liposomal prodrug of mitomycin C, Proceedings of the 28th International Symposium on Controlled Release of Bioactive Materials and 4th Consumer Diversified Products Conference, San Diego, CA, June 23–27, 2001.

127. Iyengar, B.S. et al., Nucleotide derivatives of 2,7-diaminomitosene, *J. Med. Chem.*, 31, 1612, 1988.

128. Iyengar, B.S., Dorr, R.T., and Remers, W.A., Additional nucleotide derivatives of mitosenes. Synthesis and activity against parenteral and multidrug resistant L1210 leukemia, *J. Med. Chem.*, 34, 1947, 1991.

129. Sami, S.M. et al., Preparation and antitumor activity of new mitomycin A analogs, *J. Med. Chem.*, 30, 168, 1987.

130. Bradner, W.T., Remers, W.A., and Vyas, D.M., Structure-activity comparison of mitomycin C and mitomycin A analogs, *Anticancer Res.*, 9, 1095, 1989.

131. Bradner, W.T. et al., Antitumor activity and toxicity in animals of BMY-25282, a new mitomycin derivative, *Cancer Res.*, 45, 6475, 1985.

132. Remers, W.A., Mitomycins, in *Anticancer Agents Based on Natural Product Models*, Cassady, J.M and Douros, J.D., Eds., Academic Press, New York, 1980, Chapter 4.

133. Orlamans, E.O. et al., Synthesis, mechanism of action, and biological evaluation of mitosenes, *J. Med. Chem.*, 32, 1612, 1989.

134. Weiss, M.J. et al., The mitomycin antibiotics. Synthetic studies. XXII. Antibacterial structure-activity relationships in the indoloquinone series, *J. Med. Chem.*, 11,742, 1968.

135. Allen, Jr., G.R., Poletto, J.F., and Weiss, M.J., Mitomycin antibiotics. Synthetic studies. III. Related indoloquinones, active antibacterial agents, *J. Am. Chem. Soc.*, 86, 3878, 1964.

136. Allen, Jr., G.R., Poletto, J.F., and Weiss, M.J., Mitomycin antibiotics, Synthetic studies. XVII. Indolo-quinones with variants at the 2 position, *J. Org. Chem.*, 30, 2897, 1965.

137. Poletto, J.F., Allen, Jr., G.R., and Weiss, M.J., Mitomycin antibiotics. Synthetic studies. XIX. Synthesis of indoloquinone analogs with certain C-3 variants, J. Med. Chem., 10, 95, 1967.

138. Leadbetter, G. et al., Mitomycin antibiotics. Synthesis of 1-substituted 7-methoxymitosenes, *J. Org. Chem.*, 39, 3508, 1974.

139. Hirata, T., Yamada, Y., and Matsui, M., Synthetic studies on mitomycins. Synthesis of aziridinopyr-rolo[1,2-a]indoles, *Tetrahedron Lett.*, 19, 1969.

140. Siuta, G.J., Franck, R.W., and Kempton, R.J., Studies directed toward a mitomycin synthesis, *J. Org. Chem.*, 39, 3739, 1974.

141. Kametani, T. et al., Photooxygenation of 9-keto-7-methoxy-6-methyl-9H-pyrrolo[1,2-a]indole as a synthetic approach to mitomycins, *Heterocycles*, 4, 1637, 1976.

142. Nakatsubo, F. et al., Synthetic studies toward mitomycins. 2. Total synthesis of dl-porfiromycin, *J. Am. Chem. Soc.*, 99, 8115, 1977.

143. Fukuyama, T. et al., Synthetic studies toward mitomycins. III. Total synthesis of mitomycins A and C, *Tetrahedron Lett.*, 4295, 1977.

144. Fukuyama, T. and Yang, L., Practical synthesis of (±)–mitomycin C, *J. Am. Chem. Soc.*, 111, 8303, 1989.

145. Danishefsky, S., Chemical explorations driven by an enchantment with mitomycinoids — a twenty year account, *Synlett*, 475, 1995.

146. Benbow, J.W., McClure, K.F., and Danishefsky, S.J., Intramolecular cycloaddition reactions of dienyl nitroso compounds: application to the synthesis of mitomycin K, *J. Am. Chem. Soc.*, 115, 12305, 1993.

147. McClure, K.F. and Danishefsky, S.J., A novel Heck arylation reaction: rapid access to congeners of FR 900482, *J. Am. Chem. Soc.*, 115, 6094, 1993.

148. Moertel, C.G., Reitemeir, R.J., and Hahn, R.G., Mitomycin C therapy in advanced gastrointestinal cancer. *J. Am. Med. Assoc.*, 204, 1045, 1968.

149. Doll, D.C., Weiss, R.B., and Issel, B.F., Mitomycin: ten years after approval for marketing, *J. Clin. Oncol.*, 3, 276, 1985.

150. Miller, T.P., McMahon, L.J., and Livingston, R.B., Extensive adenocarcinoma and large cell undif-ferentiated carcinoma of the lung treated with 5-FU, vincristine, and mitomycin C. *Cancer Treat. Rep.*, 64, 1241, 1980.

151. Crooke, S.T. and Bradner, W.T., Mitomycin C: a review. *Cancer Treat. Rev.*, 3, 121, 1976.

152. Early, K. et al., Mitomycin C in the treatment of metastatic transitional cell carcinoma of urinary bladder. *Cancer*, 31, 1150, 1973.

153. Baker, L.H., Opipari, M.I., and Izbicki, R.M., Phase II study of mitomycin C, vincristine, and bleomycin in advanced squamous cell carcinoma of the uterine cervix, *Cancer*, 38, 2222, 1976.

154. MacDonald, J. et al., 5-Fluorouracil (5-FU), mitomycin C (MMC), and Adriamycin (ADR) — FAM: a new combination chemotherapy program for advanced gastric carcinoma (abstract C-11), *Proc. Am. Soc. Clin. Oncol.*, 17, 244, 1976.

155. Dorr, R.T. and Von Hoff, D.D., *Cancer Chemotherapy Handbook*, Appleton and Lange, Norwalk, CT, 1994, p. 719.

156. Kaplan, M.A. and Vyas, D.M., Crystalline 7-(dimethylaminomethylene)amino-9a-methoxymitosane, Ger. Offen. DE 3,605,523, 1986.
157. Bradner, W.T. et al., Antitumor activity and toxicity in animals of BMY-25282, a new mitomycin derivative, *Cancer Res.*, 45, 6475, 1985.
158. Bradner, W.T. et al., Antitumor activity and toxicity in animals of RR-150 (7-cysteaminomitosane), a new mitomycin derivative, *Cancer Res.*, 44, 5619, 1984.
159. Senter, P.D. et al., Reassignment of the structure of the antitumor agent RR-150, *J. Antibiot.*, 41, 199, 1988.
160. Vyas, D.M. et al., Novel disulfide mitosanes as antitumor agents, *Recent Adv. Chemother, Proc. Int. Congr. Chemother 14th, Anticancer Sect. 1*, 517, 1985.
161. Bradner, W.T. et al., Antitumor activity and toxicity in animals of N-7-[2-(4-nitrophenyldithio)ethyl]mitomycin C (BMY-25067), *Invest. New Drugs*, 8, S1-S7, 1990.
162. De Boer, C. et al., Porfiromycin, a new antibiotic, *Antimicrobial. Agents Annual*, 1960, Plenum Press, New York, 1961, pp. 17-22.

24 Staurosporines and Structurally Related Indolocarbazoles as Antitumor Agents

Michelle Prudhomme

CONTENTS

I. Introduction 499
II. Mechanisms of Action 500
    A. Staurosporine and UCN-01 500
    B. K-252a...502
    C. Rebeccamycin 503
    D. AT2433-A1 and B1 503
III. Total Syntheses of the Natural Products 503
IV. Development of Drugs from the Natural Products 506
V. Clinical Applications...511
VI. Conclusions 512
References 512

I. INTRODUCTION

Bacteria and primitive organisms are important sources of various compounds of biological interest. Among the diversity of bacterial metabolites are maleimido- or maleamido-indolocarbazoles. Some of them do not bear any sugar moiety (arcyriaflavins, tjipanazole J, K-252c, BE-13793C), and others have a carbohydrate moiety linked to only one indole nitrogen (rebeccamycin, AT2433-A1 and B1) or to both indole nitrogens (staurosporines, K-252a and b) (Figure 24.1).[1–6] Staurosporine was isolated from *Streptomyces staurosporeus*[7,8] in 1977, and 3 yr later arcyriaflavins were obtained from the fruiting bodies of slime molds *Arcyria denudata*.[9] Later, more than 30 microbial staurosporine analogs were isolated from bacterial cultures. Indolocarbazoles have attracted considerable interest since the discovery in 1986 of the inhibitory properties of staurosporine toward protein kinase C (PKC), a kinase involved in many signal transduction pathways leading to a variety of cell responses, such as cell proliferation, differentiation, and gene expression.[10] K-252a and analogs bearing a five-membered ring sugar moiety have been isolated from *Nocardiopsis* and *Streptomyces* species, and they are also PKC inhibitors.[11–13] Staurosporine is a nonselective ATP-competitive kinase inhibitor that inhibits the different PKC isoenzymes, cyclin-dependent kinases (CDKs), and tyrosine kinases involved in cell proliferation.[14–17] More recently, it has been shown that UCN-01, a 7-hydroxy staurosporine isolated from cultures of a *Streptomyces* strain,[18] abrogates cell cycle arrest caused by DNA damage. This results in premature activation of mitosis in DNA-damaged cells, which leads to cell death.[19,20]

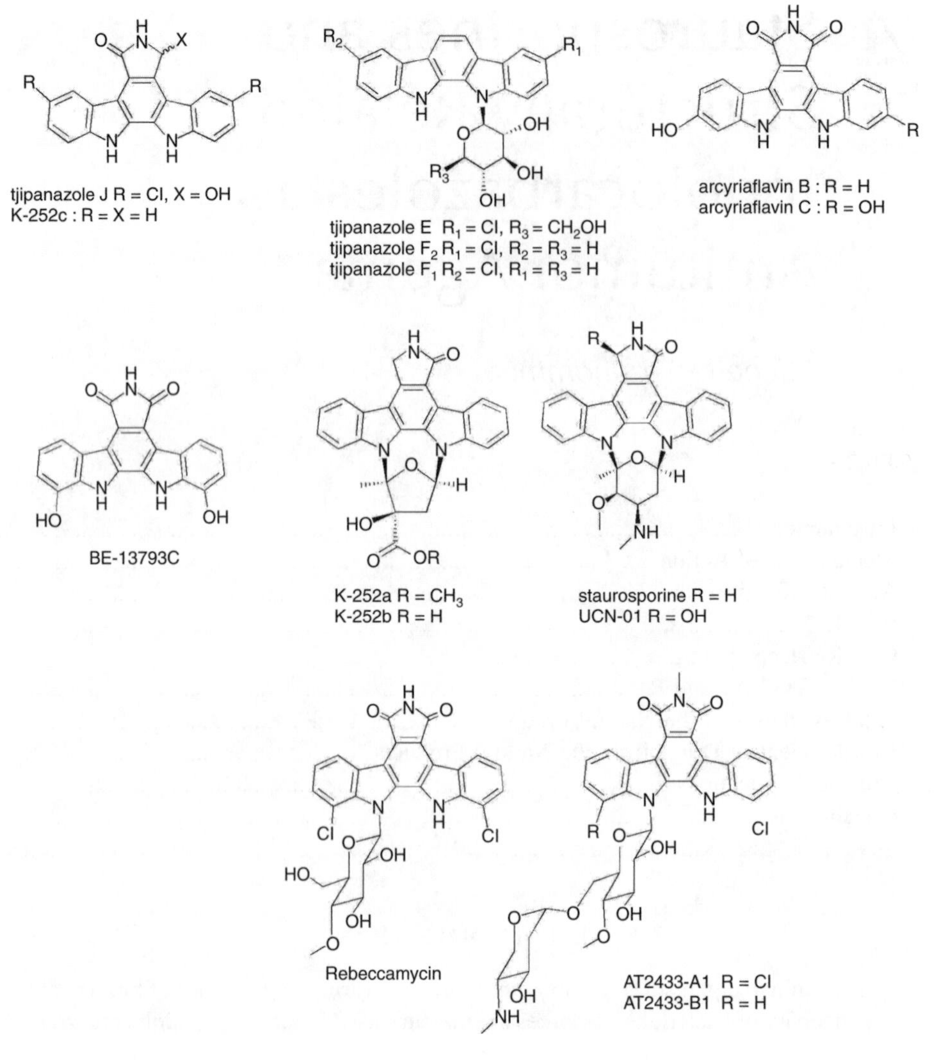

tjipanazole J R = Cl, X = OH
K-252c : R = X = H

tjipanazole E R₁ = Cl, R₃ = CH₂OH
tjipanazole F₂ R₁ = Cl, R₂ = R₃ = H
tjipanazole F₁ R₂ = Cl, R₁ = R₃ = H

arcyriaflavin B : R = H
arcyriaflavin C : R = OH

BE-13793C

K-252a R = CH₃
K-252b R = H

staurosporine R = H
UCN-01 R = OH

Rebeccamycin

AT2433-A1 R = Cl
AT2433-B1 R = H

FIGURE 24.1 Indolocarbazoles from natural sources.

The cytotoxicity of rebeccamycin and BE-13793C is linked to their capacity to inhibit topoisomerases, nuclear enzymes that control the topological state of DNA and are involved in various vital processes such as replication and transcription. Rebeccamycin is only a topoisomerase I inhibitor, whereas BE-13793C inhibits both topoisomerase I and II.[21,22] The antibiotics, AT2433-A1 and B1, structurally related to rebeccamycin, do not inhibit topoisomerases, and their cytotoxicity may be a consequence of their tight intercalative binding to DNA.[23] Because of the promising biological activities of indolocarbazoles, large structure–activity relationship studies were carried out, which led to the discovery of more potent and more selective molecules. In this chapter, only indolocarbazole compounds bearing a sugar moiety are described.

II. MECHANISMS OF ACTION

A. STAUROSPORINE AND UCN-01

Staurosporine, initially described as a PKC inhibitor, proved to be a potent and nonselective adenosine triphosphate–competitive kinase inhibitor. It strongly inhibits serine/threonine kinases

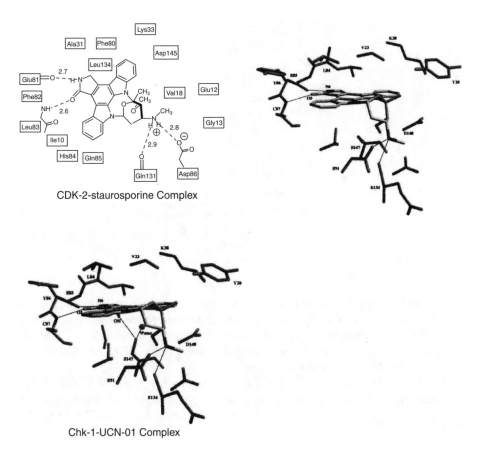

CDK-2-staurosporine Complex

Chk-1-UCN-01 Complex

FIGURE 24.2 (See color insert following page 304.) Crystal structures of staurosporine and UCN-01 complexes with kinases.

such as PKC, PKA, PKG, CDKs, CaMK, MLCK, S6-kinase, ERK-1, and aurora kinases, as well as tyrosine kinases such as PDGFR, c- and v-Src, EGFR, and many others.[14,24,25] The majority of them are extremely sensitive to staurosporine, with IC_{50} values in the low nanomolar range. The crystal structure of staurosporine in complex with the cyclin-dependent kinase CDK2 provides insight into the interactions with a variety of kinases. The carbonyl and the NH of the lactam heterocycle seem to be always involved in hydrogen bonds with amino acid residues of all kinases (Figure 24.2).[26] Staurosporine has also been cocrystallized with cAMP-dependent protein kinase PKA and Src tyrosine kinase CSK.[27,28]

More interesting is UCN-01, which exhibits greater selectivity toward protein kinases. UCN-01 inhibits "classic" PKC isoenzymes but has no effect against the "atypical" PKC isoforms, which are only activated on their regulatory domain by phosphatidylserine and phosphatidylinositol-3,4,5-triphosphate.[29] UCN-01 inhibits the growth of various human tumor cell lines. In contrast to staurosporine, UCN-01 is a relatively nonpotent direct inhibitor of CDK1 and CDK2, and it induces preferential G1-phase accumulation and reduction of cellular CDK2 activity by preventing the phosphorylation of Thr160 residue.[30–32] Moreover, UCN-01 abrogates G2 and S checkpoints.[33,34] In the cell cycle of normal cells, three principal checkpoints can be activated in response to DNA damage in the G1, S, and G2 phases. These checkpoints control the ability of cells to arrest the cell cycle, allowing time to repair the DNA. The G1 checkpoint allows DNA repair before DNA replication, the S checkpoint is involved in replication fork regulation, and the G2 checkpoint allows DNA repair before mitosis. The G1 checkpoint is dependent on the activity of p53. The *p53* gene is mutated in nearly half of all human malignancies and in large proportions of premalignant

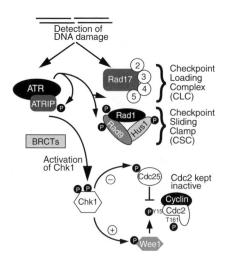

FIGURE 24.3 Activation of Chk1.

precursor lesions.[35-37] In p53 mutated cancer cells, the G1 checkpoint is lacking, and the G2 checkpoint provides cancer cells with the opportunity to repair their DNA after damage. The combination of a DNA-damaging agent with molecules that can abrogate S and G2 arrest would enhance cytotoxicity selectively in the tumor cells. The arrest in the G2 phase after DNA damage seems to be mediated by Chk1 kinase inhibition.

In the presence of DNA damage, Chk1 is activated by phosphorylation by ATM or ATR (Figure 24.3). Chk1 then phosphorylates and inactivates the Cdc25C phosphatase, thereby preventing activation of the cyclinB1/CDK1 mitotic complex through phosphorylation of Thr-14 and Tyr-15 residues controlled by Wee-1 kinase and Cdc25C phosphatase. Because Wee-1 kinase is relatively resistant to UCN-01,[14,38] this staurosporine analog should inhibit one or several kinases that phosphorylates Cdc25C on the Ser-16 residue (Chk1, Chk2, and cTAK1).[39] It has been shown that UCN-01 inhibits Chk1 and cTAK1 with IC_{50} values of 11 and 27 nM, respectively.[33] Initially, it was suggested that Chk2 was weakly inhibited by UCN-01 (IC_{50} 1040 nM). A more recent report has shown that UCN-01 efficiently inhibits both Chk1 and Chk2 kinase activity immunoprecipitated from human colon carcinoma HCT116 cells (IC_{50} about 10 nM); however, inhibition of recombinant Chk2 requires much higher concentrations of UCN-01 (IC_{50} about 300 nM).[40] In p53 mutant cells, low concentrations of UCN-01 caused S phase cells to progress to G2 before undergoing mitosis and death, whereas high concentrations caused a direct S-to-M premature mitosis.[34] X-ray structures of Chk1 in complex with staurosporine and UCN-01 have been determined. Both compounds bind in the ATP-binding pocket of the kinase (Figure 24.2).[41] The selectivity of UCN-01 toward Chk1 over cyclin-dependent kinases can be a result of the interaction of the hydroxyl group on the lactam heterocycle with the ATP-binding pocket. The inhibition activities of UCN-01 and staurosporine (Ki, app) against Chk1, CDK1, CDK2, and CDK4 have been determined. The Ki, app values for UCN-01 in nanomoles are 5.6, 95, 30, and 3600, respectively, and for staurosporine, the values are 7.8, 1.0, 2.9, and 41, respectively.[41] An additional target for the apoptosis-modulating activity of UCN-01 has recently been identified. UCN-01 inhibits 3-phosphoinositide-dependent protein kinase 1 (PDK1), which plays a central role in the regulation of the Akt-PKB survival pathway by phosphorylating Akt on Thr208 residue.[42] The PDK1-Akt survival pathway represents a new and attractive target for cancer chemotherapy.

B. K-252A

K-252a was originally identified as a PKC inhibitor, but its effects are wide-ranging and include tyrosine kinase receptor TrkA and MLK3 inhibition and activation of PI3K and MEK signaling pathways. It also induces cell cycle arrest and apoptosis by inhibiting CDK1 and Cdc25C.[11,43-45]

C. REBECCAMYCIN

Rebeccamycin was isolated in 1987 from cultures of *Saccharothrix aerocolonigenes*.[46] Its antipro-liferative activity against various tumor cell lines is linked to its ability to inhibit topoisomerase I.[47] Although rebeccamycin is structurally related to staurosporine, the biological targets of these microbial metabolites are quite different. Rebeccamycin is inactive against PKC, PKA, and topoi-somerase II but inhibits topoisomerase I,[21] whereas staurosporine is a nonselective kinase inhibitor without activity against topoisomerases. Like camptothecin, rebeccamycin is a topoisomerase I poison, stabilizing the enzyme–DNA interaction in the form of a covalent intermediate called "cleavable complex" and preventing the religation of the cleaved DNA strand.[48] More potent and soluble rebeccamycin analogs were prepared either by semisynthesis from rebeccamycin or by total synthesis.[49-51]

D. AT2433-A1 AND B1

AT2433-A1 and B1 (Figure 24.1) have been isolated from cultures of *Actinomadura melliaura*, with two other molecules bearing a primary amine instead of the methylamino group on the deoxysugar moiety.[52] Their antitumor activity was evaluated against transplantable mouse leukemia P388. AT2433-A1 was at least eight times more potent than AT2433-B1. Their cytotoxicity may be a consequence of their tight intercalative binding to DNA, preferentially to GC-rich sequences, but contrary to rebeccamycin, these disaccharide indolocarbazoles have no inhibitory effect on topoi-somerase I.[53,54] In contrast to AT2433-B1, its isomer, JDC-277, bearing an axial hydroxyl group instead of an equatorial *N*-methylamino group on the deoxysugar moiety, stimulates topoisomerase I-mediated DNA cleavage mainly at TG sites, showing that sequence-selective DNA interaction and topoisomerase I inhibition depends on the stereochemistry, and the substituents on the disac-charide unit.

III. TOTAL SYNTHESES OF THE NATURAL PRODUCTS

Because of the small quantities generally available from the natural sources and because of the possibility to have access to various analogs, total syntheses of natural indolocarbazoles have triggered considerable efforts.[55] Early work centered on aglycon construction.[56-63] The first total synthesis of staurosporine was achieved in 1995 by Danishefsky et al. in 27 steps (Figure 24.4), using sequential indole-Grignard additions to *N*-(benzyloxymethyl)-dibromomaleimide, followed by glycosylation with a 1,2-anhydrosugar, and then photocyclization. Coupling of the sugar moiety with the second indole nitrogen was performed from the exo-glycal intermediate, and after removal of the protective groups, the last step was the reduction of a carbonyl group of the imide hetero-cycle.[64,65] In 1997, Wood et al.[66,67] reported the synthesis of staurosporine and several natural compounds such as TAN-1030A, K-252a, RK286, and MLR-52 from the same intermediate obtained by cyclofuranosylation of the lactam aglycone in the presence of camphorsulfonic acid. A ring expansion approach was developed that allowed the transformation of a furanosylated indolocarbazole to a pyranosylated derivative (Figure 24.4).

The 23-step stereocontrolled total synthesis of K-252a was described in 1999 starting from indole-3-acetic acid (Figure 24.5).[68] The glycosylation was carried out before the introduction of the second indole moiety provided by tryptamine. The carboxylate substituent of the sugar was introduced via the addition of hydrogen cyanide on a carbonyl group. Four total syntheses of rebeccamycin have been reported (Figure 24.6).[69-71] Kaneko et al. proposed two synthetic approaches to the chlorinated indolocarbazole aglycone, to which a α-bromo sugar was attached via the Koenigs–Knorr method. The chlorinated aglycone was obtained from *N*-benzyloxymethyl-dibromomaleimide and either chloroindole or 7-7-dichloroindigo. Danishefsky's method used a bis-indolylmaleimide aglycone that was coupled to an anhydro sugar before photocyclization. Faul et al. performed the glycosylation of 7-chloro-indole-3-acetamide using the anhydro sugar described

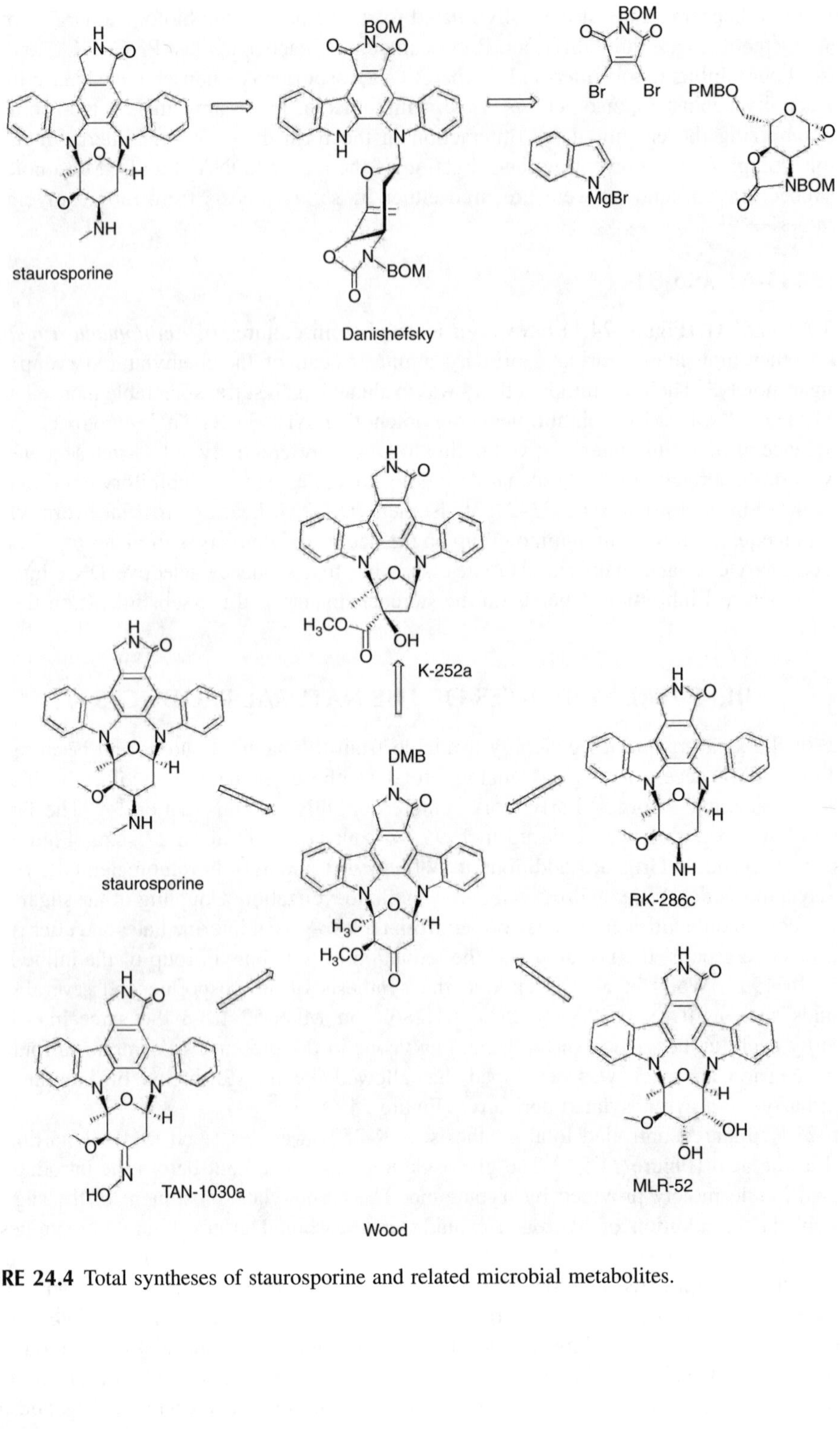

FIGURE 24.4 Total syntheses of staurosporine and related microbial metabolites.

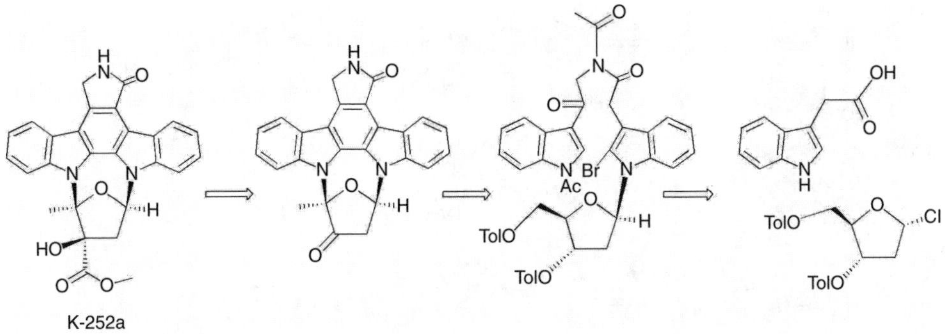

FIGURE 24.5 Total synthesis of K-252a by Kobayashi et al.

by Danishefsky before condensation with methyl 7-chloroindole-3-glyoxylate. The oxidative cyclization of the bisindolylmaleimide was carried out using palladium(II) triflate.

The synthesis of tjipanazole E was achieved by Bonjouklian et al. (Figure 24.7).[72] Tjipanazole D, obtained by condensation of *p*-chlorophenylhydrazine and 1,2-cyclohexanedione followed by a Fischer indolization, was further glycosylated using α-bromo-tetraacetylglucose. Tjipanazoles F1 and F2 were synthesized by Gilbert et al. (Figure 24.7).[73,74] Tjipanazole F2 was prepared from 1,2-bis(3-indolyl)ethane via a Mannich cyclization, followed by sequential bromination, glycosylation, oxidation, and halogen exchange. Tjipanazole F1 was synthesized via the same sequence of reactions from 1-(5-chloro-indolyl)-2-indolyl-ethane; the halogen atom was introduced at the beginning of the synthesis using 5-chloro-indole. Chisholm and Van Vranken described in 2000 the first total synthesis of AT2433-A1 and B1 via the glycosylation of the indoline intermediate prepared from *N*-methyl-dibromomaleimide and 7-chloro-indolylmagnesium bromide and indolylmagnesium bromide, respectively (Figure 24.7).[75]

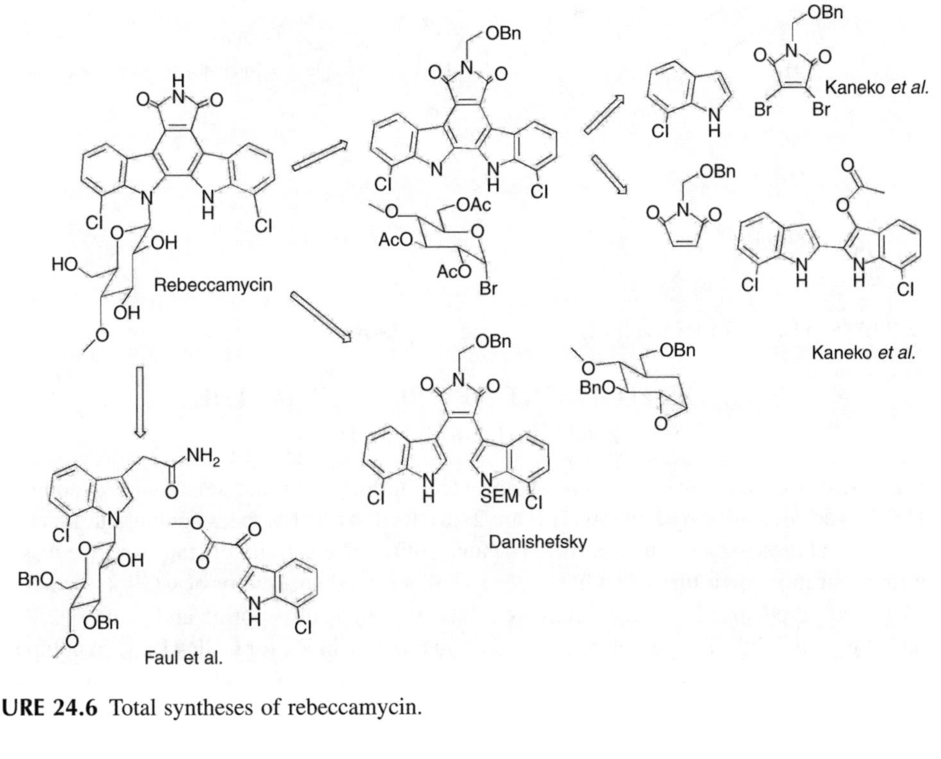

FIGURE 24.6 Total syntheses of rebeccamycin.

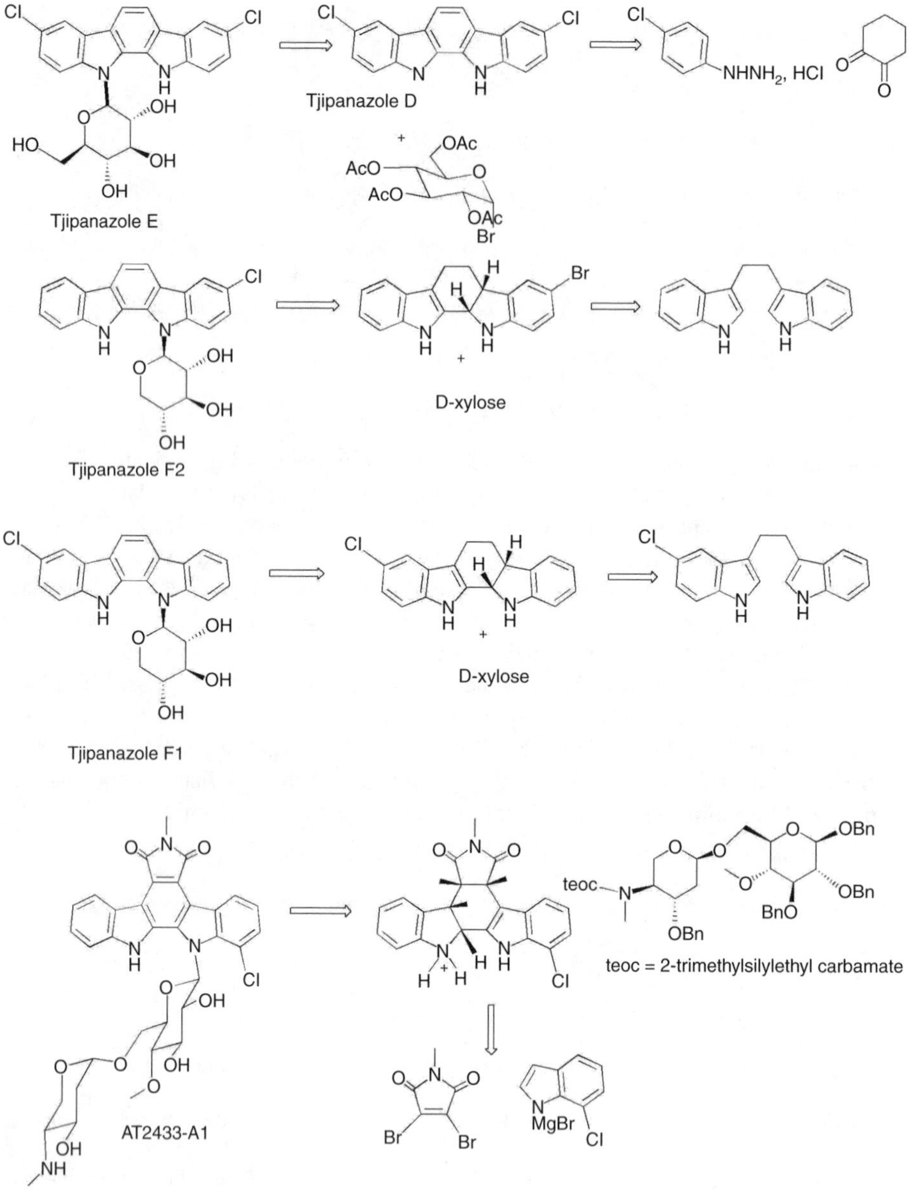

FIGURE 24.7 Syntheses of tjipanazoles E, F1, F2, and AT2433-A1.

IV. DEVELOPMENT OF DRUGS FROM THE NATURAL PRODUCTS

Structural analogs of staurosporine were synthesized, aiming at higher selectivity. Among them, CGP 41 251 was developed by Novartis (Figure 24.8). CGP 41 251 was less potent but much more selective toward kinases than staurosporine. The antiproliferative activity of staurosporine was three to nine times stronger than that of CGP 41 251. The increased specificity of CGP 41 251 has not resulted in a cell-type-specific antiproliferative effect. *In vivo*, staurosporine and CGP 41 251 were active in the growth inhibition model, with a higher therapeutic index for CGP 41 251 in comparison to staurosporine.[76,77] Like staurosporine, K-252a and UCN-01, CGP 41 251 has no effect on topoisomerase I.[78]

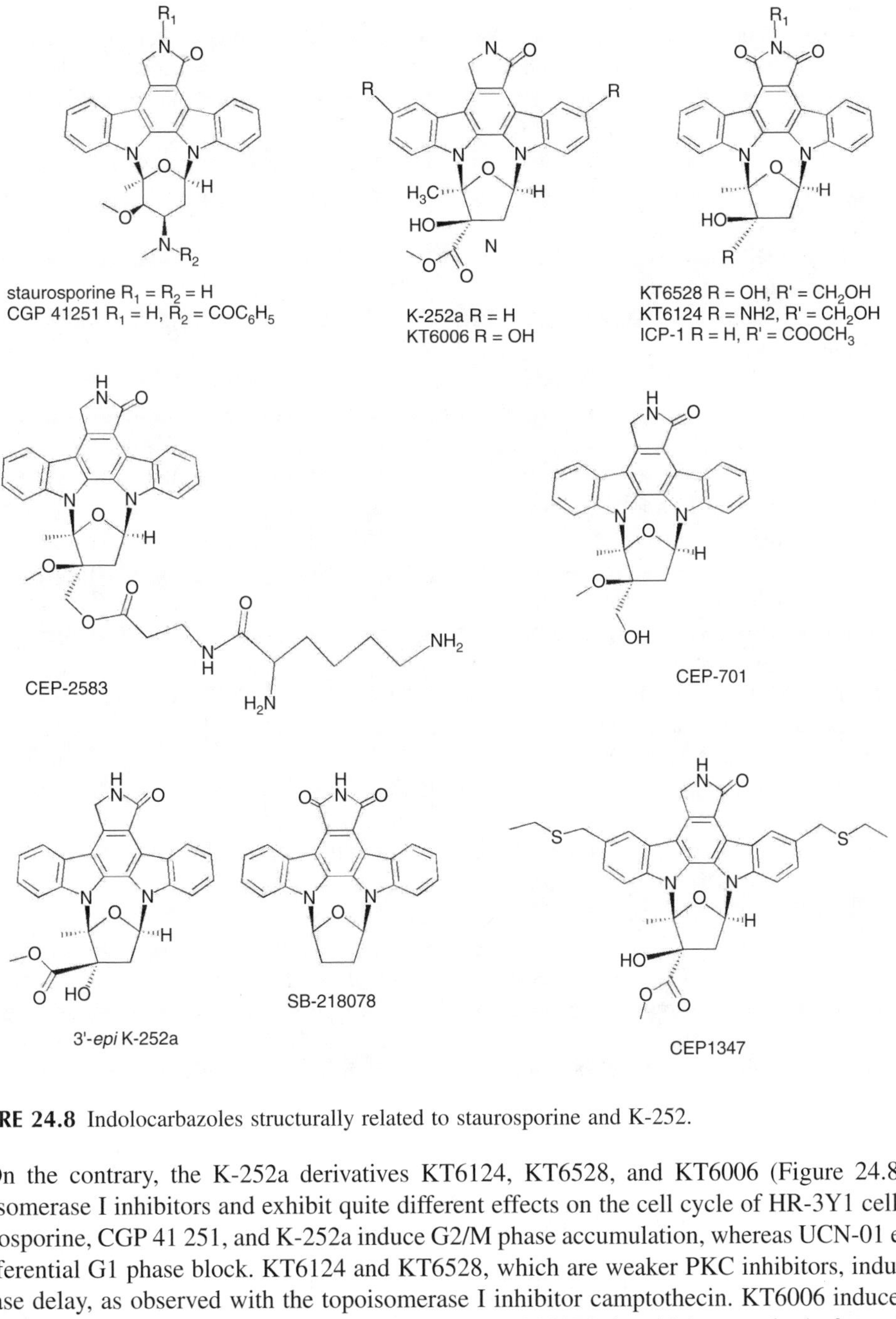

FIGURE 24.8 Indolocarbazoles structurally related to staurosporine and K-252.

On the contrary, the K-252a derivatives KT6124, KT6528, and KT6006 (Figure 24.8) are topoisomerase I inhibitors and exhibit quite different effects on the cell cycle of HR-3Y1 cells.[78,79] Staurosporine, CGP 41 251, and K-252a induce G2/M phase accumulation, whereas UCN-01 exerts a preferential G1 phase block. KT6124 and KT6528, which are weaker PKC inhibitors, induce an S phase delay, as observed with the topoisomerase I inhibitor camptothecin. KT6006 induces not only the S phase delay but also G2/M phase block. SB-218078, in which a tetrahydrofuranyl unit replaces the sugar unit (Figure 24.8), was found to enhance the cytotoxicity of DNA-damaging agents by preventing G2 arrest. As with UCN-01, SB-218078 is a potent inhibitor of human Chk1 kinase (Ki,app = 15 nM),[41] and it abrogates G2 arrest caused by either γ-irradiation or a topoisomerase I inhibitor.[80]

Avid binding of UCN-01 to human plasma proteins, not predicted from the animal studies, limits access to the tumor.[81,82] A screen of related indolocarbazoles showed that ICP-1 (Figure 24.8), a K-252a analog, was also able to abrogate S and G2 phase arrest and to enhance cytotoxicity induced by cisplatin only in p53 defective cells.[83] In contrast to UCN-01, checkpoint abrogation by ICP-1 was only slightly inhibited by human plasma. CEP-2583 and CEP-701 (Figure 24.8), tyrosine kinase receptor inhibitors developed by Cephalon, are currently undergoing phase II clinical trials.[84,85] This group has recently obtained, by semisynthesis from K-252a, the 3-*epi* diastereoisomer (Figure 24.8). Inverting the 3-alcohol resulted in a 20-n*M* inhibitor of VEGFR2 and a 1-n*M* inhibitor of TrkA tyrosine kinase compared with 43 n*M* and 13 n*M* for K-252a, respectively.[86]

CEP1347 is a semisynthetic K-252a derivative (Figure 24.8) bearing two methylthioethyl substituents in the 3 and 9 positions on the planar indolocarbazole. Both K-252a and CP1347 are neuroprotective compounds, and they mediate neuroprotective effects through the activation of neurotrophic signaling pathways. Both compounds inhibit JNK (*c*-Jun N-terminal kinase) pathway activation through similar mechanisms. Unlike K-252a, however, CP1347 does not inhibit the majority of common kinases such as PKC, PKA, MLCK, and PI-3 kinase, but it directly inhibits PAKs, a family of serine/threonine kinases, in a quite selective manner.[87–89]

BE-13793C (Figure 24.1), a topoisomerase I inhibitor, was isolated from cultures of *Streptoverticillium* species.[90] Modifications were carried out to enhance the solubility and to obtain compounds exhibiting more potent antitumor activities *in vivo*. Glycosylated derivatives ED-110 and NB-506 were prepared, as well as analogs such as J-107088 bearing the hydroxyl groups in various positions on the indolocarbazole framework (Figure 24.9).[91–94] ED-110 and NB-506 stimulate DNA cleavage by topoisomerase I without any effect on topoisomerase II. The replacement of the NHCHO substituent of NB-506 with a longer side chain [NHCH(CH$_2$OH)$_2$] reinforces the interaction with DNA without affecting the intercalative binding process.[95,96] A NB-506 regioisomer, with the two hydroxyl groups at positions 2 and 10 instead of 1 and 11, has lost its capacity to intercalate into DNA but remains an extremely potent topoisomerase I poison, showing that intercalation into DNA is not required for inhibition of topoisomerase I by indolocarbazoles. NB-506 and its regioisomer stimulate cleavage at different sites on DNA. The 2,10 isomer is up to 100-fold more toxic to tumor cells than the parent NB-506.[97] NB-506 induces apoptosis in leukemia cells via alterations of mitochondrial and caspases functions.[98] The metabolism of NB-506 has been studied, both *in vitro* and *in vivo*, in animals and humans (Figure 24.10).[99,100] In rodent plasma, NB-506 was converted to its deformylated derivative ED-501, whereas in dog and human plasma, it was converted to its anhydride ED-551, an inactive metabolite. It has been suggested that the conversion of NB-506 into anhydride ED-551 is catalyzed by metal ions with the formation of a complex in which the metal ion interacts with the three carbonyl groups. The replacement of the NHCHO group by a NHCH(CH$_2$OH)$_2$ substituent (compound D, Figure 24.10) significantly reduces the rate of metabolization into the inactive anhydride form.[101] During the course of clinical trials of NB-506, Banyu Pharmaceuticals continued to synthesize NB-506 analogs, and from among several hundred compounds, J-107088 (Figure 24.9), which showed greater inhibitory activity than NB-506 toward topoisomerase I, has recently replaced NB-506 in the clinical trials.[93]

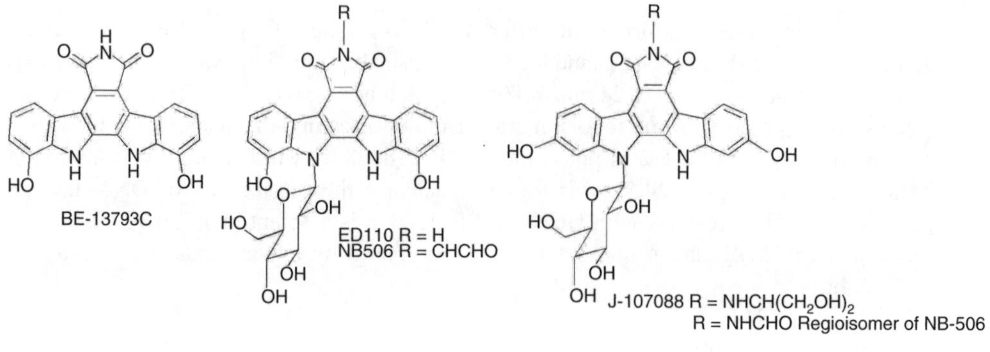

FIGURE 24.9 Chemical structures of BE-13793C, semi-synthetic derivatives, and synthetic analogs.

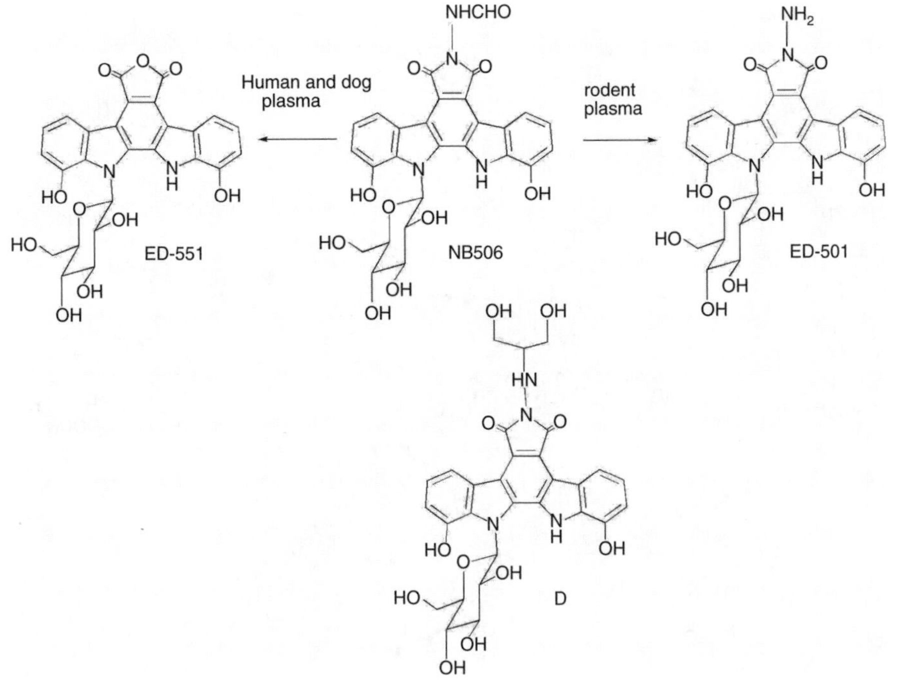

FIGURE 24.10 Metabolism of NB-506.

Large structure–activity relationship studies have also been carried out on rebeccamycin analogs prepared either by semisynthesis or by total synthesis.[50,51,102] Biosynthetic studies revealed that rebeccamycin is from two tryptophan, one glucose, and one methionine molecules.[102] Addition of 6-, 5-, or 4-fluorotryptophan to the fermentation broths resulted in rebeccamycin analogs bearing, instead of chlorine atoms, fluorine atoms in the 2 and 10, 3 and 9, and 4 and 8, positions, respectively, on the indolocarbazole unit. Maximal topoisomerase I–induced cytotoxicity is imparted by 3,9-dihalogenated compounds. The 2,10-dihalogenated compounds exhibit less topoisomerase I selective cytotoxicity, and the 1,11-disubstituted analogs have the least topoisomerase I–dependent cytotoxicity. These observations may be explained by interaction with other cytotoxic targets such as DNA or kinases. The 4,8-dihalogenated compounds are the least cytotoxic and the least active toward topoisomerase I or other targets. The methylation of the 4-hydroxyl group of the sugar moiety has little effect on topoisomerase I inhibition but results in increased cytotoxicity, very likely because of better penetration into the cells. The semisynthetic analog synthesis program developed by Bristol-Myers culminated in the discovery of the 6-*N*-diethylaminoethyl derivative, a water-soluble analog of rebeccamycin.[49]

NSC 655649, a tartrate salt, entered phase I studies (Figure 24.11). This compound has a broad spectrum of activity against solid tumors and, in contrast to rebeccamycin, inhibits topoisomerase II. NCS 655649 is currently undergoing phase II clinical trials for the treatment of hepatobiliary and gall bladder cancers.[102] Interestingly, the introduction of an amino group at the 6-position on the sugar unit (Figure 24.11) strongly enhances the capacity of the drugs to interact with DNA but almost abolishes their poisoning effect on topoisomerase I, showing that DNA and topoisomerase I represent two independent targets that can both be used for the development of antitumor rebeccamycin derivatives.[103]

The main differences between the structures of rebeccamycin and staurosporine are the sugar moiety linked to one indole nitrogen in rebeccamycin and to both indole nitrogens in staurosporine and the imide function in the upper heterocycle of rebeccamycin instead of an amide function in

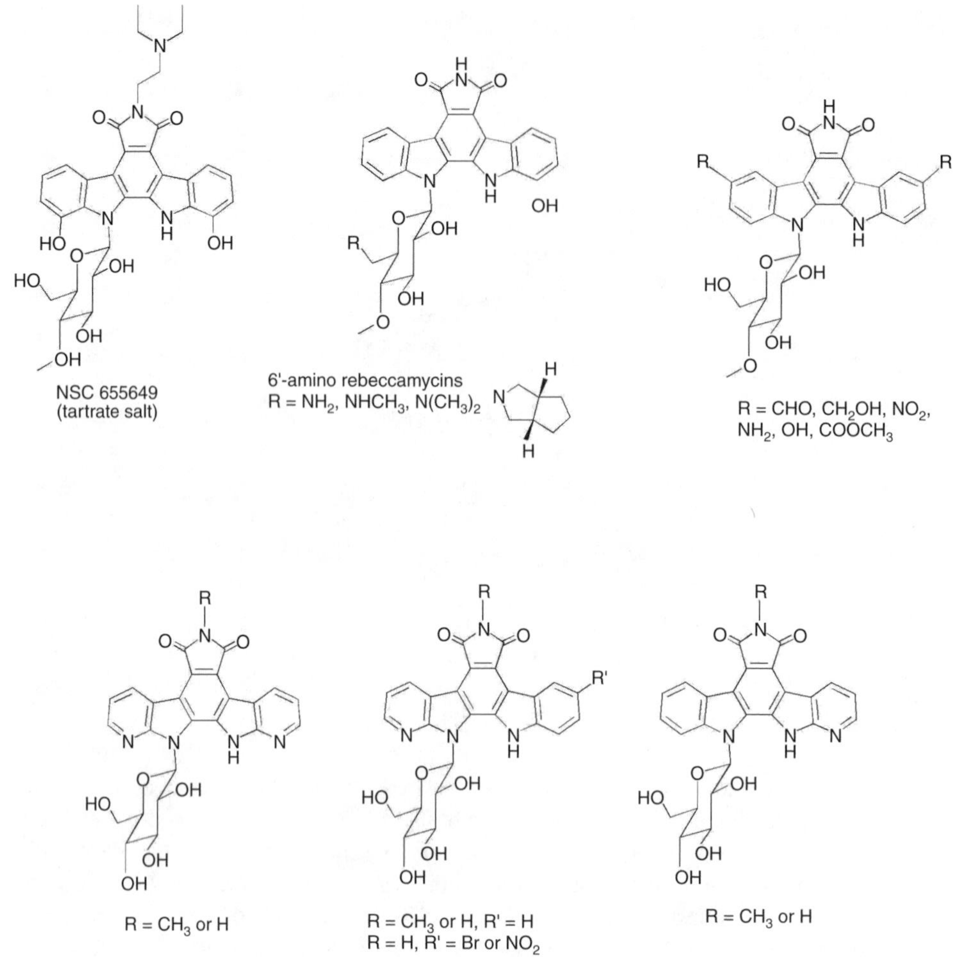

FIGURE 24.11 Semi-synthetic rebeccamycin analogs and aza-rebeccamycins obtained by total synthesis.

staurosporine. These differences should be responsible for their enzyme selectivities. Therefore, structural modifications were performed on rebeccamycin to obtain staurosporine analogs, with the sugar moiety linked to both indole nitrogens (Figure 24.12).[104,105] The anti-topoisomerase I activities of the unsubstituted imide and rebeccamycin are in the same range, whereas the activity of the regioisomeric mixture of amides is about 10-fold weaker. There is no doubt that the carbonyl group that distinguishes these compounds must be important for both DNA binding and topoisomerase I inhibition. None of these bridged compounds inhibits PKC. The two cationic molecules, bearing a *N,N*-diethylamino chain at the imide nitrogen or an amino substituent on the sugar moiety, exhibit enhanced affinity for DNA, and this translates at the biological level into a 10-fold higher cytotoxic index; however, they do not inhibit topoisomerase I, in contrast to the parent uncharged derivative (R′ = R″ = OH), which stimulates DNA single-strand breaks by topoisomerase I.[106] In contrast to rebeccamycin and dechlorinated rebeccamycin, which exhibit nonselective cytotoxicity toward various tumor cell lines, some compounds in this bridged series, such as 3,9-dinitro, 3,9-dihydroxy, and 3,9-diformyl compounds, exhibit a marked selectivity, indicating targets that are differently expressed in the various tumor cell lines.

Substitutions in 3,9-positions on the indolocarbazole framework of rebeccamycin have been carried out (Figure 24.11).[107] Depending on the structural modifications, the rebeccamycin derivatives may or may not exhibit selectivity toward the various tumor cell lines tested. 3,9-Substituents

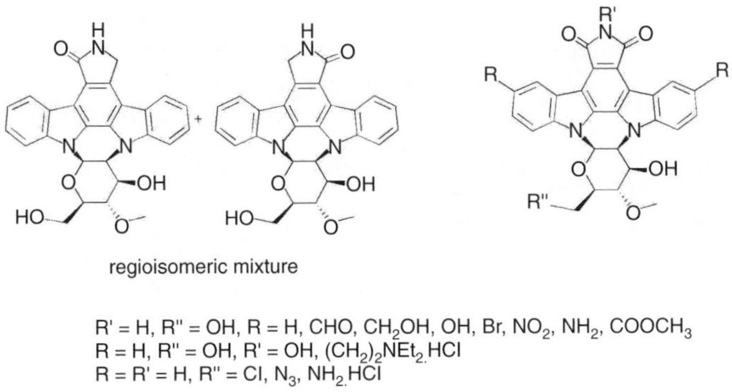

R' = H, R" = OH, R = H, CHO, CH$_2$OH, OH, Br, NO$_2$, NH$_2$, COOCH$_3$
R = H, R" = OH, R' = OH, (CH$_2$)$_2$NEt$_2$.HCl
R = R' = H, R" = Cl, N$_3$, NH$_2$.HCl

FIGURE 24.12 Staurosporine analogs from rebeccamycin.

can enhance or abolish the cytotoxicity, but they can also induce selectivity. The 3,9-dihydroxy derivative is a DNA-binding topoisomerase I poison, but it also inhibits CDK1/cyclinB and CDK5/p25. It induces accumulation of L1210 cells in the G2+M phases, which seems to be correlated with CDK1/cyclinB inhibition. This multitarget drug represents a new lead for subsequent drug design in the rebeccamycin series.

Another interesting series is that of the 7-aza-rebeccamycins obtained by total synthesis.[108,109] Surprisingly, DNA binding experiments in the monoazaindole series showed major differences between the compounds bearing the carbohydrate moiety on either the azaindole moiety or the indole unit. Compounds with the sugar attached to the indole exhibit higher affinity for DNA and stronger topoisomerase I inhibition than the parent, dechlorinated rebeccamycin. When the sugar is linked to the azaindole, however, DNA binding properties are abolished and topoisomerase I inhibitory properties are strongly weakened. Even the introduction of a bromo or a nitro group is not sufficient to restore the affinity for DNA; however, these substitutions induced marked topoisomerase I inhibition. Both monoaza or diaza rebeccamycins showed a selective action toward certain cell lines. Their cytotoxic profiles are quite different from those of the parent non-aza compounds, which are nonselective and display similar cytotoxicity against all the tumor cell lines, but like most of rebeccamycin derivatives, they induce a G2+M arrest. It can be concluded from these studies that if topoisomerase I remains the first target for rebeccamycin derivatives, they certainly have other biological targets. The first results show that kinases that control the cell cycle should also be targets for rebeccamycin analogs.

V. CLINICAL APPLICATIONS

Though there is no indolocarbazole on the market at present, some of them are undergoing clinical trials. Phase I and II trials of UCN-01 have been reported.[110] In a phase I study, where the doses were administered by continuous infusion for 72 h, a prolonged half-life of 600 h was observed. The side effects were various, and among them, hyperglycemia needed insulin treatment. However, avid plasma protein binding of UCN-01 dictated a change in dose escalation and administration schedules. Phase I trials with shorter infusions are being completed. In a phase II study, one patient with melanoma achieved a partial response and encouraging results were obtained in a patient with refractory anaplastic large cell lymphoma.[111–113] The maximal tolerated and recommended phase II dose was 42.5 mg/m^2 per day for 3 d. A phase II clinical trial of the staurosporine analog CGP41251 (PKC412) has been reported. The patients with lymphoproliferative disorders were treated at three oral dose levels of 25, 150, and 225 mg/day for 14 d. The results show that the treatment is well tolerated and reduces the tumor load in chronic B-cell malignancies.[114] Rebeccamycin derivative NCS 655649 has entered into phase I and II studies. This drug may be safely given at a maximal tolerated dose of 572 mg/m^2 in both single-dose and multiple-dose formats.

Antitumor activity was observed in two heavily pretreated ovarian cancer patients and in one patient with a soft tissue sarcoma refractory to etoposide and doxorubicin. Preliminary data of a phase I study demonstrated antitumor activity in colorectal carcinoma. A phase II trial evaluated the efficacy of NSC 655649 in patients with advanced previously minimally treated metastatic colorectal cancer. At 500 mg/m^2 once every 3 wk, the treatment was inactive.[115–118] A phase II study of NCS-655649 in metastatic renal cell cancer showed a modest antitumor activity of this drug in patients with advanced this cancer.[119]

VI. CONCLUSIONS

Indolocarbazoles represent a promising class of antitumor agents. Minimal structural modifications are able to modify the biological targets and induce selectivity toward tumor cell lines. Staurosporine is a nonselective kinase inhibitor, whereas the main target for rebeccamycin is topoisomerase I. Rebeccamycin exhibits no selectivity toward the various tumor cell lines, whereas 7-aza-rebeccamycins in which an indole moiety is replaced by a 7-azaindole unit are highly selective. Other structural modifications on the rebeccamycin framework such as substitutions in 3,9-positions may induce selectivity toward the tumor cell lines and strong inhibitory properties toward kinases involved in the progression of the cell cycle. No indolocarbazole has yet been launched on the market, but a staurosporine analog UCN-01 and a rebeccamycin derivative NCS 655649 have entered into phase I and phase II trials. UCN-01 and a K-252a analog, ICP-1, abrogate the DNA-damage induced checkpoints to cell-cycle progression in the G2 and in the S phases. This is probably the most important point because, in this manner, indolocarbazoles may exhibit highly selective activity against tumor cells without effect on normal cells, thereby achieving the ultimate goal for antitumor drugs.

REFERENCES

1. Prudhomme, M., Indolocarbazoles as anti-cancer agents, *Curr. Pharm. Des.*, 3, 265, 1997, and references provided therein.
2. Williams, D.E. et al., Holyrines A and B, possible intermediates in staurosporine biosynthesis produced in culture by a marine actinomycete obtained from the North Atlantic Ocean, *Tetrahedron Lett.*, 40, 7171, 1999.
3. Schupp, P. et al., Staurosporine derivatives from the ascidian *Eudistoma toealensis* and predatory flatworm *Pseudoceros* sp., *J. Nat. Prod.*, 62, 959, 1999.
4. Cantrell, C.H. et al., A new staurosporine analog from the prosobranch mollusk *Coriocella nigra*, *Nat. Prod. Lett.*, 14, 39, 1999.
5. Hernandez, L.M.C. et al., 4-*N*-methyl-5-hydroxystaurosporine and 5-hydroxystaurosporine, new indolocarbazole alkaloïds from a marine *Micromonospora* sp. strain, *J. Antibiot.*, 53, 895, 2000.
6. Schupp, P., Proksch, P. and Wray, V., Further new staurosporine derivatives from the ascidian *Eudistoma toealensis* and its predatory flatworm *Pseudoceros* sp., *J. Nat. Prod.*, 65, 295, 2002.
7. Omura, S. et al., A new alkaloid AM-2282 of *Streptomyces* origin. Taxonomy, fermentation, isolation and preliminary characterization, *J. Antibiot.*, 30, 275, 1977.
8. Furusaki, A. et al., X-ray crystal structure of staurosporine: a new alkaloid from a *Streptomyces* strain, *J. Chem. Soc. Chem. Commun.*, 800, 1978.
9. Steglich, W. et al., Indole pigments from the fruiting bodies of the slime mold *Arcyria denudata*, *Angew. Chem. Int. Ed. Engl.*, 19, 459, 1980.
10. Tamaoki, T. et al., Staurosporine, a potent inhibitor of phospholipid/Ca^{++} dependent protein kinase, *Biochem. Biophys. Res. Commun.*, 135, 397, 1986.
11. Kase, H., Iwahashi, K. and Matsuda, Y., K-252a, a potent inhibitor of protein kinase C from microbial origin, *J. Antibiot.*, 39, 1059, 1986.
12. Cai, Y. et al., The staurosporine producing strain *Streptomycetes longisporoflavus* produces metabolites related to K-252a. Proposal for biosynthetic intermediates of K-252a, *J. Antibiot.*, 49, 1060, 1996.

13. Yasuzawa, T. et al., The structures of the novel protein kinase C inhibitors K-252a, b, c, and d, *J. Antibiot.*, 39, 1072, 1986.
14. Gray, N. et al., ATP-directed inhibitors of cyclin-dependent kinases, *Curr. Med. Chem.*, 6, 859, 1999.
15. Hoehn, P., et al., 3-Demethoxy-3-hydroxystaurosporine, a novel staurosporine analog produced by a blocked mutant, *J. Antibiot.*, 48, 300, 1995.
16. Rialet, V. and Meijer, L., A new screening test for antimitotic compounds using the universal M phase-specific protein kinase, p34cdc2/cyclin Bcdc13, affinity-immobilized on p13suc1-coated microtitration plates, *Anticancer Res.*, 11, 1581, 1991.
17. Al-Obeidi, F.A. and Lam, K.S., Development of inhibitors for protein tyrosine kinases, *Oncogene*, 19, 5690, 2000.
18. Takahashi, I. et al., UCN-01 and UCN-02, new selective inhibitors of protein kinase C II. Purification, physico-chemical properties, structural determination and biological activities, *J. Antibiot.*, 42, 571, 1989.
19. Jackson, J.R. et al., An indolocarbazole inhibitor of human checkpoint kinase (Chk1) abrogates cell cycle arrest caused by DNA damage, *Cancer Res.*, 60, 566, 2000.
20. Yu, L. et al., UCN-01 abrogates G_2 arrest through a Cdc2-dependent pathway that is associated with inactivation of the Wee1Hu kinase and activation of the Cdc25C phosphatase, *J. Biol. Chem.*, 273, 33455, 1998.
21. Rodrigues Pereira, E. et al., Structure-activity relationships in a series of substituted indolocarbazoles: topoisomerase I and protein kinase C inhibition and antitumoral and antimicrobial properties, *J. Med. Chem.*, 39, 4471, 1996.
22. Kojiri, K. et al., A new antitumor substance, BE-13793C, produced by a Streptomycete, taxonomy, fermentation, isolation, structure determination and biological activity, *J. Antibiot.*, 44, 723, 1991.
23. Facompré, M. et al., DNA binding and topoisomerase I poisoning activities of novel disaccharide indolocarbazoles, *Mol. Pharmacol.*, 62, 1215, 2002.
24. Meggio, F. et al., Different susceptibility of protein kinases to staurosporine inhibition. Kinetic studies and molecular bases for the resistance of protein kinase CK2, *Eur. J. Biochem.*, 234, 317, 1995.
25. Mahedevan, D.; Bearss, D.J. and Vankayalapati, H., Structure-based design of novel anti-cancer agents targeting aurora kinases, *Curr. Med. Chem.-Anti-Cancer Agents*, 3, 25, 2003.
26. Lawrie, A.M. et al., Protein kinase inhibition by staurosporine revealed in details of the molecular interaction with CDK2, *Nat. Struct. Biol.*, 4, 796, 1997.
27. Prade, L. et al., Staurosporine-induced conformational changes of cAMP-dependent protein kinase catalytic subunit explain inhibitory potential, *Structure*, 5, 1627, 1997.
28. Lamers, M.B.A.C. et al., Structure of the protein tyrosine kinase domain of C-terminal Src kinase (CSK) in complex with staurosporine, *J. Mol. Biol.*, 285, 713, 1999.
29. Roy, K.K. and Sausville, E.A., Early development of cyclin dependent kinase modulators, *Curr. Pharm. Des.*, 7, 1669, 2001.
30. Akiyama, T. et al., G1-checkpoint function including a cyclin-dependent kinase 2 regulatory pathway as potential determinant of 7-hydroxystaurosporine (UCN-01)-induced apoptosis and G1-phase accumulation, *Jpn. J. Cancer Res.*, 90, 1364, 1999.
31. Sausville, E.A., Cyclin-dependent kinases: novel targets for cancer treatment, in Perry, M. Ed., 1999 *Educational Book.* American Society for Clinical Oncology, Alexandria, VA, 1999, 9.
32. Abe, S. et al., UCN-01 (7-hydroxystaurosporine) inhibits *in vivo* growth of human cancer cells through selective perturbation of G1 phase checkpoint machinery, *Jpn. J. Cancer Res.*, 92, 537, 2001.
33. Busby, E.C. et al., The radiosensitizing agent 7-hydroxystaurosporine (UCN-01) inhibits the DNA damage checkpoint kinase hChk1, *Cancer Res.*, 60, 2108, 2000.
34. Kohn, E.A. et al., Abrogation of the S phase DNA damage checkpoint results in S phase progression or premature mitosis depending on the concentration of 7-hydroxystaurosporine and the kinetics of Cdc25C activation, *J. Biol. Chem.*, 277, 26553, 2002.
35. Zhou, B.-B.S. and Elledge, S.J., The DNA damage response: putting checkpoints in perspective, *Nature*, 408, 433, 2000.
36. Cuddihy, A.R. and O'Connell, M.J., Cell-cycle responses to DNA damage in G2, *Int. Rev. Cytol.*, 222, 99, 2003.
37. Nyberg, K.A. et al., Toward maintaining the genome: DNA damage and replication checkpoints, *Annu. Rev. Genet.*, 36, 617, 2002.

38. Graves, P.R. et al., The Chk1 protein kinase and the Cdc25C regulatory pathways are targets of the anticancer agent UCN-01, *J. Biol. Chem.*, 275, 5600, 2000.

39. Tenzer, A. and Pruschy, M., Potentiation of DNA-damage-induced cytotoxicity by G2 checkpoint abrogators, *Curr. Med. Chem.-Anti-Cancer Agents*, 3, 35, 2003.

40. Yu, Q. et al., Inhibition of Chk2 activity and radiation-induced p53 elevation by the cell cycle checkpoint abrogator 7-hydroxystaurosporine (UCN-01), *Proc. Am. Assoc. Cancer Res.*, 42, 800, 2001.

41. Zhao, B. et al., Structural basis for Chk1 inhibition by UCN-01, *J. Biol. Chem.*, 277, 46609, 2002.

42. Sato, T.; Fujita, N. and Tsuruo, T., Interference with PDK1-Akt survival signaling pathway by UCN-01 (7-hydroxystaurosporine), *Oncogene*, 21, 1727, 2002.

43. Roux, P.P. et al., K-252a and CEP1347 are neuroprotective compounds that inhibit mixed-lineage kinase-3 and induce activation of Akt and ERK, *J. Biol. Chem.*, 277, 49473, 2002.

44. Angeles, T.S., Kinetics of trkA tyrosine kinase activity and inhibition by K-252a., *Arch. Biochem. Biophys.*, 349, 267, 1998.

45. Chin, L.S. et al., K-252a induces cell cycle arrest and apoptosis by inhibiting Cdc2 and Cdc25c, *Cancer Invest.*, 17, 391, 1999.

46. Bush, J.A. et al., Production and biological activity of rebeccamycin, a novel antitumor agent, *J. Antibiot.*, 40, 668, 1987.

47. Marminon, C. et al., Syntheses and antiproliferative activities of 7-azarebeccamycin analogs bearing one 7-azaindole moiety, *J. Med. Chem.*, 46, 609, 2003.

48. Bailly, C. et al., DNA cleavage by topoisomerase I in the presence of indolocarbazole derivatives of rebeccamycin, *Biochemistry*, 36, 3917, 1997.

49. Kaneko, T. et al., Water soluble derivatives of rebeccamycin, *J. Antibiot.*, 43, 125, 1990.

50. Prudhomme, M., Recent developments of rebeccamycin analogs as topoisomerase I inhibitors and antitumor agents, *Curr. Med. Chem.*, 7, 1189, 2000.

51. Prudhomme, M., Rebeccamycin analogs as anti-cancer agents, *Eur. J. Med. Chem.*, 38, 123, 2003.

52. Matson, J.A. et al., AT2433-A1, AT2433-A2, AT2433-B1, and AT2433-B2 novel antitumor antibiotic compounds produced by *Actinomadura melliaura*. Taxonomy, fermentation, isolation and biological properties, *J. Antibiot.*, 42, 1547, 1989.

53. Carrasco, C. et al., DNA sequence recognition by the indolocarbazole antibiotic AT2433-A1 and its diastereoisomer, *Nucleic Acids Res.*, 30, 1774, 2002.

54. Facompré, M. et al., DNA binding and topoisomerase I poisoning activities of novel disaccharide indolocarbazoles, *Mol. Pharmacol.*, 62, 1215, 2002.

55. Knölker, H-J. and Reddy, K.R., Isolation and synthesis of biologically active carbazole alkaloïds, *Chem. Rev.*, 102, 4303, 2002.

56. Sarstedt, B. and Winterfeldt, E., Reactions with indole derivatives XLVIII, a simple synthesis of staurosporin aglycon, *Heterocycles*, 20, 469, 1983.

57. Magnus, P.D., Exon, C. and Sear, N.L., Indole-2,3-quinodimethanes. Synthesis of indolocarbazoles for the synthesis of the fused dimeric indole alkaloïd staurosporinone, *Tetrahedron*, 39, 3725, 1983.

58. Joyce, R.P., Gainor, J.A. and Weinreb, S.M., Synthesis of the aromatic and monosaccharide moieties of staurosporine, *J. Org. Chem.*, 52, 1177, 1987.

59. Brenner, M. et al., Synthesis of arcyriarubin B and related bisindolylmaleimides, *Tetrahedron*, 44, 2887, 1988.

60. Hughes, I., Nolan, W.P. and Raphael, R.A., Synthesis of the indolo[2,3-*a*]carbazole natural products staurosporinone and arcyriaflavin B, *J. Chem. Soc. Perkin Trans. I*, 2475, 1990.

61. Moody, C.J. et al., Synthesis of the staurosporin aglycon, *J. Org. Chem.*, 57, 2105, 1992.

62. Mc Combie, S.W. et al., Indolocarbazoles. 1. Total synthesis and protein kinase inhibiting characteristics of compounds related to K-252c, *Bioorg. Med. Chem. Lett.*, 3, 1537, 1993.

63. Link, J.T., Gallant, M. and Danishefsky, S.J., The first synthesis of a fully functionalized core structure of staurosporine: sequential indolyl glycosidation by endo and exo glycals, *J. Am. Chem. Soc.*, 115, 3782, 1993.

64. Link, J.T., Raghavan, S. and Danishefsky, S.J., First total synthesis of staurosporine and *ent*-staurosporine, *J. Am. Chem. Soc.*, 117, 552, 1995.

65. Link, J.T. et al., Staurosporine and *ent*-staurosporine: the first total syntheses, prospects for a regioselective approach, and activity profiles, *J. Am. Chem. Soc.*, 118, 2825, 1996.

66. Wood, J.L. et al., Design and Implementation of an efficient synthetic approach to furanosylated indolocarbazoles: total synthesis of (+)- and (-)-K252a, *J. Am. Chem. Soc.,* 119, 9641, 1997.

67. Wood, J.L. et al., Design and Implementation of an efficient synthetic approach to pyranosylated indolocarbazoles: total synthesis of (+)-RK286c, (+)-MLR-52, (+)-staurosporine and (-)-TAN-1030a, *J. Am. Chem. Soc.,* 119, 9652, 1997.

68. Kobayashi, Y.; Fujimoto, T. and Fukuyama, T., Stereocontrolled total synthesis of (+)-K252a, *J. Am. Chem. Soc.,* 121, 6501, 1999.

69. Kaneko, T. et al., Two synthetic approaches to rebeccamycin, *Tetrahedron Lett.,* 26, 4015, 1985.

70. Gallant, M., Link, J.T. and Danishefsky, S.J., A stereoselective synthesis of indole-b -*N*-glycosides: an application to the synthesis of rebeccamycin, *J. Org. Chem.,* 58, 343, 1993.

71. Faul, M.M., Winneroski, L.L. and Krumrich, C.A., Synthesis of rebeccamycin and 11-dechlororebeccamycin, *J. Org. Chem.,* 64, 2465, 1999.

72. Bonjouklian, R. et al., Tjipanazoles, new antifungal agents from the blue-green alga *Tolypothrix tjipanasensis, Tetrahedron,* 37, 7739, 1991.

73. Gilbert, E.J. and Van Vranken, D.L., Control of dissymmetry in the synthesis of (+)-tjipanazole F2, *J. Am. Chem. Soc.,* 118, 5500, 1996.

74. Gilbert, E.J., Ziller, J.W. and Van Vranken, D.L., Cyclizations of unsymmetrical bis-1,2-(3-indolyl)ethanes: synthesis of (-)-tjipanazole F1, *Tetrahedron,* 53, 16553, 1997.

75. Chisholm, J.D. and Van Vranken, D.L., Regiocontrolled synthesis of the antitumor antibiotic AT2433-A1, *J. Org. Chem.,* 65, 7541, 2000.

76. Meyer, T. et al., A derivative of staurosporine (CGP 41 251) shows selectivity for protein kinase C inhibition and *in vitro* anti-proliferative as well as *in vivo* anti-tumor activity, *Int. J. Cancer,* 43, 851,1989.

77. Hoehn, P. et al, 3-Demethoxy-3-hydroxystaurosporine, a novel staurosporine analog produced by a blocked mutant, *J. Antibiot.,* 48, 300, 1995.

78. Akinaga, S., Diverse effects of indolocarbazole compounds on the cell cycle progression of *ras*-transformed rat fibroblast cells, *J. Antibiot.,* 46, 1767, 1993.

79. Yamashita, Y. et al., Induction of mammalian DNA topoisomerase I mediated DNA cleavage by antitumor indolocarbazole derivatives, *Biochemistry,* 31, 12069, 1992.

80. Jackson J.R. et al., An indolocarbazole inhibitor of human checkpoint kinase (Chk1) abrogates cell cycle arrest caused by DNA damage, *Cancer Res.,* 60, 566, 2000.

81. Fuse, E. et al., Unpredicted clinical pharmacology of UCN-01 caused by specific binding to human α_1-acid glycoprotein, *Cancer Res.,* 58, 3248, 1998.

82. Sausville, E. et al., Phase I trial of 72-hour continuous infusion of UCN-01 in patients with refractory neoplasms, *J. Clin. Oncol.,* 19, 2319, 2001.

83. Eastman, A. et al., A novel indolocarbazole, ICP-1, abrogates DNA damage-induced cell cycle arrest and enhances cytotoxicity: similarities and differences to the cell cycle checkpoint abrogator UCN-01, *Molecular Cancer Therapeutics,* 1, 1067, 2002.

84. Traxler, P. et al., Tyrosine kinase inhibitors: from rational design to clinical trials, *Med. Res. Rev.,* 21, 499, 2001.

85. Laird, A.D. and Cherrington, J.M., Small molecule tyrosine kinase inhibitors: clinical development of anticancer agents, *Expert Opin. Investig. Drugs,* 12, 51, 2003.

86. Gingrich, D.E. and Hudkins, R.L., Synthesis and kinase inhibitory activity of 3-(S)-*epi*-K-252a, *Bioorg. Med. Chem. Lett.,* 12, 2829, 2002.

87. Maroney, A.C. et al., CEP1347 (KT7515), a semi-synthetic inhibitor of the mixed lineage kinase family, *J. Biol. Chem.,* 276, 25302, 2001.

88. Roux, P.P. et al., K252a and CP1347 are neuroprotective compounds that inhibit mixed-lineage kinase-3 and induce activation of Akt and ERK, *J. Biol. Chem.,* 277, 49473, 2002.

89. Nheu, T.V. et al., The K252a derivatives, inhibitors for the PAK/MLK kinase family selectively block the growth of RAS transformants, *Cancer J.,* 8, 238, 2002.

90. Kojiri, K. et al., A new antitumor substance, BE-13793C, produced by a *Streptomycete.* Taxonomy, fermentation, isolation, structure determination and biological activity, *J. Antibiot.,* 44, 723, 1991.

91. Tanaka, S. et al., A new indolopyrrolocarbazole antitumor substance, ED-110, a derivative of BE-13793C, *J. Antibiot.,* 45, 1797, 1992.

92. Kanzawa, F. et al., Antitumor activities of a new indolocarbazole substance, NB-506, and establishment of NB-506-resistant cell lines, SBC-3/NB, *Cancer Res.*, 55, 2806, 1995.

93. Yoshinari, T. et al., Mode of action of a new indolocarbazole anticancer agent, J-107088, targeting topoisomerase I, *Cancer Res.*, 59, 4271, 1999.

94. Ohkubo, M. et al., Synthesis and biological activities of NB-506 analogs modified at the glucose group, *Bioorg. Med. Chem. Lett.*, 10, 419, 2000.

95. Yoshinari, T. et al., Induction of topoisomerase I-mediated DNA cleavage by a new indolocarbazole, ED-110, *Cancer Res.*, 53, 490, 1993.

96. Bailly, C. et al., Substitution at the F-ring *N*-imide of the indolocarbazole antitumor drug NB-506 increases the cytotoxicity, DNA binding, and topoisomerase I inhibition activities, *J. Med. Chem.*, 42, 2927, 1999.

97. Bailly, C. et al., Intercalation into DNA is not required for inhibition of topoisomerase I by indolocarbazole antitumor agents, *Cancer Res.*, 59, 2853, 1999.

98. Facompré, M. et al., Apoptotic response of HL-60 human leukemia cells to the antitumor drug NB-506, a glycosylated indolocarbazole inhibitor of topoisomerase I, *Biochem. Pharmacol.*, 61, 299, 2001.

99. Takenaga, N. et al., *In vitro* metabolism of a new anticancer agent, 6-*N*-formylamino-12,13-dihydro-1,11-dihydroxy-13-(β-D-glucopyranosyl)-5*H*-indolo[2,3-*a*]pyrrolo[3,4-*c*]-5,7(6*H*)-dione (NB-506), in mice, rats, dogs and humans, *Drug Metabol. Disposition*, 27, 213, 1999.

100. Takenaga, N. et al., *In vivo* metabolism of a new anticancer agent, 6-*N*-formylamino- 12,13-dihydro-1,11-dihydroxy-13-(β-D-glucopyranosyl)-5*H*-indolo[2,3-*a*]pyrrolo[3,4-*c*]-5,7(6*H*)-dione (NB-506), in rats and dogs: pharmacokinetics, isolation, identification, and quantification of metabolites, *Drug Metabol. Disposition*, 27, 205, 1999.

101. Goossens, J.-F. et al., Plasma stability of two glycosyl indolocarbazole antitumor agents, *Biochem. Pharmacol.*, 65, 25, 2003.

102. Long, B.H. et al, Discovery of antitumor indolocarbazoles: rebeccamycin, NSC 655649, and fluoroindolocarbazoles, *Curr. Med. Chem.-Anti-Cancer Agents*, 2, 255, 2002.

103. Anizon, F. et al., Rebeccamycin analogs bearing amine substituents or other groups on the sugar moiety, *Bioorg. Med. Chem.*, 11, 3709, 2003.

104. Anizon, F. et al., Synthesis, biochemical and biological evaluation of staurosporine analogs from the microbial metabolite rebeccamycin, *Bioorg. Med. Chem.*, 6, 1597, 1998.

105. Marminon, C. et al., Syntheses and antiproliferative activities of new rebeccamycin derivatives with the sugar unit linked to both indole nitrogen, *J. Med. Chem.*, 45, 1330, 2002.

106. Facompré, M. et al., DNA targeting of two new antitumor rebeccamycin derivatives, *Eur. J. Med. Chem.*, 37, 925, 2002.

107. Moreau, P. et al., Semi-synthesis, topoisomerase I and kinases inhibitory properties, and antiproliferative activities of new rebeccamycin derivatives, *Bioorg. Med. Chem.*, 11, 4871, 2003.

108. Marminon, C. et al., Syntheses and antiproliferative activities of 7-azarebeccamycin analogs bearing one 7-azaindole moiety, *J. Med. Chem.*, 46, 609, 2003.

109. Marminon, C. et al., Syntheses and antiproliferative activities of rebeccamycin analogs bearing two 7-azaindole moieties, *Bioorg. Med. Chem.*, 11, 679, 2003.

110. Dancey, J. and Sausville, E.A., Issues and progress with protein kinase inhibitors for cancer treatment, *Nat. Rev. Drug Discovery*, 2, 297, 2003.

111. Walder, S., Perspectives for cancer therapies with CDK2 inhibitors, *Drug Resistance Update*, 4, 347, 2001.

112. Senderowicz, A.M., Preclinical and clinical development of CDK inhibitors, Proceedings 1st International Symposium on Signal Transduction Modulators in Cancer Therapy, Amsterdam, 2002, p 41.

113. Sausville, E.A. et al., Signal transduction-directed cancer treatments, *Ann. Rev. Pharmacol. Toxicol.*, 43, 199, 2003.

114. Ganeshaguru, V.A. et al., A novel treatment approach for low grade lymphoproliferative disorders using PKC412 (CGP41251), an inhibitor of protein kinase C, *Hematol. J.*, 3, 131, 2002.

115. Merchant, J. et al., Phase I clinical and pharmacokinetic study of NSC 655649, a rebeccamycin analog, given in both single-dose and multiple dose-formats, *Clin. Cancer Res.*, 8, 2193, 2002.

116. Tolcher, A.W. et al., Phase I and pharmacokinetic study of NSC 655649, a rebeccamycin analog with topoisomerase inhibitory properties, *J. Clin. Oncol.*, 19, 2937, 2001.

117. Dowlati, A. et al., Phase I and pharmacokinetic study of rebeccamycin analog NSC 655649 given daily for five consecutive days, *J. Clin. Oncol.*, 19, 2309, 2001.
118. Goel, S. et al., A phase II study of rebeccamycin analog NSC 655649 in patients with metastatic colorectal cancer, *Invest. New Drugs*, 21, 103, 2003.
119. Hussain, M. et al., A phase II study of rebeccamycin analog (NSC-655649) in metastatic renal cell cancer, *Invest. New Drugs*, 21, 465, 2003.

25 Combinatorial Biosynthesis of Anticancer Natural Products

Michael G. Thomas, Kathryn A. Bixby, and Ben Shen

CONTENTS

I. Introduction 519
II. Requirements for Combinatorial Biosynthesis...520
III. Examples of Combinatorial Biosynthesis of Anticancer Natural Products....................521
    A. Daunorubicin and Doxorubicin (Type II Polyketide Synthase) 521
        1. Biosynthetic Gene Cluster and Biosynthetic Mechanisms................................521
        2. Combinatorial Biosynthesis..525
    B. Mithramycin (Type II PKS) 525
        1. Biosynthetic Gene Cluster and Biosynthetic Mechanisms................................526
        2. Combinatorial Biosynthesis..526
    C. Ansamycins (Type I PKS)..528
        1. Biosynthetic Gene Clusters and Biosynthetic Mechanisms 529
        2. Combinatorial Biosynthesis..529
    D. Epothilones (Hybrid PKS-NRPS) 530
        1. Biosynthetic Gene Cluster and Biosynthetic Mechanisms................................530
        2. Combinatorial Biosynthesis..531
    E. Leinamycin (Hybrid NRPS-PKS Featuring AT-Less PKS)..................................532
        1. Biosynthetic Gene Clusters and Biosynthetic Mechanisms 532
        2. Combinatorial Biosynthesis..533
    F. Enediynes (Iterative Type I PKS)...533
        1. Biosynthetic Gene Clusters and Biosynthetic Mechanisms 534
        2. Combinatorial Biosynthesis..534
    G. Aminocoumarins (Mixed Biosynthesis Featuring Atypical NRPS).....................535
        1. Biosynthetic Gene Clusters and Biosynthetic Mechanisms 536
        2. Combinatorial Biosynthesis..537
IV. Summary, Challenges, and Prospects 538
Acknowledgments 539
References 539

I. INTRODUCTION

Natural products have an excellent track record as sources of anticancer drugs, with 60% of today's available anticancer drugs originating from natural products, their derivatives, or mimics.[1] Traditional natural product drug discovery approaches depend primarily on bioassay-guided screening for lead identification and chemical synthesis and medicinal chemistry for lead development. Two historical shortcomings of such endeavors are the small quantities of materials available from nature and the difficulty of their chemical total synthesis and structural derivatization. Total synthesis has limited practical value for the preparation of complex natural products. Analog generation by chemical modification remains problematic because of the rich variety of functional groups that are characteristic of natural products and often requires multiple protection and deprotection steps, making the process quite difficult and laborious. If these hurdles of traditional natural products drug discovery can be overcome, we will have many more successful drugs derived from natural products.

Combinatorial biosynthesis offers promising solutions to both of those problems. Specific structural alteration in the presence of abundant functional groups can often be achieved by rational engineering of the biosynthetic machinery, and the target molecules can therefore be produced by a recombinant organism that is amenable for large-scale fermentation, providing sufficient quantities of natural products and their structural analogs by mass production, thereby lowering the production cost and reducing the environmental concerns associated with conventional chemical syntheses.[1–18]

The field of combinatorial biosynthesis started to become a reality with the seminal work of Sir David Hopwood and colleagues in the 1980s. The heterologous expression of the actinorhodin biosynthetic gene cluster[19] provided a basis for novel metabolite production as well as introducing the concept of "hybrid" natural products.[20] Since this work, the definition of combinatorial biosynthesis as a field has come to embrace more "traditional" methodologies of metabolic engineering and all aspects of molecular biological manipulations used to alter the production of the natural product of interest. Therefore, in this chapter we include targeted gene disruption/replacement, heterologous expression, hybrid pathway construction, enzyme and pathway evolution, and combinations of all these techniques under the umbrella of combinatorial biosynthesis. Page limitations prevent us from summarizing all of the work on combinatorial biosynthesis of anticancer natural products, and so we have chosen a selected set of natural products to showcase the effectiveness and potential utilities of combinatorial biosynthesis in the production and development of novel natural product-based anticancer drugs. Although significant progress has also been made in plant natural products by cell culture technology[21–23] and fungal natural products by combinatorial biosynthesis methods,[24–26] we have limited the scope of this chapter to bacterial natural products.

II. REQUIREMENTS FOR COMBINATORIAL BIOSYNTHESIS

Four prerequisites must be satisfied before any attempt to implement combinatorial biosynthesis strategies for natural product structural diversity. They are the availability of the gene clusters encoding the production of a particular natural product or family of natural products, genetic and biochemical characterizations of the biosynthetic machinery for the targeted natural products to a degree that the combinatorial biosynthesis principles can be rationally applied to engineer the novel analogs, expedient genetic systems for *in vivo* manipulation of genes governing the production of the target molecules in their native producers or heterologous hosts, and production of the natural products or their engineered analogs to levels that are appropriate for detection, isolation, and structural and biological characterization.[27]

Although each of these requirements is essential to realize the full potential of combinatorial biosynthesis, establishing an expedient genetic system for *in vivo* manipulation of the biosynthetic machinery of the targeted metabolites is of paramount importance. This requires the development or engineering of the necessary vectors (e.g., plasmids, cosmids, and *Escherichia coli–Streptomyces* artificial chromosomes) and hosts (e.g., *Streptomyces coelicolor* CH999 and *Streptomyces lividans* K4-114), mechanisms for the introduction of DNA into the host (e.g., polyethylene glycol- or CaCl$_2$-assisted transformation, electroporation, and *E. coli–Streptomyces* conjugation), precursor availability (e.g., methylmalonyl–CoA, propionyl–CoA, and nonproteinogenic amino acids), posttranslational modifications (e.g., phosphopantetheinylation), cofactor availability (e.g., coenzyme F$_{420}$), and regulatory issues (e.g., butyrolactone production). Empirical circumstances may dictate that a genetic system be developed for each natural product; however, a more desirable goal would be the establishment of a limited number of heterologous hosts whereby designed genetic engineering experiments can be carried out for any natural product or family of natural products.

III. EXAMPLES OF COMBINATORIAL BIOSYNTHESIS OF ANTICANCER NATURAL PRODUCTS

Selected in this section are only a few examples of combinatorial biosynthesis of anticancer natural products, illustrating the principle and practice of combinatorial biosynthesis of anticancer natural products. We have based our choice of the natural products on the following three criteria: the biosynthetic machinery and fundamental chemistry and biochemistry that governs their biosynthesis, the combinatorial biosynthesis strategies that have been applied to their engineering, and the clinical significance or potential of the natural products as anticancer drugs. Readers are directed to Table 25.1 for a more comprehensive summary of combinatorial biosynthesis of anticancer natural products of microbial origin covering the literature up to early 2004.

A. DAUNORUBICIN AND DOXORUBICIN (TYPE II POLYKETIDE SYNTHASE)

Daunorubicin (DNR) and doxorubicin (DXR), members of the anthracycline family of antitumor antibiotics, have been widely used clinically in various cancer chemotherapies, and DXR has been considered one of the most successful anticancer drugs ever developed.[28] The drugs intercalate into DNA noncovalently, thus preventing DNA template function as well as stabilizing the DNA–topoisomerase II cleavable complex.[29–31] The stabilization of this complex leads to DNA strand scission. Although anthracyclines have been shown to have many different effects on cellular physiology, this DNA strand-breakage is generally accepted as their primary activity.[29–31]

DNR was first isolated from a *Streptomyces peucetius* strain and was subsequently found in a number of other *Streptomyces* species. In contrast, DXR was isolated only from mutant strains that were derived from the DNR-producing *S. peucetius* wild-type strain upon chemical mutagenesis.[32] This was somewhat surprising because DXR has an additional HO- group at C-14 compared to DNR. DNR and DXR are structurally characterized by a napthacenequinone aglycon that is glycosylated with a deoxy aminosugar, L-daunosamine, at C-7-O (Figure 25.1A). Isotopic labeling experiments established that the DNR and DXR aglycones are derived from one propionate and nine acetates, whereas the L-daunosamine moiety is derived from L-glucose.[28,33]

1. Biosynthetic Gene Cluster and Biosynthetic Mechanisms

The Hutchinson group began isolating and sequencing the DNR/DXR biosynthetic gene clusters in 1989 from *S. peucetius* and *S. peucetius* subsp. *caesius*, with the latter strain producing DXR.[34] This effort continued for the next 10 yr and was complemented by parallel studies by the Strohl group on the DNR biosynthetic gene cluster from *Streptomyces* sp. strain C5,[28] leading to the eventual cloning, sequencing, and characterization of the entire DNR/DXR cluster. Central to the cluster are the *dps* genes that encode a type II polyketide synthase (PKS) for aglycone biosynthesis, the *dnm* genes that encode enzymes for TDP-L-daunosamine biosynthesis, and the *dnr* and *dox* genes for additional modifications of the DNR/DXR aglycone (Figure 25.1A and 25.1B). Many of these genes have been exploited successfully for the production of novel DNR/DXR analogs and other metabolites by combinatorial biosynthesis methods.[28,35,36]

DpsA-G and DpsY together catalyzes the assembly of the DNR/DXR aglycone backbone from one unit of propionyl-CoA and nine units of malonyl-CoA (Figure 25.1A). The process is initiated by DpsCDG to form the first intermediate of β-ketovaleryl-*S*-DpsG, with DpsC selecting for propionyl-CoA as a starter unit.[35] Further elongation of β-ketovaleryl-*S*-DpsG by DpsAB affords the linear decaketide intermediate, and DpsAB together dedicate the number of the malonyl-CoA extender units added to the growing acyl-*S*-DpsG intermediate.[37] The resultant reactive linear decaketide intermediate undergoes a series of facile transformations including regiospecific reduction (by DpsE) and multiple intramolecular cyclizations (by DpsFY), yielding the first isolatable biosynthetic intermediate, 12-deoxyaklanonic acid,[33,38–41] a key intermediate found in the biosynthesis of many other anthracyclines such as rhodomycin and aclacinomycin A.[42] Additional modifications of 12-deoxyaklanonic acid by

TABLE 25.1
Anticancer Natural Products[a]

Anticancer Antibiotic	Producing Organism	Biosynthetic Gene Cluster (Whole or Partial)	Heterologous Expression	Combinatorial Biosynthesis
Aclacinomycin	*Streptomyces galilaeus* ATCC 31615	182	66	66,183
Actinomycin D	*Streptomyces chrysomallus*	184	185	186
Ansamitocin	*Actinosynnema pretiosum*	70	187	81
Aureothin	*Streptomyces thioluteus* HKI-227	188	188	—
Bleomycin	*Streptomyces verticillus* ATCC15003	189-191	107,192-195	—
Borrelidin	*Streptomyces parvulus* Tü4055	196	—	—
C-1027	*Streptomyces globisporus*	131	—	131
Calicheamicin	*Micromonaspora echinospora* ssp. *calichensis*	130	120	—
Clorobiocin	*Streptomyces roseochromogenes* DS 12976	143	197-199	152-154
Coumermycin A$_1$	*Streptomyces rishiriensis* DSM 40489	144	197,200	150,151,201
Daunorubicin/Doxorubicin	*Streptomyces peucetius* (spp *casius*, dxr producer)	28,35,36,48	37,202,203	43,183,203
Dynemicin	*Micromonospora chersina*	128	—	—
Elloramycin	*Streptomyces olivaceus* Tü2353	204-206	204,205,207,208	209,210
Epothilones	*Sorangium cellulosum* So ce90	96-98	97,101,107,108,211	211-213
Esperamicin	*Actinomadura verrucosospora*	128	—	—
Fredericamycin	*Streptomyces griseus*	214	214	214
Geldanamycin	*Streptomyces hygroscopicus var. geldanus* NRRL 3602	79	—	—
Gilvocarcin V	*Streptomyces griseoflavus* Gö 3592	215	215	—
Hedamycin	*Streptomyces griseoruber* (ATCC15422)	216	—	—
Jadomycin	*Streptomyces venezuelae* ISP5230	217-222	203,223	203,220
Landomycin A	*Streptomyces cyanogenus* S136 (A)	224	225	226,227

Compound	Organism			
Leinamycin	*Streptomyces atroolivaceus* S-140	114,115	—	115
Maduropeptin	*Actinomadura madurae*	128	—	—
Mithramycin	*Streptomyces argillaceus*	57-59,63,228	66,229-231	51,65,66,227,232,233
Mitomycin C	*Streptomyces lavendulae* NRRL 2564	234,235	236,237	—
Macrotetrolide	*Streptomyces griseus* DSM40695	238,239	238	238,240,241
Neocarzinostatin	*Streptomyces carzinostaticus*	128,136	—	—
Nogalamycin	*Streptomyces nogalater*	242,243	66,244	66,183,245
Novobiocin	*Streptomyces spheroides* NCIB 11891	142,246	246-249	150,250
Oleandomycin	*Streptomyces antibioticus* ATCC 11891	251-259	252,253,255,259-261	258-260,262-264
Puromycin	*Streptomyces alboniger*	265	266	—
Rebeccamycin	*Saccharothrix (Lechevalieri) aerocolonigenes* ATCC 39243	267,268	269	269,305
Rifamycin B	*Amylcolatopsis mediterranei* S699	69,72,78	187	81,270,271
Saframycin	*Myxococcus xanthus*	272,273	—	—
Simocyclinone D8	*Streptomyces antibioticus* Tu6040	145,274	—	145
Sparsomycin	*Streptomyces sparsogenes*	275,276	—	—
Staurosporine	*Streptomyces* sp. TP-A0274	277	277	—
Tetracenomycin C	*Streptomyces glaucescens*	278,279	208,278,280-288	48,67,201,203,282,286,289-292
Urdamycin	*Streptomyces fradiae* Tü2717	293-296	293	226,227,297
				294,295,298,299
Valanimycin	*Streptomyces viridifaciens* MG456-hF10	300	300	—
Vicenistatin	*Streptomyces halstedii*	301	301	—

[a] Numbers in final three columns are reference numbers.

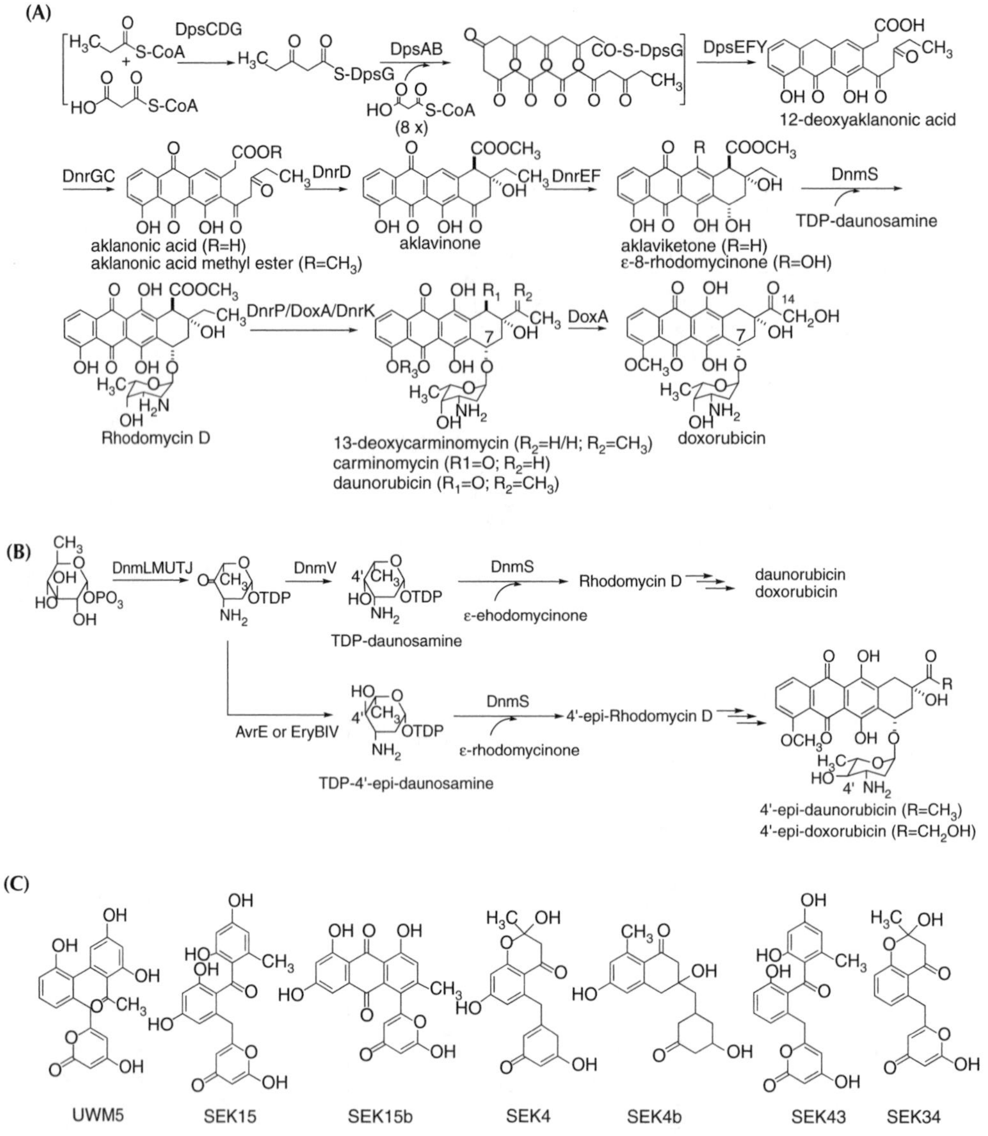

FIGURE 25.1 (A) Structures and biosynthesis of daunorubicin and doxorubicin. (B) Structures and biosynthesis of TDP-daunosamine and engineered biosynthesis of 4′-epi-daunorubicin and 4′-epi-doxorubicin. (C) Structures of novel metabolites produced by hybrid type II PKSs containing subunits of the daunorubicin/doxorubicin type II PKS.

the so-called tailoring enzymes that act after the Dps PKS enzymes finally furnish the myriad of functional groups characteristic of aromatic polyketide metabolites such as DNR/DXR. These fundamental genetic, biochemical, and chemical characterizations of the DNR/DXR biosynthetic machinery set the stage to formulate strategies, design experiments, and develop tools for combinatorial biosynthesis. The choice of the starter unit (as dedicated by DpsC), the number of the elongation units (as controlled by DpsAB), the regiochemistry of reduction and cyclization of the nascent linear polyketide (as determined by DpsEFY), and the multitude of post-PKS modifications (as furnished by DnrCDEFGKP and DoxA) represent outstanding opportunities for combinatorial biosynthesis and have all been successfully exploited to engineer structural diversity of aromatic polyketides such as DNR/DXR (Figure 25.1A and 25.1B).

The biosynthesis of TDP-L-daunosamine proceeds via six catalytic steps, starting from D-glucose-1-phosphate, and DnmLMJVUT are sufficient to catalyze this process.[43] Subsequent attachment of TDP-L-daunosamine to ε-rhodomycinone by the DnmS glycosyltransferase yields the first glycosylated product rhodomycin D (Figure 25.1B).[35] Many of the deoxy sugars found in natural products are biosynthesized by analogous pathways and often share common intermediates, and the glycosyltransferases show significant flexibility in TDP-deoxy sugar recognition and transfer. Thus, complementing aglycone engineering, the deoxy sugar biosynthetic pathways and the corresponding glycosyltransferases provide another dimension for natural production structural diversity by combinatorial biosynthesis.

2. Combinatorial Biosynthesis

One of the most informative experiments for determining the role a particular enzyme plays in the biosynthesis of a metabolite is the inactivation of the gene encoding an enzyme of interest, followed by isolation and structural characterization of the biosynthetic intermediate accumulated by the mutant strains. Thus, using either targeted gene disruption or random mutagenesis to inactivate genes, the roles that many, if not all, of the encoded enzymes play in biosynthesis can be deciphered. An additional benefit of this approach is that the intermediates accumulated by the mutants greatly facilitate the structural–activity relationship analysis of the natural products, and some of these intermediates may have biological activity distinct from the parent natural product.

Although the intermediates of the DNR/DXR pathway have not proven fruitful in generating compounds with improved biological activities, the thorough analysis of this pathway serves an excellent example illustrating the potential and effectiveness of identifying biosynthetic intermediates via targeted gene disruption. The details of these studies have been reviewed elsewhere,[27,28,35] and numerous compounds have been identified, including aklaviketone,[44] aklavinone,[45] ε-rhodomycinone,[45,46] and rhodomycin D[47,48] (Figure 25.1A). Although many of these intermediates were isolated from strains that were generated by random mutagenesis, the availability and characterization of the DNR/DXR biosynthetic gene cluster provided an alternative to do so rationally by targeted gene disruption.

Gene inactivation can be combined with gene complementation with homologs from other pathways to provide alternative functions. An excellent example of this is the conversion of the DNR/DXR-producing *S. peucetius* ATCC 29050 strain into an 4-epi-DNR/4-epi-DXR-producing strain.[49] Hutchinson and colleagues recognized that the stereochemistry of the 4-OH of TDP-L-daunosamine is controlled by DnmV, and DnmV homologs from the avermectin (AveE for TDP-oleandrose biosynthesis) or erythromycin (EryBIV for TDP-mycarose biosynthesis) pathways catalyze the analogous chemistry but with opposite stereochemistry. They reasoned that substitution of *dnmV* with *avrE* or *eryBIV* might afford the biosynthesis of TDP-4-epi-daunosamine, which could eventually lead to the production of 4-epi-DNR or 4-epi-DXR if the DnmS glycosyltransferase and other downstream enzymes can recognize and process the 4-epi-diastereomers of the biosynthetic intermediates. Remarkably, complementation of the Δ*dnmV* mutant with either *avrE* or *eryBIV* indeed resulted in the production of 4-epi-DNR and 4-epi-DXR, providing a fermentation alternative to these important anticancer drugs that are currently manufactured by low-yielding semisynthesis methods (Figure 25.1B).[49]

The DNR/DXR type II PKS has also been extensively exploited by combinatorial biosynthesis methods with other type II PKS systems such as actinorhodin,[36,50] tetracenomycin (TCM),[37] and jadomycin,[37] leading to the production of numerous novel polyketide metabolites (Figure 25.1C). Although these studies do not focus on generating new DNR/DXR analogs, they do highlight the feasibility and versatility of producing novel aromatic polyketide molecular scaffolds based on the type II PKS machinery.

B. Mithramycin (Type II PKS)

Mithramycin (MTM), isolated from *Streptomyces argillaceus* and several other *Streptomyces* species,[51] is a member of the aureolic acid group of anticancer drugs that includes the chromomycins, olivomycins, chromocyclomycins, durhamycins, and UCH9.[52] Structurally, members of the aureolic acid group of anticancer drugs (with the exception of chromocyclomycin) share a similar aglycone with two alkyl substitutions at C3 and C7 but differ in their glycosylation pattern at C6-O and C2-O. MTM has proven clinically important for the treatment of various cancers.[53] The anticancer activity of MTM is presumed to be a result of a Mg^{2+}-MTM complex binding to GC-rich DNA and disrupting DNA-dependent RNA synthesis,[52] in addition to inducing the surrounding AT-rich sequences to be sensitive to DNase II attack.[54,55]

1. Biosynthetic Gene Cluster and Biosynthetic Mechanisms

Although the polyketide origin of the aureolic acid group of anticancer natural products was established by Rosazzas and coworkers by incorporating isotope-labeled precursors into chromomycins A3,[56] it was the reinterpretation of these data and subsequent cloning and sequencing of the MTM and chromomycins biosynthetic gene clusters by Salas, Rohr, and coworkers that unambiguously established the biosynthetic pathway for this family of metabolites, involving a single linear dekaketide as a key intermediate.[52,57-61] Key features of the MTM clusters include genes encoding the MTM type II PKS and associated enzymes that catalyze the formation of the tetracyclic decaketide intermediate premithramycinone from 10 units of malonyl-CoA, genes encoding deoxysugar biosynthesis enzymes that catalyze the biosynthesis of the five deoxy sugars and their attachment at C2-O and C6-O of the MTM aglycone, and genes encoding various tailoring enzymes that catalyze modifications of the MTM aglycone, including the unique ring fission oxidation, to finally afford the characteristic tricyclic aglycone for this group of anticancer natural products (Figure 25.2A).[52] In contrast to the DNR/DXR type II PKS that requires the DpsCD to select propionyl-CoA as a special starter unit, the MTM type II PKS does not contain the DpsCD homologs; it instead derives the acetate starter unit from malonyl-CoA, a common feature believed to be true for all type II PKSs that catalyze the biosynthesis of polyketide natural products with acetate as a starter unit.[6]

One of the most fascinating aspects for MTM biosynthesis is the interplay between the glycosyltransferases and the tailoring enzymes. MTM contains a trisaccharide (D-olivose-D-oliose-D-mycarose) at C2-O (C-12a-O in premithramycinone) and a disaccharide (D-olivose-D-olivose) at C6-O (C8-O of premithramycinone), but only four glycosyltransferases (MtmGI/II/III/IV) have been identified within the MTM cluster. The timing of the glycosylation steps was deciphered by a series of gene disruption experiments followed by structural determination of the accumulated intermediate(s).[52,62,63] Thus, D-olivose is attached to premithramycinone at C12a-O by MtmGIV, which is followed by D-oliose attachment to the D-olivose by MtmGIII. The addition of D-mycarose to complete the trisaccharide moiety is still an open question, with either MtmGIII catalyzing the addition of both D-oliose and D-mycarose to D-olivose or MtmGIV catalyzing D-mycarose addition. The disaccharide (D-olivose-D-olivose) attached to premithramycinone at C8-O is synthesized first by the action of MtmGI or MtmGII and subsequently attached by either of these enzymes. Once the five sugar moieties are attached, forming premithramycin B, ring D is oxidatively cleaved by MtmOIV to finally afford MTM (Figure 25.2A).

2. Combinatorial Biosynthesis

Salas, Rohr, and coworkers have extensively characterized the MTM biosynthetic pathway by targeted gene disruption.[58,60,62-64] The resultant mutants accumulated a number of novel MTM analogs including 4-demethylpremithramycinone, premithramycinone, premithramycin A1, premithramycin A2, 9-demethylpremithramycin A3, premithramycin A3, 7-demethyl-mithramycin, premithramycin B, 4A-keto-D-olivosyl-premithramycin A2, 4A-keto-D-olivosyl-9-demethyl-premithramycin A2, and 4A-keto-D-olivosyl-demycarosyl-mithramicin (Figure 25.2A and 25.2B).

By mixing and matching genes from the MTM pathway with genes from various other biosynthetic pathways in either the MTM-producing *Streptomyces agrillaceus* or other polyketide-producing

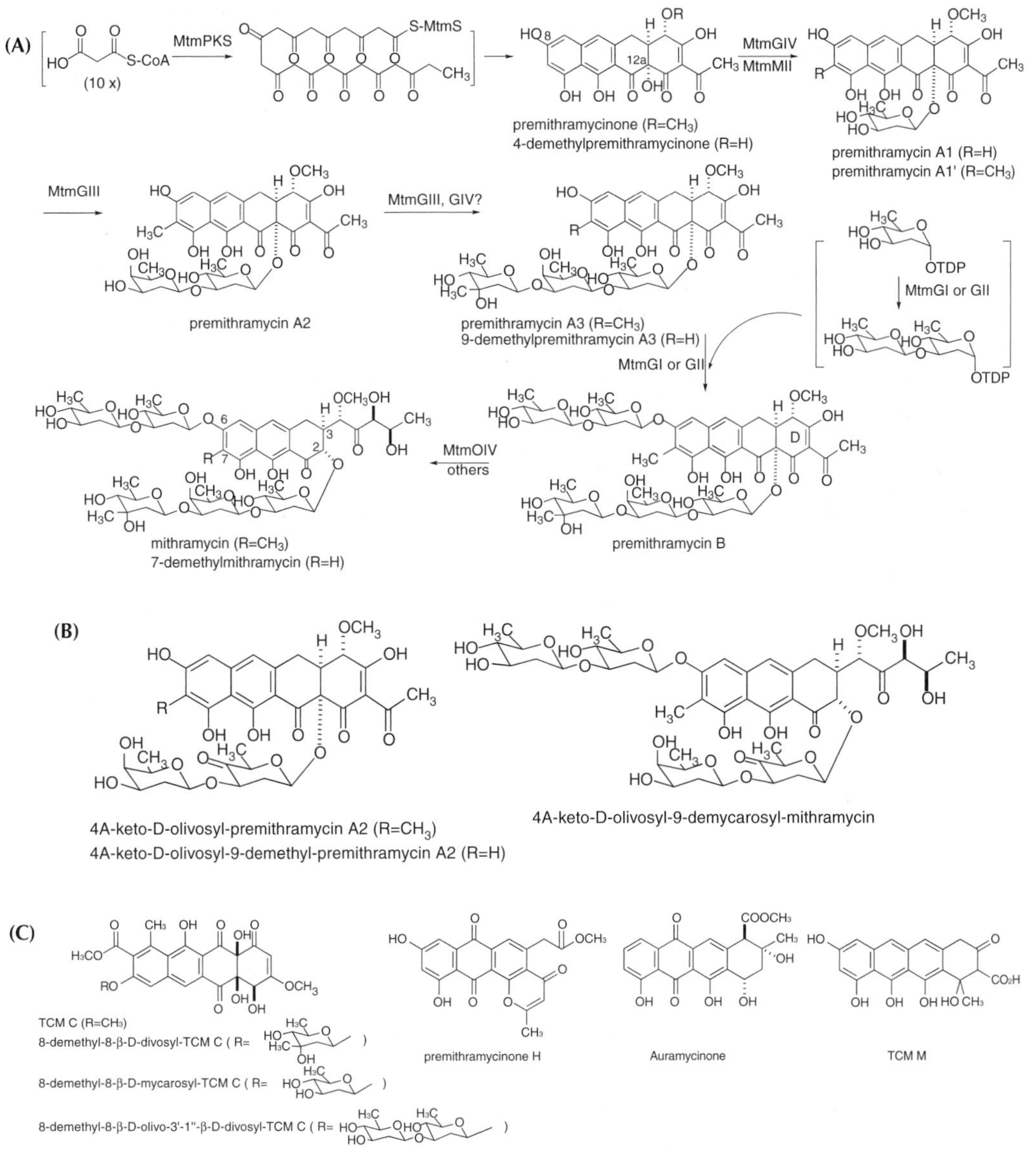

FIGURE 25.2 (A) Structures and biosynthesis of mithramycin. (B) Structures of novel mithramycin analogs produced by engineering the mitramycin pathway. (C) Structures of novel metabolites produced by mixing and matching genes from the mithramycin, tetracenomycin, and elloramycin pathways.

Streptomyces strains, Salas, Rohr, and coworkers have also systematically explored the potential of using the MTM biosynthetic machinery for the production of novel metabolites by combinatorial biosynthesis methods (Figure 25.2C). Introduction of the *tcmGH* genes from the TCM pathway, both of which encode oxygenases, into an *S. agrillaceus* Δ*mtmD* mutant resulted in the production of premithramycinone H.[65] Expression of the minimal MTM type II PKS (*mtmPKS*) along with a cyclase (*mtmX*) in an *Streptomyces galilaeus* mutant strain produced auramycinone, an intermediate in aclacinomycin biosynthesis,[66] while introduction of the same set of genes into *Streptomyces glaucescens* yielded TCM M.[67] Production of the anthracycline antibiotic elloramycin in *S. agrillaceus* resulted in the glycosylation of the elloramycin aglycon with sugars from the MTM biosynthetic pathway, leading to the production of hybrid metabolites such as 8-β-D-olivosyl-tetracenomycin C, 8-demethyl-8-β-D-mycarosyl-tetracenomycin C, and 8-demethyl-8-β-D-olivo-3-1-β-D-olivosyl-tetracenomycin C.[68] These studies highlight how enzymatic components from pathways producing different anticancer natural products can be mixed and matched to further extend the biosynthetic potential of the characterized biosynthetic machineries.

C. ANSAMYCINS (TYPE I PKS)

The ansamycins all include an aliphatic ansa bridge, giving the antibiotic family their name, and are divided into two structural groups based on whether their macrocycle contains a benzenic or naphthalenic chromophore. The benzoquinone ansamycins include the anticancer drugs geldanamycin, herbimycin A, maytansine, and ansamitocin P-3, whereas the napthoquinone ansamycins include rifamycin B, the most well-known and best characterized member of this antibiotic family[69] (Figure 25.3A). All of the ansamycins contain a 3-amino-5-hydroxybenzoic acid (AHBA) moiety that is oxidized to the benzoquinone or converted to the naphthoquinone via cyclization and oxidation with other moieties within the molecule. On the basis of results from feeding experiments with isotope-labeled precursors, the biosynthesis of both the benzoquinone and naphthoquinone ansamycins can be predicted to involve a type I PKS specifying AHBA as the starter unit.[69–72]

Although none of the benzoquinone ansamycins are currently in clinical use, the promising anticancer activities they possess have spurred interest in the development of analogs of these compounds that lack the dose-limiting toxicity in humans.[73] The biological target of these molecules is Hsp-90, an important protein chaperone in mammalian cells.[74] Geldanamycin has been shown to bind to the ATP-binding site on Hsp-90, thereby inhibiting the ATP-dependent activity of this chaperone.[75,76] In the absence of Hsp-90 function, associated protein kinases that are essential

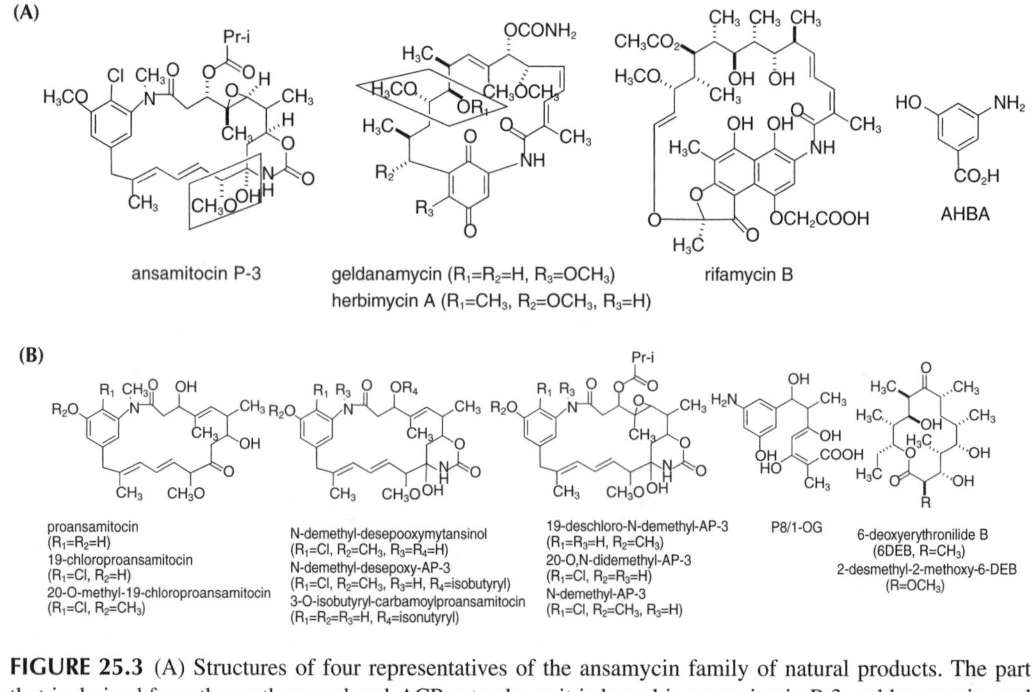

(A)

ansamitocin P-3

geldanamycin (R₁=R₂=H, R₃=OCH₃)
herbimycin A (R₁=CH₃, R₂=OCH₃, R₃=H)

rifamycin B

AHBA

(B)

proansamitocin
(R₁=R₂=H)
19-chloroproansamitocin
(R₁=Cl, R₂=H)
20-O-methyl-19-chloroproansamitocin
(R₁=Cl, R₂=CH₃)

N-demethyl-desepooxymytansinol
(R₁=Cl, R₂=CH₃, R₃=R₄=H)
N-demethyl-desepoxy-AP-3
(R₁=Cl, R₂=CH₃, R₃=H, R₄=isobutyryl)
3-O-isobutyryl-carbamoylproansamitocin
(R₁=R₂=R₃=H, R₄=isonutyryl)

19-deschloro-N-demethyl-AP-3
(R₁=R₃=H, R₂=CH₃)
20-O,N-didemethyl-AP-3
(R₁=Cl, R₂=R₃=H)
N-demethyl-AP-3
(R₁=Cl, R₂=CH₃, R₃=H)

P8/1-OG

6-deoxyerythronilide B
(6DEB, R=CH₃)
2-desmethyl-2-methoxy-6-DEB
(R=OCH₃)

FIGURE 25.3 (A) Structures of four representatives of the ansamycin family of natural products. The part that is derived from the methoxymalonyl-ACP extender unit is boxed in ansamitocin P-3, geldanamycin, and herbimycin A. (B) Structures of novel ansamitocin P-3 biosynthetic intermediates produced by engineering the ansamitocin P-3, erythromycin, and FK520 pathways.

components of signal transduction pathways cannot function properly. The failure of these protein kinases to function can result in a cytostatic effect on cancer cells.[73] Importantly, Hsp-90 and one or more of its protein kinase associates are overproduced in certain types of human cancers, indicating that Hsp-90 is a promising target for the development of anticancer drugs.

1. Biosynthetic Gene Clusters and Biosynthetic Mechanisms

The cloning and sequencing of the genes encoding the biosynthesis of the AHBA moiety of the ansamycins by Floss and coworkers[77] gave a unique handle to specifically clone the ansamycin biosynthetic gene clusters from their producing organisms. The first ansamycin biosynthetic gene cluster sequenced was the archetype of this family of molecules, rifamycin B, which was cloned from two different strains of *Amycolatopsis mediterranei*.[69,78] Bioinformatic analysis of the cloned clusters revealed that the rifamycin backbone is assembled by the Rif PKS (RifA-E), a type I PKS that specifies for AHBA as a starter unit (by RifA) and that uses two malonyl-CoA and eight methylmalonyl-CoA as extender units (by RfiA-E). Intramolecular cyclization of the full-length polyketide intermediate (by RifF), followed by subsequent modifications of the resultant macrolactam intermediate by the tailoring enzymes, finally affords the rifamycins.

The insights gained from the rifamycin B biosynthetic gene cluster, particularly the genes encoding the AHBA biosynthetic enzymes (RifG-N), greatly facilitated the cloning and sequencing of the ansamitocin[70] and geldanamycin[79] biosynthetic gene clusters from *Actinosynnema pretiosum* and *Streptomyces hygroscopicus* NRRL 3602, respectively. The Asm and Gdm type I PKS consist of not only a loading module specific for AHBA (by AsnA and GdmA1), similar to the Rif PKS, but also extending modules specific for the unusual extender unit of methoxymalonyl-acyl carrier protein (ACP), a finding that is consistent with their structures (Figure 25.3A). A dedicated set of genes encoding the biosynthesis and activation of the rare methoxymalonyl-ACP extender unit was also identified within both clusters.

2. Combinatorial Biosynthesis

Floss and coworkers have meticulously inactivated each of the candidate genes, encoding the tailoring steps for ansamitocin P-3 biosynthesis in *A. pretiosum* followed by isolation and structural elucidation of the accumulated metabolites.[80] These studies not only allowed them to identify the functions for many of the gene products and to establish the order of the tailoring steps catalyzed by these enzymes but also demonstrated the feasibility of producing novel ansamitocin analogs by combinatorial biosynthesis in the native producing organisms (Figure 25.3B). Alternatively, production of an ansamycin polyketide intermediate has also been achieved by expressing a hybrid biosynthetic pathway in *E. coli*.[81]

Khosla and coworkers introduced the following components from various biosynthetic machineries into a single *E. coli* strain: the *sfp* phosphopantetheinyl transferase gene from *Bacillus subtilis* to ensure posttranslation modification of PKS or nonribosomal peptide synthetase (NRPS) modules,[82,83] the *rifH/K/L/M/N* genes from the rifamycin producing *A. mediterranei* and the *asm23/47* genes from the ansamcyin P-3 producing *A. pertiosum* to produce the AHBA starter unit,[81] the *pccB* and *accA1* gene from *Streptomyces coelicolor* to provide the methylmalonyl-CoA extender units,[84] and the *rifA* gene from *A. mediterranei* to catalyze the initiation and the first three cycles of polyketide elongation for a tetraketide intermediate in ansamycin biosynthesis.[81,85] To overcome the difficulty of producing the four-module RifA as an intact protein, specific polypeptide linkers from 6-deoxyerythronolide B synthase (DEBS), the archetype type I PKS catalyzes the biosynthesis of erythromycins from *Saccharopolyspora erythraea*,[86–88] were fused to a divided RifA synthase to facilitate interpolypeptide transfer of the growing biosynthetic intermediate.[81] The resultant *E. coli* recombinant strain indeed produced the expected tetraketide intermediate, 2,6-dimethyl-3,5,7-trihydroxy-7-(3-amino-5-hydroxyphenyl) 2,4-heptadienoic acid (P8/1-OG) (Figure 25.3B). This work is one of the best examples of rational construction of a hybrid pathway from various biosynthetic machineries and represents an important first step toward the reconstitution of the ansamycin biosynthetic machinery in a fast-growing, genetically tractable host for engineered production of novel ansamycins.

Complementary to ansamycin production in a heterologous host, parts of the ansamycin biosynthetic machinery have also been combined with other polyketide biosynthetic machinery for the production of novel structures. For example, Floss and coworkers introduced the *asm13-17* genes from *A. pertiosum*, which encode the biosynthesis of the unusual methoxymalonyl-ACP extender unit, into a recombinant *S. lividans* strain. The latter carried an engineered DEBS whose methylmalonyl-specific PKS module-6 has been replaced by the methoxymalonyl-specific PKS module of the FK520 pathway from *S. hygroscopicus*.[79] Production of methoxymalonyl-ACP by the Asm13–17 enzymes, and incorporation of this unusual extender unit by the engineered DEBS, resulted in the production of the anticipated 2-desmethyl-2-methoxy-6-deoxyerythronolide B (Figure 25.3B), serving as another example of producing novel structures by combining features from multiple natural product biosynthetic pathways.[89] With the growing list of gene clusters sequenced for the ansamycin family of metabolites, the extensive genetic and biochemical groundwork conducted by Floss and coworkers on rifamycin biosynthesis,[90] and the enormous success at reprogramming type I PKS such as DEBS,[15] it is not difficult to envisage engineering additional novel ansamycin analogs by combinatorial biosynthesis methods in the near future.

D. EPOTHILONES (HYBRID PKS-NRPS)

The epothilones (EPOs) were first isolated from culture extracts of the myxobacterium *Sorangium cellulosum* strain So ce90, exhibiting antifungal and cytotoxic activities, and were subsequently rediscovered in a target-driven screening for molecules that stabilize tubulin by a mechanism similar to the anticancer drug paclitaxel (Taxol).[91,92] By stabilizing tubulin, the EPOs induce cell cycle arrest, leading to apoptosis, and most significantly the EPOs are active against paclitaxel-resistant cell lines, as well as cell lines resistant to other anticancer agents.[92] EPO B showed the most promising *in vitro* activity; however, this activity was markedly reduced when analyzed *in vivo*. In addition to the two-major EPOs (A and B), numerous minor analogs have also been isolated and characterized, including most notably EPO C and D (Figure 25.4A). EPO D has better activity than EPO B and paclitaxel and did not result in the same acute toxic effects as EPO B.[93] Several variants of the EPOs are currently in various stages of clinical trials.[94]

The EPOs contain a methylthiazolyl moiety attached to a 16-membered macrolactone. The main structural variations among the EPOs are the modification at C12/C13 (a double bond, as in EPO C and D vs. an epoxide as in EPO A and B) and the substitution at C12 (H for EPO A and C and CH$_3$ for EPO B and D). Feeding experiments with isotopic-labeled precursors have established a hybrid peptide–polyketide origin for EPO biosynthesis with the EPO backbone specifically derived from acetate, propionate, methionine, and cysteine.[95] These results are consistent with the hypothesis that EPO biosynthetic machinery is characterized by a hybrid nonribosomal peptide synthetase (NRPS)-PKS system (Figure 25.4A).[96–98]

1. Biosynthetic Gene Cluster and Biosynthetic Mechanisms

Molnar and coworkers and Julien and coworkers independently cloned and sequenced the EPO biosynthetic gene cluster from *S. cellulosum* strain So ce90[96] and *S. cellulosum* strain SMP44,[97] respectively. The gene nomenclature proposed by Julien and coworkers is used here.[97,98] Central to the EPO cluster are five PKS genes (*epoA-E*) encoding nine PKS modules, one NRPS gene (*epoB*) encoding one NRPS module, and *epoK* encoding a cytochrome P-450 monooxygenase. EpoA-F constitutes a hybrid PKS/NRPS/PKS megasynthase that activates, condenses, and modifies five malonyl, four methylmalonyl, and one cysteinyl unit to form the macrolactone core of EPO C. Module 4 of the Epo hybrid PKS/NRPS/PKS megasynthase apparently can accept both malonyl-CoA or methylmalonyl-CoA as extender units, resulting in the production of EPO D, in addition to EPO C. Further oxidation of EPO C and D by EpoK finally affords EPO A or B (Figure 25.4A).

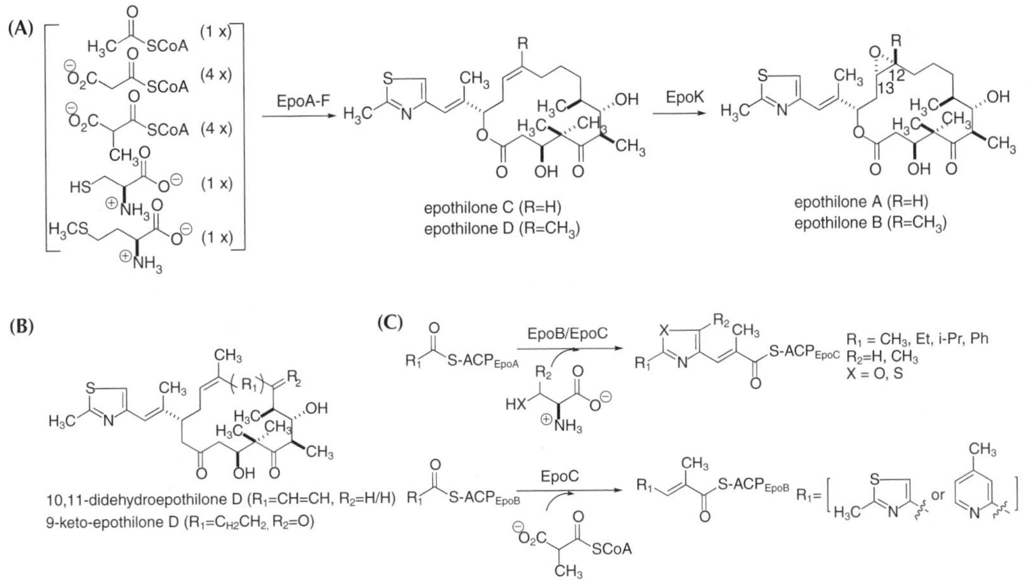

FIGURE 25.4 (A) Structures and biosynthesis of the epothilones. (B) Structures of novel epothilones produced by engineering the epothilone pathway. (C) Structures of novel analogs of epothilone biosynthesis intermediates produced by purified and/or engineered epothilone type I PKS enzymes.

2. Combinatorial Biosynthesis

Combinatorial biosynthesis of EPOs serves as an excellent example to illustrate overexpression and manipulation of an entire biosynthetic pathway in a more amenable heterologous host than its native producer. The impetus for this work was the finding that the natural EPO-producing myxobacterial strains are slow growing and that methods and tools for the genetic manipulation of these strains have yet to be fully developed. Furthermore, EPO D, the derivative that shows the greatest therapeutic index among the EPOs,[99,100] is only a minor component produced by the native strains. Functional expression of the EPO biosynthetic pathway into a faster-growing host that is genetically more tractable than the native producer therefore could be advantageous for both titer improvement of the natural products and rational engineering of the EPO pathway to selectively produce the desired EPO analog.

Julien and coworkers first cloned the entire EPO cluster (*epoA-K*) into two compatible expression vectors and introduced them into *S. coelicolor* CH999, an engineered *Streptomyces* host extensively used for combinatorial biosynthesis.[97] The resulting recombinant strain produced moderate amounts (50–100 µg/L) of EPOs, far less than required for industrial scale-up; however, this system could be amenable to yield improvement via traditional and contemporary means. The researchers then moved the entire EPO cluster into the well-studied myxobacterium *Myxococcus xanthus*.[101] Optimization of fermentation conditions for the resultant recombinant strain has subsequently led to EPO titer exceeding 20 mg/L, an acceptable yield for industrial scale production by fermentation.[93,101–103] Although these studies successfully afforded the production of EPO A and B in *S. coelicolor* CH999 and *M. xanthus*, Julien and coworkers also constructed recombinant strains that lack a functional *epoK*.[98,101] Because EpoK catalyzes the epoxidation of EPO C and D, the absence of a functional EpoK enabled the latter strains to produce EPO C and D exclusively.

As a proof of principle that rational engineering of the hybrid Epo NRPS-PKS megasynthase could potentially introduce additional structural diversities into the EPO hybrid peptide–polyketide backbone, Carney, Katz, and coworkers inactivated several domains within the EpoD PKS subunits. They took advantage of the genetic tools available in *M. xanthus* and introduced the engineered hybrid Epo NRPS-PKS into *M. xanthus* to screen for the production of novel EPO analogs.[93] Although inactivation of the enoylreductase domain of module 5 of EpoD yielded 10,11-didehydroepothilone C and D, double inactivation of the ketoreductase domain of module 6 of EpoD and EpoK resulted in the production of 9-keto-epothilone D (Figure 25.4B).[104]

Complementary to these *in vivo* studies, Walsh and coworkers have performed a number of elegant *in vitro* studies to probe the enzymatic flexibility of EpoB and EpoC to accept alternative substrates for EPO biosynthesis[105–108] (Figure 25.4C). They first showed that EpoB and EpoC are quite flexible, accepting various acyl-ACP_{EpoA} as substrates, including propionyl-, isobutyryl-, and benzoyl-ACP_{EpoA}, resulting in the production of a series of 2-alkylthiazolylmethylacrylyl-ACP_{EpoC} derivatives. They then revealed that EpoB can activate and modify not only the natural L-cysteine substrate but also L-serine and L-threonine, with each subsequently being incorporated and modified by the reconstituted ACP_{EpoA}/EpoB/EpoC complex.[107] They finally demonstrated that EpoA or EpoB can be replaced by proteins from the rapamycin, enterobactin, and yersiniabactin pathways, with efficient catalysis made possible by manipulating protein–protein recognition components at the PKS/NRPS and NRPS/PKS interfaces.[106] These studies demonstrated the potential to engineer additional EPO structural diversity by applying combinatorial biosynthesis methods to the hybrid Epo NRPS-PKS megasynthase.

E. LEINAMYCIN (HYBRID NRPS-PKS FEATURING AT-LESS PKS)

Leinamycin (LNM), produced by several *Streptomyces* species,[109,110] contains an unusual 1,3-dioxo-1,2-dithiolane moiety that is spiro-fused to a thiazole-containing 18-membered lactam ring (Figure 25.5), a molecular architecture that has not been seen in any other natural product. LNM shows potent antitumor activity *in vitro* and in various tumor models *in vivo*.[111,112] Most significantly LNM is active against tumors that are resistant to clinically important anticancer drugs, such as cisplatin, DOX, MTM C, or cyclophosphamide.[110] LNM causes alkylative DNA cleavage in the presence of thiol agents as cofactors, and this process is mediated by an episulfonium ion intermediate.[113] However, development of LNM into a clinical anticancer drug has so far been hampered by its *in vivo* instability.[111,112]

1. Biosynthetic Gene Clusters and Biosynthetic Mechanisms

Shen and coworkers developed a polymerase chain reaction method to clone genes encoding thiazole-forming NRPS and applied it to the cloning and localization of the LNM biosynthetic gene cluster from *Streptomyces atroolivaceus*.[114] Sequencing and subsequent characterization of the LNM cluster revealed that the LNM macrolactam backbone is derived from one molecule each of D-alanine and L-cysteine and six molecules of malonyl-CoA, with the two CH_3 groups at C6 and C4 derived from methionine. Strikingly, the biosynthesis of the LNM backbone from the carboxylic acid and amino acid precursors is catalyzed by a hybrid NRPS (LnmQP) and AT-less type I PKS (LnmGIJ) megasynthase, representing a novel architecture for both NRPS and PKS.[115,116] The biosynthetic origin of the 1,3-dioxo-1,2-dithiolane moiety, however, remains to be established.

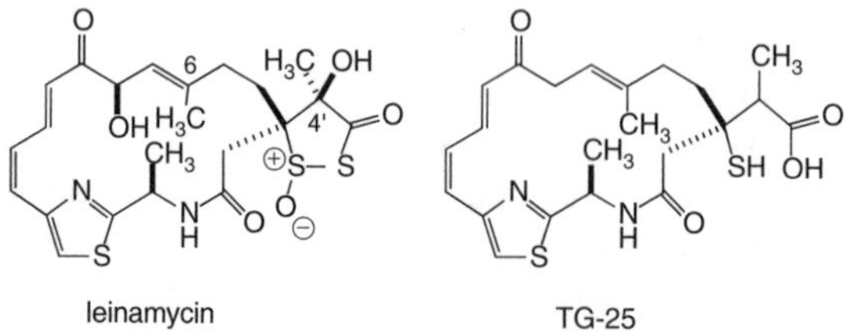

leinamycin   TG-25

FIGURE 25.5 Structures of leinamycin and TG-25. TG-25 was produced by combinatorial biosynthesis of the leinamycin pathway.

2. Combinatorial Biosynthesis

To produce LNM analogs by combinatorial biosynthesis methods, Shen and coworkers first developed an expedient genetic system for *in vivo* manipulation of LNM biosynthesis in *S. atroolivaceus*.[114] Systematic inactivation of genes from the sequenced region not only defined the precise boundaries of the LNM cluster but also generated a series of mutants that accumulated numerous novel metabolites.[116] For example, inactivation of *lnmE* abolished LNM production, but the Δ*lnmE* mutant strain produced a new compound TG-25 (Figure 25.5) (Yun, B.-S., Tang, G.-L., Cheng, Y.-Q., and Shen, B., unpublished data). TG-25 represents a novel structural variant of the LNM scaffold, featuring a free –SH group. Remarkably, although episulfonium ion formation from LNM requires a reductive environment, needing reduced thiols, preliminary studies indicated that TG-25 can generate a similar episulfonium ion under an oxidative environment, requiring O_2. TG-25 could therefore potentially be used as anticancer drug via oxidative activation *in situ* in tumors under high reactive oxygen species while remaining relatively nontoxic to normal tissues.

F. ENEDIYNES (ITERATIVE TYPE I PKS)

Since the unveiling of the neocarzinostatin (NCS) chromophore structure in 1985,[117] the enediyne family has grown steadily.[118] Over 20 enediyne natural products are currently known, and they are classified into two subcategories according to the enediyne core structures. Members of the 9-membered enediyne core subcategory are chromoproteins consisting of an apo-protein and the enediyne chromophore, including C-1027 from *Streptomyces globisporus*, NCS from *Streptomyces carzinostaticus*, and maduropeptin (MAD) from *Actinomadura madurea*. The 10-membered enediyne core subcategory includes calicheamicin (CAL) from *Micromonospora echinospora* spp. *calichenisis*, dynemicin (DYN) from *Micromonospora chersina* sp. nov. No. M965-1, and esperamicin (ESP) from *Actinomadura verrucosospora* (Figure 25.6). Although the 10-membered enediynes are all isolated as discrete small molecules, the gene product of *calC* from the CAL biosynthetic gene cluster confers CAL resistance *in vivo*.[119] Although no binding between CalC and CAL has been experimentally demonstrated, CalC apparently confers resistance by a self-sacrificing mechanism.[120] This finding raised the interesting question of whether self-sacrificing is a general mechanism for all 10-membered enediynes and whether there is an evolutionary relationship (mechanistic or structural) between CalC and the apo-proteins of the nine-membered

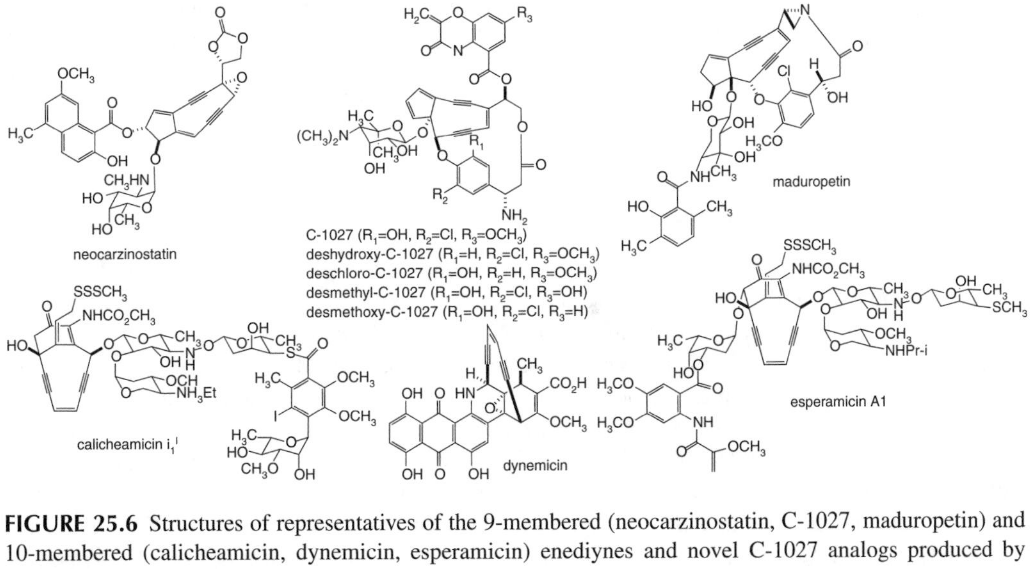

FIGURE 25.6 Structures of representatives of the 9-membered (neocarzinostatin, C-1027, maduropetin) and 10-membered (calicheamicin, dynemicin, esperamicin) enediynes and novel C-1027 analogs produced by engineering the C-1027 pathway.

enediyne chromoproteins, even through no sequence homology exists between CalC and the nine-membered enediyne apo-proteins.

The enediynes are extremely potent anticancer agents with a unique molecular architecture. As a family, they are the most potent, highly active anticancer agents ever discovered. All members of this family contain a unit consisting of two acetylenic groups conjugated to a double bond or incipient double bond within a 9- or 10-membered ring (i.e., the enediyne core). As a consequence of this structural feature, these compounds share a common mechanism of action: The enediyne core undergoes an electronic rearrangement (Bergman or Myers rearrangement) to form a transient benzenoid diradical, which is positioned in the minor groove of DNA so as to damage DNA by abstracting hydrogen atoms from deoxyriboses on both strands. Reaction of the resulting deoxyribose carbon-centered radicals with molecular oxygen initiates a process that results in both single-stranded and double-stranded DNA cleavage.[118,121,122]

Although the natural enediynes have seen limited use as clinical drugs mainly because of substantial toxicity, various polymer-based delivery systems or monoclonal antibody (mAB)-enediyne conjugates have shown great clinical success or promise in anticancer chemotherapy.[119,123–125] For example, the poly(styrene-co-maleic acid)-conjugated NCS (SMANCS®) was approved in Japan in 1993 and has been marketed since 1994 for use against hepatoma.[122] A CD33 mAB-CAL conjugate (Mylotarg®) was approved by the FDA in 2000 for the treatment of acute myeloid leukemia.[124,126] Several anti-hepatoma mAB-C-1027 conjugates have also been prepared and displayed high tumor specificity with a strong inhibitory effect on the growth of established tumor xenografts.[125] These examples clearly demonstrate that the enediynes can be developed into powerful drugs when their extremely potent cytotoxicity is harnessed and specifically delivered into the target tumor cells.

1. Biosynthetic Gene Clusters and Biosynthetic Mechanisms

Feeding experiments with isotope-labeled precursors for NCS, DYN, and ESP biosynthesis clearly established that the enediyne cores of both the 9- and 10-membered subcategories are derived from the head-to-tail condensation of (at minimum) eight acetate units.[9,127–129] This indicated a PKS mechanism of synthesis, but the exact mechanism of polyketide biosynthesis remained controversial until the Shen and Thorson groups simultaneously published the cloning, sequencing, and characterization of the CAL[130] and C-1027[131] biosynthetic gene clusters, respectively. Surprisingly, both gene clusters encode an iterative type I PKS, enediyne PKS, which is predicted to synthesize a linear polyketide intermediate. This linear intermediate is presumably modified by a series of enediyne PKS-associated enzymes to generate the enediyne core scaffolds. Both gene clusters also encode the necessary enzymes for synthesis and attachment of the appropriate deoxy aminosugar and other peripheral moieties. These studies laid the foundation for combinatorial biosynthesis of this important family of anticancer natural products.

2. Combinatorial Biosynthesis

To set the stage to apply combinatorial biosynthesis methods to the enediyne biosynthetic machinery for generating novel anticancer drugs, the Shen and Thorson groups first developed genetic systems for *S. globisporus* (C-1027), *S. carzinostaticus* (NCS), and *M. calichensis* (CAL).[128,130–132] With these tools in hand, Shen and coworkers have systematically inactivated many of the genes identified within the C-1027 and NCS genes gene clusters, unveiling many mechanistic details into enediyne biosynthesis.[128,131–136] On the basis of these studies, they have demonstrated both yield improvement for C-1027 and production of novel C-1027 analogs by combinatorial biosynthesis methods. For example, overexpression of *sgcA1* and its flanking genes in the wild-type *S. globisporus* has resulted in two- to fourfold increase in C-1027 production,[135] and inactivation of *sgcC1*, *sgcC3*, *sgcD3*, and *sgcD4* has afforded *S. globisporus* recombinant strains that produced four novel C-1027 analogs[131] (Liu, W., Ju, J., Shen, B., unpublished data).

Genetic and biochemical characterization of the C-1027 biosynthetic gene cluster has previously predicted that SgcC1 and SgcC3 catalyze the hydroxylation and chlorination steps in β-amino acid moiety biosynthesis, while SgcD3 and SgcD4 are responsible for the hydroxylation and methylation steps for the benzoxazolinate moiety.[131] Mutants of corresponding genes indeed accumulated the predicted new metabolites: deshydroxy-C-1027 (ΔsgcC3), deschloro-C-1027 (ΔsgcC3), desmethyl-C-1027 (ΔsgcD4), and desmethoxy-C-1027 (ΔsgcD3) (Figure 25.6). These results clearly demonstrated the feasibility and effectiveness to produce novel enediyne analogs by combinatorial biosynthesis methods. Strikingly, deshydroxy-C-1027 is at least fivefold more stable than C-1027 in respect to undergoing the Bergman cyclization, a property that could be potentially explored in C-1027 engineering.[131] The latter result indicates that chemical and physical properties of C-1027 could be fine-tuned by varying the peripheral moieties of the enediyne core, supporting the wisdom of diversifying the C-1027 molecular scaffold for drug discovery.

Expedient methods to access and clone enediyne gene clusters have also been developed, and the growing list of enediyne clusters available will surely accelerate the pace of applying combinatorial biosynthesis methods to the enediyne biosynthetic machinery for novel anticancer drug discovery. Farnet and coworkers have developed a high-throughput genome scanning method to detect and analyze gene clusters involved in natural product biosynthesis. This method was applied successfully to uncover numerous biosynthetic pathways encoding enediyne biosynthesis in variety of actinomycetes, including many strains that were not known previously as enediyne producers.[132] Inspired by the biosynthetic potential for enediyne natural products, as evidenced by the DNA sequence, Farnet and coworkers demonstrated that enediyne production in these organisms can be induced on optimization of their fermentation conditions. This genomics-guided approach for discovering and expressing cryptic metabolic pathways provides an outstanding opportunity to fully exploit the biosynthetic potential of microorganisms that we now know is much greater than previously appreciated.[137,138]

Complementary to the genome scanning method, the Shen and Thorson groups developed a polymerase chain reaction method to quickly access additional enediyne gene clusters on the basis of the emerging paradigm for enediyne biosynthesis featuring the enediyne PKS. This method was applied to clone the MAD, DYN, and ESP enediyne PKS loci from their respective producing organisms.[128] A phylogenetic analysis of these *bona fide* enediyne PKS genes has revealed a clear genotypic distinction between 9- and 10-membered enediyne core subcategories (i.e., C-1027, NCS, MAD vs. CAL, DYN, ESP). This finding is consistent with the postulation that the minimal enediyne PKS helps define the structural divergence of the enediyne cores and could potentially be used to predict unknown enediyne structures on the basis of their enediyne PKS sequences or to guide rational construction of hybrid enediyne biosynthetic pathways for novel enediyne production.[128]

G. Aminocoumarins (Mixed Biosynthesis Featuring Atypical NRPS)

The aminocoumarin antibiotics, novobiocin from *Streptomyces spheroides*, coumermycin A$_1$ from *Streptomyces risluriensis*, clorobiocin from *Streptomyces roseochromogenes*, and simocyclinone D8 from *Streptomyces antibioticus*, were initially identified for their antibiotic activity against Gram-positive bacteria. The interest in their role in the treatment of cancer is a result of their ability to potentiate the effects of several anticancer drugs such as etoposide and teniposide.[139,140] In addition, novobiocin was recently found to reverse drug resistance that was caused by a specific multidrug efflux mechanism in breast cancer cells, indicating a model for anticancer drug augmentation.[141]

The aminocoumarin antibiotics share a 3-amino-4,7-dihydroxycoumarin moiety that is highly decorated. Novobiocin and clorobiocin share a prenylated 4-hydroxybenzoate moiety but differ in two respects. The CH$_3$ at C-8 and the carbamoyl moieties at C-3-O in novobiocin are replaced by Cl and a 5-methyl-2-pyrrolecarbonyl moieties in clorobiocin, respectively. Coumermycin A$_1$ contains two coumarin-noviose moieties joined by a 3-methyl-2,4-dicarboxylpyrrole, with noviose deoxysugars containing 5-methyl-2-pyrrolecarbonyl substitutions at C-3-O. Finally, simocyclinone D8 is a hybrid natural product with an angucyclinone polyketide ring system connected to

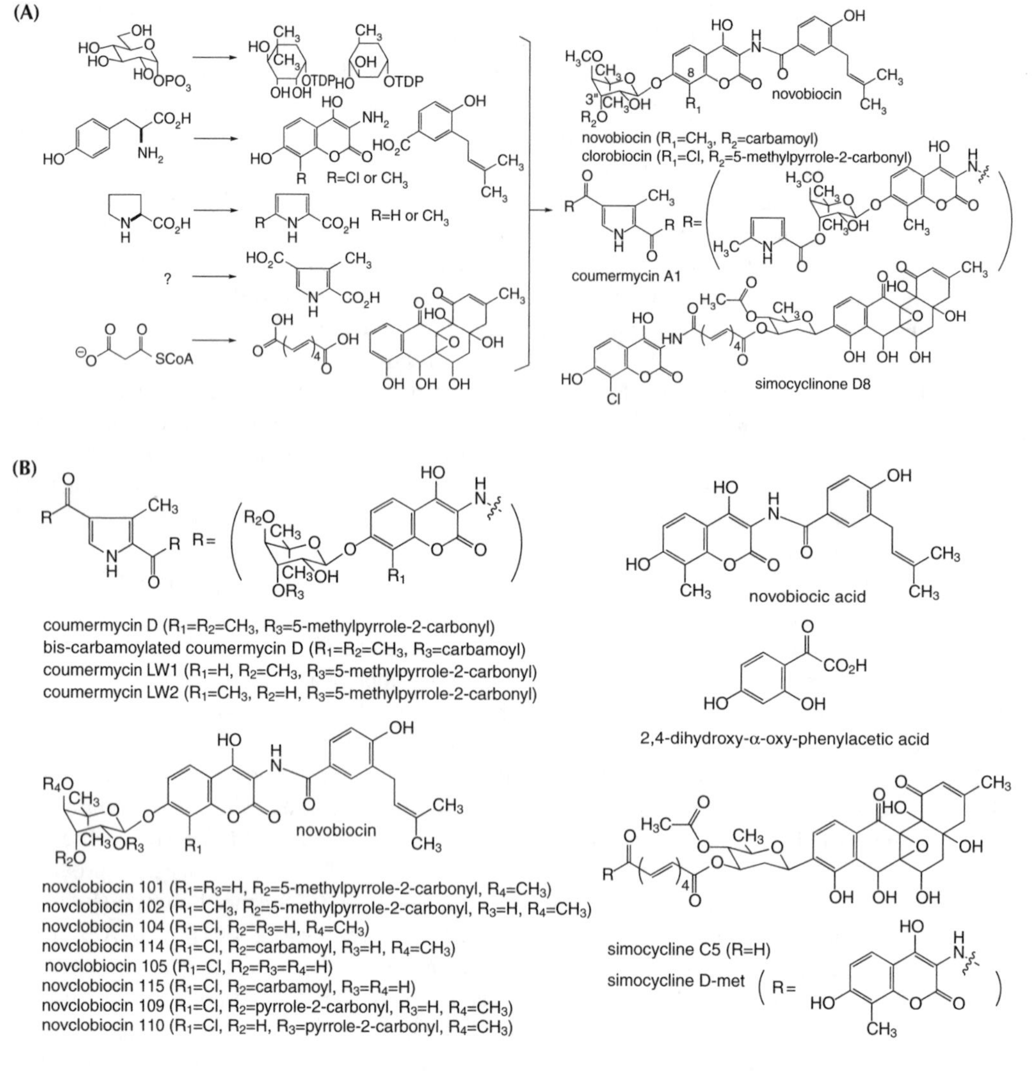
(A)
novobiocin
novobiocin (R1=CH3, R2=carbamoyl)
clorobiocin (R1=Cl, R2=5-methylpyrrole-2-carbonyl)
R=Cl or CH3
R=H or CH3
coumermycin A1
simocyclinone D8
(B)
coumermycin D (R1=R2=CH3, R3=5-methylpyrrole-2-carbonyl)
bis-carbamoylated coumermycin D (R1=R2=CH3, R3=carbamoyl)
coumermycin LW1 (R1=H, R2=CH3, R3=5-methylpyrrole-2-carbonyl)
coumermycin LW2 (R1=CH3, R2=H, R3=5-methylpyrrole-2-carbonyl)
novobiocic acid
2,4-dihydroxy-α-oxy-phenylacetic acid
novobiocin
novclobiocin 101 (R1=R3=H, R2=5-methylpyrrole-2-carbonyl, R4=CH3)
novclobiocin 102 (R1=CH3, R2=5-methylpyrrole-2-carbonyl, R3=H, R4=CH3)
novclobiocin 104 (R1=Cl, R2=R3=H, R4=CH3)
novclobiocin 114 (R1=Cl, R2=carbamoyl, R3=H, R4=CH3)
novclobiocin 105 (R1=Cl, R2=R3=R4=H)
novclobiocin 115 (R1=Cl, R2=carbamoyl, R3=R4=H)
novclobiocin 109 (R1=Cl, R2=pyrrole-2-carbonyl, R3=H, R4=CH3)
novclobiocin 110 (R1=Cl, R2=H, R3=pyrrole-2-carbonyl, R4=CH3)
simocycline C5 (R=H)
simocycline D-met

FIGURE 25.7 (A) Structures and biosynthesis of the aminocoumarins novobiocin, clorobiocin, coumermycin A1, and simocyclinone D8. (B) Structures of novel aminocoumarins and/or biosynthetic intermediates produced by combinatorial biosynthesis of the four aminocoumarin pathways.

the 3-amino-4,7-dihydroxycoumarin moiety by an octatetraene dicarboxylic acid and a deoxy sugar olivose (Figure 25.7A).

1. Biosynthetic Gene Clusters and Biosynthetic Mechanisms

Heide and coworkers have cloned and sequenced the biosynthetic gene clusters for novobiocin,[142] clorobiocin,[143] coumermycin A_1,[144] and simocylcinone D8,[145] setting an excellent stage to study aminocoumarin biosynthesis by a comparative genomics approach and to develop new aminocoumarins by combinatorial biosynthesis methods. The enzymes encoded by these clusters can be divided into three groups that are involved in the biosynthesis of the core moieties of the antibiotics, including TDP-noviose and TDP-olivose from L-glucose-1-phosphate, aminocoumarin from L-tyrosine, 5-methyl-2-pyrrolecarboxylic acid from L-proline, prenylated 4-hydroxybenzoate from

L-tyrosine, angucyclinone and the octatetraene dicarboxylic acid from acetates; the assembly of the various moieties into the characteristic molecular scaffolds of the aminocoumarin family of antibiotics; and tailoring steps such as carbamoylation, glycosylation, chlorination, and methylation (Figure 25.7A).

Although the biosynthetic machineries for the deoxy sugar and polyketide moieties of the aminocoumarin family of antibiotics are similar to those known for other natural products in *Streptomyces*, those for the aminocoumarin, 5-methyl-2-pyrrolecarboxylate, hydroxybenzoate moieties from the corresponding amino acid precursors are certainly unique, and are characterized by a family of atypical NRPS enzymes. In contrast to NRPSs known for nonribosomal peptide biosynthesis, these atypical NRPS enzymes function to tether the proteinogenic amino acid precursors to peptidyl carrier protein as a series of modifications occur to generate the aminocoumarin, 5-methyl-2-pyrrole-carboxylate, or hydroxybenzoate moiety.[142–145] Tethering a proteinogenic amino acid as an aminoacyl-S–peptidyl carrier protein for modification by tailoring enzymes has been speculated as a general strategy to sequester, and thereby divert, a fraction of the proteinogenic amino acid to provide building blocks for secondary metabolite biosynthesis.[146–149] Thus, it is not apparent how these atypical NRPS could be mixed and matched with other NRPS enzymes in combinatorial biosynthesis. Rather, the tailoring steps have turned out to be the most rewarding targets of the aminocoumarin biosynthetic machinery for structural diversities by combinatorial biosynthetic methods.

2. Combinatorial Biosynthesis

Given their mixed biosynthetic origin, targeted gene inactivation has been instrumental in deciphering the molecular logic of how the aminocoumarin family of antibiotics is assembled. These studies established the roles many of the enzymes play in their respective biosynthetic pathways, resulted in the isolation of a series of mutants and the accumulated biosynthetic intermediates, and set the stage to engineer novel metabolites by mixing and matching biosynthetic components among the four aminocoumarin gene clusters. Mutants generated and biosynthetic intermediates isolated could also be exploited by combinatorial biosynthetic methods such as directed biosynthesis and chemoenzymatic synthesis to further expand the structural diversity of the aminocoumarins.

Leading the effort of combinatorial biosynthesis of the aminocoumarins, Heide and coworkers have generated more than 50 new analogs (Figure 25.7B). Thus, inactivation of the *couN3* and *couN4* genes of the coumermycin A1 pathway yielded *S. rishiriensis* mutants that accumulated coumermycin D,[150] and that of *couO* and *couP* resulted in mutants that produced coumermycin LW1 and coumermycin LW2, respectively.[151] Introduction of *novN*, encoding the carbamoyltransferase from the novobiocin biosynthetic pathway, into the Δ*couN3* mutant strain generated the novel aminocoumarin, bis-carbamoylated coumermycin D.[150] Investigations into the function of genes involved in chlorobiocin biosynthesis by *S. roseochromogenes* var *oscitans* DS12.986 resulted in mutants that produced novclobiocins 104 and 105,[152] novclobiocin 109,[153] and novclobiocin 101.[154] Complementation of a Δ*clo-hal* mutant with the *novO* methyl transferase gene from the novobiocin biosynthetic gene cluster resulted in the formation of novclorobicin 102.[154] Expression of *novH-N* in *S. lividans* TK24 resulted in the production of novobiocic acid — the aglycone of novobiocin, and expression of *novD-K* afforded 2,4-dihydroxy-α-oxy-phenylacetic acid, demonstrating the feasibility to manipulate aminocoumarin biosynthesis in heterologous hosts.[155]

As a final example of combinatorial biosynthesis of aminocoumarin antibiotics, Heide and coworkers combined various combinatorial biosynthetic methods, including directed biosynthesis and mutational and chemoenzymatic synthesis for novel aminocoumarin production (Figure 25.7B). Thus, inactivation of *simJ1* of the simocyclinone pathway yielded an *S. antibioticus* mutant strain that accumulated simocyclinone C5. Fermentation of the Δ*simJ1* mutant strain in the presence of the aminocoumarin intermediate for novobiocin biosynthesis resulted in the production of the novel simocyclinone derivative simocyclinone D-met.[155] Alternatively, various 3-unsubstituted aminocoumarin derivatives were isolated from mutants of the clorobiocin producer *S. roseochromogenes*, and they were readily converted to the corresponding 3-carbamoylated products *in vitro*, using the purified recombinant NovN carbamoyltransferase as exemplified by novclobiocin 104 and 105 to novclobiocin 114 and 115, respectively.[156] Similar efforts in combining biochemical, genetic, and synthetic approaches to take advantage of the substrate promiscuity of CloL, NovL, and CouL resulted in the production of an additional 32 new aminocoumarins.[157]

IV. SUMMARY, CHALLENGES, AND PROSPECTS

Recent advances in cloning, and thus rapid access to biosynthetic gene clusters; fundamental understanding of secondary metabolic pathways; and emerging recombinant DNA technologies have fueled the exponential growth of the field of combinatorial biosynthesis in the last decade. Combinatorial biosynthesis has advanced from a limited set of proof-of-principle experiments into a more mature scientific discipline. With the ever-growing list of gene clusters for natural product biosynthesis becoming available, the continuous advancement of recombinant DNA technologies, and the accelerated pace in our understanding of the fundamental chemistry, biochemistry, and genetics governing natural product biosynthesis, combinatorial biosynthesis will surely play an increasing role in accessing natural products and generating natural product diversity for anticancer drug discovery.[10,11]

We have presented several examples of how combinatorial biosynthesis has been used to introduce structural diversity into anticancer natural products. We have by no means been complete in our summary. Rather, we have tried to highlight examples that show the enormous potential that combinatorial biosynthesis gives a researcher to develop or extend the metabolic capabilities of so-called "secondary metabolism." The techniques used in combinatorial biosynthesis are not necessarily new. Rather, the most significant advancement in the field comes from the sequencing of natural product biosynthetic clusters at a breathtaking pace. With this information in hand, a researcher has a blueprint for how a natural product is biosynthesized and also has an understanding of how the blueprint can be altered to build new structural diversity into a natural product.

Although there has been much success, there are still a number of hurdles that need to be overcome before the full potential of combinatorial biosynthesis can be reached. As stated previously, the importance of developing genetic systems for the manipulation of the biosynthetic pathways cannot be stressed enough. Although it is possible that genetic techniques can be developed for each organism that produces a natural product of interest, it would be more desirable to work toward the development of a series of genetically tractable universal heterologous hosts that can produce functional natural product biosynthetic pathways. This has been accomplished with some success for natural products biosynthesized by the actinomycetes. As shown by examples in this chapter, *S. coelicolor* and *S. lividans* have proven useful for the heterologous production of functional biosynthetic components of noncognate pathways from a variety of actinomycetes and also for the epothilone pathway from a myxobacterium. Development of metabolically engineered *E. coli* and *M. xanthus* strains for the production of complicated natural products[84,101,158] indicates that these bacteria can also be developed into general hosts for natural product production. Although not discussed, development and demonstration of *Bacillus subtilis* as a heterologous host to express large biosynthetic gene clusters from related bacteria have also been reported.[159,160]

These strains, however, may not be the heterologous hosts of choice for natural product biosynthetic pathways from more distantly related organisms. It is essential that heterologous hosts be developed to express gene clusters from cyanobacteria, pseudomonads, photorhabdus, and other bacterial genera known to be rich in metabolic potential. In addition, the development of strains for the heterologous production of natural products produced by filamentous fungi, plants, and animals will require the appropriate heterologous hosts to be developed. In addition to cultured organisms, harnessing the metabolic potential of uncultured microorganisms from sources such as soil (metagenomic DNA[161–165]) and marine waters[166–168] will also push the need for appropriate heterologous hosts. Many of these ecology niches are known to be rich sources of anticancer natural products.

The development of these heterologous hosts requires the parallel development of the necessary vectors for the cloning, mobilization, and expression of the heterologous genes. The generation of bacterial artificial chromosomes makes it possible to clone entire biosynthetic gene clusters onto a single plasmid.[169] New bacterial artificial chromosome derivates that replicate in *E. coli* and that can be moved by conjugation or transformation into *Streptomyces*[170–172] or *Bacillus*[173] pave the way for producing a natural product in more than one heterologous host using a single clone. The development of a universal bacterial artificial chromosome vector for mobilization into a series of engineered organisms would provide the ultimate vehicle for screening potential heterologous hosts.

Although the enabling recombinant DNA technologies are necessary, a better understanding of the fundamental chemistry, biochemistry, and genetics governing the biosynthetic machinery is critical to direct the engineering of the enzymes to have increased or altered substrate specificity or better catalytic efficiency. Knowledge of the essential domains needed for the appropriate protein–protein interactions between biosynthetic enzymes will enable more efficient catalysis between hybrid constructs. Biochemical and mechanistic characterization of natural product biosynthetic machinery requires the enzymes of interest to be overproduced, purified in their functional forms, and studied *in vitro*, posing great challenges for which we so far have few solutions.

Although the approaches outlined above focus on the rational engineering of natural products biosynthesis, mechanisms must also be investigated for protein and pathway evolution. Nature has evolved a stunning diversity of natural products through gene-shuffling processes over millions of years. With the recent advances in gene[174,175] and genome[176,177] shuffling, a future goal will be to develop genetic screens and selection that enable us to "evolve" new natural product biosynthetic pathways from the large library of natural product biosynthetic gene clusters that DNA sequencing has uncovered.

Finally, combinatorial biosynthesis alone will not reach the full potential of natural product structural diversity. It will be the combination of this approach with other established and emerging technologies[178–181] that will ultimately yield the maximal structural diversity of natural products for anticancer drug discovery.[302–304]

ACKNOWLEDGMENTS

This work was supported in part by grant IRG-58-011-46-07 from the Amercian Cancer Society to MGT. Studies on natural product biosynthesis in Shen laboratory were supported in part by an IRG from American Cancer Society and the School of Medicine, University of California, Davis; the Searle Scholars Program/the Chicago Community Trust; the University of California BioSTAR Program and Kosan Biosciences, Hayward, CA (Bio99-10045); and National Institutes of Health grants CA78747, CA94426, CA106150. BS is the recipient of National Science Foundation CAREER Award MCB9733938 and National Institutes of Health Independent Scientist Award AI51689.

REFERENCES

1. Newman, D. J., Cragg, G. M., and Snader, K. M., Natural products as sources of new drugs over the period 1981-2002, *J. Nat. Prod.* 66, 1022, 2003.
2. Cane, D. E., Walsh, C. T., and Khosla, C., Harnessing the biosynthetic code: combinations, permutations, and mutations, *Science* 282, 63, 1998.
3. Shen, B., Aromatic Polyketide Biosynthesis, in *Topics in current chemistry: biosynthesis*, Leeper, F. J. and Vederas, J. C. Springer, Berlin, 2000, pp. 1.
4. Staunton, J. and Weissman, K. J., Polyketide biosynthesis: a millennium review, *Nat. Prod. Rep.* 18, 380, 2001.
5. Staunton, J. and Wilkinson, B., Combinatorial biosynthesis of polyketides and nonribosomal peptides, *Curr. Opin. Chem. Biol.* 5, 159, 2001.

6. Du, L. and Shen, B., Biosynthesis of hybrid peptide-polyketide natural products, *Curr. Opin. Drug Discov. Devel.* 4, 215, 2001.
7. Walsh, C. T., Combinatorial biosynthesis of antibiotics: challenges and opportunities, *ChemBioChem* 3, 124, 2002.
8. Mootz, H., Schwarzer, D., and Marahiel, M., Ways of assembling complex natural products on modular nonribosomal peptide synthetases, *ChemBioChem* 3, 490, 2002.
9. Shen, B., Polyketide biosynthesis beyond the type I, II and III polyketide synthase paradigms, *Curr. Opin. Chem. Biol.* 7, 285, 2003.
10. Walsh, C. T., Polyketide and nonribosomal peptide antibiotics: modularity and versatility, *Science* 303, 1805, 2004.
11. Shen, B., Accessing natural products by combinatorial biosynthesis, *Sci. STKE* 2004, pe14, 2004.
12. Strohl, W. R., Biochemical engineering of natural product biosynthesis pathways, *Metab. Eng.* 3, 4, 2001.
13. Kennedy, J. and Hutchinson, C. R., Nurturing nature: engineering new antibiotics, *Nat. Biotechnol.* 17, 538, 1999.
14. Khosla, C., Combinatorial chemistry and biology: an opportunity of engineers, *Curr. Opin. Biotechnol.* 7, 219, 1996.
15. Khosla, C., Harnessing the biosynthetic potential of modular polyketide synthases, *Chem. Rev.* 97, 2577, 1997.
16. Kerwin, S., Toward bioengineering anticancer drugs, *Chem. Biol.* 9, 956, 2002.
17. Hutchinson, C. R. et al., Drug discovery and development through the genetic engineering of antibiotic-producing microorganisms, *J. Med. Chem.* 32, 929, 1989.
18. Hutchinson, C. R. and Fujii, I., Polyketide synthase gene manipulation: a structure-function approach in engineering novel antibiotics, *Annu. Rev. Microbiol.* 49, 201, 1995.
19. Malipartida, F. and Hopwood, D., Molecular cloning of the whole biosynthetic pathway of a *Streptomyces* antibiotic and its expression in a heterologous host, *Nature* 309, 462, 1984.
20. Hopwood, D. A. et al., Production of "hybrid" antibiotics by genetic engineering, *Nature* 314, 642, 1985.
21. Kutney, J. P., Plant cell culture combined with chemistry: a powerful route to complex natural products, *Acc. Chem. Res.* 26, 559, 1993.
22. Roberts, S. C. and Shuler, M. L., Large-scale plant cell culture, *Curr. Opin. Biotechnol.* 8, 154, 1997.
23. Bourgaud, F. et al., Production of plant secondary metabolites: a historical perspective, *Plant Sci.* 161, 839, 2001.
24. Kennedy, J. et al., Modulation of polyketide synthase activity by accessory proteins during lovastatin biosynthesis, *Science* 284, 1368, 1999.
25. Miao, V. et al., Genetic approaches to harvesting lichen products, *Trends Biotechnol.* 19, 349, 2001.
26. Nicholson, T. P. et al., Design and utility of oligonucleotide gene probes for fungal polyketide synthases, *Chem Biol* 8, 157, 2001.
27. Hutchinson, C. R., Antibiotics from genetically engineered microorganisms, in *Biotechnology of antibiotics*, 2 ed., Strohl, W. R. Marcel Dekker, New York, 1997, pp. 683.
28. Strohl, W. R. et al., Anthracyclines, in *Biotechnology of antibiotics*, 2 ed., Strohl, W. R. Marcel Dekker, New York, 1997, pp. 577.
29. Cullinane, C., van Rosmalen, A., and Phillips, D. R., Does adriamycin induce interstrand cross-links in DNA? *Biochemistry* 33, 4632, 1994.
30. Chen, A. Y. and Liu, L. F., DNA topoisomerases: essential enzymes and lethal targets, *Annu. Rev. Pharmacol. Toxicol.* 34, 191, 1994.
31. Drlica, K. and Franco, R. J., Inhibitors of DNA topoisomerases, *Biochemistry* 27, 2253, 1988.
32. Arcamone, F. et al., Adriamycin, 14-hydroxydaunorubicin, a new antitumor antibiotic from *S. peucetius* var. *caesius*, *Biotechnol. Bioeng.* 11, 1101, 1969.
33. Hutchinson, C. R., Anthracyclines, in *Biochemistry and genetics of commercially important antibiotics* Butterworths, Boston, 1995, pp. 331.
34. Otten, S. L., Stutzman-Engwall, K. J., and Hutchinson, C. R., Cloning and expression of daunorubicin biosynthesis genes from *Streptomyces peucetius* and *S. peucetius* subsp. *caesius*, *J. Bacteriol.* 172, 3427, 1990.
35. Hutchinson, C. R. and Colombo, A. L., Genetic engineering of doxorubicin production in *Streptomyces peucetius*: a review, *J. Ind. Microbiol. Biotechnol.* 23, 647, 1999.

36. Rawlings, B. J., Biosynthesis of polyketides (other than actinomycete macrolides), *Nat. Prod. Rev.* 16, 425, 1999.

37. Meurer, G. and Hutchinson, C. R., Daunorubicin type II polyketide synthase enzymes DpsA and DpsB determine neither the choice of starter unit nor the cyclization pattern of aromatic polyketides, *J. Am. Chem. Soc.* 117, 5899, 1995.

38. Grimm, A. et al., Characterization of the *Streptomyces peucetius* ATCC 29050 genes encoding doxorubicin polyketide synthase, *Gene* 151, 1, 1994.

39. Ye, J. et al., Isolation and sequence analysis of polyketide synthase genes from the daunomycin-producing *Streptomyces* sp. strain C5, *J. Bacteriol.* 176, 6270, 1994.

40. Rajgarhia, V. B. and Strohl, W. R., Minimal *Streptomyces* sp. strain C5 daunorubicin polyketide biosynthesis genes required for aklanonic acid biosynthesis, *J. Bacteriol.* 179, 2690, 1997.

41. Gerlitz, M. et al., Effect of daunorubicin dpsH gene on the choice of starter unit and cyclization pattern reveals that type II polyketide synthetases can be unfaithful yet intriguing, *J. Am. Chem. Soc.* 119, 7392, 1997.

42. Kantola, J. et al., Expanding the scope of aromatic polyketides by combinatorial biosynthesis, *Comb. Chem. High Throughput Screen.* 6, 501, 2003.

43. Olano, C. et al., A two-plasmid system for the glycosylation of polyketide antibiotics: bioconversion of epsilon-rhodomycinone to rhodomycin D, *Chem. Biol.* 6, 845, 1999.

44. Eckardt, K. et al., Biosynthesis of anthracyclinones: isolation of a new early cyclization product aklaviketone, *J Antibiot (Tokyo)* 41, 788, 1988.

45. Grein, A., Antitumor anthracyclines produced by *Streptomyces peucetius*, *Adv. Appl. Microbiol.* 32, 203, 1987.

46. McGuire, J. C. et al., Biosynthesis of daunorubicin glycosides: role of epsilon-rhodomycinone, *Antimicrob. Agents Chemother.* 18, 454, 1980.

47. Cassinelli, G. et al., 13-Deoxycarminomycin, a new biosynthetic anthracycline, *J. Nat. Prod.* 48, 435, 1985.

48. Hutchinson, C. R., Biosynthetic studies of daunorubicin and tetracenomycin C, *Chem. Rev.* 97, 2525, 1997.

49. Madduri, K. et al., Production of the antitumor drug epirubicin (4-epidoxorubicin) and its precursor by a genetically engineered strain of *Streptomyces peucetius*, *Nat. Biotechnol.* 16, 69, 1998.

50. Hopwood, D. A., Genetic contributions to understanding polyketide synthases, *Chem. Rev.* 97, 2465, 1997.

51. Remsing, L. L. et al., Mithramycin SK, a novel antitumor drug with improved therapeutic index, mithramycin SA, and demycarosyl-mithramycin SK: three new products generated in the mithramycin producer *Streptomyces argillaceus* through combinatorial biosynthesis, *J. Am. Chem. Soc.* 125, 5745, 2003.

52. Rohr, J., Mendez, C., and Salas, J. A., The biosynthesis of aureolic acid group antibiotics, *Bioorg. Chem.* 27, 41, 1999.

53. Hughes, S., Peel-White, A. L., and Peterson, C. K., Paget's disease of bone — current thinking and management, *J. Manipulative Physiol. Ther.* 15, 242, 1992.

54. Cons, B. M. and Fox, K. R., Effects of the antitumor antibiotic mithramycin on the structure of repetitive DNA regions adjacent to its GC-rich binding site, *Biochemistry* 30, 6314, 1991.

55. Cons, B. M. and Fox, K. R., The GC-selective ligand mithramycin alters the structure of (AT)n sequences flanking its binding sites, *FEBS Lett.* 264, 100, 1990.

56. Montanari, A. and Rosazza, J. P., Biogenesis of chromomycin A3 by *Streptomyces griseus*, *J. Antibiot.* 43, 883, 1990.

57. Lombo, F. et al., Characterization of *Streptomyces argillaceus* genes encoding a polyketide synthase involved in the biosynthesis of the antitumor mithramycin, *Gene* 172, 87, 1996.

58. Lombo, F. et al., Cloning and insertional inactivation of *Streptomyces argillaceus* genes involved in the earliest steps of biosynthesis of the sugar moieties of the antitumor polyketide mithramycin, *J. Bacteriol.* 179, 3354, 1997.

59. Prado, L. et al., Analysis of two chromosomal regions adjacent to genes for a type II polyketide synthase involved in the biosynthesis of the antitumor polyketide mithramycin in *Streptomyces argillaceus*, *Mol. Gen. Genet.* 261, 216, 1999.

60. Gonzalez, A. et al., The *mtmVUC* genes of the mithramycin gene cluster in *Streptomyces argillaceus* are involved in the biosynthesis of the sugar moieties, *Mol. Gen. Genet.* 264, 827, 2001.

61. Menendez, N. et al., Biosynthesis of the antitumor chromomycin A3 in *Streptomyces griseus*: analysis of the gene cluster and rational design of novel chromomycin analogs, *Chem. Biol.* 11, 21, 2004.
62. Prado, L. et al., Oxidative cleavage of premithramycin is one of the last steps in the biosynthesis of the antitumor drug mithramycin, *Chem. Biol.* 6, 19, 1999.
63. Fernandez, E. et al., Identification of two genes from *Streptomyces argillaceus* encoding glycosyltransferases involved in transfer of a disaccharide during biosynthesis of the antitumor drug mithramycin, *J. Bacteriol.* 180, 4929, 1998.
64. Fernadez Lozano, M. et al., Characterization of two polyketide methyltransferases involved in the biosynthesis of the antitumor drug mithramycin by *Streptomyces argillaceus*, *J. Biol. Chem.* 275, 3065, 2000.
65. Lombo, F. et al., The novel hybrid antitumor compound premithramycinone H provides indirect evidence for a tricyclic intermediate of the biosynthesis of the aurolic acid antibiotic mithramycin, *Angew. Chem. Int. Ed.* 39, 796, 2000.
66. Kantola, J. et al., Folding of the polyketide chain is not dictated by minimal polyketide synthase in the biosynthesis of mithramycin and anthracycline, *Chem. Biol.* 4, 751, 1997.
67. Kunzel, E. et al., Tetracenomycin M, a novel genetically engineered tetracenomycin resulting from a combination of mithramycin and tetracenomycin biosynthetic genes, *Chem.-A Eur. J.* 3, 1675, 1997.
68. Wohlert, S.-E. et al., Novel hybrid tetracenomycins through combinatorial biosynthesis using a glycosyltransferase encoded by the elm genes in cosmid 16F4 and which shows a broad sugar substrate specificity, *J. Am. Chem. Soc.* 120, 10596, 1998.
69. August, P. R. et al., Biosynthesis of the ansamycin antibiotic rifamycin: deductions from the molecular analysis of the rif biosynthetic gene cluster of *Amycolatopsis mediterranei* S699, *Chem. Biol.* 5, 69, 1998.
70. Yu, T. W. et al., The biosynthetic gene cluster of the maytansinoid antitumor agent ansamitocin from *Actinosynnema pretiosum*, *Proc. Natl. Acad. Sci. USA* 99, 7968, 2002.
71. Hatano, K. et al., Biosynthetic origin of aminobenzenoid nucleus (C7N-unit) of ansamitocin, a group of novel maytansinoid antibiotics, *J. Antibiot.* 35, 1415, 1982.
72. Tang, L. et al., Characterization of the enzymatic domains in the modular polyketide synthase involved in rifamycin B biosynthesis by *Amycolatopsis mediterranei*, *Gene* 216, 255, 1998.
73. Reider, P. J. and Roland, D. M., in *The Alkaloids*, Pelletier, S. W. Academic Press, New York, 1984, pp. 71.
74. Richter, K. and Buchner, J., Hsp90: chaperoning signal transduction, *J. Cell Physiol.* 188, 281, 2001.
75. Whitesell, L. et al., Inhibition of heat shock protein HSP90-pp60v-src heteroprotein complex formation by benzoquinone ansamycins: essential role for stress proteins in oncogenic transformation, *Proc. Natl. Acad. Sci. USA* 91, 8324, 1994.
76. Blagosklonny, M. V., Hsp-90-associated oncoproteins: multiple targets of geldanamycin and its analogs, *Leukemia* 16, 455, 2002.
77. Kim, C. G. et al., 3-Amino-5-hydroxybenzoic acid synthase, the terminal enzyme in the formation of the precursor of mC7N units in rifamycin and related antibiotics, *J. Biol. Chem.* 273, 6030, 1998.
78. Schupp, T. et al., Cloning and sequence analysis of the putative rifamycin polyketide synthase gene cluster from *Amycolatopsis mediterranei*, *FEMS Microbiol. Lett.* 159, 201, 1998.
79. Rascher, A. et al., Cloning and characterization of a gene cluster for geldanamycin production in *Streptomyces hygroscopicus* NRRL 3602, *FEMS Microbiol. Lett.* 218, 223, 2003.
80. Spiteller, P. et al., The post-polyketide synthase modification steps in the biosynthesis of the antitumor agent ansamitocin by *Actinosynnema pretiosum*, *J. Am. Chem. Soc.* 125, 14236, 2003.
81. Watanabe, K. et al., Engineered biosynthesis of an ansamycin polyketide precursor in *Escherichia coli*, *Proc. Natl. Acad. Sci. USA* 100, 9774, 2003.
82. Lambalot, R. H. et al., A new enzyme superfamily - the phosphopantetheinyl transferases, *Chem. Biol.* 3, 923, 1996.
83. Walsh, C. T. et al., Post-translational modification of polyketide and nonribosomal peptide synthases, *Curr. Opin. Chem. Biol.* 1, 309, 1997.
84. Pfeifer, B. et al., Biosynthesis of complex polyketides in a metabolically engineered strain of *E. coli*, *Science* 291, 1790, 2001.

85. Ghisalba, O. and Nuesch, J., A genetic approach to the biosynthesis of the rifamycin-chromophore in *Nocardia mediterranei*. IV. Identification of 3-amino-5-hydroxybenzoic acid as a direct precursor of the seven-carbon amino starter-unit, *J. Antibiot.* 34, 64, 1981.

86. Tsuji, S., Cane, D., and Khosla, C., Selective protein-protein interactions direct channeling of intermediates between polyketide synthase modules, *Biochemistry* 40, 2326, 2001.

87. Wu, N., Cane, D. E., and Khosla, C., Quantitative analysis of the relative contributions of donor acyl carrier proteins, acceptor ketosynthases, and linker regions to intermodular transfer of intermediates in hybrid polyketide synthases, *Biochemistry* 41, 5056, 2002.

88. Wu, N. et al., Assessing the balance between protein-protein interactions and enzyme-substrate interactions in the channeling of intermediates between polyketide synthase modules, *J. Am. Chem. Soc.* 123, 6465, 2001.

89. Kato, Y. et al., Functional expression of genes involved in the biosynthesis of the novel polyketide chain extension unit, methoxymalonyl-acyl carrier protein, and engineered biosynthesis of 2-desmethyl-2-methoxy-6-deoxyerythronolide B, *J. Am. Chem. Soc.* 124, 5268, 2002.

90. Floss, H. G. and Yu, T. W., Lessons from the rifamycin biosynthetic gene cluster, *Curr. Opin. Chem. Biol.* 3, 592, 1999.

91. He, L., Orr, G. A., and Horwitz, S. B., Novel molecules that interact with microtubules and have functional activity similar to Taxol, *Drug Discov. Today* 6, 1153, 2001.

92. Hoefle, G. et al., Antibiotics from gliding bacteria. 77. Epothilone A and B — novel 16-membered macrolides with cytotoxic activity: isolation, crystal structure, and conformation in solution., *Angew. Chem. Int. Ed.* 35, 1567, 1996.

93. Arslanian, R. L. et al., Large-scale isolation and crystallization of epothilone D from *Myxococcus xanthus* cultures, *J. Nat. Prod.* 65, 570, 2002.

94. McCarthy, A. A., Kosan biosciences. Better chemistry through genetics, *Chem. Biol.* 9, 849, 2002.

95. McAllister, K. A. et al., Biochemical and molecular analyses of the *Streptococcus pneumoniae* acyl carrier protein synthase, an enzyme essential for fatty acid biosynthesis, *J. Biol. Chem.* 275, 30864, 2000.

96. Molnar, I. et al., The biosynthetic gene cluster for the microtubule-stabilizing agents epothilones A and B from *Sorangium cellulosum* So ce90, *Chem. Biol.* 7, 97, 2000.

97. Tang, L. et al., Cloning and heterologous expression of the epothilone gene cluster, *Science* 287, 640, 2000.

98. Julien, B. et al., Isolation and characterization of the epothilone biosynthetic gene cluster from *Sorangium cellulosum*, *Gene* 249, 153, 2000.

99. Chou, T. C. et al., Desoxyepothilone B: an efficacious microtubule-targeted antitumor agent with a promising *in vivo* profile relative to epothilone B, *Proc. Natl. Acad. Sci. USA* 95, 9642, 1998.

100. Chou, T. C. et al., Desoxyepothilone B is curative against human tumor xenografts that are refractory to paclitaxel, *Proc. Natl. Acad. Sci. USA* 95, 15798, 1998.

101. Julien, B. and Shah, S., Heterologous expression of epothilone biosynthetic genes in *Myxococcus xanthus*, *Antimicrob. Agents Chemother.* 46, 2772, 2002.

102. Frykman, S. et al., Modulation of epothilone analog production through media design, *J. Ind. Microbiol. Biotechnol.* 28, 17, 2002.

103. Lau, J. et al., Optimizing the heterologous production of epothilone D in *Myxococcus xanthus*, *Biotechnol. Bioeng.* 78, 280, 2002.

104. Starks, C. M. et al., Isolation and characterization of new epothilone analogues from recombinant *Myxococcus xanthus* fermentations, *J. Nat. Prod.* 66, 1313, 2003.

105. Chen, H. et al., Epothilone biosynthesis: assembly of the methylthiazolylcarboxy starter unit on the EpoB subunit, *Chem. Biol.* 8, 899, 2001.

106. O'Connor, S. E., Chen, H., and Walsh, C. T., Enzymatic assembly of epothilones: the EpoC subunit and reconstitution of the EpoA-ACP/B/C polyketide and nonribosomal peptide interfaces, *Biochemistry* 41, 5685, 2002.

107. Schneider, T. L., Shen, B., and Walsh, C. T., Oxidase domains in epothilone and bleomycin biosynthesis: thiazoline to thiazole oxidation during chain elongation, *Biochemistry* 42, 9722, 2003.

108. Walsh, C. T., O'Connor, S. E., and Schneider, T. L., Polyketide-nonribosomal peptide epothilone antitumor agents: the EpoA, B, C subunits, *J. Ind. Microbiol. Biotechnol.* 30, 448, 2003.

109. Hara, M. et al., DC 107, a novel antitumor antibiotic produced by a *Streptomyces* sp., *J. Antibiot.* 42, 333, 1989.

110. Hara, M. et al., Leinamycin, a new antitumor antibiotic from *Streptomyces*: producing organism, fermentation and isolation, *J. Antibiot.* 42, 1768, 1989.

111. Kanda, Y. et al., Synthesis and antitumor activity of leinamycin derivatives: modifications of C-8 hydroxy and C-9 keto groups, *Bioorg. Med. Chem. Lett.* 8, 909, 1998.

112. Kanda, Y. et al., Synthesis and antitumor activity of novel thioester derivatives of leinamycin, *J. Med. Chem.* 42, 1330, 1999.

113. Gates, K. S., Mechanisms of DNA damage by leinamycin, *Chem. Res. Toxicol.* 13, 953, 2000.

114. Cheng, Y. Q., Tang, G. L., and Shen, B., Identification and localization of the gene cluster encoding biosynthesis of the antitumor macrolactam leinamycin in *Streptomyces atroolivaceus* S-140, *J. Bacteriol.* 184, 7013, 2002.

115. Cheng, Y. Q., Tang, G. L., and Shen, B., Type I polyketide synthase requiring a discrete acyltransferase for polyketide biosynthesis, *Proc. Natl. Acad. Sci. USA* 100, 3149, 2003.

116. Tang, G. L., Cheng, Y. Q., and Shen, B., Leinamycin biosynthesis revealing unprecedented architectural complexity for a hybrid polyketide synthase and nonribosomal peptide synthetase, *Chem. Biol.* 11, 33, 2004.

117. Edo, K. et al., The structure of neocarzinostatin chromophore possessing a novel bicyclo[7,3,0]dodec-adiyne system, *Tetrahedron Lett.* 26, 331, 1985.

118. Shen, B., Liu, W., and Nonaka, K., Enediyne natural products: biosynthesis and prospect towards engineering novel antitumor agents, *Curr. Med. Chem.* 10, 2317, 2003.

119. Thorson, J. S. et al., Understanding and exploiting nature's chemical arsenal: the past, present and future of calicheamicin research, *Curr. Pharm. Des.* 6, 1841, 2000.

120. Biggins, J. B., Onwueme, K. C., and Thorson, J. S., Resistance to enediyne antitumor antibiotics by CalC self-sacrifice, *Science* 301, 1537, 2003.

121. Doyle, T. W., *Enediyne antibiotics as antitumor agents*. Marcel Dekker, New York, 1995.

122. Maeda, H., Edo, K., and Ishida, N., *Neocarzinostatin: The past, present, and future of an anticancer drug.* Springer, New York, 1997.

123. Jones, G. B. and Fouad, F. S., Designed enediyne antitumor agents, *Curr. Pharm. Des.* 8, 2415, 2002.

124. Sievers, E. L. et al., Selective ablation of acute myeloid leukemia using antibody-targeted chemotherapy: a phase I study of an anti-CD33 calicheamicin immunoconjugate. *Blood* 93, 3678, 1999.

125. Brukner, I., C-1027 Taiho Pharmaceutical Co. Ltd., *Curr. Opin. Oncologic, Endocrine Metabol. Invest. Drugs* 2, 344, 2002.

126. Bross, P. F. et al., Approval summary: gemtuzumab ozogamicin in relapsed acute myeloid leukemia, *Clin. Cancer Res.* 7, 1490, 2001.

127. Thorson, J. S. et al., Enediyne biosynthesis and self-resistance: a progress report, *Bioorg. Chem.* 27, 172, 1999.

128. Liu, W. et al., Rapid PCR amplification of minimal enediyne polyketide synthase cassettes leads to a predictive familial classification model, *Proc. Natl. Acad. Sci. USA* 100, 11959, 2003.

129. Hensens, O. D., Giner, J. L., and Golberg, I. H., Biosynthesis of NCS Chrom A, the chromophore of the antitumor antibiotic neocarzinostatin, *J. Am. Chem. Soc.* 111, 3295, 1989.

130. Ahlert, J. et al., The calicheamicin gene cluster and its iterative type I enediyne PKS, *Science* 297, 1173, 2002.

131. Liu, W. et al., Biosynthesis of the enediyne antitumor antibiotic C-1027, *Science* 297, 1170, 2002.

132. Zazopoulos, E. et al., A genomics-guided approach for discovering and expressing cryptic metabolic pathways, *Nat. Biotechnol.* 21, 187, 2003.

133. Liu, W. and Shen, B., Genes for production of the enediyne antitumor antibiotic C-1027 in *Streptomyces globisporus* are clustered with the *cagA* gene that encodes the C-1027 apoprotein, *Antimicrob. Agents Chemother.* 44, 382, 2000.

134. Christenson, S. D. et al., A novel 4-methylideneimidazole-5-one-containing tyrosine aminomutase in enediyne antitumor antibiotic C-1027 biosynthesis, *J. Am. Chem. Soc.* 125, 6062, 2003.

135. Murrell, J. M., Liu, W., and Shen, B., Biochemical characterization of the SgcA1 alpha-D-glucopyranosyl-1-phosphate thymidylyltransferase from the enediyne antitumor antibiotic C-1027 biosynthetic pathway and overexpression of *sgcA1* in *Streptomyces globisporus* to improve C-1027 production, *J. Nat. Prod.* 67, 206, 2004.

136. Liu, W. et al., The neocarzinostatin biosynthetic gene cluster from *Streptomyces carzinostaticus* ATCC15944 inolving two iterative type I polyketide synthase, *Chem. Biol.*, 12, 293–302, 2005.

137. Omura, S. et al., Genome sequence of an industrial microorganism *Streptomyces avermitilis*: deducing the ability of producing secondary metabolites, *Proc. Natl. Acad. Sci. USA* 98, 12215, 2001.

138. Bentley, S. D. et al., Complete genome sequence of the model actinomycete *Streptomyces coelicolor* A3(2), *Nature* 417, 141, 2002.

139. Rappa, G., Lorico, A., and Sartorelli, A. C., Potentiation by novobiocin of the cytotoxic activity of etoposide (VP-16) and teniposide (VM-26), *Int. J. Cancer* 51, 780, 1992.

140. Lorico, A., Rappa, G., and Sartorelli, A. C., Novobiocin-induced accumulation of etoposide (VP-16) in WEHI-3B D+ leukemia cells, *Int. J. Cancer* 52, 903, 1992.

141. Shiozawa, K. et al., Reversal of breast cancer resistance protein (BCRP/ABCG2)-mediated drug resistance by novobiocin, a coumermycin antibiotic, *Int. J. Cancer* 108, 146, 2004.

142. Steffensky, M. et al., Identification of the novobiocin biosynthetic gene cluster of *Streptomyces spheroides* NCIB 11891, *Antimicrob. Agents Chemother.* 44, 1214, 2000.

143. Pojer, F., Li, S. M., and Heide, L., Molecular cloning and sequence analysis of the clorobiocin biosynthetic gene cluster: new insights into the biosynthesis of aminocoumarin antibiotics, *Microbiology* 148, 3901, 2002.

144. Wang, Z. X., Li, S. M., and Heide, L., Identification of the coumermycin A(1) biosynthetic gene cluster of *Streptomyces rishiriensis* DSM 40489, *Antimicrob. Agents Chemother.* 44, 3040, 2000.

145. Galm, U. et al., Cloning and analysis of the simocyclinone biosynthetic gene cluster of *Streptomyces antibioticus* Tü 6040, *Arch. Microbiol.* 178, 102, 2002.

146. Chen, H. et al., Formation of beta-hydroxy histidine in the biosynthesis of nikkomycin antibiotics, *Chem. Biol.* 9, 103, 2002.

147. Chen, H. et al., Aminoacyl-*S*-enzyme intermediates in β-hydroxylations and α,β-desaturations of amino acids in peptide antibiotics, *Biochemistry* 40, 11651, 2001.

148. Chen, H. and Walsh, C. T., Coumarin formation in novobiocin biosynthesis: beta-hydroxylation of the aminoacyl enzyme tyrosyl-*S*-NovH by a cytochrome P450 NovI, *Chem. Biol.* 74, 1, 2001.

149. Thomas, M. G., Burkart, M. D., and Walsh, C. T., Conversion of L-proline to pyrrolyl-2-carboxyl-*S*-PCP during undecylprodigiosin and pyoluteorin biosynthesis, *Chem. Biol.* 9, 171, 2002.

150. Xu, H. et al., Genetic analysis of the biosynthesis of the pyrrole and carbamoyl moieties of coumermycin A1 and novobiocin, *Mol. Genet. Genom.* 268, 387, 2002.

151. Li, S. M. et al., Methyltransferase genes in *Streptomyces rishiriensis*: new coumermycin derivatives from gene-inactivation experiments, *Microbiology* 148, 3317, 2002.

152. Xu, H. et al., CloN2, a novel acyltransferase involved in the attachment of the pyrrole-2-carboxyl moiety to the deoxysugar of clorobiocin, *Microbiology* 149, 2183, 2003.

153. Westrich, L., Heide, L., and Li, S. M., CloN6, a novel methyltransferase catalysing the methylation of the pyrrole-2-carboxyl moiety of clorobiocin, *ChemBioChem* 4, 768, 2003.

154. Eustaquio, A. S. et al., Clorobiocin biosynthesis in streptomyces. Identification of the halogenase and generation of structural analogs, *Chem. Biol.* 10, 279, 2003.

155. Merchant, J. et al., Phase I clinical and pharmacokinetic study of NSC 655649, a rebeccamycin analogue, giving in both single-dose and multiple-dose formats, *Clin. Cancer Res.* 8, 2193, 2002.

156. Xu, H., Heide, L., and Li, S. M., New aminocoumarin antibiotics formed by a combined mutational and chemoenzymatic approach utilizing the carbamoyltransferase NovN, *Chem. Biol.* 11, 655, 2004.

157. Galm, U. et al., *In vitro* and *in vivo* production of new aminocoumarins by a combined biochemical, genetic, and synthetic approach, *Chem. Biol.* 11, 173, 2004.

158. Pfeifer, B. and Khosla, C., Biosynthesis of polyketides in heterologous hosts, *Micro. Mol. Biol. Rev.* 65, 106, 2001.

159. Doekel, S., Eppelmann, K., and Marahiel, M. A., Heterologous expression of nonribosomal peptide synthetases in *B. subtilis*: construction of a bi-functional *B. subtilis/E. coli* shuttle vector system, *FEMS Microbiol. Lett.* 216, 185, 2002.

160. Eppelmann, K., Doekel, S., and Marahiel, M. A., Engineered biosynthesis of the peptide antibiotic bacitracin in the surrogate host *Bacillus subtilis*, *J. Biol. Chem.* 276, 34824, 2001.

161. Schloss, P. D. and Handelsman, J., Biotechnological prospects from metagenomics, *Curr. Opin. Biotechnol.* 14, 303, 2003.

162. Gillespie, D. E. et al., Isolation of antibiotics turbomycin A and B from a metagenomic library of soil microbial DNA, *Appl. Environ. Microbiol.* 68, 4301, 2002.

163. Rondon, M. R. et al., Cloning the soil metagenome: a strategy for accessing the genetic and functional diversity of uncultured microorganisms, *Appl. Environ. Microbiol.* 66, 2541, 2000.

164. Rondon, M. R., Goodman, R. M., and Handelsman, J., The Earth's bounty: assessing and accessing soil microbial diversity, *Trends Biotechnol.* 17, 403, 1999.

165. Handelsman, J. et al., Molecular biological access to the chemistry of unknown soil microbes: a new frontier for natural products, *Chem. Biol.* 5, R245, 1998.

166. Stein, J. L. et al., Characterization of uncultivated prokaryotes: isolation and analysis of a 40-kilobase-pair genome fragment from a planktonic marine archaeon, *J. Bacteriol.* 178, 591, 1996.

167. Schleper, C. et al., Genomic analysis reveals chromosomal variation in natural populations of the uncultured psychrophilic archaeon *Cenarchaeum symbiosum*, *J. Bacteriol.* 180, 5003, 1998.

168. Beja, O. et al., Construction and analysis of bacterial artificial chromosome libraries from a marine microbial assemblage, *Environ. Microbiol.* 2, 516, 2000.

169. Shizuya, H. et al., Cloning and stable maintenance of 300-kilobase-pair fragments of human DNA in *Escherichia coli* using an F-factor-based vector, *Proc. Natl. Acad. Sci. USA* 89, 8794, 1992.

170. Sosio, M. et al., Artificial chromosomes for antibiotic-producing actinomycetes, *Nat. Biotechnol.* 18, 343, 2000.

171. Sosio, M., Bossi, E., and Donadio, S., Assembly of large genomic segments in artificial chromosomes by homologous recombination in *Escherichia coli*, *Nucl. Acids Res.* 29, E37, 2001.

172. Alduina, R. et al., Artificial chromosome libraries of *Streptomyces coelicolor* A3(2) and *Planobispora rosea*, *FEMS Microbiol. Lett.* 218, 181, 2003.

173. Handelsman, J. et al., Cloning the metagenome: culture-independent access to the diversity and functions of the uncultured microbial world, in *Methods in Microbiology*, Wren, G. and Dorrell, N. Academic Press, New York, 2002, pp. 241.

174. Stemmer, W. P., DNA shuffling by random fragmentation and reassembly: *in vitro* recombination for molecular evolution, *Proc. Natl. Acad. Sci. USA* 91, 10747, 1994.

175. Stemmer, W. P., Rapid evolution of a protein *in vitro* by DNA shuffling, *Nature* 370, 389, 1994.

176. Zhang, Y. X. et al., Genome shuffling leads to rapid phenotypic improvement in bacteria, *Nature* 415, 644, 2002.

177. Patnaik, R. et al., Genome shuffling of *Lactobacillus* for improved acid tolerance, *Nat. Biotechnol.* 20, 707, 2002.

178. Barton, W. A. et al., Structure, mechanism and engineering of a nucleotidylyltransferase as a first step toward glycorandomization, *Nat. Struct. Biol.* 8, 545, 2001.

179. Barton, W. A. et al., Expanding pyrimidine diphosphosugar libraries via structure-based nucleotidylyltransferase engineering, *Proc. Natl. Acad. Sci. USA* 99, 13397, 2002.

180. Fu, X. et al., Antibiotic optimization via *in vitro* glycorandomization, *Nat. Biotechnol.* 21, 1467, 2003.

181. Thorson, J. S. et al., Structure-based enzyme engineering and its impact on *in vitro* glycorandomization, *ChemBioChem* 5, 16, 2004.

182. Raty, K. et al., Cloning and characterization of *Streptomyces galilaeus* aclacinomycins polyketide synthase (PKS) cluster, *Gene* 293, 115, 2002.

183. Kantola, J. et al., Elucidation of anthracyclinone biosynthesis by stepwise cloning of genes for anthracyclines from three different *Streptomyces* spp., *Microbiology* 146 (Pt 1), 155, 2000.

184. Schauwecker, F. et al., Molecular cloning of the actinomycin synthetase gene cluster from *Streptomyces chrysomallus* and functional heterologous expression of the gene encoding actinomycin synthetase II, *J. Bacteriol.* 180, 2468, 1998.

185. Pfennig, F., Schauwecker, F., and Keller, U., Molecular characterization of the genes of actinomycin synthetase I and of a 4-methyl-3-hydroxyanthranilic acid carrier protein involved in the assembly of the acylpeptide chain of actinomycin in *Streptomyces*, *J. Biol. Chem.* 274, 12508, 1999.

186. Schauwecker, F. et al., Construction and *in vitro* analysis of a new bi-modular polypeptide synthetase for synthesis of N-methylated acyl peptides, *Chem. Biol.* 7, 287, 2000.

187. Yu, T. W. et al., Mutational analysis and reconstituted expression of the biosynthetic genes involved in the formation of 3-amino-5-hydroxybenzoic acid, the starter unit of rifamycin biosynthesis in *Amycolatopsis mediterranei* S699, *J. Biol. Chem.* 276, 12546, 2001.

188. He, J. and Hertweck, C., Iteration as programmed event during polyketide assembly; molecular analysis of the aureothin biosynthesis gene cluster, *Chem. Biol.* 10, 1225, 2003.

189. Du, L. et al., The biosynthetic gene cluster for the antitumor drug bleomycin from *Streptomyces verticillus* ATCC15003 supporting functional interactions between nonribosomal peptide synthetases and a polyketide synthase, *Chem. Biol.* 7, 623, 2000.

190. Shen, B. et al., The biosynthetic gene cluster for the anticancer drug bleomycin from *Streptomyces verticillus* ATCC15003 as a model for hybrid peptide-polyketide natural product biosynthesis, *J. Ind. Microbiol. Biotechnol.* 27, 378, 2001.

191. Shen, B. et al., Cloning and characterization of the bleomycin biosynthetic gene cluster from *Streptomyces verticillus* ATCC15003, *J. Nat. Prod.* 65, 422, 2002.

192. Sanchez, C. et al., Cloning and characterization of a phosphopantetheinyl transferase from *Streptomyces verticillus* ATCC15003, the producer of the hybrid peptide-polyketide antitumor drug bleomycin, *Chem. Biol.* 8, 725, 2001.

193. Du, L. and Shen, B., Identification and characterization of a type II peptidyl carrier protein from the bleomycin producer *Streptomyces verticillus* ATCC 15003, *Chem. Biol.* 6, 507, 1999.

194. Du, L. et al., An oxidation domain in the BlmIII non-ribosomal peptide synthetase probably catalyzing thiazole formation in the biosynthesis of the anti-tumor drug bleomycin in *Streptomyces verticillus* ATCC15003, *FEMS Microbiol. Lett.* 189, 171, 2000.

195. Du, L. et al., BlmIII and BlmIV nonribosomal peptide synthetase-catalyzed biosynthesis of the bleomycin bithiazole moiety involving both in cis and in trans aminoacylation, *Biochemistry* 42, 9731, 2003.

196. Olano, C. et al., Biosynthesis of the angiogenesis inhibitor borrelidin by *Streptomyces parvulus* Tü4055: cluster analysis and assignment of functions, *Chem. Biol.* 11, 87, 2004.

197. Schmutz, E. et al., Resistance genes of aminocoumarin producers: two type II topoisomerase genes confer resistance against coumermycin A1 and clorobiocin, *Antimicrob. Agents Chemother.* 47, 869, 2003.

198. Pojer, F. et al., CloR, a bifunctional non-heme iron oxygenase involved in clorobiocin biosynthesis, *J. Biol. Chem.* 278, 30661, 2003.

199. Pojer, F. et al., CloQ, a prenyltransferase involved in clorobiocin biosynthesis, *Proc. Natl. Acad. Sci. USA* 100, 2316, 2003.

200. Schmutz, E. et al., An unusual amide synthetase (CouL) from the coumermycin A1 biosynthetic gene cluster from *Streptomyces rishiriensis* DSM 40489, *Eur. J. Biochem.* 270, 4413, 2003.

201. Shen, B. and Hutchinson, C. R., TetracenomycinF2 cyclase: intramolecular aldol condensation in the biosynthesis of tetracenomycin C in *Streptomyces glaucens*, *Biochemistry* 32, 11149, 1993.

202. Rajgarhia, V. B., Priestley, N. D., and Strohl, W. R., The product of *dpsC* confers starter unit fidelity upon the daunorubicin polyketide synthase of *Streptomyces* sp. strain C5, *Metab. Eng.* 3, 49, 2001.

203. Meurer, G. et al., Iterative type II polyketide synthases, cyclases and ketoreductases exhibit context-dependent behavior in the biosynthesis of linear and angular decapolyketides, *Chem. Biol.* 4, 433, 1997.

204. Decker, H. et al., Identification of *Streptomyces olivaceus* Tü 2353 genes involved in the production of the polyketide elloramycin, *Gene* 166, 121, 1995.

205. Blanco, G. et al., Identification of a sugar flexible glycosyltransferase from *Streptomyces olivaceus*, the producer of the antitumor polyketide elloramycin, *Chem. Biol.* 8, 253, 2001.

206. Rafanan, E. R., Jr. et al., Cloning, sequencing, and heterologous expression of the *elmGHIJ* genes involved in the biosynthesis of the polyketide antibiotic elloramycin from *Streptomyces olivaceus* Tü 2353, *J. Nat. Prod.* 64, 444, 2001.

207. Patallo, E. P. et al., Deoxysugar methylation during biosynthesis of the antitumor polyketide elloramycin by *Streptomyces olivaceus*. Characterization of three methyltransferase genes, *J. Biol. Chem.* 276, 18765, 2001.

208. Summers, R. G. et al., Sequencing and mutagenesis of genes from erythromycin biosynthetic gene cluster of *Saccharopolyspora erythraea* that are involved in L-mycarose and D-desomine production, *Microbiology* 143, 3251, 1997.

209. Rodriguez, L. et al., Generation of hybrid elloramycin analogs by combinatorial biosynthesis using genes from anthracycline-type and macrolide biosynthetic pathways, *J. Mol. Microbiol. Biotechnol.* 2, 271, 2000.

210. Fischer, C. et al., Digitoxosyltetracenomycin C and glucosyltetracenomycin C, two novel elloramycin analogues obtained by exploring the sugar donor substrate specificity of glycosyltransferase ElmGT, *J. Nat. Prod.* 65, 1685, 2002.

211. Arslanian, R. L. et al., A new cytotoxic epothilone from modified polyketide synthases heterologously expressed in *Myxococcus xanthus*, *J. Nat. Prod.* 65, 1061, 2002.

212. Schneider, T. L., Walsh, C. T., and O'Connor, S. E., Utilization of alternate substrates by the first three modules of the epothilone synthetase assembly line, *J. Am. Chem. Soc.* 124, 11272, 2002.

213. O'Connor, S. E., Walsh, C. T., and Liu, F., Biosynthesis of epothilone intermediates with alternate starter units: engineering polyketide-nonribosomal interfaces, *Angew. Chem. Int. Ed.* 42, 3917, 2003.

214. Wendt-Pienkowski, E. et al., Cloning, sequencing, and analysis of the fredericamycin biosynthetic gene cluster from *Streptomyces griseus*, Submitted, 2005.

215. Fischer, C., Lipata, F., and Rohr, J., The complete gene cluster of the antitumor agent gilvocarcin V and its implication for the biosynthesis of the gilvocarcins, *J. Am. Chem. Soc.* 125, 7818, 2003.

216. Bililign, T. et al., The hedamycin locus implicates a novel aromatic PKS priming mechanism, *Chem. Biol.* 11, 959–969, 2004.

217. Han, L. et al., Cloning and characterization of polyketide synthase genes for jadomycin B biosynthesis in *Streptomyces venezuelae* ISP5230, *Microbiology* 140, 3379, 1994.

218. Han, L. et al., An acyl-coenzyme A carboxylase encoding gene associated with jadomycin biosynthesis in *Streptomyces venezuelae* ISP5230, *Microbiology* 146, 903, 2000.

219. Yang, K., Han, L., and Vining, L. C., Regulation of jadomycin B production in *Streptomyces venezuelae* ISP5230: involvement of a repressor gene, *jadR2*, *J. Bacteriol.* 177, 6111, 1995.

220. Yang, K. et al., Accumulation of the angucycline antibiotic rabelomycin after disruption of an oxygenase gene in the jadomycin B biosynthetic gene cluster of *Streptomyces venezuelae*, *Microbiology* 142 (Pt 1), 123, 1996.

221. Wang, L., McVey, J., and Vining, L. C., Cloning and functional analysis of a phosphopantetheinyl transferase superfamily gene associated with jadomycin biosynthesis in *Streptomyces venezuelae* ISP5230, *Microbiology* 147, 1535, 2001.

222. Wang, L., White, R. L., and Vining, L. C., Biosynthesis of the dideoxysugar component of jadomycin B: genes in the jad cluster of *Streptomyces venezuelae* ISP5230 for L-digitoxose assembly and transfer to the angucycline aglycone, *Microbiology* 148, 1091, 2002.

223. Wohlert, S. et al., Insights about the biosynthesis of the avermectin deoxysugar L-oleandrose through heterologous expression of *Streptomyces avermitilis* deoxysugar genes in *Streptomyces lividans*, *Chem. Biol.* 8, 681, 2001.

224. Westrich, L. et al., Cloning and characterization of a gene cluster from *Streptomyces cyanogenus* S136 probably involved in landomycin biosynthesis, *FEMS Microbiol. Lett.* 170, 381, 1999.

225. von Mulert, U. et al., Expression of the landomycin biosynthetic gene cluster in a PKS mutant of *Streptomyces fradiae* is dependent on the coexpression of a putative transcriptional activator gene, *FEMS Microbiol. Lett.* 230, 91, 2004.

226. Trefzer, A. et al., Elucidation of the function of two glycosyltransferase genes (*lanGT1* and *lanGT4*) involved in landomycin biosynthesis and generation of new oligosaccharide antibiotics, *Chem. Biol.* 8, 1239, 2001.

227. Trefzer, A. et al., Rationally designed glycosylated premithramycins: hybrid aromatic polyketides using genes from three different biosynthetic pathways, *J. Am. Chem. Soc.* 124, 6056, 2002.

228. Lombo, F. et al., The mithramycin gene cluster of *Streptomyces argillaceus* contains a positive regulatory gene and two repeated DNA sequences that are located at both ends of the cluster, *J. Bacteriol.* 181, 642, 1999.

229. Blanco, G. et al., Deciphering the biosynthetic origin of the aglycone of the aureolic acid group of anti-tumor agents, *Chem. Biol.* 3, 193, 1996.

230. Fernandez, E. et al., An ABC transporter is essential for resistance to the antitumor agent mithramycin in the producer *Streptomyces argillaceus*, *Mol. Gen. Genet.* 251, 692, 1996.

231. Lozano, M. J. et al., Characterization of two polyketide methyltransferases involved in the biosynthesis of the antitumor drug mithramycin by *Streptomyces argillaceus*, *J. Biol. Chem.* 275, 3065, 2000.

232. Rix, U. et al., Modification of post-PKS tailoring steps through combinatorial biosynthesis, *Nat. Prod. Rep.* 19, 542, 2002.

233. Remsing, L. L. et al., Ketopremithramycins and ketomithramycins, four new aureolic acid-type compounds obtained upon inactivation of two genes involved in the biosynthesis of the deoxysugar moieties of the antitumor drug mithramycin by *Streptomyces argillaceus*, reveal novel insights into post-PKS tailoring steps of the mithramycin biosynthetic pathway, *J. Am. Chem. Soc.* 124, 1606, 2002.

234. Mao, Y., Varoglu, M., and Sherman, D. H., Molecular characterization and analysis of the biosynthetic gene cluster for the antitumor antibiotic mitomycin C from *Streptomyces lavendulae* NRRL 2564, *Chem. Biol.* 6, 251, 1999.

235. Mao, Y., Varoglu, M., and Sherman, D. H., Genetic localization and molecular characterization of two key genes (*mitAB*) required for biosynthesis of the antitumor antibiotic mitomycin C, *J. Bacteriol.* 181, 2199, 1999.

236. Sheldon, P. J. et al., Mitomycin resistance in *Streptomyces lavendulae* includes a novel drug-binding-protein-dependent export system, *J. Bacteriol.* 181, 2507, 1999.

237. He, M., Sheldon, P. J., and Sherman, D. H., Characterization of a quinone reductase activity for the mitomycin C binding protein (MRD): Functional switching from a drug-activating enzyme to a drug-binding protein, *Proc. Natl. Acad. Sci. USA* 98, 926, 2001.

238. Kwon, H. J. et al., Cloning and heterologous expression of the macrotetrolide biosynthetic gene cluster revealed a novel polyketide synthase that lacks an acyl carrier protein, *J. Am. Chem. Soc.* 123, 3385, 2001.

239. Shen, B. and Kwon, H. J., Macrotetrolide biosynthesis: A novel type II polyketide synthase, *Chem. Rec.* 2, 389, 2002.

240. Smith, W. C., Xiang, L., and Shen, B., Genetic localization and molecular characterization of the *nonS* gene required for macrotetrolide biosynthesis in *Streptomyces griseus* DSM40695, *Antimicrob. Agents Chemother.* 44, 1809, 2000.

241. Kwon, H. J. et al., C-O bond formation by polyketide synthases, *Science* 297, 1327, 2002.

242. Ylihonko, K. et al., A gene cluster involved in nogalamycin biosynthesis from *Streptomyces nogalater*: sequence analysis and complementation of early-block mutations in the anthracycline pathway, *Mol. Gen. Genet.* 251, 113, 1996.

243. Torkkell, S. et al., The entire nogalamycin biosynthetic gene cluster of *Streptomyces nogalater*: characterization of a 20-kb DNA region and generation of hybrid structures, *Mol. Genet. Genom.* 266, 276, 2001.

244. Ylihonko, K. et al., Production of hybrid anthracycline antibiotics by heterologous expression of *Streptomyces nogalater* nogalamycin biosynthesis genes, *Microbiology* 142 (Pt 8), 1965, 1996.

245. Kunnari, T. et al., Isolation and characterization of 8-demethoxy steffimycins and generation of 2,8-demethoxy steffimycins in *Streptomyces steffisburgensis* by the nogalamycin biosynthesis genes, *J. Antibiot.* 50, 496, 1997.

246. Steffensky, M., Li, S. M., and Heide, L., Cloning, overexpression, and purification of novobiocic acid synthetase from *Streptomyces spheroides* NCIMB 11891, *J. Biol. Chem.* 275, 21754, 2000.

247. Eustaquio, A. S. et al., Novobiocin biosynthesis: inactivation of the putative regulatory gene *novE* and heterologous expression of genes involved in aminocoumarin ring formation, *Arch. Microbiol.* 180, 25, 2003.

248. Freel Meyers, C. L. et al., Initial characterization of novobiocic acid noviosyl transferase activity of NovM in biosynthesis of the antibiotic novobiocin, *Biochemistry* 42, 4179, 2003.

249. Albermann, C. et al., Substrate specificity of NovM: implications for Novobiocin biosynthesis and glycorandomization, *Org. Lett.* 5, 933, 2003.

250. Freel Meyers, C. L. et al., Characterization of NovP and NovN: Completion of novobiocin biosynthesis by sequential tailoring of the noviosyl ring, *Angew. Chem. Int. Ed.* 43, 67, 2004.

251. Swan, D. G. et al., Characterisation of a *Streptomyces antibioticus* gene encoding a type I polyketide synthase which has an unusual coding sequence, *Mol. Gen. Genet.* 242, 358, 1994.

252. Rodriguez, A. M. et al., A cytochrome P450-like gene possibly involved in oleandomycin biosynthesis by *Streptomyces antibioticus*, *FEMS Microbiol. Lett.* 127, 117, 1995.

253. Olano, C. et al., A second ABC transporter is involved in oleandomycin resistance and its secretion by *Streptomyces antibioticus*, *Mol. Microbiol.* 16, 333, 1995.

254. Quiros, L. M. and Salas, J. A., Biosynthesis of the macrolide oleandomycin by *Streptomyces antibioticus*. Purification and kinetic characterization of an oleandomycin glucosyltransferase, *J. Biol. Chem.* 270, 18234, 1995.

255. Quiros, L. M. et al., Two glycosyltransferases and a glycosidase are involved in oleandomycin modification during its biosynthesis by *Streptomyces antibioticus*, *Mol. Microbiol.* 28, 1177, 1998.

256. Olano, C. et al., Analysis of a *Streptomyces antibioticus* chromosomal region involved in oleandomycin biosynthesis, which encodes two glycosyltransferases responsible for glycosylation of the macrolactone ring, *Mol. Gen. Genet.* 259, 299, 1998.

257. Shah, S. et al., Cloning, characterization and heterologous expression of a polyketide synthase and P-450 oxidase involved in the biosynthesis of the antibiotic oleandomycin, *J. Antibiot.* 53, 502, 2000.

258. Tang, L., Fu, H., and McDaniel, R., Formation of functional heterologous complexes using subunits from the picromycin, erythromycin and oleandomycin polyketide synthases, *Chem. Biol.* 7, 77, 2000.

259. Aguirrezabalaga, I. et al., Identification and expression of genes involved in biosynthesis of L-oleandrose and its intermediate L-olivose in the oleandomycin producer *Streptomyces antibioticus*, *Antimicrob. Agents Chemother.* 44, 1266, 2000.

260. Doumith, M. et al., Interspecies complementation in *Saccharopolyspora erythraea*: elucidation of the function of *oleP1*, *oleG1* and *oleG2* from the oleandomycin biosynthetic gene cluster of *Streptomyces antibioticus* and generation of new erythromycin derivatives, *Mol. Microbiol.* 34, 1039, 1999.

261. Rodriguez, L. et al., Functional analysis of OleY L-oleandrosyl 3-O-methyltransferase of the oleandomycin biosynthetic pathway in *Streptomyces antibioticus*, *J. Bacteriol.* 183, 5358, 2001.

262. Long, P. F. et al., Engineering specificity of starter unit selection by the erythromycin-producing polyketide synthase, *Mol. Microbiol.* 43, 1215, 2002.

263. Gaisser, S. et al., Parallel pathways for oxidation of 14-membered polyketide macrolactones in *Saccharopolyspora erythraea*, *Mol. Microbiol.* 44, 771, 2002.

264. Rodriguez, L. et al., Engineering deoxysugar biosynthetic pathways from antibiotic-producing microorganisms. A tool to produce novel glycosylated bioactive compounds, *Chem. Biol.* 9, 721, 2002.

265. Tercero, J. A. et al., The biosynthetic pathway of the aminonucleoside antibiotic puromycin, as deduced from the molecular analysis of the pur cluster of *Streptomyces alboniger*, *J. Biol. Chem.* 271, 1579, 1996.

266. Tercero, J. A., Espinosa, J. C., and Jimenez, A., Expression of the Streptomyces alboniger pur cluster in *Streptomyces lividans* is dependent on the *bldA*-encoded tRNALeu, *FEBS Lett.* 421, 221, 1998.

267. Sanchez, C. et al., The biosynthetic gene cluster for the antitumor rebeccamycin: characterization and generation of indolocarbazole derivatives, *Chem. Biol.* 9, 519, 2002.

268. Onaka, H. et al., Characterization of the biosynthetic gene cluster of rebeccamycin from *Lechevalieria aerocolonigenes* ATCC 39243, *Biosci. Biotechnol. Biochem.* 67, 127, 2003.

269. Hyun, C. G. et al., The biosynthesis of indolocarbazoles in a heterologous *E. coli* host, *ChemBioChem* 4, 114, 2003.

270. Watanabe, K. et al., Understanding the substrate specificity of polyketide synthatse modules by generating hybrid multimodular synthases, *J. Biol. Chem.* 278, 42020, 2003.

271. Hu, Z. et al., A host-vector system for analysis and manipulation of rifamycin polyketide biosynthesis in *Amycolatopsis mediterranei*, *Microbiology* 145, 2335, 1999.

272. Pospiech, A. et al., A new *Myxococcus xanthus* gene cluster for the biosynthesis of the antibiotic saframycin Mx1 encoding a peptide synthetase, *Microbiology* 141, 1793, 1995.

273. Pospiech, A., Bietenhader, J., and Schupp, T., Two multifunctional peptide synthetases and an O-methyltransferase are involved in the biosynthesis of the DNA-binding antibiotic and antitumour agent saframycin Mx1 from *Myxococcus xanthus*, *Microbiology* 142, 741, 1996.

274. Trefzer, A. et al., Biosynthetic gene cluster of simocyclinone, a natural multihybrid antibiotic, *Antimicrob. Agents Chemother.* 46, 1174, 2002.

275. Parry, R. J. and Hoyt, J. C., Purification and preliminary characterization of (E)-3-(2,4-dioxo-6-methyl-5-pyrimidinyl)acrylic acid synthase, an enzyme involved in biosynthesis of the antitumor agent sparsomycin, *J. Bacteriol.* 179, 1385, 1997.

276. Lazaro, E. et al., Characterization of sparsomycin resistance in *Streptomyces sparsogenes*, *Antimicrob. Agents Chemother.* 46, 2914, 2002.

277. Onaka, H. et al., Cloning of the staurosporine biosynthetic gene cluster from *Streptomyces* sp. TP-A0274 and its heterologous expression in *Streptomyces lividans*, *J. Antibiot.* 55, 1063, 2002.

278. Motamedi, H. and Hutchinson, C. R., Cloning and heterologous expression of a gene cluster for the biosynthesis of tetracenomycin C, the anthracycline antitumor antibiotic of *Streptomyces glaucescens*, *Proc. Natl. Acad. Sci. USA* 84, 4445, 1987.

279. Bibb, M. J. et al., Analysis of the nucleotide sequence of the *Streptomyces glaucescens tcmI* genes provides key information about the enzymology of polyketide antibiotic biosynthesis, *EMBO J.* 8, 2727, 1989.

280. Decker, H., Motamedi, H., and Hutchinson, C. R., Nucleotide sequences and heterologous expression of *tcmG* and *tcmP*, biosynthetic genes for tetracenomycin C in *Streptomyces glaucescens*, *J. Bacteriol.* 175, 3876, 1993.

281. McDaniel, R. et al., Engineered biosynthesis of novel polyketides, *Science* 262, 1546, 1993.
282. Zawada, R. J. and Khosla, C., Heterologous expression, purification, reconstitution and kinetic analysis of an extended type II polyketide synthase, *Chem. Biol.* 6, 607, 1999.
283. Shen, B. et al., Purification and characterization of the acyl carrier protein of the *Streptomyces glaucescens* tetracenomycin C polyketide synthase, *J. Bacteriol.* 174, 3818, 1992.
284. Summers, R. G. et al., The tcmVI region of the tetracenomycin C biosynthetic gene cluster of *Streptomyces glaucescens* encodes the tetracenomycin F1 monooxygenase, tetracenomycin F2 cyclase, and, most likely, a second cyclase, *J. Bacteriol.* 175, 7571, 1993.
285. Shen, B. and Hutchinson, C. R., Triple hydroxylation of tetracenomycin A2 to tetracenomycin C in *Streptomyces glaucescens*. Overexpression of the *tcmG* gene in *Streptomyces lividans* and characterization of the tetracenomycin A2 oxygenase, *J. Biol. Chem.* 269, 30726, 1994.
286. McDaniel, R. et al., Engineered biosynthesis of novel polyketides: influence of a downstream enzyme on the catalytic specificity of a minimal aromatic polyketide synthase, *Proc. Natl. Acad. Sci. USA* 91, 11542, 1994.
287. Bao, W., Wendt-Pienkowski, E., and Hutchinson, C. R., Reconstitution of the iterative type II polyketide synthase for tetracenomycin F2 biosynthesis, *Biochemistry* 37, 8132, 1998.
288. Kulowski, K. et al., Functional characterization of the *jadI* gene as a cyclase forming angucyclinones., *J. Am. Chem. Soc.* 121, 1786, 1999.
289. Khosla, C. et al., Genetic construction and functional analysis of hybrid polyketide synthases containing heterologous acyl carrier proteins, *J. Bacteriol.* 175, 2197, 1993.
290. Kramer, P. J. et al., Rational design and engineered biosynthesis of a novel 18-carbon aromatic polyketide, *J. Am. Chem. Soc.* 119, 635, 1997.
291. Fu, H. et al., Engineered biosynthesis of novel polyketides: stereochemical course of two reactions catalyzed by a polyketide synthase, *Biochemistry* 33, 9321, 1994.
292. Fu, H. et al., Engineered biosynthesis of novel polyketides: regiospecific methylation of an unnatural substrate by the *tcmO* O-methyltransferase, *Biochemistry* 35, 6527, 1996.
293. Decker, H. and Haag, S., Cloning and characterization of a polyketide synthase gene from *Streptomyces fradiae* Tü2717, which carries the genes for biosynthesis of the angucycline antibiotic urdamycin A and a gene probably involved in its oxygenation, *J. Bacteriol.* 177, 6126, 1995.
294. Hoffmeister, D. et al., The NDP-sugar co-substrate concentration and the enzyme expression level influence the substrate specificity of glycosyltransferases: cloning and characterization of deoxysugar biosynthetic genes of the urdamycin biosynthetic gene cluster, *Chem. Biol.* 7, 821, 2000.
295. Trefzer, A. et al., Function of glycosyltransferase genes involved in urdamycin A biosynthesis, *Chem. Biol.* 7, 133, 2000.
296. Faust, B. et al., Two new tailoring enzymes, a glycosyltransferase and an oxygenase, involved in biosynthesis of the angucycline antibiotic urdamycin A in *Streptomyces fradiae* Tü2717, *Microbiology* 146 (Pt 1), 147, 2000.
297. Hoffmeister, D. et al., Engineered urdamycin glycosyltransferases are broadened and altered in substrate specificity, *Chem. Biol.* 9, 287, 2002.
298. Kunzel, E. et al., Inactivation of *urdGT2* gene, which encodes a glycosyltransferase responsible for the C-glycosyltransfer of activated D-olivose, leads to the formation of the novel urdamycins I, J, and K., *J. Am. Chem. Soc.* 121, 11058, 1999.
299. Rix, U. et al., Urdamycin L: a novel metabolic shunt product that provides evidence for the role of the *urdM* gene in the urdamycin A biosynthetic pathway of *Streptomyces fradiae* Tü 2717, *ChemBioChem* 4, 109, 2003.
300. Garg, R. P. et al., Molecular characterization and analysis of the biosynthetic gene cluster for the azoxy antibiotic valanimycin, *Mol. Microbiol.* 46, 505, 2002.
301. Ogasawara, Y. et al., Cloning, sequencing, and functional analysis of the biosynthetic gene cluster of macrolactam antibiotic vicenistatin in *Streptomyces halstedii*, *Chem. Biol.* 11, 79, 2004.
302. Clardy, J. and Walsh, C. T., Lessons from natural molecules, *Nature,* 432, 7–15, 2004.
303. Galm, U. et al., Antitumor antibiotics: Bleomycin, enediynes, and mitomycin, *Chem. Rev.,* 105, 739–758, 2005.
304. Chen, J. and Stubbe, J., Bleomycins: Towards better therapeutics, *Nat. Rev. Cancer,* 5, 102–112, 2005.
305. Sanchez, C. et al., Combinatorial biosynthesis of antitumor indolocarbazole compounds, *Proc. Natl. Acad. Sci. USA,* 102, 461–466, 2005.

26 Developments and Future Trends in Anticancer Natural Products Drug Discovery*

David J. Newman and Gordon M. Cragg

CONTENTS

I. Introduction 554
II. Plant Sources...554
    A. Combretastatins...554
    B. Flavopiridol...554
    C. Olomucine, Roscovitine, and the Purvalanols 558
    D. Benzopyrans and Privileged Structures...558
    E. Microbial Associates and Endophytic Fungi 559
    F. Old Molecules, New Uses..559
        1. Bruceantin 559
        2. Triterpenoid Acids 559
        3. β-Lapachone 560
III. Marine Sources 560
    A. Microbial Associates...560
        1. Dolastatins...560
        2. Ecteinascidins 560
        3. Jaspamide/Jasplakinolide..562
        4. Geodiamolides and Chondramides...562
        5. Pederins, Mycalamides, and Related Structures 562
        6. Manzamines 563
    B. Independent Microbes 563
        1. Salinosporamide..563
    C. Marine Invertebrates...564
        1. NVP-LAQ824 and Psammaplin 564
IV. Terrestrial Microbes 565
    A. Leptomycins..565
    B. Palmarumycin CP$_1$...565
V. Plant–Microbe–Marine Interface 566
    A. Indigo and the Indirubins 566
VI. Conclusion..568
References 568

---

* Not including those products discussed in other chapters in this volume.

I. INTRODUCTION

Important, naturally derived chemotypes currently in clinical use or development as anticancer agents have been discussed in earlier chapters, and the purpose of this chapter is to provide an overview — by no means exhaustive — of developments and future trends in the generation of novel agents from nature. For reference, a list of approved antitumor agents (as of the end of 2002) is shown in Table 26.1, using the data from Newman et al.[1]

The organization of the chapter is according to the putative sources of agents where these are known or assumed, but the chapter ends with a discussion, together with examples, of what appears to be an evolving trend about the actual source organisms. As more is learned about the genetics of a "producing organism," frequently it is realized that the actual organism from which the original material was isolated and identified is either a host for microbial biosynthetic machinery or may perform a tailoring function on materials produced by a microbe.

II. PLANT SOURCES

On the basis of recent experience using the National Cancer Institute (NCI) 60–cell line screen, the potential for the discovery of a new cytotoxic chemotype from plants, comparable to the taxanes or camptothecins, appears to be relatively low. In spite of work covering vascular plants from extensive collections in the tropics over the last 18+ yrs, no novel chemotypes have, as yet, been reported. Nevertheless, a number of variations on known chemotypes or their sources have been reported, including camptothecin derivatives, materials that resemble bleomycin in their activities at the molecular level, and compounds, such as the jatrophane esters, that have now been shown to have paclitaxel-like activities.

A. COMBRETASTATINS

What is evident however, is that at least one of the classes of natural products currently in clinical trials, the combretastatins, either as the original compound or as its phosphate ester (combretastatin A4 phosphate), has generated a significant amount of interest as a basic chemotype (Chapter 3). At present, there are multiple variations on the CA4 basic structure (trimethoxy-stilbene) either in clinical trials or entering various stages of preclinical development as a drug candidate. The compounds range from very close chemical analogs to topologically equivalent molecules from a receptor perspective, but significantly different when viewed in two dimensions, though resembling the basic molecular structure of CA4 when carefully examined (1–7). The structures of a significant number of these natural product mimics are given in the recent review of Cragg and Newman, which should be consulted by the interested reader.[2]

B. FLAVOPIRIDOL

Approximately contemporaneous with the initial work on CA4, Flavopiridol (8) was prepared by the Indian subsidiary of Hoechst (now Sanofi-Aventis) following the isolation and synthesis of the plant-derived natural product, rohitukine (9). It was the most active of approximately 100 analogs made when assayed against cyclin-dependent kinases (CDKs) and showed about 100-fold more selectivity compared with its activity versus tyrosine kinases. Although it showed roughly comparable activity (IC_{50} 100–400 nM) against CDKs, it turned out to be the first compound at NCI identified as a potential antitumor agent that subsequently was proven to be a relatively specific CDK inhibitor.[3] The initial report[4] on its CDK2 inhibitory activity was made in 1994, followed by data demonstrating antitumor activity, provided by Czech et al.,[5] in 1995. Flavopiridol is currently in phase III clinical trials as an inhibitor of cyclin-dependent kinase 2 (CDK2), both as a single agent and as a modulator in combination with other agents, particularly paclitaxel and cisplatinum. It has been reported to lead to partial or complete remissions in a number of phase I patients, leading

TABLE 26.1
All Anticancer Drugs (1940s–2002)

Generic Name	Year	Source[a]
OCT-43	1999	B
Alemtuzumab	2001	B
Celmoleukin	1992	B
Denileukin diftitox	1999	B
Interferon alfa2a	1986	B
Interferon, gamma-1a	1992	B
Interleukin-2	1989	B
Pegaspargase	1994	B
Rituximab	1997	B
Tasonermin	1999	B
Teceleukin	1992	B
Trastuzumab	1998	B
BEC	1989	N
Aclarubicin	1981	N
Actinomycin D	Pre-1981	N
Angiotensin II	1994	N
Arglabin	1999	N
Asparaginase	Pre-1981	N
Bleomycin	Pre-1981	N
Daunomycin	Pre-1981	N
Doxorubicin	Pre-1981	N
Masoprocol	1992	N
Mithramycin	Pre-1981	N
Mitomycin C	Pre-1981	N
Paclitaxel	1993	N
Pentostatin	1992	N
Peplomycin	1981	N
Solamargine	1987	N
Streptozocin	Pre-1981	N
Testosterone	Pre-1981	N
Vinblastine	Pre-1981	N
Vincristine	Pre-1981	N
Alitretinoin	1999	ND
Amrubicin HCl	2002	ND
Cladribine	1993	ND
Cytarabine ocfosfate	1993	ND
Docetaxel	1995	ND
Dromostanolone	Pre-1981	ND
Elliptinium acetate	1983	ND
Epirubicin HCl	1984	ND
Estramustine	Pre-1981	ND
Ethinyl estradiol	Pre-1981	ND
Etoposide	Pre-1981	ND
Etoposide phosphate[b]	1996	ND
Exemestane	1999	ND
Fluoxymesterone	Pre-1981	ND
Formestane	1993	ND
Fulvestrant	2002	ND
Gemtuzumab ozogamicin	2000	ND
Hydroxyprogesterone	Pre-1981	ND

TABLE 26.1 (Continued)
All Anticancer Drugs (1940s–2002)

Generic Name	Year	Source[a]
Idarubicin hydrochloride	1990	ND
Irinotecan hydrochloride	1994	ND
Medroxyprogesterone acetate	Pre-1981	ND
Megesterol acetate	Pre-1981	ND
Methylprednisolone	Pre-1981	ND
Methyltestosterone	Pre-1981	ND
Miltefosine	1993	ND
Mitobronitol	Pre-1981	ND
Pirarubicin	1988	ND
Prednisolone	Pre-1981	ND
Prednisone	Pre-1981	ND
Teniposide	Pre-1981	ND
Testolactone	Pre-1981	ND
Topotecan HCl	1996	ND
Triamcinolone	Pre-1981	ND
Triptorelin	1986	ND
Valrubicin	1999	ND
Vindesine	Pre-1981	ND
Vinorelbine	1989	ND
Zinostatin stimalamer	1994	ND
Aminoglutethimide	1981	S
Amsacrine	1987	S
Arsenic trioxide	2000	S
Bisantrene hydrochloride	1990	S
Busulfan	Pre-1981	S
Camostat mesylate	1985	S
Carboplatin	1986	S
Carmustine	Pre-1981	S
Chlorambucil	Pre-1981	S
Chlortrianisene	Pre-1981	S
Cis-diamminedichloroplatinum	Pre-1981	S
Cyclophosphamide	Pre-1981	S
Dacarbazine	Pre-1981	S
Diethylstilbestrol	Pre-1981	S
Flutamide	1983	S
Fotemustine	1989	S
Heptaplatin /SK-2053R	1999	S
Hexamethylmelamine	Pre-1981	S
Hydroxyurea	Pre-1981	S
Ifosfamide	Pre-1981	S
Levamisole	Pre-1981	S
Lobaplatin	1998	S
Lomustine	Pre-1981	S
Lonidamine	1987	S
Mechlorethanamine	Pre-1981	S
Melphalan	Pre-1981	S
Mitotane	Pre-1981	S

TABLE 26.1 (Continued)
All Anticancer Drugs (1940s–2002)

Generic Name	Year	Source[a]
Mustine hydrochloride	Pre-1981	S
Nedaplatin	1995	S
Nilutamide	1987	S
Nimustine hydrochloride	Pre-1981	S
Oxaliplatin	1996	S
Pipobroman	Pre-1981	S
Porfimer sodium	1993	S
Procarbazine	Pre-1981	S
Ranimustine	1987	S
Sobuzoxane	1994	S
Thiotepa	Pre-1981	S
Triethylenemelamine	Pre-1981	S
Uracil mustard	Pre-1981	S
Zoledronic acid	2000	S
Aminogluethimide	Pre-1981	S*
Capecitabine	1998	S*
Carmofur	1981	S*
Cytosine arabinoside	Pre-1981	S*
Doxifluridine	1987	S*
Enocitabine	1983	S*
Floxuridine	Pre-1981	S*
Fludarabine phosphate	1991	S*
Fluorouracil	Pre-1981	S*
Gemcitabine HCl	1995	S*
Goserelin acetate	Pre-1981	S*
Leuprolide	Pre-1981	S*
Mercaptopurine	Pre-1981	S*
Methotrexate	Pre-1981	S*
Mitoxantrone HCI	1984	S*
Tamoxifen	Pre-1981	S*
Thioguanine	Pre-1981	S*
Bexarotene	2000	S*/NM
Raltitrexed	1996	S*/NM
Temozolomide	1999	S*/NM
Anastrozole	1995	S/NM
Bicalutamide	1995	S/NM
Camostat mesylate	1985	S/NM
Fadrozole HCl	1995	S/NM
Gefitinib	2002	S/NM
Imatinib mesilate	2001	S/NM
Letrazole	1996	S/NM
Toremifene	1989	S/NM
Bcg live	1990	V
Melanoma theraccine	2001	V

[a] Definitions as Newman et al.[1]

[b] Prodrug (not counted).

to phase II studies in patients with paclitaxel-resistant tumors. There have been a number of relatively recent reports of this agent and the various combinations with other drugs and drug candidates — a significant number of which are either natural products or are derived from natural products.[6-8]

C. OLOMUCINE, ROSCOVITINE, AND THE PURVALANOLS

Continuing on the theme of plant-derived products being leads to cell cycle modulators, Meijer, one of the early researchers in pharmacologic intervention in the cell cycle,[9] demonstrated, using the simple starfish oocyte model of cell division, that substituted purines, particularly 6-dimethylaminopurine and isopentenyladenine (from *Castanea* sp.) were found to inhibit the mitotic histone H1 kinase, which is now better known as CDK1/cyclin B, at the 50–100 μM level *in vitro*. On searching for other purine-derived compounds, the group tested a plant secondary metabolite, subsequently named olomucine (**10**), that was originally isolated from the cotyledons of the radish but had been synthesized in Australia in 1986 by Parker et al.[10] This material disproved the current dogma about adenosine triphosphate binding sites because it demonstrated an improved efficacy (IC_{50} of 7 μM) and selectivity for CDKs and, to some extent, MAP kinases by direct competition with adenosine triphosphate. Further development of this series led to roscovitine (**11**), and then following synthesis of a very focused library via combinatorial chemistry techniques, the purvalanols, which were even more potent, with IC_{50} values in the 4–40-nM range (compared with 450 nM for roscovitine).[11]

The (R)-isomer of roscovitine was selected for clinical development under the code CYC202 under license to Cyclacel in Dundee, Scotland, and currently is in phase II clinical trials in Europe. As with other signal transduction inhibitors in clinical trials, sequential treatment with cytotoxins is also being used or considered. Though some beneficial effects are seen with the signal transduction inhibitors, complete or partial responses tend to only be demonstrated when sequential treatments of signal transduction inhibitor/cytotoxin are used.

D. BENZOPYRANS AND PRIVILEGED STRUCTURES

This type of chemical modification of an existing basic "privileged structure" is documented in the combinatorial chemistry literature, particularly in the case of the formally plant-derived base structure, the benzopyran. The term, "privileged structures," was originally introduced by Evans et al.[12] for benzazepines and benzdiazepines because of their ability to bind to a multitude of unrelated classes of peptide receptors with high affinity. The term was later extended by Mason et al.[13] to other structural classes such as benzamidines, biphenyltetrazoles, spiropiperidines, indoles, and benzylpiperidines and was reviewed by Patchett and Nargund in 2000.[14] In a series of elegant papers, Nicolaou et al.[15-17] dramatically extended the concept to include benzopyrans, as a result of the researchers' hypothesis: "We were particularly intrigued by the possibility that using scaffolds of natural origin, which presumably have undergone evolutionary selection over time, might confer favorable bioactivities and bioavailabilities to library members."[15] A search of the literature showed almost 4000 2,2-dimethyl-2H-benzopyran moieties (**12**); a slight structural modification whereby the pyran ring is reduced (**13**) identified another 8000 structures. Examples of the multiplicity of bioactive derivatives of this general structure are given in detail in Figure 2 of the first paper in the series,[15] which should be consulted by the interested reader.

Using a technique of solid-phase combinatorial chemistry that used a selenenyl bromide substituted polystyrene resin,[18] the group was able to generate what were effectively "libraries from libraries" by using a massively parallel synthetic technique on the pyrans, which were produced from the seleno-linked libraries by epoxidation of the pyran double bond, followed by the opening of the epoxide by nucelophilic attack. These libraries, and subsequent derivatives based on the benzopyrans, were then tested against a variety of pharmacologic screens, and the leads were then reoptimized, using the screening data.

To date (mid-2004), four publications have come from these initial libraries, with activities in the inhibition of NADH:ubiquinone oxidoreductase giving rise to benzopyrans with IC_{50} values in the 18–55-nM range and having predominately cytostatic effects on tumor lines, thus indicating a possible potential in chemoprevention. The interested reader should consult Figure 4 and Figure 5 in the relevant publication for more information on the actual structures and activities in both types of assay.[19] This report, relevant to antitumor studies, was then followed by others in different disease states covering antibiotics, energy metabolism, and genetic dissection of a cell pathway.[20–22]

E. MICROBIAL ASSOCIATES AND ENDOPHYTIC FUNGI

As mentioned in the introduction, there is now a realization that materials that were often thought to be from — in this case — plants are actually produced in large part by another organism, and the plant host may only perform some final tailoring of the molecule. Such is the case with maytansine since, as reported by Floss (Chapter 17), the actual source of the base structure is a bacterium, with perhaps the modification leading to maytansine being performed by the plant itself. There are other examples given later in the chapter that continue on this theme.

In addition to the above example of what may be a simple transfer of material to the host from a microbe, there are some extremely interesting examples in the recent review by Strobel et al.[23] of materials that are produced by plant epiphytic microbes, predominately from the fungi. Among the organisms reported are a series of fungal epiphytes from the well-known genus *Pestalotiopsis* and a report that *Pestalotiopsis micromonospora*, a very common epiphytic fungus, produces paclitaxel at low levels. This followed on from the initial report, which was greeted with a fair amount of skepticism, that the novel fungus *Taxomyces andreanae,* isolated from *Taxus brevifolia,* produced paclitaxel at very low levels. Since the original report in 1993, many further examples of paclitaxel production at low levels have been reported in the scientific literature, not just from *Taxus* spp., but from trees of many other genera. These findings have led Strobel to suggest that paclitaxel production by epiphytic fungi is a protective mechanism for the plant host against plant pathogens, such as *Pythium* spp. and *Phytophora* spp., as paclitaxel is known to be very active as a fungicide against these particular pathogens.

F. OLD MOLECULES, NEW USES

1. Bruceantin

An example of an "old" drug of the same vintage as Taxol® and camptothecin, and having a possibility of revival, is bruceantin (**14**), which was first isolated from a tree, *Brucea antidysenterica*, used in Ethiopia for the treatment of "cancer." Activity was observed in animal models bearing a range of tumors, but no objective responses were observed in clinical trials, and further development was terminated. Interest has been revived by the observation of significant activity against panels of leukemia, lymphoma, and myeloma cell lines, as well as in animal models bearing early and advanced stages of the same cancers. This activity has been associated with the downregulation of a key oncoprotein (c-*myc*), and these data are being presented as strong evidence supporting the development of bruceantin as an agent for the treatment of hematological malignancies.[24]

2. Triterpenoid Acids

Triterpenoid acids such as oleanolic (**15**) and ursolic acid, which are common plant constituents, are known to possess weak antiinflammatory and antitumor activities. Attempts to synthesize new analogs having increased potencies have led to the synthesis of 2-cyano-3,12-dioxoolean-1,9-dien-28-oic acid (**16**) and its methyl ester, which have potent *in vitro* and *in vivo* antitumor activity against a wide range of tumors, including breast carcinomas, leukemias, and pancreatic carcinomas.

Of particular interest is the significant activity of 2-cyano-3,12-dioxoolean-1,9-dien-28-oic acid against epithelial ovarian carcinoma cell lines, including lines that were resistant to clinically used agents such as cisplatin. Because epithelial ovarian carcinoma is the leading cause of death from gynecologic cancers, further evaluation of 2-cyano-3,12-dioxoolean-1,9-dien-28-oic acid in the treatment of these cancers is being pursued.[25]

3. β-Lapachone

The relatively simple naphthoquinone β-lapachol (**17**) is a well-known compound obtained from the bark of the lapacho tree, *Tabebuia avellanedae*, and other species of the same genus native to South America, and it, together with other components of the plant, have extensive usage as an ethnobotanical treatment in the Amazonian region. Although advanced to clinical status by the NCI in the 1970s, this compound showed unacceptable levels of toxicity and was dropped.

Recently, however, a close relative, β-lapachone (**18**), has shown very interesting molecular target activity, with induction of apoptosis in transformed cells being one of the mechanisms of action. What is of prime import, however, is that one of the possible routes of apoptotic induction is via the transcription factor E2F1[26] and appears to only function in transformed cells. In contrast, no comparable effect was shown in nontransformed cells at comparable concentrations of the agent. Evidence of further involvement in the transcription processes was the recent demonstration by Choi et al.[27] that this agent induced activation of caspase-3, inhibition of NF-κB, and subsequent downregulation of bcl-2. At present, β-lapachone is in phase I clinical trials in the United States under the aegis of ArQule, who licensed it from the originators, Cyclis Pharmaceuticals. Further information on the background of these agents is given in the review by Ravelo et al.[28]

III. MARINE SOURCES

A. MICROBIAL ASSOCIATES

Perhaps it is in this area that the interplay of microbes and a host invertebrate/alga is becoming a significant part of the overall equation. For many years, it has been speculated that a number of potent marine-derived molecules that were active as antitumor agents might be the product of interactions between the organism from which the agent was isolated and associated microbes.

1. Dolastatins

The driving force behind such suppositions was the number of reports in the marine natural product literature over the last 10–15 yr, where the structures of compounds isolated from an invertebrate or, in an increasing number of cases, from multiple different invertebrate phyla in distinct geo-graphical areas of the oceans were similar to each other or to metabolites from terrestrial bacteria. Some relevant examples are the dolastatins from *Dolabella auricularia* and simplostatins, and now dolastatins, from the cyanophyte, *Simploca* sp., on which the nudibranch feeds (Chapter 11).

2. Ecteinascidins

When the structure of Et743 was established by the groups of Rinehart and Wright in 1990 (Chapter 12), it became obvious that this compound had a base structure[29] that was well known from microbial sources on land, from invertebrates in the ocean, and from a marine bacterium. Thus, Et743 resembled the naphyridinomycins and saframycins from terrestrial Streptomycetes; the renieramycins from oceanic sponges, with the latest variation, renieramycin J being recently reported by Oku et al.[30]; jorumycin from the nudibranch, *Jorunna funebris*[31]; and cyanosafracin B from a marine-derived *Pseudomonas fluorescens*. The ecteinascidins all contain a fourth

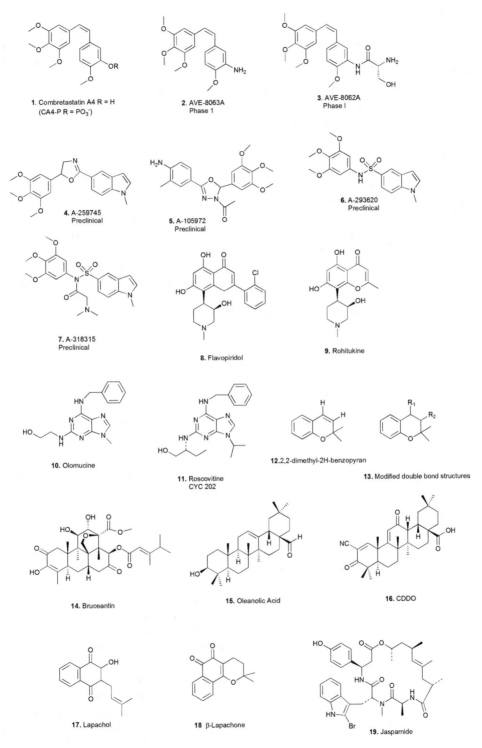

1. Combretastatin A4 R = H
(CA4-P R = PO₃⁻)

2. AVE-8063A
Phase 1

3. AVE-8062A
Phase I

4. A-259745
Preclinical

5. A-105972
Preclinical

6. A-293620
Preclinical

7. A-318315
Preclinical

8. Flavopiridol

9. Rohitukine

10. Olomucine

11. Roscovitine
CYC 202

12. 2,2-dimethyl-2H-benzopyran

13. Modified double bond structures

14. Bruceantin

15. Oleanolic Acid

16. CDDO

17. Lapachol

18 β-Lapachone

19. Jaspamide

STRUCTURES 1–19

exocyclic isoquinoline group when compared to their putative siblings, together with a bridging cysteinyl-derived grouping.

At present, there is work ongoing on identifying the endosymbionts of *Ecteinascidia turbinata*, with the ultimate aim of identifying, cloning, and expressing the putative Et743 biosynthetic genes if the potential producing organism, known currently as *Candidatus* Endoecteinascidia frumentensis, is not amenable to cultivation.[32]

Further examples of the potential involvement of microbes in the production of sponge-isolated secondary metabolites with antitumor potential (either directly or as a warhead on a delivery system) include, but are not limited to, the following.

3. Jaspamide/Jasplakinolide

Jaspamide (**19**) is the major metabolite[33] of the Pacific sponge, *Dorypleres splendens* (also identified as a *Jaspis* sp.), and was also reported as jasplakinolide when it was isolated from a Fijian-derived sample of *Jaspis johnstoni*.[34] In addition, modified jaspamides were also reported from the sponge *Jaspis splendens* collected in Vanuatu.[35] Jaspamide was initially reported to be a potent insecticide and antifungal agent,[33,34] but subsequent work by NCI scientists demonstrated that it induces actin polymerization *in vitro*, leading to disruption of the actin skeleton.[36,37] However, in spite of many attempts at NCI, no reproducible *in vivo* antitumor activity could be demonstrated because the therapeutic index was very close to unity. In this respect, it is similar to the effects seen with other actin inhibitors such as the cucurbitacins, latrunculins, and cytochalasins. However, with a suitable delivery system, these molecules may well be of utility.

4. Geodiamolides and Chondramides

The possible involvement of microbes as a source of these compounds was raised by the reports of the structures of the closely related geodiamolides,[38] from a *Geodia* sp. in 1987 and from other taxonomically distant sponges in following years.[39] The potential for microbial involvement was reinforced by the reports from the Reichenbach group in 1995, of the isolation and purification of similar molecules, the chondramides (**20**), with an 18-membered macrolide ring from a terrestrial myxobacterium.[40] Analogous to the jaspamides, their first reported activities were as antifungal compounds, but subsequent investigation demonstrated their actin-stabilizing activity.[41]

5. Pederins, Mycalamides, and Related Structures

An early entrant to the potential for microbial involvement was the recognition in the late 1980s that the highly cytotoxic sponge metabolites known as the mycalamides[42] and subsequently other derivatives of the basic structure, the onnamides, theopederins, and the ring-opened form known as irciniastatin A (also reported later as psymberin),[43,44] all contained the basic ring structure of the toxin from the *Paederus* beetle, pederin (**21**). These molecules (which currently number over 30[44]) are very interesting from the perspective of potential antitumor drug leads, but development has been bedeviled by sourcing problems, even though a formal synthesis of mycalamide A (**22**) was published in 2004 by Trost et al.[45]

Recently, however, work by Kellner[46] on the source of pederin from the *Paederus sabaeus* beetle from Turkey demonstrated that the actual source was a putative pseudomonad. This work was extended from a molecular genetic aspect by Piel et al.,[47] demonstrating a symbiosis island in the microbe that indicates the potential for horizontal gene transfer to other organisms and, most important, the presence in the microbe of a PKS-PS cluster that formally codes for the biosynthetic pathway for pederin.[48] If this organism can either be cultured or the cluster expressed in another host, then the sourcing problems of this class of molecules may well be on the way to being solved, allowing further investigation of their potential.

6. Manzamines

Although one could always make a circumstantial case that a microbe or microbes were involved in the biosynthesis of either the agents listed above or of precursors that might then be further host-modified (analogous to maytansine; see Section IIE above), it was not until 2003 that there was actual hard evidence for production of the same compound by an associated microbe and by the host invertebrate. An Indonesian sponge, collected from approximately 100 m depth, using rebreather collection techniques, produced a series of antitumor compounds known as the manzamines; on culturing the associated microbes, a *Micromonospora* sp. was purified, which on fermentation in selected media, produced the base molecule, manzamine A, and a derivative, 8-hydroxymanzamine A, both of which were isolable from the host sponge. Thus, for the first time, the thesis mentioned above was proven and the potential for production of this agent by fermentation was demonstrated.[49,50]

On the basis of these observations and data, it is probably correct to assume that each of these series of compounds has a common genetic ancestor (a microbe). Thus, it behooves natural products chemists and biologists to look beyond the "nominal sources" and to actually track down, as far as is possible, the actual progenitor. If a series of compounds is going to be useful as a drug entity, or lead thereto, fermentation is still the easiest production technique on any significant scale above a small laboratory preparation level.

B. Independent Microbes

Although pioneering work was done in the middle of the last century in isolating and culturing marine microbes from a microbiological aspect, it was not until the 1980s that significant time and effort was devoted to the isolation, identification, and fermentation of marine microbes (or microbes from marine sources) in order to evaluate their potential to produce secondary metabolites with antitumor activity that would be comparable to those known since the 1940s from their terrestrial equivalents.

1. Salinosporamide

In retrospect, what probably had the most effect on the isolation techniques was the realization that the media used for work with terrestrial microbes was much too rich in carbon for their marine-based cousins. Once a series of suitable media and storage/recovery techniques was established, in large part because of the work of Fenical and Jensen and their colleagues at The Scripps Institute of Oceanography, the field opened up, with reports of active and chemically unique molecules being published at frequent intervals. What is of prime import is that as a result of their systematic searching for marine bacteria, the Fenical group recently described a totally new genus[51] within the order *Actinomycetales*, for which they have proposed the genus name of *Salinospora*.[52]

This bacterium, on fermentation in suitable media in seawater, produced the omuralide (**23**) derivative salinosporamide A (**24**). Omuralide (**23**) is the spontaneous reaction product from the terrestrial Streptomycete metabolite, lactacystin (**25**), first reported in 1991 by Omura et al.[53] and synthesized, together with analogs, by Corey's group.[54] Subsequently, lactacystin (actually omuralide) was found to be a specific inhibitor of the 20S proteosome, modifying a subunit-specific threonine.[55] That such an inhibitor has drug potential is demonstrated by the now approved antitumor agent, Velcade®, a synthetic peptide boronate, so the search for other molecules with similar activities is worthwhile. Earlier work on natural products and synthetic agents has been covered by Kisselev and Goldberg.[56]

Salinosporamide itself, which differs from lactacystin/omuralide in terms of its substitution pattern around the pyrrolid-2-one ring, is also a potent proteosome inhibitor and is currently under preclinical development by Nereus Pharmaceuticals in San Diego, California, with a recent publication from Corey's group giving details of a simple synthetic route to the compound.[57]

TABLE 26.2
Status of Marine-Derived Natural Products in Clinical and Preclinical Trials
(Not discussed in other chapters)

Name	Source	Status (Disease)	Comment
ES-285 (Spisulosine)	*Spisula polynyma*	Phase I (Cancer)	*Rho*-GTP inhibitor
KRN-7000	*Agelas mauritianus*	Phase I (Cancer)	An Agelasphin derivative
Squalamine	*Squalus acanthias*	Phase II (Cancer)	Antiangiogenic activity as well
NVP-LAQ824	Synthetic	Phase I (Cancer)	Derived from Psammaplin, Trichostatin, and Trapoxin structures
Laulimalide	*Cacospongia mycofijiensis*	Preclinical (Cancer)	Synthesized by a variety of investigators
Vitilevuamide	*Didemnum cucliferum* and *Polysyncraton lithostrotum*	Preclinical (Cancer)	
Diazonamide	*Diazona angulata*	Preclinical (Cancer)	Synthesized and new structure elucidated.
Eleutherobin	*Eleutherobia* sp	Preclinical (Cancer)	Synthesized and derivatives made by combi-chem; can be produced by aquaculture
Sarcodictyin	*Sarcodictyon roseum*	Preclinical (Cancer) (derivatives)	Combi-chem synthesis performed around structure
Peloruside A	*Mycale hentscheli*	Preclinical (cancer)	
Salicylihalimides A	*Haliclona* sp	Preclinical (cancer)	First marine Vo-ATPase inhibitor. Similar materials from microbes, synthesized
Thiocoraline	*Micromonospora marina*	Preclinical (Cancer)	DNA polymerase inhibitor
Ascididemnin		Preclinical (Cancer)	Reductive DNA-cleaving agents
Variolins	*Kirkpatrickia variolosa*	Preclinical (Cancer)	CDK inhibitors

C. Marine Invertebrates

At present, there is a considerable number of formally marine invertebrate–derived molecules that are in preclinical studies that have not been covered in other chapters in this book. Table 26.2 lists a selection of those that are in preclinical development, and we will comment on one of them at a greater length below. The citations to the compounds in Table 26.2, together with descriptions of their sources and comments on their activities as antitumor agents, and so forth, are given in detail in the 2004 review by Newman and Cragg, which should be consulted for further information.[58]

1. NVP-LAQ824 and Psammaplin

One compound listed in Table 26.2 is in fact a totally synthetic molecule but was derived from information from natural products, from both a marine invertebrate and from terrestrial microbes. This is the Novartis agent known currently as NVP-LAQ824. The basic structure (**26**) was derived from work with both natural products and synthetic derivatives. The marine natural product, psammaplin A (**27**), which had originally been identified in 1987 by the groups of Schmitz[59] at the University of Oklahoma and Crews[60] at the University of California, Santa Cruz, was screened, together with congeners and with the microbial products trapoxin B (**28**) and trichostatin A (**29**), for their activity as histone deacetylase inhibitors by Novartis (then Ciba-Giegy) as part of a National Cooperative Drug Discovery Group program. The psammaplins were found to be extremely potent histone deacetylase inhibitors.

The synthetic path from the psammaplin, trapoxin, and trichostatin structures to the compound now known as NVP-LAQ824 was described in three papers.[61–63] These papers, and in particular the review by Remiszewski,[63] should be consulted for the chemical rationales that led from these natural products to the current clinical candidate. This agent demonstrated significant *in vitro* activity against cells derived from human multiple myeloma patients, inducing apoptotic signaling and also exhibiting histone deacetylase inhibition. At present, the material is in a phase I clinical trial against hematologic malignancies[64] at the Dana-Farber Cancer Institute.

IV. TERRESTRIAL MICROBES

At this time, most agents in advanced preclinical development as direct antitumor agents are not from terrestrial microbes, but what is becoming more important is the use of older molecules as the basis for treatments with newer agents, as well as the use of the old molecule as a lead structure for combinatorial chemistry (see Section IIF above). We will give two examples, one from each category.

A. LEPTOMYCINS

Reports have been coming out in the biological literature of the utility of the well-known microbial products, the leptomycins, and the effect that they have on tumor cells when combined with specific molecular target inhibitors. The effect of the fungal metabolite, leptomycin B (**30**), on the cell cycle was first described by Yoshida et al.,[65] when they demonstrated that this compound induced elongation of the cells of the fission yeast *Schizosaccharomyces pombe* and also stopped proliferation of rat 3Y1 cells at both G1 and G2. Removal of the leptomycin then produced tetraploid cells that could proliferate without passage through the M phase. Subsequent probing of the genetic machinery of a leptomycin-resistant strain of *S. pombe* demonstrated that the probable target of leptomycin in this organism and, by inference, in human cells was located in the regulatory cascade of *crm1*, a gene that was required for maintenance of higher-order chromosome structures, and hence cell growth in the yeast.[66] Confirmation of *crm1* as the target of leptomycin B in both the yeast and in human cells was provided by the subsequent identification of a cysteine residue (Cys-529) in this protein, as the site of covalent binding resulting from Michael addition of the cysteinyl–SH group across the α, β-unsaturated δ-lactone that is common to all of the leptomycin-like molecules. This led to a total inhibition of its function as part of a trimeric complex with RanGTP and a protein containing a nuclear export sequence whose function was to move proteins out of the nucleus into the cytoplasm.[67–69]

In 2001, Vigneri and Wang[70] reported that leptomycin B, in conjunction with the synthetic PTK inhibitor STI571 (Gleevec®), stopped egress of the oncoprotein *bcr-abl* from the nucleus of CML cells and induced apoptosis. Using target mutants, the researchers confirmed that the effect was caused by inhibition of *crm1*, which stopped the export of the *bcr-abl*/RanGTP/protein complex and led to the sequestration of the *bcr-abl* within the nucleus and induction of apoptosis. This finding had a potential clinical significance, as a fair proportion of patients treated with Gleevec® become refractory in a relatively short period of time. Work on this approach is ongoing, with over 80 citations to the initial report in the literature. An updated review by Yoshida's group on the potential for nuclear transport inhibitors was recently published, covering leptomycins and other similar agents.[71]

B. PALMARUMYCIN CP$_1$

Palmarumycin CP$_1$ (**31**) from the terrestrial fungus *Coniothyrium*, first reported by Krohn et al.[72] in 1994, was used by Wipf at the University of Pittsburgh as a base structure to synthesize a variety of compounds containing the base naphthoquinone–spiroketal structure. This study was part of a search for compounds that inhibited the thioredoxin–thioredoxin reductase system at similar levels to the activity demonstrated by the fungal metabolite, pleurotin (**32**), whose *in vitro* and *in vivo* activities in this and in related pathways were recently reported by Welsh et al.[73] In a series of recent papers, Wipf and his colleagues have presented evidence demonstrating that a simple modification of the base CP_1 structure gives compounds S11 and S12 (**33, 34**) with excellent enzyme inhibitory activity and *in vitro* cytotoxicities, comparable to those of the natural products.[74–76]

V. PLANT–MICROBE–MARINE INTERFACE

Finally, there is one class of compounds that has recently come into play as molecular target inhibitors that have a very long provenance as chemical entities — in some cases as mixtures and in others as single compounds — and whose listed activities range from dyestuffs, through a traditional Chinese medical treatment for leukemia, to very specific inhibition of an essential cellular enzyme cascade.

A. INDIGO AND THE INDIRUBINS

The compounds in question are all based on the simple heterocycle, indole. The essential aminoacid tryptophan is formed from the chorismate pathway, via cyclization of anthranilate, and indole is formed from tryptophan catabolism. If indole is hydroxylated in the 3 position, presumably by a suitable cytochrome P_{450}, then it is tautomeric with the 3 keto analog, indoxyl. Various levels of oxidation then lead to the mixture of indigo, indirubin, and their isomers that is commonly used as the source of indigo dyestuffs for either clothing or as war paint (the ancient Celtic "woad" being a mixture from the plant *Isatis tinctora*[77]).

Although usually thought of as being plant products, indigo and the indirubins have been reported from four nominally independent sources: a variety of plants,[77] a number of marine mollusks (usually belonging to the *Muricidae* family of gastropods),[78] natural or recombinant bacteria,[79] and human urine.[80] It is highly probable, however, that in all cases, the indirubins and indigo are the terminal oxidation products of tryptophan/indole catabolism — perhaps even a method of removing excess indole from the organism.

However, irrespective of the reason for the production, the indirubins have been identified as the major active ingredient of the traditional Chinese medical recipe known as Danggui Longhui Wan, used to treat chronic myelogenous leukemia[81,82] in China for many years. What is of import from both a natural product and a pharmacological perspective is the recognition by Meijer's group that the indirubins as a class were both inhibitors of several CDKs and also potent inhibitors of glycogen synthase kinase-3 (GSK-3).

Thus, in a recent paper, Meijer et al.[83] described the early work on this enzyme and then demonstrated that simple indirubins, including a brominated version (**35**) that had never been isolated from natural sources before their work with the mollusk *Hexaplex trunculus* and its chemically modified oxime derivative (**36**), were potent and selective inhibitors of both variants of GSK-3, with IC_{50} values of 22 and 5 nM, respectively. There was at least a fivefold specificity versus CDK1/Cyclin B or CDK/p25 and significantly more specificity against a wide range of other kinases. A slightly later paper from the same group gave full details of the chemistry involved and established structure–activity relationships using x-ray crystallography and molecular modeling techniques.[84]

By using immobilized indirubins, Meijer's group was able to conclusively identify GSK-3 as the target of these molecules in cell lyzates, to demonstrate that they had a significant effect *in vivo* on *Xenopus* development — consonant with the suggested mechanism, and to derive x-ray crystallographic evidence for the binding of the agents at the active site (adenosine triphosphate binding site) of GSK-3. They also demonstrated that the activation of the enzyme via phosphorylation by an as-yet-unidentified kinase on two specific tyrosine residues (GSK-3 is a Ser/Thr kinase) is blocked when indirubins are bound, perhaps because of a conformational shift in the bound versus free protein.

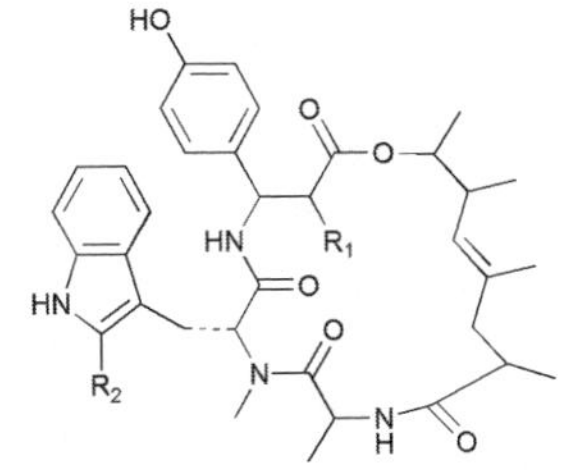

20. Chondramides
(R$_1$ and R$_2$ include Cl,
OCH$_3$ and H)

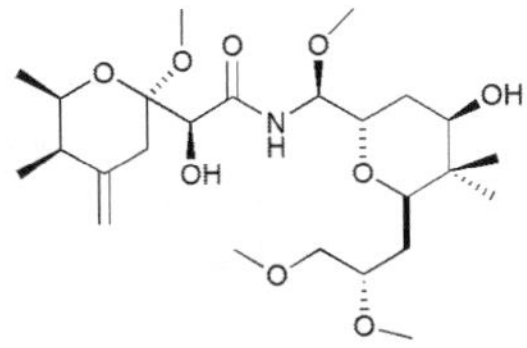

21. Pederin

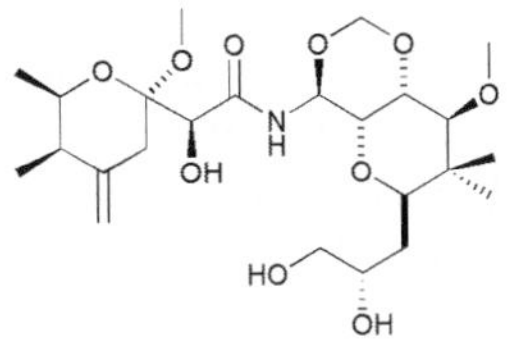

22. Mycalamide A

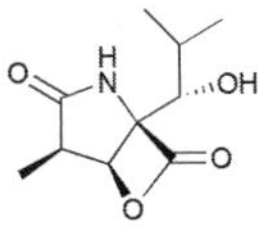

23. Omuralide

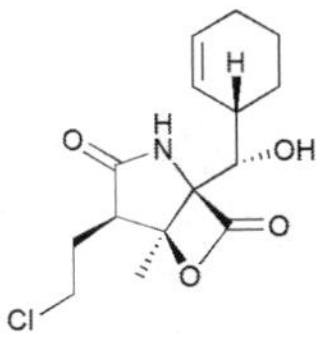

24. Salinosporamide A

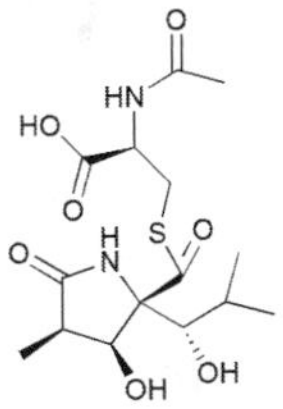

25. Lactacystin

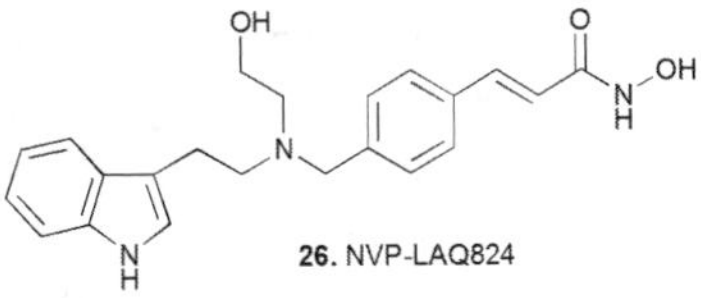

26. NVP-LAQ824

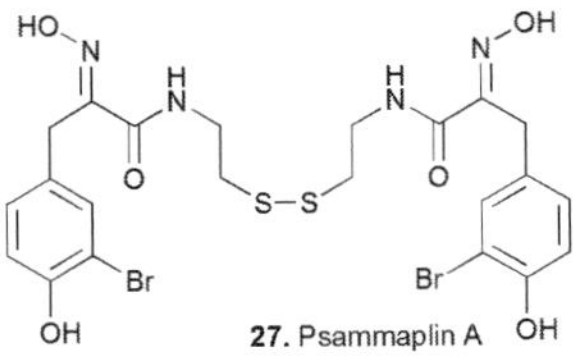

27. Psammaplin A

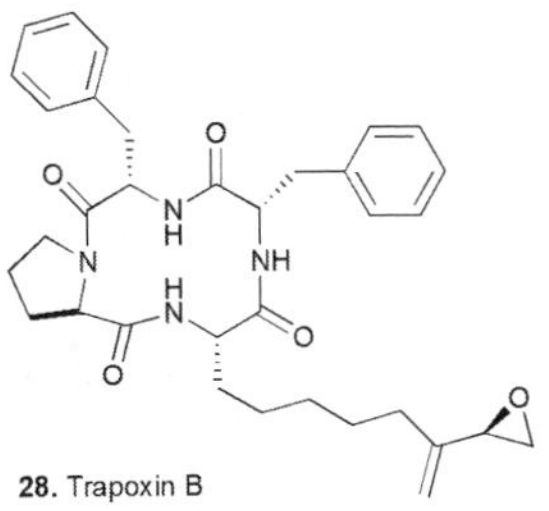

28. Trapoxin B

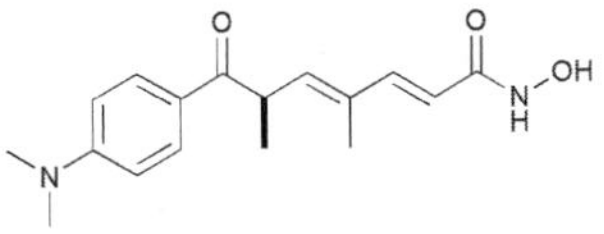

29. Trichostatin A

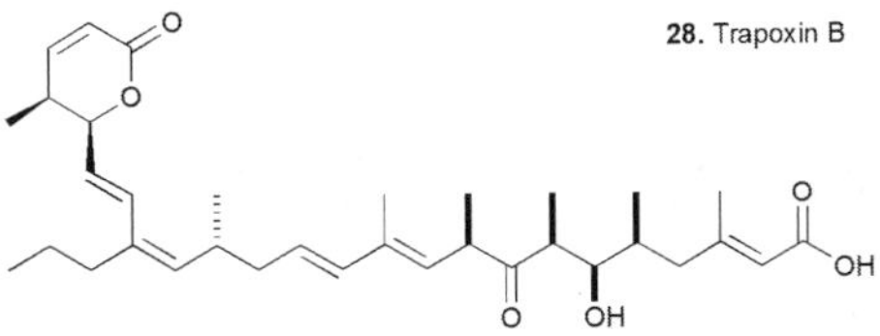

30. Leptomycin B

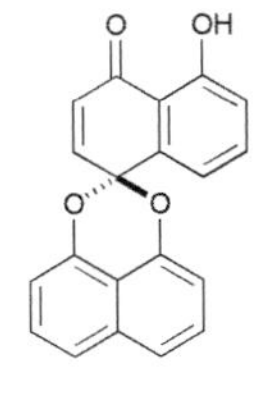

31. Palmarumycin CP$_1$

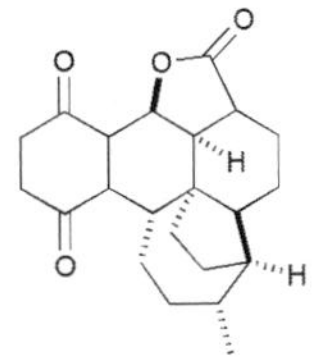

32. Pleurotin

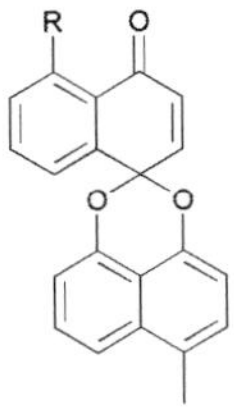

33. S11 R = H
34. S12 R = OH

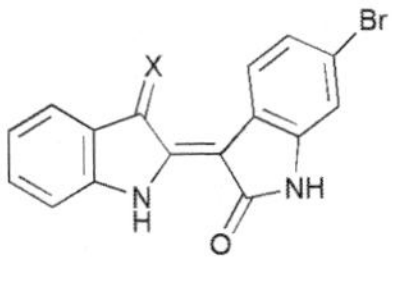

35. Bromoindirubin X = O
36. Oxime of bromoindirubin X = N-OH

STRUCTURES 20–36

These papers were then followed by another from the same group in which they demonstrated that the indirubins actually had yet another, independent action at the cell receptor level. By using the same basic suite of compounds, they demonstrated that they serve as ligands for the "orphan receptor" known as the aryl hydrocarbon receptor (AhR). As yet, no natural ligands have been identified for AhR, even though, contrary to earlier beliefs, it has existed for over 450 million years (though there have been proposals that indole-containing compounds are among the natural ligands[85]). From a series of elegant biochemical and biological studies, including the use of cells that were null–null mutants for the AhR, the researchers were able to differentiate between the cytostatic effects of some indirubins following their activation of the AhR from kinase inhibition, which appears to be the main mechanism underlying the cytotoxicity of these molecules.[86]

VI. CONCLUSION

Although one generally thinks of the organism from which a compound is isolated and purified as being the source of the natural product, it is becoming more and more difficult to actually state that this is so. From the information presented in the case of plants and marine invertebrates, it can be seen that in many instances, microbes (perhaps from all three kingdoms) are intimately involved at some stage in the production of those secondary metabolites that are being evaluated for their potential as drug leads or entities.

We deliberately did not give any details of the work now ongoing in studies of the metagenomes[87] of either soils or marine sediments, because, as yet, no compounds have been reported in the open literature that are acting as leads to new antitumor agents. However, because the microbial diversity is many orders of magnitude greater than any other source of biological diversity, such agents will be forthcoming.

REFERENCES

1. Newman, D. J., Cragg, G. M., and Snader, K. M., Natural products as sources of new drugs over the period 1981-2002, *J. Nat. Prod.*, 66, 1022, 2003.
2. Cragg, G. M. and Newman, D. J., A tale of two tumor targets: Topoisomerase I and tubulin. The Wall and Wani contribution to cancer chemotherapy, *J. Nat. Prod.*, 67, 232, 2004.
3. Sielecki, T. et al., Cyclin-dependent kinase inhibitors: Useful targets in cell cycle regulation, *J. Med. Chem.*, 43, 1, 2000.
4. Losiewicz, M. D. et al., Potent inhibition of cdc2 kinase activity by the flavanoid, l86-8275, *Biochem. Biophys. Res. Comm.*, 201, 589, 1994.
5. Czech, J. et al., Anti-tumoral activity of flavone l 86-8275, *Int. J. Oncol.*, 6, 31, 1995.
6. Dai, Y. and Grant, S., Cyclin-dependent kinase inhibitors, *Curr. Opin. Pharmacol.*, 3, 362, 2003.
7. Dancey, J. and Sausville, E. A., Issues and progress with protein kinase inhibitors for cancer treatment, *Nature Rev. Drug Disc.*, 2, 296, 2003.
8. Senderowicz, A. M., Small-molecule cyclin-dependent kinase modulators, *Oncogene*, 22, 6609, 2003.
9. Meijer, L. and Raymond, E., Roscovitine and other purines as kinase inhibitors. From starfish oocytes to clinical trials, *Acc. Chem. Res.*, 36, 417, 2003.
10. Parker, C. W., Entsch, B., and Letham, D., Inhibitors of two enzymes which metabolize cytokinins, *Phytochemistry*, 25, 303, 1986.
11. Chang, Y. T. et al., Synthesis and application of functionally diverse 2,6,9-trisubstituted purine libraries as CDK inhibitors, *Chem. Biol.*, 6, 361, 1999.
12. Evans, B. E. et al., Methods for drug discovery: Development of potent, selective, orally effective cholecystokinin antagonists, *J. Med. Chem.*, 31, 2235, 1988.
13. Mason, J. S. et al., New 4-point pharmacophore method for molecular similarity and diversity applications: Overview of the method and applications, including a novel approach to the design of combinatorial libraries containing privileged sub-structures, *J. Med. Chem.*, 42, 3251, 1999.

14. Patchett, A. A. and Nargund, R. P., Privileged structures — an update, in *Ann. Repts. Med. Chem.,* Doherty, A. M., Ed., Academic Press, San Diego, CA, 2000, Vol. 35, pp 289–298.

15. Nicolaou, K. C. et al., Natural product-like combinatorial libraries based on privileged structures. 1. General principles and solid-phase synthesis of benzopyrans, *J. Am. Chem. Soc.*, 122, 9939, 2000.

16. Nicolaou, K. C. et al., Natural product-like combinatorial libraries based on privileged structures. 2. Construction of a 10,000-membered benzopyran library by directed split-and-pool chemistry using nanokans and optical encoding, *J. Am. Chem. Soc.*, 122, 9954, 2000.

17. Nicolaou, K. C. et al., Natural product-like combinatorial libraries based on privileged structures. 3. The "libraries from libraries" principle for diversity enhancement of bezopyran libraries, *J. Am. Chem. Soc.*, 122, 9968, 2000.

18. Nicolaou, K. C. et al., Polymer-supported selenium reagents for organic synthesis, *Chem. Comm.*, 1947, 1998.

19. Nicolaou, K. C. et al., Combinatorial synthesis of novel and potent inhibitors of NADH: Ubiquinone oxidoreductase, *Chem. Biol.*, 7, 979, 2000.

20. Nicolaou, K. C. et al., Discovery of novel antibacterial agents active against methicillin-resistant *Staphylococcus aureus* from combinatorial benzopyran libraries, *ChemBioChem*, 460, 2001.

21. Nicolaou, K. C. et al., Discovery and optimization of non-steroidal *FXR* agonists from natural product-like libraries, *Org. Biomol. Chem.*, 1, 908, 2003.

22. Downes, M. et al., A chemical, genetic, and structural analysis of the nuclear bile acid receptor *FXR*, *Mol. Cell*, 11, 1079, 2003.

23. Strobel, G. et al., Natural products from endophytic microorganisms, *J. Nat. Prod.*, 67, 257, 2004.

24. Cuendet, M. and Pezzuto, J. M., Antitumor activity of bruceantin. An old drug with new promise, *J. Nat. Prod.*, 67, 269, 2004.

25. Melichar, B. et al., Growth-inhibitory effect of a novel synthetic triterpenoid, 2-cyano-3,12-dioxoalean-1,9-dien-28-oic acid, on ovarian carcinoma cell lines not dependent on peroxisome proliferators-activated receptor-γ expression, *Gynecol. Oncol.*, 93, 149, 2004.

26. Li, Y. et al., Selective killing of cancer cells by β-lapachone: Direct checkpoint activation as a strategy against cancer, *Proc. Natl. Acad. Sci., USA*, 100, 2674, 2003.

27. Choi, B. T., Cheong, J., and Choi, Y. H., β-lapachone-induced apoptosis is associated with activation of caspase-3 and inactivation of NF-κB in human colon cancer HCT-116 cells, *Anti-Cancer Drugs*, 14, 845, 2003.

28. Ravelo, A. G. et al., Recent studies on natural products as anticancer agents, *Curr. Topics Med. Chem.*, 4, 241, 2004.

29. Scott, J. D. and Williams, R. M., Chemistry and biology of the tetrahydroisoquinoline antitumor antibiotics, *Chem. Rev.*, 102, 1669, 2002.

30. Oku, N. et al., Renieramycin J, a highly cytotoxic tetrahydroisoquinoline alkaloid, from a marine sponge *Neopetrosia* sp., *J. Nat. Prod.*, 66, 1136, 2003.

31. Fontana, A. et al., A new antitumor isoquinoline alkaloid from the marine nudibranch *Jorunna funebris*, *Tetrahedron*, 56, 7305, 2000.

32. Moss, C. et al., Intracellular bacteria associated with the ascidian *Ecteinascidia turbinata*: Phylogenetic and *in situ* hybridization analysis, *Mar. Biol.*, 143, 99, 2003.

33. Zabriskie, T. M. et al., Jaspamide, a modified peptide from a Jaspis sponge, with insecticidal and antifungal activity, *J. Am. Chem. Soc.*, 108, 3123, 1986.

34. Crews, P., Manes, L. V., and Boehler, M., Jasplakinolide, a cyclodepsipeptide from the marine sponge, *Jaspis* sp., *Tetrahedron Lett.*, 27, 2797, 1986.

35. Zampella, A. et al., New jaspamide derivatives from the marine sponge *Jaspis splendans* collected in Vanuatu, *J. Nat. Prod.*, 62, 332, 1999.

36. Bubb, M. R. et al., Jasplakinolide, a cytotoxic natural product, induces actin polymerization and competitively inhibits the binding of phalloidin to f-actin, *J. Biol. Chem.*, 269, 14869, 1994.

37. Senderowicz, A. M. et al., Jasplakinolide's inhibition of the growth of prostate carcinoma cells *in vitro* with disruption of the actin cytoskeleton, *J. Nat. Can. Inst.*, 87, 46, 1995.

38. Chan, W. R. et al., Stereostructures of geodiamolides A and B, novel cyclodepsipeptides from the marine sponge *Geodia* sp., *J. Org. Chem.*, 52, 3091, 1987.

39. Coleman, J. E., Van Soest, R. W. M., and Andersen, R. J., New geodiamolides from the sponge *Cymbastela* sp. collected in Papua New Guinea, *J. Nat. Prod.*, 62, 1137, 1999.

40. Kunze, B. et al., Chondramides A-D, new antifungal and cytostatic depsipeptides from *Chondromyces crocatus* (myxobacteria). Production, physico-chemical and biological properties, *J. Antibiotics*, 48, 1262, 1995.

41. Sasse, F. et al., The chondramides: Cytostatic agents from myxobacteria acting on the actin cytoskeleton, *J. Nat. Can. Inst.*, 90, 1559, 1998.

42. Perry, N. B. et al., Mycalamide A, an antiviral compound from a New Zealand sponge of the genus *Mycale*, *J. Am. Chem. Soc.*, 110, 4850, 1988.

43. Pettit, G. R. et al., Antineoplastic agents. 520. Isolation and structure of irciniastatins A and B from the Indo-Pacific marine sponge *Ircinia ramosa*, *J. Med. Chem.*, 47, 1149, 2004.

44. Cichewicz, R. H., Valeriote, F. A., and Crews, P., Psymberin, a potent sponge-derived cytotoxin from *Psammocinia* distantly related to the pederin family, *Org. Lett.*, 6, 1951, 2004.

45. Trost, B. M., Yang, H., and Probst, G. D., A formal synthesis of (-)-mycalamide A, *J. Am. Chem. Soc.*, 126, 48, 2004.

46. Kellner, R. L. L., Molecular identification of an endosymbiotic bacterium associated with pederin biosynthesis in *Paederus sabaeus* (coleoptera: Staphylinidae), *Insect Biochem. Mol. Biol.*, 32, 389, 2002.

47. Piel, J., Hofer, I., and Hui, D., Evidence for a symbiosis island involved in horizontal acquisition of pederin biosynthetic capabilities by the bacterial symbiont of *Paederus fuscipes* beetles, *J. Bact.*, 186, 1280, 2004.

48. Piel, J., A polyketide synthase-peptide synthase gene cluster from an uncultured bacterial symbiont of *Paederus* beetles, *Proc. Natl. Acad. Sci. USA*, 99, 14002, 2002.

49. Kasanah, N. et al., Biotransformation and biosynthetic studies of the manzamine alkaloids, *Abs. Pap. 6th Int. Mar. Biotech. Conf.*, Abs S14, 2003.

50. Yousaf, M. et al., Solving limited supplies of marine pharmaceuticals through the rational and high-throughput modification of high yielding marine natural producer scaffolds, *Abs. Pap. 6th Int. Mar. Biotech. Conf.*, Abs. S14, 2003.

51. Mincer, T. J. et al., Widespread and persistent populations of a major new marine actinomycete taxon in ocean sediments, *Appl. Environ. Microbiol.*, 68, 5005, 2002.

52. Feling, R. H. et al., Salinosporamide A; a highly cytotoxic proteosome inhibitor from a novel microbial source, a marine bacterium of the new genus *Salinospora*, *Angew. Chem. Int. Ed.*, 42, 355, 2003.

53. Omura, S. et al., Lactacystin, a novel microbial metabolite, induces neuritogenesis of neuroblastoma cells, *J. Antibiot.*, 44, 113, 1991.

54. Corey, E. J. and Li, W. D., Total synthesis and biological activity of lactacystin, omuralide and analogs, *Chem. Pharm. Bull.*, 47, 1, 1999.

55. Fenteany, G. et al., Inhibition of proteosome activities and subunit-specific amino-terminal threonine modification by lactacystin, *Science*, 268, 726, 1995.

56. Kisselev, A. F. and Goldberg, A. L., Proteosome inhibitors: From research tools to drug candidates, *Chem. Biol.*, 8, 739, 2001.

57. Reddy, L. J., Saravanan, P., and Corey, E. J., A simple stereocontrolled synthesis of salinosporamide A, *J. Am. Chem. Soc.*, 126, 6230, 2004.

58. Newman, D. J. and Cragg, G. M., Marine natural products and related compounds in clinical and advanced preclinical trials, *J. Nat. Prod.*, 67, 1216, 2004.

59. Arabshahi, L. and Schmitz, F. J., Brominated tyrosine metabolites from an unidentified sponge, *J. Org. Chem.*, 52, 3584, 1987.

60. Quinoa, E. and Crews, P., Phenolic constituents of *Psammaplysilla*, *Tet. Lett.*, 28, 3229, 1987.

61. Pina, I. C. et al., Psammaplins from the sponge *Pseudoceritina purpurea*: Inhibition of both histone deacetylase and DNA methyltransferase, *J. Org. Chem.*, 68, 3866, 2003.

62. Remiszewski, S. W. et al., N-hydroxy-3-phenyl-2-propenamides as novel inhibitors of human histone deacetylase with *in vivo* antitumor activity: Discovery of (2e)-n-hydroxy-3-[4-[[2-hydroxyethyl)[2-(1h-indol-3-yl)ethyl]amino]methyl]-phenyl]-2-propenamide (NVP-LAQ824), *J. Med. Chem.*, 46, 4609, 2003.

63. Remiszewski, S. W., The discovery of NVP-LAQ824: From concept to clinic, *Curr. Med. Chem.*, 10, 2393, 2003.

64. Catley, L. et al., NVP-LAQ824 is a potent novel histone deacetylase inhibitor with significant activity against multiple myeloma, *Blood*, 102, 2615, 2003.

65. Yoshida, M. et al., Effects of leptomycin B on the cell cycle of fibroblasts and fission yeast cells, *Exp. Cell Res.*, 187, 150, 1990.

66. Nishi, K. et al., Leptomycin B targets a regulatory cascade of *crm*1, a fission yeast nuclear protein, involved in control of higher order chromosome structure and gene expression, *J. Biol. Chem.*, 269, 6320, 1994.

67. Kudo, N. et al., Leptomycin B inhibition of signal-mediated nuclear export by direct binding to *crm*1, *Exp. Cell Res.*, 242, 540, 1998.

68. Kudo, N. et al., Leptomycin B inactivates *crm*1/exportin 1 by covalent modification at a cysteine residue in the central conserved region, *Proc. Natl. Acad. Sci., USA*, 96, 9112, 1999.

69. Henderson, B. R. and Eleftheriou, A., A comparison of the activity, sequence specificity, and *crm*1-dependence of different nuclear export signals, *Exp. Cell Res.*, 256, 213, 2000.

70. Vigneri, P. and Wang, J. Y. J., Induction of apoptosis in chronic myelogenous leukemia cells through nuclear entrapment of *bcr-abl* tyrosine kinase, *Nat. Med.*, 7, 228, 2001.

71. Yashiroda, Y. and Yoshida, M., Nucleo-cytoplasmic transport of proteins as a target for therapeutic drugs, *Curr. Med. Chem.*, 10, 741, 2003.

72. Krohn, K. et al., Palmarumycins C1-C16 from *Coniothyrium*: Isolation, structure elucidation, and biological activity, *Liebigs Ann. Chem.*, 1099, 1994.

73. Welsh, S. J. et al., The thioredoxin redox inhibitors 1-methylpropyl 2-imidazoyl disulfide and pleurotin inhibit hypoxia-induced factor 1-α and vascular endothelial growth factor formation, *Mol. Can. Therap.*, 2, 235, 2003.

74. Wipf, P. et al., New inhibitors of the thioredoxin-thioredoxin reductase system based on a naphthaquinone spiroketal natural product lead, *Bioorg. Med. Chem. Lett.*, 11, 2637, 2001.

75. Lazo, J. et al., Antimitotic actions of a novel analog of the fungal metabolite palmarumycin CP₁, *J. Pharmacol. Exp. Ther.*, 296, 364, 2001.

76. Wipf, P. et al., Natural product based inhibitors of the thioredoxin-thioredoxin reductase system, *Org. Biomol. Chem.*, 2, 1651, 2004.

77. Maugard, T. et al., Identification of an indigo precursor from leaves of *Isatis tinctoria* (Woad), *Phytochemistry*, 58, 897, 2001.

78. Cooksey, C. J., Tyrian purple: 6,6′-dibromoindigo and related compounds, *Molecules*, 6, 736, 2001.

79. MacNeil, I. A. et al., Expression and isolation of antimicrobial small molecules from soil DNA libraries, *J. Mol. Microbiol. Biotechnol.*, 3, 301, 2001.

80. Adachi, J. et al., Indirubin and indigo are potent aryl hydrocarbon receptor ligands present in human urine, *J. Biol. Chem.*, 276, 31475, 2001.

81. Chen, D. and Xie, J., Chemical constituents of a traditional Chinese medicine Qing Dai, *Zhongcaoyao*, 15, 534, 1984.

82. Xiao, Z. et al., Indirubin and mesoindigo in the treatment of chronic myelogenous leukemia in China, *Leuk. Lymphoma*, 43, 1763, 2002.

83. Meijer, L. et al., GSK-3-selective inhibitors derived from Tyrian purple indirubins, *Chem. Biol.*, 10, 1255, 2003.

84. Polychronopoulos, P. et al., Structural basis for the synthesis of indirubins as potent and selective inhibitors of glycogen synthase kinase-3 and cyclin-dependent kinases, *J. Med. Chem.*, 47, 935, 2004.

85. Denison, M. S. and Nagy, S. R., Activation of the aryl hydrocarbon receptor by structurally diverse exogenous and endogenous chemicals, *Annu. Rev. Pharmacol. Toxicol.*, 43, 309, 2003.

86. Knockaert, M. et al., Independent actions on cyclin-dependent kinases and aryl hydrocarbon receptor mediates the antiproliferative effects of indirubins, *Oncogene*, 23, 4400, 2004.

87. Schloss, P. D. and Handelsman, J., Biotechnological prospects from metagenomics, *Curr. Opin. Biotechnol.*, 14, 303, 2003.

Index

A

Aclacinomycin (ACM), clinical activity, 315
Actinomycins
  analogs
    direct biosynthesis, 287
    partial synthesis, 287–288
    total synthesis, 288–289
  biological action mechanism, 284–285
  clinical applications, 292
  conformation, 283–284
  history, 281, 282f, 282t
  in vitro antitumor activity, 285
  in vivo antitumor activity/toxicity, 286
  separation and nomenclature, 282–283
  structure-activity relationships, 289
    chromophoric analogs, 291
    natural actinomycin variants, 290–291, 290t
    peptidic analogs, 291–292
  structures, 283
  synthesis, 286–287
Adriamycin. *See* Doxorubicin
Aminocoumarins, as combinatorial biosynthesis example, 535–538
Ansamitocins. *See* Maytansinoids
Ansamycins, 339
  as combinatorial biosynthesis example, 528–530
  See also Benzoquinone ansamycins
Anthracyclines, 299–300
  clinical activity, 313
    Aclacinomycin (ACM), 315
    Daunorubicin, 313
    Doxorubicin, 313–314
    Epirubicin, 314
    Idarubicin, 314–315
    liposomal anthracyclines, 315–316
    Sabarubicin, 315
  drugs development
    Daunorubicin, 301
    Doxorubicin, 301
    second-generation antitumor anthracyclines, 301–302
  mechanism of action, 302
    cardiotoxicity, 304
    DNA complex and Topoisomerase II poisoning, 302–303
    resistance (natural/acquired) to, 303
  medicinal chemistry
analogs modified in the Aglycone Moiety, 305–307
analogs modified in the Sugar Moiety, 308–312
  synthesis
    Doxorubicin, 304
    epirubicin, 304
    idarubicin, 304–305

Anti-cancer research
  biosynthetic techniques, 1
  combinatorial chemistry approach, 1
  See also Natural products
Antimitotic agents, 171, 267
Aplidin
  background, 223–224
  mechanism of action, 224–227
  preclinical drug development
    chemical synthesis, 229
    clinical studies, 229–230
    in vivo/in vitro testing, 227, 228t
    toxicology, 227–228, 228t
  See also Ecteinascidin 743 (ET-743/Yondelis)
Aspirin, 1
AVE8062, 34

B

Benzopyrans and privileged structures, 558–559
Benzoquinone ansamycins, 339–341, 340t
  clinical applications, 353–354
  mechanism of action, 340, 342–344
  medicinal chemistry, 350–353
  synthesis, 344–350
β-lapachone, 560
Bleomycins, 357–358
  antitumor agents (development of), 365–368
  clinical applications, 373–374
  mechanisms of action, 358
    DNA degradation, 358–361
    protein synthesis inhibition, 363–365
    RNA degradation, 361–363
  medicinal chemistry
    BLM combinatorial library, 372–373
    study of individual "amino acid" constituents, 370–372
  synthesis, 368–369
BN-80915 (Diflomotecan), 17
BNP-1350 (Karenitecin), 17
Bruceantin, 559
Bryostatins, 2, 137, 147
  biological activities, 138–140
  chemical syntheses, 140
  clinical trials, 141
    combination therapy, 144t
    monotherapy, 142t–143t
  future sources
    actual sources, 145
    aquaculture, 141, 145
    chemical analogs, 146–147

C

CA1P (OXI4503), preclinical studies, 37
CA4P, 34
 in combination therapy, 37
 effects on tumor blood flow, 34–35
 tumor response to (as a single agent), 35–37
Camptosar (irinotecan), 5, 6f, 16
Camptothecin (CPT) and analogs, 2, 5, 17
 clinically useful analog development, 15
  analogs in development, 16-17
  irinotecan (Camptosar), 16
  toptecan (Hcamtin), 1516
 cytotoxic mechanism, 7f
 medicinal chemistry, 1213
  A/B/C/D ring analogs, 13, 13f
  conjugates, 15
  E-ring lactone, 14
  water soluble/insoluble analogs, 14–15
 other biochemical effects, 89
 SCTI locus suppressors, 9
 preclinical studies, 34–37
 semisynthetic methods, 12, 12f
 structure, 6f
 synthetic studies, 9
  asymmetric synthesis, 9–11, 10f, 11f, 12f
  racemic CPT sythesis, 9
CC-1065 analogs and conjugates with polyamides,
  383–385, 407–408
 bisalkylators (dimers), 400–401
 cellular and pharmacological studies
  in vitro studies, 403–406
  in vivo studies, 406–407
 clinical studies, 407
 drug DNA complexes (structural characterization),
  402–403
 drug-DNA interactions, 385–386
 interaction with chromatin, 392–394
 modification in the pharmacophore unit
  analogs, 389–392
  CPI analogs, 386–387
  enantiomers of CPI analogs, 387–389
 polyamide conjugates/related sequence-directed
  structures, 394–400
Chondrmides, 562
CKD-602, 16
Combinatorial biosynthesis (anticancer natural products),
  519–520
 challenges/future directions, 538–539
 examples, 521, 522t–523t
  aminocoumarins, 535–538
  daunorubicin (DNR) and doxorubicin (DXR), 521,
   524–525
  enediynes, 533–535
  epothilones, 530–532
  leinamycin, 532–533
  mithramycin, 525–528
 requirements, 520
Combretastatins
 A-1 (CA-1), 24, 25f, 554
 A-4 (CA-4), 24, 25f

biochemical/biological mechanism of action, 25–27,
  28t–30t
 derivatives and synthetic analogs, 33–34
 folk use of plant family, 23–24, 39–40
 major groups, 24, 25f
  A-series (cis-stilbenes), 24
  B-series (diaryl-ethylenes), 24
  C-series (quinone), 24
  D-series (macrocyclic lactones), 24
 synthesis, 27, 31–33, 32f, 33f
 vasular targeting (therapeutic intervention), 27
CPT. *See* Camptothecin (CPT) and analogs
Cryptophycins, 151–152, 166
 action mechanism, 153
  cellular mechanism/cryptophycin 52 (LY355703),
   154
  effects of cryptophycin 52 (LY355703) on micrtubule
   polymerization, 154–155
 comparison (crytophycins 52/55 and 55-glycinate),
  164–165
 cryptophysin 52 (LY355703)
  clinical evaluation, 166
  in vivo antitumor activity, 165–166
 isolation and characterization, 152–153
 SARS, 158–160
  β-epoxide region/fragment A, 160C/D ester bond and
   C6/C7 positions, 162–164
  D-chlorotyrosine group/fragment B, 161–162
  phenyl group/fragment A, 160-161
 synthesis/cryptophycin 52 (LY355703), 156–158

D

Daunorubicin, 301
 clinical activity, 313
 as combinatorial biosynthesis example, 521, 524–525
Diflomotecan (BN-80915), 17
Digitoxin, 1
Discodermolide, 111–112, 171–172
 biological activity
  antitumor properties, 179–183
  immunosuppressive properties, 175, 179
 chemistry, 174
 clinical investigations, 187
 natural sources, 172
 structure-activity studies, 183–187
 structures, 173, 176–178
 synthesis, 174
  of analogs, 175
Diterpenoids, 93
Docetaxel (Taxotere), 92–93
 clinical applications, 112–113
Dolastins, 191, 560
 bioactivity/mechanisms of action, 198–202
 clinical status, 206–208
 origins/isolation/structure, 191–194
 structural modifications and SAR, 202–20
 structure/synthesis, 194–198
Doxorubicin, 301
 clinical activity, 313–314

as combinatorial biosynthesis example, 521, 524–525
  synthesis, 304
DX-8951f, 16

E

E7389, 241–242, 259–262
  background, 242–243
  in vitro characterization, 257–258
  in vivo characterization, 258–259
  structure-activity relationship, 243–244
  structure-activity relationship (structurally simplified
    analogs), 248–257
  synthesis of HB analogs, 257, 258f
  total synthesis/compound supply, 244–248
  See also Halichondrin
Ecteinascidin 743 (ET-743/Yondelis), 560, 562
  background, 216–217
  chemical synthesis, 220–222
  clinical studies, 221
    ovarian cancer, 223
    safety profile, 223
    in soft tissue sarcoma (STS), 221, 223
  mechanism of action, 217–218
  preclinical drug development
    in vitro/in vivo testing, 218, 219t, 220t
    toxicology, 218, 220
  See also Aplidin; Kahalalide F
Endophytic fungi and microbial associates, 559
Enediynes, 451, 464–465
  biosynthesis, 455–456
  cellular mechanisms of action, 458–459
  clinical experience, 463–464
  as combinatorial biosynthesis example, 533–535
  history, 451–453
  molecular mechanisms of action, 456–458
  preclinical studies, 459–463
  structural classes and producing organisms, 454–455
Epirubicin
  clinical activity, 314
  synthesis, 304
Epothilones, 111–112, 413–415, 441–442
  clinical applications, 440–441
  as combinatorial biosynthesis example, 530–532
  mechanism of action, 420–429
  natural occurrences, 415–420
  preclinical studies, 438–440
  semisynthesis, 430–435
  structure-activity relationships, 435–438
  synthesis, 429–430
ET-743. *See* Ecteinascidin 743
Etopophos, 81
Etoposide, 2
Etoposide development. *See* Podophyllotoxins and analogs
Exatecan mesylate dihydrate (DX-8951f), 16

F

Flavopiridol, 554, 558
Folk medicine
  Catharanthus roseus/Vinca rosea, 123

Cephalotaxus plant family, 47
Combretaceae plant family, 23–24, 39–40
  indigo/indirubins, 566–568
Podophyllum plant family, 72

G

GA (geldanamycin). *See* Benzoquinone ansamycins
Geodiamolides, 562
Gimatecan (ST1481), 17
GL-331, 82

H

Halichondrin, 2, 269–262
  See also E7389
Hemiasterlin, 267–269, 278–279
  See also HTI-286
HHT. *See* Homoharringtonine and related compounds
Homoharringtonine and related compounds,
    47–48, 65
  clinical applications
    acute promylocytic leukemia (APL), 64
    Chinese clinical trials, 63
    myelodysplastic syndrome (MDS), 64
    Phase I studies (U.S.), 63
    Phase I studies/solid tumors, 63–64
    Phase III studies/acute leukemia patients, 64
    Phase II studies/CML patients, 64–65
  clinical drug development, 63
  mechanism of action, 56–58
  medicinal chemistry, 58–60, 59t, 60t, 62t, 63
  structures, 48–50, 48f, 49f
  synthesis
    of acyl groups, 50–52
    total, 52–56
HTI-286, 267, 278–279
  biological activity, 277
  natural product lead structures, 267–269
  natural products biological activity, 269, 271
  preclinical and clinical development, 277–278
  synthesis, 271–275
  synthesis of analogs and SAR, 275–276
Hycamptin (topotecan), 5, 6f, 15–16
  structure, 8f

I

Idarubicin
  clinical activity, 314–315
  synthesis, 304–305
Indigo/indirubins, 566–568
Indolocarbazoles. *See* Staurosporines/structurally related
    indolocarbazoles
Irinotecan (Camptosar), 5, 6f, 16

J

Jaspamide/jasplakinolide, 562

K

Kahalalide F
  background, 230–231
  drug development
    clinical trials, 234
    in vitro/in vivo testing, 233, 233t
    preclinical toxicology, 233–234
    synthesis, 234
  mechanism of action, 231–233
  See also Ecteinascidin 743 (ET-743/Yondelis)
Karenitecin (BNP-1350), 17

L

Leinamycin, as combinatorial biosynthesis example,
    532–533
Leptomycins, 565
LY3555703. *See* Cryptophycins

M

Macrocyclic depsipeptide. *See* Cryptophycins
Manzamines, 563
Marine sponges, 278
Maytansinoids, 321
  biological activity/mechanism of action, 323–324
  biosynthesis, 330–333
  medicinal chemistry, 324–327
  natural occurrences, 321–323
  preclinical/clinical developments, 328–329
  total synthesis, 324
Microtubule stabilization (by antimitotic agents),
    171
Mithramycin, 525–528
Mitomycins, 475–479
  analogs and derivatives
    conjugated derivatives, 485–487
    structural analogs and derivatives, 484–485
  clinical applications, 489–490
  mechanism of action, 479–484
  structure-activity relationship, 487–489
  synthesis, 489
Morphine, 1
Mycalamides, 562

N

Natural products
  as lead compounds for pharmaceutical research, 1,
    278–279
  reason for importance in research
    biological and ecological rationale for novel bioactive
      secondary metabolites, 1
    historical source of drugs, 1, 23–24, 39–40
    production of otherwise inaccessible drugs, 2
    templates for future drug design, 2
Natural products as anti-cancer agents, 1, 554, 555t–557t,
    568
  marine sources/independent microbes, 563
    salinospramide, 563

marine sources/invertebrates, 564, 564t
    NVP-LAQ824 and Psammaplin, 564–565
  marine sources/microbial associates, 560
    dolastins, 560
    ecteinascidins, 560, 562
    geodiamolides and chondramides, 562
    jaspamide/jasplakinolide, 562
    manzamines, 563
    pederins/mycalamides/related structures, 562
  plant sources, 554
    benzopyrans and privileged structures, 558–559
    combretastatins, 554
    Flavopiridol, 554, 558
    microbial associates/endophytic fungi, 559
    olomucine/roscovotine/purvalanol, 558
  plant sources/"old" molecules
    β-lapachone, 560
    bruceantin, 559
    triterpenoid acids, 559–560
  plant-microbe-marine interface, 566
    indigo/indirubins, 566–568
  terrestrial agents, Palmarumycin CP$_1$, 565–566
  terrestrial microbes, 565
    leptomycins, 565
9-AC (IDEC-132), 1617
9-NC, 17
NK, 611, 81
NVP-LAQ824, 564–565
NX211, 16

O

Olomucine, 558
Oncovin (vincristine), 91
OXI4503 (CA1P), preclinical studies, 37

P

Paclitaxel, 2
Paclitaxel. *See* Taxol and analogs
Palmarumycin CP$_1$, 565–566
Pederins, 562
PharmaMar compounds, 216
Phleomycins, 365–366
Podophyllotoxins and analogs, 71–72
  clinical applications, 84
  etoposide and teniposide development, 73
  future perspectives, 85
  history, 72
  mechanisms of action, 74
    DNA topoisomerase II inhibition, 74–75
    other antineoplastic mechanisms, 75
    tubulin polymerization inhibition, 74
  structure, 72–73, 72f, 73t
  structure-activity relationships, 75
    molecular area-oriented analog synthesis, 76–79, 76t
    representative analogs, 81–82
    SAR models, 8081
  synthesis, 82–83
Psammaplin, 564–565
Purvalanols, 558

Q

Quinine, 1

R

Rebeccamycin, 503
Roscovotine, 558

S

Sabarubicin, clinical activity, 315
Salinospramide, 563
SCTI genetic locus, and CPT toxicity, 9
ST1481 (Gimatecan), 17
Staurosporines/structurally related indolocarbazoles, 499–500, 512
  clinical applications, 511–512
  drugs (from natural products), 506–511
  mechanisms of action
    AT2433-A1 and B1, 503
    rebeccamycin, 503
    staurosporine and K-252a, 502
    staurosporine and UNC-01, 500–502
  synthesis, 503, 504f, 505–506

T

Taxol and analogs, 89–90, 113, 414
  analogs in clinical trials, 113
  biosynthesis/bioproduction, 93–95
  clinical applications, 112–113
  history
    clinical development, 91
    discovery, 90
    preclinical development, 90–91
    supply crisis, 92
    Taxotere (docetaxel), 92–93
  mechanism of action, 95–96
  medicinal chemistry, 96
    prodrugs of taxol, 102
    Ring A modifications, 96–97
    Ring B modifications, 97–98
    Ring C modifications, 98–100
    Ring D modifications, 100
    SAR summary, 104
    side chain analogs, 101–102
    targeted analogs, 102–104

other natural products/similar mechanism, 111–112
  synthetic studies
    semisynthetic methods, 105–107
    total synthesis, 107–109
    taxol-tubulin interaction, 109–111
    and tubulin polymerization, 91, 91f
Taxotere (docetaxel), 92–93
Teniposide, 2
Teniposide development. *See* Podophyllotoxins
        and analogs
TOP-53, 82
Topoisomerase I (topo I), 5
  DNA relaxation mechanism, 7f
  poisoning of, 6, 8
Topotecan (Hycamtin), 5, 6f, 15–16
  structure, 8f
Triterpenoid acids, 559–560

U

U.S. National Cancer Institute
  anticancer drug leads investigations, 191
    60-cell line screen, 554
  terrestrial plants (exploratory survey), 1, 23, 90

V

Velban (vinblastine), 91
Vinblastine (Velban), 91
Vinca alkaloids, 123, 132
  clinical applications, 131–132
    Vinblastine, 2, 131
    Vincristine, 1, 131
    Vinflunine, 5, 131
    Vinorelbine, 4, 131
  discovery, 123–124
  mechanism of action, 124–125
  medicinal chemistry, 128–131
  semi and total synthesis, 125–127
Vincristine (Oncovin), 91

Y

Yondelis™. *See* Ecteinascidin 743

Z

ZD-6126, 34